AF892974

Lecture Notes in Mechanical Engineering

Lecture Notes in Mechanical Engineering (LNME) publishes the latest developments in Mechanical Engineering—quickly, informally and with high quality. Original research reported in proceedings and post-proceedings represents the core of LNME. Volumes published in LNME embrace all aspects, subfields and new challenges of mechanical engineering. Topics in the series include:

- Engineering Design
- Machinery and Machine Elements
- Mechanical Structures and Stress Analysis
- Automotive Engineering
- Engine Technology
- Aerospace Technology and Astronautics
- Nanotechnology and Microengineering
- Control, Robotics, Mechatronics
- MEMS
- Theoretical and Applied Mechanics
- Dynamical Systems, Control
- Fluid Mechanics
- Engineering Thermodynamics, Heat and Mass Transfer
- Manufacturing
- Precision Engineering, Instrumentation, Measurement
- Materials Engineering
- Tribology and Surface Technology

To submit a proposal or request further information, please contact the Springer Editor of your location:

China: Ms. Ella Zhang at ella.zhang@springer.com
India: Priya Vyas at priya.vyas@springer.com
Rest of Asia, Australia, New Zealand: Swati Meherishi at swati.meherishi@springer.com
All other countries: Dr. Leontina Di Cecco at Leontina.dicecco@springer.com

To submit a proposal for a monograph, please check our Springer Tracts in Mechanical Engineering at https://link.springer.com/bookseries/11693 or contact Leontina.dicecco@springer.com

Indexed by SCOPUS. All books published in the series are submitted for consideration in Web of Science.

Faiz Ahmad · Hussain H. Al-Kayiem ·
William Pao King Soon
Editors

ICPER 2020

Proceedings of the 7th International Conference on Production, Energy and Reliability

Springer

Editors
Faiz Ahmad
Universiti Teknologi Petronas
Ipoh, Malaysia

Hussain H. Al-Kayiem
Universiti Teknologi Petronas
Ipoh, Malaysia

William Pao King Soon
Universiti Teknologi Petronas
Ipoh, Malaysia

ISSN 2195-4356 ISSN 2195-4364 (electronic)
Lecture Notes in Mechanical Engineering
ISBN 978-981-19-1938-1 ISBN 978-981-19-1939-8 (eBook)
https://doi.org/10.1007/978-981-19-1939-8

This Springer imprint is published by the registered company Springer Nature Singapore Pte Ltd.
The registered company address is: 152 Beach Road, #21-01/04 Gateway East, Singapore 189721, Singapore

Contents

Thermo-Fluids

Turbulent Flow Estimation by Wavelet Transform

A. B. Osman, Mark Ovinis, Ahmed Muftah Montaser Mihoob, Abdalellah O. Mohmmed, and Shafriza Nisha Basah

Abstract Turbulent flow estimation from an image sequence is challenging due to the lack of dedicated flow measurement techniques. Existing techniques estimate flowrate with high uncertainty. In this paper, a new technique based on discrete wavelet transform (DWT) is proposed. Wavelets have the advantage of decomposing flow signals into numerous levels and remove input signal noise. The flow signals are first decomposed using DWT into multiple levels, then, the wavelet coefficients are correlated by the Fast Fourier Transform (FFT) based algorithm to determine the velocity field. This wavelet-based algorithm is named as DWT-FFT. DWT-FFT was evaluated first using synthetic signals and then applied for turbulent flow estimation. The accuracy of DWT-FFT was compared to classical algorithms including direct cross correlation (DCC) and direct implementation of FFT. DWT-FFT estimated the flow with an error of 0.7%, outperforming both DCC and FFT which estimated with an error of 7.14% and 12.2% respectively.

Keywords Oil spill · Cross correlation · Flow measurement

1 Introduction

The use of optical sensors for flow measurements is attractive because of the non-invasiveness of the technique. In particular, optical techniques have proven to be, by far, the most accurate technique for estimating flow rate in the Macondo well spill [1]. However, estimating the velocity field of a turbulent flow from optical data is

A. B. Osman · M. Ovinis (✉) · A. M. M. Mihoob
Mechanical Engineering Department, Universiti Teknologi PETRONAS, 32610 Seri Iskandar, Perak, Malaysia
e-mail: mark_ovinis@utp.edu.my; anotood@yahoo.com

A. O. Mohmmed
Department of Mechanical Engineering, Middle East College, Knowledge Oasis Muscat, Al Rusayl, Muscat, Oman

S. Nisha Basah
School of Mechatronic Engineering, Universiti Malaysia Perlis, 01000 Kangar, Perlis, Malaysia

F. Ahmad et al. (eds.), *ICPER 2020*, Lecture Notes in Mechanical Engineering,
https://doi.org/10.1007/978-981-19-1939-8_1

challenging because of the complex nature of the flow, which includes multi-scale behaviour in both time and space direction. Several optical techniques have been used for estimating the flow rate of oil spill estimation as reviewed in [1, 2], which includes particle image velocimetry (PIV), feature tracking velocimetry (FTV), large eddy velocimetry (LEV), and optical plume velocimetry (OPV). PIV is a common technique usually used for experimental studies and has extensive applications in fluid dynamics. However, PIV failed to estimate accurately the flow rate of the oil spill in case of Macondo well incident [1, 2]. The spatial cross correlation algorithm and the inadequate sizes of interrogation windows usually used in PIV, limited this technique to capture the fast flow of oil spill. Both FTV and LEV techniques, which are based on visible tracking of existing features in the fluid flow space, poorly estimated the flow rate. The bad quality of spill videos led to the inability of these techniques to track visible features such as coherent structures.

The most accurate optical technique used for oil spill flow rate estimation is OPV. This technique was developed by Crone et al. [3], for estimating the flow rate of underwater black smoker, which has similar flow properties to that of the underwater spill. OPV estimates the flow rate based on temporal averaging, in which the turbulent flow variation over time is considered. This could be the main reason behind its better accuracy over other techniques. Moreover, the temporal cross correlation of OPV technique is less sensitive to bias, and requires less computational time [3]. As a comparison, both PIV and OPV use cross correlation algorithm as a step to estimate the flow. However, PIV implements cross correlation spatially, while the OPV is a temporal based on cross correlation, taking only two points in the turbulent flow space. Therefore, PIV is not ideal for capturing turbulent flow due to the variability of the flow over time. This is because of its low temporal resolution (i.e. PIV estimates the velocity from two consecutive images only) and loss of spatial resolution, due to the use of interrogation window for correlation. Therefore, OPV has the advantage of good temporal and spatial resolution as compared to PIV. The proposed technique described in this study comprises the advantages of OPV technique in terms of its temporal and spatial resolution, as well as considering the multi-scale behaviour of the turbulent flow by decomposing into many levels.

The velocity can be estimated by dividing distance over time. To estimate the velocity field using a temporal averaging technique, four steps are required: (i) setting a distance between two points in the flow space, (ii) extracting the signal from these two points (v) cross correlating the two signals, and (iv) detecting the peak correlation coefficients. The time is the time that corresponds to the peak of these coefficients. The most important step for velocity estimation is the cross correlation step. Several algorithms have been used for cross correlation coefficients estimation, and they include, direct cross correlation (DCC), Fast Fourier Transform (FFT) cross correlation, and average square difference function. The DCC algorithm is based on signal amplitude in time domain, while the FFT algorithm is based on frequency information of the signal. However, transforming a signal from time to frequency domain is a trade-off between time information and frequency information.

On the other hand, the wavelet transforms have the ability to decompose a signal/image into many levels and provide both time–frequency information. The wavelet transforms are applied in several works related to fluid dynamics. These includes investigation of self-similarity and intermittency behaviour in turbulent signals [4, 5], and identification of flow structure [6]. The relation between wavelets and turbulence are discussed by Farge et al. [7]. The application of wavelet transform to turbulent signals are reviewed by Farge [8], as well as using of wavelet for fluid flow simulation [9]. However, in most of these works, the velocity field is already known and wavelet is utilized to understand the turbulent flow behaviour. Fluid motion estimated from image sequences by Dérian et al. [10, 11], combined a differential based algorithm with wavelet expansion. This combination improved the accuracy and the performance of motion estimation algorithm.

In this paper, a new technique based on combining FFT cross correlation and discrete wavelet transform is proposed and named as the DWT-FFT algorithm. The rest of the paper includes material and methods which illustrate the signal simulation model, experimental video data, as well as the developed wavelet-based algorithm, followed by results and discussion and finally the conclusion.

2 Material and Method

2.1 Signal Simulation

A turbulent signal exhibits noisy random behaviour over time [12]. In this work, a random signal is simulated to investigate the accuracy of wavelet based technique. By shifting the simulated signal with known time delay, one can use the DWT-FFT algorithm to estimate the known delay value. The two simulated and shifted signals are generated using Eqs. 1 and 2. The first signal, $S_1(n)$ is a random signal with a length of n, and the second signal $S_2(n)$ is generated by delaying S_1 by a value of d. A random noise $N(n)$ with a certain signal to noise (SNR) is then added to both S_1 and S_2.

$$S_1(n) = s(n) + N(n) \tag{1}$$

$$S_2(n) = s(n - d) + N(n) \tag{2}$$

The magnitude of SNR is calculated as the ratio of signal mean μ, to its standard deviation σ, as shown in Eq. 3:

$$SNR = \frac{\mu}{\sigma} \tag{3}$$

The first signal was delayed using Eq. 1 by a value of d and will be used as ground-truth for evaluating the proposed method. Therefore, the estimation of time delay between the two signals S_1 and S_2 require the following steps:

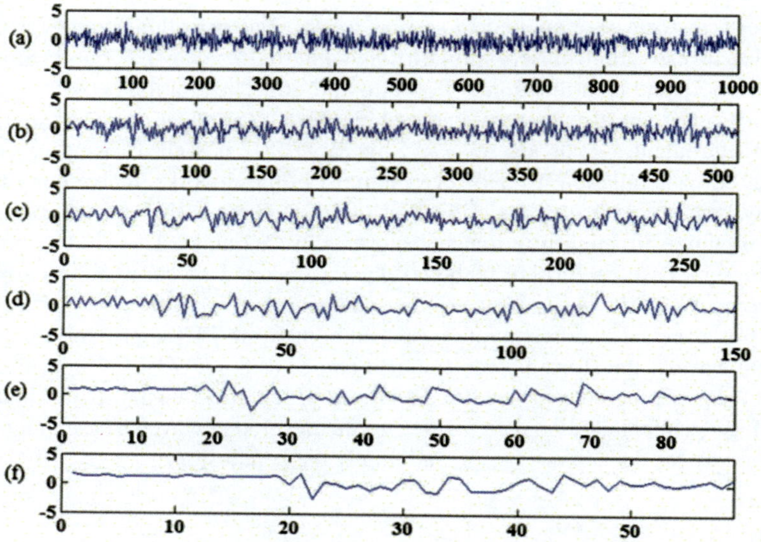

Fig. 1 A sample of the simulated random signal with the reconstructed signals: **a** original signal, **b** level-1, **c** level-2, **d** level-3, **e** level-4 and **f** level-5 signals

- Decompose both signals, S_1 and S_2, into several levels using discrete wavelet transform into approximate and detail information.
- Reconstruct the signal at each level from approximate signal.
- Calculate the time delay between the two signals reconstructed signals for each levels separately using FFT cross correlation algorithm.
- Add noise to the reconstructed signals and estimate the time delay.

In this work, the first signal length is simulated with length of 1000, and delayed by 300 to generate the second signals. Then, the first step to validate our developed algorithm in this paper is to find this delay value. For each level of wavelet, the time delay will be estimated to find out the optimum level of accuracy estimating the delay of random signals. Figure 1 shows a sample of simulated random signals with the reconstructed signals in five levels of wavelet decomposition.

2.2 *Turbulent Jet Flow Image Sequence*

In order to evaluate the accuracy of the DWT-FFT algorithm, we used an image sequence of a turbulent buoyant jet of Crone et al. [18],who developed an experimental rig to simulate a turbulent buoyant jet in order to develop the OPV technique [3]. The Crone video is available at the website of (http://www.fluidcontinuity.org/research/opv/). Detailed information of this video is summarized in Table1:

Table 1 Video parameters of turbulent buoyant jet

No	Description	Value
1	Video length	30 s
2	Number of frames	994
3	Scale	1 m = 858 pixels
4	Actual flow rate	4.2 pixel/frame
5	Nozzle diameter	18 mm

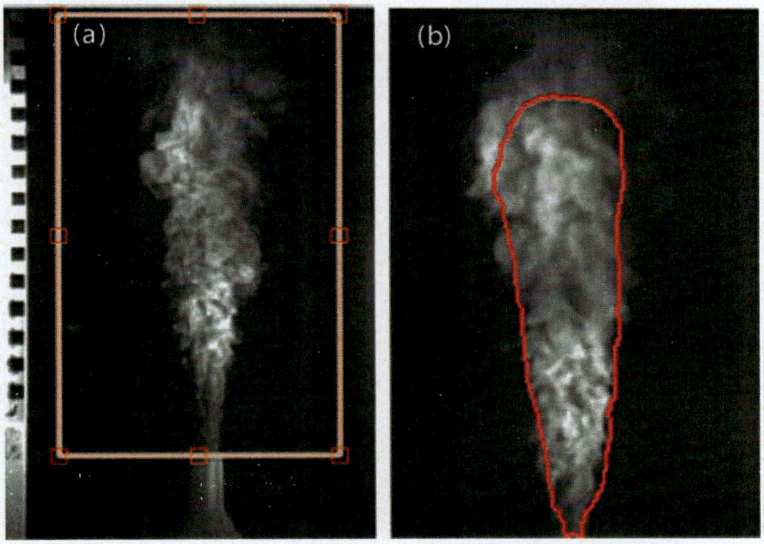

Fig. 2 Sample of turbulent buoyant jet image include **a** original image extracted from Crone video data [3], and **b** cropped image with segmented region

Figure 2, shows a sample of jet flow image which was extracted from the video. The overall video was converted into image sequence, then the flow region is cropped (i.e. rectangular region in Fig. 2(a)) for each image and saved in a 3D matrix in order to estimate its velocity. In order to find the velocity field at the jet flow region, the flow region is segmented as shown by red colour region in Fig. 2b.

2.3 Velocity Field Estimation

The overall methodology for estimating the velocity field from image sequence using the DWT-FFT algorithm is described in Fig. 3.

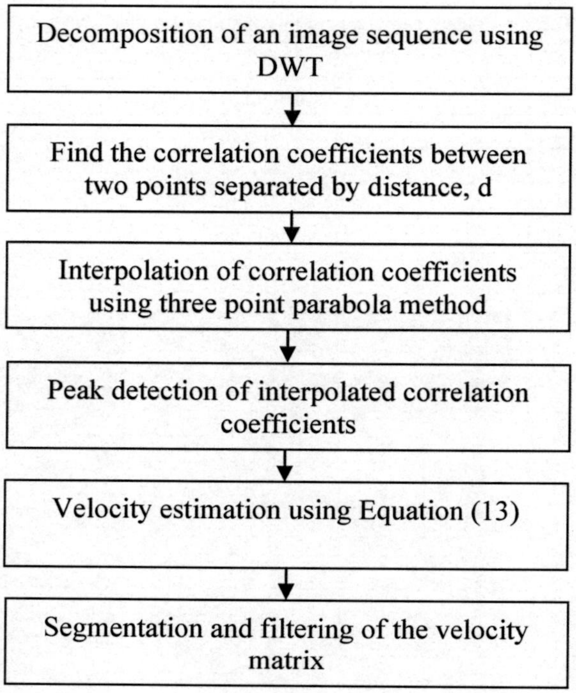

Fig. 3 Steps for estimating the velocity field of jet based on DWT-FFT algorithm

2.4 Discrete Wavelet Transform

The wavelet transform is an extension of the windowed Fourier transform, which has the advantage of presenting any signal in both time–frequency domain. The Fourier transform includes only the frequency information in a small window size. DWT is a common approach in signal/image processing field to decompose any signal/image into multi-levels, in order to analysis the signal/image in both time and frequency domain simultaneously.

If $S(t)$ is a time domain signal, the DWT of this signal $D(j, k)$ can be given by the following equation:

$$D(j,k) = \sum_t S(t)\psi_{j,k}(t) \tag{4}$$

where $\psi_{j,k}(t)$ is the wavelet coefficient associated with the signal $S(t)$, and j, k are the wavelet coefficients.

The discrete wavelet phase, $S_P(j, k)$ can be written as:

$$S_p(j,k) = D(j,k)^2 \tag{5}$$

The remaining part of the time series signal in DWT, *r(j,k)* can be given by:

$$r(j,k) = \sum s(t)\phi_{j,k}(t) \tag{6}$$

where $\phi_{j,k}(t)$ denotes the scaling function of the wavelet

$$\phi_{j,k} = 2^{-1/2}\phi_{j,k}(2^{-j}t - k) \tag{7}$$

Then the inverse of DWT can be given by:

$$S(t) = r(j,k)\phi_{j,k}(t) + \sum_j \sum_k D(i,k)\psi_{j,k}(t) \tag{8}$$

2.5 *Correlation Coefficients and Time Delay Estimation*

The second step for velocity estimation is the estimation of a time delay between the signals extracted from decomposed image sequence. The delay estimation requires three steps: estimation of correlation coefficients, interpolation, and peak detection of obtained coefficients. To find the correlation coefficients, FFT cross correlation algorithm was applied. First, both signals were transformed in a frequency domain using Fourier transform, then the transformed data multiplied and the inverse found to determine the correlation coefficients. Figure 4 illustrates the steps for finding correlation coefficients using FFT cross correlation algorithm:

Also, the FFT cross correlation algorithm can be formulated by the following Equation:

$$C = ifft(fft(S_1) \cdot conj(fft(S_2)) \tag{9}$$

where C is the correlation coefficients, S_1, S_2 are simulated and delayed signals.

The Fourier Transform of signal S is given by:

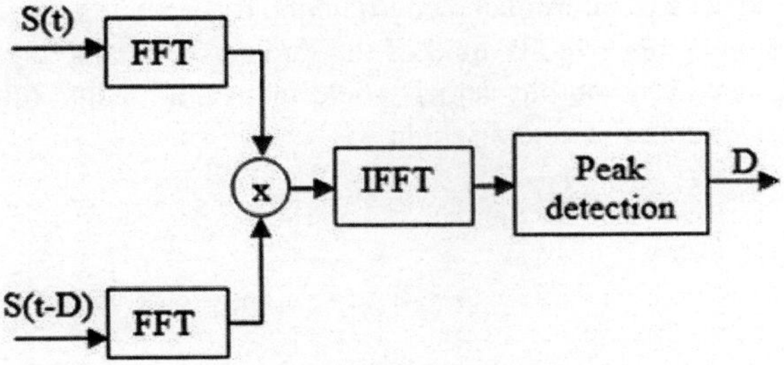

Fig. 4 Correlation coefficients calculation between two signals using FFT-correlation method

$$S_i(k) = \sum_{n=0}^{N-1} S_i(n) \cdot e^{-j\frac{2\pi}{N}n\kappa} \quad (10)$$

and the inverse of FFT is given by using the following equation:

$$S_i(j) = \sum_{k=0}^{N-1} S_i(k) \cdot e^{-j\frac{2\pi}{N}n\kappa} \quad (11)$$

For evaluating the robustness of the DWT-FFT algorithm, two algorithms were selected for comparison purpose – FFT cross correlation algorithm without embedding it in the wavelet domain, and the direct cross correlation (DCC) proposed by Crone et al. [3].

The correlation coefficients $C_{i,j,l}$ can be obtained using DCC algorithm and formulated as:

$$C_{i,j,l} \sum_{k=1}^{N-1} S_{i+d,j,k} \cdot S_{i,j,k+1} \quad (12)$$

where S is the extracted signal, and $_{I,j,l}$ are the number of rows, columns, and a number of images used to estimate the velocity respectively.

Since the correlation coefficients are obtained either by DCC or FFT algorithm, its peak detection will lead to a real time delay value. To find the complete time value with fraction, an interpolation procedure is required. Therefore, a parabola interpolation is applied to the obtained correlation coefficients. Then, the required time delay is the time value that corresponds to the peak of interpolated correlation coefficients. Once the time value is obtained, the velocity u can be easily estimated by the following Equation:

$$u = \frac{d}{\tau} \quad (13)$$

where d is a distance between two points in the turbulent flow space which was taken as 5 pixels as recommended by Crone et al. [3], and τ is the detected time value from the peak of interpolated coefficients. By repeating the same steps for time estimation, the overall velocity field of the jet can be obtained. Finally, due to the presence of noise in the output velocity field matrix, a median filter of size 10 by 10 is applied to smooth the velocity field.

3 Results and Discussion

3.1 Simulation Results

DWT-FFT algorithm is applied to estimate a known time delay between two random signals. A delay value of d = 300 is fixed between the two signals to be used as ground-truth for evaluating this technique. Then, the robustness of the algorithm is evaluated by adding noise to those signals. The range of noise level was measured by the signal-to-noise ratio (SNR) to be between −60 and 60 dB.

For implementing the DWT-FFT algorithm, five levels of wavelet decomposition were found to be enough to obtain reasonably accurate results. Figure 5 shows the relative errors in the estimated time delays for the five wavelet decomposition levels with various level of SNR. Both signal decomposition level and noise level affected directly the accuracy of the algorithm in time delay estimation. More wavelet decomposition level increases the error in time delay estimation. For example, for SNR = −60 dB, the error increased from 0.3% to 22.3% when estimated from level-1 and level-5 signals respectively. Increasing the decomposing signal using DWT usually leads to reduction in the signal length by factor two for each level as shown in Fig. 1. This resulted in uncorrelated signals and inaccurate time value. The second observation is that by increasing the noise level in the signal, the error increases and vice-versa.

To evaluate the stability of the DWT-FFT algorithm against random noise in which the noise power is constant, the time delay value was estimated 10 times for each SNR values and a summary is presented in Table 2. By increasing the SNR, the delay error reduced, as well as a reduction in the variation in delays was observed (i.e. presented by standard deviation).

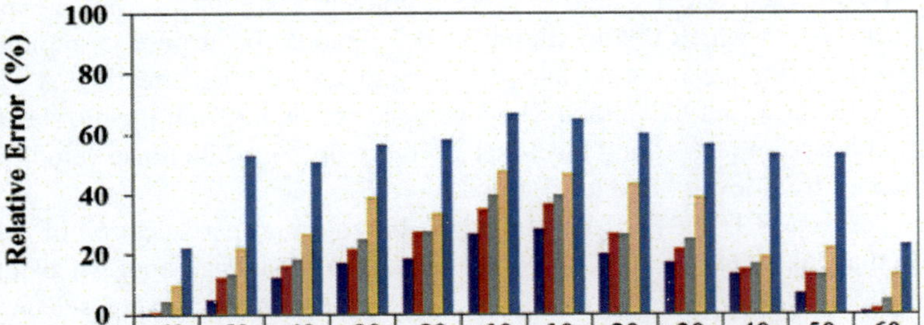

	-60	-50	-40	-30	-20	-10	10	20	30	40	50	60
Level-1	0.3	5	12.3	17	18.3	26.7	28.3	20	17	13.3	6.7	0.7
Level-2	0.7	12.3	16	21.7	27.3	35	36.7	26.7	21.7	15	13.3	1.7
Level-3	4.7	13.7	18.3	25	27.7	40	40	26.7	25	16.7	13.3	5
Level-4	9.7	22	26.7	39	33.3	47.7	46.7	43.3	39	19	22	13.3
Level-5	22.3	53	50.7	56.7	58.3	66.7	65	60	56.7	53.3	53.3	23.3

Fig. 5 Errors in estimated time delay for 5 wavelet decomposition with various SNR level

Table 2 Estimated time delay (Average, standard deviation, and error) for 10 runs in each case of SNR

Error (%)	RMSE (%)	Standard Deviation	Average of delay	SNR
59.1	10.08	31.89	122.61	5
39.7	7.27	22.99	180.82	10
28.4	8.99	28.43	21.86	20
26.7	6.27	19.84	219.97	30
23	3.4	10.75	231.07	40
19.9	5.31	16.79	240.31	50
16.9	5.7	18.01	249.23	60

3.2 *Turbulent Jet Flow Results*

The DWT-FFT algorithm is evaluated with real experimental image sequence of turbulent jet flow. These image sequences were extracted from Crone's video. The velocity field for each wavelet levels was estimated using the DWT-FFT algorithm. Figure 6 shows the estimated jet velocity field for four wavelet decomposition levels. The obtained velocity field was scaled from zero to the magnitude of the actual velocity i.e. 4.2 pixel/frame (refer to Table.1). Only the jet flow region is presented and its background filled by white colour.

Therefore, by increasing the wavelet decomposition, noisy velocity field was obtained. The artifact errors therefore increased (see Fig. 6c and d). This is due to the loss of energy from one level to the next during wavelet decomposition and reconstruction procedures. Increasing the number of decomposition levels usually increase the error in signal reconstruction. A good representation of the flow field can be obtained by first and second levels of decomposition as shown in Fig. 6a and b. The obtained velocity field in these two levels is similar to what is expected from jet theory, in which the maximum velocity should be at pure jet region near to the nozzle, while the velocity starts decaying in stream-wise direction. A smooth velocity field was observed at the fully developed region far from the nozzle area in the estimation from level-2 and level-3 (i.e. Figure 6b, c), while noise velocity field was obtained from level-3 and level-4.

The accuracy of DWT-FFT algorithm was based on proper selection of wavelet function that can be used for decomposing the turbulent signals. Four wavelet functions namely Daubechies, Symlet, Coiflet and Haar wavelets were selected for comparison purpose. Figure 7 shows the velocity field obtained by applying these different wavelets at level-2.

Almost similar velocity field was obtained using the four wavelets with small difference. Although there was some similarities, the quantification of nozzle velocity showed some variation by the use of those wavelets.

Figure 8 shows the error in nozzle velocity estimation using the different wavelets for three level of decomposition. The best result was obtained by Coiflet wavelet in

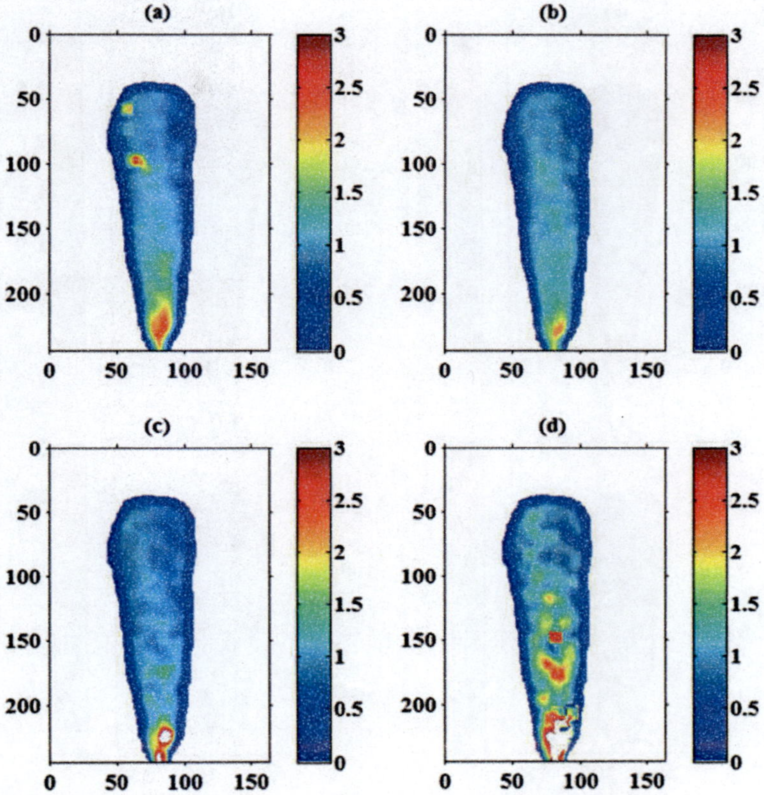

Fig. 6 Velocity field for three levels of wavelet decomposition using Daubechies wavelet function at: **a** Level-1 **b** Level-2 **c** Level-3 and **d** Level-4

the first level of decomposition with an error of 0.7%. Higher errors were obtained when the nozzle velocity was estimated from level-2 and level-3.

One of the important factors that affect the accuracy of DWT-FFT technique is the vanishing moment of wavelet function used. To investigate the effect of vanishing moment in the nozzle velocity estimation, the Coiflet wavelet with standard vanishing moment ranged from 1 to 5 was applied. Figure 9 illustrates the relative error estimated in nozzle velocity estimation as a relation to the vanishing moment. By increasing the vanishing moment from 1 to 5, the error reduced by a percentage of 73%. The same conclusion obtained by Dérian et al. [11] when Daubechies wavelet was used with different vanishing moment for motion estimation. Therefore, the best vanishing moment when Coiflet function was used in nozzle velocity estimation is equal to 5.

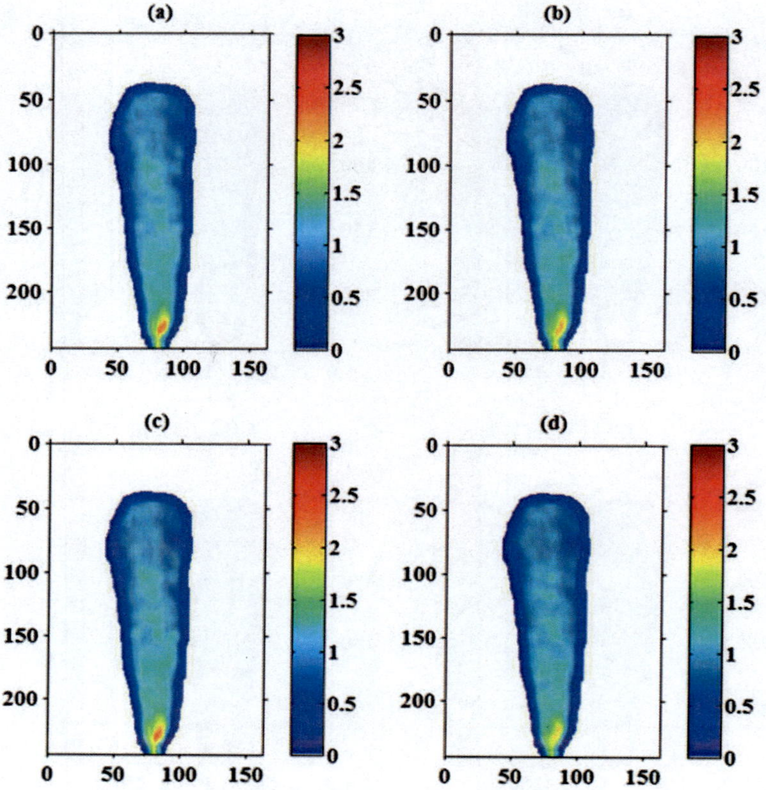

Fig. 7 Jet velocity field obtained using different wavelet functions: **a** Daubechies, **b** Symlet, **c** Coiflet and **d** Haar wavelet

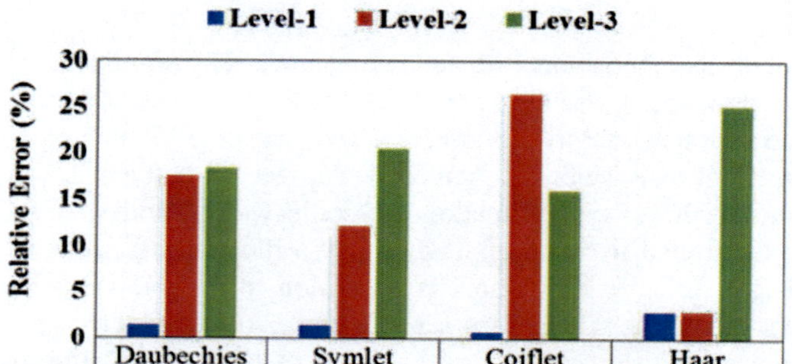

	Daubechies	Symlet	Coiflet	Haar
Level-1	1.4	1.4	0.7	2.9
Level-2	17.4	12.1	26.4	2.9
Level-3	18.3	20.5	16	25.1

Fig. 8 Relative errors in nozzle velocity estimation using different wavelet functions at various decomposition levels

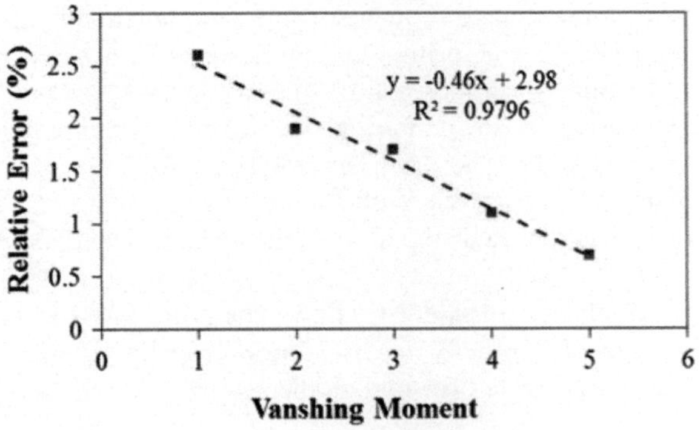

Fig. 9 The vanishing moment effect in nozzle velocity estimation

Table 3 Comparison between various algorithms

No	Algorithm	Estimated velocity	Relative error (%)
1	DCC	3.90	7.14
2	FFT-correlation	3.69	12.2
3	DWT-FFT	4.17	0.7

3.3 Comparison to Classical Algorithms

Since the main objective of this work is to develop an alternative technique for accurately estimating the nozzle velocity, the best result obtained using the developed DWT-FFT algorithm is based on using Coiflet wavelet with vanishing moment of 5 and by estimating the velocity field from the reconstructed signals of the first level of wavelet decomposition.

The accuracy of DWT-FFT algorithm was compared to two other classical algorithms namely DCC and FFT algorithms. Both algorithms were applied to estimate the velocity field of the jet flow and the nozzle velocity. Table 3 shows a comparison between the results obtained by using the three algorithms. The actual nozzle velocity, 4.2 pixel/frame, obtained by Crone et al. [3] was used as the ground truth for validation. The accuracy DWT-FFT algorithm is better than the accuracy of DCC and FFT algorithm.

4 Conclusion

In this paper, a new technique was proposed for turbulent flow estimation from image sequence. The proposed technique is based on wavelet transform, which is named as DWT-FFT. Various parameters were investigated to test the accuracy of the DWT-FFT including wavelet decomposition level, wavelet function, and the vanishing moment of wavelet function. The DWT-FFT algorithm was validated using both simulated signals and image sequence. Both simulated and experimental results showed that the DWT-FFT algorithm was more sensitive to the wavelet decomposition level and the signal noise. Increasing the decomposing level increased the error in time delay estimation. Also increasing the wavelet decomposition levels resulted in a noisy velocity field. The optimum condition for the developed algorithm was achieved using Coiflet wavelet with a vanishing moment of 5. The DWT-FFT algorithm was the most robust of the algorithms and estimated the nozzle velocity with an error of 0.7%. For future work, experimental work will be conducted by building a rig to simulate a turbulent jet flow. Then, the DWT-FFT algorithm will be applied to estimate the velocity field, and more investigation will be conducted on the effect of changing nozzle flow rate on the accuracy of the algorithm.

Acknowledgements The authors would like to express their appreciation to Universiti Teknologi PETRONAS for supporting this work under YUTP 015LC0-180.

References

1. McNutt MK, Camilli R, Crone TJ, Guthrie GD, Hsieh PA, Ryerson TB et al (2012) Review of flow rate estimates of the Deepwater Horizon oil spill. Proc Natl Acad Sci 109:20260–20267
2. Lehr B, Aliseda A, Bommer P, Espina P, Flores O, Lasheras J et al (2010) Deepwater horizon release estimate of rate by PIV. Report to the US Dept of interior
3. Crone TJ, McDuff RE, Wilcock WS (2008) Optical plume velocimetry: a new flow measurement technique for use in seafloor hydrothermal systems. Exp Fluids 45:899–915
4. Xu G, Wan B, Zhang W (2006) Application of wavelet multiresolution analysis to the study of self-similarity and intermittency of plasma turbulence. Rev Sci Instrum 77:083505
5. Ruppert-Felsot J, Farge M, Petitjeans P (2009) Wavelet tools to study intermittency: application to vortex bursting. J Fluid Mech 636:427–453
6. Hui L (1998) Flow structure identification of a turbulent shear flow with use of wavelet statistics
7. Farge M, Kevlahan N, Perrier V, Goirand E (1996) Wavelets and turbulence. Proc IEEE 84:639–669
8. Farge M (1992) Wavelet transforms and their applications to turbulence. Annu Rev Fluid Mech 24:395–458
9. Kim T, Thürey N, James D, Gross M (2008) Wavelet turbulence for fluid simulation. ACM Trans Graph (TOG) 50
10. Dérian P, Héas P, Herzet C, Mémin É (2012) "Wavelet-based fluid motion estimation. Scale space and variational methods in computer vision. Springer, pp 737–748
11. Dérian P (2012) Wavelets and fluid motion estimation. Université Rennes 1, 2012
12. Ferré-Giné J, Rallo R, Arenas A, Giralt F (1997) Extraction of structures from turbulent signals. Artif Intell Eng 11:413–419

Influence of Swirl Flow Pattern in Single Tube Burner on Turbulent Flame Blow-Off Limit

Hasanain A. Abdul wahhab

Abstract Blow-off limits for a premixed turbulent flame are influenced by swirl flow pattern. The objective of the current paper is to present experimental investigations and numerical simulation for identifying and evaluating the influence of changing the swirl intensity by increasing the mixture flow rate on the blow-off limits at different equivalence ratio values (LPG-air mixtures). Both, experimental and numerical observations for swirling turbulent flames at different conditions close to blow-off and after blow-off were discussed. The numerical analysis was carried out using ANSYS 17.0 FLUENT Premixed Model, also called Turbulent Flame Speed Closure (TFSC) model. By varying the flow rate of the mixture, the blow-off velocity in lean and rich mixing sides was determined. It has been realized that when the mixture of unburned gases flow approaches to the combustion zone, a sudden expansion of the flow field occurs; also, the inner and outer flow diverges due to the swirl momentum. Also, the mean axial velocity decreases with the increase of expansion distance, as it depends on the conservation of mass law. The results elucidated that increasing swirl intensity of unburned gas increases the flame separation distance, also the blow-off limit in the lean fuel side needs relatively to fewer values from the velocities compared to the rich fuel side.

1 Introduction

In the past few years, improving premixed combustion operations in gas turbine and aircraft engines have become a significant challenge in order to reduce NOx emissions providing a high mixing quality and low flame temperature. Moreover, the design of modern high performance propulsion systems suffers from major problem of combustion instability [1, 2]. Combustion instability generates an important issue concerning with self sustained combustion oscillations in the combustion chamber, as the pressure fluctuations with unstable rise in temperature are the cause of this closed loop coupling. More so, all these problem can induce unstable operations

H. A. Abdul wahhab (✉)
Department of Mechanical Engineering, University of Technology-Iraq, Baghdad, Iraq
e-mail: 20085@uotechnology.edu.iq

F. Ahmad et al. (eds.), *ICPER 2020*, Lecture Notes in Mechanical Engineering,
https://doi.org/10.1007/978-981-19-1939-8_2

caused to increased heat transfer at combustion chamber walls, finally may lead to total blow off [3, 4]. Also, many studies tested all the ways to determine the low typical frequency oscillations which cause large mechanical vibrations in the combustion system that may ultimately result in blow off [5–7]. The use of swirl combustors in gas turbine combustion systems is one of the best improvements to its ability to stabilize flame within wide ranges of equivalence ratios [8, 9]. Therefore, the design and development of combustion systems in gas turbines are required entirely to encounter numerous basic design elements, like a high efficiency of combustion that covers broad working limits, low greenhouse gas emissions, and low limits of flame stability [10–12]. Shanbhogue et al. [13] demonstrated that lean-premixed turbulent flames are more susceptible to blow-off. Also, they confirmed that blow-off limit of lean turbulent flames uaually starts by generation of local swirls along the combustor region where the local momentum rate exceeds the critical limit. Steinberg et al. [14] analyzed the shape of flame front for different phases in a combustor region; it is an important feature that controls the uniformity of the temperature in the burnt gases at the combustor outlet. Chan et al. [2] studied the weak swirl to investigate in stabilize propagating open premixed turbulent flames, for a range of the swirl numbers between 0.05 and 0.3, their experiments used a jet burner. They noted that the measured tangential component for flame velocity within the turbulent zone (at downstream of the burner edge) was close to zero for the entire radius of the burner, which they concluded that the flame itself was unaffected by the swirl. Rajasegar et al. [7] tested new geometrical adaptations to generic high swirl-center-fan burner to change the swirl number in order to study blow-off limit. They explained that the critical value of swirl number required for the occurrence of blow-off, depended on: Reynolds number for fuel–air flow, equivalence ratio, and swirl design features. Varying the total reactants flows from 400 to 1650 SLPM showed that increased blow-off flame behaviour. This effect was attributed to the influence of the sudden pressure drop between unburned and burned gases across the flame front which increases with total flow. Zhang et al. [15] prepared blow-off experiments of turbulent flame using $CH_4/H_2/CO$ blends on a high-pressure, high-vortex burner designed to match gas turbine conditions. Their results appeared that hydrogen controls the lean blow-off limits, and many other physically meaningful variables (laminar burning velocity, turbulent flame speed, thermal diffusivity) are simply functions of percentage of hydrogen. Generally, many techniques have been used to develop traditional deigns of burner for ensuring gas turbines steady working. These methods are either the development of aerodynamics and then turbulence in the swirling flow field in the chamber of combustion to expand stability, or using the substituted fuels for reducing the combustion impact upon the surroundings [16–18]. On the other hand, there is little information on the impact of swirl patterns for turbulent flames on the limits near blow-off [19]. Therefore, interest in observing the mechanism of the flame blow-off has increased because of the need to operate disturbed flame incinerators under different conditions: for different fuels, the application of available swirling techniques, close to the blow-off limits [20–22]. There is a limited research available upon the swirling spray flames behaviour throughout the blow-off; Dawson et al. [23] studied the cases of spray flames for premixed and non-premixed combustion in the similar swirl burner at circumstances near the blow-off limit, and many experimental investigations concentrated upon the study of swirl flames with the use of different flow channels, throughout the turbulent intensity gradient and its influence upon the distribution of the velocity of premixed flame [24, 25]. Based on previous experiments for researchers [4, 26, 27], Four analysis and diagnostic techniques were used to analyze the combustion of a medium burner array. First, a dynamic mode analysis based on the high combustion speed determined by the quality of a stable and turbulent flame [28, 29]. Second, a lean blow off limits provided additional measure of flame stability [30]. Third, the emission measurements are using a gas chromatography; mass spectrometry; and a gas analyzer to found flame combustion efficiency [31, 32]. Fourth, flame temperature measurements indicated thermal performance of the flame array [6, 28]. The review of literature showed that numerous researchers have investigated the flame structure influences from the local extinctions perspective for different fuels. But, the effects of a swirl flow patterns on blow-off limits haven't been widely studied. Therefore in the present work, the influence varying the swirl flow pattern in a single tube burner on blow-off limits was investigated. For that, swirl flames for the premixed of LPG and air has been experimented, also the present paper introduces the results of the simulation of turbulence intensity influencing on blow-off limits. The simulation was conducted utilizing ANSYS 17.0 FLUENT Premixed Model, also named "The model of turbulent flame speed closure (TFSC).

2 Experimental Implementation

2.1 Combustor Set Up and Diagnostics

The experiments were conducted by apparatus consisting of a single tube burner system and an optical system. Via this single tube burner, the premixed swirl turbulent flame was prepared; fuel gas Liquid Petroleum gas (LPG) (mixture of; Propane, n-Butane, and Isobutene) and the flow rates of air were measured. Recording the blow off limits for swirl flame front images needed for this study was carried out via the optical system. The components of apparatus are demonstrated in Fig. 1.The system of a single swirl burner includes copper tube at 80 mm diameter and at 980 mm length, before 50 mm from the end edge of the burner, four copper strips are setup inside the tube, these strips are fixed by helical location to generate a swirl flow pattern for the premixed gases while identifying the angle between the plate's axes at 90°.This type of combustion process is a premixing combustion, LPG was premixed with air before reaching the edge burner, and to prevent the interference of the ambient air near the burner edge in the combustion process, the combustion region was surrounded with a nitrogen container (the combustion region was covered by transparent polycarbonate container. This container was filled by the nitrogen gas during the experiments; it was used to block the surrounding air to join in combustion operation). Table 1 shows geometrical features of combustor region and operational conditions of the experiments. Two gas flow meters were used to measure the flow rates for LPG and air, with metered individually (±1.1% accuracy) during the experiments, these reactant flow meters at the exit of the mass flow controllers were calibrated using a RITTER Drum type (enable a measuring accuracy of ±0.2% at standard flow rate and approximately ±0.5% over the whole measuring range). Air and LPG were fed to the combustor zone at ambient conditions of 1 atm and 303 K; a wide range of equivalent ratios has been tested, where the flow rates of both gas and air were controlled to obtain the volumetric mixing ratios. Flame temperature measurements were made using PEAK Sensors type S (Platinum 90%, Rhodium10%) high temperature (normally used in oxidising atmosphere, sensitivity is between 6 and 12 μV/°C, accuracy ± 1.0), measurement elements for platinum thermocouples probe were insulated, and the thermocouples were mounted within several spaces of the combustion field with a horizontal positioning accuracy of 0.1 mm. The temperature sensors were distributed to the right and left of the burner centre line with a spacing distance of 10 mm, at the height Y = 30 mm above the burner edge. A GRAPHTEC data logger was used to acquire the flame temperature signals with an accuracy of 0.1 °C ± 0.8%. Thermocouple probes were linked to data logger and saved in computer.

The optical system comprises a laser source (He–Ne) having low power (0.21mWatt) and (713 nm) wavelength, and a set of lenses. A digital camera setup "Phantom VEO 440" with (100 ~ 180 fps) speed was used in the experiments.

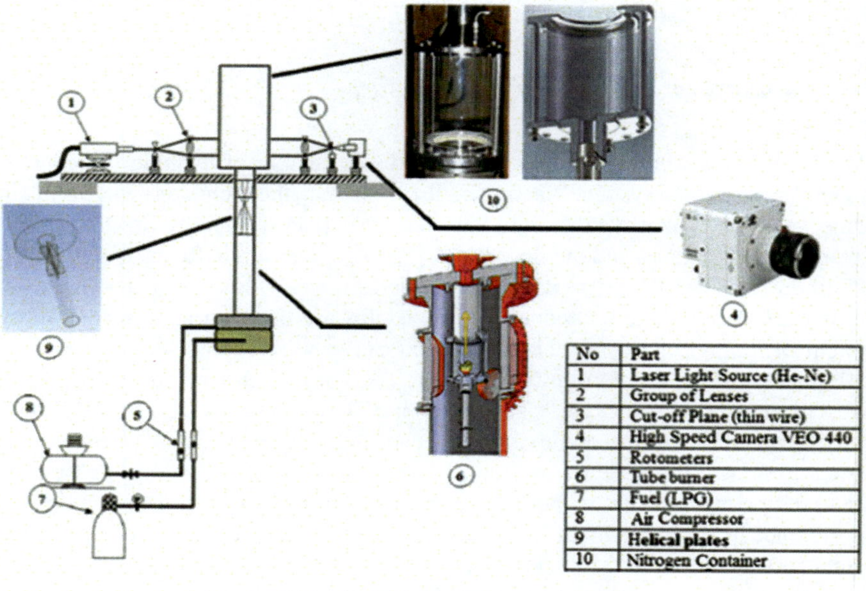

Fig. 1 Experimental setup schematic diagram (including the combustion and optical systems)

Table 1 Geometrical features of combustor region and operational conditions of the experiments

		Symbol	Value
Geometry	Burner tube diameter	D	80 mm
	Burner tube length	L	980 mm
	Number of helical strips	–	4
	Nitrogen container diameter		280 mm
	Nitrogen container height		350 mm
Operational conditions	Fuel flow rate	V_f	110–460 SLPM
	Air flow rate	V_a	390–1860 SLPM
	Equivalence ratio	φ	0.54–1.37
	Temperature of mixture	T_{in}	303 K
Fuel (LPG)	Propane	–	70% w/w
	n-Butane	–	17.5% w/w
	Isobutane	–	12.5% w/w

Images were acquired a resolution of (1280 × 720 dpi). The camera was located on a stationary base, in the test rig front at an angle of (90°) for the laser path. Also, the optical alignment alongside the section of test rig section was inspected and corroborated. Through the swirl flow pattern experiments, the top end of the single tube burner remained open. The images of blow-off limits for the swirl turbulent flame with a high contrast have been obtained with changing the LPG-air flow rates for wide ranges of equivalence ratios ($0.54 < \varphi < 1.37$). The device was utilized for the blend ignition looks like a lighter of cigarette and was positioned close the burner open end. Also, the taken images by camera were utilized for determining the geometric factors of the swirl turbulent flames in different blow-off distances. To determine the blow-off limit for a given swirl flow pattern, the composition of blend was steadily advanced up to the instant if the blow-off distance would not be able to move up from the edge of burner.

2.2 Digital Image Processing

The image processing technique has been widely used in the study of a premixed turbulent flame. The main purpose of the image processing technique is detection of the flame boundaries. Particularly from the edges of the flames which are not visibly observable, thereby making it difficult to identify the edges of the flames by means of a simple detection technique because of the low contrast ratio when those borders. Therefore, it is very necessary to apply digital image processing techniques to the raw images before they can be analyzed. In order for the image to be process effectively, a MATLAB software program was developed and used to extract information from the high-speed video data recorded. The algorithm considered each image as a graphics object where each element of a matrix contains three color intensities, the red, green, and blue (RGB) values. The RGB image is processed to a greyscale image with 8 bit depth using rgb 2-Gy function in MATLAB. Each pixel is represented with a numerical value from 0 (black) to 255 (white), where the light spots are black and flame front contours are white in rgb 2-Gy image. Several measures was undertaken to determine the basic parameters of the swirl flame, these steps include the following: background extraction and noise reduction, binary Image and thresholding operation (the thresholds were around 0.65–0.88, which made possible to reduce light structures connected to the image), and morphological operations on binary Image. Figure 2 shows the samples of images under processing. The software program is divided into three basic stages: flame structure identification; flame parameters verification, and tracking stable and blow-off flames. The MATLAB program code developed was based on an automatic video film to images analysis for measuring flame parameters such as the flame spatial location, growth factor, stability limits, and blow-off velocity. Figure 3 shows a flame structure data and main parameters.

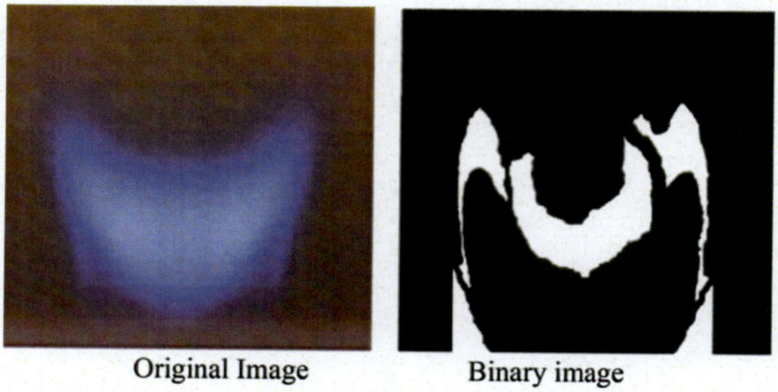

Fig. 2 The samples of images under processing

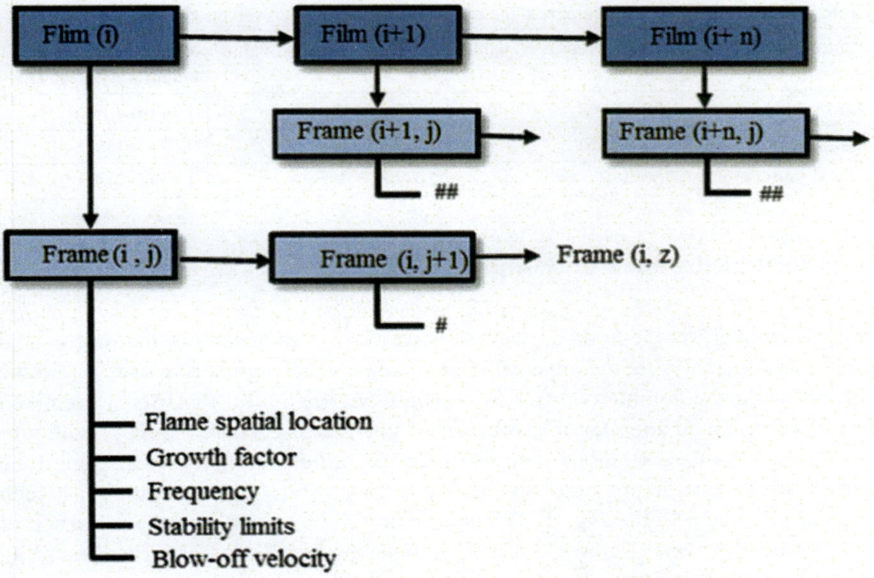

Fig. 3 A flame structure data and main parameters

2.3 Uncertainty Analysis

The uncertainty analysis performed on the experimental data in order to evaluate the uncertainties in the measurements. In every measurement, the total uncertainty can be obtained by combining the accuracy of the measuring and the sensory instrument. Addition of individual errors together, however can result in a worse case uncertainty value. The error components are unlikely to be at their maximum value and same polarity at the same time. A more realistic method is to use the root-sum-of-squares method, that is, by taking the square root of the sum of the individual errors [33].

$$\varepsilon_{\text{total}} = \sqrt{(\varepsilon_{\text{sensor}})^2 + (\varepsilon_{\text{instrument}})^2} \tag{1}$$

$$\text{Min. \& Max.Uncer.}(\%) = 100 \times \left(\frac{\varepsilon_{\text{total}}}{\text{Min. \& Max.Reading}} \right) \tag{2}$$

where, ε_{total} is the total uncertainty. The uncertainty percentage to the maximum and minimum values of recorded data was calculated from: air flow rate, fuel flow rate, flame temperature. Table 2 shows the uncertainty percentage error of the experimental yield values measured.

Table 2 The uncertainty percentage to the max. and min. values of recorded data

Parameter	Min. for recorded value	Max. for recorded value	Min. and Max. uncertainty (%) error
Air flow rate	390 SLPM	1860 SLPM	0.44–0.40%
Fuel flow rate	110 SLPM	460 SLPM	0.19–0.13%
Flame temperature	766 K	2167 K	0.26–0.21%

3 Computational Modeling

Numerical analysis methods are an effective tool for representing the mechanical problems and analysing the physical phenomena for the engineering uses [8]. In the current work, the commercial (CFD) software (ANSYS 17.0 FLUENT) Premixed Model, also named a model of "Turbulent Flame Speed Closure (TFSC)" with pre-processing was used to model and simulate the turbulence flame in a single tube burner. There exist many steps that achieved for performing the simulations; such steps are depicted in the Fig. 4. This simulation is interested into the influence of swirl flow pattern on the blow-off turbulent flame through a vertical burner according to the research objectives.

Fig. 4 Flowchart of simulation steps in ANSYS 17

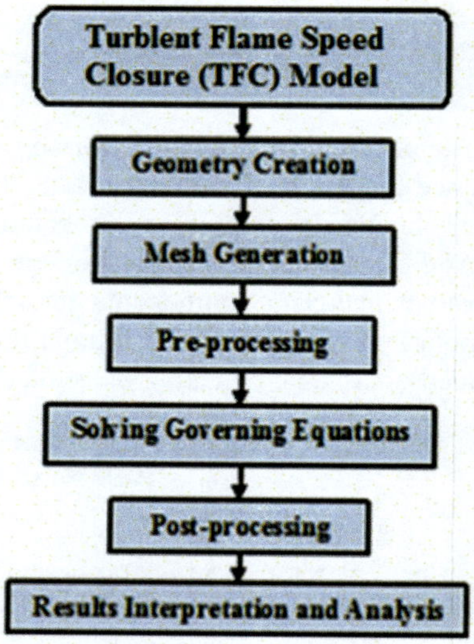

3.1 Premixed Turbulent Flame Model

A three-dimensional model of a premixed for a swirl single tube burner was prepared in ANSYS software 17.0. A CFD model was employed for investigating the blow-off limits for a premixed turbulent flame at changing different swirl flow patterns: dynamical premixed fuel LPG/air flow. The first step in numerical simulation was to make a computational domain. The dimensions of premixed swirl turbulent flame were received from the experimental setup. The geometrical model as well as the mesh of swirl single tube burner is shown in Fig. 5. The swirl combustion for turbulent flame comprises of (2) zones. The flow in the external zone is swirled, meaning that the velocity of flow possesses a tangential and axial component. The tangential component of velocity is induced via four swirl plates; having an outlet angle of (90°) to axial direction. The internal zone velocity merely possesses an axial component. The internal flow crosses throughout a vertical channel (when the four plate's edges did not extend through it), that increases the turbulence. The highly significant parameters of simulations are manifested in Table 3. The CFD model has been established at several assumptions;

- Unsteady, 1D, turbulent flame in the external zone.
- Swirl flow of fuel/gas mixture in internal zone, it is assumed that fuel gas and air are mixed together homogeneously, that the fuel gas to air ratio is of the order of unity, and evaporation phenomena can be neglected.
- There is no reaction between fuel gas and air before combustion process. And,
- It is assumed that both gases (fuel gas and air) are treated as two interacting continua and their densities remain constant.

3.2 Turbulent Flame Speed Closure Model Theory (TFSC)

The normal configuration of a premixed turbulent flame in can be represented in three dimensions by several hypothetical models. These models include all states with unstable propagation and restricted non-constant burner heat. The adiabatic equations for pre-mixed turbulent flame in physical space can be analyzed and transform these states into an active advance space [23–25]. All information, including analytical equations, is included in the user guide of Fluent [20] and guide to the theory of Fluent [19].Such model applies to the "free reaction zone" combustion system, in which the ability to determine the smallest turbulent flow scales of flame thickness as it penetrates into the flame zone is clearly shown. At the same time, it does not include detailed reaction chemistry. In this case, the reaction progression variable in the combustion modelling is f, described as the sum of ingredients types; it is normalized with the sum of the types of ingredients in a chemical equilibrium. This means that $f = 0$ denotes the unburned state and $f = 1$ represents the wholly blend of burnt fuel and air. For modelling an anterior spread of flame, the transport equation is solved for the reaction progression variable. The text should be formatted as follows:

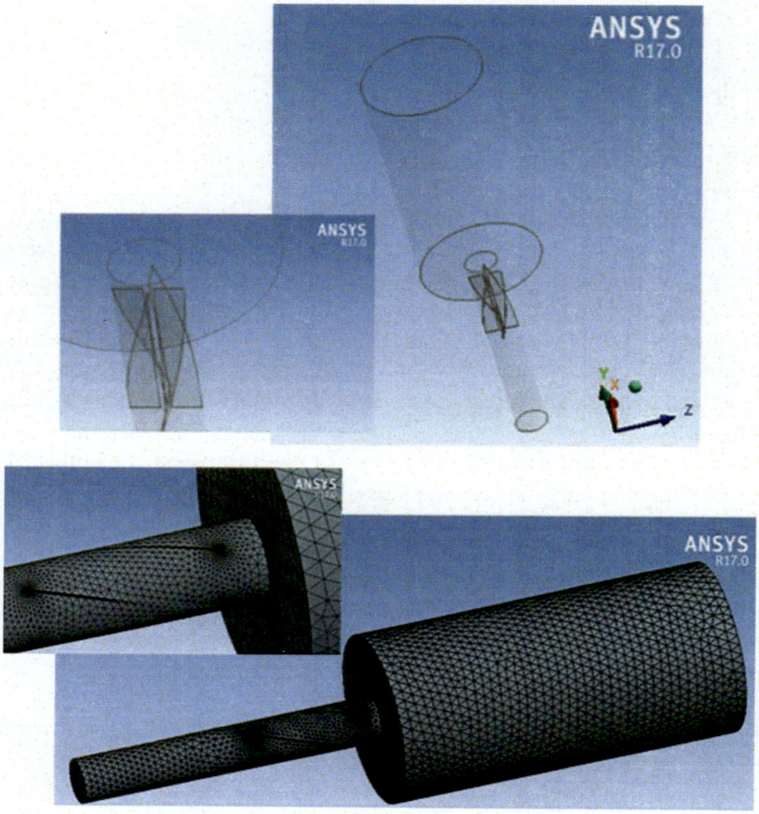

Fig. 5 Computational domain and geometrical mesh model of single tube burner

$$\frac{\partial}{\partial t}(\rho f) + \nabla.(\rho \vec{v} f) = \nabla.\left(\frac{\mu}{Ns_t}\nabla f\right) + \rho\tau \tag{3}$$

This transform formula for f defines the temporal as well as the spatial development of reaction progression within a turbulent field of flow. Where, f is the average reaction progression variable $\vec{v}$ is the vector of the complete velocity, $\mu$is the dynamic viscosity, and Ns_t is the turbulent Schmidt-no.

$$\tau = \rho_u S_t |\nabla f| \tag{4}$$

The mean reaction rate (τ) in the source term is modelled as in Eq. 2, where ρ_u and S_t are the unburned blend density and the speed of the turbulent flame, respectively. Also, the Zimont model for S_t was employed, that is depended upon the supposition of the balance free scale turbulence in a laminar flame. As a consequence, the speed of turbulent flame is merely reliant upon the big scale turbulence parameters. In this

Table 3 Important parameters of simulation model

		Value
Geometric dimensions of the combustion model		
Internal zone diameter		80 mm
External zone diameter		280 mm
External zone height		350 mm
Fuel	LPG	Propane 70% w/w n-Butane 17.5% w/w Isobutane 12.5% w/w
Inlet boundary conditions	Air	Fuel
Density	1.184 kg/m^3	0.884 kg/m^3
Axial velociry	8.5–14.4 m/s	10.4–16.6 m/s
Radial velocity	–	–
Temperature (K)	303 K	303 K
Turbulent kinetic energy (m2/s2)	2.4–3.86	1.61–2.34

case, it can be deduced that the local displacement turbulent flame speed of the flow that enters the flame critical edge (or point of anchoring) was equivalent to the local turbulent displacement flame speed (S_t).

$$S_t = S_l + k\acute{u} \tag{5}$$

where, k is an experimental constant that derived from the experiments, and ($\acute{u}$) being consider ably greater than S_l.

4 Results and Discussion

4.1 Flame Dynamics

1000 images of subsequent data were used to study of flame dynamic analyses, which is more than enough to a convergence study is to be conducted on this particular data set, the data are read in first with the first 200 binary frames, then the first 400, until 1000 is reached. All these data sets are treated independently and the statistics are compared to determine whether the growth factor statistics converge at 1000 image sets or not. The growth factor was adopted to determine the extent of separation of the swirl flame from the burner edge. The lean blow off limit is assessed by reducing the LPG flow rate while keeping the air flow rate constant. The stable swirl flame and blow-off swirl flame of the single swirl burner under premixed operating conditions are shown in Fig. 6. Initial flame stability analyses were performed under lean mixing conditions at $\varphi = 0.58$ and 0.98, as denoted by Fig. 6a, b, it can observe the modes (F1, F2) with the most stable growth factor (closest to 0) and the highest energy content, is located at the mean frequency. Also, the simultaneous spatial structure of states (a, b) in the figure, represents the time averaged flame position. The M-shaped flame exhibits with a very stability. The single swirl burner at $\varphi = 1.23$ exhibits a blow-off swirl flame (F3) started in the recirculation zone with discernible oscillations in the outer shear layer, as governed by state (c) in Fig. 6. Generally, as equivalence ratios decreases, flames begin to lift off one by one until all the flames were completely lifted from the burner edge. More so, with decreasing flame speed, flame blow off propagates across the single swirl burner. Conversely, as flow conditions were gradually reverted to a more stable operating condition, the flame self propagated across the entire swirl due to the counter rotating vortex pattern. A single solitary flame always relit all the burner elements through cross burner flow interaction, demonstrating that the flames are mutually supported by one another.

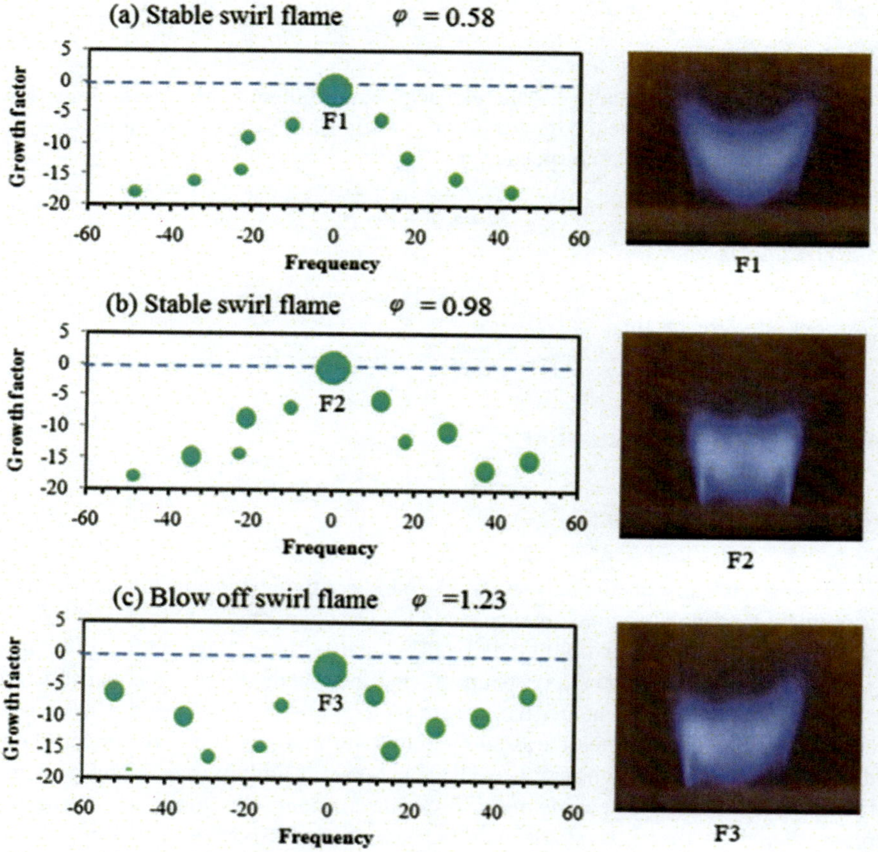

Fig. 6 Stable single swirl flame; **a** $\varphi = 0.58$; **b** $\varphi = 0.98$; and **c** blow-off swirl flame $\varphi = 1.23$

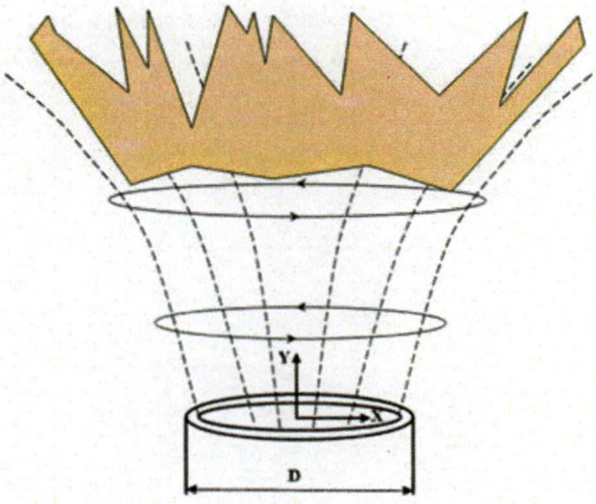

Fig. 7 Swirl flow pattern of turbulent flame

Figure 7 appears the schematic flow field made via swirl pattern, when the mixture of unburned gases flow approaches to the combustion zone, a sudden expansion of the flow field occurs; also the inner and outer flow diverges due to the swirl momentum. Also, the mean axial velocity decreases with the increase of expansion distance, as it depends on the conservation of mass law. In the combustion reacting case, the flame front is uniform, where the value of axial velocity is equivalent, but the opposed direction of the velocity of turbulent flame. Furthermore, if the flow of blend through the edge of burner approaches to the combustion zone, a shear layer is forming between inlet flow and existed gases in the zone of combustion.

Experiments and numerical analysis have shown that increasing the unburned gases velocity increases the swirl intensity, is due to increase the tangential velocity component in relation to the axial velocity component. Finally, the flame separation distance up to the burner edge increase with increasing unburned gases velocities, as shown in Fig. 8. The comparison between CFD simulation results, and experimental measurements of change the unburned gases velocity with the flame separation distance from the burner edge shows a good agreement at equivalence ratio equal $\varphi = 1.02$. The mean percentage of difference between, the experimental measurement and CFD simulation is 4.8%.

The effect of increasing the swirl intensity was tested. Figure 9 elucidates an obvious discrepancy of the front location of turbulent flame for various unburned gases flow rates when the mixture at equivalence ratio (φ) is equal to 1.02. From the images (a–e), it can be inferred that as the velocity (Uf) of unburned gases raises at (19.4–29.8 m/s), the flame separation distance from the burner exit increases at (0–156 mm). Also, a clear idea of the effect of unburned gases velocity on the blow-off limits is through increasing the turbulence intensity, the contours of burned gas volume fraction from (A–E) appears in Fig. 9 for the case of Uf = 19.5 to 29.8 m/s. It is clear that the mixture flow rate increases the turbulence of swirl flow turbulence at the nozzle edge of burner and thus reduces the stability of flame at same equivalence ratio.

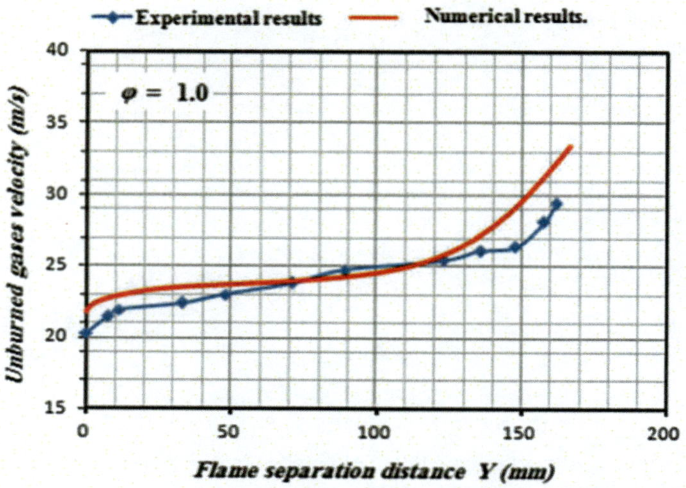

Fig. 8 Profile of flame separation distance with unburned gases velocity at equivalence ratio = 1.0

Figure 10 presents the comparison between CFD simulation results and experimental measurements of the blow-off limit for the different percentages of fuel–air mixtures on lean and rich combustion sides. Generally, if the fuel-gas flowrate rises, the blow-off velocity rises, especially for the flames of rich fuel. Also, the positive correlation observed between blow-off velocity and swirl intensity is vigorous. Somewhat similar trend is illustrated in the equivalence ratio blow-off curves, where the blow-off velocity seems to present a little difference with the equivalence ratios 0.54–1.37. The mean percentage of difference between, the numerical results and the experimental measurement is 11.9% for the lean combustion side, while the percentage mean difference for same variable is 14.8% in the rich combustion side.

4.2 Flame Temperature

Temperature measurements were made to determine the thermal performance of the single swirl burner. Figure 11 presents maximum temperatures at the flame surface under various equivalence values; lean and rich. Maximum temperatures of the single swirl burner agree very well with both of numerical values of flame temperature and adiabatic flame temperature. At rich equivalence ratio, however, the experimental flame temperature deviates from the adiabatic flame temperature possibly due to the greatly accelerated heat transfer when accounting for higher heat release. Stronger flame in single swirl burner may likely raise the overall thermal conversion efficiency to exceed Ideal values since interaction across identical flames can increase total heat

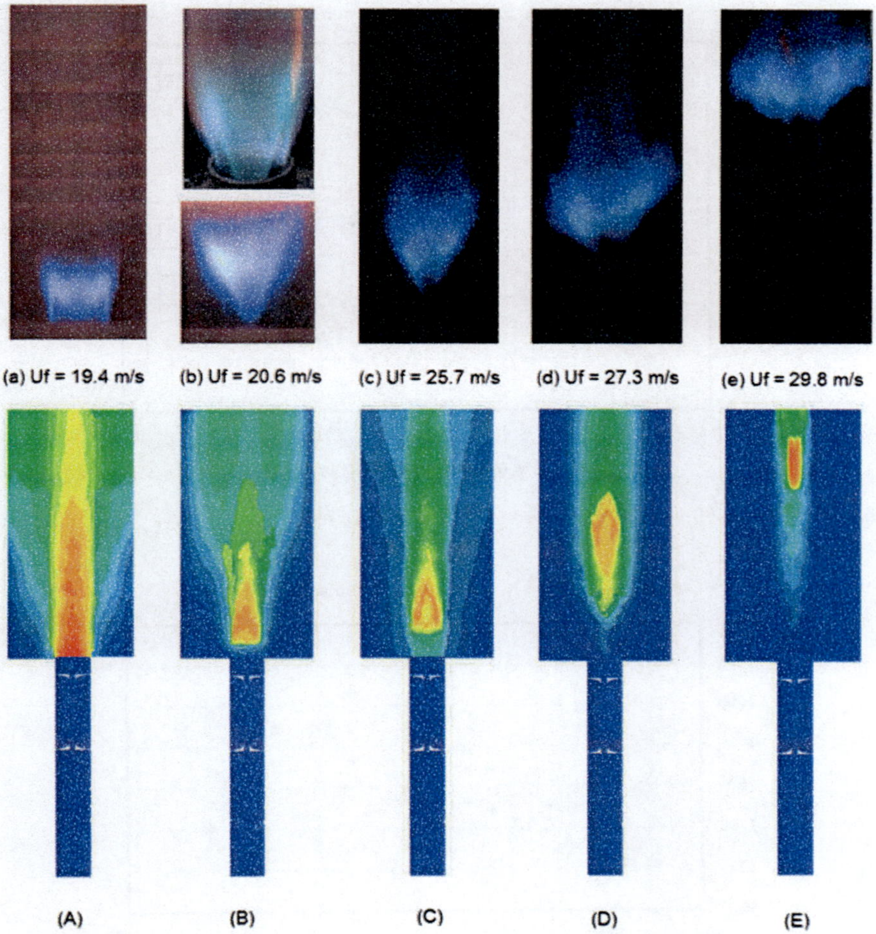

Fig. 9 Images and contours of burned gas volume fraction for turbulent flames at different flow rates

release rate by 30% over an isolated flame [6]. Figure 12 shows spatial temperature profiles along the center line of combustor zone, these values of spatially temperature profiles can be very useful in evaluating regular combustion.

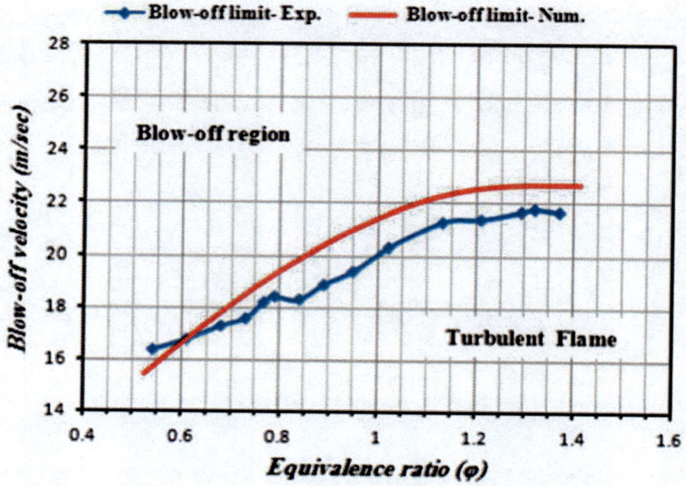

Fig. 10 The blow-off velocity profile versus the equivalence ratio

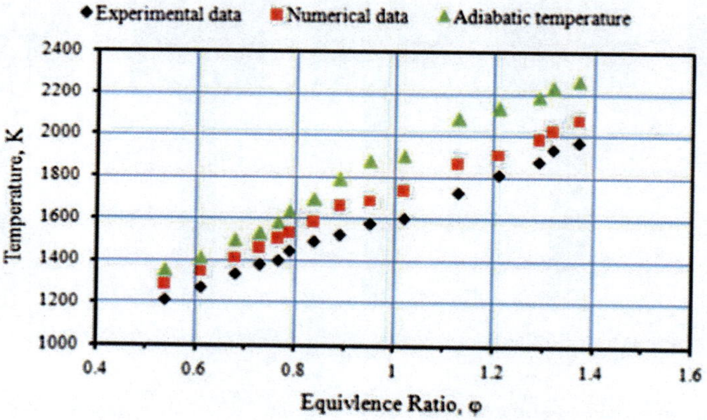

Fig. 11 Maximum flame temperatures of the single swirl burner at various equivalence ratio

5 Conclusions

The influence of variation in the swirl pattern flow intensity for a turbulent flame upon the blow-off limits in a single tube burner was studied throughout experimental tests and computational simulation. Experiments were conducted by a single tube burner system. The turbulent flame speed closure (TFC) model by using ANSYS 17.0 FLUENT Premixed Model was used for simulating the turbulent flame of LPG and air mixture, under the influence of the swirl pattern flow. The outcomes were

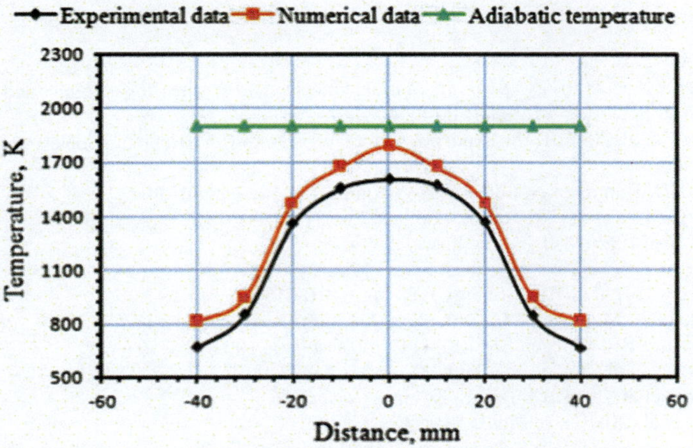

Fig. 12 Temperature profile along the burner center line: LPG/air mixture, $\varphi = 1.02$, Y = 30 mm above the burner surface

compared with the numerical and experimental data from other works in literature [7, 10, 12, 22, 34, 35], and a good agreement was obtained. The subsequent conclusions can be drawn from the analysis of the results:

- For the flow field produced by the swirl pattern, when the flow of the unburned gases enters the zone of combustion, there's a rise in the diameter, named an abrupt expansion.
- The results appeared that increasing the swirl intensity of the unburned gas increases the flame separation distance.
- The blow off velocity values increased with increasing the LPG content in the mixture.

It's strongly recommended to envisage the turbulent flames within a single swirl burner with the swirl pattern flow. Further studies could continue on elucidating the exact physical basis for the observed the correction of blow-off turbulent flame. All of the tests presented in this work were for atmospheric flames. If gas turbine applications are to be targeted, then future work must consider elevated pressures and temperatures.

Acknowledgements The author is obliged to the University of Technology, Baghdad in Iraq, for providing the Centre of Automotive and Energy Generation.

References

1. Rajasegar R et al (2017) ASME international mechanical engineering congress and exposition. American Society of Mechanical Engineers
2. Farisco F et al (2019) Turbo expo: power for land, sea, and air. American Society of Mechanical Engineers
3. Carter CD, Hammack S, Lee T (2016) Combust Flame 168:66–74
4. Mitsingas CM et al (2019) Proc Combust Inst 37:1479–1487
5. Prieler R et al (2018) J Energy Inst 91:369–388
6. Rajasegar R et al (2019) Combust Flame 199:324–337
7. Rajasegar R et al (2018) J Energy Eng 144:04018013
8. Ateshkadi A, McDonell VG, Samuelsen G (2000) Proc Combust Inst 28:1281–1288
9. Chowdhury BR, Cetegen BM (2018) Combust Flame 190:302–316
10. Marinov S et al (2012) Flow Turbul Combust 89:73–95
11. Beerer D et al (2014) J Eng Gas Turbines Power 136
12. Boxx I et al (2013) Exp Fluids 54:1532
13. Shanbhogue SJ, Husain S, Lieuwen T (2009) Prog Energy Combust Sci 35:98–120
14. Steinberg AM, Arndt CM, Meier W Proc Combust Inst 34:3117–3125
15. Zhang Q (2008) Georgia institute of technology
16. Masri A, Gounder J (2010) Combust Sci Technol 182:702–715
17. Masri AR, Kalt PA, Barlow RS (2004) Combust Flame 137:1–37
18. Meares S et al (2015) Proc Combust Inst 35:3813–3820
19. Steinberg A et al (2011) Proc Combust Inst 33:1663–1672
20. Meier W et al (2010) Exp Fluids 49:865–882
21. Yuan R et al (2015) Proc Combust Inst 35:1649–1656
22. Yuan R (2015) University of Cambridge
23. Dawson J et al (2011) 33:1559–1566
24. Sadanandan R, Stöhr M, Meier W (2009) Combust Explos Shock Waves 45:518–529
25. Zhang J, Liu C, Yang T (2017) Sci Sin Phys Mech Astron 47:070007
26. Fan A, Zhang H, Wan J (2017) Energy 123:252–259
27. Gong F, Huang Y, Huang X (2015) Int J Heat Mass Transf 82:406–418
28. Michaels D, Shanbhogue S, Ghoniem A (2017) Combust Flame 176:151–161
29. Chowdhury BR, Cetegen BM (2017) Combust Flame 178:311–328
30. Yan Y et al (2018) Int J Hydrogen Energy 43:7581–7590
31. Tong Y et al (2018) Appl Therm Eng 131:507–521
32. Tran S, Cummings R, Sahni O (2017) Comput Fluids 158:221–235
33. Bevington PR, Robinson DK (2003) Data reduction and error analysis. McGraw-Hill, New York
34. Stöhr M et al (2011) Proc Combust Inst 33(2):2953–2960
35. Ali ABS et al (2016) Int J Energy Environ 7:397

Influence of Life Green Wall and Roof Shedding on the Internal Thermal Condition of Buildings: Case Study in Malaysia

Hussain H. Al-Kayiem, Marwan Effendy, Tri W. B. Riyadi, Jundika C. Kurnia, and Akelesh Morganadus

Abstract Accumulation of heat inside the buildings is mainly caused by solar radiation, mostly from the roof and the outdoor hot weather. Life Green Wall (LGW) is a greenery system that acts as a solution to beat inside building heat accumulation in urban and rural islands. On the other hand, the roof's shedding thrashes the solar radiation impact on the interior thermal conditions. This study aims to design, implement, and evaluate the effectiveness of the LGW and roof shedding on the thermal conditions inside a low-cost building through experimental investigation. Also, the energy consumption for air-conditioning was measured and evaluated. The case study was handled in the postgraduate office in the solar site at Universiti Teknologi PETRONAS - Malaysia. The thermal conditions inside the office, before and after installing a set of PV panels on the roof and then with LGW were monitored, analyzed to compare to evaluate the influence of the rooftop shedding and LGW on the interior thermal conditions. The implementation of PV panels has reduced the power consumption by 9.7%, which is about 3.42 kWh/day. LGW, on the sides of the building and the shedding, together, had brought down the energy consumption by almost 18.3%. Results show that rooftop shedding and LGW on the sides led to significant improvement in the interior thermal conditions and considerable energy saving.

1 Introduction

Life Green Wall (LGW), which is also known as a vertical garden, is a wall that is covered by greenery. It is defined as a greening vertical layer such as walls, facades, blind walls and partition walls, and the main intention is to grow the plant on the

H. H. Al-Kayiem (✉) · J. C. Kurnia · A. Morganadus
Mechanical Engineering Department, Universiti Teknologi PETRONAS, 32610 Seri Iskandar, Perak, Malaysia
e-mail: hussain_kayiem@utp.edu.my

M. Effendy · T. W. B. Riyadi
Mechanical Engineering Department, Universitas Muhammadiyah Surakarta, 57102 Surakarta, Indonesia

F. Ahmad et al. (eds.), *ICPER 2020*, Lecture Notes in Mechanical Engineering,
https://doi.org/10.1007/978-981-19-1939-8_3

wall of buildings [1]. LGW has been adapted to beat urban heat island issues where the temperature in the urban areas is higher due to increased construction and human activities. The idea of LGW is originated way back to 500 BC, which started from the Hanging Garden of Babylon. The Roman and Greek empire also adopted this idea. This concept has evolved more from the seventeenth and eighteenth centuries in the UK and Central Europe. This idea had bigger adaptability among the Americans and Europeans in the nineteenth century, where mostly word climbers were used as the vegetation mode [1]. The large areas covered with huge building structures and the lack of greeneries have altered climatic characteristics [2–6].

LGW is vegetation that is planted in a vertical direction on a building's wall. The concept of LGW is related to all surfaces such as walls, facades, blind walls, etc. [7]. LGWs are also known in various other names. Those include green walls, vertical gardens, green vertical systems, bio-shades, etc. [8]. Some studies described LGW as direct greening systems and indirect greening systems. Although various names are used, the concept is similar, providing great benefits in fighting the urban heat island issue. The classification of LGWs is outlined in Table 1.

Table 1 Classification of life green wall [1]

Direct Green Facade	Indirect Green Facade	Living Wall
 **Plants are rooted in the soil and it climbs on facade and covers an elevation.**	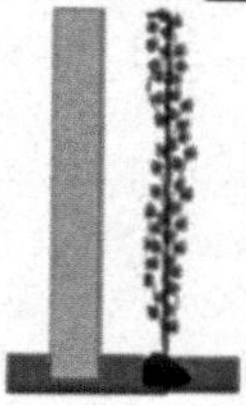 **Plant does not support directly on the facade rather there is an additional support for the plant to grow**	 **Plants are pre-vegetated and directly attached to walls with structural frame or wall.**

Different classification of the green wall has different effects in reduction of temperature [8]. Several studies have shown that the living wall is the most efficient system in reducing surface temperature and ambient temperature. In a study carried out in the Netherlands, living walls reduced temperature up to 5 °C while direct and indirect green facades reduced the temperature by 1.21 °C and 2.71 °C, respectively [9]. These results proved that the efficiency of the LGW is best as living walls continued by indirect green facade, and last is the direct green facade.

The previous investigation reported large Benefits of the installation of LGW. LGW does not only reduce temperature; instead, it possesses many economic and environmental benefits, too [8]. This solution is one key solution to beat the urban heat island effects [1]. This solution is way sustainable and effective as well as eco-friendly. Table 2 summarizes the benefits of LGW, which were investigated in previous studies.

Heat gain in a room happens because various sources include conduction through exterior walls, roofs, and glass. Even solar radiation through glass, lighting and even people are reasons for heat gain to occur in a room. These heat gains are contributors to an increase in indoor temperature. Accordingly, more energy is required to achieve a comfortable indoor environment which is also known as thermal comfort. This increased high cooling load inside the office leads to higher power consumption. As factors to these problems are from greenery reduction, the solutions could be invented by applying a contrasting act to maximize greenery as much as possible in an urbanized environment [21].

Table 2 Benefits of Life Green Wall based on previous studies

Benefits	The author that supports the claim
It is resulting in temperature reduction, which reduces cooling energy demands and energy consumptions	Safikhani et al. [8], Eumorfopoulou [10], Wong et al. [11], Alexandari et al. [4], Perini et al. [9], Jaafar [12], Perez et al. [13, 14], Onish [15]
It reduces sound penetration and noise disturbance in the insides of buildings	Wong et al. [16], Li et al. [17] and Wong et al. [18]
Provides aesthetic benefits	White et al. [19]
Environmental benefits	Safikhani et al. [8] and Donahue [20]

The new postgraduate (PG) office building at the solar research site in Universiti Teknologi PETRONAS (UTP) was constructed using low-cost materials. This office required a high cooling load to achieve its thermal comfort due to solar radiation from the exterior walls, as well as the roof of the office. This leads to higher energy consumption to achieve a comfortable indoor room temperature. This project focuses on improving the thermal condition of the building by using the LGW concept, which leads to lower power consumption. This paper is aimed to evaluate the thermal condition of the low-cost building with the presence of LGW. Besides, the contribution of LGW in reducing the cooling load of the low-cost building is evaluated. Four different cases have been considered to investigate the influence of the rooftop covering and the installation of LGWs on the sides of the experimental prototype.

2 Experimental Methodology

This study is conducted via experimental investigation within 28 -weeks period of time. The LGW constructed for this building is an indirect green façade.

2.1 Cases of Study

Four different cases have been investigated. The experimental conditions of each case are shown in Table 3. Each case is repeated for five days, and the mean is considered.

Table 3 Experiment cases

Case	Condition	Remarks
Case 1	Base case	To study the thermal condition of the room without any ventilation
Case 2	Air-conditioned ventilation	To study the thermal condition of the room when the air conditioner is turned on to 25 °C
Case 3	With covered roof by solar photovoltaic panels	To study and compare the thermal condition after installing PV panels provides shading effects for the roof
Case 4	Life Green Wall and rooftop shading	To study and compare the thermal conditions of the room after the installation of LGW

Case 1 is considered as the benchmark reference to compare the other changes in the other cases. For each case, the experiment is conducted from 8:00 a.m. till 6:00 p.m. for five consecutive days. The air-conditioned units in cases 2, 3 and 4 are set to 25 °C. This setting is selected as it is the recommended standard for a comfortable working atmosphere.

2.2 *Experimental Setup*

The experimental setup is an office used as a working and officing area for the postgraduate students working in the solar site. It consists of 8 working stations, two air conditioners and an internet server. The size of the room is around 5-m-wide and 7.25-m-long. The LGW is installed over three walls of the office, which are subjected to direct solar radiation, named wall A, wall B and wall C. outlines of the office and the LGW is shown in Fig. 1.

The office is equipped with 11 thermocouples, three solarimeters and two units to measure the power consumptions of the two air conditioners. To acquire the relevant measurement data which permits comprehensive analysis of each case. Some thermocouples are attached to the external surface of the walls, and some are installed inside the PG office (Fig. 2).

Locations and details of the measuring instruments are described in Table 4.

2.3 *Evaluation Parameter*

After the experiment is conducted, the collected data will be used to evaluate three parameters: heat gain, energy consumption and heat ratio.

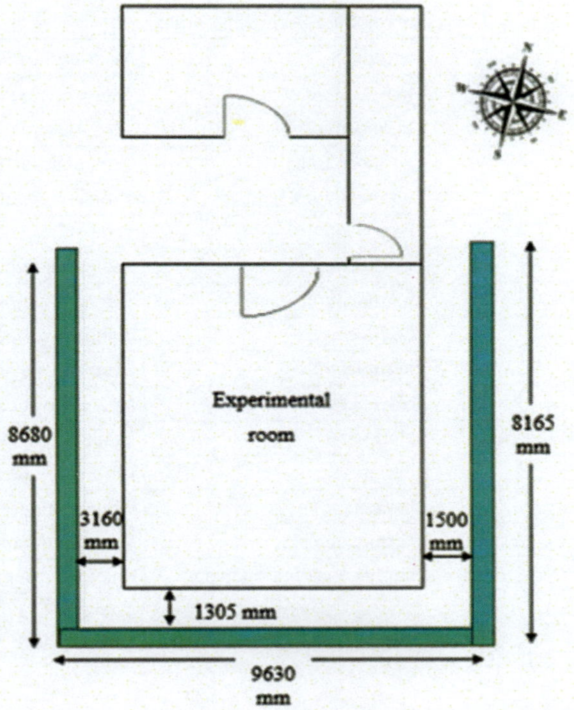

Fig. 1 Experimental office layout

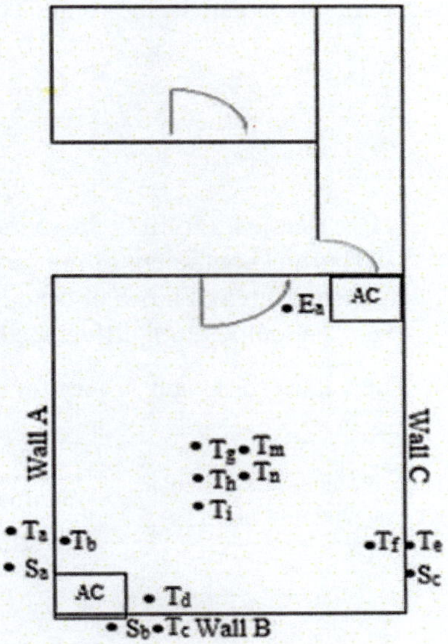

Fig. 2 Top view of the experimental setup showing the locations of the measuring point

Table 4 Indicators for the experimental setup

Indicators	Instrument	Measured value
T_a	Thermocouple	Temperature of exterior wall A
T_b	Thermocouple	Temperature of interior wall A
T_c	Thermocouple	Temperature of exterior wall B
T_d	Thermocouple	Temperature of interior wall B
T_e	Thermocouple	Temperature of exterior wall C
T_f	Thermocouple	Temperature of interior wall C
T_g	Thermocouple	Temperature of room at 100 cm
T_h	Thermocouple	Temperature of room at 150 cm
T_i	Thermocouple	Temperature of room at 200 cm
S_a	Solarimeter	Irradiance at exterior wall A
S_b	Solarimeter	Irradiance at exterior wall B
S_c	Solarimeter	Irradiance at exterior wall C
T_m	Thermocouple	Temperature of roof
T_n	Thermocouple	Temperature of roof
E_a	Energy consumption meter	Energy consumption of air conditioning unit

2.3.1 Heat Gain

The heat is gain is evaluated for every wall and roof, and the sum of these parameters is the total heat gain of the room. The heat gain is calculated using Eq. (1).

$$Q_{gain} = U \times A \times \Delta T \tag{1}$$

where;

Q_{gain}: Heat gain from the particular surface (Watt).
U: Thermal coefficient of the surface (W/m^{2+}K).
A: Surface area of the wall or top roof (m^2).
ΔT: The temperature difference between the internal and external surface.

The values for U and A vary for each wall and roof are listed in Table 5.

Table 5 Values of particulars for heat gain formula

Wall/Roof	Thermal coefficient, U (W/m^{2+}K)	Area of surface, A (m^2)
Wall A	4.853	22.92
Wall B	4.853	13.80
Wall C	4.853	22.92
Roof	0.290	35.96

2.3.2 Energy Consumption Per Day

The energy consumption every hour from 8:00 am. till 6:00 pm. over five days for each case was recorded. The values are then cumulated, and the average consumption of energy by the air-conditioned unit per day is calculated.

3 Results and Discussion

The improvement in the thermal conditions is predicted relative to the bare case 1. The improvements are evaluated in terms of heat gain by the office's interior, the air temperature inside the office and the saving in energy consumed for the air conditioning.

3.1 Heat Gain Analysis

The transient state of heat gain analysis in the four cases is shown in Fig. 3. From the total heat gain of the room, the value of heat gain falls between -2 till 0.5 kW. The negative value is due to the heat gain trapped inside the building causing heat flow in the opposite direction.

For the other case, the heat gain in case 4 has a slight reduction compared to case 2 (with cooling) and case 3 (with covered roof). Case 3 shows a slightly high heat gain due to the higher ambient temperature during the experiment period discussed in section B.

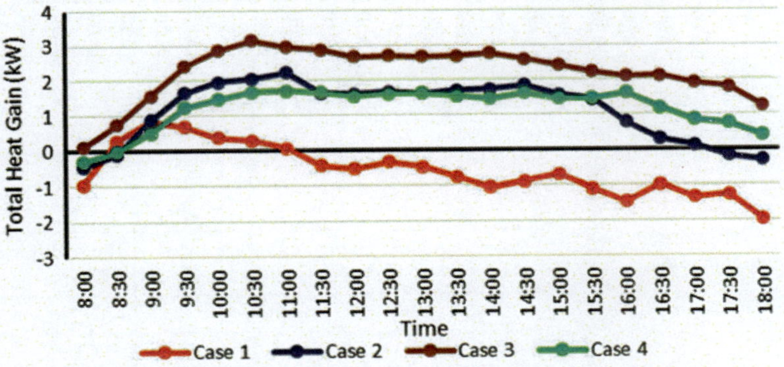

Fig. 3 Transient behavior of the predicted heat gain of the experimental

3.2 Temperature Analysis

Temperature analysis includes the average room temperature (Fig. 4) and the difference between ambient and indoor temperature (Table 6). Based on Fig. 4, the average room temperature is remarkably highest during case 1 since heat is trapped in the room due to the absence of cooling ventilation. The temperature is extremely high, ranging from 34 °C and increasing to 43 °C in the afternoon. The temperature is significantly higher than the ambient temperature, where the maximum ambient temperature during case 1 was only about 32 °C. This is the reason for the negative heat gain value in section A discussed earlier.

In case 2, case 3 and case 4, the indoor temperature is successfully reduced for each case, showing a clear advantage of the shading effects, especially by the LGW where case 4 experiment marks the lowest room temperature, which is about 25 °C.

From Table 6, despite having the highest average ambient temperature, the room temperature is reduced by 4.69 °C from the ambient temperature with the presence of LGW. The value indicates that LGW can improve the thermal condition by 98% compared to the bare wall condition.

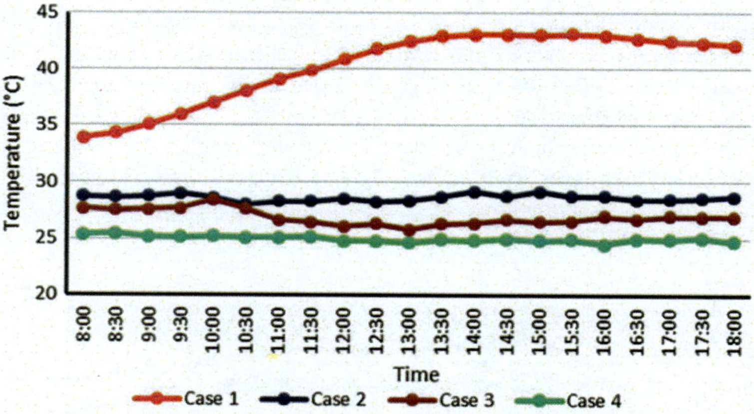

Fig. 4 Transient behavior of the room temperature. (mean of 5 days measurements for each case)

Table 6 Constant values for heat gain prediction

Case	Ambient temp. (°C)	Average room temp. (°C)	Temp. difference (°C)
1	30.00	40.26	−10.26
2	28.82	28.74	0.09
3	29.56	24.91	4.64
4	31.63	26.94	4.69

Table 7 Analysis of mean energy consumption per day

Case	Total energy consumption (kWh/day)	Reduction of energy consumption (kWh/day)	Percentage reduction (%)
2	18.66	Reference point	Reference point
3	16.86	1.8	9.65
4	15.24	3.42	18.33

In case 3, after the shading effects on the roof, the value also has been decreased by 4.64 °C. The value in case 4 will be higher if the ambient temperature value is lower during case 4.

Comparing case 3 and case 4 with just the cooling ventilation in case 2, LGW is proven to improve the room temperature by 4.69 °C. This case is similar to Perini's study, where the indirect green façade has been proven to reduce by 2.71 °C. In a nutshell, LGW has been proven to improve the thermal condition inside the office.

3.3 Energy Consumption Analysis

During cases 2, 3 and 4, two air conditioner units were used to remove the office's heat gain. The energy consumption readings were recorded, and the mean energy consumption per day is evaluated, as shown in Table 7.

Based on the total energy consumption per day, it is noticed that originally the room consumed up to 18.66 kWh per day to remove the heat gain in the room. The implementation of PV panels on the rooftop has reduced the power consumption by 9.65%, which is about 3.42 kWh/day. LGW has proved to be cost-saving since it could reduce energy consumption by almost 18.33% per day. Similar to this, a study conducted in Dubai has proven that the implementation of LGW has reduced the air conditioning energy demand by up to 20% [22]. These values are expected to reduce further once the plants grow and provide 100% shading towards the office.

4 Conclusion

To be adapted to the low-cost building, an indirect green façade measure is used, which is slightly more effective than a green façade and also, the walls were far lower in cost compared to having a direct green façade. The implementation of LGW shows a reduction of temperature up to 4.7 °C, which is about 98% improvisation compared to the bare wall condition.

Since the thermal conditions have been improved, LGW has reduced the cooling load by the air conditioning units, where a decrease in the amount of energy consumption by up to 18.3% is achieved. It is about 3.42 kWh per day. This could save up to 54 kWh per month, which is a considerable amount of energy, which leads to cost-saving.

Life Green Wall is an effective measure to withstand the solar radiation problem in the postgraduate office at the new solar site. This study has proven that LGW does not necessarily need to be built expensively, yet a simple, low-cost solution could also serve a similar purpose.

It is recommended to carry out numerical simulation of the studied cases to develop a validated procedure applied to larger buildings and more complicated architecture.

Acknowledgements The authors wish to acknowledge both Universiti Teknologi PETRONAS (UTP)—Malaysia and Universitas Muhammadiyah Surakarta (UMS)—Indonesia for supporting the research under International Matching Research Grant. Particular acknowledgment is due to UTP for the financial support to publish the article under research grant RIPHEN, CS: 015MD0-063.

References

1. Besir AB, Cuce E (2018) Green roofs and facades: a comprehensive review. Renew Sustain Energy Rev 82:915–939
2. Pugh TAM, MacKenzie AR, Whyatt JD, Hewitt CN (2012) Effectiveness of green infrastructure for improvement of air quality in urban street canyons. Environ Sci Technol 46(14):7692–7699
3. Arnfield AJ (2003) Two decades of urban climate research: a review of turbulence, exchanges of energy and water, and the urban heat island. Int J Climatol 23(1):1–26
4. Eleftheria AJP (2008) Temperature decreases in an urban canyon due to green walls and green roofs in diverse climates. Build Environ 43(4):480–493
5. Tzoulas K, Korpela K, Vennc S, Yli-Pelkonen V, Ka´zmierczak A, Niemela J, James P (2007) Promoting ecosystem and human health in urban areas using green infrastructure: a literature review. Landsc Urban Plan 81(3):167–178
6. Sheweka S, Magdy AN (2011) The living walls as an approach for a healthy urban environment. Energy Procedia 6:592–599
7. Manso M, Castro-Gomes J (2015) Green wall systems: a review of their characteristics. Renew Sustain Energy Rev 41:863–871
8. Safikhani T, Abdullah AM, Remaz D, Baharvand MO (2014) A review of energy characteristic of vertical greenery systems. Renew Sustain Energy Rev 40:450–462
9. Katia Perini MO, Haas EM, Raiteri R (2011) Greening the building envelope, façade greening and living wall systems 1(1):1–8
10. Eumorfopoulou E, Kontoleon KN (2009) Experimental approach to the contribution of plant-covered walls to the thermal behaviour of building envelopes. Build Environ 44:1024–1038
11. Wong NH, Alex Tan AYK, Tan PY, Wong NC (2009) Energy simulation of vertical greenery systems. Energy Build 41(12):1401–1408
12. Jaafara BIS, Rebab MN, Rasidia MH (2013) Impact of vertical greenery system on internal building corridors in the tropic. Procedia Soc Behav Sci 105:558–568
13. Pérez G, Rincóna L, Vila A, González JM, Cabeza LF (2011) Green vertical systems for buildings as passive systems for energy savings. Appl Energy 88:4854–4859
14. Pérez G, Martorell JC, Cabeza LF (2014) Vertical Greenery Systems (VGS) for energy saving in buildings: a review. Renew Sustain Energy Rev 39:139–165

15. Onishi A, Cao X, Ito T, Shi F, Imur H (2010) Evaluating the potential for urban heat-island mitigation by greening parking lots. Urban For Urban Green 9(4):323–332
16. Wong NH, Tan AK, Tan PY, Chiang K, Wong NC (2010) Acoustics evaluation of vertical greenery systems for building walls. Build Environ 45:411–420
17. Li HN, Chau CK, Tang SK (2010) Can surrounding greenery reduce noise annoyance at home? Sci Total Environ 408(20):4376–4384
18. Wong NH, Tan YK, Chen Y, Sekar K, Tan PY, Chan D, Chiang K, Wong NC (2010) Thermal evaluation of vertical greenery systems for building walls. Build Environ 45:663–672
19. White EV, Gatersleben B (2011) Greenery on residential buildings: does it affect preferences and perceptions of beauty? J Environ Psychol 31:89–98
20. Donahue JD (2011) An empirical analysis of the relationships between tree cover, air quality and crime in urban areas. 2011, thesis, Georgetown University, Washington DC
21. Perini K, Rosasco P (2013) Cost–benefit analysis for green façades and living wall systems. Build Environ 70:110–121
22. Haggag M, Hassan A, Elmasry S (2014) Experimental study on reduced heat gain through green façades in a high heat load climate. Energy Build 82:668–674

Production of Volatile and Char During Pyrolysis and Combustion in Domestic Wastewater Sludge by Using Thermogravimetric Analysis

M. S. Zakaria, Suhaimi Hassan, and Mohd Faizairi Mohd Nor

Abstract Rapid increase in world population and economic growth, improvements in living standards after the industrialization of the world brought about high demand of energy. In the last two decades, extensive work has been done for alternative and sustainable sources and technologies to meet up current and future energy demand such as solar energy, wind energy, hydro and biofuels energy. One of the viable biomass sources available in Malaysia is Domestic wastewater sludge. Further study on this biomass is required to promote as the biomass fuel. This study focused on the production of volatile and char yield during pyrolysis and combustion with particle size range between 0.4 mm and 2.0 mm and heating rate (5, 10, 15, 20, 30, 50 and 100 K/min) by using thermogravimetric analysis. Based on the finding, the total volatile shows a tendency to decrease when the heating rate increase with the highest value was 84.51%. However, the highest char was produced at the highest heating rate was 23.80%. The lowest heating rate is favourable since produce high volatile matter content and low char content.

1 Introduction

Energy demands in Malaysia are increasing sharply due to the growth of population and economy. Malaysia's population in 2007 hit 27.4 million and expected to reach 33.4 million and 37.4 million by the year of 2020 and 2030, respectively [1]. Until today, fossil fuel has been the main source for electricity generation and is heavily utilized over the past few decades. The overdependence on this single fuel source creates a huge challenge in the energy and power sector in this country. It is known that non-renewable energy such as fossil fuel is not sustainable. In Malaysia, the oil reserve is expected to deplete in around 20 years' time while natural gas may still be sustainable for the next 36 years [2]. The impact to the environment, climate change and pollution due to fossil fuel burning have also become serious issues that caught

M. S. Zakaria (✉) · S. Hassan · M. F. M. Nor
Department of Mechanical Engineering, Universiti Teknologi PETRONAS, 32610 Seri Iskandar, Perak, Malaysia
e-mail: mohd_16005855@utp.edu.my

F. Ahmad et al. (eds.), *ICPER 2020*, Lecture Notes in Mechanical Engineering,
https://doi.org/10.1007/978-981-19-1939-8_4

the attention of the government and authorities. The combustion of petroleum, coal and natural gas produce substantial amount of gaseous emissions such as Carbon Dioxide (CO2), Nitrogen Oxide (NOx), and Sulphur Dioxide (SO_2). The increase of CO_2 gas emission based on data recorded from 1980 until 2006, reveals the impact of fossil fuel consumption to the environment, which has reached a worrying state [3].

Domestic Wastewater Sludge (DWS), as one of renewable energy has a large potential as an alternative fuel due to its economic viability and benefits to the environment. One of the viable options to solve these problems is by converting this waste material into a source of energy through Thermochemical Conversion (TCC) technologies. In the earlier works, the feasibility of DWS to become an alternative solid fuel for power generation application has been investigated and its potential has been proven [4–9].

In order to recover energy from biomass waste material through TCC process, an in-depth understanding of the thermal conversion involving the decomposition process of DWS during the elevation of temperature is essential. The knowledge of kinetic parameters such as ash, volatile matter and char content should be properly studied. In order to investigate the kinetic analysis of DWS, thermogravimetric kinetic studies are important. Different operational condition and properties of materials are the most important variables that can influence the result of the experiment. Variation of experimental condition can give clues for optimization and reaction mechanism of sample in the thermal decomposition process.

2 Methodology

Proximate analysis is important to determine the relative percentage of the combustible and non-combustible components of the solid fuel. These components are measured when the samples are heated under controlled conditions of temperature using thermogravimetric method. The proximate analysis involves the use of a Thermogravimetric Analyzer (TGA-Perkin Elmer PYRIS1TGA). The TGA test procedures were carried out in accordance to the ASTM E1131-98 Standard, where the changes in sample weight is used as a basis to determine the sample's response to changes in temperature. TGA procedure is used to measure the weight changes in a biomass sample as a function of temperature or time in controlled oxygen and nitrogen environment, and these measurements not only improve the production but also simplify the interpretation of the result [10–13]. A heating procedure as listed in Table 1 was then administered in the TGA to remove the moisture content and volatile matter and measure the remaining fixed carbon and ash in the samples. In this study, there are 7 heating rates were studied which are 5, 10, 15, 20, 30, 50 and 100K/min.

Table 1 Heating procedure for TGA analysis

Temperature (°C)	Time (min)	Heating rate (K/min)	Gas
30–110	–	5	N2
110–600	–	5	N2
600–600	10	–	N2
600–850	–	5	N2

3 Results

3.1 Derivative Thermogravimetric (DTG)

Table 2 lists the DTG profiles of DWS at different heating rates. In devolatilization stage, temperature range of DTG is broader as heating rate was increased. DTG curve start at temperature 110 °C for heating rate 5–20 K/min. However, devolatilization stage for heating rate 30, 50 and 100 K/min start at 120 °C, 145 °C and 150 °C, respectively. The final temperature shows an increment from 430 °C to 550 °C on rising from 5 K/min to 100 K/min heating rate. Peak DTG (*DTGmax*) and the corresponding temperature (*Tp*) also tends to increase to the higher value.

Based on Table 2, *DTGmax* for 5 K/min is −3.26 %/min. *DTGmax* rises sharply to −59.83 %/min (in absolute value) when 100 K/min heating rate is used. The increment is approximately 20 times higher, corresponding to the increasing of heating rate. The relation of *DTGmax* and heating rate can be clearly seen from linear fit plotted in Fig. 1, where the heating rate is directly proportional to *DTGmax*. The increasing value of *DTGmax* attributed to the increasing value of maximum decomposition rate of DWS. This reveals that greater thermal energy is provided to improve heat transferred from ambient to DWS material during pyrolysis.

The relation of *Tp* to the heating rate is also shown in Fig. 1. Based on Fig. 1, it can be observed that *Tp* slightly increase when heating rate is increased. As a result, the curvature of curve shifted to the higher temperature. The maximum rate of decomposition took place at temperature 276.75 °C for 5 K/min and 341.84 °C for

Table 2 DTG profiles of DWS pyrolysis at different heating rates

Heating Rates (K/min)	*Trange* (°C)	*DTGmax* (%/min)	*Tp* (°C)	Total volatile (%)	Char (%)
5	110–430	−3.26	276.75	84.51	15.49
10	110–450	−6.29	294.52	83.97	17.87
15	110–480	−9.36	306.35	82.13	18.02
20	110–500	−12.69	312.10	80.72	19.28
30	120–510	−17.77	334.12	78.58	21.42
50	145–530	−29.19	341.78	77.90	21.46
100	150–550	−59.83	341.84	76.20	23.80

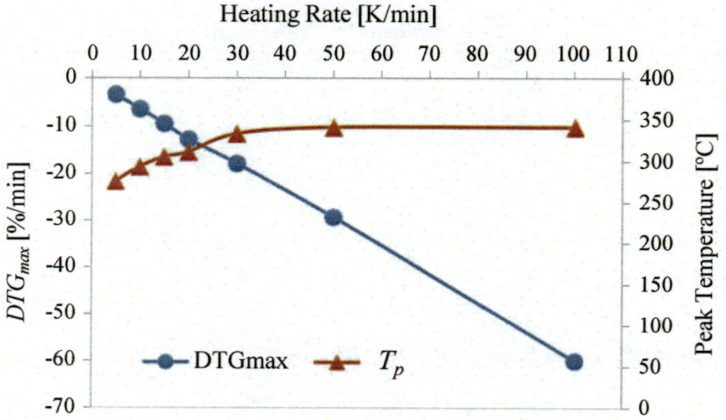

Fig. 1 Maximum decomposition rate, DTGmax and the corresponding temperature, *Tp* as a function of heating rate

100 K/min respectively. The reason is due to the short retention time when higher heating rate is applied. Decomposition process is delayed, thus higher temperature is required for DWS to decompose. Despite this, at lower heating rate, DWS starts to decompose at maximum rate once reaching low temperature. This indicates that DWS is easier to decompose in lower heating rate than that in the higher heating rate [14–17].

Similarly, Mohammed et al. [18] claimed that increasing heating rate implies high thermal energy was equipped to the system instantaneously. Significantly, this is the factor that is hindering the system to obtain equilibrium temperature within the furnace and material. Additionally, at higher heating rate, limitation of heat transfer caused the equilibrium between temperature in furnace and sample takes longer time to be attained [19–22]. Besides, materials react in a short time when higher heating rate is practiced. This causes the shifting of temperature to the higher value in order to complete the decomposition process.

3.2 *Total Volatile and Char*

Volatile and char are the two important substances that should be taken into consideration in order to examine the influence of heating rate to DWS decomposition process. Apparently, total volatile reduces as a result of increased heating rate. Conversely, char formation increase is influenced by the increasing of heating rate. The reduction of total volatile is from 84.51% to 76.2% is in accordance to the heating rate increment. On the other hand, char production increase from 15.59% to 23.8%. Despite the increment of char production is very little especially when changing the heating rate from 10 K/min to 15 K/min (0.15% difference) and 30 K/min to 50 K/min (0.04%

difference), it can be seen that heating rate has significant influence to the total char formation. The relation of heating rate to the total volatile and char can be clearly interpreted as presented in Fig. 2.

Lower heating rate leads to an efficient heat transfer because the process is gradually progressed. As a result, the effectiveness of decomposition process is enhanced, thus increase the liberation of volatile product. In contrast, when carrying out the experiment using higher heating rate, shorter time is given for the material to acquire the heat at a certain temperature. Heat transfer occurs poorly, hence release less amount of volatile product and oppositely form large amount of char residue at the end of the decomposition process. Studies carried out by Altun et al. [13] on the pyrolysis of Silopi Asphaltite by using TGA also agreed that the reason for volatile increase when lower heating rate was used is due to the effectiveness of heat transfer which occurred when temperature is elevated. The material's particle cracked efficiently resulted more volatile release when weight loss took place.

Ounas et al. [17] defined that the reason for small amount of volatile release during higher heating rate can be related to the heat transfer limitation, where the temperature gap between ambient in furnace and sample's particle is large. Other than the presence of non-equilibrium state, the existence of temperature gradient may also differ the temperature of particle's surface and between its inner parts. Therefore, the decomposition and release of volatile process might be affected, where the decomposition rate is found to occur rapidly than the volatile released. This result is in line with the study conducted by Mani et al. [9] for pyrolysis of wheat straw and Idris et al. [16] for oil palm biomass respectively. However, studies done by Ceylan et al. [10] for hazelnut husk found that heating rate seems not much affected by the yielding of volatile matter.

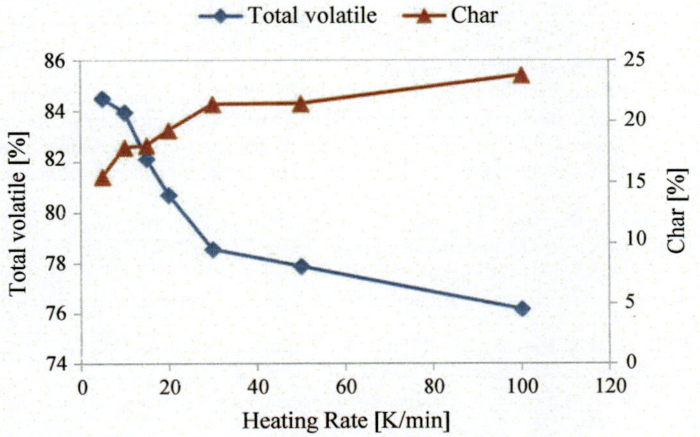

Fig. 2 Total volatile and char yields at every heating rate

4 Conclusion

The findings in this study proved that the thermal decomposition process of DWS is thoroughly influenced on experimental conditions such as heating rate and particle size of sample in pyrolysis and combustion environment. The selection of the appropriate heating rate together with the most representative particle size is crucial to be taken into consideration in order to ensure the efficiency of thermal degradation process for later application in TCC technology. According to the finding, it is suggested that the selection of small particle size together with relatively low heating rate is required to meet the optimum outcome of decomposition process.

Acknowledgements This research was funded by a grant from Yayasan Universiti Teknologi PETRONAS (FRGS Grant).

References

1. Pang Y, Shen S, Chen Y (2017) High temperature steam gasification of corn straw pellets in downdraft gasifier: preparation of hydrogen-rich gas. Waste Biomass Valorization 10(7):127
2. Musabbikhah M, Saptoadi H, Subarmono S, Arif Wibisono M (2015) Modelling and optimisation of coconut shell drying and carbonisation using multi-response taguchi method with multi-response signal-to-noise procdure. Int J Eng Trans C: Asp 28(3):464–472
3. Bidabadi M, Noroozi MJ (2015) Effect of heat recirculation in biomass flame stability within a cylindrical micro-combustor (technical note). Int J Eng Trans C: Asp 28(3):454–459
4. Manaf LA, Samah MAA, Zukki NIM (2009) Municipal solid waste management in Malaysia: practices and challenges. Waste Manag 29(11):2902–2906
5. Muyibi SA, Ambali AR, Eissa GS (2007) The impact of economic development on water pollution: trends and policy actions in Malaysia. Water Resour Manag 22:485–508
6. Ahmad S, Kadir MZAA, Shafie S (2011) Current perspective of the renewable energy development in Malaysia. Renew Sustain Energy 15:897–904
7. Shamsul NS, Kamarudin SK, Rahman NA, Kofli NT (2011) An overview on the production of bio-methanol as potential renewable energy. Renew Sustain Energy 33:578–588
8. Supatata N, Buates J, Hariyanont P (2013) Characterization of fuel briquettes made from sewage sludge mixed with water hyacinth and sewage sludge mixed with sedge. Int J Environ Sci 4:179–181
9. Tiara T, Agustina TE, Faizal M (2018) The effect of air fuel ratio and temperature on syngas composition and calorific value produced from downdraft gasifier of rubber wood-coal. Int J Eng Trans C: Asp 31(9):1480–1486
10. Ceylan E (2015) A Pagelication of microwaves for sewage sludge conditioning. Water Resour 39:4749–4754
11. Jaafar MZ, Kheng W, Kamaruddin N (2003) Greener energy solutions for a sustainable future: issues and challenges for Malaysia. Energy Policy 31:1061–1072
12. Molla AH (2012) A potential resource for bioconversion of domestic wastewater sludge. Bioresour Technol 85:263–272
13. Altun P, Bonin G, Prone A, Massiani C (1999) Distribution, movement and plant availability of trace metals in soils amended with sewage sludge composts: application to low metal loadings. Sci Total Environ 161–179
14. Manara P, Zabaniotou A (2012) Towards sewage sludge based biofuels via thermochemical conversion – a review. Renew Sustain Energy Rev 16:2566–2582

15. Domínguez A, Menéndez JA, Inguanzo M, Pís JJ (2006) Production of bio-fuels by high temperature pyrolysis of sewage sludge using conventional and microwave heating. Bioresour Technol 97:1185–1193
16. Idris A, Hassan S, Nor MFM, Fadhil M (2014) Thermogravimetric kinetic behavior of Malaysian Poultry Processing Dewatered Sludge (PPDS) during combustion. Appl Mech Mater 695:433–437
17. Ounas K, Winter WT, Stipanovic AJ (2015) A modulated-TGA approach to the kinetics of lignocellulosic biomass pyrolysis/combustion. Polym Degrad Stab 97:1606–1615
18. Mohammed M, Casal D, Pevida C, Pis JJ, Rubiera F (2010) Thermal behaviour and kinetics of coal / biomass blends during co-combustion. Bioresour Technol 101:5601–5608
19. Mekhilef S, Saidur R, Safari A, Mustaffa WESB (2011) Biomass energy in Malaysia: current state and prospects. Renew Sustain Energy Rev 15(7):3360–3370
20. Abdalla A (2013) Development of fuel briquettes from biomass residue. Universiti Teknologi Petronas
21. Alam MZ, Muyibi SA, Wahid R (2008) Statistical optimization of process conditions for cellulase production by liquid state bioconversion of domestic wastewater sludge. Biores Technol 99(11):4709–4716
22. Ibrahim ABA, Aris MS, Chin YS (2012) Development of fuel briquettes from dewatered poultry sludge. In: International conference on future electrical power and energy system lecture notes in information technology, vol 9, pp 469–476

Char and Ash Forming Elements of Tropical Beach Seawrack

Jezreel Deyto Tuquero and Shaharin Anwar Sulaiman

Abstract The utilization of energy from renewable sources had been steadily increasing. Bioenergy's contribution to the final energy consumption was estimated at 12%, while traditional biomass was at 6.9%. The ever-increasing biomass utilization through direct combustion and other thermochemical conversions will increase the generation of ash and char. The chemical composition of biomass ash is as diverse as its origin and source. Seawrack or beach-cast seagrass (*Enhalus Acoroides* (L.f.) Royle deposited along the shorelines had been the focus of this paper. However, investigation on seawrack as feedstock has not yet been conducted. Data and knowledge on the viability of beach-cast seagrass gasification is still lacking. The absence of data would further result in continued disregard of a potential energy resource. This paper aims to probe and determine the elemental composition of beach-cast seagrass char and ash to further understand its performance during the gasification process. EDX analysis on the ash residue was performed to identify its surface elemental composition. FTIR spectroscopy and chemical analysis were conducted to identify the organic compounds present in the beach-cast seagrass sample. Results of the EDX analysis of beach-cast seagrass ash showed the major and minor elements in the beach-cast seagrass ash in decreasing order of availability are O (53.18%) > C (29.63%) > Ca (9.23%) > Na (2.59%) > Cl (2.24%) > Mg (2.12%) > S (0.55%). The Chlorine content of beach-cast seagrass is very minimal, which indicates that corrosion and fouling may not occur during gasification. As well as the concentration of Ca and Mg is beneficial in the improvement of the product gas and reduction of tar formation during biomass gasification. The presence of lignin, known as the main component responsible for char formation, was confirmed through FTIR spectroscopy. Further,

J. D. Tuquero · S. A. Sulaiman (✉)
Department of Mechanical Engineering, Universiti Teknologi Petronas, 32610 Seri Iskandar, Perak, Malaysia
e-mail: shaharin@utp.edu.my

J. D. Tuquero
e-mail: jezreel_16000583@utp.edu.my

J. D. Tuquero
Electrical Engineering Department, Palawan State University, 5300 Puerto Princesa City, Philippines

F. Ahmad et al. (eds.), *ICPER 2020*, Lecture Notes in Mechanical Engineering,
https://doi.org/10.1007/978-981-19-1939-8_6

chemical analysis validated that beach-cast seagrass *Enhalus Acoroides* comprises lignin at 14.93%, cellulose at 74.11%, and hemicellulose at 21.04%.

1 Introduction

Despite the challenges brought by the Covid crisis, renewable energy has remarkably been robust. Data showed that from January 2020 to October 2020, the auctioned renewable capacity was 15% higher than for the same period during the preceding year [1]. During the ten years (2007–2017), the combined renewables grew annually by 3% vis-à-vis the total final energy consumption. As of 2019, the world's total renewable power capacity was at 2,588 GW, and the existing bio-power capacity was at 139 GW. On the other hand, traditional biomass (wood, charcoal, cow dung), as fuel for cooking and heating needs, took a share of 6.9% in the global final energy consumption. Despite the meager energy supply share from biomass in the global energy, bioenergy contributed an estimated 12% (45.2 exajoules) to the total final energy consumption in 2018 [2]. The current utilization of biomass through direct combustion is estimated to be around 95–97%. With this quantity, the amount of biomass ash produced worldwide annually is estimated to be approximately 480 million tons which are comparable to that of coal ash at 780 million tons per year at present [3].

With the increasing demand for biofuels, seaweeds or macroalgae and seagrasses have recently emerged as viable feedstock for biofuels [4]. In keeping with the 2 °C Scenario (2DS), the International Energy Agency underscored the essential role of biomass in the decarbonization options in the coming years. Based on projections, the energy coming from biomass will be at 17% of the final energy demand by 2060. For bioenergy to significantly impact the decarbonization of energy systems, the supply of biomass feedstock should be increased fivefold compared to current usage. The continuous work on the evaluation of new biomass feedstock like marine biomass is imperative [5].

Beach-cast seagrasses or wrack are dislodged or drifted seagrasses that accumulate along the coastlines over time. Annually, as part of its natural cycle, seagrasses shed their leaves or blades. These blades settle at the bottom of the sea, for which, during strong winds, waves, typhoons, and high tides are tossed and accumulate along beaches [6, 7]. At the Kingston coast in Southern Australia, from 2003 to 2013, it was recorded that the annual average collected wrack was at 12,900 m^3 [6]. The Philippines has wide seagrass beds that usually thrive in sandy and soft-bottom areas in shallow waters near the shore. Seagrasses are known to have very high productivity as these can multiply without cultivation and fertilizers. Fourteen species of seagrasses are in the Philippines, considered the second highest worldwide, next only to Australia [8]. There is a possibility of an increase in the global distribution of seagrasses due to the changing climate that causes the rise in sea level, resulting in increased seagrasses to thrive on [9].

Enhalus acoroides (L.f.) Royle, the species of interest in this study, is the largest species of seagrass in the Philippines that reaches a height of more than a meter. *Enhalus acoroides* thrive in the tropics. It grows in Kenya, Seychelles, Madagascar, Saudi Arabia, Yemen, India, Sri Lanka, Andaman Islands, Malaysia, Kampuchea, Vietnam, Ryukyu Islands, Indonesia, Papua New Guinea, Mariana Islands, Caroline Islands, Queensland, and New Caledonia [10].

Biomass feedstocks vary in physical, chemical, and biological properties as they come from various sources [11]. Several studies on gasification of different biomass feedstock have shown that the composition of syngas, thermal degradation, calorific values, and reactivity are all influenced by the feedstock properties [12]. Ash in biomass can significantly affect the conversion process if it contains alkali metals such as potassium or halides such as chlorine, as these components can cause agglomeration. Ash is an inorganic solid residue after the solid fuel is completely burned. Ash is composed of silica, aluminum, iron, and calcium; small amounts of magnesium, titanium, sodium, and potassium may also be present. Ash obtained during gasification is not necessarily inherent from the biomass itself. Often during collection and multiple handling, rocks, dirt, and other impurities contaminate biomass. On the other hand, char residue during gasification contains volatiles, ash, and fixed carbon. It was very reactive and highly porous, and it did not cake [13]. Knowing the ash and char composition of biomass is necessary to understand the dynamics that may affect the conversion process. As well as aid in the determination of possible disposal or valorization of ash and char, such as soil ameliorant, activated carbon, and graphene/carbon fibers.

In the study of 141 varieties of biomass, there exists a high variability of the chemical composition of ash among biomass groups and sub-groups. Wood and woody biomass ash are rich in calcium (Ca), magnesium (Mg), manganese (Mn), and sulfur (S). Herbaceous and agricultural biomass ash has a high amount of chlorine (Cl), potassium (K), sodium (Na), phosphorus (P), and silicon (Si). Aquatic biomass ash is enriched with Cl, Na, and S. Calcium oxide (CaO) content in biomass ash varies in the interval of 1.0–83.5%. A high amount of CaO concentration is usually present in wood stems, trunks, and large poplar branches, spruce, birch, and maple. Low CaO concentration is typical in rice husks, some algae, coconut shells, and corn cobs. For agricultural and herbaceous biomass, Ca comes from P fertilizers. The sodium oxide (Na_2O) content in biomass ash varies in the interval of 0.1–28.7%. Sodium content in biomass ash is an indication of biomass species that grow in a saline environment. The chlorine content in biomass ash varies in the interval of 0.01–18.2%. Compared to solid fossil fuel ashes, Cl in biomass ash is higher in content. An extremely high concentration of Cl is present in some salt-tolerant biomass like thistle and algae. High Cl concentrations are found in crops that are cultivated with fertilizers. Wood generally has a very low Cl concentration. Chlorine is the most problematic of the elements found in biomass since it causes deposition, corrosion, and fouling during thermochemical conversion. High S (plus Cl) concentrations in some biomasses are due to chemical contaminations, fertilizers, insecticides, lacquer, plastics, rubber, dyestuff, or preservatives [3].

While biofuel production from marine biomass has been the subject of research over the years, gasification of marine biomass has not been exhaustively explored. This observation was noted by Brennan and Owende [14] due to limited available data related to gasification of marine biomass. More importantly, based on the survey of related literature conducted by the researcher, no studies have yet been conducted on thermochemical conversion through downdraft gasification of beach-cast seagrass (*Enhalus Acoroides* (L.f.) Royle) as biomass feedstock. It is in this context that this paper will attempt to explore and understand the composition of ash and char in beach-cast seagrass.

2 Methodology

The beach-cast seagrass biomass feedstock used for the experiments was collected from a tropical beach in Puerto Princesa City, Palawan, Philippines. The feedstock was washed in seawater to remove the sand and dirt particles before drying under the sun. Before the gasification experiment, the feedstock was subjected to further drying in a convection oven at 105 °C for 24 h and crushed into smaller bits. The mass of the feedstock that was fed to the gasifier reactor was at 60 g per run. A laboratory-scale electrically heated downdraft gasifier system with air as a gasifying agent was employed in the experiments. Before loading the feedstock into the reactor, the reactor was heated incrementally by 100 °C until it reached the desired temperature settings of 700, 800, and 900 °C. Once the reactor reached the desired temperature setting, the gasifying agent (air) coming from a compressor was fed into the reactor at flow rates of 1.5, 2.5, and 3.5 L/min. Ash and char residues were collected every after the experiment and stored in a sealed container. Energy-dispersive X-ray spectroscopy (EDX) analysis on the ash residue was performed to identify its surface elemental composition. The equipment used for this purpose was a Phenom XL Scanning Electron Microscope (SEM). A Fourier transform infrared (FTIR) spectroscopy was performed using a Perkin Elmer FT-TR Spectrometer Frontier employing Attenuated Total Reflectance (ATR) accessory to identify the organic compounds present in a sample. The chemical compositions of beach-cast seagrass were determined by following the standards outlined in TAPPI test methods and standard practices.

3 Result and Discussion

3.1 Ash and Char Yield

It was observed that the amount of ash and char collected ranges between 29 and 53% of the 60 g initial mass of feedstock that was fed into the reactor. The highest collection of ash and char was recorded when the gasification temperature was set at 800 °C with an airflow rate (AFR) of 2.5 L/min. The lowest collection (only ash) that was recorded was at a temperature of 900 °C with an AFR of 3.5 L/min. Figure 1 shows that the residue yields decreased as the temperature and AFR gets higher. It was also observed at higher temperature and AFR that the char decreased while the ash increased.

3.2 *Elemental Composition of Beach-Cast Seagrass Ash*

Table 1 shows the list of the elements in the combined map and elements count (Fig. 2) that were identified through EDX analysis of the beach-cast seagrass ash. Results

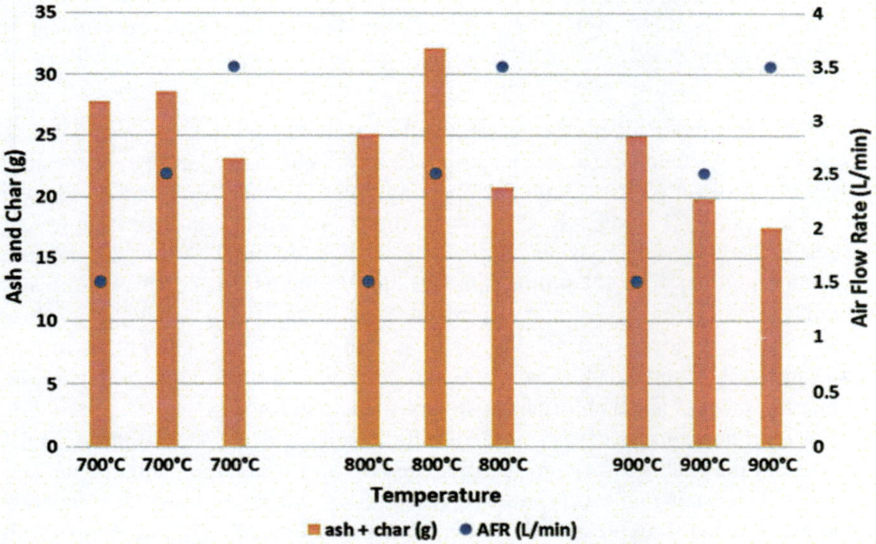

Fig. 1 Ash and char yield at varying gasification temperature

Table 1 Elemental composition of beach-cast seagrass ash obtained from EDX analysis

Element symbol	Element name	Atomic (%)	Weight (%)
O	Oxygen	53.18	47.20
C	Carbon	29.63	19.74
Ca	Calcium	9.23	20.53
Na	Sodium	2.59	3.31
Cl	Chlorine	2.24	4.41
Mg	Magnesium	2.12	2.86
S	Sulfur	0.55	0.98

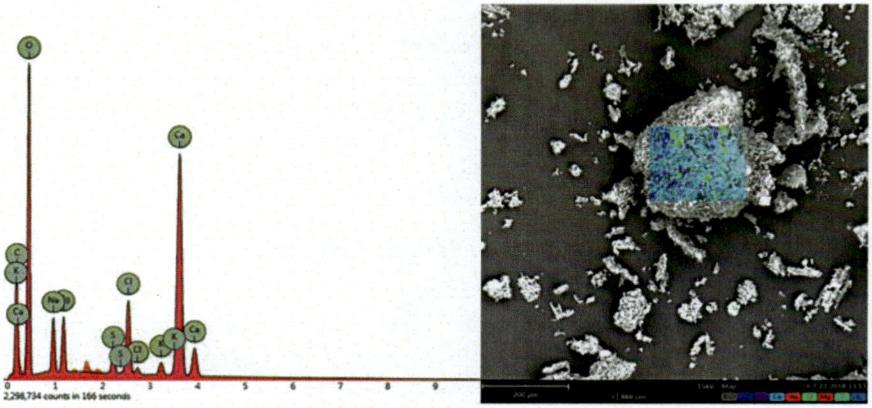

Fig. 2 Elements count and combined map during the energy dispersive X-ray Spectroscopy of beach-cast seagrass ash

showed that the major and minor elements in the beach-cast ash are oxygen (53.18%), carbon (29.63%), calcium (9.23%), sodium (2.59%), chlorine (2.24%), magnesium (2.12%), and sulfur (0.55%) with a decreasing order of availability expressed as O > C > Ca > Na > Cl > Mg > S. Chlorine, sodium, and sulfur contents are inherent for beach-cast seagrass, considering that it is an aquatic biomass that thrived in a saline environment. The calcium and magnesium elements that comprise lime and dolomite minerals may have come from the contaminants (e.g., shells) that cling to the seagrass blades. One notable result is that the chlorine content of beach-cast seagrass is minimal. Otherwise, this could pose a serious concern during the gasification process since chlorine causes corrosion and fouling.

Some implications can be gained from the results with respect to the disposal and reuse of the beach-cast seagrass ash related to carbon, calcium, and magnesium contents. The carbon in the ash can serve as an indicator in determining the efficiency of the conversion through gasification [15]. One possible application of biomass ash is to reuse it as fuel but may require some post- treatments [16]. Ca and Mg often used as catalysts to improve the product gas and reduce the tar formation during biomass gasification [17]. Also, high ash and low combustible sulfur along with high CaO contents in biomass ash can be beneficial in the mitigation of sulfur oxide (SOx) emission [18].

3.3 Char Forming Composition

Results of the FTIR spectroscopy (Table 2 and Fig. 3) conducted on beach-cast seagrass *Enhalus Acoroides* samples had revealed the presence of lignin, cellulose,

Table 2 Peak assignments in the infrared spectrum of beach-cast seagrass

Frequencies (cm^{-1})	Structure/Compound type	Bonds
3286.45	Hydroxyl	O–H stretch
2919.88	Alkane	– C–H stretch
2850.00		
1611.27	Carbonyl	– C = O stretch
1412.38		– C–O stretch
1332.99		– C = O bend
1241.42	Alkane	– C–H bend
1140.38	Carbonyl	– C–O stretch
1088.66		
1044.07		
1024.48		
962.03		
874.58	Carboxyl	– COOH bend
836.33	Alkane	– C–H bend
814.80		
767.90		
714.25		
659.81		

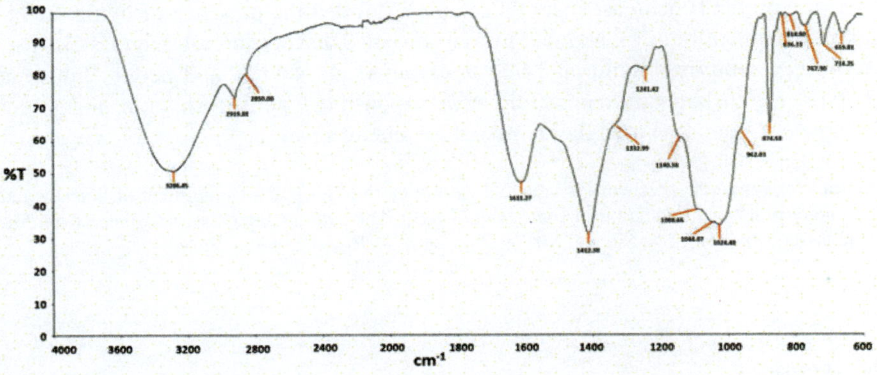

Fig. 3 FTIR spectrum of beach-cast seagrass

and hemicellulose. The O–H stretching vibration that peaked at 3283.45 cm^{-1} corresponds to the hydroxyl group of alcohols and phenols. The two peaks at 2919.88 cm^{-1} and 2850.00 cm^{-1} fall under the C–H stretching vibration in the alkanes. This C–H stretching and bending frequencies belong to the methyl and methylene groups in alkanes that are present in hydrocarbons [19]. Lignin is the main contributor in the formation of hydrocarbons at the spectral regions of 3200 to 2900 cm^{-1} while cellulose contributes to the formation of aldehydes at 2860 to 2770 cm^{-1}. Moreover, hydrocarbons and aldehydes are formed from hemicellulose [18]. Also, hydrocarbons and aldehydes were formed from hemicellulose. The results of chemical analyses on beach-cast seagrass confirmed the presence of lignin at 14.93%, cellulose at 74.11%, and hemicellulose at 21.04%.

In the pyrolysis of lignocellulosic materials, lignin which comprises the resinous part of the seagrass blades is the main component of char production [20]. Further, in this investigation, it was observed that char was no longer present at gasification temperatures 800 °C with an AFR of 3.5 L/min. and 900 °C with an AFR of 2.5 and 3.5 L/min. Gasification conditions at higher temperatures and higher AFR revealed that the lignin underwent complete degradation.

4 Conclusions

Results of the EDX analysis showed that the major and minor elements in the beach-cast seagrass ash are oxygen, carbon, calcium, sodium, chlorine, magnesium, and sulfur. Chlorine content of beach-cast seagrass is very minimal which is a good indicator that corrosion and fouling may not occur during gasification Also, the concentration of Ca and Mg in the beach-cast seagrass is beneficial in the thermo-chemical conversion process. Ca and Mg are often us ed as catalysts to improve the product gas and reduce tar formation during biomass gasification. FTIR spectroscopy on the beach-cast seagrass sample exhibited the presence of lignin, cellulose, and hemicellulose. Chemical analysis revealed that beach-cast seagrass *Enhalus Acoroides* comprises lignin at 14.93%, cellulose at 74.11%, and hemicellulose at 21.04%. Lignin being the main component responsible for char formation, underwent complete degradation at higher temperature and airflow rate.

Acknowledgements This work is being supported by the Philippine Department of Science and Technology-Science Education Institute (DOST-SEI). The authors appreciate assistance from Mr. Muddasser, Mr. Hafif, Ms. Siti and Mr. Mior of Biomass Research Laboratory.

References

1. IEA (2020) Renewables 2020, IEA, Paris. https://www.iea.org/reports/renewables-2020
2. REN21 (2020) Renewables 2020 Global Status Report
3. Vassilev SV, Vassileva CG, Song Y-C, Li W-Y, Feng J (2017) Ash contents and ash-forming elements of biomass and their significance for solid biofuel combustion. Fuel 208:377–409
4. Marquez GPB, Santiañez WJE, Trono GC, Montaño MNE, Araki H, Takeuchi H, Hasegawa T (2014) Seaweed biomass of the Philippines: sustainable feedstock for biogas production. Renew Sustain Energy Rev 38:1056–1068

5. Brown A, Le Feuvre P (2017) Technology roadmap: delivering sustainable energy. International Energy Agency
6. Primary Industries and Regions South Australia (PIRSA) (2014) South Australian Beach-Cast Seagrass and Marine Algae Fishery Assessment. Accessed from https://www.environment.gov.au
7. Marquez GP, Reichardt WT, Azanza RV, Klocke M, Montano MN (2013) Thalassic biogas production from sea wrack biomass using different microbial seeds: cow manure, marine sediment and sea wrack-associated microflora. Bioresour Technol 133:612–617
8. Department of Environment and Natural Resources (DENR), B.F.A.R. of D.A., Department of the Interior and Local Government and Coastal Resource Management Project (2001) Managing Coastal Habitats and Marine Protected Areas. Cebu City, Philippines
9. Short F, Carruthers T, Dennison W, Waycott M (2007) Global seagrass distribution and diversity: a bioregional model. J Exp Mar Biol Ecol 350(1–2):3–20
10. Meñez EG, Phillips RC, Calumpon HP (1983) Seagrasses from the Philippines. Smithsonian Contributions to the MarineSciences. Smithsonian Institution, p 40
11. Capareda SC (2014) Introduction to biomass energy conversions. Taylor and Francis Group
12. Sikarwar VS, Zhao M, Fennell PS, Shah N, Anthony EJ (2017) Progress in biofuel production from gasification. Prog Energy Combust Sci 61:189–248
13. Basu P (2010) Biomass gasification and pyrolysis practical design. Elsevier Inc.
14. Brennan L, Owende P (2010) Biofuels from microalgae—a review of technologies for production, processing, and extractions of biofuels and co-products. Renew Sustain Energy Rev 14(2):557–577
15. Huang S, Wu S, Wu Y, Gao J (2017) Structure characteristics and gasification activity of residual carbon from updraft fixed-bed biomass gasification ash. Energy Convers Manage 136:108–118
16. Pels JR, Nie DSD, Kiel JHA (2005) Utilization of ashes from biomass combustion and gasification. In: 14th European biomass conference and exhibition
17. Herman AP, Yusup S, Shahbaz M, Patrick DO (2016) Bottom ash characterization and its catalytic potential in biomass gasification. Procedia Eng 148:432–436
18. Plis A, Lasek J, Skawińska A, Kopczyński M (2014) Thermo-chemical properties of biomass from Posidonia oceanica. Chem Papers 68(7)
19. Silverstein RM, Webster FX, Kiemle DJ (2005) Spectrometric identification of organic compounds, 7th edn. Wiley, New York
20. Órfão JJM, Antunes FJA, Figueiredo JL (1999) Pyrolysis kinetics of lignocellulosic materials-three independent reaction models. Fuel 78:349–358

Analysis on Part Load Performance of EC Operations at District Cooling Plant

Mohd Amin Abd Majid, Masdi Muhammad, Takashi Sasaki, and Ainul Amar Mokhtar

Abstract Electric chillers (EC) that are installed at District cooling (DC) plants are designed for charging thermal energy storage (TES) during off peak period. The stored chilled water in the TES is used to support the demand of chilled water during peak period. This is the case of EC at Universiti Teknologi PETRONAS (UTP) DC plant due to insufficient supply of chilled water (CW) from steam absorption chillers (SAC) and TES. At the same time, the EC were also operated during the day time to support the demand of CW during peak timeand this was not per design. This led to EC were being operated at partial load. This study is on partial load operations of EC at the plant during year 2016 and 2017 covering four EC, each with 325 RT capacity. The main function of the ECs was to charge the TES during off peak period. Cases arise EC were also operated during the day. This resulted to EC were not being operated at full load. They were operated at partial load. This study is to determine the part load performance of EC operated under these conditions. Historical data was used for the study. Findings indicate most of the time EC were operated at about 10 h during the day.

Keywords Electric chillers · Chilled water · Thermal energy storage · COP

1 Introduction

Chillers are one of the largest energy consumers within a building and this has a big impact on operational cost. Therefore, it is important to monitor chillers operation. At UTP DC plant during understudy period there were four electric chillers (EC) each with 325 RT capacity, used to charge thermal energy storage (TES) during off peak period. The chilled water stored in the thermal storage was then used to support the chilled water demand during peak period. Thermal storage facility main

M. A. Abd Majid · M. Muhammad (✉) · A. Amar Mokhtar
Universiti Teknologi PETRONAS, Seri Iskandar, Malaysia
e-mail: masdimuhammad@utp.edu.my

T. Sasaki
Hitachi Asia (Malaysia) Sdn Bhd, Kuala Lumpur, Malaysia

F. Ahmad et al. (eds.), *ICPER 2020*, Lecture Notes in Mechanical Engineering,
https://doi.org/10.1007/978-981-19-1939-8_7

purpose to provide support of chilled water demand during peak period to support the steam absorption chillers (SAC). However, at UTP DC, when the supply of chilled water from SAC and TES was insufficient EC were operated to provide the support. When this occurred the EC also consumed electricity during the day. This is not as per design of EC installation. At the same time when this happened the EC were not being operated at full load. They were operated at partial load. This study is to determine the part load performance of EC operated under these conditions.

2 UTP GDC Plant

UTP district cooling (DC) plant is a cogeneration plant. The plant produces power and chilled using natural gas as fuel. Chilled water is generated by steam absorption chillers (SAC) and electric chillers (EC). The SAC were driven by steam from heat recovery steam generators while EC were driven by electricity from the gas turbines [1]. The system function whereby during the day the SAC support the requirements of chilled water. EC were operated to charge the TES at night. The TES was used to support the chilled water requirements during the day to support SAC. When insufficient supply of chilled water form SAC and TES occurred during peak time, EC were operated to supplement the requirements.

COP

The paper [2] reviews national standards and guidelines for the energy performance of chiller systems which provide cooling energy in commercial buildings but their operation causes the major proportion of electricity consumption in the building sector. The standards reviewed are currently adopted in nine places (in alphabetical order): Australia, California, Canada, China, Chinese Taipei, the EU, Hong Kong, New Zealand and the USA. Under standard rating conditions at full load operation, the minimum required coefficient of performance (COP) ranges from 2.40 to 3.06 for air-cooled chillers and from 3.80 to 6.39 for water-cooled chillers. Some standards state also the COP requirements at part load operation with integrated part load values or seasonal energy efficiency ratios. A classification scheme is provided in the EU and Chinese standards which helps labelling and certifying high efficient chiller products. Criteria for enhancing energy ratings of chiller systems are suggested.

EMSD [3] highlighted on the advantage of installing VSD to air con systems. Most compressors are equipped with induction motor with fixed speed and run at full power since it is designed to handle maximum load. While VSD save energy whenever electric motors run at less than full power. For motor equipped with VSD, the power utilisation by the motor will varies with the cube of motor speed. This means that a reduction of speed by 20% will result in reduction of power by nearly half i.e. 50%. Since most HVAC equipment seldom run at full power, significant energy savings can be made for HVAC system equipped with VSD.

3 Methodlogy

The adopted methodology are as follows

i. Operating data of the four EC, ECA, ECB, ECC and ECD were acquired from UTP GDC. The data cover from June 2016 to January 2017.
ii. The data was used to analyse COP, COP vs RT produced and cost of per unit RT.
iii. Comparative study was then done.

4 Operation of ECs

The four ECS each with capacity 325 RT were used to charge thermal storage system during off peak period i.e. at night. The thermal storage system was 10,000 RTh capacity. During the day time the thermal storage was discharged to support the chilled water demand by the university. A typical chilled water demand during the day is as shown in Fig. 1. (The plot is for 14/2/19). The chilled water demand is met by two-unit steam absorption chillers and supported by chilled water from thermal storage. Most of the days. insufficient supply of chilled water occurred. The available Electric chillers (EC) were used to provide the additional requirements. The scope of the study is on the extent of EC support of chilled water during the days.

5 Results and Discussion

Tables 1, 2, 3 and 4 show operating data during the study period for ECA, ECB, ECC and ECD respectively.

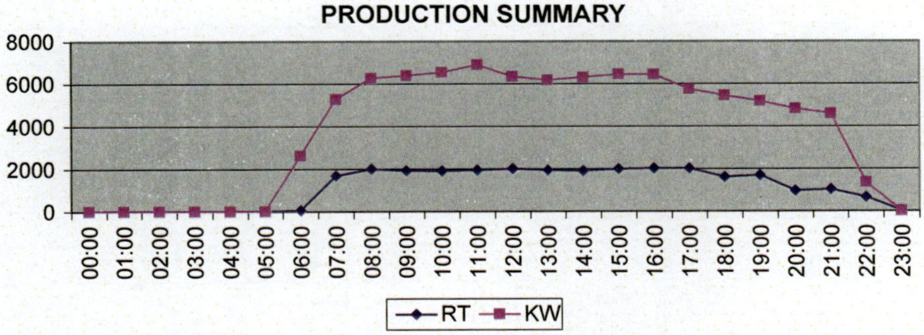

Fig. 1 Hourly demand of chilled water within normal operation

Table 1 ECA Operation data from June 2014 to February 2017

Month	EC0121A						
	Heat summation (RT)	Electric power (KW)	RT/KW	RT/KW RM0.4210/KW	RT/KW RM0.3720/KW	%differ	C.O.P
Jun-14	243.09	162.6	1.49	0.63	0.56	0.12	5.36
Jul-14	234.76	149.1	1.57	0.66	0.59	0.12	5.58
Aug-14	134.7	77.6	1.74	0.73	0.65	0.12	5.75
Sep-14	195.53	116.7	1.67	0.71	0.62	0.12	5.91
Oct-14	215	129.8	1.66	0.7	0.62	0.12	5.85
Nov-14	194.3	118.2	1.64	0.69	0.61	0.12	5.94
Dec-14	195.6	116.8	1.67	0.71	0.62	0.12	5.94
Jan-15	170.7	102	1.67	0.7	0.62	0.12	5.94
Feb-15	212.1	136.4	1.56	0.66	0.58	0.12	5.94
Mar-15	208.9	133.3	1.57	0.66	0.58	0.12	5.94
Apr-15	200.1	136.4	1.47	0.62	0.55	0.12	5.94
May-15	174.7	133.9	1.3	0.55	0.49	0.12	5.94
Jun-15	173.9	134.1	1.3	0.55	0.48	0.12	4.6
Jul-15	160.0.1	126.7	1.26	0.53	0.47	0.12	4.6
Aug-15	192.4	131.1	1.46	0.61	0.54	0.12	4.6
Sep-15	154.7	85.8	1.8	0.76	0.67	0.12	4.6
Oct-15	112.8	72.5	1.56	0.65	0.55	0.12	4.6
Nov-15	136.66	99.57	1.37	0.58	0.67	0.12	4.6
Dec-15	184.4	136.9	1.35	0.57	0.58	0.12	4.6
Jan-16	175.1	128.5	1.36	0.57	0.51	0.12	5.14
Feb-16	202.2	118.5	1.71	0.72	0.5	0.12	6.22
Mar-16	221.8	161.4	1.37	0.58	0.51	0.12	4.22
Apr-16	282.1	152.2	1.85	0.78	0.63	0.12	5.44
May-16	227.6	143.7	1.58	0.67	0.51	0.12	4.22
Jun-16	131.4	118.1	1.11	0.47	0.69	0.12	4.17
Jul-16	145.6	118.5	1.23	0.52	0.59	0.12	4.22
Aug-16	169.4	139.6	1.21	0.52	0.41	0.12	4.17
Sep-16	108.8	88.3	1.23	0.54	0.46	0.12	4.22
Oct-16	118.4	95.6	1.24	0.53	0.46	0.12	4.17
Nov-16	170	131.7	1.29	0.51	0.45	0.12	4.22
Dec-16	169	134.5	1.26	0.54	0.46	0.12	4.17
Jan-17	101.4	84.5	1.2	0.53	0.46	0.12	4.22
Feb-17	6.07	76.5	0.08	0.03	0.03	0.12	4.17

Table 2 ECB operating data from June 2014 to February 2017

Month	EC0121B						
	Heat summation (RT)	Electric power (kW)	RT/kW	RT/kW RM0.4210/kW	RT/kW RM0.3720/kW	%differ	C.O.P
Jun-14	181.6	134.6	0.83	0.38	0.35	0.12	3.23
Jul-14	134.88	133.63	0.87	0.41	0.36	0.12	3.35
Aug-14	111.41	107.42	1.04	0.44	0.38	0.12	3.5
Sep-14	133.06	146.14	0.34	0.4	0.35	0.12	3.32
Oct-14	135.62	146.54	0.83	0.38	0.34	0.12	3.35
Nov-14	144.37	150.01	0.36	0.41	0.36	0.12	3.36
Dec-14	134.46	140.78	0.36	0.4	0.36	0.12	3.36
Jan-15	121.74	133.77	0.81	0.38	0.34	0.12	3.22
Feb-15	135.13	147.22	0.82	0.38	0.34	0.12	3.23
Mar-15	146.06	165.13	0.8	0.38	0.33	0.12	3.21
Apr-15	116.11	131.34	0.88	0.37	0.33	0.12	3.16
May-15	141.81	165.53	0.66	0.36	0.32	0.12	3.03
Jun-15	124.88	146.82	0.65	0.36	0.32	0.12	3.01
Jul-15	126.47	141.22	0.8	0.38	0.33	0.12	3.08
Aug-15	130.78	142.33	0.82	0.38	0.34	0.12	3.24
Sep-15	175.13	183.43	0.82	0.38	0.34	0.12	3.23
Oct-15	176.31	182.13	0.82	0.38	0.34	0.12	3.22
Nov-15	172.2	178.41	0.87	0.41	0.36	0.12	3.4
Dec-15	163.75	166.37	1.02	0.43	0.38	0.12	3.58
Jan-16	180.23	182.23	0.88	0.42	0.37	0.12	3.5
Feb-16	208.66	203.35	1.03	0.43	0.38	0.12	3.62
Mar-16	188.76	136.04	0.87	0.41	0.36	0.12	3.42
Apr-16	166.63	175.8	0.35	0.4	0.35	0.12	3.37
May-16	154.75	165.24	0.34	0.38	0.35	0.12	3.31
Jun-16	131.11	153.64	0.82	0.35	0.31	0.12	3.1
Jul-16	118.42	123.13	0.87	0.41	0.36	0.12	3.41
Aug-16	122.74	128	0.36	0.4	0.36	0.12	3.34
Sep-16	134.47	145.03	0.83	0.38	0.34	0.12	3.18
Oct-16	137.77	146.03	0.83	0.38	0.35	0.12	3.2
Nov-16	156.53	181.7	0.87	0.37	0.32	0.12	3.12
Dec-16	144.87	153.3	0.34	0.4	0.35	0.12	3.24
Jan-17	118	128.8	0.82	0.38	0.34	0.12	3.18
Feb-17	17.07	111.78	0.15	0.06	0.06	0.12	0.76

Table 3 Operating data: ECC from June 2016 to February 2017

Month	EC0121C						
	Heat summation (RT)	Electric power (kW)	RT/kW	RT/kW RM0.4210/kW	RT/kW RM0.3720/kW	%differ	C.O.P
Jun-14	162.76	138.28	0.82	0.35	0.31	0.12	2.34
Jul-14	64.03	114.38	0.82	0.34	0.3	0.12	2.8
Aug-14	57.5	71.13	0.81	0.34	0.3	0.12	2.72
Sep-14	53.38	72.31	0.81	0.34	0.3	0.12	3
Oct-14	81.8	37.25	0.64	0.35	0.31	0.12	3.12
Nov-14	111.46	128.07	0.87	0.37	0.32	0.12	3.08
Dec-14	134.4	160.4	0.64	0.35	0.31	0.12	3.04
Jan-15	114.44	136.61	0.64	0.35	0.31	0.12	2.88
Feb-15	136.24	163.47	0.83	0.35	0.31	0.12	2.36
Mar-15	142.83	173.32	0.82	0.35	0.31	0.12	2.83
Apr-15	131.55	157.66	0.83	0.35	0.31	0.12	2.88
May-15	128.41	156.47	0.82	0.34	0.3	0.12	2.88
Jun-15	144.81	181.17	0.8	0.34	0.3	0.12	2.64
Jul-15	123.88	151.65	0.82	0.34	0.3	0.12	2.8
Aug-15	117.87	143.73	0.82	0.35	0.3	0.12	2.81
Sep-15	155.66	182.33	0.65	0.36	0.32	0.12	3
Oct-15	142.7	172.33	0.83	0.35	0.31	0.12	2.82
Nov-15	154.81	177.75	0.87	0.37	0.32	0.12	3.23
Dec-15	130.66	133.43	0.34	0.4	0.35	0.12	3.44
Jan-16	122.37	133.56	0.88	0.37	0.33	0.12	3.18
Feb-16	146.73	161.33	0.32	0.33	0.34	0.12	3.34
Mar-16	153.78	188.3	0.81	0.34	0.3	0.12	2.83
Apr-16	161.46	136.15	0.82	0.35	0.31	0.12	2.34
May-16	113.66	144.3	0.73	0.33	0.28	0.12	3.05
Jun-16	143.25	165.14	0.77	0.33	0.28	0.12	2.73
Jul-16	35.55	114.45	0.83	0.35	0.31	0.12	2.82
Aug-16	121.8	143.3	0.81	0.34	0.3	0.12	2.66
Sep-16	110.83	136.3	0.81	0.34	0.3	0.12	2.82
Oct-16	110.03	132.52	0.83	0.35	0.31	0.12	2.35
Nov-16	147.37	175.63	0.64	0.35	0.31	0.12	3
Dec-16	36.46	123.52	0.78	0.33	0.28	0.12	2.7
Jan-17	32.13	54.13	0.53	0.25	0.22	0.12	2.36
Feb-17	0	17.38	0	0	0	0	0

Table 4 ECD operating data from November 15 to February 17

Month	ECO121D						
	Heat summation (RT)	Electric power (kW)	RT/kW	RT/kW RM0.4210/kW	RT/kW RM0.3720/kW	%differ	C.O.P
Nov-15	162	192.60	0.84	0.35	0.31	0.12	3
Dec-15	154	189	0.82	0.34	0.30	0.12	3
Jan-16	173	217.70	0.80	0.34	0.30	0.12	3
Feb-16	124	156	0.80	0.34	0.30	0.12	3
Mar-16	204	225	0.91	0.38	0.34	0.12	3
Apr-16	174	187	0.94	0.39	0.35	0.12	3
May-16	192	228	0.84	0.35	0.31	0.12	3
Jun-16	182	232	0.79	0.33	0.29	0.12	3
Jul-16	154	196	0.79	0.33	0.29	0.12	3
Aug-16	152	187	0.81	0.34	0.30	0.12	3
Sep-16	156	195	0.80	0.34	0.30	0.12	3
Oct-16	160	201	0.79	0.33	0.30	0.12	3
Nov-16	178	219	0.81	0.34	0.30	0.12	3
Dec-16	149	192	0.77	0.33	0.29	0.12	3
Jan-17	122	161	0.76	0.32	0.28	0.12	3
Feb-17	0.07	20	0.00	0.00	0.00	0.12	0

6 Summarised Results for ECA

i. The COP of ECA varies subsantially throughout the years from Jun 2014 to Feb 2017. The annual average COP are 5.67,5.2, 4.64 and 4.5 for 2014, 2015, 2016 and 2017 respectively. Overall average is 5. The average annual COP seems decreasing with years. This is above the range specified by [2].

ii. RT per kW varies from 0.08 to 1.85 with average of 1.41

iii. Each RT cost varies from RM0.03 to RM0.78 for rate RM0.421 per kWh and RM0.02 to RM0.69 for rate RM 0.372 per kWh.

7 Summarised Results for ECB

i. The average annual COP of ECB were; 3.36, 3.22, 3.32, 3.16 for 2014, 2015, 2016 and 2017 respectively. Overall average is 3.27. The trend also shows decreasing with the exception for year 2016 which was higher than the value for 2015. This is above the range specified by [2]

ii. RT per kW varies from 0.15 to 1.04 with average of 0.91

iii. Each RT cost varies from RM0.06 to RM0.44 for rate RM0.421 per kWh and RM0.06 to RM0.39 for rate RM0.372 per kWh.

8 Summarised Results for ECC

i. The average annual COP of C were; 2.97, 3.0, 3.12 and 2.36 for 2014, 2015, 2016 and 2017 respectively. Overall average is 2.86. The trend also shows decreasing with the exception for year 2016 which was higher than the value for 2015. This is within the range specified by [2] which specified that air cooled chillers COP range should be between 2.4 and 3.09.
ii. RT per kW varies from 0.59 to 0.94 with average of 0.83
iii. Each RT cost varies from RM0.25 to RM0.40 for rate RM0.421 per kWh and RM0.22 to RM0.35 for rate RM 0.372 per kWh.

9 Summarised Results for EC D

i. The average annual COP of ECC, were; 2.91, 2.88, and 2.52 for 2015, 2016 and 2017 respectively. Overall average is 2.77. The trend also shows decreasing
ii. RT per kW varies from 0.76 to 0.94 with average of 0.82. The average is within the range specified by [2]
iii. Each RT cost varies from RM0.32 to RM0.39 for rate RM0.421 per kWh and RM0.28 to RM0.35 for rate RM 0.372 per kWh.

Plots of COP, COP vs Heat summation and COP vs electric powers are included in Figs. 2, 3 and 4 respectively.

Tables 5, 6 and 7 show operation hours for ECA, ECB, ECC and ECD, during selected months of study period.

Plot of operating hours daily for chillers during the months of March 2016, June 2016 and September 2016 for chillers A, B, C and D are as shown in Figs. 5, 6 and 7 respectively.

10 Discussion

1. The analysis as shown by Tables 4, 5, 6 and 7 and Figs. 4, 5, 6 and 7 indicate that the four ECs at the UTP DC plant namely ECA, ECB, ECC and ECD were being operated for many hours during the day. Many of the days the ECs were operated up to 10 h, indicating insufficient chilled water from the thermal storage to support the chilled water requirements. Hence the ECs were being operated to top up the chilled water requirements. This is outside the scope the system was being designed.

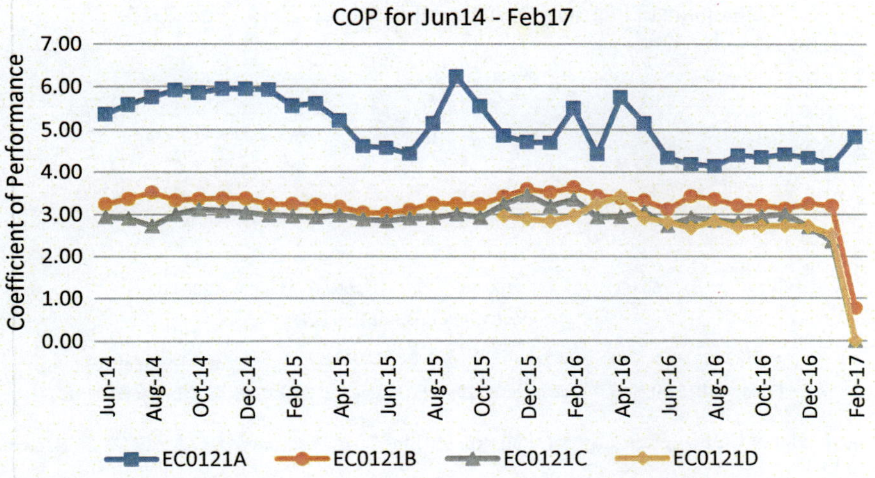

Fig. 2 Plot of COP for June 2014–February 2017 for ECA, ECB, ECC and ECD

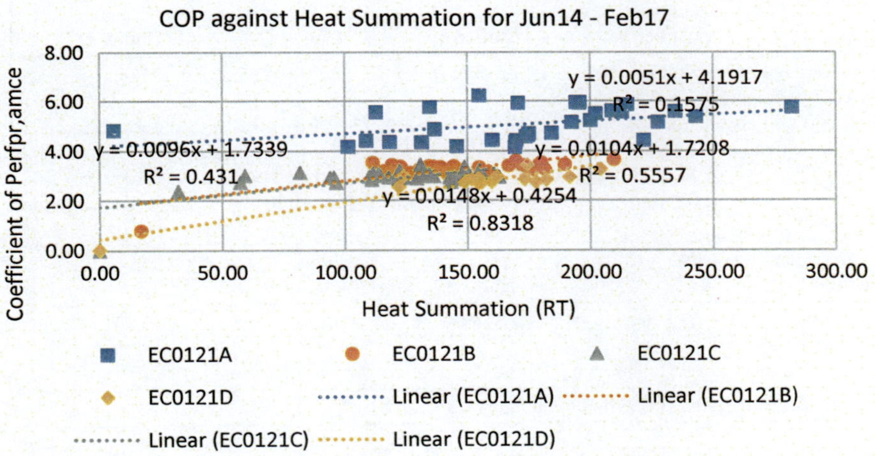

Fig. 3 Plot of COP vs heat summation (RT) for June 2014–February 2017 for ECA, ECB, ECC and ECD

2. From the evaluated values of COP ECA, ECB, ECC and ECD, it is noted that the chillers were not operated at full capacity. Hence based on [4] the power utilized was not optimized. Thus to optimise all the four chillers should be equipped with variable frequency drive (VFD).
3. The total day time operating hours for Chillers ECA, ECB, ECC and ECD for March2016, June 2016 and Sept 2016 (245 + 175 + 205 + 233 + 193 + 184

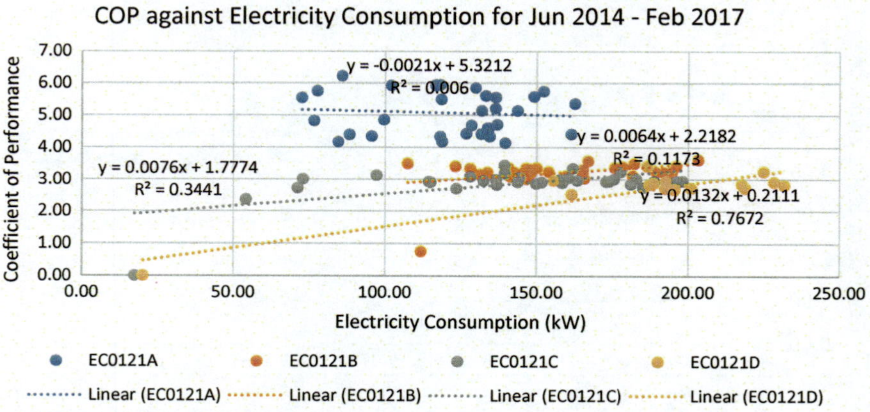

Fig. 4 Plot of COP vs electricity consumption for June 2014–February 2017 for ECA, ECB, ECC and ECD

Table 5 Hours during the day-time operation for ECA, ECB, EC C and ECD operated during March 2016

	Chiller A	Chiller B	Chiller C	Chiller D
No. of days EC were operated to support chilled water demand	31(17 × 10 h, 1 × 9 h, 4 × 8 h, 1 × 7 h, 1 × 6, 2 × 5, 1 × 4 h, 4 × 2 h) 17 days operated for 10 h	27 (3 × 0, 2 × 2, 4 × 1, 1 × 3, 2 × 4, 4 × 5, 3 × 6, 1 × 8, 0 × 9, 11 × 10) 11 days operated for 10 h	30(0 × 1, 5 × 2,2 × 3,0 × 4,3 × 5,5 × 6,3 × 7,3 × 8,0 × 9, 9 × 10) 9 days operated for 10 h	31 (1 × 0, 1 × 1, 1 × 2,0 × 3,1 × 4,2 × 5,5 × 6,2 × 7,3 × 8,2 × 9,13 × 10) 13 days operated for 10 h
Total operating hours	245	175	205	233
Total RTh generated by ECs	61,890	37,189	36,538	56,241
Total operating hours is 858 h generating 191,858 RTh				

+ 182 + 203 + 165 + 201 + 172 + 204) was 2362 h. This equivalent to 787 h monthly average. Average for each chiller was 197 h.

4. The data indicate the four chillers were operated at part load all the time This is not efficient. In order to improve the performance of the chillers VFD should be fitted to the four chillers. This will allow the chillers to perform more efficiently during part load conditions. Retrofitting of VFD can often lead to energy saving of around 20% [3–5].

The ECs were operated between 2 and 10 h during the day.
Analysis of electric rate (Fig. 8 and Table 8).

Table 6 Hours during the day time for ECA, ECB, ECC and ECD operated during June 2016

	Chiller A	Chiller B	Chiller C	Chiller D
No. of days EC were operated to support chilled water demand	193(0/0, 2/1, 1/2, 2/3, 0/4, 2/5, 2/6, 1/7, 1/8, 0,9,14/10)-14 days operated for 10 h	184(4/0, 1/1, 2/2, 2/3, ¾, 1/5, 0/6, 2/7, 2/8, 4/9, 9/10)-9 days operated for 10 h	182(1/0,2/1,2/2, 5/3, 4/4, 0/5, 2/6, 1/7, 2/8, 0/9,11/10)-11 days operated for 10 h	203(2/0,1/1, 3/2, 2/3, 0/4, 2/5,2/6, 1/7, 1/8, 1/9, 15/10)-15 days operated for 10 h
Total operating hours	193	184	182	203
Total RTh generated by electric chillers	42,833	38,851	37,246	47,703
Total monthly daytime operating hours is 762 h generating 166,133 RTh				

Table 7 Hours during the day time for ECA, ECB, ECC and ECD operated during September 2016

	Chiller A	Chiller B	Chiller C	Chiller D
No. of days EC were operated to support chilled water demand	165(8/0,1/1,3/2,0/3,0/4,3/5,0/6,0/7,3/8,1/9,118/10) 18 days operating for 10 h	201(1/0, 1/1, ½, 2/3, 2/4, 4/5, 3/6, 4/7, 3/8, 2/9, 9/10) 9 days operating for 10 h	172(3/0, 0/1,1/2, 5/3, ¼, 5/5/3/6, 1/7, 3/8, 3/9,5/10) 5 days operating for 10 h	204(3/0, 1/1, 2/2, 0/3, ¼, 2/5,0/6, 4/7, 5/8, 3/9, 9/10) 9daysoperating for 10 h
Total operating hours	165	201	172	204
Total RTh generated by EC	34,755	39,836	32,259	42,421
Total operating hours for four EC is 742 h generating 119,271 RTh				

i. Strong linear correlation is observed (except for ECA). Electric rate (average) of ECA was lower percentage than other ECs.
ii. Average electric rate (kWh/RTh) varies from 0.6386 to 1.1847.

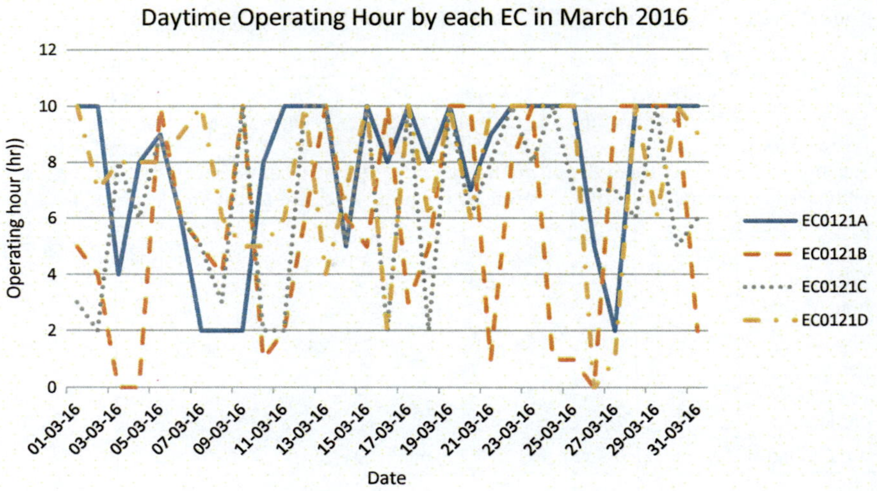

Fig. 5 Daytime operating hours for ECA, ECB, ECC, ECD during March 2016

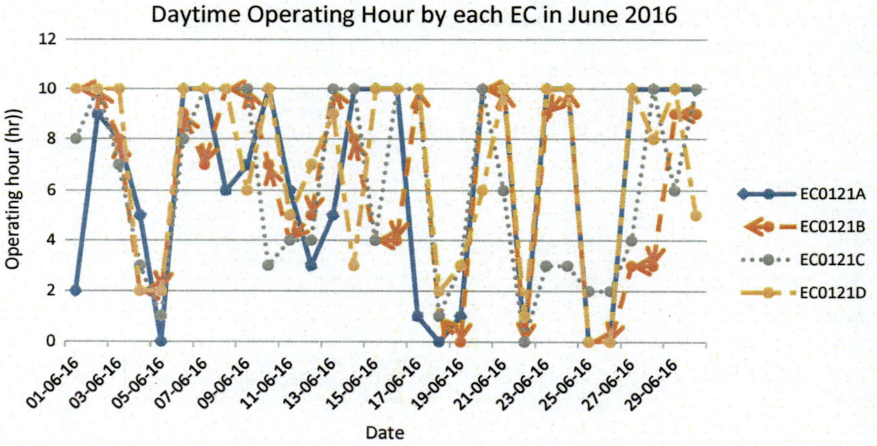

Fig. 6 Day time operating hours of ECA, ECB, ECC, ECD during June 2016

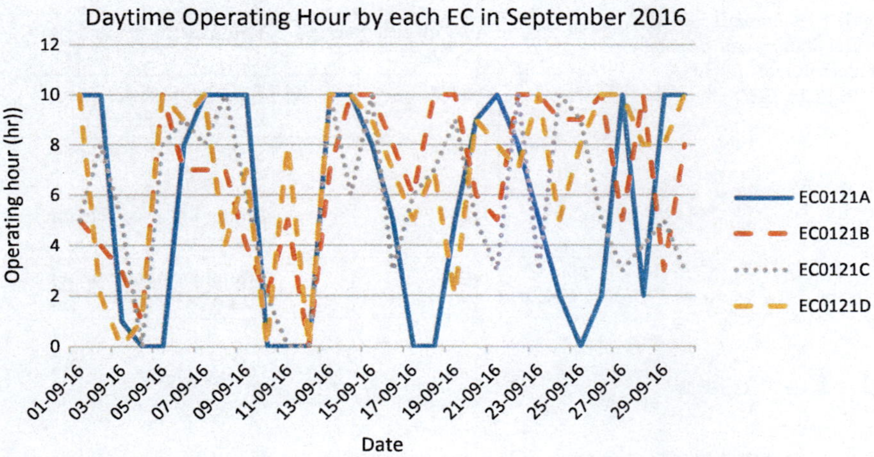

Fig. 7 Daytime operating hours of ECA, ECB, ECC and ECD during September 2016

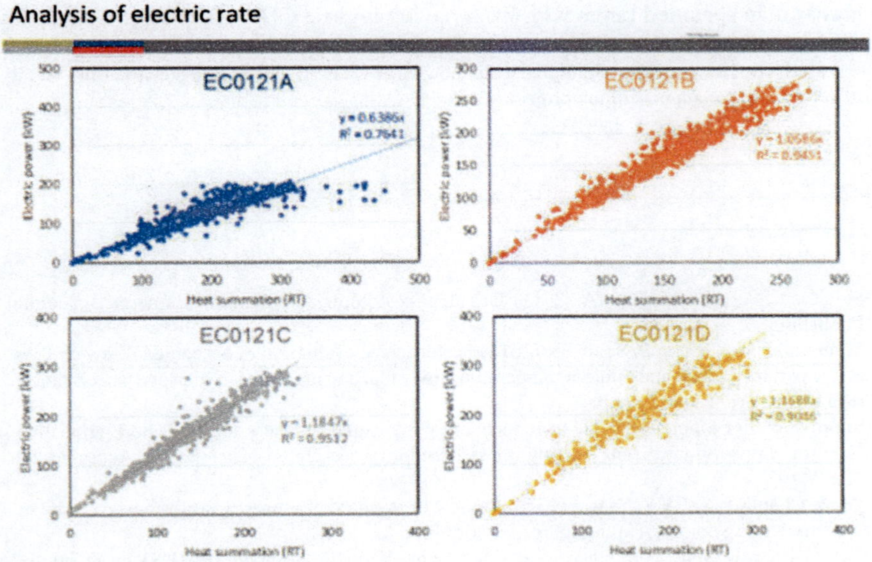

Fig. 8 Correlation analysis of electric rate for ECA, ECB, ECC and ECD

Table 8 Summarise the results of correlation analysis of electric rates for ECA, ECB, ECC, ECD

EC	AVG electric rate (kWh/RTh)	Remarks
A	0.6386	Correlation difference increases in high RT
B	1.0586	Strong linear correlation is observed
C	1.1847	Strong linear correlation is observed
D	1.1688	Strong linear correlation is observed

11 Conclusion

The study noted that all the four ECs were operated more than 10 h during the day. The ECs were operated during the day to support the demand chilled water during the day. Even though the ECs were operated with partial load all the COP are within or higher than specified range with ECA having highest COP of 5.

Acknowledgements Authors would like to acknowledge UTP, HITACHI and Newton Link Grant (British Council) for support of the project.

References

1. Sulaiman SA, Abd Majid MA (2011) Gas district cooling in Malaysia. Lambert Academic Publishing
2. Manimaran W, Chacko S, Soori PK (2014) Performance evaluation of air cooled screw chillers at low part load ratios and outdoor temperatures in Dubai and measures to improve performance. Int J Smart Grid Clean Energy
3. Improve chiller efficiency, How to reduce energy consumption of a chiller, The Engineering Mindset, https://www.theengineeringmindset.com/improve-chiller-efficiency/1. Accessed 22 Mar 2019
4. Yu W, Chan KT, Si RKY, Yang J (2014) Review of standards for energy performance of chiller systems serving commercial buildings. Energy Proc **61**
5. EMSD Energy Land, Appliances and Equipment, Variable speed drives (VS) in air conditioning systems. https://www.emsd.gov.hk/energyland/en/appAndEquip/equipment/airconditioning/vsd.html. Accessed 22 Mar 2019

Computational Analysis of Air Velocity in a Solar Chimney with an Extended Ground Thermal Absorption Medium

Iylia E. Jamil, Hussain H. Al-Kayiem, and Y. S. Chin

Abstract The bulk size of the solar collector is one of the main setbacks in adopting solar chimneys for power generation. The objective of this paper is to investigate the possibility of extending the absorbing medium outside the canopy. The paper presents numerical investigation procedure and simulation results for solar chimney enhancement with extended ground thermal absorbing assembly with no cover. Solar radiation is an important factor in influencing the ambient conditions outside the system. The paper presents numerical investigation procedure and simulation results for solar chimney enhancement with extended ground thermal absorbing assembly with no cover. Through CFD, the velocity and temperature inside the solar chimney are considered performance indicators. The novel, simple, easily implemented, and low-cost approach of adding the extended ground material around the collector increases the temperature in the region before the inlet to the collector and improves the flow inside the system. Results show an increase of up to 8.8% in air velocity at the base of the chimney model with the proposed extended absorbing area, which is around a 29% increase in the air's power at the chimney base.

1 Introduction

Due to the limited resource of conventional energy, the research and development of renewable energy technology have expanded over the decades. The solar chimney power plant (SCPP) is an alternative method for electrical power generation utilizing solar as a resource. An SCPP is a solar thermal system which converts solar energy into electricity. It typically consists of three main components: a solar collector, a chimney tower and a wind turbine. The working principle of the SCPP is simple;

I. E. Jamil (✉) · H. H. Al-Kayiem
Solar Thermal Advanced Research Center [STARC], Universiti Teknologi PETRONAS, 32610 Seri Iskandar, Perak, Malaysia
e-mail: iylia_0008404@utp.edu.my

Y. S. Chin
Department of Engineering and Design, University of Chichester, West Sussex, UK

F. Ahmad et al. (eds.), *ICPER 2020*, Lecture Notes in Mechanical Engineering,
https://doi.org/10.1007/978-981-19-1939-8_8

solar insolation passes through the semi-transparent collector, which heats up the air within the collector passage.

Additionally, it creates a greenhouse effect between the ground and the collector, which further heats the air underneath the collector. The increased air temperature causes buoyancy, which triggers the stack effect. Consequently, continuous updraft is created as hot air leaves through the chimney and ambient air enters from the collector peripheral inlet. This flow provides the kinetic energy that runs through the wind turbine installed at the chimney base. The amount of power due to kinetic energy in the wind stream is $P_{wind} = {}^{1}\!/_{2} m v_{chimney}^2$

The first pilot solar chimney system was built in Manzanares, Spain, in 1982. With a collector radius of 122.0 m, chimney height and diameter of 194.0 and 10.0 m, respectively, the plant was designed to produce a 50 kW electrical power output. From this point, many studies revolve around the Manzanares plant to investigate the defining parameters of the SCPP efficiency. One investigation on analytical and numerical models that consider the influence of ambient conditions and structural parameters on the power output was performed by Bernardes et al. [1]. The results show that the power output increases by increasing chimney height, collector area and transmittance.

To widen the research, the development of a small-scale prototype of the SCPP has increased in popularity up to this day. In Kota, India, a prototype of 8.0 m in height and 12.0 m collector diameter has been built [2]. The experiment shows that solar intensity has an effect on air velocity and ambient air temperature. Kasaeian [3] presented the effects of ambient conditions on the performance of SCPP through a scaled model and dimensional analysis. Roozbeh [4] also proved that increase in solar radiation would increase the power output of the system. In another research, Jiakuan [5] showed that the power output of the power generating system increases with global solar radiation intensity, collector area and chimney height.

In Universiti Teknologi PETRONAS, a small-scale pilot plant has been built with a 6.0 m collector diameter and 6.3 m chimney height. From this model, a series of investigations have been published. Al-Azawiey et al. [6, 7] presented the influence of the collector geometry on SCPP performance. A larger collector area shows a significant increase in the air velocity inside the chimney. In terms of height, the lower height of the collector will provide higher overall efficiency. An optimum collector height-to-diameter ratio is also suggested for designing future models. To target the hours with low solar radiation, Azeemuddin et al. [8], Al-Azawiey et al. [9] and Aurybi et al. [10] have all studied various types of energy storage mediums and assisted thermal energy sources, in addition to solar, that can sustain the SCPP performance throughout the day. Jamil et al. [11] presented a rudimentary analysis for solar chimney enhancement by considering the solar radiation effect on the surrounding environment of the collector, in particular the ambient air temperature. SCPP was simulated as a 2D numerical domain, and the velocity and temperature inside the solar chimney as the ambient temperature of the air outside are varied from 307 to 314 K. At the highest ambient temperature, results show an increase in air velocity of 0.55 m/s at ground level and 1.85 m/s at the height of 1 m along with the chimney. Further, an investigation of the study to include the ambient domain is proposed.

The conclusion from the above literature confirms that ambient conditions are a ruling factor in the performance of SCPP. Despite extensive works that have been done so far, to this date, no work has been published to investigate the effect of incoming air temperature on the system performance, independent of solar irradiance. This paper presents a new method of enhancing the SCPP performance by adding thermal absorbing material on the ground in the immediate region before the air inlet into the system. This extension theoretically increases ambient air temperature near the inlet and increases the air temperature that goes into the collector. The main aim of the numerical simulation in the present work is to allow flow visualization for qualitative evaluation and to measure the velocity of the flow at various locations.

2 Computational Implementation

The numerical simulation is carried out for three cases. Case 1 is a conventional SC model. Case 2 and Case 3 are enhanced SC models with extended thermal absorption medium around the collector region, as shown in Fig. 1. Experimental data has been

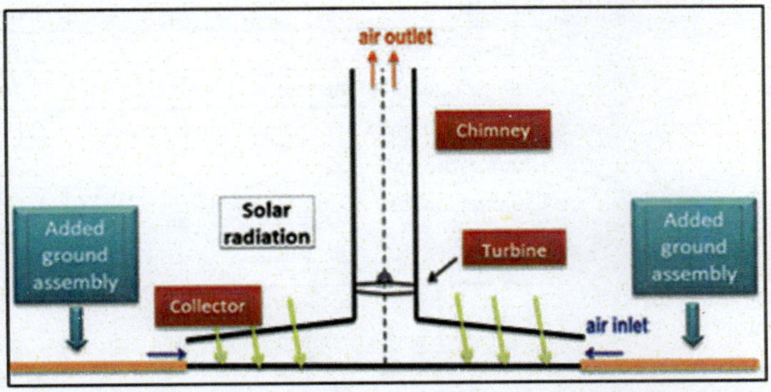

Fig. 1 Schematic diagram of solar chimney power plant with added ground assembly for thermal absorption

collected from the experimental measurement of utilizing the experimental SC model in the solar site in UTP. The experimental investigation results have been used to validate the numerical results obtained from the simulation presented in this paper.

2.1 System Modelling

The SC is modeled using according to the geometry shown in Table 1. All cases are modeled according to the geometry described.

The collector consists of the collector cover i.e. the canopy and the ground. The collector roof, or the canopy, is modeled at two inclination angles, as shown in Fig. 2. The inner canopy radius of 1.3 m slopes at 20° from the horizontal. The outer canopy extends from the inner canopy and slopes at 5°. The inclination angles are designed as such to enhance and channel buoyant air stream to the chimney and avoid the abruptness of air convergence.

The study involves creating an ambient domain to visualize the airflow going into the system, specifically over the extended ground medium in Case 1 and Case 2. The flow and the temperature distribution at immediate regions in the extended ground can be observed by creating a far-field boundary. The ambient domain is created as a rectangular box around the SC model. The ambient domain extends 2 m to the right from the collector inlet and 2 m above the chimney outlet. By comparison, the simulation with a larger ambient domain is also performed. It shows that the influence on the accuracy of numerical simulation results for the objective of the presented work is negligible.

Table 1 Geometry of base model

Parameter	Dimension
Collector radius	3.30 m
Collector height at the inlet	0.10 m
Collector angle of inclination from horizontal	20°, 5°
Chimney height	14.0 m
Chimney radius	0.076 m

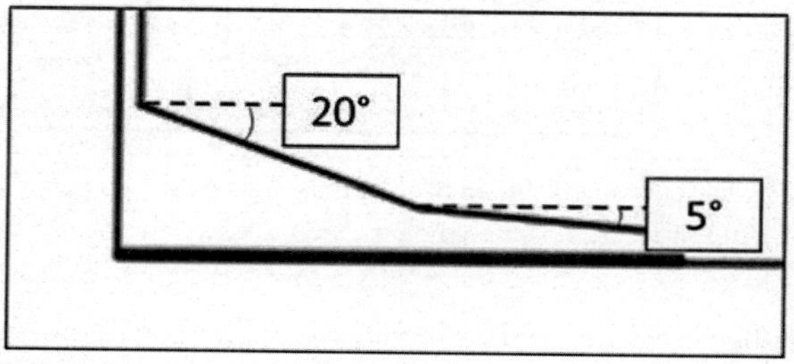

Fig. 2 Canopy slop angle of the collector roof

2.2 Numerical Solution

The fluid motion inside the model is solved through the conservation equation of mass, momentum, and energy in ANSYS Fluent software. Simulation is carried out assuming steady-state, turbulent, 2-D flow.

The incompressible flow assumption is justified for the model since the maximum Mach number is much smaller than 0.1. The buoyancy effects are considered by applying the Boussinesq model in (1) to allow a variation of density as a function of temperature. This model treats density as a constant value in all solved equations, except for the momentum equation's buoyancy term. This model accounts for the full, bouncy effect, where the density varies with temperature, and the flow is motivated by the force of gravity, which influences the change in density.

$$(\rho - \rho_o)g \approx -\rho_o\beta(T - T_o)g(\rho) = \rho_o(1 - \beta\Delta T) \tag{1}$$

Equation (1) is valid when $(1 - \beta\Delta T) \ll 1$.

The radiation transfer equation was solved by Discrete Ordinates (DO) model. Heat transfer in the SCPP system involves all three modes: conduction, convection, and radiation. Radiation heat transfer mainly occurs in the collector, which is covered by different types of semi-transparent material. The canopy material is nearly transparent for incident solar radiation but partly opaque for infrared radiation from the ground. In the present simulations, the discrete ordinate (DO) radiation model was adopted to solve the radiative transfer equation. It is the only model that can be used to model semi-transparent walls of various types. The DO model can compute non-gray radiation using a gray band model and works well across a full optical thickness range.

In a 2D numerical simulation of the solar chimney, incident solar radiation on the ground through the semi-transparent collector is commonly treated as an internal heat source or heat flux. [12].

2.3 Boundary Conditions and Material Selection

The boundary conditions for the model are presented in Fig. 3. Creating a symmetry wall along the center of the SC allows a reduction of the computational model's physical size, thus reducing the computing time. The inlet to the system is the edge of the domain parallel to the collector inlet, while the outlet is the edge of the domain above the chimney outlet. Both the inlet and outlet are set at zero gauge pressure.

The collector thermal conditions are set as coupled, with an internal emissivity of 0.9 and heat generation rate of 8 W/m^3. The solar radiation intensity is set at 800 W/m^2 based on experimental values. The collector is treated as a semi-transparent wall, where the Discrete Ordinates model will solve the radiation. At the base of the model are two separate components. The ground under the collector is a zone set as

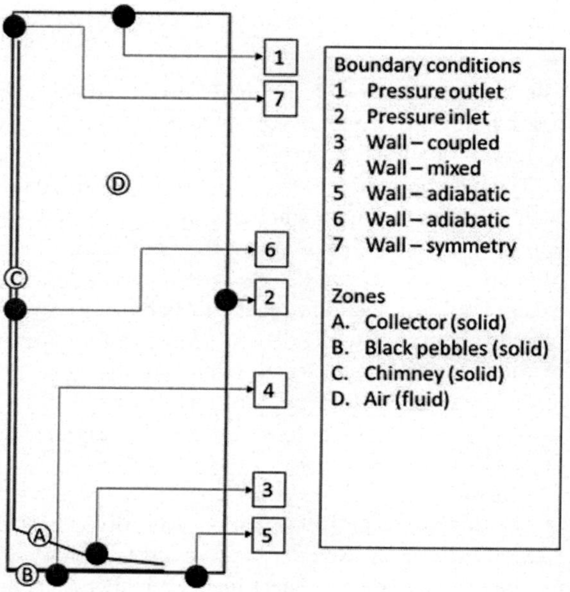

Fig. 3 Main zones in the 2D model of the solar chimney (A, B, C and D) and the boundary condition at each surface

black-painted pebbles. The temperature of the zone is 320 K, with a heat transfer coefficient of 8 W/m^2·K. Outside the collector, the absorbing area is the normal ground set as an adiabatic wall.

The material selected for the canopy is Perspex sheet with a thickness of 4 mm. A transmission coefficient of 0.92 [10] allows solar irradiation to penetrate through the collector to create a greenhouse effect. The ground is covered with a thermal absorption medium. The medium chosen for this set-up is black-painted pebbles, based on a reported experiment [9]. Such a type of ground has absorptivity of 0.9 and emissivity of 0.95. Black painted pebbles with a thickness of 10 cm can increase the solar intensity absorption of the ground. This is applied to all cases in the simulation. For Case 2 and Case 3, an extension of the ground medium is created outside the canopy for an extra radius of 1 and 2 m, respectively. A summary of the material properties used in the simulation is presented in Table 2.

"SIMPLE" algorithm was applied to solve the calculations. A structured grid was adopted, with the grid near the walls being refined adaptively. The second-order upwind scheme was chosen for the convective terms. Simulation results using different grid distributions and different grid refinement strategies indicated that the grids are sufficiently fine to obtain grid-independent solutions. The simulation model is built with 111,884 elements and 112,108 nodes using quadrilateral elements.

Table 2 Material properties for simulation

Physical property	Collector	Ground	Chimney
Material	Perspex	Black pebble	PVC
Absorption coefficient	0.06	0.9	0.04
Transmission coefficient	0.92	0	0
Density (kg/m^3)	2700	2640	833
Specific heat (J/kg·K)	840	820	1170
Thermal conductivity (W/m·K)	0.78	1.73	0.19
Emissivity	0.9	0.9	0.91

3 Results and Discussion

The results of air velocity and temperature obtained from the simulation at various parts of the system are presented in this section.

3.1 Validation

To validate the model, the results for the velocity at the chimney base and the air temperature difference inside the system from the numerical simulation are compared against experimental data. Table 3 shows the summary of the comparison of the result.

As the simulation is conducted based on constant ambient conditions, the data referred to from the experiment is taken at the closest values. Several parameters like ambient wind and humidity are not taken into account for the simulation. This coarse assumption may be more or less influence the accuracy of the simulation results compared to the practical working conditions outside the SC but will not significantly influence the main objective of this paper. The model is accepted for further investigation based on the percentage difference between the experiment and simulation data.

Table 3 Comparison of data between simulation and experiment

Parameter	Experiment	Simulation	Difference (%)
Velocity at chimney base	1.256 m/s	1.102 m/s	12
Δ T	3.8 K	4.1 K	7.9

3.2 Velocity Analysis

Velocity results are taken at several points inside the system. At the chimney, the velocity at the base is shown in Fig. 4. Velocity values are plotted against the chimney's radius at the respective locations, from radius 0 m at the center to radius 0.15 m at the chimney wall. The velocity approaches zero at the wall of the chimney and is maximum at the center. At the chimney base, the maximum velocity for Case 1 is 1.02 m/s, which increases by only 1.25%. For Case 3, the velocity is 1.1 m/s which indicates an 8.8% improvement from the conventional model.

Figure 5 shows the plot of velocity at the center of the chimney along with its height. The average velocity for all three cases is 0.95, 0.97 and 1.08 m/s, respectively, for Case 1, Case 2 and Case 3. It is observed in all cases that as the flow enters the

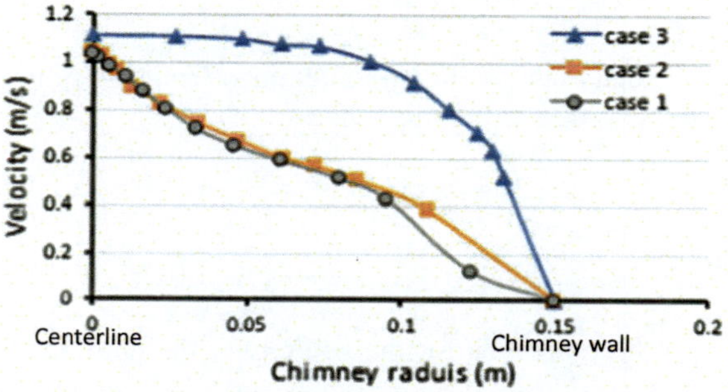

Fig. 4 Velocity profile along chimney radius at chimney base

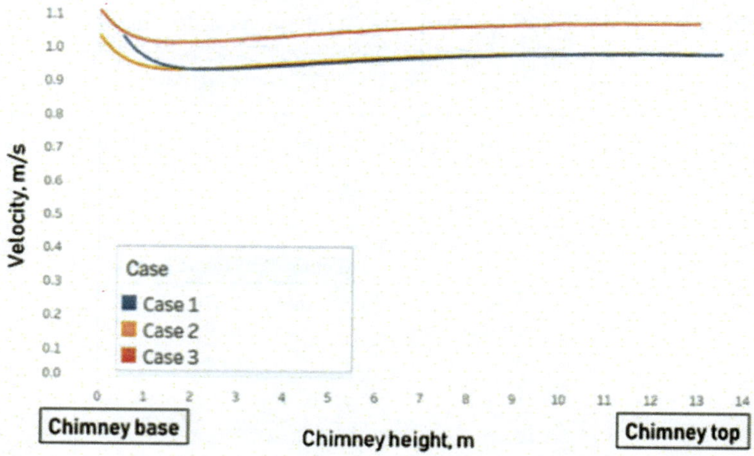

Fig. 5 Velocity along chimney height

chimney, there is a slight dip in the velocity at the first 1 m height before the velocity picks up again to match the velocity at the entrance.

At the inlet into the solar chimney, the flow velocity can be seen in Fig. 6, the plot from the base of the inlet to the canopy roof. The average velocity of airflow at the inlet for Case 1 is 1.46 m/s, for Case 2 is 1.47 m/s, and for Case 3 is 1.51 m/s. In all cases, the velocity approaches zero as the air flows under the roof of the collector.

Figure 7 shows the air velocity under the collector from the entrance to a distance of 2 m from the entrance. At all points under the collector, the velocity of Case 3 shows the highest value throughout, while the trend that the velocity increases as the airflow towards the chimney are true for all cases.

Flow visualization can be observed in Fig. 8, taken from the results for Case 3 at various solar chimney locations. At the collector inlet, there is a surge of airflow going into the system. The maximum velocity observed within the domain is visible at local points near the collector inlet. This is tied to the pressure drop at the collector inlet, which can be seen in the pressure contour shown in Fig. 9.

3.3 Temperature Analysis

With the inclusion of the ambient domain in the simulation, the extended ground assembly temperature is visible for observation, as shown in Fig. 10. The argument

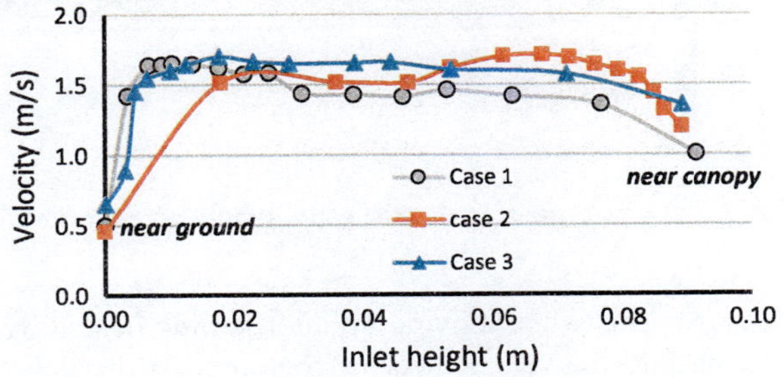

Fig. 6 Velocity at the collector inlet

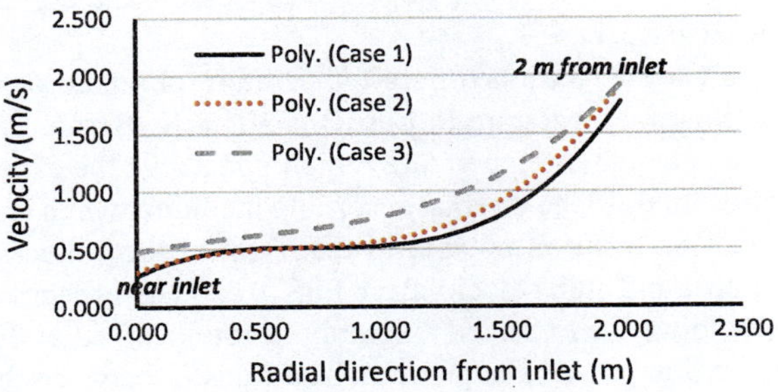

Fig. 7 Velocity along collector radius

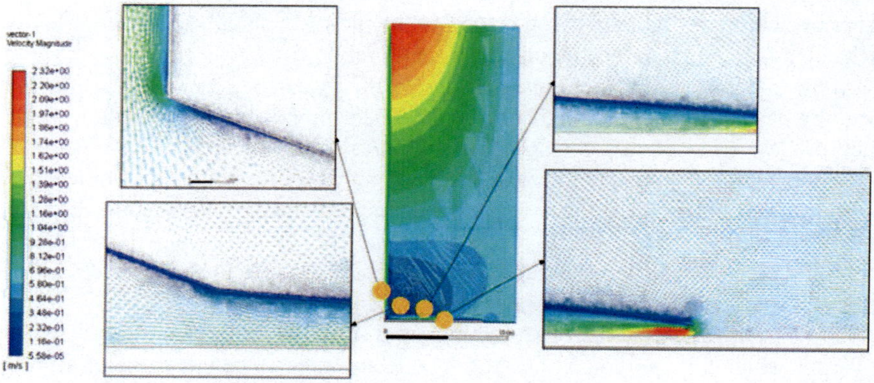

Fig. 8 Airflow velocity vector at various locations in the domain for Case 3

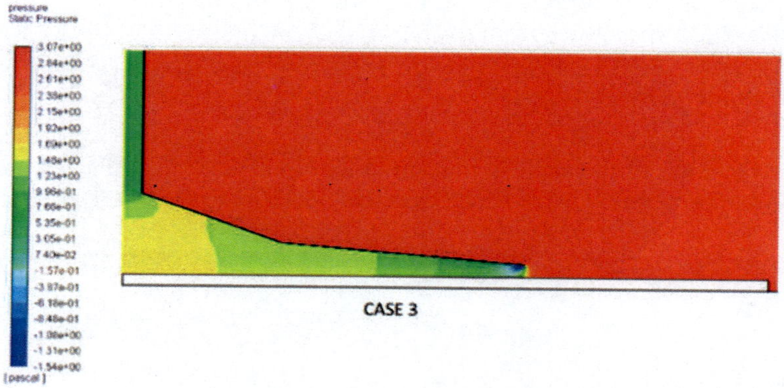

Fig. 9 Pressure contour around the collector inlet area for Case 3

that air particles are moving up after getting heated by the ground is discussed by looking into the velocity vector contours. Air particles are excited and get kinetic energy near the extended ground and move towards the inlet of the collector. The particle's velocity with extended areas, case 2 and case 3, are higher compared to the normal case, case 1.

In Case 2, the maximum temperature observed at the ground is 302.7 K. The maximum temperature in Case 3 is 305.8 K. Before the entrance, it is evident that there is a higher temperature region just above the extended ground assembly. This corresponds to the thermal absorbing medium, which is set up with a heat generation rate. This result is notable in that right at the collector inlet, which shows lower pressure and high velocity from Figs. 9 and 10, respectively, the temperature increase seen above the extended assembly is not observed at this location. The air surge in this region prevents convection heat transfer between the ground and the air stream from taking place.

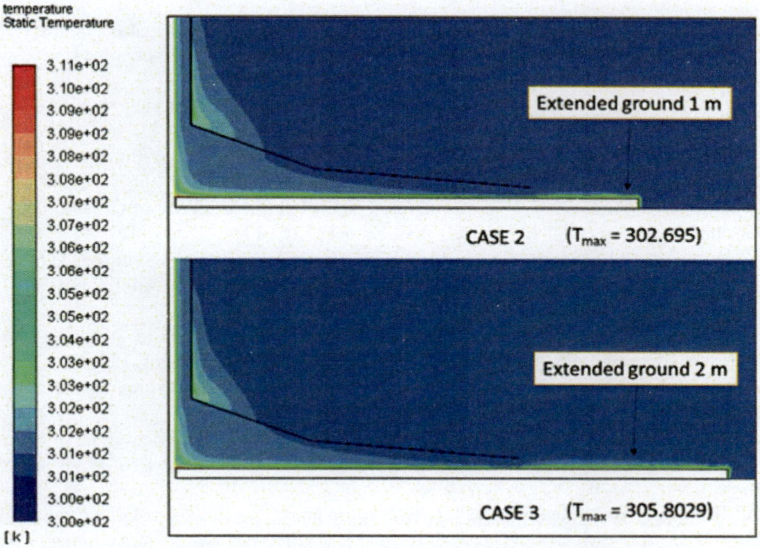

Fig. 10 Temperature contour around the collector inlet area for Case 2 and Case 3

4 Conclusion

The solar chimney is simulated by the validated numerical procedure. The far-field approach is adopted in the current study to avoid the unjustified assumption at the inlet and outlet of the system. Results show a notable increase in air velocity at all locations inside the solar chimney when the extended thermal absorbing ground is considered. This investigation demonstrated the possibility of reducing the canopy area and extending the ground surrounding the collector inlet covered by black-painted pebbles. The findings are recommended to be adopted in the practical application of the solar chimney for power generation, where it imposes cost reduction as the canopy area is reduced.

Acknowledgements The authors acknowledge Universiti Teknologi PETRONAS for financial support to present the paper utilizing YUTP FRG with CS: 015LC0-404.

References

1. Bernardes AS, Weinrebe VA (2003) Thermal and technical analyses of solar chimneys. Sol Energy 75:511–514
2. Lal S, Kaushik SC, Hans R (2016) Experimental investigation and CFD simulation studies of a laboratory scale solar chimney for power generation. Sustain Energy Technol Assess 13:13–22

3. Kasaeian A, Amirifard M, Ahmadi MH, Kasaeian F (2017) Investigation of the effects of ambient temperature and dimensional parameters on the performance of solar chimney power plants. Int J Low-Carbon Technol 12:335–348
4. Sangi R, Amidpour M, Hosseinizadeh B (2011) Modeling and numerical simulation of solar chimney power plants. Sol Energy 85(5):829–838
5. Zhou X, Yang J, Xiao B, Hou G (2007) Simulation of a pilot solar chimney thermal power generating equipment. Renewe Energy 32(10):1637–1644
6. Al-Azawiey SS, Al-Kayiem HH, Hassan SB (2017) On the influence of collector size on the solar chimneys performance. MATEC Web Conf 131:02011
7. Al-Azawiey SS, Al-Kayiem HH, Hassan SB (2016) Investigation on the influence of collector height on the performance of solar chimney power plant. ARPN J Eng Appl Sci 11(20):12197–12201
8. Azeemuddin I, Al-Kayiem HH, Gilani SI (2013) Simulation of a collector using waste heat energy in a solar chimney power plant system. WIT Trans Ecol Environ 179:933
9. Al-Azawiey SS, Hassan SB (2016) Heat absorption properties of ground material for solar chimney power plants. Int J Energy Prod Manag 1(4):403–418
10. Aurybi MA, Al-Kayiem HH, Gilani SIU, Ismaeel AA (2017) Numerical assessment of solar updraft power plant integrated with external heat sources. WIT Trans Ecol Environ 226:657–666
11. Jamil IE, Sing CY, Al-Kayiem HH (2020) Rudimentary analysis for solar chimney enhancement. IOP Conf Ser Mater Sci Eng 863(1); 5th UTP-UMP-uaf symposium on energy systems, SES 2019 Kuantan. Malaysia, pp 1–2. Article number 012063
12. Inc. A 2016 ANSYS Fluent Theory Guide v17.1 ANSYS 17.1 Doc

Preliminary Numerical Evaluation of Temperature Fluctuation in Two Different Converging Mixing-Tee Junctions

William Pao and Hazizam Zamri

Abstract MLNG-Tiga of Bintulu decided to change their aging mixing-tee configuration in one of their natural gas regenerative heat cycle piping. In the original converging T-junction, hot gas is flowed in the main pipe with cold gas intersecting midway, resulted in thermal mixing. Even though the new design still utilized the concept of converging T-junction, the hot and cold gas are routed to collide with each other before exiting as mixed gas. Since the thermal mixing efficiency is not known, the relative performance of these two mixing-Tee could not be evaluated. The present study focused on the evaluation of thermal mixing of two different T-junctions, namely an intersecting Tee and a colliding Tee, both operated within limited parameters, with the intention to understand the thermal performance. The influence of different configurations of inlets and outlet was investigated numerically using the standard K-ε turbulence model with Liquified Natural Gas (LNG) as the working fluid. Literature survey revealed that a lot of research on T-junction has been conducted on geometric factors such as diameter ratios and different fluid properties, surprisingly, a comparative study on the effects of different flow configuration is yet to be found. The present research showed that due to different impact configurations, the colliding tee offered better thermal mixing compared to the intersecting tee, resulting in higher level of turbulent. Consequently, colliding tee will experience higher temperature fluctuation at 0.25D downstream of the mixing intersection, with possibility of shortening of its service life.

1 Introduction

The Malaysia Liquified Natural Gas (MLNG) plant in Bintulu receives its natural gas feedstock from offshore waters of Sarawak. Amongst the many complex process trains in MLNG, one of them is the natural gas regenerative cycle, which serves to remove moisture from the feedstock. In one part of the regenerative cycle, hot and

W. Pao (✉) · H. Zamri
Mechanical Engineering Department, Universiti Teknologi PETRONAS, 32610 Seri Iskandar, Perak, Malaysia
e-mail: William.Pao@UTP.edu.my

F. Ahmad et al. (eds.), *ICPER 2020*, Lecture Notes in Mechanical Engineering,
https://doi.org/10.1007/978-981-19-1939-8_9

cold gas are mixed in the T-junction before the thermally mixed natural gas is sent downstream for further processing. As the piping aged and repeated detection of through cracks in the mixing junction, MLNG decided to rehabilitate the system by partly redesigning the most problematic section, i.e. the mixing T-junction.

In the original mixing junction design, see Fig. 1a, hot gas is flowed in the horizontal main pipe while cold gas is injected from the vertical branch pipe and mixed in the T-junction. In the newly proposed design, see Fig. 1b, hot and cold gas are set to collide with each other and mixed in the T-junction before existing through branch pipe. Even though both flow configurations are categorized as converging T-junction in the literature because of the merging of the flow stream, the thermal mixing behaviour is likely to be very different in between these two T-junctions.

Without prior knowledge of thermal mixing performance of mixing-tee, it is impossible to inform scientifically if the 'renovation' would result in better performance. The objective of this research is to numerically compare the thermal mixing behaviour between two flow configurations of converging T-junction.

2 Literature Review

The T-junction is a significant component of the structure of the pipeline. It can differ in form and is mainly used to divide or combine pipeline flow. The T-connection can, however, be used for the mixing of two liquids, liquid and gas or two gasses. Thermal fatigue in pipeline is a major issue in the gas industry [1]. Thermal mixing is a situation where streams of hot and cold flows mixed, causing temperature fluctuations and exchange of energy between hot and cold streams, resulting in a single stream with averaged temperature. These temperature fluctuations, often in a random pattern, are significant contributors to cyclic thermal stresses, which can lead to eventual pipe failure in the long term [2].

The issue of thermal fatigue has been the interest of researchers since decades ago. Previous work spans out from the 1980's where experimental investigations were carried out. Experimental investigations were conducted by [3–5], to investigate mixing of two different fluid streams at a mixing tee. McFarland and Landy [3]

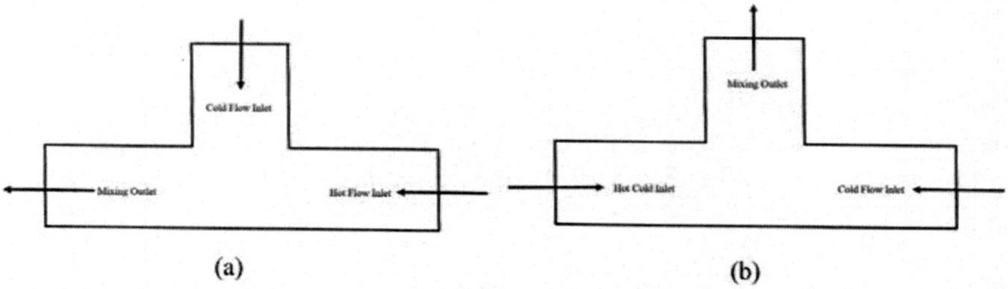

Fig. 1 **a** Original intersecting Tee and **b** new design colliding Tee

discussed the data obtained from steady and transient experiments and suggested good mixing requirements of a mixing tee. Maruyama et al. [4, 5] described analytical procedures to optimize mixing conditions and concluded that degree of mixing is highly dependent on geometrical features of the tee. Researchers such as in the works of [7–9] have also conducted research to find best possible methods to extend life of piping systems by reducing the thermal fatigue. Ogawa et al. [10] studied the influence of upstream elbows of the mixing tee on temperature and velocity distributions. The measurements were taken using thermocouples and particle image velocimetry (PIV). Their results showed the temperature fluctuation intensity was larger with the elbow and concluded that this was due to the vortex generation behind the branch pipe jet. Ayhan and Sokmen [11] investigated the velocity and temperature fluctuations numerically by comparing Reynold's Averaged Navier–Stokes (RANS) and Large Eddy Simulations (LES) models. They further discovered that the frequency range 2–5 Hz possesses the most energy and identify this as the main culprit frequency contributed to thermal fatigue.

In the experimental study of Zhou et al. [12], temporary or continuous cyclic mixing processes were investigated by flowing different fluids of different temperatures into the mixing tee. The cold flow was kept constant at 20 °C while hot flow was varied in between 240 °C and 160 °C. Measurements were taken using thermocouples similar to [10] at a sampling rate of 100 Hz. Their investigation showed that inner wall areas of the pipes are prone to material damage due to thermal fatigue and could be identified with the developed spectrum patterns. The change in temperature difference has significant influence on the thermal fatigue downstream of the tee. Chen et al. [13] tried to correlate the outlet mixing temperature by varying the inlet temperatures and flow rates. Their work concluded that lower main pipe flow rate will lead to better mixing effects, and turbulence effects could be improved with a higher flow rate from the branch pipe.

To further improve CFD results of the particular filed of investigation and to more accurately evaluate thermal fatigue in mixing tees, Qian et al. [1] has introduced the application of the dynamic Smagorinsky model (DSM) that is to be used with the Large Eddy Simulation (LES) sub-grid scale (SGS) turbulence model. His simulations show that temperature fluctuations intensity is remarkably close to experimental results. However, the model required high computational capability and may not be feasible. Zhou [14] presented a mixing envelope diagram which can be utilized as an operation guideline to reduce thermal fatigue at the mixing tee. Mixing quality was assessed by Zughbi et al. [15] where they recommended mixing tees with inclined angle produced better mixing quality than mixing tee with 90° right angle.

Lee et al. [16] used Large Eddy Simulations to investigate temperature fluctuations and the structural response from the mixing tee. They concluded that the differential temperature between hot and cold stream is the single most dominant factor contributed to thermal fatigue. This conclusion is similar to the findings of Zhou et al. [12]. Selvam et al. [17] investigated thermal mixing of flows in horizontal T-junction with low branch velocities. Constant flow rates with different temperature ranges between hot and cold flows were conducted. His results showed that unstable stratified flow was subjected to high oscillations in temperatures. Their experiments were in the frequency range of 0.2–2 Hz, with no dominant frequency found. This, however, was not inline with the conclusions made by [11, 14], of which thermal mixing occurred at dominant frequency in the range of 2–5 Hz. Kickhofel et al. [18] experimentally investigated the mixing tee with thermally driven density stratification. The experiment was similar to [17], where low flow rates were fixed. The experimental outcome agreed with [12, 16], that large differential temperature leads to strong turbulence and large local thermal gradients.

The literature survey revealed that there were many investigations of thermal mixing using T-junction, mostly focusing on flow rates, variation of hot and cold temperature inlets. Surprising, all the published paper focused on the mixing-tee configuration as shown in Fig. 1a; no paper is found on mixing-tee with flow configuration as shown in Fig. 1b. This is a research gap by which the present research is intended to fill.

3 Methodology

3.1 Modelling of Non-isothermal Gas Flow in CFD

To analyse the thermal fluid flow problem, it is required to follow the correct methodology of CFD. The pre-requisite for computational fluid dynamic simulation is pre-processor, solvers and post-processor. The flow chart in Fig. 2 will be used as the basis.

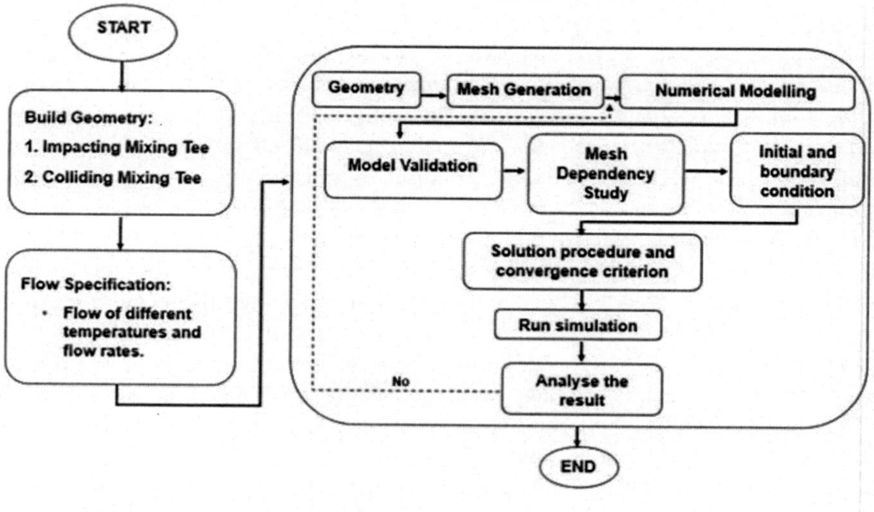

Fig. 2 Methodology of modelling

3.2 DOE and Model Geometry

Design Of Experiment (DOE) is adopted to ensure a structured variation of influencing factors can be evaluated with minimum time of the simulations. In this paper, only two factors are considered, i.e. temperature and flow rate. A two-level design of low and high values has been selected in Table 1 based on the input from MLNG engineers. Altogether, 16 cases for different factors combination has to be analysed.

The mixing-tee geometries are shown in Fig. 3 and the corresponding data in Table 2.

3.3 Numerical Model

To solve the simulation in the mixing tees, a set of governing equations, turbulence models and boundary equations as appropriate were used to investigate the thermal mixing [20].

The governing equations are as follows:

Continuity equation:

$$\nabla \cdot (\rho \mathbf{v}) = 0 \tag{1}$$

Table 1 Design of experiment for the present research

Factor	Level	Cold flow	Hot flow
Temperature (°C)	Low	21	120
	High	120	360
Flow rate (Kg/S)	Low	1.67	3.31
	High	9.07	10.71

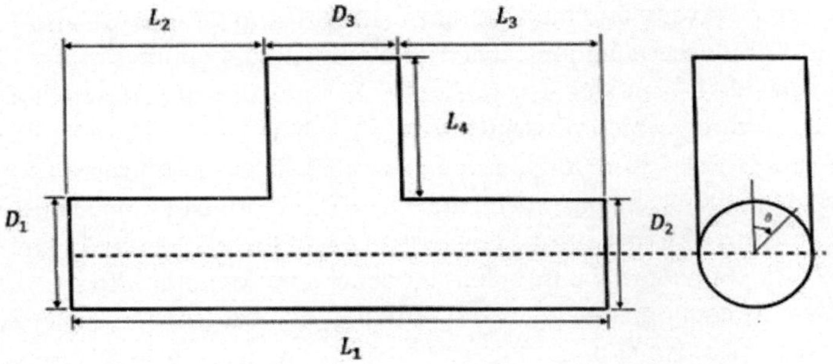

Fig. 3 Schematic diagram of mixing tees

Table 2 Geometric parameters for mixing-tee

Geometric parameter	Impacting mixing tee	Flow configuration	Colliding mixing tee	Flow
Diameter, D_1(mm)	304.8	Outlet	304.8	Cold inlet
Diameter, D_2(mm)	304.8	Hot inlet	304.8	Hot inlet
Diameter, D_3(mm)	304.8	Cold inlet	304.8	Outlet
Length, L_1(mm)	2200.0	–	2400.0	–
Length, L_2(mm)	1047.6	–	1047.6	–
Length, L_3(mm)	847.6	–	1047.6	–
Length, L_4(mm)	444.0	–	1500.0	–

Momentum equation:

$$\frac{\partial}{\partial t}(\rho \mathbf{v}) + \nabla \cdot (\rho \mathbf{v} \otimes \mathbf{v}) = -\nabla p + \nabla\left(\mu^{*}\nabla \cdot \mathbf{v}\right) + \rho \mathbf{g} \tag{2}$$

where, $\mu* =$ effective viscosity = $\mu + \mu_t$.

Energy equation:

$$\nabla \cdot \left(\rho C_p \mathbf{v} T\right) = \nabla \cdot \left(\lambda^{*} \nabla T\right) \tag{3}$$

where, $\lambda* =$ effective conductivity $= \lambda + \lambda_t$.

The turbulence induced viscosity and conductivity, μ_t and λ_t respectively are to be evaluated based on the turbulence model adopted in the current simulation.

The Reynolds-Averaged Navier–Stokes (RANS) model utilizes viscosity terms to close the turbulence equations. The turbulence model applied to the simulation was the Realizable K-ε model. The realizable K-ε model was selected due to a good performance in flows which possesses strong gradients in temperature or pressure, separation and recirculation due to turbulence [19]. Two variables are solved which are the turbulent kinetic energy, k, and the rate of dissipation of kinetic energy, ε. The technique has proved to offer good convergence and the reduction of computational effort. Further detail about the turbulent model can be found in [20].

The transport equations were solved utilizing the academic package of ANSYS Fluent Version 19.2. No slip as well as adiabatic conditions were given to the wall. The boundary conditions at inlets were set up as mass flow inlets based on Fig. 5. The outlet boundary condition was set as the outflow. The transport equations were discretized utilizing the second order upwind scheme, similar to [2]. Pressure values were interpolated through the standard scheme and the SIMPLE algorithm was used to solve transport equations. The equations were set to be considered converged when residual values were below 1×10^{-6}.

Although there are some arguments in the use of the K-ε model due to accuracy issues especially when compared to Large Eddy Simulation (LES), for the purpose of this study it is adequate to study the thermal behaviours of the flow as stated in [21] where both models predict similar mixing behaviours at the same areas in his study. [22] has also stated that unsteady-state turbulence models including URANS are suitable for resolving temperature oscillations. Due to limitations of computational capability as well as the academic package of ANSYS Fluent which offers limited use, the current simulation was run using a computer with Intel® Core™ i7-7700HQ CPU and 8 GB of RAM.

3.4 *Mesh Convergence Test*

There are a few mesh types to be applied to a geometry in finite volume solvers such as tetrahedral, hexahedral and ply meshes [23]. To discretize the fluid model, a combination of tetrahedral and hexahedral meshes were used to minimize error.

The inlet branches and outlet branches of the mixing tee are set to have hexahedral mesh which gives a smooth outlook and reduces the numerical diffusion error [24]. However, to achieve a good and smooth mesh transition, the intersecting and mixing areas of the T-junction were discretized using tetrahedral mesh. These areas are finer in its discretization so as to capture the non-linear behaviour of thermal mixing. Reason for using hybrid mesh combination is to ensure the overall quality of the mesh skewness was kept in between 0 and 0.8 and its orthogonal quality within 0.2–1.0 according to the mesh metrics.

In addition to this, inflation with smooth transitions were applied near the vicinity of the walls to capture accurately near-wall flow patterns. To better assure the stability in computational use and results quality, a mesh independency study was performed. The mesh in the current study were generated within the pre-built meshing module in ANSYS. As the K–ε turbulence model was adopted, refinement of the grid before simulation is crucial. To measure this, it is important to assure the y + value for the generated mesh is less than 1. Figure 4 showed the details of different mesh refinements and its parameters. The maximum pressure was quantified from 5 different element sizes. As the results showed in Fig. 5, the differences quantified in total pressure in Mesh #4 and #5 was less than 1%, Mesh #4 was considered a reasonable choice and used in the present study unless otherwise stated.

Mesh	Element Size (Mm)	Number of Elements	Mesh Visualization on Outlet
1	30	48428	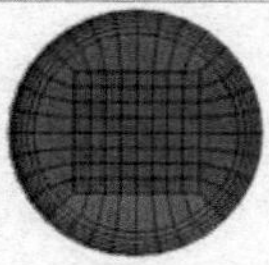
2	25	75651	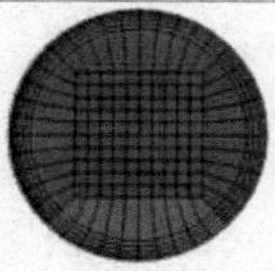
3	20	115739	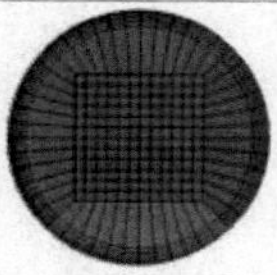
4	15	272062	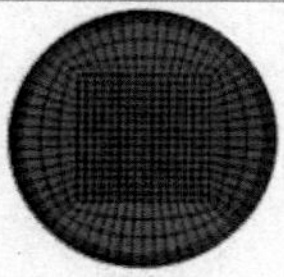
5	10	454805	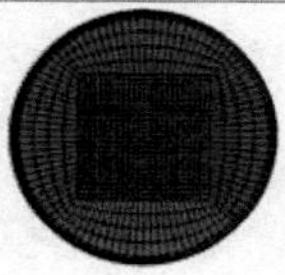

Fig. 4 Mesh dependency tests

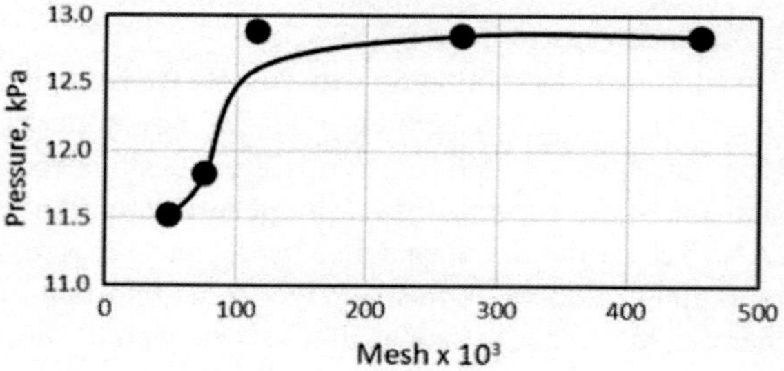

Fig. 5 Pressure convergence versus number of elements

4 Results and Discussions

4.1 Model Validation

The present CFD model was first validated using an example taken from Kuhn et al. [24], which was also using the commercial package of ANSYS Fluent. The algorithm used for pressure and velocity coupling was PRESTO, with a second order bounded differencing scheme (BCD) to approximate the diffusive and convective parts of the discretized equations in the model. The mesh volume was significantly higher (1,900,800 elements) in their research compared to the present study. Large Eddy Simulation was used in their research to predict the fluid temperature and velocity fields. As of the geometry of the T-junction, similar diameters were used for the main and branch pipe.

For the inlet boundary condition, cold fluid entered the main pipe via the side-arm with a velocity of 0.85 m/s and temperature of 8°C. The hot fluid flowed inside the main pipe with velocity of 2.55 m/s and temperature of 83 °C. The working fluid used in the simulations were water. Their simulation was directly validated against experimental results, which is not available to the present authors.

Figure 6 showed velocity contour of present model compared with [24]. It should be noted that in the present study, limitations to computational facilities and of FLUENT academic version, the contour is acceptable even though it could be better.

As pointed out in the literature, the most accurate turbulent model for thermal mixing is LES, even though many authors still preferred K–ε model. With limited computational capability of the present research, LES simulations were not investigated. However, it is an interest to know the departure of K–ε from LES, to allow one to establish an intuition of sense of error. Figure 7 compared the temperature fluctuation of the current study against a similar example taken from [11] using LES turbulence model. There is a notable difference in capability of each turbulence models in the analysis of thermal mixing. LES is able simulate closer-to-reality turbulent, while K-ε only showcased the relative region of turbulent intensity. From current application and objective point of view, K-ε is sufficiently refined.

Fig. 6 Comparison of captured velocity field between **a** Kuhn et al. [24] and **b** present model

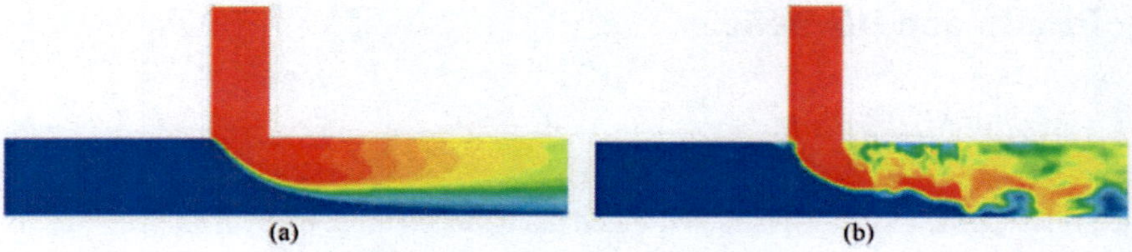

Fig. 7 Comparison of temperature fluctuation between **a** present model and **b** LES model [11]

4.2 Spatial Temperature Distribution

Figure 8 shows a schematic diagram of the colliding mixing tee, and Fig. 9a, cross-sectional temperature distribution at different sampling points along the mixing outlet. The variation of the mean temperature at centerline 0d denotes that this is the mixing region for both fluids. The further away from the mixing outlet, the mean temperature becomes lower until reaching an average temperature of 502 K. Figure 9 also compared the cross-sectional temperature distribution between the present model and the result by Dimonfe et al. [6].

Figure 10 shows temperature distribution for the intersecting T-junction and Fig. 11a cross-sectional temperature distribution at different sampling points along the mixing outlet. The sampling points were the same as in Fig. 9 for the mixing outlet.

A qualitative comparison between Figs. 8 and 10 indicated that thermal stratification, where the gas is not mixed thermally, is likely to form in intersecting T-junction. A qualitative comparison between Fig. 11a, b also indicated there is a stark difference

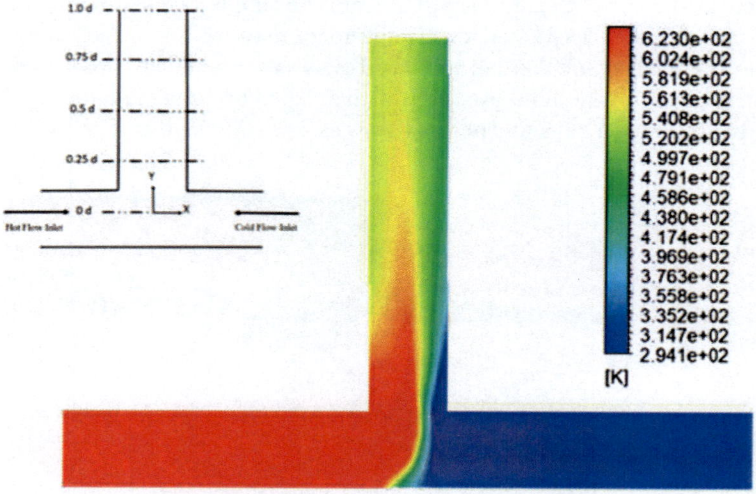

Fig. 8 Temperature distribution of colliding T-junction

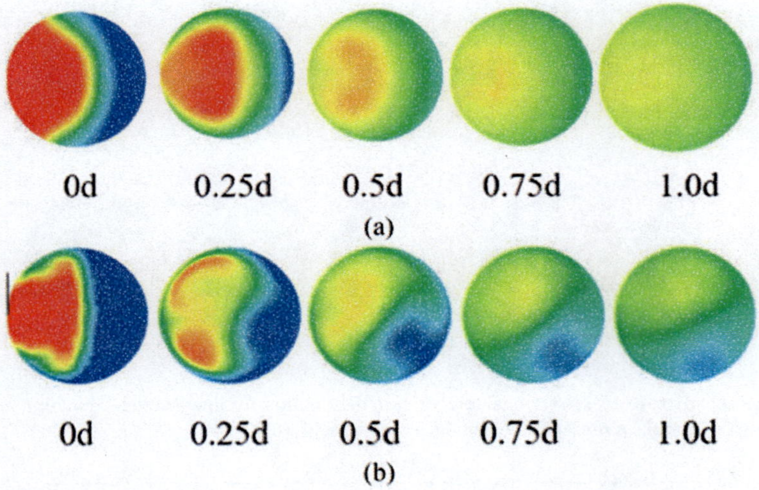

Fig. 9 Comparison of cross-sectional temperature distribution for colliding T-junction at five different cross-sections **a** present model and **b** Dimofte et al. [6]

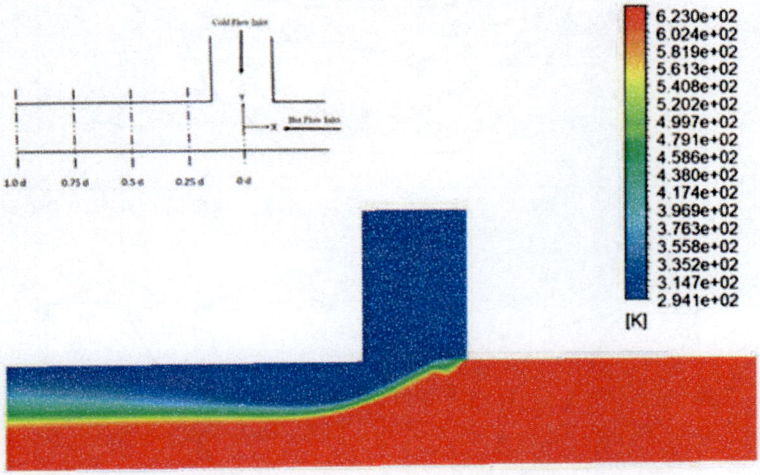

Fig. 10 Temperature distribution of intersecting T-junction

in cross-sectional temperature distribution between the present model and Dimofte et al. [6], which required further investigation.

Spatial temperature distributions around the circumference of the mixing outlet, at different *d*, were also investigated, where the sampling points were taken at 45° increment as shown in Fig. 12. The change in temperature difference along the circumference of the sections shows the degree of mixing taking place. In the study

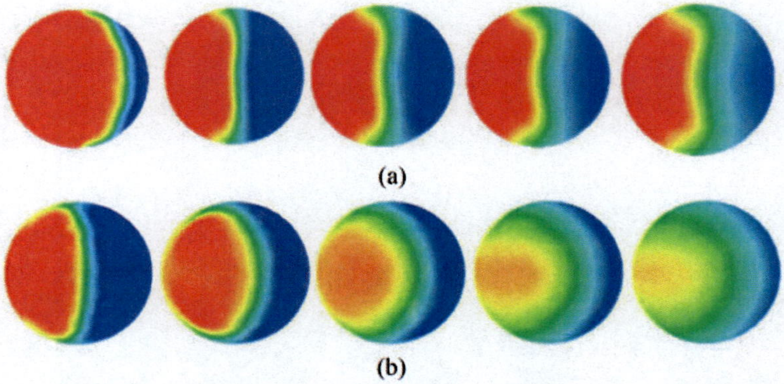

Fig. 11 Comparison of cross-sectional temperature distribution for intersecting T-junction at five different cross-sections **a** present model and **b** Dimofte et al. [6]

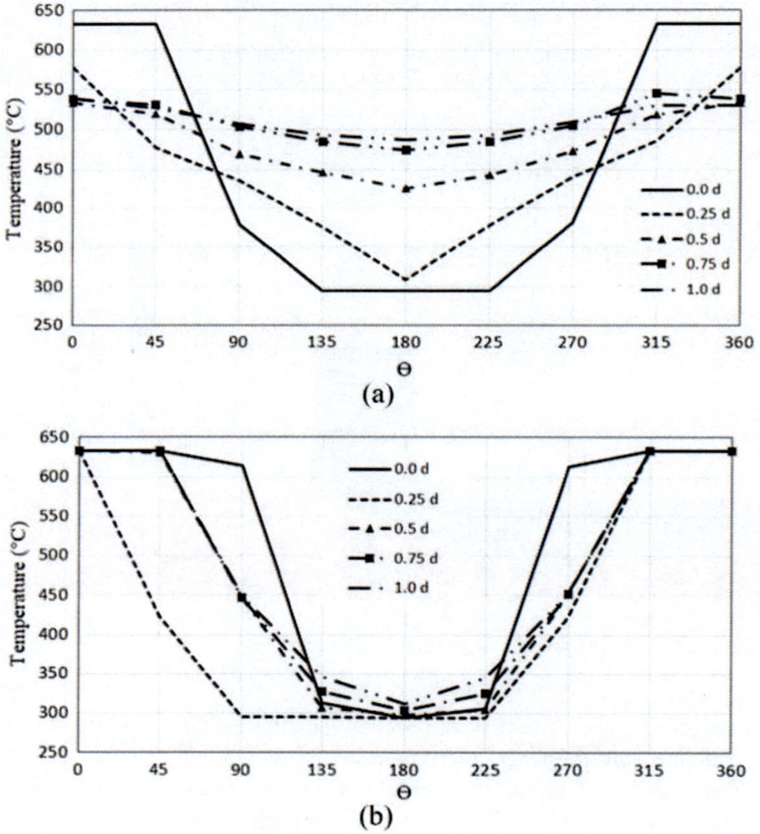

Fig. 12 Circumferential temperature distribution at different angles for **a** intersecting and **b** colliding T-junction

of thermal mixing in [13], benchmarks were formulated to evaluate the thermal mixing quality based on the temperature difference between maximum and minimum readings at each section. Good mixing will result in a difference of less than 278 K, medium quality in between 279 and 281 K while bad mixing will result in a difference of larger than 282 K. In Fig. 12a, at *0d*, the temperature difference exceeded 339 K and is considered bad mixing, while *0.25d, 0.5d, 0.75d* and *d*, featured good mixing quality as temperature differences were below 278 K. This demonstrated that in intersecting mixing-tee, thermal mixing occurs from 0.25d onwards.

In contrast to Fig. 12a, b showed an unsymmetric circumferential temperature profile at *0.25d*, a region where severe thermal stratification occurred. It was also noticed that the temperature difference along the mixing outlet remained significant. Using the benchmark grading from [13], all these sections had bad mixing quality.

4.3 Temporal Temperature Fluctuation

A transient simulation was performed with both cases of mixing tees for a total flow time of 10 s. The first 5 milli-second of the simulation were removed as no mixing occurred during that timeframe. Only in-plane averaged temperature at 0, 0.25d and 0.5d are presented and compared in Fig. 13. It is immediately apparent that the

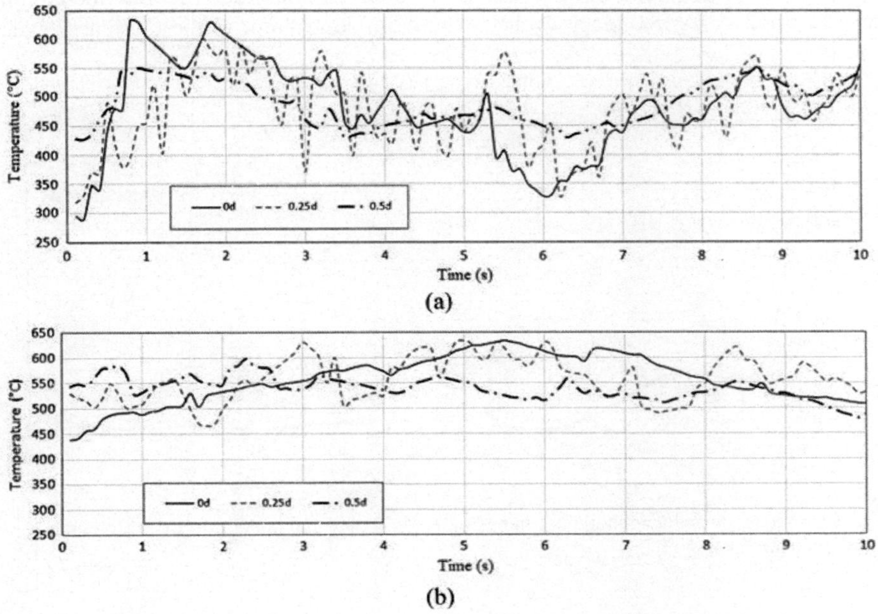

Fig. 13 Temperature fluctuation versus time at different mixing outlet locations for **a** colliding intersecting and **b** intersecting junction

colliding junction had a more volatile temperature fluctuation over a wide temperature range of 300–640 K in comparison to intersecting junction, where the temperature fluctuation is milder with a narrower temperature range of 480–640 K. It was noticed that in both cases, temperature fluctuations at the 0d section is significantly less compared to other sections of the tee. In both cases, the temperature fluctuation was observed the highest in *0.25d* section. The thermal fluctuations occurred throughout the simulation with no sign of dying down. At *0.5d*, temperature fluctuation started to damp-out.

The comparison in intensity of both mixing tees for temperature fluctuations are notably different. As in all sections of the mixing branch, the colliding mixing tee features greater intensity of fluctuations. The impacting mixing tee also features temperature fluctuations although at a lower intensity. Based on reports prior to the investigation, pipe structural failures of the impacting mixing tee have happened a few times before. Based on this, it is assumed that with a higher intensity of fluctuation, the lifespan of the colliding mixing tee would be significantly less, and the occurrence of pipe failures are very likely.

In order to reveal more information about Fig. 13, the temperature versus time data was subjected to Fast Fourier transform to decompose its magnitude and frequency, as shown in Fig. 14. The most significant and the first frequency occurred at 4.8 Hz, within the range reported in [11, 14, 17] with a magnitude of 190 and 240 K differential temperature, for intersecting and colliding T-junction, respectively. For intersecting T-junction, the smaller fluctuating frequency did not exceed 100 Hz; but the colliding T-junction ran up to 800 Hz. This small amplitude thermal fluctuation maybe small in its frequency cycle, is the main contributor to the thermal fatigue failure.

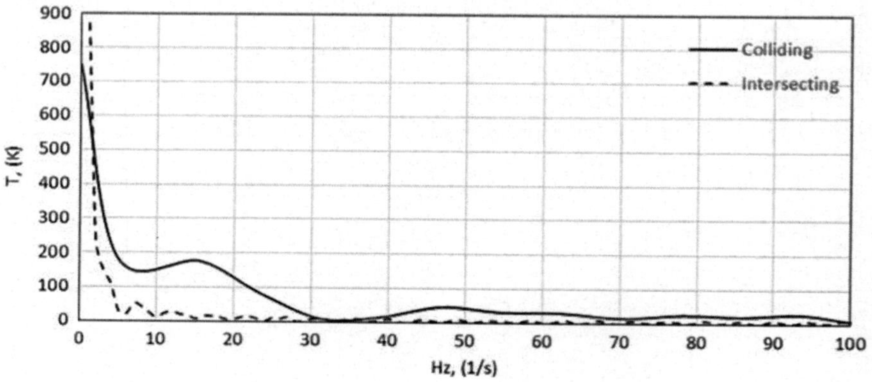

Fig. 14 Frequency and magnitude decomposition of colliding and intersecting T-junction temperature fluctuation using FFT

5 Conclusions

The present research compared numerically the thermal mixing behaviour of two mixing converging T-junctions with different flow configuration, namely an intersecting and colliding T-junction. Literature survey revealed that there is a technical gap in the reporting of colliding T-junction where not much information could be found. Validation of present CFD model suggested that the results were of sufficient accuracy for preliminary qualitative analysis.

The different flow configurations of the mixing tees resulted in different thermal mixing behaviour. It is concluded that the thermal mixing quality of the colliding mixing tee is significantly better in comparison to the intersecting mixing tee. Thermal stratification behaviour did not show in colliding T-junction. However, the advantage in using colliding mixing-tee is over-masked by its severe temperature amplitude fluctuation and intensity, a major cause of thermal fatigue. Consequently, a better thermal mixing in colliding mixing-tee may have a traded-off with shorter structural life cycle.

Acknowledgements This research is sponsored by Ministry of Education Malaysia through the Fundamental Research Grant Scheme FRGS/1/2019/TK03/UTP/02/10 and PETRONAS through YUTP-Fundamental Research Grant Cycle 2019.

References

1. Qian S, Kanamaru S, Kasahara N (2015) High-accuracy CFD prediction methods for fluid and structure temperature fluctuations at T-junction for thermal fatigue evaluation. Nucl Eng Des 288:98–109
2. Naik-Nimbalkar VS, Patwardhan AW, Banerjee I, Padmakumar G, Vaidyanathan G (2010) Thermal mixing in T-junctions. Chem Eng Sci 65(22):5901–5911
3. McFarland BL, Landy DG (1980) Turbulent mixing of two streams in a 90 0 tee. In: AIChE symposium series, vol 76
4. Maruyama T, Mizushina T, Watanabe F (1982) Turbulent mixing of two fluid streams at an oblique branch. Int Chem Eng 22(2):287–294
5. Maruyama T, Suzuki S, Mizushina T (1981) Pipeline mixing between two fluid streams meeting at a T-junction. Int Chem Eng 21(2):205–212
6. Dimofte E, Popescu F, Ion IV (2016) Numerical modelling of mixing fluids at different temperatures. In: 5th international conference of thermal equipment, renewable energy and rural development, p 35
7. Faidy C (2003) Thermal fatigue in mixing tees: a step by step simplified procedure. In: Proceedings of 11th international conference on nuclear engineering (ICONE-11). Tokyo, Japan
8. Hu LW, Kazimi MS (2003) Large eddy simulation of water coolant thermal striping in a mixing tee junction. In: Proceedings of the 10th international topical meeting on nuclear reactor thermal hydraulics (NURETH-10). Seoul, Korea
9. Westin J et al (2008) High-cycle thermal fatigue in mixing tees: Large-Eddy Simulations compared to a new validation experiment. In: 16th international conference on nuclear engineering, pp 515–525

10. Ogawa H et al (2005) Experimental study on fluid mixing phenomena in T-pipe junction with upstream elbow. In: The 11th international topical meeting on nuclear reactor thermal-hydraulics (NURETH). Avignon, France, pp 2–6
11. Ayhan H, Sökmen CN (2012) CFD modeling of thermal mixing in a T-junction geometry using LES model. Nucl Eng Des 253:183–191
12. Zhou M, Kulenovic R, Laurien E (2018) Experimental investigation on the thermal mixing characteristics at a 90° T-Junction with varied temperature differences. Appl Therm Eng 128:1359–1371
13. Chen MS et al (2014) Experimental observations of thermal mixing characteristics in T-junction piping. Nucl Eng Des 276:107–114
14. Zhou M, Kulenovic R, Laurien E (2019) Advanced flow pattern for describing tangential flow oscillation in thermal-mixing pipe flow at a horizontal T-Junction. Int J Therm Sci 136:328–336
15. Zughbi HD et al (2003) Mixing in pipelines with side and opposed tees. Ind Eng Chem Res 42(21):5333–5344
16. Lee JI et al (2009) Numerical analysis of thermal striping induced high cycle thermal fatigue in a mixing tee. Nucl Eng Des 239(5):833–839
17. Selvam PK et al (2017) Thermal mixing of flows in horizontal T-junctions with low branch velocities. Nucl Eng Des 322:32–54
18. Kickhofel J et al (2016) T-junction cross-flow mixing with thermally driven density stratification. Nucl Eng Des 309:23–39
19. Lin CH, Ferng YM (2016) Investigating thermal mixing and reverse flow characteristics in a T-junction using CFD methodology. Appl Therm Eng 102:733–741
20. ANSYS Fluent 2016 Fluent 16.1 User's Guide, Fluent Inc., Lebanon, USA
21. Cozewith C et al (1991) Computer simulation of tee mixers for nonreactive and reactive flows. Ind Eng Chem Res 30(1):270–275
22. Forney LJ, Monclova LA (1994) Numerical simulation of pipeline tee mixers: comparison with data. In: AIChE symposium series, vol 90, No 299, pp 141–143
23. Ban S et al (2018) Numerical simulation of two-phase slug flow regime in horizontal pipe and its validation. Int J Numer Meth Heat Fluid Flow 28(6):1279–1314
24. Kuhn S et al (2010) Computational study of conjugate heat transfer in T-junctions. Nucl Eng Des 240(6):1548–1557

Examining Combustion Characteristics of Water-In-Palm Oil Biodiesel Emulsified Fuels for Micro-Explosion

N. H. Amran, Z. A. Abdul Karim, M. N. I. Martang, and I. B. Dalha

Abstract Micro-explosions during the combustion of palm oil water-in-biodiesel emulsified fuel caused simultaneous reduction in both NOx and smoke from the diesel engine exhaust. The existence of water particles in the palm oil biodiesel emulsified fuel cause micro-explosions due to the difference in volatilities and boiling temperatures of the water and biodiesel as the water particles reached their superheated limit first before the biodiesel. In addition, water lowers the adiabatic flame temperature and therefore acts as a heat sink during the combustion process. Ascertaining the occurrences of micro-explosion in the combustion chamber by direct visualisation is a difficult task, however, it can be made possible by observing the trends of the combustion phases. The secondary atomisation caused by the micro-explosion can be depicted on the combustion characteristics curve by determining the heat release rate of combustion. This study aimed to ascertain the occurrences of micro-explosion and reduction in flame temperature by analysing the combustion phases of water-in-palm oil biodiesel in a diesel engine with different blends of emulsified fuels. Emulsified biodiesel fuels with 11% and 13% water by volume blended with palm oil biodiesel were used as fuel in an unmodified single-cylinder direct injection engine at 0%, 25%, 50%, 75% and 100% loads while maintaining the engine speeds at 2000 rpm. WiBE 8-8-13 shows the highest peak for both in-cylinder pressure and heat release rate with 3.87% and 23.51% increase respectively, as compared to B10 fuel at high load condition. The emission characteristics recorded by WiBE 8-8-11 shows lowest emission of NOx by 52.83% but increase the percentage of CO by 43.75% compare B10 fuel.

N. H. Amran (✉) · Z. A. Abdul Karim
Department of Mechanical Engineering, Universiti Teknologi PETRONAS, 32610 Seri Iskandar, Perak, Malaysia
e-mail: hazira.amran@gmail.com

M. N. I. Martang
Sime Darby Plantation Academy, Carey Island, 42960 Selangor, Malaysia

I. B. Dalha
Faculty of Engineering, Ahmadu Bello University, Samaru 1045, Zaria, Nigeria

F. Ahmad et al. (eds.), *ICPER 2020*, Lecture Notes in Mechanical Engineering,
https://doi.org/10.1007/978-981-19-1939-8_10

1 Introduction

Rapidly scaling up in energy demand of conventional fuels in developed countries, particularly in the automobile sector coupled with rising greenhouse gases (GHG) emissions product from the burning of fossil fuels are the signals to develop renewable energy sources with lesser effects on the environment [1]. For instance, the annually International Energy Agency (IEA) Report projected fastest pace of global energy consumption for this decade rose by 2.3% [2]. Given in 2018, natural gas accounted over almost 45% of its total in cumulative energy consumption. IEA reported that ascending in market of all fuels, with fossil fuels accounting about approximately 70% towards the development. A dramatic spike in consumption has been shown in carbon dioxide emissions from global combustion of conventional fuels and manufacturing activities. In addition, an estimated 36.57 billion tonnes of carbon dioxide were emitted globally throughout 2018 [3].

Malaysia, Vietnam, Indonesia and Thailand, among other Southeast Asia countries have abundance of natural resources including oil palms which is palm oil biodiesel became popular for replacement petroleum-based diesel in transportation [4]. The greatest benefit of biodiesel is its renewability property, which domestically manufactured through transesterification process from plant oils, animal fats or waste cooking oil [5–7]. Biodiesel originated from palm oil methyl ester (POME) is a less polluting fuel alternatively fuel to diesel, emitting 62% lesser greenhouse gases (GHG) emissions than fossil fuels; and it outperforms soybean by 40%, rapeseed and sunflower biodiesel by 45% and 58%, respectively [8]. In addition to being less toxic and harmful, it is biologically degradable and feasible as an engine fuel without requiring any modification. In line with biodiesel drive, the National Biofuel Policy (NBP) came into force in 2006 with stated objectives to reduce the heavy reliance on fossil fuels while utilizing qualities employed by palm oil biodiesel especially its renewability and sustainability [9]. Furthermore, Malaysia has consistently developed high quality biodiesel at competitive prices that comply with international biodiesel requirements (ASTM D 6271 and EN 14214) [10]. Malaysia does seems to have a promising rising growth of 490million litres in 2017 for biodiesel trend, which is expected to hit 815 million litres in 2027, indicating a 66% increase [11].

Recent research has focused on incorporating water into biodiesel as an emulsified fuel that can be utilized in an existing diesel engine [12, 13]. Emulsified biodiesel fuel exhibits promising potential in reducing both formation of NO_x and particulate matters simultaneously with improvement in combustion efficiency [14, 15]. Water particles in the emulsion is claimed to cause two phenomena. First, a secondary atomisation of the fuel known as micro-explosion due to different volatility between water and oil particles [16, 17]. Water particles in the emulsion reaching their superheated limit first, and evaporating before the diesel oil [18]. Micro-explosion improves air–fuel mixing which reduces the formation of soot, hence lower emissions of particulate matters. Second, the reduction in the flame temperature during combustion due to the water acting as a heat sink, resulting in the lowering NOx and PM emissions level at the same time [19]. The micro-explosion phenomena reported by Vellaiyan and

Amirthagadeswaran [20] is primarily related to the disintegration of water droplets that create very fine vapour particles which eventually lead in secondary atomization. In secondary atomization, the contact surface area between tiny particles in the fuel and air increases significantly, thereby enhances the efficiency of combustion. Similarly, the fuel molecules are broken down by vapour particles allowing the fuel to completely burn. Vellaiyan and Amirthagadeswaran [21] also investigated the effects of water in diesel with 5%, 10%, 15% and 20% volume of water. Findings revealed a significant reduction in harmful emission, concentrating primarily on NOx and soot particles despite offsetting on the engine performance. In 2017, Cahyono et al. [22] reported on the impacts of water in fuel on performance and NOx emission in diesel engine. Experimental results discovered remarkable reduction of NOx emission up to 50.5% but conversely reduction in engine performance. Furthermore, the researcher Vigneswaran et al. [23] studied the efficiency, combustion and emission in unmodified diesel engine running with water in diesel (W/D) emulsion along with additive concentrations of 5%, 10%, 15% and 20% and compared the results with neat diesel as baseline. W/D and W/D with 10% additive emulsified fuels improved 2.96% and 7% in BTE, respectively when compared to diesel. W/D emulsion exhibits the lowest NOx production up to 27.96% compared to W/D with 5% additive, while W/D with 10% additive showed only slight increase in NOx.

Although extensive studies on emulsified fuels are available, the analysis of combustion characteristics within the combustion chamber in relation to micro-explosion event is still incomplete and inadequate. The aim of this research work is to ascertain the micro-explosions phenomena when the engine runs on water-in-biodiesel (WiBE) fuels by examining the combustion characteristics inside the combustion chamber, thus assist in correlating the emission control strategy for compliance with emission regulations. In the current study, two primary combustion characteristics were considered, i.e. in-cylinder pressure (ICP) traces and rate of heat release (HRR) analysis with variance in crank angle. For all emulsified fuels tested, combustion characteristics including ignition delay, timing of injection, and peak cylinder pressure having a significant impact on diesel engine performance and emission levels. The heat release model is based on Thermodynamics' first law, which is given as follows:

$$\frac{dQ}{d\theta} = \frac{\gamma}{\gamma - 1} p \frac{dV}{d\theta} + \frac{1}{\gamma - 1} V \frac{dP}{d\theta} \tag{1}$$

and,

$$\gamma = \frac{C_p}{C_v} \tag{2}$$

where $dQ/d\theta$ is heat release rate (kJ/deg), P is the in-cylinder pressure (kPa), V is the cylinder volume (m^3), γ is the specific heats ratio, θ is crank angle (deg), C_p is specific heat at constant pressure, C_v is specific heat at constant volume.

2 Description of Experimental Setup

The materials and experimental procedure in this study includes water-in-biodiesel (WiBE) fuels preparation, engine testing and exhaust measurement testing are presented in this section.

2.1 *Fuel Preparations*

This research investigated the occurrence of micro-explosion in WiBE fuels, which is base fuel B10 fuel containing 10% palm oil methyl ester (POME) blended with 11% and 13% water from the total emulsion volume. Stable emulsions of water-in-biodiesel were prepared with the presence of surfactant. During emulsification, surfactants help to maintain the surface tension of oil and water. The optimal HLB values of 7 and 8 were obtained by using two surfactants, Span 80 and Tween 85. A set of 12 samples of WiBE fuels were prepared with three dissimilar dosages of surfactant of 7, 8 and 9%. Table 1 shows the compositions of the WiBE fuels that are used in this series of investigation.

The WiBE fuels are blended using Hielscher's UP100H Ultrasonic Homogenizer operated at 70% amplitude and 0.5 cycle for 20 min. Firstly, biodiesel B10 and both surfactants Span 80 and Tween 85 are placed in the same container and undergoes direct sonication for 5 min. Next, volumetric percentage of distilled water is added into the solution, the two immiscible liquids experienced vigorous mixing by the ultrasonic wave for a length of 15 min until it turns into milky white emulsion.

2.2 *Properties of the Test Fuels*

Table 2 shows the result for density, viscosity, calorific value and sauter mean diameter (SMD) of the test fuels. Table 3 shows the comparison between physchochemical properties of biodiesel requirements from multiple countries.

Table 1 Composition of WiBE fuels

	Sample ID	Percentage of diesel (%)	Percentage of biodiesel (%)	Percentage of water (%)	Percentage of surfactant (%)		
					Tween 85	Span 80	
	Neat diesel	100	–	–	–	–	
	B10 fuel	90	10	–	–	–	
Sample ID	Composition	HLB value	Surfactant dosage (%)	Distilled water (ml)	Surfactant (ml)		B10 fuel (ml)
					Tween 85	Span 80	
WiBE-1	7-7-11	7	7	110.00	3.10	4.60	882.30
WiBE-2	7-7-13			130.00	3.67	5.43	860.90
WiBE-3	7-8-11		8	110.00	3.55	5.25	881.20
WiBE-4	7-8-13			130.00	4.19	6.21	859.60
WiBE-5	7-9-11		9	110.00	3.99	5.91	880.10
WiBE-6	7-9-13			130.00	4.72	6.98	858.30
WiBE-7	8-7-11	8	7	110.00	4.25	3.45	882.30
WiBE-8	8-7-13			130.00	5.03	4.07	860.90
WiBE-9	8-8-11		8	110.00	4.86	3.94	881.20
WiBE-10	8-8-13			130.00	5.74	4.66	859.60
WiBE-11	8-9-11		9	110.00	5.47	4.43	880.10
WiBE-12	8-9-13			130.00	6.46	5.24	858.30

2.3 Engine Set-Up

A Yanmar L100V single cylinder, 4-stroke, air cooled test rig diesel engine is being used for this study. Figure 2 depicts the engine test rig's configuration. The tests are conducted at a continuous speed of 2000 rpm with five separate loads varying from minimum to maximum of 0%, 25%, 50%, 75% and 100%. The eddy current dynamometer mounted to the engine determines the engine load and speed, while a rotary style encoder is utilized to measure crank angle. In-cylinder pressure is measured using a Kistler water-cooled pressure transducer mounted in the engine cylinder head. Thermocouples measures air and exhaust gas temperatures. The airflow rate was measured with a hot-wire airflow meter and the air pressure was measured with a pressure sensor. Before running the engine using the WiBE fuels, the engine is operated for about 20 min to stabilize the engine with neat diesel. After the engine attained a steady state, it then shifted to B10 fuel and finally the WiBE fuels. The engine was switched back to neat diesel operation after each test and ran for about 15 min before being shut down. The KANE AUTOPLUS 5–2 was used to quantify NOx, CO and HC emissions, whereas smoke meter is utilized to measure smoke opacity. For accuracy, each sample data is recorded twice. All technical specifications of the engine test bed and emission analyser are shown in Tables 4 and 5, while the schematic layout of the experimental setup and image of the engine test bed are presented in Figs. 1 and 2, respectively.

3 Results and Discussion

3.1 Combustion Characteristics

The following sections discuss the combustion characteristics, i.e. in-cylinder pressure trends and heat release rate curves of four WiBE blends under two loading conditions of 25% and 75% load of constant 8% surfactant dosage.

3.1.1 In-Cylinder Pressure

The variation of in-cylinder pressure (ICP) for four WiBE fuels under two loading conditions of 25% and 75% load of constant 8% surfactant dosage are depicted in Fig. 3. Peak cylinder pressure corresponds to the potentiality of fuel to blend well with air and burn during pre-mixed combustion phase [25]. From the results plotted in Fig. 3a, at low loading condition, the ICP trends exhibited by all fuels are almost the same and comparable with insignificant difference during the combustion stroke. For WiBE with 11% water content (7-8-**11** and 8-8-**11**) showed insignificant drop

Table 2 Laboratory characterization properties of test fuels

Sample ID	Composition	Density at 25 °C (kg/m^3)	Viscosity at 40 °C (mm^2/s)	Calorific value (MJ/kg)	Sauter mean diameter (SMD) (μm)
WiBE-1	7-7-11	856.6	3.462	40.291	2.11
WiBE-2	7-7-13	866.1	3.869	39.647	1.58
WiBE-3	7-8-11	859.2	4.107	39.825	2.23
WiBE-4	7-8-13	861.8	4.217	38.733	2.10
WiBE-5	7-9-11	854.8	3.641	39.077	2.09
WiBE-6	7-9-13	859.2	4.183	38.919	1.59
WiBE-7	8-7-11	851.5	3.504	40.105	1.67
WiBE-8	8-7-13	859.4	3.963	38.543	1.63
WiBE-9	8-8-11	858.7	3.894	39.664	2.45
WiBE-10	8-8-13	862.2	3.911	38.623	2.13
WiBE-11	8-9-11	844.8	3.754	42.268	2.10
WiBE-12	8-9-13	850.8	4.348	41.637	1.63

Table 3 Biodiesel requirements from multiple countries [3, 24]

Parameters	USA (ASTM)	Malaysia	France (general official)	Italy (UNI)	Germany (DIN)	Austria (ON)	Korea
Density at 15 °C g/cm^3	0.8–0.9	0.8783	0.87–0.89	0.86–0.90	0.875–0.89	0.85–0.89	0.86–0.89
Viscosity at 40 °C mm^2/s	1.9–6.0	4.415	3.5–5.0	3.5–5.0	3.5–5.0	3.5–5.0	1.9–5.5
Flash point (°C)	130	182	100	100	100	100	>120
Pour Point (°C)	–	15	−10	0–5	–	–	–
Cetane Number, CN	≥47	56	≥49	–	≥49	≥49	–

Table 4 Technical specifications of the test bed diesel engine

Description	Specifications
Model	Yanmar L100V
Type	Single cylinder, 4-stroke, air cooled
Bore x Stroke (mm)	86 × 75
Displacement (cm^3)	435
Compression ratio	20.0 ± 0.3
Fuel injection timing	13°CA BTDC
Continuous rate output	3600 rpm/6.2 kW
Maximum rate output	3600 rpm/6.8 kW

of less than 1% in peak cylinder pressure traces compared to the B10 fuel. Further increase in water percentage, i.e. 13% water content (7–8-**13** and 8-8-**13**) exhibit 3.74% higher in peak pressure compared to the B10 fuel. This could be due to the physical properties of water of having higher density as the volume of water increase in the emulsion, thus triggering auto-ignition of the fuel [26–28]. Auto-ignition occurs owing to the volume of water content in the emulsion triggers more combustible air–fuel mixing [29]. On top of that, micro-explosion phenomenon could be attributed to the substantial peak growth of the emulsified fuels which delivers better atomization of air–fuel mixing subsequently led to higher in-cylinder pressure [29]. There is insignificant ignition delay (ID) showed at lower engine load. Next, Fig. 3b portrays two peaks of cylinder pressure at high loading condition of 75%. The increase of area under the pressure curves as the load increased signify more work is done at higher engine load. The addition of 11% of water in the emulsions achieved 5.55% increment during first peak ICP while reduction of 4.61% during

Table 5 Technical specifications of gas analyser

Description	Specifications	
Model	Kane AUTOplus 5–2	
Type	Automotive exhaust gas analyser	
Carbon Monoxide, CO	In ambient air	0
	Certified calibration gas	3.51%
Hydrocarbon, HC	In ambient air	0
	Certified calibration gas	1068 ppm
Oxygen, O_2	In ambient air	20.9%
	Certified calibration gas	0
Carbon Dioxide, CO_2	In ambient air	0
	Certified calibration gas	13.44%
Nitric Oxides, NO	In ambient air	0
	Certified calibration gas	1018 ppm
Measured flow rate	≥0.75 l/min	
Measured pump suction	≥200 mbar	

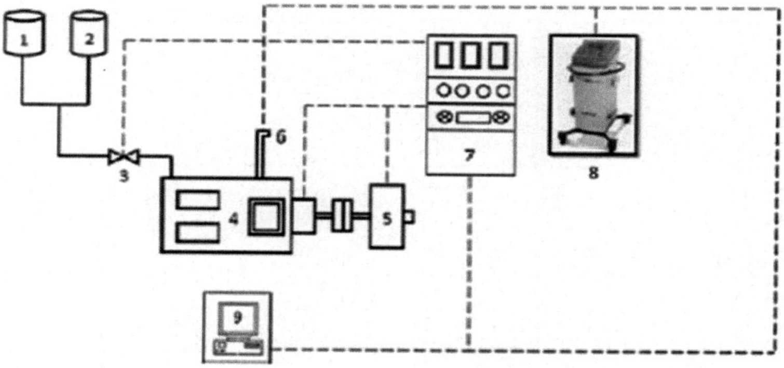

Fig. 1 Schematic layout for engine test rig. 1, 2. Fuel tank 3. Fuel flow meter 4. Diesel engine 5. Eddy current dynamometer 6. Exhaust hose 7. Engine Control Unit 8. Emission Analyser 9. PC

second peak ICP in comparison with B10 fuel. Inclusion of water as emulsified fuel lowers cylinder pressure by absorbing flame temperature in the combustion zone [23]. In contrast, for emulsion with 13% water content, both first and second peak show increment in ICP by 2.62% and 3.87% respectively as compared to B10 fuel. Expansion of second peak occurred due to more heat generated during combustion process. Heat vigorously increase could be due micro-explosion of water particles into fine droplets during combustion, as well as extra force acting on the piston surface [21]. As acknowledged, both WiBE of 13% water content developed higher peak as compared to the B10 fuel. This phenomenon can be associated with rapid

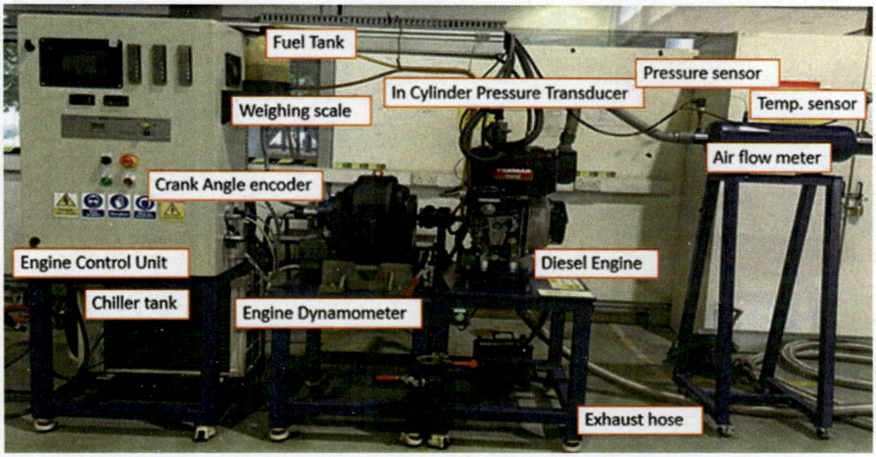

Fig. 2 Experimental test rig Yanmar L100V diesel engine part details

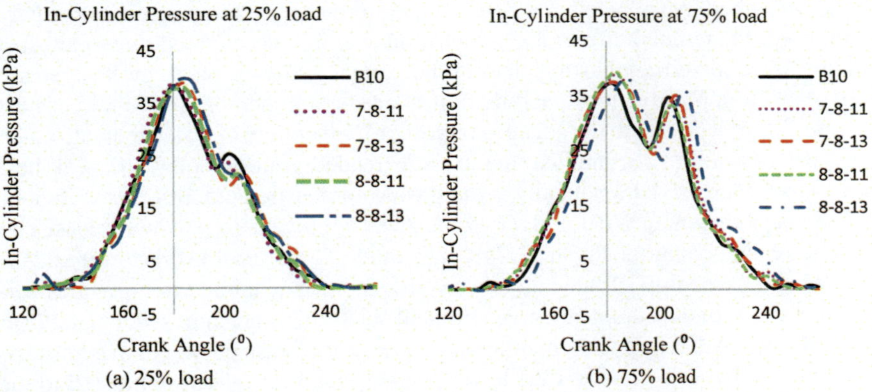

Fig. 3 In-cylinder pressure of 8% surfactant dosage

burning and completely uncontrolled combustion due to relatively high fuel supply at higher engine load [21]. More fuel is set to release into the chamber at higher loads, the oxygen molecule in the WiBE fuel continues to increase the combustion reaction and further accelerates the cylinder pressure.

3.1.2 Heat Release Rate, HRR

Figure 4 presents heat release rate (HRR) plots at different crank angle of B10 fuel and four WiBE fuels at two different loading condition of 25% and 75% load of 8% surfactant dosage. Both plots revealed significant difference during the ignition

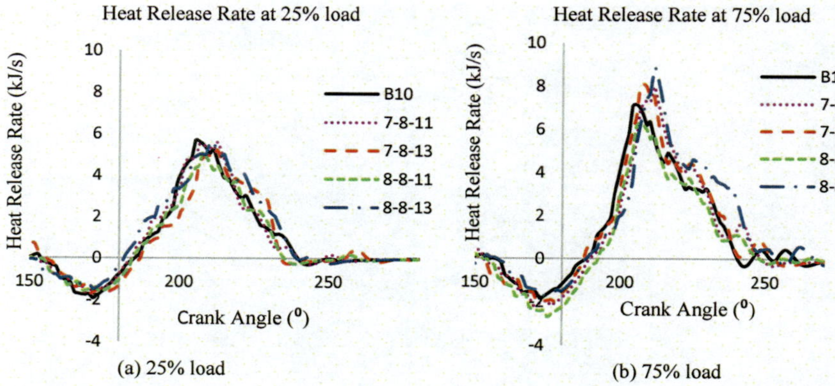

Fig. 4 Heat release rate of 8% surfactant dosage

delay period and combustion peak. During the fuel injection period, a clear negative heat release rate can be observed. The reason is that during the initial compression stroke and when fuel which is at a lower temperature than the in-cylinder temperature is injected, the temperature in the combustion chamber decelerated due to air–fuel mixing at pre-mixed combustion phase. When auto ignition too place, the heat release rate grows to a positive level, and there is a substantial increase in the rate of heat release [21, 25]. Owing to the negative impact on temperature of water content, which caused more fuel to accumulate due to the extended ignition delay, the peak heat release rate increased dramatically at the start of combustion [25, 30]. Also, obvious difference is shown by WiBE 8-8-11 which gives significant drop of heat release rate at both loading conditions with 11.24% falls at 25% load and 14.00% lower at 75% load as compared to B10 fuel and other WiBE fuel. As exhibited in Fig. 4a, at low engine load, heat development curves from all WiBE fuels are almost comparable to the B10 fuel. All WiBE fuels manifested lower heat release curve compared to the B10 fuel. WiBE 8-8-11 shows the lowest fall of peak heat release compared to the B10 fuel and other WiBE fuels with 11.24% decrease. Meanwhile, at high loading condition as observed in Fig. 4b, there are abnormalities to the both WiBE fuels with water content of 11% and 13%. Addition of 11% water attributed to longest ID period but gives lowest peak of HRR. Conversely, the emulsifieds fuels with 13% water content produced shortest ID period but exhibits advanced heat release. Good combustion inside combustion chamber indicates by higher heat release rate [23]. From the Fig. 4, it was clear that WiBE 8-8-13 had highest heat release rate at high loading conditions, participating in better and complete combustion within combustion chamber.

3.2 *Exhaust Emission*

The following sections discuss the results of exhaust gas emission measurements from four WiBE fuels under loads of 0%, 25%, 50%, 75% and 100%, at constant engine speed of 2000 rpm. Due to unavoidable breakdown of the smoke meter, the discussion presented in this article is only on the emissions of NOx, CO and CO_2.

3.2.1 *Nitrogen Oxides, NO_x*

The trends of NOx emission for B10 fuel and four WiBE fuels under various loads (0%, 25%, 50%, 75% and 100%) and constant speed of 2000 rpm are shown in Fig. 5. The amount of NOx steadily rises as the engine load increase for all fuels as presented in the plot. At higher load, the amount of heat generated is higher due to the increase burning of fuel led to high temperature inside the combustion zone. As is well documented, NOx formation is primarily temperature dependent and increases as temperature increase in the combustion chamber [31]. However, all WiBE fuels show significant drop of NOx emission with WiBE 8-8-11 exhibited 46.68% lowest as compared to the B10 fuel. The decline in all WiBE fuels can be attributed to the incorporation of water in the emulsion that lowered the peak flame temperature during combustion [12], as the presence of water acts as heat sink. At higher loads, significant NOx reduction is evidence due to the occurrences of micro-explosion of WiBE fuels which further improves combustion and hence, lowers the emissions of NOx.

3.2.2 Carbon Monoxide, CO

Figure 6 exhibits the emission of CO for B10 fuel and four WiBE fuels under varying loads of 0%, 25%, 50%, 75% and 100% at constant engine speed of 2000 rpm. As

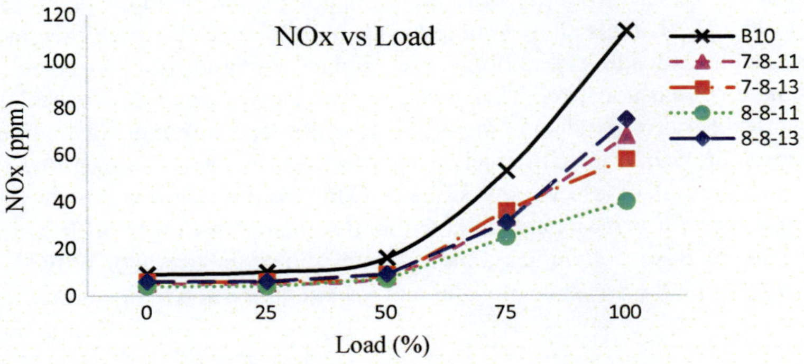

Fig. 5 NOx emission against loading condition

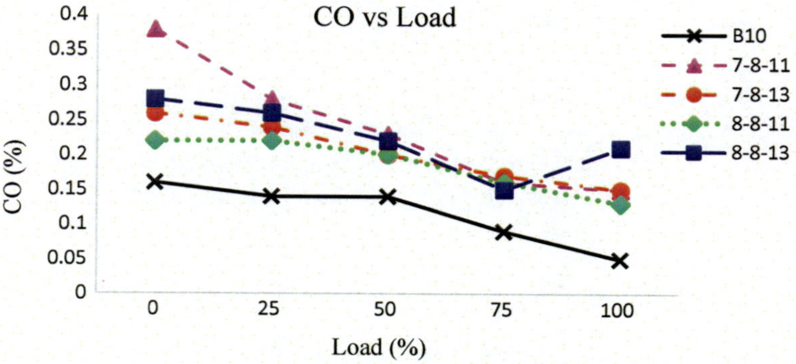

Fig. 6 CO emission against loading condition

seen in the plot, general trends of all fuels demonstrate that CO emissions decrease gradually with engine loads. At low loads, there is relatively high production of CO due to lower combustion temperature affected by the water in the fuel which makes it difficult to oxidize from CO to CO_2 [14, 28]. As the engine loads increase, the temperature in the combustion chamber increase and accelerates the oxidation of CO to CO_2 resulting in lower magnitude formation of CO [31]. However, there are irregularities for WiBE fuel sample 8-8-13, at 75% engine load, CO emissions drastically increased. The plot also shows that all WiBE fuels emit higher CO compared to B10 fuel. This could be due to the incorporation of water in WiBE which lowers the combustion temperature, thus slightly increase CO formation From the plot, in Fig. 6, WiBE 8-8-13 showed lowest CO emission compare to others emulsion by 43.5%.

3.2.3 *Carbon Dioxide, CO_2*

Generally, CO_2 formed from the complete combustion of fuel in air in which CO_2 formation is inversely proportional to the formation of CO. From the plot representation in Fig. 7, all fuels show similar trends as B10 fuel, in which increment of CO_2 emissions as loading conditions arise. Higher magnitude of CO_2 means closer to complete combustion [32]. This shows that enhancement of air–fuel mixing at full load levels correspond to complete combustion and a tremendous amount of CO_2 formation [29]. All WiBE fuels except for WiBE 7-8-11 showed lower CO_2 emissions as a result of lesser conversion of CO to CO_2 due to lower overall combustion temperature. It is reasonable to note that the change in CO_2 of all WiBE when compared to B10 is not significant, indicating that the combustion of WiBE fuels is complete, despite the presence of water and lower combustion temperatures.

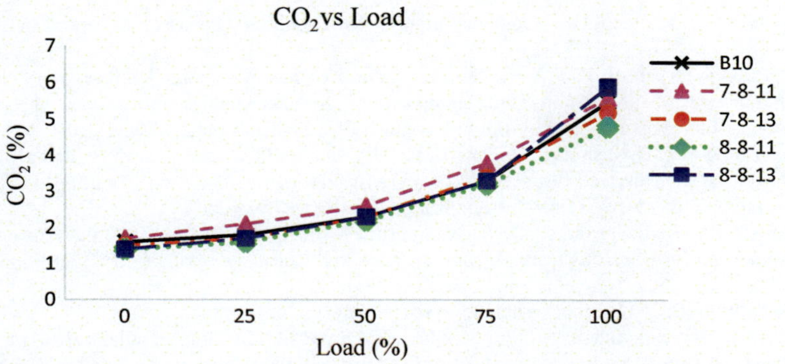

Fig. 7 CO_2 emission against loading condition

4 Conclusion

The aim of the work to ascertain the micro-explosion effects of water-in-biodiesel fuels by examining the combustion characteristics has been successfully achieved. From the analyses, several conclusions can be drawn:

1. WiBE fuels caused the occurrences of micro-explosion at higher engine loads as observed in the ICP trends and HRR curves.
2. Both ICP trends and HRR curves displayed significant differences due to the different combustion characteristics during the pre-mixed combustion and the uncontrolled combustion phases.
3. All WiBE exhibited significant reduction in NOx, but increase in CO due to the heat sink effects which lowers the flame temperature. While, insignificant change in the emissions of CO_2.

Acknowledgements The authors acknowledge the Department of Mechanical Engineering, Universiti Teknologi PETRONAS for providing laboratory facilities to carry out this research work and The Murata Science Foundation FY2020 for the financial support.

References

1. Ministry of Plantation Industries and Commodities Malaysia (2006) The National Biofuel Policy, pp 1–13
2. International Energy Agency (2019) Global energy demand rose by 2.3% in 2018, its fastest pace in the last decade-news, pp 1–5
3. International Energy Agency (2018) Global Energy & CO_2 Status Report, pp 1–11
4. Yusoff MNAM, Zulkifli NWM, Sukiman NL, Chyuan OH, Hassan MH, Hasnul MH, Zulkifli MSA, Abbas MM, Zakaria MZ, Zulfattah M (2020) Sustainability of palm biodiesel in transportation: a review on biofuel standard, policy and international collaboration between Malaysia and Colombia. Bioenergy Res, pp 1–18

5. Malaysian Investment Development Authority (MIDA) 2020 Our way : Advanced biofuel, pp 1–12
6. Rezania S, Oryani B, Park J, Hashemi B, Yadav KK, Kwon EE, Hur J, Cho J (2019) Review on transesterification of non-edible sources for biodiesel production with a focus on economic aspects, fuel properties and by-product applications. Energy Convers Manag 20:1–15
7. Mansir N, Teo SH, Rashid U, Saiman MI, Tan YP, Alsultan GA, Taufiq-Yap YH (2018) Modified waste egg shell derived bifunctional catalyst for biodiesel production from high FFA waste cooking oil-A review. Renew Sustain Energy Rev 82:3645–3655
8. Rahyla RS, Radin Firdaus RB, Purwaningrum F (2017) Upgrading of Malaysian palm oil biofuel industry: lessons learned from the USA and Germany's policies. Cogent Food Agric 3(1):1–11
9. Wahab AG (2019) Biofuels Annual Report (Malaysia), pp 1–10
10. Masjuki HH, Kalam MA, Mofijur M, Shahabuddin M (2013) Biofuel: policy, standardization and recommendation for sustainable future energy supply. Energy Procedia 42:577–586
11. Singh D, Sharma D, Soni SL, Sharma S, Kumar Sharma P, Jhalani A (2020) A review on feedstocks, production processes, and yield for different generations of biodiesel. Fuel 262:1–15
12. Ithnin AM, Yahya WJ, Azrin MA, Ramlan NA, Abdul Kadir H, Sidik NAC, Koga T (2018) Emulsifier-free water-in-diesel emulsion fuel: its stability behaviour, engine performance and exhaust emission. Fuel 215:454–462
13. Hasannuddin AK, Yahya WJ, Sarah S, Ithnin AM, Syahrullail S, Sugeng DA, Razak IFA, Abd Fatah AY, Aqma WS, Rahman AHA, Ramlan NA (2018) Performance, emissions and carbon deposit characteristics of diesel engine operating on emulsion fuel. Energy 142:496–506
14. Ithnin AM, Ahmad MA, Bakar MAA, Rajoo S, Yahya WJ (2015) Combustion performance and emission analysis of diesel engine fuelled with water-in-diesel emulsion fuel made from low-grade diesel fuel. Energy Convers Manag 90:375–382
15. Ithnin AM, Noge H, Kadir HA, Jazair W (2014) An overview of utilizing water-in-diesel emulsion fuel in diesel engine and its potential research study. J Energy Inst 87(4):273–288
16. Yahaya Khan M, Abdul Karim ZA, Aziz ARA, Tan IM (2016) Experimental study on influence of surfactant dosage on micro explosion occurrence in water in diesel emulsion. Appl Mech Mater 819:287–291
17. Mondal PK, Mandal BK (2019) A comprehensive review on the feasibility of using water emulsified diesel as a CI engine fuel. Fuel 237:937–960
18. Khan MY, Abdul Karim ZA, Aziz ARA, Tan IM (2014) Experimental investigation of microexplosion occurrence in water in diesel emulsion droplets during the leidenfrost effect. Energy Fuels 28(11):7079–7084
19. Yahaya Khan M, Abdul Karim ZA, Aziz ARA, Tan IM (2017) A case study on the influence of selected parameters on microexplosion behavior of water in biodiesel emulsion droplets. J Energy Resour Technol 139(2):1–10
20. Vellaiyan S, Amirthagadeswaran KS (2016) The role of water-in-diesel emulsion and its additives on diesel engine performance and emission levels: a retrospective review. Alexandria Eng J 55(3):2463–2472
21. Vellaiyan S, Amirthagadeswaran KS (2016) Influence of water-in-diesel emulsion fuel and compression ratio on combustion, performance and emission characteristics of diesel engine. J Sustain Energy Eng 3(3):238–253
22. Cahyono B, Fathallah AZM, Semin, Pahlevi N (2017) Effect of water fuel emulsion on performance and NOx emissions of diesel engine. Int J Marine Eng Innov Res 1(4):221–232
23. Vigneswaran R, Annamalai K, Dhinesh B, Krishnamoorthy R (2018) Experimental investigation of unmodified diesel engine performance, combustion and emission with multipurpose additive along with water-in-diesel emulsion fuel. Energy Convers Manag 172:370–380
24. Malaysian Palm Oil Board (MPOB) (2007) Biodiesel solutions:fuel properties of normal and Low Pour Point Palm Biodiesel pp 1–8
25. Rajasekar E, Selvi S (2014) Review of combustion characteristics of CI engines fueled with biodiesel. Renew Sustain Energy Rev 35:390–399

26. Tesfa B, Mishra R, Zhang C, Gu F, Ball AD (2013) Combustion and performance characteristics of CI (compression ignition) engine running with biodiesel. Energy 51:101–115
27. Maawa WN, Rizalman M, Yusri IM, Febrina R (2019) Effect of emulsified palm oil biodiesel-diesel blends to the combustion characteristics of compression ignition engine. IOP Conf Series Mater Sci Eng 469(1):1–11
28. Jhalani A, Sharma D, Soni SL, Sharma PK, Sharma S (2019) A comprehensive review on water-emulsified diesel fuel: chemistry, engine performance and exhaust emissions. Environ Sci Pollut Res 26(5):4570–4587
29. Vellaiyan S (2019) Enhancement in combustion, performance, and emission characteristics of a biodiesel-fueled diesel engine by using water emulsion and nanoadditive. Renew Energy 145:2108–2120
30. Vellaiyan S, Amirthagadeswaran KS, Vijayakumar S, Varun B (2014) Emission characteristics of diesel engine using water-in- diesel emulsified fuel and its CFD analysis. Int J Appl Environ Sci 9(5):973–6077
31. Mondal PK, Mandal BK (2019) A comparative study on the performance and emissions from a CI engine fuelled with water emulsified diesel prepared by mechanical homogenization and ultrasonic dispersion method. Energy Rep 5:639–648
32. Mohsin R, Majid ZA, Shihnan AH, Nasri NS, Sharer Z (2014) Effect of biodiesel blends on engine performance and exhaust emission for diesel dual fuel engine. Energy Convers Manag 88:821–828

Effects of Nozzle Diameter and Number of Carbon Atoms in Fuel on Flame Quenching in Counter Burner

Ahmed H. Mola, Hasanain A. Abdul Wahhab, and Zainab H. Naji

Abstract Size of burner nozzle is influencing the quenching of premixed flame. This paper describes an experimental study of quenching of premixed counter flame. Flames of Methane, Propane, LPG with air mixtures propagating in counter burner of a different nozzles' diameter have been experimentally investigated. Influence of counter burner nozzles diameters and fuel (Methane, Propane, LPG)/air mixing ratio on the flame burning velocity and quenching diameter has been determined. Schlieren technique was used to analyze the effect of changing nozzles with diameters of 9, 12, 16, 18 and 23 mm on the quenching diameter. Significant results are obtained in the area of counter flame. Quenching diameter in the counter flow depend basically on the nozzle diameter and velocity of fuel and air. For both lean and rich mixtures, quenching diameters are almost the same for top and bottom flame disc and increased with increasing the nozzles diameter. The quenching diameter is greater than the quenching distance for mixtures with the equivalence ratio $0.46 < \varphi < 1.57$. The existence of this difference is probably caused by the dead space. Also, the increase number of carbon atoms in fuel increased the quenching diameter.©

Nomenclature

Latin symbols

d_q Quenching diameter, in (mm).
d_c Disc Flame diameter, in (mm).
d_n Nozzle diameter, in (mm)
Su Burning velocity, in (cm s^{-1})
Pe Peclet number

A. H. Mola · H. A. Abdul Wahhab (✉) · Z. H. Naji
Department of Mechanical Engineering, University of Technology, Baghdad, Iraq
e-mail: 20085@uotechnology.edu.iq

F. Ahmad et al. (eds.), *ICPER 2020*, Lecture Notes in Mechanical Engineering,
https://doi.org/10.1007/978-981-19-1939-8_11

Greek symbols

η The thermal diffusivity

1 Introduction

The matter of flame quenching behaviour is necessary for the design of many combustion tools such as ramjets, after-burners, pilot flames among [1, 2]. The problem is how to uphold the flame from being rehabilitated out of the combustion chamber or away from burner edge because of high velocity of the gas against that of the flame. The quenching term in combustion process means extinguishing a flame. Flame quenching occurs at the walls of combustion chambers and burners [2]. The minimum distance of flame separation cannot be attained is named a quenching diameter *dq*. According to this definition, quenching diameter value depends on type of fuel, mixing ratio and directions of flame propagation (design of burner edges) [3–5]. So, flame quenching study by the wall is very important for understanding the combustion operation near the burner wall [6, 7]. The quenching diameter and flame characteristics escort to quenching conditions were studied for premixed flames in methane/air mixtures [8] and other studies in propane/air mixtures [9]. The quenching diameter depends on the direction of flame front on burners, upwards or downwards limits. The perfect diffusion flame and its stretch are the physical factors responsible for the difference between two directions of propagation.

In this research, the work scope was to determine influence of changing of nozzle diameter and number of carbon atoms in fuel on quenching diameter in counter burner for downward and upward propagating flames in fuel/air mixtures. The research has been carried out experimentally and Schlieren is the main technique utilized for capturing the flame data. Optical system consists of laser source (He–Ne) with low power 0.21 mWatt and wave length of 713 nm, and a group of lenses is used. A digital camera setup, Phantom VEO 440, Image processing algorithm was developed to extract information from the recorded images.

2 Experimental Setup

The experiments were conducted by apparatus consists of counter burner system and optical system. By the counter burner the mushroom-shaped tented flame is prepared; fuel gas (Methane, Propane, LPG) and air flow rates are measured. Recording the mushroom-shaped tented flame front images required for the study is performed by the optical system. The apparatus components are illustrated in Fig. 1. Counter burner system includes two counter copper tubes, everyone from them end edge

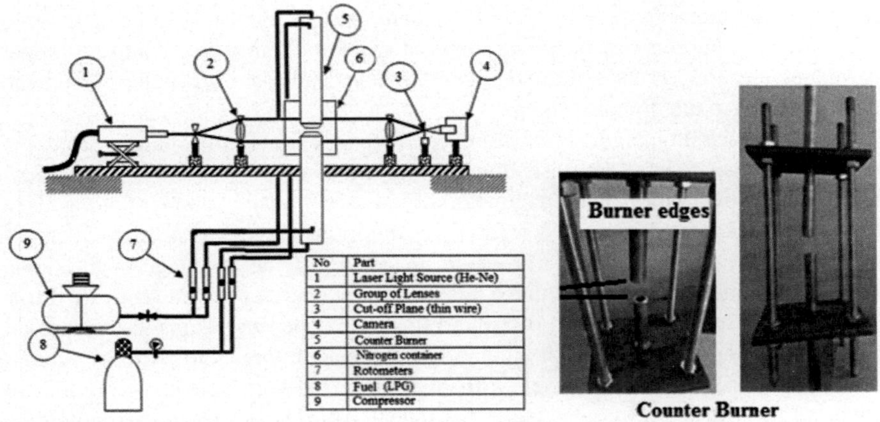

Fig. 1 Schematic view of the experimental setup

with different nozzle diameters: 9, 12, 16, 18 and 23 mm (with changing nozzle angle), these burners fixed by vertical frame from steel while identified the distance between burners edges at 20 mm. This type of combustion process is premixing combustion, fuel (Methane, Propane, LPG) was premixed with air before arrived the edge burner, and to prevent the interference of the ambient air near burner edge in the combustion process, the combustion region was surrounded with Nitrogen container (the combustion region was covered by transparent polycarbonate container, this container was filled by the nitrogen gas during experiments, it was used to block surrounding air to join in combustion operation). Four gas flowmeters were used to measurement flow rates for fuel gas and air; during experiments a wide range of equivalent ratios have been tested, where the flow rates of both gas and air were controlled to obtain the volumetric mixing ratios and the equivalent ratios were determined.

Optical system consists of laser source (He–Ne) low power 0.21mWatt and wave length of 713 nm, and a group of lenses. A digital camera setup "Phantom VEO 440" with 100 ~ 180 fps speed was used in the experiments. Images were acquired a resolution of 1280 × 720 dpi. The camera was positioned on a fixed base, in front of the test rig at angle is 90° for the laser path. The optical alignment along the test rig section were also checked and confirmed. During Counter flame experiments, the top and bottom end of the burner remained open. The opposite configuration was during upward flame propagation. Images for flame disc with high contrast have been obtained with changing the nozzle diameters for burner's edges and CH_4, C_3H_8, LPG–air for wide ranges of equivalence ratios within range of $0.46 < \varphi < 1.57$. The device used for mixture ignition resembled a cigarette lighter and was located near the open end of the burner. Images obtained from the first of them were used to determine the burning velocity. Also, images taken with the camera were used to determine the geometrical parameters of the flames in different nozzle diameters. To determine the quenching diameter for a given nozzle diameter, d_n, the limit mixture composition was gradually approached-up to the moment when the flame disc could no longer travel between the top and bottom burner edge. The dead space was determined from the images from the digital camera.

Image processing algorithm was developed to extract information from the recorded images. The program is divided into three stages: flame front identification; and verification of flame parameters. First stage, image subtraction was used to reduce background noise. Therefore, a calibration image was acquired without combustion; it was subtracted from each image to provide an image of the flame front with reduced background noise. Second stage, each image was converted to a binary image using a threshold of around 0.8–0.85. The output image replaces all pixels in the input image with luminance greater than a threshold with the value 1 (white) and replaces all other pixels with the value 0 (black), where it was possible to suppress light structures connected to the image. Third stage, some morphological functions were used to improve the quality of the image and the flame front shape. Finally, images with the flame front were prepared for the next step quantitative analysis such as measuring the diameter of the disc flame. Depended on most commonly method to determination the burning velocity, measured the amount of gas mixture consumed per second and divided on the area of flame disc [10]:

$$Su = \frac{4Q_{mix}}{\pi d_c^2} \tag{1}$$

where, Su is the burning velocity cm/s, d_c is flame disc diameter (mm), Q_{mix} is unburned gases volumetric flow (cm^3/s). Also, the quenching diameter is associated with parameters of the flammable mixture by the critical Peclet number (Pe) [11]:

$$Pe = \frac{Su.d_q}{\eta} \tag{2}$$

where, η is the thermal diffusivity of the unburned mixture and d_q is the quenching characteristic diameter.

3 Results and Discussion

Experiments began with recorded the counter flame images for high range of equivalence ratio in the different of the nozzle diameters9, 12, 14, 18 and 23 mm. Then, their diameter was systematically increased until the counter flame was able to propagate in them. The burning velocities Su as well as a function of the equivalence ratio for rich and lean sides for methane, Propane, LPG/air mixing at different nozzle diameters are presented in Fig. 2. The results demonstrate that the burning velocity increased with increasing nozzle diameter.

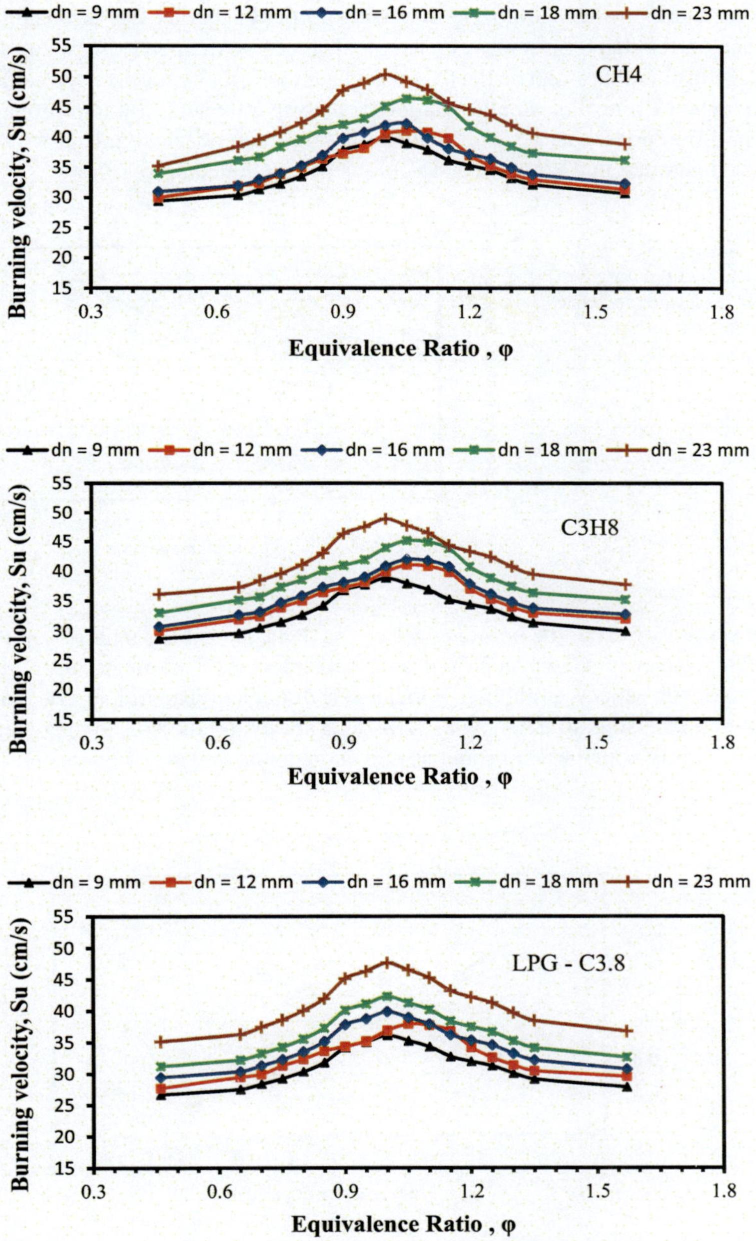

Fig. 2 Burning velocity as function of the equivalence ratio for a different nozzle angle

Figure 3 shows that the nozzle diameter of counter burner has an insignificant influence on the flame burning velocity, *Su*. The nozzle diameter of the burner also has an inconsiderable effect on the flame propagation velocity for the nozzle diameter greater than 12 mm. For smaller values of these parameters, a flame propagation velocity difference is visible. This velocity is smaller in nozzle diameter 9 mm than in the higher value in 23 mm for a stoichiometric methane/air mixture.

Fig. 3 Burning velocity *Su* as a function of the nozzle diameter for a stoichiometric methane/air mixture

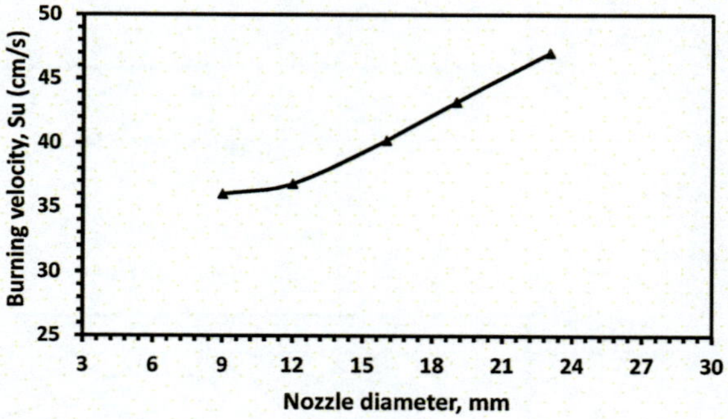

The cooling effect of the burner edges on the flame is visible for narrower nozzles edge. This effect is shown in Fig. 4 as a dependency of the dead space and the nozzle diameter ratio as a function of the nozzle diameter (changing nozzle diameter depend on increasing nozzles angle). The dead space and the nozzle diameter ratio increase slightly with the decreasing nozzle diameter up to 9 mm. For smaller nozzle

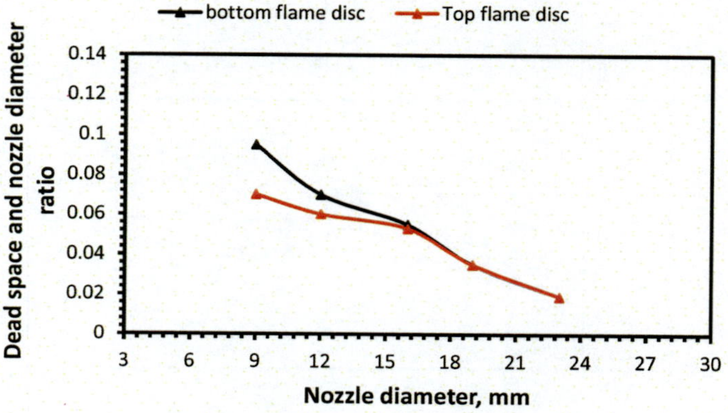

Fig. 4 Dead space and nozzle diameter ratio as a function of nozzle diameter

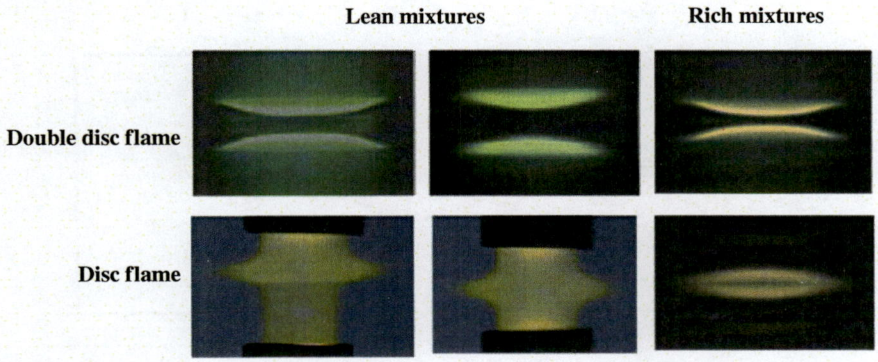

Fig. 5 Images of the disc and double disc flames in lean and rich mixtures in nozzle diameter = 16 mm

diameter, this increase is sharper and is accompanied by a decrease in the laminar burning velocity and a possible lowering of the local temperature of a chemical reaction. On the other hand, the dead space and nozzle diameter ratio at small nozzle diameter 9 mm was higher value for the bottom flame disc from the top disc flame.

From images recorded by the digital camera, differences between the propagating limit of disc and double disc counter flames in lean and rich mixtures in nozzle diameter $d_n = 16$ mm can be seen in Fig. 5. The differences are in the flame colour and in the extent of the dead space. Lean disc and double disc flames are blue, the rich ones show a bright blue. The quenching diameters as a function of the mixture concentration are shown in Fig. 6. One can see that for lean and rich mixtures, the quenching diameters are almost increased with increasing nozzles diameter. While, the increase number of carbon atoms in fuel (1 atom in methane, 3 atoms in propane, and 3.8 atoms in LPG) increased the quenching diameter. Also, the existence of a difference between the quenching distance and the quenching diameter is probably caused by the dead space.

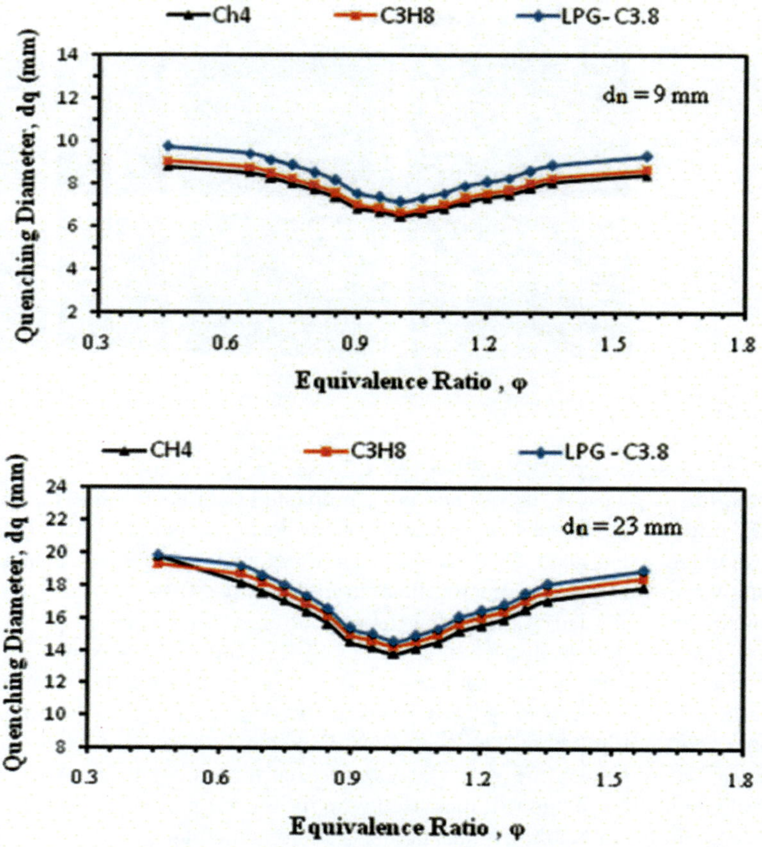

Fig. 6 Experimentally determined quenching diameter as a function of the equivalence ratio

4 Conclusions

This paper describes several experimental investigations considering the various quenching limits for premixed counter flames. Flames in Methane, Propane, LPG/air mixtures propagating in counter burner of a different nozzle diameter have been experimentally investigated. Influence of the changing nozzle diameters for counter burner, methane, propane, LPG/air mixing ratio on the flame burning velocity and quenching diameter has been determined. Schlieren technique was used to analyze the effect of nozzles diameter, 9, 12, 16, 18, and 23 mm on the quenching diameter. Significant results were obtained in the area of counter flame. Quenching diameter in the counter flow depend basically on the value of diameter and velocity of fuel and air. For both lean and rich mixtures, quenching diameters are almost the same for top and bottom flame disc and is increasing with increasing the nozzles diameter. The quenching diameter is greater than the quenching distance for mixtures with the equivalence ratio $0.46 < \varphi < 1.57$. The existence of this difference is probably caused by the dead space. Also, the increase number of carbon atoms in fuel increased the quenching diameter.

Acknowledgements The authors are obliged to the University of Technology, Baghdad in Iraq, for providing the technical assistance available in the Centre of Automotive and Energy Generation.

References

1. Song ZB, Ding XW, Yu JL, Chen YZ (2006) Combust Explos Shock Waves 42(3):268–276
2. Abdul Wahhab HA, Aziz AR, El-adawy M, Ismael MA, Firmansya (2018) Int J Eng Tech **7**:230–233
3. Kim NL, Maruta K (2006) Combust Flame 146:283–301
4. Zamashchikov VV (2004) Combust Explos Shock Waves 40(5):545–552
5. Maruta K, Kataoka T, Kim NI, Minaev S, Fursenko R (2005) Proc Combust Inst 30:2429–2436
6. Jarosinski J, Podfilipski J, Fodemski T (2002) Combust Sci Technol 174:167–187
7. Law CK (2006) Combustion physics. Cambridge University Press, New York
8. Sotton J, Boust B, Labuda A, Bellenoue M (2005) Combust Sci Technol 177(7):141–162
9. Bellenoue M, Kageyama T, Labuda SA, Sotton J (2003) Exper Therm Fluid Sci 27:323–331
10. Sher E, Heywood JB, Hacohen J (1992) Combust Sci Technol 83:323–325
11. Maxime K, Sergei L, Julien S, Marc B (2009) Proceedings. 22nd ICDERS 1:1–4

Determining the Influencing Factors for Micro-Explosion of Emulsified Biodiesel Fuel

Syed Muhammad Syed Masharuddin, Zainal Ambri Abdul Karim, Mior Azman Meor Said, Nur Hazira Amran, and Mhadi A. Ismeal

Abstract Emulsion of palm oil biodiesel can reduce the usage of diesel, promotes complete combustion and reduce both nitrogen oxides and particulate matter. The current researchers found that hydrophilic-lipophilic 7, 8, and 9, surfactant dosage 7% and 9%, and water content 11% and 13% are optimum blends as emulsified fuels for the diesel engine. However, there was a gap for surfactant dosage 8% on the optimum blends. Hence, this study aims to ascertain the emulsion properties meeting the diesel fuels standard followed by visualization technique with Liedenfrost to record the evolution leading to micro-explosion. The results on all the emulsions did not show clear water after 90 days up to 114 days of visual observation indicating stable emulsion. The viscosity and density values with 13% water content were higher than 11%. While the calorific values and sauter mean diameter of 11% water content were higher than 13%. Micro-explosion observed for all single droplet emulsion exhibited coalescence of the small water particles into the larger diameter water particles as the main factor for the occurrences of micro-explosion.

Keywords Water-in-biodiesel emulsion · Palm oil · Direct ultrasonic homogenizer · Leidenfrost technique · Micro-explosion

---

S. M. Syed Masharuddin (✉) · Z. A. Abdul Karim · M. A. Meor Said
Centre for Automotive Research and Electric Mobility (CAREM), Universiti Teknologi PETRONAS, 32610 Seri Iskandar, Perak, Malaysia
e-mail: syed_19000318@utp.edu.my

N. H. Amran
Mechanical Engineering Department, Universiti Teknologi PETRONAS, 32610 Seri Iskandar, Perak, Malaysia

M. A. Ismeal
Mechanical Engineering Department, University of Bahri, Khartoum Bahri, Khartoum, Sudan

F. Ahmad et al. (eds.), *ICPER 2020*, Lecture Notes in Mechanical Engineering,
https://doi.org/10.1007/978-981-19-1939-8_12

1 Introduction

A mixture of diesel and palm oil methyl ester as the alternative fuel for an internal combustion engine can be recommended as a national biodiesel feedstock [1]. Biodiesel can be used continuously as a substitute for fossil energy sources in Malaysia and renewable energy resources [2–5].

An emulsion with the addition of water in diesel as alternative fuel is a promising strategy that can decrease simultaneously the emissions of both particulate matter (PM) and nitrogen oxides (NOx) [5–11]. Fuel modification technique cost less than modification of diesel engine. Aiyshah [5] stated that Euro 5 compliance diesel with 6.5% of water content showed 45% reduction of NOx and 8.56% reduction of fuel consumption. All samples of water-in-biodiesel emulsions involving in the micro-explosion result in lower fuel consumption compared to fuel without water content [5, 6]. Water-in-biodiesel fuel in the form of emulsion caused the break-up of the fuel droplets by the phenomenon known as micro-explosion [7, 8]. As the emulsion droplet is heated, water encapsulated in the fuel vaporized, reached superheat temperatures and then exploded into microdroplets, also called secondary atomization. These microdroplets expand the fuel surface contact area, subsequently enhancing the combustion of the fuel [6, 9–11]. Micro-explosion will not happen when the water content is zero in the fuel [12, 13]. Likewise, the presence of water will lower the adiabatic flame temperature as it acts as a heat sink [9]. Another phenomenon that occurs when the emulsion droplet temperature is lower than the boiling point of water is puffing, which is the ejection of child droplets from the emulsion [14, 15].

Since both water and diesel are immiscible liquids, it is crucial in forming water-in-diesel emulsions with the presence of surfactant [6, 9, 10]. The surfactants are classified based on hydrophilic-lipophilic balance (HLB) due to their lyophobic and lyophilic groups[9, 16] and function to absorb at the interfaces as a surface-active agent between immiscible liquids establishing a stable emulsion [16, 17]. Generally, an HLB value of 9 and below tends to be suitable for forming water-in-fuel emulsions [3]. Even though, there are a variety of surfactants available, hydrophilic Tween 80 and lipophilic Span 80 surfactants are opted by most researchers [2–4, 8, 9, 18–23]. Also, researchers claimed that an emulsion mixed with two surfactants of the different hydrophilic-lipophilic properties would offer better emulsion stability rather than stabilizing with a single surfactant [17]. Researchers using ultrasonic homogenizer obtained smaller water particles size and more evenly distributed water particles compared to mechanical homogenizer [2, 9, 24–27]. Lin and Chen [26] preferred ultrasonic homogenizer due to low carbon monoxide (CO) emission and low fuel consumption rate when they obtained intensive micro-explosion. Chen [28] stated that micro-explosion could be divided into low, medium, and high-intensity micro-explosion based on the strength and timing of the micro-explosion. Avulapati [29, 30] reported that micro-explosion form coalescence water particles size more than 150–300 μm promote strong micro-explosion and water particles size in-between 80 μm and 150 μm promote weak micro-explosion.

Table 1 Parameter composition of water-in-palm oil derived biodiesel emulsions [25]

Item	Composition	Optimum composition
HLB value	6, 7, 8, 9	7 and 8
Surfactant dosage % of water volume	7, 9, 11	7 and 9
Water content %	9, 11, 13, 15	11 and 13

An extensive review of the past investigation works on water-in-diesel emulsions found that petroleum-derived diesel was commonly used for the emulsions. Literature relating to biodiesel emulsions, explicitly palm oil methyl ester biodiesel emulsions and production method using ultrasonic homogenization can be found from Abdul Karim et al. [25]. The study identified the optimum composition of emulsion with regards to stability, water particle sizes and distribution in the emulsion as shown in Table 1.

However, there was a gap for emulsion blends for the surfactant dosage of 8%. The effects of water content, surfactant dosage and HLB values on emulsion characteristics based on utilizing biodiesel B5 (5% palm oil methyl ester and 95% pure diesel) are still uncertain. Moreover, the optimum water-in-palm oil derived biodiesel emulsion that meets the specifications for a diesel engine is as yet unsure. Therefore, this paper aimed at ascertaining the emulsion properties meeting the diesel fuel standard followed by visualization technique with Liedenfrost to record the evolution leading to micro-explosion and determining the influencing factors for micro-explosion.

2 Methodology

The following sections describe the production, composition, and characterization of water-in-palm oil derived biodiesel emulsions (WiPE) for the current study.

2.1 Emulsion Preparation

The production of emulsions for the current study follows the procedures described in Ref. [25]. Figure 1a shows the command process of water-in-palm oil derived biodiesel emulsions. Firstly, ultrasonic homogenizer command I, which is the direct sonification of biodiesel B5 & surfactant for 3 min using an ultrasonic homogenizer. Followed by ultrasonic homogenizer command II where, distilled water is added to the solution and being direct sonication for another 10 min. The time taken for this process of WiPE is based on 100 ml per sample. The time taken may vary for different volumes of the WiPE samples to ensure the mixture blends well.

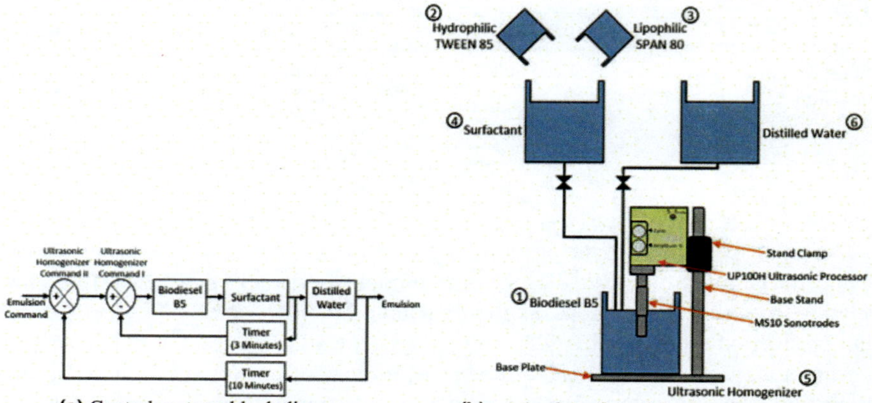

(a) Control systems block diagram **(b)** production of emulsions equipment setup

Fig. 1 Processes involved in the production of emulsions

Table 2 Composition of water-in-palm oil derived biodiesel emulsions for the current study

Sample ID	HLB	Surfactant dosage % of water volume	Water content %	Volume in ml			
				Biodiesel B5	TWEEN 85	SPAN 80	Distilled water
7-8-11	7	8	11	88.12	0.355	0.525	11
7-8-13	7	8	13	85.96	0.419	0.621	13
8-8-11	8	8	11	88.12	0.486	0.394	11
8-8-13	8	8	13	85.96	0.574	0.466	13

Note: sample ID refers to HLB value, followed by surfactant dosage, and water content

Figure 1b shows the emulsion blends setup consist of water, oil, and two surfactants with the different hydrophilic-lipophilic property. The cycle and amplitude % of the ultrasonic homogenizer were 0.7 and 50, respectively for all blends.

The composition for water-in-biodiesel emulsions with 8% surfactant dosage based on the emulsion blends gap discussed previously is shown in Table 2. The determination of the defined HLB values can be found in Khan [31].

2.2 *Laboratory Characterization of WiPE*

The physical characterizations investigations are conducted which included density at 25 °C using DA-645 Density Meter, viscosity at 40 °C using ARG-2 Rheometer, calorific value using Bomb Calorimeter, emulsion stability by visual observation water particle sizes and measuring the sedimentation height of the untouched emulsions in a clear glass until it was unstable, microscopy observation using Olympus BX51 for water particle sizes calculated based on the sauter mean diameter (SMD), and ultimate analysis of the emulsions using CHNS analyzer.

2.3 Micro-Explosion Analysis of WiPE

Figure 2 shows the experimental set up for visualization of micro-explosion by a single emulsion droplet [29, 32]. The technique includes the study of the image sequence towards micro-explosion using software called ImageJ (open-source image processing program) captured by high-speed camera Miro M310 at resolution 512 × 384 px, sample rate 950 fps, and exposure time 1000 μs, and temperature against time using NI cRIO-9063 with module 9211. The time taken is obtained by the image number divided by the sample rate. After the calibration using a suitable scale yielded, every 60 px is equal to 1 mm. Meanwhile, the fuel droplet temperature was obtained using wired thermocouple and the aluminium hotplate temperature was maintained at 500 ± 2 °C throughout the entire experiment. The parent droplet begins approximately 3.7 mm for all emulsion samples with the temperature begins around 90 °C. The aluminium hotplate receives the heat from the ceramic type heating element to create the Leidenfrost effect [33] to avoid the emulsions droplet from rapid evaporates. All temperatures can be monitor using the LabView software connected to National Instruments (NI) cRIO-9063 module 9211.

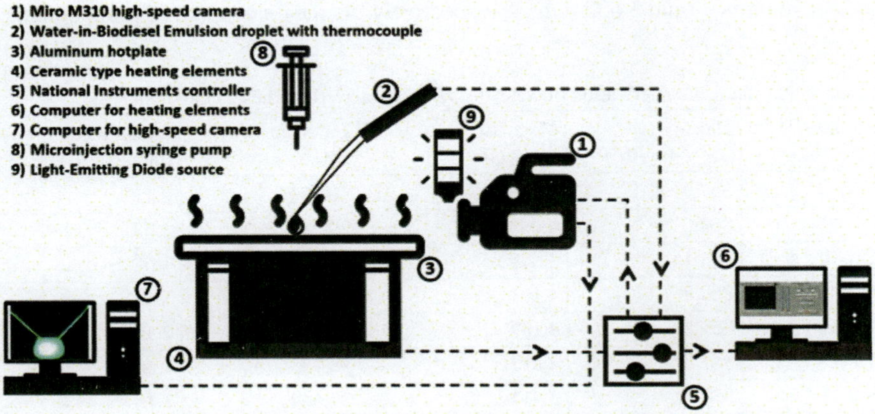

Fig. 2 Setup for Leidenfrost technique of water-in-palm oil derived biodiesel emulsions

3 Results and Discussion

Table 3 shows the result of density, viscosity, calorific value, and SMD. Unbold sample ID refers to the previous study and bold sample ID refers to current research. All values obtained are following the standard of Water/Diesel Fuel Regulation by France and Italy.

3.1 Physical Characterization of Water-In-Palm Oil Derived Biodiesel Emulsions

The effects of water content percentage on the density of the blends are exhibited in Fig. 3a. Although the emulsion samples of the current study, which is surfactant dosage 8 exhibited higher density values compared to surfactant dosage of 7 and 9 except 7-7-13 which is slightly higher by 0.249%, the percentage differences are only between 0.151% and 0.809%. It is observed that the minimum and maximum density values of the emulsion samples are 851.8 kg/m^3 and 864.0 kg/m^3, respectively. A trend can be observed that the density of the emulsions generally increases with an increase in the water content in the mixture due to the higher density of water compared to the biodiesel as agreed by many researchers [21, 23, 34, 35]. This trend is observed for all the 12 emulsion samples.

The viscosity values increase with the addition of water and also showed in increasing trend with increase in HLB [4, 14, 19, 21, 23], as shown in Fig. 3b. It is observed that the minimum and maximum viscosity values of the emulsion samples were 4.565 mm^2/s and 3.635 mm^2/s, respectively. As the percentage of water content

Table 3 Physical characterization of water-in-palm oil derived biodiesel emulsions

Sample ID	Density at 25 °C (kg/m^3)	Viscosity at 40 °C (mm^2/s)	Calorific value (MJ/kg)	Sauter mean diameter (μm)
7-7-11	857.9	3.635	40.291	2.11
7-7-13	864.0	4.063	39.647	1.58
8-7-11	855.1	3.679	40.105	1.67
8-7-13	860.8	4.161	38.543	1.63
7-8-11	**859.2**	**4.107**	**39.825**	**2.10**
7-8-13	**861.8**	**4.217**	**38.733**	**1.60**
8-8-11	**858.7**	**3.911**	**39.664**	**2.18**
8-8-13	**862.2**	**3.918**	**38.623**	**1.66**
7-9-11	857.0	3.823	39.077	2.09
7-9-13	860.5	4.392	38.919	1.59
8-9-11	851.8	3.942	42.268	2.10
8-9-13	856.5	4.565	41.637	1.63

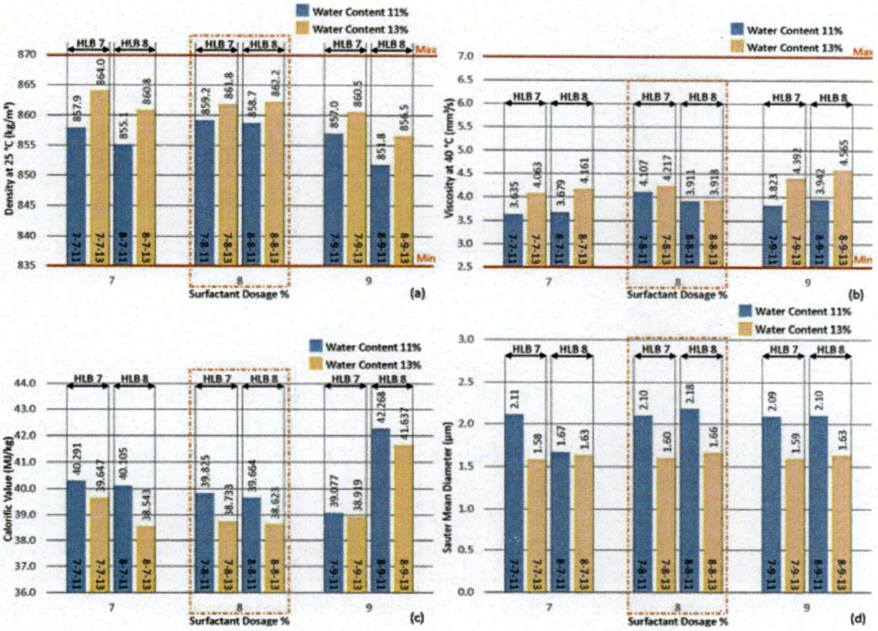

Fig. 3 Physical properties of 12 emulsion samples **a** density, **b** viscosity, **c** calorific values and **d** SMD

increases, more complex micellar structures are formed, thus resulting in increased viscosity. Higher viscosity means it needs higher collision energy to break through the surfactant layer to produce separation or coalescence. The lower viscosity of WiPE may contribute to smooth coalescence and caused the earlier micro-explosion [36].

The presence of water leads to a lower heating value of WiPE, as shown in Fig. 3c, which can contribute to a slight reduction in engine performance when used as fuel. However, it is envisaged that due to the occurrences of micro-explosion of the WiPE, it will strongly enhance the engine performance with both low emissions and low fuel consumption. This is the results of increases increase in combustion rate and air–fuel mixing enhancement [18, 37]. Smaller calorific value gains from the higher water content [2, 4, 21].

The measured water particle diameters in the emulsion exhibited in Fig. 3d show maximum and minimum SMD of 2.18 μm and 1.58 μm, respectively. It was observed that SMD at 11% water content was higher than at 13%. The SMD of 13% water content was very similar to SMD of surfactant 7 and 9. This SMD shows the consistency of producing small particle sizes when using ultrasonic homogenizer and proved that the blending technique gave some impact to the size of water particles [3, 9, 17]. HLB values do not seem to have any effect on the emulsion SMD.

Table 4 Ultimate analysis of emulsion samples

Sample name	CHNS			
	Carbon (%)	Hydrogen (%)	Nitrogen (%)	Sulfur (%)
WiPE 7-8-11	72.580	11.542	0.720	0.031
WiPE 7-8-13	70.990	11.851	0.450	0.028
WiPE 8-8-11	72.160	11.973	0.380	0.021
WiPE 8-8-13	68.810	12.227	0.280	0.019

3.2 Ultimate Analysis

The ultimate analysis of the four WiPE with 8% surfactant dosage is shown in Table 4. The analysis shows that the carbon, nitrogen, and sulfur content were higher for WiPE with a water content of 11% compared to 13%, which can be attributed to the influence of surfactant along with lesser water. The hydrogen content was lesser for WiPE with a water content of 11% compared to 13%. Increasing the water content will resulting in reduction of carbon, nitrogen and sulfur, meanwhile, increment of hydrogen. This indicates the existent of nitrogen and sulfur is highly likely from the derivatives used for the palm oil derived biodiesel. In addition, HLB 7 shows more percentage of nitrogen and sulfur compared to HLB 8, lipophilic SPAN 80 contributes more nitrogen and sulfur compared to hydrophilic TWEEN 85. Overall, all sulfur percentage of the emulsion samples are within the limit of the standards that is 0.031% for Italy and France standards.

3.3 Emulsion Stability—Visual Observation

In ensuring the best WiPE is produced, the stability of the samples is measured. The stability of the emulsion is important to ensure it is compatible with the engine. The samples were kept motionless in a glass bottle (Fig. 4b) to observe their physical changes. The periodical check of 3 days was made to measure the percentage of water/oil layer separation. All emulsions of water-in-palm oil derived biodiesel are stable with an intact emulsion volume percentage of 60–69% for 114 days. After 114 days, the WiPE gradually became unstable in the following order 7-8-13, 7-8-11, 8-8-13, and 8-8-11. Clear oil and creaming at 11% water content was higher than at 13% and sedimentation at 11% water content was lesser than at 13%. The surfactant dosage promotes the optimum stability of WiPE [4]. A combination of two types of surfactant (i.e. hydrophilic and lipophilic) prolongs the separation over time [17].

Water-lover surfactant (hydrophilic property) shows longer separation behavior, which is HLB 8 compared to HLB 7 that contains more oil-lover surfactant (lipophilic

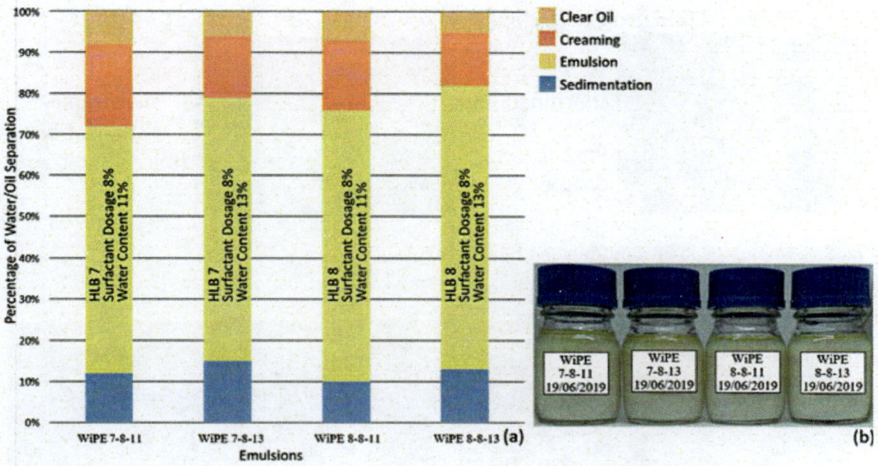

Fig. 4 **a** Percentage of water/oil separation emulsion samples and **b** visual observation of emulsion samples

property). More surfactant of hydrophilic property (TWEEN 85) added to the emulsions provide more desirable stability compared to the surfactant of lipophilic property (SPAN 80). In terms of stability, surfactants with HLB 8 maintain stable without a clear water layer longer than HLB 7.

3.4 Puffing and Micro-Explosion Phenomena

In the occurrence of puffing and micro-explosion phenomena, Fig. 5 shows the sample image of (a) puffing, and (b) micro-explosion. In Fig. 5a shows two separated droplets, the bigger one is called parent droplet the main sample droplet meanwhile the smaller one is called child droplet due to occasional discharged of small water particles which has not merged, reached the surface of the droplet and exit the parent droplet. Following Fig. 5b shows finer droplets exposed to atmosphere after the water particles inside the parent droplet changed to vapour and erupted, which is called micro-explosion due to secondary atomization.

3.5 Sequence Image Towards Micro-Explosion

For each of the emulsion, Fig. 6 shows the images sequence of emulsion droplet towards micro-explosion from initial to final which is as soon as the WiPE droplet (a) 7-8-11, (b) 7-8-13, (c) 8-8-11, and (d) 8-8-13 touching the hotplate surface until micro-explosion occurred. For coalescent, by the usage of the microscope view

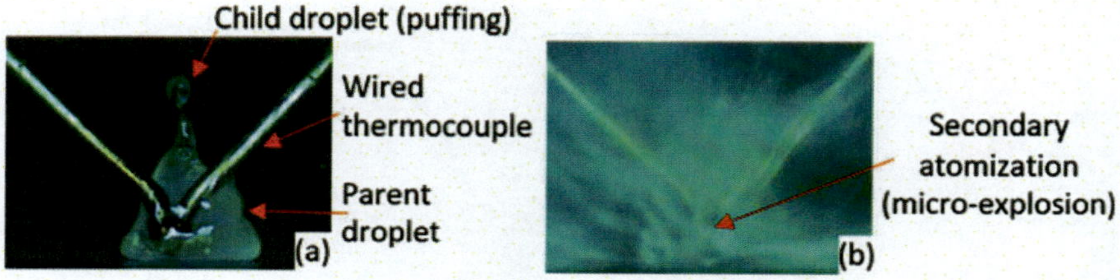

Fig. 5 Sample image of **a** puffing and **b** micro-explosion

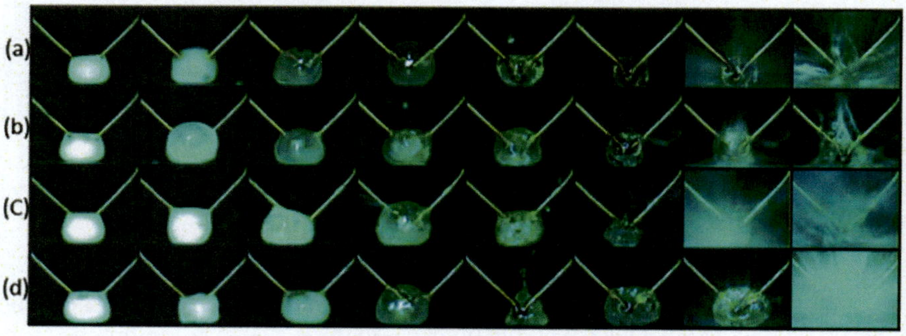

Fig. 6 Images sequence of emulsion droplet towards micro-explosion

resulting the appearance of multiple water particles in Fig. 6 shows around the diameter size of (a) 2.10 μm, (b) 1.60 μm, (c) 2.18 μm, and (d) 1.66 μm, initially, respectively. Later under the heat absorption, the homogenous emulsion starts to separate calmly. Sometimes, the motion of the droplet was violent, the droplet starts to bounce and dance, while the shape changes from round to oval, expanding and retracting for several moments. This is due to coalescence of the small water particles into the larger diameter water particles, which can be visually observed as the main factor for the occurrences of micro-explosion. The parent droplet begins approximately 3.7 mm for all emulsion samples with the temperature begins around 90 °C. Few things can be observed, the emulsion samples from milky appearance turned clearer and all emulsions were having both puffing and micro-explosion. The micro-explosion occurred at varying temperature for each WiPE samples from as low as 157.17 °C up to 187.50 °C. The time taken towards micro-explosion varied from 5.836 s to 7.631 s.

3.6 WiPE Droplet Temperature

Figure 7a shows the droplet temperature as monitored using National Instruments (NI) from the moment the WiPE 7-8-11 droplet touches the hotplate surface until just when the micro-explosion occurred with a total duration of 7.631 s. At 3.863 s, the temperature taken was 139.31 °C and at 5.708 s was 154.55 °C. Figure 7c shows the temperature value for WiPE 8-8-11 of 122.35 °C at 3.103 s, increased drastically to 139.36 °C at 4.417 s, which is 17.01 °C increment. The micro-explosion occurred at 5.836 s, with a temperature of 166.60 °C. WiPE 8-8-11 showed the earliest time for micro-explosion occurrence followed by 8-8-13, 7-8-13, and 7-8-11, respectively. For the same surfactant dosage of 8% and water content of 11%, the results show the time taken and the temperature required to obtain micro-explosion was lower at HLB 8 compared to HLB 7. These indicate higher hydrophilic properties contribute more to micro-explosion.

Based on Fig. 7b, twice a peak increment of temperature was observed. At 0.645 s the temperature increased from 97.50 °C to 110.94 °C, which is 13.44 °C and at 2.736 s the temperature increased from 135.32 °C to 148.71 °C, which is an increment of 13.39 °C. The temperature of micro-explosion was 186.60 °C at 6.432 s. The temperature observed in Fig. 7d obtained micro-explosion at 6.187 s with a temperature of 157.17 °C, the lowest temperature required for micro-explosion occurrence than 8-8-11, 7-8-13 and 7-8-11, respectively.

The composition for surfactant dosage and water content are the same, which are 8% and 13% respectively, HLB 8 shows the micro-explosion occurred at an earlier time and lower temperature compared to HLB 7. These indicate both water content of 11% and 13% having better micro-explosion with the more hydrophilic property.

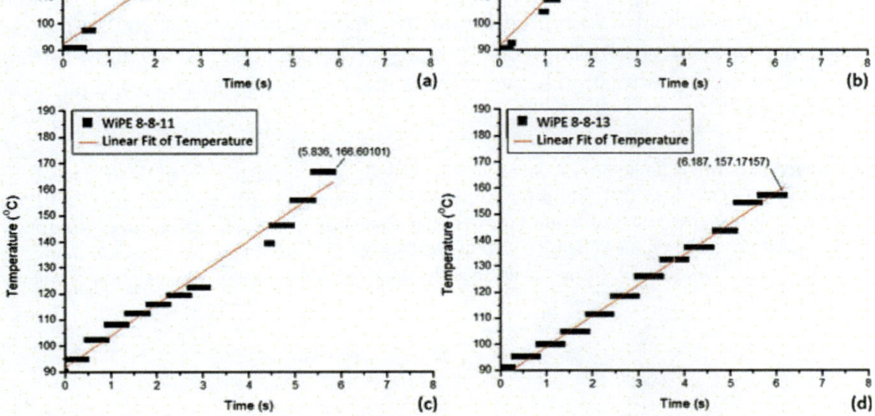

Fig. 7 Droplet temperature of WiPE **a** 7-8-11, **b** 7-8-13, **c** 8-8-11 and **d** 8-8-13

Water-lover surfactant (hydrophilic property) shows lesser time leads to micro-explosion, which is HLB 8 compared to HLB 7 that contains more oil-lover surfactant (lipophilic property). More surfactant of hydrophilic property (TWEEN 85) added to the emulsions provide more desirable micro-explosion compared to the surfactant of lipophilic property (SPAN 80). In terms of micro-explosion, surfactant with HLB 8 shows viscosity prove to cause coalescence faster and obtain micro-explosion at a lower temperature.

4 Conclusion

The results on the emulsion stability of all the WiPE did not show clear water (separation) after 114 days of visual observation. Thus, the emulsions are considered as stable. For physical properties, the viscosity and density value at 13% water content were higher than at 11%. On the other hand, the calorific values and SMD at 11% water content were higher than at 13%. For ultimate analysis, the carbon, nitrogen, and sulfur percentage of 11% water content were higher than 13%. Meanwhile, the hydrogen percentage with 13% water content was higher than 11%.

WiPE 8-8-11 is the best composition, which is HLB 8, surfactant dosage (8% of water volume), and water content 11%, which is claimed to be easier to coalescence and micro-explosion. After the observation of micro-explosion for all WiPE droplets, the factors influencing the occurrence of micro-explosion has been established.

1. HLB 8 contain higher concentration of hydrophilic. HLB 8 obtain micro-explosion earlier compared to HLB 7.
2. Water content 11% resulting larger SMD. SMD not less than 2.2 μm, larger SMD easy to coalescence. At HLB 8, 11% water content obtain micro-explosion earlier compared to 13% water content.

Since all the WiPE samples have been produced and characterization has been done, it is recommended to reproduce all the stable emulsions for engine performance measurements and exhaust emission tests. These emulsions will allow further investigation in the correlation between the actual engine performance and physical properties of WiPE.

Acknowledgements The authors would like to acknowledge the support given by the Universiti Teknologi PETRONAS to the Centre for Automotive Research and Electric Mobility (CAREM) in performing this research and Ministry of Higher Education for the financial support through Fundamental Research Grant Scheme (FRGS Cost Centre: 0153AB-L67). Also, special thanks to Mohamed Syafiq Sallehuddin for his assistance in experimental set up for visualization of micro-explosion.

References

1. Lia Laila LO (2017) Kaji Eksperimen Angka Asam dan Viskositas Biodiesel Berbahan Baku Minyak Kelapa Sawit dari PT Smart Tbk. J Teknol Proses Dan Inov Ind 2(1):27–31
2. Patil H, Waghmare J (2017) Biodiesel-water emulsions: an alternative approach for conventional fuels. Int Res J Eng Technol 4(7):1200–1204
3. Debnath BK, Saha UK, Sahoo N (2015) A comprehensive review on the application of emulsions as an alternative fuel for diesel engines. Renew Sustain Energy Rev 42:196–211
4. Kapadia H, Brahmbhatt H, Dabhi Y, Chourasia S (2019) Investigation of emulsion and effect on emission in CI engine by using diesel and bio-diesel fuel: a review. Egyptian J Pet 28(4):323–337
5. Mazlan NA, Yahya WJ, Ithnin AM, Kadir HA, Ramlan NA, Sugeng DA, Muhammad Adib AR, Koga T, Mamat R, Che Sidik NA (2018) Effects of different water percentages in non-surfactant emulsion fuel on performance and exhaust emissions of a light-duty truck. J Clean Prod 179:559–566
6. Watanabe H, Suzuki Y, Harada T, Matsushita Y, Aoki H, Miura T (2010) An experimental investigation of the breakup characteristics of secondary atomization of emulsified fuel droplet. Energy 35(2):806–813
7. Crookes RJ, Kiannejad F, Nazha MAA (1997) Systematic assessment of combustion characteristics of biofuels and emulsions with water for use as diesel engine fuels. Energy Convers Manag 38(15–17):1785–1795
8. Debnath BK, Sahoo N, Saha UK (2013) Adjusting the operating characteristics to improve the performance of an emulsified palm oil methyl ester run diesel engine. Energy Convers Manag 69:191–198
9. Ithnin AM, Noge H, Kadir HA, Yahya WJ (2014) An overview of utilizing water-in-diesel emulsion fuel in diesel engine and its potential research study. J Energy Inst 87(4):273–288
10. Morozumi Y, Saito Y (2010) Effect of physical properties on microexplosion occurrence in water-in-oil emulsion droplets. Energy Fuels 24:1854–1859
11. Tran T, Ghojel JI (2005) Impact of introducing water into the combustion chamber of diesel engines on emissions-an overview. Asia Pacific Conf Combust 237438547:33–36
12. Khan MY, Abdul Karim ZA, Aziz ARA, Tan IM (2014) Experimental investigation of microexplosion occurrence in water in diesel emulsion droplets during the leidenfrost effect. Energy and Fuels 28:7079–7084
13. Cen C, Wu H, Lee C, Fan L, Liu F (2019) International Journal of Heat and Mass Transfer Experimental investigation on the sputtering and micro-explosion of emulsion fuel droplets during impact on a heated surface. Int J Heat Mass Transf 132:130–137
14. Yahaya Khan M, Abdul Karim ZA, Aziz ARA, Heikal MR, Crua C (2017) Puffing and microexplosion behavior of water in pure diesel emulsion droplets during leidenfrost effect. Combust Sci Technol 189(7):1186–1197
15. Kim H, Won J, Wook S (2018) Evaporation of a single emulsion fuel droplet in elevated temperature and pressure conditions. Fuel 226:172–180
16. Scarpete D (2013) Diesel-water emulsion, an alternative fuel to reduce diesel engine emissions. A Review. Mach Technol Mater (7):13–16
17. Won J, Baek SW, Kim H, Lee H (2019) The viscosity and combustion characteristics of single-droplet water-diesel emulsion. Energies 12:1–12
18. Vellaiyan S, Amirthagadeswaran KS (2016) Influence of water-in-diesel emulsion fuel and compression ratio on combustion, performance and emission characteristics of diesel engine. J Sustain Energy Eng. 3(3):238–253
19. Bukkarapu SkR, Jyothi Y, Raju LS, Babu GC, Narayanan K (2017) Review on current trends in water in oil emulsions. Int J Mech Eng Technol 8(4):359–371
20. Singh R, Sharma S, Gangacharyulu D (2016) The effect of nano-additive blended water-diesel emulsion fuel on CI engine performance and emissions—a review. IJCS Proc Int Conf Adv Emerg Technol 4:23–27

21. Reham SS, Masjuki HH, Kalam MA, Shancita I, Rizwanul Fattah IM, Ruhul AM (2015) Study on stability, fuel properties, engine combustion, performance and emission characteristics of biofuel emulsion. Renew Sustain Energy Rev 52:1566–1579
22. Aziz A, Jusoh A, Mamat R, Abdullah AA (2013) Effect of water content and Tween 80 to the stability of emulsified biodiesel. Appl Mech Mater 465–466:191–195
23. Lin CY, Lin SA (2007) Effects of emulsification variables on fuel properties of two- and three-phase biodiesel emulsions. Fuel 86(1–2):210–217
24. Abdul Karim ZA, Tee D, Khan MY, Hagos FY (2018) Investigation of water-in-biodiesel emulsion characteristics produced by ultrasonic homogenizer. MATEC Web Conf 225(01012):1–7
25. Abdul Karim ZA, Kaur E, Syed Masharuddin SM, Khan MY, Hagos FY (2020) The characteristics of water-in-biodiesel emulsions produced using ultrasonic homogenizer. Alexandria Eng J 59(1):227–237
26. Lin CY, Chen LW (2008) Comparison of fuel properties and emission characteristics of two- and three-phase emulsions prepared by ultrasonically vibrating and mechanically homogenizing emulsification methods. Fuel 87(10–11):2154–2161
27. Taurozzi JS, Hackley VA, Wiesner MR (2012) Preparation of nanoparticle dispersions from powdered material using ultrasonic disruption preparation of nanoparticle dispersions from powdered material using ultrasonic disruption. NIST Spec Publ 1200-2(1.1):1–14
28. Chen G, Li Y, Lan C, Lin H, Chao Y (2017) Micro-explosion and burning characteristics of a single droplet of pyrolytic oil from castor seeds. Appl Therm Eng 114:1053–1063
29. Avulapati MM, Ganippa LC, Xia J, Megaritis A (2016) Puffing and micro-explosion of diesel-biodiesel-ethanol blends. Fuel 166:59–66
30. Avulapati MM, Megaritis T, Xia J, Ganippa LC (2019) Experimental understanding on the dynamics of micro-explosion and puffing in ternary emulsion droplets. Fuel 239:1284–1292
31. Khan MY, Abdul Karim ZA, Aziz ARA (2016) An experimental investigation of microexplosion in bio fuel emulsion. ARPN J Eng Appl Sci 11(24):14303–14307
32. Zhu M, Zhang Z, Zhang Y, Liu P, Zhang D (2017) An experimental investigation into the ignition and combustion characteristics of single droplets of biochar water slurry fuels in air. Appl Energy 185(2):2160–2167
33. Khan MY, Abdul Karim ZA, Aziz ARA, Tan IM (2017) A case study on the influence of selected parameters on microexplosion behavior of water in biodiesel emulsion droplets. J Energy Resour Technol 022203:1–10
34. Ghannam MT, Selim MYE (2009) Stability behavior of water-in-diesel fuel emulsion. Pet Sci Technol 27(4):396–411
35. Siegmund SC, Storbeck ML, Cross JB, Fogler HS (1980) Physical properties of water in fuel oil emulsions (density and bulk compressibility). J Chem Eng Data 25:72–74
36. Chen RH, Wang WC, Chen YW (2016) Like-drop collisions of biodiesel and emulsion diesel. Eur J Mech B/Fluids 60:62–69
37. Vellaiyan S, Amirthagadeswaran KS (2016) The role of water-in-diesel emulsion and its additives on diesel engine performance and emission levels: A retrospective review. Alexandria Eng J 55(3):2463–2472

Experimental Investigation of Puffing/Microexplosion in Fuel Droplets and Effects of Injection Modes on It

Zuhaib Nissar, A. Rashid A. Aziz, Morgan R. Heikal, and Mhadi A. Ismael

Abstract Puffing/microexplosion potentially increases atomization and enables better air–fuel mixing in sprays, thereby improving combustion. In this paper, we have investigated the physics leading to puffing and microexplosion in fuel droplets and identified factors such as droplet detachment modes and droplet-air interaction that influence puffing/microexplosion. A microsyringe is used to inject droplets as small as 100 μm radius in a control volume chamber maintained at 500 °C and atmospheric pressure. Droplet dynamics are captured using high-speed camera coupled with a long-distance microscopic lens using backlight illumination technique as the droplet traverses and undergoes puffing and microexplosion. In-house image processing codes are used to track droplet motion and compute its geometrical parameters, which enable us to decipher microexplosion dynamics. During this study, two significant factors were identified to cause puffing/microexplosion; thinning of droplet shell and interfacial instabilities. A general trend of increasing puffing times was observed with increasing droplet radii. The modes of detachment do affect the instabilities at the droplet surface which further amplify them and causes early puffing. Furthermore, the rate of bubble growth was amplified for the droplets traversing at higher Reynolds number.

1 Introduction

The use of fuels like water in diesel emulsion (WIDE) as a fuel source is attractive due to its ability to reduce NOx and soot particles, its use in combustion engines also obviates the environmental concerns of greenhouse gas emission and carbon sequestration. The addition of water lowers the local flame temperature which

Z. Nissar (✉) · A. R. A. Aziz · M. R. Heikal · M. A. Ismael
Center for Automotive Research and Electric Mobility (CAREM), Universiti Teknologi PETRONAS, 32610 Seri Iskandar, Perak, Malaysia
e-mail: zuhaib_16000073@utp.edu.my

M. R. Heikal
Advanced Engineering Centre, School of Computing, Engineering and Mathematics, Sir Harry Ricardo Laboratories, University of Brighton, Brighton BN2 4GJ, UK

F. Ahmad et al. (eds.), *ICPER 2020*, Lecture Notes in Mechanical Engineering,
https://doi.org/10.1007/978-981-19-1939-8_13

reduces NOx emissions [1, 2]. As water evaporates, it supplies hydroxy (OH) radicals to the combustion process thereby reducing soot formation [3]. These emulsions increased brake thermal efficiency by 6%, with a 30% reduction in NOx and particulate emissions, and 70% reduction in unburnt hydrocarbons [4, 5]. It has been found that immiscible emulsions like water in diesel undergo a phenomenon called puffing/microexplosion due to difference in the volatilities of their components. Microexplosions can be beneficial to secondary atomization and enhance penetration depth of fuel sprays in internal combustion engines.

Microexplosion phenomenon is caused due to the effect of diffusional entrapment of volatile component, i.e. water, inside an emulsion droplet [6]. When WIDE droplet is heated under high-temperature conditions, evaporation of diesel at the surface leads to the development of boundary layer. In contrary to this, droplet core has a higher concentration of water which heats up gradually to surpass its boiling point and become superheated. According to thermodynamics there is a maximum superheat limit a liquid can accumulate called critical limit. This condition is metastable until the superheated water attains the critical limit. At nucleation temperature, the metastable state breaks, and formation of a new vapour phase takes place in the form of small vapour bubbles. If nucleation temperature coincides with the critical limit, it is called homogeneous nucleation. Otherwise, nucleation which occurs before the critical limit is called heterogeneous nucleation [7]. Bubbles which have radius less than that of their critical radii will collapse. These bubbles combine to form a giant bubble. With intense surge in internal pressure, the bubble pinches through the droplet surface, which causes ebullition of an emulsion droplet. If the phenomenon is extensive, it is called microexplosion, while as, if it is limited to a portion of an emulsion droplet, it is called puffing [8, 9]. These minion droplets lead to enhancement in the net surface area which in turn accelerates the gasification process and may help in homogenizing charge in internal combustion engines [10]. Different models have been suggested to estimate the times to puffing, also called microexplosion delay time which is based on the penetration of heat inside the droplet [11]. However, the significant gap between the predicted times to puffing and the experimental values has been left unaddressed.

In the view of above considerations, we have investigated the physics of puffing/microexplosion in WIDE droplets. We have enquired the role of droplet generation modes and interaction between droplet and air on the microexplosion delay time. Details of the experimental setup are described in the next section. The analysis performed on the experimental data is presented in Sect. 3. The results of the experimentally observed puffing/microexplosion in WIDE droplets are discussed in Sect. 4. The results are summarized in Sect. 5.

2 Working Principles

Shadowgraphy mainly deals with the optics of inhomogeneous medium [12]. The inhomogeneities and disturbances may occur due to the motion of bulk matter, turbulence, and thermal convection. These disturbances may lead to small changes in the density gradient and with it the refractive index. Refractive index n is given by

$$n = \frac{c_0}{c} \tag{2.1}$$

where $c_0 = 3 \times 10^8 m/s$ is the universal speed limit,c is the speed of light in the medium.

A well-known Gladstone–Dale relation is used to link the density of fluid ρ and its refractive index.

$$\frac{n-1}{\rho} = k, \tag{2.2}$$

where k is the Gladstone–Dale coefficient.

If a geometrical light ray traverses in the z-direction, the optical inhomogeneities refract or bend light rays in proportion to the gradients of refractive index in an x–y plane (wave-front) [12]. The resulting ray curvature is given by

$$\frac{\partial^2 x}{\partial z^2} = \frac{1}{n}\frac{\partial n}{\partial x}; \frac{\partial^2 y}{\partial z^2} = \frac{1}{n}\frac{\partial n}{\partial y} \tag{2.3}$$

Integrating along the optical axis z, components of the angular ray deflection in the x and y directions can be written as:

$$\varepsilon_x = \frac{L}{n_0}\frac{\partial n}{\partial x}; \varepsilon_y = \frac{L}{n_0}\frac{\partial n}{\partial y} \tag{2.4}$$

where L is the optical path length,n_0 is the refractive index of the surrounding medium.

In shadowgraphy, illuminance level in the image responds to the second spatial derivative of refractive index, i.e. $\frac{\partial^2 n}{\partial x^2}$ and $\frac{\partial^2 n}{\partial y^2}$. It displays the ray displacement resulting from the deflection and is thus a mere shadow. This can be seen in Fig. 1. However, it allows us to visualise the most noticeable features of a subject with little changes in illumination.

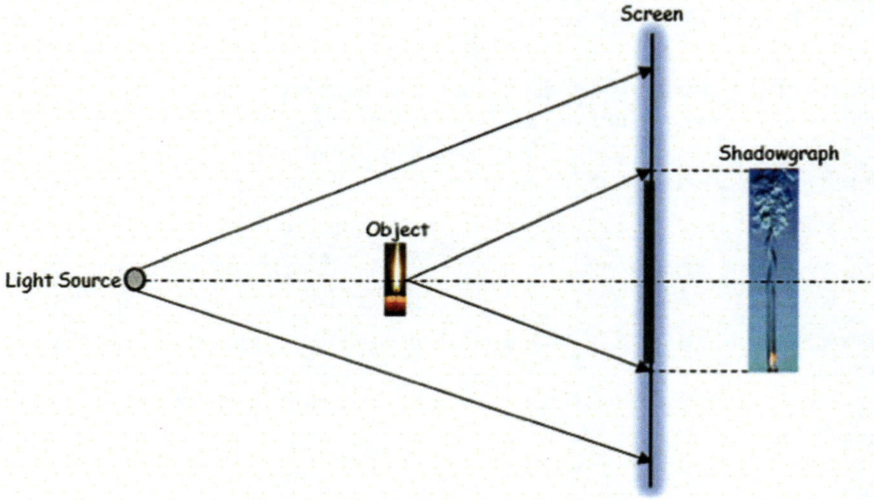

Fig. 1 Schematic showing the principles of Shadowgraphy

3 Experimental Procedure

Experiments are carried out at Flow Visualization Laboratory, Universiti Teknologi PETRONAS, Malaysia. Primary purpose of these experiments is to gather decipher puffing/microexplosion dynamics in order to relate them with different approaches to model the phenomenon. The setup consists of a constant volume combustion chamber, which is maintained at a constant ambient temperature of 500 °C and atmospheric pressure. The schematic of the experimental setup is given in Fig. 2. The cylindrical chamber has an opening for fuel injection and is mounted with a ceramic electric heater. The heater is connected to a PID controller wired with K-type thermocouple in order to measure and ascertain a specific steady-state gas temperature at the centre of the chamber. The heater is insulated with fibreglass wool to prevent heat losses and the chamber is assembled with borosilicate windows using flanges in order to provide optical access for the flow visualisation.

A lipophilic surfactant (Span 80—1% volume of water) is used as an emulsifier to prepare 15% (v/v) water in diesel emulsion. The mixture is mechanically stirred for 10 min at 1000 rpm to homogenise the emulsion. The experiment involves injection of free-falling emulsion droplets using a micro-pump UMP3 with Micro4 Controller. This controller would allow us to monitor injection time and amount of fuel and subsequently produce droplets of regular size. The needle is positioned at the inlet once the steady-state temperature is achieved inside the chamber. This is done to prevent any heat transfer at the needle-tip before injection.

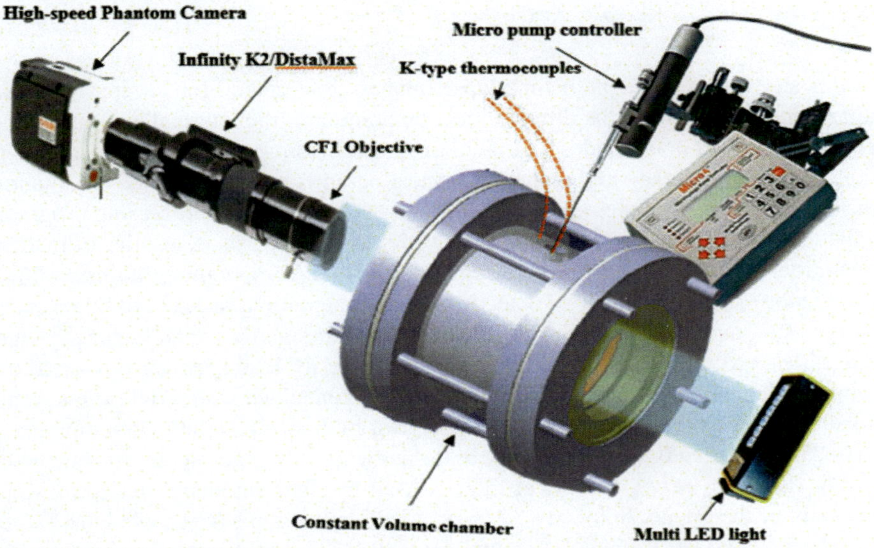

Fig. 2 Schematic of the combustion chamber showing the optical accesses and the long-distance microscope imaging system. The Multi LED light used to provide the back illumination. Reprinted from Energies, Volume 11, No. 9, Ismael et al., Investigation of Puffing and Micro-Explosion of Water-in-Diesel Emulsion Spray Using Shadow Imaging, P 2281, Copyright MDPI (2018)

The optical setup consisted of a Multi LED (LT-V8-15) (Tokyo, Japan) for back illumination, high-speed video camera (Phantom Miro M310) (Wayne, NJ, USA) coupled with a long-distance microscope (Infinity K2) (Boulder, CO, USA) and zoom lens (CF1 objective) (Boulder, CO, USA). The camera is adjusted to a resolution of 256×800 pixel^2, the frame rate of 12,000 fps and exposure time of 2 μs. The image scale factor is calculated using the measurements from the diameter of the syringe needle, which is known to us as 0.3 mm. The uncertainties in the microexplosion delay time are associated with the least count of the optical setup (1/12,000 s). The uncertainty for the measurement of the droplet diameter is related to the detection of the droplet perimeter, which is estimated to be ± 37 μm. Droplets with diameters smaller than 50 μm are ignored due to the large experimental uncertainty. The uncertainty in velocity measurements is estimated based on the reports and assessments published earlier, according to which the global uncertainty is less than 2% even under supersonic applications [13].

4 Analysis

In this study, different sets of time-lapse images are prepared for different droplet diameters, starting from the droplet injection point until the aftermaths of microexplosion. Using in-house codes for digital image processing, motion and trajectory of the droplets are tracked and their geometrical parameters are evaluated simultaneously. The stack of images are imported to a Java-based image processing program called ImageJ for pre-processing [14]. Pre-processing consists of a series of matrix operations, namely, image contrast enhancement, filter operations, thresholding, morphology operations, feature detection, and segmentation. ImageJ considers each image as a graphics object where each element of the matrix contains three colour intensities, red, green, and blue (RGB) values. The RGB images are processed to 8-bit greyscale images in which each pixel is represented with numerical value from a list of $2^8 = 256$ values. This value is assigned by linearly scaling from min–max to 0–255, such that 0 and 255 usually represent white and black spots, respectively.

For detecting edges, the Sobel filter is used to highlight sharp changes in the intensity of the image. In this, two 3×3 convolution kernels (as shown below) are used to generate vertical and horizontal derivatives. The final image is produced by combining the two derivatives using the square root of the sum of the squares [15, 16].

$$\begin{bmatrix} 1 & 2 & 1 \\ 0 & 0 & 0 \\ -1 & -2 & -1 \end{bmatrix} \begin{bmatrix} 1 & 0 & -1 \\ 2 & 0 & -2 \\ 1 & 0 & -1 \end{bmatrix}$$

This is followed by a thresholding operation. Thresholding is a type of image segmentation that isolates objects by converting a grey-scale image into a binary image. In this operation, the two levels are assigned to pixels that are below or above the specified threshold value. Pixels with luminance greater than this specified threshold value are assigned value 1 (white) and value 0 (black) if otherwise. Figure 3 demonstrates the order of the aforementioned operations performed for the time-lapsed images of emulsion droplets.

During post-processing, the periphery of the droplet is captured from the binary image and tracked for the entire stack of images using in-house codes. This allows us to measure different geometrical parameters of the droplet such as area, radius, droplet centroid, perimeter, shape factor and aspect ratio, thereby enabling us to ascertain droplet dynamics.

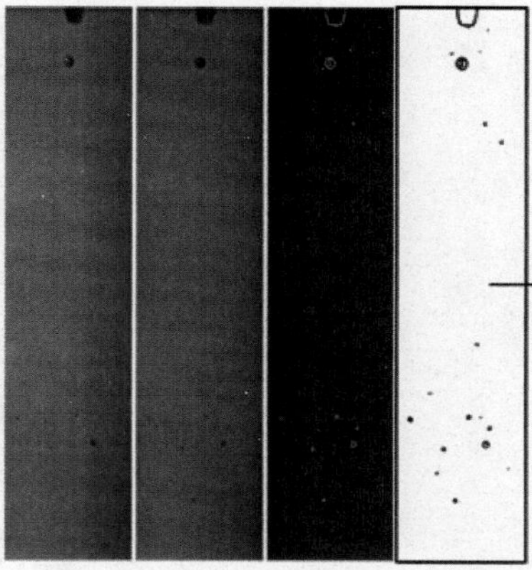

Fig. 3 Chronology of image processing operations, where **a** original image, **b** 8-bit grayscale image, **c** Sobel filter for edge detection, **d** binary image using threshold feature

5 Results and Discussion

In this section, results of investigation during droplet detachment and after it are presented and dynamics of droplets leading to microexplosion/puffing are explained based on the evolution of droplet parameters such as radius, motion, geometry and shape factor. The significant results are provided and discussed as follows:

5.1 Effects of Droplet Detachment

It is observed that microexplosion is sensitively dependent on the modes through which these droplets are detached which is also supportive to the experiments by Wang et al. [2]. As the droplet is detached from the syringe tip, its mode of generation and preliminary deformation induce instabilities at the surface. This appears to effect microexplosion delay time as will be seen later. Two modes of detachment are identified during experiments, stable and unstable. Stable mode is characterized by uniform striations or vertical ribs along the droplet crown exhibiting symmetry and reduced shape deformations at the surface after detachment, while in unstable mode there are either irregular striations or no striations at all along the crown, the contours are very asymmetric and deformations are intense after the detachment. During detachment these rib-like structures undergo buckling leading to atomization, thus forming child droplets and a bigger droplet near the syringe tip (Fig. 4).

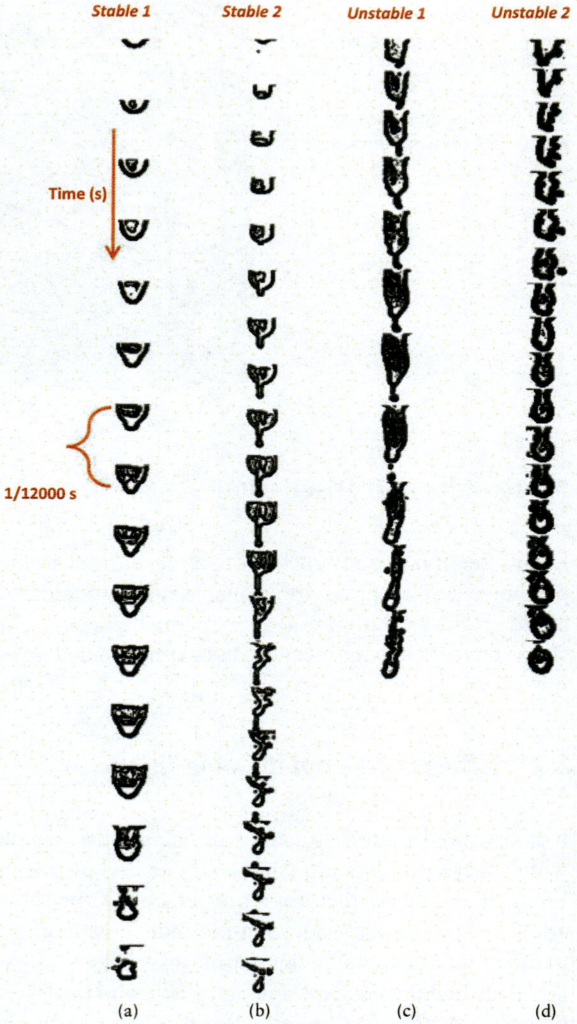

Fig. 4 Image sequences showing different droplet detachment modes at ambient temperature of 500 °C and 101.325 kPa. Time interval between frames is 1/12,000 s; **a** and **b** refer to stable modes, **c** and **d** refer to unstable modes

5.2 *Droplet Dynamics*

As the droplet undergoes free fall, heat from the ambient air penetrates inside the droplet causing internal nucleation. During the experiments, it is observed that internal nucleation occurred faster during unstable droplet detachment mode. It is hard to assume homogeneous nucleation as there are many factors contributing to the formation of bubble nucleus inside the droplet much of which are probabilistic in nature. This also includes cavitation due to in-syringe flow effects. Oza and Sinnamon [17, 18] called it an internal mode of nucleation. There were test runs

Fig. 5 Image showing two bubble nuclei being formed inside the droplet

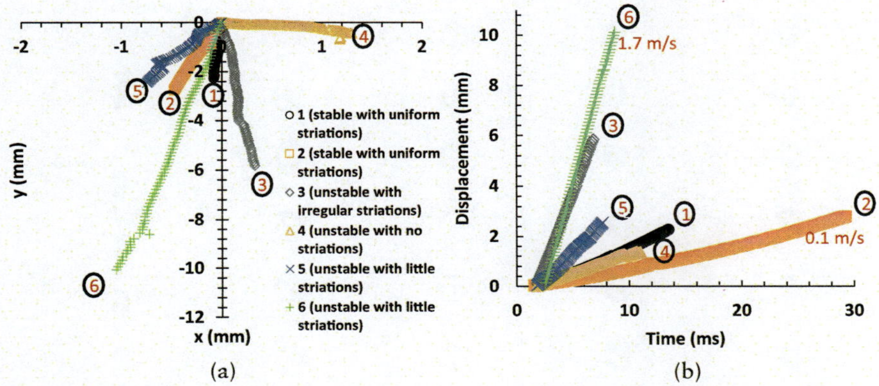

Fig. 6 **a** Trajectories of droplets during free fall inside the constant volume chamber at 500 °C and 101.325 kPa; **b** plots describing the motion of droplets

in which the nucleation seems to have occurred prior to droplet detachment which is also confirmed in the literature. Analysis of these effects requires investigation of in-syringe flow effects (Figs. 5 and 6).

Figure 7 shows the genesis and growth of droplet due to nucleation and bubble expansion. A bubble nucleus forms at 17/6000 s which expands due to bubble pressure leading to thinning of liquid shell around the droplet, and at 62/6000 s the droplet undergoes puffing giving away child droplets. It is observed that a single bubble nucleus forms close to the droplet centre in most cases, however, droplets containing binary bubble nuclei are observed as well (see Fig. 5). These observations are consistent with the experiments by Wang et al. [2].

Using in-house codes for image processing, motion and dynamics of the droplets is traced. Figure 6a shows the trajectory of droplets right from the tip of the syringe till the point where they undergo puffing. In order to estimate the average velocity of droplets, displacement of the droplet centroid is plotted against time. In Fig. 6b velocities of these droplets range from 0.1 m/s to 1.7 m/s. Based on the values of initial droplet diameter $2R_{d0}$ and their velocity V, droplet Reynolds number $Re = \frac{2\rho_{air}\cdot V\cdot R_{d0}}{\mu_{air}}$ is calculated. Properties of air are evaluated for $T_g = 500°\text{C}$; dynamic viscosity of air $\mu_{air} = 35.47 \times 10^{-6}\text{Pa/s}$ and its density $\rho_{air} = 0.4567\text{kg/m}^3$.

Figure 8 shows variations of droplet diameter with time, which gives us insights towards puffing regardless of droplet diameter. Time is normalised with the thermal diffusion time scale (τ) of the emulsion droplet, where $\tau = R_0^2/\kappa, \kappa = 8.7335 \times 10^{-8}\text{m}^2/\text{s}$ is the thermal diffusivity of the emulsion mixture whose value is evaluated for initial droplet temperature $T_0 = 40°\text{C}$. One can see that bubble expansion inside

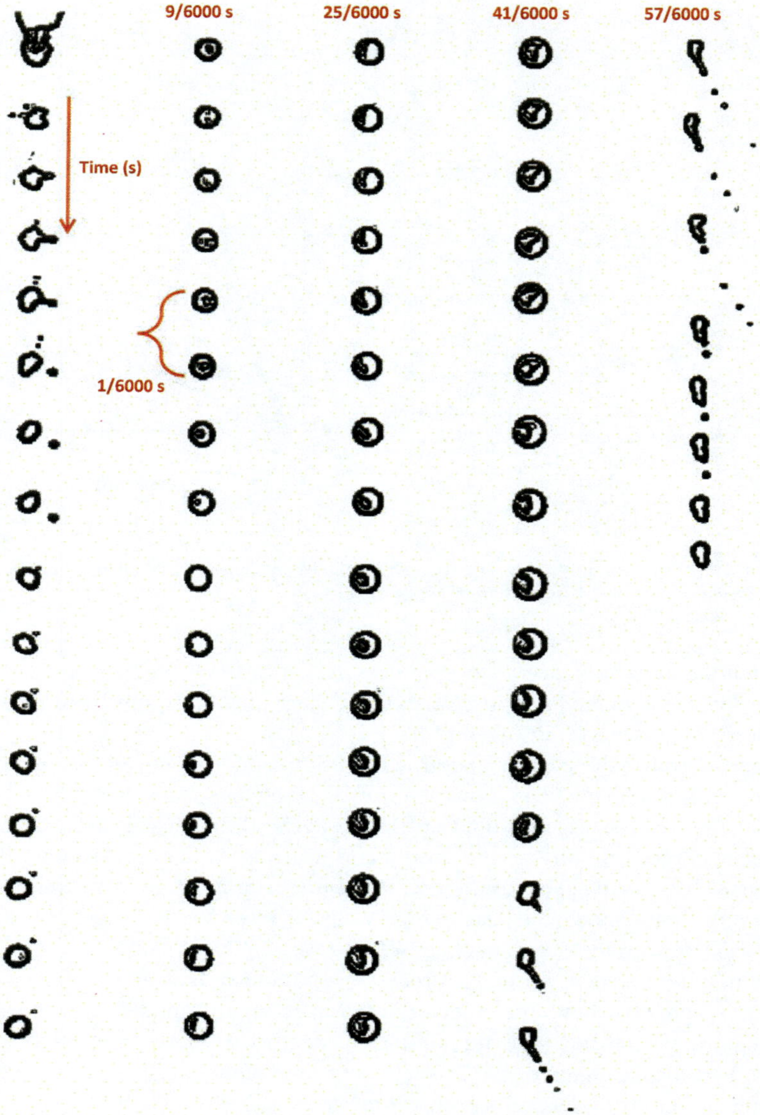

Fig. 7 Image sequence showing droplet of $2R_{d0} = 200$ μm undergoing bubble expansion and puffing. Time interval between frames is 1/6000 s

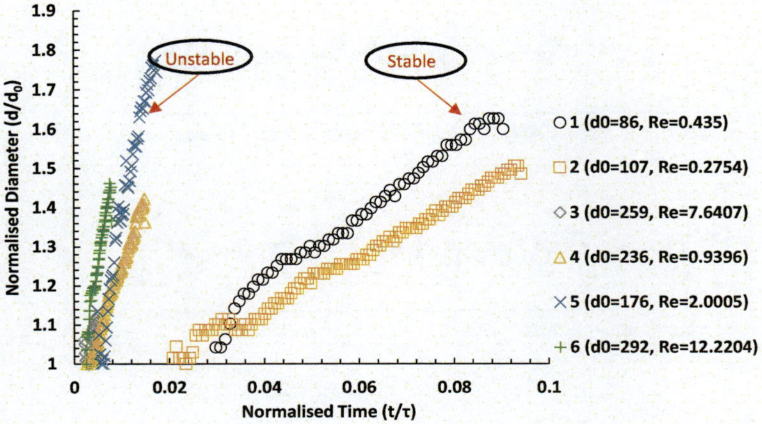

Fig. 8 History of droplets undergoing bubble expansion. Normalized diameter vs normalized time, where $\tau = R_0^2/\kappa$. Details of droplet detachment mode, initial diameter d_0 (μm) and Reynolds number Re are given on the right

the droplet causes an increase in the droplet diameter thereby causing thinning of the droplet shell and eventually leading to puffing/microexplosion. It is observed that the timescale of puffing is much less than the diffusion timescale. Two regimes of droplet expansion are recognized which are mainly related to the modes of droplet detachment and also the magnitude of droplet Reynolds number. In case of unstable detachment, puffing is found to occur quite earlier while as in stable mode the same effect is prolonged. Also, while comparing curves 1 and 2, it is observed that droplets falling at higher Reynolds numbers appear to have higher rates of droplet expansion. Similar trend could be seen for the unstable regime. This can be partly attributed to increasing rates of heat transfer due to convection and internal circulation at higher Reynolds numbers.

To investigate further on these modes, evolution of shape factors of the droplets with time were examined. The results are presented in Fig. 9, where (1), (2) are observed as stable modes while the rest refer to unstable modes of generation. It can be seen from these figures that there are very little fluctuations in the amplitude in case of stable mode causing longer times to puffing. In unstable mode, these fluctuations appear to be intense causing shorter times to puffing. Also, it is observed that droplet takes longer to detach from the syringe tip in case of stable mode.

It is deciphered that the modes of droplet detachment and motion of the droplet induce instabilities at the droplet surface which give rise to droplet oscillations and undulations thereby causing wrinkling at the droplet surface. This allows the bubble to reach the droplet surface due to intense bubble pressure and cause thinning of the liquid shell at the surface, finally punching through it much earlier than anticipated. This is consistent with the results in [19] in which experimental values of times to puffing were always shorter than the predicted values. Based on this, two factors were identified to cause puffing/microexplosion; thinning of the liquid shell and

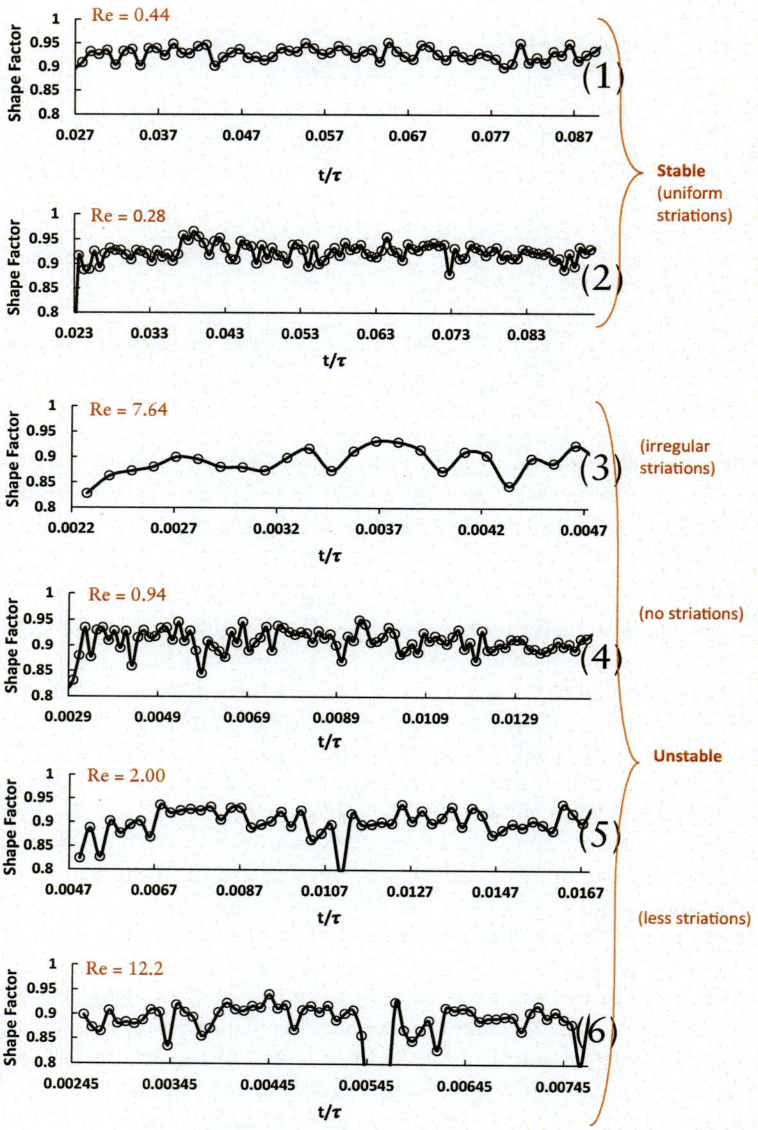

Fig. 9 Fluctuations in the shape factor of the droplet with time; **1**–**6** are in the same order as in the previous figure

interfacial instabilities. The modes of detachment and Reynolds number affect these instabilities which can further amplify them and cause early puffing. Furthermore, the rate of bubble growth is amplified for the droplets traversing at higher Reynolds numbers.

6 Conclusion

Droplets of water in diesel emulsion were visualised using backlight illumination technique as they traversed hot ambient. The events of puffing/microexplosion were captured at 12,000 fps with the help of high-speed camera which was coupled with a microscopic lens. Using in house codes for image processing, dynamics of the droplets were traced, and their geometrical parameters were evaluated. Bubble nucleation was observed shortly after the bubble had detached from the microsyringe. As the bubble increased in size, it caused swelling of the droplet and eventually led to puffing. Thinning of the liquid shell and droplet oscillations were found to be the primary cause which led to puffing/microexplosion. A general trend of increasing puffing times with increasing droplet diameters was observed, although in all cases. Two regimes of droplet puffing were observed as the normalised values of droplet diameter and time were compared. The modes of detachment were found to affect the instabilities at droplet surfaces which can lead to early puffing. Furthermore, the rate of bubble growth was amplified for the droplets at higher Reynolds numbers.

Acknowledgements The current work was carried out under the Centre for Automotive Research and Electric Mobility (CAREM), Universiti Teknologi PETRONAS (UTP), supported by the ministry of higher education Fundamental Research Grant Scheme (FRGS) (Grant FRGS/1/2017/TK10/UTP/01/2) and UTP Graduate Assistant (GA) studentship.

References

1. Law CK, Lee CH, Srinivasan N (1980) Combustion characteristics of water-in-oil emulsion droplets. Combust Flame 37:125–143. https://doi.org/10.1016/0010-2180(80)90080-2
2. Wang CH, Liu XQ, Law CK (1984) Combustion and microexplosion of freely falling multicomponent droplets. Combust Flame 56(2):175–197. https://doi.org/10.1016/0010-2180(84)90036-1
3. Glassman I, Yetter RA, Glumac NG (2015) Combustion. Academic Press
4. Elsanusi OA, Roy MM, Sidhu MS (2017) Experimental investigation on a diesel engine fueled by diesel-biodiesel blends and their emulsions at various engine operating conditions. Appl Energy 203:582–593. https://doi.org/10.1016/j.apenergy.2017.06.052
5. Attia AMA, Kulchitskiy AR (2014) Influence of the structure of water-in-fuel emulsion on diesel engine performance. Fuel 116:703–708. https://doi.org/10.1016/j.fuel.2013.08.057
6. Miglani A, Basu S, Kumar R (2014) Insight into instabilities in burning droplets. Phys Fluids 26(3):032101. https://doi.org/10.1063/1.4866866
7. Faghri A, Zhang Y (2006) Transport phenomena in multiphase systems. Academic Press, Boston

8. Watanabe H, Okazaki K (2013) Visualization of secondary atomization in emulsified-fuel spray flow by shadow imaging. Proc Combust Inst 34(1):1651–1658. https://doi.org/10.1016/j.proci.2012.07.005
9. Ismael MA, Heikal MR, Aziz ARA, El-Adawy M, Nissar Z, Baharom MB, Zainal EZA, Firmansyah, Crua C (2018) Investigation of puffing and micro-explosion of water-in-diesel emulsion spray using shadow imaging. Energies 11(9):2281. https://doi.org/10.3390/en11092281
10. Law CK (2006) Combustion physics. Cambridge University Press, Cambridge
11. Sazhin SS, Rybdylova O, Crua C, Heikal M, Ismael MA, Nissar Z, Aziz ARBA (2019) A simple model for puffing/micro-explosions in water-fuel emulsion droplets. Int J Heat Mass Transf 131:815–821. https://doi.org/10.1016/j.ijheatmasstransfer.2018.11.065
12. Settles GS (2001) Schlieren and shadowgraph techniques. Springer, Berlin, Heidelberg
13. Robbe C, Nsiampa N, Oukara A, Papy A (2014) Quantification of the uncertainties of high-speed camera measurements. Int J Metrol Quality Eng 5(2):201(1–9). https://doi.org/10.1051/ijmqe/2014007
14. Schindelin J, Arganda-Carreras I, Frise E, Kaynig V, Longair M, Pietzsch T, Preibisch S, Rueden C, Saalfeld S, Schmid B, Tinevez J-Y, White DJ, Hartenstein V, Eliceiri K, Tomancak P, Cardona A (2012) Fiji: an open-source platform for biological-image analysis. Nat Methods 9(7):676–682. https://doi.org/10.1038/nmeth.2019
15. Sobel I, Feldman G (1973) A 3x3 isotropic gradient operator for image processing. In: Duda R, Hart P (eds.) Pattern classification and scene analysis. John Wiley & Sons, pp 271–272
16. Anoraganingrum D (1999) Cell segmentation with median filter and mathematical morphology operation. IEEE. https://doi.org/10.1109/iciap.1999.797734
17. Oza RD, Sinnamon JF (1983) An experimental and analytical study of flash-boiling fuel injection. SAE International. https://doi.org/10.4271/830590
18. Oza RD (1984) On the mechanism of flashing injection of initially subcooled fuels. J Fluids Eng 106(1):105–109. https://doi.org/10.1115/1.3242383
19. Nissar Z, Rybdylova O, Sazhin SS, Heikal M, Aziz ARBA, Ismael MA (2020) A model for puffing/microexplosions in water/fuel emulsion droplets. Int J Heat Mass Transf 149:119–208. https://doi.org/10.1016/j.ijheatmasstransfer.2019.119208

Development of Model and Program for Prediction of Gas Turbine Design Point Unknown Parameters

Zainal Ambri Abdul Karim, Aklilu Tesfamichael Baheta, and Milad Fakour Dabbaghi Milad

Abstract With the aid of design parameters and application of thermodynamic laws and relationships, in general theoretical performance analysis of gas turbines can be implemented. However, available design point parameters are limited, and this has made analysis of gas turbine difficult. Moreover, Manufacturer of gas turbines usually provide certain essential information for general understanding due to business concern. Thus, the aim of this study is to develop a model and program for prediction of unknown performance parameters of a single shaft gas turbine at the design point. In this study, gas turbine components have been modelled and a program was developed in engineering equation solver. A graphical user interface is also developed so that the program is user friendly. To predict the unknown parameters manufacturers' available parameters from catalogue namely, pressure ratio, cycle efficiency, exhaust gas flow, exhaust gas temperature, and power output are used as input. Single shaft gas turbine was simulated with certain realistic assumptions. Results of the analysis were compared with previous work that was done in MATHCAD Program and showed good agreement.

Nomenclature

T_1	Inlet air temperature (K)
T_2	Exit air temperature (K)
P_{ratio}	Pressure ratio between compressor blading region
X_{H_2O}	Water content of the inlet air ($kgH_2O/kgAir$)
X_{CH_4}	Pure methane air ratio
C_P (Air)	Constant pressure specific heat capacity of dry air (kJ/kg)

A. T. Baheta (✉)
College of Electrical and Mechanical Engineering Department, Sustainable Energy Center of Excellence, Addis Ababa Science and Technology University, 16417 Addis Ababa, Ethiopia
e-mail: aklilu.tesfamichael@aastu.edu.et

Z. A. Abdul Karim · M. F. D. Milad
Mechanical Engineering Department, Universiti Teknologi Petronas, Bandar Siri Iskandar, 32610 Perak, Malaysia

F. Ahmad et al. (eds.), *ICPER 2020*, Lecture Notes in Mechanical Engineering,
https://doi.org/10.1007/978-981-19-1939-8_14

C_P (H_2O)	Constant pressure specific heat capacity of water (kJ/kg)
$\dot{m}_{air}$	Mass flowrate of air
R_{H_2O}	Gas constant of water
R_{Air}	Gas constant of dry air
η_{el}	Electrical Efficiency
η_{mech}	Mechanical efficiency
η_p	Polytropic efficiency
γ	Specific heats ratio

Subscripts

C	Compressor
T	Turbine
2	Compressor inlet
3	Compressor exit
4	Turbine inlet
5	Turbine exit

1 Introduction

Gas Turbines are one of the most widely used group of internal combustion engines that have been used in various industries such as oil & gas, process plants, and aviation, as well domestic and smaller related industries [1]. In energy systems, they are claimed as a crucial part or component in generating power which can give a vast contribution in order to fulfil the growth in power demands. Since 19's, gas turbines have been subject to continuous development in various industrial application with electric power generation being one of the major focus areas [2]. There are several factors that have impact on gas turbine performance such as installation location, size, component efficiency, configuration, operation, and type of fuel used. Hence, these factors have to be considered and analysed in order to get the expected gas turbine performance and performance verification.

As it is always important for gas turbine powered electrical plants to provide maximum output with increasing demand of energy requirement. The performance analysis of gas turbine becomes an issue to the industries for both its design and its operations. Therefore, it is obligatory to perform modelling of gas turbine engine for estimating the performance parameters not only at the design but also at off-design conditions.

Gas turbine performance simulation starts with design point data. However, gas turbines manufacture usually provide certain essential information for the general understanding due to business concern. The data available in gas turbine catalogue are often with certain exclusion of important parameters, this has resulted difficulty to carry out the full range theoretical performance analysis of gas turbine. According to the manufacturers catalogues, they only provide five important parameters namely, thermal efficiency, mass flowrate and temperature of exhaust gases, compression ratio and power output. Moreover, these available data are not enough parameters in order to carry out the full analysis at the design point of gas turbine.

Models have been developed for different purpose, for instance, based on first principles mathematical model of a laboratory SR-30 engine was developed using the state variable method by Singh et al. [3]. They also developed a simulator with a graphical user interface in the Simulink environment. Lee et al. also [4] developed gas turbine model for prediction of gas turbine performance. A stage-stacking method was used for the compressor, and a stage-by-stage model was adopted in the turbine model. Furthermore, accurate and reliable method to predict the full-range performance of gas turbines are requirements for the research development and assessment on a gas turbine based system such as: the combined cycle with Carbon Dioxide bottoming cycle [5], a solar hybrid combined cycle system, the combined cooling, heating and power system [6] and solid oxide fuel cell/gas turbine hybrid system [7]. Therefore, availability of such data and parameters, which allow gas turbines user to carry out comparison of varies gas turbines' working performance is needed.

There are few available software's which can determine the unknown parameters however, they are expensive, and their source code are not accessible and hence they are not suitable for modification. Thus, the main objectives of this study are to model single shaft gas turbine using a thermodynamic approach for prediction of hidden performance parameters and to develop a program with a graphical user interface (GUI) using engineering equation solver (EES) software.

2 Thermodynamic Modelling

Performance prediction a gas turbine involves modeling of each component and assumptions. Figure 1 shows the main components considered in the model of the gas turbine. In the modeling each component considered as a black box and one-dimensional thermodynamic model is applied under steady state conditions. The model is a component by component list that describes process characteristics and then they are mathematically linked with mass and energy balance principles as they are mechanically linked in the gas turbine to be simulated.

The mathematical model for each engine component in the gas turbine is included as follows. Compressor exit temperature calculation is solved by solving simultaneously Eqs. (1) and (2) [5].

$$\eta_{p,c} = \frac{\int_{T2}^{T3}\left[\frac{C_p(Air).T+X_{H_2O}.C_p(H_2O).T}{R_{air}+X_{H_2O}.R_{H_2O}.T}\right]dT}{\ln(P_{ratio})} \tag{1}$$

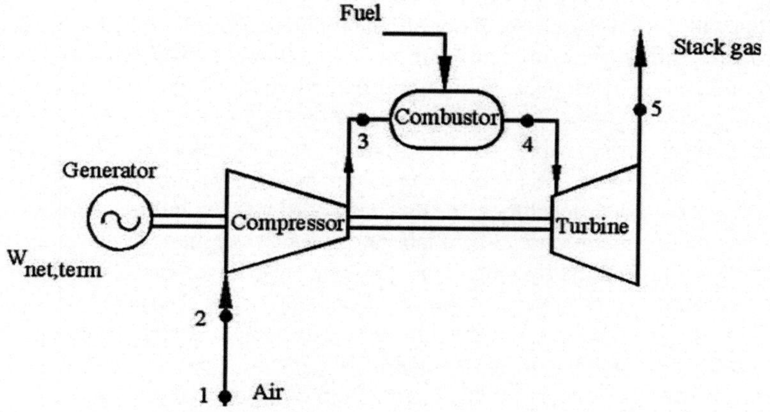

Fig. 1 Schematic diagram of a single shaft gas turbine

$$\frac{T_3}{T_2} = \left(\frac{P_3}{P_2}\right)^{\frac{\gamma-1}{\gamma.\eta_{p,c}}} \tag{2}$$

Similarly, the turbine inlet temperature value can be found by solving simultaneously Eqs. (3) and (4) for known exit temperature [5, 8].

$$\eta_{p,t} = \frac{\int_{T4}^{T3}\left[\frac{C_p(Air).T+X_{H_2O}.C_p(H_2O).T+X_{CH_4}.C_p(CH_4).T}{(R_{air}+X_{H_2O}.R_{H_2O}+X_{CH_4}.R_{CH_4}).T}\right]dT}{\ln(P_{ratio})} \tag{3}$$

$$\frac{T_4}{T_3} = \left(\frac{P_4}{P_3}\right)^{\left(\frac{\gamma-1}{\gamma}\times\eta_{p,t}\right)} \tag{4}$$

Then, the specific power consumption by the compressor can be calculated from Eq. (5) [8].

$$w_c = \int_{T2}^{T3}\left[C_p(Air).T + X_{H_2O}.C_p(H_2O).T\right]dT \tag{5}$$

whereas, the turbine specific power output is given as [8]

$$w_t = \int_{T4}^{T3}\left[C_p(Air).T + X_{H_2O}.C_p(H_2O).T + X_{CH_4}.C_p(CH_4).T\right]dT \tag{6}$$

Heat extraction, which reduces the "mixed combustion temperature" to "mixed turbine inlet temperature". This temperature difference corresponds to the turbine inlet temperature as defined by ISO 2314 standards. In order to calculate the mixed combustor temperature Eq. (7) can be used [2].

$$T_{mix} = \frac{\int_{T2}^{T5} \left[C_p(Air).T + X_{H_2O}.C_p(H_2O).T\right] dT}{LHV - \int_{T1}^{T5} \left(C_p(CH_4).T\right) dT} \tag{7}$$

The pressure at the compressor blading inlet is denoted as P_{CB1} while in Eq. (8) P_{CB2} means the outlet pressure of the compressor blading.

Pressure ratio correction

$$P_{CB1} = P_1.(1 - P_{Inletpressureloss}) \tag{8}$$

$$P_{CB2} = P_1.P_{ratio}(1 + P_{diffuserpressureloss}) \tag{9}$$

$$P_{ratio} = \frac{P_{CB2}}{P_{CB1}} \tag{10}$$

The gas turbine net specific power output is given as

$$P_{GT} = P_{turbine} - P_{compressor} \tag{11}$$

The gas turbine net power out is given by Eq. (12).

$$P_{el} = \eta_{el}.\eta_{mech}.\dot{m}_{air}.P_{GT} \tag{12}$$

In this study the subroutine was written in EES environment to develop the simulation model. Water composition of inlet air to the compressor is measurable under the ambient conditions. Furthermore, once the water content is obtained, the compressor and the turbine unknown parameters (compressor exit and turbine inlet) are obtained iteratively until the initial assumed values become close to the predicted values within a given percentage error. After these values are predicted then the remaining unknown parameters are evaluated. The availability of a Graphical User Interface (GUI) is one advantages of the EES software. Hence, a user-friendly GUI was developed for the simulation model. It will help the user to add input parameters for the EES simulation model, run the model and it will also display all the results. This will make the model simple to change the inputs, run the program, see the results and in general user friendly.

3 Results and Discussion

Several assumptions and conditions are used in order to develop a thermodynamic model for a gas turbine engine. In the analysis the engine is assumed to be run at ISO conditions, i.e., 15 °C, 60% relative humidity, and sea level. Pure methane was taken as the fuel and all the working fluids of the system such as air, water, and fuel are assumed ideal gas conditions and the remaining assumptions are shown in Table 1. All the thermodynamic properties were taken from the internal library of the EES.

In order to validate the results obtained from EES model, the specification of Siemens SGT6-8000H gas turbine data used in MATHCAD simulation model is show in Table 2 [2] were used as input for our model simulation. The results from the MATHCAD model were used as a reference value for validation process, since the actual data of the gas turbine are not available in open literature.

Tables 1 and 2 values with other assumptions have been used as input to run the simulation model developed in EES environment and the results of the simulation model are shown in Table 3. The table also shows the MATCAD simulation results and as can be seen the results obtained from our model agrees with the MATCAD simulation results with an error of less than 1.21%. The simulation gives the unknown parameters, namely, air mass flow rate, fuel mass flow rate, compressor exit temperature, turbine inlet temperature, and compressor and turbine efficiencies.

To make the simulation model user friendly a graphical user interface has been developed as shown in Fig. 2. The figure represents the GUI of the EES model, which consist of a schematic view of the gas turbine that is analyzed in the EES program. As can be seen, it describes clearly all the gas turbine parameters, input values,

Table 1 Assumption parameters in EES

Assumption parameters	Value
Inlet pressure loss	1%
Outlet pressure Loss	1.3%
Internal pressure loss	6.5%
Compressor diffuser pressure loss	2%
Mechanical efficiency	99.5%
Electrical efficiency	98.6%
Low heating value of methane	50 MJ/kg

Table 2 Catalogue parameters for Siemens SGT6-8000H [2]

Catalogue parameters	Value
Electrical power output	274 MW
Overall thermal efficiency	40.16%
Exhaust mass flowrate	600 kg/s
Exhaust temperature of flue gas	620 °C
Pressure ratio of the compressor	20

Table 3 Results comparison between Mathcad and EES

Variable	Mathcad program	EES program	Error (%)
Gas turbine model	SGT6-8000H	SGT6-8000H	–
Inlet air mass flow rate (Kg/s)	586.355	589.81	0.59
Dry air mass flow rate (Kg/s)	582.645	589.74	1.21
Fuel mass flow rate (Kg/s)	13.645	13. 50	1.06
Compressor exit temperature (K)	721.95	722.737	0.11
Turbine inlet Temperature (K)	1601.65	1604.442	0.17
Polytropic efficiency of compressor	92%	91.1%	0.97
Polytropic efficiency of turbine	85.99%	86%	0.011
Isentropic efficiency of compressor	88.26%	87.57%	0.78
Isentropic efficiency of turbine	89.53%	89.54%	0.011

assumptions values and the output parameters. The user can change the turbine data given by manufacturer and other input data, assumptions such as ambient conditions in the EES model (via graphical interface) then press the calculation tab to run the program. Once the calculation process finished, the corresponding results will be displayed in the output section of the GUI and these values are also shown in the schematic diagram of the gas turbine at their corresponding state points.

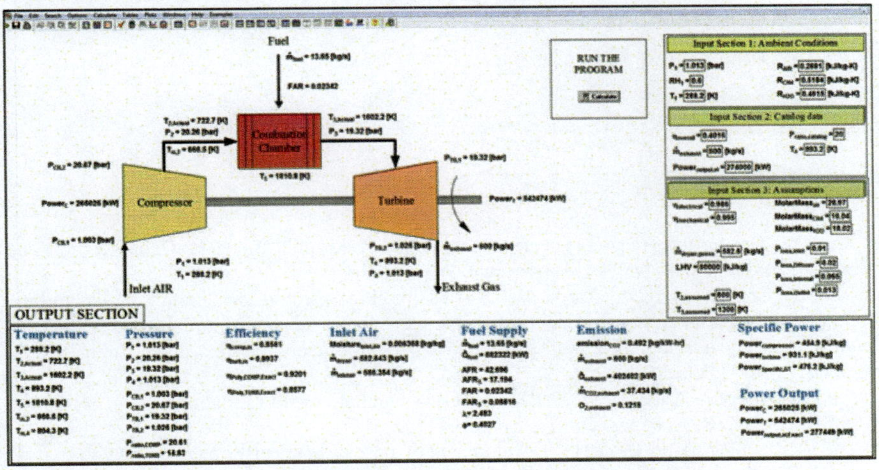

Fig. 2 Graphical user interface (GUI) of the EES Model

4 Conclusion

In order to evaluate the gas turbine off design performance, it is important to know the specification of the turbine at the design point. Nevertheless, the parameters of gas turbine at the design point reported by manufacturer are limited and most of the required information are kept confidential. Hence, this project has developed a model that can predict all the design point data using the manufacturers' available catalogues data as input. The model was developed based on basic thermodynamics concepts. Engineering equation solver environment was used to develop the program for prediction and comparison of the results. Furthermore, the program with a graphical user interface is developed to illustrate the various outputs once the users feed in the assumed values and inputs parameters. The results of the simulation model were compared with previous results done using MATCHAD and comparison showed that good agreements.

The simulation tool can be used for different purposes including for the industry as a simulation tool for performance prediction in the analysis of the gas turbine model during the selection, installation, and/or maintenance phases. It can also be used in the educational sector for teaching of thermodynamics course. This model is user-friendly, and the program can be easily understood and used to analyze a gas turbine performance for various scenarios. However, the program needs to be extended so that it will include off design performance of the gas turbine.

Acknowledgements The authors would like to thank Universiti Teknologi PETRONAS (UTP) for providing all the necessary supports to conduct this study and approval of the grant no. 015LCO-066 to carry out the research.

References

1. Soares C, Gas turbines in simple cycle and combined cycle applications. In: Dennis R (ed.) The gas turbine handbook. NETL, Morgantown, WV
2. Ranasinghe C, Noor H, Jayasuriya J (2014) A simplified method for determining gas turbine performance parameters based upon available catalogue data. In: ASME Turbo Expo 2014: turbine technical conference and exposition. American Society of Mechanical Engineers, pp V006T08A002–V006T08A002
3. Singh R, Maity A, Nataraj PSV (2018) Modeling, simulation and validation of mini SR-30 gas turbine engine. IFAC Papers OnLine 51–1:554–559
4. Lee JJ, Kang DW, Kim TS (2011) Development of a gas turbine performance analysis program and its application. Energy 36(8):5274–5285
5. Saravanamuttoo GFCRHIH, Cohen H (2001) Gas turbine theory
6. Erdem HH, Sevilgen SH (2006) Case study: effect of ambient temperature on the electricity production and fuel consumption of a simple cycle gas turbine in Turkey. Appl Therm Eng 26(2):320–326
7. Pathirathna KAB (2013) Gas turbine thermodynamic and performance analysis methods using available catalog data, ed.

8. Wettstein HE (2007) Lecture on gas turbine cycle analysis aimed at application to catalog data from commercially available machines. In: ASME Turbo Expo 2007: power for land, sea, and air. American Society of Mechanical Engineers, pp 475–480

Numerical Investigation on the Effect of Vibration on Cooling Enhancement of Heat Sinks

Ambgaha Hewage Dona Kalpani Rasangika, Mohammad Shakir Nasif, William Pao, and Rafat Al-Waked

Abstract Heat sinks are commonly used in electronic industry to improve the cooling process and different technologies are adopted together with conventional heat sinks in order to enhance heat transfer rate. One of the methods is applying forced vibration on heated heat sink. This study applies the oscillating frequency between 20 Hz and 2000 Hz to 3D rectangular fin heat sinks under constant Reynold number and non-dimensional amplitude. Different heat sinks were considered with fin height between 0.0127 and 0.0203 m. The results of numerical simulation indicated that increase in oscillating frequency increases the Nusselt number while increase in fin height leads to a decrease in the Nusselt number. The maximum Nusselt number was obtained when the fin height was minimum with frequency at the highest level.

1 Introduction

Miniaturization of the electronic industry has led to a dramatic increase in heat generation per unit volume of electronic components and in the past two decades, the failure rate of the electronic component has exponentially increased with increasing temperature. Unless the operating temperature is adequately controlled, the heat generated will affect the functionality and reliability of the electronic equipment. The fan-based electronic cooling method is mostly used because of the high performance to cost ratio. Heat sinks are commonly used with fan-based cooling technology to enhance the cooling process of electronic components.

When air blows through the hot surface, a thermal boundary layer is generated due to difference in temperature between the inlet air and hot surface. This thermal boundary layer resists the heat dissipation of the heat sinks. Vibration of the hot

A. H. D. Kalpani Rasangika · M. S. Nasif (✉) · W. Pao
Department of Mechanical Engineering, Universiti Teknologi PETRONAS, Seri Iskandar, 32610 Perak, Malaysia
e-mail: mohammad.nasif@utp.edu.my

R. Al-Waked
Department of Mechanical and Maintenance Engineering, German Jordanian University, Amman, Jordan

F. Ahmad et al. (eds.), *ICPER 2020*, Lecture Notes in Mechanical Engineering,
https://doi.org/10.1007/978-981-19-1939-8_15

surface can be used to periodically distort this boundary layer and achieve higher heat transfer rate. Improvement of heat transfer rate depends on the vibrational conditions, inlet air velocity and size, shape, orientation, material and surface roughness of the heat sink.

Gururatna and Xianchang [1] numerically investigated horizontal vibration effect on 2D square pin fin. The vibration frequency was within the range of 50–1000 Hz while amplitude of the vibration kept constant. Heat transfer rate was enhanced up to 500 Hz frequency level and beyond the 500 Hz it decreases rapidly.

Park et al. [2] experimentally investigated the vibration effect on thermal resistance of isolated aluminium fins. The thermal resistance of two isolated fins was calculated under vibration and non-vibration conditions. Results show that the values of thermal resistance are low under vibration condition as compared to non-vibration condition. The vibration frequency was also found to be inversely related to thermal resistance as higher frequencies resulted in lower thermal resistance as vice versa. The minimum thermal resistance value was obtained at the resonance frequency of each fin. This is because at resonance frequency, the maximum amplitude is obtained. This led to a conclusion that amplitude has a dominant effect on heat transfer enhancement of vibrating fin.

Influence of vertical oscillations on force convection heat transfer coefficient was experimentally investigated by Waheb and Jamal [3]. The Test model was aluminium block which consists of a substrate and two fins. Under constant amplitude, vibrational frequency, input power, airflow velocity was varied and the effect of these parameters on force convection heat transfer coefficient was analysed. The results show that vibration has high impact on heat transfer coefficient at high power and low inlet velocity ranges. However, the improvement of coefficient depends on the vibrational intensity, inlet velocity and supplied power level.

Kadhim and Mery [4] experimentally investigated the influence of vertical vibration on natural heat transfer rate of copper sinusoidal heated surface with different orientations (horizontal, vertical and facing downward Position). In this study, vibration frequency (5–25 Hz), amplitude (3–5 mm), surface heat (250–1500 W/m^2), Rayleigh Number, vibrational Reynold number and Prandtl number were considered as parameters. Result of the experiment shows that the vibration has positive influence on heat transfer rate and vibrational mean Nusselt number. When amplitude of vibration is increased vibrational mean Nusselt number increases for every case. However, the improvement of these values depends on the Reynold number, Rayleigh number and orientation of the sinusoidal surface.

Dey and Chakrborty [5] investigated the vibration effect on enhancement of convective cooling numerically. In their study, 3D oscillating fin (0.5 $\times$ 1 $\times$ 0.01) inches with copper substrate (1 $\times$ 1) inches was considered. A 20 W, uniformly distributed power supply was connected to the base of the substrate. Vibration frequency (2–2000 Hz), non-dimensional vibration amplitude, Reynold number (100–1000) were considered as parameters. Numerical results indicated that oscillating fin improves the heat transfer process as compared to the static fin due to the formation of tip-leakage vertices.

From the review of literature, it can be clearly seen that vibration enhances heat transfer of the surface by disturbing the thermal boundary layer. However, Dey and Chakrborty [5] did not consider the effect of the fin height on heat transfer enhancement of the oscillating fin. The scope of this study is to evaluate the vibration effect on heat transfer enhancement of single 3D oscillating fin, to understand the vibration effect on heat transfer enhancement associated with oscillating fin which subjected to the cosinusoidal vibration for a range of frequencies with constant amplitude and Reynold number. In addition, the effect of oscillating fin height on convective heat transfer enhancement has been investigated for range of frequencies. Also, vibration frequency and height of the fin were considered as parameters. The range of the oscillation frequency and simulation conditions have been obtained from the research done by Dey and Chakrborty [5].

2 Model Description and Governing Equation

A schematic of the computational domain is represented in Fig. 1a. The size of the rectangular fin, substrate and fluid domain are (0.0127 × 0.0254 × 0.00025 m), (0.0254 × 0.0254 m) and (0.0508 × 0.0381 × 0.0254 m) respectively. The velocity inlet and pressure outlet were considered with ambient conditions of air. In the x_1 direction, uniform velocity has been employed for flow inlet and transverse vibration (perpendicular to the fin length) was applied on the vertical rectangular fin. The flow was assumed as incompressible and laminar flow. Also, fluid properties are assumed to be constant and effect of the body forces were neglected. Hence governing equation can be expressed as [6–8]:

$$\text{Continuity equation} = \frac{\partial u_i}{\partial x_i} \tag{1}$$

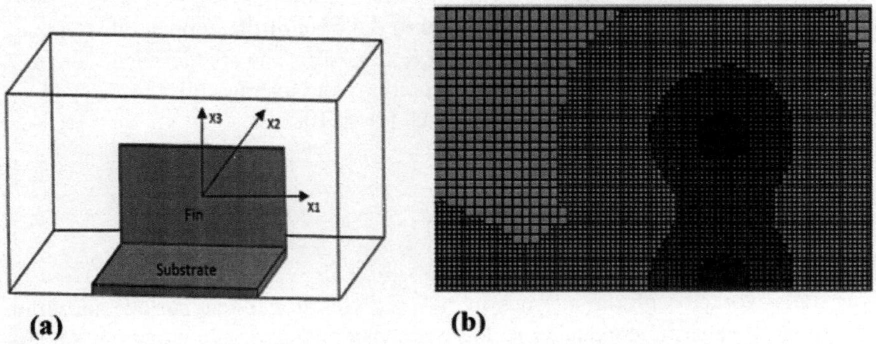

Fig. 1 **a** Physical geometry, **b** computational domain

$$\text{Momentumequationforithdirection}: \rho\frac{\partial u_i}{\partial t} + \rho\frac{\partial(u_i u_j)}{\partial x_j} = -\frac{\partial P}{\partial x_i} + \mu\frac{\partial}{\partial x_j}\left(\frac{\partial u_i}{\partial x_j}\right) \tag{2}$$

$$\text{Energy conservation equation} = \rho c_p\frac{\partial T}{\partial t} + \rho c_p\frac{\partial(u_j T)}{\partial x_j} = \frac{\partial}{\partial x_j}\left(k\frac{\partial T}{\partial x_j}\right) \tag{3}$$

Here the velocity in the ith direction is represented by the u_i and ρ, μ and P are the density, viscosity and hydrostatic pressure of the fluid respectively. Also, k is thermal conductivity, c_p is constant specific heat and q'' represents the rate of heat generation. For fully developed conditions, on the outlet surface,

$$\frac{\partial u_1}{\partial x_i} = 0, \frac{\partial u_2}{\partial x_i} = 0, \frac{\partial u_3}{\partial x_i} = 0, \frac{\partial \theta}{\partial x_i} = 0 \tag{4}$$

where u_1, u_2, u_3 are velocities in each direction and θ is represents the dimensionless temperature corresponding to the ambient temperature. The walls of the oscillating fin and substrate have no slip conditions and the fin motion is governed by the following time dependent velocity equation.

$$V = V_m \cos(2\pi f t) \tag{5}$$

$$V_m = 2\pi f \cdot \left(\frac{a_f}{H}\right) \cdot r \tag{6}$$

where V_m is dimensional maximum oscillating velocity of the oscillating fin, a_f is maximum tip amplitude of the oscillating fin, H is height of the fin, r represents the distance between the considered point on the fin and root of the fin and f is frequency of the oscillating fin. Considering oscillatory velocity field, the point of the fin which connected to the substrate do not oscillate while tip of the fin oscillates with the maximum amplitude. The boundary conditions used in this study are close to the actual operating conditions of the electronic cooling systems. The uniformly distributed 20 W power input was supplied to the base of the copper substrate and temperature of the inlet air was set to 300 K. Based on the characteristic length L (length of the fin which parallel to flow direction) and average inlet velocity ($U =$ 2 m/s), the Reynold number was nearly equal to the 1000.

3 Numerical Method

This study worked with the interaction between vibrational heat surface and fluid. As a result, the thermal field and the flow are considered as a time moving boundary problem. To simulate the vibration, User Define Function was generated using C platform and it was compiled to the solver. Here, the area average Nusselt number was used to measure the heat transfer enhancement of oscillating fin.

The numerical technique which was considered in this study is based on previous research work [5]. Figure 1b illustrates the mesh configuration of the simulation model. The hexahedral grid was considered with 355,622 fluid elements and for capturing of the thermal boundary layer, local biased mesh was considered around the substrate and oscillating fin. Oscillating fin with substrate was numerically solved using finite-volume solver which is available in the ANSYS FLUENT. The pressure velocity coupling was achieved using SIMPLE scheme. Second order upwind scheme was used for solving pressure, momentum and energy equations, while transient formation was obtained using second order implicit scheme. The time step size of each case depends on the frequency of the oscillation and in the current study, the time step was considered as $\tau/8$ for all subsequent cases. Here τ represents the period of oscillation of the fin. The criteria for convergence have been specified as 10^{-6} for all residual variables.

4 Results and Discussion

4.1 Validation of Numerical Results

By comparing the results with the numerical results from Dey and Chakrborty [5], the computational methodology is thus validated. The validation was performed with range vibrational frequency under constant dimensionless amplitude ($\frac{a_f}{H}$) of 15. Figure 2 represents the comparison of Nusselt number improvement obtained in this study and that proposed by Dey and Chakrborty [5]. The results obtained from the present computational study makes a good argument with the previous numerical results. The maximum percentage deviation from the previous numerical results was found to be 3.5%. thus, it is concluded that the computational results of this study have a good agreement with the previous numerical results [5].

Computations of this study were initially performed under non vibration conditions. Then piezo actuator type sinusoidal vibration is applied to the fin with different oscillating frequencies under constant Re number of 1000, fin height of 0.0127 m and maximum tip amplitude of 0.0024 m. The static fin generates thermal boundary layer on each side of the fin which resulted to minimal heat dissipation from the fin surface. This developed boundary layer is expected to rupture using oscillation motion which applied on the fin mentioned by Eq. (5).

The vibrational characteristics plays pivotal role in thermal performance of vibration fin. Moreover, this effect may be depending on the dimensions of the fin. Hence, to understand the effect of fin heights on thermal performance of a vibrating fin at various frequency levels, simulations were performed with different heat sinks at range of vibrational frequencies.

Figure 3 shows Nusselt number variation with the height of the oscillating fin at different frequency levels.

Oscillating fin height was changed from 0.0127 to 0.0203 m while Reynolds number and maximum tip amplitude were kept constant at 1000 and 0.0024 m, respectively. It is clearly seen that increasing fin height leads to decrease the Nusselt number of the vibrating fin. However, as the frequency increases, percentage reduction in Nusselt number decreases. With an increase in height of the oscillating fin from 0.0127 m to 0.0203 m, the maximum percentage reduction in Nusselt number was found to be 8.9% at lowest frequency level. The increase in fin height affects the dissemination of convective flow and interferes in the oscillation of the fin. This

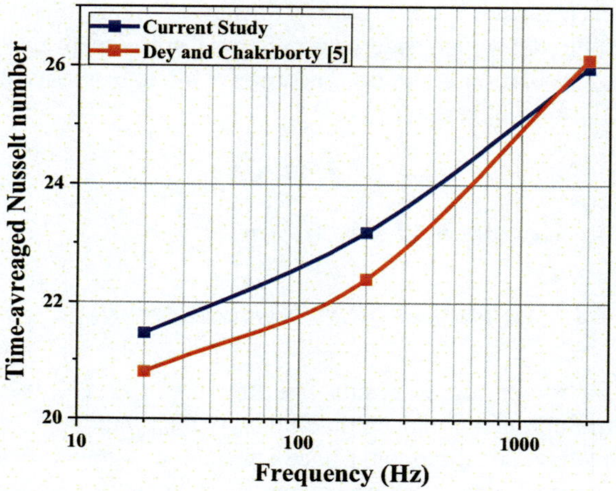

Fig. 2 Comparison of numerical results with the numerical results of Dey and Chakrboty [5]

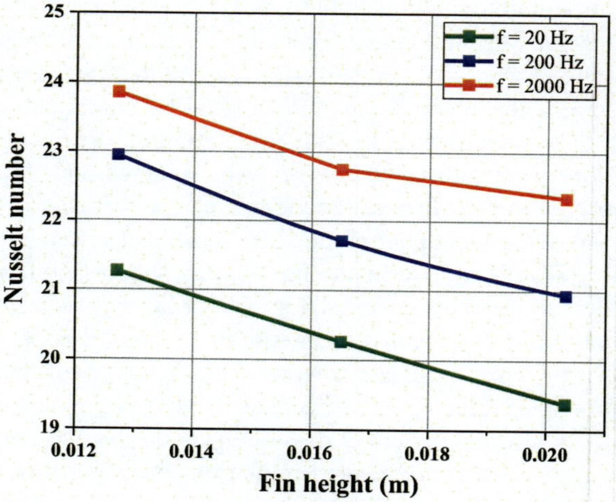

Fig. 3 Effect of the fin height on the Nusselt number

may be attributed to the effect of the height of the oscillating fin, where the vibration was unable to significantly affect the boundary layer due to large fin height. However, when frequency increases, vibrational energy which propagated between the solid and air interface become strong and it induced higher convective flow. Thus, increasing fin height at higher frequency has lower percentage reduction in Nusselt number as compared to lower frequency range.

Figure 4a, b and c shows temperature profile on a plane which passes through the tip of the fin for different frequency levels. It can be clearly seen that with changing oscillating frequency from 20 to 2000 Hz, the disturbance around the fin on either side spreads out. This phenomenon leads to higher heat dissipation from the oscillating fin with increasing frequency.

5 Conclusion

The vibration effect on heat transfer enhancement of the 3D oscillating fins was investigated using numerical method. The vibration frequency and fin height are considered as parameters while Reynold number and non-dimensional amplitude is kept constant. The area average Nusselt number increases as the oscillating frequency is increased in all considered cases. When frequency increases from 20 to 2000 Hz at maximum tip amplitude of 0.0024 m, Nusselt number was increased by 12.2%. Considering the Nusselt number improvement, it is concluded that the heat transfer rate of hot surface improves as the oscillating frequency increases due to increase

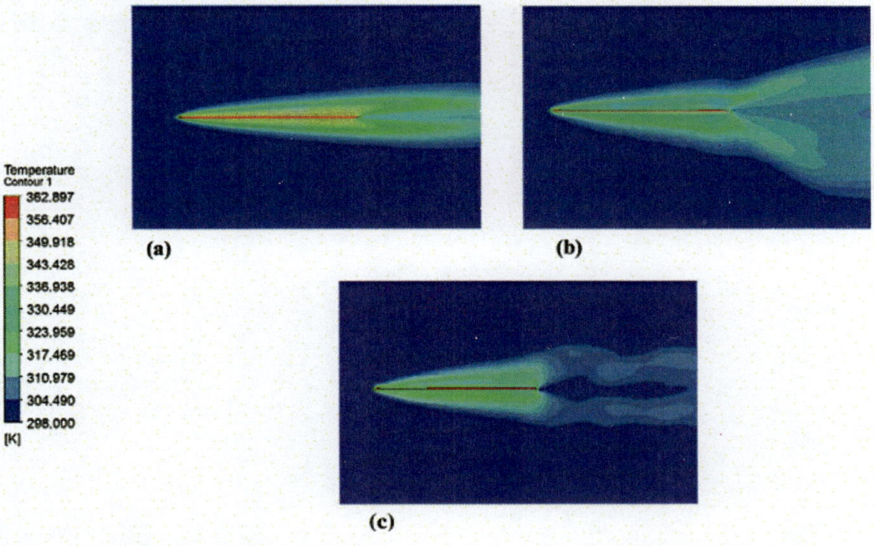

Fig. 4 Temperature profiles on a fin tip plane: **a** $f = 20$ Hz, **b** $f = 200$ Hz, **c** $f = 2000$ Hz

of the oscillating speed. However, when fin height is increased, the Nusselt number decreased in all investigated cases and maximum reduction of Nusselt number was found to be 8.9%. It can be concluded that best utilization of fin height of oscillating fin is when the fin height is short. In every frequency case, increasing of fin height of oscillating fin smoothly decrease the Nusselt number.

Acknowledgements The authors appreciate the financial support received from the Ministry of Education Malaysia through the FRGS 015MA0-032.

References

1. Gururatana S, Li X (2013) Heat transfer enhancement of small scale heat sinks using vibrating pin fin. 10(8):801–809
2. Park KT, Kim DK, Lee JW, Lee MG, Kim HJ (2011) Study on vibration-assisted heat transfer for isolated fins. 290:2364–2368
3. Waheb J (2016) Experimental study of the effect of vertical oscillation on forced convection heat transfer coefficient of vertical channel. May 2016
4. Kadhim ZK (2016) Influence of vibration on free convection heat transfer from sinusoidal influence of vibration on free convection heat transfer from sinusoidal surface. July 2016
5. Dey S, Chakrborty D (2009) Enhancement of convective cooling using oscillating fins ☆. Int Commun Heat Mass Transf 36(5):508–512
6. Fu W, Tong B (2002) Numerical investigation of heat transfer from a heated oscillating cylinder in a cross flow. 45:3033–3043
7. Fu W (2006) Effects of a vibrational heat surface on natural convection in a vertical channel flow. 49:1340–1349
8. Huang WFÆC (2008) Limitation enhancement of a mixed convection by a vibrational heat surface in a vertical channel. 717–725

Effect of Stator Core Materials on the Performance Characteristics of a Free Piston Linear Generator Engine

Wasiu B. Ayandotun, Abdul-Rashid B. A. Aziz, Salah E. M. Elfakki, Ezrann Z. Abidin, Ahmed T. Raheem, and Farith I. Harith

Abstract This paper presents a type of energy conversion device, a free-piston linear generator (FPLG) engine which can generate electrical energy when a free piston engine (FPE) is coupled to a generator. An FPLG engine prototype is designed, fabricated, and developed to operate both in combustion and expansion mode at the Center for Automotive and Electric Mobility (CAREM). The generator unit consists of different components. Stator core is one of the components which is known to contribute to the generating capacity of the engine. So, experimental tests were conducted using two stator core materials - Perspex stator core, GU(P) and metal stator core, GU(M) materials, to study the material effect on the engine performance. The operating pressure of the working fluid, compressed air, was maintained at 8 bar pressure, while 33.3 Hz and 66.6 Hz were the operating frequencies set for the intake valve opening and closing respectively. Based on the experimental data, the motion characteristics and output performance of the engine were discussed. The results shows that the stator core materials have effect on engine performance. Compared with the other material, GU(M) material is found to have a shorter translator stroke and higher cogging force. However, translator velocity that resulted at the operating condition is higher with GU(P) than with the GU(M). The peak power and indicated efficiency were found to be 423.4 W and 64.2%, respectively when GU(M) material is used. Therefore, the power generation of an FPLG engine can improve if suitable stator core material is chosen for the generator unit.

W. B. Ayandotun · A.-R. B. A. Aziz (✉) · S. E. M. Elfakki · E. Z. Abidin · A. T. Raheem · F. I. Harith
Center for Automotive Research and Electric Mobility (CAREM), Universiti Teknologi PETRONAS, 32610 Perak, Seri Iskandar, Malaysia
e-mail: rashid@utp.edu.my

Federal Polytechnic, Zamfara State, PMB 1012, Kaura Namoda, Nigeria

F. Ahmad et al. (eds.), *ICPER 2020*, Lecture Notes in Mechanical Engineering,
https://doi.org/10.1007/978-981-19-1939-8_16

1 Introduction

The crisis of fossil fuels shortage and global warming coupled with the demand for low fuel consumption technology and above all stringent emission legislations have motivated researchers all over the world to improve existing conventional engines by developing alternative fuels and also by investigating new energy conversion devices [1–3]. One of the devices/technologies being researched is a free-piston linear generator (FPLG) engine. This is a new engine that has drawn the interests of a lot of researchers all over the world because of its unique features [4]. Contrary to the conventional engines, an FPLG engine is a 'crankless' internal combustion engine with no restriction of piston motion inside the cylinder by a crankshaft. It has few moving parts, thus making it a compact engine with low maintenance costs and reduced frictional losses. The overview of the background study of the engine has been presented by Mikalsen and Roskilly [1]. These potential advantages make FPE engines be considered as a promising alternative to conventional engines and therefore provide wide areas of applications for its use.

In the 1990s, the researchers at the Technical University Dresden developed the first free piston expander which operates on the pressurized working fluid to expand through the motion of a linear piston [5]. Since then the free piston concept has been applied to the development of free piston expander in recent times. Hou et al. investigated the performance characteristics of a free piston expander linear generator for small scale organic Rankine cycle. In addition to this, they analysed the keys factors that affect on the engine performance. They employed compressed air as the working fluid with intake pressure, external load resistance, and operating frequency varying from 2.0 bar to 3.0 bar, 10 Ω to 50 Ω, and 1.5 Hz to 2.5 Hz respectively. The results of the tests show that intake pressure and external load resistance have a great impact on actual stroke, velocity, and exhaust temperature. With increasing intake pressure and external load resistance, actual stroke and velocity increased while exhaust temperature decreased. Also increasing trend in the indicated efficiency was recorded with the increase in intake pressure and external load resistance. The maximum voltage and power output generated by the engine are 44.4 V and 111.6 W, respectively to achieve maximum indicated efficiency of 81% [6]. Li et al. proposed a free piston expander-linear generator which has it application in recovering waste heat from vehicle engine. The conceptual design and 3D simulation of the model were analyzed using the computational fluid dynamics method. The simulation was conducted at a working frequency and intake pressure of 3 Hz and 0.2 MPa respectively. The results revealed that the FPE can attain indicated efficiency of 66.2% and maximum power output of 22.7 W [7]. Weiss et al. designed and tested a FPE operating at a low temperature source for small scale waste heat recovery. The model developed was investigate under different conditions. The results of the tests revealed that reducing piston mass leads to a reduction in the piston stroke length and increase in the operating frequency also increases power output. 25.6 mW and 88–90% of the power output and energy recovery efficiency could be attained respectively [8].

Xu et al. studied the effect parameters such as in-cylinder pressure, motion characteristics and output performance of a free linear expander-linear generator prototype at different valve timings. They investigated further the effect of external load resistance on motion characteristics and output performance with cam plate A and B. The results of their study showed that motion characteristics and output performance are significantly influenced by the intake duration and exhaust duration. Indicated efficiency showed a decreasing trend with respect to increasing external load resistance. However, the maximum indicated efficiency of 66.2% was attained [9]. Zhang et al. designed and investigated experimentally a double acting free piston expander incorporating slider-based/outlet control scheme to replace the throttling valve of the transcritical CO_2 refrigeration cycle. The results show that the engine can operate optimally at frequency range of 10–17 Hz to achieve isentropic efficiency of 62% using pV diagram analysis [10]. Li et al. studied the effects of intake duration, intake pressure, theoretical stroke, and external load resistance on performance and efficiency using a single-piston FPE-LG prototype for small-scale ORC application. The results revealed that the peak output power and energy conversion efficiency can attain maximum value of 32.5 W and 30.8% respectively [11].

In a similar research conducted by Wang et al., a dual-piston air-driven free piston linear expander was used to experimentally study the relation between system operation characteristics such as piston motion, output voltage, operation frequency, and energy conversion efficiency. The output of the study showed that output voltage is sensitive to piston velocity. 55% energy conversion efficiency could be reached at a driven pressure of 3.79 bar [12]. Tian et al. experimentally studied the operation characteristics and output performances of a FPE-LG engine based on variable factors such as intake pressure, operation frequency, and external load resistance. The results of the tests revealed that at an operation frequency of 1.0 Hz and external load resistance of 20 ohms, the maximum velocity and peak power output achieved are 0.69 m/s and 96 W, respectively. However, with an operating frequency of 2.0 Hz and intake pressure of 2.6 bar, the energy conversion efficiency rose up to 45.82% [13].

Researchers have shown a considerable interest in FPE particularly when in expansion application. Several research works have been conducted on FPE linear expander as mentioned in previous literature. The advantages of FPE-LG engine such as compact structure, operation flexibility and output power produced have necessitated various applications of technology. Application such as small-scale organic Rankine cycle (ORC), which can transform waste heat from exhaust of an IC engine into electricity and other uses. However, in this research, a prototype of the novel technology designed and developed at the Centre for Automotive and Electric Mobility (CAREM) was operated using an air test rig to perform a preliminary test to understand the performance response of the engine to either of the stator core materials when used in a generator unit assembly.

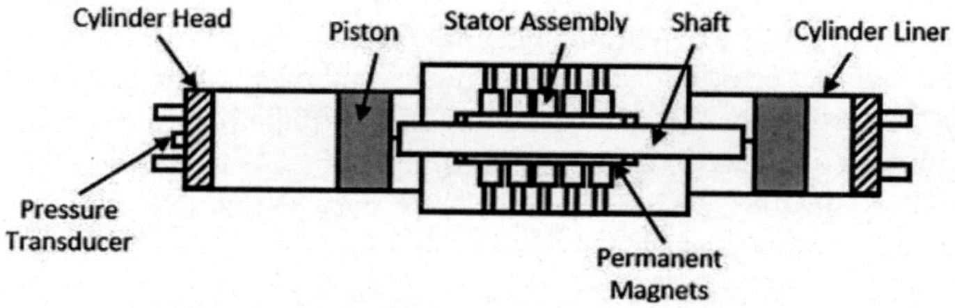

Fig. 1 Components of a free piston linear generator (FPLG) engine

2 Experimental Procedure

2.1 Description of FPLG Engine

The experimental test rig (FPLG) is designed and developed using locally available resources and materials for both the components and control. It primarily consists of two free piston expanders and a tubular linear generator. The FPLG engine consists of a piston rod (shaft) and two pistons connected to each end of the shaft as shown in Fig. 1. The pistons are then enclosed inside cylinder liners. Positioned between the two cylinders is a tubular linear generator, which consists of a translator and a stator. Mounted on the translator shaft are series of bonded permanent magnets, while the stator housed the wound coils which are connected to a full-wave rectifier bridge. The position of the translator is electronically measured and controlled using a linear encoder and magnetic strip as it moves back and forth. The in-cylinder pressure during the expansion process was measured using a pressure transducer together with a charge amplifier. The operation of the FPLG engine is controlled by the National Instruments (NI) controller based on the signal sends and receives. Data for parameters such as pressure, displacement and the output current are logged using NI data acquisition system. The engine is also equipped with both cooling and lubricating systems for lubricating of moving parts and removal of dissipated heat, respectively. The air supply system basically consists of an air compressor, pressure regulators, an air receiver tank, manifolds, and electromagnet valves.

2.2 Working Principle and Experimental Procedure

The FPLG engine can operate two modes—expansion and combustion modes. Figure 2 illustrates the working process of the FPLG engine in expansion mode, and Table 1 showing the detail specification of the engine. Electromagnetic valves are connected to the either side of the cylinder head. When the system is initialised to operate, the valves are opened to allow 8 bar compressed air from the receiver tank to inject into the right cylinder for the set opening duration. With this, the translator is driven from its Top Dead Center (TDC) towards its Bottom Dead Center (BDC) i.e.

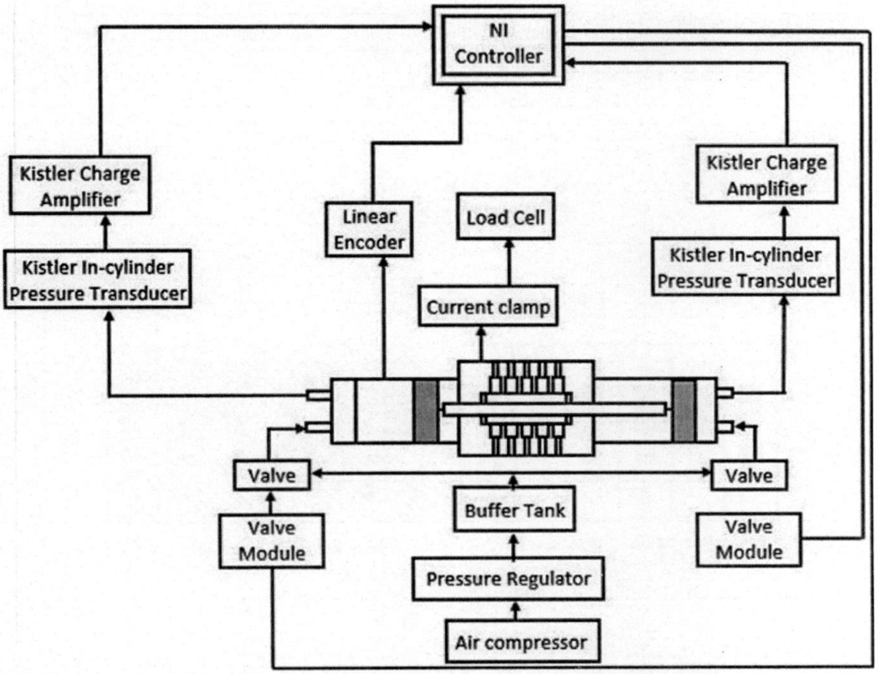

Fig. 2 A schematic diagram of the experimental setup

Table 1 Engine specification parameters of the FPLG engine

Engine Specifications	Value
Cylinder bore (mm)	56
Maximum stroke (mm)	84
Cylinder displacement (cc)	221
Number of cylinder liner	2
Operating frequency (Hz)	10
Moving mass (kg)	7
Intake valve opening time (ms)	5–100

from the right to the left till the expansion process lasts. While the right cylinder is at this position (BDC), the valves changed the flow direction. Meanwhile, the piston is at TDC for the left cylinder. The valves are opened to inject compressed air into the cylinder through the ports, thereby driving the piston from TDC towards BDC (i.e. from the left to the right) until the expansion process completes. At this position, the air in the left cylinder is driven out by the movement of the left piston, and the valves at the left side then change the flow direction. The system repeats the working processes all over again while the compressed air continues to drive the translator

Table 2 Experimental conditions

Experimental conditions	Value
Aspect Ratio (AR)	1.5
Air intake pressure (bar)	8
Valve opening frequency, VOF (Hz)	33.3
Valve closing frequency, VCF (Hz)	66.6
External load resistance (ohms)	4.7

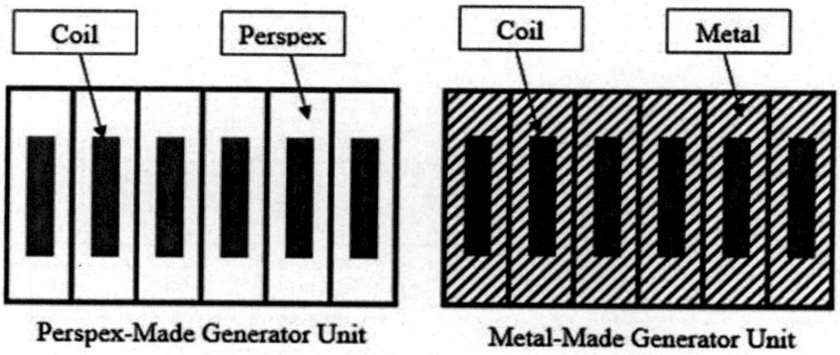

Fig. 3 Stator core arrangements **a** with Perspex and **b** with metal

back and forth. With the linear motion of the translator, the LG then transforms the kinetic energy into electricity that is consumed by an external load. For this test, the engine was first coupled with the generator unit made of Perspex, GU(P) and then operated and repeated thrice at the set experimental condition (see Table 2). Thereafter the generator unit was replaced with the generator unit made of metal, GU(M) and equally operated at the set condition. The arrangement of the generator unit core arrangement vis-à-vis with the coils is illustrated in Fig. 3a, b for GU(P) and GU(M), respectively.

3 Experimental Results and Discussion

3.1 Motion Profile

Figure 4 shows the dynamic linear motion profile of FPLG engine as a function of time. The variation in the stroke length is an attribute of a FPLG engine because of the absence of a crankshaft to constrain piston motion. In other words, the top dead centers (TDC) and bottom dead centers (BDC) are not fixed. In this case, the middle of a stroke is defined as the displacement origin of coordinates and the limits of travel, known as mechanical TDC and BDC, are 48 mm towards the TDC and −48 mm towards the BDC. During operation of the FPLG engine, the displacement of travel of the translator occurs between the mechanical TDC and BDC. So, the

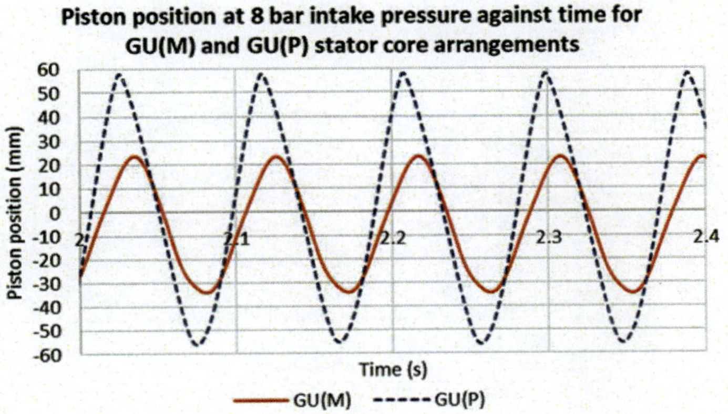

Fig. 4 Translator motion profile against time

actual stroke of the translator is the distance between the TDC and BDC. From the figure, the frequency of operation of GU(M) and GU(P) is the same though their amplitudes are different. Different TDC position can be observed with the cycles. However, the cycle to cycle variation calculated for GU(M) and GU(P) are 0.13% and 3.74%, respectively which indicates that the engine performed stably for all the tests. From the motion profiles, the GU(M) profile shows a shorter expansion and exhaust positions of 35 mm and 23 mm, respectively than with GU(P) which had full expansion and exhaust positions. So, it can be said that the stator core material has effect on the motion control of an FPLG engine.

3.2 *Velocity Profile*

During the operations of the FPLG engine, the position of the translator was measured and based on this parameter, the speed/velocity of the translator profile was obtained. Figure 5 illustrates the velocity profiles of the translator motion from BDC to TDC i.e., clockwise from the left side position at a driven pressure of 8 bar using different stator core materials. The curve was processed using a moving average of 5. The piston motion from left to right shows the velocity of the translator during exhaust stroke while from right side position depicts the expansion stroke. The figure clearly shows that the translator travelled at a low velocity and acceleration when approaching TDC and BDC during exhaust and expansion strokes. However, while departing TDC and BDC, the piston showed a faster velocity and acceleration. Moreover, the maximum velocity of the translator is found to occur at the middle of a stroke. For the GU(P), the translator is found to run at high speed of 3.5 m/s due to less resistance induced by the stator core on the translator motion. With this, the translator could travel full stroke between 48 mm (0 mm BTDC) and -48 mm (96 mm BTDC). However for GU(M), the translator travelled at a lesser speed of

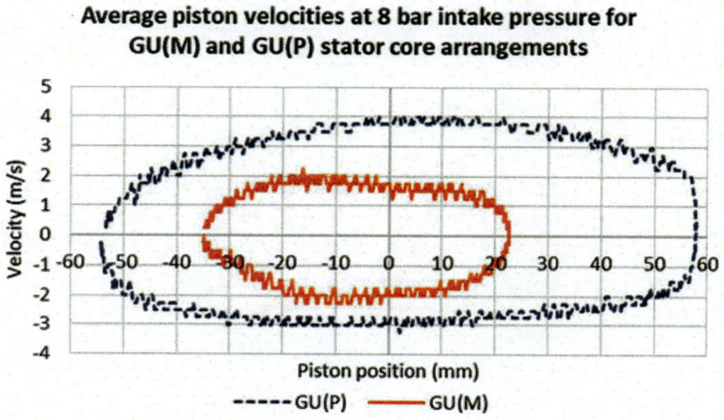

Fig. 5 Translator velocity against piston position

2.5 m/s compared to the GU(P). This is due to the high resistance (cogging force) induced by the stator core material on the translator motion. Hence, shorter translator stroke between 23 mm (25 mm BTDC) and −35 mm (83 mm BTDC) was obtained. The velocity profile produced by the translator is similar to an ellipse shape. This is the trending profile of a FPLG engine and similar to the research results in Refs. [4, 12].

3.3 Pressure–Volume Diagram

The pressure–volume (p–V) diagram for the FPLG engine working in expansion mode under different stator core materials arrangements is presented in Fig. 6. The high/low pressures recorded for GU(M) and GU(P) are 7.28/1.00 bar and 5.64/1 bar respectively. The area under the curve shows the workdone by the compressed air on the piston and is usually expressed as indicated workdone. From the curves, the indicated workdone calculated for the GU(M) and GU(P), respectively are 127 J and 104 J, respectively.

3.4 Power (Root Mean Square) and Peak Power

Figures 7 and 8 illustrates the variation of the root mean square of power, P(rms) and peak power, P(peak) output of the FPLG engine for different stator core materials when operated at a frequency of 33.3 Hz and intake pressure of 8 bar. Generally, it can be observed that the peak power occurred at the middle portion of a stroke. This is because highest velocity of the translator is recorded at this position. For the GU(P), it can be observed that power output was produced throughout the stroke length.

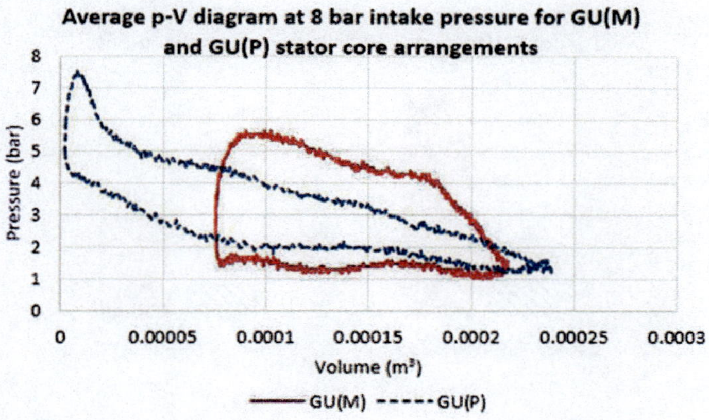

Fig. 6 In-cylinder pressure against volume

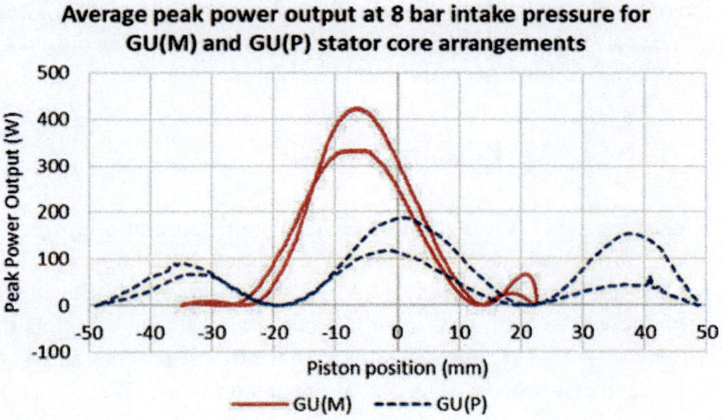

Fig. 7 Power output against piston position

However, for the GU(M), the translator motion was impeded owing to the stator core materials the generator unit is composed of. At the set operation condition, the peak power attained by the GU(M) and GU(P) are 423.4 W and 189.6 W, respectively. 120.2 W and 62.9 W are the calculated and corresponding Prms generated for the GU(M) and GU(P) respectively.

3.5 Work-Electric Conversion and Indicated Efficiencies

At a set operating condition and stable engine run, a FPLG engine can generate a peak power output and average power output based on valve opening and closing

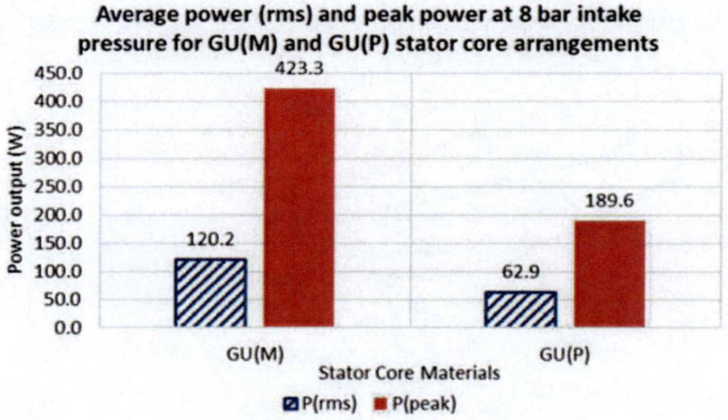

Fig. 8 Power (rms) and peak power of GU(M) and GU(P) stator core

timings, which of course controls the back and forth motion of a piston. The electric energy output (E_{output}) can be calculated by integrating the power output (P_{output}) with respect to time as follows according to [11, 12]:

$$E_{output} = \int P_{output} dt \tag{1}$$

Figure 9 illustrates the P–V diagram of GU(M) stator core at 8 bar pressure for the actual working processes and ideal working processes. In the figure, $1 \to 2 \to 3$ depicts the actual intake process, and $1' \to 2' \to 3'$ represents the ideal intake process. $3 \to 4$ shows the actual expansion process and $3' \to 4$ depicts the ideal expansion process. $4 \to 5 \to 1$ represents the actual exhaust process and $4 \to 5' \to 1'$ shows the ideal exhaust process. The actual expansion work ($W_{exp-actual}$) is then

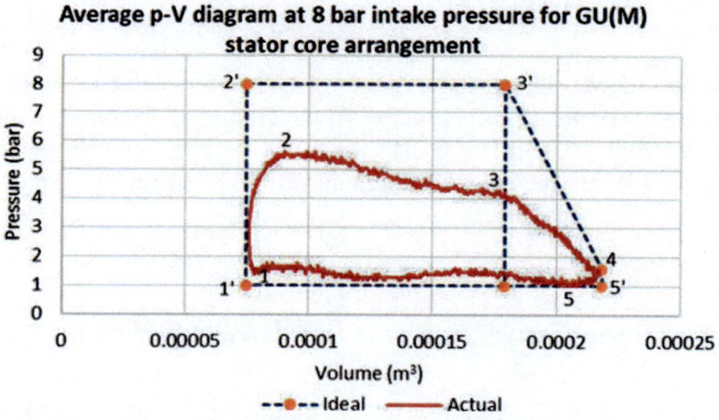

Fig. 9 The P–V diagaram of GU(M) stator core

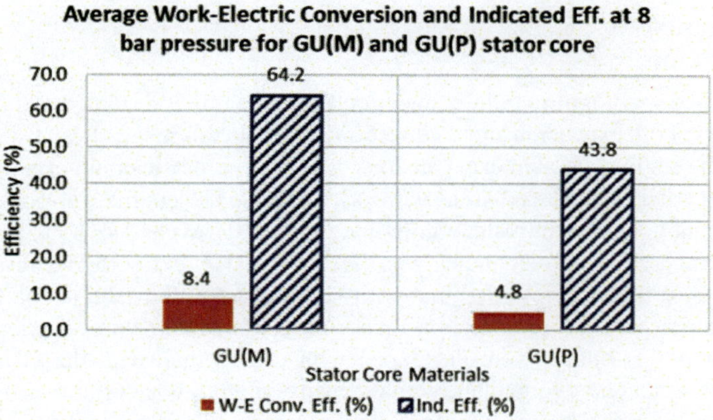

Fig. 10 Work-electric conv. and indicated efficiencies of GU(M) and GU(P) stator core

calculated by integrating the actual P–V diagram as follows [11]:

$$W_{exp-actual} = \int Pdv \tag{2}$$

According to Refs. [11], the work-energy (W-E) conversion efficiency is defined as the ratio of the electric energy output to the actual expansion work and is represented as follows:

$$\eta_{w-exp} = \frac{E_{output}}{W_{exp-actual}} \tag{3}$$

Based on the above equations and the test data, the η_{w-exp} for the GU(M) and GU(P) is calculated as 8.4% and 4.8% respectively and illustrated in Fig. 10.

Also on the basis of p–V diagram using Fig. 9, the indicated efficiency ($\eta_{indicated}$) is defined as as the ratio of the area bounded by the actual working process (A_{actual}) of the experimental p–V diagram to the area bounded by the ideal working process (A_{ideal}) and computed as follows [10, 14]:

$$\eta_{indicated} = \frac{A_{actual}}{A_{ideal}} \tag{4}$$

Based on the above equations and the test data, the $\eta_{indicated}$ for GU(M) and GU(P) is calculated as 64.2% and 43.8% respectively and displayed as shown in Fig. 10.

4 Conclusion

A two-stroke spark ignition direct injection FPLG engine is designed and fabricated to operate in both expansion and combustion modes. In this research, an experimental study of the engine operated in expansion mode is presented to conduct experimental tests using 8 bar compressed air as the working fluid. The engine was operated with two different generator unit stator core materials—GU(M) and GU(P) at a constant valve opening/closing frequencies were set at 33.3 Hz and 66.6 Hz, respectively. The results of the tests revealed that with GU(M) stator core, the piston translates at a shorter stroke during operation than with GU(P) stator core. Higher velocity and lower power output were developed with GU(P) than with the GU(M). The relation between energy input and actual expansion work were established and used to compute the work-energy conversion efficiency and indicated efficiency. GU(M) and GU(P) can attain 8.4% and 4.8% of w-e conversion efficiency respectively, and 64.2% and 43.8% indicated efficiency for GU(M) and GU(P) respectively. The type of stator core material the generator unit is assembled with affects engine output performance.

The experimental results presented in this paper will provide the guidance for further improvement on the power output generation of the FPLG engine by employing better stator core materials. With this, further experimental research will be conducted on the stator core materials with the hope to improve engine performance characteristics.

Acknowledgements This study was supported financially by the Petroleum Research Fund (PRF) Grant (0153AB-A34) under the project titled "Development of a concept-demonstrator free piston linear generator (FPLG) powered by flexi-fuel" and the Centre for Automotive Research and Electric Mobility (CAREM), Universiti Teknologi PETRONAS.

References

1. Mikalsen R, Roskilly AP (2007) Appl Therm Eng 27:2339–2352
2. Bielaczyc P, Woodburn J, Szczotka A (2014) Appl Energy 117:134–141
3. Chandra R, Vijay VK, Subbarao PMV, Khura TK (2011) Appl Energy 88:3969–3977
4. Hanipah MR, Mikalsen R, Roskilly AP (2015) Appl Therm Eng 75:493–503
5. Heyl, P and Quack, H, "Free piston expander-compressor for CO2-design, applications and results," in *Proceedings of the 20th international congress of refrigeration*, Sydney, 1999. doi.
6. Hou X, Zhang H, Xu Y, Tian Y, Zhao T, Li J, Yu F (2018) Appl Therm Eng 144:209–218
7. Li G, Zhang H, Yang F, Song S, Chang Y, Yu F, Wang J-F, Yao B (2016) Energies 9:300
8. Weiss, L W. Journal of Mechanical Design. 2010 132
9. Xu Y, Tong L, Zhang H, Hou X, Yang F, Yu F, Li J, Zhao T, Li J, Zhang M (2018) Appl Therm Eng 142:555–565
10. Zhang B, Peng X, He Z, Xing Z, Shu P (2007) Appl Therm Eng 27:1629–1636
11. Li J, Zhang H, Tian Y, Hou X, Xu Y, Zhao T, Wu Y (2019) Appl Therm Eng 152:751–761

12. Wang Y, Chen L, Jia B, Roskilly AP (2017) Appl Energy 195:93–99
13. Tian Y, Zhang H, Li G, Hou X, Yu F, Yang F, Yang Y, Liu Y (2017) Appl Therm Eng 127:184–193
14. Hou X, Zhang H, Yu F, Liu H, Yang F, Xu Y, Tian Y, Li G (2017) Appl Energy 208:1297–1307

Spray Characteristics and Droplet Micro-Explosion of Water in Diesel Emulsion

Mhadi A. Ismael, A. Rashid A. Aziza, Morgan Heikal, Cyril Crua, Ezrann Z. A. Zainal, Salah E. Mohammed, and Zuhib Nisar

Abstract This paper highlights the potential usefulness of water-in-diesel emulsions when exposed to high-temperature, thereby improving the air–fuel mixing process utilising micro-explosion resulting from multi-component fuel combustion. Although the high injection pressure significantly affects the spray characteristics and fuel mixing process, observation of micro-explosion in the spray is rarely addressed. In this study, an emulsion made with 5%, 10%, and 15% of water by volume is investigated at different injection pressure (500, 750 and 1000 bar). The experiments are conducted in a constant volume chamber under evaporative conditions. A high-speed camera coupled with a long-distance microscope was used to magnify and visualise the spray droplets micro-explosion in the spray. The raw images of spray were then analysed using a purpose-built image processing algorithm to identify both spray penetration length and dispersion angle. Our measurements indicated that the differences in water content (10 and 15%w) suggest an increase in both spray penetration length and cone angle due to fuel evaporation and micro-explosion. Furthermore, the spray breakup enhances with the increase in water content.

M. A. Ismael (✉) · A. R. A. Aziza (✉) · E. Z. . A. Zainal · S. E. Mohammed · Z. Nisar
Centre for Automotive Research and Electric Mobility, Universiti Teknologi PETRONAS, Perak, Malaysia
e-mail: Mhados1@gmail.com

A. R. A. Aziza
e-mail: rashid@utp.edu.my

M. A. Ismael · S. E. Mohammed
Mechanical Engineering Department, University of Bahri, Bahri, Khartoum, Sudan

M. Heikal · C. Crua
Advanced Engineering Centre, University of Brighton, Brighton BN2 4GJ, UK

F. Ahmad et al. (eds.), *ICPER 2020*, Lecture Notes in Mechanical Engineering,
https://doi.org/10.1007/978-981-19-1939-8_17

1 Introduction

In the new global warming, diesel fuel has become a central issue for emission pollutants. As a fuel base solution, much of the current literature on the use of water in diesel engines have shown a remarkable reduction in nitrogen oxides (NOx) and particulate matter (PM) [1]. The NOx reduction can be explained by the presence of water in the diesel fuel, which leads to absorbing the heat of combustion, consequentially, lower exhaust gas temperature [2]. Several techniques have been suggested to identify the use of water in the engine cylinder. Although the direct injection method strategies facilitate the control of water to fuel ratio [3], the main challenge faced by many researchers is that the engine was suffering from wear and degraded combustion [3]. The water emulsified diesel is found to be the better option of these. The multi-component fuel with different boiling points leads to secondary atomisation utilising micro-explosion phenomena [4]. The micro-explosion is the dispersion of droplets into tiny droplets due to the continuous boiling of the water inside the oil, which leads to the disruption of the primary drops and causes secondary atomisation. As a result, the secondary drops evaporated faster and mixed with the ambient air [5].

The literature has recently emerged that offers contradictory findings on the brake specific fuel consumption (BSFC) and brake thermal efficacy (BTH) regarding the emulsified fuel before injection into a diesel engine. Some researchers [6] reported an increase in BSFC due to the lower calorific value of the emulsions. This can be attributed to replacing the amount of diesel with the equal amount of water added in the emulsion, while others showed a significant increase in BTH due to their better combustion characteristics [7]. On the other hand, contradictory to the previous findings, some authors [8] observed a remarkable reduction in BSFC and increased thermal efficiency in water in diesel emulsion compared to that of diesel fuel. The decrease in BSFC may be attributed to the distribution of dispersed water droplets in the base fuel, as it can affect the occurrence and strength of micro-explosion [9]. In line with this, it has been reported that an increase of the engine effective efficiency up to 20% and 25% reduction of NOx emission in the case of large dispersed water droplet size [10]. However, with a small dispersed water droplet size, the BSFC reduced by 5.55% compared to 3.89% in the case of large dispersed droplet size [8].

All these contradictory findings are due to the complicated combustion behaviours of emulsified fuels. Micro-explosion has been identified as a significant contributing factor to the combustion of emulsified fuels. Several factors can explain this observation inside an engine. For example, an observation of a longer ignition delay and a shorter combustion duration in the case of water in diesel emulsion fuel than neat diesel. Some researchers have observed an increase in the ignition delay [11]. The longer ignition delay can be attributed to the lower volatility and higher viscosity of the dispersed phase [12], while the shorter combustion reaction may be contributed by the effects of micro-explosion phenomena [7]. The longer ignition delay is often favourable for fuel burns simultaneously to enhance the pressure and heat release. In addition, the distribution of water dispersed size was found to be related to the amount of water in the emulsion [13]. The dispersed droplet size increased with the increase of water content. The optimum water content in the emulsion to get a high strength of micro-explosion and high stability of the emulsion was 10% [14]. Whilst with 10% water content, a slight reduction in combustion pressure was found [15] and increased by 5% at full engine load with15% water content [16]. On the other hand, in spite of many difficulties about the role of micro-explosion, however, single droplet experiments are often used to facilitate the observation of this behaviour, and it well reported that the water content and its distribution are among the most important factors [17].

However, the dispersed droplet size distribution during the spray has been classified depending on injection pressure in an operating engine. In this trend, the effect of fuel injection equipment on the distribution of water droplet size was carried out by some authors [18]. A significant decrease in the water droplet size was found at high injection pressure (1500 bar). However, some water content in the emulsion was evaporated during the injection. Furthermore, direct evidence of a micro-explosion in the spray combustion under real engine conditions has rarely been captured in the chamber reactor experiments. Together, these studies indicate the need to understand water effects in the fuel on spray penetration length and its dispersion angle. Besides those characteristics, spray droplets micro-explosion is sorely missed.

Therefore, this study set out to assess the effect of water content and high injection pressure on spray characteristics. For this purpose, an optical constant volume combustion chamber was developed to visualise the spray images. A high-speed camera was used to capture the full spray characteristics. The current study will clarify several aspects of existing micro-explosion during the spray.

2 Method and Procedure

The fuel used in the current study was an emulsion of 5%w, 10%w and 15%w water (by volume) and the neat diesel as the base fuel. The span 80 was used as an emulsifier agent. The emulsion preparation and its properties are discussed in the authors' previous work [18].

The experiments were carried out in an evaporating condition using a common rail injection system. The injection pressure was varied to 500, 750 and 1000 bar. The schematic of the experiment setup is shown in Fig. 1, which is composed of a constant volume chamber, high-pressure fuel injection system, shadowgraph imaging system and data acquisition system.

2.1 Constant Volume Chamber

A constant volume chamber was developed and used for spray visualisation (Fig. 1). The chamber has a cylindrical structure with two optically accessible windows. The

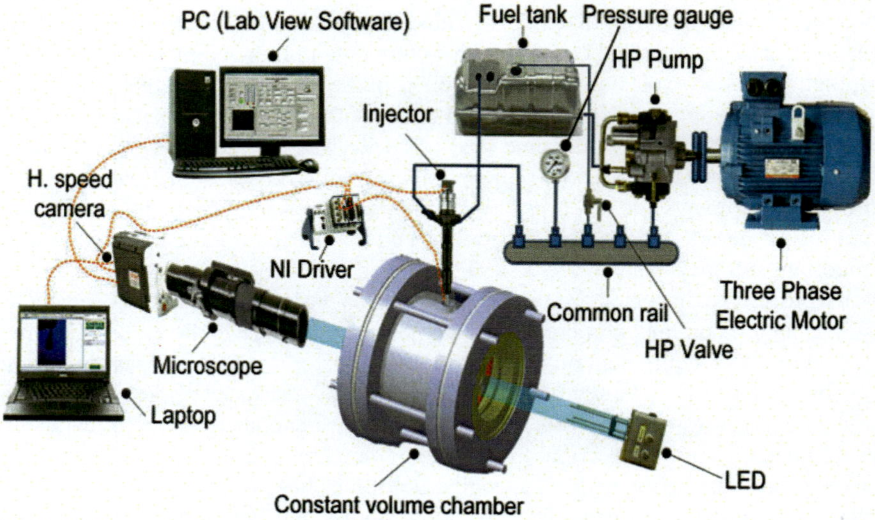

Fig. 1 Schematic diagram of the experimental system setup

diameter of the window is 100 mm, the spray penetration can reach 60 mm, excluding the length of the injector tip nozzle and the bottom of the glass window. A ceramic electric heater was mounted on the cylinder wall to heat the chamber. Two thermocouples (K-type) were used to measure the temperatures on the chamber surfaces and gas temperature located in the canter inside the chamber. The chamber wall temperature was set to 900 K. The surrounding gas (air) temperatures were measured at the axial distance of the injector tip along the spray axis, and the temperature was 873 $\pm$ 3 K.

2.2 *Optical System and Resolution*

Two different arrangements of optical systems were conducted to acquire details of the entire spray and spray boundary at high injection pressure. In this study, the first optical setting, a high-speed video camera (Phantom Miro M310) and Multi LED (LT-V8-15) for illumination are positioned opposite each other to visualise the shadow of the full spray. The camera was set to a resolution of 64 $\times$ 296 pixels and 74, 074 fps. Based on the calibration scale, the final optical scale is 0.24 mm/pixel. The visual scale was 0.028 mm/pixel. The experimental condition is shown in Table 1.

Table 1 Experimental conditions

Test fuels (v/v)	Ambient temperature (K)	Injection pressure (bar)	Injection duration (ms)	Nozzle diameter (mm)
Net diesel, 5%w 10%w, 15%w	873	500, 750, 1000	2	0.2

2.3 *Spray Measurements*

The evaporative raw images captured by the high-speed Phantom Miro (M310) camera were then analysed using a purpose-built MATLAB image processing algorithm to detect the spray penetration and spray dispersed angle simultaneously automatically. The global image thresholding method is often used for spray image segmentation [19]. An automatic thresholding algorithm was used to segment the spray boundary from the spray shadow (Fig. 2). The spray tip penetration was defined as the distance from the injector nozzle orifice to the spray boundary along with the axial distance. The dispersed spray angle was defined as the angle that corresponds to the triangle whose area is equal to the projected area of the spray boundary from the injection nozzle to the spray half of spray tip penetration. The spray angle measured using the following equation:

$$\theta = 2arctan\left[\frac{A}{(S/2)^2}\right](1) \quad (1)$$

where A is the projected area, and S is the penetration length. In order to represent the statistical analysis, the mean, median and standard deviation were calculated for both the spray penetration length and dispersion angle. In each case, three videos were processed together, and median, mean and standard deviation were calculated. The accuracy of measurements of the spray tip penetration was estimated to be $\pm$ 0. 3 mm. The end of injection was not defined well. Therefore a cutoff at frame 40 was used in all the cases (t = 526.5 μs).

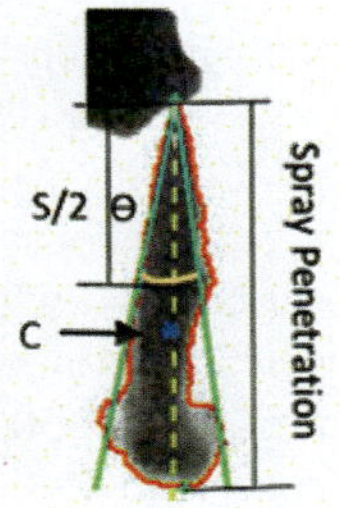

Fig. 2 A Snapshot of the Matlab image processing shows the definition of spray penetration length and dispersion angle measurement

3 Results and Discussion

This section presents the effect of both the water concertation and injection pressure of water in diesel emulsion on spray tip penetration and sprays cone angle. We have also discussed the spray droplets puffing and micro-explosion during the injection.

3.1 Effect of Injection Pressure and Water Content on the Spray Penetration

Figure 3 shows the images for the structure of spray tip penetration for 10% water content injected at different injection pressures (500, 750 and 1000 bar). As shown in these images, the injection pressure significantly influenced the spray penetration length as expected [38, 40, 53]. The quantitative analysis of the spray penetration length for different water content, i.e. 5%, 10% and 15% and injection pressure (500, 750, and 1000 bar) are plotted in Fig. 4. It should be noted that the neat diesel was not used for the comparison because it showed very similar results to 5% water content. As shown in the figure, with the increase in the injection pressure and water content, the spray penetration lengths are increased due to more spray momentum and spray breakup caused by the aerodynamic shear stress. The effect of water content on spray penetration is clearly observed for 500 bar injection pressure (Fig. 4). However, with the increase of injection pressure to 750 bar, the differences between the spray penetration lengths significantly decreased. No significant differences were found between 5%, 10% and 15% water content at 1000 bar injection pressure. This indicated that the effect of high injection pressure (750 bar upward) mitigated the influence of the viscosity of the emulsion and droplet micro-explosion.

3.2 Effect of Injection Pressure and Water Content on the Spray Dispersion Angle

To further examine the role of the injection pressure on emulsified fuels, the dispersion angles for different injection pressures on different water content i.e. 5, 10 and 15%w are given in Fig. 5. The spray started and developed with an angle between 35° and 47°, as the spray continued, the dispersion angle decreased and appeared to approach constant values between 15° to 13°. This was consistent with the previous studies [24] which found the spray dispersion angle for completely atomised diesel spray was independent of injection pressure. It can be seen that at low injection pressure i.e. 500 bar, the water content had little effect on the spray dispersion angle compared to the injection sprays at 750 and 1000 bar. As discussed earlier, this was thought to be due to enhanced micro-explosion at low injection pressure during the spray. At low injection pressure, the spray velocity, which was determined by

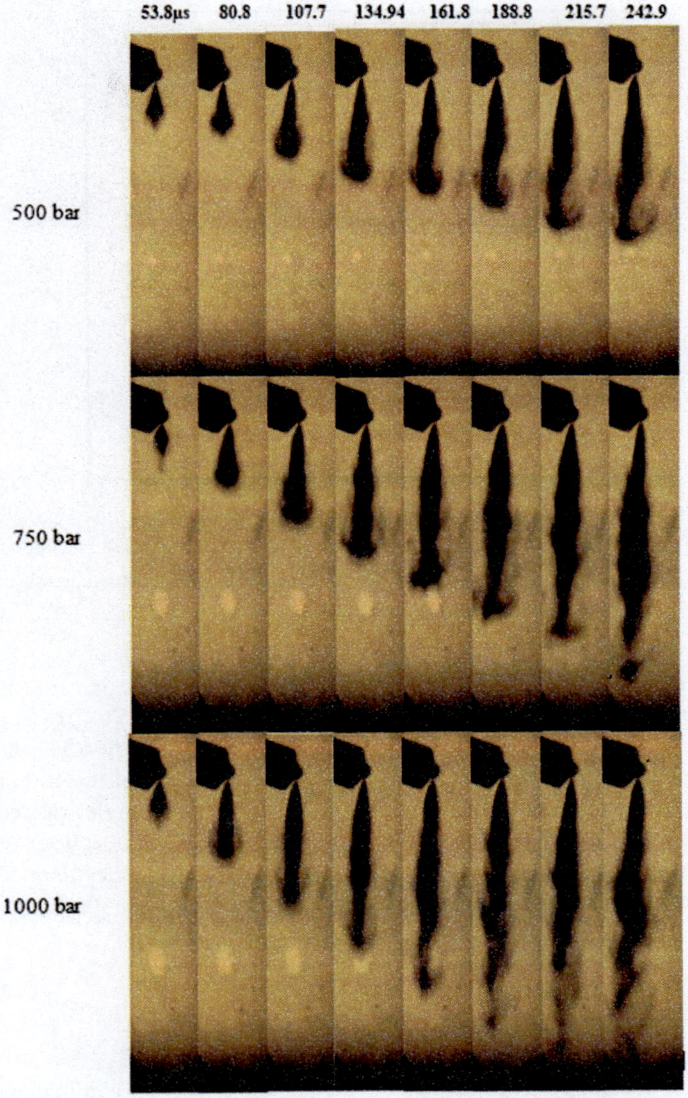

Fig. 3 Images of spray structure of 10% water content at different injection pressure (500, 750 and 1000 bar) at a temperature of 873 K

fuel injection velocity, became low. Hence, the momentum exchanged between the emulsion spray and the surrounding ambient temperature became significant at low injection pressure [25]. A combined qualitative and quantitative analysis is given in Figs. 4 and 5, respectively. It is possible to hypothesise that an increase of the cone angle resulted from droplet breakup caused by micro-explosion. With the increase

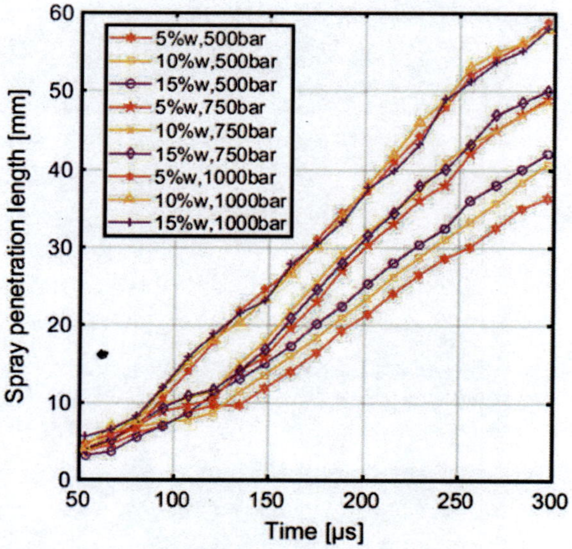

Fig. 4 Spray penetration for 5, 10 and 15% water content at different injection pressure (500, 750 and 1000 bar) at a temperature of 873 K

of water content in the emulsion, the intensity of micro-explosion increased. It's shown clearly in Fig. 6, with the increase of water content from 5 to 10% at an injection pressure of 500 bar, the dispersion spray angle increased by 4.3% and 16.4% in the case of 15% water content. However, with the increase of injection pressure to 750 bar, the effect of micro-explosion on spray dispersion angle was reduced to 0.4% and 1.7%, respectively. Interestingly, with the increase of injection pressure to 1000 bar, no differences were found between 10 and 15% water content compared to 5% water content.

4 Conclusion

An experimental study was conducted to examine the effect of water in diesel emulsion (5%w, 10%w and 15%w by volume) on spray characteristics and spray droplet behaviour injected into a high-temperature environment. The measurements of the spray summarised in the following:

1. Both the spray tip penetration and spray angle were increased with the increase of water content from 0 to 15%. Spray cone angle increased by 16.4% in case of 15% water content and 500 bar injection pressure.
2. For 5% water content, the spray characteristics were similar to those of 0% due to the loss of some water through the injection system.

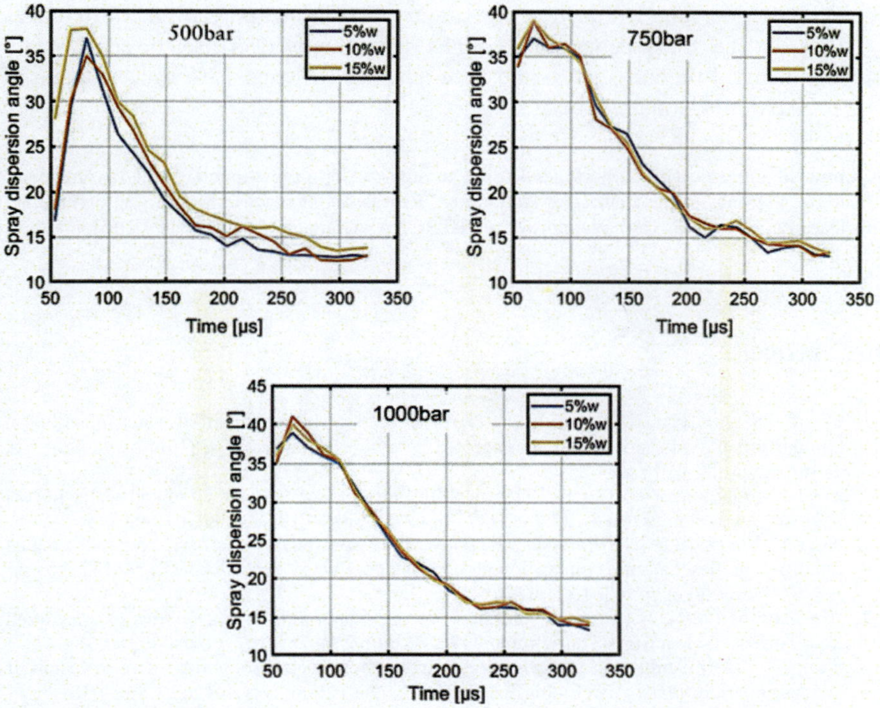

Fig. 5 Spray dispersion angle of 5, 10 and 15% water content at different injection pressure (500, 750 and 1000 bar) at a temperature of 873 K

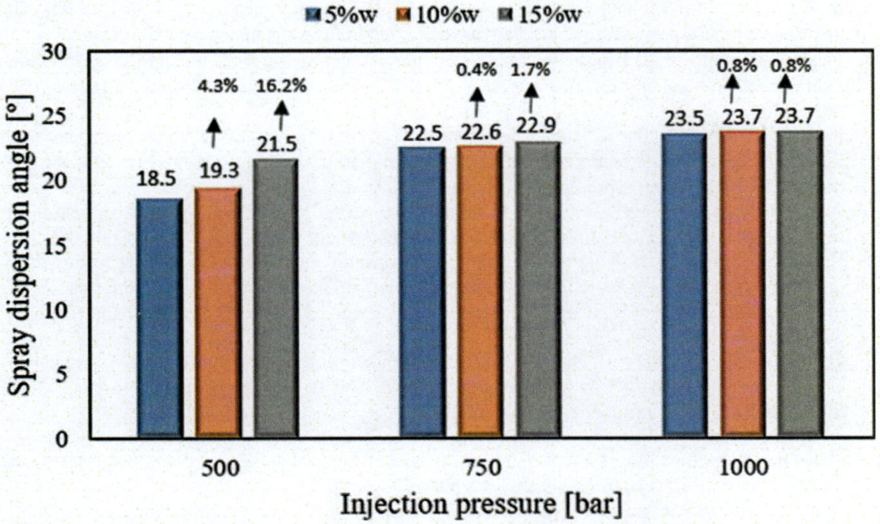

Fig. 6 Spray dispersion angle for 5%, 10% and 15% water content at different injection pressure

3. With the increase of injection pressure to 750 bar and onward, the differences between the spray penetration lengths significantly decrease. In contrast, no significant differences are found between 5%, 10% and 15% water content at 1000 bar injection pressure.

Acknowledgements The authors would like to acknowledge the support given by Universiti Teknologi Petronas to the Centre for Automotive Research and Electric Mobility in performing this research. The research was supported by PETRONAS Company Grant [YUTP_015LC0-133].

References

1. Ma X, Zhang F, Han K, Zhu Z, Liu Y (2014) Effects of intake manifold water injection on combustion and emissions of diesel engine. Energy Procedia 61:777–781. https://doi.org/10.1016/j.egypro.2014.11.963
2. Yokev N, Greenberg JB (2018) Linear stability analysis of laminar premixed water-in-fuel emulsion spray flames. Fuel 222:733–742 https://doi.org/10.1016/j.fuel.2018.02.113
3. Nour M, Kosaka H, Abdel-Rahman AK, Bady M (2016) Effect of water injection into exhaust manifold on diesel engine combustion and emissions. Energy Procedia 100:178–187. https://doi.org/10.1016/j.egypro.2016.10.162
4. Abu-Zaid M (2004) An experimental study of the evaporation characteristics of emulsified liquid droplets. Heat Mass Transf 40(1):737–741. https://doi.org/10.1007/s00231-003-0473-5
5. Shinjo J, Xia J, Ganippa LC, Megaritis A (2014) Physics of puffing and microexplosion of emulsion fuel droplets. Phys Fluids 26(10):103302. https://doi.org/10.1063/1.4897918
6. Elsanusi OA, Roy MM, Sidhu MS (2017) Experimental investigation on a diesel engine fueled by diesel-biodiesel blends and their emulsions at various engine operating conditions. Appl Energy 203:582–593. https://doi.org/10.1016/j.apenergy.2017.06.052
7. Doudou A, Garbinato B, Guerraoui R (2015) Preparation of water-biodiesel emulsion fuels with CNT & alumina nano-additives and their impact on the diesel engine operation. SAE Tech Pap 01(0904)https://doi.org/10.4271/2015-01-0904.Copyright
8. Ithnin AM et al (2018) Emulsifier-free water-in-diesel emulsion fuel: Its stability behaviour, engine performance and exhaust emission. Fuel 215:454–462. https://doi.org/10.1016/j.fuel.2017.11.061
9. Mura E, Josset C, Loubar K, Huchet G, Bellettre J (2010) Effect of dispersed water droplet size in microexplosion phenomenon for water in oil emulsion. At. Sprays 20(9):791–799. https://doi.org/10.1615/AtomizSpr.v20.i9.40
10. Attia AMA, Kulchitskiy AR (2014) Influence of the structure of water-in-fuel emulsion on diesel engine performance. Fuel 116:703–708. https://doi.org/10.1016/j.fuel.2013.08.057
11. Reham SS, Masjuki HH, Kalam MA, Shancita I, Rizwanul Fattah IM, Ruhul AM (2015) Study on stability, fuel properties, engine combustion, performance and emission characteristics of biofuel emulsion. Renew Sustain Energy Rev 52:1566–1579. https://doi.org/10.1016/j.rser.2015.08.013
12. Ogunkoya D, Li S, Rojas OJ, Fang T (2015) Performance, combustion, and emissions in a diesel engine operated with fuel-in-water emulsions based on lignin. Appl Energy 154:851–861. https://doi.org/10.1016/j.apenergy.2015.05.036
13. Noor El-Din MR, El-Hamouly SH, Mohamed HM, Mishrif MR, Ragab AM (2013) Water-in-diesel fuel nanoemulsions: Preparation, stability and physical properties. Egypt J Pet 22(4):517–530.https://doi.org/10.1016/j.ejpe.2013.11.006
14. Mura E, Calabria R, Califano V, Massoli P, Bellettre J (2014) Emulsion droplet micro-explosion: Analysis of two experimental approaches. Exp. Therm. Fluid Sci. 56:69–74. https://doi.org/10.1016/j.expthermflusci.2013.11.020

15. Baskar P, Senthil Kumar A (2017) Experimental investigation on performance characteristics of a diesel engine using diesel-water emulsion with oxygen enriched air. Alex Eng J 56(1):137–146.https://doi.org/10.1016/j.aej.2016.09.014
16. Lin CY, Chen LW (2006) Emulsification characteristics of three- and two-phase emulsions prepared by the ultrasonic emulsification method. Fuel Process Technol 87(4):309–317. https://doi.org/10.1016/j.fuproc.2005.08.014
17. Ismael MA, Heikal MR, Aziz AA, Crua C (2018) The effect of fuel injection equipment of water-in-diesel emulsions on micro-explosion behaviour. EnergS 11(7):1650. https://doi.org/10.3390/en11071650
18. Ismael MA, Heikal MR, Aziz A. RA, Syah F, Zainal AEZ, Crua C (2018) The effect of fuel injection equipment on the dispersed phase of water-in-diesel emulsions. Appl. Energy 222:762–771.https://doi.org/10.1016/j.apenergy.2018.03.070
19. Wang Z et al (2018) Experimental investigation on spray, evaporation and combustion characteristics of ethanol-diesel, water-emulsified diesel and neat diesel fuels. Fuel 231(April):438–448. https://doi.org/10.1016/j.fuel.2018.05.129
20. Hays RB (2013) Utilisation of the healthcare system for authentic early experience placements. Rural Remote Health 13(1):240–251. https://doi.org/10.1016/j.enconman.2016.10.063
21. Mohammed S, Iffa E (2010) Spray characteristic comparisons of compressed natural gas and hydrogen fuel using digital imaging. J Appl Sci 1–5
22. Yu J, Hillamo H, Sarjovaara T, Hulkkonen T (2011) experimental investigation on low pressure gas jet characteristics by tracer-based PLIF technique. J Phys 1–14
23. Semin(2012) Injector nozzle spray on compressed natural gas engines: A technical review. Int Rev Mech Eng 6(5):5013
24. Wang Z et al (2018) Effects of water content on evaporation and combustion characteristics of water emulsified diesel spray. Appl Energy 226(1037):397–407. https://doi.org/10.1016/j.apenergy.2018.06.023
25. Klein-Douwel RJH, Frijters PJM, Somers LMT, de Boer WA, Baert RSG (2007) Macroscopic diesel fuel spray shadowgraphy using high speed digital imaging in a high pressure cell. Fuel 86(12–13):1994–2007. https://doi.org/10.1016/j.fuel.2006.11.039

Study the Thermal Efficiencies for Household Compressors

Loauy Abd Al-Azez Mahdi and Hasanen M. Hussen

Abstract This study focuses on the efficiencies of household compressor in order to understand the behaviour and clarify performance. The study covers the compressors range of swept volume (2.24–11.15) cm^3 depending on manufacturing data at 54.4 °C condenser temperature and variable evaporator temperatures with range (−35 to −10) °C. The coefficient of performance (C.O.P) value for all the ranges of swept volume is over unity for evaporator temperature −25 to −10 °C, and less than unity when the evaporator temperature go down −30 °C, this must take more care from the designer and the manufacturer in order to improve the performance of household compressors. The volumetric efficiency which found approximately between (40 and 50%) when the compressors working according to ASHRAE Technical committe8.9 at evaporator temperature −23 °C and condenser temperature 54.4 °C, but when the evaporator temperature vary for the range (−35 to −10) °C the maximum value be at −10 °C. The polytropic and isentropic efficiency which appear lower than 50% for all ranges of capacities, that means the performance be lower and the losses must be pointed and working hardly to decrease it. Finally, the second law efficiency which is lower than 40% which means low performance and need more work to improve it.

Keywords Performance · Compressors · Manufacturing · Data · Efficiencies

Nomenclature

ASHRAE	American Society of Heating, Refrigerating and Air-Conditioning Engineers
C.O.P	Coefficient of performance
E.E.R	Energy Efficiency Ratio

L. Abd Al-Azez Mahdi (✉) · H. M. Hussen
U.O.T.-Iraq, Baghdad, Iraq
e-mail: louayalkaesy@gmail.com; 20035@uotechnology.edu.iq

H. M. Hussen
e-mail: drhasanenalsaedi@gmal.com; 20046@uotechnology.edu.iq

F. Ahmad et al. (eds.), *ICPER 2020*, Lecture Notes in Mechanical Engineering,
https://doi.org/10.1007/978-981-19-1939-8_18

h_1	Refrigerant inlet enthalpy, kJ/kg
h_2	Refrigerant outlet enthalpy, kJ/kg
h_{2s}	Refrigerant reverse outlet enthalpy, kJ/kg
$\dot{m}_r$	Refrigerant mass flow rate, kg/h
N_c	Number intake strokes per unit time or compressor rotation rate
v_1	Specific volume of the refrigerant at inlet, m^3/kg
Q_e	Cooling capacity, W
S	Swept volume, cm^3
T_e	Evaporator Temperature, °C
W	Power consumption, W
W_c	Mechanical power available for compression, W
w_p	Polytropic work, kJ/kg
W_e	Mechanical power output, W
η_v	Volumetric Efficiency
η_e	Electric Efficiency
η_{is}	Isentropic Efficiency
η_p	Polytropic Efficiency
η_{II}	Second law Efficiency
η_m	Mechanical Efficiency
$_{1,2}$	Subscripts for inlet and outlet to compressor

1 Introduction

The efficiencies are the important parameters, which guide the engineers to understanding the performance and done the improvement. Efficiencies depends on the inlet and outlet conditions are more general and suitable for comparison and modelling the compressors. The electrical, volumetric, polytropic, isentropic, and second law efficiencies of the household compressors are the identity of the compressors working, so the clarify of its behaviours are very important. This study done in order to complete the aim of household compressors in Ref. [1].

2 Literature Review

The researchers which pay attention to the subject of general performance specially the coefficient of performance and the efficiencies of refrigerators and freezers systems are;

- Shaffer and Lee [2] studied the energy losses of hermetic refrigeration compressor. They were separated the mechanical losses, valve losses, suction gas heating losses, leakage and heat transfer losses and study the effects on the EER.

They experimentally testing a standard ¼ horse power hermetic compressor and compare with manufacturer data. They found the motor efficiency to be 70%. Their advice is to reduce the valve bounce and reducing the suction gas heating.

- Duggan et al. [3] used experimental methods for calorimeter compressor test, and compared with flow rater test in order to know the more efficient method. The flow rater test was more accurate than calorimeter.
- Mackensen et al. [4] applied a physics-based method to represent the compressor performance in refrigeration systems with limited experimental data. The method takes the positive displacement compressors, semi-hermetic and open type. Various manufacturers data for compressors were obtained. A volumetric efficiency expression was found to the refrigerant mass flow rates based on the polytropic compression process with clearance volume. The maximum average means errors of 3.7% for reciprocating compressor, 2.3% for scroll compressor, and 0.6% for screw compressor were found for overall performance of the model.
- Possamai and Todescat [5] applied the ideal Carnot refrigeration and the performance of vapor compression refrigeration system to represent high efficiency compressors. The study foxed on the compressor efficiency that can reach 50% of the ideal efficiency. The losses divided into thermodynamic losses, friction losses, and electric losses. The electric efficiency reaches 87.3%, while the friction efficiency 87.6%. The study recommended reducing the friction losses in order to improve the overall compressor efficiency.
- Rigola et al. [6] used advanced numerical simulation model to study the thermal and dynamic optimization of hermetic reciprocating compressors depending on lump thermal conductance approach. The study shows that the electric efficiency slightly above 90% at r.p.m. 2900–3000 and the torque was the minimum. The results also showed the effect of discharge valve diameter and suction valve mufflers on the coefficient of performance (C.O.P.) and volumetric efficiency.
- Ribas et al. [7] reviewed the researches that deal with the numerical analysis for household compressors and divided it into Integral numerical group, Hybrid numerical group, and Differential numerical group. The review divided and locate the sources of losses as follow: 49% superheating, 25% suction, 22% discharge, and 4% leakage. They advise to take more care for gas super heating to reduce the large losses and to improvement the compressor performance. The magnitude of electric efficiency in the review was found to be between 87 and 88%.
- Negrao et al. [8, 9] simulate the unsteady behavior of mass flow rate and power of reciprocating compressors in order to found a semi-empirical mathematical model. Calorimeter data sets of two compressors are used to fitted the thermodynamic equations model. The ratio of actual to ideal volumetric efficiency is linearly dependent on the pressure ratio, and the power is linear function of mass flow rate and compressor isentropic work were found.

The previous literatures do not explore the performance of wide capacity range household compressors. The literatures done for one or two size of household compressors. Therefore, the aim of this work is to clarify the performance of household compressors for a wide range of swept volume (2.42–11.15) cm^3 and evaporator temperature (−35 to −10) °C at a condenser temperature 54.4 °C by study C.O.P, thermal, mechanical, and electric efficiencies.

3 Generalized Approach for C.O.P and Efficiencies

- The performance of compressor is determined by thermodynamic term called coefficient of performance [10]:

$$\text{C.O.P} = \frac{\text{Q}_\text{e}}{\text{W}} \tag{1}$$

The refrigeration capacity and the power consumption are taken from manufacture data in order to find the COP.

- The ratio between the actual flow rate inlet to compressor and the displacement rate of the compressor is represent the volumetric efficiency is [11, 10].

$$\eta_v = \frac{\dot{m}_\text{r} V_1}{N_c S} \tag{2}$$

The reduction in volumetric efficiency related to re-expansion of gas in clearance volume, pressure drops in inlet and outlet valves, pressure drops in the passages like inlet and outlet Muffler, leakage from compression chamber between the cylinder and piston, internal superheat due to motor heat dissipated and the friction, pressure ratio, compressor r.p.m, and refrigerant gas temperature. The volumetric efficiency is effected the capacity.

- The ratio of actual power consumption to the power consumption which mean the isentropic efficiency is needed to done the adiabatic and reversible process operating between the actual inlet and outlet refrigerant conditions [10–12]:

$$\eta_{is} = \frac{\dot{m}_\text{r}(h_{2s} - h_1)}{m_r(h_2 - h_1)} \tag{3}$$

The isentropic efficiency is decreased due to mechanical friction in the bearing, pressure drops in tubes and passages, heat transfer to the refrigerant, the percent of the clearance volume, pressure ratio, compressor r.p.m., and the super heat of the refrigerant gas inlet.

- The polytropic efficiency is the ratio between the power consumption need for reversible polytropic process for the refrigerant to the actual power consumption [11, 12]:

$$\eta_p = \frac{\dot{m}_r w_p}{W} \tag{4}$$

$$w_p = \int_{p_1}^{p_2} v dp = P_1 V_1 \frac{n}{n-1}\left[\left(\frac{P_2}{P_1}\right)^{\frac{n-1}{n}} - 1\right] \tag{5}$$

n: polytropic coefficient or polytropic index based on:

$$n = \frac{ln\left(\frac{P_2}{P_1}\right)}{ln\left(\frac{v_1}{v_2}\right)} \tag{6}$$

The polytropic index effected by friction and heat.

- The second law efficiency

$$\eta_{\text{II}} = \frac{C.O.P}{C.O.P_{carnnot}} \tag{7}$$

Which indicate the useful power required in order to done a work.

- Electric energy is supplied to an electric motor to be converted into mechanical work. This give use to an electrical efficiency:

$$\eta_e = \frac{w_e}{W} \tag{8}$$

The electric efficiency slightly above 90% at r.p.m. 2900–3000 and the torque was the minimum [7, 13].

- The mechanical efficiency is an index of the friction losses. In the household compressors the direct driven for the machine mean approximately no losses, so it is at range 95 to 98% if the r.p.m. 3000 for reciprocating compressor [12]

$$\eta_m = \frac{w_c}{W} \tag{9}$$

4 Results and Discussion

Figure 1 shows the relation between the C.O.P and swept volume. The C.O.P value is high if the evaporator temperature is high and drop when the evaporator temperature decreases for constant swept volume, while for constant evaporator temperature with different swept volume the value increase for swept volume range (2.5 to 7) cm^3 and above that it is remain constant. for the ranges of working swept volume the magnitude is over unity when the compressor working at evaporator temperature −25 to −10 °C and the swept volume more than 5 cm^3, and the C.O.P less than unity for evaporator temperature (−35 and −30) °C. This point leads to discuss

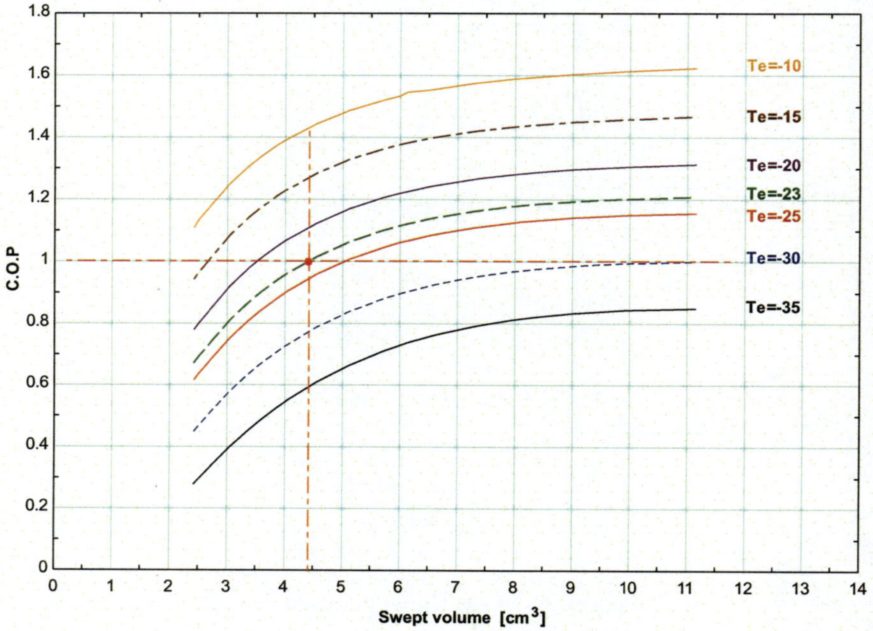

Fig. 1 The relation between C.O.P and swept volume for −35 to −10 °C evaporator temperature

the standard which dependable to build the refrigeration systems and specially the working evaporator temperature, because if we build new refrigeration system at different evaporator temperature from ASHRAE Technical committee [14] may be reduce the power consumption and improve the working performance.

The volumetric efficiency is plotted in Fig. 2 with swept volume. The value of the volumetric efficiency is low when the evaporator temperature is low and it is increase when the evaporator temperature increases for constant swept volume. For constant evaporator temperature the value of the volumetric efficiency remains constant, but when the evaporator temperature (−35 and −30) °C the value is lower 30%, when the evaporator increases above that the value be more than 40%. For standard evaporator temperature −23 °C the value is approximately 50% which agree with the value obtain by Groll [15] for the household compressors the volumetric efficiency for household compressors be in range 40 to 70% for high pressure ratio. The value reach 95% for evaporator temperature −10 °C. The value of volumetric efficiency at −23 °C may push the designer and manufacturer to review and may be change the standard working temperature for refrigeration system or do more researches to improve the compressor in order to increase the volumetric efficiency.

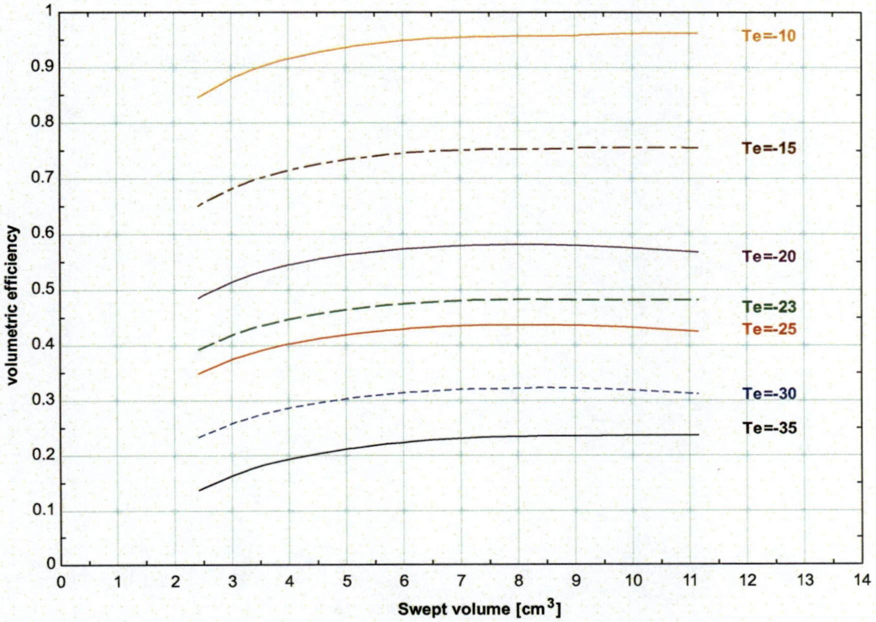

Fig. 2 The relation between volumetric efficiency and swept volume for −35 to −10 °C evaporator temperature

Figures 3 and 4 represent the polytropic and isentropic respectively with the swept volume. The polytropic and isentropic efficiency seem identical and the differences is due the index number between the two efficiencies. The figures show lower performance for the ranges of household compressor which less than 40% for all compressors working lower than 4.5 cm^3 swept volume at all evaporator temperature and around 40% for the swept volume larger than 5 cm^3, these values also agree with Groll [15] the isentropic efficiency for the household compressors be in range 40 to 60%, which backup strongly the lower efficiency for the household compressors that need to encourage the manufacturer to support the researches to work hardly and find the sectors of big losses in the compressors and done amendment for it.

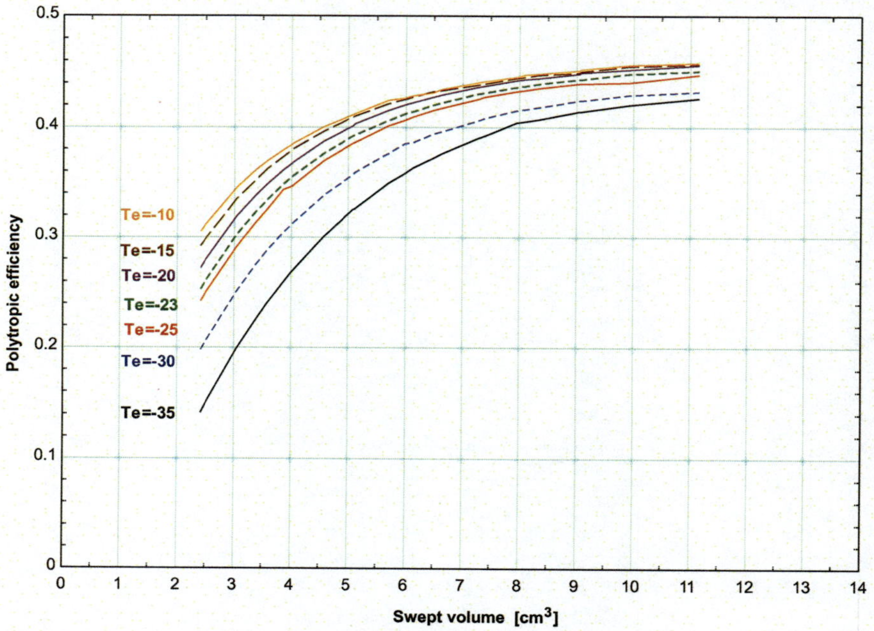

Fig. 3 The relation between polytropic efficiency and swept volume for −35 to −10 °C evaporator temperature

Figure 5 represented the relation between the second low efficiency and the swept volume, the value is increase when the evaporator temperature increases for constant swept volume. For constant evaporator temperature the values are low when the swept volume less than 7 cm^3 and above that the second low efficiency value increase from 30% to less than 40%.

5 Conclusion

The thermal efficiencies found low when working under ASHRAE Technical Committee 8.9 standard for the sizes of swept volume (2.42 to 11.15) cm^3 for constant condenser temperature 54.4 °C and variable evaporator temperature (−35 to −10) °C, and the main results points are:

- for lower swept volume compressors less than 7 cm^3 the C.O.P, volumetric, isentropic, polytropic, and second law efficiency values are low and that is strong effect on the performance of the vapor compression refrigeration system.

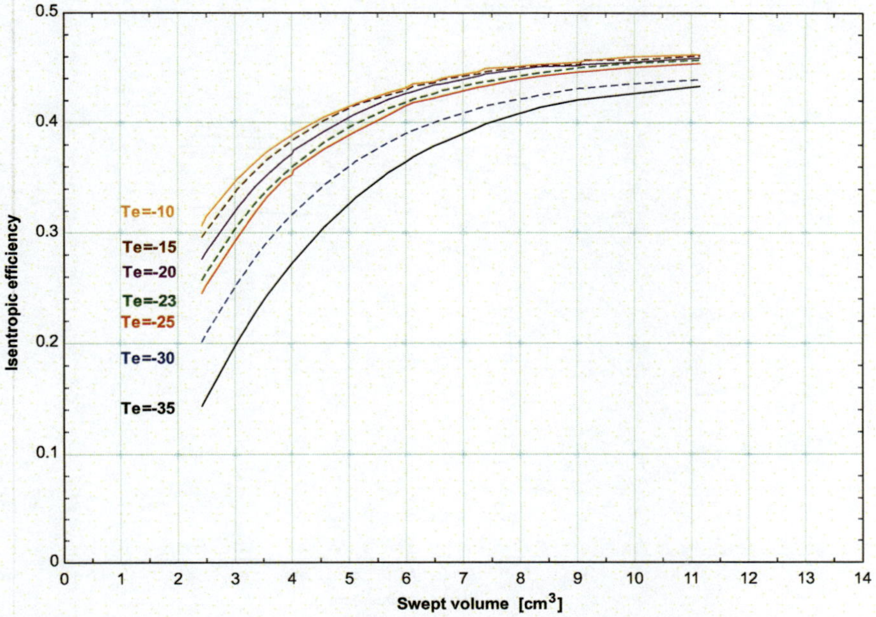

Fig. 4 The relation between isentropic efficiency and swept volume for −35 to −10 °C evaporator temperature

- The evaporator working temperature should be above −23 °C in order to increase the compressors performance (C.O.P) more than unity.
- The volumetric efficiency is very low less 50% when the system working according to ASHRAE Technical committee 8.9 at evaporator temperature −23 °C and condenser temperature 54.4 °C, which effect hardly on the refrigerant mass flow rate and decrease it, due to high pressure difference between the condenser and evaporator which effect on the cooling capacity and decrease it.
- The isentropic, polytropic, and second law efficiency around 40%.
- The summation of decreases in the efficiencies will effect to reduce the household compressors output which mean less cooling capacity and dissipated more power.

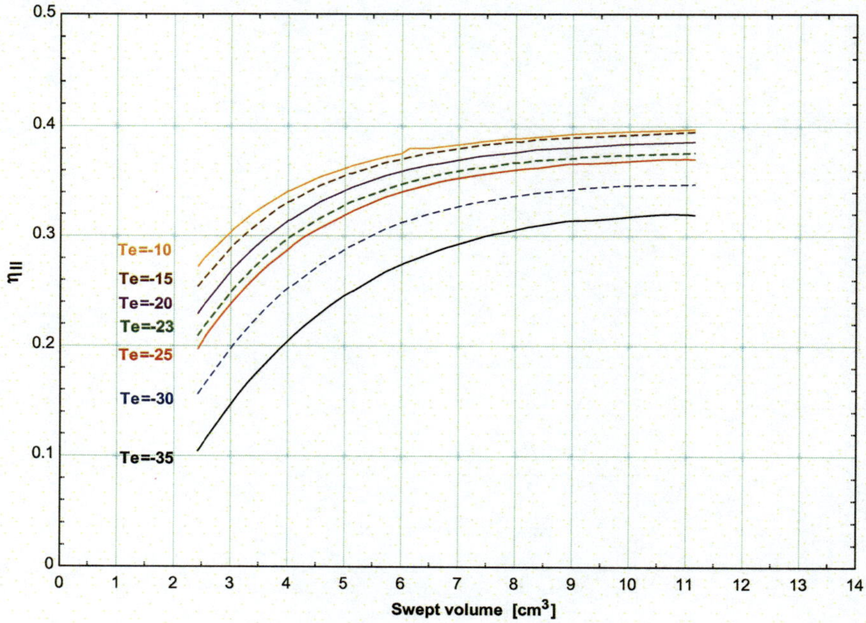

Fig. 5 The relation between second law efficiency and swept volume for −35 to −10 °C evaporator temperature

References

1. Abd AlAzez Mahdi L, Esmaael Habib E, Abd Almunam L (2016) Prediction of performance equations for household compressors depending on manufacturing data for refrigerators and freezers. Appl Mech Mater 818:184–209
2. Shaffer RW, Lee WD (1976) Energy consumption in hermetic refrigerator compressors. In: International compressor engineering conference, Purdue University
3. Duggan MG, Hundy CF, Lawson S (1988) Refrigeration compressor performance using calorimeter and flow rater techniques. In: International compressor engineering conference, Purdue University
4. Mackensen, Klein, Reindl (2002) Characterization of refrigeration system compressor performance. In: International compressor engineering conference, Purdue University
5. Possamai FC, Todescat ML (2004) A review of household compressor energy performance. In: International compressor engineering conference, Purdue University
6. Rigola J, Perez-Segarra CD, Oliva A (2005) Parametric studies on hermetic reciprocating compressors. Int J Refrig 28:253–266
7. Ribas FA, Deschamps CJ, Fagotti F, Morriesen A, Dutra T (2008) Thermal analysis of reciprocating compressors- a critical review. In: International compressor engineering conference, Purdue University
8. Negrao COR, Erthal RH, Andrade DEV, Silya LW (2010) An algebraic model for transient simulation of reciprocating compressors. In: International compressor engineering conference, Purdue University

9. Negrao COR, Erthal RH, Andrade DEV, Silya LW (2011) A semi-empirical model for the unsteady-state simulation of reciprocating compressors for household refrigeration application. Appl Therm Eng 31:1114–1124
10. Threlkeld JL (1962) Thermal environmental engineering. Prentice-Hall, Englewood Cliffs, NJ
11. Ueno K, Bye RE, Hunter KS (2003) Compressor efficiency definitions. University of Colorado, May 12
12. Domanski P (1984) Computer modeling of the vapor compression cycle with constant flow area expansion device. In: NBS building science series 155
13. ASHRAE/ANSI Standard 23-1993 Methods of testing for rating positive displacement refrigerant compressors and condensing units. American Society of Heating, Refrigerating and Air-Conditioning Engineers, Inc., Atlanta
14. ASHRAE Technical Committee 8.9 - Residential Refrigerators and Freezers (2011)
15. Groll E (2010) Overview of compressor technology and their analyses. In: Short course on introduction to compressors, Ray W. Herrick Laboratories, Purdue University, West Lafayette, IN, July 10–11

Investigation of Heat Transfer Enhancement of a Planar Latent Heat Thermal Energy Storage

Luthfi A. F. Haryoko, Jundika C. Kurnia, and Mohd A. Abd Majid

Abstract Latent heat thermal energy storage (LHTES) has been recognized as one of the the best options of thermal energy storage (TES) options due to its higher energy storage capacity. It has been used in various industrial application and building air-conditioning systems. Meanwhile, fractal flow design has been widely adopted to improve heat transfer performance inside LHTES. The fractal flow design is expected to give better heat transfer efficiency as compared to conventional S-shaped pattern. Nevertheless, limited studies has been reporting the performance of planar LHTES with fractal fluid passages. This study is therefore conducted to investigate the heat transfer performance of planar LHTES with various fluid passage design. An experimental set-up with various fluid passage designs such as conventional serpent-shaped (S), diamond fractal-shaped (D), and rectangular fractal-shaped (R) is developed. The experimental condition is carried out under variation of heat transfer fluid (HTF) inlet temperature at a range of 70–90°C (above PCM melting temperature) with constant mass flow rate of 0.01 kg/s. The results revealed that the R-shaped design has the highest overall TES efficiency compared to D-shaped (14.97% lower) and S-shaped (83% lower) design respectively. This study can provide a basic reference for fluid passage configuration in planar LHTES.

1 Introduction

For decades, renewable energy as a clean energy sources have been considered by global community as promising candidate to replace fossil fuel. Accordingly, numerous research and development has been conducted and reported. Some of the most preferred renewable energy sources are solar and wind energy due to its abundant availability. Nevertheless, their adoptability is significantly low due to its intermittent nature: It is not available at all time. This generates discrepancy between energy supply and demand. Therefore, an energy storage is needed to cover

L. A. F. Haryoko · J. C. Kurnia (✉) · M. A. Abd Majid
Department of Mechanical Engineering, Universiti Teknologi PETRONAS, 32610 Bandar Seri Iskandar, Perak Darul Ridzuan, Malaysia
e-mail: jundika.kurnia@utp.edu.my

F. Ahmad et al. (eds.), *ICPER 2020*, Lecture Notes in Mechanical Engineering,
https://doi.org/10.1007/978-981-19-1939-8_19

the hole between energy supply and demand [1]. Thermal energy storage (TES) is one of preferable options as compared to other types of energy storage due to its relatively simple working mechanism. There are three types of TES, sensible heat thermal energy storage (SHTES), latent heat thermal energy storage (LHTES) and chemical energy storage [2]. Among TES, LHTES has received considerable attention due to its large storage capability with minor increase in temperature, thus reducing heat loss to the environment [3, 4]. This behaviour is primarily associated to the phase change material (PCM) high latent heat of fusion. However, PCM has major disadvantage in terms of its low thermal conductivity.

To overcome this issue, comprehensive studies have been conducted to improve the thermal performance of LHTES and numerous enhancement method have been developed and evaluated. One of the most common method that has been widely used because of its simple design, ease of manufacture and minimum construction cost is enlarging the heat transfer area [5]. Fractal design shaped is one of the mechanisms that can be implemented to improve thermal performance of LHTES during melting and solidification process of PCM. The design profile is widely found in respiratory system, and blood vessel structure of plants and animals. This unique flow passages has an intrinsic advantage of reducing flow resistance and high heat transfer capability [6]. Senn and Poulikakos [7] investigated laminar thermal hydraulic performance in fractal and serpentine flow microchannel nets by using numerical approach. The results showed that the fractal channel has 60% lower pressure drop at the same configurations compared to the serpentine flow channel. Chen et al. [8] developed a model to analyse the heat transfer and pressure drop characteristics in fractal minichannel heat sink. They indicated that the fractal channel has significant results in both heat transfer and pressure drop as compared to the conventional S-shaped. Zhang et al. [9] investigated the thermal and flow behaviour of bifurcations and bends in fractal microchannel networks by using numerical simulations. The outcomes observed that the aspect ratio and flow rate are the major component to determine heat transfer efficiency and flow distribution along the channel. It is because of fractal channel with a smaller aspect ratio of 0.333 was confirmed to have better overall heat transfer performance compared with the other microchannel networks as reported in the literature. Zheng et al. [10] studied the effect of fractal net fins to the melting performance of a TES unit. The results revealed that the fractal net fins considerably improved the melting heat transfer characteristics of the selected PCM in a TES unit. Li et al. [11] investigated the effect of fin structure (Koch-fractal fin) on the solidification mechanism of a LHTES unit. The results shown that the heat rate for the discharging process on LHTES unit with Koch-fractal fins is better compared to radial fins.

Notwithstanding the auspicious heat transfer escalation predicted by fractal channel in planar LHTES, based on author's best knowledge, no study investigating effect of fractal channel on heat transfer performance of planar LHTES has been conducted and published. Hence, the objective of this study to investigate the heat transfer performance in various design of flow channels of planar LHTES. The effect of HTF inlet temperature are evaluated and discussed based on the experimental condition. The results are presented to show the overall TES efficiency, charging, and discharging efficiency and serve as a design guideline for flow channel arrangement in planar LHTES.

2 Experimental Methodology

To assess the thermal storage performance of planar LHTES using fractal flow channel with PCM as a storage medium, experimental test rig was prepared. Prior to the experimental setup, each TES container with specific HTF flow channel were filled up with melted PCM based on the geometrical aspect of the HTF. This includes the weight of the melted PCM and the area of the storage unit. The experimental setup consists of three coiled tube with three meranti wood as the storage container: the inner diameter is made of 15 mm with 1 mm thickness copper tube (ASTM B280) and the total length of the coiled tube is 12 m. The coiled tube will be constructed into three different types of fractal channel such as D-Shaped, R-Shaped, and S-Shaped as shown in Fig. 1. The area of fluid passage including its container are 50 cm × 30 cm.

The experimental setup and layout of the test rig are displayed in Figs. 2 and 3 with details of geometrical and operating conditions tabulated in Table 1. The heat transfer fluid in this study is tap water whilst the phase change material used

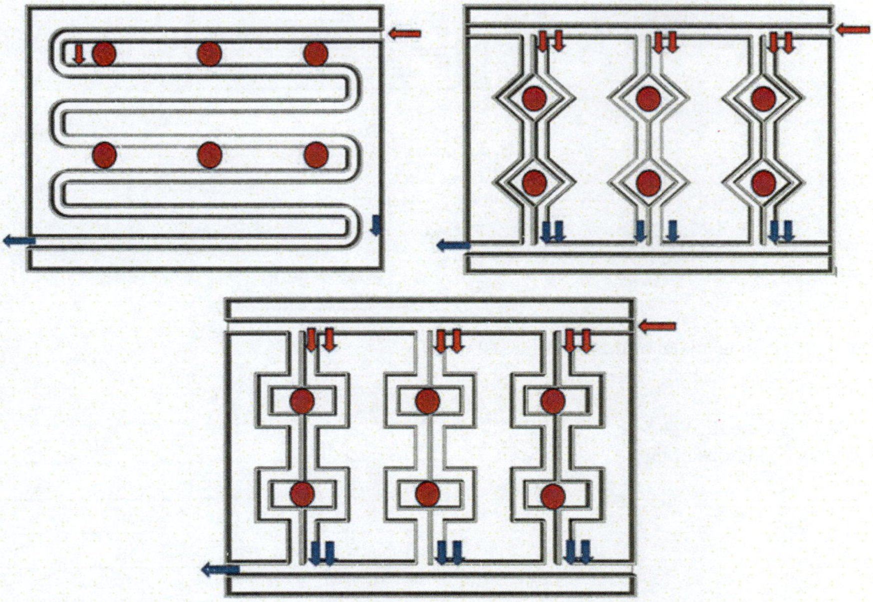

Fig. 1 The design specification drawings based on fractal design profile in horizontal planar orientation with thermocouples positioning

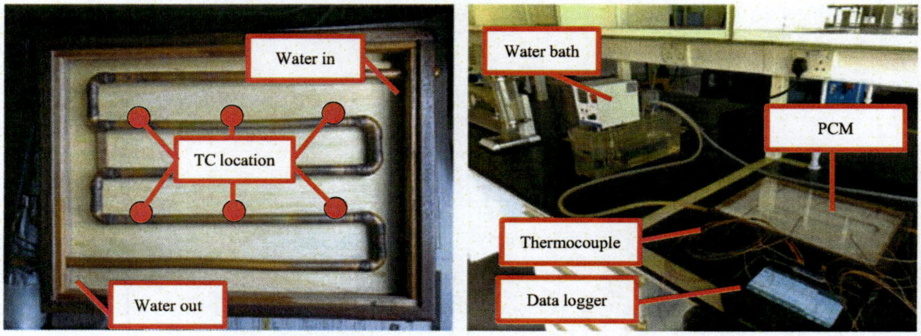

Fig. 2 Experimental setup

is technical grade parrafin wax (n-Octadecane). The thermophysical properties of the utilized PCM is taken to be similar to the previous study [12] and is summarized in Table 2.

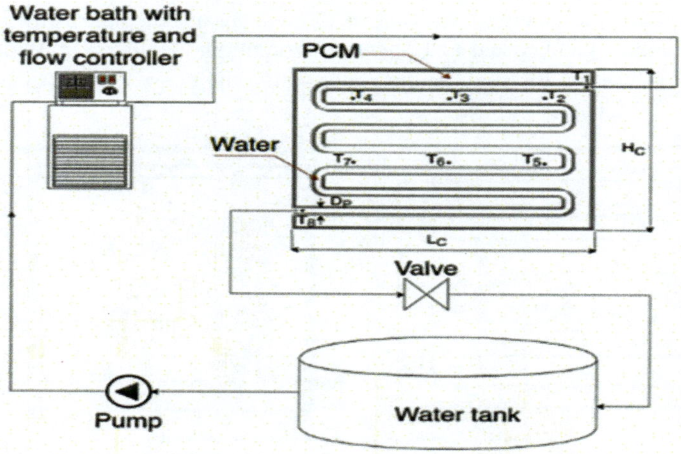

Fig. 3 Schematic layout of the experimental test rig

Table 1 Geometrical and operating conditions of the study

Parameters	Value	Unit
H_C	0.3	m
L_C	0.5	m
D_P	0.015	m
T_{in} (hot water)	70,80,90	°C
T_{in} (cold water)	22,23,24,25	°C

Table 2 Thermophysical properties of PCM used in the study

Parameters	Value	Unit
Melting point, T_m	59.9	°C
Solid–solid phase change temperature, T_{ss}	43.9	°C
Latent heat of melting, L_m	190	kJ/kg
Specific heat of Solid, $C_{p(s)}$	2.0	kJ/kg.K
Specific heat of Liquid, $C_{p(l)}$	2.15	kJ/kg.K

A total of 8 thermocouples type-K are used to measure the temperature and are connected to datalogger (OMEGA OM-DAQXL) with 16 channels input to record the data. Water inlet to the energy storage is supplied by water bath (Thermo Cool Tech 320) at controlled temperature and flow rate. Prior to the experimental runs, calibration of the thermocouples is conducted to ensure the accuracy of the measured data. Two-point calibration is utilized to assess the collected data from thermocouple with a linear relation. Boiling and melting point are captured from the highest and lowest point of temperature calibration. The reading of thermocouples for boiling and melting point are put into boiling water and ice cubes. Temperature result will be compared with a standard thermometer. For consistency throughout the experimental study, thermocouples are paired and labelled with the data logger. The equation of thermocouple calibration is described below:

$$T_{\text{actual}} = T_{\text{detected}} \frac{BoilingPointReading - MeltingPointReading}{100}. \quad (1)$$

Once calibration is conducted, experiment is carried out as follows: (i) Water bath is begun with constant flow rate of 0.01 kg/s, (ii) The hot water (70, 80 and 90 °C) is provided to the LHTES until the temperature of designated location in the LHTES has reached equilibrium at the provided temperature of HTF, (iii) Once the TES is fully charged, the water is substituted with a room temperature water and provided to the LHTES in the same way for LHTES discharging, (iv) The cold water (22–25°C) is provided until the LHTES temperature at the specific location has reached equilibrium at the provided temperature of HTF. During the experiment, HTF inlet, PCM inlet and outlet temperature is documented using the data logger. To ensure validity of the experimental outcome, each experimental test was repeated three times. While for the sampling time, the reading was taken within 10 min interval on each temperature profile.

However, all experimental tests have some degree of uncertainty that may come from a variety of sources. The process of evaluating the uncertainty associated with a measurement result is often called as measurement uncertainty. There are two measurement uncertainty including random and systematic uncertainty, depends on each case of the experimental study measurement. Since this study is validated by repeating the measurement of each experimental test, hence average and standard deviation is necessary to be evaluated. The equations of average (mean) and standard deviation are expressed as follow.

Table 3 Experimental design matrix and results

$T_{i,HTF}$ (°C)	PCM mass (kg)	$T_{e,PCM}$ (°C)	$T_{m,PCM}$ (°C)	$T_{m,PCM}$ (°C)	Q_{stored} (kJ)	$Q_{supplied}$ (kJ)	$Q_{released}$ (kJ)	$\eta_{charging}$ (%)	$\eta_{discharging}$ (%)	η_{TES} (%)
	2.3 PCM_S	61.33		32.00	572.38	4022.00	401.86	14.23	70.20	10.00
70	2.7 PCM_D	60.67		22.67	718.50	1055.00	527.44	68.10	73.40	50.00
	3.0 PCM_R	64.67		22.00	828.12	854.76	502.32	96.90	60.70	58.80
	2.3 PCM_S	75.67		22.67	686.20	1611.60	527.44	42.60	76.90	32.70
80	2.7 PCM_D	74.83	60	21.33	807.90	2492.40	527.44	32.44	65.28	21.20
	3.0 PCM_R	79.67		27.67	890.85	1359.50	703.25	65.50	78.90	51.70
	2.3 PCM_S	86.00		21.00	744.97	1816.56	552.55	41.00	74.17	30.41
90	2.7 PCM_D	86.00		21.00	874.53	1816.56	753.48	48.14	86.15	41.48
	3.0 PCM_R	89.76		25.67	967.93	1009.20	602.78	95.90	62.28	59.72

$$Average(mean) = \frac{x_1 + x_2 + \cdots + x_1}{N}, \tag{2}$$

$$s = \sqrt{\frac{\sum \delta x_i^2}{(N-1)}} \tag{3}$$

The performance of the studied TES is examined by following a method previously utilized by Li et al. [13] which is based on energy analysis. In the method, the studied energy storage performance is presented in several parameters including overall TES efficiency, charging efficiency, and discharging efficiency which are defined as follows,

$$\eta_{\text{charging}} = \frac{Q_{\text{stored}}}{Q_{\text{supplied}}}, \tag{4}$$

$$\eta_{\text{discharging}} = \frac{Q_{\text{discharged}}}{Q_{\text{stored}}}, \tag{5}$$

$$\eta_{\text{overall}} = \frac{Q_{\text{discharged}}}{Q_{\text{supplied}}}. \tag{6}$$

Here, Q_{supplied}, Q_{stored} and $Q_{\text{discharged}}$ are the energy supplied from the HTF to the energy storage during charging, energy stored inside the energy storage and energy discharged from the energy storage during discharging. These are defined as [14, 15]

$$Q_{\text{stored}} = \int_{T_i}^{T_m} m_{pcm} c_{p,pcm} dT + m_{pcm} \beta L + \int_{T_m}^{T_f} m_{pcm} c_{p,pcm} dT, \tag{7}$$

$$Q_{\text{supplied}} = \dot{m}_{htf} \int_0^{t_{\text{charge}}} (T_{in} - T_{out}) dt, \tag{8}$$

$$Q_{\text{discharged}} = \dot{m}_{htf} \int_0^{t_{\text{discharge}}} (T_{out} - T_{in}) dt. \tag{9}$$

where m_{pcm} is the mass of PCM, $\dot{m}_{htf}$ is the mass flow rate of HTF, L is the specific latent heat of PCM, $c_{p,pcm}$ and $c_{p,htf}$ are the specific heat of HTF and PCM, respectively. T_{in} is the HTF inlet temperature, T_{out} is the outlet HTF temperature, t_{charge} is the charging time, $t_{\text{discharge}}$ is the discharging time, T_i and T_f are the initial and final temperature of PCM, respectively. The PCM liquid fraction, β is assumed to be 1 at the end of the charging.

3 Results and Discussion

3.1 *Overall TES Efficiency, Charging Efficiency and Discharging Efficiency*

For overall TES efficiency, R-shaped design is found to have the highest TES efficiency in melting and solidification process followed by D-shaped (14.97% lower) and S-shaped (83% lower) at the initial temperature 70 °C. It is because of the fractal design profile could enhanced heat transfer due to thermal mixing distribution at bifurcation point. Hence, heat transfer efficiency of R-shaped is higher as compared to others fractal channel due to higher surface area in contact with PCM.

For charging efficiency, R-shaped profile is found as the best configurations for all investigated inlet temperature. R-shaped design is found to have the highest charging efficiency followed by D-shaped (29.70% lower) and S-shaped (85.30% lower) at the initial temperature 70 °C. PCM mass is greatly influenced the amount of energy storage capacity and shorten the time needed to reach equilibrium status.

For discharging efficiency, D-shaped channel is observed as the best configurations for all inlet temperature. D-shaped design is found to have the highest discharging efficiency followed by S-shaped (3.20% lower) and R-shaped (12.70% lower) at the initial temperature 70 °C. PCM mass is not greatly influenced in solidification phase.

3.2 *Charging and Discharging Phase in LHTES*

R-shaped with advantageous of fractal design has showed faster melting and solidification time to complete by reaching inlet HTF temperature $T_{i(HTF)}$ after 360 min (6 h) as compared to other design configurations; D-shaped reachs it at 620 min (10.33 h) and S-shaped at 980 min (16.33 h). This results was observed at inlet HTF temperature of 70 °C. However, for different inlet temperature different pattern was observed: longer time to reach steady state for S-shaped and D-shaped and shorter time for R-shaped at 80 °C and maintained at 90 °C. Overall, fractal designs shows faster charging discharging process as compared to the conventional S shaped design.

4 Conclusion

An experimental investigation on the heat transfer performance of a latent heat thermal energy storage using fractal channel with PCM as a storage medium has been conducted. An experimental setup of LHTES has been prepared and calibrated to obtain temperature data for calculation and analysis of the performance of the studied LHTES. Several different of fractal channel are evaluated to examine their influence on the behaviour of LHTES. To examine the influence of fractal channel in investigating the performance of TES, a variation of hot and cold-water inlet temperature is applied. On storage container aspect, each of LHTES container are filled up with melted PCM based on the geometrical condition of the HTF flow channel. This method is recommended to easily evaluate heat transfer performance in LHTES. The results indicate that the R-shaped profile has the highest overall TES efficiency as compared to other passage design. The effect of flow rate and PCM mass will be evaluated in our next study to comprehensively assess the performance of the LHTES.

Acknowledgements Authors would like to acknowledge sponsorship of Newton Link Grant No 015ME0-015 to attend the conference.

References

1. Kurnia JC, Sasmito AP (2018) Numerical investigation of heat transfer performance of a rotating latent heat thermal energy storage. Appl Energy 227:542–554
2. Sharma A, Tyagi VV, Chen CR, Buddhi D (2009) Review on thermal energy storage with phase change materials and applications. Renew Sustain Energy Rev 13(2):318–345
3. Abhat A (1983) Low temperature latent heat thermal energy storage: Heat storage materials. Sol Energy 30(4):313–332
4. Lacroix M (1993) Numerical simulation of a shell-and-tube latent heat thermal energy storage unit. Sol Energy 50(4):357–367
5. Agyenim F, Hewitt N, Eames P and Smyth M 2010 A review of materials, heat transfer and phase change problem formulation for latent heat thermal energy storage systems (LHTESS) *Renewable and Sustainable Energy Reviews* **14** 2 615–28
6. Huang Z, Hwang Y, Aute V and Radermacher R 2016 Review of Fractal Heat Exchangers 10
7. Senn S M and Poulikakos D 2004 Laminar mixing, heat transfer and pressure drop in tree-like microchannel nets and their application for thermal management in polymer electrolyte fuel cells *Journal of Power Sources* **130** 1–2 178–91
8. Chen Y, Zhang C, Shi M and Yang Y 2009 Thermal and hydrodynamic characteristics of constructal tree-shaped minichannel heat sink *AIChE Journal*
9. Zhang C, Lian Y, Hsu C-H, Teng J, Liu S, Chang Y-J, Greif R (2015) Investigations of thermal and flow behavior of bifurcations and bends in fractal-like microchannel networks: Secondary flow and recirculation flow. Int J Heat Mass Transf 85:723–731
10. Zheng J, Yu C, Chen T, Yu Y, Wang F (2019) Optimization of the Melting Performance of a Thermal Energy Storage Unit with Fractal Net Fins Processes 7(1):42
11. Li J, Huang Y, Zhang C, Liu X (2019) Numerical study on the solidification performance of a latent heat storage unit with Koch-fractal fin. Fractals 27(07):1950108
12. Regin AF, Solanki SC, Saini JS (2006) Latent heat thermal energy storage using cylindrical capsule: Numerical and experimental investigations. Renewable Energy 31(13):2025–2041
13. Li G (2016) Sensible heat thermal storage energy and exergy performance evaluations. Renew Sustain Energy Rev 53:897–923
14. Pielichowska K, Pielichowski K (2014) Phase change materials for thermal energy storage. Prog Mater Sci 65:67–123
15. Mahfuz M H, Anisur M R, Kibria M A, Saidur R and Metselaar I H S C 2014 Performance investigation of thermal energy storage system with Phase Change Material (PCM) for solar water heating application *International Communications in Heat and Mass Transfer* **57** 132–9

Performance Evaluation of Thermal Energy Storage with Phase Change Material in Smooth and Expanded Pipe Flow Using Numerical Simulation

B. W. Riyandwita, J. C. Kurnia, S. Hastuty, and M. A. Barrinaya

Abstract Energy storage is needed in the renewable energy system due to the fluctuation nature of the energy source. Thermal energy storage with phase change material (PCM) has been proven to have higher performance due to its latent heat properties. In this study, heat transfer performance comparison of two geometric pipe profiles is performed to evaluate the effect of higher heat transfer area at same PCM volume using Computational Fluid Dynamics (CFD). Pressure drop in the heat transfer fluid (HTF) side is also evaluated. Base model of the smooth pipe is taken from existing experiment and a new pipe is modified by expanding the outer diameter of HTF side 50% bigger than the base model, resulting 14.9% higher heat transfer area. It is shown that at Reynolds number of 4162, there is only slightly higher heat transfer performance represent by slightly higher average liquid fraction and slightly lower outlet temperature in the expanded pipe compare to the smooth one over time. Meanwhile, the pressure drop of the HTF is increased by around 29%. For current Reynolds number, increment of heat transfer area is not enough to increase the heat transfer performance. Other key factor related to the heat transfer performance need to be investigated.

1 Introduction

Energy demands for application in households, public and industrial sectors may vary on time bases, whether it is within hours, days or even weeks. To overcome these demands, energy storage system is introduced. Several energy storage type are available such as mechanical, electrical, chemical, thermal and biological energy

---

B. W. Riyandwita (✉) · S. Hastuty · M. A. Barrinaya
Department of Mechanical Engineering, Faculty of Industrial Technology, Universitas Pertamina, Jakarta 12220, Indonesia
e-mail: byan.wr@universitaspertamina.ac.id

J. C. Kurnia
Department of Mechanical Engineering, Universiti Teknologi PETRONAS, 32610 Bandar Seri Iskandar, Perak Darul Ridzuan, Malaysia

F. Ahmad et al. (eds.), *ICPER 2020*, Lecture Notes in Mechanical Engineering,
https://doi.org/10.1007/978-981-19-1939-8_20

storages. The most common energy storage for those demands is thermal energy storage (TES) systems that is embedded as a system with the energy sources.

There are two types of TES, that is sensible and latent heat storage. Sensible heat storage is able to store the energy by increasing or reducing the temperature of the storage medium. It is depends on the specific heat and volume of the medium. Commonly use material for the medium is rock, ground or water. Meanwhile, latent heat storage is able to store the energy through phase change. Phase change materials (PCMs) use in latent heat storage usually perform the energy storage by changing from solid to liquid. The specific heat of solidification or vaporization is important for latent heat storage design. Both sensible and latent TES also may occur in the same storage material [1]. Combination of sensible and latent thermal energy storage has been shown to increase the energy storage capacity [2–5].

Application of PCM in cylindrical geometries are considered most promising due to its minimum volume of the PCM. The first configuration is where the PCM fills the shell and the HTF flows through the inner pipe [6–8]. The second model is the reverse of model one where the PCM fills the inner pipe and HTF flow through the outer one [7, 9]. The third model is multi-tube similar to shell and tube configuration with multiple tube [8]. Niyas and Muthukumar have compared all the models with the same PCM volume and smooth pipe configuration. It was shown that the third configuration reduce the charging time about 22.38% and 59.32% when compared with model one and two [7].

Due to the fact that the thermal conductivity of the PCM is relatively lower that the HTF, low heat transfer process is expected in the PCM side. That is why most study focused on how to increase the heat transfer in the PCM side with finned tube in different configurations [10–12]. There is no study to yet that focused on HTF side enhancement with conjunction of thermal energy storage in order to maximize its performance.

In this study, two geometrical profiles of smooth and expanded pipe is compared. Effect on increment of the heat transfer area between HTF and PCM is investigated. Similar to heat exchanger application, pressure drop in the HTF is also importance in the overall performance, so pressure drop of the HTF side is also observed.

2 Mathematical Modeling

Numerical investigation of heat transfer process is performed using Computational Fluid Dynamics (CFD) method by axisymmetric two-dimensional domain of circular pipe with PCM wall as a thermal energy storage. Two geometry profiles for the pipe, smooth and expanded pipe is compared with the same PCM wall volume and evaluate its performance. For this study, solidification/melting model is performed to capture the latent thermal energy storage process in the PCM. The formulation of the model is based on enthalpy-porosity method. Water is employed as the heat transfer fluid (HTF) and paraffin wax (n-octadecane) is selected for the PCM material.

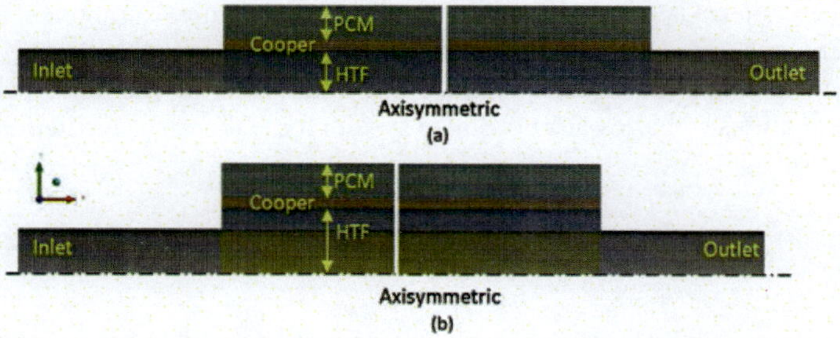

Fig. 1 Two-dimensional axisymmetric CFD domain for **a** smooth and **b** expanded pipe

2.1 Grid Generation and Boundary Conditions

The base model for this study is taken from the study of Lacroix [6]. The base model of smooth pipe consists of two concentric tubes. The inside tube is made of copper with inner diameter of 0.0127 m, outer diameter of 0.0158 m and length of 1.0 m. The outside tube for PCM is made with Plexiglas with outer diameter of 0.0258 m. The expanded pipe is a modification of the smooth one, with inside tube inner diameter of 0.01905 m (1.5 times larger than the base model), outer diameter of 0.0222 m and length of 0.7661 m (shorter than the base model) while the thickness and the total volume of the PCM is maintained at 0.005 m and 3.2686×10^{-4} m^3, respectively. The heat transfer area interface between the HTF and PCM increase by 14.9% for the expanded pipe.

In this study, hexagonal mesh with inflation layer mesh suitable for enhanced wall function is generated for the grid as shown in Fig. 1. The number of hexagonal element for the smooth and expanded pipe is 112,000 and 113,112, respectively.

The inlet and outlet region are extended 0.127 m to avoid reversed flow and entrance effect during the simulations. The Reynolds numbers of the flow is around 4162 (flow rate of 0.0315 kg/m^3). The initial temperature is 295.7 K and the inlet HTF temperature is 305.7 K. The outlet of HTF is set as pressure outlet boundary condition at atmospheric pressure. All the walls are set as no-slip adiabatic conditions.

2.2 Numerical Method and Governing Equations

For this investigation, commercial ANSYS Fluent package is used. SIMPLE algorithm with double precision is selected. PRESTO scheme was selected for the pressure correction equation and second order upwind scheme is used form momentum, energy, turbulent kinetic energy and dissipation equations with transient assumption. For transient analysis, time step of 0.1 s is selected in this study. The realizable $k - \varepsilon$ turbulence model with enhanced wall function is used for turbulent modeling.

Convergence criteria is set to be 10^{-6} for continuity, velocities components, k and ε. The convergence criteria for energy equation is set to be 10^{-7}.

Two-dimensional governing equations for continuity, momentum, energy and turbulent equations in Cartesian coordinate system for HTF can be written as follows:

Continuity:

$$\frac{\partial \rho_{htf}}{\partial t} + \frac{\partial}{\partial x_j}(\rho_{htf} u_j) = 0 \quad (1)$$

Momentum:

$$\frac{\partial}{\partial t}(\rho_{htf} u_j) + \frac{\partial}{\partial x_j}(\rho_{htf} u_j u_k) = -\frac{\partial p}{\partial x_j} + \frac{\partial}{\partial x_j}\tau + \rho_{htf} g \quad (2)$$

Energy:

$$\frac{\partial}{\partial t}(\rho_{htf} T) + \frac{\partial}{\partial x_j}(\rho_{htf} u_j T) = \frac{\partial}{\partial x_j}\left(\frac{\lambda_{htf}}{c_{p,htf}}\frac{\partial T}{\partial x_j}\right) \quad (3)$$

Turbulent kinetic energy:

$$\frac{\partial}{\partial t}(\rho_{htf} k) + \frac{\partial}{\partial x_j}(\rho_{htf} k u_j) = \frac{\partial}{\partial x_j}\left[\left(\mu_{htf} + \frac{\mu_t}{\sigma_k}\right)\frac{\partial k}{\partial x_j}\right] + G_k - \rho_{htf}\varepsilon \quad (4)$$

Turbulent dissipation rate

$$\frac{\partial}{\partial t}(\rho_{htf}\varepsilon) + \frac{\partial}{\partial x_j}(\rho_{htf}\varepsilon u_j) = \frac{\partial}{\partial x_j}\left[\left(\mu_{htf} + \frac{\mu_t}{\sigma_\varepsilon}\right)\frac{\partial \varepsilon}{\partial x_j}\right] + \rho_{htf} C_1 S\varepsilon - \rho_{htf} C_2 \frac{\varepsilon^2}{k + \sqrt{\nu\varepsilon}} \quad (5)$$

where ρ_{htf} is water density, x_j is directional position, p is pressure, τ is viscous stress tensor, g is gravity acceleration and it is set in positive x direction to avoid gravitational buoyancy effect on the PCM, T is temperature, λ_{htf} is thermal conductivity of water, $c_{p,htf}$ is water specific heat (4200 J/kgK), μ_{htf} is water dynamic viscosity, $C_1 = \max\left[0.43, \frac{\eta}{\eta+5}\right], \eta = S\frac{k}{\varepsilon}$, $S = \sqrt{2S_{ij}S_{ij}}$, $S_{ij} = \frac{1}{2}\left(\frac{u_i}{x_j} + \frac{u_j}{x_i}\right)$, $\mu_t = \rho C_\mu \frac{k^2}{\varepsilon}$. G_k represents the turbulence kinetic energy production due to the gradients of mean velocity and calculated as $G_k = \mu_t S^2$. The constant of $C_2 = 1.9$, $\sigma_k = 1.0$ and $\sigma_\varepsilon = 1.2$ are selected. C_μ is a relation of the rates of mean strain and rotation, the angular velocity, and the turbulence fields (k and ε) [13].

Meanwhile, the governing equations for continuity, momentum, energy and turbulent equations in Cartesian coordinate system for PCM can be written as:

Continuity:

$$\frac{\partial \rho_{pcm}}{\partial t} + \frac{\partial}{\partial x_j}(\rho_{pcm} u_j) = 0 \quad (6)$$

Momentum:

$$\frac{\partial}{\partial t}(\rho_{pcm}u_k) + \frac{\partial}{\partial x_j}(\rho_{pcm}u_j u_k) = -\frac{\partial p}{\partial x_j} + \frac{\partial}{\partial x_j}\tau + \rho_{pcm}g + S_M \tag{7}$$

Energy:

$$\frac{\partial}{\partial t}(\rho_{pcm}H_{pcm}) + \frac{\partial}{\partial x_j}(\rho_{pcm}u_j H_{pcm}) = \frac{\partial}{\partial x_j}\left(\lambda_{pcm}\frac{\partial T}{\partial x_j}\right) \tag{8}$$

Turbulent kinetic energy:

$$\frac{\partial}{\partial t}(\rho_{pcm}k) + \frac{\partial}{\partial x_j}(\rho_{pcm}ku_j) = \frac{\partial}{\partial x_j}\left[\left(\mu_{pcm} + \frac{\mu_t}{\sigma_k}\right)\frac{\partial k}{\partial x_j}\right] + G_k - \rho_{pcm}\varepsilon + S_k \tag{9}$$

Turbulent dissipation rate

$$\frac{\partial}{\partial t}(\rho_{pcm}\varepsilon) + \frac{\partial}{\partial x_j}(\rho_{pcm}\varepsilon u_j) = \frac{\partial}{\partial x_j}\left[\left(\mu_{pcm} + \frac{\mu_t}{\sigma_\varepsilon}\right)\frac{\partial \varepsilon}{\partial x_j}\right] + \rho_{pcm}C_1 S\varepsilon - \rho_{pcm}C_2\frac{\varepsilon^2}{k + \sqrt{\nu\varepsilon}} + S_\varepsilon \tag{10}$$

where ρ_{pcm} is paraffin density, λ_{pcm} is thermal conductivity of paraffin, $c_{p,pcm}$ is paraffin specific heat (2130 J/kgK), μ_{pcm} is paraffin dynamic viscosity,$S_M = \frac{(1-\beta)^2}{(\beta^3+\epsilon)}A_{\text{mush}}u_k$, $S_k = \frac{(1-\beta)^2}{(\beta^3+\epsilon)}A_{\text{mush}}k$ and $S_\varepsilon = \frac{(1-\beta)^2}{(\beta^3+\epsilon)}A_{\text{mush}}\varepsilon$. Where ϵ is a small number (0.001) to prevent division by zero, A_{mush} is the mushy zone constant (10^5)[13]. For energy equation in PCM, $H_{pcm} = h_{pcm} + \Delta H_{pcm}$ with sensible heat $h_{pcm} = h_{pcm,ref} + \int_{T_{ref}}^{T} c_{p,pcm}$. Latent heat of PCM, $\Delta H_{pcm} = \beta L$, where L is specific latent heat of paraffin (243.5 kJ/kg [6]) and β is a function of liquid fraction.

$$\beta = \begin{cases} 0 & \text{if} T < T_s \\ \frac{T-T_s}{T_s-T_l} & \text{if} T_s < T < T_l \\ 1 & \text{if} T > T_l \end{cases} \tag{11}$$

where for this study, solidus temperature, $T_s = 300.15\text{K}$ and liquidus temperature, $T_l = 300.65\text{K}$[14].

For the cooper, only heat conduction occurred. The energy equation for the cooper is as follow

$$\frac{\partial}{\partial t}(\rho_{cooper}T) + \frac{\partial}{\partial x_j}(\rho_{cooper}u_j T) = \frac{\partial}{\partial x_j}\left(\frac{\lambda_{cooper}}{c_{p,cooper}}\frac{\partial T}{\partial x_j}\right) \tag{12}$$

where ρ_{cooper} is cooper density (8978 kg/m^3), λ_{cooper} is cooper thermal conductivity (387.6 w/mK), and $c_{p,cooper}$ is cooper specific heat at constant pressure (381 J/kgK).

2.3 Constitutive Relations

To capture the effect of temperature change on thermal properties, temperature dependent water properties of density, viscosity and thermal conductivity from Kays et al. [15] are used as follow

$$\rho_{htf} = 753.2 + 1.88T - 3.57 \times 10^{-3}T^2 \tag{13}$$

$$\lambda_{htf} = -0.5981 + 6.53 \times 10^{-3} - 8.354 \times 10^{-6}T^2 \tag{14}$$

$$\mu_{htf} = \left(2.591 \times 10^{-5}\right) \times 10^{\frac{238.3}{T-143.2}} \tag{15}$$

The paraffin thermal properties of density, viscosity and thermal conductivity are as follow [6, 16]

$$\rho_{pcm} = \frac{774}{9 \times 10^{-4}(T - T_l) + 1} \tag{16}$$

$$\lambda_{pcm} = \begin{cases} 0.358 \text{ if } T < T_s \\ 0.148 \text{ if } T > T_l \end{cases} \tag{17}$$

$$\mu_{pcm} = 0.001 \times e^{\left(-4.649 + \frac{1.79 \times 10^{-3}}{T}\right)} \tag{18}$$

To employ the some of the properties, user-defined functions (UDF) is compiled and loaded in the calculation.

3 Results and Discussion

To evaluate the heat transfer performance of PCM latent TES due to HTF side geometrical effect, two different geometrical feature is compared. Prior to investigate it, validation of the simulation is conducted. Validation is performed by comparing the temperature data at two locations from existing experiment and modeling of Lacroix [6]. Figure 2 show the comparison of those temperature locations. T1 is the temperature of the first location and T2 is the temperature of the second one. With origin coordinate (0,0) at cross section point between inlet and axis line, the first location is at $x = 0.637$ m and $y = 0.0178$ m, while the second one is at $x = 1.077$ m and $y = 0.0168$ m. As can be seen in Fig. 2, the result from current simulation shows relatively good agreement with the results from experiment and modeling of Lacroix [6].

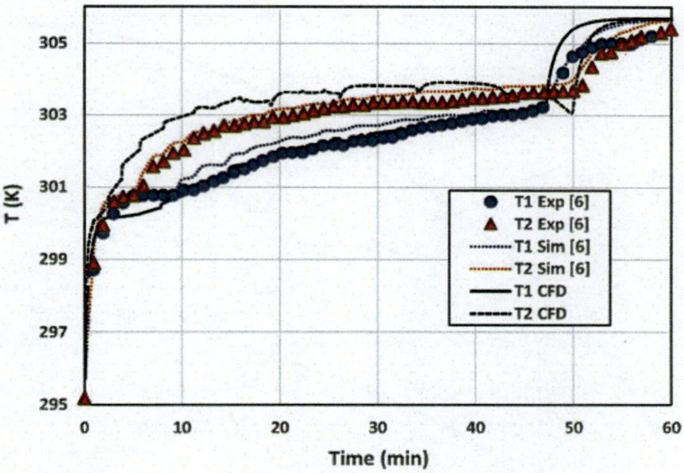

Fig. 2 Comparison of temperature variation at two location from simulation of current study and Lacroix [6]

Average liquid fraction, β_{ave}, is average value of the liquid fraction of the PCM in outer tube. It is showing how much paraffin has been melted due to heat transfer from the HTF. Comparison of average liquid fraction of PCM in smooth and expanded pipe flow is shown in Fig. 3. It is shown in the figure that the average liquid fraction over time for the expanded pipe is almost similar to the one with the smooth pipe. Slightly higher value is observed at higher time period, meaning that the PCM with expanded pipe configuration has slightly higher performance compare to the smooth one over time. This is confirmed with the outlet temperature comparison in Fig. 4. It is shown that the outlet temperature is slightly lower in the longer time period, meaning that more heat has been absorbed in the expanded pipe compare to the smooth one over time.

Figure 5 shows the comparison of HTF pressure drop due to different geometric profile. It is show that around 29% difference resulted due to the different geometric profile. Insignificant heat transfer enhancement by the expanded pipe is traded with the gain of pressure drop. From current study it is seen that expanded pipe is unable to get significant heat transfer performance although as mentioned previously that the heat transfer area is increased by 14,9%. For more comprehensive evaluation, variation of the expanded ratio (ratio of expanded to smooth inner diameter), mass flow rate and inlet temperature need to be performed.

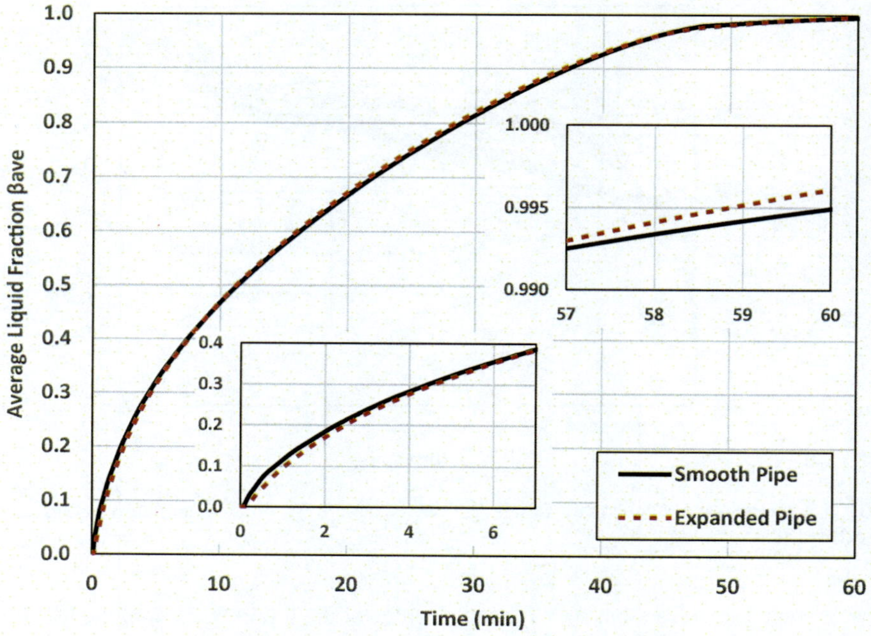

Fig. 3 Comparison of average liquid fraction of PCM in smooth and expanded pipe flow

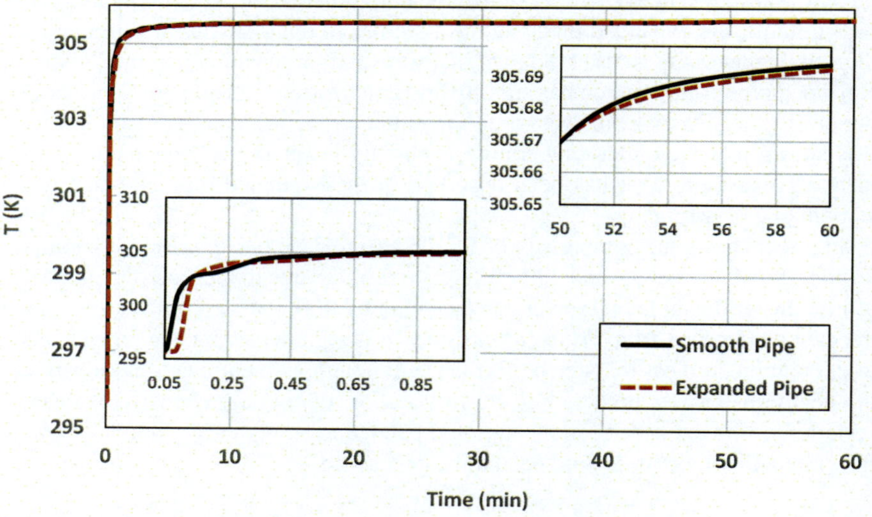

Fig. 4 Comparison of HTF outlet temperature in smooth and expanded pipe flow

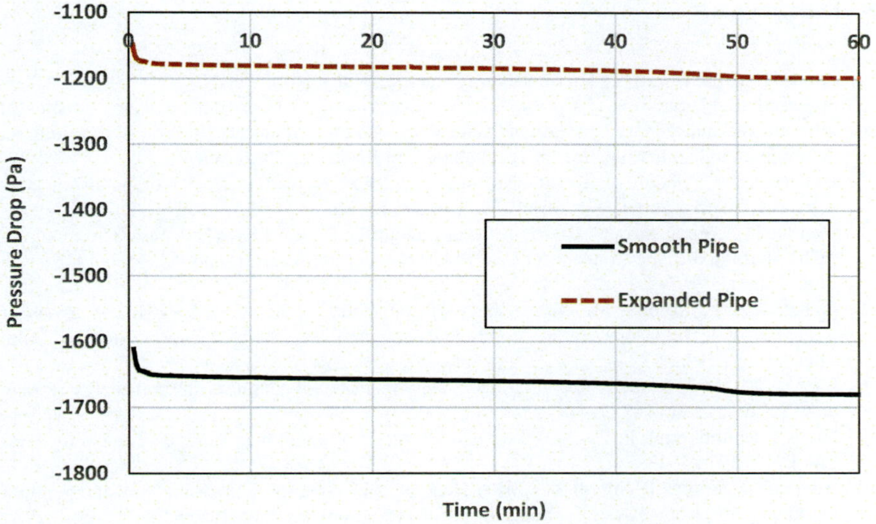

Fig. 5 Comparison of HTF pressure drop in smooth and expanded pipe flow

4 Conclusions

Computational Fluid Dynamics of heat transfer performance for latent thermal energy storage using PCM is performed. Comparison of two geometric pipe profiles is conducted to see the effect of higher heat transfer area at the same PCM volume to its heat transfer performance. Pressure drop in the heat transfer fluid (HTF) side is also evaluated. Smooth and expanded pipe flow is compared with PCM wall. It is shown that at Reynolds number of 4162, there is almost no noticeable heat transfer performance in the expanded pipe. Slightly higher average liquid fraction and slightly lower outlet temperature are observed in the expanded pipe compare to the smooth one. Meanwhile, the pressure drop increment is more noticeable than the heat transfer performance. For current Reynolds number, increment of heat transfer area is not enough to increase the heat transfer area. Evaluation on other variable such as variation of mass flow rate and inlet temperature might be useful get a better insight on the heat transfer phenomena of expanded pipe with PCM wall. From this study, it can be seen that enhancement of the HTF side proof to be more challenging that PCM side because of the pressure drop factor.

Acknowledgements This work is performed under 2019 join collaborative research grant by Universitas Pertamina and Universiti Teknologi PETRONAS.

References

1. Dincer I, Rosen M (2013) Thermal energy storage. Wiley, Hoboken, N.J
2. Abdelsalam M, Sarafraz P, Cotton J, Lightstone M (2017) Heat transfer characteristics of a hybrid thermal energy storage tank with Phase Change Materials (PCMs) during indirect charging using isothermal coil heat exchanger. Sol Energy 157:462–476
3. Underwood C, Shepherd T, Bull S, Joyce S (2018) Hybrid thermal storage using coil-encapsulated phase change materials. Energy and Buildings 159:357–369
4. Galione P, Pérez-Segarra C, Rodríguez I, Lehmkuhl O, Rigola J (2014) A New Thermocline-PCM Thermal Storage Concept for CSP Plants. Numerical Analysis and Perspectives. Energy Procedia 49:790–799
5. Geissbühler L, Kolman M, Zanganeh G, Haselbacher A, Steinfeld A (2016) Analysis of industrial-scale high-temperature combined sensible/latent thermal energy storage. Appl Therm Eng 101:657–668
6. Lacroix M (1993) Numerical simulation of a shell-and-tube latent heat thermal energy storage unit. Sol Energy 50(4):357–367
7. Niyas P, Muthukumar P (2013) Performance analysis of latent heat storage systems. Int J Sci Eng Res 4(12):74–79
8. Agyenim F, Eames P, Smyth M (2010) Heat transfer enhancement in medium temperature thermal energy storage system using a multi-tube heat transfer array. Renew Energy 35(1):198–207
9. Jesumathy SP, Udayakumar M, Suresh S (2012) Heat transfer characteristics in latent heat storage system using paraffin wax. J Mech Sci Technol 26(3):959–965
10. Kurnia JC, Sasmito AP, Jangam SV, Mujumdar AS (2013) Improved design for heat transfer performance of a novel phase change material (PCM) thermal energy storage (TES). Appl Therm Eng 50(1):896–907
11. Sciacovelli A, Gagliardi F, Verda V (2015) Maximization of performance of a PCM latent heat storage system with innovative fins. Appl Energy 137:707–715
12. Liu Z, Sun X, Ma C (2005) Experimental investigations on the characteristics of melting processes of stearic acid in an annulus and its thermal conductivity enhancement by fins. Energy Convers Manag 46(6):959–969
13. Ansys 2019 FLUENT 19.0: Theory Guide, Ansys. Inc.
14. Kurnia JC, Sasmito AP (2017) Numerical investigation of heat transfer performance of a rotating latent heat thermal energy storage. Appl Energy 227:542–554
15. Kays W, Crawford M, Weigand B (2005) Convective heat and mass transport, 4th edn. Singapore, McGraw Hill
16. Reid RC, Prausnitz JM, Poling BE (1987) The properties of gases and liquids, 4th edn. Singapore, McGraw Hill

Experimental Evaluation of Drag Reduction in Pipe Flow Using Streamlined Turbulence Modifiers

Hussain H. Al-Kayiem, A. R. Othman, Ali B. Abed, and Ali M. Tukkee

Abstract Fluid Transportation is sustained by high power pumping to overcome the drag created by the shear stress, with considerable cost and energy consumption. In the current paper, a new technique is proposed, which meets the industry's desires to reduce drag and save pumping energy by mounting small streamlined protrusions in the pipe's short segments' inner wall. The protrusions are functioning as turbulence modifiers by changing the turbulence structure in the flow near the wall. Series of experimental measurements were carried out using two configurations of TMs in 0.0572-m-diameter GI pipe, with water flow covering Reynolds number range from 55,000 to 110,000. The current study focuses on the effect of the population of the TMs on drag reduction. The results demonstrated drag reduction in the low population case, which has 30 TMs separated by 4.8 mm, by approximately 6.3%. The high population case showed an increase in drag. However, it is possible to push the boundary for drag reduction efficiency limit by further investigation of configurations of TMs. It is highly recommended to investigate the technique with a different population and more geometries of the turbulence modifiers.

Keywords Drag reduction · Riblet · Turbulence structure · Turbulence modifier

1 Introduction

Pipelines infrastructure plays an important role in the modern lifestyle operation and economic functions of any developed or developed country. In 2016 alone, Americans have reportedly consumed 29% of natural gas and 40% of oil as part of the total energy consumption in the world, where every day, pipelines transport an average amount of 19.63 million barrels [1]. With the ever-growing demand for

H. H. Al-Kayiem (✉) · A. M. Tukkee
Solar Thermal Advanced Research Centre (STARC), Universiti Teknologi PETRONAS, 32610 Seri Iskandar, Perak, Malaysia
e-mail: hussain_kayiem@utp.edu.my

H. H. Al-Kayiem · A. R. Othman · A. B. Abed
Mechanical Engineering Department, Universiti Teknologi PETRONAS, 32610 Seri Iskandar, Perak, Malaysia

F. Ahmad et al. (eds.), *ICPER 2020*, Lecture Notes in Mechanical Engineering,
https://doi.org/10.1007/978-981-19-1939-8_21

energy consumption, the need to construct more pipelines to transport the fuels and feedstocks is expected to increase as well. The importance of pipelines is not just limited to the oil and gas industry but also marked its significance in most industries around the world. According to a study conducted by Dena, a German energy agency [2], the energy consumption by the industries in the European Community (EC) is reportedly 951 TWh (Trillion Watt-hour) in total in the year 2000. More than half of this energy, approximately 65%, was contributed by electric motors-driven machines such as pumps to transport fluids across Europe.

The high consumption of energy by pumps in pipeline transportation has been a vital issue across the industries as it increases the operating costs and hazardous impact to the environment. The fundamental cause of high energy consumption in a pipe is a high-pressure drop of fluid in the pipe due to high drag encountered. The significance of drag reduction cannot be underestimated. Hence, many pieces of research have been done intensively for the past seven decades to optimize the drag reduction in the pipeline. Tom is one of the first researchers to discover the drag reduction capabilities of polymer additives in fluid flow, which shows a promising 80% drag reduction [3]. Ever since his discovery, additives in fluid flow to achieve drag reduction have been widely applied in industries.

In contrast, additives' improvement of drag reduction performance became one of the most intensively studied topics. Despite the high effectiveness of additives in drag reduction, this method's most challenging issue is the extra steps required to undergo removal processes of the additives, which nullify the total energy and cost savings [4]. In addition to that issue, adding additives in the fluid may alter the physical and chemical properties of the fluid, which makes this method infeasible in industries such as the food industry and pharmaceutical industries [4]. As a result, researchers have been prompted to switch their attention to the study of non-additive methods.

Non-additives methods are divided into active control and passive control. Active methods introduce external energy input to the system to alter the fluid flow's turbulence structure near the wall by controlling the flow separation, such as the oscillating wall method. Passive methods are steady and require no energy in controlling the turbulence structure of fluid flow by utilizing roughness elements of the in-contact surface [5]. The most popular passive methods are riblets and real shark skin, as shown in Fig. 1.

Active control methods are not favored in industrial practices due to extra energy input needed, nullifying the total savings and requires complex installation on existing pipeline facilities, making them infeasible yet. In comparison, passive control methods appear to be a strong candidate as a long-term solution to drag reduction. The general performance of drag reduction by riblets is within the range of 5–13%, which is still much lower than additives methods' performance. The failure of the passive-control method in delivering competitive drag reduction performance results from a poor understanding of its drag reduction mechanism in turbulent structure, which then hinders the development of optimum geometry for the in-contact surface of pipelines [4]. Drag force encountered by a turbulent fluid flow in a smooth pipe is high-skin-frictional drag due to momentum exchange of eddies, resulting

Fig. 1 Shark-skin [5]

in high-pressure drop [6]. An attempt to study three-dimensional trapezoidal-blade riblets compared to other existing shapes of riblets has indicated 9.9% of drag reduction performance achieved by blade riblets that are higher than other shapes of riblets [7].

The common geometries of riblet structures investigated are V-shaped, U-shaped, L-shaped with a broad range of 5–11% of drag reduction performance [8–10]. The early works were conducted by [11] and [12]. They proved the capability of the V-groove riblet in producing drag reduction of 8%. Flow over segmented trapezoidal blade riblets (L-shaped) was investigated by [13] and produced a drag reduction of 7.29%. Martin and Bhushan [14] found that the highest drag reduction performance achieved by segmented L-shaped riblet was 5.2%. Despite being closer to the shape of the real shark-skin scale in terms of segmented feature, the drag reduction performance of segmented riblet is worse than the continuous riblet. It is suggested that the segmented gaps on real shark skin may not help in reducing drag encountered but function to eject the contaminant particles on shark skin [13, 14]. One of the limitations imposed by the riblet structure is introducing protrusions into the flow system, which increases the wetted surface area, despite achieving higher turbulent drag than skin-frictional drag overall. This serves as one of the factors that hinder the progress of drag reduction performance by riblet structure.

In addition to reducing the wetted area, the design structure of streamlining of riblets would undoubtedly smoothen the flow over the riblets and reduce the drag opposite the flow. From the results reported by [15], it has been found that a streamlined-shaped wing promotes laminar flow along the streamwise direction of flow. In their simulation of streamlined energy promotors investigated in a rectangular channel, [16] reported a drag reduction of 7.0%. They concluded that the streamlined-riblet structure improves the drag reduction performance by possessing a lower wetted surface area and lowering the velocity gradient of flow due to streamlined features while retaining the drag-reducing capabilities of conventional riblet configuration.

The literature recommends streamlined riblets, but no experimental data is available in the literature to support the recommendation. As such, it is essential to incorporate the streamlining feature in the design of riblet geometry. Therefore, the current research objective is to investigate the streamlined riblets' contribution, named Turbulence Modifiers (TMs), to the drag reduction in pipe flow. Three cases of riblets structures have been studied experimentally. One case is a clean pipe without riblets. This case which is the bare pipe, is considered a benchmark case. With different riblets and different populations, the other two cases are case 1 with a low population and case 2 with a high population. The investigation covered a large turbulence range of Re, from 55,000 to 110,000 in 0.0572-m-diameter GI pipe.

2 Experimental Method

This study is conducted via experimental investigation utilizing a flow loop installed in the advanced fluid research Lab in the Mechanical department—UTP. The original system was modified to meet the requirement of the current investigations.

2.1 Flow Loop

The flow loop consists of three flow lines with 2, 3- and 4-inch diameters. Originally, the 2-inch flow line is made of PVC, then replaced with GI pipe in the current investigation as the friction losses are not visible enough in the PVC pipe. The image of the flow loop is shown in Fig. 2.

The test pipe is made from Galvanized Iron (GI) of 0.0572 m inner diameter and 0.0036 m wall thickness, 10 m length. It consists of four sections, two sections of 5 m length, connected by pipe union and the test section of 1 m length connected

Fig. 2 Three lines flow loop. The one in the model, with a 2-inch diameter, is used in the current measurements

by a flange that could be easily dismantled and re-assembled to change different configurations of energy promotors. Water is supplied to the test pipe by a centrifugal pump type EBARA 3 M 50-125/2.2 with a flow rate range of 24–60 m^3/h and ahead of 8–19 m. the water supply system is shown in Fig. 3. At the end of the test section, water is conveyed by a returning PVC pipe and discharged to a tank and then recirculated by the pump to the test section in a mode of close circulation loop. However, to provide an idea about the uncertainty in the measurement, the accuracy and specifications of the instruments are given in Table 1.

2.2 Data Acquisition and Instruments

The flow loop is supported by measuring instruments connected to a data acquisition system to permit online monitoring, recording and control. The data acquisition consists of PC with in-house developed software to manage the operation of the

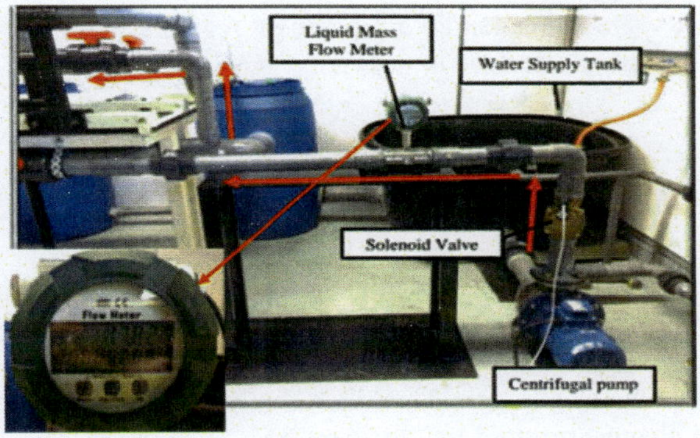

Fig. 3 The water supply system and flow rate measuring instrument

Table 1 Specifications and accuracies of the measuring instruments

Measurement and control device	Model	Specification
Liquid flow meter	FLT liquid turbine flow meter	Measuring range: 2–40 m^3/h Accuracy: ± 0.5% of the reading
Pressure gauge	Digital type WINTERS DPG215	Measuring range: 0–160 psi Uncertainty: ± 0.2 psi Accuracy: ± 1.6% of the reading
Differential Pressure Transmitter (DPT)	High-performance differential pressure transmitter EJX110A	Accuracy: ± 1% of the reading

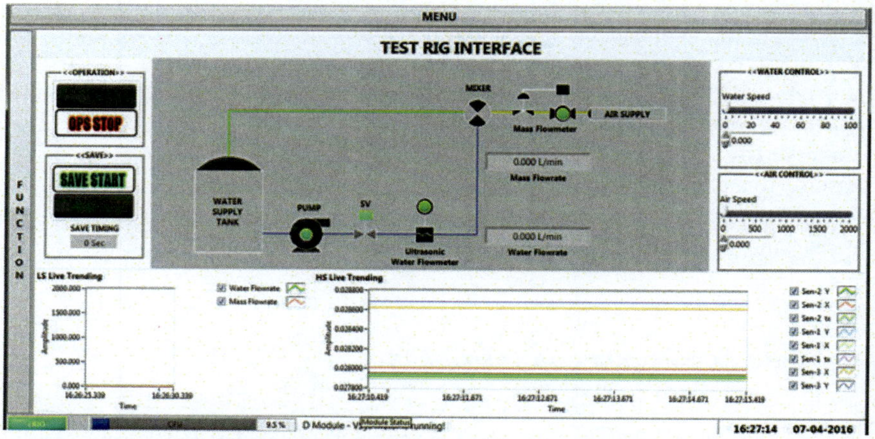

Fig. 4 Data acquisition outline as appearing in the PC interface of control and measurement collection from the measuring instruments

system. A check valve and solenoid arrangement regulate the water flow rate. The test rig-instrumentations interface is shown in Fig. 4.

2.3 *Turbulence Modifiers*

The production of the TMs with sufficient accuracy is a real challenge as there is a need for high precision machines such as Wire Cut EDM to achieve the best TMs with a streamlined shape. To meet the shape of the streamlined TMs, CNC machine is the best way to achieve the desired configuration with sufficient accuracy. Another challenge is the changeability of the different TMs in the same test section. This is achieved by grooving the inlet section of the pipe at the test section. The pipe's inner surface is grooved by 2 mm in the radial direction and 20 mm in the axial direction using CNC machine. The ring of the TMs, shown schematically in Fig. 5a, is then insert and fitted inside the groove, as shown in Fig. 5b.

The modifiers are installed at a distance larger than 1.78 m downstream of the control valve to ensure fully developed flow.

Three experimental cases have been investigated. One case is a bare case for benchmarking, and two different cases have been considered with a different population of TMs. The shape of the TM is the same in both cases, but the number of TMs in each ring is differing. The particulars of each case, with schematic drawing, are outlined in Table 2.

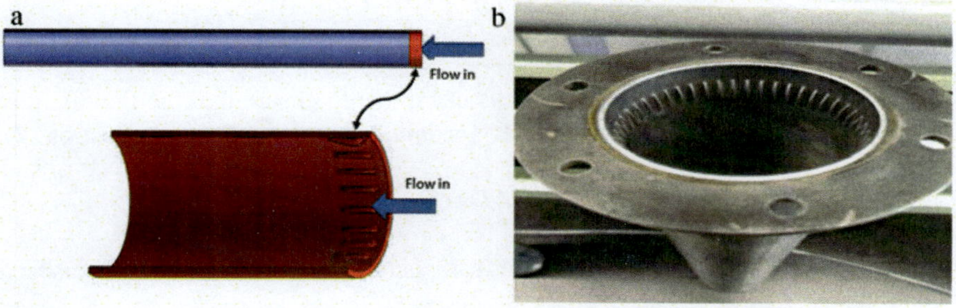

Fig. 5 **a** Schematics of the turbulence modifier installation inside the pipe, at the inlet section. **b** Turbulence modifier installation inside the pipe, at the inlet section

Table 2 Morphology of the turbulence modifiers and cases

Case	Morphologies					Figure
	Width, W_{EP} (mm)	Length, L_{EP} (mm)	Height, h_{EP} (mm)	Population per periphral	Spacig (mm)	
Bare case	–	–	–	–	–	–
Case 1	0.6	10	1.25	30	4.8	
Case 2	0.6	10	1.25	48	2.8	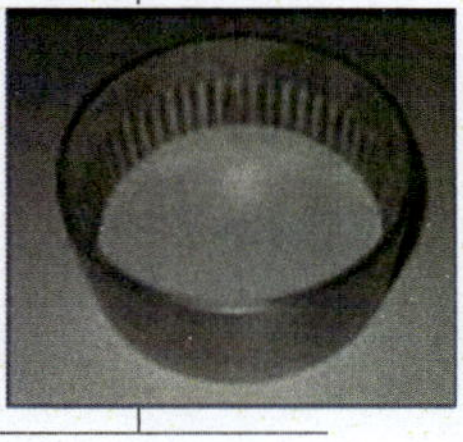

2.4 Evaluation Parameter

The main parameter of concern is the pressure drop which is the indicator for drag reduction. The drag reduction is predicted as:

$$DR\% = \frac{(\Delta p/L)_{with\,TM} - (\Delta p/L)_{smooth}}{(\Delta p/L)_{smooth}} \times 100 \quad (1)$$

The entrance length was calculated as in Eq. (2) to ensure that the flow is fully developed and is not affected by the valve installed upstream of the test Section (2).

$$l_e - 4.4D \times \mathrm{Re}^{1/6} \quad (2)$$

The pressure drop is predicted theoretically using the well know Darcy Weisbach equation.

$$\Delta p = f\frac{L}{D}\frac{v^2}{2}\rho \quad (3)$$

where, $L = 2$ m, D = 0.0572 m, v = Q/A, ρ = 999 kg/m^3. The frictional factor, f is estimated as Re and the pipe roughness, ε, which is taken as 0.16 mm according to the new GI pipe standards, using Colebrook Eq. (4).

$$1/f^{0.5} = -2.0\log\left(\frac{\varepsilon/D}{3.7} + \frac{2.51}{\mathrm{Re}\,f^{0.5}}\right) \quad (4)$$

As Colebrook is implicit, where f appears on both sides, it is solved by iteration with the first guess determined by Haaland friction Eq. (5).

$$1/f^{0.5} = -1.8\log\left[\frac{6.9}{\mathrm{Re}} + \left(\frac{\varepsilon/D}{3.7}\right)^{1.11}\right] \quad (5)$$

3 Results and Discussion

The improvement in the thermal conditions is predicted relative to the bare case 1. The improvements are evaluated in terms of heat gain by the office's interior, the air temperature inside the office and the saving in energy consumed for the air conditioning.

3.1 Experimental Justification

The reliability of the developed setup, to provide accurate results, is justified by comparing the pressure drop's measurement results with theoretical ones covering the entire range of Re, from 50,000 to 110,000 for a bare pipe without any energy

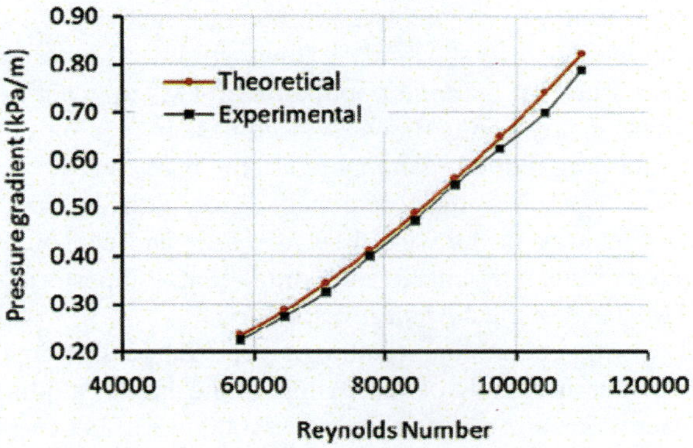

Fig. 6 Comparison between the measured and predicted pressure gradient of bare pipe

promotors inserts. The measurement data of the pressure gradient is directly obtained from the DPT reading divided over 2 m. The pressure gradient's theoretical values are predicted using the well-known design formulas of Haaland, Eq. (5) to get a first guess of the friction factor and Colebrook, Eq. 4, to estimate the final friction factor by iteration. The predicted friction factor is then used to evaluate the pressure gradient using the well-known Darcy-Weisbach equation. The comparison results of the experimental and theoretical pressure gradient are shown in Fig. 6. The predicted values show higher slightly higher estimation than the measured pressure gradient over the entire range of Re. the mean square error percentage is 0.75%. This justifies the accuracy and suitability of the test setup for such investigation.

3.2 Analysis of Turbulence Modifier Population

Two cases of the population have experimented. Case 1 is with a high population as there are TMs over the internal peripheral of the pipe. The spacing between the TMs is 4.8 mm. Case 2 is with the higher population as the spacing is 2.8 mm. Figure 7 shows the measurement results of the pressure gradient in the bare case. The maximum drag reduction reached 6.3% for the streamlined TMs conflagration of case 1 where the population is less separated by 4.8 mm, as shown in Fig. 3.9. This reduction is considered in good agreement with the previously published literature that the range of reduction is confined between 8 and 13% [16].

As the population increased, the drag is increased by around 7.0%. This non-consequent influence could be attributed to insight analysis of the flow near the wall. In the low population case, the vortices near the wall are lifted from the very close region to the buffer zone of the turbulence profile. The near-wall vortices are not interacting with each other. They just have a smaller sone near the wall, between the TMs and just moving up. This is acting as a positive influence to reduce the drag overall tested Re range, from 50,000 to 110,000. In the case of a large population, the presence of the TMs close to each other causes the vortices to interact with each other leading to higher local turbulence intensity. This will negatively influence and increase the drag shear on the flowing fluid. Another issue that may also require to be discussed that the high population of TMs may act as a flow cross-section reducer. Then, the hydraulic diameter becomes smaller, and the friction increase is causing larger drag reduction compared to the bare case.

In this sense, the researchers agree with the conclusion of [17] that a smaller morphology in the wetted surface area and capable of slowing down the vortices interactions may improve the drag reduction performance of riblet structures or the TMs used in the current investigation.

However, another feature is realized in this investigation that the deviation, whether increasing or decreasing, of the drag, is tending to be higher as Re increases. At low Re, around 55,000 up to 70,000, the effect of the TMs is insignificant. Above 70,000 Re, the influence of the TMs becomes more obvious. The reason that at low Re, the turbulence vortices are small, and the insertion of TMs is not effective to lift them from the near wall to the core of the velocity profile. This fact is existing in the investigated range of Re and the morphologies of the TMs. Further configurations and morphologies of the TMs are essential to understanding the turbulence change mechanism and the consequences of drag reduction.

4 Conclusion

Drag reduction may be achieved by various fixed fixtures installed inside the pipe's inner surface, which is a passive drag reduction technique. This is the industry's desire as these passive control methods appear to be a strong aspirant as a long-term solution to drag reduction despite drag reduction by riblets within the range of 5% to 13%, which are still much lower than drag additives methods. The current investigation experimented with a new type of riblets which are functioning as turbulence modifiers. The population of the TMs is the focus of the study. It is concluded that the population of the riblets, or TMs, is a strong contributor to the performance of

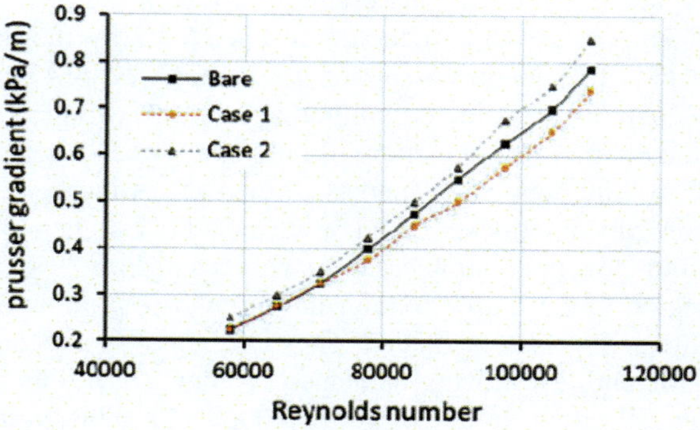

Fig. 7 Measurement results of the pressure gradient for bare pipe and two cases of TMs population

drag reduction by turbulence modification. The results indicate a significant fact that a high population may increase the drag, while a low population reduces the drag.

It is highly recommended to carry out a more experimental and numerical investigation on the effect of various morphologies on restricting the turbulence near the wall of pipes and the consequences on drag reduction.

Acknowledgements The authors express their acknowledgment to Universiti Teknologi PETRONAS for supporting the research work technically and financially under research grant YUTP CS-015LC0-026.

References

1. The Importance of Pipeline Infrastructure (2017) https://consumerenergyalliance.org/2017/05/importance-pipeline-infrastructure/. Accessed Jan 2020
2. Vogelesang H (2008) An introduction to energy consumption in pumps. https://doi.org/10.1016/S0262-1762(07)70434-0
3. Toms B (1948) Proceedings of 1st international congress rheology, Holland, Vol II, 1949, p 135
4. Abdulbari HA et al. Going against the flow—A review of non-additive means of drag
5. Joshi S, Gujarathi Y (2014) A review on active and passive flow control techniques. Int J Recent Technol Mech Elect Eng 15:16–22
6. Dean B, Bhushan B (2010) Shark-skin surfaces for fluid-drag reduction in turbulent flow: a review. Philos Trans Royal Soci London A: Math Phys Eng Sci 368:4775–4806
7. Bechert D et al (1997) Experiments on drag-reducing surfaces and their optimization with an adjustable geometry. J Fluid Mech 338:59–87
8. Liu KN et al (1990) Drag reduction in pipes lined with riblets. AIAA J 28(10):545–551. https://doi.org/10.2514/3.10459
9. Chu DC, Karniadakis GE (1993) A direct numerical simulation of laminar and turbulent flow over riblet-mounted surfaces. J Fluid Mech 250(1). [Published online by Cambridge University Press: 26 April 2006]. https://doi.org/10.1017/S0022112093001363
10. Jung YC, Bhushan B (2010) Biomimetic structures for fluid drag reduction in laminar and turbulent flows. J Phys Condens Matter Inst Phys J 22(3):035104. https://doi.org/10.1088/0953-8984/22/3/035104
11. Walsh MJ, Weinstein LM (1978) Drag and heat transfer on surfaces with small longitudinal fins. In: 11th fluid and plasma dynamics conference, american institute of aeronautics and astronautics, July 10–12. https://doi.org/10.2514/6.1978-1161
12. Walsh MJ (1980). Drag characteristics of V-groove and transverse curvature riblets, viscose drag reduction. AIAA168–184. https://doi.org/10.2514/5.9781600865466.0168.0184
13. Bechert DW, Bruse M, Hage W (2000) Experiments with three-dimensional riblets as an idealized model of shark skin. Exp Fluids 28(5):403–412. https://doi.org/10.1007/s003480050
14. Martin S, Bhushan B (2016) Fluid flow analysis of continuous and segmented riblet structures. RSC Adv 6(13):10962–10978. https://doi.org/10.1039/C5RA20944G
15. Bokhorst EV, de Kat R, Elsinga GE, Lentink D (2015) Feather roughness reduces flow separation during low Reynolds number glides of swifts. J Exp Biol 218(20):3179–3191. https://doi.org/10.1242/jeb.121426
16. Al-Kayiem HH, Lim DC, Kurnia JC (2020) Large eddy simulation of near wall turbulent flow over streamlined riblet-structured surface for drag reduction in a rectangular channel. Thermal Sci 59–59 [online]. https://doi.org/10.2298/TSCI190909059A

17. Caram J, Ahmed A (1991) Effect of riblets on turbulence in the wake of an airfoil. AIAA J 29(11):1769–1770

Numerical Investigation of Laminar Convective Heat Transfer of Nanofluid in Parallel Cooling Channels

Luthfi A. F. Haryoko, Agus P. Sasmito, and Jundika C. Kurnia

Abstract Cooling electronic component has become indispensable part of electronic devices. A good cooling method provides assurance that the electrical component less likely experience failure. One of the major methods of cooling electronic component is applying nanofluid in channels for heat transfer purpose. Conventional parallel channel design tends to have low pressure drop which indicates less pumping power for the cooling fluid. However, the drawback of the design is that the heat distribution is not sufficiently uniform. The main objective of the research is to make some modifications on the previous parallel channel and propose new parallel designs by using computational fluid dynamics. The flow behaviour and heat transfer performance of a Newtonin nanofluid are investigated under laminar flow condition. Taguchi method is used to determine the best combination of key parameters on this study. The results indicate that the proposed design shows better heat transfer performance compared to the conventional parallel channel. The addition of nanoparticles into the base fluid gives substantial effect on the heat transfer performance of the proposed cooling channel. Optimum operating parameters is obtained with the objective function of minimizing pressure drop and maximizing uniformity of heat distribution.

1 Introduction

Electronic devices have become inseparable part of our daily life. Various electronic appliances like, computer, generators, transformers, transmitters, semiconductors and so on have been crucial in our activities. During operation, some portion of electrical energy supplied to these devices are converted into heat, creating

L. A. F. Haryoko · J. C. Kurnia
Department of Mechanical Engineering, Universiti Teknologi PETRONAS, Bandar Seri Iskandar, 32610 Perak Darul Ridzuan, Malaysia

A. P. Sasmito (✉)
Department of Mining and Materials Engineering, McGill University, 3450 University, Frank Dawson Adams Bldg., Montreal, QC 327, Canada
e-mail: agus.sasmito@mcgill.ca

F. Ahmad et al. (eds.), *ICPER 2020*, Lecture Notes in Mechanical Engineering,
https://doi.org/10.1007/978-981-19-1939-8_22

unwelcomed thermal stresses. This thermal stresses, occuring especially at interface, decrease device lifetime drastically [1]. Such circumstance causes the study of heat transfer becoming very important that many researchers have been funded to enhance the heat transfer strategies for cooling applications of electrical devices. Accordingly, numerous cooling methods have been introduced in electronic industries to maintain the temperature of the microchip low[2].

Although water is considered as one of the best medium to carry thermal energy during heat transfer process, the introduction of nanofluid, which is a base fluid containing nanometer-sized particles, have open new possibility of having better heat transfer. Implementation of nanofluid has been recommended based on the reports of higher thermal conductivity compared to base fluid [3]. Research from Hwang et al. [4], Lee et al. [5], Eastman et al. [6] and Yu et al. [7] indicated that nanofluid have better thermal conductivity than base fluid. This thermal conductivity is widely regarded as the results of Brownian motion of nanoparticles which leads to reduction in thermal boundary layer [8]. What is more, nanofluid is environmentally friendly [9]. Sajid and Ali [10] highlighted the application of nanofluid in electronic cooling device applications. The examples of nanofluids suspensions are water-Al_2O_3, water-CuO, water-SiC and others. To improvise the heat transfer performance of nanofluid, several studies of manipulating the key parameters like types of nanofluid, concentration of nanofluid and types of channels of nanofluid have been carried out and the results show that most of them have big impacts on the heat transfer of nanofluid.

Apart from studies investigating on parameters like types and concentration of nanofluid, there is a another dominant parameter dictating the performance of electronic cooling system, which is the geometrical design of nanofluid cooling channels. It is observed that due to the low pressure drop, parallel channels have better potential than coiled based channel after some modifications. Speaking of parallel-channel design, it consists of two manifolds which are wide channels that connected to multiple parallel channels [11]. The manifold acts as a distributor or collector which allows the cooling fluid flows in parallel manner and causing pressure difference [12–15]. There are two types of manifold structure, which are bifurcation and consecutive types [16, 17]. For the bifurcation-type manifold, due to its branched channel in the form of tree where the length of channels gradually decreases, the fluid flow in it will yield high pressure drop. This is not suitable for applications requiring low pressure drop [18]. Meanwhile, consecutive-type manifold branches into small holes which reside on the side wall of manifold for which low pressure drop can be achieved [19]. The investigation of pressure drop and flow distribution in one of the interesting channel design which is parallel channel design with Z-type arrangement has been carried out by Wang [20]. The study takes friction and inertia terms into account for the comparison of Z-type arrangement with the U-type arrangement. The concluding remark from the study is that the geometrical structure with characteristics parameters and flow condition should be properly adjusted in order to improve uniformity and achieve an optimal design of the Z-type arrangement. The study of addition of ribs on microchannel conducted by Esmaili et al. [21] has influential impacts in heat transfer. The study shows the comparison of flat channels and ribbed channel where ribbed channels yield higher pressure drop but a significant increase in heat transfer coefficient. Moreover, the final analysis of the result indicated that the addition of ribs on microchannel has better advantage over the penalty of having higher pressure drop. By configuring the characteristics parameters, inclined-ribbed channel with angle 45° achieves the best heat transfer performance among all the inclined-ribbed channel. The investigation of enhancing flow uniformity in parallel mini-channels with fin-pinned inlet header is conducted by Song et al. [22]. Two pin fin arrangements of trapezoidal and rectangular header shapes were included in the study. The outcome of the study is that pin fins add very minimal pressure drop but improves heat sink thermal performance.

According to the previous studies, cooling electrical components with nanofluid has been one of the effective solutions to achieve higher cooling performance. Nevertheless, different cooling channels have different heat transfer performance. As previously discussed, parallel channel design exhibits low pressure drop which shows its potential to be widely used in cooling application. However, this design has a drawback of having non-uniform heat distribution. Thus, parallel channel design should be further innovated with modifications. Therefore, the aim of this study is to investigate the heat transfer performance of nanofluid in various kinds of parallel cooling channels using numerical prediction and to assess the best combination of key parameters for the lowest pressure drop and the highest uniformity of heat distribution using Taguchi method. Taguchi method is a well-known method used in design experiment plan by statistically minimising the test runs to determine the best condition of parameters with different levels. This method is especially useful when it comes to configuration of key parameters or characteristics dimensions of certain geometry. The study is expected to serve as a guideline to design efficient cooling channels in electronic devices.

2 Mathematical Model

As shown in Fig. 2, cooling channel and the solid separator are the two major components of the mathematical model in this study. This combined model represents the occurrence of a conjugate heat transfer for nanofluid, and heat flux generated from electronic chip. The heat generated from electronic microchip is represented by a prescribed heat flux at the bottom of the solid separator. The heat transfer occurs through the solid separator by conduction and then to the nanofluid by convection. Referring to Fig. 1, the area of the microchip is based on an Intel microprocessor with model number of i7-3700 K, which is 37.5 × 37.5 mm. Additionally, the accuracy of the results is furtherly firmed by keeping the area of the solid separator as constant whereas the design will not exceed the boundary of the separator and the total length of all designs have low difference. Eight channel designs were investigated in which the first design was taken as a datum of parallel design. The other seven designs are parallel designs with certain modifications.

General information

Type	CPU / Microprocessor
Market segment	Desktop
Family	Intel Core i7
Model number	i7-3770K
CPU part numbers	CM8063701211700 is an OEM/tray microprocessor CM8063701211701 is an OEM/tray microprocessor BX80637I73770K is a boxed processor with fan and heatsink (English version) BXC80637I73770K is a boxed processor with fan and heatsink (Chinese version)
Frequency	**3500 MHz**
Maximum turbo frequency	3900 MHz (1 or 2 cores) 3800 MHz (3 cores) 3700 MHz (4 cores)
Low power frequency	1600 MHz
Bus speed	**5 GT/s DMI**
Clock multiplier	**35**
Package	1155-land Flip-Chip Land Grid Array
Socket	**Socket 1155 / H2 / LGA1155**
Size	1.48" x 1.48"
Weight	1oz / 27.1g (CPU) 13.1oz / 372.6g (box)

Fig. 1 Technical specification of microprocessor i7-3700 K

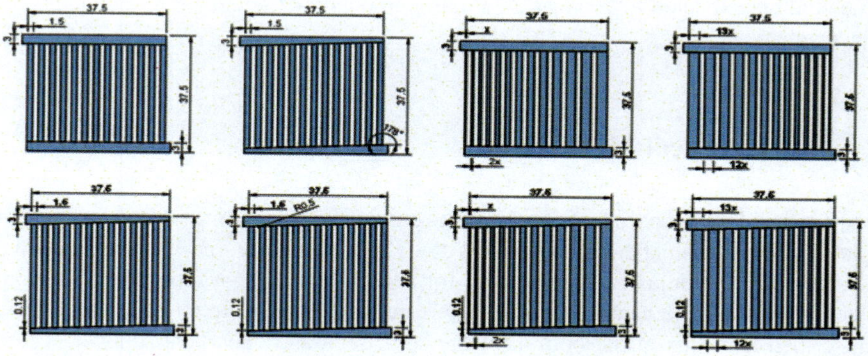

Fig. 2 Dimensions of the parallel channel designs (where x = 3/14)

2.1 Governing Equations

The governing equations of heat conduction by the wall of solid separator is described by,

$$k_s \nabla^2 T = 0, \tag{1}$$

where k_s is the heat conductivity of solid separator and T is the temperature.

Meanwhile for the cooling channel, the fluid flow and convective heat transfer are considered by using the conservation equation of mass, momentum, and energy.

$$\nabla.\left(\rho_{nf} u\right) = 0, \tag{2}$$

$$\nabla.\left(\rho_{nf} u \otimes u\right) = -\nabla p + \nabla.[u_{nf}([16]\nabla u + \left(\nabla u\right)^T], \tag{3}$$

$$\nabla.\left(\rho_{nf} c_{p,nf} u T\right) = -\nabla.\left(k_{nf} \nabla T\right). \tag{4}$$

where ρ_f is fluid density, u is fluid velocity, p is pressure, u_f is dynamic viscosity of the fluid, $c_{p,f}$ is specific heat of the fluid and k_f is thermal conductivity of the fluid.

2.2 Constitutive Relations

Water and water-based nanofluids are the main medium used in this study. For the thermal conductivity, according to Sasmito et al. [23], the static part involves Maxwell's theory while the dynamic part involves Brownian motion of nanoparticles which both contribute to the thermal conductivity. The Maxwell's theory is applied in this modelling as the average speed of the motion of the molecules of two regions may be changing from time to time. Meanwhile, the Brownian motion involves the random collision between nanoparticles with base fluid molecules. As per the kinetic theory of particles, the temperature rises due to the motion. However the Brownian motion decreases along with the increase in the bulk viscosity.

The heat transfer performance of the different designs of cooling channel is compared in a kind of analysing term and the term is defined as equation below,

$$Figure\,of\,Merit,\,FoM = \frac{W}{W_{pump}} \tag{5}$$

where W_{pump} is the pumping power required to drive the fluid flow through the channel. Figure of merit can be explained as a numerical quantity that measures the overall efficiency of a system.

2.3 Boundary Conditions

The boundary conditions for the flow inside the cooling channel are prescribed as follows:

- Inlet: The inlet of the cooling channel is defined with a fixed constant initial temperature with various mass flow rate which represent the Reynolds number of 250, 500, 750, 1000, 1250 and 1500 which are in laminar flow condition.

$$\dot{\mathrm{m}} = \dot{\mathrm{m}}_{in}, \; T = T_{in} \tag{6}$$

- Outlet: Pressure and streamwise gradient of the temperature is set to be zero.

$$p = p_{out}, \mathrm{n}. \nabla T = 0 \tag{7}$$

- Bottom wall of solid separator: A constant heat flux of a typical sample of microchip is prescribed at the base of the solid separator (whereas z = 1.5×10^{-3} m).

$$n.k_s \nabla T = Q_{mc} \tag{8}$$

- Flow channel and solid separator interface: No slip condition for velocity on this interface as the distance between the wall will be zero.

$$u = 0 \tag{9}$$

- Flow channel walls: No slip and adiabatic conditions are set.

$$u = 0, \; n.k_w \nabla T = 0 \tag{10}$$

- Solid separator walls: Adiabatic conditions are set.

$$n.k_s \nabla T = 0 \tag{11}$$

3 Numerical Methodology

All the geometry designs were drawn in AutoCAD software and later imported into ANSYS Workbench 16.2 for meshing and labelling. All the dimensions were based on actual dimension which illustrated a real cooling channel design for cooling electronic component. Computational Fluid Dynamics (CFD) software, ANSYS Fluent 16.2 was used to solve the mathematical model for nanofluid in parallel cooling channel and to implement boundary conditions and constitutive relations needed. Five different amounts of meshing were run to ensure mesh independent result. The pressure drop and temperature outlet of flow channel were compared to ensure the mesh was suitable to obtain accurate results. 4.2×10^5 amount of meshing can obtain the best result, which meshing quality in terms of average orthogonal quality and skewness are around 0.75 and 0.25 respectively. This meshing quality is convincing in getting accurate results.

Table 1 Levels of independent variables for the L9 orthogonal array

Parameters	Level 1	Level 2	Level 3
Type of nanofluid	Al_2O_3-water	CuO-water	TiO_2-water
Concentration of nanofluid (%)	0.5	1	2
Reynolds number	250	750	1500
Heat flux (W/m^2K)	10,000	20,000	50,000

To take into account the thermo-physical properties of the nanofluid materials such as aluminium oxide—water, copper oxide—water and titanium oxide—water, C language code user-defined functions (UDF) is compiled and incorporated to the model. The numerical model was solved with Semi-Implicit Pressure-Linked Equation (SIMPLE) algorithm, least square cell based for gradient, second-order upwind for pressure, first-order upwind for momentum and energy. Meanwhile for the convergence criteria for relative residuals of 10^{-6}, around 100–250 iterations were needed to run each simulation. Each run took 5–15 min on a workstation with a quad-core processor (2.80 GHz) and 8 GB of RAM.

In this study, four parameters determining the performance of the nanofluid in parallel cooling channel are evaluated, i.e., type of nanofluid, concentration of nanofluid, Reynolds number and heat flux. Four parameters are studied for each level as tabulated in Table 1. Taguchi method with the help of using Minitab software is applied in analysing the required runs for the simulation [24]. The best combination of parameters that will be used in this study is L9 orthogonal array. The L9 orthogonal array is meant for understanding the effect of 4 independent factors each having 3 factor level values. This array assumes that there is no interaction between any two factors. Taguchi method is used to control sensitivity of each parameter and determine the perfect combination of the design factors for parallel cooling channel (Table 2).

Table 2 Experimental design matrix

Run No	Type of nanofluid	Concentration of nanofluid (%)	Reynolds number	Heat flux (W/m^2K)
1	Al_2O_3-water	0.5	250	10,000
2	CuO-water	1	250	20,000
3	TiO_2-water	2	250	50,000
4	CuO-water	0.5	750	50,000
5	TiO_2-water	1	750	10,000
6	Al_2O_3-water	2	750	20,000
7	TiO_2-water	0.5	1500	20,000
8	Al_2O_3-water	1	1500	50,000
9	CuO-water	2	1500	10,000

4 Results and Discussion

The numerical simulations were carried out for eight channel designs under different mass flow rates which represents different laminar Reynolds number ranging from 250 to 1500. After deciding the best channel design in terms of pressure drop and uniformity of heat distribution, Taguchi method was carried out on the best design with different key parameters such as type of nanofluid, concentration of nanofluid, mass flow rate and heat flux. The best condition of achieving the lowest pressure drop and the lowest standard deviation of temperature which represents the best uniformity of heat distribution were analysed with *S/N* ratio graph.

4.1 Model Validation

In this study, the numerical model is validated with empirical correlation from Cengel [25]. Figure 3 shows good agreement between numerical model and empirical correlation of the outlet temperature with regards to the Reynolds number. It was found that the relative error for the outlet temperature and Reynolds number is less than 5%. As the Reynolds number increases, the outlet temperature will decrease by a few points. This phenomenon illustrates that the Reynolds number has a significant contribution for the temperature distribution inside the studied model. However, both studies give the stable trend on each *Re* and outlet temperature starting from *Re* of 250–1500. This result gives strong justification that the developed model correctly account for heat transfer distribution of nanofluid in parallel cooling channels.

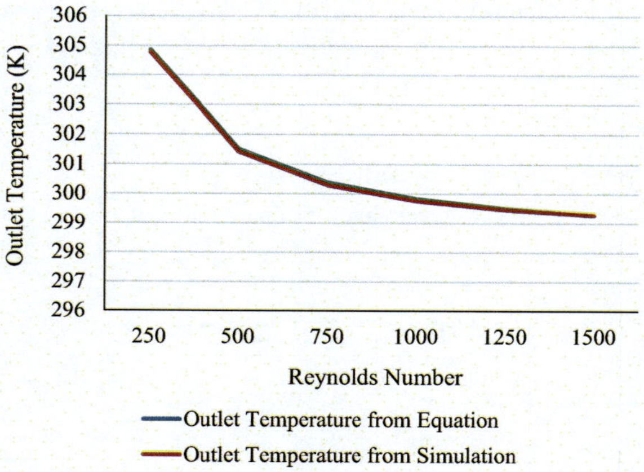

Fig. 3 Outlet temperature under different Reynolds number—comparison between present model and correlation proposed by Cengel [25]

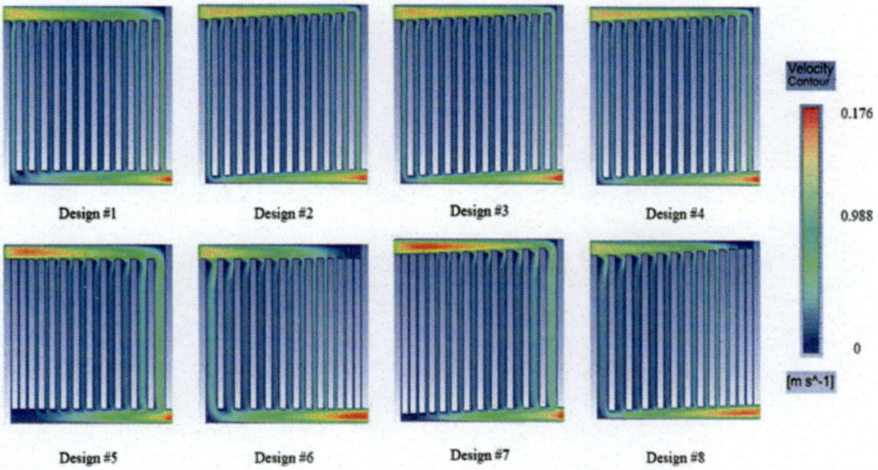

Fig. 4 Velocity contour of different channel designs where water is the cooling fluid

4.2 Effect of Different Channel Designs

4.2.1 Water

The velocity contour of different channel designs is presented in Fig. 4. It is observed that each of parallel channel gives different fluid flow distribution with water as a cooling fluid. Each of design shows the highest velocity flow for parallel channel is located at the inlet and outlet area. Nevertheless, there is a considerable difference observed on different parallel channels where the velocity flow in design #5, #6, #7 and #8 could get faster on the top of the smaller parallel channels. Figure 5 displays the temperature profile at the surface of cooling channel of different channel designs. It is observed that each of parallel channel gives different temperature distribution with water as a cooling design. There are no significant differences for the temperature profile in design #1, #2, #3 and #4 while others design show major contributions to the temperature profile. It can be seen that the highest temperature contour in design #5, #6, #7 and #8 is from the highest velocity flow which is located at the top of the smaller channel. It proves that the velocity distribution could make a positive impact on the convective heat transfer in parallel cooling channel.

4.2.2 Nanofluid

The velocity contour of different channel designs with nanofluid as working fluid is illustrated in Fig. 6. It is observed that each of parallel channel gives different fluid flow distribution with 0.5% concentration of Al_2O_3-water as a cooling fluid. Each of design shows the highest velocity flow for parallel channel is located at the inlet and

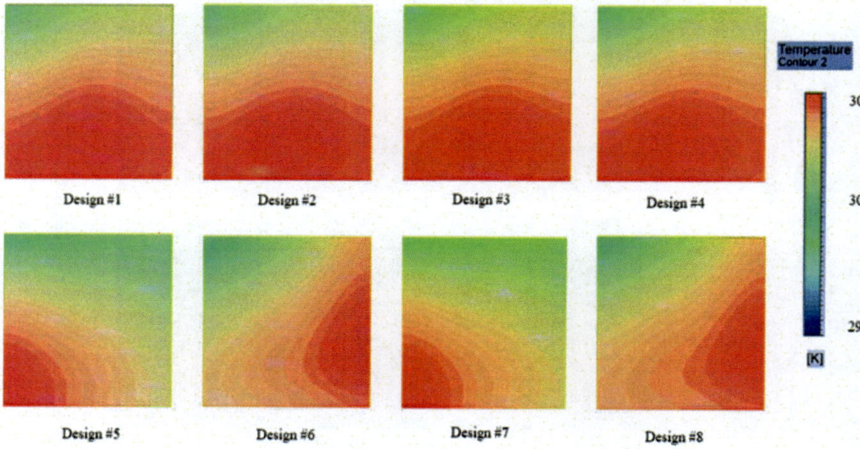

Fig. 5 Temperature profile at the surface of cooling channel of different channel designs where water is the cooling fluid

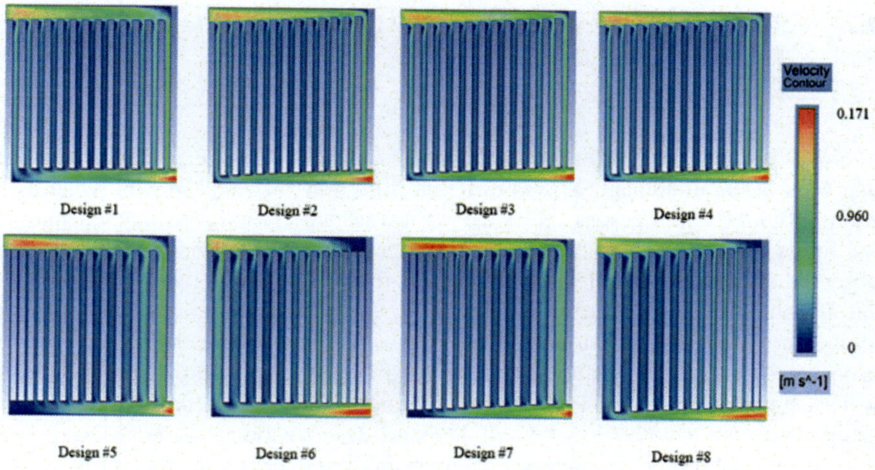

Fig. 6 Velocity contour of different channel designs where 0.5% concentration of Al_2O_3-water nanofluid is the cooling fluid

outlet condition. However, there is an interesting part on parallel channel where the velocity profile is not strongly changed by the supplementary nanoparticle suspension, especially at low concentrations. This is because of the nanofluid suspension does not considerably alter viscosity of the fluid [23]. Despites of that nanofluid effect, the velocity distribution in design #5, #6, #7 and #8 have a substantial performance since the velocity flow could get faster on the upper of the smaller channels. Figure 7 presents the temperature profile at the surface of cooling channel of different

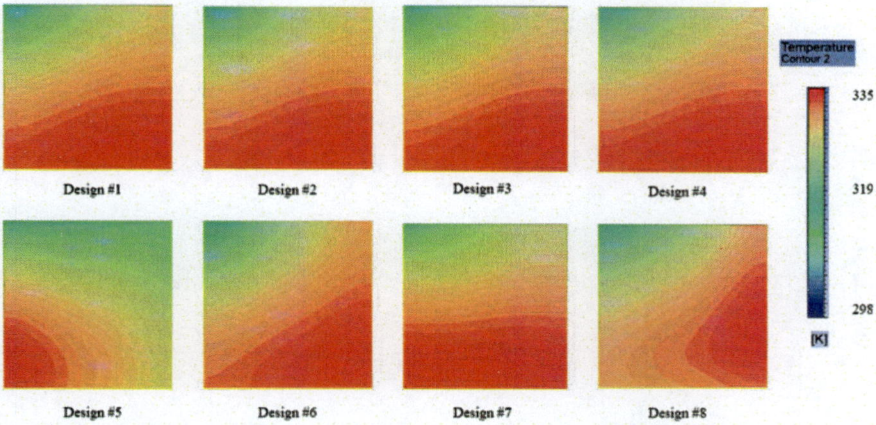

Fig. 7 Temperature profile at the surface of cooling channel of different channel designs where 0.5% concentration of Al_2O_3-water nanofluid is the cooling fluid

channel designs. It is observed that each of parallel channel gives different temperature distribution with 0.5% concentration of Al_2O_3-water as a cooling fluid. This is due to nanoparticles concentration could enhance the thermal conductivity of the water. There are no substantial differences for the temperature profile in design #1, #2, #3 and #4 while others design indicates major contributions to the temperature profile. Hence, the temperature profile in design #5, #6, #7 and #8 is more uniformly distributed across the channel due to the velocity distribution on these designs are the highest one. It confirms that the velocity distribution with the additional of nanoparticle concentration has a strong correlation with the convective heat transfer in parallel cooling channel.

4.3 Optimization

Previously, we have discussed the effect of different channel design on the heat transfer performance of parallel cooling channel with and without nanofluid. In this section, we will further evaluate the best combination of operating parameters to generate the minimum pressure drop and maximum uniformity of heat distribution. The signal to noise (*S/N*) ratio for each parameter is displayed in Fig. 8 with the options is set to "smaller is better". According to the both *S/N* ratio, it is observed that the optimum combination of key parameters for the minimum pressure drop is condition of *Re* 250, 1% concentration of CuO-water nanofluid under 50,000 W/m^2K of heat flux. Hence, this condition works the best for electronic microchip which generally generates 50,000 W/m^2K of heat flux. Nonetheless, the optimum combination of key parameters for the maximum uniformity of heat distribution is condition of *Re* 1500, 1% concentration of Al_2O_3-water nanofluid under 10,000 W/m^2K of heat

a

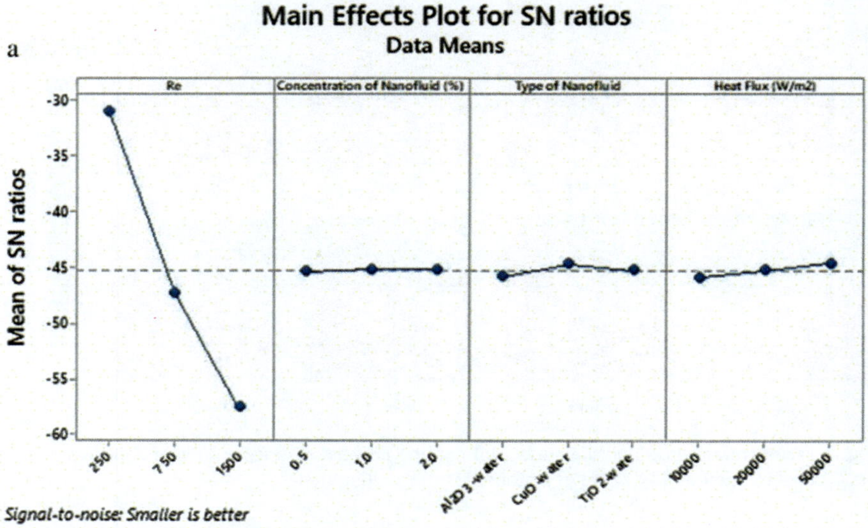

b

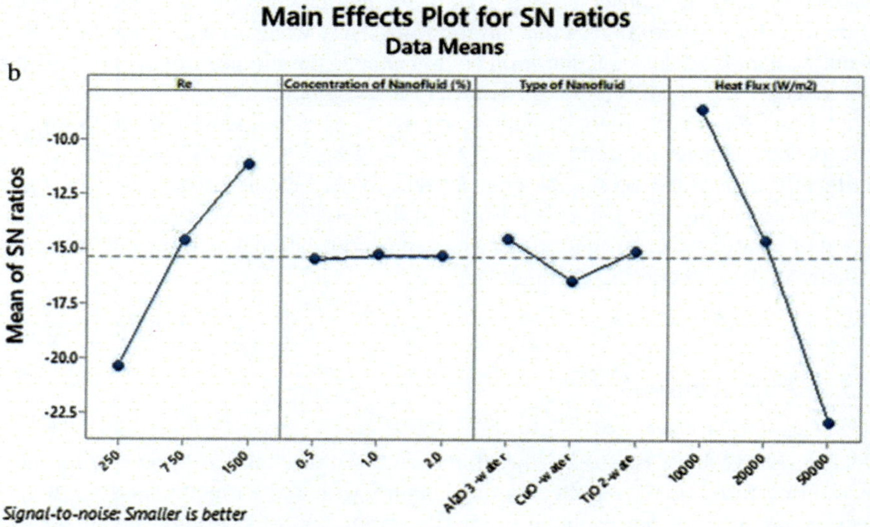

Fig. 8 *S/N* response graph of various operating parameters with respect to **a** pressure drop and **b** standard deviation of temperature

flux. This result indicates that the maximum uniformity of heat distribution achieved when the heat flux is in the minimum condition. For the heat flux of 10,000 W/m^{2}K, it is normally generated by minimum density electronic devices or fuel cell. In other words, the specific options of 10,000 W/m^{2}K in heat flux is still hard to achieve uniform heat distribution for electronic cooling microchip.

5 Conclusion

Numerical investigation of laminar convective heat transfer of nanofluid in parallel cooling channels has been conducted. Various parallel channel design for heat transfer performance of nanofluid as a cooling fluid have been evaluated and discussed. Taguchi method is used to determine the best configurations of key parameters to achieve the minimum pressure drop and maximum uniformity of heat distribution. The results indicate that the conventional parallel channel still yields the minimum pressure drop compared to the other seven modified parallel channels. However, the design #5, #6, #7 and #8 are the best configurations to achieve the maximum convective heat transfer. The addition of nanoparticles on the base fluid on these designs results in better convective heat transfer. Meanwhile, the combination of key parameters to achieve the minimum pressure drop is *Re* 250, 1% concentration of CuO-water nanofluid with heat flux of 50,000 W/m^2K and the key parameters to achieve the maximum uniformity of heat distribution is *Re* 1500, 1% concentration of Al_2O_3-water nanofluid with heat flux of 10,000 W/m^2K.

Acknowledgements The authors gratefully acknowledge the facility and financial support from the Ministry of Higher Education Malaysia through Fundamental Research Grant Scheme (FRGS) no FRGS/1/2018/TK07/UTP/02/8 (CS: 015MA0-028).

References

1. Schelling PK, Shi L, Goodson KE (2005) Managing heat for electronics. Mater Today 8(6):30–35
2. Murshed SMS (2016) Introductory chapter: electronics cooling—An overview. Electronics Cooling
3. Venerus DC et al (2010) Viscosity measurements on colloidal dispersions (Nanofluids) for heat transfer applications. Appl Rheol 20(4)
4. Hwang Y, Lee JK, Lee CH, Jung YM, Cheong SI, Lee CG, Ku BC, Jang SP (2007) Stability and thermal conductivity characteristics of nanofluids. Thermochim Acta 455(1):70–74
5. Lee JH, Hwang KS, Jang SP, Lee BH, Kim JH, Choi SUS, Choi CJ (2008) Effective viscosities and thermal conductivities of aqueous nanofluids containing low volume concentrations of Al2O3 nanoparticles. Int J Heat Mass Transf 51(11):2651–2656
6. Eastman JA, Choi SUS, Li S, Yu W, Thompson LJ (2001) Anomalously increased effective thermal conductivities of ethylene glycol-based nanofluids containing copper nanoparticles. Appl Phys Lett 78(6):718–720
7. Yu W, Xie H, Chen L, Li Y (2009) Investigation of thermal conductivity and viscosity of ethylene glycol based ZnO nanofluid. Thermochim Acta 491(1):92–96
8. Ali HM, Sajid MU, Arshad A (2017) Heat transfer applications of TiO_2 nanofluids. Application of Titanium Dioxide
9. Dharmalingam R, Sivagnanaprabhu KK, Kumar BS, Thirumalai R (2014) Nano materials and nanofluids: an innovative technology study for new paradigms for technology enhancement. Procedia Eng 97:1434–1441
10. Sajid MU, Ali HM (2019) Recent advances in application of nanofluids in heat transfer devices: a critical review. Renew Sustain Energy Rev 103:556–592

11. Ghani IA, Che Sidik NA, Kamaruzzaman, Jazair Yahya NW, Mahian O (2017) The effect of manifold zone parameters on hydrothermal performance of micro-channel HeatSink: a review Int J Heat Mass Transf 109:1143–1161
12. Abubakar SB, Sidik NAC, Ahmad AS (2016) The use of Fe_3O_4-H_2O_4 nanofluid for heat transfer enhancement in rectangular Microchannel Heatsink 23(10)
13. Bajura RA, Jones EH (1976) Flow distribution manifolds. J Fluids Eng 98(4):654–665
14. Acrivos A, Babcock BD, Pigford RL (1959) Flow distributions in manifolds. Chem Eng Sci 10(1–2):112–124
15. Razali AA, Sadikin A (2015) A CFD simulation study on pressure drop and velocity across single flow microchannel heat sink 8(10)
16. Amador C, Gavriilidis A, Angeli P (2004) Flow distribution in different microreactor scale-out geometries and the effect of manufacturing tolerances and channel blockage. Chem Eng J 101(1):379–390
17. Liu H, Li P, Lew JV (2010) CFD study on flow distribution uniformity in fuel distributors having multiple structural bifurcations of flow channels. Int J Hydrogen Energy 35(17):9186–9198
18. Wang J (2011) Theory of flow distribution in manifolds. Chem Eng J 168(3):1331–1345
19. Manoj Siva V, Pattamatta A, Das SK (2014) Effect of flow maldistribution on the thermal performance of parallel microchannel cooling systems. Int J Heat Mass Transf 73:424–428
20. Wang J (2010) Pressure drop and flow distribution in parallel-channel configurations of fuel cells: Z-type arrangement. Int J Hydrogen Energy 35(11):5498–5509
21. Esmaili Q, Ranjbar AA, Porkhial S (2018) Experimental analysis of heat transfer in ribbed microchannel. Int J Therm Sci 130:140–147
22. Song JY, Hah S, Kim D, Kim SM (2019) Enhanced flow uniformity in parallel mini-channels with pin-finned inlet header. Appl Thermal Eng 152:718–733
23. Sasmito A, Kurnia J, Mujumdar A (2011) Numerical evaluation of laminar heat transfer enhancement in nanofluid flow in coiled square tubes. Nanoscale Res Lett 6(1):376
24. Kong T. Taguchi methods in experimental design 10
25. Cengel YA (2004) Heat transfer: a practical approach, 2nd edn. McGraw-Hill, New York

Modeling and Validation of Engine Cylinder Pressure During Starting of Free-Piston Linear Generator by Brushless Motor Commutation

Saiful A. Zulkifli, Mohd Syaifuddin Mohd, and Abd Rashid Abd Aziz

Abstract This paper describes mechanical modeling of a specific configuration of free-piston linear engine-generator (FPLG) during starting, by brushless commutation of linear motor. To account for piston blow-by, or air leakage through piston rings at low starting speeds, the model incorporates an air mass transfer algorithm for the air compression-expansion process in the engine cylinders. The model is verified with experimental data. Pressure profiles obtained during experimental validation of the FPLG are discussed, for different motoring conditions. The improved model shows a reduction of 28% in compression pressure compared to the ideal case. In single-stroke tests, the airflow corrective factor value of $C = 3.7 \times 10^{-7}$ is found to give the closest profile to experimental pressure for the right cylinder and $C = 5.5 \times 10^{-7}$ for the left cylinder, while $C = 4.1 \times 10^{-7}$ best matches the data of cyclic motoring tests. Results show that by increasing motoring voltage, the piston's amplitude remains unchanged, since increasing motoring energy leads to better compression-expansion process, more effective air spring and a pressure profile closer to the lossless system.

1 Introduction

A two-stroke, dual-cylinder free-piston engine (FPE) is coupled to a linear electric generator to produce a free-piston linear engine-generator (FPLG), for use as onboard powerhouse in series hybrid electric vehicle (HEV) or as portable power generator [1, 2]. A practical method to start the linear engine is to operate the machine as

S. A. Zulkifli (✉)
Electrical and Electronics Engineering Department, Universiti Teknologi PETRONAS, 32610 Seri Iskandar, Malaysia
e-mail: saifulazrin_mz@utp.edu.my

M. S. Mohd · A. R. Abd Aziz
Mechanical Engineering Department, Universiti Teknologi PETRONAS, 32610 Seri Iskandar, Malaysia

S. A. Zulkifli · M. S. Mohd · A. R. Abd Aziz
Center for Automotive Research and Electric Mobility (CAREM), 32610 Seri Iskandar, Malaysia

F. Ahmad et al. (eds.), *ICPER 2020*, Lecture Notes in Mechanical Engineering,
https://doi.org/10.1007/978-981-19-1939-8_23

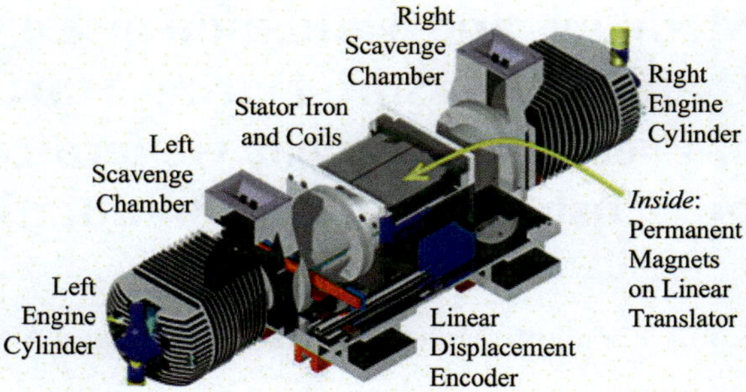

Fig. 1 A free-piston linear engine-generator (FPLG) prototype at *UTP*

a linear motor to reciprocate the piston-magnet-translator assembly between two cylinders to initiate combustion. However, due to the very large peak compression force during starting and limited current rating of stator coils, the translator cannot be moved end-to-end in a single stroke. A strategy has been proposed which utilizes the air-spring quality of the engine cylinders prior to combustion [3].

This work describes a mechanical model of the FPLG, consisting of an improved compression-expansion model representing the engine cylinders during starting—it incorporates air leakage through the piston rings at low piston speed. The model is verified with experimental data and pressure profiles are investigated for different starting conditions. Modeling and simulation are implemented on *MATLAB Simulink* while field experimentation is carried out using *National Instruments'* embedded controller and data acquisition and *LabView* Real-Time software. Figure 1 shows the FPLG prototype used in this study: a 5-kW system jointly developed by *Universiti Teknologi PETRONAS* (UTP), *Universiti Malaya* and *Universiti Kebangsaan Malaysia* [4, 5].

2 Resonance Strategy and Mechanical Modeling of FPLG Prototype

Mechanical aspects of FPLG starting and its modeling are discussed in this section.

2.1 Mechanical Resonance Strategy

In the absence of combustion, engine cylinders exhibit an air-spring behavior, so that at sufficient piston speed, air inside the cylinder compresses and expands as the piston moves into and out of the cylinder, absorbing and dissipating energy respectively. Thus, the cylinders behave as ordinary mechanical springs capable of storing and delivering energy, effectively creating a bounce phenomenon at each end of the stroke. Figure 2 shows how a dual-cylinder configuration of FPLG can

Fig. 2 Spring-mass representation and mechanical resonance process

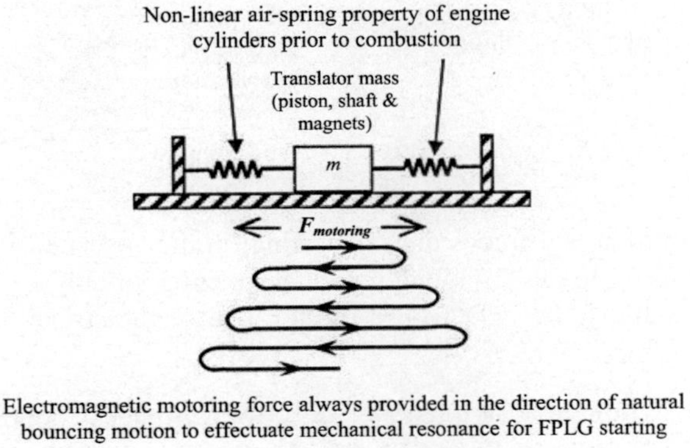

be likened to a spring-mass-spring system: a mass in the center sandwiched by two springs attached to a fixed reference [3].

If very little energy is lost in the bounce process, it is possible to apply motoring force of sufficient magnitude to reciprocate the piston assembly in small amplitudes initially. Over time, its amplitude and speed will grow—due to resonance—to achieve the final required stroke length (69 mm), speed (3–5 Hz cyclic frequency) and compression pressure (about 7–9 bars). However, there is one fundamental problem: at low starting speeds, the air-spring characteristic of an engine cylinder is heavily affected by piston speed. This is due to air leakage around the piston rings, referred to as piston blow-by. The slower the piston speed, the more is the quantity of air that leaks through, so that it becomes possible to manually push the piston assembly from end to end, although slowly due to the large compression force.

This is in contrast to ordinary mechanical spring, whose spring force depends only on displacement and not on speed of moving mass. For the FPLG, the dependence on piston speed due to piston blow-by affects the cylinders' effectiveness in the reciprocation and starting process [3]. Thus, the lower the motoring force, the lower is piston speed and the less effective is the bounce process as more energy is lost during the bounce. At the final required starting frequency of 3–5 Hz, the piston is still operating in the low-speed region where compression-expansion process is much affected by speed. Effectiveness of the air spring is thus very important during starting and is assessed in this work through analysis of the engine cylinder pressure profiles, via simulation and field experimentation.

2.2 *Mechanical Modeling of FPLG*

Modeling of FPLG requires identifying mechanical forces during starting. In the present work, motoring force appears as a mechanical force contributing to the total net force—the electrical process of current injection responsible for creating the force is not considered. Figure 3 shows a simplified free-body diagram indicating

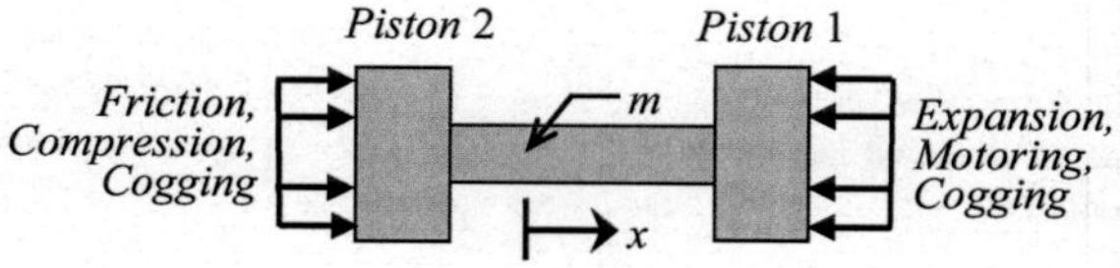

Fig. 3 FPLG free-body diagram during starting process

translational forces during starting: in the absence of combustion, the translator is subject to the following forces, neglecting vibration, as it moves from the right-end of the stroke (TDC 1) to the left-end of the stroke (TDC 2) (from Cylinder 1 to Cylinder 2) [3, 6, 7].

1. Compression force $F_{compression}$ of Cylinder 2, opposing translator motion,
2. Expansion force $F_{expansion}$ (suction force for the initial stroke) of Cylinder 1, assisting translator motion (opposing motion for the initial stroke),
3. Friction force f between piston ring and cylinder liner and between translator shaft and linear bearing,
4. Magnetic cogging force $F_{cogging}$ pushing or pulling on the translator, and
5. Electrical motoring force $F_{motoring}$ which should be in the same direction as translator motion in order to effectuate a successful resonating strategy.

Figure 4 shows a schematic diagram indicating the intake and exhaust ports, and a graph of the mechanical forces versus displacement, for the 69-mm total stroke length. Compression and expansion forces are due to air pressure acting on the piston surface. Magnetic cogging force results from static interaction between the permanent magnets' magnetic field and the iron-cored stator. Depending on translator position, cogging force may be positive or negative, thus resisting or assisting translator motion. Figure 5 shows the cogging force over the entire stroke, obtained via finite-element analysis performed by the *Universiti Malaya* team. Due to symmetry of FPLG design, the profile is symmetrical with respect to the reference center position. It has zero values at certain positions of the stroke, around which are stable rest positions in the absence of external force.

Instantaneous values of friction cannot be obtained accurately, due to hardware limitation. For the purpose of the present analysis, a fixed value of 230 N is used, acquired from a relatively simple but reliable set of experiments. Adjustments to this value are made in the later stages of validation and analysis. Cogging and friction are static forces with no dependency on translator speed, while compression and expansion forces are dynamic—have a heavy speed dependency for low cyclic speed. For an ideal gas, both compression and expansion processes are governed by the same relationship between pressure and volume inside an engine cylinder when the exhaust port is closed.

$$P_1 V_1^k = P_2 V_2^k \tag{1}$$

where P is pressure, V is volume and subscripts *1* and *2* denote instantaneous values of pressure or volume at different times or displacements. The constant k

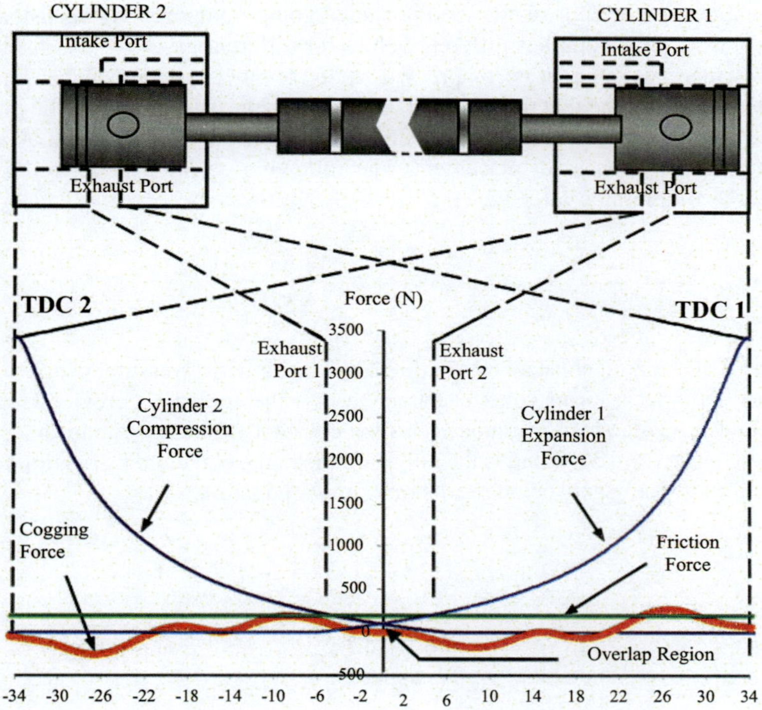

Fig. 4 FPLG schematic and mechanical forces versus displacement

Fig. 5 Magnetic cogging force versus displacement

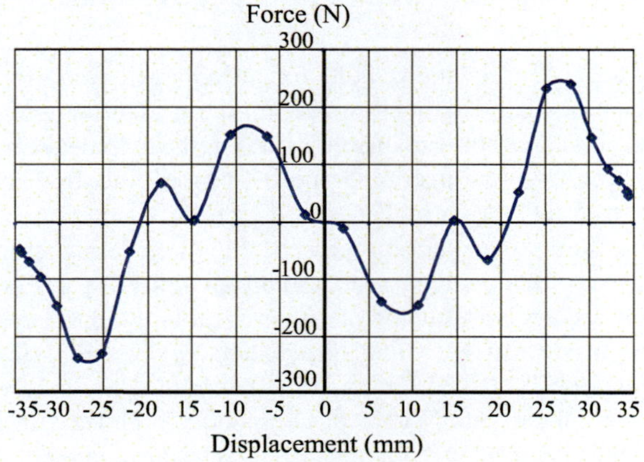

is the adiabatic constant of the medium undergoing compression-expansion—air in the present case—and has different values for different gases[1] [3]. Consider the case in which one cylinder is compressed while the other cylinder's exhaust port is already open. Assume that the resultant compression pressure acting on the piston's surface is uniform across the piston's surface area and can thus be taken as a one-dimensional function of displacement. The following equation is derived for the cylinder's compression force [2, 8]:

$$F_{compression}(x) = K_1 \cdot \left(\frac{K_2}{x+l}\right)^k \tag{2}$$

where K_1 and K_2 are constants determined by atmospheric pressure, piston surface area and cylinder trapped volume. Parameter l is the equivalent crevice length of the cylinder head and is thus another system constant, while x is the instantaneous piston distance from TDC and is the only variable in the equation. Derivation for the cylinder undergoing expansion results in a similar equation.

$$F_{expansion}(x) = K_3 \cdot \left(\frac{K_4}{x+l}\right)^k. \tag{3}$$

From the expressions of $F_{compression}$ and $F_{expansion}$ above, non-linearity of these forces with respect to displacement is apparent.

2.3 Improved Model for Compression-Expansion Force

The above equations hold for an adiabatic and isentropic process in which no heat or mass is gained or lost. In the present system which involves piston rings, air compression-expansion inside engine cylinders cannot be considered isentropic because at low operating speeds, piston blow-by occurs: air leakage through the piston rings from higher-pressured to lower-pressured region. Thus, the ideal relationship above is not sufficient for FPLG modeling during starting. The compression-expansion model must incorporate an air mass transfer algorithm to account for the air leakage, resulting in less-than-ideal in-cylinder pressure, as shown in Fig. 6.

In order to determine the actual energy stored during the compression stroke, the amount of air leakage must be determined accurately. The major leakage path is through the piston ring gap. The amount of air leakage at any given speed and load is controlled primarily by the airflow resistance in the flow path between the combustion chamber and scavenging chamber. The amount of air leakage is also inversely proportional to the operating speed, so that as speed increases, the time period of

[1] Several factors affect slightly the value of k, which are assumed negligible and thus ignored in the present modeling of FPLG. The constant value of k used in this study is 1.38.

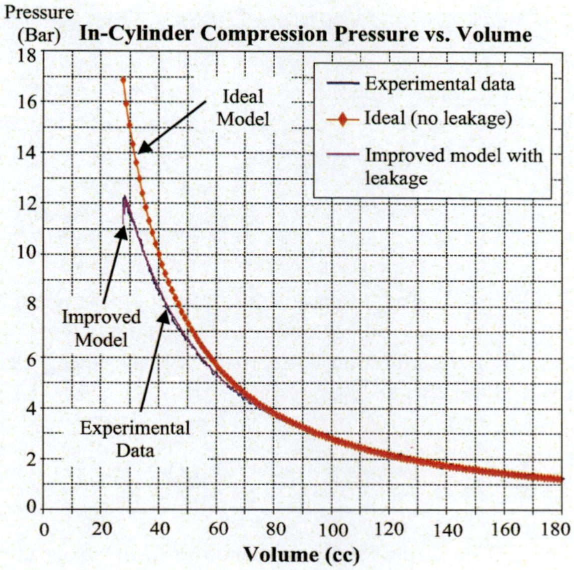

Fig. 6 Comparison between experimental data, ideal model and improved compression-expansion model

half cycle decreases. Thus, the amount of air leakage can be calculated as the flow of gas through a nominal cross-sectional passage of a given area. By incorporating an airflow corrective factor, C and time period, Δt, the effect of operating speed on air leakage can be included.

$$m_{loss} = Q * \rho * C * \Delta t \tag{4}$$

m_{loss}	mass of air leakage
Q	volumetric flow rate
ρ	calculated gas density
C	airflow corrective factor
Δt	time period of half cycle

This improved model has been correlated and validated with experimental data. Figure 6 also shows simulation results of the final validated model, compared to experimental data of cranking the FPLG at 440 cycles/minute, along with the ideal case without leakage. The improved model shows a difference (reduction) of 28% in compression pressure compared to the ideal case. Since force is pressure multiplied by piston area, this reduction translates to a difference of more than 2 kN. Thus, if leakage were not accounted for, simulation results would not be valid.

2.4 Dynamic Equation During Starting Process

For mechanical equation of FPLG during starting, Newton's second law of motion is used. Let m be the total mass of the moving translator (mass of shaft, pistons and magnets) and a its acceleration.

$$\sum F_x = ma_x = m\frac{d^2x}{dt^2} \tag{5}$$

During starting, motion is possible along a single axis only (x), ignoring vibration along other axes. Incorporating the mechanical forces described above, we thus have the following equation to represent FPLG during starting, for a translator motion from TDC1 to TDC2 (Fig. 4).

$$\sum F_x = ma_x = m\frac{d^2x}{dt^2} = F_{mot}(\mathrm{x}) + F_{exp}(x) - F_{comp}(x) - F_{cog}(x) - f \tag{6}$$

F_{mot} (x) represents motoring force resulting from current injection into the FPLG coils. If we let the motoring force to be constant-magnitude, we have F_{mot} $(x) = F_{mot}$. From (2) and (3), assuming adiabatic and isentropic process and $x = 0$ at TDC 2 (Fig. 4), F_{comp} (x) and F_{exp} (x) are given by [9, 10]:

$$F_{comp}(x) = K_1 \cdot \left(\frac{K_2}{x + l}\right)^k \tag{7}$$

$$F_{exp}(x) = K_3.\left(\frac{K_4}{L - x + l}\right)^k \tag{8}$$

where K_1, K_2, K_3, K_4, L and l are all system constants. L is the total stroke length from TDC2 to TDC1. Thus, the dynamic equation for FPLG becomes:

$$F_{mot} + F_{exp}(x) - F_{comp}(x) - F_{cog}(x) - f = m\frac{d^2x}{dt^2} \tag{9}$$

$$F_{mot} + K_3.\left(\frac{K_4}{L - x + l}\right)^k - K_1 \cdot \left(\frac{K_2}{x + l}\right)^k - F_{cog}(x) - f = m\frac{d^2x}{dt^2} \tag{10}$$

Considering non-linearity of F_{cog} (x), $F_{comp}(x) = K_1 \cdot \left(\frac{K_2}{x+l}\right)^k$ and $F_{exp}(x) = K_3.\left(\frac{K_4}{L-x+l}\right)^k$, it is not possible to solve the above equation for x in closed algebraic form; even worse if velocity dependency of compression-expansion due to air leakage is incorporated into F_{comp} (x) and F_{exp} (x) and if the motoring force is not constant but having some relation to other system parameters.

3 Electrical Motoring for Starting of FPLG and Brushless Linear Motor Operation

To provide for the force to reciprocate the piston assembly, the FPLG is operated as a brushless, permanent-magnet linear motor. Current is injected into the stator coils creating a magnetic field which interacts with the existing magnetic field of the permanent magnets on the translator shaft. A motor force is created which then pushes on the translator shaft in a certain magnitude and linear direction depending on the relative position of the permanent magnets with respect to the fixed stator coils [11, 12]. Figure 7 shows a schematic diagram depicting production of the linear motoring force. The hardware responsible for current injection consists of a common type of motor drive used in rotating brushless motor application: 3-phase, full-wave MOSFET voltage-source inverter bridge. Figure 7 shows the complete electrical circuit consisting of the inverter bridge, controller, gate driver and a 3-phase, Y-connected coils of FPLG [3, 13].

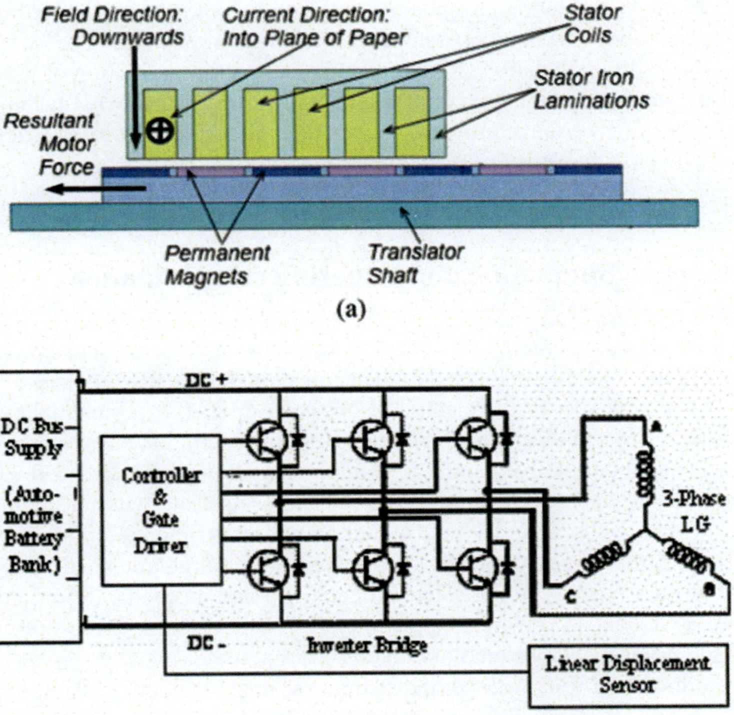

Fig. 7 **a** Interaction of magnetic fields to produce linear motoring force and **b** electrical circuit consisting of MOSFET inverter bridge, controller & gate driver and 3-phase FPLG

Commutation concerns the process of current injection: proper steering of current into the right coils at the right time by switching on and off certain drive transistors, based on translator position and electromagnetic K_V (or K_F) profiles of the three phases of FPLG. This relative position is critical to the effective motoring and starting of FPLG, as interaction between the two fields is different at different positions along the stroke of the translator. Two distinct commutation techniques are possible, namely, rectangular and sinusoidal commutation. Rectangular or block commutation consists of discrete on/off switching of the transistors at fixed, finite and pre-determined positions of the translator. Two variations of rectangular commutation studied in this work are 6-step and square-wave [3, 12].

Six-step commutation is the simplest and most commonly used in 3-phase brushless motor operation; at any one time, two out of the three phases are energized so that current is positive in one phase and negative in another, while the remaining third phase is floating (not energized.) Square-wave commutation offers slightly higher current levels with the same DC bus voltage since all three phases are energized, with two phases having one current direction while the remaining phase has the opposite current flow.

Sinusoidal commutation also involves energization of certain pairs of transistors based on displacement but it performs the switching at a very high rate (several kHz), using pulse width modulation (PWM) technique, to allow regulation of the injected current by varying the duty cycle of the energized transistors. This technique can be used to achieve certain control objectives such as current control or extended for motion and velocity control of the moving translator.

4 Motoring Simulation and Experimental Validation

The FPLG mechanical model discussed above is combined with an electrical model of FPLG to form an integrated electromechanical model for starting investigation of FPLG. After inclusion of the electrical subsystem model, experimental validation of both electrical and mechanical models using low excitation energy (low DC bus voltage—single automotive battery) are implemented in motionless coil switching tests, generated emf tests and single-stroke motoring tests without compression. Details of the electrical subsystem modeling and experimental validation can be found in [3]. Good correspondence is observed between simulation and experimental results but improvement of the electrical back emf and mechanical friction models are required to enhance the models' validity.

Further experimentation and simulation are implemented with higher DC bus voltage in motoring tests with compression, to validate the compression-expansion model and to understand FPLG system behavior when subjected to different energization level, besides to further analyze and refine the individual models. Single-stroke and cyclic motoring tests are carried out. Details of the test results, discussion and model refinement are provided in [3]. The following is a summary of the results.

4.1 Single-Stroke Motoring

Figure 8 shows pressure curves of single-stroke motoring with 6-step and square-wave commutation, using 24-V bus voltage (2 standard automotive batteries). Both exhibit generally unsmooth profile, due to current injection switching between coils and also cogging force. Peak pressure values are about the same, although they occur at slightly different locations—at the cutoff positions for current injection (27.5 mm for 6-step and 25 mm for square-wave). This and the slightly different shape of the pressure trace with respect to positions of stagger and kinks are solely due to the different switching positions between 6-step and square-wave commutation. Otherwise, all else are quite similar—at the DC bus level of 24 V, one commutation technique has no conceivable advantage over the other.

Figure 9 below shows validation of compression-expansion model for single-stroke motoring. The right and left graphs show compression pressure of the right cylinder and suction pressure of the left cylinder, respectively. The first curve of each graph is experimental pressure while the other curves are simulation profiles using different values of leakage factor, or airflow corrective factor represented by C in Eq. (4). In the development of the improved compression-expansion model that includes the air leakage effect, this constant parameter, C, which is closely related to the effective leakage area around the piston rings, is introduced. It is a very critical and sensitive parameter in the compression-expansion model which significantly affects simulation accuracy. Essentially, at any specific piston speed, the larger the value of the corrective factor, the more is the amount of air leakage.

Preliminary validation tests have found a value of $C = 4.5 \times 10^{-7}$ to give simulation results closest to experimental data. However, the value did not consistently

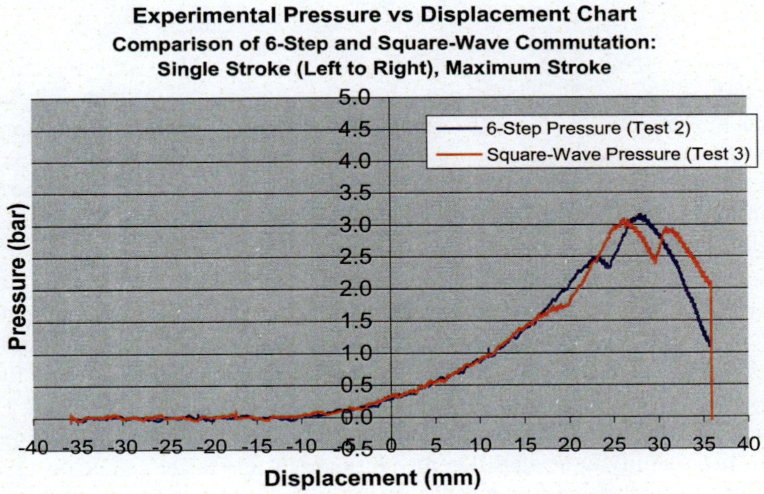

Fig. 8 Single-stroke motoring test for comparison of 6-step and square-wave commutation

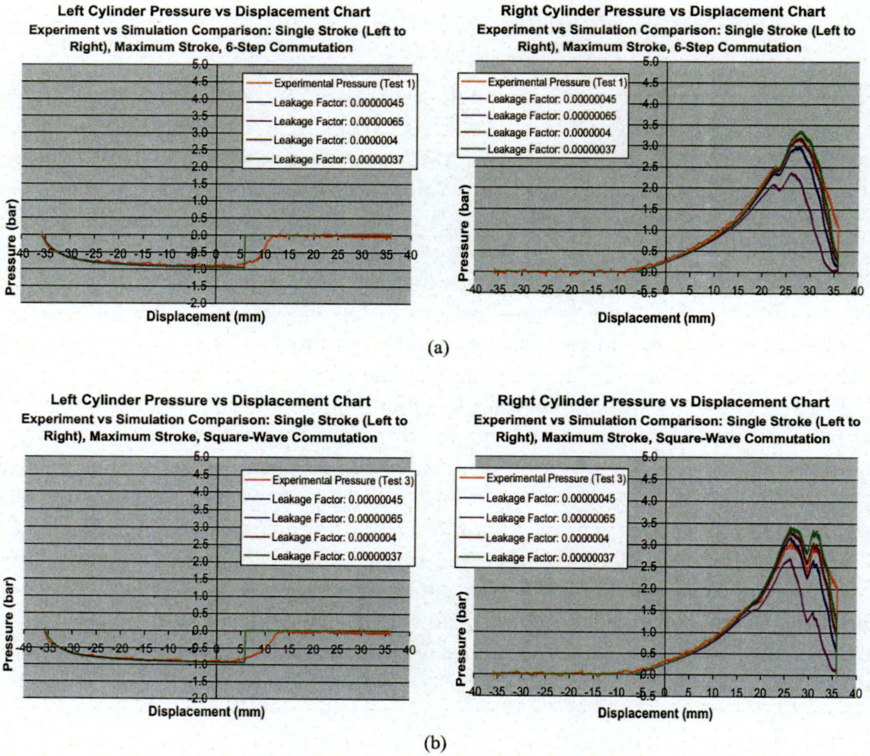

Fig. 9 Validation of compression-expansion model using single-stroke motoring with **a** 6-step commutation and **b** square-wave commutation

deliver comparatively accurate results for all the tests, which were implemented across a small range of motoring speed. In the present validation stage, it is discovered that the corrective factor, also termed leakage factor, is not a constant parameter of operation, as it should be according to the design of the compression-expansion model. In other words, the value discovered earlier does not give accurate results now; instead, it needs to be tuned to obtain the best simulation pressure profiles for the right cylinder.

As shown in the graphs, different values of leakage factor are used in the simulation. The trend is justifiable, indicating general validity of the compression model: the larger the leakage factor, the lower is the resultant compression pressure due to more pressure leakage. In the end, the value of $C = 3.7 \times 10^{-7}$ is found to give the closest profile to the experimental pressure. Graphs on the left side of Fig. 9 show suction pressure profiles of the left cylinder as the translator moves from left to right. Invariability to using different values of leakage factor can be observed in these graphs. Another weakness of the compression model is also seen: the experimental increase from negative pressure to atmospheric when the left cylinder port starts to open, is far from the sudden jump of the simulation profile (the infinite slope of simulation is expected as the mathematical model is a simple on–off relay for the port open/closure.) Thus, in future work, a finer model incorporating second-order fluid dynamic behavior would be needed to improve the model's accuracy.

4.2 Cyclic Motoring

Figure 10 shows cyclic motoring tests using a 24-V DC bus. Pressure leakage effect can be clearly seen from the difference in upward (compression) stroke and downward (expansion) stroke; if the situation were ideal (no leakage), these pressure curves should overlap. Pressure trace for upward (compression) stroke shows an overlap between the two commutation techniques; thus, even with higher motoring energy of square-wave commutation, the resultant compression pressure is no different from that of 6-step commutation.

Figure 11 shows validation of the compression-expansion model for cyclic motoring. As before, the parameter that requires tuning is the leakage factor, C. If the compression model is good, the value should be close to, if not equal to the previous value found for single-stroke motoring. As seen in Fig. 11, different values were tested and the same expectation of the relation between leakage factor and pressure profile is observed: a bigger leakage factor leads to a lower pressure profile. Note that only 6-step commutation is used in this validation process. The value of $C = 3.7 \times 10^{-7}$ found for the right cylinder in the previous single-stroke test works well with the present cyclic motoring case, as is the value of $C = 5.5 \times 10^{-7}$ found for the left cylinder. However, the value of $C = 4.1 \times 10^{-7}$ seems to match the data even better for the right cylinder.

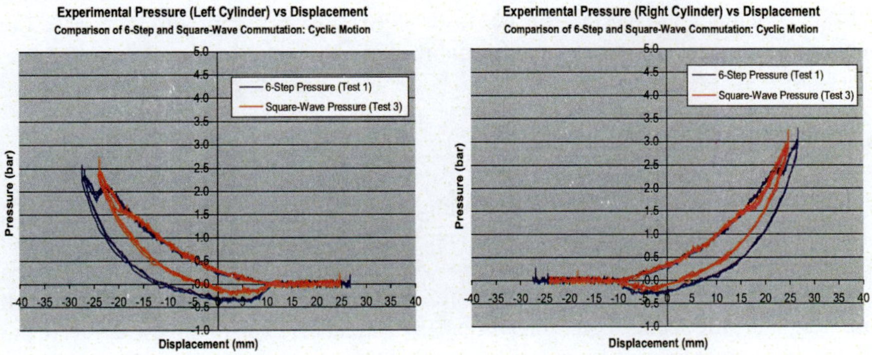

(a) (b)

Fig. 10 Cyclic motoring test for comparison of 6-step and square-wave commutation: **a** left cylinder pressure and **b** right cylinder pressure

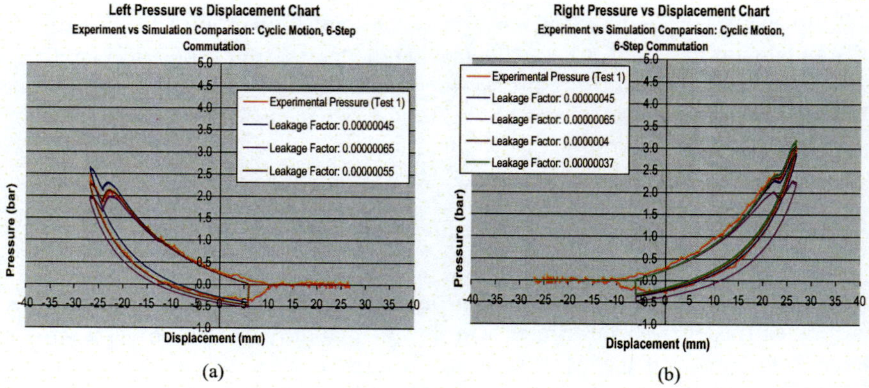

Fig. 11 Validation of compression-expansion model with 6-step commutation and cyclic motoring: **a** left cylinder pressure and **b** right cylinder pressure

4.3 Final Validation with Higher DC Bus Voltage

After considerable model refinement involving airflow corrective factor (leakage factor), friction and several electrical parameters,[2] the models are tested with higher DC bus voltage: 36 V and 48 V. Firstly, the two techniques of 6-step and square-wave commutation are tested experimentally with cyclic motoring to compare their performance with a higher DC bus, since there was no conceivable difference with 24-V excitation and single-stroke motoring (Fig. 8). Figure 12 provides experimental comparison for cyclic motoring with 48 V. A very minor gain in the pressure profiles is observed: the forward compression curve of square-wave shows a slightly higher pressure profile compared to 6-step. Also, the square-wave profile seems slightly 'tighter', indicating less air leakage during the bounce process. Nonetheless, the difference is insignificant that further testing of FPLG motoring will be carried out with 6-step commutation only, to test viability of FPLG starting strategy.

4.3.1 Single-Stroke Motoring.

Figure 13 compares experimental and simulation results using 24 V, 36 V and 48 V DC bus (2, 3 and 4 automotive batteries), for single-stroke motoring. A uniform increase in experimental pressure profiles with increase in bus voltage can clearly be seen, even in the displacement region before reaching peak pressure. Although each simulation profile seems lower than its experimental curve (up to 20% lower as in the 36 V bus) the general contours are captured in the simulation results (for example,

[2] Full details of model validation, refinement and analysis of the FPLG system under various DC bus excitation are provided in [3].

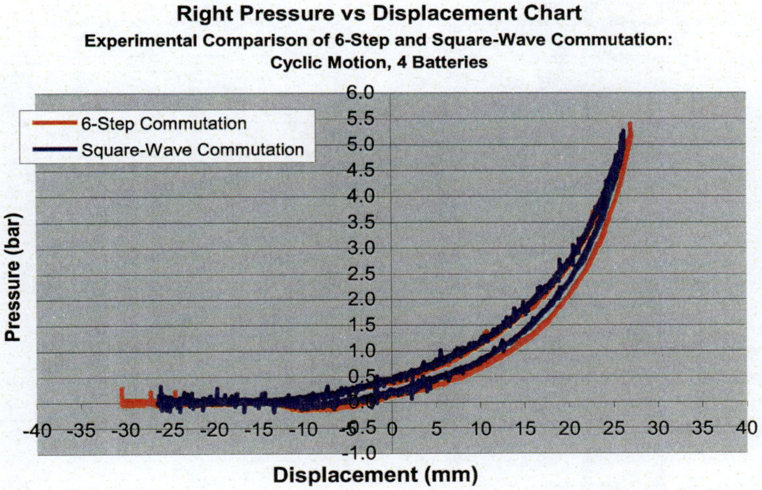

Fig. 12 Experimental comparison of 6-step and square-wave with cyclic motoring and 48 V

the slight dip in the 24 V profile) and the same increasing trend with increasing DC bus voltage is observed.

Both experimental and simulation profiles are smoother compared to the 24 V profiles: with more motoring energy, the FPLG can better overcome the forces of cogging and friction, resulting in smoother profiles. Both experimental and simulation pressure show upward and downward strokes that are closer to each other than the 24 V curves, which is also due to the higher motoring energy. This is due to reduced air leakage in the compression-expansion process.

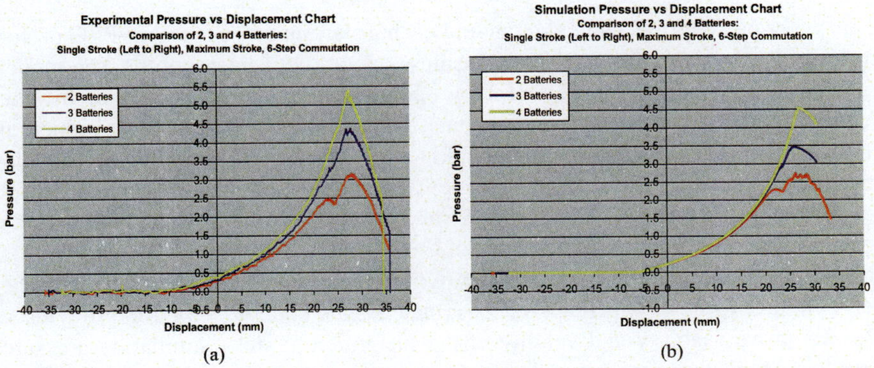

Fig. 13 Right-cylinder pressure profiles with single-stroke motoring, 6-step commutation and various bus voltage: **a** experimental and **b** simulation results

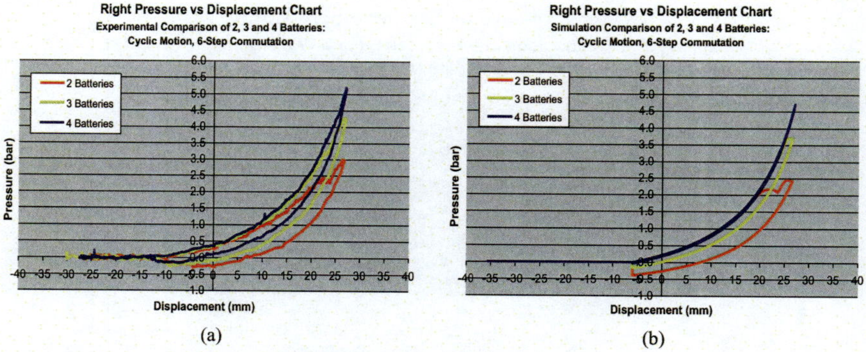

Fig. 14 Right-cylinder pressure profiles with cyclic motoring, 6-step commutation and various bus voltage: **a** experimental and **b** simulation results

4.3.2 Cyclic Motoring

Figure 14 compares experimental and simulation results using 12 V, 36 V and 48 V DC bus (2, 3 and 4 automotive batteries), for cyclic motoring. As in the single-stroke case, the experimental pressure profiles continue to become 'tighter' (upward and downward strokes are closer to each other), as expected. In the simulation pressure graph with 48 V bus (4 batteries), the upward and downward strokes are so close that they appear to be a single profile, inferring very little air leakage during the compression-expansion process.

5 Conclusion

This paper presents mechanical modeling during starting process of an FPLG prototype, by mechanical resonance and brushless operation of linear motor. The model incorporates pressure leakage effect at low operating speed due to piston blow-by. The model is verified against experimental data. The improved model shows a reduction of 28% in compression pressure compared to the ideal case. Simulation runs to validate the integrated FPLG electromechanical model are implemented, using different levels of DC bus voltage, in single-stroke and cyclic motoring tests. In single-stroke tests, the airflow corrective factor value of $C = 3.7 \times 10^{-7}$ is found to give the closest profile to experimental pressure for the right cylinder and $C = 5.5 \times 10^{-7}$ for the left cylinder, while $C = 4.1 \times 10^{-7}$ best matches the data of cyclic motoring tests. Results prove usability and validity of the integrated model - simulation pressure profiles exhibit the same general contours as experimental pressure profiles. They show the same increasing trend in increasing the DC bus voltage. Although cyclic amplitude is not increased with increased bus voltage, the additional energy is spent in increasing the air spring's effectiveness of the engine cylinder. Higher motoring force leads to higher velocity which in turn leads to less pressure leakage, resulting in better compression and expansion performance and hence a more effective air spring character of the engine cylinder.

Acknowledgements This research work was supported by the Ministry of Science, Technology and Innovation (MOSTI), Malaysia under IRPA grant 03-99-02-0001 PR0025/04-01. Contributions from the previous FPLG team members from UTP, *Universiti Malaya* and *Universiti Kebangsaan Malaysia* are highly appreciated.

References

1. Annen KD, Stickler DB and Woodroffe J (2002) Miniature Internal Combustion Engine (MICE) for portable electric power. In: Proceedings of the 23rd army science conference, Florida
2. Hoff E, Brennvall JE, Nilssen R, Norum L (2004) High power linear electric machine - made possible by gas springs. In: Proceedings of the Nordic workshop on power and industrial electronics, Norway
3. Zulkifli SA (2007) Modeling, simulation and implementation of rectangular commutation for starting of free-piston linear generator. Thesis, Universiti Teknologi *PETRONAS*, Malaysia, M.Sc
4. Nor KM, Arof H, Wijono (2004) Design of a 5kW tubular permanent magnet linear generator. In: Proceedings of the international power engineering conference, Bristol, pp 528–532
5. Arof H, Eid AM, Nor KM (2004) On the issues of starting and cogging force reduction of a tubular permanent magnet linear generator. In: Proceedings of the Australasian universities power engineering conference (AUPEC2004), Brisbane
6. Cawthorne WR (1999) Optimization of a brushless permanent magnet linear alternator for use with a linear internal combustion engine, Ph.D. Thesis. West Virginia University, Morgontown
7. Houdyschell D (2000) A diesel two-stroke linear engine, Master's Thesis, West Virginia University, Morgontown
8. Aichlmayr HT (2002) Design considerations, modeling and analysis of micro-homogeneous charge compression ignition combustion free-piston engines, Ph.D. Thesis, University of Minnesota
9. Nemecek P, Sindelka M, Vysoky O (2004) Modeling and control of linear combustion engine. In: Proceedings of the IFAC symposium on advances in automotive control, pp 320–325
10. Vysoky O (2003) linear combustion engine - control and application. In: 10th anniversary of the foundation of the faculty transportation sciences, Prague
11. Fazal I, Karsiti MN, Zulkifli SA, Ibrahim, T, Rama Rao KS (2010) Modeling and simulation of a moving-coil linear generator. In: Proceedings of the international conference on intelligent and advanced systems (ICIAS 2010), Kuala Lumpur, Malaysia
12. Abidin EZZ, Ibrahim AA, Aziz ARA, Zulkifli SA (2015) Effect of motoring voltage on compression ratio of a free-piston linear generator engine. J Mech Eng Sci 8:1393–1400
13. Zulkifli SA, Mohd S, Saad N, Aziz ARA (2014) Impact of motor size & efficiency on acceleration, fuel consumption & emissions of split-axle through-the-road parallel hybrid electric vehicle. Appl Mech Mater 663:498–503

Influence of Twisted Tape Inserts with Perforation on Heat Transfer and Pressure Drop Inside Circular Tube: Numerical and Experimental Investigation

Sahira Hasan Ibrahim and Hasanain A. Abdul Wahhab

Abstract Active heat transfer enhancement by inserts is attributed to imposed turbulence in the pipe flow. This work describes a numerical and experimental analysis to study the effect of twisted tape inserts as turbulence promoters to enhance heat transfer inside a circular tube heated with a constant heat flux. An experimental setup has been manufactured, and the aluminium pipe chosen of 1500 mm length, with aspect ratio, L/D = 60 which is heated from the outer surface with constant heat flux. Four different twisted tapes of constant width of 20 mm and twisted ratio of 3.15, with different central-circular holes of 2, 4, 6, and 8 mm diameter were used to investigate the effect of tape perforation on heat transfer coefficient and fraction factor at Reynolds numbers of range 4.5–12.5 $\times$ 10^3. The numerical analysis was carried out using ANSYS 19.0 FLUENT with k-ε Turbulent Model. The results show that the larger values of average Nusselt number were obtained for ordinary twisted tape is about 1.96 times more than the plain tube at Reynolds numbers of 12.5 $\times$ 10^3.

1 Introduction

For decades, interest in developing techniques of increasing heat transfer rates, and using them widely in the design and manufacture of heat exchangers, has been increased [1]. Among the various augmentation techniques, Heat exchanger are used in different processes ranging from conversion radiation and recovery of thermal energy in various industrial, commercial and domestic applications, some common examples include steam generation and condensation in power and cogeneration plants sensible heating and cooling in thermal pressing of chemical, pharmaceutical and agricultural products, fluid heating in manufacturing and waste heat recovery etc. [2, 3]. The need of enhance the thermal performance of heat exchangers and

S. Hasan Ibrahim (✉) · H. A. Abdul Wahhab
Department of Mechanical Engineering, University of Technology, Baghdad, Iraq
e-mail: 20063@uotechnology.edu.iq

F. Ahmad et al. (eds.), *ICPER 2020*, Lecture Notes in Mechanical Engineering,
https://doi.org/10.1007/978-981-19-1939-8_24

their production cost reduction lead to continuous development in use many techniques termed as "Heat Transfer Augmentation", where the heat augmentation techniques improve convective heat transfer by reducing the thermal resistance in a heat exchanger [4, 5], also these techniques were classified either as passive or active, the passive techniques does not need any external power during operation, so the additional power needed to increase heat transfer is taken from the stored power in the system such as twisted tapes, extended or treated surfaces. While active techniques need to external power such as surface vibration and mechanical aids [6]. Twisted tape is a type of passive heat transfer augmentation techniques which has been used as heat transfer enhancing devices [4]. The important effects induced by the tapes are swirl flow which improves fluid mixing, helically twisting motion which offers an effectively larger flow partitioning and blockage of the tube flow cross section which leads to a higher flow velocity [7, 8]. He et al. [9] tested single twisted tape insert in circular tube, they noticed that heat transfer rate increased significantly than tube without inserts, and as twist ratio decreases, the heat transfer rate and friction factor become larger. Several experiments executed for the single twisted tape insert with many different twist ratios to enhancement heat transfer and describes to relationships for Nusselt number and friction factor [10–12]. Bhattacharyya et al. [13] studied the enhance of heat transfer for laminar flow in circular tubes using twisted tape inserts and rough surfaces, this investigation was carried out experimentally. They confirmed major conclusion that the twisted tapes test in conjunction with ribs perform significantly better than the individual use of tapes, and these reported results are useful for the design of solar thermal heaters and heat exchangers. Subsequently, many researchers simulated heat transfer system problems with setup the twisted tapes inserts in circular tubes, the numerical simulation were conducted utilizing ANSYS FLUENT [14–16].

The review of literature showed that numerous researchers have investigated heat transfer enhancement by twisted tapes inserts in the pipe flow. Generally, twisted tapes inserts are help to increase fluid velocity near the inner surface of pipe due to dividing the flow within the pipe. But in some of applications, these twisted tapes inserts may contribute to excessive flow turbulence, which would eventually lead to a state of thermal instability. Therefore, it requires continuous development of the ideas proposed in the past and finding solutions to the most important obstacles that prevent their commercial application. The objective of the present work is to investigate numerically and experimentally for the effect of using twisted tape inserts as turbulence promoters to enhance heat transfer, also study to the effect twisted tapes perforation by a central- circular holes on heat transfer coefficient and friction factor inside the pipe, and compared the performance with normal twisted tape.

2 Experimental Implementation

The experimental tests were carried out in circular tube, which was heated from outer surface with constant heat flux. The test rig included; test section, heating system, and measuring instruments. Figure 1 shows the schematic of experiment set up. The test section consists of circular tube and twisted tapes. The circular tube was made of T-305 aluminium of 25 mm outer diameter, 1500 mm length, and 3 mm thickness with L/D about 60. Twisted aluminium tapes at dimensions 20 × 1750 mm were used for manufacturing six inserters at the shape of twisted tapes with twisted ratio Y/D = 3.15 and perforated with different circular holes of 2, 4, 6, and 8 mm diameter, as shown in the Fig. 2, These were made by warped the tape around circular pipe and then drag to the required Y/D. Water was used as the working fluid, due to high heat capacity which enables easy control of temperature, and conventionality of using it as cooling fluid in many heat exchangers. Water flow rates was controlled by ball valve which supply required flow rates of water through the test section from a feed water storage tank of 1.5 m^3 capacity. Heating system (steel tape heater) was used for heating the outer surface of the circular tube with uniform surface heat flux 39.5 KW/m^2, this heater was manufactured from steel electric tape with high resistance. The electrical tape was wrapped around the tube; the heater was covered with a double layer of mica tape of 0.8 mm thickness for electrical insulation. Varaic, TDGC2-1P voltage regulator of 25 A was used to controlled the power supplied by heater.

Ten thermocouple wires type K were used to measured the local surface temperatures along the test section, as shown in Fig. 3. Each thermocouple was fixed on the pipe wall after cleaning the wall, and a good adhesive material was used to fixed the thermocouples head, and then connected to a digital thermometer through a selector switch. There were additional two thermocouple wires type K were used to measured the water inlet and exit temperature, all thermocouples calibrated with a mercury thermometer.

3 Computational Modeling

Numerical analysis methods are an effective tool for representing the mechanical problems and analysing the physical phenomena for the engineering uses [17]. In the current work, the commercial (CFD) software (ANSYS 17.0 FLUENT) by the k-ε model was selected to solve the considered cases because the effect of swirl on turbulence is included in this model, which enhances the accuracy of swirling flows. There exist many steps that achieved for performing the simulations. This simulation is interested into the influence of twisted tape inserts with perforation on heat transfer and pressure drop inside circular tube according to the research objectives.

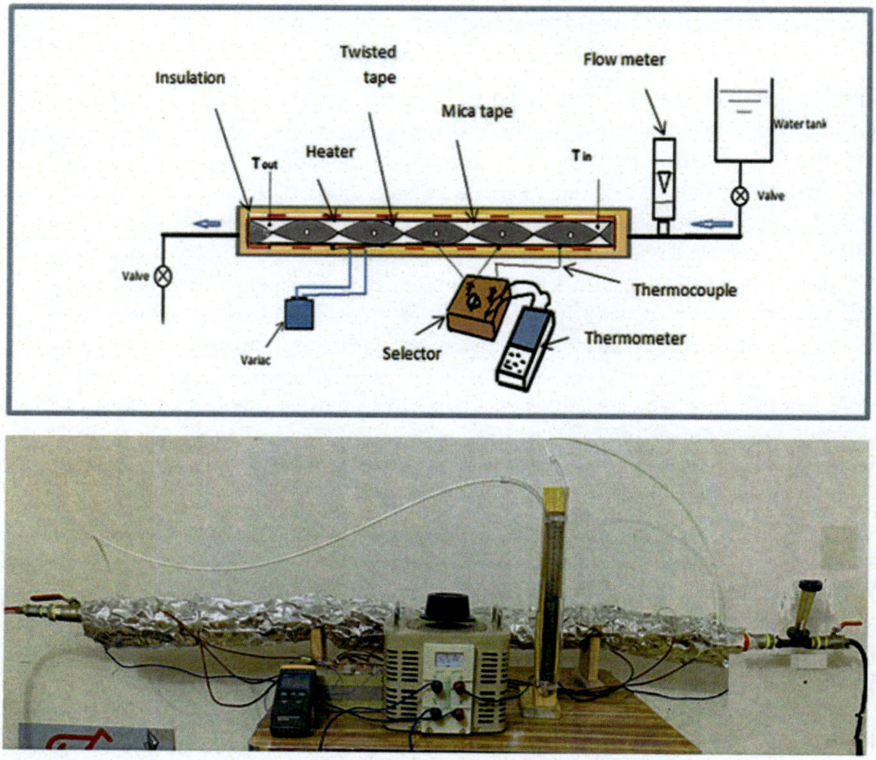

Fig. 1 Schematic diagram and photo of the experimental setup

Fig. 2 Twisted tapes with different holes diameters

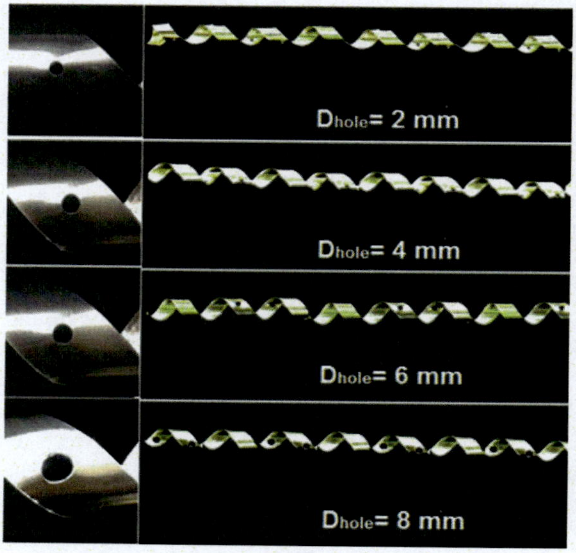

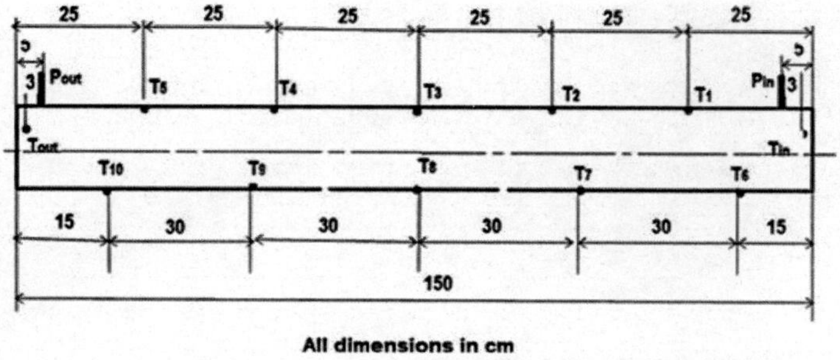

Fig. 3 Schematic of pipe dimensions with thermocouples distribution on pipe wall

Table 1 Important parameters of simulation

Dimension of Inserts	Hole diameter of Inserts (mm)	Nodes in meshing	Elements in meshing
Twisted tape;	2	1,728,233	1,374,139
Width, 20 mm	4	1,734,534	1,394,532
Thickness, 3 mm	6	1,767,832	1,421,369
Length before twisting, 1750 mm Length after twisting, 1500 mm	8	1,789,432	1,445,632

3.1 Description of Geometry and Modeling

To improving in style inserts in the pipes of heat exchanger increased a lot of ideas in field of heat transfer enhancement. Hence, in this study a twisted tape inserts with perforations in pipe heat exchanger is simulated. Analysis was carried by twisted tape inserts with varying diameter perforations and keeping tape dimensions constant. Different holes diameter in twisted tape inserts taken into consideration are shown in the Table 1. Figure 4 displays the geometry and mesh of twisted tape insert and geometry of inner pipe. All dimensions kept same as experimental investigation. Geometry and meshing simulation is performed on CFD software package of ANSYS 18.0, sizing of grid is performed by command body sizing with 1 mm for pipe. A detail of nodes and elements of meshing is given in the Table 1.

3.2 **Governing Equation**

In this numerical study, 3D governing equations are used with flow specification of turbulent flow. In this study twisted tape inserts are used in order to increase the turbulence which in turns will result in increased rate of heat transfer. While

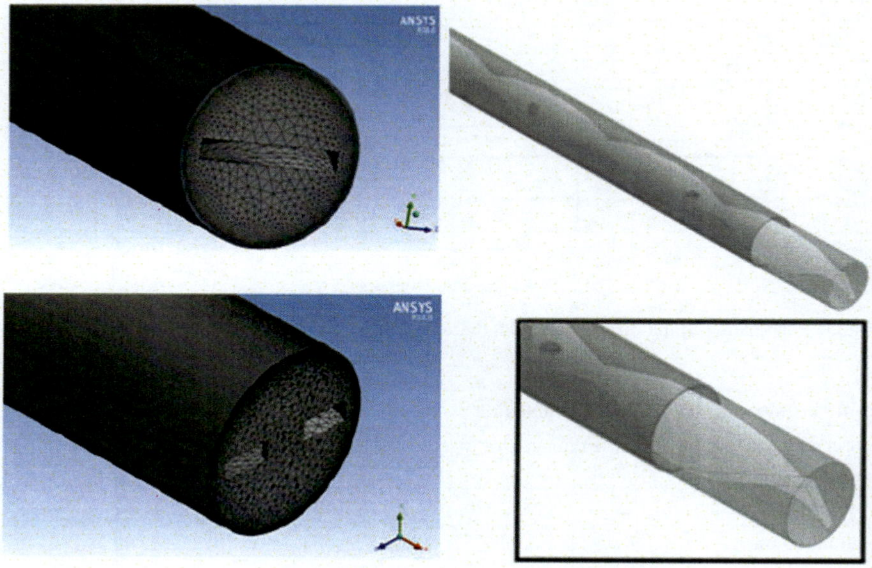

Fig. 4 Geometry and mesh of twisted tape inserts, with perforations inside the pipe

the perforations helps to increased flow homogeneity. The k-ε turbulence model is selected in which two equations one for turbulent kinetic energy and one for dissipation rate is solved. Other than equation of k-ε model following are the governing equations which are used during the analysis [18].

$$\frac{\partial}{\partial z_j}(\rho \overline{u_j}) = 0 \tag{1}$$

$$\overline{u_j}\frac{\partial}{\partial z_j}(\rho \overline{u_i}) = B - \frac{\partial p}{\partial z_i} + \frac{\partial}{\partial z_j}\left(\mu\left(\frac{\partial \overline{u_j}}{\partial z_i} + \frac{\partial \overline{u_i}}{\partial z_j}\right) - \rho \overline{u_i u_j}\right) \tag{2}$$

$$\overline{u_j}\frac{\partial}{\partial z_j}(\rho cp \overline{T}) = H + \frac{\partial}{\partial z_j}\left(\rho cp \alpha \frac{\partial T}{\partial z_j} - \rho cp \overline{u_j T}\right) \tag{3}$$

These governing equations are utilized by FLUENT package of ANSYS. To obtain results nearest to preciseness the solution is made up to convergence where a variable seems to be constant against the iterations.

4 Results and Discussion

4.1 Experimental Results

The experimental results has been reported and discussed for the steady turbulent flow inside horizontal pipe which heated by a constant heat flux, (39.23 k W/m^2) applied on the outer surface of the pipe. The twisted tape inserts considered experimentally have pitch to width ratio (y/w) of 3.15 with a central circular holes of different diameters of 2, 4, 6, and 8 mm respectively, and ordinary twisted tape has been examined to find out the effect of tape perforation on heat transfer rate and friction factor for water flow at Reynolds number of range between (4.5–12.5 × 10^3).

Figure 5a shows average Nusselt number variation with Reynolds number for plain tube without tape inserts, which it is increased from 45.2 to 90.5 as Reynolds number increased from 4.5 × 10^3 to 12.5 × 10^3. This variation was compared with previous empirical correlation of Dittus Boelter [4] for heat transfer inside smooth tube, as shown in this figure there is a good agreement and the maximum division is 14%. Figure 5b shows the friction factor variation with Reynolds number for plain tube which is increased from 0.041 to 0.03 as Reynolds number increased from 4.5 × 10^3 to 12.5 × 10^3. This variation was compared with previous empirical correlation of Blacius [6] for the flow inside smooth tube, as shown in this figure there is a good agreement and the maximum division is 8%.

Figure 6a shows the comparison between average Nusselt number for plain tube and ordinary twisted tape of twist ratio (y/w) is 3.15, as shows the average Nusselt number was proportionally with Reynolds number for the range of (4.5–12.5 × 10^3). For this case Nusselt number for pipe fitted with twisted tape is higher than the plain tube by average percentage 94% at Reynolds number range (4.5–12.5 × 10^3). While Fig. 6b shows the comparison between fraction factor for plain tube and ordinary twisted tape of twist ratio (y/w) is 3.5, as shows the fraction factor is decreasing with Reynolds number increasing for the range of 4.5 × 10^3–12.5 × 10^3. For this case friction factor for tube fitted with twisted tape is higher than the plain tube by 2.7 times at the same Reynolds number range.

Figure 7a represents the comparison between average Nusselt number for ordinary twisted tape and the perforated twisted tapes inserts with circular holes of 2, 4, 6, 8 mm diameter, respectively. the same above behaviour was observed for all tapes the average Nusselt number was proportionally with Reynolds number for the range of (4.5–12.5 × 10^3) and the average Nusselt number decreasing with increasing the hole diameter for practical Reynolds number and its deceased by 5.1, 10, 16, 23%, then the ordinary twisted tape at Reynolds number of 7.8 × 10^3. So the highest heat transfer rate was obtained for ordinary twisted tape. Figure 7b shows the comparison between friction factor for ordinary twisted tape and the perforated twisted tapes with circular holes of 2, 4, 6, 8 mm diameter, respectively. The same above behaviour was observed for all tapes the fraction factor was proportionally with Reynolds number for the range of (4.5–12.5 × 10^3) and the friction factor decreasing with increasing

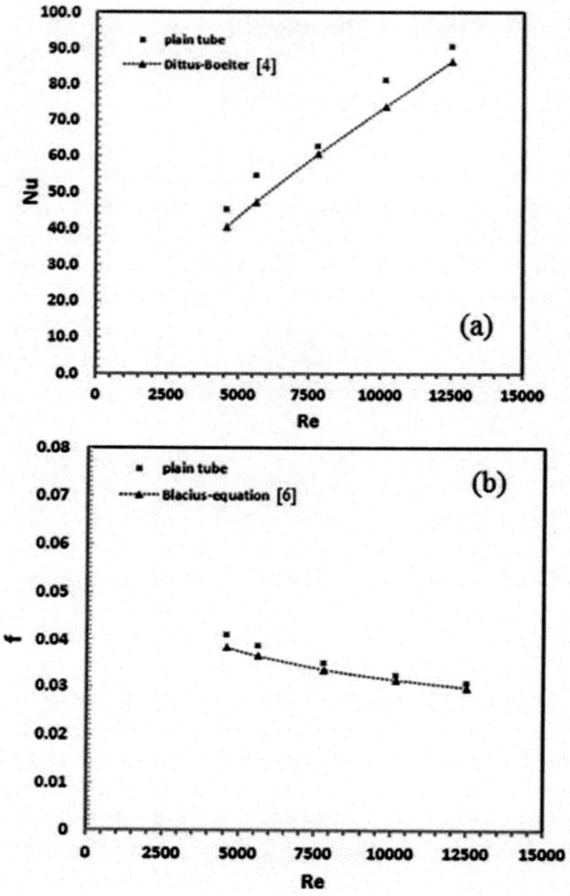

Fig. 5 Comparison of results; **a** Nusselt number variation with Reynolds number, **b** Friction factor variation with Reynolds number

the hole diameter for practical Reynolds number and its deceased by 6, 14, 24.5, 34% then the ordinary twisted tape at Reynolds number of 7.8×10^3.

4.2 Numerical Results

Figure 8 shows the temperature contours with different cases of twisted tapes inserts inside one of heat exchanger pipes. Case (a) for twisted tape insert with y/w ratio is 3.15, cases (b)–(e) for twisted tape inserts with perforations of 2, 4, 6, 8 mm diameter, respectively. It can be seen that the increasing of the perforation diameter for twisted tape, the temperature decreases at the same water flow conditions. In the position of 1200 mm, The water temperatures are found for twisted tape insert with perforations of 2, 4, 6, 8 mm diameter are 373.6 k, 368.4 k, 352.5 k and 337.43 k, respectively.

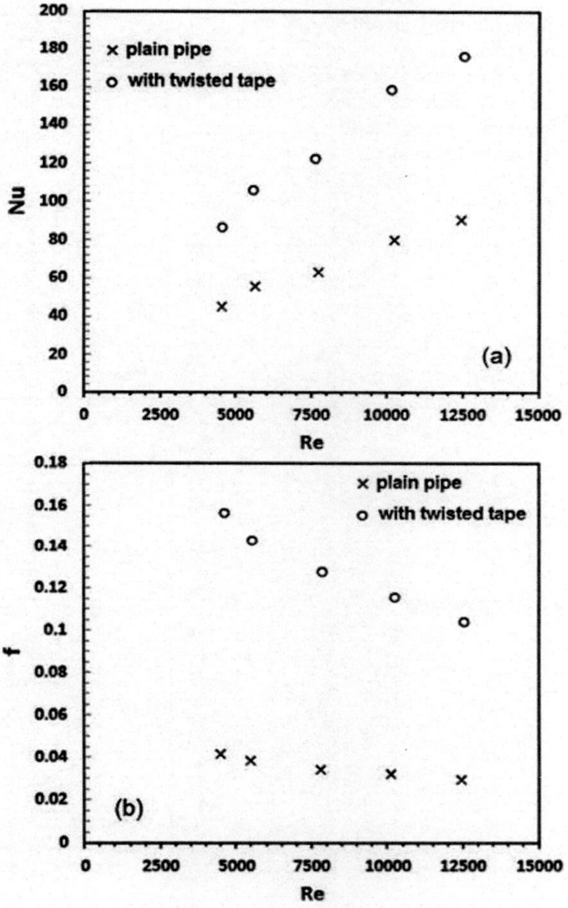

Fig. 6 Comparison of results for plain pipe and pipe with twisted tape; **a** Nusselt number variation with Reynolds number, **b** Friction factor variation with Reynolds number

Figure 9 shows the velocity contours with different cases of twisted tapes inserts inside the pipe. Case (a) for twisted tape insert with y/w ratio is 3.15, cases (b) and (c) for twisted tape inserts with perforations of 2, and 8 mm diameter, respectively. It can be noticed that the water flow in the pipe with ordinary twisted tape is vortex flow, because of vortex behaviour caused by twisted tape in a pipe, this increases the affected heat transfer region, thereby heat transfer rate increases, pressure drop also increases. And this pressure drop varies according to Reynolds number and the twisted ratio of twisted tape. Whilst in the state of twisted tape inserts with the perforations is also vortex flow, but clearly, the vortex intensity decreased with increasing holes diameter. So, these holes provide improving bulk water mixing between the central region and the near wall region, resulting in disruption of the thermal boundary layer.

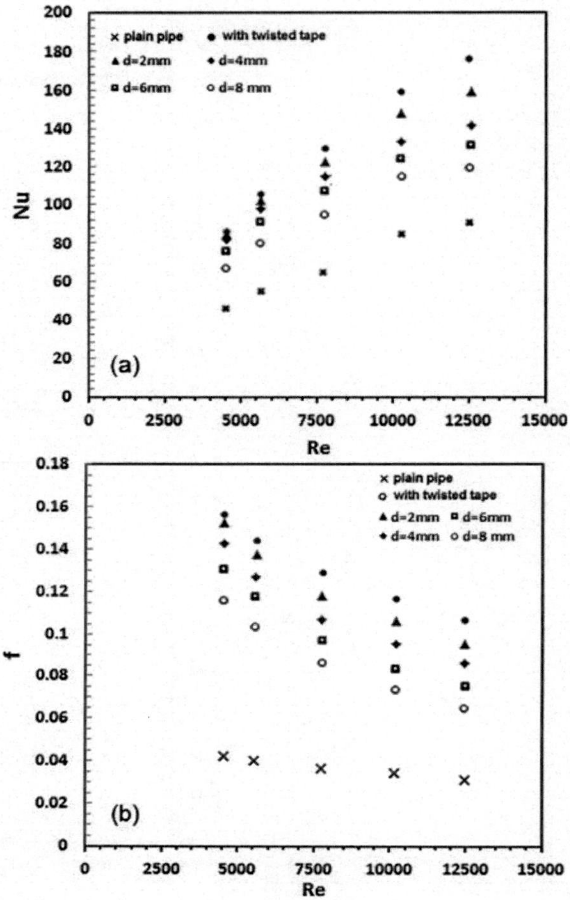

Fig. 7 Comparison of results for ordinary twisted tape and perforated tapes of 2, 4, 6, 8 mm diameter; **a** Nusselt number variation with Reynolds number, **b** Friction factor variation with Reynolds number

5 Conclusions

This work describes a numerical and experimental analysis to study the effect of twisted tape inserts as turbulence promoters to enhance heat transfer inside a circular tube heated with a constant heat flux. The characteristics of heat transfer (Nusselt number) and friction factor for different twisted tapes with perforations 2, 4, 6, 8 mm inserted inside a horizontal pipe, have been studied. Accordingly, the following conclusions are drawn:

- The behaviour was observed for all tapes the average Nusselt number was proportionally with Reynolds number for the range of (4.5–12.5 $\times$ 10^3) and the average Nusselt number decreasing with increasing the hole diameter for practical Reynolds number and its deceased by 5.1, 10, 16, 23%, then the ordinary

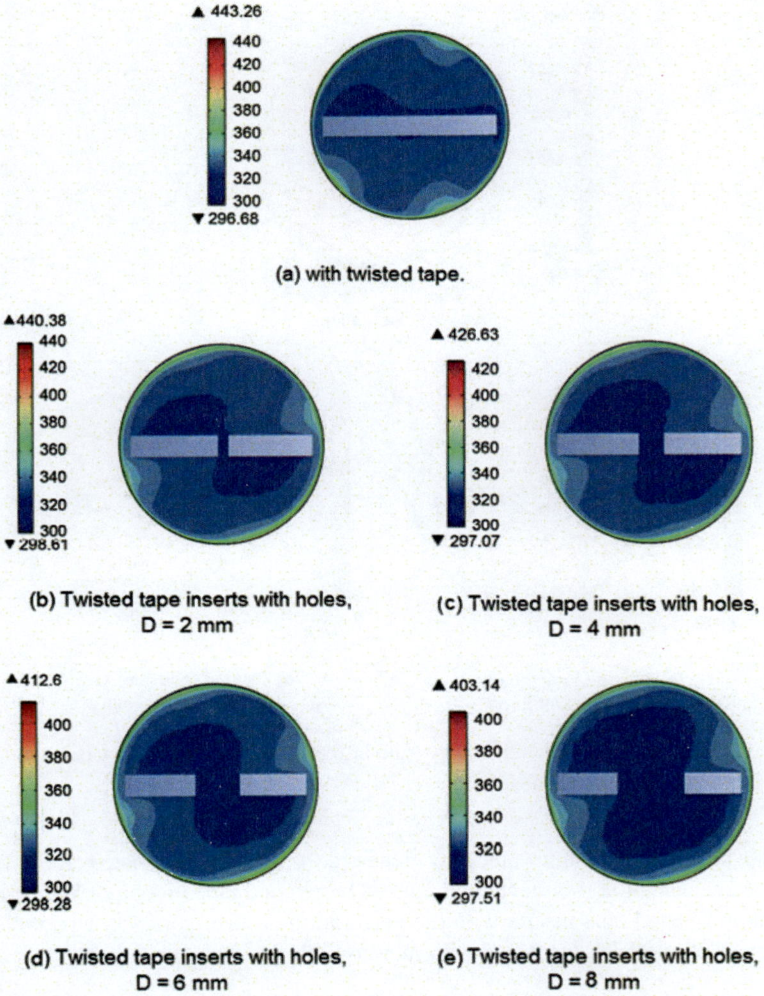

Fig. 8 Temperature contours of **a** with twisted tape inserts, **b–e** for twisted tapes inserts with different perforations of 2, 4, 6, and 8 mm, respectively

twisted tape at Reynolds number of 7.8 × 103. So the highest heat transfer rate was obtained for ordinary twisted tape.

- The behaviour was observed for all tapes the fraction factor was proportionally with Reynolds number for the range of (4.5–12.5 × 10^3) and the friction factor decreasing with increasing the hole diameter for practical Reynolds number.
- Numerical results appeared that the increasing of the perforation diameter for twisted tape, the temperature decreases.

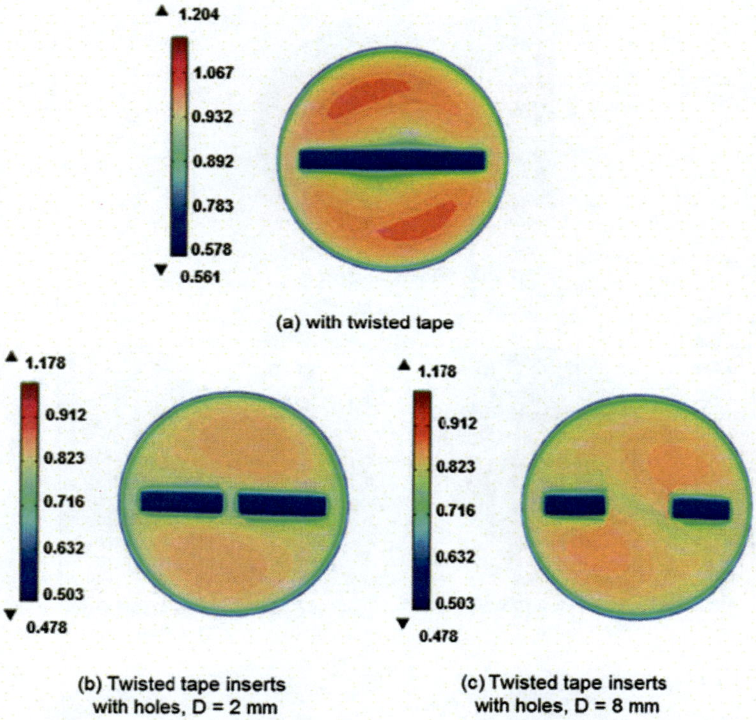

Fig. 9 Velocity contours of **a** with twisted tape inserts, **b** and **c** for twisted tapes inserts with perforations of 2, and 8 mm, respectively

- It can be noticed that the water flow in the pipe with ordinary twisted tape is vortex flow, because of vortex behaviour caused by twisted tape in a pipe, Whilst in the state of twisted tape inserts with the perforations is also vortex flow, but clearly, the vortex intensity decreased with increasing holes diameter.

Acknowledgements The authors are obliged to the University of Technology, Baghdad in Iraq, for providing the Centre of Automotive and Energy Generation.

References

1. Manglik R, EBergles A (1993) Transaction of ASME. J Heat Transf 115:890–896
2. Watching N, Smith E, Pongjet P (2006) 2nd international conference on environment, Bangkok, Thailand, pp 21–23
3. Paresh AJ, Patil VH (2005) International conference on science, technology and management, University of Delhi, Delhi, India, pp 1277–1284

4. Al-Kayiem HH, Muna SK, Saud TT (2020) Chapter in the book: inverse heat conduction and heat exchangers. Published by Intec open, London, UK
5. Moorthy CH, Srinivas V, Bharadwajan K (2017) Inter J Mech Prod Eng Res Dev 7:195–206
6. Sivakumar K, Ratan K, Murall S, Prakash S, Thanigaivel V, Sryakumalar T (2016) Int J Chem Sci 14:439–445
7. Eiamsa-ard S, Wongcharee K, Eiamsa-ard P, C Thianpong (2010) Exp Ther Fluid Sci 34:1151–1161
8. Ajaj HZ (2012) Anba J Eng Sci 5:139–145
9. He Y, Men Y, Zhao Y, Lu H, Ding Y (2009) Appl Ther Eng 29:1965–1972
10. Bhattacharyya S, Benim AC, Chattopadhyay H, Banerjee A (2018) Exp Heat Trans 32:411–425
11. Onkar H, Rahul P, Jyoti K, Gargi K, Mayura M (2016) Int J Curr Eng Tech 277–286
12. Gawande KR, Deshmukh AV (2017) IRA-Int J Tech Eng 13:34–42
13. Bhattacharyya S, Saha S, Saha SK (2013) Exp Ther Fluid Sci 44:727–735
14. Bhattacharyya S, Chattopadhyay H, Benim AC (2017) Int J 17:390–396
15. Pourrajabi M, Nezhad AR, Pourrajabi MR (2013) Int Sym Adv Sci Tech. Iran: 7th SASTech
16. Saleh FA (2018) Int J Mech Mechatron Eng 19:1–14
17. Sidhappa KA, Sonal SH (2016) Int J Innov Eng Res Technol 3:34
18. Incopera FP, Dewitt DP (1986) Fundamentals of heat transfer, 4th Ed. 154

The Effect of Load and Injection Timing on the Emitted Pollutants Concentrations by Engine Fuelled by Biofuel

Hayder A. Dhahad and Miqdam T. Chaichan

Abstract Diesel engines have won attention because they are one of the most important ways to provide people with reassurance, whether through electricity generation or transportation. However, Diesel engines emit many dangerous pollutants that affect health and the environment. To reduce these effects, researchers adopted alternative fuel types of diesel such as biodiesel. In this study, a diesel engine was tested and operated using biodiesel produced from rapeseed and the emitted pollutants were measured and compared to their counterparts for Iraqi commercial diesel fuel. The results showed a decrease in the CO concentrations when using biodiesel, with a slight increase in the HC concentrations, and these concentrations increased by delaying the injection timing. NOx concentrations increased with increasing engine load and with advancing injection timing. The resulting concentrations when using Biodiesel are higher due to the availability of oxygen. The smoke number increased at an advanced injection timing and decreased by delaying this timing, as it increases for biodiesel more than for diesel. At low loads, the smoke number of diesels is higher for all the studied injection times. The previous results contradicted many of the previous studies published in the literature because the fuel used (Iraqi diesel) is a fuel with a high sulphur content while the aforementioned studies adopt diesel with an ultra-low sulphur content.

1 Introduction

Internal combustion engines, whether diesel or gasoline, occupy a great place in public life, especially in the field of transportation and electricity production. The crowded streets with cars, buses, and various types of construction vehicles, in addition to trains in major cities, is evidence of the importance of these engines [1,

H. A. Dhahad (✉)
Department of Mechanical Engineering, University of Technology, Baghdad, Iraq
e-mail: hayder_abed2002@yahoo.com

M. T. Chaichan
Energy and Renewable Energies Technology Center, University of Technology, Baghdad, Iraq

F. Ahmad et al. (eds.), *ICPER 2020*, Lecture Notes in Mechanical Engineering,
https://doi.org/10.1007/978-981-19-1939-8_25

2]. Unfortunately, the burning of diesel and gasoline fuels produces gases that are harmful to the environment and health [3]. The escalating concentrations of these pollutants, especially those that enhance global warming, have called for a global pause to confront the dangers of pollution due to these engines [4, 5]. Many countries of the world have begun to set strict laws and instructions to reduce pollutants emitted from internal combustion engines [6]. These strict regulations and guidelines pressed manufacturers and researchers to find promising ways to reduce engine exhaust pollutants [7]. Some researchers have proposed new methods of fuel combustion in engines such as homogeneous charge compression ignition (HCCI) [8], pre-mixed charge compression ignition (PCCI) and the use of a low-temperature combustion (LTC) method [9, 10], nanoadditives [11], among others. One way to reduce engine exhaust pollutants is to use alternative fuels for diesel and gasoline. Researchers suggested many types of fuels, for example, for gasoline, they suggested hydrogen [12], liquefied petroleum gas [13], natural gas [14], and alcohols of all kinds such as ethanol, methanol and others [15, 16]. As for diesel, the most important alternative that has proven highly effective is biodiesel [17, 18]. One of the most important and closest fuels to diesel is biodiesel; it is a promising alternative as it is a clean, renewable fuel. Biodiesel consists of methyl fatty acid or ethyl esters [19]. Biodiesel is made from vegetable oils, animal fats, or waste oil used in cooking or industry [20]. Studies have proven that biodiesel can be added to petroleum diesel fuel in any proportion, and therefore diesel fuel can be partially or completely replaced [21, 22]. Biodiesel has several impacts on diesel engine emissions. In Europe, rapeseed is commonly used in the production of biodiesel (rapeseed methyl ester RME). The introduction of this ester as a biofuel revitalized the rapeseed fields that have become a common sight on this continent. Biodiesel has many benefits, including: Reducing the concentrations of pollutants emitted from the engine, the most important of which are unburned hydrocarbons (THC), carbon monoxide (CO) and particulate matter [23].

Iraq suffers from the aging of its oil refineries and their consumption due to forty years of harsh wars and a fatal economic blockade. Iraqi diesel is considered one of the worst types of diesel in the world due to its high Sulphur content ranging from 1 to 2.5% by weight [24]. These poor fuels cause a high emission of smoke in addition to the rest of the emissions [25]. Replacing this fuel with biodiesel means ridding the Iraqi environment of hundreds of thousands of tons of harmful emissions annually. From here, the study presents a proposal to use an alternative to Iraqi diesel fuel, which is a type of biodiesel produced from rapeseed. This paper examines the impact of combustion characteristics on the emitted engine pollutants of Iraqi diesel and RME for a wide range of injection pressures, injection timings and loads using measured in-cylinder pressure data for combustion analyses.

2 Experimental Setup

2.1 The Tests Rig

The tests were conducted with a Ford Motor Company high speed direct injection (HSDI) diesel. The engine is equipped with a modern injection system from Delphi Diesel Systems Ltd (DDS). This system is controlled by a controller that can control various operating performance parameters. Table 1 lists the used engine specifications.

The engine is supplied by several thermocouples to measure several parameters such as intake air, exhaust gas, and cooling water. Also, the engine is supplied by a pressure transducer to measure the pressure inside the combustion chamber.

2.2 Regulated Gases Measurements

The Horiba MEXA-7170DEGR gas analyser instrument was used to measure the CO, CO_2, THC and NOx gases emitted. Carbon dioxide and carbon dioxide emissions are measured using the non-dispersive infrared (NDIR) method, nitrogen oxides emissions are measured using the chemical luminescence method, and unburned hydrocarbons (HC) are measured by flame ionization detection technique. AVL 415S

Table 1 The used engine specifications

Parameter	Symbol	Value
Cylinders' number	Number	4
Bore	mm	86
Stroke	mm	86
Swept volume	cc	1998.23 cc
Compression ratio	–	18.2:1
Cylinder pressure (Maximum)	bar	150
Piston design	–	Central bowl in piston
Nominal bowl volume	cc	21.7
Oil type	–	15W40
Water temperature	°C	70
Maximum turbine speed	rpm	180 000
Engine speed (Idle)	rpm	750 ± 5
No load speed (Maximum)	rpm	4800 ± 50

Table 2 The used fuels specifications

Parameter	Units	Diesel	Biodiesel (RME)
Chemical formula		C14H26.18	C18.96H35.29O2
Cetane number	–	49	53.97
Density at 15 °C	(kg/m^3)	829.1	8821.55
Viscosity at 40 °C	(cSt)	2.647	4.428
LCV	(MJ/kg)	42.73	37.6
Sulphur	(ppm)	10,000	5
Total aromatics	(wt%)	25.4	00
Carbon	(wt%)	84.41	77.12
Hydrogen	(wt%)	12.65	12.04
Oxygen	(wt%)	00	10.93
Cloud point	°C	−11	−17
Flash point	°C	69	70
Ash content	(wt%)	<0.015	<0.005

smoke meter used to evaluate engine soot. AVL 415Scounts the Filter Smoke Number (FSN). A photoelectric measuring head fixed inside this meter is used to determine the filter paper darkening resulted from a known volume of the exhaust gas sample that is passing through it. The scale used automatically changes the sample size, and can also changes the effective length, to reach the most appropriate concentration of soot, so this scale is accurate to 0.002 FSN. The smoke number is measured at concentrations from 0 to 10, as 0 represents the white paper filter and 10 represents exactly the blackness of the paper.

2.3 *Fuel Used*

Iraqi diesel with high sulphur content and locally produced was used in the trials for comparison. Shell Global Solutions provided the tested rapeseed methyl ester (RME). Table 2 lists the used fuels specifications.

3 Results and Discussions

Figure 1 shows the carbon monoxide concentrations when the engine is operating at high and low loads. In general, CO emissions decrease with increasing load, as a result of the observed increase in the combustion chamber temperatures. Controlling the equivalence ratio is the best way to control the levels of carbon monoxide emitted. When the engine is running at low load, the temperatures inside the combustion chamber are low, which results in unstable combustion reactions. Also, when the

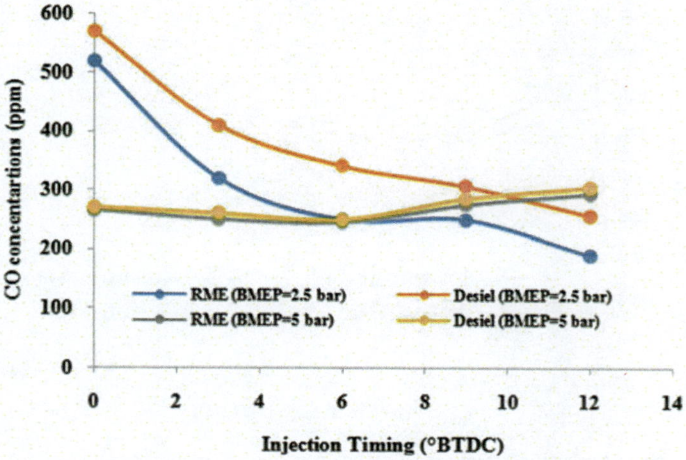

Fig. 1 The effect of injection timing on CO concentrations at variable loads

engine is running at high load, working at a rich equivalence ratio causes a reduction in the percentage of oxygen inside the combustion chamber, which means that the chances of converting fuel carbon into carbon dioxide are reduced, which causes some of the fuel not to be completely burned. The presence of oxygen in the chemical biodiesel formulation at a high percentage causes the combustion to improve, whether in rich or weak equivalent ratios. Delayed injection timing at a low engine load increases CO emission due to incomplete combustion [26]. This incomplete combustion occurs in the expansion phase at a low temperature. As for advanced injection, this effect did not appear as the combustion chamber temperature is high, which means a change in the conditions of this pollutant formation.

Figure 2 illustrates the HC variation due to changing engine injection timing unburned hydrocarbon concentrations at different engine loads. HC concentrations decrease with the increase in engine load, as increased load causes an increase in the combustion temperature inside the combustion chamber. The results show that the HC emissions of biodiesel at high loading are high most of the injection times and are higher than that of the diesel fuel counterpart. HC concentrations increase as a result of flame suppression and the high viscosity of RME plays an important role in this regard. Also, the high viscosity of biodiesel causes wetting of the cylinder walls, causing incomplete combustion. Advanced injection causes an increase in HC concentrations due to a decrease in the charge density at the time of injection which increases the wetting of the wall.

Figure 3 shows the variability of NOx concentrations at variable loads and injection timings. Delaying the timing of the injection causes a decrease in the NOx concentrations and is one of the main methods of controlling these concentrations. Delay in injection timing causes the temperature inside the combustion chamber to decrease due to combustion during the expansion stroke. Also, a high load causes an increase in these concentrations compared to a low load. Nitrogen in the air needs

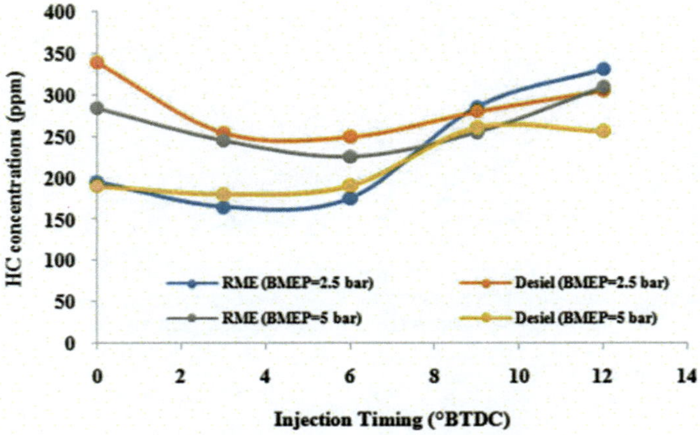

Fig. 2 The effect of injection timing on HC concentrations at variable loads

to have available concentrations of oxygen for the reaction and to high temperatures and sufficient time for the reaction to take place [27]. All these things together are provided by working with advanced timing and high load with biodiesel. As the figure's curves reveal, NOx concentrations for RME are higher than NOx emitted by diesel for all the tested injection timings due to the availability of oxygen in its reactions.

Figure 4 depicts the smoke number variation due to changing in load and injection timing. RME fuel smoke concentrations are always lower than diesel fuel, when the engine is running at high loads. When working at low loads, these concentrations of the two tested fuels converged with a slight increase in favor of biodiesel. The decrease in the smoke number when working with biodiesel can be referred to high

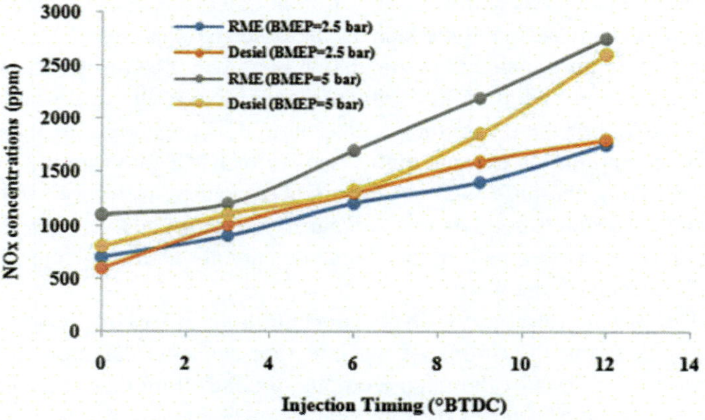

Fig. 3 NOx concentration variation due to changing in injection timing at variable load

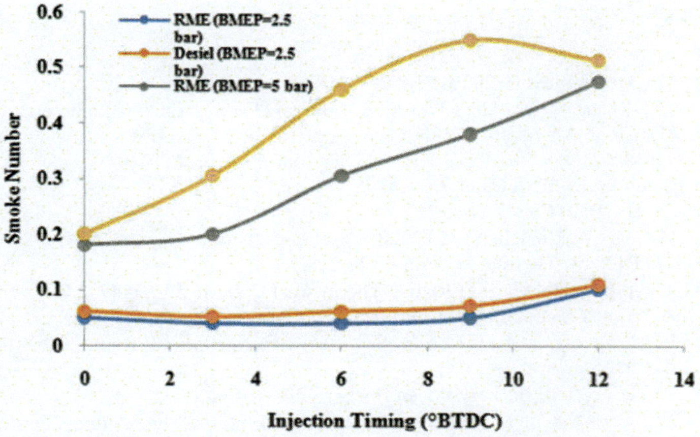

Fig. 4 The effect of injection timing on smoke number at variable loads

oxygen content in it [28, 29]. The improvement is most pronounced at higher engine loads. Also, the decrease in aromatic substances in the biodiesel formulation causes a decrease in the formation of soot, as aromatization in particular tends to increase the tendency to form soot due to the length of the hydrocarbon chain [21, 30]. Delaying injection timing generates much more heat during the pre-mixed phase. This increase in generated heat causes reduction in soot concentrations. Also, the duration of the proliferation phase decreases with delaying the timing of the injection causing the period of soot formation to be reduced.

4 Conclusions

Biodiesel is an alternative fuel to petro-diesel due to its distinct properties, as it is of plant or animal origins in addition to the presence of a high percentage of oxygen in its chemical composition. In this study, the use of biodiesel produced from rapeseed was examined and the pollutants emitted from the engine were compared with the emissions from an Iraqi diesel fuel. The results of the study showed that NOx biodiesel concentrations are higher than their diesel counterparts due to the availability of oxygen. The CO pollutants are lower for all loads and injection timings studied. The RME engine emits higher HC concentrations when advancing the injection timing and vice versa when delaying. The smoke number is low at low loads and increases with increasing load. This number is higher for the case of diesel use at all the studied injection times.

Acknowledgements The authors are obliged to the University of Technology, Baghdad in Iraq, for providing the Centre of Automotive and Energy Generation.

References

1. Hayder AD, Mohammed F (2020) Fuel 279:118384
2. Chaichan MT, Kazem HA, Abed TA (2018) Environ Dev Sustain 20(2):589–603
3. Hayder AD, Sinan AA (2019) ARPN J Eng Appl Sci 14:1478–1489
4. Olabi AG, Maizak D, Wilberforce T (2020) Sci Total Environ:141249.
5. Meloni E, Palma V (2020) Catalysts 10:745
6. Giechaskiel B (2020) Catalysts 10:587
7. Hayder AD, Ali SA, Chaichan MT (2020) Case Stud Thermal Eng 20:100651
8. Chaichan MT (2018) Thermal Sci Eng Progr 7:45–53
9. Fathi M, Ganji DD, Jahanian O (2020) J Therm Anal Calorim 139:2447–2454
10. Atmanli A, Yilmaz N (2020) Fuel 260:116357
11. Magaril E, Magaril R, Al-Kayiem HH, Skvortsova E, Anisimov I, Rada EC (2019) Sustainability 11(7):2165
12. Acar C, Dincer I (2020) Int J Hydrogen Energy 45:3396–3406
13. Ahn S, Cho J, Baek S, Kim K, Ko J, Myung C L and Park S (2020) Int J Autom Tech 21:451–457
14. Xu Y, Zhang Y, Gong J, Su S, Wei Z (2020) Fuel 266:116957
15. Chaichan MT, Kadhum AH, Al-Amiery AA (2017) Case Stud Thermal Eng 10:611–620
16. Khunathorncharoenwong N, Charoensuppanimit P, Assabumrungrat S and Kim-Lohsoontorn P (2020) Int J Hydrogen Energy
17. Ekaab NS, Hamza NH, Chaichan MT (2018) Case Stud Thermal Eng 13:100381
18. Thamer I, Al-Kayiem HH (2012) J Appl Sci 12(24): 2604–2609
19. Chaichan MT (2017) Int J Hydrogen Energy 43:5415–5435
20. Rathnam VM, Modak JM, Madras G (2020) Fuel 275:117998
21. Dhahad HA, Chaichan MT, Megaritis T (2019) Energy 181:1036–1050
22. Al-Kayiem HH, Abdul Wahhab HA, Magaril E, Aziz ARA (2018) AIP Conf Proc 2035:020008
23. Rimkus A, Matijošius J, Manoj Rayapureddy S (2020) Energies 13:2398
24. Dhahad HA, Chaichan MT (2020) Thermal Sci Eng Progr 100:535
25. Poisson C, Boucher S, Selby D, Ross SP, Jindal C, Efird JT, Bith-Melander (2020) Int J Environ Res Public Health 17:3299
26. Peirce DM, Alozie NSI, Hatherill DW, Ganippa LC (2013) Energy Fuels 27:3838–3852
27. Rounce P, Tsolakis A, York APE (2012) Fuel 96:90–99
28. Sergey U, Harald V, Jørgen BN, Erik H (2013) Fuel Process Technol 106:350–358
29. Hamza NH, Ekaab NS, Chaichan MT (2020) Alexandria Uni Eng 59:1717–1724
30. Di Y, Diming L, Zhiyuan H, Piqiang T (2011) SAE Paper

Experimental Investigation on Sub-Cooled Degree Effect on the Flow Boiling in a Microchannel Heat Sink

Qahtan A. Al-Nakeeb, Ekhlas M. Fayyadh, and Moayed R. Hasan

Abstract The effects of degree inlet sub-cooled, heat fluxes, and mass fluxes on the flow boiling characteristics (heat transfer and pressure drop) in single rectangular microchannel was investigated at operation system pressure of 1 bar. The microchannel was shaped by machining 300 μm wide × 700 μm deep groove (i.e. a hydraulic diameter 420 μm) into the surface of brass block having length of 60 mm. The working fluid was deionized water. The experiments were conducted over range of mass flux from 300 to 600 kg/m^2.sec, and heat flux (based on wall) from 5.4 to 376.5 kW/m^2 at the degree of sub-cooled (20 K and 35 K). The results found that the pressure drop and heat transfer values are dependent on the degree of inlet degree sub-cooled over the range of heat fluxes tested, wherever lower values of pressure drop and decreasing of heat transfer coefficient values were noted at higher degree of inlet degree of sub-cooled. However, the dominate mechanism of heat transfer is the nucleate and forced convection boiling for present work. The results of boiling pressure drop and heat transfer coefficient were compared with different micro scale correlation; these correlations predicted an acceptable reasonably well for the flow boiling characteristics.

Keywords Microchannel · Single phase flow · Two phase flow · Sub-cooled flow boiling · Heat transfer coefficient · And pressure drop

Q. A. Al-Nakeeb (✉) · E. M. Fayyadh · M. R. Hasan
Mechanical Engineering Department, University of Technology, Baghdad-Iraq, Iraq
e-mail: 20077@uotechnology.edu.iq

E. M. Fayyadh
e-mail: 20084@uotechnology.edu.iq

M. R. Hasan
e-mail: 20002@uotechnology.edu.iq

F. Ahmad et al. (eds.), *ICPER 2020*, Lecture Notes in Mechanical Engineering,
https://doi.org/10.1007/978-981-19-1939-8_26

1 Introduction

The earlier several years ago has seen unexpected increasing a computer processor efficiency that has been induced, for the largest part of the computer; through relentless pursuit of partial minimize components in the processor itself. Computer processors are not limited to the significant increase in dissipation of heat per unit surface area and per unit volume. In addition, for these systems, heat dissipation rates have risen to stages which difficult to control by using traditional techniques for cooling. In recent years, many cooling schemes have been produced which capitalize on the advantages of phase change phenomenon to attain the required effectiveness of cooling. These include thermosyphons for pool boiling and flow boiling through a channel. A special type of channel flow boiling systems is the two-phase micro-channel heat sinks. For instance in some applications, such as missiles, avionics, and compact computers, these kind of features have made heat dissipation devices a prime contender for small, lightweight cooling systems. Finally, the realistic advantages of flow boiling through the micro-channel heat sinks have drawn significant attention to many issues relating to their characteristics of phase change [1].

Flow boiling heat transfer characteristics in a multi microchannel heat sink was studied by Qu and Mudawar [2] by using de-ionized water represented a working substance. The test heat sink contained 21 rectangular multi-microchannels grooved by dimensions, 231 μm in width and 713 μm in depth. The heat sink was made of free-oxygen copper. The experiment was conducted at mass flux with ranges of 135–402 kg/m^2.sec, inlet temperatures are 30 °C and 60 °C and an outlet pressure is 1.17 bar. It was found that the coefficient of heat transfer for saturation water flow boiling depends on mass flux and weak dependency on heat flux at moderate and high heat flux. So, the dominate heat transfer coefficient for deionized water flow through microchannels heat sink is forced convective boiling mechanism.

Liu and Garimella [3] explored water flow boiling heat transfer in two configurations of microchannnels (275 × 636 μm^2) and (406 × 1063 μm^2). The experiment conducted at inlet temperature of 67–95 °C, and range of mass fluxes with (221–1283 kg/m^2.sec). It was found that the degree of sub-cooling and velocity had a slightly effect on the onset nucleation boiling in the boiling curve.

Agostini et al. [4] investigated the effect of sub-cooled and pressure on flow boiling in microchannel using refrigerant R245fa as a working fluid flow. The microchannel consisted of 67 channels with 223 μm wide, 680 μm high and 80 μm separating wall thickness. The experiment were carried out at rang of mass flux from 282 up to 1501 kg/m^2.sec and base heat flux from 3.7 up to 190 W/cm^2 at degree of sub-cooled (0–19 K) and the saturation pressure from 141 to 273 kPa. The results showed that the coefficient of heat transfer increased with increased the heat flux values and independent of the mass flux and thermodynamic quality. Meanwhile, the results proved that there is not much downstream affected of different level of sub-cooled on heat transfer coefficient for saturation boiling region. In contrast, the values of heat transfer coefficient decreased with increased sub-cooled degree.

Studying the heat transfer characteristics for two phase heat sink at low temperature by using indirect refrigeration cooling system in rectangular microchannels was done by Mudawar and Lee [5]. The authours conducted the experiments at range of the hydraulic diameter 175.7, 200, 334, and 415.9 μm with a different range of working fluid temperature from −30 °C to 0 °C using a pre-cooled of the fluid working substance (Hydrofluoroether (HFE)). The microchannel was made of oxygen free copper blocks. It was found that increasing sub-cooled will cause the decreasing of the two phase pressure drop. Also, increasing two phase cooling performance of microchannel by decreasing the hydraulic diameter as a result of increasing the wetted area and the mass velocity. Additionally, [6] extended their work to study the effect of decreasing working fluid temperature on the onset of boiling. It was found that the onset of boiling will be delay; simultaneously it reduces the size of bubble and coalescence effects that leads to enhance the performance of cooling system. Harirchian and Garimella [7] studied the influence of mass flux (225–1420 kg/m^2.sec) and the microchannel sizes on the flow boiling patterns by using high speed camera. Multichannel was grooved on the silicon heat sink. 5 K of sub-cooled of Fluorinert (F-77) at the inlet of the channels, seven different width from 100 μm to 5850 μm, and all the depth is 400 μm are studied. The heat transfer data showed that flow regimes with microchannel width 400 μm and over are the same with nucleate boiling dominate over the range of heat fluxes, while in channels with narrower width, convective boiling dominates. Five flow patterns have been recognized, specifically: bubbly, slug, churn, wispy-annular and annular flows. Bubble nucleation at the wall is suppressing at low heat flux with the smaller sizes of microchannel.

However, in a highly degree of sub-cooled, the heating microchannel surfaces collapsed the coalescing bubbles into many fine bubbles, this regime called Microbubble Emission Boiling (MEB). Koichi Suzuki et al. [8] conducted the sub-cooled of water in the horizontal rectangular multichannel heat sink, which consisted of 11 channel having hydraulic diameter 155 μm. This heat sink was made of copper block. Maximum pressure in the channel was 25 mm Hg and the outlet channel was opening to atmosphere. At the higher heat flux, the microchannels were completely covered with large coalescing bubbles which the coalescing bubble was covered a whole heating surface. The length of the heating unit and the amount of heating are playing a significant role to create a microbubble emission boiling through the channels.

Morever, Galvis and Culham [9] investigated flow boiling features in two different single copper microchannels that have the identical length and aspect ratio nonetheless with hydraulic diameters 217 μm and 419 μm. The experiments were performed by a range of mass fluxes (350–1373) kg/m^2sec, also heat fluxes (31.7–1414) kW/m^2 at two different inlet water sub cooling (50 and 70) K. It was reported that the nucleate boiling mechanism was the dominant mechanism of heat transfer in a smaller microchannel and vice versa with the bigger microchannel where the convective boiling mechanism possibly will dominate over mechanism of nucleate boiling.

The influence of mass and heat fluxes on the coefficient of heat transfer has been studied by Mohamed et al. [10]. The authors conducted experiment in a single microchannel with 1 mm width and 0.39 mm height which was made of the oxygen copper block surface. The operation parameters are inlet temperature of 89 °C, mass flux from 200 up to 800 kg/m^2.sec and heat flux with range of 56–865 kW/m^2. High speed and high resolution camera was used to capture flow pattern. At all mass fluxes, the unstable flow boiling were observed at boiling incipience. At low value of mass and heat fluxes, the heat transfer coefficient depends on the heat fluxes only. Visualization method proved that the bubbles nucleate were rapidly growth without departure from the channel surface. However, elongated bubbles that formed were leading instability with a great fluctuation.

Yin et al. [11] investigated the flow boiling with high aspect ratio in michrochannels having hydraulic diameter of 571 μm. The experiments were carried out at inlet water temperature of 65 °C, ranges of mass flux 261–961 kg/m^2.sec, and range of heat fluxes 631–987 kW/m^2. It was observed that nucleate boiling dominated through the high aspect ratio microchannels i.e. the heat transfer coefficient generally depends on the heat flux instead of the mass flux. The bubble confinement Influences on the heat transfer through the microchannel, which it is significant in small and moderate aspect ratio of microchannels. In contrast, the confinement bubbles did not occur at large aspect ratio of microchannel.

Ramesh et al. [12] studied experimentally the effect of sub-cooled on the heat transfer and pressure drop through a single microchannel with dimension of (1 mm wide × 0.49 mm deep × 40 mm long). Three de-ionized water inlet temperature were selected (30, 50, and 70 °C) for the range of heat fluxes 260–1100 kW/m^2, and mass fluxes range 528–1188 kg/m^2.sec. The results shows that there are three different heat transfer regimes, the single phase heat transfer, stable sub-cooled boiling heat transfer, unstable sub-cooled boiling heat transfer. Additionally, the experimental studying indicated that the sub-cooled heat transfer coefficient depends on the heat flux and practically independent on the mass flux (nucleate flow boiling was dominated heat transfer mechanism).

Suha and Ekhlas [13] carried out experiment in single microchannel to study the influence of inlet degree sub-cooled on flow boiling. Heat sink was made of oxygen free copper with hydraulic diameter of 0.3 mm. The operation conditions of experimental work for mass flux and heat flux ranges are (78–800 kW/m^2) and (1700 and 2100 kg/m^2.sec) respectively for two inlet degrees of sub-cooled 21 K and 31 K. The results have shown that increasing mass flux leads to increasing the heat transfer coefficient. Also, the convective flow boiling is dominated mechanism. In addition, heat transfer coefficient was increasing with increasing degree of sub-cooled.

Vivian and Karayiannis [14] studied experimentally the influence of inlet sub-cooled on the flow boiling parameters in multi-microchannel using HFE-7200 as working fluid. The microchannel consisted of 44 channels with 0.36 mm in width, 0.7 mm in depth and 0.1 mm separating wall thickness. The experimental work was done at mass flux of 200 kg/m^2.sec and wall heat flux within the range from 25.9 to 180.7 kW/m^2 at inlet sub-cooled degree of (5, 10 and 20 K) and system pressure of 1 bar. The experimental results were shown that decreased in pressure drop when the sub-cooled degree is increased. Moreover, the highly inlet sub-cooled degree produced a lower average heat transfer coefficient and less effect on the local heat transfer coefficient.

However, previous literature review of flow boiling heat transfer, one can observe that the dominant heat transfer mechanisms and influences of heat and mass fluxes on heat transfer flow boiling characteristics through microchannel are not obvious. Actually, there is a discrepancy on the dominant mechanisms of heat transfer. Some researchers have stated that the nucleate boiling mechanism is a dominant heat transfer mechanism. In contrast, other researchers have decided that the convective boiling mechanism is the dominant heat transfer mechanisms in microchannels. To understand the characteristics and mechanism(s) of flow-boiling heat transfer in microchannels, the above review shows that further research is still needed. Consequently, one of the main purposes of the present work is to study the effect of inlet degree of sub-cooled (20 K and 35 K) for heat flux range (5.4–376.5 kW/m^2) and mass fluxes (300, 400, 500, and 600 kg/m^2.sec) on the flow boiling heat transfer and pressure drop in single micrchannel with hydraulic diameter of $D_h = 420\ \mu m$, using de-ionized water as working fluid. At the same time, validated correlations are needed for the wide use of multichannel heat sinks in cooling high heat flux devices. The results of the current experimental analysis have also been used to test proposed correlations that predict heat transfer rates.

2 Experiment Devices and Data Reduction

2.1 Experiment Setup

The experimental system comprises of de-ionized water reservoir, a micro-gear pump, a pre-heater, the test section, a sub-cooler, and in-line filter. The rotameter model (LZB-3WBF) is used for the experiment works, which calibrated with an accuracy of ± 0.64 ml/min. A chiller unit is used for cooling purposes in sub-cooler and condenser. Figure 1 and Plate 1 are represented the schematic diagram and experimental facility set-up respectively. Strong boiling for around 1 h was used for the degassing of de-ionized water in liquid tank. To relief gases that are non-condensable to the environment, the valve that positioned on the top of the condenser was opened. For separating all particles from water a (5 μm) filter was installed before micro-gear pump. Later on, the degassed water was pumping into the test section by a micro-gear pump. Pre-heater was used to control the working fluid inlet temperature.

The test section consists of brass block with a single microchannel on its upper surface, cartridge heaters, polycarbonate housing and top cover plate. Single rectangular microchannel was drilled on the upper surface of the brass block using a milling machine at a feeding rate of (10 mm/min.). The brass block has overall dimensions 12 mm width, 60 mm length, i.e. (12 × 60 mm^2 as a base surface area) and 60 mm in height. The nominal dimensions of the microchannel are 0.3 mm width, 0.7 mm depth. As see Fig. 2, two cartridge heaters of 100 W, heating power each were inserted

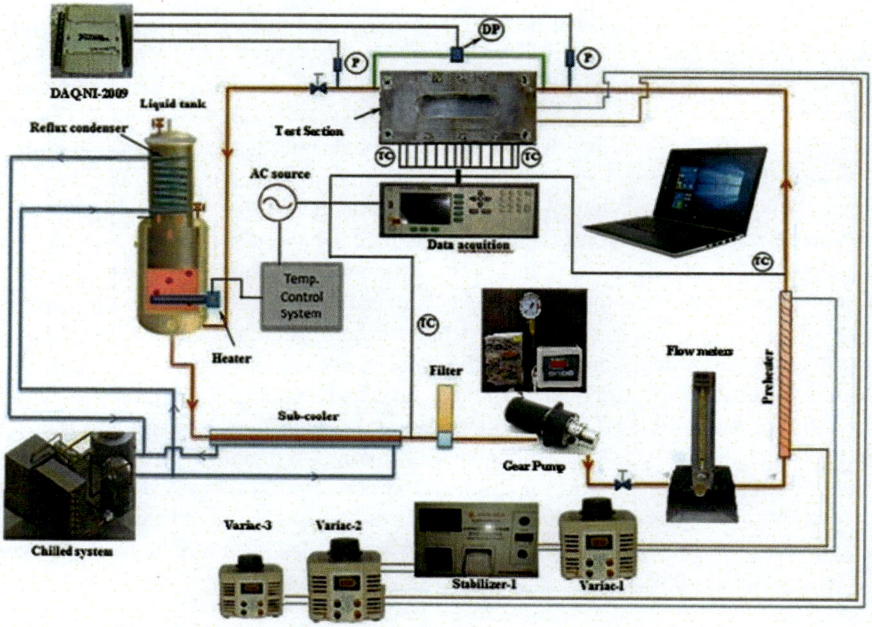

Fig. 1 Schematic diagram of the experimental facility

horizontally at the bottom of the brass block to supply the heating power to the test section. Four K-type thermocouples were implanted along the vertical centerline of the metal block at 14.7 mm equidistance to measure the heat fluxes. Seven thermocouples type-K were implanted at 1 mm underneath the channel and distributed laterally direction of the channel with a distance equal to 9.67 mm. All thermocouples diameter that implanted at the plane center of the brass block are 0.5 mm. The accuracy of thermocouples calibration were about ± 0.5 K The brass block was place into the polycarbonate housing and was sealed using O-rings as seen in Fig. 2. The housing of polycarbonate comprises of the inlet manifold and outlet manifold, inlet -outlet sub channels. Thermocouples type K with diameter of 1 mm was introduced at inlet-out manifolds to measure inlet–outlet temperature of fluid. It's calibrated with an accuracy of ± 0.5 K. The pressure of fluid at inlet -outlet manifolds were measured using absolute pressure transducers model (MPX5500DP). It's calibrated with an accuracy of ± 4.27 kPa %. Pressure drop across the test section was measured directly using differential pressure transducer (MPX4250DP). It was calibrated with an accuracy of ± 2.34 kPa. The dimensions of the inlet/outlet manifold and subchannels are (20 mm wide, 10 mmheight and 10 mmdepth), while the sub channels are (2 mm wide, 0.7 mm depth and 7 mmlength). To pass thermocouples wires from brass block through polycarbonate housing, a number of holes were drilled with diameter of (0.9 mm) in it. Layer of Polycarbonate having thickness 6 mm was sandwiched between the top polycarbonate cover plate and top-surface of the housing. To prevent

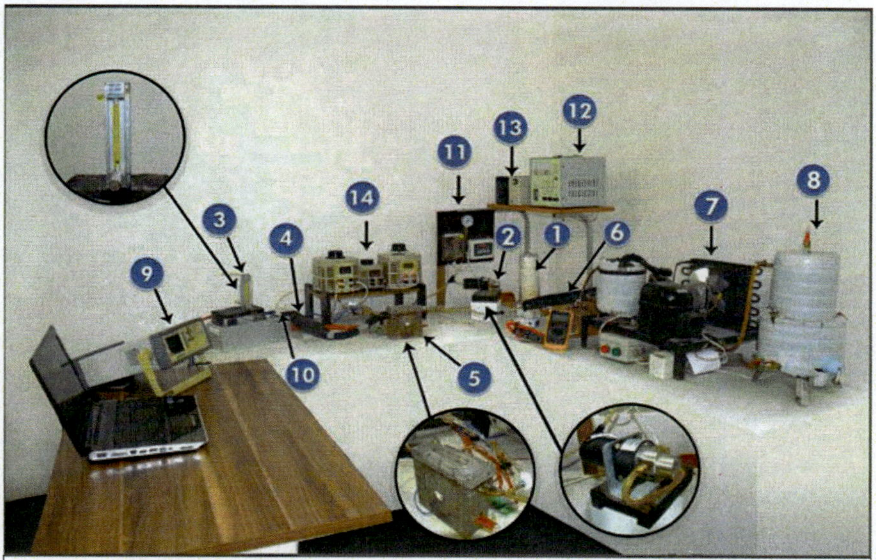

1: Filter	2: Micro-Gear Pump	3: Flowmeter	4: Pre-Heater	5: Test-section	6: Cooler	7: Chiller Unit
8: De-Gasses	9: Data Logger	10: DAQ-NI-6009	11: Gear Pump Controller	12: AC-Stabilizer	13: Unit Power Supply	14: Variacs

Plate 1 Experimental facility

any leakage from upper-surface of the housing, a layer of polycarbonate was locked by using the O-ring established in the housing at its top surface. As shown in Plate 2, the top cover plate has a visualization window of similar dimension as the block microchannel including the sub-channels. The data of temperature were recorded using data logger (Applent AT 4532x) reading. Bus-power multifunction DAQ USB (model NI USB-6009) was used as interface device to compiler the pressure transduce signals (0–5 DC–Volt) from the pressure transduce gauges to the LabVIEW software. The data were reading and recording through NI USB-2009 and LabView software at time interval of 1 ms. The experiments were conducted by constant flow rate and increasing the heat flux gradually by increasing the electrical heating power. The data were saved for 3 min and then the data results used in the data reduction process.

2.2 Data Reduction

2.2.1 Single Phase Data Reduction

For single phase flow, the net pressure drops through microchannel ΔP_{ch} is given by:

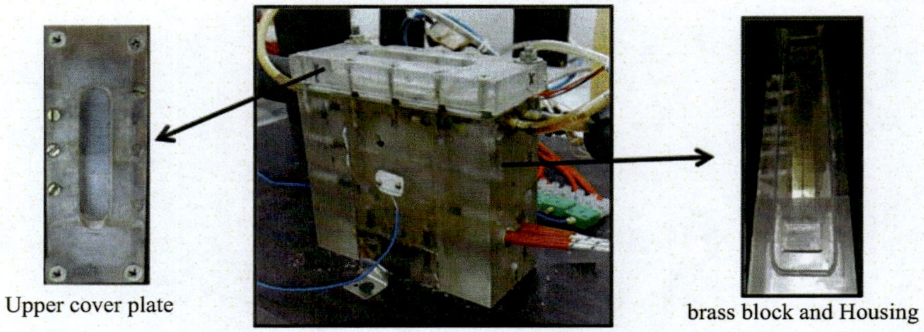

Plate 2 Test section

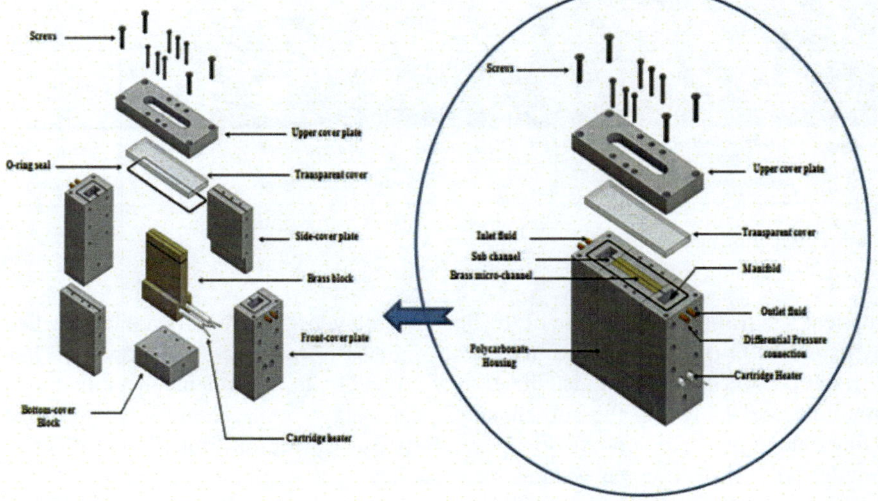

Fig. 2 Test section design and construction

$$\Delta P_{ch} = \Delta P_m - \Delta P_{loss} \tag{1}$$

Due to the inlet sub-channel and the outlet sub-channel, the contribution of losses at the maximum Reynolds number (Re = 2322) are 0.42 percent and 0.42 percent, respectively. For a lower Reynolds number, these values are considered lower. So the pressure losses due to sudden contraction and expansion were calculated as given by equation below:

$$\Delta P_{loss} = \Delta P_{sch,i} + \Delta P_{sch,o} + \Delta P_{sc} + \Delta P_{se} \tag{2}$$

The impact of the pressure losses at the highest value of Reynolds number (Re = 2322) as a result of the inlet subchannel, and the outlet subchannel were 0.42%, and 0.42%, respectively. These values will be lower for lower Reynolds number. So, losses of pressure due to sudden contraction and sudden expansion were calculated as given by equation below:

$$\Delta P_{loss} = (K_{sc1} + K_{se2}).\frac{V_{sch}^2 . \rho}{2} + (K_{sc2} + K_{se1}).\frac{V_{ch}^2 . \rho}{2} \quad (3)$$

In the above equation (K_{sc1}, andK_{sc2}) is the sudden contraction loss coefficients from the manifold to the subchannel and from the subchannel to the microchannel respectively, while(K_{se1}, andK_{se2}) is the sudden expansion loss coefficients from the microchannel to the subchannel and from the subchannel to the manifold. The values of loss coefficients (K_{sc1}, K_{sc2} K_{se1} and K_{se2}) are determined as (0.5, 0.47, 0.72 and 0.81) respectively [15].

The heat flux added to the test section $q_b^{''}$ was calculated as:

$$q_b^{''} = \frac{P - Q_{loss}}{A_b} \quad (4)$$

$$P = IV \quad (5)$$

where *P* is the heating power Watt unit, *I* and *V* are the electrical current and electrical voltage in Amperes unit and Volts unit, respectively, and A_b is the base area of the heat sink's surface which calculated with width base *W* and heated length (L_{ch}) of base block is given as $A_b = W.L_{ch}$. Here, Q_{loss} is the heat loss to the ambient, which evaluated by heating the test section without pumping the fluid (dry-test). Once the test section became steady, the heating power was documented. The process was repeated; finally the heating loss was correlated with temperature difference between the test section and the environment for different magnitudes of power input and system temperature. This method was adopted by many researchers [16] and [13].

The local heat transfer coefficient ($h_{sp}(z)$) is determined as following:

$$h_{sp}(z) = \frac{q_b^{''} W}{(T_W(z) - T_f(z))(W_{ch} + 2H_{ch})} \quad (6)$$

where *W* is the width of metal-block, Also W_{ch}, andH_{ch} are respectively, the width and height of the channel. The local thermocouples were inserted at distance (*th*) of 1 mm beneath the channel. Therefore, the values of thermocouples reading ($T_{tc}(z)$) were amended using the one-dimensional heat conduction formula to evaluated the value of inner surface of microchannel temperature ($T_w(z)$), as expressed by equation below:

$$T_W(z) = T_{tc}(z) - \frac{q_b^{''} th}{K_s} \quad (7)$$

By assuming the uniform boundary condition the locally fluid temperature ($T_f(z)$) can be calculated by equation below:

$$T_f(z) = T_i + \frac{q_b^{''} W.z}{\dot{m} C_p} \quad (8)$$

where (T_i, C_p, and $\dot{m}$) are inlet fluid temperature, specific heat of the fluid, mass flow rate and z is the axial location of the channel length.

The average Nusselt number is given by

$$Nu = \frac{1}{L_{ch}} \int_0^{L_{ch}} \frac{h_{sp}(z).D_h}{K_l} dz \quad (9)$$

2.2.2 Two-Phase Flow Data Reduction

Sub-cooled deionized water is supplied into the microchannel. So, water maintains a single phase liquid state until thermodynamic equilibrium quality approaches to zero. For this, the length of microchannel can be separated into two zones: an upstream sub-cooled inlet zone (single phase region) and downstream saturated zone; zero dryness fraction location represented the dividing point between the two zones. Consequently, the channel pressure drop is divided into the two regions, single-phase pressure drop region, ΔP_{sp} and a two-phase pressure drop region, ΔP_{tp}. Therefore, the net two-phase pressure drop through the channel can be express as:

$$\Delta P_{tp} = \Delta P_{ch} - \Delta P_{sp} \quad (10)$$

The single-phase pressure drop through the single phase region can be evaluated from equation as below:

$$\Delta P_{sp} = \frac{2 f_{app} G^2}{\rho_l D_h} L_{sp} \quad (11)$$

$$L_{sp} = \frac{\dot{m} C_p (T_{sat} - T_i)}{q_b^{''} W} \quad (12)$$

T_{sat} can be estimating by using the value of inlet pressure, assuming that the pressure drop across the sub-cooled zone is relatively small; this is realistic regard to a slight values of pressure drop through the microchannel. For developing flow, apparent friction factor (f_{app}) can be calculated using the following equation by Shah and London [17].

$$f_{app} = \frac{3.44}{Re(L^*)^{1/2}} + \frac{(f_{FD}Re) + \frac{K(\infty)}{4L^*} - \frac{3.44}{(L^*)^{1/2}}}{Re\left(1 + C(L^*)^{-0.2}\right)} \tag{13}$$

where L^* is the dimensionless axial distance $\left(L_{sp}/ReD_h\right)$, while $K(\infty)$ and C are constant depend on the aspect ratio. Regarding to the geometry utilized in the present work, these values are 1.7784×10^{-4}, and 1.1962. Also, Poiseuille number (f_{FD} Re) for fully developing single phase fluid flow was formulated by Shah and London [17].

$$(f_{FD}Re) = 24\left(1 - 1.35534\beta + 1.9467\beta^2 - 1.7012\beta^3 + 0.9653\beta^4 - 0.2537\beta^5\right) \tag{14}$$

In the above equations β is the channel aspect ratio.

At the two-phase region, the locally of heat transfer coefficient can be computing by the equation below according to energy balance equation:

$$h_{tp}(z) = \frac{q_b'' W}{(T_w(z) - T_{sat(z)})(W_{ch} + 2H_{ch})} \tag{15}$$

where $T_{sat(z)}$ defines as the locally fluid saturation temperature, and can be evaluated base on the local pressure which flowing in the channel.

The thermodynamic vapor quality at any axial position $x(z)$ was calculated as given by equation below:

$$x(z) = \frac{i(z) - i_l(z)}{i_{lg}(z)} \tag{16}$$

where $i_l(z)$, and $i_{lg}(z)$ are the liquid specific enthalpy and the latent heat of vaporization respectively which evaluated at the local pressure. Equation above covers the concept of vapor quality to sub-cooled liquid flow, for this case; it will assume negative value. The local enthalpy of the fluid, $i(z)$, at a location z through the channel is represented by energy balance.

$$i(z) = i_i + \frac{q_b'' W}{\dot{m}} \tag{17}$$

with i_i the specific enthalpy at the entrance to the channel. To asses fluid property in Eq. (16) and to evaluate the locally temperature of the two phase flow, an assumption for the pressure gradient must be assumed. In this study, the local pressure through the channel is calculated by assuming the linearly pressure drop with axial length of the micro-channel as reported by [18] and [13].

$$P_{sp(z)} = P_i - \left(\frac{z}{L_{sp}}\right)\Delta P_{sp} \quad (18)$$

$$P_{tp}(z) = P_{sat}\left(L_{sp}\right) - \left(\frac{z - L_{sp}}{L_{ch} - L_{sp}}\right)\Delta P_{tp} \quad (19)$$

Experimental results were attained at 1 atm. system pressure over the range of parameters: mass fluxes (300–600 kg/ m^2.sec) and heat fluxes (5.4–376.5 kW/m^2). The analysis of uncertainty was done using the approach clarified by Coleman and Steel [19] and the values were 10.2–15.7%, 11.4–15.9% and 2.2–50% respectively for the values of heat flux, coefficient of heat transfer and fanning friction factor. The higher value of the uncertainty of the friction factor was up to (50%), as a consequence of the errors in the flow rate and pressure drop measuring instruments through the lower values of Reynolds number.

3 Result and Discussion

3.1 Single Phase Experimental Setup Validation

The experimentation of single phase was performed to verify the experimental system before boiling flow experiments. Figure 3a shows the comparison of experimental result for friction factor with the correlation of Shah and London [17] correlation

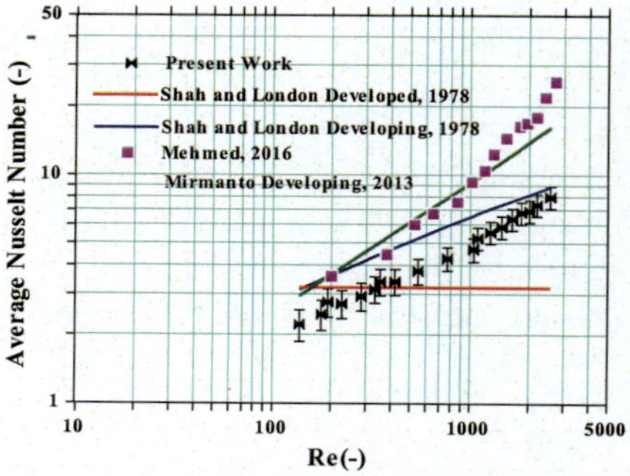

[b] Average Nusselt number comparisons with conventional and micro-scale flow correlations.

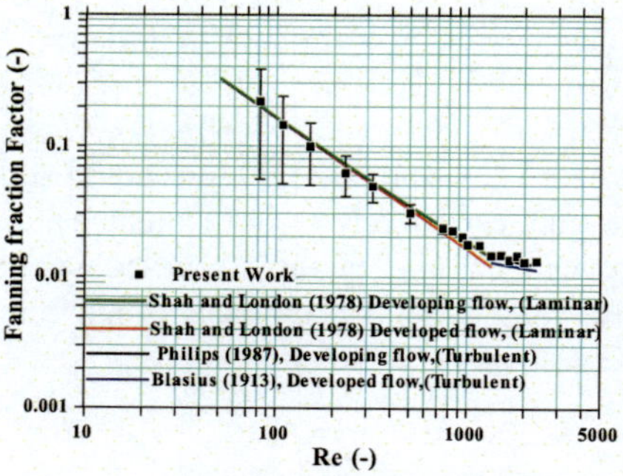

[a] Experimental fanning fraction factor results compared with fluid flow correlations.

Fig. 3 **a** Experimental fanning fraction factor results compared with fluid flow correlations. **b** Average Nusselt number comparisons with conventional and micro-scale flow correlations for developing flow (Eq. 13) and fully developed flow ($f_{FD} = C/Re$) as showed in Table A1, Appendix A. The constant C is equal to 16.3935 for the geometry used in the current study. The figure shows that there is a good promising with the correlations and the deviation is within the experimental uncertainty. Figure 3b shows the experimental Nusselt number ($Nu_{exp} = h.D_h/k_f$) compared with the predictions correlation from the correlation of Shah and London [17], and experimental correlation of Mirmanto [20], and experimental result of Mehmed [10], which it defined Table A2, Appendix A. It is clear that the experimental results prove a comparable tendency, values of Nusselt number growths with Reynolds number but the values are lower than the prediction from the correlation and previous researchers' results.

3.2 *Two-Phase Results and Discussion*

3.2.1 Boiling Curve

The influence of mass flux on the boiling curve plotted consistent to degree of inlet sub-cooled (20 K and 35 K) at system pressure of 1 atm. can be presented in Fig. 4. Data points on this curve was taken at steady state conditions and show the plot of the heat flux based on wall channel versus temperature difference between the temperature measured by the thermocouple located near the channel outlet and saturation temperature. As shown, for a certain degree sub-cooled, at low heat fluxes and different mass fluxes, the slopes of all boiling curves are constant, indicated the single phase of heat transfer. As the heat flux increases, exhibits a sudden change in slope from its single phase, indicating the onset of nucleate boiling had commenced at that location. Also from the figure below, shows that the onset of nucleate boiling is delay as an increasing of mass flux at a certain inlet degree of sub-cooled.

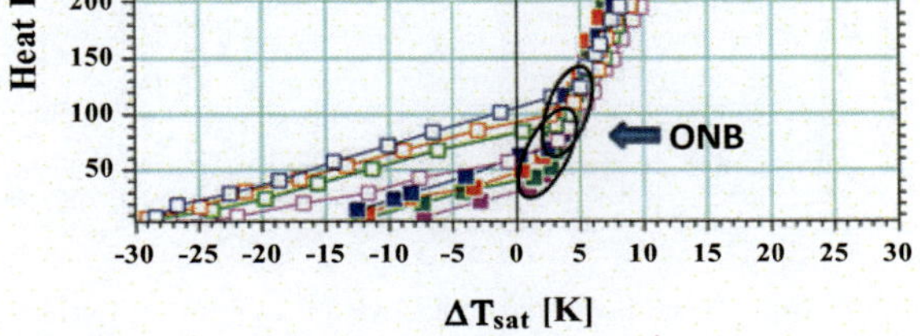

Fig. 4 The effect of inlet degree of sub-cooled (20 K and 35 K) on the Boiling Curve

Increasing of degree of sub-cooled leads to increasing in degree of superheated. Also, it delayed the onset of boiling Lee and Mudawar [6]. As shown in figure above, the lower inlet temperature required highly amount of heat flux to reach the boiling incipience than higher inlet temperature fluid flow, the values wall superheated was higher at the higher inlet sub-cooled than the lower inlet sub-cooled for along the range of heat fluxes for the certain mass flux. Therefore, onset of nucleate boiling (ONB) was occurred earlier in a low inlet sub-cooled degree Zhang [21].

3.2.2 Heat Transfer Result

The existence of local temperature sensors helps to calculate the local coefficients of heat transfer. Figure 5a, b show the variation of the local heat transfer coefficient with heat flux and thermodynamic equilibrium respectively for the range of mass fluxes 300–600 kg/m^{2}sec at two degree of inlet sub-cooled (20 K and 35 K). These figures view at the position of thermocouple location in (Z/L = 0.98), which associated to the highest value of saturated boiling and the larger vapor thermodynamic quality. However, water at the inlet of microchannel is at a sub-cooled condition then progressivity was heated up on the length of the test section. So, the flow boiling is separated in two zones [2]: Sub-cooled and saturated zone. In sub-cooled (quality (x) < 0), the heat flux range is within low and moderate, whereas in saturated zone (x > 0), the heat flux input is high. As shown in figures, the local heat transfer coefficient shows a same behavior with respect to the thermodynamic quality and wall heat flux for any given mass flux and inlet degree of sub-cooed. Where, for low to moderate heat flux at sub-cooled region, x < 0 the local heat transfer coefficient values are small, then followed by sudden increase in the local heat transfer coefficient until the vapor quality reaches zero (x = 0) at wall heat fluxes (64.2, 77.9, 93.8 and 116.7 kW/m^2) at 20 K degree of sub-cooled and (98.29, 127.1, 157.2, and 225.3 kW/m^2) at 35 K degree of sub-cooled for mass fluxes of 300, 400, 500 and

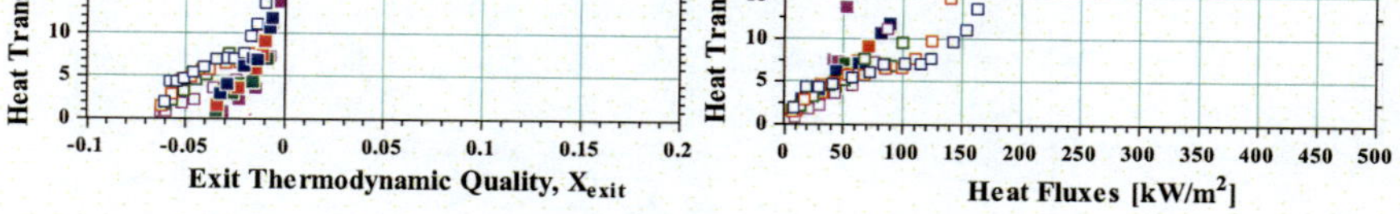

Fig. 5 Sub-cooled flow boiling heat transfer coefficient versus **a** exit thermodynamic quality and **b** wall heat flux for two different inlet degree of sub-cooled

600 kg/m^2.sec respectively. For vapor quality greater than zero ($x > 0$), saturated region, the local heat transfer coefficient begins decreasing with increasing heat flux and vapor quality. At higher heat fluxes, the local saturated heat transfer coefficient becomes largely insensitive to heat flux for the range tested. This tendency cannot support the mechanism of nucleate boiling (the values of heat transfer coefficient increases with increases the values of heat flux). Furthermore, a trend of local heat transfer coefficient was reported by previous studies of flow boiling in microchannels which documented that the convective boiling is the prevailing heat transfer process. For instance, Qu and Mudawer [2] stated that the coefficient of heat transfer increased with increasing mass flux values but it decreased when the vapor thermodynamic quality increased. It was settled that the convective boiling mechanism is the dominant heat transfer mechanism. Additionally, Lee and Garimella [22] found out that the value of coefficient of heat transfer improved with heat flux at low and moderate heat flux inputs, while it reduced with heat flux at higher heat flux input value. The authors reported that at low and moderate heat flux input, nucleate boiling dominant the boiling process, while convective boiling became the dominant mechanism of heat transfer at greater heat flux input. However, they did not show any mass flux influence on the coefficient of heat transfer as an indication. These proofs produce the conlcusion which the nucleate and forced convection boiling mechanism is the dominat mechanism for the current work. At the low mass flux, nucleate boiling starting earlier, in addition the higher heat transfer coefficient occurs at the higher mass fluxes, higher mass fluxes required a larger amount of heat for boiling incipience.

However, the higher amount of coefficient of heat transfer happened at the lower value degree of sub-cooled. As a result of the lower inlet sub-cooled have a higher frequencies of bubble departure rate and shorter waiting time for bubble creation cycle, this is because the bulk liquid temperature required a small time to reached the necessarily wall superheated for bubble nucleation. Vice versa, at the higher inlet sub-cooled degree as reported by Lee and Karayiannis [14].

The influence of the heat flux on the average heat transfer coefficient along the microchannel with range of mass fluxes 300,400,500, and 600 kg/m^2sec and degree of inlet subcooled (20 K and 35 K) can be illustrated in Fig. 6. Generally, for any given value of inlet subcooled the average heat transfer reveals a similar behavior with respect to the heat flux, that the higher values of average heat transfer coefficient occurs at a lower mass fluxes through the two-phase flow region. However, at the lower inlet degree of sub-cooled the average heat transfer coefficient along the channel is higher by 150%-200% than the average heat transfer coefficient at the higher inlet degree of sub-cooled and for all values of mass flux, as presented in Fig. 6.

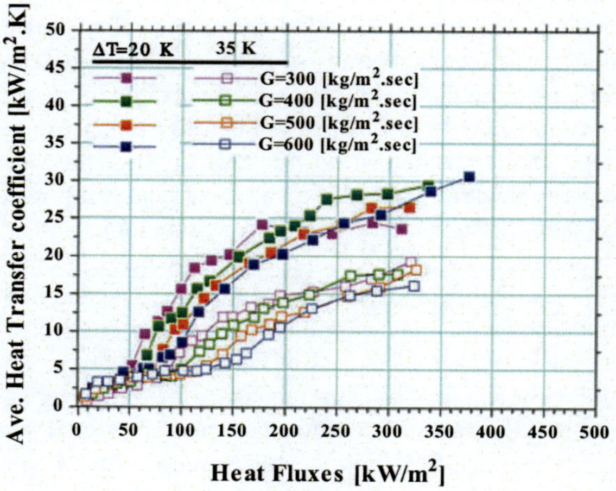

Fig. 6 The effect inlet degree of sub-cooled (20 K and 35 K) on the Average heat transfer coefficient at different mass and wall heat flux

3.2.3 Pressure Drop Results

Considering a pressure drop for the fluid flowing (de-ionized water) through the microchannel is a very significant parameter, because this parameter related with the power losses that helping the judging the effectiveness of the heat sink. So, the effect of wall heat flux on pressure drop at range of mass fluxes (300–600) kg/m^{2}sec and a degree of sub-cooled (20 K and 35 K) can be shown in Fig. 7. As shown in the figure, which the values of pressure drop along the microchannel with heat flux follows for all range of mass fluxes a same trend. However, for given value of mass flux and degree of sub-cooled, the trend curve of the pressure drop can be divided into two regions. The first region is the single phase region, where the pressure drop is decreased slightly with increasing wall heat flux as a result of the relatively decreasing

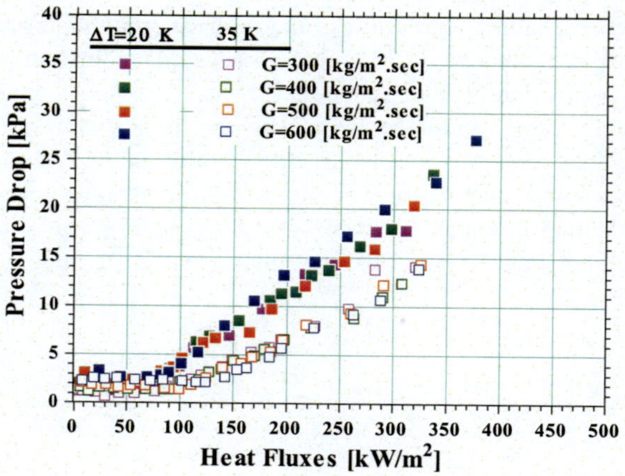

Fig. 7 The effect of inlet degree of sub-cooled (20 K and 35 K) on pressure drop microchannel for different mass and heat flux

in liquid viscosity. Increasing wall temperature causes a reduction in liquid viscosity. The effect of the inlet sub-cooled on the pressure drop is very small in single phase region due to the pressure drop of fluid flow depended on the viscosity. So, the fluid temperature for the lower inlet sub-cooled degree is greater than the higher inlet degree of sub-cooled, this trend is the same for all mass flux range. While the second region is the two phase region which it begins at a nucleate flow boiling. The values of pressure drop increasing sharply with increasing the wall heat fluxes. Furthermore, the pressure drop increased with increasing in the mass fluxes as a result of an increasing the acceleration pressure drop, since the inertia force defeat the evaporation momentum force besides increasing the frictional pressure drop also. However, increasing in the degree of sub-cooled will decrease the channel pressure drop. This was vigorous bubble in the channel condensate by higher degree of inlet sub-cooled. Also, it is contributed a small void fraction in the channel. The same versa, at the lower degree of sub-cooled. two phase microchannel pressure drop is highly dependent on the void fraction as reported by Lee and Karayiannis [14] and Zhang [21].

4 Prediction of Correlations

The heat transfer coefficient and pressure drop resulting from tests realized for this study is compared to the heat transfer coefficient and pressure drop predicted by correlations which taken from the open literature listed in Appendix B. The experimental conditions which conducted in theses correlation, were mass fluxes with (300 up to 600) kg/m^2.sec and heat fluxes range 5.4–376.5 kW/m^2 at degree of inlet sub-cooling (20 K and 35 K). The comparison between the experimental data and the selected correlation are based on the mean absolute error (MAE) defined by equation for heat transfer coefficient as given by:

$$MAE = \frac{1}{N}\sum\left|\frac{h_{exp.-h_{pred}}}{h_{exp}}\right| * 100\% \tag{20}$$

Also, for pressure drop as follows:

$$MAE = \frac{1}{N}\sum\left|\frac{P_{exp.-P_{pred}}}{p_{exp}}\right| * 100\% \tag{21}$$

4.1 Heat Transfer Coefficient Prediction

Present experimental data are compared with heat transfer coefficient suggested by Kandlikar and Balasubramanian [23], Sun and Mishima [24], and Li and Wu [25]. The comparsion of the Kandlikar and Balasubramanian [23] correlation with experimental results is depicted in Fig. 8. In The first Fig. 8a was found that the experimental data over predicted with a MAE value of 34.6%, at higher heat flux. This correlation was used to predict the coefficient of heat transfer where the flow was dominated by nucleate boiling (i.e. the heat transfer coefficient increases with increases heat flux). While the others correlation's for Kandlikar and Balasubramanian [23] was used when the flow was dominated by convective boiling. Which provides well predicted within a MAE value of 25%, as shown in Fig. 8b.

Correlation of Sun and Mishima [24] was intended for small to microchannel with diameter ranging 0.21–6.05 mm for different liquid including water and founded on 2505 flow boiling data sets. In this correlation, the convective boiling component was taken into consideration by introducing the liquid Webber number. As shown in Fig. 9, the correlation is predicting very well with the experimental data with a MAE value of 20.46%. Finally, the experimental data of the present study is compared with the correlation of Li and Wu [25]. This correlation based on the 3744 data point with twelve different liquids including de–ionized water at ranges of hydraulic diameters (0.148–3.25 mm).The correlation considered the effect of surface tension by using the Bond number. As demonstrated in Fig. 10 that the correlation deviates little from experimental data with MAE value of 14.75%. It can be proved from these prediction correlations that the dominated mechanism of the heat transfer in the current study is convective flow boiling dominated mechanism.

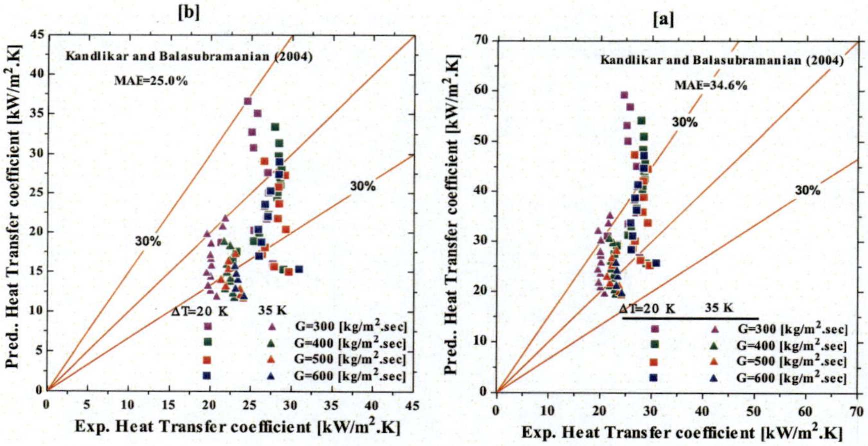

Fig. 8 Comparison of experimental heat transfer coefficient data with **a** nucleate boiling mechanism correlation and **b** convection boiling mechanism correlation of Kandlikar and Balasubramanian [23], (20 K and 35 K) inlet temperature sub-cooled

Fig. 9 Comparison of experimental heat transfer coefficient data with correlation of Sun and Mishima [24], at 20 K and 35 K inlet temperature sub-cooled

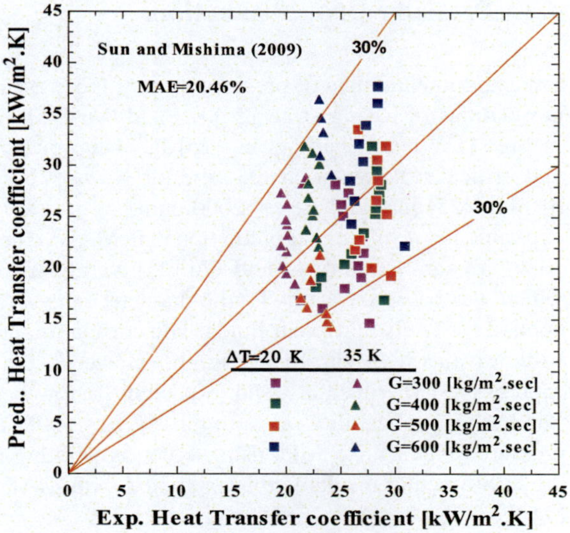

Fig. 10 Comparison of experimental heat transfer coefficient data with correlation of Li and Wu [25], at (20 K and 35 K) inlet temperature sub-cooled

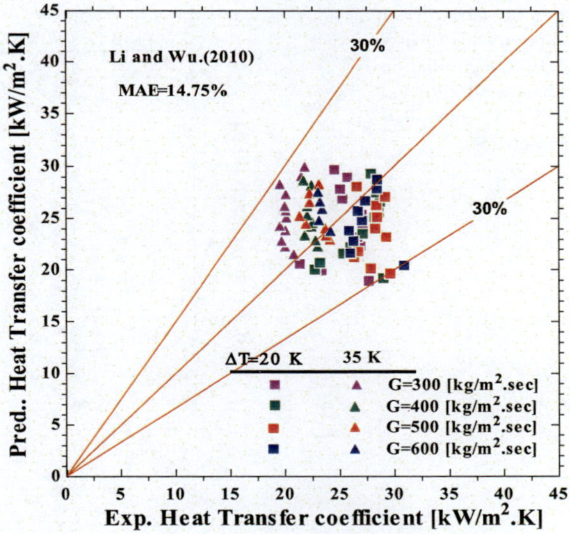

4.2 Pressure Drop Prediction*:*

The experimental data of pressure drop of the present study are compared with the correlations of Zhang at al. [26], Li and Wu [25], and Yan et al. [27]. As shown in Fig. 11 Where Zhang et al. [26] has been proposed correlation for horizontal and vertical rectangular channel and tubes which based on hydraulic diameter rang (0.146–6.25) mm and $Re_f \leq 2000$ and $Re_g \leq 2000$. This correlation predicts the experimental results under prediction with MAE value of 35.8%, as shown in Fig. 11. Also, the correlation of Li and Wu [25] were predicted for the experimental data, which was based on the hydraulic diameter ranges from 0.148 up to 3.25 mm and applied for twelve different fluids. The correlation predicts poor results with MAE value of 39.1% as shown in Fig. 12. However, the correlation doesn't correctly capture the experimental trend, especially at low heat flux. The correlation for J. Yan has an applicability range of mass fluxes 6000–10,000 kg/m^{2}ses at the system pressure range of 3–5 MPa using water as working fluid. The correlation predicts the experimental results within predicted with MAE value of 31.1%, as shown in Fig. 13.

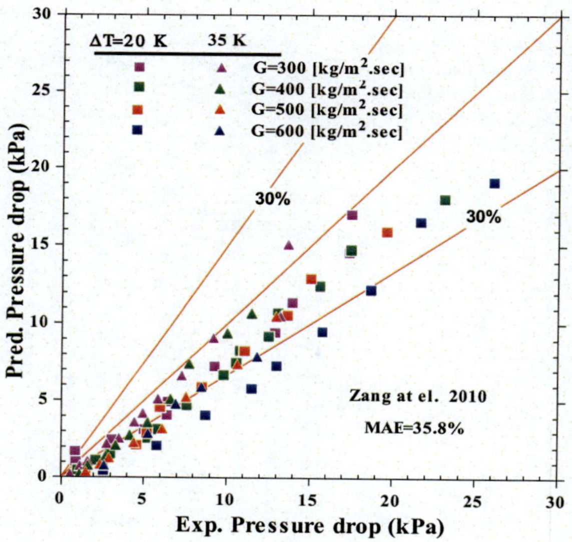

Fig. 11 Comparison of experimental Pressure drop data with correlation of Zhang et al. [26], at (20 K and 35 K) inlet degree of sub-cooled

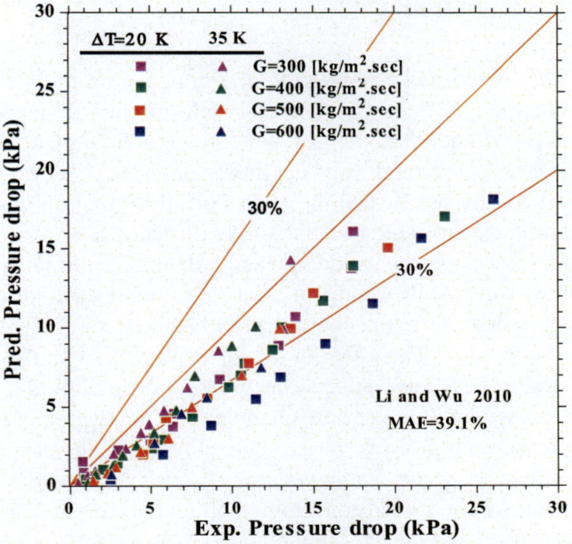

Fig. 12 Comparison of experimental Pressure drop data with Li and Wu [25] correlation, at (20 K and 35 K) inlet degree of sub-cooled

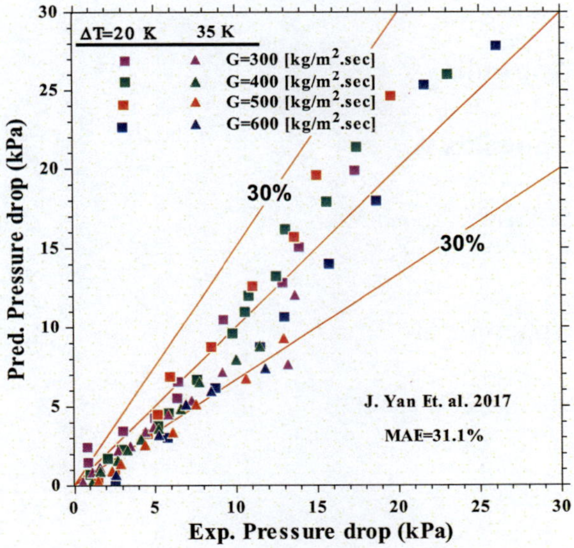

Fig. 13 Comparison of experimental Pressure drop data with Yan [27] correlation, at (20 K and 35 K) inlet degree of sub-cooled

5 Conclusion

The experimental works of different degree of inlet sub-cooled flow boing in a brass single microchannel heat sink utilizing de-ionized water were accomplished at (20 K and 35 K) degree sub-cooled with range of mass fluxes 300, 400, 500, and 600 kg/m^2.sec and wall heat fluxes 5.6–376.5 kW/m^2. Markedly, at low heat flux and different mass fluxes, the slopes of all boiling curves at a certain inlet sub-cooled are constant, indicated the single phase of heat transfer. As the value of heat flux increases, shows a sudden change in slope from its single-phase flow to two-phase flow zone. At the same time, the onset of nucleate boiling is delay as an increasing of mass flux. Also, the coefficient of heat transfer increasing with increasing the value of mass flux. The effect of the inlet sub-cooled on the pressure drop is very small in single phase region due to the pressure drop of fluid flow depended on the viscosity. For the two phase region, changing in inlet degree of sub-cooled are affected on the pressure drop of two phase state. Obviously, increasing in the degree of sub-cooled will decrease the channel pressure drop. The dominate heat transfer mechanism is the nucleate and forced convection boiling for present work. The microscale correlations could predict the experimental data reasonably well.

Appendix

Appendix A

See Tables A1 and A2.

Appendix B

See Tables B1 and B2.

Table A1 Single phase fluid flow validation equations

	Flow Regime		Equations	References
Fanning friction factor	Laminar flow	Developing flow	$f_{app} = \frac{3.44}{Re(L^*)^{1/2}} + \frac{(f_{FD}Re) + \frac{K(\infty)}{4L^*} - \frac{3.44}{(L^*)^{1/2}}}{Re\left(1 + C(L^*)^{-0.2}\right)}$	Shah and London [17]
		Developed flow	$(f_{FD}Re) =$ $24\left(1 - 1.35534\beta + 1.9467\beta^2 - 1.7012\beta^3 + 0.9653\beta^4 - 0.2537\beta^5\right)$	
	Turbulent flow	Developing flow	$f_{appturb.} = \left(0.0929 + \frac{1.0161D_h}{L}\right)Re^{*\left(-0.268 - \frac{0.3193D_h}{L}\right)}$ Where: $Re^* = Re\left(2/3 + \frac{11}{24}\beta(2 - \beta)\right)$	Phillis [28]
		Developed flow	$f_{FD} = 0.079Re^{-0.25}$	Blasius (2013)

Table A2 Single phase heat transfer fluid flow validation equations

	Flow Regime		Equations	References
Nusselt No	Laminar Flow	Developing flow	$Nu = 0.775L{*}_t^{(-1/3)}(f\,Re)_{FD}^{(1/3)}$ $L{*}_t = \frac{L}{RePrD_h}$ $L{*}_t$ is a dimensionless channel	Shah and London [17]
		Developed flow	$Nu =$ $8.235(1 - 10.6044\beta + 61.1755\beta^2 - 155.1803\beta^3 + 176.9203\beta^4 - 72.9236\beta^5)$	
		Developing flow	$Nu = Re^{0.283}Pr^{-0.513}L{*}_t^{-0.309}$	Mirmanto [20]

Table B1 Flow boiling heat transfer correlations

Applicability range	Correlations	Reference
G = 13 to 8179 kg/m^{2}s, D = 4–32 mm Water, R-11, R-12, R-22, R-13, R-134a, R-152a	For Re < 1600 $h_{tp,NBD} =$ $0.6683Co^{-0.2}(1-x)^{0.8}h_L + 1058Bo^{0.7}(1-x)^{0.8}F_{fl}h_L$ $h_{tp,CBD} =$ $1.136Co^{-0.9}(1-x)^{0.8}h_L + 667.2Bo^{0.7}(1-x)^{0.8}F_{fl}h_L$ $h_L = \frac{Nuk}{D_h}$ Whare: $h_{tp,NBD}$: Two-phase heat transfer coefficient for nucleate boiling dominated $h_{tp,CBD}$: Two-phase heat transfer coefficient for convective boiling dominated F_{fl}: fluid surface parameter = 1 for water	Kandlikar and Balasubramanian [23]
D_h=0.21–6.05 mm	$h_{tp} = \frac{6Re_L^{1.05}Bo^{0.54}K_l}{We_L^{0.191}(\rho_L/\rho_g)^{0.142}D_h}$	Sun and Mishima [24]
Based on 3744 date points. twelve different fluids including water, D_h=0.148–3.25 mm	$h_{tp} = 334Bo^{0.3}\left(BdRe_L^{0.36}\right)^{0.4}\frac{K_L}{D_h}$ $Bd = \frac{g\Delta\rho D_h^2}{\sigma}$	Li and Wu [25]

Table B2 Flow boiling pressure drop correlations

Applicability range	Correlations	Reference
P = 3–5 MPa, G = 6000–10,000 kg/m^2. Sec and q''=7.5–12.5 MW/m^2	$\Phi^2 = 1 + 2250 Bo^{1.5} Ja^{-1.43}\left(\frac{\rho_l}{\rho_g}\right)^{0.2}$	Yan et al. [27]
D_h=0.146–6.25 mm, Horizontal and Vertical, Rectangular channel and tubes, $Re_f \leq 2000$, $Re_f \geq 2000$,	$\Delta P_a = \frac{G^2}{\rho_L}\left[\frac{X_e^2}{\alpha_e}\frac{\rho_L}{\rho_g} + \frac{(1-X_e)^2}{(1-\alpha_e)} - 1\right]$ $\alpha_e = \frac{1}{1+0.28X_M}$ $X_M = \left(\frac{1-X_e}{X_e}\right)^{0.64}\left(\frac{\rho_L}{\rho_g}\right)^{0.36}\left(\frac{\mu_L}{\mu_g}\right)^{0.07}$ $\Delta P_f = \frac{L_{tp}}{X_e}\int_0^x \frac{2 f_L G^2}{D_h \rho_L}\Phi_L^2 dx$ $\Phi_L^2 = 1 + \frac{C}{X} + \frac{1}{X^2}$ $X = \left(\frac{f_L}{f_g}\right)^{0.5}\left(\frac{\rho_L}{\rho_g}\right)^{0.5}\left(\frac{1-x_e}{x_e}\right)$ $C = 2\left(1 - e^{(-0.358/L_a)}\right)$ $L_a = [\sigma/g\Delta\rho]^{0.5}/D_h$	Zhang et al. [26]
D_h=0.148–3.25 mm, twelve different fluids, 769 data points	$\Delta P_a = \frac{G^2}{\rho_L}\left[\frac{X_e^2}{\alpha_e}\frac{\rho_L}{\rho_g} + \frac{(1-X_e)^2}{(1-\alpha_e)} - 1\right]$ $\alpha_e = \frac{1}{1+0.28X_M}$ $X_M = \left(\frac{1-X_e}{X_e}\right)^{0.64}\left(\frac{\rho_L}{\rho_g}\right)^{0.36}\left(\frac{\mu_L}{\mu_g}\right)^{0.07}$ $\Delta P_f = \frac{L_{tp}}{X_e}\int_0^x \frac{2 f_L G^2}{D_h \rho_L}\Phi_L^2 dx$ $\Phi_L^2 = 1 + \frac{C}{X} + \frac{1}{X^2}$ $X = \left(\frac{f_L}{f_g}\right)^{0.5}\left(\frac{\rho_L}{\rho_g}\right)^{0.5}\left(\frac{1-x_e}{x_e}\right)$ $C =$ $\begin{cases} 11.9 Bd Re^{0.5} Bd \leq 1.5 \\ 109.4\left(Bd Re^{0.5}\right)^{-0.56} 1.5 < Bd \leq 11 \end{cases}$ $B_d = g\Delta\rho D_h^2/\sigma$	Li and Wu [25]

References

1. Yarin LP, Mosyak A, Hetsroni G(2008) Fluid flow, heat transfer and boiling in micro-channels. Springer Science & Business Media, Sep 19
2. Qu W, Mudawar I (2003) Flow boiling heat transfer in two-phase micro-channel heat sinks—I. Experimental investigation and assessment of correlation methods. International journal of heat and mass transfer 46(15):2755–2771, Jul 1

3. Liu D, Garimella SV (2007) Flow boiling heat transfer in microchannels. J Heat Transf 129(10):1321
4. Agostini B, Thome JR, Fabbri M, Michel B, Calmi D, Kloter U (2008) High heat flux flow boiling in silicon multi-microchannels–Part II: heat transfer characteristics of refrigerant R245fa. Int J Heat Mass Transf 51(21–22):5415–5425, Oct 1
5. Lee J, Mudawar I (2008) Fluid flow and heat transfer characteristics of low temperature two-phase micro-channel heat sinks–Part 2. Subcooled boiling pressure drop and heat transfer. Int J Heat Mass Transf 51(17–18):4327–41, 1 Aug
6. Lee J, Mudawar I (2008) Fluid flow and heat transfer characteristics of low temperature two-phase micro-channel heat sinks–Part 1: experimental methods and flow visualization results. Int J Heat Mass Transf 51(17–18):4315–4326, 1 Aug
7. Harirchian T, Garimella SV (2009) Effects of channel dimension, heat flux, and mass flux on flow boiling regimes in microchannels. Int J Multiph Flow 35(4):349–362, 1 Apr
8. Suzuki K, Nomura T, Hong C, Yuki K (2009) Subcooled flow boiling with microbubble emission in a michrochannel. In: International conference on micro/nanoscale heat transfer, Jan 1, Vol 43901, pp 83–89
9. Galvis E, Culham R (2012) Measurements and flow pattern visualizations of two-phase flow boiling in single channel microevaporators. Int J Multiph Flow 1(42):52–61, Jun
10. Karayiannis TG, Mahmoud MM, Özdemir MR (2016) Flow boiling heat transfer in a rectangular copper microchannel 2(2):761–773
11. Yin L, Xu R, Jiang P, Cai H, Jia L (2017) Subcooled flow boiling of water in a large aspect ratio microchannel. Int J Heat Mass Transf 1(112):1081–1089, Sep
12. Ramesh B, Jayaramu P, Gedupudi S (2019) Subcooled flow boiling of water in a copper microchannel: experimental investigation and assessment of predictive methods. Int Commun Heat Mass Transf 1(103):24–30, Apr
13. Mohammed SA, Fayyadh EM (2020). Experimental investigation of heat transfer and flow characteristics in different inlet subcooled flow boiling in microchannel. In: IOP conference series: materials science and engineering, Vol 671, No 1, p 012023. IOP Publishing,Jan
14. Lee VY, Karayiannis TG (2020) Effect of inlet subcooling on flow boiling in microchannels. Appl Therm Eng 25(181):115966, Nov
15. Shaughnessy EJ, Katz IM, Schaffer JP (2005) Introduction to fluid mechanics. Oxford University Press, New York
16. Koşar A, Kuo CJ, Peles Y (2005) Boiling heat transfer in rectangular microchannels with reentrant cavities. Int J Heat Mass Transf 48(23–24):4867–4886, 1 Nov
17. Shah RK, London AL (1978) Laminar flow forced convection in ducts, Supplement 1
18. Mirmanto M (2014) Heat transfer coefficient calculated using a linear pressure gradient assumption and measurement for flow boiling in microchannels. Int J Heat Mass Transf 1(79):269–278, Dec
19. Coleman HW, Steele WG (2018) Experimentation, validation, and uncertainty analysis for engineers. John Wiley & Sons, May 8
20. Karayiannis TG, Lewis JS, Kenning DB. Single-phase flow and flow boiling of water in horizontal rectangular microchannels (Doctoral dissertation, Brunel University School of Engineering and Design PhD Theses)
21. Zhang S, Yuan W, Tang Y, Chen J, Li Z (2016) Enhanced flow boiling in an interconnected microchannel net at different inlet subcooling. Appl Therm Eng 5(104):659–667, Jul
22. Lee PS, Garimella SV (2008) Saturated flow boiling heat transfer and pressure drop in silicon microchannel arrays. Int J Heat Mass Transf 51(3–4):789–806, 1 Feb
23. Kandlikar SG, Balasubramanian P (2004) An extension of the flow boiling correlation to transition, laminar, and deep laminar flows in minichannels and microchannels. Heat Transfer Eng 25(3):86–93, 1 Apr
24. Sun L, Mishima K (2009) Evaluation analysis of prediction methods for two-phase flow pressure drop in mini-channels. Int J Multiph Flow 35(1):47–54
25. Li W, Wu Z (2010) A general correlation for adiabatic two-phase pressure drop in micro/mini-channels. Int J Heat Mass Transf 53(13–14):2732–2739, 1 Jun

26. Zhang W, Hibiki T, Mishima K (2010) Correlations of two-phase frictional pressure drop and void fraction in mini-channel. Int J Heat Mass Transf 53(1–3):453–465, 15 Jan
27. Yan J, Guo P, Bi Q, Liu Z, Zhang Q, Yang Z (2017) Pressure drop for highly subcooled water flow boiling under high heat and mass fluxes. Appl Therm Eng 1(124):1061–1074, Sep
28. Phillips RJ (1987) Forced-convection, liquid-cooled, microchannel heat sinks (Doctoral dissertation, Massachusetts Institute of Technology)

Comparative Assessment of Friction Head Losses in Pipe Flow

Saeed A. Madodi, Hasanain A. Abdul Wahhab, and Nabeel Sameer Mahmoud

Abstract In spite of widely investigated and reported determination of frictional losses due to turbulent pipe flow, it is still demonstrating inconsistency. This work is presenting comparisons of estimated frictional losses results in study turbulent pipe flow as measured experimentally, estimated by empirical relations and predicted numerically. The Wilson frictional losses have been evaluated at inclination angles of 0°, 45° and 90°, different pipe materials, different inside diameters and at various flow rates ranging from 1.8 to 5.1 m^3/s covering range of Reynolds number from 1925 to 4100. The results appeared that the head loss is strongly dependent on pipe inclination angle and this behaviour increases with its increase, for different pipe material and with different pipe diameter. Numerical results shown turbulence intensity increased with the inclination of pipe, particularly, its start to become unstable near at the inlet, with velocity profiles becoming more turbulence. The minimum head loss per unit length was estimated for three of the inclination angles (0°, 45°, and 90°), using this type of data, the feasibility of a piping system can be assessed. The numerical results and experimental data were compared and a good agreement was obtained.

1 Introduction

The single phase flow occurs due to the difference in pressure between one region and another and as a result, the fluid particles move from the high pressure region to the low pressure region, thus possessing a certain kinetic energy for a specific velocity, where the fluid velocity is directly proportional to the amount of pressure drop. And it depends on the dimensions and shape of the space through which the fluid passes [1]. While in the two-phase flow, the principle differs with regard to the movement of the carried material, as the kinetic energy of the fluid is used and part of

S. A. Madodi (✉)
Technical Institute/Kut, Middle Technical University, Baghdad, Iraq
e-mail: saeedabbas@mtu.edu.iq

H. A. Abdul Wahhab · N. S. Mahmoud
Mechanical Engineering Department, University of Technology, Baghdad, Iraq

F. Ahmad et al. (eds.), *ICPER 2020*, Lecture Notes in Mechanical Engineering,
https://doi.org/10.1007/978-981-19-1939-8_27

it is given to the portable material that is to be transported from one place to another or given a kinetic energy to be used for certain purposes [2, 3]. Therefore, two-phase flow is widely used in most engineering disciplines, as researchers have noticed since the beginning modern industry this phenomenon has many applications, some of which are applied in power plants and factories producing chemicals in addition to many other applications [4–6]. On the other hand, the study of flow field in the curve pipes are important as the straight pipes in production systems; chemical, biochemical, mechanical, civil, nuclear, and engineering as helically coiled heat exchangers, cooling channels…etc. The curvature in pipes (changing direction of the fluid flow) influences on flow patterns and wall shear stress. As it is caused increase loss energy such as an increase in friction losses and pressure drop. This behaviour is same with change the direction of the pipe or changing in flow inclination angle [7, 8]. The fluid flow through pipes is resisted by viscous shear stresses within the fluid and the turbulence that occurs along the internal walls of the pipe, created by the roughness of the pipe material. This resistance is usually known as pipe friction and is measured is feet or meters head of the fluid, thus the term head loss is also used to express the resistance to flow [9]. The resistance increases for flow rate values in transition and turbulent regions. Studies elsewhere have shown that high velocities produce high resistances to flow in pipes and hence head losses values for particular type of surface roughness [10].

Many researchers ensured that for any flow of fluid in a pipe exhibiting some form of roughness; and the head loss due to friction was produced, Mousavi and Hooman [10], Qasim and Khudheyer [11]. Pressure loss in steady pipe flow is calculated using the Darcy-Weisbach equation; this equation includes the Darcy friction factor [12]. The exact solution of the Darcy friction factor in turbulent flow is got by looking at the Moody diagram [6] or by solving it from the Colebrook equation [2]. If the Darcy friction factor must be known only once, the Moody diagrams good. In past years, this diagram was rather laborious to program in a computer code and doesn't offer any advantages [12–14]. In order to avoid the iteration of the Darcy friction factor from the Colebrook equation, new equations are deduced. In this study three equations to replace the Colebrook equation are validated, these three are Blasius, Swamee-Jain, and Haaland equations [15]. The research is included experimental and numerical investigations to steady turbulent flow, in which the pipes are used circular, and the solved Darcy friction factors are compared to the solution of the Colebrook equation. Also, it was evaluate friction head losses and pressure drop at the angles inclination pipes (0°, 45° and 90°), for different material types and inside diameters but to same length, various flow rates on the Wilson friction loss calculations using water as the process fluid at 25 °C.

2 Experimental Procedure

An experimental study of steady water flows inside two different pipe materials (Cast. Iron and PVC) with two different inside diameter (12.7 and 25.4 mm) and installing in three directions: horizontal (at angle 0°), inclination (at angle 45°), and vertical (at angle 90°) was performed. Figure 1 showed a schematic of the experimental setup. Also, the experimental setup used consists of a Centrifugal pump, Water Tank, set of flow rate measuring instruments (Rotameters), a scaled board on which U-tubes Manometers are assembled for the hydrostatic pressure readings, eight experimental pipes and a vales for flow control.

As depicted in figure, pressure measurement ports were placed on pipes at sixteen points at the ends of each test pipes (0 cm and 125 cm respectively). For the flow to be fully developed according to relation (L = 60 D) from entrance length (it is a distance from pipe inlet to a downstream location where fully developed flow is established) [7]. After ensuring that the flow between two ports is arrived to fully developed, the pressure drop values recorded between two ports in each pipe were used to determine the friction losses for the fully developed pipe flow investigation. The head loss between each two ports on pipe was obtained from the difference between both water port heights. The head friction losses of the pipe flow determined through the relation given in Eq. (1) were compared with pressure drop ($\Delta p/\gamma$) at the

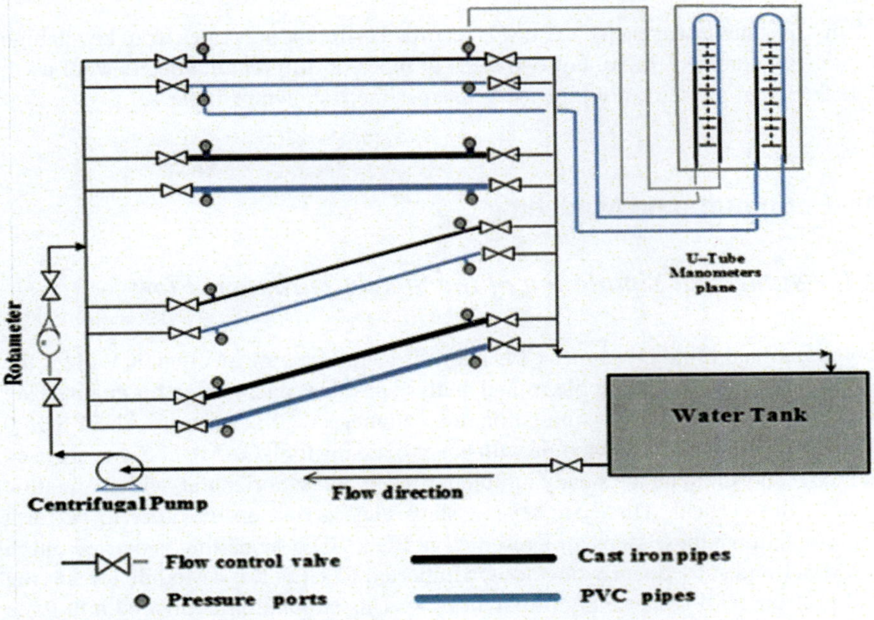

Fig. 1 A schematic of the experimental setup

Table 1 The physical properties of water and technical properties of pipes

Physical properties of water		
	Mean temperature, T	25 °C
	Density, ρ	997.13 kg/m^3
	Dynamic viscosity, μ	0.000891 kg/m.s
Technical properties of Cast. Iron pipe		
	Pipe diameter, D Pipe length, L	12.7 – 25.4 mm 1250 mm
	Roughness, Φ	0.15 mm
Technical properties of PVC pipe		
	Pipe diameter, D Pipe length, L	12.7–25.4 mm 1250 mm
	Roughness, Φ	0.0015 mm

given flow rates each in experiments. The physical properties of water in experiments were specified according to the average water temperature statistically measured in experiments, Table 1.

$$H_f = \frac{\Delta p}{\gamma} = f\frac{L}{D}\frac{V^2}{2g} \tag{1}$$

In experiments for each case, it was recorded different flow rates from low to high value with a control valve. For two sizes of pipes of different diameters were used. The technical properties of pipe used in experiment shown in Table 1.

3 Computational Modelling

3.1 Numerical Simulation of the Steady Turbulent Flow

Computational Fluid Dynamics CFD analysis techniques are an efficient tool for the represent of mechanical problems and analyse physical phenomena for engineering applications. In the present simulation, the commercial CFD software ANSYS 18.0 FLUENT Turbulent Flow Module with pre-processing tool, GAMBIT 2.4.6, are used to model and simulate the steady turbulent flow in pipes; horizontal, with inclination at 45°, and vertical. There are several steps that carried out in order to perform the simulations; these steps are presented in Fig. 2. The modelling is carried out in GAMBIT and subsequently the mesh is imported FLUENT, k-ε Module for solving and post processing as steady turbulent flow. This simulation is divided into three modellers according to the research objectives. In the first modeller, the dynamics of steady turbulent flow is explained. The k-ε model is used to simulate the turbulent flow

Fig. 2 Simulation steps in ANSYS 18.0

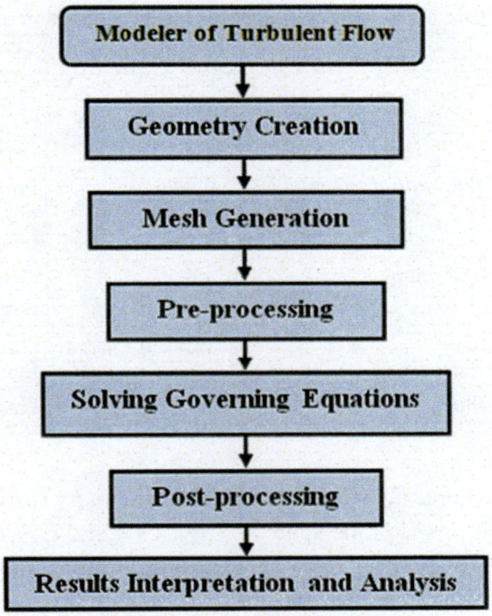

in horizontal two sizes of pipes. In second modeller, the dynamics of turbulent flow through inclination with 45° is explained. In third modeller, includes flow modelling in vertical pipes.

3.2 *Governing Equations*

The fluid flow at the chemical and mechanical plants and inside of the dynamic systems can be considered in batches because the inflow of fluid inside the system varies with different; his physical properties, momentum value, and the pipe wall roughness within the system. The loss head at the flow channels of the hydrodynamic system can be determined using standard k-ε model:

k-Transport equation:

$$\rho \frac{D_k}{D_t} = \frac{\partial}{\partial x_i}\left[\left(\mu + \frac{\mu_t}{\sigma_k}\right)\frac{\partial k}{\partial x_i}\right] + \mu_t S^2 - \rho\varepsilon;\ S = \sqrt{2S_{ij}S_{ij}} \tag{2}$$

ε-Transport equation:

$$\rho \frac{D_\varepsilon}{D_t} = \frac{\partial}{\partial x_i}\left[\left(\mu + \frac{\mu_t}{\sigma_\varepsilon}\right)\frac{\partial \varepsilon}{\partial x_i}\right] + \frac{\varepsilon}{k}\left(C_{\varepsilon 1}\mu_t S^2 - \rho C_{\varepsilon 2}\varepsilon\right) \tag{3}$$

Table 2 Boundary conditions for the simulation of turbulent flow model

Component	Boundary type	Value
Horizontal pipe	Wall	Pipe wall roughness depend on the materials properties (Cast iron and PVC), the flow assumed no-slip
Pipe with inclination at 45°	Wall	Pipe wall roughness depend on the materials properties (Cast iron and PVC), the flow assumed no-slip
Vertical pipe	Wall	Pipe wall roughness depend on the materials properties (Cast iron and PVC), the flow assumed no-slip
Pipe inlet	Velocity inlet	Inlet parameters depending on ambient condition
Pipe outlet	Pressure outlet	Pressure difference

where, μ_t and ε are turbulent viscosity (production) and dissipation, respectively. k is inverse time scale. σ_k, σ_ε, $C_{\varepsilon 1}$, and $C_{\varepsilon 2}$ are empirical constants determined from benchmark experiments of simple flows using water.

3.3 Boundary Conditions and Simulation Parameter Cases

This simulation employed turbulent model which helps to determine the various velocities, pressure drop, and turbulence kinetic energy of the piping system from simulation results following the technical properties of three models with different pipe inclination; horizontal, with 45°, and vertical. The boundary conditions used in the simulation are presented in Table 2.

Figure 3 shows the 3D geometrical model and mesh for three position of piping system. Then the flow information for the fluid is exchanged periodically. This approach can save computation time and has enough accuracy in dealing with many complex non-linear problems. The pressure based type solver was used. Nodes and elements of fine mesh for all models represented in Table 3.

4 Results and Discussion

Figure 4 represents the relationship between pressure drop and Reynolds number of different cases for turbulent flow where it is noticed through the graph that the pressure drop is strongly dependent on pipe inclination angle and this behaviour increases with its increase, for different pipe material and with different pipe diameter. Because the amount of pressure drops in the pipe, it depends on the velocity of fluid flowing inside it. The reason is that there are two forces that affect the movement of the fluid inside the pipe, which are momentum and shear forces, both of which represent

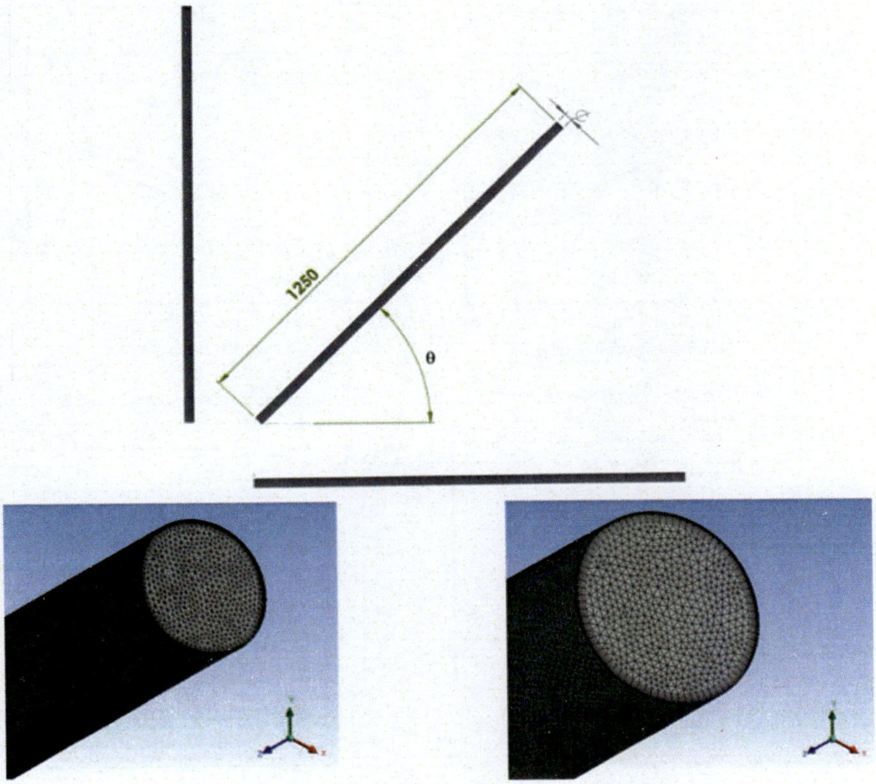

Fig. 3 Computational domain and geometrical mesh model of piping system

Table 3 Nodes and elements for three models of piping system

Model	Dimensions	Nodes	Elements
Horizontal	Pipe diameter 12.7 mm, pipe length 1250 mm	4,183,466	3,333,974
Inclination at 45°	Pipe diameter 25.4 mm, pipe length 1250 mm	7,893,458	6,662,694
Vertical			

energy that moves from the fluid towards the walls of the pipe and generates one force, which is the flow impedance force. This effect increases with the velocity of the flow. More so, the increasing of pipe inclination angle and its wall roughness was increased pressure drop rates.

According to Fig. 5, firstly for any pipe material, and with different pipe diameter, the head loss per unit length is strongly dependent on pipe slope when the pipe inclination is 90° (vertical), the head loss per unit length is significantly higher than for flow in sloping pipes at 45° and horizontal for the same diameter. Secondly,

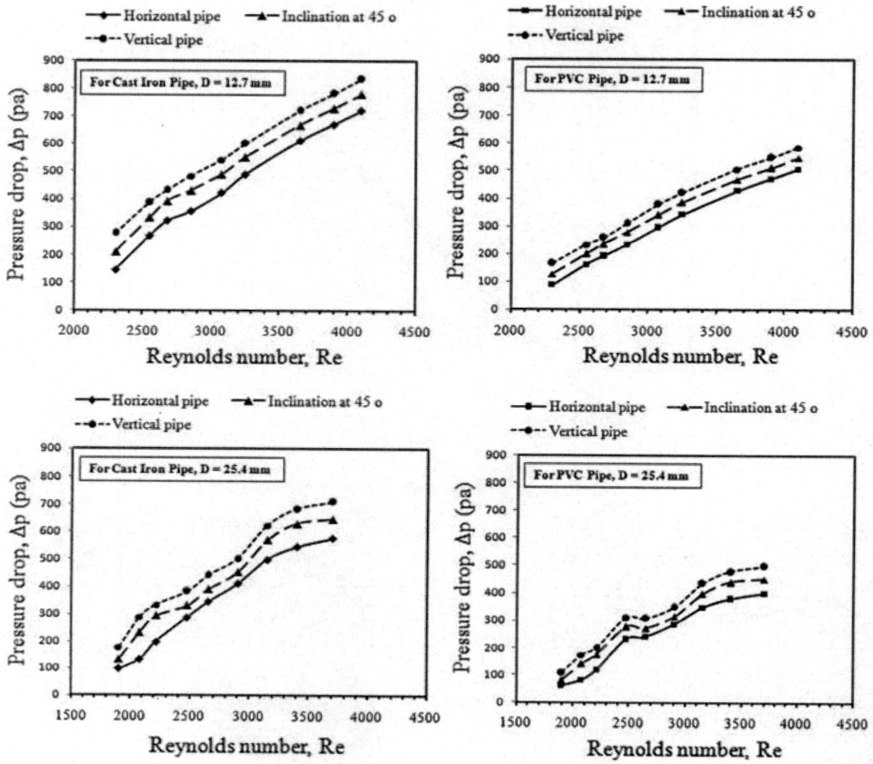

Fig. 4 Pressure drop as a function of Reynolds number for different pipe materials (Cast Iron and PVC) and pipe diameter (12.7 and 25.4 mm)

Head loss per unit length is also as strongly dependent on pipe material when the roughness of pipe wall is high value (increase friction and shear intensity with high Reynolds numbers). Thirdly, based on figure, head loss per unit length was lower for the larger pipe diameter. In fact, for a given pipe inclination and pipe wall roughness, the head loss per unit length increases with decreasing discharge below a critical Reynolds number. For each pipe material type, pipe inclination, and pipe diameter, the lowest possible head loss per unit length can be estimated from the data. This information can be used to predict the best type of materials that can be used in a pipe of known slope. The minimum head loss per unit length was estimated for three of the inclination angles (0, 45°, and 90°), using this type of data, the feasibility of a piping system can be assessed. If the inlet and outlet heads, pipe inclination, and pipe diameter are known, the maximum length of bypass pipe possible can be determined.

Investigation focused on turbulence intensity profiles and the extent to which it is affected by variation of flow direction. The simulation contours of k-ε model for higher the Reynolds number 4100 in a Cast Iron pipe, D = 25.4 mm at a different

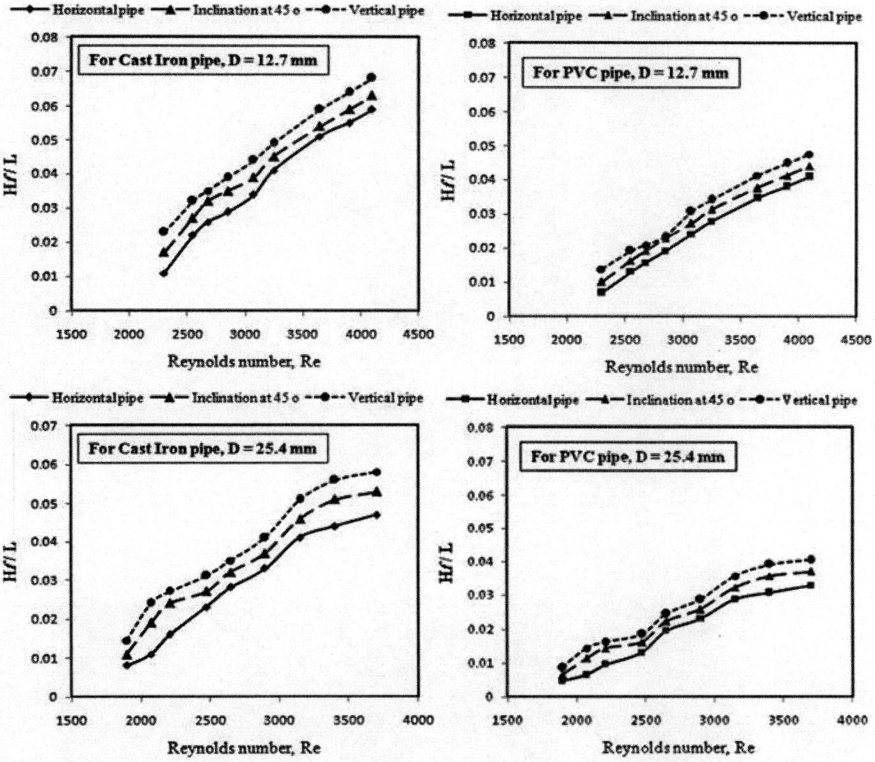

Fig. 5 Head losses as a function of Reynolds number for different pipe materials (Cast Iron and PVC) and pipe diameter (12.7 and 25.4 mm)

inclination angles 0, 45°, and 90°, indicated in Fig. 6. For all flow geometries considered here, the turbulence intensity increased with the inclination of pipe and starts to become unstable near at the inlet flow, with velocity profiles becoming more inflectional in nature. A turbulence profiles continues to slow through the front flow gradually until to pipe end, but this behaviour is deceleration with increase a inclination angle for pipe.

Pressure distribution at different inclination angles of the flow cycle, as simulated by all the models, are presented at pipe diameter 25.4 mm measured from the inlet in Fig. 7. The profiles predicted by three variations of the k-ε model are almost the same throughout the cycle. From the contours of pressure, it is observed that the pressure is maximum value at the inlet and decreases as a water approaches and passes through pipe to the end (as shown by the red and blue colour).

The comparison between numerical data and experimental measurements of the variation of pressure drop as a function of Reynolds number for a Cast Iron pipe and pipe diameter 25.4 mm was presented in Fig. 8. The results of pressure drop for three models at different inclination angles 0, 45°, and 90° shows a good agreement

Fig. 6 Contours of turbulence intensity in a Cast Iron pipe, D = 25.4 mm at a different inclination angles 0, 45°, and 90°

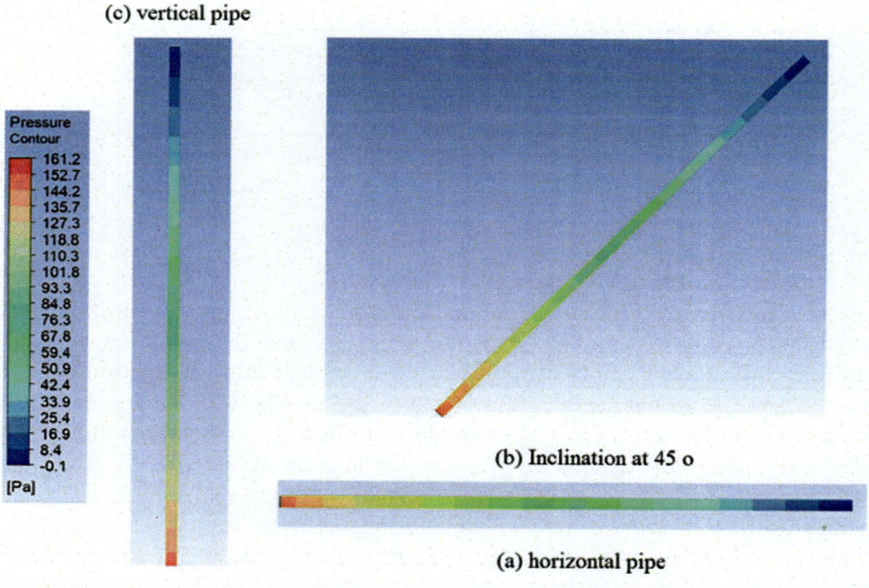

Fig. 7 Contours of pressure distribution in a Cast Iron pipe, D = 25.4 mm, at a different inclination angles 0, 45°, and 90°

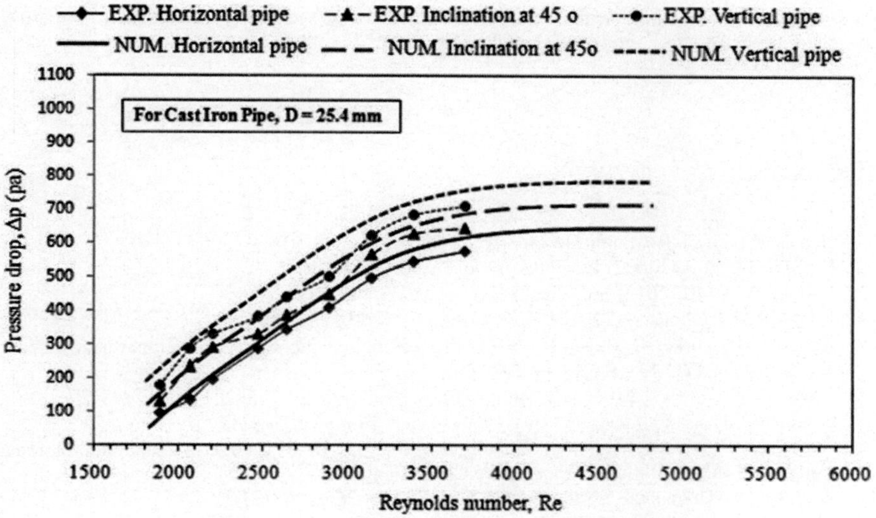

Fig. 8 Ccomparison between numerical data and experimental measurements

at the Reynolds number range 1925 to 3800. The mean percentage of differences between the CFD simulation and the experimental measurement is 2.8%, 3.8%, and 6.2%, respectively. More so, the outcomes were compared with the numerical and experimental data from other works in literature [4, 11], and a good agreement was obtained.

5 Conclusion

The influence of variation in the pipe inclination angle of 0°, 45°, and 90°, different pipe materials, different inside diameters in the turbulent flow was studied through experimental tests and computational simulation. Using this type of studies, the feasibility of a piping system can be assessed. If the inlet and outlet heads, pipe inclination, and pipe diameter are known. The maximum length of bypass pipe possible can be determined. The results appeared that the head loss is strongly dependent on pipe inclination angle and this behaviour increases with its increase, for different pipe material and with different pipe diameter. Also, head loss is also as strongly dependent on pipe material when the roughness of pipe wall is high value. Numerical results shown turbulence intensity increased with the inclination of pipe and it starts to become unstable near at the inlet flow, with velocity profiles becoming more turbulence. The numerical results and experimental data were compared and a good agreement was obtained.

Acknowledgements Authors are obliged to the University of Technology, Baghdad, for providing the centre for automotive and the energy generation.

References

1. Hansa VS, Sainib RP, Sainic JS (2010) Sol Energy 84:898–911
2. Kim D, Ghajar AJ (2002) Exp Therm Fluid Sci 25:659–676
3. Lips S, Meyer PJ (2012) Int J Heat Mass Transf 55:405–412
4. Bhagwat SM, Ghajar AJ (2014) Int J Multiph Flow 59:188–205
5. Adelaja AO, Dirker J, Meyer JP (2014) Proceedings of the 15th international heat transfer conference, (IHTC-15), Kyoto, pp 1–14
6. Ghajar AJ, Tang CC (2012) Adv Multiph Flow Heat Transf 4:175–201
7. Shoukri M, Hassan I, Gerges I (2003) Can J Chem Eng 81:205–211
8. Adebayo AR, Abdulraheem A, AlShammari AT (2013) In: SPE Nigeria annual international conference and exhibition. Society of Petroleum Engineers
9. Ebrahimi A, Khamehchi E (2015) J Nat Gas Sci Eng 26:1306–1316
10. Mousavi SS, Hooman K (2006) Energy Convers Manag 47:2011–2019
11. Qasim SM, Khudheyer SM (2008) J Eng Dev 16:142–157
12. Nasiruddin, Kamran Siddiqui MH, Int J Heat Fluid Flow28:318–328
13. Mokhtari M, Barzegar Gerdroodbary M, Yeganeh R, Fallah K (2017) Eng Sci Technol Int J 20:1106–1114
14. Gajusingh ST, Shaikh N, Siddiqui K (2010) Exp Therm Fluid Sci 34:590–602
15. Wong TT, Leung CW, Li ZY, Tao WQ (2003) Int J Heat Mass Transf 46:4629–4638

Numerical Investigation on the Effect of Vertically Oriented CPC on Spillage Loses in Evacuated Tube-Based Power Tower Receiver

Ibrahim A. Hussain, Syed I. Gilani, Hussain H. Al-Kayiem, and Mohamad Z. Abdullah

Abstract The existing design of the evacuated tube-based solar power tower (ET-SPT) receiver suffers greatly from optical losses due to optical spillage; where some reflected radiation escape through the gaps between the receiver tubes, and some miss the absorber due to difference in the diameters of the glass cover and the metallic absorber tube. It is aimed in the current work to minimize the concentrated solar power spillage losses at the receiver. Potential of integrating compound parabolic concentrator in the ET-SPT receiver design was investigated. A proposed methodology of a W-shape cavity, compound parabolic concentrator (CPC) coupled with an evacuated tubular absorber, with 40 and 70 mm tube and glass diameters, respectively was designed and numerically analysed for optical performance using ray tracing software TracePro. Parameters considered in the investigation of the vertically oriented receiver design (CPC + Vacuum tube) comprise of glass tube and absorber diameters, distance between the tubes, and the angular distance between the farthest heliostats to the east and the west of the tower. Simulation results have shown that flux on target (receiver) with CPC is about 3 times more than that of ETR without CPC. Significant improvement in the receiver optical performance is achieved by minimizing the amount of spillage loses from 40 to 11% with an average receiver optical efficiency of 75%.

1 Introduction

Solar power tower (SPT) or central receiver system (CRS) is a renewable energy electricity generation facility [1]. It comprises a collection of solar-tracking mirrors known as 'heliostats', an energy receiver, and a power conversion system. A typical SPT system is shown in Fig. 1. Heliostats track, reflect and concentrate the incident

I. A. Hussain (✉) · S. I. Gilani · H. H. Al-Kayiem · M. Z. Abdullah
Department of Mechanical Engineering, Universiti Teknologi PETRONAS, Seri Iskandar, Malaysia
e-mail: ibrahim_18000592@utp.edu.my

I. A. Hussain
Department of Mechanical Engineering, Ahmadu Bello University, Zaria, Nigeria

F. Ahmad et al. (eds.), *ICPER 2020*, Lecture Notes in Mechanical Engineering,
https://doi.org/10.1007/978-981-19-1939-8_28

Fig. 1 Solar power tower system [2]

solar radiation on the energy receiver which is on top of a tower. The energy receiver is the most vital component of the SPT system. The receiver is designed to absorb the highly concentrated solar radiation at high temperature and transport most of this energy to a primary working-fluid (i.e. receiver coolant) flowing through the receiver tubes. This could be the working fluid directly employed in the power-conversion system or a separate heat-transfer fluid. Depending on the receiver type, the heat-absorbing surfaces of the receiver consist of panels of parallel tubes around or enclosed within the receiver.

1.1 Conventional SPT Receivers and Their Challenges

The two most widely used receivers in SPT systems are external and cavity receiver. External and cavity are tubular type receivers (Fig. 2), and they both utilize liquids as heat transfer fluid [3]. What differentiates these two types of receiver is the location of the heat-absorbing tubes. In external receiver, the heat absorbing tubes are located on the outer surface of the cylindrical or rectangular receiver. Whereas in cavity receiver, the tubes are enclosed within the inner surface of the receiver. Volumetric receiver is another type of SPT receiver. It is not a tubular type of receiver and not commonly used in commercial SPT systems.

Optical as well as thermal losses affect the performance of nearly all types of receivers currently employed by SPT systems [4–6]. In external receiver design, the tubes are unprotected from the surrounding atmosphere, so the receiver is prone to high thermal losses. External receiver has the advantage of having a large acceptance angle of up to 360°; this implies it can accept solar radiation from all directions. While in cavity receiver design, the thermal losses are controlled by enclosing the heat-absorbing tubes within the receiver cavity, but the receiver is limited to an acceptance angle of 60–120°.

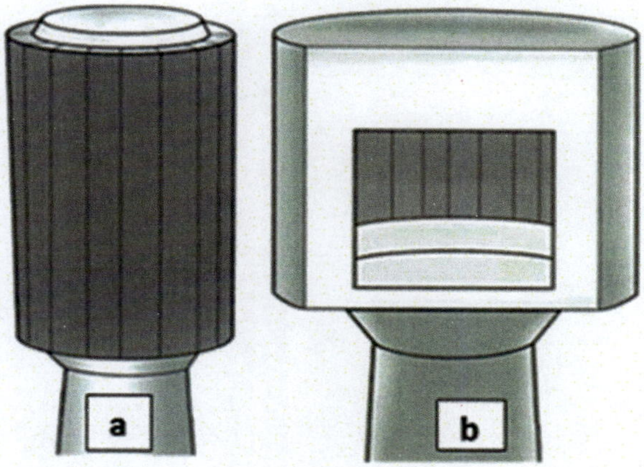

Fig. 2 SPT external **a** and cavity **b** receivers [5]

1.2 Evacuated Tube-Based Solar Power Tower (ET-SPT)

Evacuated tube collectors are mostly used as stand-alone systems for the production of domestic hot water [7], integrated with either parabolic trough concentrator [8] or compound parabolic concentrator [9] for hot water and low pressure steam generation. But evacuated tube absorbers integrated with compound parabolic concentrator exhibits better performance than other concentrators [10] especially for medium temperature range applications.

ETR is seldom used for SPT systems. The temperature ranges at which the commercial SPT plants operate is much higher than the working temperature of ETR. But for much smaller capacity SPT that requires low pressure steam in the mid temperature range (200–300 °C) for their operations, the use of conventional SPT receivers is uneconomical. Moreover, conventional SPT receivers have the disadvantages mentioned earlier in the previous section.

In order to minimize the thermal loses which are inherent in the conventional SPT receiver, evacuated tube was first proposed by [6] as power tower receiver. This is due to its good thermal performance, relatively low cost, convenient installation and easy transportability [11]. Moreover, ETR can be utilized around the receiver wall in such a way that a 360° heliostat field can be employed. Figure 3 shows evacuated tube receiver employed in SPT system. ET-SPT receiver technology is proved to be thermally efficient, due to the presence of vacuum envelope between the glass and the metallic tubes in ETR; which highly reduces the convection loses. But the challenge with ET-SPT receiver is high optical loses due to spillage of concentrated solar radiation within the receiver [6].

Fig. 3 Evacuated tube-based SPT receiver [12]

1.2.1 Spillage Losses in ET-SPT Receiver

Concentrated solar radiation spillage loses in ET-SPT receiver are categorized into two: spillage around the receiver and spillage within the receiver tubes. Spillage around the receiver is the solar radiation outside the reach of the receiver; this occurs when the rays from the heliostat are not reflected on the receiver aperture. It is either by the sides, on top or below the receiver aperture. While the spillage within the receiver tubes is the concentrated solar radiation that escapes through the gaps between tubes or penetrates the glass cover but miss the absorber; this also occurs as a result of diameter difference between the glass cover and the heat absorbing tube. Therefore, these two types of spillage loses make the total loses due to spillage at the ET-SPT receiver.

ET-based SPT is an emerging technology, as such, not much work is found about it in the literature [13]. A model of a central receiver system based on evacuated tubes receiver was developed and investigated by [14]. The simulated result shows good thermal performance, but high cosine and spillage losses of concentrated solar radiation were reported.

In an attempt to reduce the spillage loses within the receiver by reducing the amount of concentrated solar radiation escaping through the gaps between the receiver tubes, a zigzag arrangement of ETR tubes to maximize the heat capturing capability was also used in another investigation by [12]. An improved receiver

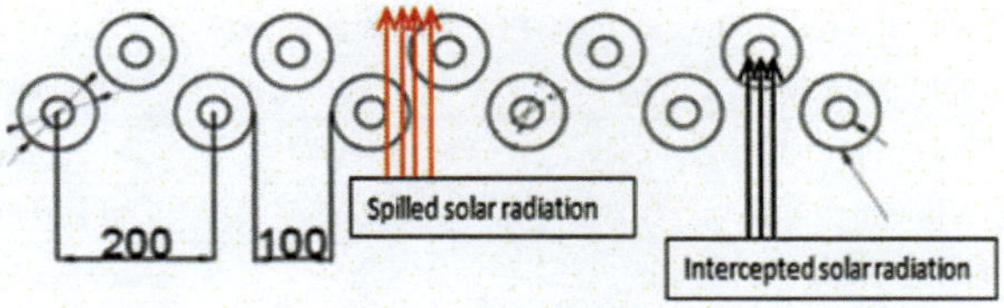

Fig. 4 Crossection of receiver tubes in zigzag arrangement showing intercepted and spilled solar radiation [12]

thermal performance is reported, but 40% of the concentrated solar radiation is reported as spillage losses within the receiver tubes. This shows that the spillage losses are not eliminated even with the new arrangement of the receiver tubes. Figure 4 shows how concentrated solar radiation passes through the gaps and through the glass cover even with zigzag arrangement of the receiver tubes.

In another work by the same authors [12], PV and back flat reflectors were integrated to ET-SPT receiver, and were able to recapture 19.8% and 20%, respectively of the spilled concentrated solar radiation within the receiver. One of the factors that affect the energy collection at SPT receiver is the spillage loses of concentrated solar radiation; which if minimized can greatly enhance the amount of energy absorbed [15].

Based on the literature, spillage of concentrated solar power within the receiver is an issue which requires great attention to improve the performance of the ET-SPT receiver. No doubt that one of the methods or the most promising method is the re-design and arrangement of the ETR at the top of the tower.

Accordingly, in an attempt to minimize these loses and improve the receiver optical performance, a proposed design of a non-imaging secondary concentrator integrated with evacuated tube receiver to intercept maximum amount of solar radiation from the heliostat field and deliver it to the absorber is presented. The proposed technique is modeled and simulated using a ray-tracing software TracePro to predict some performance parameters of a solar power system.

2 Method

2.1 System Description

The system comprises of heliostats, tower and newly designed receiver (CPC + ETR). The new receiver which consists of a W-shaped cavity compound parabolic concentrator with an evacuated tubular absorber is mounted on top of the tower. The parameters utilized for the geometrical design of both the field and the CPC are shown in Table 1.

Table 1 Heliostat field and CPC design parameters

Field parameters	Value (m)	CPC parameters	Value
Tower optical height	9.25	Acceptance half angle (θ)	45°
Heliostat width	1.25	Absorber diameter	40 mm
Heliostat height	2.5	Glass cover diameter	70 mm
Heliostat height above the ground	1.5	Absorber optical length	2.32 m

2.2 Geometrical Modeling of the Proposed CPC

Two most important things in CPC geometrical design are the acceptance half-angle (θ_a) and the CPC aperture. The acceptance half-angle was determined by the farthest heliostats to the east and west of the tower; which in turn determines the concentration ratio of the CPC. The absorber length is the same as that of the reflector. Therefore, the maximum concentration ratio (C_r) is given by Eq. 1 [16].

$$C_r = \frac{1}{\sin\theta_a} \tag{1}$$

Since the evacuated tube is to be utilized as the absorber, the height of the reflector and the radius of the glass cover were obtained using Eqs. 2 and 3 [16, 17].

$$h = \frac{a}{2}\left[\frac{1}{\sin\theta_a \tan\theta_a} + \frac{1}{2} + \frac{1}{\pi \sin\theta_a}\right] \tag{2}$$

$$R = \sqrt{2}r \tag{3}$$

where 'R' is the radius of the glass cover, 'r' the radius of the absorber, 'a' the perimeter of the absorber, and 'h' is the full height of the CPC.

The reflector upper portion is generated using the relation given by [18] in Eq. 4. The relationship is the same for the ideal non-imaging cusp reflector.

$$\begin{aligned} x &= r\sin\vartheta - \rho(\vartheta)\cos\vartheta \\ y &= -r\cos\vartheta - \rho(\vartheta)\sin\vartheta \end{aligned} \tag{4}$$

where '$\rho(\vartheta)$' is the distance between two points 'jk' as shown in Fig. 5a, one point on the curve of the reflector and a corresponding point on the tangent of the absorber [18, 19]. String method is used to determine the distance '$\rho(\vartheta)$' for both the parabola and the region below the absorber using the relation given by McIntire [16].

$$\rho(\vartheta) = r\frac{\left[\vartheta + \theta_a + \frac{\pi}{2} - 2\alpha - \cos(\vartheta - \theta_a)\right]}{[1 + \sin(\vartheta - \theta_a)]} \quad \text{for}\,\theta_a + \frac{\pi}{2} \le \vartheta \le \frac{3\pi}{2} + \theta_a \tag{5}$$

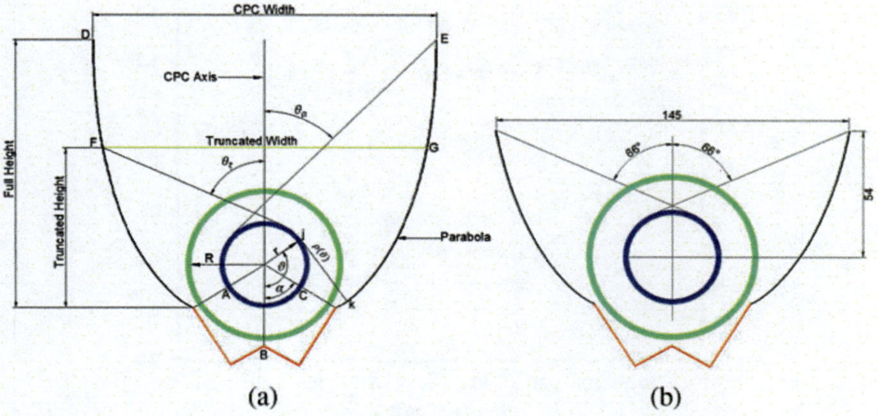

Fig. 5 A W-shape cavity CPC with an evacuated tubular absorber. **a** Full height. **b** Truncated

$$\rho(\vartheta) = r(\vartheta - \alpha), \text{ for } 2\sigma \leq \vartheta \leq \theta_a + \frac{\pi}{2} \tag{6}$$

The region below the absorber is designed by shifting the origin of the involute as in the case of ideal non-imaging reflector through angle 'α' to points A and C; this implies that the top outer part of the 'W' is the starting point of the reflector curve. For the 'W' shape cavity at the bottom of the reflector, its external section is tangent to the glass cover through the points (−r tan(σ), − r) and (r tan (σ), − r), where $\sigma = \cos^{-1}$ (r/(R) and α = (tanσ − 2σ), while the internal section is the imaginary line from the upper tangent extended through the center [16].

2.2.1 Effect of Truncation on the Concentrator

The height of the reflector considerably determines the manufacturing cost of the concentrator due to the cost of the materials. The upper portion of the reflector is almost perpendicular to the concentrator aperture; it can be truncated without making much impact on the performance of the concentrator. Truncation increases the optical efficiency of CPC by increasing its acceptance angle [20], at the same its effect on the thermal efficiency is very minimal. According to [21], optical efficiency is predicted to increase by over 20% with truncation of about 50% of the reflector, and the decrease in thermal efficiency is about 1%. Truncation of the concentrator brings about savings in material cost, an increase in acceptance angle, and field view of the absorber [10], but with a slight decrease in concentration ratio. The decrease in the CPC width occurs rapidly at around 50% truncation; this is shown in Fig. 6. Note that the width is proportional to the concentration ratio. Therefore, the CPC utilized in this work is truncated at 50% of its full height measured from the center of

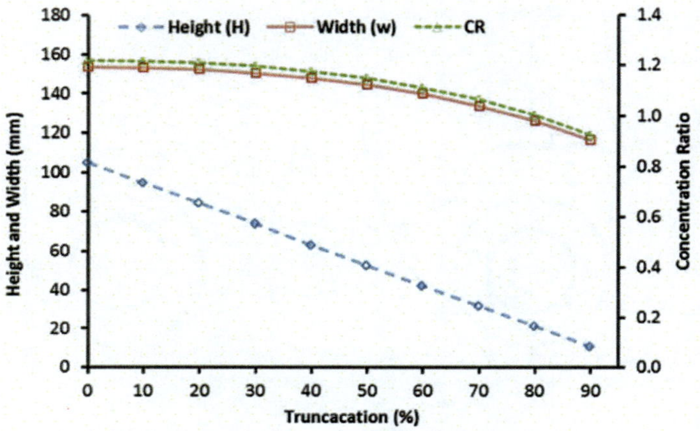

Fig. 6 Effect of truncation on height, width and concentration ratio

the absorber; this truncation increases the CPC acceptance angle for direct radiation to 66° as shown in Fig. 5b, and with only 6% decrease in the concentration ratio.

2.3 Ray Tracing Simulations

The 3D model of the CPC was drawn using AutoCAD, and then imported into TracePro. TracePro is a Monte Carlo ray tracing software for optical and illumination analysis. Surface and material properties of different components that made up the CPC were defined in the TracePro. Solar emulator is used to typically imitate the sun, where parallel sun and 10,000 numbers of rays were assumed for the simulation. Other sun's properties were defined together with coordinates of the location and that of the system. The coordinates of the system which comprises of heliostat base, tower base, heliostat, and tower were retrieved from the 3D model. After the system was set up, the ray-tracing was carried out for a typical day in December. The flow diagram for the simulation is shown in Fig. 7.

The heliostat was positioned due north of the tower. To keep the reflected rays on the target (receiver), relationships for azimuth elevation sun-tracking method are used to determine the azimuth and elevation angles of the heliostat. Parameters considered are the horizontal distance from the tower, the height of the tower, latitude of the location, hour angle, Sun's azimuth, and elevation angles for each hour. Heliostat azimuth is considered positive due east and negative due west of the tower. The target position is above the heliostat; therefore, it is considered positive, and the latitude of the location is 4.384°N. The analysis was carried out with a single heliostat and a single ETR coupled with CPC.

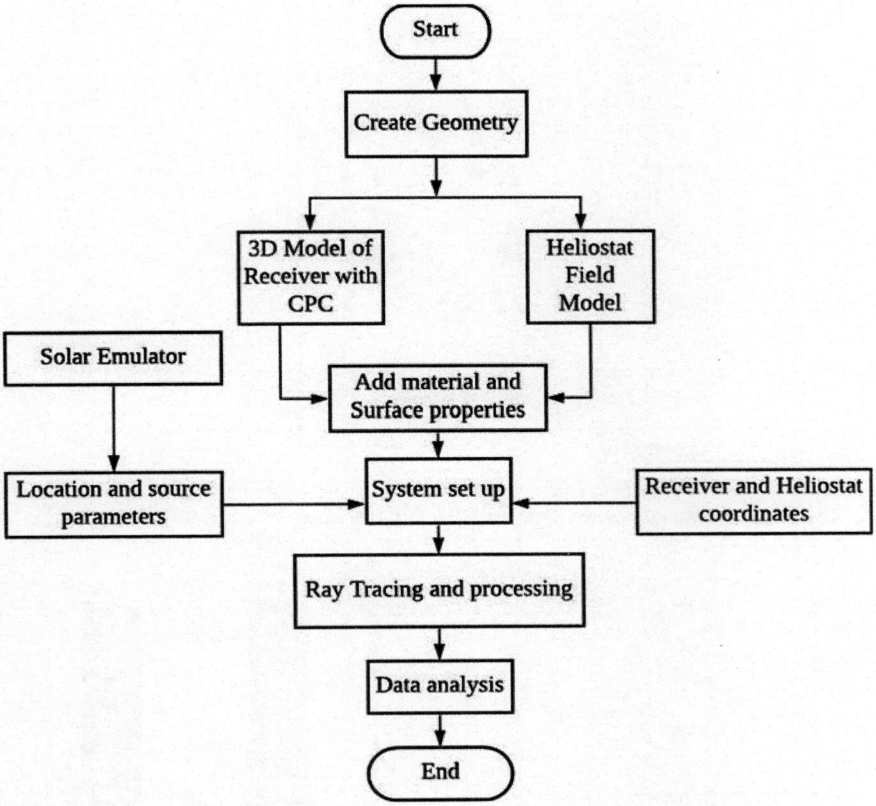

Fig. 7 Flow diagram for the simulation process

3 Results and Discussions

Ray tracing simulation of the newly designed receiver was carried out using TracePro to determine its optical performance. Figure 8 shows the diagram for the ray tracing. The heliostat intercepts and reflects the solar radiation to the target (receiver) which is 9.5 m away from the heliostat and at 8.25 m above the center of the heliostat.

The optical performance of the modified cavity CPC was carried out using TracePro. To further verify the CPC performance, a simulation was also carried out using Tonatiuh so that the results can be compared. The simulation for both optical software was carried out for eight hours, between 10 o'clock and 5 o'clock in December. There is a good agreement between the results of the hourly power absorbed as shown in Fig. 9a. Though the difference between the power absorbed at 2 o'clock is significant for the Tonatiuh and the TracePro but it was compensated by the difference in the power absorbed at the hours before solar noon. This is clearly seen in the total power absorbed at the end of the simulation hours. The difference

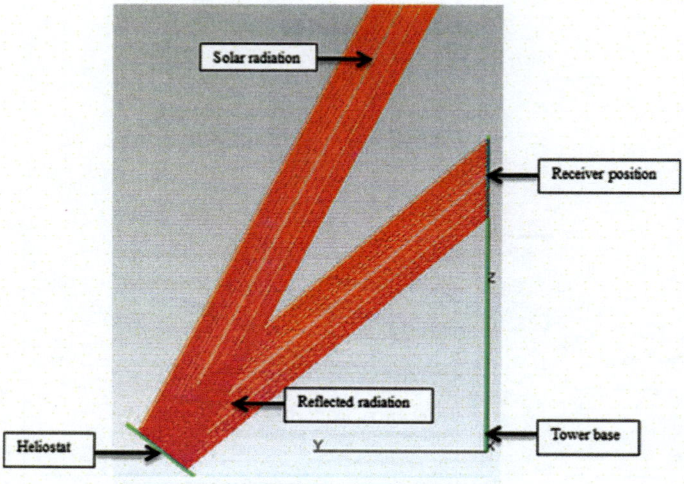

Fig. 8 Ray tracing diagram of one of the heliostat in SPT system

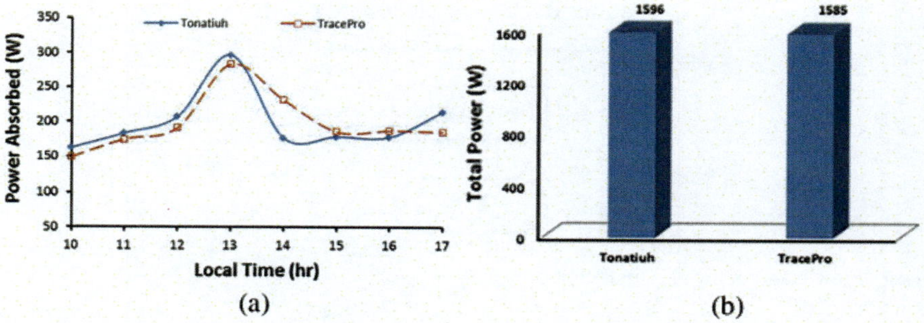

Fig. 9 **a** Intercepted power at the absorber and **b** Total absorbed power

between the total absorbed power at the end of the simulation day is <1% as shown in Fig. 9b.

3.1 Analysis of the Power on the Receiver with and Without CPC

Figure 10 shows the maximum solar power recorded at the absorber for each hour during the eight hours recording of data. Minimum hourly value was recorded at 10 o'clock, while maximum value was recorded at around 1 o'clock for receiver with and without CPC. The upward and downward trend of the amount of radiant

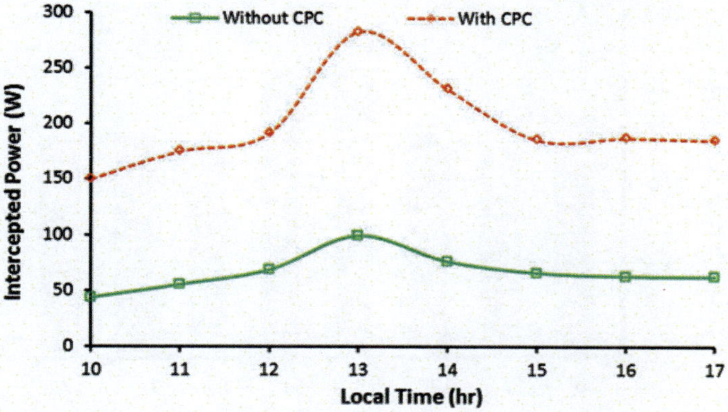

Fig. 10 Maximum hourly intercepted solar power

flux recorded at the absorber with and without CPC before and after the solar noon is related to the amount of direct normal irradiance; which increases and decreases before and after the solar noon, respectively.

3.2 *Recuperative Spillage Loses*

Hourly spillage of concentrated solar power recaptured with the new receiver is shown in Fig. 11. This is the power that would have been lost if CPC is not integrated with the ETR. It is the difference between the solar radiation absorbed by the receiver with and without CPC. It can be deduced from the figure that recaptured spillage loses and the amount of intercepted solar power at the receiver have a direct relationship. The more the incident solar radiation at the receiver aperture the higher the likelihood of some of the radiation escaping between the gaps and through the tubes. But with the presence of CPC, this radiation is captured and reflected on to the absorber.

3.3 *CPC Efficiency*

Hourly optical efficiency of the non-imaging concentrator is shown in Fig. 12; it is determined by the amount of the total solar power recorded at the absorber to the one recorded at the receiver aperture. All the parameters that determine the efficiency of the reflector such as multiple reflections, gap losses, reflectivity of the concentrator etc. are taken into account during the simulation process. Concentrator optical efficiency increases with increase in the amount of solar power at the receiver because the higher the power at the receiver aperture the higher the tendency to recapture more power that tends to escape.

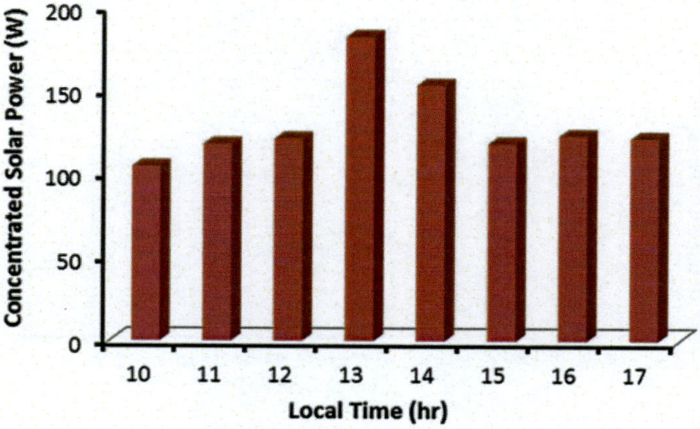

Fig. 11 Hourly recaptured spillage loses using CPC

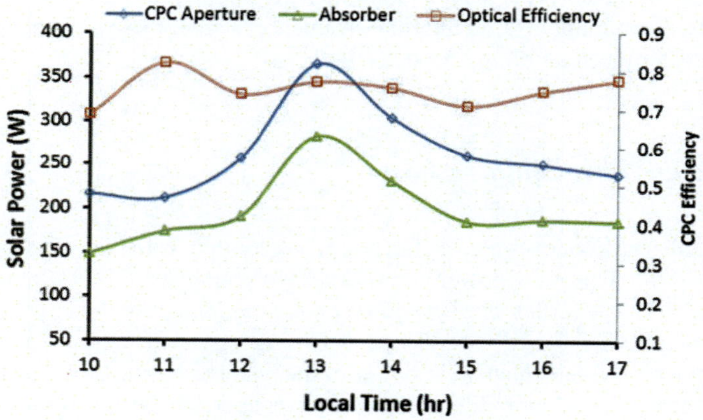

Fig. 12 Hourly CPC optical efficiency

The difference in the optical performance of the receiver (without CPC) used in Ref. [12] and the new receiver (with CPC) design is shown in Fig. 13. Only about 46% of the total concentrated solar power at the receiver aperture is absorbed in the current receiver design, while about 75% of the total concentrated power at the receiver aperture is absorbed in the new receiver design. The increase in the amount of solar power absorbed is due to the presence of the secondary concentrator.

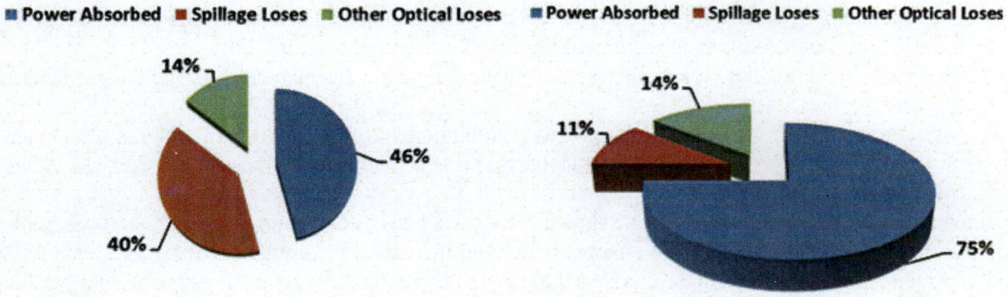

(a) (b)

Fig. 13 **a** Receiver without CPC. **b** Receiver with CPC

4 Conclusion

Ray tracing simulation of a vertically oriented ET-SPT receiver integrated with compound parabolic concentrator was carried out using TracePro to determine the optical performance of the receiver. The simulation was performed for 8 h in December. The newly designed receiver which comprises of evacuated tubes and CPC has shown a better optical performance than the receiver without CPC. It accomplished a reduction of spillage loses from 40 to 11% and increased optical efficiency with 29%. This was achieved by minimizing the amount of spillage losses (concentrated solar radiation that escape between the receiver tubes and solar radiation that penetrate the tube but missed the absorber) through recapturing and redirecting it back to the absorber.

Acknowledgements The authors acknowledge Universiti Teknologi PETRONAS (UTP), Malaysia for providing financial support under YUTP-FRGS research grant (cost center: 015LCO-209) to conduct this research.

References

1. Cruz N, Salhi S, Redondo J, Álvarez J, Berenguel M, Ortigosa P (2018) Hector, a new methodology for continuous and pattern-free heliostat field optimization. Appl Energy 225:1123–1131
2. Islam MT, Huda N, Abdullah A, Saidur R (2018) A comprehensive review of state-of-the-art concentrating solar power (CSP) technologies: Current status and research trends. Renew Sustain Energy Rev 91:987–1018
3. Behar O, Khellaf A, Mohammedi K (2013) A review of studies on central receiver solar thermal power plants. Renew Sustain Energy Rev 23:12–39
4. Ho CK, Iverson BD (2014) Review of high-temperature central receiver designs for concentrating solar power. Renew Sustain Energy Rev 29:835–846
5. Ho CK, McPheeters CO, Sharps PR (2018) Hybrid CSP/PV receivers: converting optical spillage to electricity. In: AIP conference proceedings, p 170006

6. Ali BH, Gilani S, Al-Kayiem HH (2016)Mathematical modeling of a developed central receiver based on evacuated solar tubes. In: MATEC web of conferences, p 02005
7. Morrison G, Budihardjo I, Behnia M (2004) Water-in-glass evacuated tube solar water heaters. Sol Energy 76:135–140
8. Mohammad ST, Al-Kayiem HH, Assadi MK, Sabir O, Khlief AK (2018) An integrated program of a stand-alone parabolic trough solar thermal power plant: Code description and test. Case Stud Therm Eng 12:26–37
9. Madduri A, Loeder D, Beutler N, He M, Sanders S (2012) Concentrated evacuated tubes for solar-thermal energy generation using stirling engine. IEEE Energytech 2012:1–6
10. Akhter J, Gilani SI, Al-Kayiem HH, Ali M (2019) Optical performance analysis of single flow through and concentric tube receiver coupled with a modified CPC collector under different configurations. Energies 12:4147
11. Sabiha M, Saidur R, Mekhilef S, Mahian O (2015) Progress and latest developments of evacuated tube solar collectors. Renew Sustain Energy Rev 51:1038–1054
12. Khlief AK, Gilani SI, Al-Kayiem HH, Mohammad ST (2018)Design a new receiver for the central tower of solar energy. In: MATEC web of conferences, p 02009
13. Ali BH, Gilani S, Al-Kayiem HH (2014)Investigation of evacuated tube collector performance at high temperature mode using TRNSYS simulation model. In: Applied mechanics and materials, pp 155–160
14. Khlief AK, Gilani SI, Al-Kayiem HH, Mohammad ST (2017) Design and modelling of novel evacuated receiver tube of concentration solar power system by using heliostat stadium arrangement. In: MATEC web of conferences, p 02002
15. García L, Burisch M, Sanchez M (2015) Spillage estimation in a heliostats field for solar field optimization, pp 1269–1276
16. McIntire W (1980) New reflector design which avoids losses through gaps between tubular absorbers and reflectors. Sol Energy 25:215–220
17. Rabl A, Goodman N, Winston R (1979) Practical design considerations for CPC solar collectors. Sol Energy 22:373–381
18. Akhter J, Gilani S, Ali M (2018)Design and optical modeling of a low-profile stationary concentrating solar collector for medium temperature heat supply. J Telecommun Electron Comput Eng (JTEC) 10:65–71
19. Kim YS, Balkoski K, Jiang L, Winston R (2013) Efficient stationary solar thermal collector systems operating at a medium-temperature range. Appl Energy 111:1071–1079
20. Carvalho M, Collares-Pereira M, Gordon J, Rabl A (1985) Truncation of CPC solar collectors and its effect on energy collection. Sol Energy 35:393–399
21. Ustaoglu A, Okajima J, Zhang X-R, Maruyama S (2018) Truncation effects in an evacuated compound parabolic and involute concentrator with experimental and analytical investigations. Appl Therm Eng 138:433–445

Numerical Analysis to Study the Effect of Foam Thickness on The Thermal-Hydraulic Performance of The Metal Foam Heat Exchanger

Noor M. Jasim, Ekhlas M. Fayyadh, and Moayed Razoki Hasan

Abstract In this study the effects of key parameters including the Reynolds number and metal foam thicknesses on the heat transfer and pressure drop are examined. A numerical simulation is performed for 5 PPI copper foam with constant porosity (0.89) that wrapped array of heat exchanger with different dimensionless foam thickness 0.2, 0.43, 0.6, 0.8, and 0.91 for range of Reynolds number of (1311–6555). Two-dimensional simulations using the thermal non-equilibrium energy model and the Darcy-Forchheimer flow model are performed on metal foam heat exchangers. The governing equations are resolved using the Star-CCM + program, Version 2019.1, a commercial multiphysics finite-volume solver. Hot tubes have a constant temperature while cold air passes at a constant input temperature of 25 °C. It is found that within the range of Reynolds number and the same level of compactness, the heat transfer and pressure drop is increased with increasing Reynolds number. However, it was found that metal foam with a thickness of 0.43 has the highest area goodness factor ratio at a high Reynolds number, while metal foam with a thickness layer of 0.6 is the highest area goodness factor ratio at a low Reynolds number.

Keywords Copper metal foam · Area goodness factor · Foam layer thickness · High porosity · Staggered tube bundle

N. M. Jasim (✉) · E. M. Fayyadh · M. Razoki Hasan
Mechanical Engineering Department, University of Kufa, Kufa, Iraq
e-mail: noorm.alhasnawi@uokufa.edu.iq

E. M. Fayyadh
e-mail: 20084@uotechnology.edu.iq

M. Razoki Hasan
e-mail: 20002@uotechnology.edu.iq

N. M. Jasim
Mechanical Engineering Department, University of Technology, Bagdad, Iraq

F. Ahmad et al. (eds.), *ICPER 2020*, Lecture Notes in Mechanical Engineering,
https://doi.org/10.1007/978-981-19-1939-8_29

Nomenclature

a_{sf}	The solid–fluid specific surface area, $\mathbf{m^2}$	A	Cross sectional area, $\mathbf{m^2}$
$A.G.F$	Area Goodness Factor	$A.G.F.R$	Area Goodness Factor ratio
A_f	Frontal area of the core, $A_r = b/a$	A_r	Channel aspect ratio, $A_r = b/a$
b	Height of the rectangular channel, m	c_p	Specific heat of fluid-phase, **J/kg·°K**
C_f	Inertia drag coefficient	d_f	Fiber diameter, **m**
d_p	Pore diameter, **m**	D	Diameter of the tube, **m**
D_F	Diameter of the Foam, **m**	f	friction factor
g	Gravity acceleration, m/s^2	h_{sf}	Solid-to-fluid interfacial specific heat transfer coefficient, Wm^{-2} K^{-1}
J	The Colbrun Factor	k	Thermal conductivity, $\mathbf{Wm^{-1}}$ $\mathbf{°K^{-1}}$
K	Permeability of the porous medium, $\mathbf{m^2}$	L	Channel/core length, **m**
$\dot{m}$	Mass flow rate, **kg/s**	N_{row}	Number of Pipe in row
Nu	Nusselt number	p	Dimensional pressure, Pa
PPI	Pore per inch	Pr	Prandtl Number, $Pr = \nu_f / \alpha$
Re	Reynolds number, $Re = u_{in}\ a/\nu_f$	Re_d	Pore Reynolds number, $Re_p = \rho_f\ u_p\ d / \mu_f$
S_T	Transversal distance	S_L	Longitudinal distance
t	Time, **sec**	T	Dimensional temperature, °**K**
u, v	Dimensional velocity components, **m/s**	x, y	Dimensional coordinates, **m**
Greek symbols			
ρ	Density, $\mathbf{kg/m^3}$	ε	Foam porosity
μ	Dynamic viscosity, $\mathbf{N{\cdot}s/m^2}$	κ	Solid to fluid-phase thermal conductivity ratio
μ_e	Effective dynamic viscosity, $\mathbf{N{\cdot}s/m^2}$	ω	Pore density, PPI
ν_f	Kinematic viscosity, $\mathbf{m^2/s}$	χ	Foams tortuosity
Subscripts			
f	Fluid phase of metal foam	fe	fluid effective
F	Foam material	L	Longitudinal pitch
p	Pipe/pore	s	solid phase of metal foam
se	Solid effective	T	Transversal pitch

1 Introduction

High porosity metal foam is one of the modern materials that was used to develop the performance of heat exchangers because of its excellent mechanical properties in terms of permeability of the flow through it and low fluid resistance [1]. On the other hand, it has good thermal properties because it has a high heat transfer coefficient. In addition to other features such as being relatively lightweight and good stiffness [2]. Many researchers, such as Boomsma et al. [3], Klett [4], and Bhattacharya and Mahajan [5] indicated that when using foam for enhancement performance of heat exchanger, improvement of the heat transfer is accompanied an increasing of the pressure drop.

However, the effect of pore density on the performance heat exchanger was investigated by Liu et al. [6]. It was found that an increase of pore density increases both pressure drop and the Nusselt number of the foam-filled tube, particularly when the conductivity ratio k_s/k_f is very large. Huisseune et al. [2] have confirmed that increasing the density of the pores increases the surface area of heat transfer and pressure drop. Furthermore, the tortuosity flow also leads to augment mixing of fluid flow inside the voids resulting in an increase in turbulence and an increase in the coefficient of heat transfer for fixed high porosity of open-cell metal foam [2].

The permeability property of foams has an important role in the study of fluid flow through it, which describes the possibility, ease of fluid flow and the magnitude of the resistance of the solid matrix to flow. The phenomenon of thermal insulation described by Kang Chen et al. [7], when the permeability of the medium is low, the air with slow velocity does not pass easily, so mass flow and the heat transfer are very little. However, Xiao [8] experimentally demonstrated that the porous copper with high porosity and low pore density has high permeability. Also, Rahmati et al. [9] noted that the low permeability, that means a low Darcy number, makes the foam layer similar to the layer of the solid body so that it prevents flowing through it. An increasing the foam area leads to the increase in heat removal from foam, so greater amount of heat that transfer to the air flow, this was also clarified and proven by Mehrdad's [10]. The thickness of the foam has the same effect as the density of the pores, where the thick foam has low penetration. However, greater the thickness of the foam layer, the surface area will be larger, which leads to increase the resistance of fluid flow, i.e. a decrease the amount of air and the forced thermal transfer, T'Joen et al. [11], Kang Chen et al. [7] and Yue Wang et al. [12]. In applications of a foam-covered heat exchanger, at high permeability, a high Darcy number, the liquid flows faster so that heat transfer is greatly increased, and thermal performance is improved through high thermal conductivity, as demonstrated by Saman Al-Rashidi et al. [13]. Bernando et al. [14] explained that the heat transfer is higher when the amount of foam is lower, at low velocity, but in the case of high velocity, the heat is higher at larger thickness. This refers to the tube heat exchanger that is coated with foam which was completely filled the flow channel. Therefore, when increasing the thickness ratio and Reynolds number, an increase in the thermo-hydraulic performance coefficient of the exchanger. Chambio and Hooman [15], has also demonstrated experimentally that the' efficiency of heat exchangers of transfer heat is large when the foam thickness is larger. However, Note that this thermal feature is accompanied by a rise in the pumping power.

Foam materials are among the novel materials used to improve the performance of heat exchangers. Therefore, this numerical study was prepared to estimate the thermal and dynamic effect of copper foams of various specifications and their ability to improve the performance of heat exchangers so as to increase the heat transfer with the minimum pumping capacity. The layers of high porosity copper foam (89%) with a pore density of 5PPI and variable dimensional thickness ($0.5(D_F - D)/D$) of 0.2, 0.43, 0.6, 0.8, and 0.91 at range of Reynolds number of (1311–6555) were used to wrapped heated tubes in a staggered arrangement that is cooled with air.

2 Physical Model and Governing Equations

Figure 1 illustrates a staggered three-row tube bundle in which configuration is characterized by the tube with diameter D, the transversal pitch S_T and longitudinal pitch S_L. However, the size of the heat exchanger is similar to that of conventional air-cooled finned tube heat exchanger by Kays and London [16]. The geometrical details are listed in Table 1.

Five samples are listed in Table 2 having different thickness (i.e. five different foam heights) with constant porosity. It was modeled as a homogenous and isotropic

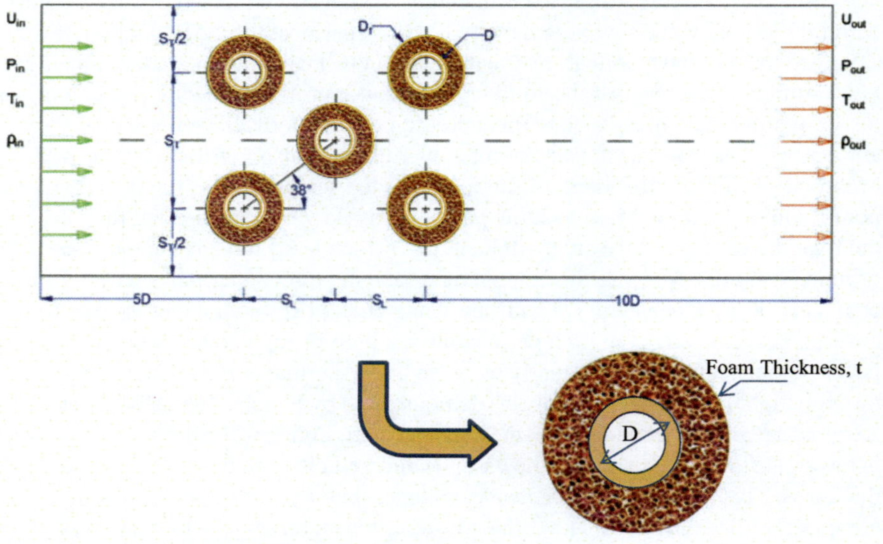

Fig. 1 The dimension of physical model

Table 1 Dimension of the staggered pipe heat exchanger

Item	Heat exchangers description
Tube diameter, D (mm)	20
Dimensionless fin height, $0.5(D_{\mathrm{Fin}} - D)/D$	0.43
Dimensionless foam thickness, $0.5(D_F - D)/D$	0.2, 0.43, 0.6, 0.8, 0.91
Foam thickness, t (mm)	4, 8.6, 12, 16, 18.2
Transversal pitch, S_T (mm)	69.2
Longitudinal pitch, S_L (mm)	44.5
Tube numbers (N)	5
Row number (N_{row})	3
H.X. width: L2 (mm)	200
H.X. height: L3 (mm)	138.4

Table 2 Five samples different thickness for foam-wrapped tubes

Dimensionless Foam Thickness 0.2, Sample 1	Dimensionless Foam Thickness 0.43, Sample 2	Dimensionless Foam Thickness 0.6, Sample 3	Dimensionless Foam Thickness 0.8, Sample 4	Dimensionless Foam Thickness 0.91, Sample 5
				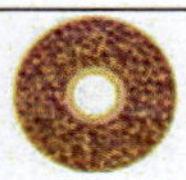

porous media with a high porosity 0.89. Then, it was wrapped on tubes with a dimensional thickness of $(0.5(D_F - D)/D)$.

It is possible to represent the computational geometry by a symmetrical shape in flow direction [17, 18], as shown in Fig. 2. Moreover, the boundary conditions are presented in Table 3 and Fig. 2. However, the upstream distance from entry air to the center of the first-row about five times of tube diameter D, while downstream distance is ten times of tube diameter D [19]. The tube length can be assumed long sufficient so that the end impacts can be neglected and as a consequence the airflow

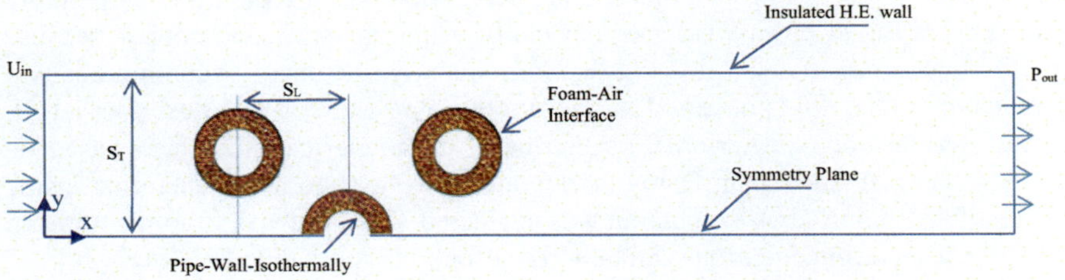

Fig. 2 The computational field of the problem

Table 3 The boundary conditions and input data of the numerical model

Boundary conditions		Input data	
At $\left\{\begin{matrix} x = 0 \\ 0 \le y \le S_T \end{matrix}\right\}$	$u = u_{\text{in}}$ $v = 0$ $T = T_{\text{in}}$	Inlet velocity	$u_{\text{in}} = 1, 2, 3, 4$ & 5
At $\left\{\begin{matrix} x = L \\ 0 \le y \le S_T \end{matrix}\right\}$	$v = \frac{\partial u}{\partial x} = 0$ $\frac{\partial T_f}{\partial x} = \frac{\partial T_s}{\partial x} = 0$	Inlet temperature	$T_{\text{in}} = 25$ °C
At the wall boundary $\left\{\begin{matrix} x = 0,\ L \\ y = S_T \end{matrix}\right\}$	$u = v = 0$ $T_f = T_s = 0$	Pipe temperature	$T_{\text{pipe}} = 100$ °C
At $\left\{\begin{matrix} x = 0,\ L \\ y = 0 \end{matrix}\right\}$	Symmetric plane		
At the pipe boundary	$u = v = 0$ $T_f = T_s = T_{\text{pipe}}$		

domain can be regarded as to be two-dimensional [21]. The heat exchange system is characterized by the height $2S_T$ [20].

The pipe wall temperature (T_p) is assumed to be uniform and constant, (wall temperature will be higher than the ambient temperature), and this tube will be simulate tube in an air cooled condenser where the tube side fluid undergoes phase change. While the flow of air represented as the cross-flow at a uniform, fixed velocity with a constant temperature of 25 °C at the entrance of the duct. The airflow inside the rectangular duct can be incompressible, viscous, fully-developed turbulent flow. Odabaee and Hooman [19, 22], assumed the local thermal equilibrium (LTE) condition exit between the solid and fluid phases to investigate the effect of porous materials on heat transfer in their study. Alvandifar et al. [23], also simulated the forced convection through metal foam heat exchanger with the assumption of LTE. This assumption gives an inaccurate perception because the fluid temperature differs from the steel temperature, especially when using air and copper because the thermal conductivity ratio is high. Consequently, the energy equation has two forms, one for the fluid phase and another for the solid phase, which describes LTNE (local thermal non-equilibrium). Furthermore, the numerical study was based on the Forchheimer–Brinkman Darcy model, and there was no heat generation inside the medium, and viscous dissipation was neglected. Then the mass, momentum and energy conservation equations can be written for solid and fluid as follows [24]:

- Mass conservation equation:

$$\left(\frac{\partial u}{\partial x}+\frac{\partial u}{\partial y}\right)=0 \tag{1}$$

- X-Momentum conservation:

$$\frac{\rho_f}{\varepsilon^2}\left(u\frac{\partial u}{\partial x}+v\frac{\partial u}{\partial y}\right)=-\frac{\partial p}{\partial x}-I\frac{\mu u}{K}-I\frac{\rho_f C_f}{\sqrt{K}}\left|\vec{V}\right|u+\frac{\mu}{\varepsilon}\left(\frac{\partial^2 u}{\partial x^2}+\frac{\partial^2 u}{\partial y^2}\right) \tag{2}$$

- Y-Momentum conservation:

$$\frac{\rho_f}{\varepsilon^2}\left(u\frac{\partial v}{\partial x}+v\frac{\partial v}{\partial y}\right)=-\frac{\partial p}{\partial y}-I\frac{\mu v}{K}-I\frac{\rho_f C_f}{\sqrt{K}}\left|\vec{V}\right|v+\frac{\mu}{\varepsilon}\left(\frac{\partial^2 v}{\partial x^2}+\frac{\partial^2 v}{\partial y^2}\right) \tag{3}$$

- Energy conservation:

  For the fluid:

$$(\rho\, C_p)_f\,(u\frac{\partial T_f}{\partial x}+v\frac{\partial T_f}{\partial Y})=\frac{\partial}{\partial x}(k_{fe}\frac{\partial T_f}{\partial x})+\frac{\partial}{\partial y}(k_{fe}\frac{\partial T_f}{\partial y})+h_{sf}\,a_{sf}\,(T_s-T_f) \tag{4}$$

  For the solid:

$$0=\frac{\partial}{\partial x}(k_{se}\frac{\partial T_s}{\partial x})+\frac{\partial}{\partial y}(k_{se}\frac{\partial T_s}{\partial y})+h_{sf}\,a_{sf}\,(T_f-T_s) \tag{5}$$

where I in the momentum equations is equal to 1 for foam region, and equal to 0 for fluid region. The characteristics of the fluid Cp, μ, and ρ are specific heat, the dynamic viscosity, and density respectively. while Boomsma and Poulikakos [25] models are utilized to estimate the effective thermal conductivity of a fluid and a solid k_{fe} & k_{se}. Moreover, the porosity of the media is ε. Whilst, the airflow resistance (as a drag form) and flow permeabilities are C_f & K respectively. Calmidi's models [26] were used to estimate these two variables as:

$$K=0.00073\,d_p^2(1-\varepsilon)^{-0.224}\left(\frac{d_f}{d_p}\right)^{-1.11} \tag{6}$$

$$C_f=0.00212(1-\varepsilon)^{-0.132}\left(\frac{d_f}{d_p}\right)^{-1.63} \tag{7}$$

The pore d_p and fiber diameter d_f can be estimated as functions of pore density ω and foam porosity ε, respectively, Calmidi [26]:

$$d_p=\frac{0.0254}{\omega} \tag{8}$$

$$d_f = 1.18\, d_p \sqrt{\frac{(1-\varepsilon)}{3\pi}} \frac{1}{1 - e^{-(1-\varepsilon)/0.04}} \tag{9}$$

Also, h_{sf} is interfacial heat transfer coefficient of solid/fluid field of metal foam, Lu et al. submitted mathematical model to calculate [5] as:

$$Nu_{sf} = \frac{h_{sf} d}{k_f} = \begin{Bmatrix} 0.76 \text{Re}_d^{0.4} \text{Pr}^{0.37}, \ (1 \le \text{Re}_d \le 40) \\ 0.52 \text{Re}_d^{0.5} \text{Pr}^{0.37}, \ (40 \le \text{Re}_d \le 10^3) \\ 0.26 \text{Re}_d^{0.6} \text{Pr}^{0.37}, \ (10^3 \le \text{Re}_d \le 2 \times 10^5) \end{Bmatrix} \tag{10}$$

$$d = \left(1 - e^{-(1-\varepsilon)/0.04}\right) d_f \tag{11}$$

Finally, Fourie and Du Plessis submitted mathematical model to calculate a_{sf} [27] as:

$$a_{sf} = \frac{3}{d_c}(3-\chi)(\chi-1) \tag{12}$$

$$d_c = \frac{2 * d_p}{(3-\chi)} \tag{13}$$

$$\chi = 2 + 2\cos\left(\frac{4\pi}{3} + \frac{1}{3}\cos^{-1}(2\varepsilon - 1)\right) \tag{14}$$

The most important parameters for analyzing the performance of heat transfer through a heat exchanger are the friction coefficient (*f*), and the thermal transfer coefficient (*h*) & (Nu). Thus:

$$h = \frac{Q}{2\pi * R * (T_w - T_{f,b})} \tag{15}$$

$$Nu = \frac{h * D}{k_f} \tag{16}$$

where k_f fluid thermal conductivities coefficient, T_w is the temperature of the solid's pipe (constant wall temperature), $T_{\text{f,b}}$ is the bulk temperature of the inlet and outlet fluid, *D*, Pipe diameter. The factor of friction of a staggered heat exchanger can be described in terms of Reynolds number as follows:

$$f = \frac{2 * \Delta p}{N_{\text{row}} * \rho * U_{\text{in}}^2} \tag{17}$$

where ΔP is the difference of pressure upstream and downstream, and N_{row}, Number of pipes, while the Reynolds number base on pipe diameter is:

$$\text{Re}_D = \frac{\rho * U_{in} * D}{\mu} \tag{18}$$

3 Numerical Solution

In this study, the numerical analysis was performed using commercial Computational Fluid Dynamics (CFD) software (Simcenter STAR-CCM+, Version 2019.1). The control volume method is used to solve the equations of mathematical model, and by using the numerical method (the finite-volume method) with arranging the staggard grid. The momentum and energy equations have convective coefficients, which the scheme of second upwind was used to estimate and determine. The velocity of flow and field pressure can be associated by utilizing the SIMPLE algorithm as described by [28]. The coupled set of governing equations is solved using pressure correction for divergent velocity field. The under-relaxation factor is of value in the range (0.8) which is having been selected as the best or most appropriate in the equation of pressure correction. Also, note that the correction fields of temperature and velocity are updated. On the other hand, the numerical solution becomes converge when the deviation of iterations is less than 10^{-5}. Additionally, the thermo-physical properties of air under the temperature of 298 K are selected. The grid system is applied in represented of the computational model [29]. A polyhedral mesh was created with irregular size of 1 mm. At the wall boundary, the mesh is refinement with prism mesh layers, and also at the interface boundaries for foam/fluid in two fields. The schematic of mesh is presented in Fig. 3.

For each layer thickness of foam, Grid independence calculations were performed within four set of grid sizes at $\text{Re}_\text{D} = 5666$ as shown in Table 4. This table displays

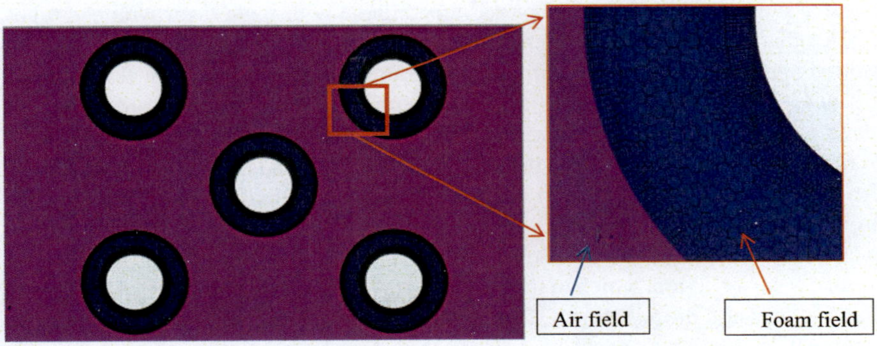

Fig. 3 The generated mesh of the problem

Table 4 Nusselt number shifting depending on the grid size at different foam thickness

$0.5(D_F - D)/D$	Grid no	1	2	3	4
0.2	No. of Cells	78,977	88,511	105,657	116,302
	Nu	194.054	197.290	201.525	201.946
0.43	No. of Cells	65,617	83,448	101,784	118,607
	Nu	290.782	291.256	298.972	299.547
0.6	No. of Cells	60,905	86,480	99,678	114,508
	Nu	317.661	332.354	336.236	337.371
0.8	No. of Cells	60,964	70,419	97,711	115,573
	Nu	351.807	354.543	357.980	358.166
0.91	No. of Cells	64,375	80,540	100,238	105,901
	Nu	367.141	368.345	370.629	371.029

that the difference in the value of Nusselt number between set grid No. 3 and No. 2 was within range of 2.1–0.9%, while the difference between set grid No. 4 and No. 3 was within range of 0.2–0.1%. From this one can found that the value of Nusselt number does not varied between grid No. 3 and No. 4, therefore set grid No. 3 was adopted.

To check the validity of this simulation, the predicted total heat transfer for a staggered three-row tube bundle coated with open-cell aluminum foam at range of air velocity are compared with experimental work of Chumpia [15] that have the same shape, arrangement, and foam specifications as shown in Fig. 4. This figure display that the numerical simulation agrees with the experimental work of Champia [15].

4 Results

In this section the effects of foam layer thickness and Reynolds number on the heat transfer and pressure drop of staggered tube bundle with foam layer wrapped tubes heat exchanger are evaluated and discussed. Figure 5 displays the pressure drop plotted against the Reynolds number for different dimensional foam thickness of 0.2, 0.43, 0.6, 0.8 and 0.91. As seen in figure, Generally, the pressure drop (dp) reveals a similar behaviour with respect to the Reynolds number (Re) for given value of foam thickness, where (dp) is directly proportional to the (Re). Also, from figure, the pressure drop increases with increasing foam layer thickness and the pressure drop for the foam layer can be as high as almost (10–12) times that of fin tube heat exchanger. This refers that the increase of the thickness results in an increase in the surface area of friction, which in turn an increases the pressure drop. One can observe that this pressure drop is the pressure difference between the inlet and outlet of the computational domain. So; it was due to the existence of a solid obstacle as well

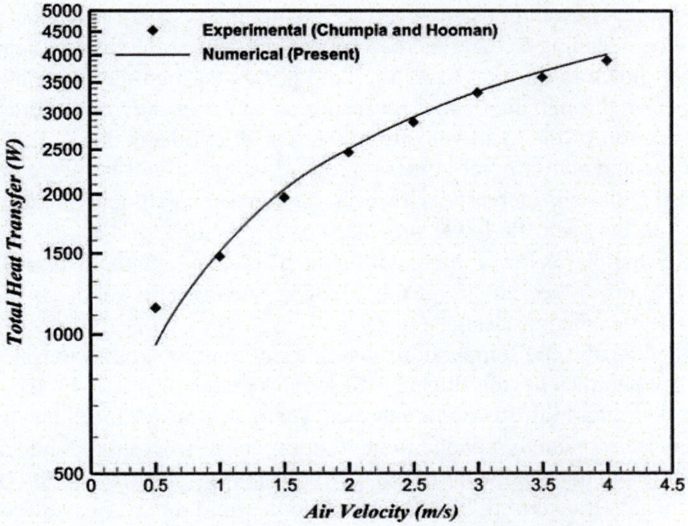

Fig. 4 Comparison of the total heat transfer computed from simulation numerically with the corresponding data measured experimentally by Chumpia [15]

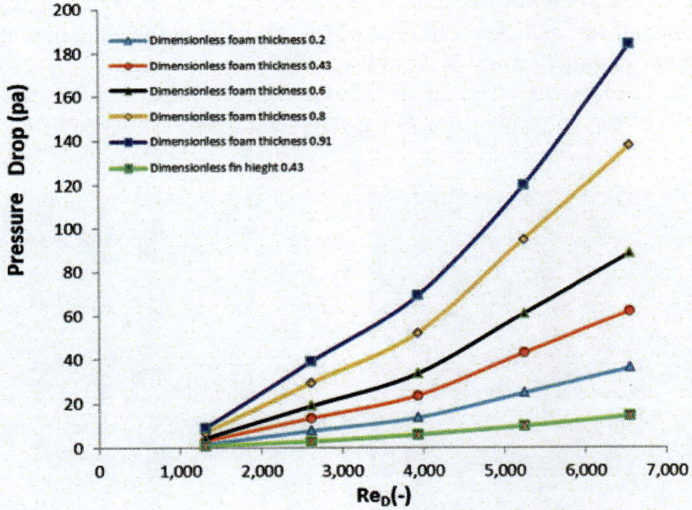

Fig. 5 Pressure drop variation with reynolds number for different dimensionless foam thickness and fin

as a foam layer. The solid obstacles results a hydraulic resistance by blocking apart of flow passage leading to local increase in the velocity of the fluid passing around the body as shown in the Fig. 6. As the fluid passes through the domain, a part of fluid that passes through the foam layer results a form drag and viscous drag. Hence, there is reduction of the local velocity at the rear of cylinders due to the blockage effect and the stagnation point at this region. This region will be decreased as the foam layer thickness is increased. However, the highest air velocity is located in the middle region between the lower tube and the upper tube, and in all cases, when the Reynolds number is low or high, within the whole computational domain, which nearly double times the velocity of the inlet free stream. This should be taken into account in heat exchanger designs.

Figure 7 describe the temperature distribution contour around the metal foam tubes for five samples having dimensionless thicknesses of 0.2, 0.43, 0.6, 0.8, and 0.91 at the low and high Reynolds number. The heat transfer mechanism consists of two main parts, namely conduction and convection, where the conduction heat transfer is through the solid walls of the matrix from the hot surface to the end of the foam, this heat is transferred, by convection, to the fluid that passes adjacent to the solid surface. According to the figure, general speaking for low and high Reynolds number, there is large change in air temperature as it moves through the tube bank (from left to right) and the highest temperature is approximately achieved at rear cylinder due to stagnation point. This temperature is decline as the foam thickness is increased, which demonstrate to be true that the best heat transfer can be obtained when the foam layer is thicker. For High Reynolds number the contour temperature is lower value than that for low Reynolds number.

Figure 8, displays the effect of foam layer thickness upon the Nusselt number within range of Reynolds numbers. For given metal foam layers, the behavior trend

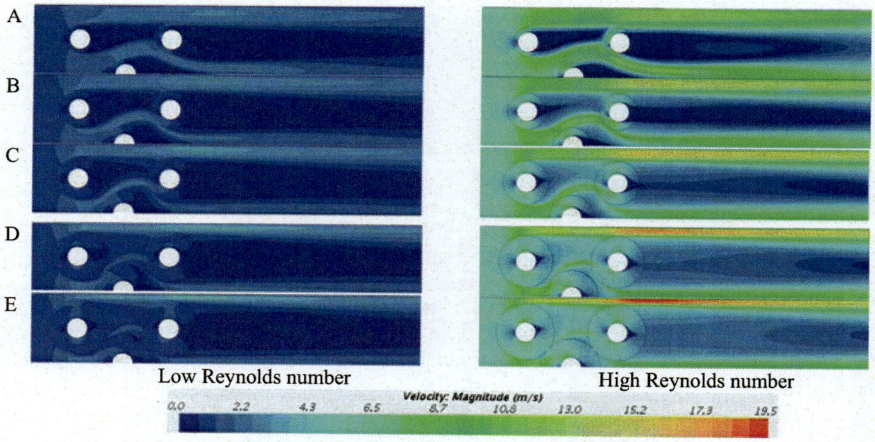

Fig. 6 Velocity contour at low and high Reynolds numbers for different foam layer thickness: A = 4 mm, B = 8.6 mm, C = 12 mm, D = 16 mm, and E = 18.2 mm

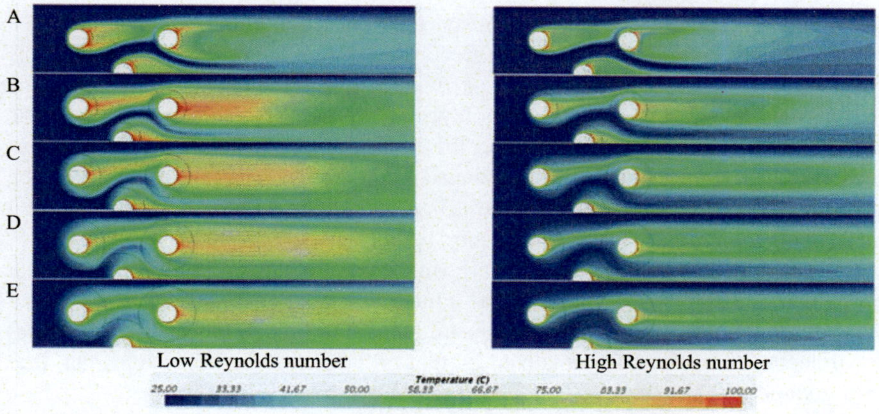

Fig. 7 The temperature contour for low and high Reynolds number and dimensionless foam thickness: A:0.2, B = 0.43, C = 0.6, D = 0.8, and E = 0.91

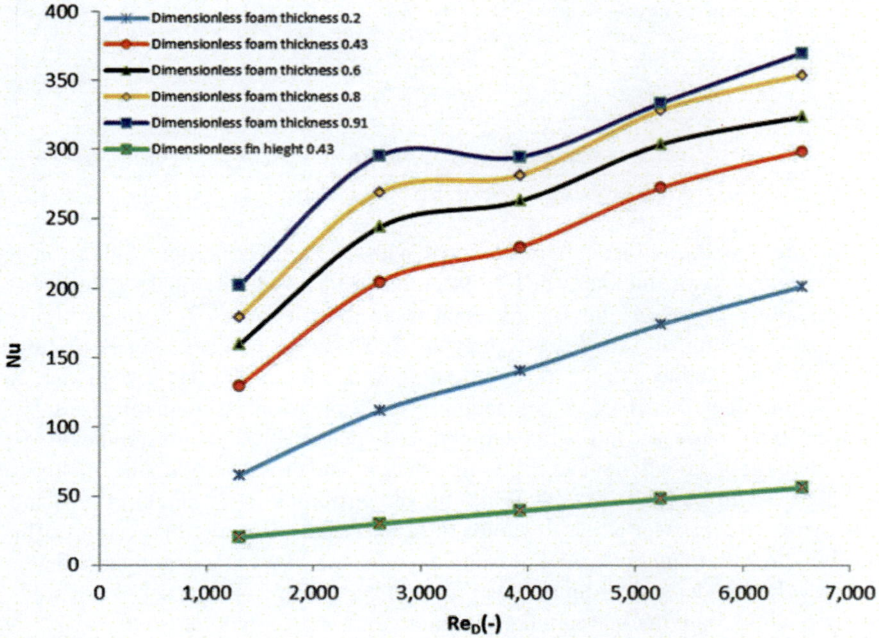

Fig. 8 Nusselt number varation with reynold number for different dimensionless foam thickness and fin

of Nusselt number will be similar with respect to Reynolds number, which is increased as the Reynolds number is increased. According to this figure, the thickness of the foam has a great influence on the heat transfer coefficient, where increasing the Nusselt number with increasing foam layer thickness, the result of increasing the area of solid surface, that is extended the surface area for heat transfer and thus increased heat transfer between solid matrix and fluid. Comparing metal foam tube banks and the fin tube bank, the *Nu* values of metal foam tube banks can be about 4 to 10 times higher than that of the fin tube bank.

For the purpose of evaluating the performance of the copper foaming coated heat exchangers relative to the conventional finned tube design, from both heat transfer and pumping power point view, the area goodness factor A.G.F. was used for metal foam heat exchanger and normalized by that of a finned-tube heat exchanger, (the area goodness factor ratio (A.G.F.R.) for all sample here, as follows:

$$\text{A.G.F.} = \frac{j}{f^{1/3}} = \frac{\text{Nu}}{\text{Re}_D * (\text{Pr} * f)^{1/3}} \tag{19}$$

where j is Colburn factor:

$$j = \frac{\text{Nu}}{\text{Re}_D * \text{Pr}^{1/3}} \tag{20}$$

$$\text{A.G.F.R.} = \frac{\text{Area Goodness factor of foam tube}}{\text{Area Goodness factor of finned tube}} = \frac{(j/f^{1/3})_{\text{Foam}}}{(j_o/f_o^{1/3})_{\text{Fin}}} \tag{21}$$

The area goodness factor is represented with a Reynolds number for range of dimensionless foam thickness in Fig. 9. As seen in this figure, at low Reynolds number, the goodness factor for all metal foam thickness are (3–5) times higher than that of the finned-tube heat exchanger. As Reynolds number is increased the ratios gradually reduces to 2.5–3.5. According to this figure, for any given value of Reynolds number, the area goodness factor for the dimensionless foam thickness 0.2, 0.43, and 0.6 is increased as the foam thickness increased till Reynolds number 4600, where thickness 0.6 will deviate from this behavior and crossover with the thickness of 0.43, i.e. will be lower than it. While for dimensionless foam thickness 0.8 and 0.91 there is a crossover at the Reynolds of (2600) approximately. Interestingly, the tube bundle with foam layer thickness of 0.6 shows the highest area goodness factor from low Reynolds number up to Reynolds number of 6555 at each set of Reynolds number, while foam thickness of 0.43 is suitable for high Reynolds number.

For understanding this fluctuation, Fig. 10 represents the relationship of area goodness factor ratio to a dimensionless thickness of layers, at a low Reynolds number 1311 and at a high Reynolds number $\text{Re}_D = 6555$. As clear from the figure, at low Reynolds number, the dimensionless thickness 0.43 allows air to pass through it; this means that the amount of transfer heat is higher than pumping power. As the thickness of foam is increased the volume of air flow is decreased due to the high

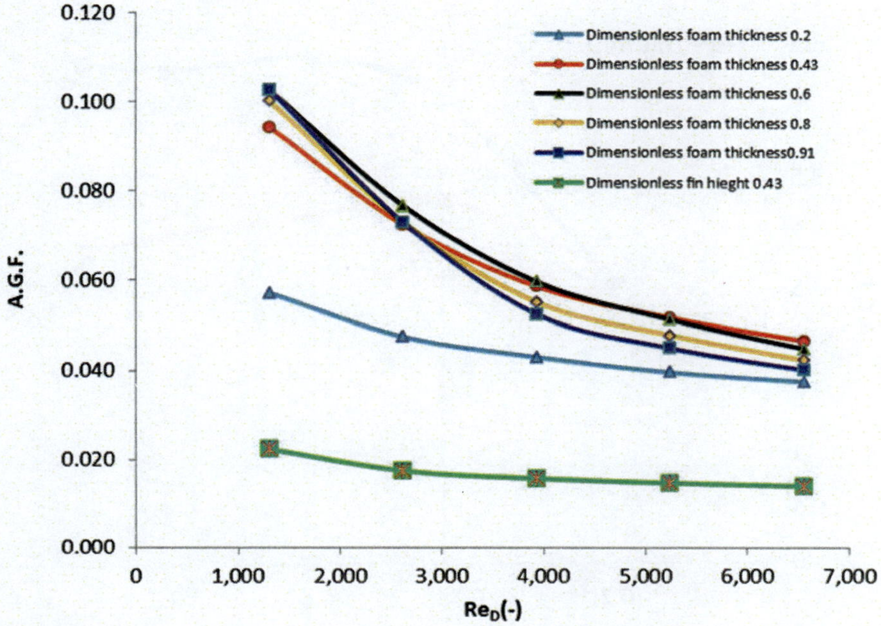

Fig. 9 Area goodness factor ration versus reynolds number for dimensionless foam thicknesses

resistance i.e. the pumping power required will be higher, so the area goodness factor ratio gradually reduced. While at a high Reynolds number, the air velocity is larger, so it can penetrate into the foam thickness from 0.43 until it reaches a dimensionless thickness 0.6, which has the highest area goodness factor ratio, then increasing thickness of foam layer more than this value, leads to require pumping power higher than the amount of heat transfer, so the area goodness factor ratio is reduced slightly.

5 Conclusion

The layers of high porosity copper open-cell metal foam (89%) with a pore density of 5PPI that has a variable dimensionless thickness of (0.2, 0.43, 0.6, 0.8, & 0.91) were used to wrapped heated tubes in a staggered arrangement that is cooled with air. Through numerical analysis using STAR CCM + software, it was found that both heat transfer and pressure drop increase, when Reynolds number and dimensionless layer thickness are increased. To determine the optimum thickness, an area goodness factor ratio was set as a standard for improving the thermal-hydraulic performance. The metal foam dimensionless layer thickness of 0.43 has the highest goodness factor ratio at a high Reynolds number, while the highest area goodness factor ratio is with a 0.6 dimensionless thickness at a low Reynolds number.

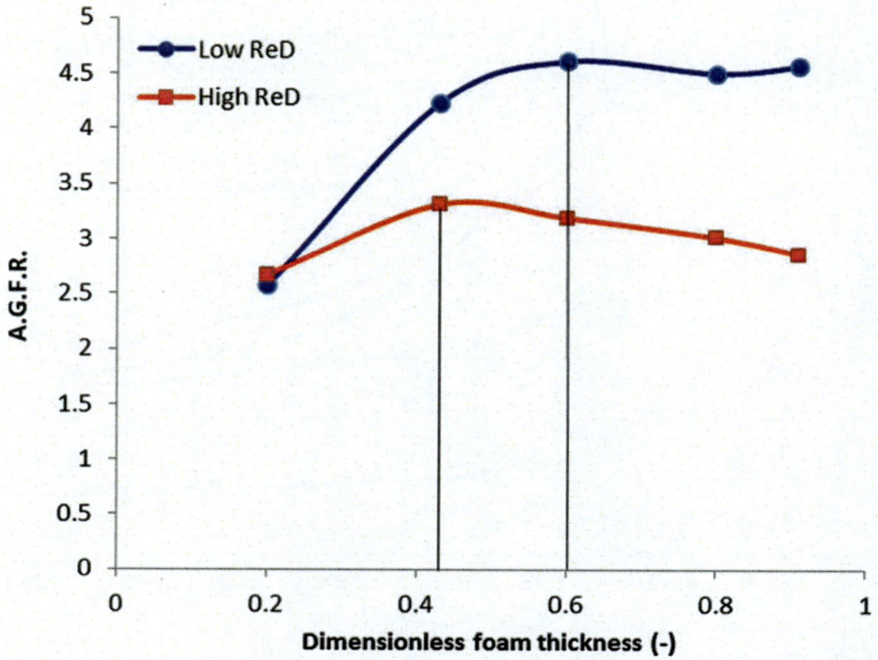

Fig. 10 Area goodness factor ratio variation with dimensionless foam thickness at low and high reynolds number

References

1. Chen Y et al (2017) Simulation and optimization of metal-foam tube banks for heat transfer enhancement of exhaust heat exchangers. In: 9th international conference on applied energy ICAE2017 21–24 cardiff energy procedia, pp 3863–3869
2. Huisseune H et al (2013) Simulation of an aluminum foam heat exchanger using the volume averaging technique. In: 8th international conference on porous metals and metallic foams, metfoam procedia materials science, vol 4, pp 353–358
3. Boomsma K et al (2003) Metal foams as compact high performance heat exchangers. Mech Mater 35:1161–1176. https://doi.org/10.1016/j.mechmat.2003.02.001
4. Bhattacharya A, Mahajan RL (2002) Finned metal foam heat sinks for electronics cooling in forced convection. J Electron Packag Copyright by ASME SEPTEMBER 124
5. Bhattacharya A, Mahajan RL (2006) Finned metal foam heat sinks for electronics cooling in forced convection. J Electron Packag 124:155–163
6. Lu W et al (2006) Thermal analysis on metal-foam filled heat exchangers Part I: metal-foam filled pipes. Int J Heat Mass Trans 49:2751–2761. https://doi.org/10.1016/j.ijheatmasstransfer.2005.12.012
7. Chen K et al (2018) Experimental investigation on enhanced thermal performance of staggered tube bundles wrapped with metallic foam. Int J Heat Mass Trans 122:459–468. https://doi.org/10.1016/j.ijheatmasstransfer.2018.01.115
8. Xiao Z (2013) Heat transfer, fluid transport and mechanical properties of porous copper manufactured by lost carbonate sintering. Ph.D. thesis of the University of Liverpool

9. Rahmati N et al (2017) Numerical study of heat transfer in turbulent cross-flow over porous-coated cylinder. In: Proceedings of the ASME fluids engineering division summer meeting FEDSM Waikoloa, Hawaii, USA
10. Taheri M (2015) Analytical and numerical modeling of fluid flow and heat transfer through open-cell metal foam heat exchangers. Ph.D. thesis University of Toronto
11. T'Joen C et al (2010) Thermo-hydraulic study of a single row heat exchanger consisting of metal foam covered round tubes. Int J Heat Mass Trans 53:3262–3274. https://doi.org/10.1016/j.ijheatmasstransfer.2010.02.055
12. Chen Y et al (2017) Simulation and optimization of metal-foam tube banks for heat transfer enhancement of exhaust heat exchangers. In: 9th international conference on applied energy, ICAE2017, 21–24 Cardiff, UK, Energy Procedia, pp 3863–3869
13. Rashidi S et al (2013) Fluid flow and forced convection heat transfer around a solid cylinder wrapped with a porous ring. Int J Heat Mass Trans 63:91–100. https://doi.org/10.1016/j.ijheatmasstransfer.2013.03.006
14. Buonomo B et al (2018) Numerical investigation on a heat exchanger in aluminum foam. In: The 73rd conference of the Italian thermal machines engineering association energy procedia, vol 148, pp 782–789. https://creativecommons.org/licenses/by-nc-nd/4.0/
15. Chumpia A (2015) Improving the performance of air-cooled condensers by using metal foams. Ph.D. thesis of The University of Queensland
16. Kays and London (1964) Compact heat exchangers. McGraw-Hill Book
17. Rollet-Mietet P et al (1999) LES and RANS of turbulent flow in tube bundles. Int J Heat Fluid Flow 20:241–254
18. Moulinec C et al (2004) Direct numerical simulation on a Cartesian mesh of the flow through a tube bundle. Int J Comput Fluid Dyn 18(1):1–14 (Taylor & Francis Ltd). https://doi.org/10.1080/1061856031000140211
19. Odabaee M et al (2011) Metal foam heat exchangers for heat transfer augmentation from a cylinder in cross-flow. https://doi.org/10.1007/s11242-010-9664-y
20. Shah RK, Sekulic DP (2003) Fundamentals of heat exchanger design. Wiley
21. Al-Sumaily GF (2014) Forced convection heat transfer from a bank of circular cylinders embedded in a porous medium. J Heat Trans Copyr by ASME 136
22. Odabaee M, Hooman K (2012) Metal foam heat exchangers for heat transfer augmentation from a tube bank. Appl Therm Eng 36:456–463. https://doi.org/10.1016/j.applthermaleng.2011.10.063
23. Alvandifar N et al (2018) Partially metal foam wrapped tube bundle as a novel generation of air cooled heat exchangers. Int J Heat Mass Trans 118:171–181. https://doi.org/10.1016/j.ijheatmasstransfer.2017.10.104
24. Al Husseny ANM (2016) Heat transfer enhancement using rotating metallic porous media. Ph.D. thesis of the University of Manchester
25. Boomsma K, Poulikakos D (2001) On the effective thermal conductivity of a three-dimensionally structured fluid-saturated metal foam. Int J Heat Mass Trans 44:827–836
26. Calmidi VV (1998) Transport phenomena in high porosity metal foams. University of Colorado
27. Fourie JG, Du Plessis JP (2002) Pressure drop modelling in cellular metallic foams. Chem Eng Sci 57(14):2781–2789
28. Patankar SV, Spalding DB (1972) A calculation procedure for heat, mass, and momentum transfer in three-dimensional parabolic flow. Int J Heat Mass Transf 15:1787–1806
29. Rhie CM, Chowt WL (1983) Numerical study of the turbulent flow past an airfoil with trailing edge separation. AIAA J 21(11)

Optimum Hydraulic Diameter of Mini-Channel Absorber Collector in Thermal Performance

Zainab H. Naji and Mohammed F. Mohammed

Abstract Numerical analyses were carried out to investigate the thermal and hydraulic performance of the modified design of solar collector (SC) based on a mini-channel (MC). In the use of mini-channel heat exchangers (HEs), it's essential to possess precise design tools to predict the transfer of heat and the drop of pressure. Numerical simulations were conducted via solving the governing equations (Continuity, Momentum and Energy equations) in a laminar regime in (3) dimensions via utilizing the FLUENT software (version 16). Propylene glycol was used as working fluid to enhance the heat transfer of the mini-channel SCs. Results manifested that the outlet temperature of Propylene glycol increased with the hydraulic diameter raising. Also, when Reynolds number (Re) was increased, the outlet temperature of water decreased for all cases in this study. The numerical simulation results elucidated the Nusselt number (Nu) and the coefficient of heat transfer increase with increasing Reynolds number and when the hydraulic diameter (D_h) decreased. The percentage increase in the case of the D_h mini-channel (2.181) is about (71%, 66.6%, 60.7% and 57%) than D_h (4) at the Re (10, 30, 50 and 70), respectively.

Keywords Mini-channel · Enhancement heat transfer · Absorber collector

Z. H. Naji (✉) · M. F. Mohammed
University of Technology, Baghdad, Iraq
e-mail: 20082@uotechnology.edu.iq

M. F. Mohammed
e-mail: 20011@uotechnology.edu.iq

F. Ahmad et al. (eds.), *ICPER 2020*, Lecture Notes in Mechanical Engineering,
https://doi.org/10.1007/978-981-19-1939-8_30

1 Introduction

The mini-channel HE collects the attributes of the high surface area per unit volume, the big coefficient of heat transfer, and the small inventory of the operating fluid [1]. To fulfil the need of the high removal of heat flux, the product of (the heat transfer coefficient and the heat transfer surface area) should be raised, and because the coefficient of heat transfer is associated with the hydraulic diameter, raising the surface area being one option. The area of heat transfer can be raised via employing MCs on the heat creating body. The behaviour of the flow of working fluid in the channels is obtained via the D_h of the channel, the cross-sectional area of the channel and its shape. For obtaining a high transfer of heat, a smaller D_h and a bigger area of the heat transfer of the channel being favoured, thus the narrow channels number with a high depth is convenient, but this disadvantage of MC increases the drop of pressure and thus needs higher power of pumping.

The slight scale channels (mini or micro) have been a popular area of study for the past (3) decades due to their high rate of heat transfer, particularly for HEs. Several surveys being conducted via various researchers on the flow and heat transfer features in the MCs. Diaz, 2008 [2] conducted a numerical analysis for the performance of a MC-based evacuated-tube solar collector and a compared it with a more standard round-tube collector. Keeping the similar free flow area of the tube for the operating fluid, the MC tube attached to a metallic absorber depicts an enhancement of approximately (5%) in the effectiveness. Also, a design with no absorber fin was analysed which depicts a borderline enhancement in the efficiency with regard to a MC tube having a thick absorber fin. The MC tube dimensions for the non-absorber-fin design were optimized for reducing the drop of pressure without considerably raising the tube cost. Gaurav Agarwal, 2010 [3] studied experimentally the thermo-hydrodynamic of concurrently evolving single-phase flow throughout a MC array that comprises (15) rectangular parallel channels (w is equal to 1.1 $\pm$ 0.02 mm, d is equal to 0.772 $\pm$ 0.005 mm, and D_h is equal to 0.907 mm), machined upon a plate of copper having a thickness of (8 mm). Re was changed from (200) to (3200), whereas the Prandtl number (Pr) was kept over a (3–4) range. The principal inferences of investigation were that the evolving flows offer too high coefficients of heat transfer in the zones of entrance, and thus of attention for the uses of mini-micro scale high heat flux removal uses. Mehta et al. [4] investigated experimentally the thermo-hydrodynamics of a thermally evolving single-phase flow of water throughout a square MC conducted within the laminar zone. It was concluded that the higher K value heads to the interface wall temperature isothermalization, while the lower approach leads to the state of a fixed heat flux. Also, Infra-Red Thermography (IRT) can be positively used to determining experimentally Nu through the initial thermal progress zone. The experimental Nu revealed a convincing agreement with the calculations. Hasan et al., [5] studied numerically the axial heat conduction in an isosceles right triangular micro-channel HE for a laminar, three dimensional, incompressible, single-phase, steady-state flow. The effective factors upon the axial heat conduction in the isosceles right triangular micro-channel HE are the thickness (t) of wall, the ratio of thermal conductivity (Kr), the volume of channel, the hydraulic diameter (D_h), and the Reynolds no. (Re). It was concluded that the axial heat conduction raised with the rise of increasing Reynolds number (Re), the thickness (t) of wall, and the channels no. but reduced with the rise of hydraulic diameter (D_h) each individually. Sahar et al., [6] studied the influence of the D_h and aspect ratio up on rectangular micro channel by using Ansys fluent 14.5. The numerical simulation was presented in a single-phase flow and heat transfer characteristics. The range of D_h (micro channel) was from (0.1) to (1 mm), and the aspect ratio was kept constant. The next case was varying the aspect ratio in the range (0.39–10 mm), whereas hydraulic diameter being reserves fixed (0.56 mm). The water was used as working fluid, and the Re range was (100–2000). The results elucidated that the effect of aspect ratio was not significant on the heat transfer. Also, the hydraulic diameter was more significant on the heat transfer coefficient. Martineza et al. [7] presented modeling and experimental work to obtain the thermal performance of the collector system with MC.

The evaluation will be completed via a coupled regime that tests at a full scale, and such test will include some interconnected enveloped cladding components linked to a joined solar thermal regime, with a heat pump. Such regime will be utilized for heating an indoor area in an office building, and provides the inland hot water to it. Nearly (20 m^2) of the energetic SCs shall be installed, and a comparatively-poor insulated (80 m^2) office shall be overhauled. Also, such test shall work to determine the figures of the working performance of collector, describe the working temperature, and authenticate the governing algorithms for avoiding the unsafe icing and blemish states at frontage level. The aims of the present investigation is to employ a numerical simulation program (Ansys 16) to study the effect of the D_h of MC solar collectors on the heat transfer characteristics and modify the thermal performance by increasing the quantity of heat transferred to the operating fluid (Propylene glycol) and the optimal suggestion.

2 Numerical Formulation

2.1 System Geometry

The MC solar collector basic geometry in the present work consists of forty nine rectangular MCs with various hydraulic diameters (2.18, 2.26, 3.07, 3.42, 3.73 and 4) mm. The width of channel is constant (4 mm), but the height varied (0.5 mm). These channels are inserted inside HE box (400 mm × 700 mm × 20 mm), as shown in Figs. 1 and 2. The working fluid (propylene glycol) flows across it with various rates of mass flow in the range (0.00201–0.0254) kg/s. The entry and exit of water into the collector are from the middle of the header for more a regular flow distribution. The coupling of velocity-pressure was affected throughout the SIMPLE algorithm (Semi Implicit Method for Pressure-Linked Equations), evolved via Patankar [8].

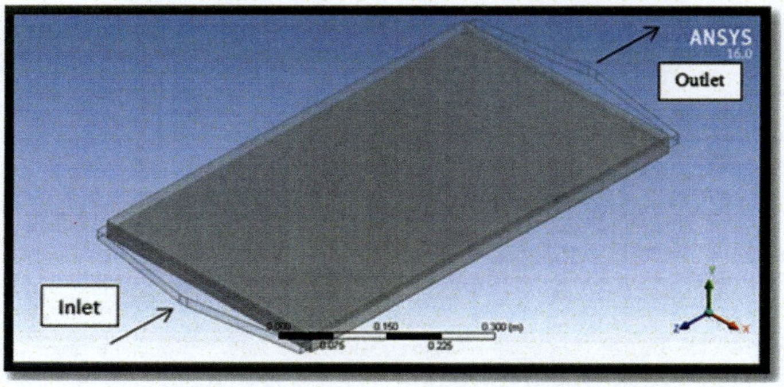

Fig. 1 The system geometry of SCs

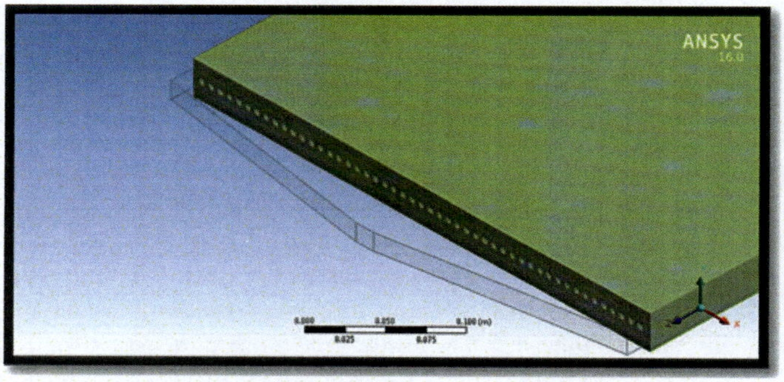

Fig. 2 The system geometry of MC solar collectors

The second-order upwind schemes being selected for the schemes of solution. The state of laminar flow was utilized. 3D mesh generations were constructed. Also, accordingly about four million cells (element) generated, as shown in Fig. 3a, b. Uniform heat flux (200 W/m^2) was used to show the effect of D_h on the MC solar collectors performance. The Numerical simulation was conducted utilizing a steady-state implicit pressure based solver that is an in-built in the viably accessible software FLUENT (Version 16) [9].

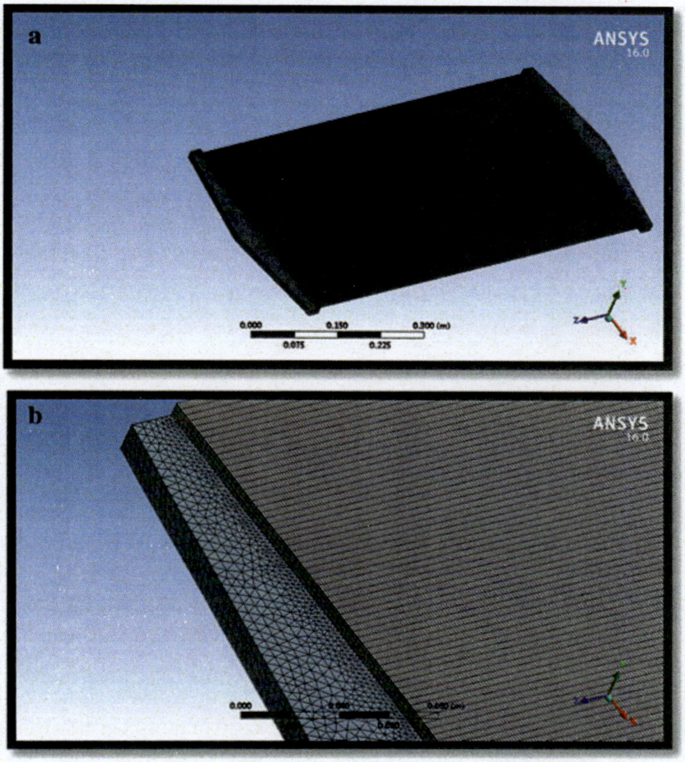

Fig. 3 **a** The mesh generation of SCs. **b** Mesh generation

2.2 *Mean Flow Equations*

In order to analyse the flow field in MC solar collectors, a solution of continuity, momentum in addition to the energy equation being needed. CFD was used to assess the hydraulic diameters influence upon the thermal performance on SC based on MC. The computational domains (two parts) are meshed with control volumes constructed around every grid employing (workbench) that's the pre-processor for FLUENT (Version 16).

The whole equations are introduced in Cartesian coordinate [10].

Continuity Equation:

$$\nabla(\rho V) = 0 \tag{1}$$

Momentum equation:

$$\frac{\partial}{\partial x_i}(\rho U_i U_j) = \frac{\partial \rho}{\partial x_i} + \frac{\partial}{\partial y_j}\left[\mu\left(\frac{\partial U_i}{\partial x_i} + \frac{\partial U_j}{\partial x_j}\right) - \overline{\rho U_l U_J}\right] \tag{2}$$

Energy equation:

$$\frac{\partial}{\partial x_i}(\rho U_i T) = \frac{\partial}{\partial x_j}\left[\frac{\mu}{P_r}\frac{\partial T}{\partial x_j} - \rho \vec{U}_l t\right]. \tag{3}$$

2.3 The Boundary Conditions and the Working Parameters

The inlet boundary condition is indicated as various mass flow rates in the range (0.00201–0.0254) kg/s depending on the Reynolds number (10–70). The inlet temperature is constant at (283 K) for various hydraulic diameters. The outflow boundary condition is implemented at outlet. The material used for the whole system is aluminium. The wall boundary conditions are utilized for bounding the solid and fluid and zones. The interface between the absorber MC and the water is characterized as a wall with a coupled state to affect the conjugate heat transfer from the absorber MC to the propylene glycol. Uniform heat flux (200 W/m^2) solar insolation is implemented at the upper surface of the HE box. Zero heat flux condition is applied at the bottom and side surfaces to affect the insulated conditions.

3 Result and Discussion

3.1 Relation Between Outlet Water Temperatures with Reynolds Number

This item will present and discuss the effect of hydraulic diameters on the thermal performance, as illustrated in Fig. 4. In general, the results depict that the outlet Propylene glycol temperature decreased when the Reynolds number increased for the all hydraulic diameters presented in this study. Also, the outlet water temperature is increased with the increase of D_h. This effect is because of the surface area increase of MC that led to absorb more heat and then transfer into propylene glycol.

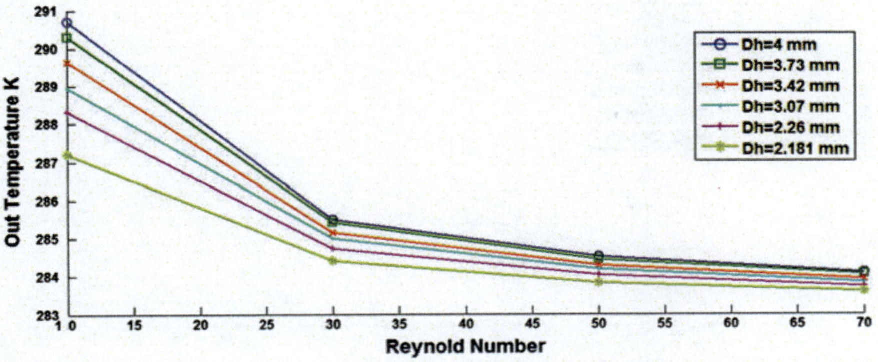

Fig. 4 The Relationship between outlet water temperatures and Reynolds number at different MC hydraulic diameters

3.2 *Effect of Heat Transfer Performance on Mini-Channel Hydraulic Diameter*

Figure 5 shows the relation between the coefficient of heat transfer and Reynolds number. The results manifested that the coefficient of heat transfer increases gradually with the increase of Reynolds number, due to the increase of the working fluid velocity that leads to a more heat transfer. Also, the coefficient of heat transfer increases when the hydraulic diameter decreases. The percentage increase in the case of hydraulic diameter MC (2.181) is about (71%, 66.6%, 60.7% and 57%) than D_h (4) at the Reynolds number (10, 30, 50 and 70), respectively. Figure 6 presents the Nusselt number for various hydraulic diameters. The results revealed that Nusselt number increases with increasing the Reynolds number and when the

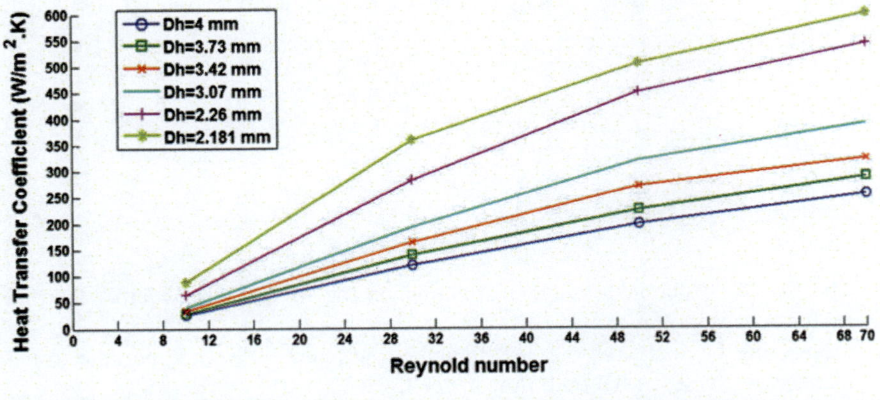

Fig. 5 The Relationship between heat transfer coefficients and Reynolds number at different MC hydraulic diameters

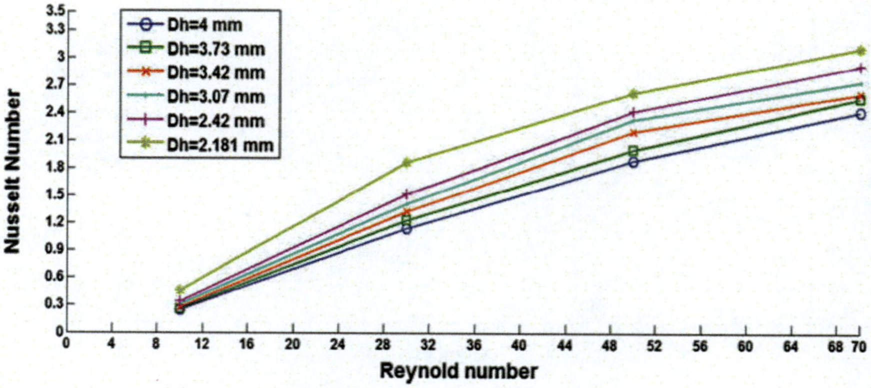

Fig. 6 The Relationship between Nusselt number and Reynolds number at different MC hydraulic diameters

D_h decreased. This effect is due to the increment of the working fluid velocity inside the MC when increasing the D_h that leads to increase the heat transfer coefficient and finally increase the Nusselt number. The behaviour of Nusselt number with Reynolds number for various hydraulic diameters is validated with [6].

3.3 *Flow and Temperature Distribution*

Figure 7 evinces the flow distribution inside the MC. Figures 8 and 9 present the temperature distribution of a MC solar collector, and the numerical simulation results elucidated the temperature increase in the direction of flow. This is due to the absorption of more heat from the MC in the direction of flow. Also, the decrease of the D_h raises the working fluid velocity that led to increase the heat transfer coefficient inside the MC.

4 Conclusions

From the numerical simulation presented in this study, the following conclusions can be drawn:

- The hydraulic diameter is more affecting on the MC solar collectors thermal performance.
- Decreasing D_h increases the heat transfer coefficient as well as Nu due to the increased velocity of the fluid flow inside the MC.
- It has been found that the D_h (4 mm) gives more outlet Propylene glycol temperature, then it is more acceptable in the heating applications.

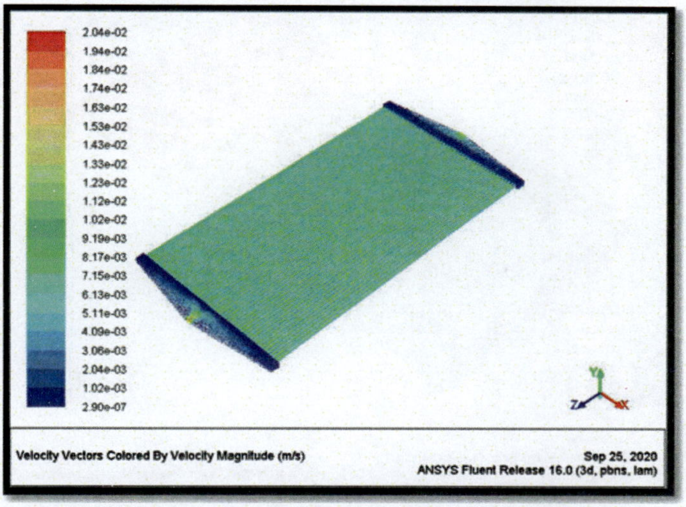

(a)

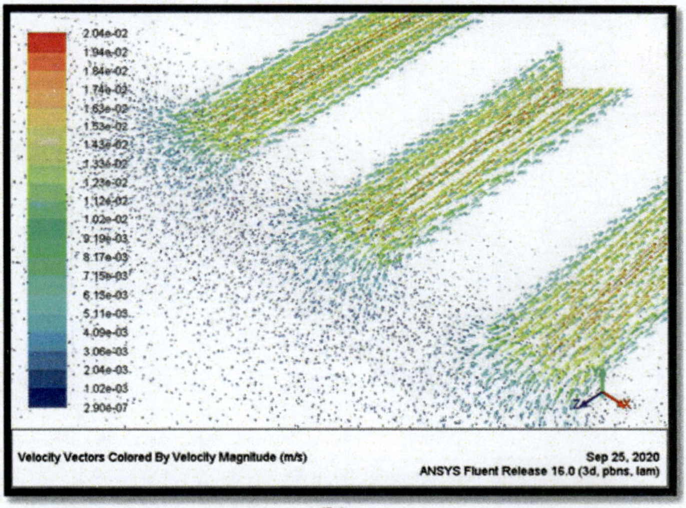

(b)

Fig. 7 The velocity vector inside the MC solar collectors

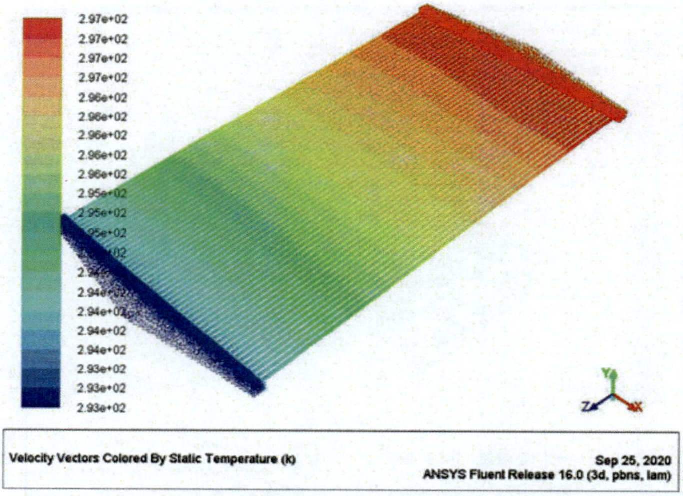

Fig. 8 The static temperature inside the MC solar collectors

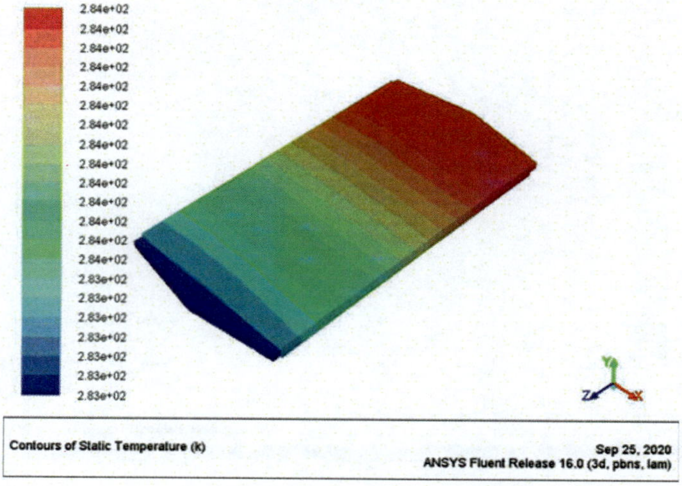

Fig. 9 The static temperature of the MC solar collectors

Acknowledgements The authors are obliged to the University of Technology, Baghdad-Iraq, Mechanical Engineering Department, Air Conditioning and Freezing Engineering Branch.

References

1. Duffie JA, Beckman WA (2006) Solar engineering of thermal processes, 3rd edn. Wiley, Interscience, New York
2. Diaz G (2008) Performance analysis and design optimization of a mini-channel evacuated-tube solar collector. Boston, Massachusetts, USA IMECE2008-67858
3. Agarwal G, Moharana MK, Khandekar S (2010) Thermo-hydrodynamics of developing flow in a rectangular mini channel array. In: 20th national and 9th international ISHMT-ASME heat and mass transfer conference
4. Mehta B, Dhyani A, Khandekar S (2011) Estimation of laminar single-phase heat transfer coefficient in the entrance region of a square mini channel using infra-red thermograph 1. In: Proceedings of the 21st national & 10thISHMT-ASME heat and mass transfer conference. IIT Madras, India
5. Ismael M, Hayder H, Hasan M, Abid GA (2014) Study of the axial heat conduction in parallel flow micro channel heat exchanger. J King Saud University Eng Sci 26:122–131
6. Sahar AM, Mahmoud MM, Wissink J, Karayiannis TG (2014) Effect of hydraulic diameter and aspect ratio on single phase flow and heat transfer in a rectangular microchannel "Conference Paper"
7. Martineza RG, Goikoleaa BA (2017) Performance assessment of an unglazed solar thermal collector for envelope retrofitting. Energy Procedia 115:361–368
8. Patankar SV, Spalding DB (1972) A calculation procedure for heat, mass and momentum transfer in three-dimensional parabolic flows. Int J Heat Mass Trans 15:1787
9. ANSYS FLUENT, Version 16.1, ANSYS
10. Frank M (2019) White. "Fluid Mechanics", 4h edn. University of Rhode Island

Numerical Analysis of Spark Ignition Engine Fueled by Iraqi Liquefied Petroleum Gas (LPG)

Mahmoud A. Mashkour and Ahmed Y. Ahmed

Abstract Sources of pollution generated from petrol and Diesel internal combustion engines have a major impact on the environment and human health. As an alternative fuel, LPG is therefore introduced to provide greater performance and lower consumption and emissions. Much of the published practical research regarding LPG, but contradictory results among them in terms of emission and consumption compared to gasoline fuel. In order to obtain the most accurate results, simulation program for Iraqi Liquefied Petroleum Gas supplied by the petrol stations was performed by MATLAB software using a quasi-dimensional, two-zone combustion model. The developed mathematical model is based on the first law of thermodynamics, equations for energy, mass conservation, equation of state, Weibe function and heat transfer from the cylinder for both burn and unburned zones. The engine parameters which can be studied in this research, in-cylinder gas pressure, in-cylinder gas temperature, thermal efficiency, specific fuel consumption, engine exhaust gas emissions and others. The simulation results show that the predicted in-cylinder gas pressure and temperature for LPG are lower than that for gasoline fuel at operating conditions $\phi = 0.8$, N = 2000 rpm, CR = 9. Also the predicted work and brake power of gasoline fuel more than LPG fuel. Furthermore, the simulation results show that the brake specific fuel consumption of LPG fuel improves by 2.25% throughout the engine speed comparing with gasoline fuel and the indicated thermal efficiency of LPG was less than that for gasoline but there is an improvement in both NO_x emission for LPG fuel by 30.3% and HC emission for LPG fuel by 0.3% comparing with gasoline fuel at the same operating conditions.

Keywords Iraqi Liquefied Petroleum Gas (LPG) · SI engine · Two-zone combustion model · Engine performance · NOx emissions

M. A. Mashkour (✉)
Mechanical Engineering Department, University of Technology - Iraq, Baghdad, Iraq
e-mail: Mahmoud.A.Mashkour@uotechnology.edu.iq

A. Y. Ahmed
Ministry of Education, Baghdad, Iraq

F. Ahmad et al. (eds.), *ICPER 2020*, Lecture Notes in Mechanical Engineering,
https://doi.org/10.1007/978-981-19-1939-8_31

Notation

a	Weibe efficiency factor
a_s	Stoichiometric molar air–fuel ratio
A_u	Area of unburned gases, m^2
A_w	Area of heat transfer, m^2
B	Bore, m
C_b	The blow by coefficient, sec^{-1}
c_{pb}	Burned specific heat at constant pressure, kJ/kg K
c_{pu}	Unburned specific heat at constant pressure, kJ/kg K
CR	Compression ratio
d_{splug}	Offset distance of the spark plug from the central axis of the cylinder, m
EGR	Fraction of exhaust gas recirculated
H.V.	Heating value of the fuel, kJ/kg
h_1	Specific enthalpy, kJ/kg
h_b	Specific enthalpy of burned gases, kJ/kg
h_g	Heat transfer coefficient, W/m^2 K
h_u	Specific enthalpy of unburned gases, kJ/kg
k	Weibe form factor
L	Connecting rod length, m
m	Mass, kg
m_{IVC}	Mass entering of the mixture when inlet valve closing, kg
$\dot{m}_1$	Mass flow rate, kg/s
N	Rotational speed, rpm
N_{cyl}	Number of cylinders
P	Pressure, bar
P_m	Motored cylinder pressure, kPa
P_r	Reference pressure, kPa
Q	Heat input, kJ
S	Cylinder stroke, m
T_b	Burned gas Temperature, K
T_g	Temperature of gases, K
T_r	Reference temperature, K
T_u	Unburned gas Temperature, K
T_w	Cylinder wall temperature, K
U	Internal energy, kJ
$\overline{U_P}$	Mean piston velocity, m/s
v	Volume, m^3
v_b	Burned specific volume, m^3/kg
V_d	Displacement volume, m^3
V_r	Reference volume, m^3
v_u	Unburned specific volume, m^3/kg
w	Engine speed, rad/s
W_b	Brake work done, J

W_i	Indicated work done, J
x_r	Residual gas fraction
γ	Polytrophic constant
θ_d	Duration of heat release, degrees
θ_{IVC}	Angle of the inlet valve closing, degrees
θ_{IVO}	Angle of the inlet valve opening, degrees
θ_s	Start of heat release, degrees
ϕ	Equivalence ratio

1 Introduction

One of the alternative sources of energy used instead of gasoline in the spark ignition (SI) engine is Iraqi Liquefied Petroleum Gas (LPG). LPG, which is a combination of butane (C_4H_{10}) and propane (C_3H_8) gases and other hydrocarbon compounds in a few proportions, is an effective fuel in general for internal combustion engines. For SI engines, LPG is an environmentally friendly fuel with potential emission advantages over gasoline [1]. Furthermore, LPG is liquefied under pressure, compressed and stored in steel tanks under pressure ranging from 1.03 to 1.24 MPa. It is used for heating, cooking, and can be used as engine fuel. Moreover as early as 1930, when a car was first powered on LPG on the eastern coast of the United States, LPG was first placed into use as a SI engine fuel. Now the advantages of contributing to energy stability, the economy and the environment by using LPG are becoming increasingly important to many governments. LPG also has a clean burning fuel and less emission, so some call it Green Fuel. Experiences say that LPG has some advantages over gasoline due to the following [1]:

1. Reduced pollution to the environment due to the cleanliness of the combustion produced by the LPG fuel engines, compared to the liquid fuel engines.
2. It reduces engine maintenance.
3. It offers faster cold starting.
4. It provides overall lower operational costs.
5. Ease of transportation of liquefied gas economically due to the major developments in building tanks.

Most studies confirm that LPG less carbon with buildup burns cleaner, less oil contamination, engine wear is reduced and life of some components of engine such as bearing and piston rings are much higher than with gasoline [2]. Moreover, as compared to gasoline in the range of (RON 91) to (RON 91), LPG has a higher octane number than (RON 105) (RON 976). The thermal efficiency of the engine can be increased by using LPG as fuel in spark ignition engines to provide better combustion in the engine, thus leading to a better fuel economy compared to gasoline. By changing the engine, the car can be converted to running on dual fuel LPG and gasoline.

Several researchers have studied the use of LPG in SI engines. Siti Sabariah et al. [3] based on a An analysis and simulation of the combustion process in the cylinder of the small SI engine running on LPG and gasoline fuel was given by the single-zone thermodynamic model. Researchers find that, relative to diesel, the LPG engine has led to lower emissions of NOx and HC. They have also shown that simulation software creation is appropriate for measuring engine output under various operating conditions. Mohsin Gatea et al. [4] conducted a computer evaluation of a four-cylinder SI engine running in petrol or dual fuel mode (LPG-PETROL). The results show that the LPG fuel improves the engine brake specific fuel consumption and brake thermal efficiency but reduces the brake power compared to gasoline fuel also the predicted maximum cylinder pressures for LPG are lower than that pressure for gasoline fuel. Dhakar et al. [5] studied the ability of using LPG in internal combustion engines. Researchers have found that the emissions from LPG are less than diesel and gasoline fuels. Also, the LPG can be used in dual fuel mode both with gasoline and diesel. Furthermore, LPG improves the emission, performance and combustion characteristics of LPG fueled engine under several operating conditions. Paczuski et al. [6] presented a report on LPG as an internal combustion engine fuel and noted that solving problems effectively would be a significant step forward in enhancing the competitiveness of LPG as an engine fuel. As well as a range of threads of advanced behavior aimed at further enhancing logistics performance, more full use of LPG benefits as an engine is presented. A comparative study by simulation of the behavior of a spark ignition engine powered by LPG and gasoline in transient regimes was performed by Nisulescu et al. [7]. Researchers have found that the values obtained by the simulation for the accumulation of pollutant emissions have been calculated in the context of the simulation.

Transient acceleration regimes for the two energy sources have resulted in emissions being decreased during the service of LPG relative to gasoline in operating conditions as follows: CO with about 20%, HC with about 27% NOx with about 68%. On the other hand, However, Aldhaidhawi et al. [8] prepared a numerical study to determine the combustion characteristics of engine emissions and the performance of petrol, LPG and ethanol running spark ignition engines. Researchers concluded that LPG and ethanol had lower effective torque and effective power than gasoline, and that both ethanol and LPG fuels were found to emit high unburned hydrocarbons compared to gasoline, with constant results recorded at all compression ratios.

It requires necessary to understand the characteristics of Iraqi LPG and to have an in-depth understanding of the mechanisms of the system so as to be able to connect it correctly and carry out programmed periodic maintenance. This paper aims to present detailed about theoretical investigation on the emission and the performance characteristics of spark ignition engine fueled of Iraqi LPG at different operating conditions and parameters.

2 Methodology

The model used for simulating LPG is a quasi-dimensional, two-zone combustion model that solves the differential equations associated with compression, combustion and expansion. Figure 1 is a schematic representation of the two regions, showing that the combustion gases are divided at the same pressure into a burned and unburned part, and that the flame propagates over the distance between the two portions [8]. The modeling is based on equations for mass conservation, equation of state, equation of energy and mass fraction burned which all programing by MATLAB program. The open system energy equation applied to the cylinder contents, is [9].

$$\frac{dQ}{d\theta} - P\frac{dV}{d\theta} = \frac{dU}{d\theta} + \frac{\dot{m}_1 h_1}{\omega} \tag{1}$$

2.1 Crank Slider Model

As a function of the crank angle, the volume of the piston cylinder can be determined from the compression ratio, the stroke, bore and connecting rod length. The geometrical parameters of the piston cylinder can be represented by the model of the crank slider shown in Fig. 1 [10].

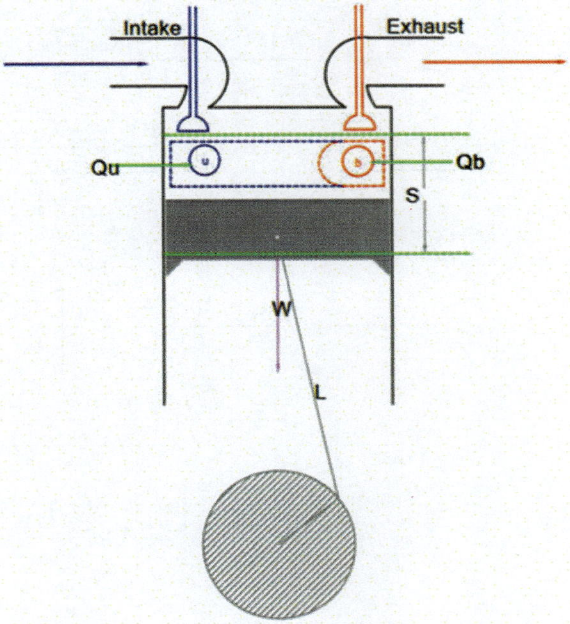

Fig. 1 Schematic of the two-zone thermodynamic system

$$V(\theta) = \frac{V_d}{CR - 1} + \frac{V_d}{2}\left[\varepsilon + 1 - cos\theta - \left(\varepsilon^2 - \sin^2\theta\right)^{1/2}\right] \tag{2}$$

where:

$$\varepsilon = \frac{2L}{S}$$

Upon differentiation

$$\frac{dV}{d\theta} = \frac{V_d}{2}\sin\theta\left[1 + \cos\theta\left(\varepsilon^2 - \sin^2\theta\right)^{\frac{-1}{2}}\right] \tag{3}$$

2.2 *Cylinder Gas Temperature and Cylinder Gas Pressure Models*

This model is derived from the two-zone energy equation. As a function of the crank angle [9], the temperature and pressure equations are derived for burned and unburned areas:

$$\frac{dT_b}{d\theta} = \frac{-hA_w(T_b - T_w)}{\omega m c_{pb} {x_b}^{1/2}} + \frac{T_b}{c_{pb}}\frac{\partial v_b}{\partial T_b}\frac{dP}{d\theta} + \frac{h_u - h_b}{x_b c_{pb}}\left[\frac{dx_b}{d\theta} - \left(x_b - {x_b}^2\right)\frac{C_b}{\omega}\right] \tag{4}$$

$$\frac{dT_u}{d\theta} = \frac{-hA_w\left(1 - {x_b}^{1/2}\right)(T_u - T_w)}{\omega m c_{pu}(1 - x_b)} + \frac{T_u}{c_{pu}}\frac{\partial v_u}{\partial T_u}\frac{dP}{d\theta} \tag{5}$$

$$A = \frac{1}{m}\left(\frac{dV}{d\theta} + \frac{VC_b}{\omega}\right) \tag{6}$$

$$B = \frac{hA_w}{\omega m}\left[\frac{1}{c_{pb}}\frac{\partial v_b}{\partial T_b}{x_b}^{1/2}(T_b - T_w) + \frac{1}{c_{pu}}\frac{\partial v_u}{\partial T_u}\left(1 - {x_b}^{1/2}\right)(T_u - T_w)\right] \tag{7}$$

$$C = -(v_b - v_u)\frac{dx_b}{d\theta} - \frac{\partial v_b}{\partial T_b}\frac{h_u - h_b}{c_{pb}}\left[\frac{dx_b}{d\theta} - \frac{\left(x_b - {x_b}^2\right)C_b}{\omega}\right] \tag{8}$$

$$D = x_b\left[\frac{T_b}{c_{pb}}\left(\frac{\partial v_b}{\partial T_b}\right)^2 + \frac{\partial v_b}{\partial p}\right] \tag{9}$$

$$E = (1 - x_b)\left[\frac{T_u}{c_{pu}}\left(\frac{\partial v_u}{\partial T_u}\right)^2 + \frac{\partial v_u}{\partial p}\right] \tag{10}$$

$$\frac{dp}{d\theta} = \frac{A + B + C}{D + E} \tag{11}$$

2.3 Combustion Model

In this model the Weibe function has been used to determine cumulative heat release fraction $x_b(\theta)$ inside combustion chamber of spark ignition engine which depends generally on the combustion chamber shape. This S-shaped curve can be represented analytically by a Weibe function as shown [9]:

$$x_b(\theta) = 1 - exp\left[-a\left(\frac{\theta - \theta_s}{\theta_d}\right)^{k+1}\right] \tag{12}$$

The heat release rate as a function of the crank angle is obtained by differentiating the function of the Weibe cumulative heat release as shown [9]:

$$\frac{dQ}{d\theta} = Q_{in}\frac{dx_b}{d\theta} \tag{13}$$

2.4 Heat Transfer Model

Convection in spark ignition engines is the main heat transfer method from the cylinder gases to the wall, with radiation being just 5%. The loss of heat to the wall is given by [11]: a Newtonian model is used.

$$\frac{dQ_w}{d\theta} = h_g A_w \left(T_g - T_w\right) \tag{14}$$

h_g represent the heat transfer coefficient which is needed in order to calculate heat loss from the cylinder. The Hohenberg correlation is used to calculated heat transfer coefficient [11]:

$$h_g = 3.26B^{-0.2}P^{0.8}w^{0.8}T^{-0.55} \tag{15}$$

$$w = \left[C_1\overline{U_P} + C_2\frac{V_d T_r}{P_r V_r}(P - P_m)\right] \tag{16}$$

where:

$$\overline{U_P} = 2SN \tag{17}$$

$C_1 = 2.28$ and $C_2 = 0$ during compression.

$C_1 = 2.28$ and $C_2 = 3.24 \times 10^{-3}$ during combustion and expansion.

The area exposed to heat transfer within the engine cylinder for both the burnt and the unburnt regions can be calculated from the following equations [11]:

$$A_w = \frac{\pi B^2}{2} + \frac{4V}{B} \tag{18}$$

$$A_b = A_w {x_b}^{1/2} \tag{19}$$

$$A_u = A_w\left(1 - {x_b}^{1/2}\right) \tag{20}$$

The motored cylinder pressure (p_m) is an important terminology affecting the transmission of heat in internal combustion engines can be defined by the equation as a polytrophic process [11]:

$$p_m = p_r\left(\frac{V_r}{V}\right)^{\gamma} \tag{21}$$

2.5 Nitrogen Oxide NO$_x$ Emission Model

For any fuel-to-air equivalence ratio, the generalized equation for the total combust reaction is given as [9]:

$$\begin{aligned} &C_a H_b O_c N_d + \frac{a_s}{\varphi}(O_2 + 3.76N_2) \rightarrow \\ &n_1 CO_2 + n_2 H_2 O + n_3 N_2 + n_4 O_2 + n_5 H_2 + n_6 CO \end{aligned} \tag{22}$$

For the formation of NOx, the extended Zeldovich mechanism was applied. The following reactions [9] describe this mechanism briefly as:

$$O + N_2 \rightleftharpoons NO + N \qquad K_1 = 3.1 \times 10^{10} \times exp\left(\frac{-160}{T}\right) \tag{23}$$

$$N + O_2 \rightleftharpoons NO + O \qquad K_2 = 6.4 \times 10^{6} \times T \times exp\left(\frac{-3125}{T}\right) \tag{24}$$

2.6 Hydrocarbons (HC) Emission Model

The unburned HC emissions which formed during a particular engine cycle of SI engine expressed as [12]:

$$HC_{formed} = [SF_{crevice}(1 - f_{ox}) + SF_{wall}(1 - f_{ox})] \times f_{ox}(1 - f_{oxexh}) \quad (25)$$

f_{ox} represents the fraction of HC oxidation within the engine cylinder is known as [12]:

$$f_{ox} = 1 - \left(\frac{P_{bulk}}{P_{max}}\right)\left(\frac{T_{HC}}{T_{HCadj}}\right) \quad (26)$$

where P_{bulk} and P_{max} are represented the bulk gas pressure, the theoretical maximum combustion pressure respectively while T_{HC}, T_{HCadj} are represented the bulk gas temperature and the adjusted log mean average temperature at the threshold of HC oxidation which are calculated from the following equations [12]:

$$T_{HC} = \frac{(T_{bulk} - T_{coolant})}{ln\left(\frac{T_{bulk}}{T_{coolant}}\right)} \quad (27)$$

$$T_{HCadj} = 1600 + 0.759 \times imep - 0.00051 \times imep^2 \quad (28)$$

Also, f_{oxexh} represents the fraction of HC oxidation within the exhaust is expressed as [12]:

$$\begin{aligned} f_{oxexh} &= 0.866 - 0.0000148 \times N - 0.0007 \times imep \\ &-0.00791 \times 0.829 \times \left(\frac{EGR}{100}\right) - 0.0000255 \times T_w \end{aligned} \quad (29)$$

As well as, $SF_{crevice}$ represents the crevice emissions index is defined as [12]:

$$SF_{crevice} = 5443 \times \left(\frac{P_{peak}}{imep}\right) \times \left(\frac{V_{crevice}}{\frac{V_d}{N_{cyl}}}\right)\left(\frac{1}{T_{coolant}}\right) \times f_{unburnd} \times f_{vapor} \times f_{mod} \quad (30)$$

The terms $f_{unburnd}$, f_{vapor} and f_{mod} are represented unburned gases, fuel vapor, and modification factor respectively which are defined form as [12]:

$$f_{unburnd} = 1 - x_r - EGR \quad (31)$$

$$f_{vapor} = \frac{1}{1 + AF_{ac}} \tag{32}$$

$$f_{mod} = 1 - 0.858 \times \left(\frac{d_{splug}}{B}\right) \tag{33}$$

Also, from the terms that remained for calculate the unburned HC emissions index (SF_{wall}) is expressed as [12]:

$$SF_{wall} = 63024 \times \left(\frac{1}{imep}\right)\left(\frac{1}{AF_{ac} \times 10^{(0.0082 \times T_{oil})} \times B}\right)\left(\frac{P_{inlet} + P_{inlet} \times CR^{\gamma}}{2}\right) \tag{34}$$

In defining the inlet pressure, Heywood and Hamrin developed a fitted equation defined as [12]:

$$P_{inlet} = 0.09875 + 0.009861 \times imep \tag{35}$$

2.7 Trapped Mass in Control Volume

The mass in the control volume is determined via an equation [9]. For $\theta_{IVC} \geq \theta \geq -360°$ crank angle through intake stroke as [9]:

$$m = \frac{V_{(\theta)}}{v_u} \tag{36}$$

For $\theta_{EVO} \geq \theta \geq \theta_{IVC}$ valve is closed as:

$$m(\theta) = m_{IVC}.exp\left[\frac{-C_b(\theta - \theta_{IVC})}{\omega}\right] \tag{37}$$

For 360°crankangle $\geq \theta \geq \theta_{EVO}$ blow down and exhaust as:

$$m = \frac{V_{(\theta)}}{v_b} \tag{38}$$

2.8 Engine Performance

Mean effective pressure (mep) means the average pressure applied to the piston in the cylinder of the engine at each power stroke and continuously adjusted during the cycle. There are two types of mean effective pressure: brake mean effective pressure (bmep), which is determined on the basis of the brake power (bp) transported by the engine, and indicated mean effective pressure (imep), which is produced in the cylinder and can be measured, both of which can be calculated as [10]:

$$imep = \frac{W_i}{V_d} \tag{39}$$

$$bmep = \frac{W_b}{V_d} \tag{40}$$

Also, can be calculated the indicated power (*ip*) inside the engine cylinder and brake power (*bp*) are given as [10]:

$$ip = \frac{imep \times V_d \times n \times N_{cylinder}}{60000} \tag{41}$$

$$bp = \frac{bmep \times V_d \times n \times N_{cylinder}}{60000} \tag{42}$$

where:

$$n = \frac{N}{2} for\, 4 - stroke\, and\, N\, for\, 2 - stroke\, enines$$

Engine friction data in SI engine are well correlated by an Eq. (43) which use in four strokes cycle, four cylinder and displacement between 845 and 2000 cm^3 [13]:

$$fmep = 0.97 + 0.15\left(\frac{N}{1000}\right) + 0.05\left(\frac{N}{1000}\right)^2 \tag{43}$$

The average effective pressure produced by the engine's useful work is called the mean effective brake pressure (bmep), which can be calculated as [13]:

$$bmep = imep - fmep \tag{44}$$

Other engine performance brake specific fuel consumption and the indicated thermal efficiency can be defined as [10]:

$$bsfc = \frac{\dot{m}_f}{bp} \tag{45}$$

$$\eta_{ITH} = \frac{ip}{\dot{m}_f \times H.V.} \tag{46}$$

3 Assumptions and Simulation Conditions

In this research, the expectations are defined as follows. The template is obtained from the four-stroke spark ignition engine without super charge and exhaust gas recirculation. The engine is used for preparing fuel in this research carburetor system. All the reactant fuel with air is completely in the system. Air with fuel mixture is presumed as ideal gas all the period. The combustion chamber design is assumed no any effect simulation procedures. The conditions for simulation and model overview are summarized in Table 1 and Fig. 2.

In this paper, the specifications of the Mercedes Benz engine were used in the internal combustion engines laboratory at university of technology which is shown in the Table 2.

4 Results and Discussion

The quasi-dimensional, two-zone combustion model was developed as the simulation model for the engine combustion modeling of four-stroke four-cylinder spark ignition (SI) engine fueled by gasoline and Iraqi LPG in MATLAB environment. The engine parameters such as in-cylinder gas pressure, in-cylinder gas temperature as a function of crank angle and the performance parameters such as brake power, brake specific fuel consumption, indicated thermal efficiency, exhaust gas emissions as a function of engine speed have been compared between gasoline and Iraqi LPG.

The variation of in-cylinder pressure with crank angle at operating condition is shown in Fig. 3 as $\varphi = 0.8$, $N = 2000$ rpm, $CR = 9$ for LPG and gasoline fuels. The simulation results show that, due to the high density of gasoline relative to LPG, the

Table 1 Simulation conditions

Parameters	Value
Pressure inside intake manifold	100 kPa
Gasoline chemical formula	C_7H_{17}
Gasoline heating value	44,430 kJ/kg
Stoichiometric A/F ratio for Gasoline	15.03
LPG chemical formula	$C_{3.34}\ H_{8.68}$
Stoichiometric A/F ratio for LPG	15.57
LPG heating value	49,160 kJ/kg
Equivalence ratio	0.8
Start of heat release	−35°
Duration of heat release	60°
The blow-by coefficient	0.8 s^{-1}
Residual gas fraction	0.1

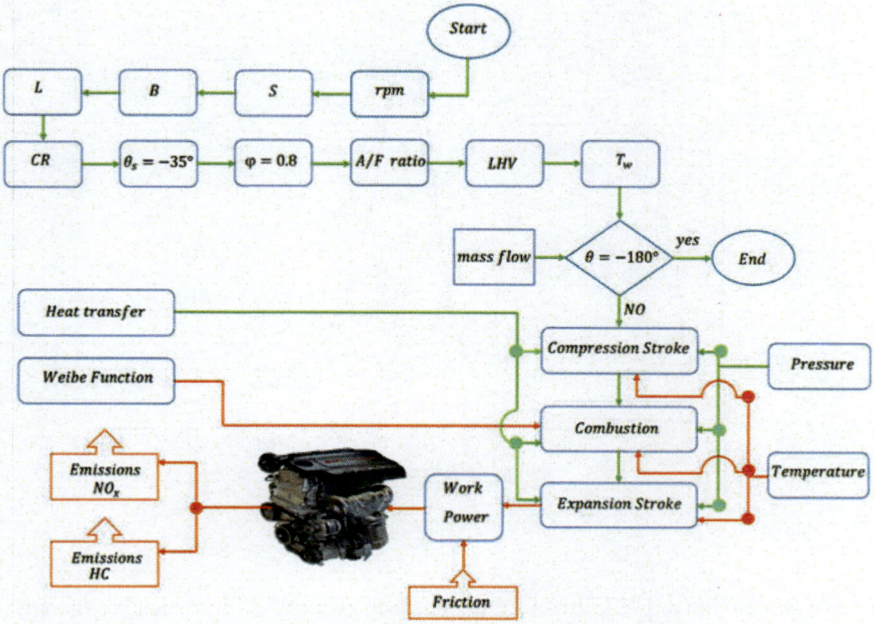

Fig. 2 Model overview

Table 2 Engine specifications

Particulars	Specifications
Engine type	Mercedes-Benz 200E
Compression ratio	9:1
Number of cylinders	4
Displacement (cc)	1997 cc 2L
Bore x Stroke (mm)	89 × 80.25
Connecting rod length (mm)	150
Number of valves per cylinder	2
Number of strokes	4
Intake valve open before TDC (degree)	11°
Intake valve close after BDC (degree)	53°
Maximum valve lift (mm)	8
Inlet valve diameter (mm)	41.2

gasoline fuel creates more in-cylinder pressure than that of LPG fuel. The quantity of air–fuel mixture fueled by LPG inside cylinder less caused it react each other less and it caused the in-cylinder pressure decreased. Furthermore due to higher content

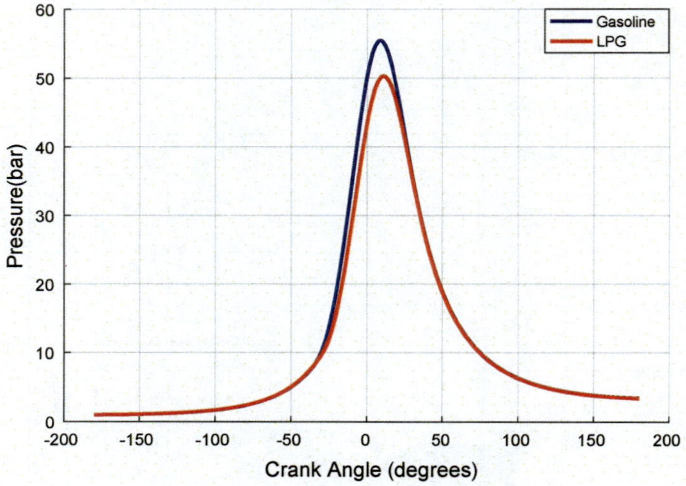

Fig. 3 Variation of in-cylinder pressure with crank angle

of carbon atoms than LPG, therefore the produced energy is higher lead to increase in-cylinder pressure.

The variation of in-cylinder gas temperature with crank angle at operating conditions is shown in Fig. 4 as $\varphi = 0.8$, $N = 2000$ rpm, $CR = 9$ for LPG and gasoline. The in-cylinder gas temperature of gasoline is higher due to more burned gas released. This is because high gasoline density could produce elevated burned gas during the

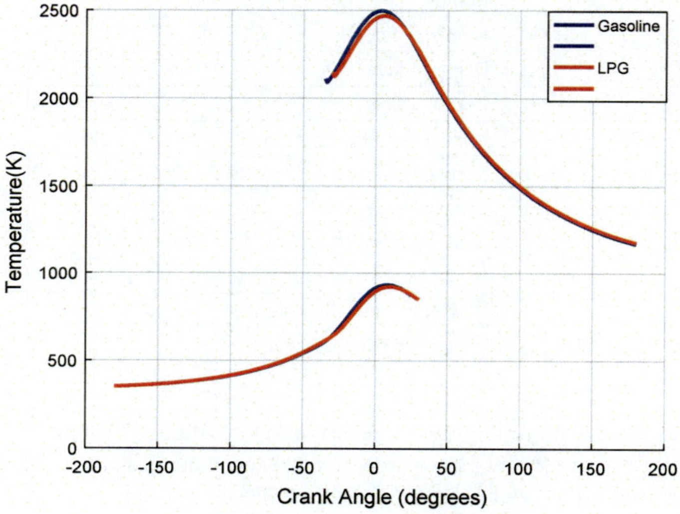

Fig. 4 Variation of in-cylinder gas temperature with crank angle

combustion process. Figure 5 represents the variation of cylinder work with crank angle at same operating condition as $\phi = 0.8$, N = 2000 rpm, CR = 9 for LPG and gasoline. The results show that the gasoline fuel produce work more than LPG fuel. This is because of the higher pressure applied on the surface area of the piston for gasoline than LPG fuel.

Figure 6 shows comparison between brake power of LPG with brake power of gasoline versus engine speed at equivalence ratio as $\phi = 0.8$, CR = 9. The results explain that the LPG fuel produce less brake power than that gasoline due to low density of LPG as its physical properties as a gaseous fuel and less work done for LPG.

Comparisons of brake specific fuel consumption for LPG and gasoline versus engine speed at $\varphi = 0.8$ and CR = 9 are shown in Fig. 7. Owing to the low heating value for gasoline fuel compared to LPG fuel and the stoichiometric air–fuel ratio for gasoline fuel less than LPG fuel, the reduction of brake specific fuel consumption for LPG compared to gasoline in the engine speed by approximately 2.25%.

Figure 8 shows comparison between indicated thermal efficiency of LPG with indicated thermal efficiency of gasoline versus engine speed at equivalence ratio $\phi = 0.8$, CR = 9. As shown in this figure, the gasoline fuel has higher indicated thermal efficiency than that for LPG fuel throughout the engine speed due to indicated mean effective pressure for LPG less than indicated mean effective pressure for gasoline also noted the difference small because stoichiometric air–fuel ratio and available energy of combustion for LPG more than gasoline.

Figure 9 shows NO_x resulted from LPG and gasoline fuels at same conditions $\phi = 0.8$, CR = 9 for a wide range of engine speed, NO_x concentrations for LPG less than gasoline by 30.3% due to high diffusion ability of LPG displacing much

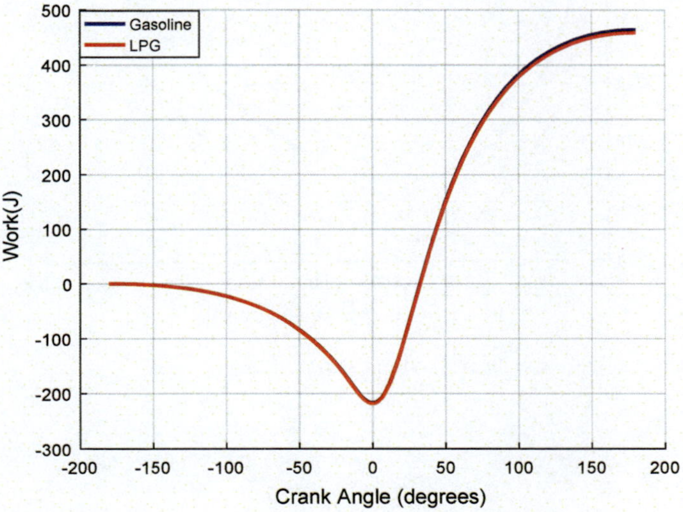

Fig. 5 Variation of cylinder work with crank angle

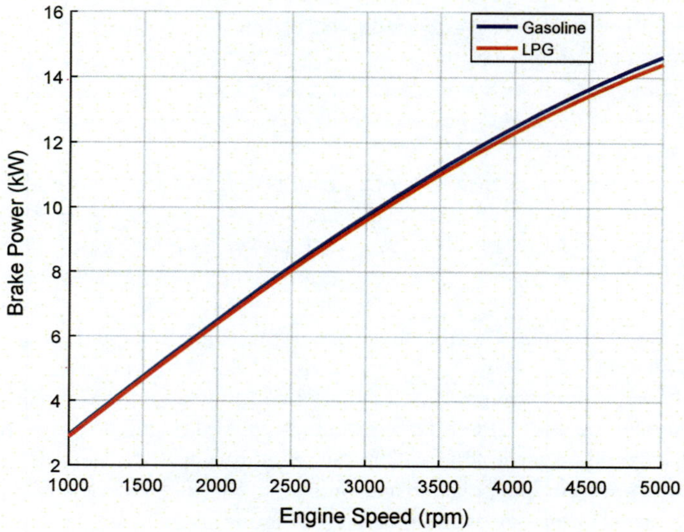

Fig. 6 Variation of brake power with engine speed

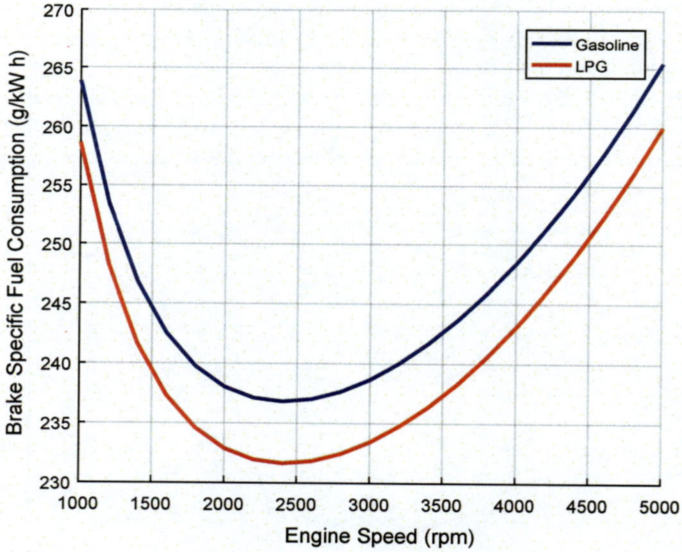

Fig. 7 Variation of brake specific fuel consumption with engine speed

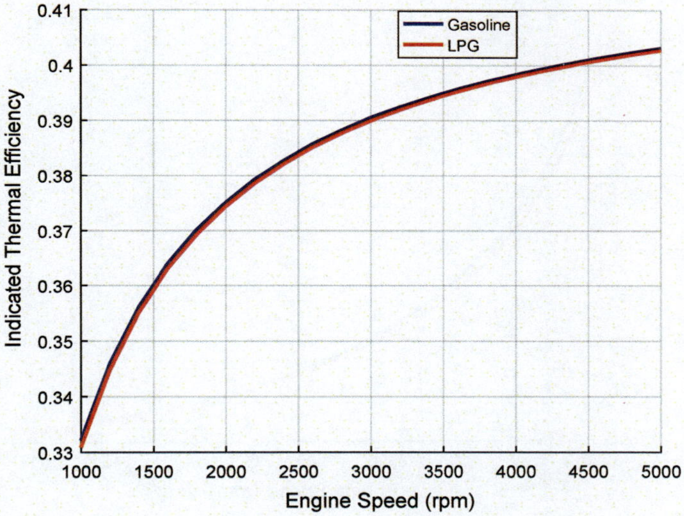

Fig. 8 Variation of indicated thermal efficiency with engine speed

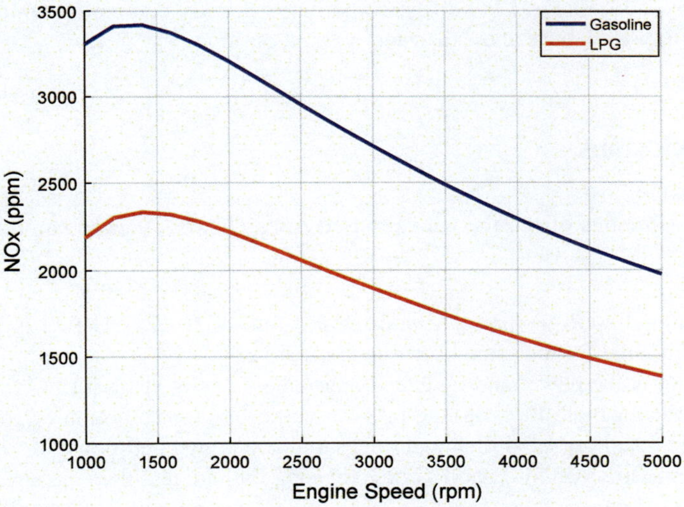

Fig. 9 Variation of NO_x with engine speed

of oxygen from air causing non-availability of oxygen to form nitrogen oxides also high temperature for combustion chamber of gasoline and high amount of oxygen lead to the formation of nitrogen oxides.

Figure 10 shows HC resulted from LPG and gasoline fuels at same condition $\phi = 0.8$, $CR = 9$ for a wide range of engine speed, HC concentrations for LPG lower than

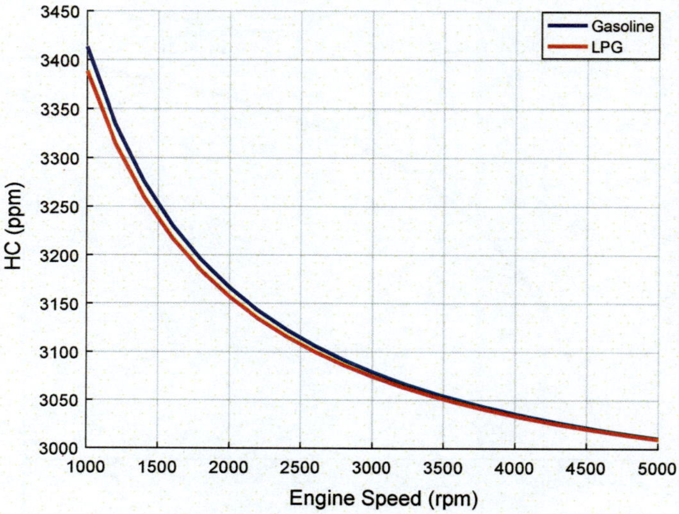

Fig. 10 Variation of HC with engine speed

gasoline by 0.3% due to LPG diffuses and burns quickly comparing with gasoline was not completely burn and exhausted without burning.

5 Conclusions

The major findings of the theoretical based research work can be summarized as follows:

1. The in-cylinder gas pressure and temperature of Iraqi LPG fuel is less than gasoline fuel for the same operating conditions.
2. Both of brake power and indicated thermal efficiency of Iraqi LPG fuel is less than the gasoline fuel, but the brake specific fuel consumption of gasoline is higher than Iraqi LPG fuel under the same operating conditions.
3. The emissions of NO_x and HC of Iraqi LPG fuel are less than the emissions of gasoline.
4. The use of Iraqi LPG as fuel for SI internal combustion engine produces good overall performance, significantly less pollutant to the environment and cheaper than gasoline fuel.

References

1. Ramadhas AS (2011) Alternative fuels for transportation. Taylor and Francies
2. Ismail A, Husaini Abu Bakar M (2019) Advanced engineering for processes and technologies. ISSN 1869–8433, Germany
3. Siti Sabariah M, Syafiqah Nabilah AS, Rosli AB, ZJunaidi Z, Mustafar MT (2019) Analysis and simulation of combustion and emission on small engine. Mater Sci Eng 469012076
4. Mohsin Gatea A, Esmail Amori K, Unis Salih H (2019) Computational assessment of four cylinders SI engine working on petrol or dual fuel mode(LPG-Petrol). Iraqi J Mech Mater Eng 19(4), Dec
5. Dhakar R, Tripathi A, Raj J (2016) Use of LPG in internal combustion engines-a state of art review. Automobile Engineering Department, ISSN-2349-5162
6. Paczuski M, Marchwiany M (2016) Liquefied Petroleum Gas (LPG) as a fuel for internal combustion engines. INTECH. https://doi.org/10.5772/61736
7. Nisulescu V, Ivan F, Iozsa D, Banca1 G (2017) Comparative analysis by simulation for behavior of a spark-ignition engine fueled with gasoline and LPG in the transient regimes. IOP Conf. Ser.: Mater Sci Eng 252:012088
8. Aldhaidhawi M, Najiand Kassy M, Subhi A (2019) Numerical study of combustion characteristic, performance and emissions of a SI engine running ongasoline, ethanol and LPG. Test Engineering and Management, January 2020. https://doi.org/10.36348/SJEAT.2019.v04i09.007
9. Ferguson CR, Kirkpatrick A (2016) Internal combustion engines: applied thermos sciences. John and Wiley Sons, Inc.
10. Pulkrabek Willard W (1997) Engineering fundamentals of the internal combustion engine. Prentice Hall, Upper Saddle River, NJ
11. Caton JA (2016) An introduction to thermodynamic cycle simulations for internal combustion engines. John Wiley & Sons, Ltd, p 950984
12. Hamrin DA, Heywood JB (1995) Modeling of engine-out hydrocarbon emissions for prototype production engines. SAE International, Warrendale, PA, SAE Technical Paper
13. Heywood JB (1988) Internal combustion engine fundamentals. McGraw-Hill Book Company, New York, NY

Numerical Investigation of LAMEE in Different Weather Conditions by Varying Solution Concentration with Air Inlet Humidity Ratio

Abbas Raza, Mohammad Shakir Nasif, and Rafat Al-Waked

Abstract LAMEE is an indirect contact air dehumidifier, and it also works as a solution regenerator. It has resolved carry over problem of desiccant solution by using a semi-permeable membrane which is permeable to water vapours only. A numerical model of cross flow configuration LAMEE has been made for investigating its performance under Malaysian and Pakistani weather conditions. In this model, desiccant solution concentration has varied from 40 to 50%, and inlet humidity ratio of air is increased twice and thrice for observing the effect on sensible as well as latent effectiveness. With desiccant solution concentration 50% increases latent effectiveness by 8.4%, while doubling inlet humidity ratio of air enhances latent effectiveness by 1.7% under Malaysian weather condition. In contrast, latent effectiveness is increased by 2.3% with 50% solution concentration and 1.24% by doubling inlet air humidity ratio under Pakistani weather condition. Sensible effectiveness does not vary significantly in all cases.

1 Introduction

Now a day's people are spending most of their time in buildings. As per fact that 90% of people time spend inside the buildings and hence thermal comfort should be achieved to maintain productivity [1]. As of 2019 building sector is the largest consumer of energy and accounting for 40% of global energy consumption along with greenhouse gases emissions [1]. According to health experts, under high humidity there is high chances for a human body to be affected from different health diseases and decay of buildings' material. For better health and productivity, indoor environment quality (IEQ) should be according to ASHRAE standards for HVAC system.

A. Raza · M. Shakir Nasif (✉)
Mechanical Engineering Department, Universiti Teknologi PETRONAS, 32610 Bandar Seri Iskandar, Perak, Malaysia
e-mail: mohammad.nasif@utp.edu.my

R. Al-Waked
Department of Mechanical and Maintenance Engineering, German Jordanian University, Amman 11180, Jordan

F. Ahmad et al. (eds.), *ICPER 2020*, Lecture Notes in Mechanical Engineering,
https://doi.org/10.1007/978-981-19-1939-8_32

Under thermal comfort conditions, the relative humidity between 30 and 60% as given in ASHRAE standards [2]. Therefore, air dehumidification plays a pivotal role in maintaining the indoor air quality and temperature of building. For air dehumidification, in conventional processes, air is dehumidified by lowering the temperature of air below its dew point, and then reheat the air for maintaining the indoor temperature of air [2]. In all this process, a huge amount of electrical energy is consumed. Furthermore, condensation yields bacteria in the cooling coils. Due to these problems, this method of air dehumidification is not a desirable way. In case of recovering heat and moisture from the exhaust air in the buildings, three types of energy recovery (ER) devices are available including air-to-air membrane energy exchangers, energy wheels and liquid-to-air membrane energy exchangers (LAMEEs) [3]. With low regeneration temperature of desiccant and no carry over problem of desiccant, LAMEE has been extensively used and prefer over other energy recovery devices. Furthermore, LAMEE used 100% fresh air in each cycle, and it provides a shield against the spread of virus especially in hospitals. Liquid-to-air membrane energy exchangers (LAMEEs) have been used as an air dehumidifier as well as desiccant solution regenerator in Liquid desiccant air conditioning systems. LAMEE used as an air dehumidifier under hot and humid weather conditions, whereas it is used to humidify the air in cold and dry weather.

The selection of desiccant made depending upon thermos-physical properties of the desiccant and cost is also a weighting factor in the selection. The suitable available desiccants are Lithium Bromide (LiBr), Lithium chloride (Licl), Magnesium chloride ($MgCl_2$) and Calcium Chloride ($CaCl_2$). Keeping the risk of crystallization of desiccant solutions under same operating conditions, MgCl2 is likely to be crystallized followed by $CaCl_2$, $LiCl_2$ and LiBr is the lowest among the desiccants [4]. There have been many developments made in LAMEE research, especially in mathematical and numerical modelling for predicting its performance. Fan et al. [5] made a numerical investigation on the temperature and moisture distribution in a crossflow LAMEE and he found out that the number of heat transfer units (NTU), the heat capacity ratio (Cr*), and the number of mass transfer units (NTUm) are key parameters in determining the performance of LAMEE.

Namvar et al. [6] made a steady-state performance analysis of a counter-crossflow. According to his findings, the effectiveness (Sensible, Latent and Total) are showing upward trend with increase in NTU and Cr* along with effectiveness is highly sensitive to Cr* at lower Cr* values. With respect to flow configurations in LAMEE, Mahmud et al. [7] found that the bottom-to-top flow configuration in RAMEE which is a combination of two or more LAMEEs and an aqueous desiccant solution in a closed loop, has high effectiveness than that of top-to-bottom flow configuration. Ge et al. [8] made different experiments for investigating water vapor diffusion resistance and modulus of elasticity of membranes as these properties are vital and effect the effectiveness of LAMEE. In his study, he found that crystals formed around the membrane pores and these crystals would enhance vapor diffusion resistance which results in lowering latent effectiveness. Moreover, he discussed about creep of membrane happens on exposing to large temperatures or stresses and stressed to take measures to avoid from high liquid pressure. Guo et al. [9] focused on regenerator part of LAMEE in his research. Using CFD, he made a numerical model and found out that response time for regeneration is highly influenced by air and solution velocity. Increasing solution velocity will enhance latent effectiveness for the regeneration of liquid desiccant, while increasing the air velocity will reduce response time of regeneration. Junze Chu et al. [10] made an experimental study of LAMEE performance by measuring its temperature fields. In his findings, he pointed out that solution temperature has negative effect on sensible, latent and total effectiveness. By increasing solution temperature increases vapor pressure of water vapors in desiccant solution and narrowing the gradient of moisture as a result latent effectiveness decrease. Moreover, increasing solution temperature will absorb less heat from the hot and humid air and decreasing sensible effectiveness.

Past research did not focus on the effect of varying inlet humidity ratio of air along with changing solution concentration under different weather conditions for the sake of optimizing LAMEEs performance. In this paper, ANSYS FLUENT CFD software is used to model heat and moisture transfer in LAMEE to predict the sensible and latent effectiveness under Malaysian weather (hot and humid) and Pakistani weather (hot and dry) conditions along with increasing inlet humidity ratio of air and increasing concentration of solution desiccant. To overcome the limitation of ANSYS FLUENT, UDF code is developed to simulate moisture transfer through membrane.

2 Modelling of LAMEE

A single channel Flat Plate LAMEE model as shown in Fig. 1. has been designed using ANSYS Fluent for the sake of simulating heat and moisture transfer across the membrane. In this model, one fluid hot and humid air having high temperature and vapour pressure while other fluid which is desiccant solution having low temperature and vapour pressure are flowing in cross flow configuration and one semi-permeable membrane, which allows only water vapours to permeate, between two fluids, has been modelled. Using Pressure based solver along with Species transport model,

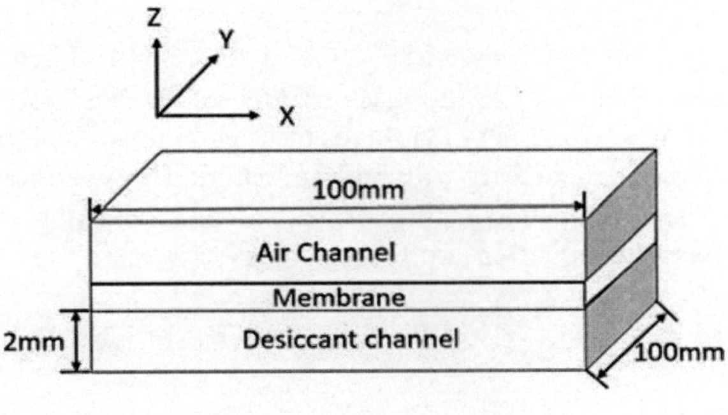

Fig. 1 Configuration of single channel Flat-Plate LAMEE

vapour mass fraction is introduced inside both fluids. For reducing the computational time of simulation, symmetry assumption of volume of fluid on both sides of energy exchanger is taken. As a matter of fact, ANSYS is unable to simulate moisture across the membrane. For this, in-house User Define Function (UDF) was written in C++ and with this in-house UDF, moisture transfer across the membrane is observed. Semi-Implicit Method for Pressure Linked Equations (SIMPLE) algorithm was adopted for calculating pressure and velocities fields. Whereas Second order upwind discretization scheme was used for momentum, energy, and species equations. Several assumptions have been taken into consideration while developing in-house UDF:

- Both fluids are laminar.
- Heat and moisture is transferred along the z-axis only.
- No heat is transferred between LAMEE and surroundings.
- Membrane allows water vapours till dynamic equilibrium state.

2.1 Governing Equations

Apart from fundamental equations of fluid flow, additional equation is required for defining species with mass diffusion. The equations for incompressible steady state air flow is given below:

$$\nabla.(\rho\Phi V) = \nabla.(r\Phi\nabla\Phi) + S\Phi \tag{1}$$

Energy transfer due to conduction and species diffusion is represented in this CFD model is given below:

$$\nabla.(\vec{V}(\rho \mathrm{E} + \mathrm{p})) = \nabla.(K\nabla T - \sum i\mathrm{h}i \cdot \vec{j} \cdot i) \tag{2}$$

The species conversion equation is given below:

$$\nabla.\left(\rho\vec{V}\mathrm{Y}i\right) = \nabla.\vec{J}i \tag{3}$$

K is thermal conductivity, T is temperature of inlet air, hi is sensible enthalpy of species i, Yi is the mass fraction of species$i$, $\vec{J}\ i$ and is the diffusion flux of species$i$. FLUENT [11] solves these equations. Heat transfer across the LAMEE can be calculated easily with built in feature. However, for moisture transfer, UDF code to be developed to enable software to model moisture transfer across membrane. The mass flux equation is represented as [12]:

$$\vec{J}i = -\rho.Di, m.\nabla Yi \tag{4}$$

Di, m is the diffusion coefficient of ith species (water vapour), J is the moisture flux of moist air, ρ is the density of air and ∇Yi is the change in mass fraction of water vapours.

Membrane surface is modelled as a thin wall owing to its small thickness and with developed UDF code, Fluent will enable species to move across the thin wall. Using Fick's law [13]

$$Jv = -\rho a.Dv, a \left.\frac{\partial Yv}{\partial z}\right|_{hs} = -\rho m.Dm \left.\frac{\partial Yv, m}{\partial z}\right|_{m} = -\rho c.Dv, c \left.\frac{\partial Yv}{\partial z}\right|_{cs} \quad (5)$$

The developed in-house UDF code was based on Fick's Law, which has given mass fraction of water vapour in the air side and mass fraction in the cold desiccant solution. The difference of mass fraction of water vapour values at air inlet and air outlet ensures the viability of UDF. The mass weighted average value of vapour mass fraction in the solution side of membrane obtained through developed in-house UDF will be used for calculating Latent effectiveness of LAMEE. With the help of this UDF, the mass weighted average value at air outlet is calculated and Latent effectiveness of this modelled LAMEE is validated against Huang et al. experimental values with an acceptable range of error.

2.2 *Mesh Sensitivity*

This CFD model is gone through mesh sensitivity analysis. On having a Grid 1(number of cells 13,472) and Grid 2 (number of cells 16,393) the percentage difference is 2.5% in the value of humidity ratio of air outlet. On further increasing the number of cells, Grid 3 (number of cells 19,205), the percentage difference is 2.3% in the value of humidity ratio of air outlet. On more refining of mesh, Grid 4 (number of cells 24,145), the percentage difference is 1.7% in humidity ratio of air outlet and 0.013% in the value of air temperature outlet as compared to Grid 3. This Grid 4 is taken for calculation as the temperature and humidity ratio of air outlet values are in acceptable range of error with Huang et al. experimental values [15].

2.3 *Membrane Modelling*

For modelling the membrane, under steady state conditions, the net heat and mass transfer across the thin wall will be equal to zero. One dimensional equations which are defining heat and mass transfer inside the membrane. The steady state equations for heat and mass [14] are given here below:

$$\frac{\partial^2 Tm}{\partial z^2} = 0 \quad (6)$$

$$\frac{\partial^2 Yi, m}{\partial z^2} = 0 \tag{7}$$

2.4 Boundary Equations

The inlet of air and desiccant solution are taken as Velocity inlet. The outlets of both fluids as Pressure outlet. Volume of both fluids are taken as symmetry. Membrane as a coupled wall.

3 LAMEE Performance

The performance of LAMEE can be gauged by many key parameters. Some of them are given below:

3.1 Sensible Effectiveness

It is the ratio of actual heat transfer to the maximum heat transfer. It is presented in a given equation.

$$€s = \frac{Tair,in - Tair,out}{Tair,in - Tsol,in} \tag{8}$$

3.2 Latent Effectiveness

It is the ratio of actual moisture transfer to the maximum moisture transfer. The equation for Latent effectiveness is given below:

$$€_L = \frac{Wair,in - Wair,out}{Wair,in - Wsol,in} \tag{9}$$

The above-mentioned equations for sensible and latent effectiveness are applicable when ($Cr^* \geq 1$). Cr^* is the ratio of heat capacity of desiccant solution to the heat capacity of air.

$$Cr* = \frac{\dot{m}_{sol}.Cp_{sol}}{\dot{m}_{air}.C_{pair}} \tag{10}$$

$\dot{m}_{sol}$ is the mass flow rate of desiccant solution flow (kg/s), Cp_{sol} is the specific heat of desiccant solution (J/kg K), $\dot{m}_{air}$ is the mass flow rate (kg/s) of air and Cp_{air} is the specific heat(J/Kg K) of air. The Eqs. 8, 9 and 10 have been taken from Abdel-Salam et al. [14].

Taking the reference values of $\dot{m}_{air}$, $\dot{m}_{sol}$,Cp_{air} from Huang et al. [15] and Cp_{sol} from Ashfin [16], the Cr* is found to be greater than 1. In this paper for calculating sensible and latent effectiveness, the above-mentioned formulas are applicable (Table 1).

It is worth mentioning that PVDF membrane is taken in this CFD model of LAMEE, along with solution desiccant is LiBr. The concentration of solution is varied at 40% and 50% respectively and concentration variation effect on LAMEE effectiveness is investigated under hot and humid as well as hot and dry weather conditions. Moreover, air inlet humidity ratio is also varied along with 40% and 50% desiccant concentration and its effect on LAMEE effectiveness is also observed.

4 Results and Discussion

4.1 Effect on Latent Effectiveness with 40% & 50% LiBr Concentration

On having the reference values of different volumetric flow rates from Huang et al. [15] along with 40% and 50% LiBr concentration, Latent effectiveness is decreasing with increasing air volumetric flow rate due to less residence time of air. However, Latent effectiveness with 50% LiBr founds to be 8.39% higher than that of 40% LiBr under same operating conditions in Malaysian weather, whereas latent effectiveness is enhanced by only 2.3% with 50% LiBr under Pakistani weather as it can be seen in Fig. 2a, b.

It is evident from the Fig. 3a, b, the variation in the water vapour mass fraction on membrane surface from air inlet which has high vapour mass to air outlet having low vapour mass is more in 50% LiBr under same operating conditions and as a result, high increase in latent effectiveness is observed with 50% LiBr solution in comparison to 40% LiBr solution.

Table 1 Membrane and other transport properties reference values [15]

Property	Value and unit
Membrane thermal conductivity	$0.5\ \mathrm{Wm^{-1}\ K^{-1}}$
Membrane density	$1780\ \mathrm{kg\ m^{-3}}$
Membrane diffusion coefficient	$3 \times 10^{-6}\ \mathrm{m^2\ s^{-1}}$
Membrane thickness	1×10^{-4} m
Inlet temperature of air	303.15 K
Inlet temperature of desiccant	298.15 K

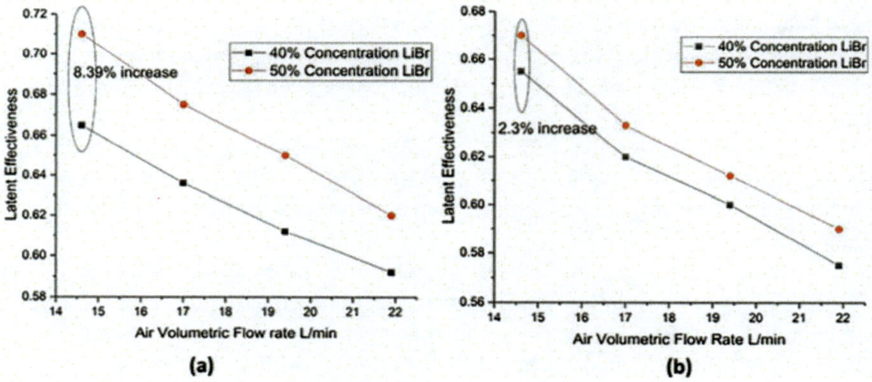

Fig. 2 **a** Latent Effectiveness with different desiccant concentration under Malaysian weather; **b** Latent Effectiveness with different desiccant concentration under Pakistani weather

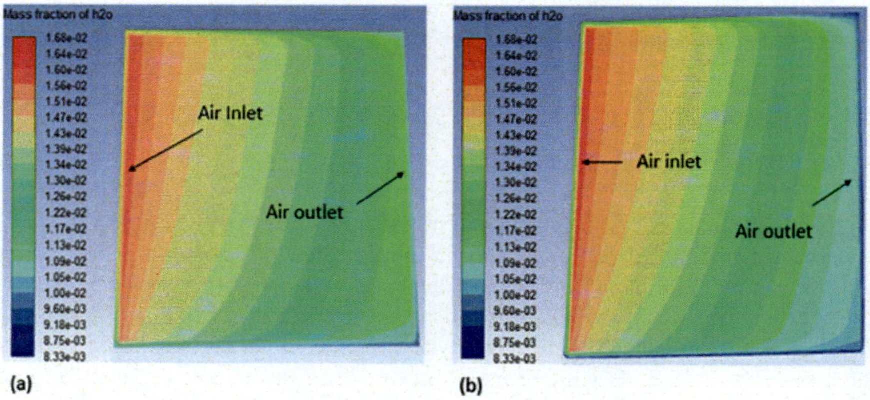

Fig. 3 **a** distribution of vapour mass fraction on membrane surface with 40% LiBr under Malaysian weather; **b** distribution of vapour mass fraction on membrane surface with 50% LiBr under Malaysian weather

From the above figures, it can be deduced that vapour mass fraction variation on membrane surface from air inlet to air outlet is slightly higher in Fig. 4b than that of Fig. 4a. and it has little bit high latent effectiveness. Moreover, this variation is small as compared to Malaysian weather conditions as the surrounding weather is hot and dry.

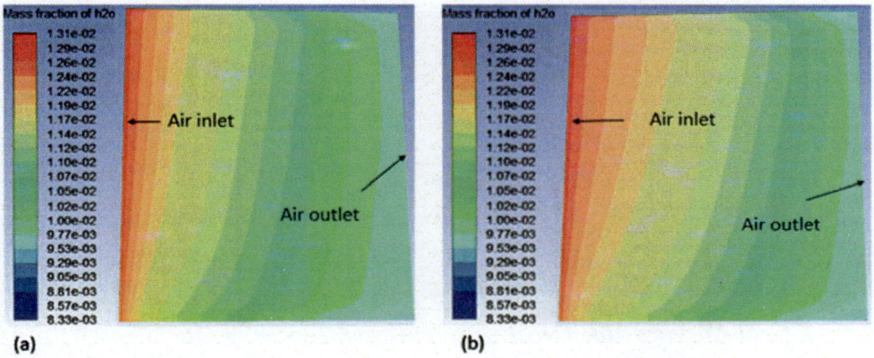

Fig. 4 **a** Distribution of vapour mass fraction on membrane surface with 40% LiBr under Pakistani weather; **b** Distribution of vapour mass fraction on membrane surface with 50% LiBr under Pakistani weather

4.2 Effect of Air Inlet Humidity Ratio with 40% LiBr Concentration

Under Malaysian weather conditions, on taking the 2 times of air inlet humidity ratio with 40% concentration of LiBr, 2% increase in Latent effectiveness is observed, and on taking 3 times of air inlet humidity ratio while keeping other parameters same, the latent effectiveness further increased by 0.9% and sensible effectiveness does not depict significant change in both cases as it can be seen in Table 2.

Likewise, taking 2 times of inlet humidity ratio with 40% solution concentration of LiBr under Pakistani weather conditions, the latent effectiveness has been increased by 1.7% and on increasing 3 times of air inlet humidity ratio, latent effectiveness is more enhanced by 0.9%, while there is no change observed in sensible effectiveness as mentioned in Table 3.

Reference values of Volumetric flow rate of air 14.6L/min and desiccant solution of 0.128 L/min is taken from Huang et al. [15] and 40% LiBr humidity ratio 0.007328 kg/kg from [16].

Table 2 LAMEE performance under Malaysian weather with 40% LiBr

Input Parameter	Outlet parameters				
Wair,inlet	Wair,outlet	Tair,outlet	Latent effectiveness	Sensible effectiveness	% increase in latent effectiveness
Kg/kg	(Kg/kg)	(K)			
0.01684	0.01051	299.55	0.655	0.72	–
0.03368	0.01607	299.55	0.668	0.719	2
0.05032	0.02134	299.55	0.674	0.72	0.9

Table 3 LAMEE performance under Pakistani weather with 40% LiBr

Input Parameter	Outlet parameters				
Wair,inlet	Wair,outlet	Tair,outlet	Latent effectiveness	Sensible effectiveness	% increase in latent effectiveness
Kg/kg	(Kg/kg)	(K)			
0.01312	0.00794	299.6	0.63	0.715	–
0.02624	0.01259	299.6	0.641	0.714	1.7
0.03936	0.01863	299.6	0.647	0.716	0.93

4.3 Effect of Air Inlet Humidity Ratio with 50% LiBr Concentration

Moreover, on increasing 2 times of inlet humidity ratio with 50% concentration of LiBr, latent effectiveness is increased by 1.7%, while on enhancing 3 times of humidity ratio of air make further 0.83% increase in latent effectiveness. Apart from it, sensible effectiveness remains same in both cases under Malaysian weather conditions as it is evident from Table 4.

Similarly, under Pakistani weather conditions, 1.24% increase in latent effectiveness is observed with 2 times of inlet humidity ratio of air and latent effectiveness is further increased by 0.9% on making 3times of inlet humidity ratio of air with 50% desiccant concentration. However, sensible effectiveness does not change in both scenarios and mentioned in Table 5.

Reference values of Volumetric flow rate of air 14.6L/min and desiccant solution of 0.128 L/min is taken from Huang et al. [15], 50% LiBr humidity ratio (0.004307 kg/kg) [16], and inlet humidity ratio of air is 0.01684 kg/kg for Malaysian weather and 0.01312 kg/kg for Pakistani weather condition.

Table 4 LAMEE performance under Malaysian weather with 50% LiBr

Input Parameter	Outlet parameters				
Wair,inlet	Wair,outlet	Tair,outlet	Latent effectiveness	Sensible effectiveness	% increase in latent effectiveness
Kg/kg	(Kg/kg)	(K)			
0.01684	0.00794	299.5	0.71	0.727	–
0.03368	0.01238‘	299.5	0.722	0.729	1.7
0.05032	0.01682‘	299.5	0.728	0.735	0.83

Table 5 LAMEE performance under Pakistani weather with 50% LiBr

Input Parameter	Outlet parameters				
Wair,inlet	Wair,outlet	Tair,outlet	Latent effectiveness	Sensible effectiveness	% increase in latent effectiveness
Kg/kg	(Kg/kg)	(K)			
0.01312	0.00744	299.57	0.644	0.715	–
0.02624	0.01193	299.6	0.652	0.71	1.24
0.03936	0.01601	299.58	0.658	0.714	0.9

5 Conclusion

CFD simulations of LAMEE have been performed with different desiccant solution concentrations under Malaysian and Pakistani weather conditions. The latent effectiveness of LAMEE is higher in Malaysian conditions, while sensible effectiveness does not vary significantly under both weather conditions. On increasing desiccant solution concentration from 40 to 50%, latent effectiveness under Malaysian weather condition is enhanced by 8.39%, while under Pakistani weather conditions, it shows 2.3% enhancement in latent effectiveness. However, sensible effectiveness depicts little variation in both weather conditions. Increasing humidity ratio of air is not an economical option for increasing latent effectiveness as maximum increase is 2% by keeping 2 times of inlet humidity ratio of air under Malaysian weather and it has shown less than 2% increase in latent effectiveness under Pakistani weather.

Acknowledgements The authors are highly indebted to the financial assistance received from Universiti Teknologi PETRONAS in carrying out this research study.

References

1. Qi R, Dong C, Zhang LZ (2020) A review of liquid desiccant air dehumidification: from system to material manipulations. Energy Build 215:109897. https://doi.org/10.1016/j.enbuild.2020.109897
2. Liu X, Qu M, Liu X, Wang L (2019) Membrane-based liquid desiccant air dehumidification: a comprehensive review on materials, components, systems and performances. Renew Sustain Energy Rev 110(September 2020):444–466. https://doi.org/10.1016/j.rser.2019.04.018
3. Ghadiri Moghaddam D, Besant RW, Simonson CJ (2014) Solution-side effectiveness for a liquid-to-air membrane energy exchanger used as a dehumidifier/regenerator. Appl Energy 113:872–882. https://doi.org/10.1016/j.apenergy.2013.08.037
4. Ge G, Abdel-Salam MRH, Besant RW, Simonson CJ (2013) Research and applications of liquid-to-air membrane energy exchangers in building HVAC systems at University of Saskatchewan: a review. Renew Sustain Energy Rev 26:464–479. https://doi.org/10.1016/j.rser.2013.04.022

5. Zhang LZ (2010) An analytical solution for heat mass transfer in a hollow fiber membrane based air-to-air heat mass exchanger. J Memb Sci 360(1–2):217–225. https://doi.org/10.1016/j.memsci.2010.05.015
6. Min J, Wang L (2013) Coupled heat and mass transfer during moisture exchange across a membrane. J Memb Sci 430:150–157. https://doi.org/10.1016/j.memsci.2012.12.018
7. Zhang LZ, Jiang Y (1999) Heat and mass transfer in a membrane-based energy recovery ventilator. J Memb Sci 163(1):29–38. https://doi.org/10.1016/S0376-7388(99)00150-7
8. Ge G et al (2014) Material properties and measurements for semi-permeable membranes used in energy exchangers. J Memb Sci 453:328–336. https://doi.org/10.1016/j.memsci.2013.11.013
9. Guo X, Wen T, Wang M, Luo Y, She X (2021) Performance investigation of a liquid desiccant regenerator with CFD technology. Appl Therm Eng 184:116055. https://doi.org/10.1016/j.applthermaleng.2020.116055
10. Chu J, Chen Z, Bai H, Zhu J (2019) Experimental study of liquid to air membrane energy exchanger (LAMEE) performance by measuring its temperature fields. Therm Sci Eng Prog 10(October 2018):163–168. https://doi.org/10.1016/j.tsep.2019.01.005
11. Al-Waked R, Nasif MS, Mostafa DB (2018) Enhancing the performance of energy recovery ventilators. Energy Convers Manag 171(May):196–210. https://doi.org/10.1016/j.enconman.2018.05.105
12. Al-Waked R, Nasif MS, Morrison G, Behnia M (2013) CFD simulation of air to air enthalpy heat exchanger. Energy Convers Manag 74:377–385. https://doi.org/10.1016/j.enconman.2013.05.038
13. Al-Waked R, Nasif MS, Morrison G, Behnia M (2015) CFD simulation of air to air enthalpy heat exchanger: variable membrane moisture resistance. Appl Therm Eng 84:301–309. https://doi.org/10.1016/j.applthermaleng.2015.03.067
14. Abdel-Salam MRH, Ge G, Fauchoux M, Besant RW, Simonson CJ (2014) State-of-the-art in liquid-to-air membrane energy exchangers (LAMEEs): A comprehensive review. Renew Sustain Energy Rev 39(November):700–728. https://doi.org/10.1016/j.rser.2014.07.022
15. Huang SM, Zhang LZ, Tang K, Pei LX (2012) Fluid flow and heat mass transfer in membrane parallel-plates channels used for liquid desiccant air dehumidification. Int J Heat Mass Transf 55(9–10):2571–2580. https://doi.org/10.1016/j.ijheatmasstransfer.2012.01.003
16. Afshin M (2010) Selection of the liquid desiccant in a run-around membrane energy exchanger. [Master′s thesis] Saskatoon, Saskatchewan, Canada: University of Saskatchewan

Utilization of Vibration to Enhance the Thermal Performance of Flat Plate Solar Collectors—A Numerical Study

Tanweer Hussain, Mohammad Shakir Nasif, Hilmi Hussin, and Nurril Ikmal Shamsul Azha

Abstract This study investigates the combioned effect of low-frequency vibrations and heat transfer fluid on the fluid outlet temperature of flat plate solar collectors (FPSC). Low-frequency vibrations in the range of 5–10 Hz with a constant amplitude of 3 mm were applied on the FPSC having water (H_2O) and Aluminum Oxide (Al_2O_3–H_2O) water nanofluid as HTF. Computational Fluid Dynamics (CFD) study was carried out to simulated FPSC and the model validated with the experimental results from the literature. The results showed a noticeable enhancement in the fluid outlet temperature of FPSC after applying vibrations. With water as HTF, the fluid outlet temperature without vibrations was 41.4 °C, while by applying low-frequency vibrations of 5 and 10 Hz, it increased to 48.7 °C and 49.5 °C respectively. While with Aluminum Oxide (Al_2O_3–H_2O) water nanofluid as HTF, the fluid outlet temperature without vibrations was 50.68 °C while by applying low-frequency vibrations of 5 and 10 Hz, it increased to 51.3 °C and 52.5 °C respectively.

1 Introduction

Solar energy has many advantages over nonrenewable energy resources and it is abundantly available without any negative effect on the environment [1]. The average Solar Energy Potential of the world is about 6000–6900 Wh/m^2 per day, while in Malaysia it is 4000–5000 Wh/m^2 per day, therefore Malaysia has low-temperature heat sources and a favorable environment for Solar Collectors [1, 2]. Due to the low-intensity of Solar Energy, the weather conditions recommend Flat Plate Solar Collectors (FPSC) for home use, because it requires low solar radiation [3]. The internal structure of FPSC is depicted in Fig. 1 [5].

Researchers are always looking for methods to improve the performance of thermal systems including solar collectors. Different methods have been investigated experimentally as well as numerically by the researchers to get the maximum

T. Hussain · M. S. Nasif (✉) · H. Hussin · N. I. S. Azha
Department of Mechanical Engineering, Universiti Teknologi PETRONAS, 32610 Seri Iskandar, Malaysia
e-mail: mohammad.nasif@utp.edu.my

F. Ahmad et al. (eds.), *ICPER 2020*, Lecture Notes in Mechanical Engineering,
https://doi.org/10.1007/978-981-19-1939-8_33

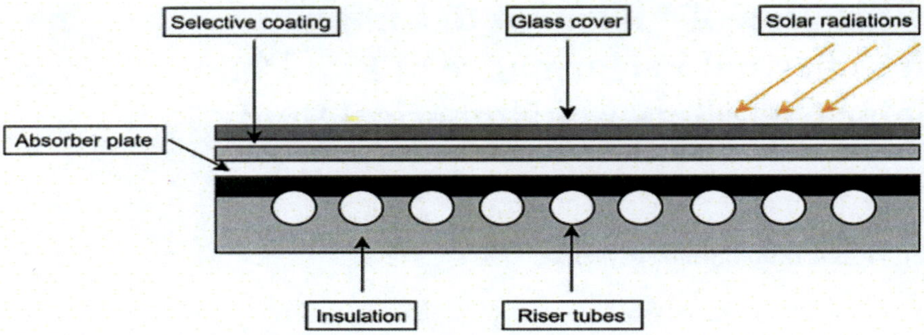

Fig. 1 Schematic diagram of FPSC

thermal performance and to improve the design of Solar Collectors. The most important parameters, which can enhance the thermal performance of FPSC are the number of tubes, tube diameter, length and width of the collector, mass flow rate of the fluid, and insulator thickness. To enhance the thermal efficiency of FPSC, the nanofluids are also being used as the Heat Transfer Fluid (HTF). Farajzadeh et al. [4] investigated the effect of nanofluids on the thermal performance of FPSC. They tested different concentrations of Al_2O_3–H_2O, TiO_2-H_2O, and a mixture of both nanofluids. With their experimental and numerical investigation, they found that a mixture of both nanofluids with an equal ratio can enhance the thermal efficiency of the FPSC up to 26% while, the efficiency was enhanced by 19% and 21% by using Al_2O_3–H_2O and TiO_2–H_2O nanofluid respectively. Hawwash et al. [5] studied the effect of double distilled water and Al_2O_3–H_2O nanofluid on the performance of FPSC, they examined six different concentrations of nanofluid within the range of 0.1–3% and observed that the optimum efficiency of FPSC can be achieved by using Al_2O_3–H_2O nanofluid up to 0.5% volume fraction. Genc et al. [6] did a numerical study to know the effect of nanofluids on the thermal performance of FPSC. They used different concentrations of nanoparticles and mass flow rates to get the maximum output. The maximum increase of 7.2% in outlet temperature was obtained with 0.004 kg/s with 3% volume concentration and maximum efficiency of 83.90% was achieved at 0.06 kg/s with a 1% volume concentration. They concluded that with lower mass flow rate, the higher thermal efficiency can be achieved by using nanofluids, and after 0.016 kg/s, the effect of nanofluids will be diminished, and the base fluid will become the working fluid. They named this flow rate as the critical flow rate of their study. Similarly, other nanofluids like Cu-H_2O, MgO-H_2O, Ag-H_2O etc. are also used in FPSC and gave a significant enhancement in the thermal performance and efficiency of FPSC.

Recently, the idea of vibrational mechanisms has been introduced to increase the performance of heat pipe Solar Collectors [7]. Vibrations destabilize the flow of fluids and results in an increase in heat transfer which is an active technique to enhance the heat transfer in Flat Plate Solar Collectors. Experimental investigation on the heat transfer enhancement due to induced vibrations was started in the 1960s [8].

The vibrational energy either can be directly supplied to the working fluid (acoustically excited systems) or the structure (mechanical excitations) [9]. Alaei et al. [10] designed a horizontal heat pipe to study the effect of low-frequency vibrations having the frequency ranges from 10 to 25 Hz and amplitude of 2 mm, and observed that by increasing the frequency in the above-mentioned range, the coefficients of heat transfer (condensation and boiling) were increased. The temperature difference was also decreased in the evaporator and condensation section with the increment of the vibrational frequency. It was also observed that the frequency does not affect a high filling ratio, and maximum thermal performance was observed with the applied frequency of 25 Hz. On the other hand, filling ratio, and maximum thermal performance was observed with the applied frequency of 25 Hz. On the other hand, Atashi et al. [11] experimentally investigated the effect of low-frequency vibrations with a frequency of 10, 15,20, and 25 Hz on boiling heat transfer by applying a vertical cylinder as a source of heat in the water pool and observed an enhancement in the boiling heat transfer by utilizing the vibrations with a decline in wall temperature which means the wall temperate is dependent of applied vibrations. The maximum enhancement ratio was observed when the frequency reached 25 Hz and the minimum was found in static condition (when there are no vibrations) for applied heat flux. By applying low-frequency vibrations, the improvement in the boiling heat transfer enhancement ratio was up to 118%.

Although the effect of vibrations to increase the thermal efficiency of solar collectors has been proved and discussed in various review papers [9, 12].

To prove this effect of vibrations in the flow tubes filled with water in a solar water heater. Cioc and Laurențiu [13] did a numerical as well as experimental study with low-frequency and low amplitude vibrations to know the effect of vibrations on heat transfer in solar water heating collectors. They filled the flow tube of the solar water heating collector with water and vibrated it with different frequencies and found a decline in the time for heat exchange at the resonance frequency (9 Hz) and an increase in the vibrating pipe temperature. Sahlani and Eidan [7] proved the heat transfer enhancement in heat pipe Evacuated Tube by using mechanical vibrations in their experimental work. They utilized low-frequency vibrations having frequencies of 2 to 10 Hz with three different volumetric flow rates of water. The increase in the coefficients of evaporation and condensation was observed with an increase in the frequency of vibrations. The thermal efficiency of the Evacuated Tube Solar Collector increased up to 78% by using vibrations when compared to the static condition with the same test conditions. In this paper, a 3-D CFD model of an FPSC is presented to investigate the effect of low-frequency vibrations with Water (H_2O) and Aluminum Oxide (Al_2O_3–H_2O) water Nanofluids as HTF on the fluid outlet temperature. A comparison between the static model and vibrating model having the frequency ranges from 5 to 10 Hz with a constant amplitude of 3 mm was investigated numerically to get the maximum fluid outlet temperature.

2 Methodology

For numerical simulations, a benchmark model of an FPSC (Fig. 2) consisting of an absorber plate having the dimensions of 1.65 m length, 0.1 m width, and 0.0005 m thickness, and a tube having 0.0125 diameter and 0.0007 m thickness.

A mesh independencestudy was carried out on the proposed CFD model, and numerical sensitivity analysis was applied to the fluid outlet temperature. Figure 3 shows the difference between the fluid outlet temperature decreased while increasing the number of elements in the mesh domain. Fluid outlet temperature results did not change beyond 217,939 elements. However, when vibration was modelled, and due to the dynamic mesh used in the vibration modelling fine mesh was used to obtain accurate results, and hence the number of elements was kept at 462,000.

The solar model validation was performed by using the fluid outlet temperature as a performance indicator. Figure 4 shows the comparative results between Gunjo et al.

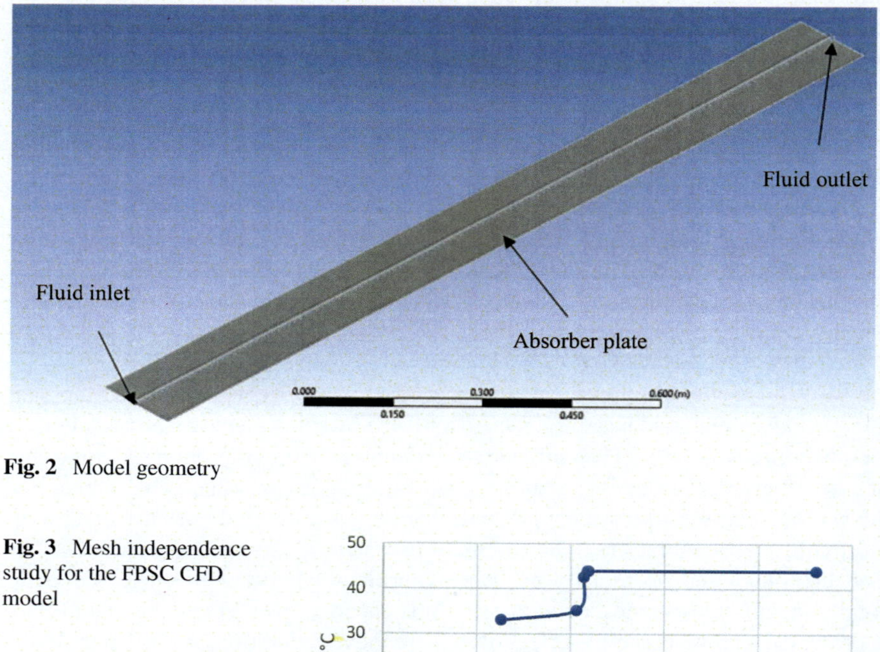

Fig. 2 Model geometry

Fig. 3 Mesh independence study for the FPSC CFD model

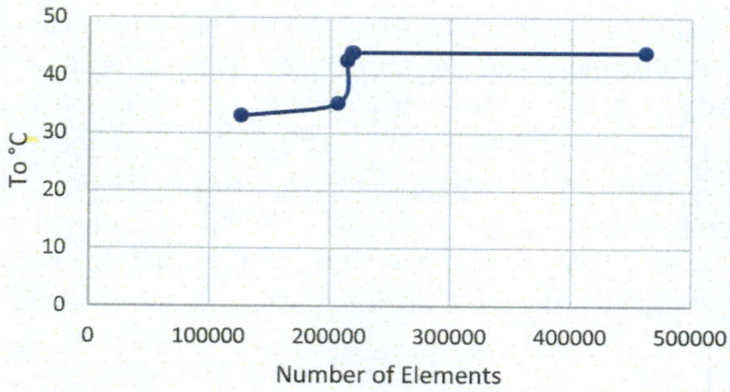

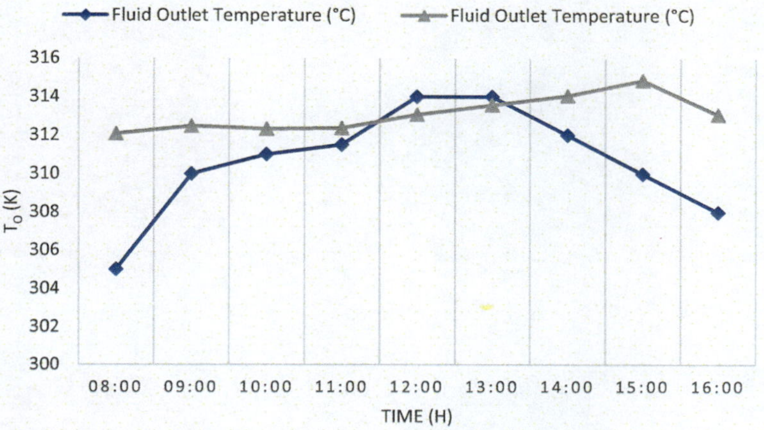

Fig. 4 Comparison between the fluid outlet temperature in this study and the experiment in the literature

[14] experimental results and the simulations of this study with maximum deviation 2.62% which is considered as good agreement.

Once the model was validated, the combined effect of low-frequency vibrations and heat transfer fluid on the fluid outlet temperature of flat plate solar collectors was studied numerically. Low-frequency vibrations in the range of 5–10 Hz with a constant amplitude of 3 mm were applied on the FPSC having Water (H_2O) and Aluminum Oxide (Al_2O_3–H_2O) water Nanofluid as HTF. The solar radition used is 750 W/m^2, fluid inlet temperature (T_{in} = 25 °C), and the mass flow rate of 0.00161 kg/s were kept constant throughout the study.

3 Results and Discussion

The validated model was used to determine the combined effect of low-frequency vibrations and heat transfer fluid on the fluid outlet temperature of flat plate solar collectors. Fluid outlet temperature was investigated by using different vibrations.

3.1 Effect of Vibration on Temperature Distribution at the Pipe Outlet

The effect of vibrations on temperature distribution at the pipe outlet was analyzed in this study. The initial fluid outlet temperature of non-vibrated FPSC while using water as HTF was 41.39 °C, while by applying vibrations, it increased 48.76 °C with 5 Hz frequency (7.37 °C increased), and it increased 49.5 °C with 10 Hz frequency (8.11 °C

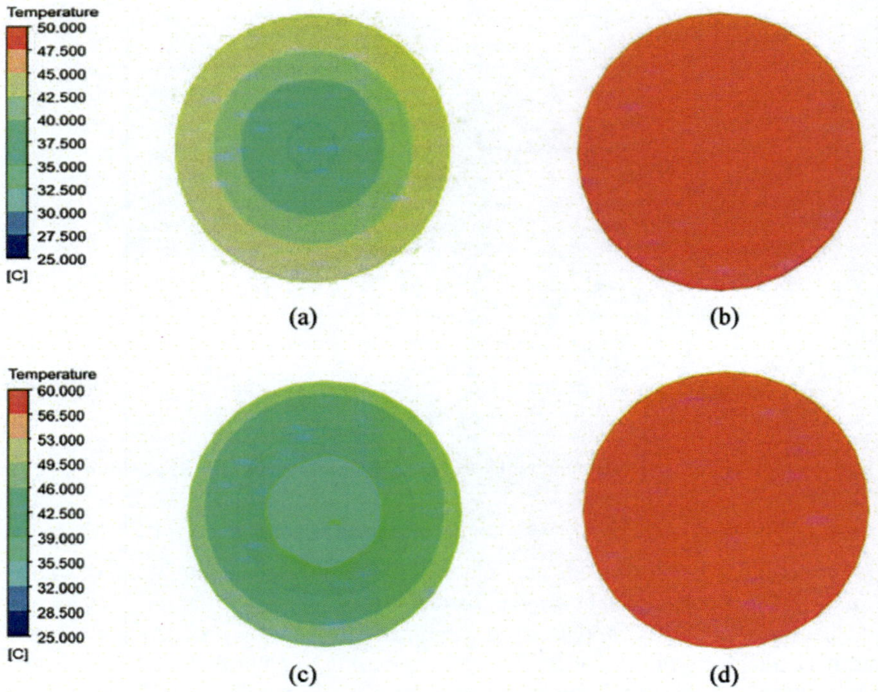

Fig. 5 Temperature contours of fluid outlet: **a** Non-vibrated system with water as HTF, **b** Vibrated system with water as HTF, **c** non-vibrated system with Aluminum Oxide (Al_2O_3–H_2O) water Nanofluids as HTF, **d** Vibrated system with (Al_2O_3–H_2O) water Nanofluids as HTF

increased). Figure 5a shows temperature contours of fluid outlet of non-vibrated system with water as HTF, while in Fig. 5b a uniform distribution of temperature can be seen throughout the pipe outlet section after applying vibrations.

Moreover, the initial fluid outlet temperature of non-vibrated FPSC by using Aluminum Oxide (Al_2O_3–H_2O) water Nanofluid as HTF was 50.68 °C, while by applying vibrations, it increased to 51.30 °C with 5 Hz frequency (0.62 °C increased) and it increased to 52.5 °C with 10 Hz frequency (1.82 °C increased). Figure 5c shows the temperature contours of the fluid at the pipe outlet with Aluminum Oxide (Al_2O_3–H_2O) water Nanofluid as HTF, while in Fig. 5d a uniform distribution of temperature can be seen throughout the pipe outlet section after applying vibrations.

Thermal performance based on fluid outlet temperature was analyzed from this study. It was observed that low-frequency vibrations can significantly enhance the fluid outlet temperature while water is being used as heat transfer fluid. While, with Aluminum Oxide (Al_2O_3–H_2O) water nanofluid as heat transfer fluid, low-frequency vibrations a small increment in the fluid outlet temperature was observed. Figure 6 shows a comparison of water and Al_2O_3–H_2O Nanofluid between the fluid outlet of the non-vibrated system with a vibrated system used in this study.

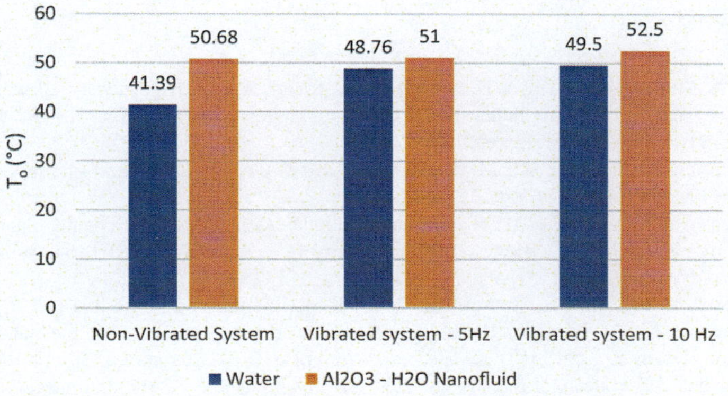

Fig. 6 Comparison of water and Al_2O_3–H_2O Nanofluid between the fluid outlet temperature of non-vibrated system with vibrated system

4 Conclusion

A CFD model of FPSC was modeled and validated with maximum deviation of 2.62%. The validated CFD model was used to investigate the combined effect of low-frequency vibrations and heat transfer fluid on the fluid outlet temperature of FPSC. Low-frequency vibrations in the range of 5–10 Hz with a constant amplitude of 3 mm were applied on the FPSC having Water (H_2O) and Aluminum Oxide (Al_2O_3–H_2O) water Nanofluid as HTF. The results showed that:

- With water as HTF, the fluid outlet temperature without vibrations was 41.39 °C, while by applying low-frequency vibrations of 5 and 10 Hz, it increased to 48.756 °C and 51 °C respectively.
- While with Aluminum Oxide (Al_2O_3–H_2O) water Nanofluid as HTF, the fluid outlet temperature without vibrations was 50.68 °C while by applying low-frequency vibrations of 5 and 10 Hz, it increased to 51.3 °C and 52.4 °C respectively

5 Conflicts of Interest

The authors declare no conflicts of interest.

Acknowledgements The authors wish to acknowledge the support of Universiti Teknologi PETRONAS for providing some of the resources to support this research.

Funding This research was funded by Yayasan Universiti Teknologi PETRONAS (YUTP), Grant Number 015LC0-108.

References

1. Jamar A, Majid ZAA, Azmi WH, Norhafana M, Razak AA (2016) A review of water heating system for solar energy applications. Int Commun Heat Mass Transf 76:178–187. https://doi.org/10.1016/j.icheatmasstransfer.2016.05.028
2. Elnugoumi MG, Bin ZA, Kh M, Almsafir M (2012) Current status and challenges of solar energy in Malaysia. A Rev 2(4):313–320
3. Sakhaei SA, Valipour MS (2019) Performance enhancement analysis of The flat plate collectors : a comprehensive review. Renew Sustain Energy Rev 102(May 2018):186–204. https://doi.org/10.1016/j.rser.2018.11.014
4. Farajzadeh E, Movahed S, Hosseini R (2018) Experimental and numerical investigations on the effect of Al2O3/TiO2[sbnd]H2O nanofluids on thermal efficiency of the flat plate solar collector. Renew Energy 118:122–130. https://doi.org/10.1016/j.renene.2017.10.102
5. Hawwash AA, Abdel Rahman AK, Nada SA, Ookawara S (2018) Numerical investigation and experimental verification of performance enhancement of flat plate solar collector using nanofluids. Appl Therm Eng 130:363–374. https://doi.org/10.1016/j.applthermaleng.2017.11.027
6. Genc AM, Ezan MA, Turgut A (2018) Thermal performance of a nanofluid-based flat plate solar collector: a transient numerical study. Appl Therm Eng 130:395–407. https://doi.org/10.1016/j.applthermaleng.2017.10.166
7. Al Sahlani A, Eidan AA (2018) Controllable vibrating system to enhance the performance of heat pipe evacuated tube solar collector. J Mech Eng Res Dev 41(3):67–73. https://doi.org/10.26480/jmerd.03.2018.67-73
8. Wong SW, Chon WY (1969) Effects of ultrasonic vibrations on heat transfer to liquids by natural convection and by boiling. AIChE J 15(2):281–288. https://doi.org/10.1002/aic.690150229
9. Franco A, Bartoli C (2019) Heat transfer enhancement due to acoustic fields: a methodological analysis. Acoustics 1(1):281–294. https://doi.org/10.3390/acoustics1010016
10. Alaei A, Hasanzadeh Kafshgari M, Atashi H (2012) A new designed heat pipe: An experimental study of the thermal performance in the presence of low-frequency vibrations. Heat Mass Transf und Stoffuebertragung 48(4):719–723. https://doi.org/10.1007/s00231-011-0908-3
11. Atashi H, Alaei A, Kafshgari MH, Aeinehvand R, Rahimi SK (2014) New pool boiling heat transfer in the presence of low-frequency vibrations into a vertical cylindrical heat source. Exp Heat Transf 27(5):428–437. https://doi.org/10.1080/08916152.2013.797940
12. Azha NIS, Hussin H, Nasif MS, Hussain T (2020) Thermal performance enhancement in flat plate solar collector solar water heater: a review. Processes 8(7). https://doi.org/10.3390/PR8070756
13. Ciocănea A, Bureea DL (2017) the influence of flow tube vibrations over the efficiency of solar water heating collectors. Energy Procedia 112:330–335. https://doi.org/10.1016/j.egypro.2017.03.1059
14. Gunjo DG, Mahanta P, Robi PS (2017) CFD and experimental investigation of flat plate solar water heating system under steady state condition. Renew Energy 106:24–36. https://doi.org/10.1016/j.renene.2016.12.041

Life Cycle Assessment and Economic Analysis of a Vapor Compression System Integrated with a Large District Cooling Plant

Mohd Amin Abd Majid, Chima Cyril Hampo, Ainul Bt Akmar, Hamdan Haji Ya, and Mazli Mustapha

Abstract In most District Cooling DC setups, Vapor Compression Chillers VCC are installed to provide Chilled Water CW for charging the Thermal Energy Storage TES tank or direct CW supply to the DC network. The operation of these VCC systems accounts for the most electrical power supplied to the entire DC plant. This strengthens the need to conduct an environmental and cost implication analysis using these systems for refrigeration purposes. A gate-to-gate Life Cycle Assessment LCA and Life cycle Cost LCC methodologies are utilised in this study for these objectives. A case study of four VCC systems installed in a large DC plant in Malaysia is examined. It is observed from the inventory results that the climatic temperature conditions had a significant influence on the monthly cooling demand, energy consumed, environmental impact and cost incurred. The result reveals that fuel choice had the most significant influence on the environmental impact contribution of the use phase. The LCA result of the considered phase (use phase) shows that the carbon footprint per refrigeration is given as 0.05 kg $CO_{2\,eq}$/RTh, while the cost per refrigeration ton of the VCC system is estimated to be MYR 0.005/RTh.

Keywords Life cycle assessment · District cooling · Global warming potential · Malaysia

Abbreviations

AC	Air Condition
CL	Cooling Load
COP	Coefficient of Performance
CW	Chilled Water
DC	District Cooling
ECC	Electric Centrifugal Chiller

M. Amin Abd Majid (✉) · C. Cyril Hampo · A. Bt Akmar · H. Haji Ya · M. Mustapha
Department of Mechanical Engineering, Universiti Teknologi PETRONAS, 32610 Bandar Seri Iskandar, Perak, Malaysia
e-mail: Mamin_amajid@utp.edu.my

F. Ahmad et al. (eds.), *ICPER 2020*, Lecture Notes in Mechanical Engineering,
https://doi.org/10.1007/978-981-19-1939-8_34

EL	Environmental Load
EP	Electric Power
FU	Functional Unit
GHG	Green House Gas
GWP	Global Warming Potentials
HVAC	Heat, Ventilation and Air Conditioning
ISO	International Standard Organization
LCA	Life Cycle Assessment
LCEL	Life Cycle Environmental Load
LCI	Life Cycle Inventory
LCIA	Life Cycle Impact Assessment
SAC	Suction Absorption Chiller
TES	Thermal Energy Storage
VCC	Vapor Centrifugal Chiller
VCS	Vapor Compression System

1 Introduction

Vapor Compression System VCS serves as the most widely used means for air-conditioning. Several industrial plants utilise large vapour-compression refrigeration systems. The VCS in district cooling plants is also used to charge thermal storage tanks during off-peak periods, i.e. at night, to take advantage of lower electricity cost. The operation of VCS accounts for the highest energy consumption in a district cooling system. Optimising the performance of energy-intensive system like the VCS will assist in mitigating against the advance environmental and economic effects.

Life Cycle Assessment LCA has been accepted as a tool to evaluate the environmental impact throughout the product life cycle. The LCA guidelines are published in ISO 14040 [1] and 14,044 [2]. The LCA assessment covers four phases: goal and scope, inventory analysis, impact assessment, and interpretation. The LCA framework is shown in Fig. 1.

This technique is used for assessing the environmental aspects associated with a product over its life cycle. The most important applications include.

1. Analysis of the contribution of the life cycle stages to the overall environmental load, usually to prioritise improvements on processes.
2. Comparisons between products or internal use Various researchers have undertaken studies on LCA for vapour compression system (VCS) had been undertaken by different researchers.

Gu et al. [3] reported an LCA study on the design of cooling and heating sources of HVAC systems using an integrated index covering total environmental impact including energy consumption, resources consumption and pollutant emission through building a life to guide the design of building energy system. Life cycle

Fig. 1 LCA Framework ISO 14040 [1]

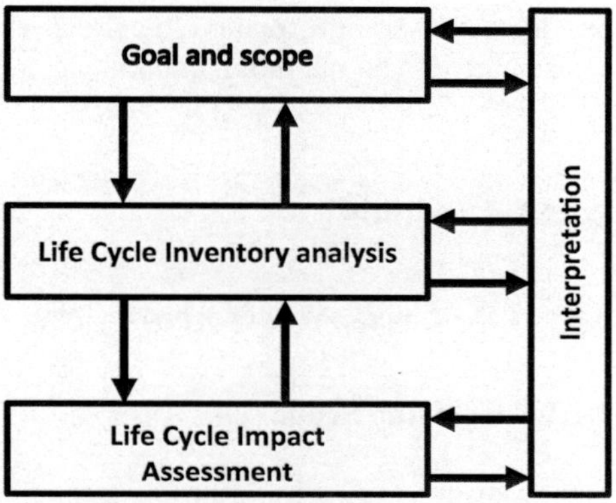

ecological load (LCEL) and economic performance of four cooling and heating sources schemes were analysed and compared, and the optimal option is proposed. Based on the findings, they noted that the system with a water-cooled electrical chiller for cooling and a gas-fired boiler for heating has the lowest LCEL and best economic performance.

Almutairi et al. [4] reported a study on VCS in Saudi Arabia. The study is on LCA and economic analysis of residential air conditioning in Saudi. The LCA study is related to residential building air conditioning. Results show that the system operation phase is responsible for the highest environmental impacts and that the type of fuel used influences the magnitude of the impacts in each region. Economic implications had more influence than environmental concerns in Saudi Arabia. The study also evaluated the relationship of economics to environmental effects, life cycle cost (LCC), and payback period (PBP) by analysing variability in input prices on the uncertainty associated with the final results. Economic analysis indicates that using a more efficient air conditioner is not attractive to the customer for the apartment and traditional house.

Cascini et al. [5] investigated the environmental impact associated with the life cycle of a commercial refrigerator system. The environmental load caused by the adoption of alternative refrigerants and different use configurations is evaluated and discussed.

Integration of VCS within the DC plant is viewed as an alternative option to address the technical and economic issues on meeting the increasing space cooling requirements to cater to population growth. In Malaysia, there are several installations of this nature. Among them are GDC UTP, Putrajaya Plant 1 and 2, and the DC of Kuala Lumpur International Airport. However, there is no specific environmental and economic study on evaluating the installation of this nature in Malaysia. This study is, however, inspired to address this gap. A gate-to-gate LCA and LCC methodology is adopted to evaluate the environmental impact and economic implication of operating these VCC systems in the DC plant in Malaysia. Findings in this study will provide necessary information to policymakers and refrigeration experts in ensuring a sustainable plant operation and management.

2 Methodology

2.1 Life Cycle Assessment of the Vapor Compression Chiller

2.1.1 Goal and Scope

This study aims to quantify the environmental impact incurred from the operation of the VCC system integrated with a large district cooling plant in Malaysia, a country with a hot, humid weather condition with a largely stable temperature. A gate-to-gate LCA methodology is used for this purpose. Thus, the environmental contributions resulting from the energy used to power the chiller system throughout its use phase are evaluated. The Umberto LCA + software is used to model the system and carry out the study's numerical analysis.

2.1.2 System Description

The DC plant is installed with four VCC systems and two TES tanks with a total holding capacity of 50,000RTh with 24.5 m in diameter and 27.8 m in height. The VCC system uses the R134a refrigerant and supplies chilled water to the Air handling Unit at 5 °C and receives at 11 °C, while it supplies cooling water at 33 °C and receives at 28 °C to and from the Cooling Tower. In this study, the ecological impact resulting from the operation of 4 VCC systems installed in a large DC plant will be considered. The electric chiller system is designed to function on a full operating strategy where two VCC plants operate at full capacity during the off-peak period (10–8 am) to charge the TES tanks. The other two VCC plants are utilised as standby chiller plants to assist the cooling load during system breakdown or scheduled maintenance of the plant and high chilled water demand scenarios. The system configuration of the DC plant setup is shown in Fig. 2 [6]. Data for analysis in this study is taken from the daily system operations of the VCC plants.

2.1.3 Functional Unit FU

The idea of FU helps provide the basis for comparison of these systems with other design alternatives in other to highlight possible and meaningful benefits. Therefore, the FU for this study is the amount of chilled water in Refrigeration Ton RT per consumed electricity.

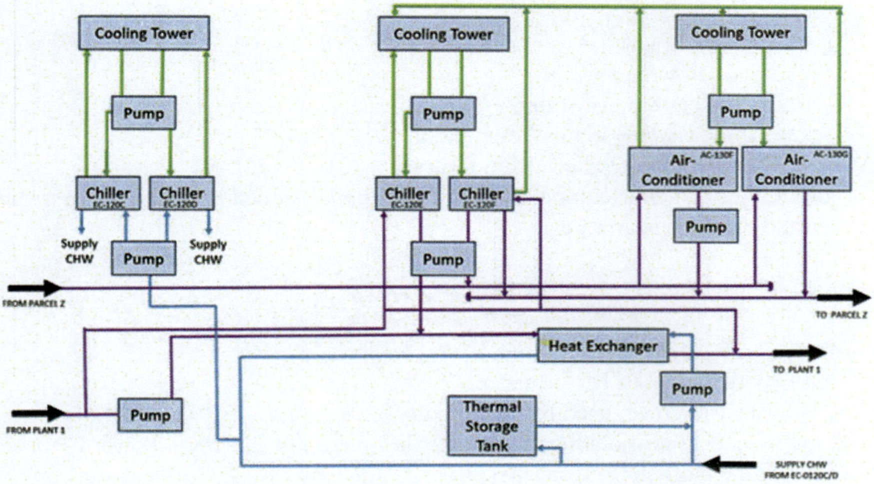

Fig. 2 Configuration of the large DC plant using block diagram [6]

2.1.4 System Boundaries

Based on the unavailability of credible data, the system boundary considered in this study covers the 25 years use-phase of the life cycle of the VCC system. The system boundary of the research is structured in such a way as to quantify the amount of electricity consumed by the VCC systems in producing chilled water and the equivalent indirect emission incurred as a result.

2.2 Life Cycle Inventory

Operation data of four VCC systems were acquired and used to define the characterisation of the system unit in terms of estimating its environmental impact. Data for this study's analysis is obtained from plant technicians, technical datasheets of the unit components, and literature. These data included minute sensor readings of the four VCC system in a full study year (2019). Parameters revealed in the data included chilled water temperatures and mass flowrates, chilled water suction and discharge pressure, and Tons of Refrigeration RT of chilled water generated per unit time. Electrical Power EP utilised to drive these systems are acquired directly from the grid electricity supply. The quantity of electrical energy used by the plant per year is assumed to be the same throughout its lifespan.

Mathematical formulation to estimate the Cooling Load satisfied per hour by the chillers is given by Eq. (1);

$$\mathrm{Qin} = \mathrm{m_w} C_p (T_{cw_o} - T_{cw_o}) \quad (1)$$

Were,
Q_{in} = Cooling Load:
MW = mass flowrate.
C_f = specific capacity of water.
T_{cwo} = Chilled Water supply Temperature.
T_{cwi} = Temperature Chilled Water return.

The COP of a vapour-compression chiller system, defined based on the first law of thermodynamics, is given as:

$$\text{COP} = Q_{in}/W_{in} \tag{2}$$

Were,
COP = Coefficient of Performance,
W_{in} = Electric Power used by the compressor.

A mathematical formulation to estimate the amount of Electrical Power EP used per hour by the chillers is given Eq. (3).

$$W_{in} = \sum_{m=1}^{m=j}\left\{\sum_{n=1}^{n=60}\frac{Q_{evap}}{\text{COP}}.3.51685\right\} \tag{3}$$

W_{in} = Electrical Power demand (kWh),
m = number of Chiller,
n = number of minutes,
j = number of considered working chillers,
Converting RT to kW: 1 refrigeration ton = 3.5168525 kilowatts.

2.3 *Life Cycle Impact Assessment*

The impact category indicator can be calculated by multiplying the LCI result with the characterisation factor. The characterisation factors for the Primary energy calculations are based on the Cumulative Energy Demand method [7], which provides an estimation of the consumption of renewable (biomass, wind, solar, geothermal, water) and non-renewable (fossil, nuclear) energy sources. The environmental characterisation factors are derived from the ReCiPe Midpoint impact assessment method [8, 9] on the Umberto LCA + software. The following impact categories listed in Table 1 have been chosen to calculate the environmental performances of the FU throughout the system life cycle stages.

Table 1 LCIA selected categories

ReCiPe midpoint	Unit
climate change, GWP100	kg CO_2-Eq
fossil depletion, FDP	kg oil-Eq
freshwater ecotoxicity, FETPinf	kg 1,4-DCB-Eq
human toxicity, HTPinf	kg 1,4-DCB-Eq
marine ecotoxicity, METPinf	kg 1,4-DCB-Eq
ozone depletion, ODPinf	kg CFC-11-Eq
photochemical oxidant formation, POFP	kg NMVOC
terrestrial ecotoxicity, TETPinf	kg 1,4-DCB-Eq
water depletion, WDP	m^3

3 Economic Data Sources and Assumptions

3.1 Operation and Maintenance Cost

The estimation of operation cost can be calculated by evaluating the energy consumption of the VCC throughout the daily operation of the annual usage of the VCC. The cost of electricity used in the calculation is based on the Tenaga Nasional Berhad TNB tariff, and the information is listed in Table 2. In this study, the TNB tariff used is tariff C2.

Using thermodynamics performance analysis with the system operation data, the equation is used to estimate the amount of used electrical energy by the VCC

Table 2 Electricity tariff category: Tenaga Nasional Berhad [10]

Tariff category	Current rates (1 Jan 2014)
Tariff B—Low voltage commercial tariff	
For the first 200 kWh (1 – 200 kWh) per month	43.5 sen/kWh
For the next kWh (201 kWh onwards) per month	50.9 sen/kWh
The minimum monthly charge is RM7.20	
Tariff C1—Medium voltage general commercial tariff	
For each kilowatt of maximum demand per month	30.3 RM/kW
For all kWh	36.5 sen/kWh
The minimum monthly charge is RM600.00	
Tariff C2—Medium voltage peak/off-peak commercial tariff	
For each kilowatt of maximum demand per month during the peak period	45.1 RM/kW
For all kWh during the peak period	36.5 sen/kWh
For all kWh during the off-peak period	22.4 sen/kWh
The minimum monthly charge is RM600.00	

throughout its hourly, daily, monthly, and annual usage. The average cost of maintenance as used throughout the year is also included in the cost of operation. Equation 4 and 5 reveals the formula used to determine the overall operation cost.

$$C_T = (T_E \times W_{in}) \tag{4}$$

$$C_O = C_T + C_M \tag{5}$$

T_E—Electricity Tariff per kWh (MYR/kWh).
W_{in}—Electrical Power demand (kWh).
C_M—Maintenance cost (MYR).
C_T—Total cost of electricity.
C_O—Operation cost of the technology (MYR).

These represent the total costs incurred over the system use phase per Rth of refrigeration generated. As mentioned before, all costs are assumed to be steady-state. Equation 6 presents the formula to calculate the LCC.

$$\text{LCC} = \frac{C_O}{Q_{in}} \tag{6}$$

LCC—Life cycle cost per refrigeration (MYR/RTH).

4 Result

4.1 *Life Cycle Inventory Analysis*

The readings of the monthly cooling load as satisfied by the four VCC are presented in Fig. 3. From the cumulative result, April and June represent the months with the highest and lowest cooling load. This result indicates that climatic weather conditions have a significant influence on refrigeration demands. The percentage contribution made by individual VCC systems in satisfying the monthly and yearly cooling needs are revealed in Figs. 3 and 4, respectively. The result shows that VCC A and B fulfilled 79% of the annual cooling load, while VCC C and D covered 21%. This indicates that VCC A and B consumed a more significant amount of electricity than VCC C and D in their overall operation.

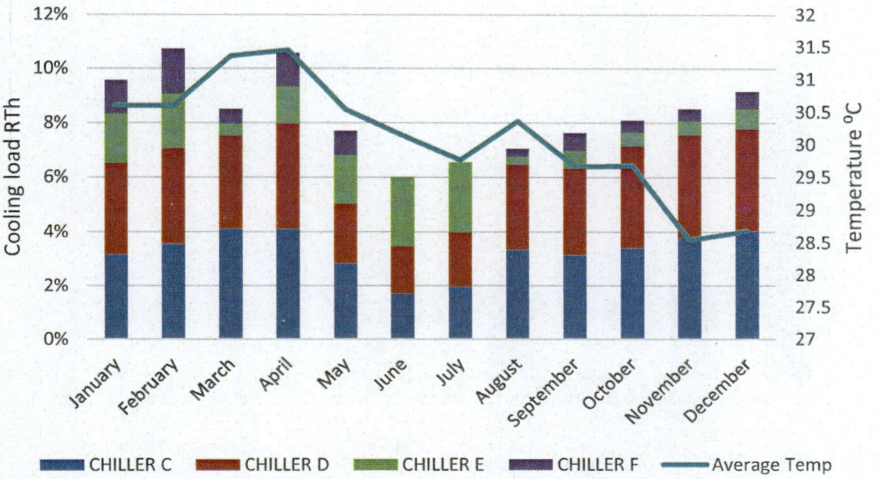

Fig. 3 Monthly Cooling Load, VCC contribution, and average climate temperature

Fig. 4 Percentage contribution of the individual VCC system in the DC plant in the study year [11]

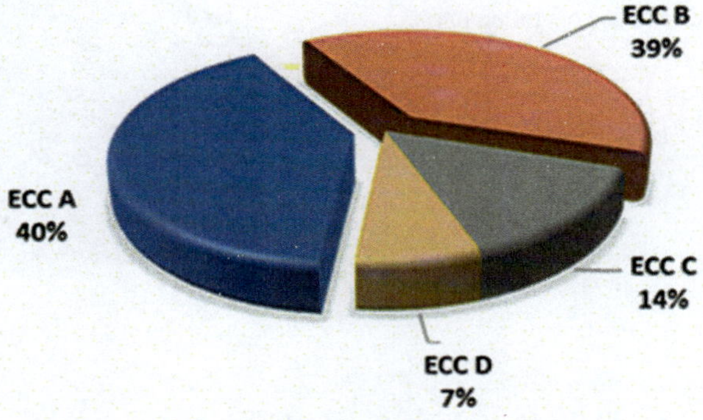

4.2 *Life Cycle Impact Assessment Results*

The result of the Life Cycle Impact Assessment per 1RTh of produced chilled water is presented in Fig. 5. The overall result of the carbon footprint per refrigeration of the VCC throughout the system use-phase is given as 0.05 kg $CO_{2\,eq}$. The figure shows that "climate change, GWP100" is the highest environmental impact influenced most by the considered use phase of the VCC. This is due to the high ratio of fossil fuel in the fuel mix of Malaysia grid generation.

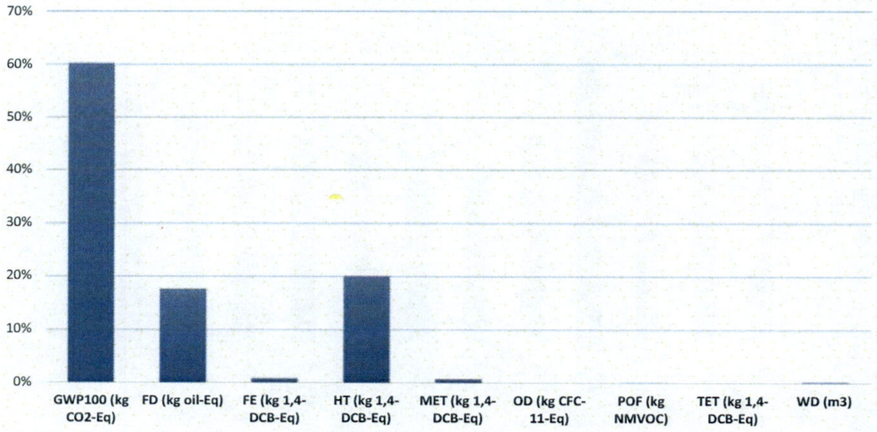

Fig. 5 Result of the life cycle impact assessment

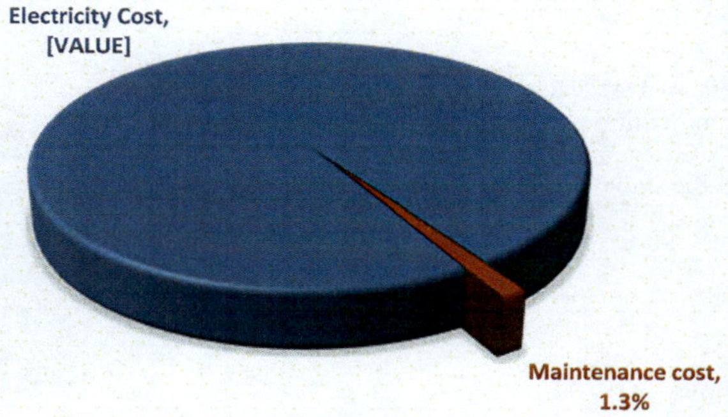

Fig. 6 Result of the contribution to the Life cycle cost of the operation phase

4.3 Result of Life Cycle Cost

From the base study result evaluation, the overall life cycle cost per refrigeration ton of the VCC system throughout its use phase is estimated to be MYR 0.005/RTh. As shown in Fig. 6, the cost of purchasing electrical power accounted for almost the entire operational cost, with an estimated contribution of 98.7% compared to the cost used for maintenance.

5 Conclusion

This paper performed a compressive analysis of the annual cooling load and the corresponding energy demand in different climatic temperature conditions experienced in each month in Malaysia. A gate-to-gate LCA and LCC methodologies are used to assess the environmental and economic implications of the operation of four VCC systems used to produce chilled water in a large DC plant installed in Malaysia. The LCA result shows that the carbon footprint per refrigeration is given as 0.05 kg $CO_{2\ eq}$/RTh, with GWP recording the highest environmental impact (60%) compared to the other chosen categories of the Recipe midpoint. This high GWP value is due to the high ratio of fossil fuel in the fuel mix of Malaysia grid generation, which drives the system throughout its 25 years of usage. The result of the cost evaluation shows that electricity cost accounted for about 98% of the entire operational cost, with the life cycle cost per refrigeration ton of the study is estimated at MYR 0.005/RTh.

Acknowledgements The authors would like to acknowledge the financial support of the YUTP Grant, Life Cycle Assessment, and economic analysis of the vapour compression cooling system integrated within District Cooling plant, YUTP Grant 015LCO-262 this research.

References

1. ISO-Norm I (2006) Environmental management—life cycle assessment—principles and framework ISO 14040: 2006. ISO: Geneva, Switzerland
2. Ryding S-O (1999) ISO 14042 Environmental management* Life cycle assessment* life cycle impact assessment. Int J Life Cycle Assess 4(6):307
3. Gu L et al (2007) Life cycle assessment for cooling and heating sources of building. In: Proceedings of 6th international conference on indoor air quality, ventilation and energy conservation in buildings: sustainable built environment
4. Almutairi KT, Greg B, Jasmina A, Salem ND (2015) Life cycle assessment and economic analysis of residential air conditioning in Saudi Arabia. Energy Build 102:370–379
5. Cascini A et al (2013) Life Cycle Assessment of a commercial refrigeration system under different use configurations. In: 18th Summer School Francesco Turco 2013. AIDI-Italian Association of Industrial Operations Professors
6. Abd Majid MA et al (2020) Analysis of a thermal energy storage tank in a large district cooling system: a case study. Processes 8(9):1158
7. Moreno Ruiz E et al (2020) Documentation of changes implemented in ecoinvent database v3. 7. ecoinvent Association. Zürich, Switzerland
8. JCR E (2010) ILCD handbook: general guide for life cycle assessment—provisions and action steps. Publications Office of the European Union, Luxembourg
9. Chomkhamsri K, Wolf M-A, Pant R (2011) International reference life cycle data system (ILCD) handbook: review schemes for life cycle assessment. In: Towards life cycle sustainability management, pp 107–117
10. Commission E (2018) Malaysia energy statistics handbook 2017. Suruhanjaya Tenaga (Energy Commission) Malaysia, Putrajaya
11. Hampo CC, Akmar AB, Majid MAA (2021) Life cycle assessment of an electric chiller integrated with a large district cooling plant. Sustainability 13(1):389

Improved Hydrocarbon Recovery Volume Estimation Model Using Integrated Asset Modelling (IAM) Approach

Mohd Razmi Ziqri Ahmad Shukri, Mohd Zaidi Jaafar, Mohamed Zamrud Zainal, and Mohd Khairi Abu Husain

Abstract An Integrated Asset Modelling for reservoir X is presented and the results were compared to the individual model. A sensitivity study was conducted to identify the most critical and sensitive geological, reservoir, production technology and facilities parameters towards hydrocarbon prediction. The individual model construction for REVEAL reservoir model, PROSPER well model, GAP network model and RESOLVE platform. This base case integrated model was used for hydrocarbon prediction including cumulative oil production and hydrocarbon recovery. The results of the sensitivity analysis are shown using a tornado chart and spider plot. Complete governing equations and the method is described in detail to permit readers to replicate all results. For the sensitivity analysis, ten parameters were selected from geological, reservoir, wells and facilities input parameters. It can be concluded that the Integrated Asset Modelling giving the similar results with the individual model and the topmost sensitive parameters are Initial Oil Column and Net To Gross (NTG) which are 70% and 30% increment respectively. The novelty of the integrated asset modelling approach is in the ability to combine reservoir, well and network models to predict the critical parameters affecting hydrocarbon volumetric prediction.

M. Razmi Ziqri Ahmad Shukri (✉) · M. Zaidi Jaafar
Faculty of Engineering, School of Chemical and Energy Engineering, Universiti Teknologi Malaysia, 80990 Johor, Malaysia
e-mail: razmiziqri@gmail.com

M. Razmi Ziqri Ahmad Shukri
PETRONAS Carigali Sdn Bhd, 50088 Kuala Lumpur, Malaysia

M. Zamrud Zainal
HALLIBURTON North Africa, Hydra, Algeria

M. Razmi Ziqri Ahmad Shukri · M. Khairi Abu Husain
Razak Faculty of Technology and Informatics, Universiti Teknologi Malaysia, 54100 Kuala Lumpur, Malaysia

F. Ahmad et al. (eds.), *ICPER 2020*, Lecture Notes in Mechanical Engineering,
https://doi.org/10.1007/978-981-19-1939-8_35

1 Introduction

In the petroleum business, all oil and gas fields worldwide operated as integrated systems including reservoirs, wells, pipeline networks and surface facilities. Oil and gas companies traditionally breaking the engineering studies into parts and study them separately. Each silo analysis depends on a variety of individual software to acquire the best recovery factor of hydrocarbon with optimum economic recovery from the assets. Due to natural integration that exists between all the elements that make up the oil and gas system, any changes that are made on one part affects every other component that is in pressure communication with it.

The concept of integration modelling involves the elimination of artificially boundary conditions by individual discipline and facilitates the construction of models that behave as close to reality as possible. For example, any change in the pipeline model will automatically change the behaviour of wells, reservoir upstream and facilities downstream. Integrated modelling can capture this behaviour while still allowing the engineers to create model separately. These allow operators to first understand the field behaviour and control them in an integrated platform, which is subsequently used to optimize hydrocarbon production and recovery.

This study will focus on the hydrocarbon recovery and sensitivity analysis on critical parameters of Field X using Integrated Asset Modelling (IAM) approach. Field X is located in offshore Sabah, approximately 43 km Northwest of Labuan and 12 km from Samarang field. Figure 1 shows the location of Field X. It is operated by PETRONAS Carigali Sdn. Bhd. Some of the data such as PETREL 100 Static, core data, PVT data, MDT/RFT data, DST well test data, well log, deviation survey and surface contour map are available for the analysis (Fig. 2).

Petroleum engineers and analyst have a problem to recognize the interaction between all system elements that affecting the overall performance of the reservoir. Some engineers in the team are working in a silo and they cannot observe the effect of many parameter changes in the whole system. This may contribute to significant loss of resources, higher operational cost, and wrong recovery estimation, hence affecting the economic analysis.

The accurate determination of hydrocarbon recovery volume is vital to decide for any investment into any project. Furthermore, it also important for the selection of production techniques, peak rate, overall management and reservoir surveillance activities through the field life.

There are 4 objectives of this paper as listed below:

1. Run integrated modelling to obtain base case results including cumulative oil production and recovery factor estimation.
2. Run low case and high case scenario using an integrated modelling approach by changing parameters of REVEAL and GAP.
3. Identify critical parameters that affect cumulative oil estimation and recovery factor.
4. Recommend the possible solution to reduce uncertainties and increase accuracy for recovery factor estimation.

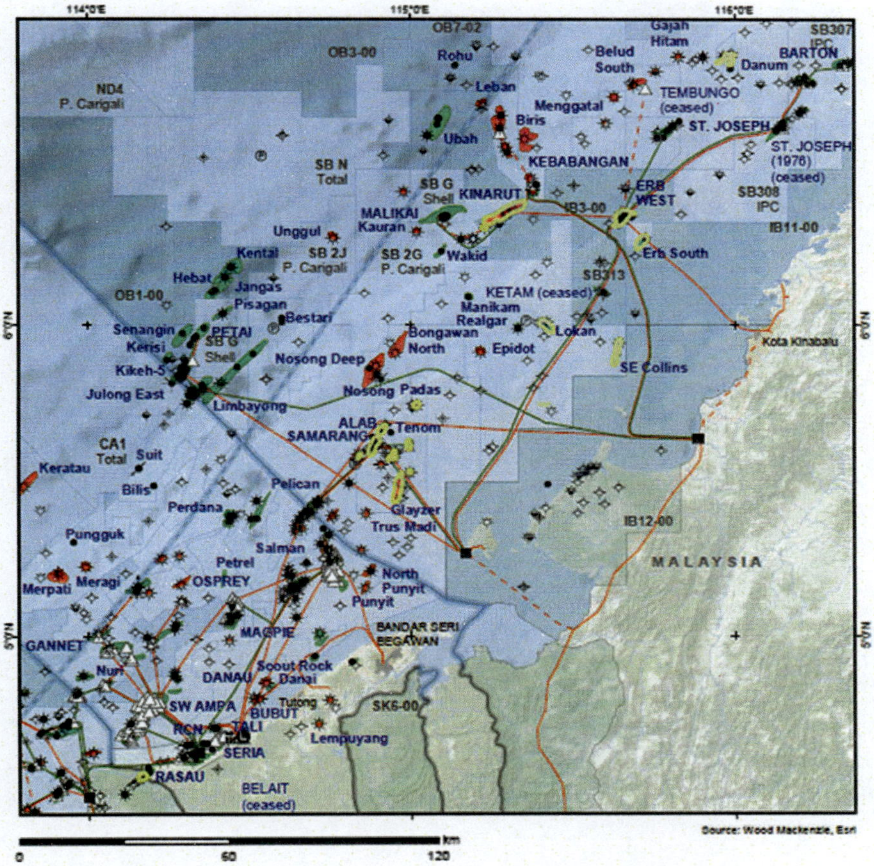

Fig. 1 Location of Field X

Fig. 2 The workflow of the integration analysis

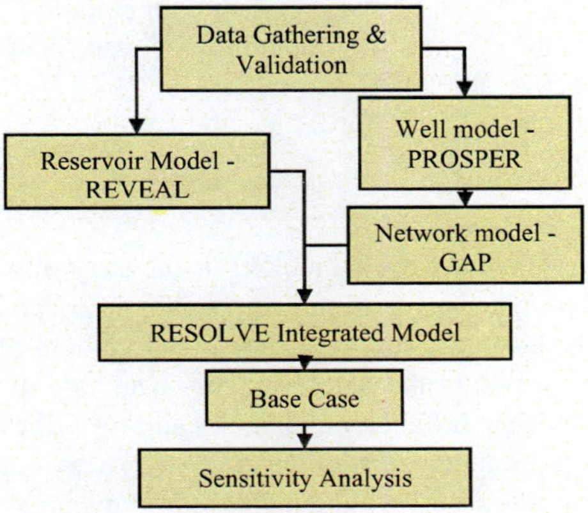

2 Volumetric Calculation, IAM and Sensitivity Analysis Overview

2.1 *Oil in Place and Recovery Factor Estimation*

An accurate estimation of hydrocarbon in place and RF is crucial in quantifying the resources and planning the overall management for the field [1].

2.1.1 Calculation of Hydrocarbon Volumes

The basic volumetric equation is outlined as:

$$STOIIP = N = \frac{V\varnothing(1 - Swc)}{Boi} \tag{1}$$

where,

V = the net bulk volume of the reservoir rock.

Ø = the porosity, or volume fraction of the rock which is porous.

Swc = the connate or irreducible water saturation .

Boi = oil formation volume factor.

In Eq. (1), the parameters Ø and Swc are normally determined by petrophysical analysis. The net bulk volume, V, is obtained by geological and fluid pressure analysis.

2.1.2 Recovery Factor and Ultimate Recovery

STOIIP can be converted into an equation for calculating the ultimate oil recovery simply by multiplying by the recovery factor (RF), which represents the fraction of recoverable oil [1].

$$\mathbf{UR} = \frac{\mathbf{V}\varnothing(1 - \mathbf{Swc})}{\mathbf{Boi}} \times \mathbf{RF} \tag{2}$$

The determination of RF is the most important single task of the Reservoir Engineer. The two main categories of hydrocarbon recovery are called primary and supplementary. Primary recovery is the volume of hydrocarbons which can be produced by utilizing the natural energy available in a reservoir. In contrast, supplementary recovery is the oil obtained by adding energy to the reservoir-fluid system. The most common type of supplementary recovery is water flooding in which water is injected into the reservoir and displaced oil towards the producing wells.

2.2 Introduction of Integrated Asset Modelling Software

The IPM software is introduced by Petroleum Experts (PETEX) – a specialist in Petroleum Engineering and Structural Geology software. The software can be used to complete hydrocarbon production system including reservoirs, wells and surface network.

The IPM software such as PVTP, MBAL, REVEAL, PROSPER, GAP and RESOLVE can be used together, allowing the analyst and engineers to design a complete integration model. It will include the reservoir, wells and complete surface system. The integrated model can be optimized before running production forecast. IPM models are very useful in the calculation of production forecasts and the optimization of field developments. These models only provide a single possible solution each time they are run.

The benefits achieved from achieving an integrated approach are [3]:

1. Improved sharing of data and cooperation of teams.
2. Lesser time to update/validate models.
3. Increased accuracy in forecasting workflows.
4. Fast decision making due to automated workflows.

This IPM added value to the project by promoting integration, communication and engagement across the discipline [4]. The structured model building approach resulted in greater confidence and accuracy in production forecast. Each engineering discipline may have a piece of the puzzle to find the best solution for achieving optimized production rates and economics. The best results can be only achieved by using the total system (reservoir to sales point) approach. It is used the production system optimization of an integrated reservoir, wells and surface networks to help identify opportunities and maximize the value of the entire production system.

The IPM model is studied based on 3 parameters as listed below:

1. Integration—To evaluate the impact of an option on the entire system
2. Quantification—To compare different options on a consistent and absolute system.
3. Validation—To identify gaps in data and deficiencies in the technical model for improvement purposes.

Several advantages of the IPM approach as listed below.

1. More effective and efficient for the field
2. Pressure interaction between subsurface and surface
3. Accounting for facilities constraint and identification of system bottlenecks and pressure
4. Simulation of a complex production system and
5. Availability of important information for the decision making

The IPM approach is used for history matching, validating and quantity checking. The reservoirs and wells are then being integrated to run forecasts and make management decision concerning scheduling and compressor selection. It is concluded that the IPM approach allowed engineers to focus on engineering analysis because of automated routine data-processing tasks. It also significantly reduced manual data entry for model updates and improves the accuracy of production allocation over traditional allocation methods.

2.3 Sensitivity Analysis

Sensitivity analysis is defined as a technique to determine how independent variable values will affect a dependent variable under a given set of assumptions. It is also known as what-if analysis.

The process of recalculating outcomes under alternatives assumptions to determine the impact of variables under sensitivity analysis can be useful for the following reasons [2]:

1. Testing the robustness of the system in the presence of uncertainties and increased understanding of the relationships between input and output variables of a system.
2. Uncertainty reduction, through the identification of model inputs that cause significant uncertainty in the output.
3. Searching for errors in the model by encountering unexpected relationships between input and outputs.

For models with many parameters, a primary sensitivity analysis can ease the calibration stage by only focusing on the sensitive parameters. Not knowing the sensitivity of parameters can result in time being uselessly spend on non-sensitive parameters.

The recent sharp decline in the price of liquid hydrocarbon drove many companies to adopt stringent austerity measures to counter economic imbalances. The profitability of any oil and gas project is a function of the factors like oil price, drilling cost, field size, field life, operating cost, rate of interest, platform cost and many more. Making a sensitivity analysis of these factors will enable to identify the key factors that control profitability and help in manoeuvring them in such situations.

3 Methodology

After addressing data gathering and validation, the construction of REVEAL, PROSPER, GAP and RESOLVE models are done. Integration of the different components into a single, full system was analysed. The integrated model will be treated as the base case model. Sensitivity analyses were done to find which critical parameters that give the most effect on hydrocarbon recovery and cumulative oil production. Based on the result, discussion on the top parameters will be done to increase accuracy hence the cumulative oil production estimation will become more robust.

3.1 The Building of the Individual Model

Before proceeding with integrated RESOLVE modelling, the individual REVEAL, PROSPER and GAP modelling were created as shown in Figs. 3, 4 and 5 respectively.

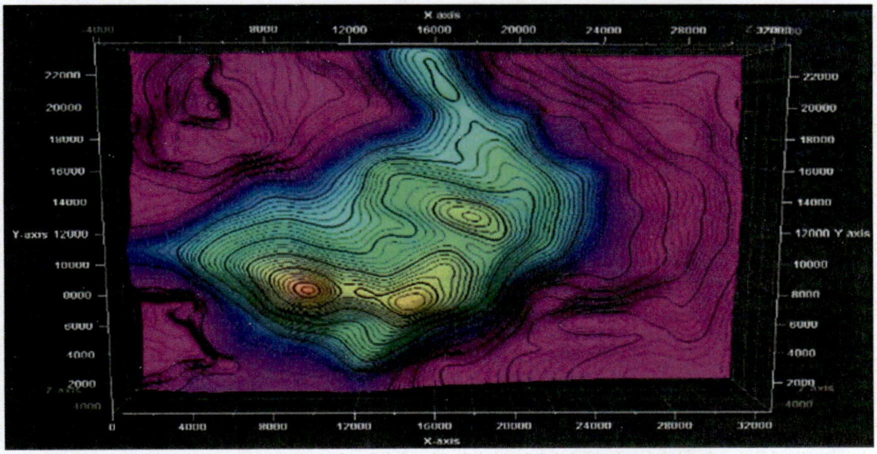

Fig. 3 Surface map of Field X

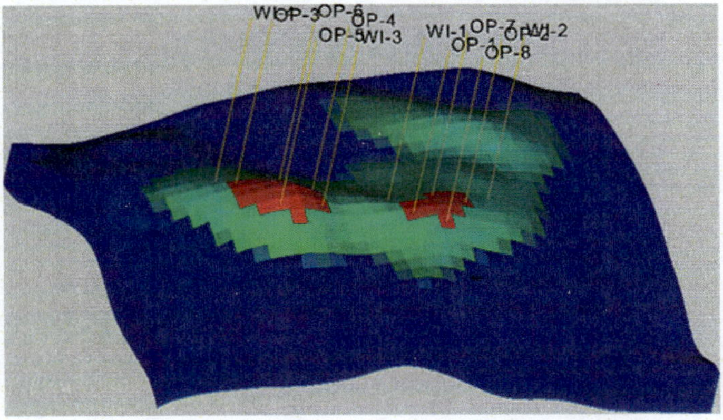

Fig. 4 Location of producer and injector wells in the REVEAL model

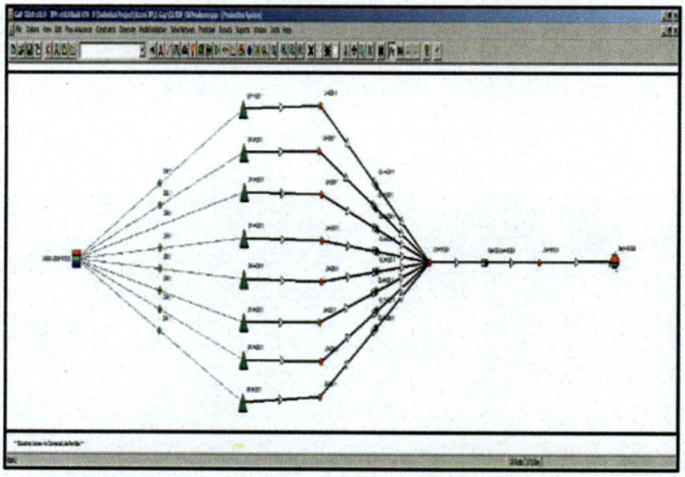

Fig. 5 GAP network for producer wells

3.2 Integrated Model Study

Once REVEAL and GAP models are ready, RESOLVE model was created and both were summoned. The RESOLVE model is shown in Fig. 6. Before proceeding with the sensitivity analysis on the parameters that give a significant effect on the

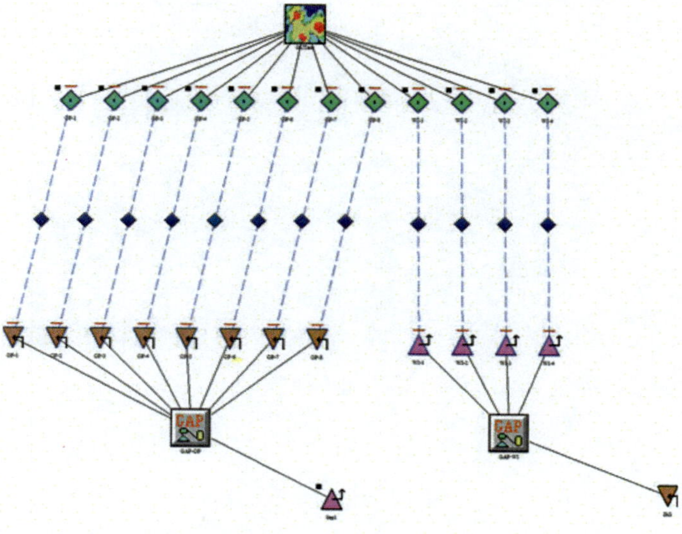

Fig. 6 Integrated base case RESOLVE model

hydrocarbon recovery, the base case needs to be run to make sure all the input parameters are in a good connection.

The results for the integrated RESOLVE base case study will be compared with individual REVEAL software results. This is very important to analyse if there are big changes in their results. The comparison will be discussed in the following chapter.

3.3 Sensitivity Analysis

Flow diagram for sensitivity analysis was created to ease to workflow as shown in Fig. 7 below.

The RESOLVE base case model was created using the most likely value input data. These values were selected based on P50 scenario. The model was run and the results such as field cumulative oil production and recovery factor were recorded. Later, this base case model will be used as the reference for the sensitivity analysis for low and high case value.

The uncertainties together with the range of low and high case are defined and will become the input parameters. The model will be run to analyse the parameters that give the most influential uncertainties. Sensitivity analysis results are then presented by using tornado plot to visualize the impact of each pre-defined parameters. To run sensitivity on RESOLVE, the Scenario Manager overview is used. The scenarios stored can be submitted to a cluster for parallel, batch processing. These scenarios can be set up using Event-Driven Scheduling (EDS) options and the result for the base case and sensitivity analysis will be shared at the Results and Discussion section. For the parameters that cannot be called using Scenario Manager, manual change can be done at an individual model and the simulation run still will be done using RESOLVE software.

After running the base case scenario study, several sensitivity cases were created to evaluate the critical parameter that gives a significant impact on cumulative oil

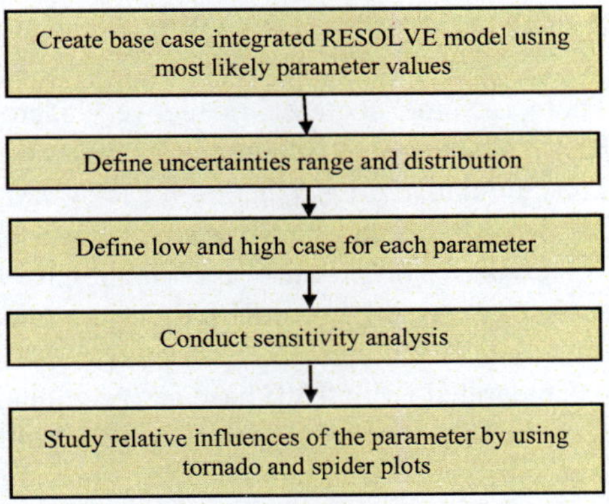

Fig. 7 Flow diagram for sensitivity analysis

Table 1 Sensitivity parameters for a simulation study

No	Parameters	Base case	
		Value	Reference
1	Separator Pressure, Psi	300	GAP
2	Water Injection Pressure, Psi	2500	GAP
3	Years of Production, Years	20	GAP
4	Oil Gravity, API	25.66	PVT, REVEAL
5	Formation GOR, SCF/BBL	500	PVT, REVEAL
6	Gas Gravity, sp.gravity	0.616	PVT, REVEAL
7	Porosity, %	Refer to Appendix	Reservoir, REVEAL
8	Permeability, mD	Refer to Appendix	Reservoir, REVEAL
9	NTG, %	57	Reservoir, REVEAL
10	Initial Oil Column, ft	268	Reservoir, REVEAL

production and oil recovery factor. Below are ten (10) selected parameters that were varied for the sensitivity analysis. The high case and low case value are a plus and minus 30% of base case value respectively as referring to Table 1. For porosity and permeability value, the sensitivity analysis will be done for each layer in a z-direction as tabulated in the Appendices.

4 Results and Discussion

There is some decrease in recovery factor and cumulative oil production from REVEAL analysis to integrated RESOLVE base case analysis as illustrated in Fig. 8. The cumulative oil production for REVEAL and integrated RESOLVE base case is 26.65 MMSTB and 25.51 MMSTB respectively. There is around 4% of reduction of cumulative oil production estimation for Field X due to the existence of well and surface constraint. From REVEAL individual software analysis alone, there is very limited well and surface consideration. This might be the reason for the reduction for integrated RESOLVE base case cumulative oil production estimation.

Generally, it was found that Field X is still not in the most optimum development scenario. This can be supported by the end trend for cumulative oil production plot in Fig. 8. This trend showed that the field still has big potential for oil to be produced.

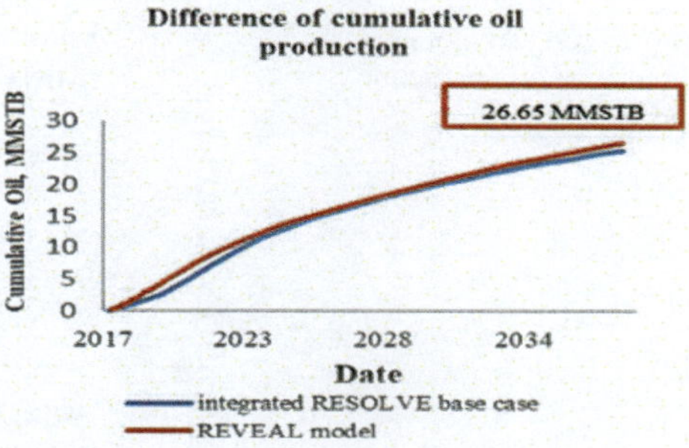

Fig. 8 The difference of cumulative oil production estimation for REVEAL and integrated RESOLVE model

To improve the base case cumulative oil production, the number of production and injector wells should be increased and the number of years of production can be added.

However, the optimization of cumulative oil production is not the main objective of this case study. The real intention of this study is to make a sensitivity study on parameters that give the highest impact on cumulative oil production estimation. The Recovery Factor (RF) for the field is only 19.8%. It is expected to have RF more than 35% since the water injection approach was applied for this field. It is noted that the base case study can be improved to get the highest cumulative oil and RF.

4.1 Sensitivity Analysis Results

The results for the sensitivity parameters are shown in Table 2. The tornado plots in Fig. 9 shows the impact of selected parameters toward cumulative oil production variation (Fig. 10).

Table 2 Sensitivity parameters for a simulation study

No	Parameters	Cumulative oil, MMSTB		
		Low case, (−30%)	Base case	High case, (+30%)
1	Separator Pressure	25.04	25.51	25.99
2	Water Injection Pressure	24.03	25.51	26.68
3	Years of Production	21.04	25.51	28.72
4	Oil Gravity	22.99	25.51	27.19
5	Formation GOR	25.51	25.51	25.51
6	Gas Gravity	25.43	25.51	25.66
7	Porosity	21.41	25.51	28.62
8	Permeability	22.70	25.51	27.75
9	NTG	17.21	25.51	33.69
10	Initial Oil Column	13.24	25.51	43.42

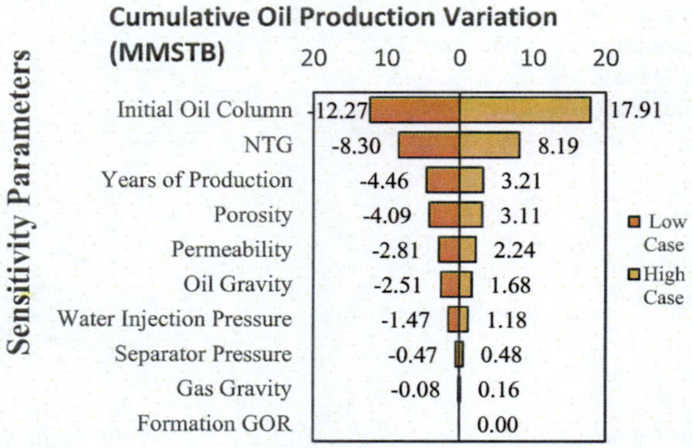

Fig. 9 Tornado plot for cumulative oil production variation

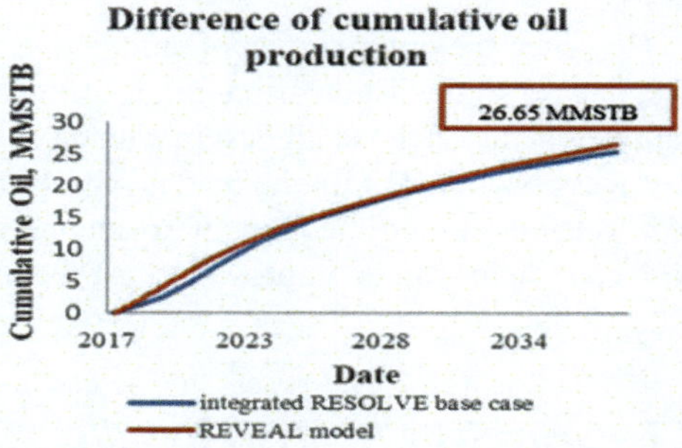

Fig. 10 Spider plot of sensitivity analysis results

Based on the Tornado and Spider Plot analysis, two parameters give the most significant effects on the base case cumulative oil production. These parameters are Initial Oil Column and Net To Gross (NTG). These parameters are considered as critical parameters for Field X cumulative oil estimation. As shown in Fig. 9, with an increase of 30% initial oil column and NTG values, the cumulative oil production will be increased by 70% and 30% respectively from the base case value.

The Initial Oil Column is very much dependent on Gas Oil Contact (GOC) and Oil Water Contact (OWC). By using the pressure gradient plot from Repeat Formation Tester (RFT) data, the accurate OWC and GOC value need to be used since it gives the highest impact on cumulative oil estimation.

Furthermore, for this case study, to make a sensitivity analysis of the initial oil column, the selected fluid contact that shifted is only OWC. This may lead to uncertainty because either GOC or both GOC and WOC can be shifted as well for the sensitivity analysis. Different detail study related to initial oil column sensitivity analysis can be done in the future.

NTG plays a significant role to determine how much oil in place for Field X reservoir. Shale cut-off, gamma-ray log reading, bulk density, porosity and permeability cut-off are the main factor to determine NTG value. All these parameters need to be evaluated carefully during petrophysical analysis to acquire the most accurate NTG value. Small changes in NTG value will give a significant impact on the cumulative oil production estimation.

5 Conclusion

Based on the integrated simulation study results, the conclusions are:

1. The base cumulative oil production estimation for Field X is 25.5 MMSTB (RF of 19.8%)
2. Two (2) parameters; Initial Oil Column, 70% and Net To Gross (NTG), 30% increment were identified as the topmost critical parameters for cumulative oil production estimation of Field X.
3. A proper data quality check and uncertainty management are required to increase accuracy for cumulative oil production estimation.

Several recommendations are made for the way forward of this study:

1. Optimization of the base case scenario is required before proceeding to any sensitivity study. The natural depletion approach should be prioritized before any secondary or tertiary depletion approach.
2. Cluster sensitivity analysis approach should be done to optimize simulation-running time.
3. Combination of PETREL dynamic simulation to represent the reservoir and GAP networking model during the integrated RESOLVE study.
4. Involvement of EXCEL analysis to see the impact of any parameter change towards economics

Acknowledgements Authors would like to acknowledge PETRONAS Carigali Sdn Bhd and Universiti Teknologi Malaysia for support and contribution in completing the research.

References

1. Dake LP (1978) Fundamentals of reservoir eng. Elsevier, The Netherlands, pp 62–103
2. Mach J, Proano E, Brown KE (1979) A nodal approach for applying systems analysis to the flowing and artificial lift oil or gas well. Soc Pet Eng
3. Madray R, Coll C, Veitch G, Chiboub C, Butter M, Azouzi S, Saada T (2008) Integrated field modelling of the MISKAR field. Soc Pet Eng
4. Ozdogan U, Keating JF, Knobles MM, Chawathe A, Seren D (2008) Recent advances and practical applications of integrated production modelling. Soc Pet Eng

Reliability

Analysis of Optimally Tuned Active Dynamic Vibration Absorber

T. V. V. L. N. Rao, Ainul Akmar Mokhtar, Masdi Muhammad, and Hamdan Haji Ya

Abstract Vibration can cause reliability problems and fatigue failures on equipment, and if resonances are involved, can result in very significant effects. A dynamic vibration absorber, eliminates or minimizes the amplitude of vibration of a structure when set to the forced vibration frequency. An active dynamic undamped and damped vibration absorber is proposed. The effects of optimum parameters of tuning for an active damped dynamic vibration absorber on the vibration reduction of a structure are discussed. Optimized design parameters of mass-spring-damper-actuator active vibration absorber proposed in this work are derived based on minimization of amplitude of resonant vibration of the structure. In an undamped dynamic vibration absorber, spring of absorber mass connected to rigid base with negative stiffness can broaden the frequency range of vibration mitigation. In a damped dynamic vibration absorber, negative stiffness of absorber mass connected with rigid base could significantly reduce the peak value of amplitude-frequency response of the primary system for lower mass ratio.

Abbreviations

b	Ratio of stiffness constant of absorber mass connected with rigid base to stiffness of absorber mass; $b = k/k_2$
c	Damping constant of absorber mass connected with rigid base
c_2	Damping of absorber mass, Ns/m
c_c	Critical damping constant, Ns/m; $c_c = 2m\omega_n$
d	Ratio of damping constant of absorber mass connected with rigid base to damping of absorber mass; $d = c/c_2$

T. V. V. L. N. Rao (✉)
Department of Mechanical Engineering, School of Engineering & Technology, The Assam Kaziranga University, Jorhat, Assam 785006, India
e-mail: tvvlnrao@gmail.com

A. Akmar Mokhtar · M. Muhammad · H. Haji Ya
Department of Mechanical Engineering, Universiti Teknologi PETRONAS, Bandar Seri Iskandar, Perak Darul Ridzuan, Malaysia

F. Ahmad et al. (eds.), *ICPER 2020*, Lecture Notes in Mechanical Engineering,
https://doi.org/10.1007/978-981-19-1939-8_36

f	Ratio of natural frequencies; $f = \omega_a / \omega_n$
F_o	Steady state force, N
k	Stiffness constant of absorber mass connected with rigid base
k_p	Proportional gain, N/m
k_d	Derivative gain, Ns/m
k_1	Stiffness of main mass, N/m
k_2	Stiffness of absorber mass, N/m
m_1	Main mass, kg
m_2	Absorber mass, kg
r_1, r_2	Nondimensional resonant frequencies of machine with absorber mass; $r_1 = \Omega_1 / \omega_2, r_2 = \Omega_2 / \omega_2$
p	Ratio of proportional gain to stiffness of absorber mass; $p = k_p / k_2$
q	Ratio of derivative gain to damping constant of absorber mass; $q = k_d / c_2$
g	Forced frequency ratio; $g = \omega / \omega_n$
X_1, X_2	Steady state response of main and absorber mass, m
δ_{st}	Static deflection, m;$\delta_{st} = F_o / k_1$
μ	Mass ratio; $\mu = m_2 / m_1$
ζ	Damping ratio;$\zeta = (c_2 + k_d) / c_c$
ω_a, ω_2	Natural frequency of absorber, rad/s; $\omega_a = \omega_2 = \sqrt{(k_2 + k_p) / m_2}$
ω_n, ω_1	Natural frequency of main mass, rad/s; $\omega_n = \omega_1 = \sqrt{k_1 / m_1}$

1 Introduction

A vibration absorber can be used to reduce the excessive vibration of structures/machines under the action of a force whose frequency of excitation coincides nearly with natural frequencies. The natural frequencies of structures/machines with dynamic vibration absorber are away from the forced/excitation frequency. Usually undamped vibration absorbers operate at constant speed [1]. The vibration amplitude of primary/main system is nonzero in a damped vibration absorber as damping is added when the frequency band is narrow for the absorber to be effective for operation. Liu and Liu [2] presented analysis of optimum damped dynamic vibration absorber. The optimized design for the minimum amplitudes of resonant vibration of a hybrid dynamic vibration absorber was presented by Cheung et al. [3]. An active dynamic vibration absorber using the PD control which does not require any feedback from the main structure was proposed by Chatterjee [4, 5]. The standalone feature of active vibration absorber makes it user-friendly as the absorber runs on the PD control of the absorber mass. The optimal PD control parameters of the standalone active absorber have shown to have accomplished better than a passive absorber [5]. Shen et al. [6] analytically studied optimal parameters of dynamic vibration absorber with negative stiffness. The negative stiffness means that the deformation is in reverse direction of the applied exterior force and the system with negative stiffness may have better vibration control under stable conditions. Results in harmonic and random excitation show that the DVA could significantly reduce the peak value of amplitude-frequency response of primary system [6]. Rao et al. [7] presented the influence of active control on undamped and damped vibration absorber.

This study investigates an undamped and damped dynamic vibration absorber with positive and negative stiffness of absorber mass connected with rigid base. Frequency range of undamped dynamic vibration absorber is determined using analytical method. Reduction in the peak values of amplitude-frequency response of the primary system for optimized design parameters of mass-spring-damper-actuator active damped dynamic vibration absorber is evaluated.

2 Analysis

The behaviour of an undamped and damped dynamic vibration absorbers with positive and negative stiffnesses of absorber mass connected with a rigid base, using analytical method is presented.

2.1 Active Undamped Dynamic Vibration Absorber

The schematic of two-degree-of freedom active undamped vibration absorber is shown in Fig. 1, wherein m_2 (a supporting/secondary mass) is attached to m_1 (primary/main mass) with springs of stiffnesses k_2 and k respectively.

Using an active control system with proportional gain to control the vibration of mass m_2, the equations of motion are expressed as

$$m_1\ddot{x}_1 + k_1 x_1 - k_2(x_2 - x_1) = F_o \sin\omega t - F_c \tag{1}$$

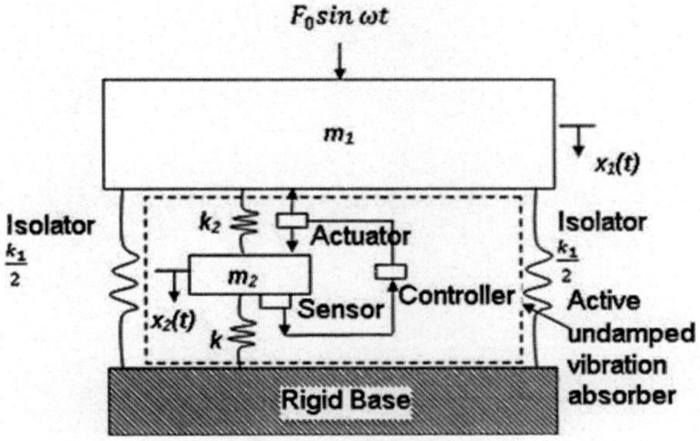

Fig. 1 Active undamped dynamic vibration absorber

$$m_2\ddot{x}_2 + k_2(x_2 - x_1) + kx_2 = F_c \tag{2}$$

where the actuator will be designed to exert a control force F_c such that $F_c = -k_p x_2$.

Using harmonic solution $x_j = X_j \sin \omega t,\ j = 1, 2$, the steady-state amplitude of masses m_1 and m_2 are obtained as

$$X_1 = \frac{\left(k_2 + k + k_p - m_2\omega^2\right)F_o}{\left(k_1 + k_2 - m_1\omega^2\right)\left(k_2 + k + k_p - m_2\omega^2\right) - k_2\left(k_2 + k_p\right)} \tag{3}$$

$$X_2 = \frac{k_2 F_o}{\left(k_1 + k_2 - m_1\omega^2\right)\left(k_2 + k + k_p - m_2\omega^2\right) - k_2\left(k_2 + k_p\right)} \tag{4}$$

Amplitude of primary/main mass (m_1) equal to zero in Eq. (3) results in

$$\omega^2 = \frac{\left(k_2 + k + k_p\right)}{m_2} \tag{5}$$

Amplitude of primary/main mass operating at its resonant frequency will be zero using the undamped vibration absorber, the natural frequencies of main mass and absorber mass coincides with operating frequency as

$$\omega^2 = \omega_1^2 = \frac{k_1}{m_1} = \omega_2^2 = \frac{\left(k_2 + k + k_p\right)}{m_2} \tag{6}$$

The two resonant frequencies Ω_1, Ω_2 introduced by dynamic vibration absorber (while eliminating the vibration at operating frequency ω) can be found by setting the denominator of Eq. (4) to zero. The roots obtained (two resonant frequencies Ω_1, Ω_2) are expressed in square of nondimensional resonant frequencies of machine with absorber mass are given as

$$r_1^2, r_2^2 = \left(1 + \frac{\mu}{2(1 + p + b)}\right) \pm \sqrt{\left(1 + \frac{\mu}{2(1 + p + b)}\right)^2 - \left(1 + \frac{\mu b}{(1 + p + b)^2}\right)} \tag{7}$$

2.2 *Optimally Tuned Active Damped Dynamic Vibration Absorber*

The vibration amplitudes the structure/machine can be reduced by active dynamic damped vibration absorber as shown in Fig. 2.

The equations of motion are given as

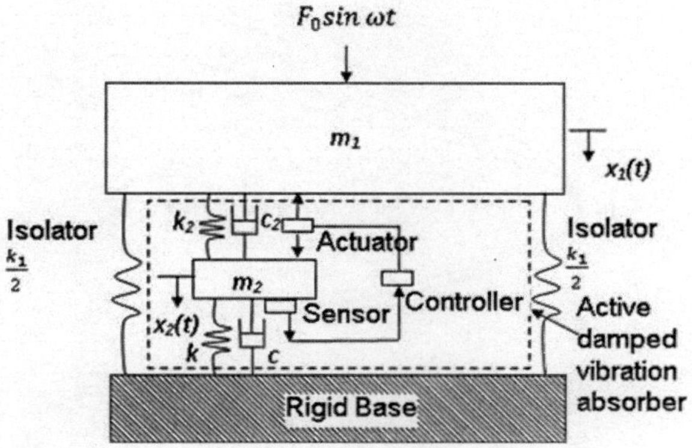

Fig. 2 Active damped dynamic vibration absorber

$$m_1\ddot{x}_1 + k_1x_1 - k_2(x_2 - x_1) - c_2(\dot{x}_2 - \dot{x}_1) = F - F_c \tag{8}$$

$$m_2\ddot{x}_2 + k_2(x_2 - x_1) + c_2(\dot{x}_2 - \dot{x}_1) + kx_2 + c\dot{x}_2 = F_c \tag{9}$$

where $F_c = -k_p x_2 - k_d \dot{x}_2$.

The general transfer function for the two degree of freedom system can be expressed as

$$\frac{X_1}{\delta_{st}} = \frac{G_{22}}{\Delta} \tag{10}$$

where

$$G_{22} = \hat{s}^2 + 2\zeta\hat{s} + f^2 \tag{11}$$

$$\Delta = \hat{s}^4 + \hat{s}^3 2\zeta\left(1 + \frac{\mu}{1+q+d}\right) + \hat{s}^2\left(f^2 + 1 + \frac{\mu f^2}{1+p+b} + \frac{4\mu d\zeta^2}{(1+q+d)^2}\right) + \hat{s}2\zeta\left(1 + \frac{(b+d)\mu f^2}{(1+p+b)(1+q+d)}\right) + f^2\left(1 + \frac{b\mu f^2}{(1+p+b)^2}\right) \tag{12}$$

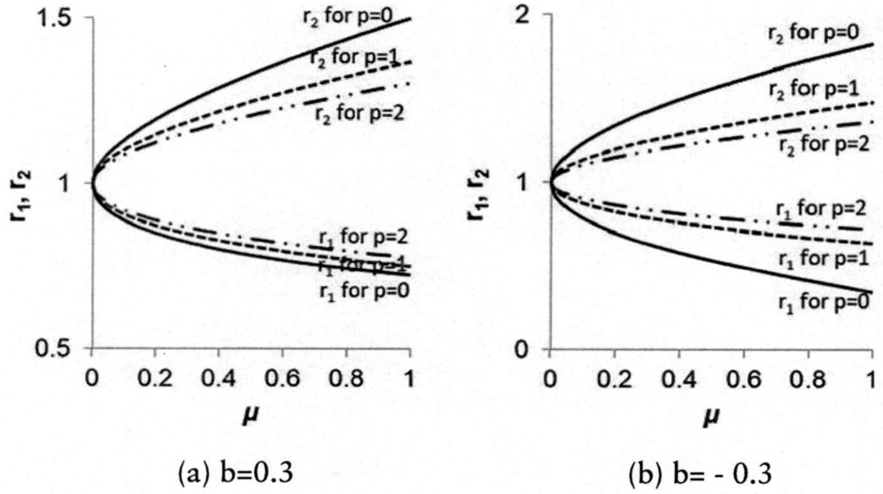

Fig. 3 Variation of r_1 and r_2

3 Results and Discussion

3.1 Active Undamped Dynamic Vibration Absorber

The parameters considered in the analysis of active undamped dynamic vibration absorber are: ratio of proportional gain to stiffness of absorber mass ($p = 0, 1, 2$); mass ratio ($\mu = 0$–1). Figure 3a–b show the variation of nondimensional resonant frequencies (r_1 and r_2) of machine with absorber mass as a function of mass ratio (μ). It can be seen that the difference between r_1 and r_2 increases with increase in mass ratio μ. Results of variations in resonant frequencies are compared for positive and negative values of ratios of stiffness constant of absorber mass connected with rigid base to stiffness of absorber mass ($b = 0.3$ and -0.3). As shown in Fig. 3b, it is of interest to note that when an absorber mass, specifically when connected to rigid base, with negative stiffness would result in a broader frequency range of vibration mitigation.

3.2 Optimally Tuned Active Damped Dynamic Vibration Absorber

The parameters considered in the analysis of active damped dynamic vibration absorber are: mass ratio ($\mu = 1/5, 1/20$); damping ratio ($\zeta = 0.1$); ratio of natural frequencies ($f = 1$); ratio of proportional gain to stiffness of absorber mass ($p =$

0,1); ratio of derivative gain to damping constant of absorber mass ($q = 0$). Results of nondimensional steady state response of main mass (X_1/δ_{st}) are compared for positive and negative values of ratios of stiffness constant of absorber mass connected with rigid base to stiffness of absorber mass (b = 0.3 and −0.3). Figure 4a–d show the variation of nondimensional steady state response of main mass (X_1/δ_{st}) as a function of forced frequency ratio (g). Lower peak values of amplitude-frequency response of the primary system are obtained for ratio of proportional gain to stiffness of absorber mass (p) of 0 (Fig. 4a, b) compared to $p = 1$ (Fig. 4c–d). An absorber satisfying the condition in which the ordinates of points of two peaks are equal is

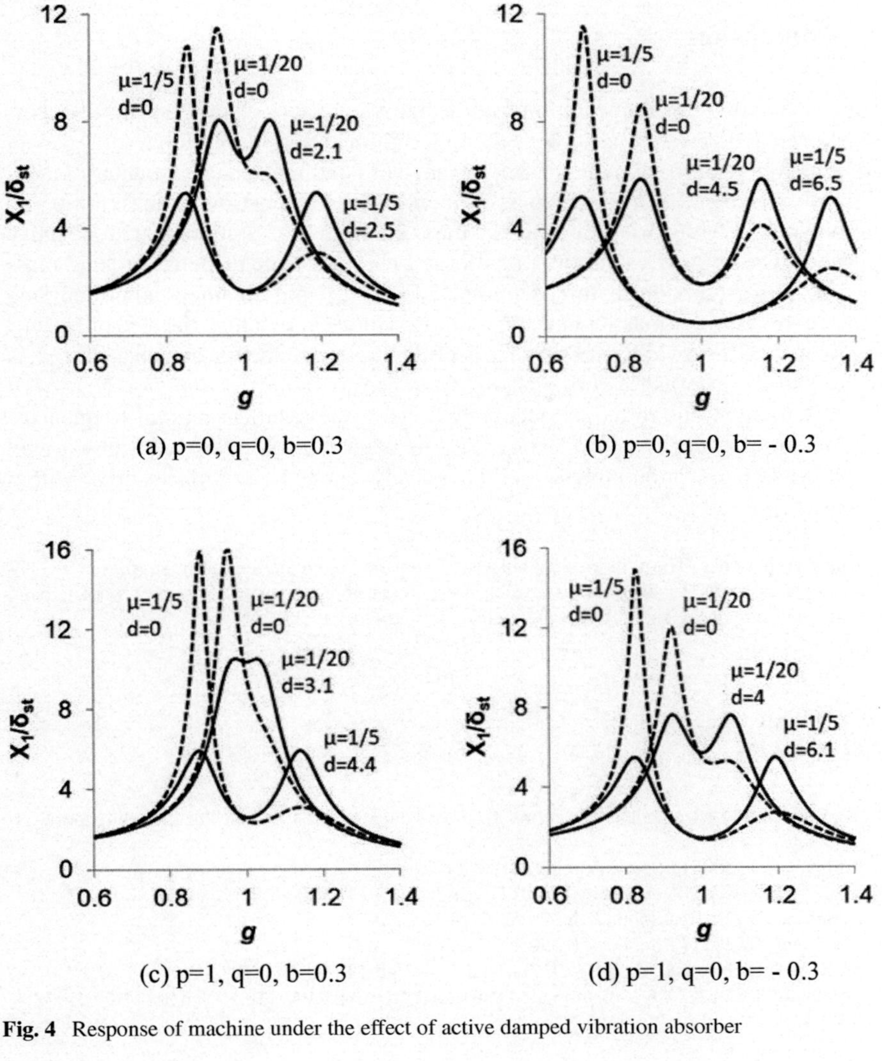

Fig. 4 Response of machine under the effect of active damped vibration absorber

optimally tuned active damped dynamic vibration absorber. The optimal values are found by making the nondimensional response as flat as possible at the two peaks. As shown in Fig. 4a–d, lower peak values of amplitude-frequency response of the primary system are obtained for mass ratio (μ) of 1/5 compared to 1/20 respectively. As shown in Fig. 4b and d, a damped dynamic vibration absorber with negative stiffness of absorber mass connected with rigid base (ratio of stiffness constant of absorber mass connected with rigid base to stiffness of absorber mass (b) of –0.3) could significantly reduce the peak value of amplitude-frequency response of the primary system for lower mass ratio (μ of 1/20).

4 Conclusion

The present study examines optimized design parameters of mass-spring-damper-actuator active vibration absorber based on minimization of amplitude of resonant vibration of the primary/main mass. The range of nondimensional resonant frequencies of machine with absorber mass (r_1 and r_2) as a function of mass ratio (μ) are broadened with spring of absorber mass connected to rigid base with negative stiffness. The optimal values of nondimensional steady state response of main mass (X_1 / δ_{st}) as a function of forced frequency ratio (g) are obtained using damping constant of absorber mass connected with rigid base. In practice, the size of m_2/m_1, and hence the bandwidth over which a useful vibration reduction can be achieved, is limited by the external factors such as space, permissible extra weight etc., as well as the limitation due to the broadening effects of the resultant natural frequencies (one natural frequency is to increase and the other is to decrease, and thus it may affect the safe operation/inclusion of the actively damped vibration absorber within the system).

Acknowledgements The authors would like to convey their appreciation for the support of YUTP-FRG Grant 2017 for funding this research. The authors appreciate the support of Madanapalle Institute of Technology and Science and Universiti Teknologi PETRONAS.

References

1. Rao SS (2011) Mechanical vibrations Pearson Education South Asia Pte Ltd, Singapore, pp 832–843
2. Liu K, Liu J (2005) J Sound Vib 284:1181–1189
3. Cheung YL, Wong WO, Cheng L (2012) J Sound Vib 331:750–766
4. Chatterjee S (2010) J Sound Vib 329:5397–5414
5. Chatterjee S (2011) J Sound Vib 330:1285–1299
6. Shen Y, Peng H, Li X, Yang S (2017) Mech Syst Signal Pr 85:193–203
7. Mokhtar RTVVLN, A A Muhammad M Hussain P, (2016) ARPN. J Eng Appl Sci 11(22):13380–13383

Optimum Cutting Time Ratio of Modified Whitworth Quick Return Mechanism

Nabile N. Swadi and Hussein M. H. Al-Khafaji

Abstract Cutting time ratio (T_r) is among the most important factors indicating the effectiveness of Whitworth quick return mechanism (WQRM). This investigation added a novel alteration that enhanced the T_r of a traditional laboratory WQRM. The modification was accomplished by integrating a group of additional links into the WQRM. This addition comprised a crank, a ground (base), a sliding block (pivot) and a slotted link. An analytical model and a computational dynamic simulation were conducted on the modified WQRM (MWQRM) to analyse the performance of its mechanism. Different values for the lengths of the new extra links were examined to validate their contribution to increasing the T_r values. Results illustrated that the MWQRM effectively enhances T_r, in which the T_r exceeds eight times for the MWQRM in comparison with the conventional one.

Keywords Modified (novel) Whitworth quick return mechanism · Cutting time ratio · Analytical method of linkage's kinematic · Computational dynamics of modified quick return mechanism

1 Introduction

In many applications, certain linkages are used to execute repetitive processes such as pushing components along a finished product or an assembly line, machining, fastening parts and packaging apparatus. The total time of these procedures affects the production process in terms of product quantity, sustainable energy (consumption) and cost. The total time is usually divided into two main parts, namely, one for the working stroke (effective) and the other for returning stroke (ineffective). Designers tend to minimize or maximize these two parts individually or totally in accordance with design requirement. Several scholars are interested in optimising time by positing algorithms and methodologies [1–5], whereas others tended to

N. N. Swadi (✉) · H. M. H. Al-Khafaji
Mechanical Department, University of Technology, Baghdad, Iraq
e-mail: 20065@uotechnology.edu.iq

F. Ahmad et al. (eds.), *ICPER 2020*, Lecture Notes in Mechanical Engineering,
https://doi.org/10.1007/978-981-19-1939-8_37

change the structure and configuration of machine (mechanisms) [6, 7] or use a control systems for actuators movement of mechanism [8–10].

Whitworth quick return mechanism (WQRM) is a common type of linkage used in repetitive industry processes. WQRM is a mechanism that extracts two strokes of reciprocating motions from a rotary motion and features different duration of time for each stroke; it is called a WQRM when the time consumed for the return stroke is less than that of the working stroke [11–13]. The ratio of working stroke time to return time is called cutting time ratio T_r. This ratio is a factor that should be taken in consideration during the design of the WQRM, and it is a kinematic property of the linkage itself that could be found firmly from the device geometry [14].

Many researchers contributed to the analysis and development of the quick return mechanism (QRM). Several scholars investigated the kinematic and dynamic performance of the QRM [15–17]. Others presented alterations and additions to the QRM, enabling the machine to be compatible with the requirements of operations, such as slider velocity [8, 18, 19], and cutting time ratio [6, 7, 11].

The current work tested the enhancement of the T_r due to the uncommon modification of conventional WQRM. The change was implemented by adding a set of links to a classical WQRM. The analytical and computational dynamic analyses were utilized for modelling and analysing the modified WQRM (MWQRM). The effectiveness of the dimensions of additional links on the T_r value was examined.

2 Configuration of MWQRM

2.1 *Unmodified WQRM*

The original laboratory WQRM and the computational model based on it are shown in Fig. 1. The mechanism has two parts. Part 1 comprises four links, namely, foundation (link 1), crank (link 2), pivoting component (link 3) and slotted bar (link 4). Part 2 involves crankshaft (link 5), connecting rod (link 6) and slider (link 7), in addition to the frame (link 1). Links 4 and 5 work as one unit. Figure 2 displays the 3D model of the computational model of unmodified WQRM.

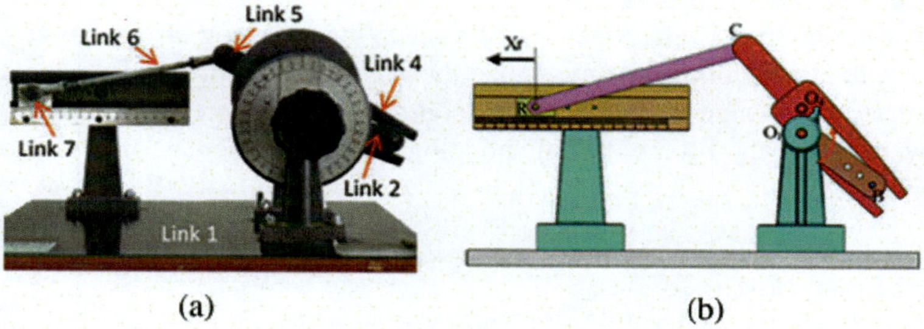

Fig. 1 Traditional laboratory WQRM **a** Original WQRM **b** Computational model

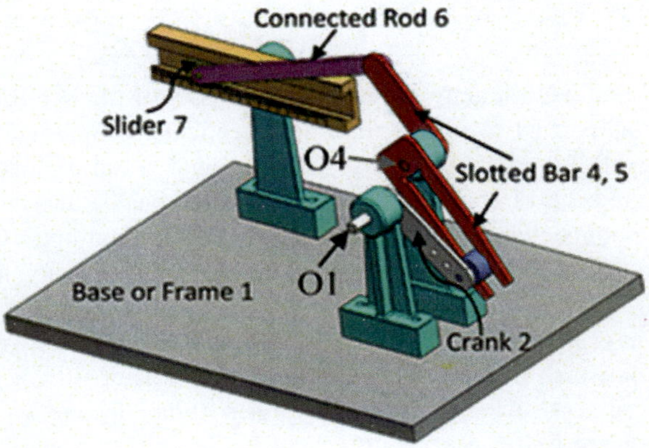

Fig. 2 3D computational model of unmodified WQRM

2.1.1 Model Validation

Experiments were performed to validate the computational model of WQRM with the real laboratory WQRM. These tests included a practical measurement of the slider's (link 7) movement (displacement X_r) of the real laboratory WQRM. Numerical solutions were performed via SOLIDWORKS software. Figure 3 presents a sample of comparisons between the measured and simulated values, which shows a good fit between the two trials.

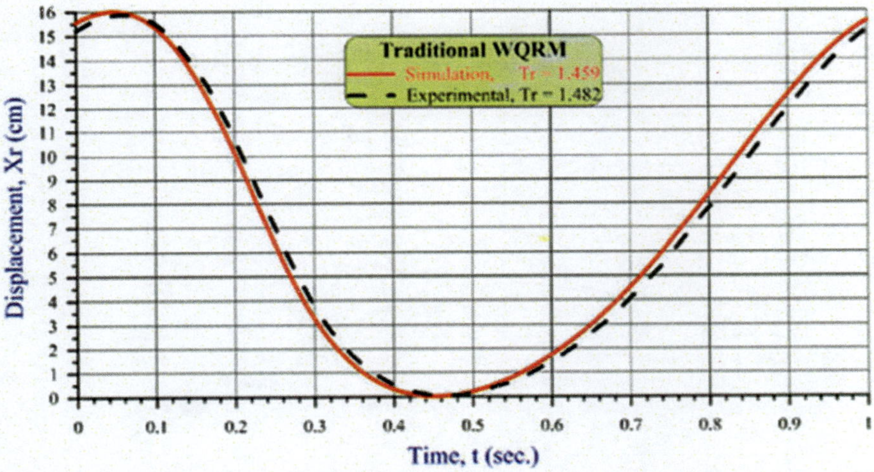

Fig. 3 Validation between experimental and simulation model

2.2 MWQRM

This study modified a computational laboratory model of the WQRM by adding a new section called part 01 to it. The modified mechanism consisted of three parts, namely, 01, 1 and 2. Figure 4 shows that the additional group 01 involves four new links, namely, crankshaft 02, pivot element 03, slotted bar 04 and base (frame 01). Extra links were connected to the standard links at a common point O1. Therefore, links 2 and 04 rotate at the same angular displacement and velocity, as explained in Figs. 4 and 5.

MWQRM was represented by two configurations in accordance with the position of frame O_{o2} relative to frame O_1.The configuration was positive r_{o1} when frame O_{o2} was below frame O_1, whereas it was negative r_{o1} when frame O_{o2} was above the frame O_1, as depicted in Fig. 5.

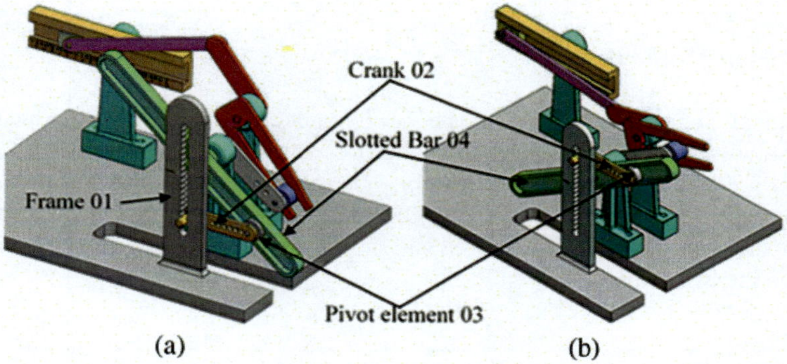

Fig. 4 Computational model of the modified WQRM **a** positive r_{o1}, **b** negative

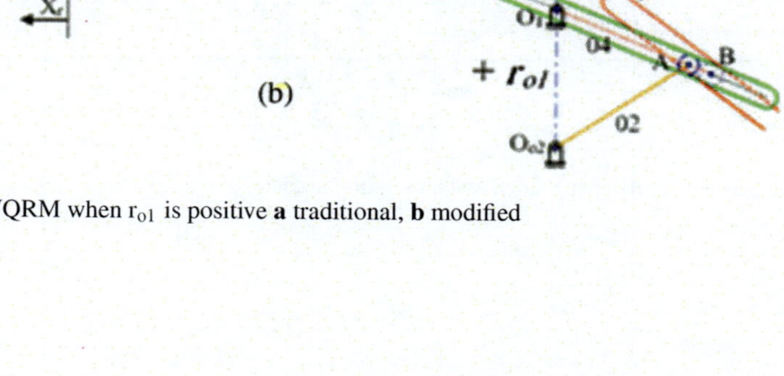

Fig. 5 WQRM when r_{o1} is positive **a** traditional, **b** modified

3 Theory

Analytical and rigid-body computational dynamic methods were conducted to calculate slider displacements and cutting time. Open-source software (GNU Octave 5.1) was utilised to build a programme for an analytical method that was based on direct vector method (closure equations), and SOLIDWORKS was used for computational dynamic simulations, Because of its accurate result consistent with the results of analytical analysis, besides it gives a good visualization of the mechanism movement [20].

All calculations were conducted with a constant angular velocity of link O2 as input for a mechanism with one complete cycle (0–360°) The rotation direction is displayed in Fig. 5.

3.1 Modelling of MWQRM

3.1.1 *Positive* r_{o1}

- Group 01

Group 01 is the additional part of the MWQRM that has a positive r_{o1}, which is shown in Fig. 5. As shown in Fig. 6, crank link O2 is connected to frame O_1 by revolute joint O_{o2} and to slotted bar (link O4) with a prismatic joint (A). The slotted bar link O4 is attached to the base with revolute joint O_1. The closure equations of this part are presented in Fig. 7, and the relations of the variables are as follows:

$$\mathbf{R_{o4}} = \mathbf{R_{o1}} + \mathbf{R_{o2}} \tag{1}$$

$$r_{o4}(i\cos\theta_{o4} + j\sin\theta_{o4}) = r_{o1}(i\cos\theta_{o1} + \sin\theta_{o1}) + r_{o2}(i\cos\theta_{o2} + j\sin\theta_{o2}) \tag{2}$$

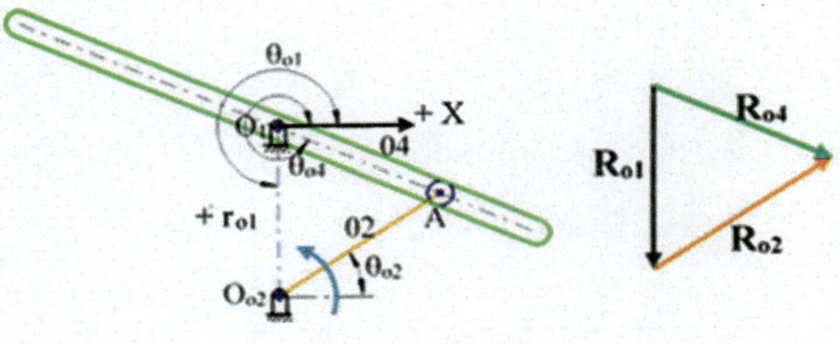

Fig. 6 Group 01 of the modified WQRM when r_{o1} positive

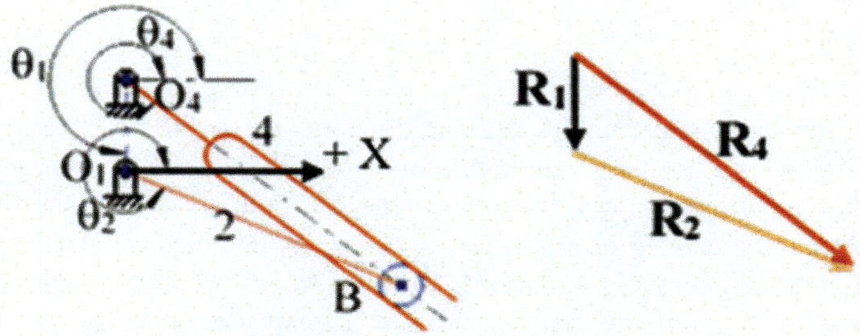

Fig. 7 Group 1 of the modified WQRM when r_{o1} positive

where:

r_{o1} = length of the link (O1– O_{o2}).

r_{o2} = length of the link O2 (O_{o2} – A).

Separating Eq. (2) into its real and imaginary parts gives

$$r_{04}\cos\theta_{04} = r_{01}\cos\theta_{01} + r_{02}\cos\theta_{02} \tag{3}$$

$$r_{04}\sin\theta_{04} = r_{01}\sin\theta_{01} + r_{02}\sin\theta_{02} \tag{4}$$

Figure 6, depicts that link r_{o1} is located at ($\theta_{o1} = 270°$) with respect to the positive X-axis. r_{o1} and r_{o2} are constant (time invariants) and known during the constant rotation of link 02 at angular velocity

$$\omega_{o2} = 60\,\text{rpm (1 rev/s) CCW}$$
$$\sin\theta_{o1} = -1 \text{ and } \cos\theta_{o1} = 0$$

θ_{o2} is known (changed from 0 to 360°), in Eqs. (3) and (4) for θ_{o4} and r_{o4} to be in terms of r_{o1}, r_{o2} and θ_{o2}:

$$r_{04}\cos\theta_{04} = r_{02}\cos\theta_{02} \tag{5}$$

$$r_{04}\sin\theta_{04} = -r_{01} + r_{02}\sin\theta_{02} \tag{6}$$

Equations (5) and (6) are manipulated to give

$$(r_{04})^2 = (r_{01})^2 + (r_{02})^2 - 2r_{01}\cdot r_{02}\sin\theta_{02} \tag{7}$$

$$r_{04} = \sqrt{r_{o1}^2 + r_{o2}^2 - 2r_{o1}.r_{o2}.\sin\theta_{o2}} \tag{8}$$

Dividing Eqs. (6) by (5) and solving for θ_{o4} yield

$$\theta_{o4} = \tan^{-1}(\frac{r_{o2}\sin\theta_{o2} - r_{o1}}{r_{o2}\cos\theta_{o2}}) \tag{9}$$

- Group 1:

Group 1 represents the second part of MWQRM. Figure 7 indicates that the group consists of crankshaft link 2 and is connected to base 1 by revolute joint O_1 and the slotted bar (link 4) with the pivot element (link 3) by a prismatic joint (B) from one side. On the other side, the slotted bar (link 4) is attached to the base with revolute joint 04.

From the vector diagram of group 1 clarified in Fig. 7, the relations are as follows:

$$R_4 = R_1 + R_2 \tag{10}$$

$$r_4(i\cos\theta_4 + j\sin\theta_4) = r_1(i\cos\theta_1 + j\sin\theta_1) + r_2(i\cos\theta_2 + j\sin\theta_2) \tag{11}$$

Equating the real and imaginary terms of Eq. (11) leads to

$$r_4\cos\theta_4 = r_1\cos\theta_1 + r_2\cos\theta_2 \tag{12}$$

$$r_4\sin\theta_4 = r_1\sin\theta_1 + r_2\sin\theta_2 \tag{13}$$

Figure 7 demonstrates that link 1 (O_4 O_1) is located at ($\theta_1 = 270°$) with respect to the positive X-axis. r_1 and r_2 are constant (time invariants). Link 2 of this part rotates with the same angular displacement and velocity of link 04 because they are connected with the same revolute joint at O1, as illustrated in Figs. 4 and 5. Therefore, $\theta_2 = \theta_{o4}$.

$$\sin\theta_1 = -1 \text{ and } \cos\theta_1 = 0$$

$$r_4\cos\theta_4 = r_2\cos\theta_{o4} \tag{14}$$

$$r_4\sin\theta_4 = -r_1 + r_2\sin\theta_{o4} \tag{15}$$

From Eqs. (14) and (15),

$$r_4 = \sqrt{r_1^2 + r_2^2 - 2r_1.r_2.\sin\theta_{o4}} \tag{16}$$

$$\theta_4 = \tan^{-1}(\frac{r_2\sin\theta_{o4} - r_1}{r_2\cos\theta_{o4}}) \tag{17}$$

- Group 2

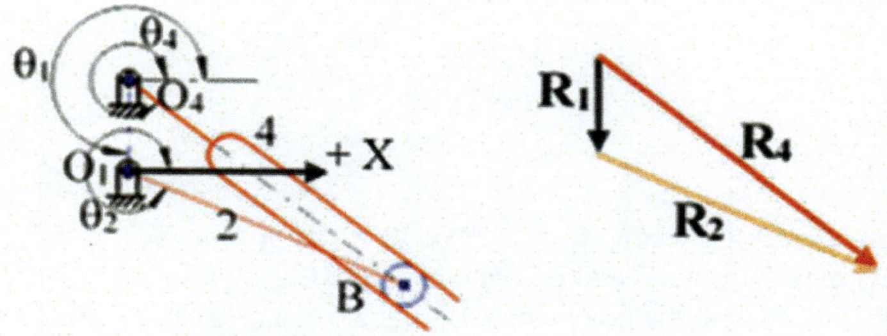

Fig. 8 Kinematic model of slider-crank mechanism (group 2 when r_{o1} is positive)

The last part of the MWQRM was group 2. It contained the end-effector's destination path of the mechanism which was the effective displacement of the cutting tool (point R). Group 2 was posited as a slider-crank mechanism, as shown in Fig. 8. Revolute joint O4 was used to link the crankshaft (link 5) to the frame, and revolute joint C was utilised to join link 5 to the connecting rod (link 6). The rod (link 6) was connected to the slider (link 7) by using revolute joint R.

The vector loop closure of this linkage shown in Fig. 8 is

$$\mathbf{R}_{11} = \mathbf{R}_5 + \mathbf{R}_6 \tag{18}$$

$$r_{11}(i\cos\theta_{11} + j\sin\theta_{11}) = r_5(i\cos\theta_5 + j\sin\theta_5) + r_6(i\cos\theta_6 + j\sin\theta_6) \tag{19}$$

When comparing imaginary and real terms,

$$r_{11}\cos\theta_{11} = r_5\cos\theta_5 + r_6\cos\theta_6 \qquad (19)$$

$$r_{11}\sin\theta_{11} = r_5\sin\theta_5 + r_6\sin\theta_6 \tag{20}$$

where

r_{11} = distance between O_4 and point R.

r_5 = length of the crankshaft (link 5).

r_6 = length of the connecting rod (link 6).

($\theta_{11} = 180°$ and $\theta_5 = \theta_4 \pm 180°$) with respect to the real positive X-axis. r_5 and r_6 are constant whereas r_{11} and θ_6 are changing with time.

sin (180) = 0 and cos (180) = - 1

$\sin\theta_5 = \sin(\theta_4 \pm 180°) = -\sin\theta_4$ and $\cos\theta_5 = \cos(\theta_4 \pm 180°) = -\cos\theta_4$

Equations (20) and (21) reduce to

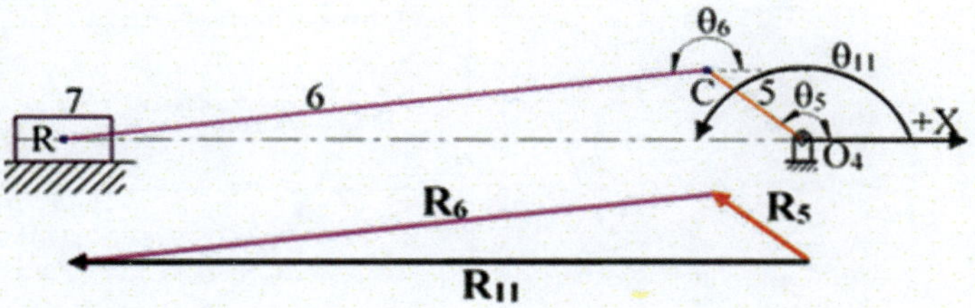

Fig. 9 The modified WQRM when r_{o1} is negative

$$-r_{11} = -r_5\cos\theta_4 + r_6\cos\theta_6 \quad (21)$$

$$0 = -r_5\sin\theta_4 + r_6\sin\theta_6 \quad (22)$$

Equation (23) was arranged to acquire

$$\theta_6 = \sin^{-1}(\frac{r_5\sin\theta_4}{r_6}) \quad (23)$$

From Eq. (22)

$$r_{11} = r_5\cos\theta_4 - r_6\cos\theta_6 \quad (24)$$

Therefore, the position of the cutting tool (point R) is

$$X_r = r_{11} - (r_6 - r_5) \quad (25)$$

3.1.2 Negative r_{o1}

The configuration of the MWQRM was named negative when the magnitude of r_{o1} was negative by shifting the point O_{o2} above point O_1, as shown in Fig. 9. The Relations derived in the previous section for positive r_{o1} was used r_{o1} negative after substituting the positive value of r_{o1} to negative in Eqs. (8) and (9).

3.2 Cutting Time Ratio (T_r)

A QRM gives a reciprocating motion in which the time taken for travel in advance-stroke is more than in the return-stroke. The difference in time is an advantage of this apparatus, and the total time of operation is the sum of return stroke and advance stroke time. The goal of designers is to reduce the total time to operate the machine.

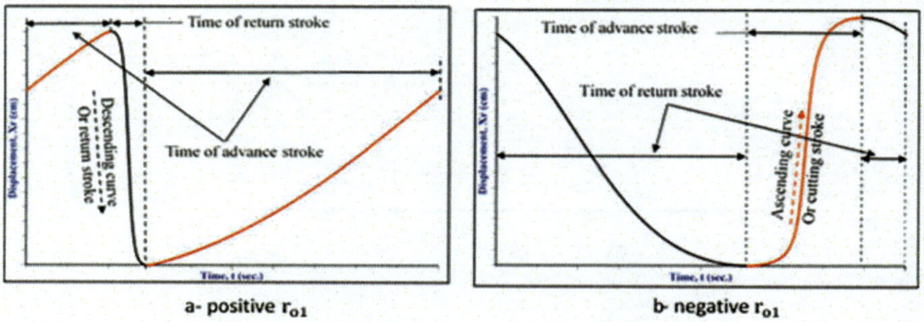

Fig. 10 A graph illustrates the time of advance-stroke and return-stroke

A reduction in the total time is influenced by several variables, such as the type of process that the mechanism performs, the progression length and the operating speed required for this process. Consequently, the trend to diminish the total time is towards decreasing the return time. The mechanism's effectiveness was measured from the time point of view by increasing the T_r which was covered by the following equation:

$$T_r = \frac{\text{time of advance stroke}}{\text{time of return stroke}} \tag{26}$$

The plots of the slider displacement X_r of the MWQWM versus the total time of operation for positive and negative r_{o1} are presented in Fig. 10. Figure 10a shows two complementary regions to the advance-stroke time, whereas the middle one is for the return stroke time. Their sum represents the total time. By contrast, in Fig. 10b, two zones exist for the negative r_{o1}; they are for return stroke and for the advance stroke. Depending on Fig. 10, the T_r formula is

$$T_r = \frac{\text{total time of ascending Xr curve}}{\text{Total time of discent Xr curve}} \tag{27}$$

4 Methodology

This work aimed to investigate the performance of WQRM after modification. Quantitative calculations were performed by using analytical equations and computational model. A programme was built using open-source software (GNU Octave 5.1) for the analytical method and SOLIDWORK software was used to implement the computational part. The same dimensions of links for the original laboratory WQRM were taken in establishing the analytical and computational models. Different dimensions of additional links were tested to study their effect on the response of the slider's

Table 1 The sets of dimensions for additional links (positive r_{o1} and r_{o2})

The dimensions of r_{o2} (mm)	The dimensions of positive r_{o1} (mm)
75	30, 37.5, 45, 52.5, 60, 67.5
60	7.5, 15, 22.5 30, 37.5, 45, 52.5
45	15, 22.5, 30,37.5

Table 2 The sets of dimensions for additional links (negative r_{o1} and r_{o2})

The dimensions of r_{o2} (mm)	The dimensions of negative r_{o1} (mm)
75	–15, –30, –45, –52.5
60	–15, –30, –45, –52.5
45	–15, –22.5, –30, –37.5

movement and the improvement of cutting time interval. The tested lengths of r_{o1} in the two configurations (positive and negative) were constrained to be less than r_{o2} ($r_{o1} < r_{o2}$). Otherwise, the behaviour of MWQRM would be changed to a different one. The sets of dimensions are listed in Tables 1 and 2. Three different length values were set for arm r_{o2}, each of which corresponds to the different dimensions for arm r_{o1}. Each set of r_{o1} was stepped 7.5 mm. All trials were conducted under constant input speed of the arm r_{o2} and one revolution.

The input constant speed was 60 rpm counter clockwise (CCW). From the tests, the displacement X_r of the slider was calculated throughout the total time consumed during the return and advanced strokes. The T_r of each trial was counted.

5 Results and Discussion

Depending on the sets of dimensions listed in Tables 1 and 2, twenty-nine (29) tests were conducted to gather data of the slider's movement that changed with time and calculate the T_r for each set. Experimental and computational simulation tests were performed on a real laboratory traditional WQRM to validate the computational solution and compute the T_r which were 1.482 and 1.459 for the real experiment and simulation, respectively.

Figures 11, 12 and 13 indicate the analytical and simulation results of the slider displacement X_r versus the total time for the MWQRM which was working with a constant angular velocity (60 rpm CCW as an input to r_{o2}). These figures clarify that the behaviour of the displacement curves for the two configurations (positive r_{o1} and negative r_{o1}) differed. The value of T_r for the positive r_{o1} was larger than that for the negative r_{o1} for the same absolute value of r_{o1} (same length). The maximum values of T_r for the positive r_{o1} were 12.293 and 12.393 from the analytical and computational solutions, respectively. The set of dimensions was $r_{o2} = 75$ mm and $r_{o1} = 67.5$ mm. The analytical and computational (SOLIDWORKS) solutions showed

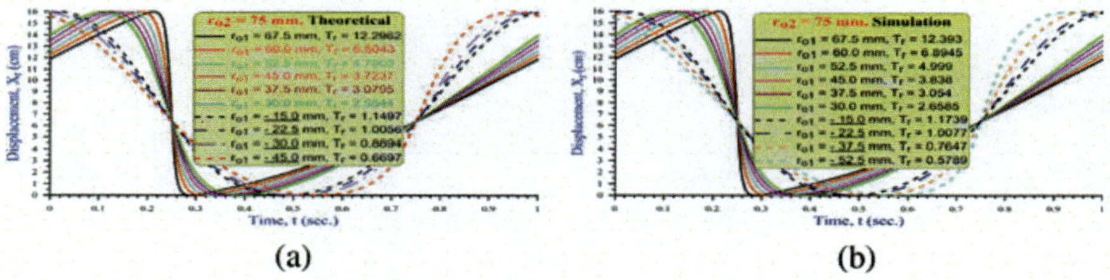

Fig. 11 Analytical X_r versus time ($r_{o2} = 75$ mm), **a** theoretical **b** simulation

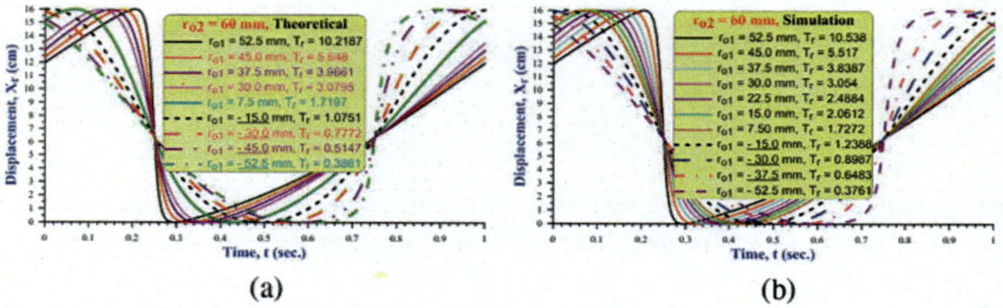

Fig. 12 Analytical X_r versus time ($r_{o2} = 60$ mm), **a** theoretical **b** simulation

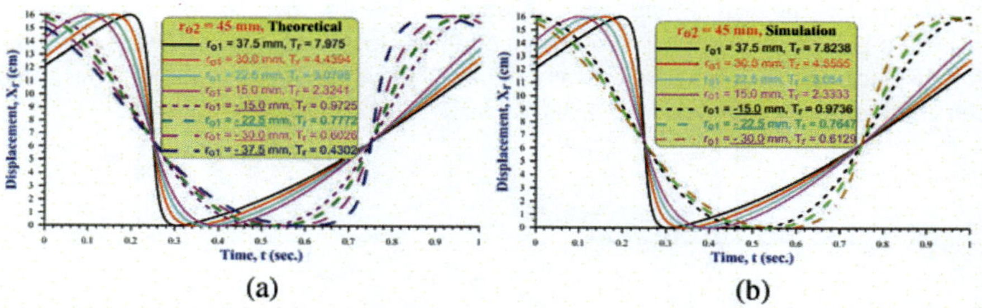

Fig. 13 Analytical X_r versus time ($r_{o2} = 45$ mm), **a** theoretical **b** simulation

good compatibility in T_r's results. The minimum and maximum absolute percentage errors between the built programme and simulation calculations of T_r ranged from 0.12% to 3.69% in the positive ro_1 condition and from 0.0297 to 15.63% in case of negative r_{o1}.

Figures 14, 15 and 16 demonstrate the relation between T_r and r_{o1} when r_{o2} was constant. T_r increased with increasing r_{o1}. The relation of T_r with r_{o2} at constant r_{o1} is shown in Tr versus r $_{o1}$ (r $_{o2}$ = 45 mm). Figure 17 a slightly decrease in T_r value occurred with increasing r_{o2} when the configuration of MWQRM was positive r_{o1}. In the negative r_{o1} configuration T_r was almost constant.

The enhancement in the cutting time ratio T_r for the positive r_{o1} MWQRM was larger than eight times from the traditional WQRM. In the negative r_{o1} MWQRM

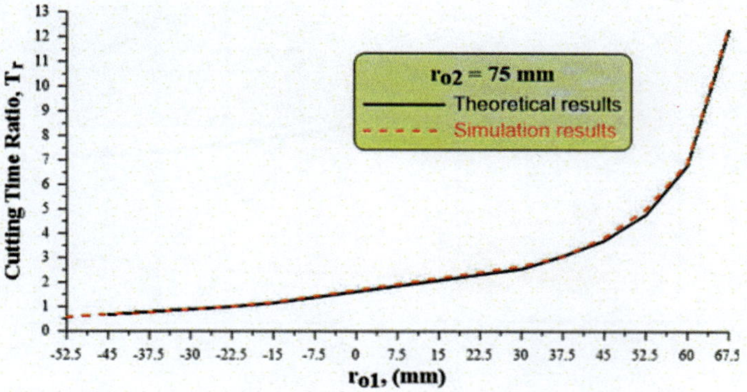

Fig. 14 T_r versus r_{o1} ($r_{o2} = 75$ mm)

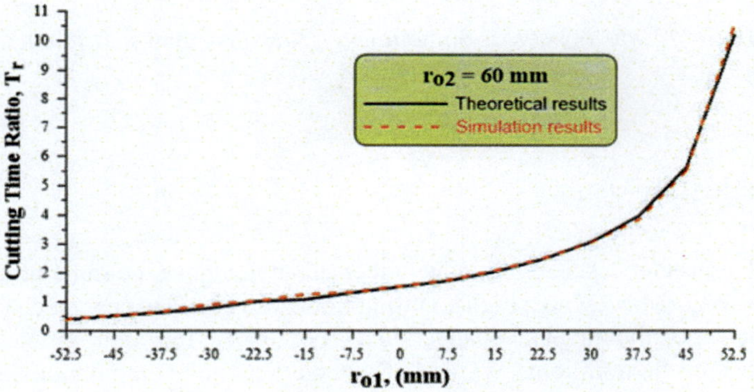

Fig. 15 T_r versus r_{o1} ($r_{o2} = 60$ mm)

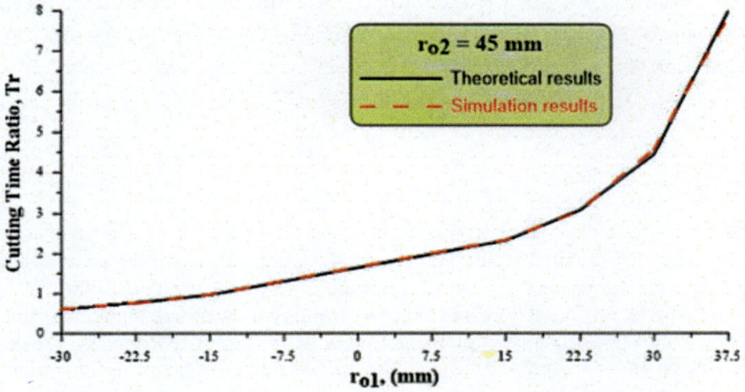

Fig. 16 T_r versus r_{o1} ($r_{o2} = 45$ mm)

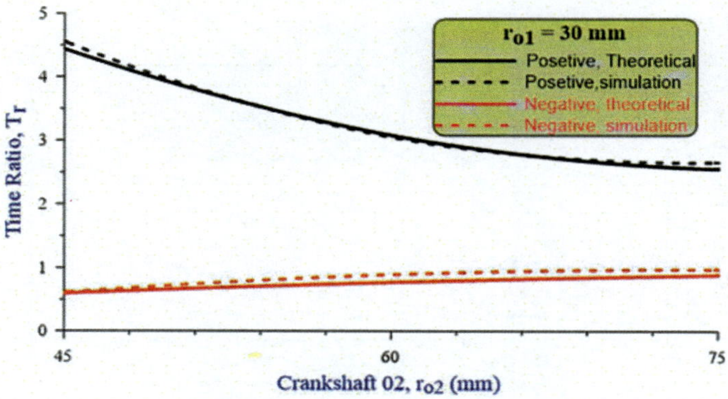

Fig. 17 T_r versus r_{o2} (r_{o1} is constant)

no increase in T_r was observed. Sometimes, it was less than that for the common WQRM.

6 Conclusion

This work presented a successful novel mechanism that gave the same function of a traditional WQRM but with a higher cutting time ratio (T_r).The new mechanism was realised by supplementing and connecting a set of links to the crank-ground revolute join (CGRJ) of the traditional WQRM. The analytical and computational solutions were applied for two configurations of MWQRM. These configurations were based on the location of the supplementary links relative to the connection join (CGRJ). The effective configuration for increasing T_r was the one when the extra links were positioned in a place under the CGRJ of WQRM. The other configuration that was over the CGRJ was inefficient for this purpose.

References

1. Saravanan R, Janakiraman V (2007) Study on reduction of machining time in CNC turning centre by genetic algorithm. Int Conf Comput Intell Multimed Appl 481–486
2. Paper R, Montiel-ross O, Melin P (2012) Methodology to optimize manufacturing time for a CNC using a high-performance implementation of ACO. Int J Adv Robot Syst Artic 9(4):1–10
3. Tapoglou N, Mehnen J, Butans J, Morar NI (2016) Online on-board optimization of cutting parameter for energy efficient CNC milling. Procedia CIRP 40:384–389
4. Park H, Qi B, Dang D, Yu D (2018) development of smart machining system for optimizing feedrates to minimize machining time. J Comput Des Eng 5(3):299–304
5. Hallmann M, Goetz S, Schleich B, Wartzack S (2019) Optimization of build time and support material quantity for the additive manufacturing of non-assembly mechanisms. Procedia CIRP 84:271–276

6. Rao LVB, Lakshminarayana K (1984) Optimal designs of RSSR crank-rocker mechanism-i. General time ratio. Mech Mach Theory 19(4/5):431–441
7. Suareo FO, Gupta KC (1988) Design of quick-returning R-S-S-R mechanisms. J Mech Transm Autom Des 110(4):423–428
8. Fung R-F, Chen K-W (1997) Constant speed control of the quick return mechanism driven by a DC motor. JSME Int J, Ser C 40(3):454–461
9. Lin F, Member S, Wai R (2001) A hybrid computed torque controller using fuzzy neural network for motor-quick-return servo mechanism. IEEE/ASME Trans Mechatron 6(1):75–89
10. Tomic M, Miloševic M, Tomic N, Pavlovic ND, Pavlovic V (2017) Remote control of the mechatronic redesigned slider-crank mechanism in service. Facta Univ Ser Mech Eng 15(2):257–268
11. Podhorodeski RP, Nokleby SB, Wittchen JD (2004) Quick-return mechanism design and analysis projects. Int J Mech Eng Educ 32(2):100–114
12. Robert LN (2008) Kinematics and dynamics of machinery, 1st ed, McGraw-Hill, USA
13. Hsieh WH, Tsai CH (2009) A study on a novel quick return mechanism. Trans Can Soc Mech Eng 33(3):139–152
14. Dicker JJ, Pennock GR, Shigley JE (2015) Theory of machines and mechanisms, 4th edn, Oxford University Press, Inc, USA
15. Vaz A, Thommen GK (2010) Modeling and simulation of the dynamics of the quick return mechanism : a bond graph approach. In: 10th national conference on industrial problems on machines and mechnical (IPRoMM 2010), vol 34, pp 23–30
16. Monková K, Monka P, Hloch S, Valíček J (2011) kinematic analysis of qquick-return mechanism in three various approaches. Teh Vjesn 18(2):295–299
17. Dong JG (2014) Optimization design on the performance of the shaper. Appl Mech Mater 472:3–7
18. Dwivedit SN (1984) Application of whitworth quick retrun mechanism for high velocity impacting press. Mech Mach Theory 19(1):51–59
19. Praneeth JSAI, Sreedharala AK, Darisi NK, Elangovan M (2014) Design of whitworth quick return mechanism using non-circular gears. Int J Mech Prod Eng 2(6):59–64
20. Swadi NN (2016) Kinematics characteristics of the internal and external geneva mechanism under the effect of multi design parameters

Performance Prognostics of Gas Turbines Using Nonlinear Filter

Shazaib Ahsan, Tamiru Alemu Lemma, Muhammad Baqir Hashmi, and Mebrahitom Asmelash Gebremariam

Abstract The Original equipment manufacturers (OEM) advice to follow a fixed cycle for an overhaul which is required to recover the deteriorations. However, there is a need to investigate prognosis-based maintenance to find optimum time for maintenance activities. This paper proposes a simultaneous state and parameter estimation technique to evaluate the performance and prognostics of gas turbine using particle filter approach based on Bayesian inference framework and isentropic efficiency. Particle Filter is reported to better foresee health status of non-linear systems guaranteeing design efficiency, improved safety and reduced maintenance cost. The component level deterioration due to fouling is investigated for compressor and turbine section. The simulation data used for prognosis is generated from industry-standard Gas Turbine Simulation Program software developed by the Netherlands Aerospace Centre. The prognosis of gas turbine has been performed by using multi-sensory data in form of isentropic efficiency. The results of prognosis provided information about the accuracy and precision of the proposed approach with 95% and 90% confidence interval. The results obtained reveal a promising prospect of the proposed method for predicting remaining useful life of deteriorating gas turbines and can support in maintenance decision making for industries using gas turbines. A parametric study is also performed to see the effect of noise on prognosis accuracy. It is concluded that with increased noise in data, the proposed approach predicts pessimistic RUL.

S. Ahsan · T. Alemu Lemma (✉) · M. Baqir Hashmi
Department of Mechanical Engineering, Universiti Teknologi PETRONAS, 32610 Seri Iskandar, Perak Darul Ridzuan, Malaysia
e-mail: tamiru.lemma@utp.edu.my

S. Ahsan
Department of Mechanical Engineering, University of Manitoba, Winnipeg, Manitoba R3T 5V6, Canada

M. Asmelash Gebremariam
Faculty of Mechanical and Manufacturing Engineering, University Malaysia Pahang, Gambang, Malaysia

Department of Energy and Petroleum Engineering, University of Stavanger, 4036 Stavanger, Norway

F. Ahmad et al. (eds.), *ICPER 2020*, Lecture Notes in Mechanical Engineering,
https://doi.org/10.1007/978-981-19-1939-8_38

Keywords Gas turbine · Transient operation · Amplitude modulated pseudo random signal · prognosis · Non-linear filter · Particle filter · Fouling

1 Introduction

The ever-growing demand for eco-friendly, sustainable and efficient power generation equipment has sparked up various challenges that need to be met by the prime moves for thermal power plants i.e. gas turbines. With increasing energy demand, these facilities are being expanded and becoming more advance and complex resulting in higher risk [1].

Although today's gas turbines are much efficient than before with efficiencies ranges approximately around 60% [2], the advantage is eclipsed by high deterioration rates due to harsh working environments. Even with a good filtration system and normal operating conditions with clean fuel, the gas path components gets fouled, damaged, corroded or eroded [3]. A deteriorated gas turbine generates more greenhouse gases that result in ecological damage and maybe led to a high penalty on the operators due to stringent environmental protocols. For safer, eco-friendly and economically feasible gas turbine operations maintenance schedule is an increasing area of research. The gas turbine these days are equipped with more advance monitoring and diagnostics tools [4–6]. The operational parameters such as gas turbine efficiency, exhaust gas temperature, specific fuel consumption or pressure drop across components are monitored continuously to detect, identify and rectify malfunctions. An illustration of a typical gas turbine is shown in Fig. 1.

There are many causes of malfunctions in context of gas turbine which can be classified into soft and hard degradation like fouling, erosion, foreign object damage, shaft misalignment, VIGV issues etc. Once the malfunction is detected there are two different tasks that need to be done i.e., diagnosis and prognosis. The two approaches (diagnosis and prognosis) can be visualized from Fig. 2 which illustrates the difference between diagnostics and prognostics with respect to failure. Prognostics is often

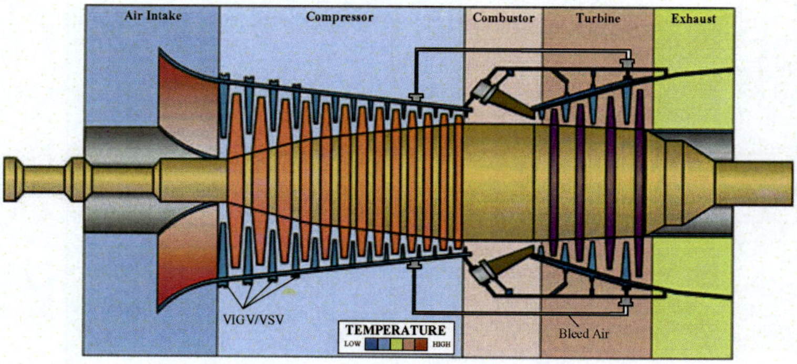

Fig. 1 Single-shaft gas turbine model

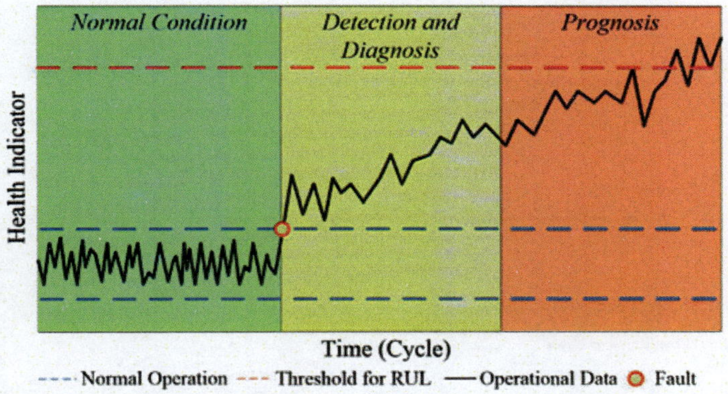

Fig. 2 Illustration of diagnosis and prognosis concept

misunderstood by diagnostics and detection which are in reality separate areas of research. Diagnostics encompasses both fault detection and identification and can be helpful in fully discerning the culprit for a particular fault and time of occurrence. Volponi [7] defines prognosis as "the ability to predict the future condition of a component and/or system of components". ISO-13381–1 defines prognosis as; "estimation of time to failure and risk for one or more incipient failure modes" [8]. The prognostics system can predict any undesirable incident earlier to its occurrence and aims to provide essential information for risk-control and management [9]. Prognostics can also enhance cost savings and operational safety.

Gas turbine engines are subjected to harsh and severe operating environments with elevated temperatures, stress and external obstructions like FOD. Hence, proper maintenance practices are required that do good condition monitoring based on available sensory data. The field of prognostics is an area of interest of many researchers in different fields which include medical sciences [10–12], engineering notably in batteries [13–15], crack propagation [13, 16, 17], bearings [18–20], helicopter gearbox [21]. In the context of gas turbines, many techniques are available for the prognosis or prediction of RUL. These includes support vector machines (SVM) [2, 9, 22, 23], Bayesian forecasting [6, 24, 25], Kalman Filters [6, 26], Particle Filter [27–30], state-space model [25], neural networks [9, 29, 31–34], statistical and computational technique of independent component analysis [35], regression methods [6, 36–38], Dempster Shafer regression [9] and exponential smoothing [6]. These methods can be categorized into groups: (i) physics-based approaches and (ii) data driven models [38–42] and (iii) hybrid models [6, 9, 23, 29]. Classification of different methods available for prognosis is shown in Fig. 3. Most of the researchers used either power output [43], temperature [44], pressure ratio [28, 29], compressor mass flow [43, 45], Vibration data [46] or efficiency [28, 29, 43, 45] as health indicator.

The gas turbines use bleed valves and VIGV to avoid surge and stall conditions hence they are considered suitable for part load applications. The problem with

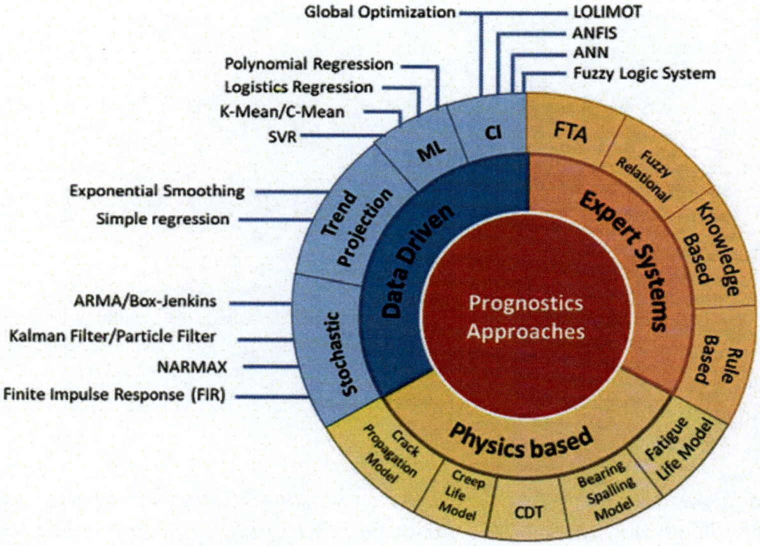

Fig. 3 Classification of Prognostics approaches

part-load operation, however, is that any performance deterioration or upset due to intrinsic/internal faults will be hidden to the analyst since the system finds its own way to compensate for internal change. Therefore, even if the malfunction is detected at an early stage, predicting any performance degradation or loss is much more difficult than the case for full load operation. The effect of performance deterioration in the compressor of the gas turbine is demonstrated in Fig. 4 with the blue continuous lines showing normal operation while red discontinued line represents deteriorated operations.

According to the actual scenarios where Gas Turbines are to be operated for supporting gird purpose, transient and dynamic operating conditions with VIGV operations are becoming an area of interest of the researchers as in the last couple of years [9, 23, 41, 47]. Furthermore, major work on prognostics of the gas turbine

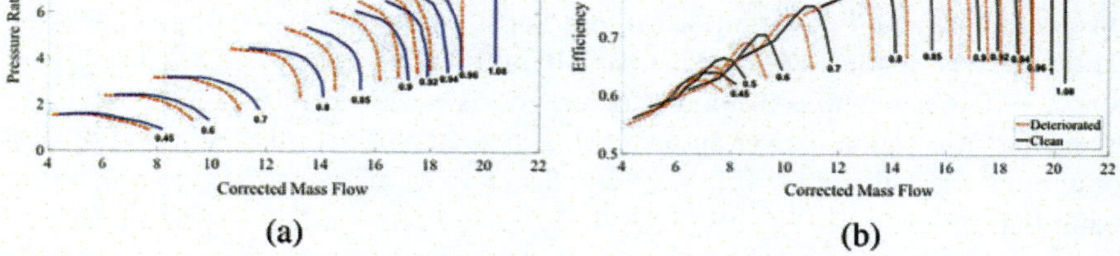

Fig. 4 Effect of performance deterioration on **a** compressor map and **b** efficiency of the compressor

is done on aero engines [9, 23, 36, 45] and still, industrial Gas Turbine for power generation purpose seems nearly untouched. To our knowledge, there is no reputed research specific to variable geometry gas turbine prognosis. This paper presents the non-linear approach of Particle Filter with simultaneous state and parameter estimation for prognosis and remaining useful life prediction of gas turbine operating at transient conditions with VIGV operations. The study has been performed on a single-shaft gas turbine engine and a two-shaft gas turbine engine. The effect of measurement noise on prognosis and remaining useful life prediction is also studied.

The organization of this paper is in the following manner: in Sect. 2, the proposed approach for prognosis i.e., Particle Filter approach is explained followed by a brief overview of the cases to be addressed using the proposed approach in Sect. 3. Later, discussion on results is presented in same section along with parametric study. At last, the conclusion drawn from the study are presented and future work is recommended.

2 Methodology

For complex systems, it is very difficult or nearly impossible to build a physics-based model. For such situations, the operational data containing input and output information is used to predict the system's future behaviour. In the present study, the method employed is the Particle Filter.

2.1 Nonlinear Filter

The real-life applications often face nonlinearities during operations and then linear models are replaced by nonlinear ones [48]. The main feature of these nonlinear filters is to approximate the nonlinear function is the state dynamic and measurement models [49]. Some of the available nonlinear filters include Unscented Kalman Filter (UKF), Extended Kalman Filter (EKF) and Particle Filter (PF). In the current study investigated the nonlinear approach of PF for Prognosis of gas turbine engines.

PF approach is a statistical technique that depends on the Bayesian approach for developing the posterior probability function based on available data. Particle Filter assumes N independent and identically distributed (i.i.d) particles according to $p(x_{0:t}|y_{1:t})$ [50], often termed as samples, the particles are simulated representing states in which a system can be at a time t.

The PF approach is also recognized as Sequential Monte Carlo technique [51, 52] which combines Bayesian learning methods with importance sampling. It is a recursive Bayesian state-estimator that employs discrete particles/samples for approximating the estimated estate's posterior distribution. PF comprises of two stages namely: prediction stage and updating stage. In the prediction stage, the system model is used to predict the state probability density function (PDF) forward from one measurement time to the next. The state PDF spreads, and the prediction generally deviates as the data is highly likely to contains measurement noise. In the updating stage, the true measurement is incorporated to amend the prediction PDF by multiplying it with the likelihood/weight. The particles in the state-space are periodically resampled to match the posterior distribution of the estimated state. As a result, the posterior is nearer in similarity to the likelihood than to the prior [53]. The general process of PF is based on state-transition function $\boldsymbol{f}(.) : \mathrm{R}^n\mathrm{x}\mathrm{R}^n \rightarrow \mathrm{R}^n$ (damage model, possibly non-linear) and the measurement-function $\boldsymbol{h}(.) : \mathrm{R}^n\mathrm{x}\mathrm{R}^n \rightarrow \mathrm{R}^n$ as given by Eqs. (1) and (2).

$$x_k = f(x_{k-1}, u_k, \theta) + w_{k-1}; \mathrm{k} = 1,2,...\mathrm{n} \tag{1}$$

$$z_k = h(x_\mathrm{k}, u_k, \theta) + v; \mathrm{k} = 1,2,...\mathrm{n} \tag{2}$$

Here, k is time step value; $\boldsymbol{w}_{k-1}$ and $\boldsymbol{v_k}$ are process and measurement white noise, respectively. They are generally assumed to be following Gaussian distribution i.e. $\boldsymbol{w}_{k-1} \sim \mathcal{N}(\boldsymbol{w}_{k-1}, 0, \boldsymbol{\Psi})$ and $\boldsymbol{v}_k \sim \mathcal{N}(\boldsymbol{v}_k, 0, \boldsymbol{\Phi})$, where $\boldsymbol{\Psi}\epsilon\mathrm{R}^{nxn}$ and $\boldsymbol{\Phi}\epsilon\mathrm{R}^{nxn}$ are covariance matrices; $\boldsymbol{x}$ is the vector of damage state; $\boldsymbol{u}_k \in \mathrm{R}^{n_u}$ is the vector or received measurements; $\boldsymbol{\theta} \in \mathrm{R}^{n_\theta}$ is a vector of model constants and $\boldsymbol{z} \in \mathrm{R}^{n_z}$ is measurement data.

For estimation of state and parameter simultaneously, a parameter vector is considered as the augmentation of state-space given by Eq. (3). $\boldsymbol{\theta}_k$ varies by a random walk given by Eq. (4). The particle filter state and measurement equation can be then rewritten for augmented state $\boldsymbol{X}_k$ as in Eq. (5).

$$\boldsymbol{X}_k = \begin{bmatrix} \boldsymbol{x}_k \\ \boldsymbol{\theta}_k \end{bmatrix} \tag{3}$$

$$\boldsymbol{\theta}_k = \boldsymbol{\theta}_{k-1} + \boldsymbol{\vartheta}_{k-1}; \text{where, } \boldsymbol{\vartheta}_{k-1} \sim \mathcal{N}(0, \boldsymbol{W_{k-1}}) \tag{4}$$

$$\begin{aligned} X_k &= f(X_{\mathrm{k}-1}, u_k, \theta) + w_{k-1}; \mathrm{k} = 1,2,...,\mathrm{n} \\ Z_k &= h(X_\mathrm{k}, u_k, \theta) + v_k; \mathrm{k} = 1,2,...,\mathrm{n} \end{aligned} \tag{5}$$

Based on the above equations, the transition pdf $p(\boldsymbol{x}_k|\boldsymbol{x}_{k-1})$ and the state pdf $p(\boldsymbol{x}_{k-1}|z_{1:k-1})$ at t_{k-1} is obtained, respectively. Using Chapman-Kolmogorov equation, Eq. (6) the prior state pdf at t_k is predicted.

$$p(\boldsymbol{x}_k|z_{1:k-1}) = \int p(\boldsymbol{x}_k|\boldsymbol{x}_{k-1}).p(\boldsymbol{x}_{k-1}|z_{1:k-1}).d\boldsymbol{x}_{k-1} \tag{6}$$

Upon the availability of new data, the prior state PDF is updated based on Bayesian rule for calculating the posterior state PDF as in Eq. (7).

$$p(\boldsymbol{x}_k|z_{1:k}) = \frac{p(z_k|\boldsymbol{x}_k).p(\boldsymbol{x}_k|z_{1:k-1})}{p(z_{k-1}|z_{1:k-1})} \tag{7}$$

The propagation of state PDF is a conceptual solution that cannot be obtained analytically due to the reason for evaluating high-dimensional complex integrals. For the reason, the sequential importance sampling algorithm is used for approximating the posterior state PDF numerically using Eq. (8) for a set of particles sampled from an importance PDF $q(\boldsymbol{x}_k|\boldsymbol{x}_{k-1}, z_{1:k})$ [54].

$$p(\boldsymbol{x}_k|z_{1:k}, \boldsymbol{\theta}) \approx \sum_{i=1}^{N_s} w_k^i.\delta(\boldsymbol{x}_k - \boldsymbol{x}_k^i) \tag{8}$$

$$\delta(\boldsymbol{x}_k - \boldsymbol{x}_k^i) = \begin{cases} 1; \boldsymbol{x}_k \neq \boldsymbol{x}_k^i \\ 0; \boldsymbol{x}_k = \boldsymbol{x}_k^i \end{cases} \tag{9}$$

where, $\delta(\boldsymbol{x}_k - \boldsymbol{x}_k^i)$ is the Dirac delta function concentrated at $\boldsymbol{x}_k^i$ [54, 55], the Dirac delta is given in Eq. (9); w_k^i is weights of $\boldsymbol{x}_k^i$, which is updated and normalized by Eq. (10).

$$w_k^i = w_{k-1}^i \frac{p(z_k|x_k^i)p(x_k^i|x_{k-1}^i)}{q(x_k^i|x_{k-1}^i, z_{1:k})}, w_k^i = w_k^i / \sum\nolimits_{i=1}^{N_s} w_k^i \tag{10}$$

Where, N_s is the number of particles, $p(z_k|\boldsymbol{x}_k^i)$ and $p(\boldsymbol{x}_k^i|\boldsymbol{x}_{k-1}^i)$ can be effectively defined by Eqs. (1) and (2) [54]. Simplifying above equation as Eq. (11).

$$w_k^i = w_{k-1}^i\, p(z_k|x_k^i), w_k^i = w_k^i / \sum\nolimits_{i=1}^{N_s} w_k^i \tag{11}$$

Following the procedure highlighted in [56], the simplified version to calculate weight or likelihood of the particles is given by Eq. (12).

$$\omega_k^i(z_k^i; z_k^m, \Sigma) = \frac{1}{\sqrt{(2\pi)^D.|\boldsymbol{\Sigma}|}}.\exp\left(-\frac{(z_k^m - z_k^i)^T.\boldsymbol{\Sigma}^{-1}.(z_k^m - z_k^i)}{2}\right) \tag{12}$$

where, z_k^m is the mean value that is considered to be measurement value; D is the number of states at time k; $\boldsymbol{\Sigma}$ is the covariance matrix. The estimated state variables are then calculated as in Eq. (13).

$$\hat{\boldsymbol{x}}_k = \sum_{i=1}^{N_s} \omega_k^i.\boldsymbol{x}_k^i \tag{13}$$

The problem associated with sequential importance sampling is the particle degeneracy. After some iterations, all except one of the particles have negligible weights. For overcoming this degeneracy issue, the resampling approach is used by mean of

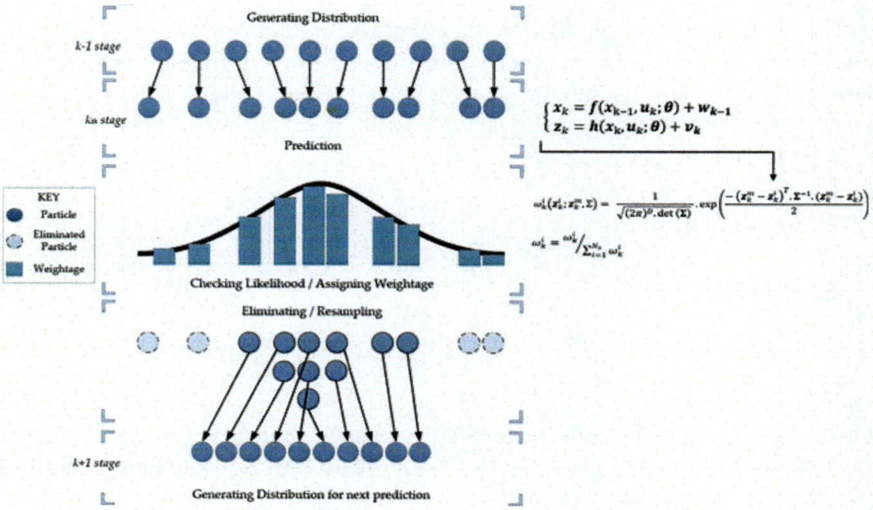

Fig. 5 Particle filtering approach [30]

which particles with low weightage at every stage are eliminated and new particles are created in the range of high weightage particles.

There are several of Particle Filter algorithms available in the literature that follow the theme illustrated in Fig. 5. The algorithms available are: (i) Generic Particle Filter, (ii) Sampling Importance Resampling, (iii) Auxiliary Sampling Importance Resampling, (iv) Regularised Particle Filter, details of which can be found in [53].

2.2 *Model Details*

2.2.1 Problem Definition

The first task in particle filter modelling and simulation is to define the combine states and parameters vector $\boldsymbol{X}$. The number of particles is also defined in these steps that directly relates to the computational power required for running particle filter. The threshold limit is also defined here along with state transition and measurement equation.

2.2.2 Prognosis

This step first initiates the particle filter approach by creating initial distribution around state and parameter vector $\boldsymbol{X}$. The state and parameter values are then predicted for every particle using state transition equation followed by measurement values prediction using measurement equation. The particle filter model equations for this study are given by Eq. (14) as the compressor fouling is proposed to be following exponential curve [57]. Upon availability of actual measurement value, likelihood is calculated for all predicted measurement values using Eq. (12). After the calculation of likelihood values, resampling is done to reduce the computational power required to process particle with low weights.

$$\begin{cases} x_k = \exp(b_k.\Delta t).x_{k-1};\ \text{state transition equation} \\ z_k = x_k;\ \text{Measurement equation} \end{cases} \tag{14}$$

2.2.3 Post-processing

After all available measurement values are processed (at time t_c), the final estimation is made using Eq. (13). Based on particle filter model equations (Eq. (14)), the trajectory is extended until the threshold limit. The point at which the threshold line is intersected by the trajectory is declared as predicted failure time, t_p. Remaining useful life (RUL) is then predicted by subtracting last available measurement time value t_c from predicted failure time t_p. The corresponding equation for RUL calculation is given by:

$$\text{RUL} = t_p - t_c \tag{15}$$

Particle filter approach is a probabilistic approach hence the PDF of RUL can be calculated in a similar manner for each particle for the certain confidence interval (CIN). The process of predicting RUL values with CIN can be illustrated as in Fig. 6. The confidence level α is selected based on the certainty level of the data. The more the confidence level is chosen, the more compact the prediction results are to the mean prediction.

3 Results and Discussion

3.1 GSP Data

For this study, two cases are considered that are namely Turbojet Engine (aero-derivative engine) and a two-shaft engine (Industrial gas turbine). Gas Turbine Simulation program (GSP) was used for simulating both gas turbines' configuration. GSP is an offline component-based gas turbine simulation software capable of steady and transient simulations developed by National Aerospace Laboratory (NLR), Netherlands. GSP allows the study related to gas turbine deterioration, control system

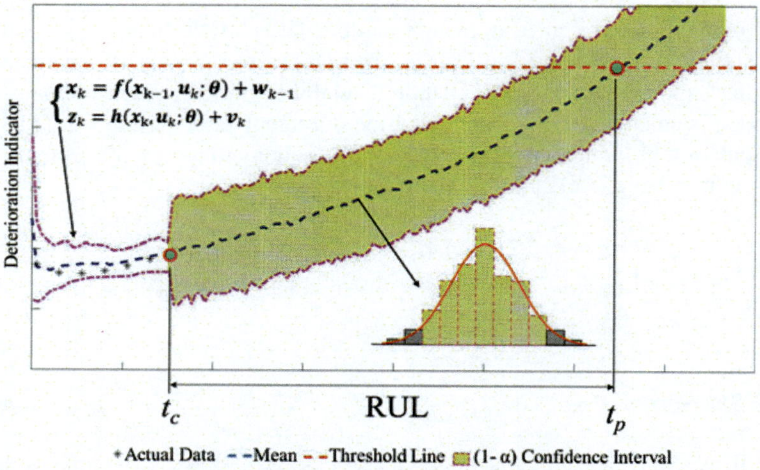

Fig. 6 Calculation of RUL with CIN

malfunctioning, ambient conditions and emissions. For this study, a transiently operating gas turbine is simulated that is induced with multi-component degradation. Both the cases are briefly described below followed by the method of generating data. The deteriorations induced are in compressor and turbine of the gas turbines more specifically fouling.

3.1.1 Case 1- Turbojet

Turbojet engines are mostly used for powering the jet engines due to their higher thrust per unit weight ratio [58]. The lower maintenance cost, high payload and longer range are the reason for its preference in the aero industry more specifically military [58]. Turbojet engine can be equipped with an afterburner that enhances the speed of the aircraft. The configuration of the Turbojet engine is given in Fig. 7a. The specifications used for turbojet modelling on GSP is given in Table 1. The respective compressor maps are also shown in Fig. 7b and c.

3.1.2 Case 2- Two-Shaft

The second case studied in this research was a turboshaft gas turbine engine. The application of the turboshaft gas turbine includes pumping stations and helicopters (e.g., GE 700-401C/-701C for Apache, Black Hawk and Sea Hawk helicopter). Turboshaft is a two-shaft configuration where the first shaft rotates the compressor while the other shaft delivers the power. The specifications used for

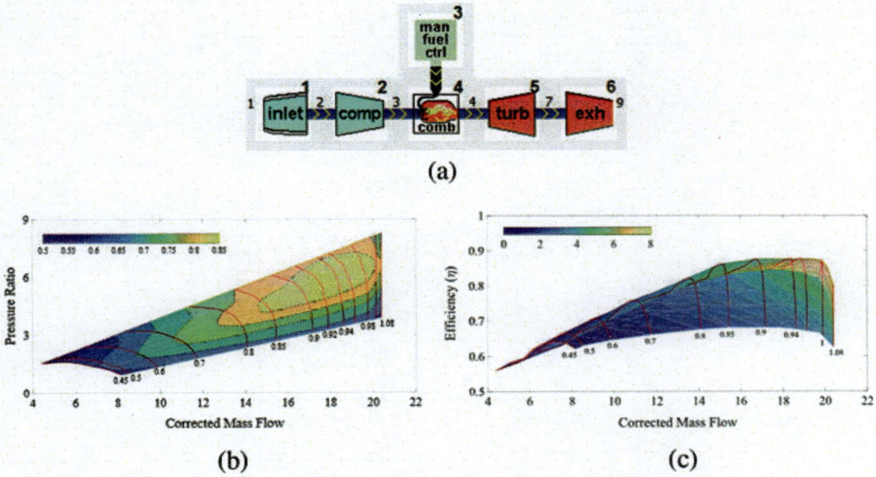

Fig. 7 Turbojet engine configuration on GSP **a** Schematic **b** PR versus Mass flow (Efficiency gradient) **c** Efficiency versus Mass Flow (PR gradient)

Table 1 Turbojet engine details

Component	Parameter	Unit	Value
Inlet	Mass flow	kg/s	19.9
Compressor	Rotor speed	RPM	16,540
	Isentropic efficiency	%	82.5
	Pressure ratio	–	6.92
Combustor	Fuel flow	kg/s	0.38
	Combustion efficiency	%	99.5
	Fuel	–	Jet A/A1, JP-8
	LHV	–	43,031
Turbine	Power	MW	5.2
	Isentropic efficiency	%	88
	Spool inertial moment	$kg.m^2$	0.7578
	Mechanical efficiency	%	99
	Thrust force	kN	14.18

Table 2 Two-Shaft Industrial gas turbine details

Component	Parameter	Unit	Value
Inlet	Mass flow	kg/s	4.6122
	Pressure ratio	–	0.988
Compressor	Rotor speed	RPM	44,700
	Isentropic efficiency	%	82.1
	Pressure ratio	–	17.5
Combustor	Fuel flow	kg/s	0.1043
	Combustion efficiency	%	98.5
	Fuel	–	Jet A/A1, JP-8
	LHV	–	43,031
Turbine	Isentropic Efficiency	%	85
	Spool Inertial Moment	$kg.m^2$	0.0603
	Mechanical Efficiency	%	99
Power Turbine	Power	MW	1.34
	Isentropic Efficiency	%	85
	Spool Inertial Moment	$kg.m^2$	8.08
	Mechanical Efficiency	%	99
	Rotor Speed	rpm	20,900

turbojet modelling on GSP is given in Table 2. Figure 8 shows the schematic and compressor maps for Two shaft engine configurations.

3.1.3 Data Generation

For the current study, GSP is used for prognosis of gas turbine working in a transient operating condition. In order to simulate transient operation, the mass flow of was modulated with time using Amplitude Modulated Pseudo-Random Binary Sequences (APRBS) proposed by Deflorian and Zaglauer [59]. Mujtaba et al. [60] have recently used APRBS of mass flow rate as the design of the experiment for model identification in pipeline leak detection study.

The selection of the excitation signal indicates the quality of the model where it is necessary to cover all significant areas of input spaces and frequencies [59]. This aspect is satisfied using the Amplitude modulated Pseudo-Random Binary Sequence (APRBS) [59]. APRBS can be defined as a sequence of step function fixed by N

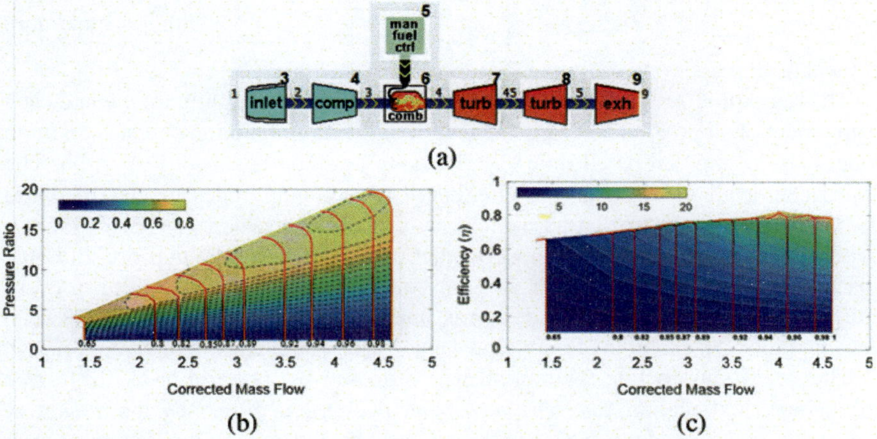

Fig. 8 Two Shaft engine configuration on GSP **a** Schematic **b** PR versus Mass flow (Efficiency gradient) **c** Efficiency versus Mass Flow (PR gradient)

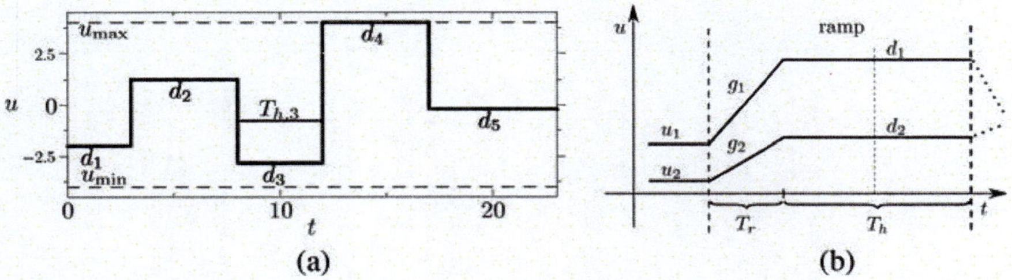

Fig. 9 **a** APRBS signal in time domain [59], **b** illustration of ramp function [59]

design point in range $d_s \in [u_{min}, u_{max}]$ and dwell time $T_{h,s}$ as shown in Fig. 9a. However, in industries, the actuators are not arbitrary fast enough to jump between signal values and rather change gradually. This is achieved by using slopes defined by dwell time $T_{h,s}$ and two consecutive design points:

$$U = \sum_{s=1}^{N} U_s\big(d_s, d_{s-1}, T_{h,s}\big) \tag{16}$$

The singular ramp function U_s consist of time discrete input sequences defined by sequence time T_s and sampling time h [61]:

$$U_s\big(d_s, d_{s-1}, T_{h,s}\big) = \{u_{k+i}\}_{i=1}^{T_s/h} \tag{17}$$

where, $k = \frac{1}{h}\sum_{j=1}^{s-1} T_j$ and $u_{k+i} = \begin{cases} d_s & , \frac{T_{r,s}}{h} \le i \le \frac{T_s}{h} \\ g.i + d_{s-1} & , 1 \le i \le \frac{T_{r,s}}{h} \end{cases}$; with $g \le$ maximum allowable slope.

The ramp time $T_{r,s}$ and dwell time $T_{h,s}$ are multiple of sampling time h and each component of slope g_j. Therefore, the length of the ramp function depends on ramp time, that depends on dwell time T_s. The ramp function is shown in Fig. 9b. The approaches available to achieve a better distribution of amplitudes are classified into the model-free and model-based method of design of experiment (DOE) [59]. Latin hypercube (LHC) is one of the most popular distributions for the purpose. In LHC, space is divided into N intervals and only one design point d_s is placed in every column and row. A good LHC design is one having maximum distance between design points. An example of mass flow value based on APRBS is shown in Fig. 10.

The deteriorations rate for fouling was proposed by Tarabrin et al. [57] to be following an exponential rate. The percentage decrease in efficiency and mass flow due to compressor and turbine deterioration were provided by [62, 63], Table 3. The introduced deterioration is in range taken from the source just mentioned and the rate was chosen to be exponentially decreasing within 2000 h of operation. The rate of deterioration for mass flow and isentropic efficiency is illustrated in Fig. 11.

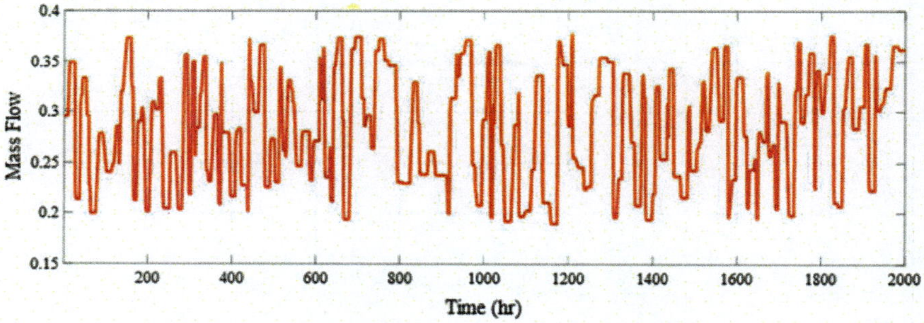

Fig. 10 Mass flow for transiently operating gas turbine

Table 3 Relationship between the faults and health parameters [62, 63]

Fault type	Mass flow change (A)	Isentropic efficiency change (B)	Ratio A: B
Compressor fouling	[0, –7.5%]	[0, –2.5%]	3:1
Compressor erosion	[0, –4%]	[0, –2%]	2:1
Turbine fouling	[0, –4%]	[0, –2%]	2:1
Turbine erosion	[0, + 4%]	[0, –2%]	2:1

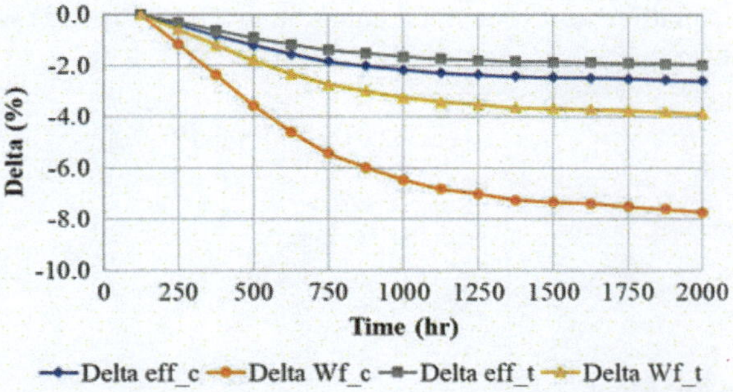

Fig. 11 Deterioration rate introduced in GSP corresponding to compressor and turbine fouling

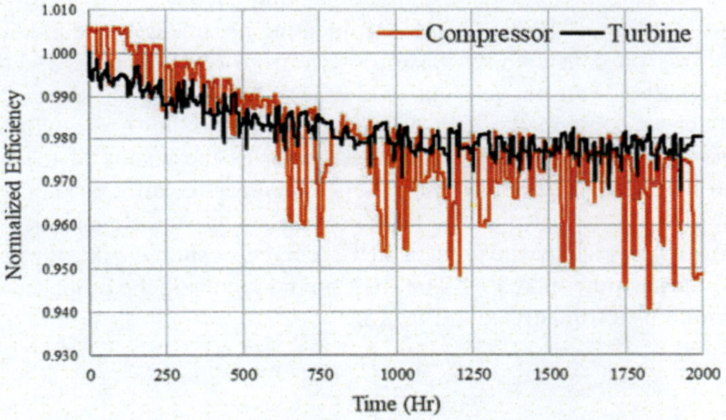

Fig. 12 The data for Prognosis with 10% of design point efficiency noise

The generated data for isentropic efficiencies after introducing multiple component faults is shown in Fig. 12. For the study, the isentropic efficiency for the compressor and turbine is normalize using Eq. (16).

$$\eta_{normalized} = \frac{\eta_{isentropic}}{\eta_{design}} \tag{16}$$

3.2 *Particle Filter Results*

For prognosis, an exponential degradation was assumed as proposed by Tarabrin et al. [57]. The threshold value for prognosis was set to 0.9 as the system is already deteriorated at this value to a great extent. The particle filter technique was used to predict the RUL of the gas turbine and study the effect of noise on prediction. For the state transition equation, an exponential equation given by Eq. (17) was considered. The measurement equation is given by Eq. (18).

$$x_k = x_{k-1}.exp(-b.t) + w_k \tag{17}$$

$$z_k = x_k + v_k \tag{18}$$

3.2.1 Case 1- Turbojet

From the data set of 2000 h of transiently operating gas turbine, 1000 h data was considered for state and parameter estimation simultaneously. A span of 100 h was taken that comprises of 10 data points for updating parameters. The prediction was first performed on data without noise directly from GSP. Figure 13 shows RUL prediction results for the above-mentioned case.

The effect of noise in the data was analysed by adding noise with a certain percentage of design efficiency. Table 4 summarizes the results of the prediction obtained for a different level of noise. The RUL values for different noise level are illustrated in Fig. 14. Analysing the RUL predictions, it can be concluded that with the addition of noise in data the predicted RUL decreases showing that the probability of failure in near future increases. The RUL prediction using the proposed approach for data with noise is illustrated in Fig. 15.

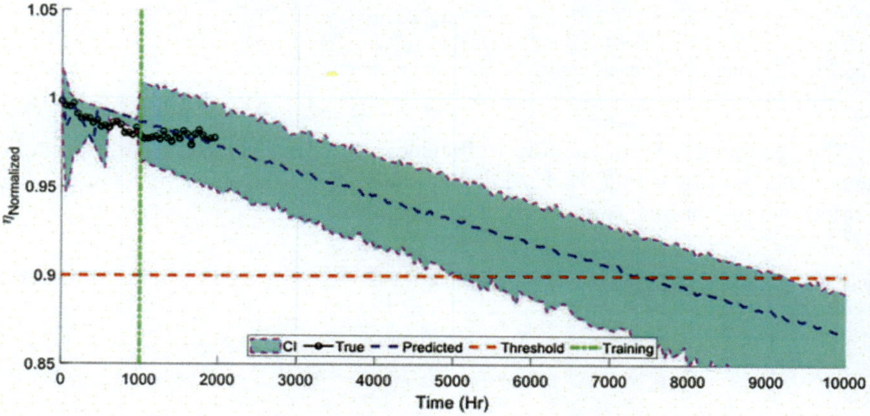

Fig. 13 Prediction results for Particle filter technique (0% design efficiency noise)

Table 4 Effect of noise on prediction using the Particle filter

Percentage noise	Failure prediction (Hr)			RUL Prediction at 1000 h (Hr)		
	Upper bound	Mean	Lower bound	Upper bound	Mean	Lower bound
0.0%	5044	7409	9250	4044	6409	8250
2.5%	3364	5917	7652	2364	4917	6652
5.0%	2362	4533	6724	1362	3533	5724
7.5%	2033	4042	5622	1033	3042	4622
10.0%	2241	3101	5900	1241	2101	4900

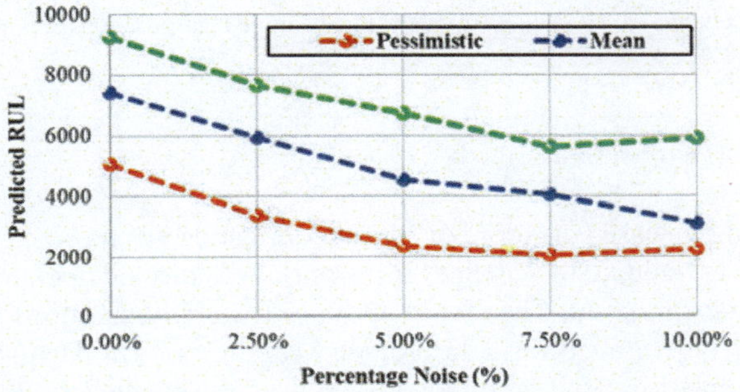

Fig. 14 Effect of noise on Prediction using the particle filter

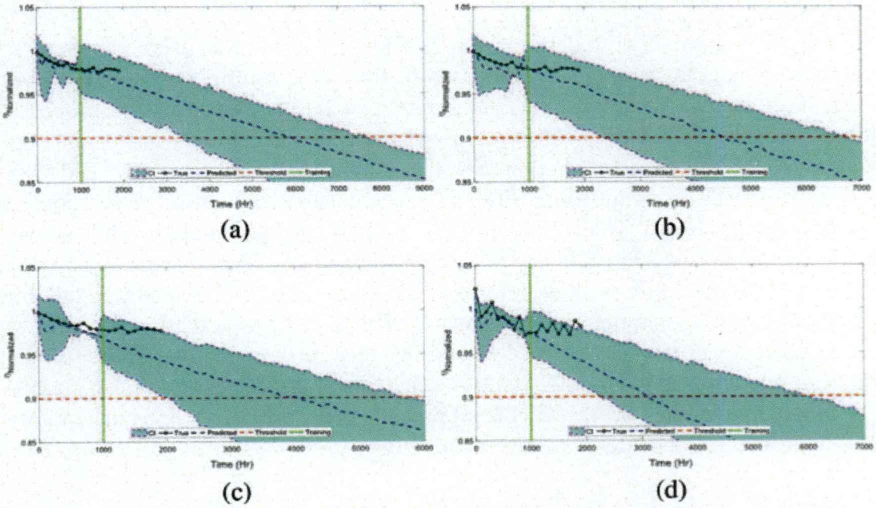

Fig. 15 Effect of noise on prediction using particle filter **a** 2.5% noise, **b** 5.0% noise, **c** 7.5% noise and **d** 10.0% noise

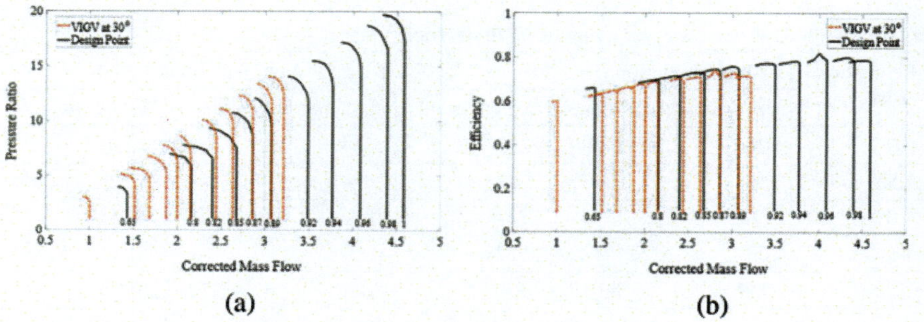

Fig. 16 Effect of change in VIGV angle for Two shaft engine configuration **a** PR versus Mass flow **b** Efficiency versus Mass Flow

3.2.2 Case 2- Two-Shaft

The Turbojet engine case was more of an aero-derivative gas turbine type. The two-shaft case was also studied to account for industrial gas turbine operating transiently away from the design point. The fact is during off-design operations the behaviour of the system changes due to variable geometry operations. The Two-shaft industrial gas turbine operated in off-design conditions was investigated in this research as they are widely used for power generation purpose in the field. The effect of variable geometry operation on compressor maps is shown in Fig. 16 at 30-degree change in VIGV angle.

For generating data with deterioration, a similar method was used employed as for Turbojet engine. Similar deteriorations of fouling at multiple component level were introduced in the two-shaft configuration of the gas turbine. The proposed prognosis approach of Particle filter was used to first estimate the state and parameters simultaneously. After the estimation of state and parameters as in Eqs. (17) and (18), the trajectory was further projected. The RUL was calculated for the data as proposed in methodology, Fig. 6. The RUL prediction using the proposed approach is shown in Fig. 17a and b for 95% CIN and 90% CIN respectively. The concept of Severity level was introduced in prediction to alert the operator and maintenance staff for the level of urgency required in maintenance. The 95% CIN prediction account for more particles to predict RUL by considering most of them as true. However, for transient operation, this big CIN value is debatable and may result in false alarms. The prediction results for 90% CIN level predicts more precise RUL as uncertainty in measurement is considered more realistically.

4 Conclusions and Recommendations

In this research, two different configurations of gas turbines were studied that are having fouling as deterioration. The aim of this work was to predict the performance

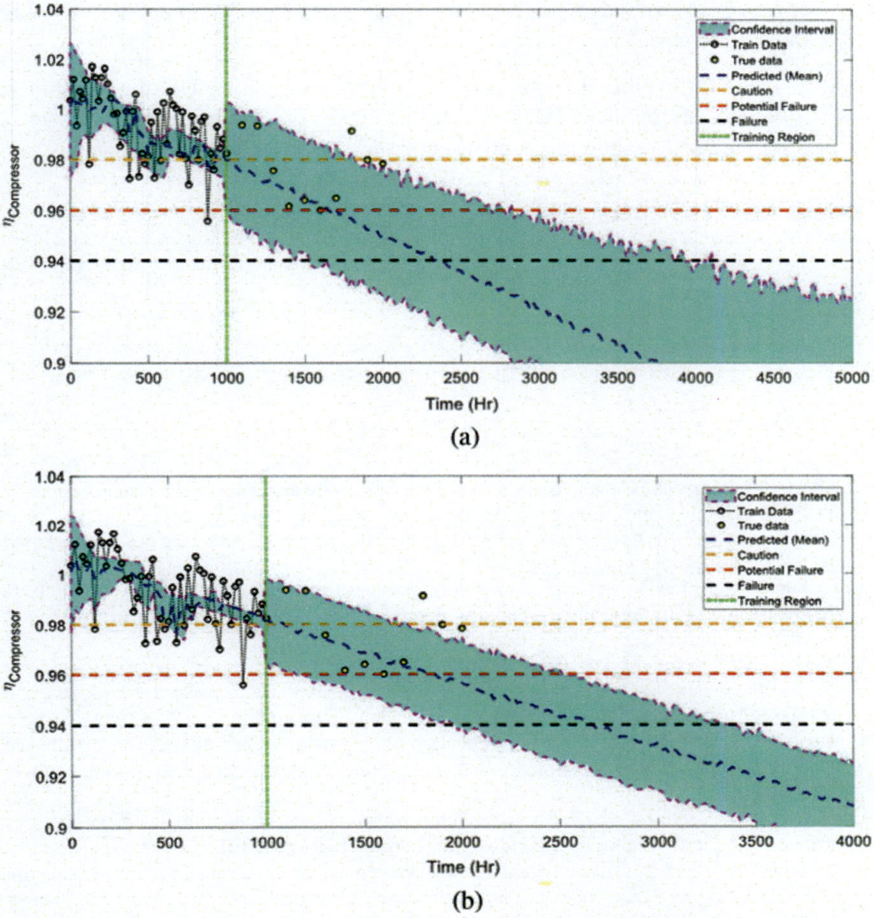

Fig. 17 Prediction results for Two-shaft engine **a** with 95% CIN **b** with 90% CIN

behaviour of transiently operated gas turbines, a phenomenon most common for power generating role gas turbines. A stochastic approach of Particle Filter employed to estimate state and parameter of deteriorating gas turbine simultaneously. The effect of different level of noise and CIN levels were also investigated. The conclusion drawn from the study are as follows:

- Particle Filter shows promising results in term of prediction RUL for gas turbines operating transiently.
- Measurement noise was found to affect RUL by predicting pessimistic RUL.
- With more CIN level chosen for study, the RUL prediction range increase. It is recommended to use low but justified value for CIN for transient operations.

- A concept of severity level has been introduced that help in early decision making for the maintenance crew.

Although Particle Filter is found good enough for prediction RUL of gas turbines, it was tested on component level. For further research, optimization approaches for prognosis-based maintenance planning can be explored. Also, prognosis and reliability analysis can be integrated for better maintenance planning.

Acknowledgements The authors are thankful to Universiti Teknologi PETRONAS for the facilities provided to conduct this research. The study is partly funded by YUTP-FRG (cost centre: 015LC0-028).

References

1. Ahsan S, Lemma TA, Gebremariam MA (2020) Reliability analysis of gas turbine engine by means of bathtub-shaped failure rate distribution. Process Saf Prog 39:e12115
2. Zhou D, Zhang H, Weng S (2015) A new gas path fault diagnostic method of gas turbine based on support vector machine. J Eng Gas Turbines Power 137(10):102605
3. Diakunchak IS (1991) Performance deterioration in industrial gas turbines. In: ASME 1991 international gas turbine and aeroengine congress and exposition. American Society of Mechanical Engineers, pp V004T10A012–V004T10A012
4. Jaw LC (2005) Recent advancements in aircraft engine health management (EHM) technologies and recommendations for the next step. In: ASME turbo expo 2005: power for land, sea, and air, 2005. American Society of Mechanical Engineers, pp 683–695
5. Davison CR, Drummond C (2010) Application of cost matrices and cost curves to enhance diagnostic health management metrics for gas turbine engines. J Eng Gas Turbines Power 132(4):041604
6. Cavarzere A, Venturini M (2012) Application of forecasting methodologies to predict gas turbine behavior over time. J Eng Gas Turbines Power 134(1):012401
7. Volponi AJ (2013) Gas turbine engine health management: past, present and future trends. In: ASME turbo expo 2013: turbine technical conference and exposition, 2013. American Society of Mechanical Engineers, pp V008T46A001–V008T46A001
8. Monitoring IC (2004) Diagnostics of machines-prognostics part 1: general guidelines. ISO13381-1:2004 (e), vol ISO/IEC Directives Part 2, IO f. S, p. 14, 2004
9. Xu J, Wang Y, Xu L (2014) PHM-oriented integrated fusion prognostics for aircraft engines based on sensor data. IEEE Sens J 14(4):1124–1132
10. Llovet JM, Brú C, Bruix J (1999) Prognosis of hepatocellular carcinoma: the BCLC staging classification. In: Seminars in liver disease, vol 19, no. 03. © 1999 by Thieme Medical Publishers, Inc, pp 329–338
11. Hamilton BE, Kellogg FS (1928) Cardiac disease in pregnancy: prognosis; medical and obstetric handling. J Am Med Assoc 91(25):1942–1944
12. Beck JR, Pauker SG (1983) The Markov process in medical prognosis. Med Decis Making 3(4):419–458
13. An D, Choi J-H, Kim NH (2013) Prognostics 101: a tutorial for particle filter-based prognostics algorithm using Matlab. Reliab Eng Syst Saf 115:161–169
14. Miao Q, Xie L, Cui H, Liang W, Pecht M (2013) Remaining useful life prediction of lithium-ion battery with unscented particle filter technique. Microelectron Reliab 53(6):805–810
15. Zhou Q, Son J, Zhou S, Mao X, Salman M (2014) Remaining useful life prediction of individual units subject to hard failure. IIE Trans 46(10):1017–1030

16. Kumar A, Saxena B, Srivastava A, Goel A (2011) Physics based prognostic health management for thermal barrier coating system. In: Proceeding. Prognostic and health management society conference, 2011, vol 2, no 6, pp 25–29
17. Kumar A, Srivastava A, Goel N, Nayak A (2010) Model based approach and algorithm for fault diagnosis and prognosis of coated gas turbine blades. In: 2010 IEEE/ASME international conference on advanced intelligent mechatronics (AIM), 2010. IEEE, pp 800–805
18. Li K, Wu J, Zhang Q, Su L, Chen P (2017) New particle filter based on GA for equipment remaining useful life prediction. Sensors 17(4):696
19. Deutsch J, He M, He D (2017) Remaining useful life prediction of hybrid ceramic bearings using an integrated deep learning and particle filter approach. Appl Sci 7(7):649
20. Orsagh R, Roemer M, Sheldon J, Klenke CJ (2004) A comprehensive prognostics approach for predicting gas turbine engine bearing life. In: ASME turbo expo 2004: power for land, sea, and air, 2004. American Society of Mechanical Engineers, pp 777–785
21. Lim R, Mba D (2014) Condition monitoring and remaining useful life prediction using switching Kalman filters. Int J Strat Eng Asset Manag 8 2(1):22–36
22. Hu W, Jiyuan L, Jianguo C, Yang G, Bo C, Liying J (2014) Fault diagnosis of gas turbine based on support vector machine. In: Presented at the control and decision conference (2014 CCDC), The 26th Chinese
23. Nieto PG, Garcia-Gonzalo E, Lasheras FS, de Cos Juez FJ (2015) Hybrid PSO–SVM-based method for forecasting of the remaining useful life for aircraft engines and evaluation of its reliability. Reliab Eng Syst Safety 138:219–231
24. Zaidan MA, Harrison RF, Mills AR, Fleming PJ (2015) Bayesian hierarchical models for aerospace gas turbine engine prognostics. Expert Syst Appl 42(1):539–553
25. Sun J, Zuo H, Wang W, Pecht MG (2012) Application of a state space modeling technique to system prognostics based on a health index for condition-based maintenance. Mech Syst Signal Process 28:585–596
26. Baptista M, Henriques EM, de Medeiros IP, Malere JP, Nascimento Jr CL, Prendinger H (2018) Remaining useful life estimation in aeronautics: combining data-driven and Kalman filtering. Reliab Eng Syst Safety
27. Wang P, Gao RX (2016) Markov nonlinear system estimation for engine performance tracking. J Eng Gas Turbines Power 138(9):091201
28. Wang P, Gao RX (2014) Particle filtering-based system degradation prediction applied to jet engines. In: Annual conference of the prognostics and health management society, vol 2014
29. Daroogheh N, Baniamerian A, Meskin N, Khorasani K (2015) A hybrid prognosis and health monitoring strategy by integrating particle filters and neural networks for gas turbine engines. In: 2015 IEEE conference on prognostics and health management (PHM), 2015. IEEE, pp 1–8
30. Ahsan S, Lemma T, Muhammad M (2019) Prognosis of gas turbine remaining useful life using particle filter approach. Materialwiss Werkstofftech 50(3):336–345
31. Tahan M, Muhammad M, Karim ZA (2017) A multi-nets ANN model for real-time performance-based automatic fault diagnosis of industrial gas turbine engines. J Brazilian Soc Mech Sci Eng 1–12
32. Rahmoune MB, Hafaifa A, Guemana M (2017) Fault diagnosis in gas turbine based on neural networks: vibrations speed application. In: Advances in acoustics and vibration. Springer, pp 1–11
33. Ghafir MA, Li YG, Wang L (2013) Creep life prediction for aero gas turbine hot section component using artificial neural networks. J Eng Gas Turbines Power 136(3):031504–031504–9
34. Rigamonti M, Baraldi P, Zio E (2016) Echo state network for the remaining useful life prediction of a turbofan engine. In: annual conference of the prognostics and health management society 2015, 2016, pp 255–270
35. Ma J, Wang Q, Li G, Zhou D (2014) Reconstruction-based fault prognosis for flue gas turbines with independent component analysis. Asia-Pac J Chem Eng 9(2):205–213
36. Li Y (2010) Gas turbine performance and health status estimation using adaptive gas path analysis. J Eng Gas Turbines Power 132(4):041701

37. Tsoutsanis E, Meskin N, Benammar M, Khorasani K (2016) A dynamic prognosis scheme for flexible operation of gas turbines. Appl Energy 164:686–701
38. Ahsan S, Lemma TA (2017) Remaining useful life prediction of gas turbine engine using autoregressive model. In: MATEC web of conferences, 2017, vol 131. EDP Sciences, p 04014
39. George V, Frank L, Michael R, Andrew H, Biqing W (2006) Intelligent fault diagnosis and prognosis for engineering systems. Wiley Inc
40. Fentaye AD, Gilani SIU-H, Baheta AT (2016) Gas turbine gas path diagnostics: a review. In: MATEC web of conferences, 2016, vol 74. EDP Sciences, p 00005
41. Hanachi H, Liu J, Banerjee A, Chen Y, Koul A (2015) A physics-based modeling approach for performance monitoring in gas turbine engines. IEEE Trans Reliab 64(1):197–205
42. Yan J (2014) Machinery prognostics and prognosis oriented maintenance management. Wiley
43. Hanachi H, Mechefske C, Liu J, Banerjee A, Chen Y (2017) Enhancement of prognostic models for short-term degradation of gas turbines. In: 2017 IEEE international conference on prognostics and health management (ICPHM), 2017. IEEE, pp 66–69
44. Marinai L, Singh R, Curnock B, Probert D (2003) Detection and prediction of the performance deterioration of a turbofan engine. In: Proceedings of the international gas turbine congress, 2003. GTSJ, pp 2–7
45. Li Y, Nilkitsaranont P (2009) Gas turbine performance prognostic for condition-based maintenance. Appl Energy 86(10):2152–2161
46. Orchard ME, Vachtsevanos GJ (2007) A particle filtering approach for on-line failure prognosis in a planetary carrier plate. Int J Fuzzy Logic Intell Syst 7(4):221–227
47. Tsoutsanis E, Meskin N, Benammar M, Khorasani K (2015) Transient gas turbine performance diagnostics through nonlinear adaptation of compressor and turbine maps. J Eng Gas Turbines Power 137(9):091201
48. Giron-Sierra JM (2017) Digital signal processing with matlab examples, vol 3. Springer
49. Gordon N, Ristic B, Arulampalam S (2004) Beyond the kalman filter: particle filters for tracking applications, vol 830, no 5,. Artech House, London, pp 1–4
50. Doucet A, De Freitas N, Gordon N (2001) An introduction to sequential Monte Carlo methods. In: Sequential Monte Carlo methods in practice. Springer, pp 3–14
51. Orchard ME, Vachtsevanos GJ (2009) A particle-filtering approach for on-line fault diagnosis and failure prognosis. Trans Inst Meas Control 31(3–4):221–246
52. Zio E, Peloni G (2011) Particle filtering prognostic estimation of the remaining useful life of nonlinear components. Reliab Eng Syst Saf 96(3):403–409
53. Arulampalam MS, Maskell S, Gordon N, Clapp T (2002) A tutorial on particle filters for online nonlinear/non-Gaussian Bayesian tracking. IEEE Trans Signal Process 50(2):174–188
54. Zhu Z, Meng Z, Cao T, Zhang Z, Dai Y (2017) Particle filter-based robust state and parameter estimation for nonlinear process systems with variable parameters. Meas Sci Technol 28(6):065003
55. Schön TB (2010) Solving nonlinear state estimation problems using particle filters-an engineering perspective. Linköping University, Linköping, Sweden, Division of Automatic Control
56. Varon LAB, Orlande HRB, Eliçabe GE (2016) Combined parameter and state estimation in the radio frequency hyperthermia treatment of cancer. Numer Heat Transfer, Part A: Appl 70(6):581–594
57. Tarabrin A, Schurovsky V, Bodrov A, Stalder J-P (1998) An analysis of axial compressor fouling and a blade cleaning method. J Turbomach 120(2):256–261
58. El-Sayed AF (2017) Aircraft propulsion and gas turbine engines. CRC Press
59. Deflorian M, Zaglauer S (2011) Design of experiments for nonlinear dynamic system identification. IFAC proceedings 44(1):13179–13184
60. Mujtaba SM, Lemma TA, Taqvi SAA, Ofei TN, Vandrangi SK (2020) Leak detection in gas mixture pipelines under transient conditions using Hammerstein model and adaptive thresholds. Processes 8(4):474
61. Deflorian M, Klöpper F, Rückert J (2010) Online dynamic black box modelling and adaptive experiment design in combustion engine calibration. IFAC proceedings 43(7):703–708

62. Qingcai Y, Li S, Cao Y, Zhao N (2016) Full and part-load performance deterioration analysis of industrial three-shaft gas turbine based on genetic algorithm. In: ASME turbo expo 2016: turbomachinery technical conference and exposition, 2016. American Society of Mechanical Engineers, pp V006T05A016–V006T05A016
63. Rahimi F, Lemma TA (2017) Gas turbine fault diagnostics using fuzzy systems. In: 2017 7th IEEE international conference on control system, computing and engineering (ICCSCE), 2017. IEEE, pp 314–319

A Comprehensive Review of Prescriptive Analytics for Naval Vessels Risk-Based Maintenance Decision-Making

Mohd Adha Mat Esa and Masdi Muhammad

Abstract Prescriptive analytics is a novel maintenance approach by recommending the maintainer, the right maintenance action at the right time through prioritization of maintenance based on the risk. It assists in improving the existing preventive maintenance approach by offering a dynamic maintenance plan based on risk or Risk-based Maintenance Scheduling. To date, there is no clear definition of Risk-based Maintenance as well as its framework, and it has not been applied extensively in naval vessel maintenance application which is unusual compared to plant maintenance. This paper reviews Risk-based Maintenance application in industries in an approach to develop prescriptive analytics for a comprehensive and dynamic naval vessels Risk-based Maintenance decision-making framework. It aims to establish an optimal maintenance scheduling to improve naval vessels availability and reduce maintenance costs while meeting stakeholders' expectations. Objectives and considerations are also defined for peculiarities application of naval vessels. The framework is then evaluated to improve the effectiveness of the current Preventive Maintenance practice and the inconsistency of the existing maintenance decision-making model using the expert judgement in Reliability Centered Maintenance.

Keywords Naval maintenance · Prescriptive analytics · Risk-based maintenance

1 Introduction

Risk-based Maintenance (RBM) Scheduling framework which consists elements of Risk Assessment and Maintenance Scheduling has been widely applied in various industries. Currently, it has yet to be extensively adopted in the domain of naval vessels maintenance [4]. The applicability of this maintenance framework's models has to be further studied to suit the maintenance of naval vessels to improve the effectiveness of Preventive Maintenance (PM), Condition Based Maintenance (CBM) and Reliability Centered Maintenance (RCM) approaches. RBM has evolved from

M. Adha Mat Esa (✉) · M. Muhammad
Faculty of Engineering, Mechanical Engineering Department, Universiti Teknologi Petronas, Seri Iskandar, Perak, Malaysia
e-mail: mohd_17005329@utp.edu.my

F. Ahmad et al. (eds.), *ICPER 2020*, Lecture Notes in Mechanical Engineering,
https://doi.org/10.1007/978-981-19-1939-8_39

utilizing expert judgment which is the limitation of RCM in relation to inconsistency issue, to utilizing probabilities and optimization techniques. This is to ensure dynamic risk quantification and maintenance scheduling [17] with maintenance cost treatment. A naval vessels maintenance scheduling framework has to be developed to maximize naval vessels' availability and minimize maintenance cost by studying the existing RBM Scheduling application and applying it to naval vessels maintenance. The maintenance scheduling framework should be reliable and comprehensive in its assessment, based on risk and other influencing factors to enable accurate and optimal maintenance decision-making (maintenance scheduling).

The idea of RBM Scheduling for ships is not new and has been accepted in principle by the maritime regulatory bodies [20]. Still, to date, the adoption of RBM Scheduling for naval vessels has not been significantly researched. Application of RBM in the maritime domain is applied only on moored floating breakwater [5] but not for naval vessels. Previously, maintenance decision-making based on risk is performed using the judgment of experts [8]. Still, it has introduced uncertainty and inconsistency as the individual experience cannot be measured and continues to evolve. In addition, this technique's main limitation is that it is time-consuming. Thus, optimization techniques which treat maintenance scheduling as an optimization problem can offer consistency and efficiency.

The existing RCM for ship application does not strictly schedule maintenance based on risk [16]. A review of the studies of RBM Scheduling development in various industries concerning all applicable models and techniques used [1] found that none of the studies acknowledges that they fit the definition of RBM Scheduling. Even though their content overlapped, they did not cite one another. Therefore, consistent terminology and critical reviews of the existing RBM studies are critical in developing an RBM Scheduling framework for naval vessels maintenance. The success factor of RBM Scheduling to reduce costs has also not been measured [2]. In addition, human error in performing maintenance task has been identified as a contributive factor in the reduction of availability [17], and this should be taken into account in the development of an RBM Scheduling application.

2 Limitations of the Current Maintenance Practice

The effectiveness of PM has been put into question due to the lack of improvement to the naval vessels' availability, apart from the significant maintenance costs and resources allocated for maintaining a complex system of naval vessels. PM schedule, as suggested by Original Equipment Manufacturer (OEM) is not an optimal maintenance approach for naval vessels. It may not serve its purpose to avoid premature failure due to the possibility of human error during routine maintenance [13]. At times, the repetition of the same failure mode or even facing worsening situations (major unplanned maintenance or breakdown maintenance) may be due to the error in maintenance or the failure to accurately identify and correctly solve the root cause of the problem.

RCM has also been introduced as a maintenance decision-making framework, to prioritize the conduct of PM, based on the criticality of the equipment. This because its failures would adversely affect the plant's availability. Thus, it becomes PM prioritization, as RCM is a reliability program rather than a PM reduction program [3]. However, most of the time, the implementation of RCM is challenging due to its complexity. It is also unlikely to be succeeded due to the following factors:

- Loss of in-house control.
- An incorrect mix of personnel performing the analysis.
- Unnecessary and costly administrative burdens.
- Fundamental RCM concepts are not understood.
- Confusion determining system functions.
- Confusion over the system's boundaries and interface.
- Divergent expectations.
- Confusion regarding the convention.
- Misunderstanding hidden failures and redundancy.
- Misunderstanding run-to-failure.
- Inappropriate component classifications.
- Instruments were not included as part of the RCM analysis.

Therefore, despite its applicability and the potential benefits of the RCM approach, it is likely that the challenges of data requirements failure and the extensive implementation and use of organizational resources have hindered the adoption of the RCM framework aboard naval vessels. Moreover, as mentioned earlier, there is also an issue of consistency in RCM decision-making using expert judgment. Multiple states of degradation have also not been considered (none of the existing techniques suggested specific maintenance actions for the application).

2.1 Room of Improvements

In addition to expert knowledge, equipment or components of historical data (past performance and failure data), the physical models of equipment performance or components degradation will be used to predict when the failure will happen in the future. Hence, appropriate maintenance actions will be scheduled Just in Time (JIT), before the failure happens. *The scheduling of m*aintenance will also depend on the criticality of the equipment or its components (which failure modes are critical and how often they may occur) to the overall naval vessels' availability and reliability for maintenance. This can be determined using the Failure Modes, Effects and Criticality (FMEAC) Analysis of the RCM approach. At the same time, the Condition Monitoring (CM) real-time data as well as the data from the PdM techniques will be continuously fed to monitor and evaluate the condition and performance of the critical equipment by updating the physical models of the equipment's performance or components' degradation.

With the consideration of historical cost data related to specific maintenance action in the previous analysis, it is possible to develop quantitative models that can be optimized using optimization techniques to determine the most cost-effective maintenance schedule and optimal life cycle plan for the naval vessels. Such a plan determines, for example:

- When to schedule minor and major maintenance activities and when to perform revisions, modifications or life extension program (over the complete life cycle).
- Which equipment or components should be maintained on the use-based (every X time-units), condition-based (using which condition), and correctively only (run-to-failure) basis.

The goal of such a plan is to optimally schedule maintenance for all the systems, such that only minimal time is needed for a ship to be out of use for maintenance activities (downtime). Maintenance should also be conducted during the best possible times while achieving the lowest possible life cycle costs (LCC) throughout the life cycle of the naval vessels. The data of failure modes and how often they occur can be used in performance prediction and assessment, to determine and prioritize the criticality of the equipment or its components. The data is established based on the vessels' failure risks and impact before the appropriate maintenance actions, e.g. PM or CBM (including CM and PdM), can be scheduled.

An improved way of performing maintenance is based on the condition of the equipment. The maintenance, where JIT is performed before the predicted failure, is based on real-time CM data or Predictive Maintenance (PdM) techniques called the Condition Based Maintenance (CBM). In the era of big data and internet of things, the development of sensors and machine learning technology such as Artificial Neural Network (ANN) will permit future application of data-driven models to support complex naval vessels maintenance decision-making and solve incompatible data standards issues between different types of machinery. Nowadays, machinery in the modern naval vessels is equipped with sensors for CM, together with the Ship Maintenance Information System (SMIS) to capture and analyze the failure data. The CM real-time data and historical failure data can be utilized using Reliability Centered Maintenance (RCM) framework, which prioritizes maintenance based on risks (failures and human errors) and costs. Inconsistency issue in manual decision making by expert knowledge can be dealt with using the advancement of computer software. However, the applicability of the big data processing technology poses challenges for the naval forces in terms of its implementation due to privacy-related risks and the availability of skilled workers and software tools [15]. As more and more influencing factors are involved in the defining equipment or its components' criticality based on the risks that needed to be included in the maintenance decision-making process, the more accurate and optimal maintenance scheduling will be made. In order to do that, the factors have to be identified, quantified, modelled and solved using available optimization techniques.

A suitable decision rule is required for multi-objective optimization approaches to suggest a specific maintenance action. It is also suggested that either optimization or expert judgment, or a combination of these should be investigated to overcome their respective limitations. Appropriate techniques (optimization) from other fields and applications are also applicable to solve the problem of maintenance scheduling (including inspection and repair times) to optimize the life-cycle cost of maintenance. A general overview of optimization algorithms is provided by Venter [23].

In line with the maintenance framework proposed by Gits [11] and Waeyenbergh and Pintelon [24], further research to explicitly link the knowledge from the technical data to the cost data is fundamentally needed. This is to find optimal maintenance decision-making for naval vessels application. Further research will also benefit from the work of Zoeteman [26], but the focus should be placed more on quantitative and technical data. There are three recent reviews of related literature [18, 18], and it is noted that the optimization maintenance scheduling, over critical equipment or its components with a clear choice between various maintenance actions (e.g., PM or CBM), has received limited attention [25]. The quantitative models in the study require certain inputs, which indicates the importance of monitoring the asset's life cycle and continuously improve the asset's life cycle plan (continuous improvement and continuous focus on LCC throughout the life cycle of naval vessels is not very common in practice). Part of this information may come from various internal (engineering division) and external sources (contractors).

3 RBM Scheduling Framework for Naval Vessels Maintenance Application

In the RBM Scheduling framework, the maintenance requirement is triggered based on the risk associated with the equipment or its component's failure mode. Thus, how maintenance is strategized and scheduled depends on the level of risk of the predicted failure. The higher the risk, the more focus should be given in strategizing and maintaining the equipment. The level of risk is determined by the frequency of failure or number of failures as well as the impact of failure. This concept is similar to the FMEAC in the RCM, but the difference is the assessment of the risks in RCM is made by experts. As argued earlier, this type of assessment could possibly introduce inconsistency and requires extensive preliminary work in strategizing the maintenance approach. In contrast, RBM is not just about strategizing the maintenance approach but dynamically schedule maintenance based on the triggered risk. The literature on how risk is quantified to prioritize maintenance, according to the level of risk in RCM, are as follows:

- *Risk Indices:* Dinovitzer [6], Klein Woud et al. [14] and Smith [21] quantified risks using risk indices for probability and likelihood of failure of individual pieces of equipment or its components and read a suitable value from an appropriate table of risk index values. It has been used in RCM to rank equipment and their components based on the level of risk associated with each component if it fails. Though this approach is simple and time-efficient, it introduces inconsistency in terms of risk assessment because it is influenced by individual experience, which cannot be measured and consumes time as experience continues to evolve over time. Therefore, the category of risk associated with component failure may need to be periodically reassessed by experts who are appropriately trained and well experienced to improve its accuracy.

- *Probabilistic approaches:* Diamantoulaki and Angelides [5], Dong and Frangopol [7], Giorgio et al. [10] and Handani et al. [12] utilized probabilistic approaches to quantify risk with a continued focus on individual pieces of equipment or components. These methodologies specifically assumed that risk is a time-dependent property. Some of the existing methods are limited to their specific applications due to their assumption; limited if failures are not random and generalized as their assumptions are not as restrictive. These methods require further development so that they may be widely applied to different pieces of naval vessels equipment or components.
- *Bayesian Method:* None of the existing methods applied for risk assessment has taken into consideration the degradation states of the equipment or components other than a binary state of failure or no failure. The Bayesian method can be applied as it takes into consideration any available prior information. The information includes information about previous failure experiences such as historical failure data, equipment or component failures or degradation model of equipment or other equipment with the same failure mechanism and the likelihood (real-time equipment parameters or failure data) of posterior information being produced (prediction of future failure) in its analysis. Hence, the method assists experts in assessing the risk and schedule maintenance more efficiently.

3.1 Prioritizing Maintenance Based on Risk

Due to the overloading of the PM work on-board naval vessels, the analysis of FMEAC in RCM is useful to strategize a suitable maintenance approach by prioritizing which PM to be focused on and at the same time screen out those PM works which is less critical based on the risk category associated with each component failure's mode. In traditional RCM, each system will be broken down to the level of its components. However, it does not categorize the systems into groups of functions, as in the risk assessment of naval vessels. It is important to note that each system on-board a naval vessel has to be categorized into three naval vessel's main functions of readiness which are float, move and fight before it can be broken down to the level of components. The weightage for each function needs to be assigned as the float function has to be the highest priority, followed by the move and lastly the fight function. Then, the risk associated with each component if it fails is assessed in terms of the frequency of failure or number of failure and the effect of failure on the equipment or component's overall functionality and performance which contributes to the naval vessel's readiness or availability, equipment downtime and cost of repair. The relationship of the function-system can be represented by a hierarchy tree, as shown

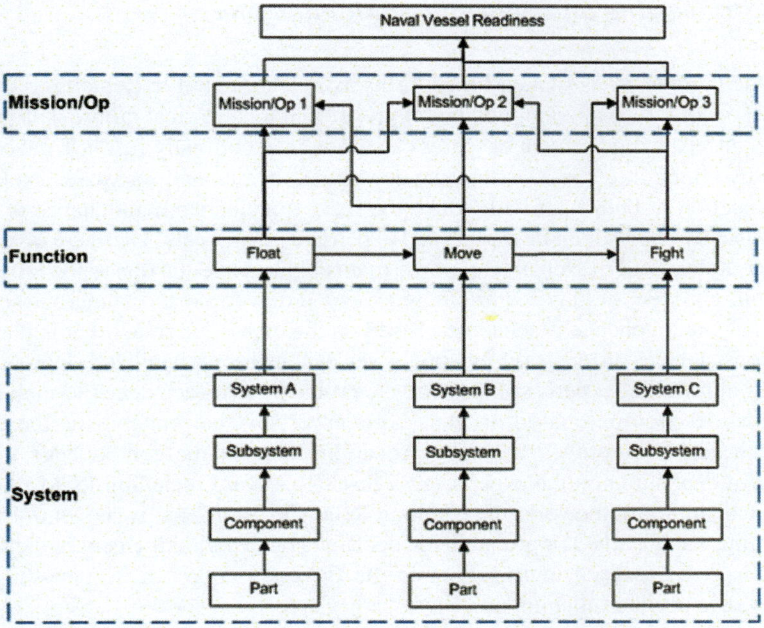

Fig. 1 Function system hierarchy tree

in Fig. 1 to represent the use of novel naval vessels maintenance risk assessment or other application domains.

Other important factors that are mandatory and cannot be compromised in the risk assessment are safety (e.g., fire, flooding or accident which can jeopardize the safety of the ship, its crew and loss of life) and environment effect. Besides that there are other considerable factors such as vessel mobility [8], affecting both the operational environment of the on-board systems, the supply of resources to the vessel; vessel motion, the marine environment, alignment of the schedule to an existing maintenance cycle for the ship and the capability of the vessel's crew to conduct maintenance at sea and the operational profile of the vessel (Usage Upkeep Cycle).

The issue of inconsistency in RCM risk assessment can be avoided by feeding data of frequency of failure, downtime, and cost of repair from the ship's maintenance information system for each component. By using the Bayesian method, historical failure data, physical model of degradation and failure or other available data can be used as prior information which will then be verified by the likelihood of real-time equipment component parameters to generate posterior information to predict accurate failure time.

3.2 Strategizing and Scheduling Maintenance

Once the component is ranked based on its associated risk, the Remaining Useful Life (RUL) of the component will be retrieved from its historical failure data. Thus, maintenance and requirement of spares can be planned optimally Just in Time before it fails by taking into consideration the downtime, the demand for spares lead time and allocation of budget. It will reduce the risk of unplanned maintenance or BM, which usually entails long downtime and expensive repair costs. These are often due to the complexity of the repair, e.g., if the main engine seizes up due to the failure of the cooling system, it may require more extended downtime and numerous spares. Previously, maintenance is scheduled based on the expert judgment that is plagued by inconsistencies in decision-making. Even when the probabilistic approach of RCM is introduced, experts still provide the probability of each decision-making in the RCM decision tree based on their experience [9]. The probabilistic the RCM approach can be improved by using probability data of the component's failure instead of depending on the experts. As a first step toward including RCM into the RBM Scheduling framework, the resulting RBM Decision Tree is shown in Fig. 2. The traditional YES/NO or probabilistic decisions by expert judgement in the RCM Decision Tree has been improved using the Bayesian approach in modelling the maintenance decision-making process.

In addition, the decision-making in RCM is limited to strategizing the maintenance approach (utilizing PM, Corrective Maintenance, CBM, periodic test, redesigning the system, or rescheduling PM) but not scheduling the maintenance. Currently, the complexity of the maintenance decision-making process requires appropriate techniques rather than just using expert judgment. Diamantoulaki and Angelides [5], Dong and Frangopol [7], Giorgio et al. [10] and Handani et al. [12] utilized the optimization technique to schedule maintenance. This technique often incorporates a detailed treatment of maintenance costs, though it does not specifically schedule maintenance activities [10]. Multi-Objective Optimization is among the optimization techniques that can help achieve an optimal decision in scheduling maintenance. Dong and Frangopol [7] performed maintenance scheduling using Multi-Objective Optimization where it can optimize inspection and repair times of the ship's structural details to minimize inspection costs, repair costs and risks. It also captures all of the factors influencing maintenance scheduling as constraints or as part of the cost function.

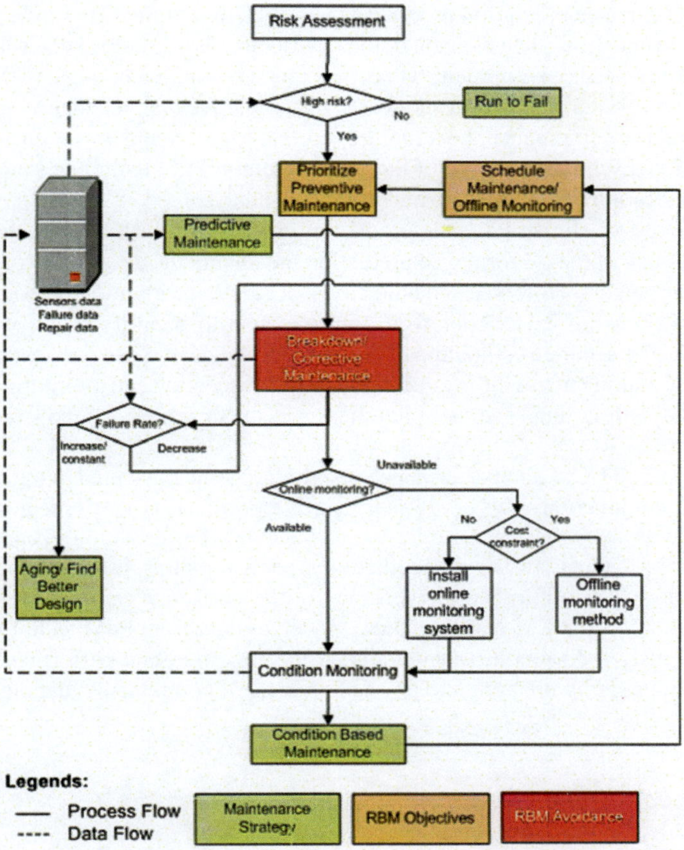

Fig. 2 RBM decision diagram

4 Other Considerations

4.1 Performance Prediction

Performance of in-service equipment can be continuously monitored. By using historical data and validated by real-time CM data as well as data from PdM techniques, it can help us to prioritize and optimize maintenance resources. This concept of maintenance, which is based on the condition of in-service equipment, is called CBM. CBM will be performed when the need arises, after one or more indicators show that the equipment is going to fail or that equipment performance is deteriorating. This approach is cost-saving compared to the routine or time-based preventive maintenance because the tasks are only performed when warranted.

The performance prediction process is the foundation of the Life Cycle Management (LCM) and the critical element of the PdM techniques to help in determining the condition of in-service equipment. This allows for a prediction of when maintenance should be performed. The beauty of most of the predictive maintenance techniques is that they require equipment to be running at normal capacity and do not interfere with the operational schedules. Some of the more common PdM techniques are thermal imaging, vibration analysis, sonic and ultrasonic analysis.

According to Soliman [22], the prediction process depends, to a greater extent, on the accuracy of the performance prediction model and the descriptors of its probabilistic parameters. However, in some cases, accurate information on some model parameters does not exist. Therefore, future efforts to quantify these parameters are crucial. The accurate estimation of the characteristics of these parameters can be achieved by making use of the available structures' inspection and monitoring results, and this will help to improve the reliability assessment process of the structures or equipment.

The proposed LCM and performance updating scheme presented in the study can be used in conjunction with other intervention optimization and scheduling techniques. The use of Structural Health Monitoring (SHM) information in updating the parameters of load and resistance prediction models for ships, is another interesting topic which will also improve the accuracy of the performance prediction. Therefore, in this study, the CM real-time data, as well as data from PdM techniques, will be continuously fed to monitor and validate the condition and performance of the critical equipment by updating the physical models of components' degradation.

4.2 Probability of Detection

More researchers are coming to realize that in many cases, especially on ageing equipment, when irreversible changes occur on the microstructural level, is it considerably more dangerous, and failure may occur suddenly. For this reason, the Probability of Detection (PoD) has been introduced as an indicator of the sensitivity level for monitoring or inspection methods to allow the detection of flaws at this pre-defect stage. The PoD has been successfully used for the probabilistic inspection optimization for civil, marine, and aerospace structures [19, 19, 19, 19] and has been widely used to assess the capacity of the inspection method to detect cracks (minimum detectable crack size). The accurate estimation of the PoD parameters of CM and PdM is crucial for maintenance decision-making. More research is needed to establish the PoD for different CM and PdM techniques. Other models for quantifying the inspection, as well as CM and PdM's effectiveness, are required. The approach presented in the study uses a novel inspection accuracy parameter. However, more research is still needed to quantify the descriptors of this parameter for different inspection and monitoring methods.

4.3 Performance Assessment

The monitoring activities, to assess the equipment's condition and performance, yield a large amount of data, whose analyses and interpretation may require significant financial resources [22]. Since these monitoring activities are crucial for the maintenance of critical equipment on-board naval vessels based on their conditions, an optimal monitoring schedule and selecting methods with minimum error for defect or damage detection has to be established. The minimum errors usually arise from using multiple short-term monitoring actions compared to continuous long-term monitoring.

4.4 Effect of Maintenance

The effect of any maintenance action on the structural [22] or equipment performance is generally difficult to quantify, especially when using probabilistic performance indicators. Soliman [22] presented two types of maintenance; the one stops further damage propagation or performance degradation (e.g. PM or CBM), and the other results in the restoration of structural or equipment performance to the initial level (e.g., repair by replacement or overhaul). These two types represent the lower and upper extremes of maintenance complexity and cost (i.e., degree of performance restoration). However, the actual maintenance can yield other levels of performance restoration (e.g., 90% performance restoration) which depends on the human factor. The possibility of human error will affect the quality of restoration due to the challenging environment of maintenance on-board naval vessels and the complexity of the job [13]. Therefore, further research in the field of naval vessels maintenance is needed to:

- Establish the relationship between various maintenance actions and the associated performance restoration.
- Establish the relationship between human reliability (human error probability) of various maintenance actions.
- Incorporate these maintenance actions into the intervention optimization approaches presented in this study.

5 Conclusion

The peculiarities and complexities in the naval vessels' maintenance require optimal maintenance scheduling framework which aims to maximize availability and minimize maintenance costs while meeting the stakeholders' KPI. RBM Scheduling framework for naval vessels application with the help advancement of big data analysis tools offers a feasible solution for maintenance decision-making uncertainties previously offered by expert judgments using the RCM method. The framework utilizes a quantitative approach in reviewing the overall designed system reliability in the hierarchy tree and making maintenance decision using a ranking system as an object of maintenance action. Finally, Machine Learning will continuously assess the risk of failure by processing and analyzing the system/equipment, big data, learn the trend and generate a dynamic maintenance strategy to support in the decision-making process. With the input from the historical and condition monitoring data, the framework shall have a comprehensive and quantifiable maintenance decision support structure by allowing stakeholders value co-creation of conflicting performance requirements. In addition, the framework will also incorporate flexibility subject to ensure continuous condition monitoring of the performance using a computerized maintenance management system.

Acknowledgements The authors acknowledge the support of Universiti Teknologi Petronas in conducting this research and supporting the publication of this paper.

Funding No funding.

References

1. Arunraj N, Maiti J (2007) Risk-based maintenance—techniques and applications. J Hazard Mater 142(3):653–661
2. Baliwangi L, Ishida K, Arima H, Artana KB (2006) Optimizing ship machinery maintenance scheduling through risk analysis and life cycle cost analysis. 25th international conference on offshore mechanics and arctic engineering. American Society of Mechanical Engineers, 2006, pp 127–134
3. Bloom N (2005) Reliability-centered maintenance (RCM). McGraw-Hill
4. Cullum J, Binns J, Lonsdale M, Abbassi R, Garaniya V (2018) Risk-based maintenance scheduling with application to naval vessels and ships. Ocean Eng 148:476–485. https://doi.org/10.1016/j.oceaneng.2017.11.044
5. Diamantoulaki I, Angelides DC (2013) Risk-based maintenance scheduling using monitoring data for moored floating breakwaters. Struct Saf 41:107–118
6. Dinovitzer A, Basu R, Holt K, Bourne J, Pegg N, Hansen K, Benckhuysen J (1997) A hybrid approach to warship structural maintenance management. Discussion. Authors' closure
7. Dong Y, Frangopol DM (2015) Risk-informed life-cycle optimum inspection and maintenance of ship structures considering corrosion and fatigue. Ocean Eng 101:161–171
8. Eruguz AS, Tan T, van Houtum G-J (2015) A survey of maintenance and service logistics management: classification and research agenda from a maritime sector perspective. Beta Working Paper Technische Universiteit Eindhoven
9. Eisinger S, Rakowsky UK (2000) Modeling of uncertainties in reliability centered maintenance—a probabilistic approach. Reliab Eng Syst Saf 71:159–164
10. Giorgio M, Guida M, Pulcini G (2015) A condition-based maintenance policy for deteriorating units. An application to the cylinder liners of marine engine. Appl Stoch Models Bus Ind 31(3):339–348
11. Gits CW (1992) Design of maintenance concepts. Int J Prod Econ 24:217–226
12. Handani DW, Ishida K, Nishimura S, Hariyanto S (2011) System dynamics simulation

13. Islam R, Abbassi R, Garaniya V, Khan F (2017) Development of a human reliability assessment technique for the maintenance procedures of marine and offshore operations.https://doi.org/10.1016/j.jlp.2017.10.015
14. Klein Woud J, Smit K, Vucinic B (1997) Maintenance programme design for minimal life cycle costs and acceptable safety risks. Int Shipbuild Prog 44(437):77–100
15. Lazakis I, Raptodimos Y, Varelas T (2017) Predicting ship machinery system condition through analytical reliability tools and artificial neural networks.https://doi.org/10.1016/j.oceaneng.2017.11.017
16. Lazakis I (2011) Establishing an innovative and integrated reliability and criticality based maintenance strategy for the maritime industry. University of Strathclyde
17. Lazakis I, Turan O, Aksu S (2010) Increasing ship operational reliability through the implementation of a holistic maintenance management strategy. Ships Offshore Struct 5(4):337–357
18. Nicolai RP, Dekker R (2008) Optimal maintenance of multi-component systems: a review. In: Complex system maintenance handbook. Springer Series in Reliability Engineering, Part D: 263–286
19. Righiniotis TD (2004) Influence of management actions on fatigue reliability of a welded joint. Int J Fatigue 26(3):231–239
20. Shorten DC (2013) Marine machinery condition monitoring: why has the shipping industry been slow to adopt?. In: 10th international conference on condition monitoring, CM 2013 and machinery failure prevention technologies 2013, MFPT 2013, June 18, 2013–June 20, 2013. British Institute of Non-Destructive Testing, Krakow, Poland, pp 1037–1046
21. Smith BD (1989) Risk management in repair work decision-making. Nav Eng J 101(3):220–230
22. Soliman M (2015) Life-cycle management of civil and marine structures under fatigue and corrosion effects, (January), 414
23. Venter G (2010) Review of optimization techniques. Encyclopedia of aerospace engineering, pp 1–12. https://doi.org/10.1002/9780470686652.eae495
24. Waeyenbergh G, Pintelon L (2002) A framework for maintenance concept development. Int J Prod Econ 77:299–313
25. Wang L, Chu J, Wu J (2007) Selection of optimum maintenance strategies based on a fuzzy analytic hierarchy process. Int J Prod Econ 107:151–163
26. Zoeteman A (2004) Railway design and maintenance from a life-cycle cost perspective: a decision-support approach. PhD thesis. TRAIL research school

CFD Simulation Approach to Predict Fire-Influenced Domino Accident Propagation Pattern

Asher Ahmed Malik, Mohammad Shakir Nasif, Ainul Akmar Mokhtar, and Mohd Zahirasri Mohd Tohir

Abstract Domino effect by fire has been known for many industrial devastations in the past records. Risk-based approaches in terms of consequence and probability are being applied to avoid and mitigate such accidents. The incident radiative heat flux triggers domino effect by fire to nearby equipment. This heat flux has been calculated using semi-empirical correlations which fail to capture full fire dynamics. Fire Dynamics Simulator (FDS) predicts time varying heat flux with well-defined fire characteristics. This study uses well valid FDS software to predict heat flux to equipment in proximity. The obtained heat flux is used to estimate the escalation probability to each tank. The results of FDS are verified with the outcomes of correlations on a tank farm case study. Wind effect is induced in order to represent a realistic scenario revealing a major alteration in domino propagation pattern. The incident heat flux due to tilted flame doubles in magnitude with a four-order increase in the escalation probability for the most affected tank, intensifying its vulnerability.

1 Introduction

The domino effect is defined as " a cascade of events in which the consequence of preceding accident is increased both spatially and temporally by the following ones, thus leading to a major accident" [1]. There are three major contributors in fired domino events. First, a primary accident in a unit which can be a pool fire, tank fire, jet fire etc. Second, an escalation vector (heat radiation or fire impingement) due to

A. Ahmed Malik · M. Shakir Nasif (✉) · A. Akmar Mokhtar
Mechanical Engineering Department, Universiti Teknologi PETRONAS, 32610 Seri Iskandar, Perak, Malaysia
e-mail: mohammad.nasif@utp.edu.my

A. Ahmed Malik · M. Shakir Nasif
Centre of Advanced Process Safety, Universiti Teknologi PETRONAS, 32610 Seri Iskandar, Perak, Malaysia

M. Zahirasri Mohd Tohir
Faculty of Engineering, Department of Chemical and Environmental Engineering, Universiti Putra, UPM, 43400 Serdang, Selangor, Malaysia

F. Ahmad et al. (eds.), *ICPER 2020*, Lecture Notes in Mechanical Engineering,
https://doi.org/10.1007/978-981-19-1939-8_40

primary fire accident. Third, a secondary accident that occurs due to the physical effects of incident escalation vector from primary fire.

The severity of domino effect by fire has been identified in the past research and accident reports. The relative probabilistic analysis show that 27% of domino accident sequence is initiated by fire. Statistically, fire is recorded to be the leading specific cause for the domino accidents acquiring more than half of the causal category [2]. Most of the severe domino accidents involved flammable materials. Due to the severe nature, it has become a major concern to analyze domino effect probabilistically and consequentially.

In regards to probabilistic study, Khakzad et al. [3] introduced a methodology based on Bayesian Networks (BNs) to predict the propagation pattern of domino effect in process plant. Domino levels were defined and the escalation probability for each domino level was calculated. Khakzad [4] presented a methodology to obtain the escalation probability spatially as well as temporally based on Dynamic Bayesian Networks (DBNs). Kamil et al. [5] conducted probabilistic analysis of domino effect by utilizing time-dependent parametric equation and displayed the propagation pattern by implementing the methodology on a case study using probabilistic technique Petri-nets (PNs).

Furthermore, due to severe nature of domino effect, numerous studies were performed on the consequence modelling using Computational Fluid Dynamics (CFD) software. Jujuly et al. [6] utilized ANSYS CFX 14 to analyze domino effect due to LNG pool fire using quiescent and non-quiescent wind condition. It was concluded that wind has the potential to trigger fire-induced domino effect. The study highlighted the significance of wind magnitude in domino effect propagation. Baalisampang et al. [7] analyzed spilled fire scenarios in Floating Production Storage and Offloading (FPSO) facility. They investigated the impact of heat flux from fire to the adjacent equipment structure. It was found that incident radiant intensity of 50 kW/m^2 elevates the surface temperature of steel to around 700 °C. Hence, reducing its yield strength and modulus of elasticity to 78% and 68%, respectively. Ahmadi et al. [8] performed well validated dike and tank fire consequence modeling using FDS and obtained the incident heat flux values at tanks in close proximity with and without wind. They observed that dike pool fire had the ability to trigger domino effect whereas tank fire is incapable to initiate domino effect at 6 m/s as well as at no wind.

Although incident heat flux is predicted using advanced CFD software like FDS, it has not been utilized to obtain the escalation probability and propagation pattern of domino effect. The escalation probabilities computed using semi-empirical correlations may overestimate the escalation due to invalid assumption of flame geometry [8]. Therefore, this study takes a step forward by using the incident heat flux predicted by FDS to compute escalation probabilities to each equipment. Furthermore, propagation pattern from simulation outcomes is also presented. A comparison of predicted escalation probability and propagation pattern from FDS and correlation is provided in order to comment on the percent deviation in results. In addition, the effect of wind on escalation probability and escalation chain is investigated in order to address the limitations of semi-empirical correlations.

2 Computational Fluid Dynamics (CFD)

As discussed above, CFD is widely utilized to model the consequence of domino events in the past research. It is becoming a promising tool in the domino effect analysis and is deemed to be more accurate than the traditional empirical models [9]. CFD is a modern technique applied to solve a set of partial differential equations derived from basic laws of mass, energy, and momentum conservation. These set of equations are solved for a large quantity of control volumes inside the domain [10]. For fire analysis, CFD uses combustion model, turbulence model and radiation model to predict the fire behavior [11]. In the framework of QRA, CFD is an integral part of risk analysis studies best suited to show the impact of domino effect on the equipment. CFD simulations have become the preference for fire scenario and impact modeling as literature justifies their compatibility to experiments [12]. CFD represents the influence of damaging factor i.e., incident heat flux onto nearby equipment. Thereby highlighting the most vulnerable units using heat flux contours.

2.1 Fire Dynamics Simulator (FDS)

FDS is a CFD model of fire-driven fluid flow that solves numerically a form of Naiver-Stokes equations [13]. FDS is preferred as it is well validated software package, requires low dedicated hardware for simulation and shows reliable dynamic solutions to fire propagation [8, 13–16]. In this study, FDS is used for consequence modelling of pool fire event as it is well specialized in approximating three dimensional fire scenarios and producing transient solutions to fire growth [8, 17].

3 Methodology

The methodological steps to predict the domino propagation pattern are as follows.

3.1 Consequence Modeling Using FDS

In the present study, consequence modeling of pool fire accident is performed using FDS particularly in tank farms. The pool fire gives off heat radiation to adjacent tanks hence leading to their failure. FDS is used to obtain this time varying incident heat flux. The failed tanks may involve in further chain of fire accidents, contributing to domino propagation. More details and dimensions of the considered case study is provided in Sect. 4.

3.2 Failure Criterion

As discussed above, the primary pool fire in one of the tanks will incident the radiative heat flux to nearby tanks. Therefore, there is a need to define a failure criterion based on which the domino effect will take place. A heat flux threshold is defined for the tank failure. Any incident heat flux value exceeding 15 kW/m^2 is considered vulnerable and is the cause of domino effect via escalation. This heat flux threshold is from Cozzani et al. [18] defined by comparing the time for effective mitigation (TEM) with time to fail (TTF). TEM is site-specific, which for their supposition is 15 min. Logically, TTF must be less than TEM to contribute to domino effect. Based on varying heat fluxes and tank volumes, the TTFs were obtained from experimentally developed correlation. It was noticed that at heat flux below 15 kW/m^2, the TTFs were more than TEM (15 min) hence domino effect is unlikely and vice versa. Therefore, 15 kW/m^2 is the determined critical value which when exceeded will contribute to domino effect.

3.3 FDS Mesh Independency and Validation Study

As mentioned, the cause of escalation is due to the radiant heat from pool fire, hence in this paper the primary accident which is considered as pool fire is validated against the experiment. The pool fire experiment chosen for validation was conducted by Fire Research Institute of Japan [19]. Arabian Light crude oil with 1 m pan diameter is simulated in FDS and incident radiative heat flux is measured at length to diameter ratio equals to 3 ($L/D = 3$). The obtained radiative heat flux is sensitively analyzed and validated against experimental results. At first, mesh independency test is conducted at three suggested grid resolutions (R) which are 4, 10 and 16 [20]. Smaller value of R corresponds to coarser mesh and vice versa. The results of mesh independency test as depicted in 'Fig. 1a' show that at $R = 10$ and $R = 16$ the predicted heat flux values are nearly the same. Hence, as we increase from $R = 10$ the predicted heat flux becomes independent of the grid resolution. Therefore, to limit the computational time $R = 10$ is chosen as an optimum grid resolution for this study. 'Fig. 1b' shows the comparison of FDS predicted incident heat flux with the experiment. The predicted heat flux is in close agreement with the experiment at $R = 10$ and $R = 16$. Moreover, out of these two, $R = 10$ is a computational cost compromising grid resolution as the predicted heat flux at this value is also close to the experimental outcome.

3.4 Induction of Wind Effect

After validating the model, the wind effect on the domino escalation is investigated. Wind magnitude and direction has major influence on the domino effect propagation.

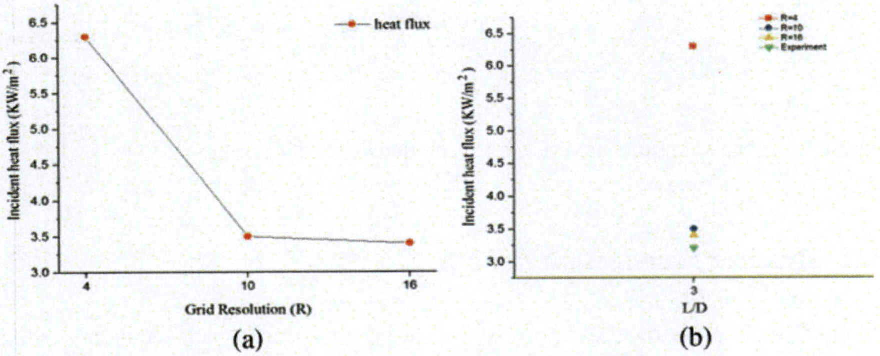

Fig. 1 Validation study **a** Mesh independency test of incident radiative heat flux at varying grid resolution (R) **b** Validation of FDS model against experiment

Its influence on incident heat flux has been investigated in the past researches using CFD in terms of heat flux contours [6, 8]. However, this study analyzes the effect of wind on escalation probability which was not being investigated.

3.5 *Probabilistic Analysis of Domino Effect*

Since this article attempts to represent the domino effect consequentially as well as probabilistically therefore, probability of escalation is additionally computed under wind and no wind scenario. The probability of escalation of fire is dependent on its damage factor caused by the incident radiative heat flux. The damage factor can be thermal dose on the affected equipment or its time to failure (TTF). In the present study, TTF of the affected equipment is the causing factor utilized to estimate the escalation probability [4]. Equations (1) and (2) are the correlations used to estimate escalation probability. They were developed from numerous simulations of experiments specifically for atmospheric storage tanks affected by heat flux from a distant fire source [18, 21].

$$\ln(\mathrm{TTF}) = -1.128\ln(\mathrm{Q}) - 2.667 \times 10^{-5}\mathrm{V} + 9.877 \tag{1}$$

$$\mathrm{Y} = 12.54 - 1.847\ln(\mathrm{TTF}) \tag{2}$$

where, TTF = time to failure (s); Q = Incident Heat Flux (kW/m^2); V = tank volume (m^3).

Using the probit value (Y), Escalation probability (EP) can be obtained using the standard form of normal distribution function [22]:

$$EP = \frac{1}{\sqrt{2\pi}} \int_{-\infty}^{Y-5} e^{-u^2}/2\,du \quad (3)$$

where ‘u’ is the substitution variable for the density function.

3.6 *Domino Propagation Pattern*

After computing the escalation probability of primary fire to each secondary tank, domino accident propagation pattern is suggested. The escalation is represented by integrating the outcomes of CFD and empirical correlations (probability models). The incident heat flux is obtained from computer software FDS under wind and no wind state. The heat flux from FDS is used in Eq. (1) to obtain time to failure and thus escalation probability is calculated using Eq. (3). Based on the incident heat flux and escalation probability, the exposure of target unit is defined as qualitative vulnerability states. Hence, this study provides an approach to predict the domino accident propagation under real scenarios using more reliable heat flux prediction technique.

4 Case Study

A tank farm case study is extracted from the previous research [5, 23]. The study consists of four equal-sized cylindrical atmospheric storage tanks containing gasoline. The maximum storage capacity for each tank is 10 tonnes. Each tank is surrounded by a dike of 10 m diameter. Pool fire is the most credible accident scenario with a diameter of 10 m, occurring at Tank 1. The bird’s eye of the tank farm is shown in ‘Fig. 2’. Asia Wind Energy Association (AWEA) suggests mean monthly wind speed in the eastern Malaysian peninsula as 3 m/s. Therefore, for this analysis wind magnitude of 3 m/s is chosen [24]. Moreover, wind directed to the East is expected to be worst scenario as it may enhance the domino escalation in the considered tank farm configuration.

5 Results and Discussion

In this section, the results of domino effect propagation under wind and no wind condition are presented. For this study, the propagation of domino accident is suggested by combining the outcomes of two major aspects of risk studies i.e., consequence and probabilistic analysis. From consequence modeling the incident

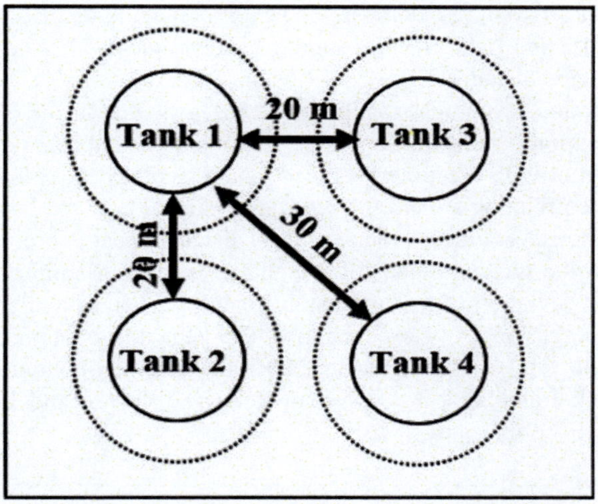

Fig. 2 Tank farm layout considered for this case study

radiative heat flux is obtained from FDS which is then utilised to compute the escalation probability to each tank. The significant results are provided and discussed as follows.

5.1 *Domino Effect Propagation (No Wind)*

As mentioned in the case study, Tank 1 is under pool fire of dimeter 20 m. The pool fire will give off heat radiations onto nearby tanks. The time varying incident radiative heat flux on each tank obtained from FDS under no wind has been depicted in 'Fig. 3'. As seen in the figure, the heat flux first reaches the peak value within few seconds and then attains a quasi-steady state throughout. The quasi-steady incident

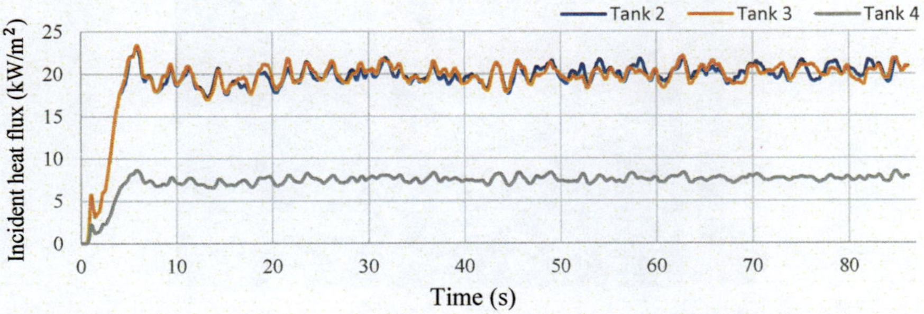

Fig. 3 Incident heat flux from Tank 1 to each tank with no wind condition

heat flux is averaged to have a comparable heat flux magnitude. Moreover, it is also noticed that Tank 2 and Tank 3 receive approximately same incident heat flux because they are at the same distances.

As a verification, a comparison of predicted heat flux from FDS and semi-empirical correlation is indicated in Table 1. As mentioned above, the incident heat flux obtained from FDS are time averaged. The calculated heat flux values from semi-empirical correlation are taken from Kamil et al. [5].

The comparison table (Table 1) shows that at distances near to fire, the FDS results agree well with the incident heat flux calculated from semi-empirical correlation with percentage deviation of approximately (0.49%) whereas a slight deviation is noticed at farther distances (9% deviation). The heat flux contour obtained from FDS is shown in 'Fig. 4'. It is noticeable from the heat flux contour that heat flux received by Tank 2 and Tank 3 is approximately 20.7 kW/m^2 and Tank 4 receives approximately 10 kW/m^2.

Table 1 Comparison of heat flux prediction by FDS and correlations under no wind condition

Incident radiative heat flux 'kW/m^2' (no wind)		
Tank1 on Fire		
	FDS	Correlation [5]
Tank2	20.4	20.5
Tank3	20.6	20.5
Tank4	10.0	11.0

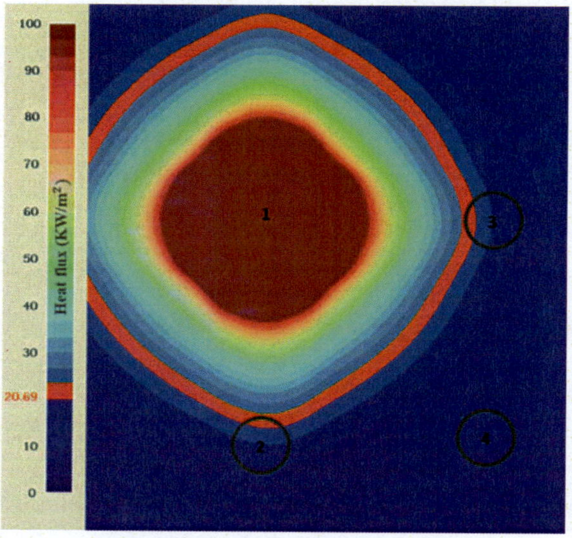

Fig. 4 Incident Heat flux contours from FDS under no wind

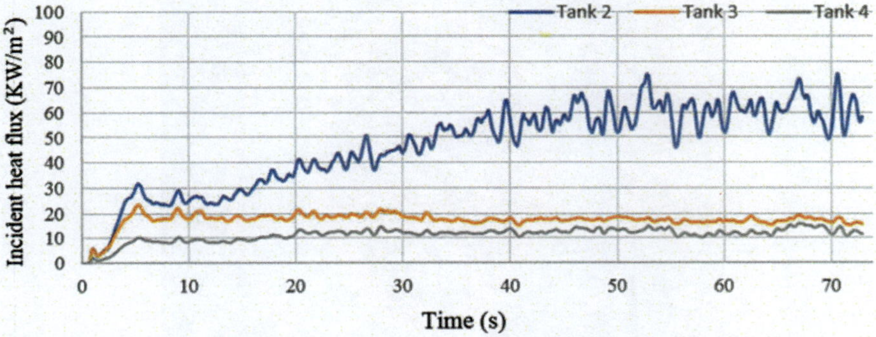

Fig. 5 Incident heat flux from Tank 1 to each tank with 3 m/s East wind condition

5.2 Domino Effect Propagation (Under Wind)

As mentioned above, the effect of 3 m/s wind speed is investigated in this paper, taking into consideration the wind from East side of the tank farm. The time varying incident radiative heat flux obtained under 3 m/s wind in the East direction wind is portrayed in 'Fig. 5'.

As seen from 'Fig. 5', the incident heat flux received by Tank 2 has shown significant increase in its value. The averaged value elevates from 20.4 kW/m^2 to 59.5 kW/m^2 which is more than twice increase. The wind magnitude and direction show considerable effect on Tank 3 and Tank 4. By comparing the heat flux with no wind state, Tank 3 has become less vulnerable, and Tank 4 might become a vulnerable unit if the threshold value descends.

The heat flux contour in 'Fig. 6' shows the variation of heat flux due to induced wind. Due to wind directed in the East, the flame tilts towards Tank 2 making it most vulnerable under high incident heat flux.

Table 2 provides the time averaged heat flux predicted by FDS at 3 m/s wind in the East direction. Tank 2 has significantly gained the incident heat flux due to flame tilt with approximately 200% increase in the received flux. This increase in heat flux will contribute to tank failure at much earlier time and continuing the domino chain to other tanks.

The escalation probabilities to each tank when Tank 1 was in pool fire under wind and no wind condition are tabulated in Table 3.

From Table 3, under no wind Tank 3 is the most vulnerable with highest escalation probability followed by Tank 2 and Tank 4. Previous study suggested the same escalation probability for Tank 2 and Tank 3 [5]. FDS predicted slightly different heat flux for the tanks at same distances. Although the difference in received heat flux and escalation probability is relatively less for Tank 2 and Tank 3 but it is not negligible specifically in the severe domino effect scenario. 3 m/s wind increased the

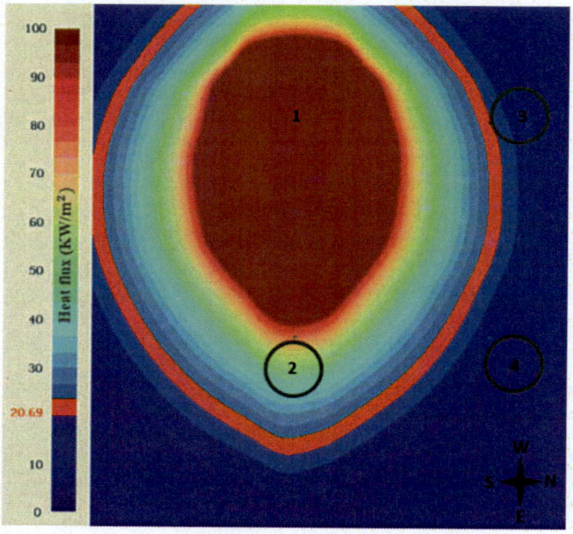

Fig. 6 Incident Heat flux contours from FDS under 3 m/s East wind

Table 2 Heat flux prediction by FDS under wind condition

Incident Radiative Heat Flux ‘kW/m^2 (3 m/s East)	
Tank1 on Fire	
	FDS
Tank2	59.5
Tank3	17
Tank4	12

Table 3 Escalation probabilities of fire propagation from primary fire under wind and no wind

Escalation Probabilities			
	No wind		
	Tank 2	Tank 3	Tank 4
Tank 1 (Fire)	5.0×10^{-6}	5.45×10^{-6}	2.0×10^{-9}
	Wind 3 m/s East		
Tank 1 (Fire)	1.4×10^{-2}	8.0×10^{-7}	1.65×10^{-8}

Least Vulnerable   Less Vulnerable   Most Vulnerable

vulnerability of Tank 2 and its escalation probability increased by 4 orders from 5. 0 $\times$ 10^{-6} to 1.4 $\times$ 10^{-2}. The propagation path under 3 m/s wind speed deviated and the domino effect escalated from Tank 1- Tank 2- Tank 3-Tank 4.

6 Conclusion

The present study evaluated the risk of domino effect in terms of consequence and probability using advanced modelling tool. FDS was utilized to predict the radiative heat flux to the nearby storage tanks. A tank farm case study was analysed to implement this verdict and heat flux from a 10 m diameter pool fire was predicted onto nearby tanks. For quiescent wind condition the heat flux values anticipated by FDS were similar to semi-empirical calculations performed by previous studies. The propagation pattern in the considered case study for no wind was suggested as Tank 1 $\rightarrow$ Tank 3 $\rightarrow$ Tank 2 $\rightarrow$ Tank 4. Wind was included to analyse its effect on incident heat flux and domino effect escalation. FDS results for non-quiescent wind condition showed a drastic change in the predicted heat flux which affected the propagation pattern. Due to wind in the East, Tank 2 which was less vulnerable under no wind received a surplus amount of flux reaching incident heat flux value from 20.4 kW/m^2 to 59.5 kW/m^2 with a four-order increase in escalation probability from 5. 0 $\times$ 10^{-6} to 1.4 $\times$ 10^{-2}.On the other hand, the heat flux received by Tank 3 is reduced to 17 kW/m^2 from 20.6 kW/m^2 due to wind tilting the flame towards Tank 2. Tank 4 which was not under the influence of domino effect at no wind nearly approached the threshold heat flux value with incident heat flux as 12 kW/m^2 increasing its vulnerability to catch domino effect. The suggested propagation pattern at 3 m/s East wind is Tank 1 $\rightarrow$ Tank 2 $\rightarrow$ Tank 3 $\rightarrow$ Tank 4.

Acknowledgements The authors would like to acknowledge the financial support received from Universiti Teknologi PETRONAS through its YUTP grant number 015LC0-159 to fund this research.

References

1. Delvosalle C (1996) Domino effect phenomena: definition, overview and classification. https://doi.org/10.1007/1-4020-3721-X_2
2. Darbra RM, Palacios A, Casal J (2010) Domino effect in chemical accidents: main features and accident sequences. J Hazard Mater 183:565–573
3. Khakzad N, Amyotre P, Khan F et al (2013) Domino effect analysis using Bayesian networks. Risk Anal 33:292–306
4. Khakzad N (2015) Application of dynamic Bayesian network to risk analysis of domino effects in chemical infrastructures. Reliab Eng Syst Saf 138:263–272
5. Kamil MZ, Taleb-Berrouane M, Khan F et al (2019) Dynamic domino effect risk assessment using Petri-nets. Process Saf Environ Prot 124:308–316

6. Masum Jujuly M, Rahman A, Ahmed S et al (2015) LNG pool fire simulation for domino effect analysis. Reliab Eng Syst Saf Epub ahead of print 2015. https://doi.org/10.1016/j.ress.2015.02.010
7. Baalisampang T, Abbassi R, Garaniya V et al (2017) Fire impact assessment in FLNG processing facilities using computational fluid dynamics (CFD). Fire Saf J 92:42–52
8. Ahmadi O, Mortazavi SB, Pasdarshahri H et al (2019) Consequence analysis of large-scale pool fire in oil storage terminal based on computational fluid dynamic (CFD). Process Saf Environ Prot 123:379–389
9. Hansen OR, Davis S, Gavelli F et al (2013) Onshore explosion studies -benefits of cfd-modelling. Chem Eng Trans 31:205–210
10. McGrattan K, Hostikka S, McDermott R et al (2019) Fire dynamics simulator technical reference guide: mathematical model. Natl Insitute Stand Technol 1:201
11. Carlsson J (1999) Fire modelling using CFD. Dep Fire Saf Eng Jörgen Carlsson Lund Inst Sci Technol Lund Univ Lund, 120
12. Kim BJ, Yoon JY, Yu GC et al (2011) Heat flow analysis of an FPSO topside model with wind effect taken into account: a wind-tunnel test and CFD simulation. Ocean Eng 38:1130–1140
13. McGrattan K, Hostikka S, Floyd J et al (2010) Sixth edition fire dynamics simulator user' s guide. NIST Spec Publ 1019. Epub ahead of print 2010. https://doi.org/10.6028/NIST.SP.1019
14. Mcdermott R, Sixth edition fire dynamics simulator technical reference guide, Vol 1 : Mathematical Model 6.3. 1
15. McGrattan K, Sixth edition fire dynamics simulator technical reference guide, Vol 2: Verification 2
16. McGrattan K, Sixth edition fire dynamics simulator technical reference guide, Vol 3: Validation 3
17. Ryder NL, Sutula JA, Schemel CF et al (2004) Consequence modeling using the fire dynamics simulator. J Hazard Mater 115:149–154
18. Cozzani V, Gubinelli G, Salzano E (2006) Escalation thresholds in the assessment of domino accidental events. J Hazard Mater 129:1–21
19. Koseki H, Kokkala M, Mulholland GW (2006) Experimental study of boilover in crude oil fires. Fire Saf Sci Proc Third Int Symp, 865–874
20. Mandloi D, Lilley D. FDS: grid refinement studies with the fire dynamics simulator code for one-room structural fires with a variety of burning items, 1–22
21. Landucci G, Gubinelli G, Antonioni G et al (2009) The assessment of the damage probability of storage tanks in domino events triggered by fire. Accid Anal Prev 41:1206–1215
22. Mannan S (2012) Lees's loss and prevention in the process industries—hazard identification, Assessment & Control
23. Khakzad N, Khan F, Amyotte P et al (2014) Risk management of domino effects considering dynamic consequence analysis. Risk Anal 34:1128–1138
24. Association AWE. No Title. https://www.asiawind.org/research-data/market-overview/malaysia/ (2019)

Demonstrating Aleatoric Uncertainty in Remaining Useful Life Prediction Using LSTM with Probabilistic Layer

Ahmad Kamal Bin Mohd Nor, Srinivasa Rao Pedapati, Masdi Muhammad, and Mohd Amin Abdul Majid

Abstract A Remaining Useful Life prediction with Aleatoric uncertainty is presented in this paper. A Long Short-Term Memory (LSTM) architecture with probabilistic layer is employed where a normal distribution layer is incorporated to produce the predicted Health Index (HI) distribution of turbofan engines. Compared to the performance of other point estimates techniques in the literature, the probabilistic LSTM achieved a competitive performance in predicting the turbofan's RUL and RUL sequence and have the advantage to express the level of uncertainty along its sequence prediction. This work is important as it reflect a real-world deep learning application where uncertainty indication is needed to evaluate prediction for important decision-making process.

Keywords Aleatoric uncertainty · Probabilistic neural network · Remaining useful life · CMAPPS · Turbofan

1 Introduction

Safety and human lives protection are an utmost priority in the aerospace industry. Sudden failure of aerospace assets during operational mode is the worst-case scenario that demand full attention from industry players and reliability practitioner. To this end, Prognostic and Health Management (PHM) has emerged as one of the key domains in reducing major accidents, enhancing asset's reliability, and improving availability while at the same time optimizing maintenance cost [1, 2].

PHM activities consist of fault detection and isolation, fault diagnostic and prognostic, i.e., predicting time of failure, as well as health management [3]. In the prognostic field, the remaining useful life (RUL) prediction, which estimates the equipment's remaining time before failure is particularly challenging [4].

Currently, 3 categories of RUL prediction techniques are widely employed: model-based, data-driven based and the hybrid of both methods [1, 3, 4]. While

A. K. Bin Mohd Nor · S. R. Pedapati (✉) · M. Muhammad · M. A. Abdul Majid
Mechanical Engineering Department, Universiti Teknologi PETRONAS, Perak 32610 Seri Iskandar, Malaysia
e-mail: srinivasa.pedapati@utp.edu.my

F. Ahmad et al. (eds.), *ICPER 2020*, Lecture Notes in Mechanical Engineering,
https://doi.org/10.1007/978-981-19-1939-8_41

being more accurate in prediction, model-based methods, which models the underlying physic properties of the system, are difficult to develop. Data-driven methods however thrive with the ever-increasing amount of data coupled with intelligent sensor and data-processing power. Amongst all, deep-learning (DL) has proven to be the most powerful in the last few years [2]. This technique essentially consists of training models with historical run-to-failure data, making it to learn the nonlinear relationships from the data in a black-box fashion [3].

As a black-box approach, the prediction of DL models is not easily interpretable. Most of the available work in the literature were developed to predict scalar point estimates without any account of uncertainty. Naturally, this hinder its application in the real-world, safety-sensitive, high investment domains where uncertainty of the estimation is vital before any costly decision-making. As an example, according to our founding, only 4 out of 282 publications [5–8] linked to the famous CMAPPS Turbofan dataset from 2008 to this day include the presentation of prediction uncertainty. However, none explain in detail the interpretation of uncertainty obtained. Literature review is done on the publications presented in a CMAPPS dataset systematic review paper [9] (106 papers) together with additional publication until July 2020 (176 papers) using the methodology presented in [9] on 9 libraries: IEEE Xplore, ScienceDirect, Springer, AIS, Wiley Online Library, Taylor & Francis, Sage Publishing, Emerald Insight and Scopus.

1.1 Uncertainty in Deep Learning.

Uncertainty in deep learning linked to the quality of input data is known as Aleatoric uncertainties [10, 11, 13–15]. This uncertainty can happen due to noise, data acquisition error or stochasticity captured in the input data, which is the usual situation encountered in the real world [10, 12–15]. This type of uncertainty cannot be reduced further by having more data if no improvement is done on the data acquisition method.

This paper analyzes and interpret the Aleatoric uncertainty of turbofan engine RUL predictions using LSTM neural network with probabilistic layer. It puts into light the randomness in the output prediction linked to deep learning that need to be considered especially for vital application's decision-making process.

2 Related Works

Uncertainty quantification work in deep learning is currently being used in many different fields, catering for specific application. Bayesian deep learning (BDL) is incorporated in convolution neural network (CNN), and Bidirectional LSTM in [6] to predict the remaining useful life of bearing with uncertainty. In [16] a probabilistic normalized convolutional neural network (NCNNs) is used on a depth completion dataset. A prognostic method with uncertainty quantification involving hydraulic mechanisms of circuit breakers using BDL coupled with a sequential Bayesian boosting algorithm is presented in [17]. In [18], a diagnostic of Diabetic retinopathy by DR|Graduate, a deep learning-based DR grading CAD system framework with uncertainty is proposed. A probabilistic Super-Resolved q-Space deep learning method is applied in [19] to measure brain tissue microstructure. Monte Carlo dropout procedure is executed in [20] to provide confidence in neural network leaf skeleton photo reconstruction. A Bayesian deep multisource learning (BDMSL) is used to diagnose glaucoma disease with uncertainty in [21]. In [22], a BDL with Variational Inference and Monte Carlo method is proposed to study the influence of inflow uncertainty on reservoir operation rules in the Three Gorges Dam of the Yangtze River. Input-dependent noise model and variational dropout is used in [23] to provide uncertainty in medical image enhancement problem. Bayesian inference is used to predict plant problem with uncertainty in [24].

3 Methodology

3.1 LSTM Architecture with Probabilistic Layer.

Regression is done using LSTM architecture whose optimum layers and hyperparameters obtained via Bayesian hyperparameter optimization [25]. 500 evaluations are done to find the best hyperparameters in the search space presented in Table 1 that minimize the validation error. The training target are HI corresponding to the normalized value of the RUL target for each cycle. A Normal distribution layer is incorporated as the output layer of the LSTM network, outputting both mean and standard deviation of the HI prediction.

LSTM architecture is increasingly used in time-series data modelling due to the capability of learning long-term dependencies [9]. LSTM employs gating mechanism to regulate the data flow and retain important memory or forget invaluable ones [26]. LSTM consists of Input Layer, Hidden Layer and Output Layer with a recurrent mechanism enabling the output of the Hidden Layer to be added with new input. This enable the modeling of sequential data.

Table 1 BayesOpt search space

Layers	LSTM hidden units	Dense layer size	Dropout rate	Mini batch size	Max epoch	Learning rate
Search space						
1 LSTM 1 Dense 1 Dropout	10 to 1000	10 to 500	0.000 to 0.500	26 to 130	10 to 2000	5e-4 to 1e-3

LSTM gating equation:

$$f_t = \sigma\left(W_f.X_t + R_f.h_{t-1} + b_f\right) \tag{1}$$

$$i_t = \sigma(W_i.X_t + R_i.h_{t-1} + b_i) \tag{2}$$

$$o_t = \sigma(W_o.X_t + R_o.h_{t-1} + b_o) \tag{3}$$

$$M_t = \varphi(W_c.X_t + R_c.h_{t-1} + b_c) \tag{4}$$

$$C_t = f_t * C_{t-1} + i_t * M_t \tag{5}$$

$$h_t = o_t * \varnothing(C_t) \tag{6}$$

W* = Input Weight

R* = Recurrent Weight

b* = bias

$\sigma, \varphi, \varnothing$ = activationfunctions

Equations (1) to (3) are Forget Gate, Input Gate and Output Gate equations. Equation (4) is used to calculate the candidate value that will be added to the new state "t" (5) together with the old state "t–1", regulated by the Forget Gate. (6) represents the final output of LSTM. "." is matrix multiplication while "" is point to point multiplication [27].

3.2 *Network Performance Assessment.*

We assess the performance of our network by calculating the Root Mean Squared Error (RMSE) and Scoring Function of the obtained RUL prediction as respectively shown in Eqs. (7) and (8), (9) and (10) [28]. The scoring function gives higher score for the same error in early prediction than late prediction. It penalizes late prediction than the early ones.

$$RMSE = \sqrt{\frac{1}{M}\sum_{i=1}^{M}\left(RUL_{truth}^{(i)} - RUL_{predicted}^{(i)}\right)^2} \tag{7}$$

$$s = M\sum_{i=1}^{M} s_i \tag{8}$$

$$s = \left\{ \begin{array}{l} e^{\frac{-d_i}{13}} - 1, d_i < 0 \\ e^{\frac{di}{10}} - 1, d_i > 0 \end{array} \right\} \quad (9)$$

$$di = \left(RUL_{predicted} - RUL_{true}\right) \quad (10)$$

3.3 Uncertainty Quantification of RUL and of HI Sequence Prediction

The LSTM will predict a sequence of HI normal distributions time series for each fleet. Then we sample 1000 points and calculate the mean,$\mu_{(i)}$ and standard deviation, $\sigma_{(i)}$ of each distribution, obtaining a sequence of HI distribution mean for each fleet. We obtain the RUL prediction by taking the first HI distribution mean in the sequence and reverse the normalization. We plot the predicted HI distribution mean sequence together with standard deviations as Upper and Lower bounds as shown in Eqs. (11) and (12), to visualize the area of uncertainty. The Aleatoric Uncertainty is represented by the distance between the predicted HI distribution mean curve and the Upper and Lower bounds. To have more detail about the uncertainty of the prediction, we plot the standard deviations versus time and observe their variation behavior.

$$Upper\ Bound_{Fleet(i)} = \mu_{(i)} + 2\sigma_{(i)} \quad (11)$$

$$Lower\ Bound_{Fleet(i)} = \mu_{(i)} - 2\sigma_{(i)} \quad (12)$$

4 Case Study

CMAPPS ((Commercial Modular Aero Propulsion System Simulation) Turbofan runto-failure datasets were published in 2008 by Nasa Prognostic Centre (PCoE) of Ames 2 Research Centre (http://ti.arc.nasa.gov/projects/dataprognostics) consisting of 4 complete sets of Training, Testing and ground truth RUL for numerous turbofan engines (Fig. 1). The simulated data was obtained by simulating a variety of operational conditions and injecting faults of varying degree degradation.

The chosen training dataset and testing data consist of 260 recorded turbofan degradations. Each Training and Testing data correspond to a turbofan whose health condition deteriorate after certain cycle [29]. Each fleet training and testing data are time series sequence comprising of Time (Cycle), 3 Operating Conditions (OC) and 21 sensors measurements as in Table 2, without any RUL information. The OC refers to different operating regimes combination of Altitude (O-42 K ft.), Throttle

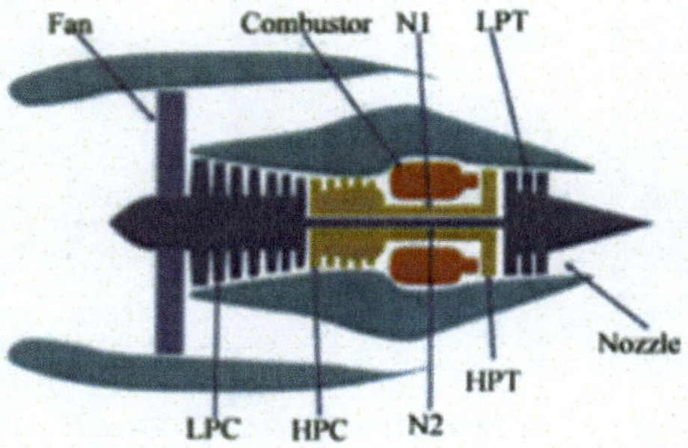

Fig. 1 Turbofan engine diagram [30]

Table 2 Sensors description

Sensor	Ref	Description	Unit
S1	T2	Total temperature fan inlet	^{0}R
S2	T24	Total temperature at LPC outlet	^{0}R
S3	T30	Total temperature at HPC outlet	^{0}R
S4	T50	Total temperature at LPT outlet	^{0}R
S5	P2	Pressure at fan inlet	psia
S6	P15	Total pressure in bypass-duct	psia
S7	P30	Total pressure at HPC outlet	psia
S8	Nf	Physical fan speed	rpm
S9	Nc	Physical core speed	rpm
S10	Epr	Engine pressure ratio (P50/P2)	N/A
S11	Ps30	Static pressure at HPC outlet	psia
S12	Phi	Ratio of fuel flow to Ps30	Pps/psi
S13	NRf	Corrected fan speed	rpm
S14	NRc	Corrected core speed	rpm
S15	BPR	Bypass ratio	N/A
S16	farB	Burner fuel–air ratio	N/A
S17	htBleed	Bleed enthalphy	N/A
S18	Nf_dmd	Demanded fan speed	rpm
S19	PCNfR_dmd	Demanded corrected fan speed	rpm
S20	W31	HPT coolant bleed	lbm/s
S21	W32	LPT coolant bleed	lbm/s

Table 3 Signal trend

Monotonic signal
2,3,4,7,8,9,11,12,13,14,15,17,20,21

Resolver Angle (20–100), and Mach Number (0–0.84) [30]. High levels of noise are incorporated, and the faults encountered are hidden by the effect of various operational conditions [30]. Only the ground truth RUL for each turbofan, are given. We thus need to do data preparation to select the appropriate sensors, visualize the sensor trends and extract each cycle target HI.

4.1 Degradation Assumption

Each fleet might be used in different operating conditions. As such the extend of degradation is different from one another. Sensors information extracted are crucial in addition to flight duration and flight conditions to model the accumulated degradation. The degradation is mainly due to wear and tear due to usage pattern that progress into component faults. The type of component fault could be single or a combination of any of these failures: Fan, Low Pressure Compressor (LPC), High Pressure Compressor (HPC), High Pressure Turbine (HPT) and Low-Pressure Turbine (LPT) (Table 1) [30].

4.2 Data Preparation

4.2.1 Sensor Selection

14 strictly monotonic sensors are selected as presented in Table 3 [28, 31]. Strictly monotonic sensors are useful as they best represent single trending degradation contrary to irregular and unchanged signals. The corresponding sensors are sensors 2,3,4,7,8,9,11,12,13,14,15,17,20,21.

4.2.2 Piece-Wise Linear Degradation

To obtain RUL sequence labels for training, piece-wise linear degradation model is employed [32, 33]. Each fleet health is thus considered stable in the beginning until the start of deterioration which is the failure start point that initiate a linear degradation until failure.

Each fleet is associated with time series corresponding to the total operational duration of the fleet and the last cycle indicate the final instant before failure. Thus initially, we model the RUL of a fleet to start with the value of the last cycle and

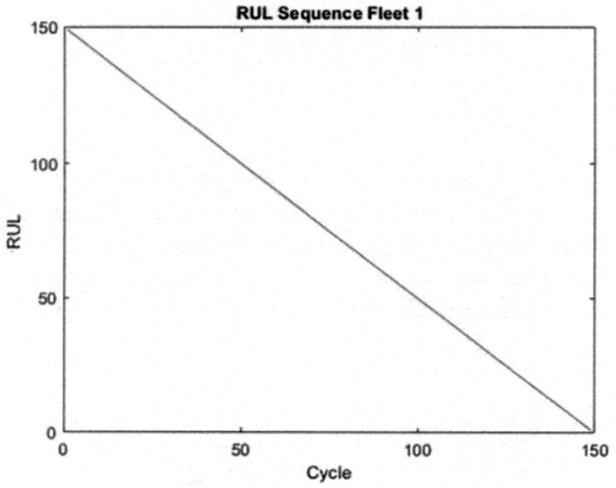

Fig. 2 Initial RUL for Fleet 1

Table 4 Mean of sensor's response means

Sensor	S2	S3	S4	S7	S11	S12	S15	S20	S21	Mean
Cycle	49	51	50	62	44	56	50	65	58	**53.8**

degrade linearly until 0. As seen in Fig. 2, Fleet 1 has a cycle that start from 1 to 149. Considering a linear degradation, Fleet 1 initially has an RUL of 149 that decreases until 0.

The failure start point in each sensor is obtained by using Cumulative Sum (CUSUM) anomaly detection technique, which returns the first index of the upper and lower cumulative sums of each sensor signal that have drifted beyond 5 standard deviations above and below a target mean, indicating the initiating points of degradation [34]. Then, the mean of the failure start points for each sensor can then be calculated. From these means, we can calculate the mean of the means of the sensor's detection, which is our failure start point. The means are presented in Table 4.

Combining the linear degradation obtained earlier and the failure start point, the transformed Fleet 1 RUL sequence is presented in Fig. 3.

4.2.3 Final RUL Quantification Process and HI Establishment

To enable generalization, we set the RUL that are above 100 to 100. We then normalize the new RUL into HI using Eq. (13) that takes the value of 1 (healthy) and 0 (failure). The piece-wise HI sequence curve is shown in Fig. 4 [35].

$$HI_{(t)} = \frac{RUL_{max} - RUL_{(t)}}{RUL_{max} - RUL_{min}} \tag{13}$$

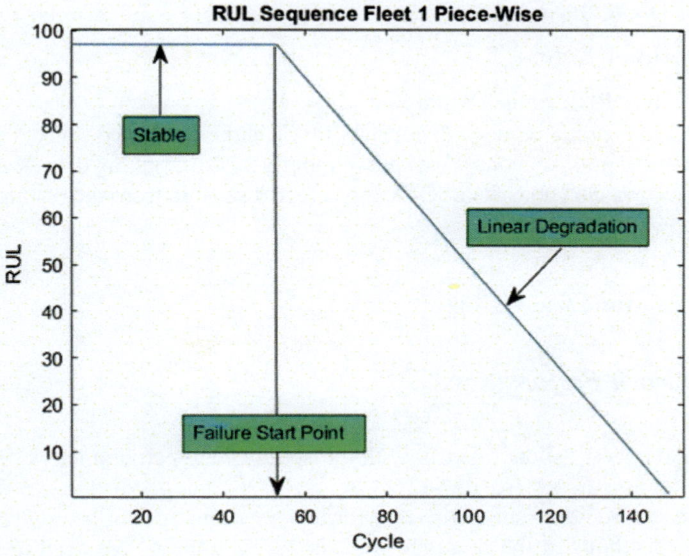

Fig. 3 Final RUL sequence Fleet 1

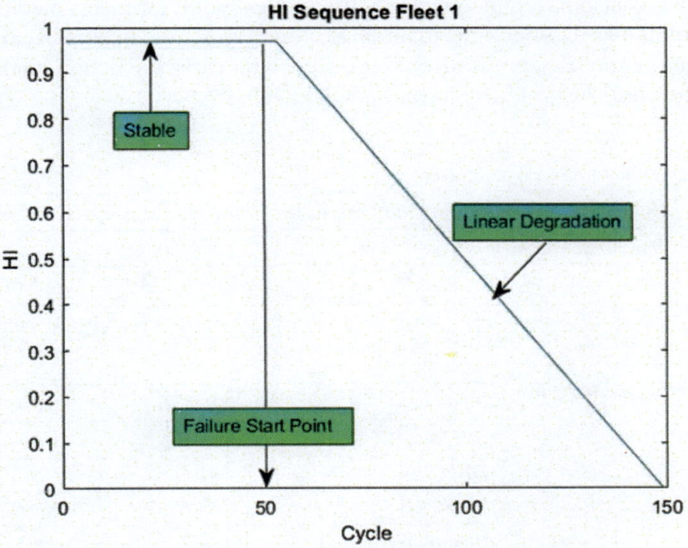

Fig. 4 Final HI Fleet 1

$RUL_{(t)}$ = RUL at time t

RUL_{max} = Health State RUL

RUL_{min} = Final RUL Before Failure

The LSTM training inputs will consist of OCs and the selected sensors while the HI will be the training response. We split randomly 20% of training data as validation data. The testing data consists of OCs and selected sensors from the testing dataset.

5 Result and Discussion

5.1 Network Performance

We obtained lowest validation error with the optimized hyperparameters as listed in Table 5.

The network achieves competitive performance compared to other point estimates data-driven methods in the literature like shown in Table 6. We need to consider that this is just a simple LSTM neural network compared to other more complicated structures such as improved Grey Model (GM) [38] and DeepCNN [37]. The training of the model is also different and more difficult than point estimates methods as the model outputs distributions rather than single point prediction. In term of early failure prediction, the model score is also competitive with early prediction amounting to 67% of the whole fleet of turbofan as shown in Table 7.

Table 5 BayesOpt optimized parameters

Layers	Hidden units	Dense size	Mini batch size	Dropout	Max epoch	Learning rate
1 LSTM	*Optimized Parameters*					
1 Dense 1 Dropout	635	331	43	0.03	14	5.613e-4

Table 6 RMSE and score of models

Method	RMSE	Score
SVM [36]	51.5	N/A
DNN [36]	41.8	N/A
MLP [37]	37.6	17,972
Proposed Method	**24.1**	**1095**
Improved GM [38]	23.0	1193
DeepCNN [37]	18.5	1287

Table 7 Early RUL prediction versus late RUL prediction

Early	Late
67%	33%

5.2 Aleatoric Uncertainty for HI Sequence Prediction

For illustration purpose, we present the HI and HI sequence result of Turbofan 2 and 10. As shown in Figs. 5 and 7, the HI mean, or RUL mean prediction of the Turbofan 2 and 10 are not far from the ground truth. We can also see from Figs. 6 and 8 that the model gives early failure degradation prediction for both fleets.

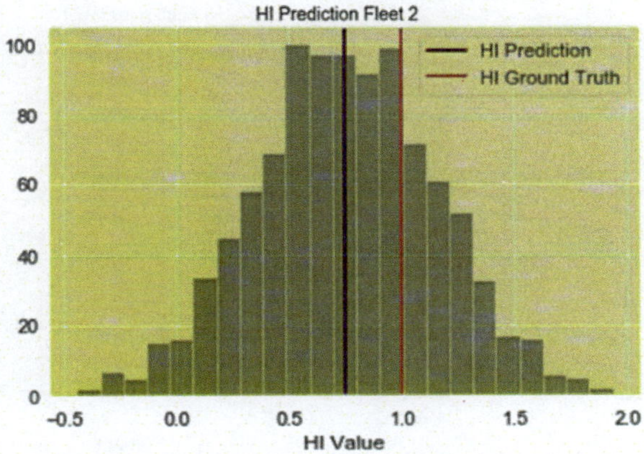

Fig. 5 HI prediction distribution Fleet 2

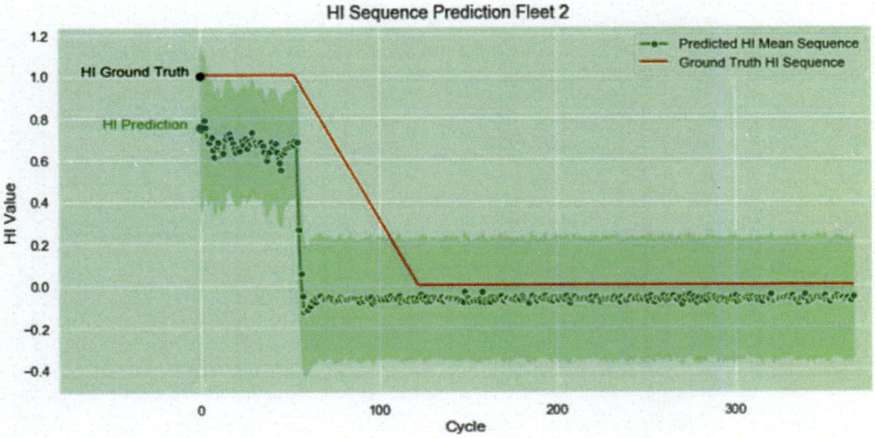

Fig. 6 HI sequence prediction Fleet 2

From the illustrations given in Figs. 6 and 8 we can see that the model gives early HI prediction for Fleet 2 and 10. We can also verify from Figs. 6 and 8 that the predicted HI sequence for these fleets follow the trend of the ground truth HI sequence.

From the standard deviation plots in Figs. 9 and 10, we can see that all cases start with a high uncertainty before decreasing to a low point. It signifies the high uncertainty associated with predicting the initial HI in each case, translated as the imprecise prediction of RUL.

Then, the uncertainty shows monotonic fluctuations for a certain period before increasing rapidly. These oscillations signify that the model predicts with a near-stable level of uncertainty on the true healthy state of the engine which is monotonic.

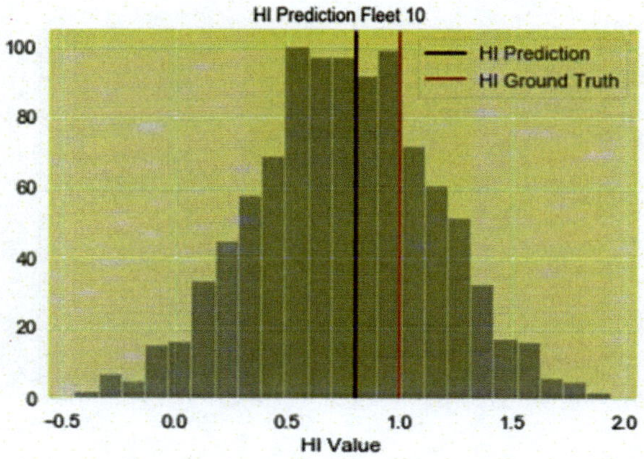

Fig. 7 HI prediction distribution Fleet 10

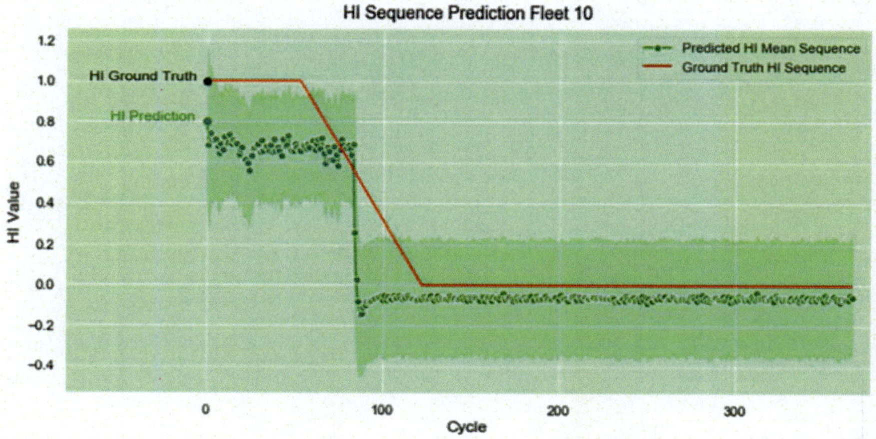

Fig. 8 HI sequence prediction Fleet 10

As shown in Figs. 6 and 8, the predicted sequence seems to follow the ground truth trend, but a gap exists between the predicted HI mean sequence and ground truth sequence, thus the oscillation. Then the sudden increase in uncertainty value indicates that the model predicted the start of deterioration as shown in Figs. 9 and 10. As we can see in Figs. 6 and 8, the HI sequence prediction decrease is steeper than the ground truth curve. Then the uncertainty decreases gradually to a small fluctuation until the end of the cycle as shown in Figs. 6 and 8 as well as 9 and 10. Along the decrease, we see in Figs. 6 and 8 that the model predicted the HI to be 0 earlier than the ground truth, thus explaining the high uncertainty before stabilization as shown in Figs. 9 and 10. Finally, like the healthy state, the model predicted with a near-stable level of uncertainty the end-of-life state of the turbofan, which is also monotonic.

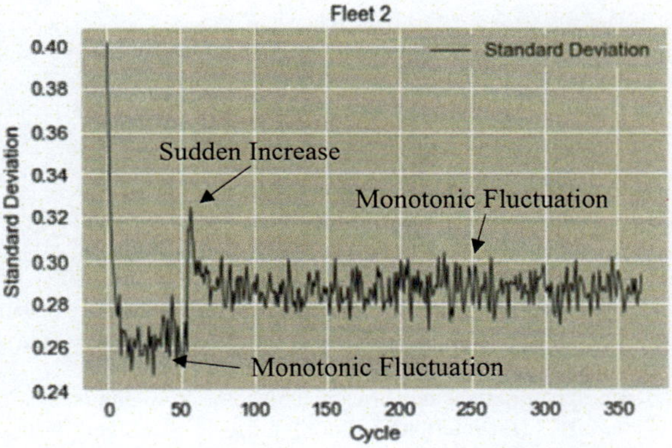

Fig. 9 HI prediction uncertainty Fleet 2

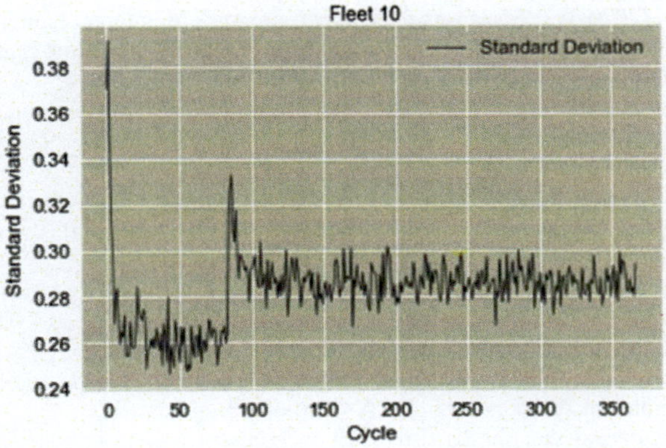

Fig. 10 HI prediction uncertainty Fleet 10

6 Conclusion

In this paper, the Aleatoric uncertainty influenced by noisy sensorial data is put into evidence in an RUL prediction problem using a probabilistic LSTM neural network. Normal distribution layer is incorporated at the output layer of the neural network to produce the predicted Health Index (HI) distribution. The network achieved a competitive performance in predicting the RUL and RUL sequence of turbofan fleets compared to other point estimate methods. By examining the bounded HI sequence prediction plot and the standard deviations, we show and interpret the level of uncertainty along the time series prediction.

Acknowledgements The authors would like to acknowledge Dr. Mohd Amin Abdul Majid and the sponsorship of Newton Link Grant No. 015ME0-015 for this conference paper.

References

1. Si X, Ren Z (2019) A data-fusion based prognostic method for complex degrading system. 2019 Prognostics and system health management conference (PHM-Qingdao), Qingdao, China, pp 1–6. https://doi.org/10.1109/PHM-Qingdao46334.2019.8943070
2. Ellefsen AL, Ushakov S, Æsøy V, Zhang H (2019) Validation of data-driven labeling approaches using a novel deep network structure for remaining useful life predictions. IEEE Access 7:71563–71575. https://doi.org/10.1109/ACCESS.2019.2920297
3. Li F et al (2018)A light gradient boosting machine for Remainning useful life estimation of aircraft engines. 2018 21st international conference on intelligent transportation systems (ITSC), Maui, HI, pp 3562–3567, https://doi.org/10.1109/ITSC.2018.8569801
4. Ge Y, Sun L, Ma J (2019) An Improved PF remaining useful life prediction method based on quantum genetics and LSTM. IEEE Access 7:160241–160247. https://doi.org/10.1109/ACCESS.2019.2951197
5. Kim M, Liu K (2020) A Bayesian deep learning framework for interval estimation of remaining useful life in complex systems by incorporating general degradation characteristics. IISE Trans,https://doi.org/10.1080/24725854.2020.1766729
6. Peng W, Ye Z, Chen N (March 2020) Bayesian deep-learning-based health prognostics toward prognostics uncertainty. IEEE Trans Industr Electron 67(3):2283–2293. https://doi.org/10.1109/TIE.2019.2907440
7. Xi X, Chen M, Zhang H, Zhou D (2018) An improved non-Markovian degradation model with long-term dependency and item-to-item uncertainty. Mech Syst Signal Process 105:467–480. https://doi.org/10.1016/j.ymssp.2017.12.017
8. Wang P, Youn BD, Hu C (2012) A generic probabilistic framework for structural health prognostics and uncertainty management. Mech Syst Signal Process Mech Syst Signal Process, 28,https://doi.org/10.1016/j.ymssp.2011.10.019
9. Zschech P, Bernien J, Heinrich K (2019) Towards a taxonomic benchmarking framework for predictive maintenance: the case of NASA's Turbofan Degradation
10. Harakeh A (2017) Adversarial Robustness of uncertainty aware deep neural networks
11. Kendal, A, Gal Y (2017) What uncertainties do we need in Bayesian deep learning for computer vision?
12. Prado A, Kausik R, Venkataramanan L (2019) Dual neural network architecture for determining epistemic and Aleatoric uncertainties

13. Gustafsson FK, Danelljan M, Schon TB (2020) Evaluating scalable Bayesian deep learning methods for robust computer vision. 2020 IEEE/CVF conference on computer vision and pattern recognition workshops (CVPRW), Seattle, WA, USA, 2020, pp 1289–1298, https://doi.org/10.1109/CVPRW50498.2020.00167
14. Le MT, Diehl F, Brunner T, Knol A (2018) Uncertainty estimation for deep neural object detectors in safety-critical applications. 2018 21st international conference on intelligent transportation systems (ITSC), Maui, HI, 2018, pp 3873–3878, https://doi.org/10.1109/ITSC.2018.8569637
15. Ghoshal B, Tucker A, Sanghera B, Wong WL (2019) Estimating uncertainty in deep learning for reporting confidence to clinicians when segmenting Nuclei image data. 2019 IEEE 32nd international symposium on computer-based medical systems (CBMS), Cordoba, Spain, 2019, pp 318–324,https://doi.org/10.1109/CBMS.2019.00072
16. Eldesokey A, Felsberg M, Holmquist K, Persson M (2020) Uncertainty-aware CNNs for depth completion: uncertainty from beginning to end. 2020 IEEE/CVF conference on computer vision and pattern recognition (CVPR), Seattle, WA, USA, 2020, pp 12011–12020, https://doi.org/10.1109/CVPR42600.2020.01203
17. Li G, Yang L, Lee C, Wang X, Rong M, A Bayesian deep learning RUL framework integrating epistemic and Aleatoric uncertainties. In: IEEE transactions on industrial electronics, https://doi.org/10.1109/TIE.2020.3009593
18. Araújo T, Aresta G, Mendonça L, Penas S, Maia C, Carneiro A, Mendonça AM, Campilho A (2020) DR|GRADUATE: uncertainty-aware deep learning-based diabetic retinopathy grading in eye fundus images. Medical Image Anal 63:101715, ISSN 1361-8415,https://doi.org/10.1016/j.media.2020.101715
19. Qin Y, Liu Z, Liu C, Li Y, Zeng X, Ye C (2021) Super-resolved q-space deep learning with uncertainty quantification. Med Image Anal 67:101885, ISSN 1361-8415,https://doi.org/10.1016/j.media.2020.101885
20. Godefroy G, Arnal B, Bossy E (2021) Compensating for visibility artefacts in photoacoustic imaging with a deep learning approach providing prediction uncertainties. Photoacoustics 21:100218, ISSN 2213-5979,https://doi.org/10.1016/j.pacs.2020.100218
21. Chai Y, Bian Y, Liu H, Li J, Xu J (2021) Glaucoma diagnosis in the Chinese context: an uncertainty information-centric Bayesian deep learning model. Inform Process Manag 58(2):102454, ISSN 0306–4573, https://doi.org/10.1016/j.ipm.2020.102454
22. Liu Y, Qin H, Zhang Z, Yao L, Wang Y, Li J, Liu G, Zhou J (2019) Deriving reservoir operation rule based on Bayesian deep learning method considering multiple uncertainties. J Hydrol 579:124207, ISSN 0022-1694,https://doi.org/10.1016/j.jhydrol.2019.124207
23. Tanno R, Worrall DE, Kaden E, Ghosh A, Grussu F, Bizzi A, Sotiropoulos SN, Criminisi A, Alexander DC (2021) Uncertainty modelling in deep learning for safer neuroimage enhancement: demonstration in diffusion MRI. NeuroImage 225:117366, ISSN 1053-8119,https://doi.org/10.1016/j.neuroimage.2020.117366
24. Hernández S, López JL (2020) Uncertainty quantification for plant disease detection using Bayesian deep learning. Appl Soft Comput 96:106597, ISSN 1568-4946,https://doi.org/10.1016/j.asoc.2020.106597
25. Ruben M (2014) BayesOpt: a Bayesian optimization library for nonlinear optimization, experimental design and bandits. J Mach Learn Res 15(1):3735–3739
26. Sherstinsky A (2018) Fundamentals of recurrent neural network (RNN) and long short-term memory (LSTM) network
27. Zheng S, Ristovski K, Farahat A, Gupta C (2017) Long short-term memory network for remaining useful life estimation, 88–95. https://doi.org/10.1109/ICPHM.2017.7998311
28. Ge Y, Wu J, Jiang X (2019) A prediction method using bayesian theory for remaining useful life. 2019 international conference on quality, reliability, risk, maintenance, and safety engineering (QR2MSE), Zhangjiajie, China, pp 856–862, https://doi.org/10.1109/QR2MSE46217.2019.9021252
29. Ramasso E, Saxena A (2014) Performance benchmarking and analysis of prognostic methods for CMAPSS datasets. Int J Prognost Health Manag 5:1–15

30. Saxena A, Goebel K, Simon D, Eklund N (2008) Damage propagation modelling for aircraft engine run-to-failure simulation. International conference on prognostics and health management.https://doi.org/10.1109/PHM.2008.4711414.M
31. Gao Y, Zhou J, Wu K, Zhao G, Hu C (2019) Construction method of turbine engine health indicator based on deep learning. 2019 prognostics and system health management conference (PHM-Qingdao), Qingdao, China, pp 1–6, https://doi.org/10.1109/PHM-Qingdao46334.2019.8943055
32. Wang Q, Zheng S, Farahat A, Serita S, Gupta C (2019) Remaining useful life estimation using functional data analysis. 2019 IEEE international conference on prognostics and health management (ICPHM), San Francisco, CA, USA, 2019, pp 1–8, https://doi.org/10.1109/ICPHM.2019.8819420
33. Li J, Li X, He D (2019) Domain adaptation remaining useful life prediction method based on AdaBN-DCNN. 2019 prognostics and system health management conference (PHM-Qingdao), Qingdao, China, pp 1-6,https://doi.org/10.1109/PHM-Qingdao46334.2019.8942857
34. "Detect small changes in mean using cumulative sum" MATLAB MathWorks® https://www.mathworks.com/help/signal/ref/cusum.html
35. Heimes F (2008) Recurrent neural networks for remaining useful life estimation. In: IEEE international conference on prognostics and health management
36. Bruneo D, De Vita F (2019) On the use of LSTM networks for predictive maintenance in smart industries. 2019 IEEE international conference on smart computing (SMARTCOMP), Washington, DC, USA, pp 241–248, https://doi.org/10.1109/SMARTCOMP.2019.00059
37. Babu GS, Zhao P, Li XL (2016) Deep convolutional neural network-based regression approach for estimation of remaining useful life. 21th international conference on database systems for advanced applications (DASFAA). Dallas, USA: Spinger, pp 214–228
38. Peng K, Pi Y, Jiao R, Dong J, Zhang K, Zhang C (2019) Remaining useful life prediction for aircraft engines based on grey model. 2019 prognostics and system health management conference (PHM-Qingdao), Qingdao, China, pp. 1–5,https://doi.org/10.1109/PHM-Qingdao46334.2019.8943000

Monitoring the Changes in State to Indicate Faults in Gas–Liquid Pipelines

Seshu Kumar Vandrangi, Tamiru Alemu Lemma, Syed Muhammad Mujtaba, and Srinivasa Rao Pedapati

Abstract The pipeline accidents due to faults (leakage and blockage) have been leading to major environmental issues. Various techniques have been established to detect faults in a pipeline. But each having pros and cons. Thus, there is a necessity for early and accurate fault detection to prevent failures. To identify the leaks in the pipelines, there is a necessity to monitor variables. A two-phase transient model is used for two different operating conditions (with leakage and without a leak) for the variable analysis. An OLGA (OiL GaS) based model is used to simulate a leakage and generate data for monitoring variables such as pressure, velocity, rate of flow, and, temperature at inlet and outlet over the flow of pipeline. The leak locations and sizes were varied, and respective variables were studied. The leak sizes were assorted for 0% (no leak), 20% and 50% leak opening, and the inlet and outlet variables were analyzed. Except for the temperature, all other variables are increasing with the movement of the leak location. The mass flow rate has been decreased 0.91 times the original mass flow rate for 20% leak opening condition. The percentage change of variable is relatively constant from the point of the leak. And, a dip of 0.77 times was observed for a 50% leak opening condition which is maintained constant as mentioned above.

Keywords Fault detection · Variables · Two phase · Pipeline accidents · Mass rate flow

1 Introduction

The first commercial oil well was drilled in Titusville, Pennsylvania, in 1859, as a result of the need for oil production. Yearly oil and gas with a value measured in millions of dollars is transported on both land and undersea, and even a minor loss might result in significant economic consequences. Even the most advanced leak detection is relatively little in terms of pipeline construction and installation costs.

S. Kumar Vandrangi · T. Alemu Lemma (✉) · S. Muhammad Mujtaba · S. Rao Pedapati
Department of Mechanical Engineering, Universiti Teknologi PETRONAS, 32610 Seri Iskandar, Perak, Malaysia
e-mail: tamiru.lemma@utp.edu.my

F. Ahmad et al. (eds.), *ICPER 2020*, Lecture Notes in Mechanical Engineering,
https://doi.org/10.1007/978-981-19-1939-8_42

Spillage is the most serious problem that may develop in a pipeline setup, and it is currently a terrible reality in the field.

The issue of fault diagnosis determination comprises initial distinguishing the error through detection, finding the origin of the issue through isolation and afterwards recognize its extent through identification. Fault diagnosis isn't just significant but also a key if a tolerant fault control framework is to be executed [1]. Fault diagnosis methods are categorized into (i) data driven-based and (ii) model-based techniques [2]. Data-driven techniques use existing complex systems data to determine occurring faults, whereas model-based techniques are naturally used if mathematical models are available. Fault detection is defined as the determination of the presence of a fault, whereas fault isolation can be defined as the determination of the kind and location of the fault [3]. The potential of a system identification technique for leak detection in gas mixture pipelines has been tested by Mujtaba et al. [4]. The recent developments in leak detection techniques in single-phase and two-phase flow pipelines can be studied through various review and research articles [5, 6]. The classification of leak detection systems based is presented in Fig. 1.

Multi-phase pipe flow through a solitary tube is a zone vital to the oil and gas industry. Multi-phase stream in pipes isn't limited to the oil business; two- and three-phase streams are experienced in numerous other engineering applications such as drainage, plants etc. [7].

The research until now has dealt with steady-state and single-phase flow analysis. But there is no proper analysis on variables affecting the leak in multi-phase flow pipelines. This current research monitors the variables through OLGA, a multi-phase

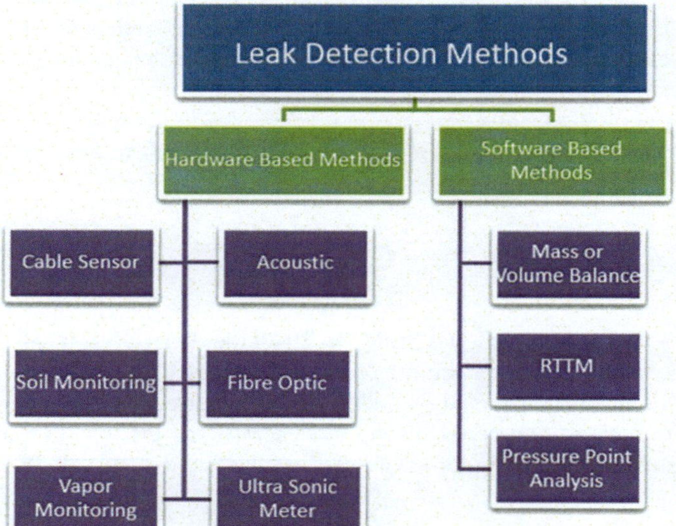

Fig. 1 Categorization of oil/gas pipeline leak detection systems based on their technicality [7]

simulator for leak analysis by inducing and modeling a leak. The research article is organized as below mentioned. The first section introduces the background study. The second section briefs about various fault detection techniques. The third section gives details on OLGA methodology and boundary conditions. The results are plotted and discussed in the fourth section, with the conclusions in the fifth section.

2 Fault Detection Techniques

In multi-phase flow systems, erroneous calculations in the design or incorrect operation can result in serious mishaps. All of these acts can be traced back to a failure to assess actual flow patterns. As a result, flow identification and prediction are critical in the calculation of flow resistance, the criterion of flow stability, and the technical application of heat transfer or pressure drop theory. Various fault detection methods are further examined.

2.1 Backpressure Technique

Scott et al. [8] have concluded that fluids distributing from a pipeline lead to an apparent loss of transportability of the pipeline, i.e. "transmission efficiency" is decreased. Critical flow can be termed as the leak occurring at a high flow rate such that sonic velocity is attained at the wall.

As shown by Scott et al. [8], the gas flow rate at the outlet in a multi-phase flow pipeline undergoing a leak is given as a function of inlet pressure and outlet pressure in the below-mentioned form

$$q_{sc} = F_{Leak}\left(F_{2-\varphi}\right)_q \left(CZTf_{sg}L_p/d^5\right)^{-0.5}\left(P_{in}^2 - P_{out}^2\right)^{0.5} \tag{1}$$

where Z is the real gas compressibility factor and "C is a constant", "f is the friction factor", and "d is the diameter of the pipe". Here the superficial single-phase conditions are denoted by "SG". The term (F_2-φ), is known as two-phase efficiency (2-φ), is defined as

$$\left(F_{2-\varphi}\right) = (dP/dx)_{SG}/(dP/dx)_{2-\varphi} \tag{2}$$

Comparing the Eqs. (1) and (2), the two-phase term creates a flow regime dependent term. It separates the leak of single-phase flow from the multi-phase flow.

When leaks are greater than 1.5″ in vertical pipes, according to Scott and Maria [7], they are relatively easy to identify. Leaks of 1 to 2 inches in horizontal pipes, on the other hand, are difficult to detect since the fluctuations in these circumstances are very small and require the use of a high precision instrument to detect. Leaks smaller than 1 inch, on the other hand, cannot be identified. Griffin and Smith [9] confirm the detection method used by Scott and Maria [7] for leaks in deep water, multi-phase pipelines.

2.2 Transient Modeling

Another approach for estimating flow rates that uses conservation equations is the RTTM ("Real-Time Transient Method") for leak detection. The difference between the predicted and measured values of the flow variable determines the leak. The above-mentioned methods are used in conjunction with flow, temperature, velocity, and pressure measurements in this method [7]. The modelling of multi-phase leaks has no virtual experiment work so far. The performance of a flowline undergoing a leak can be estimated with the advanced transient multi-phase flow simulators. Wang and John [10] introduce an LDS approach based on transient modelling in their research. On a SCADA-based data acquisition system, Fig. 2 depicts how this strategy works.

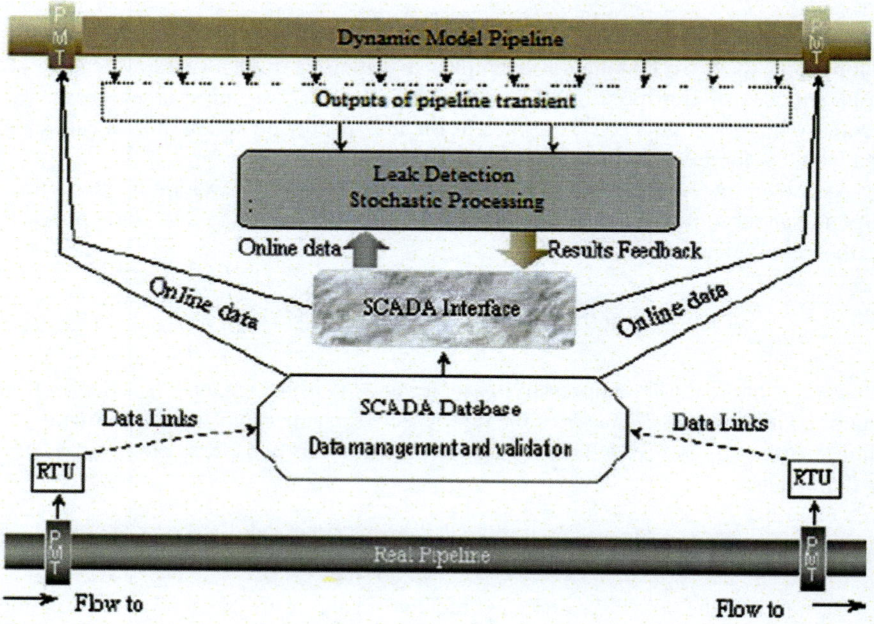

Fig. 2 The leak detection method based on SCADA [10]

2.3 OLGA Modeling

Figueiredo et al. [11] provided a single-pressure two-fluid flow model with a leak that may be used to simulate stratified-pattern two-phase flows in a horizontal pipeline with a leak. For two-phase streams, leak localization model testing is based on the "intersection of the hydraulic grade lines." Using the OLGA test equipment, pipeline reactions under zero (no) spill conditions and practical heterogeneous leak characteristics were replicated on a gas pipeline. Simulated leaks were shown to have distinct flow rate characteristics that could be identified by the ANN [12].

The SRPT, NPW approach, and pressure gradient approaches are used to create an LDS with a leak detection and localization module. This software also includes a model identification architecture to reduce "false alarms." The OLGA simulation findings are comparable to the PIPESIM results. It has been confirmed that the suggested leak detection system has good sensitivity and accuracy [13]. Aside from that, many researchers use alternative methodologies and strategies [9, 14–18].

The researchers employed a variety of leak detection technologies, and leak localization can be either inside or outside. Each method has its own set of benefits and drawbacks. Dynamic modelling, however, distinguishes out among them. The idea of operation is to detect leaks by looking for inconsistencies between measured data and generated values based on conservation equations and the fluid's EOS. A notable strength is that it may be used for leak detection and localisation, and it can handle a big volume of data quickly. A key flaw is the high computational complexity, which is costly and labour intensive.

Since hardware-based solutions are expensive, a software-based strategy that monitors the variables important for leak detection and localisation is utilised instead. Modeling a pipeline and inducing a single leak at various locations and for various leak sizes is done using OLGA software. The behaviour of the outlet variables is investigated by adjusting the leak sizes and locations.

3 OLGA Modelling Methodology

The OLGA (Oil and Gas) code was developed by IFE and SINTEF, as well as Norway. The code is a one-dimensional extended two-fluid model that comes in two flavours: steady-state point model (OLGAS) and comprehensive transient computational code (OLGA) [19]. Equations such as conservation of mass, conservation of momentum, and conservation of energy, as well as equations of state (EOS) for fluids, are employed among all the various flow modelling methodologies. In this study, a multi-phase simulator called OLGA (Schlumberger group) was utilised to model the reaction to producing several leaks of various sizes and locations. Table 1 shows the required pvt file with compositional fluid parameters, which was prepared with Multiflash software and used to model a single-phase pipe of 100 m length.

Table 1 Fluid composition used in OLGA

Components	Amount (mol) (Afebu [12])	Amount (mol) (Current Research)
N-Hexane	0.12	0.12
I-Pentane	0.72	0.72
N-Pentane	0.74	0.74
N-Butane	0.77	0.77
H_2S	0.97	0.97
Nitrogen	0.98	0.98
I-Butane	2.46	2.46
Propane	2.9	2.5
CO_2	4.09	4.09
Ethane	8.03	7.53
Methane	78.22	76.22
Water	NA	2.9

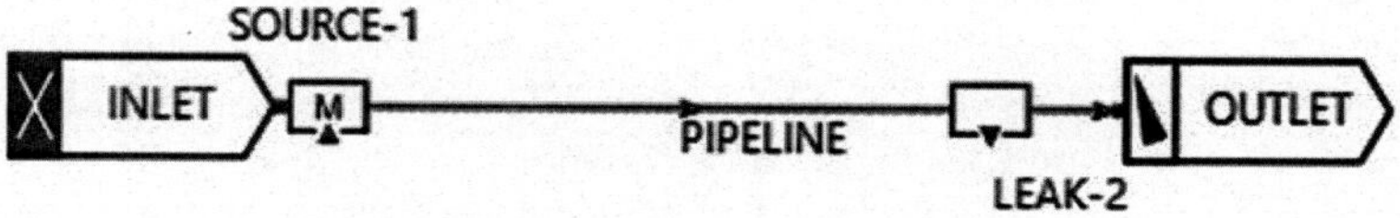

Fig. 3 The simulated OLGA model

3.1 Pipe Modeling

The pipe was fastened at one end with a mass flow source inlet node and the other end with a pressure-driven exit node. The pipe was then divided into five (5) equal segments, with leaks of various sizes (as a ratio of the nominal flow rate) forced at various points along the pipe using a valve (Leak 2), as illustrated in Fig. 3. The model's response to numerous leaks in the pipeline at the same time was also tested.

3.2 Boundary Conditions

Table 2 lists the boundary conditions utilised in the simulations of transient multiphase flow in a pipeline. For validation purposes, the pipeline was divided into five pieces. Later, it was separated into 20 parts for greater precision, and simulations were run with no leak and leak situations at various pipeline places and with variable leak sizes. Leak and no-leak circumstances will be tracked in the outlet variables.

Table 2 Boundary conditions

Component	Details
Rate of flow	10.4 kg/s
Inlet Temperature	45 °C
Inlet Pressure	29.42 bar
Outlet Temperature	27 °C
Outlet Pressure	24.5 bar
Pipe Diameter	8 In
Pipe Length	100 m
Pipe Material	Steel
Flow Rate	kg/s

4 Leak Modelling—OLGA Results

The internal pipe dynamics were characterised by profile patterns that were typical of the size and location of leaks on the pipe. Figure 4 illustrates the profile charts for the zero (no) leak condition for GT (total mass flow), UG (gas velocity), PT (pressure), and TM (temperature). The results of simulation without water for single phase flow have been validated [20]. Further simulations were performed for various other scenarios as mentioned below.

4.1 Leakage at Various Locations

Similarly, the required pvt file was extracted using Multiflash software with the inclusion of water and other fluids (Table 1). A two-phase flow pipe was modelled with a length of 100 m long with transient flow conditions. The simulations were performed using transient flow conditions, but for a simple understanding, the results

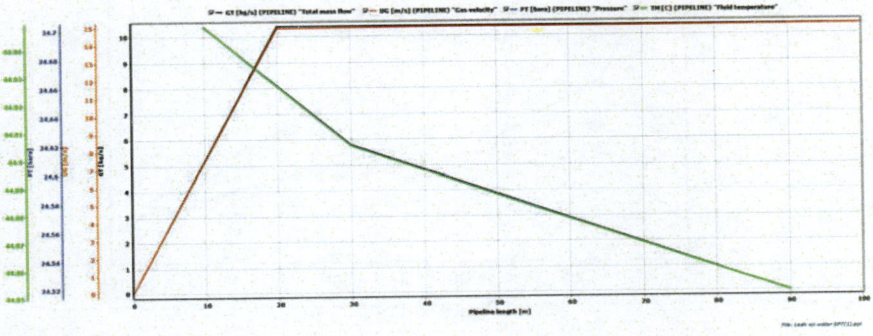

Fig. 4 Profile pattern for Zero (No) Leaks

were plotted with respect to pipeline length. The profile plots were plotted for 50% leak at 5 m, 20 m, 40 m, 95 m of the pipeline and the variables such as mass flow, pressure, temperature and gas velocity were studied as shown in Figs. 5, 6, 7 and 8. The deviation at 15 m can be observed from Figs. 5, 6, 7 and 8. Hence the leak was

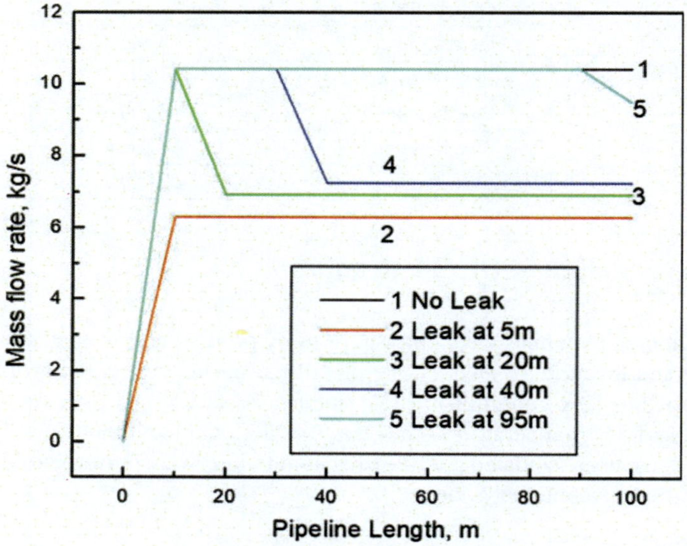

Fig. 5 Total rate of flow versus length for 50% leak opening at various leak positions

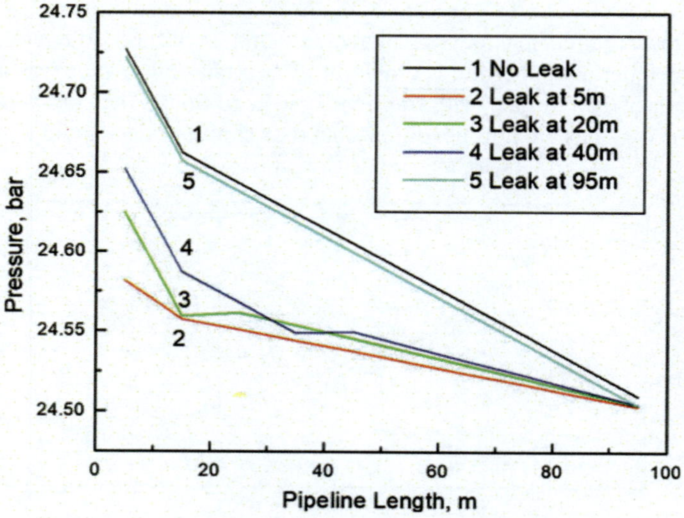

Fig. 6 Pressure plotted against length for 50% leak opening at various leak positions

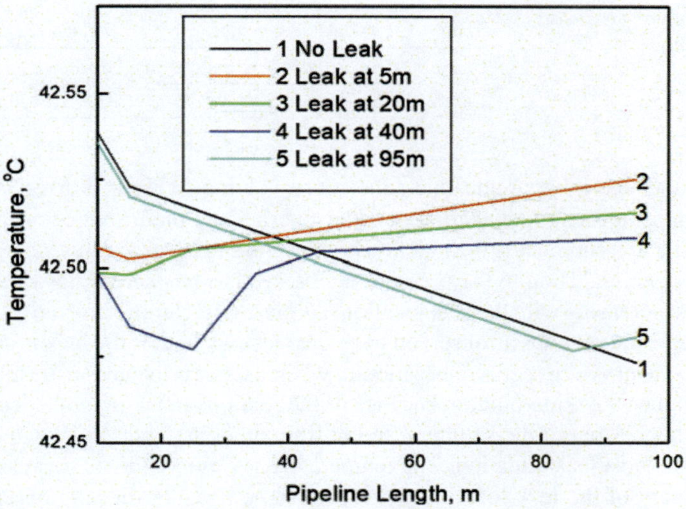

Fig. 7 Temperature plotted against length for 50% leak opening at various leak positions

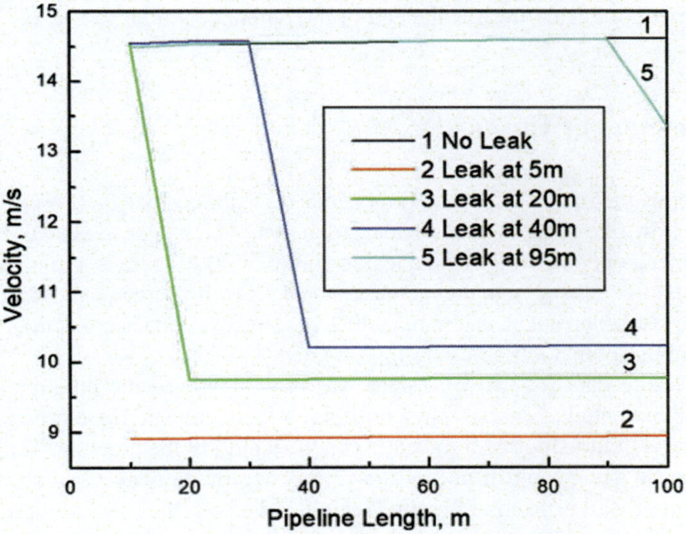

Fig. 8 Velocity plotted against length for 50% leak opening at various leak positions

placed at 15 m of the pipe length, and simulations were performed. The cases which are simulated are tabulated in the below mentioned Table 3.

Mass flow rate at various positions is plotted in Fig. 5 for 50% leak opening. The changes in flow rate can be clearly seen from the point of the leakage. As the leak location increases, the mass flow rate also observed to be increasing. Therefore,

Table 3 Different cases of leak simulations

S. no	Leak at	Leak sizes varied
Case 1	5 m, 20 m, 40 m, and, 95 m	50%
Case 2	15 m	0%, 20%, and, 50%

the more the leak is near the inlet, the more volume of fluid is lost. Indicating the importance of leak localization in order to eradicate the losses occurred due to leakage. The pressure inside the pipeline is shown in Fig. 6 for various locations with 50% leak opening. Though the pressure is observed to be decreasing with pipeline length, it is increasing with the leak location. As the leak location is moved towards the end of a pipe, the pressure is observed to be decreasing. And, if the leak is situated at either end, the pressure seems to be smooth, whereas, if it is located in the middle, the pressure values are quite random because of the sections in the pipeline design [12].

The temperature profile is quite different from other variables as shown in Fig. 7. Consider the post leak situation. The temperature is observed to be decreasing with the movement of the leak location. Here also some irregularities are observed due to the pipeline segmentation. And also, a temperature change is observed at the leak locations. When it comes to velocity profile, it is evident, as shown in Fig. 8. There is a dip in velocity at the leak locations and the velocity seems to be increasing with the movement of a leak towards the end of the pipeline, similar to Fig. 6.

4.2 *Leakages of Various Sizes*

Figure 9 plots the rate of flow data for various leak dimensions at 15 m of pipeline. The mass flow rate reduces as the leak size grows, which is as expected. However, between a leak opening of 0% and a leak opening of 20%, there is a relatively small variation in rate of flow. With the increase in leak opening percentage, the mass flow rate decreases. When a leak is less than 20% in size, it is impossible to conclude that it is present until other variables are investigated.

The pressure data for two leak sizes was plotted against the pipeline length in order to obtain detailed data, as shown in Fig. 10. As indicated, the pressure inside the pipeline is diminishing as it travels down the length of the pipeline. The pressure is found to be decreasing in proportion to the size of the leak opening, which is clear. A abrupt shift at 15 m of pipeline length, indicating the leak location, is also a common point among the observations.

The mass flow rate has been decreased 0.91 times the original mass flow rate for 20% leak opening condition. The percentage change of variable is relatively constant from the point of the leak. And, a dip of 0.77 times was observed for a 50% leak opening condition which is maintained constant as mentioned above. For 20% and

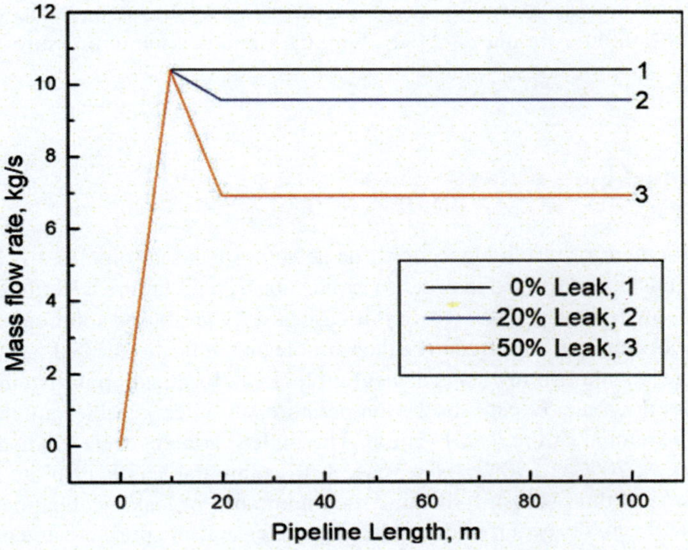

Fig. 9 Total flow rate plotted against length for different leak sizes

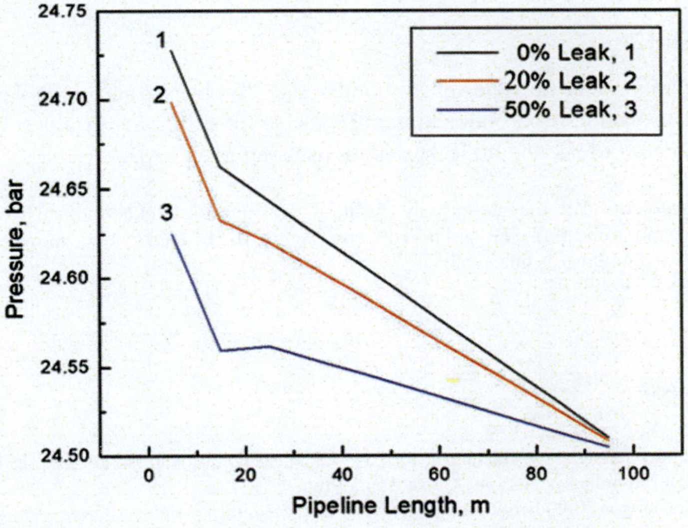

Fig. 10 Pressure plotted against length for different leak sizes

50% leak opening, the change in pressure appears to fluctuate along the length of the pipeline with minute variances. After 15 m, the rapid decline in pressure change is readily seen in both cases.

5 Conclusions

The development of a reliable system for detecting leaks in pipelines is a critical issue that must be addressed in order to avoid catastrophic failures. Hardware-based approaches are, indeed, expensive. In this case, a software-based solution is applied. The mass flow rate is discovered to be the variable that indicates the leakage; the pressure change, as in most tiny leaks, is negligible. Leak localization is also influenced by the mass flow rate. Except for the temperature, all other variables are increasing with the movement of the leak location. The outlet variables were studied for two leak openings, 20% and 50%. As per case 1, the more the leak is near the inlet, the more volume of fluid is lost. Indicating the importance of leak localization in order to eradicate the losses occurred due to leakage. In case 2, the pressure and mass flow rates are observed to be decreasing as the leak size grows. The change in mass flow rate is shown to decrease as the leak size increases, making the variable important in locating the leak.

Future Work

The simulations can be repeated for multi leak conditions and variables can be studied. The data can be obtained using OLGA for different variables. The data can be trained for leak and no leak conditions using MATLAB.

Acknowledgements We the authors are thankful to Universiti Teknologi PETRONAS for providing support under YUTP grant with cost centre 015LC0-251 in creating a research environment and guiding us financially.

References

1. Willersrud A, Blanke M, Imsland L (2015) Incident detection and isolation in drilling using analytical redundancy relations. Control Eng Pract 41:1–12
2. Zhang Y, Jiang J (2008) Bibliographical review on reconfigurable fault-tolerant control systems. Annu Rev Control 32(2):229–252
3. Isermann R, Balle P (1997) Trends in the application of model-based fault detection and diagnosis of technical processes. Control Eng Pract 5(5):709–719
4. Mujtaba SM, Lemma TA, Taqvi SAA, Ofei TN, Vandrangi SK (2020) Leak detection in gas mixture pipelines under transient conditions using hammerstein model and adaptive thresholds. Processes 8(4):474
5. Mujtaba SM, Lemma TA, Gebremariam MA (2020) Fault detection system for long-distance gas mixture pipelines using statistical features. Advances in Manufacturing Engineering. Springer, pp 287–303

6. Syed MM, Lemma TA, Vandrangi SK, Ofei TN (2020) Recent developments in model-based fault detection and diagnostics of gas pipelines under transient conditions. J Nat Gas Sci Eng, 103550
7. Scott SL, Barrufet MA (2003) Worldwide assessment of industry leak detection capabilities for single & multiphase pipelines. 2003: Offshore Technology Research Center College Station
8. Scott SL, Lei L, Jinghai Y (1999) Modeling the effects of a deepwater leak on behavior of a multiphase production flowline. In: SPE/EPA exploration and production environmental conference. 1999. Society of Petroleum Engineers
9. Griffin JM, Smith JR (2005) Field-scale evaluation of leak detection for gas-liquid two-phase flow in Deepwater production systems. In: ASME 2005 fluids engineering division summer meeting. 2005. American Society of Mechanical Engineers
10. Wang S, Carroll JJ (2006) Leak detection for gas and liquid pipelines by transient modeling. I: International oil & gas conference and exhibition in China. 2006. Society of Petroleum Engineers
11. Figueiredo AB, Sondermann CN, Patricio RA, Bodstein GC, Rachid FB (2017) A leak localization model for gas-liquid two-phase flows in nearly horizontal pipelines. In: ASME 2017 international mechanical engineering congress and exposition. 2017. American Society of Mechanical Engineers
12. Afebu K, Abbas A, Nasr G, Kadir A (2015) Integrated leak detection in gas pipelines using OLGA simulator and artificial neural networks. In: Abu Dhabi international petroleum exhibition and conference. 2015. Society of Petroleum Engineers
13. Bai Y, Zhang T, Li Y, Ruan W, Zhu K (2017) A new leak detection method for subsea pipelines. Ships Offshore Struct 12(sup1):S144–S152
14. Datta S, Sarkar S (2016) A review on different pipeline fault detection methods 41:97–106
15. Gajbhiye RN, Kam SI (2008) Leak detection in subsea pipeline: a mechanistic modeling approach with fixed pressure boundaries. In: Offshore technology conference. 2008. Offshore Technology Conference
16. Kam SI (2010) Mechanistic modeling of pipeline leak detection at fixed inlet rate. J Petrol Sci Eng 70(3–4):145–156
17. Lopezlena R, Sadovnychiy S (2019) Pipeline leak detection and location using boundary feedback estimation: case study. J Pipeline Syst Eng Pract 10(3):04019017
18. Lunger T, Karami H (2019) Leak detection in wet natural gas transportation within hilly terrain pipelines. Unconventional Resources Technology Conference (URTEC)
19. Hauge E (2007) Advanced leak detection in oil and gas pipelines using a nonlinear observer and OLGA models. Institutt for teknisk kybernetikk
20. Vandrangi SK, Lemma TA, Mujtaba SM (2021) Determination and analysis of leak estimation parameters in two-phase flow pipelines using OLGA multiphase software. Sustain Comput Inform Syst, 100564

Natural Frequency Analysis of Pipe with Different Boundary Conditions Using Analytical and Finite Element Method

Nor Azliana Badardin, Ainul Akmar Mokhtar, Nabihah Sallih, and Muhammad Harith Irfan Ishak

Abstract Vibration-induced failures in piping systems often result from the excitation of natural frequency of an individual piping span. In this paper, vibrations characteristics of pipe with different pipe end conditions have been analysed using the analytical and finite element approach. The natural frequencies of the pipe are investigated by varying the length of the pipe with fixed-free end support and fixed–fixed end support, with and without fluid. Finite element models of the pipe structure are prepared using ANSYS Workbench software. The natural frequency for fixed-free pipe is much lower than fixed–fixed pipe for both with and without fluid. The natural frequency for pipe with fluid is lower than the pipe without fluid for both boundary conditions. The results are compared with analytical method in order to validate the finite element model. A good agreement is found between the analytical and the numerical results. Linear equations that predict the first natural frequency are formulated using regression analysis. It is found that the regression equations describing the correlation of the first natural frequency on the length of the pipelines has the form of power function for all pipe end conditions. This approach can be used with subsequent dynamic analyses such as response spectrum analysis or harmonic analysis for further research.

1 Introduction

Pipes has been used in various types of industries for so many years. As any other engineering structure, piping systems are exposed to damage due to various reasons, including vibration. A pipe vibrates because of many factors. The most common factors are due to excessive pulsation, mechanical resonance and inadequate supports due to various excitation mechanism. Vibration-induced failures in piping systems often result from the excitation of a mechanical natural frequency of an individual piping span. This can cause failure to equipment and in certain situations, there is

N. Azliana Badardin (✉) · A. Akmar Mokhtar · N. Sallih · M. Harith Irfan Ishak
Mechanical Engineering Department, Universiti Teknologi PETRONAS, 32610 Seri Iskandar, Perak, Malaysia
e-mail: nor_18001934@utp.edu.my

F. Ahmad et al. (eds.), *ICPER 2020*, Lecture Notes in Mechanical Engineering,
https://doi.org/10.1007/978-981-19-1939-8_43

possibility release of containment of the process fluid that might be flammable or contain dangerous product which can causes loss and might put personnel and the environment at risk as well. Therefore, to ensure that piping systems are free from excessive vibrations, it is important to be able to calculate the mechanical natural frequencies of individual piping spans [1]. In engineering design and analysis, these parameters are essential for better understanding of the vibration problem.

Any structural system, such as a pipe, will exhibit a series of natural frequencies which depend on the distribution of mass and stiffness throughout the system. Modal analysis is usually used to find the piping natural frequencies and the associated mode shapes [2]. Obtaining the dynamic characteristics of a structure is crucial to prevent possible condition of resonance generated between pipes and other equipment. When the excitation frequency is approximately equal to the natural frequency of the system, resonant phenomena occur and small periodic driving forces can produce large amplitude vibrations that leads to premature failure of the pipe. It is necessary to carry out modal analysis prior to other dynamic analysis such as harmonic analysis and response spectrum analysis. Modal analysis using numerical method to obtain the natural frequencies of pipes or any other structures has been done by many authors. Author [2] proposed a technique to find the likelihood of occurrence on increased vibrations in pipeline by modal analysis using ANSYS Workbench software. The effect of diameter, length and wall thickness of the straight sections pipeline on the pipe natural frequencies were investigated. From the results, regression equations are obtained using the pairwise and multiple regression analysis to represent the relationship between the first natural frequency with length, diameter and the pipe wall thickness.

A modal analysis of a fluid-filled u-shaped pipeline under fluid structure interaction was done by [3] for determining the natural frequencies and mode shapes of the pipe under different fluid pressures using ANSYS Workbench software and experimental modal analysis. Experimental modal analysis was done to verify the simulation model. The results from the simulation agrees well with the results from experimental modal analysis. Liu et al. [4] carried out a modal and vibration analysis of a filter system in a petrochemical plant to demonstrate the significant effect of vibration on the structural integrity of piping. A modal analysis is performed to obtain the natural frequencies of the system. Kudus et al. [5] studies the free vibration of pipe structures exhibiting corrosion. The effect of wall thickness reduction in pipe structure on the pipe vibration characteristics is examined using the modal analysis. A commercial finite element software is used to perform the modal analysis to obtain the natural frequency and mode shapes of the baseline model. This investigation found that natural frequency of the pipe reduces significantly for damaged pipe structure and there is localisation of mode shape near the damaged region. A mathematical model was developed by author [6] for dynamics and stability of fluid-filled pipe with guided supports and compared the results with simulation model using FEA. The pipe is considered as Euler–Bernoulli beam and the equation of motion is derived by using the Hamilton's variation approach. The effects on the natural frequency of the pipe for an empty pipe, static fluid and fluid flow conditions with three different pipe support materials were investigated. ABAQUS software is used to analyse the influence of fluid velocity on natural frequencies for the pipe with fluid for four different fluid velocities by varying the valve position. Comparison was made and close agreement was found between the analytical results and FEA results.

Sharma [7] obtained the natural frequency and mode shapes of a simply supported beam by doing experiments, theoretical and numerical modal analysis. Fast Fourier Transform was utilized to convert the time domain signal obtained from the experiment to frequency-domain signal. Then finite element modal analysis was done using ANSYS software. Results from the simulation were then compared with theoretical analysis and were found to be in good agreement. Ashrafizadeh [8] investigated a failure of a high-pressure gas pipe due to crack. The stress and modal analysis using FEM was done apart from visual inspection and metallurgical evaluation. Stress analysis was done to find the areas with maximum stress caused by the gas pressure inside the pipe while modal analysis was done to evaluate the dynamic characteristic of the system due to vibration of a valve. From the natural frequencies and mode shapes of the pipe and split tee, it was found that the maximum displacement occurred at the location of peak stress where the crack was initiated. Author [9] investigated the natural frequencies of isotropic rectangular plates with different types of supports using the wave propagation method. The effect of varying the length and width of the plates on the natural frequency of the plates are investigated. Finite element method (FEM) was used for the vibrational behaviour analysis of impeller of a centrifugal pump by [10]. The effect of number of impeller blades, impeller blade thickness and the impeller disk thickness on the frequencies and the corresponding mode shapes of the impeller were observed. ANSYS software, were used for the modelling and simulation of the impeller.

This paper investigates the free vibration of a straight pipe that varies in length, with and without fluid. The effect of length and pipe end conditions on the natural frequency of the pipe is evaluated through analytical and numerical modal analysis. ANSYS Finite Element package software is used to do the numerical simulation. Regression equations are formulated by regression analysis to describe the relationship between the first natural frequency and the length, fluid filled and boundary conditions of the pipe. The results from analytical analysis and simulation using ANSYS software for the natural frequency is then compared.

2 Methodology

Obtaining the dynamic characteristics of a structure is crucial to avoid possible condition of resonance. To calculate the dynamic properties especially the natural frequencies and the mode shapes of a pipe, usually modal analysis method is used [2]. In this paper, analytical and FEA modal analysis is done to investigate the first mode of vibration of the pipe and results are compared. Only the first mode is chosen because it often defines the highest loads in a structure hence it is the most critical in the case of resonance regimes. Two different types of boundary conditions are studied which are fixed-free and fixed–fixed for both pipe with fluid and without fluid. Regression analysis is done for numerical analysis results to obtain the equation of function to express the relationship between all the parameters.

2.1 Analytical Modal Analysis

The natural frequencies for piping span can be approximated using distributed uniform beam vibration theory. Natural frequency of a straight uniform piping span can be calculated using the equation below [1, 11, 12]:

$$f_o = \frac{\lambda}{2\pi}\left(\frac{gEI}{\mu l^4}\right)^{\frac{1}{2}} \tag{1}$$

In Eq. (1),
f_o: pipe span natural frequency, Hz
λ: Frequency factor, dimensionless
g: Gravitation constant, 386 in / s^2
E: Modulus of elasticity, psi
I: Moment of inertia, in^4
l: Pipe span length, in
μ: Weight per unit length of beam (including fluid insulation), lbs/in (Equal to ρA if the weight of fluid and insulation is negligable
ρ: Pipe density, lbs/in^3
A: Pipe cross-section area, in^2

For pipe with fluid, the natural frequency can be calculated by multiplying the calculated natural frequency by the square root ratio of the weight of the pipe divided by the total weight per length of the pipe, plus the fluid and the insulation. The frequency factors for calculating the first two natural frequencies of the straight pipe are given in Table 1 while the mode shapes for fixed-free and fixed–fixed pipe configuration are illustrated in Fig. 1.

The geometric parameters and material properties used for analytical and finite element modal analysis of the pipe is taken from a published journal paper [13] shown in Table 2. To see the influence of pipe length on natural frequency of the pipe, the analysis started with 2 m length then the length is increased gradually following the configuration from [13] as shown in Fig. 2. Additional analysis was done to see the behaviour of the natural frequency for pipe length shorter than 2 m and longer than 8 m.

Table 1 Frequency Factors for idealized piping spans [1]

Piping configuration	Frequency factor	
	1st	2nd
Fixed–Free	3.52	22.4
Fixed–Fixed	22.40	61.70

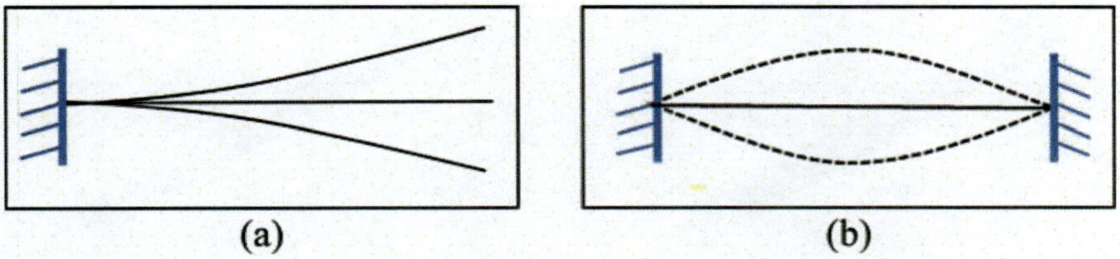

Fig. 1 Vibration mode-shapes for **a** fixed-free and **b** fixed–fixed pipe span configuration [1]

Table 2 Properties and dimensions of the pipe section

Pipe outside diameter (m)	0.16
Pipeline wall thickness (m)	0.01
Modulus of Elasticity of pipe material (GPa)	200
Pipeline Density (kg/m^3)	7850
Fluid Density (kg/m^3)	900
Poisson Ratio	0.3

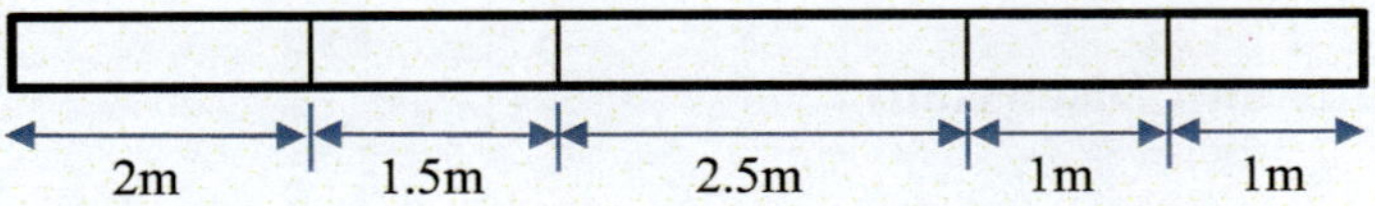

Fig. 2 Pipe configuration used in this study

2.2 Finite Element Modal Analysis

The numerical modal analysis of the pipeline for free vibration is performed using ANSYS Workbench software. The geometric parameters and material properties is taken from Table 2. The material properties assigned was the same as structural steel in the ANSYS material library. Analysis starts with 2 m pipe length and is added gradually following Fig. 2 to make the total pipe length to 3.5 m, 6 m, 7 m and 8 m. For the pipe with fluid, the study will not consider the fluid flow inside the pipe, therefore the fluid is considered as static fluid. Three-dimensional solid element for both pipe and fluid in the pipe is constructed in ANSYS Workbench and computational modal analysis is carried out to generate the pipe natural frequencies and mode shapes. Analysis using this type of element is computationally intensive, but it is the most accurate. Since the pipeline configuration is not complex, solid element is used for all the analysis. Two boundary conditions that are used for the analysis are fixed-free and fixed–fixed. Figure 3 shows the meshing of the straight pipe and fluid in pipe using solid element with their respective boundary conditions. The modal analysis was then carried out to determine the natural frequencies and the corresponding mode shapes. For this study, the modal frequency was limited to only the first mode.

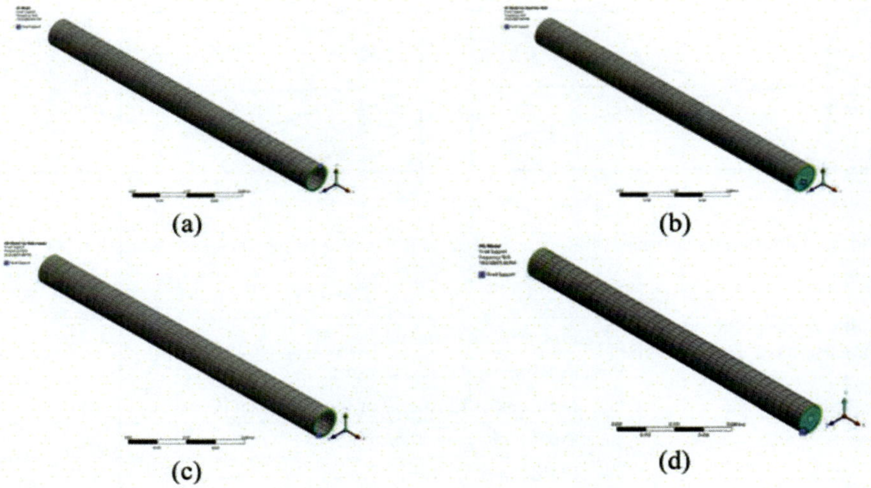

Fig. 3 Meshing of the straight pipe with their respective boundary conditions **a** fixed-free without fluid **b** fixed-free with fluid **c** fixed–fixed without fluid and **d** fixed–fixed with fluid

3 Results and Discussion

A comparison between the analytical result and numerical simulation result from ANSYS for the first natural frequency of the pipeline is carried out. For fixed-free pipe condition without fluid, the frequencies are listed in Table 3 and the natural frequencies for fixed-free pipe with fluid are listed in Table 4. For fixed–fixed pipe, the obtained natural frequencies are listed in Tables 5 and 6 for pipe with fluid and pipe without fluid respectively. Please note that in order to get more data points for regression analysis, additional modal analyses were done to see the behaviour of

Table 3 Natural frequency for fixed-free pipe conditions, without fluid

Length of pipe span (m)	1st natural frequency (Hz)		
	Analytical	Numerical	% difference
1.0	150.27	145.03	3.49
1.5	66.79	65.76	1.54
2	37.57	37.24	0.88
3.5	12.27	12.23	0.33
6	4.17	4.17	0.00
7	3.07	3.06	0.33
8	2.35	2.35	0.00
10	1.50	1.50	0.00
15	0.67	0.67	0.00
20	0.38	0.38	0.00

Table 4 Natural frequency for fixed-free pipe conditions, with fluid

Length of pipe span (m)	1st natural frequency (Hz)		
	Analytical	Numerical	% difference
1.0	128.06	124.88	2.48
1.5	56.92	56.52	0.70
2	32.02	31.99	0.09
3.5	10.45	10.50	0.48
6	3.56	3.58	0.56
7	2.61	2.63	0.77
8	2.00	2.01	0.50
10	1.28	1.29	0.78
15	0.57	0.57	0.00
20	0.32	0.32	0.00

Table 5 Natural frequency for fixed–fixed pipe conditions, without fluid

No	Length of pipe span (m)	1st natural frequency (Hz)		
		Analytical	Numerical	% difference
1	1.0	956.25	735.71	23.06
2	1.5	425.00	371.93	12.49
3	2	239.06	220.86	7.61
4	3.5	78.06	75.97	2.68
5	4	59.77	58.52	2.09
6	4.5	47.23	46.44	1.66
7	5	38.25	37.73	1.36
8	6	26.56	26.31	0.94
9	7	19.52	19.38	0.72
10	8	14.94	14.86	0.54

the natural frequency for total pipe length 1 m, 1.5 m, 10 m, 15 m and 20 m. From the analysis done for fixed-free pipe both with and without fluid it is found that the analytical and numerical results for the straight pipe are having a good agreement and the maximum difference between them is less than 5%. From the trend, it is found that the difference is increasing with the decreasing of pipe length. For fixed–fixed pipe, the maximum difference is quite high compared to fixed-free pipe for both with and without fluid. The difference decreases with the increase of the pipe length. From the analysis done it is found that the analytical and numerical results for the straight pipe starting to have a good agreement for pipe length 3.5 m and above.

Comparing pipe without fluid and with fluid inside, the natural frequency for pipe with fluid is lower than the pipe without fluid for both boundary conditions. This is due to the fluid act as damping and the effect of fluid mass which added to the

Table 6 Natural frequency for fixed–fixed pipe conditions, with fluid

No	Length of pipe span (m)	1st natural frequency (Hz)		
		Analytical	Numerical	% difference
1	1.0	814.95	636.07	21.95
2	1.5	362.20	320.80	11.43
3	2	203.74	190.22	6.64
4	3.5	66.53	65.30	1.85
5	4	51.01	50.28	1.43
6	4.5	40.31	39.89	1.02
7	5	32.65	32.41	0.73
8	6	22.64	22.60	0.18
9	7	16.63	16.64	0.06
10	8	12.73	12.76	0.24

total mass of the system that is inversely proportional to the natural frequencies [14]. The natural frequency for fixed-free pipe is much lower than fixed–fixed pipe. This is because the stiffness is much lower and the flexibility is higher compared with fixed–fixed pipe. Overall, starting from 3.5 m, the natural frequency is decreasing with the increase of the pipe length. However, the natural frequency is starting to increase when the pipe length is shorter than 3.5 m long. The difference between analytical and simulation also increase when the pipe length is shorter than 2 m for fixed-free pipe and for fixed–fixed pipe, the difference starting to increase for pipe length shorter than 3.5 m.

Mode shape describe the pattern in which the pipe will deform when vibrating at free vibration and when exited at resonant frequency. Mode shapes show how the pipe interact with other parts in the system around it when vibrating. From Tables 7 and 8, mode shape obtained by numerical simulation followed closely the shape described in Fig. 1 which is bending mode for both fixed-free end condition and fixed–fixed pipe end condition.

Regression analysis were performed using MS Excel to obtain the equations that most closely describe the numerical analysis results. A simple regression analysis was performed to determine the dependence of the value of the first frequency of natural frequency on the length of the pipeline for both fixed-free pipe and fixed–fixed pipe. From the simple regression analysis results shown in Figs. 4 and 5, we can conclude that there is dependency of the first natural frequency on the length of the pipeline. The regression equation has the form of a power function and the coefficient of determination is $R^2 \geq 0.8$ for all cases. Equations (2) and (3) describes the equation for pipeline with fixed-free support without and with fluid while Eqs. (4) and (5) describes the equations for fixed–fixed pipeline without and with fluid respectively.

$$F_{1\,fixed-freewithoutfluid} = 147l^{-1.991} \tag{1}$$

Table 7 Mode shapes for the 1st natural frequency, fixed-free pipe conditions

Pipe Length (m)	Fixed-free, without fluid		Fixed-free, with fluid	
	1st Natural Frequency	Mode shape	1st Natural Frequency	Mode shape
2	37.24	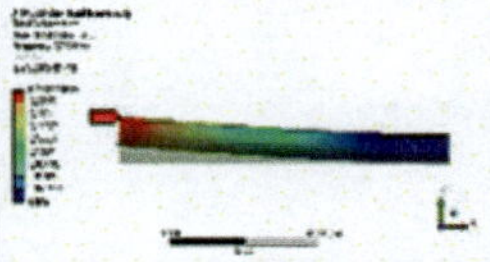	31.99	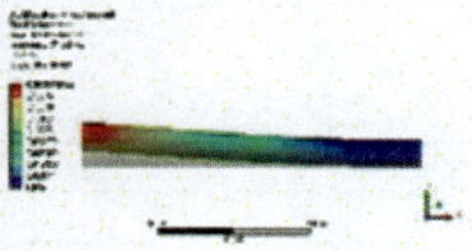
3.5	12.23	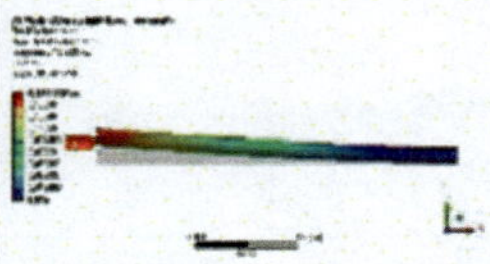	10.50	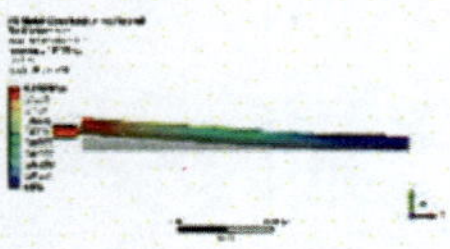
6	4.17	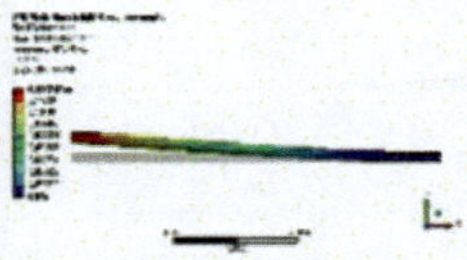	3.58	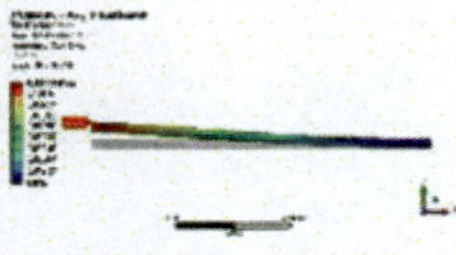
7	3.06	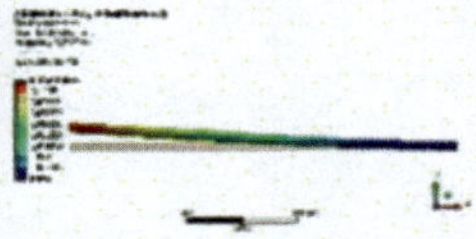	2.63	
8	2.35	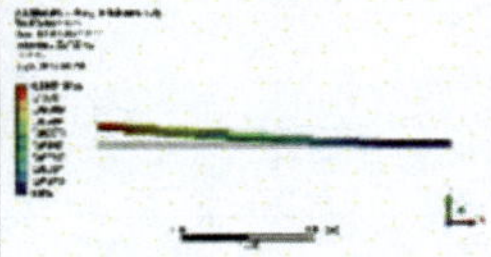	2.01	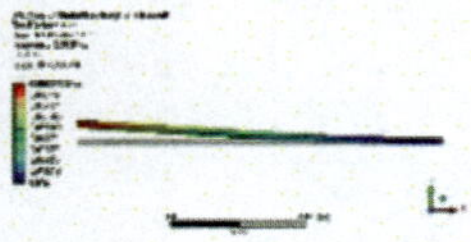

$$F_{1\,fixed-freewithfluid} = 126.47l^{-1.992} \quad (2)$$

$$F_{1\,fixed-fixedwithoutfluid} = 805.82l^{-1.927} \quad (3)$$

$$F_{1\,fixed-fixedwithfluid} = 695.54l^{-1.929} \quad (4)$$

Table 8 Mode shapes for the 1st natural frequency, fixed–fixed pipe conditions

Pipe Length (m)	Fixed–fixed, without fluid		Fixed–fixed, with fluid	
	1st Natural Frequency	Mode shape	1st Natural Frequency	Mode shape
2	220.86	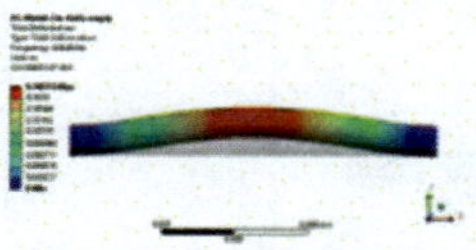	190.22	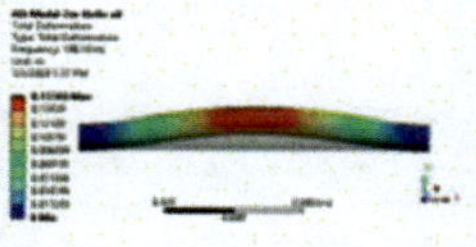
3.5	75.97	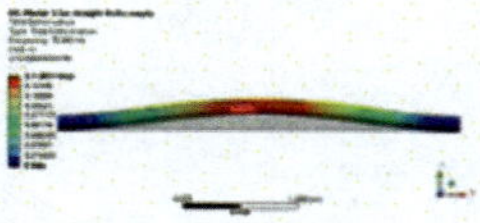	65.30	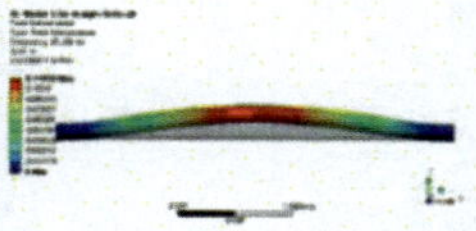
6	46.44	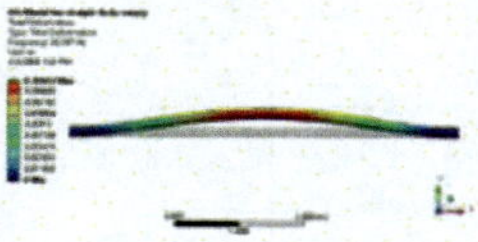	39.89	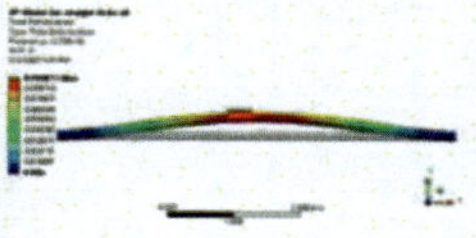
7	37.73	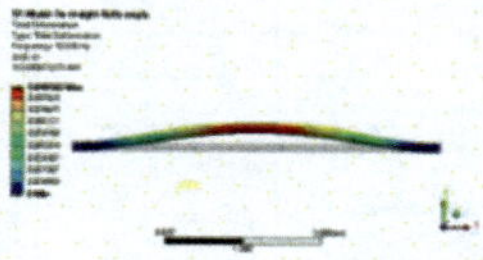	32.41	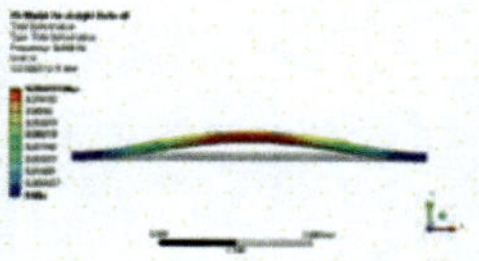
8	26.31	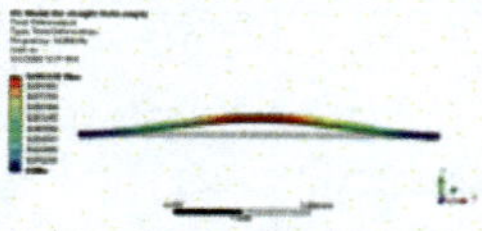	22.60	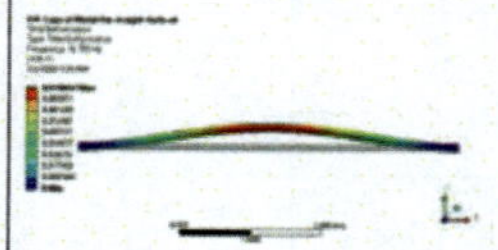

4 Conclusion

A simple approach is presented for the free vibration analysis of a straight pipe with two different types of boundary conditions. The purpose of this paper is to investigate the effects of pipe length and presence of fluid on the natural frequency using analytical and numerical analysis. From the analysis, it is found that the natural frequency calculated using analytical and numerical results for the straight pipe are having a good agreement. The natural frequency for fixed-free pipe is much lower than fixed–fixed pipe for both with and without fluid while the natural frequency for pipe with fluid is lower than the pipe without fluid for both boundary conditions. Overall, for the influence of pipe length, starting from 3.5 m, the natural frequency is decreasing with the increase of the pipe length. However the natural frequency of the pipe starting to increase when the pipe length is less than 3.5 m long.

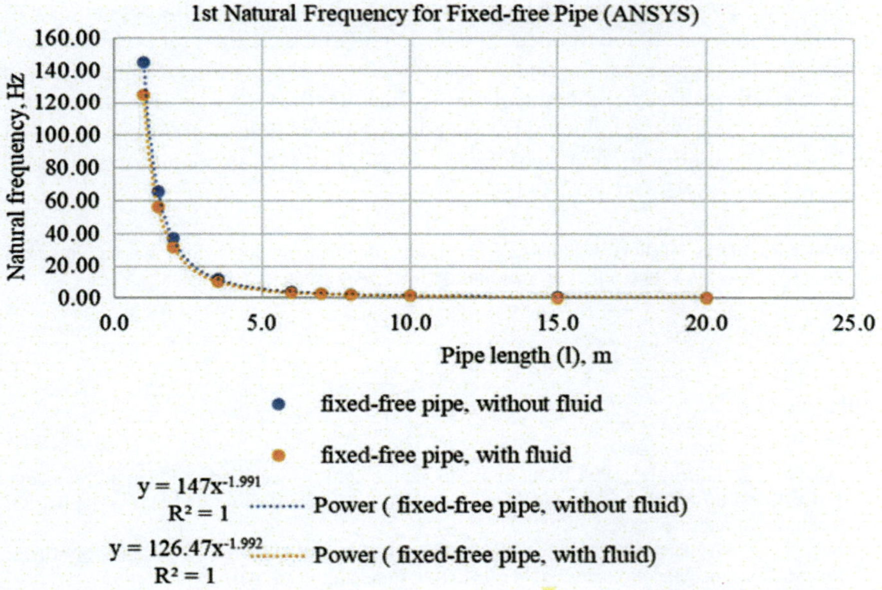

Fig. 4 Graphs of the dependencies of the first natural frequencies on length for fixed-free with and without fluid and their regression models

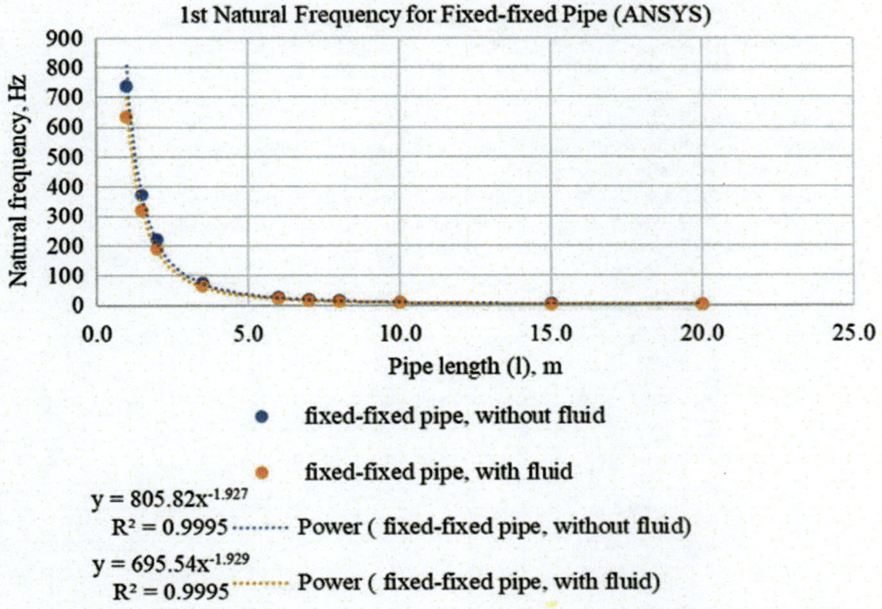

Fig. 5 Graphs of the dependencies of the first natural frequencies on length for fixed–fixed with and without fluid and their regression models

A paired regression analysis was performed as well to determine the dependence of the value of the first natural frequency on the length of the pipeline for both boundary conditions. It is found that the regression equation describing the correlation of the first natural frequency on the length of the pipelines has the form of power function for all cases. The regression model of the dependence of the first natural frequencies on the length for pipeline sections with fixed-free and fixed–fixed boundary conditions is determined as well.

Acknowledgements Authors wishing to acknowledge financial support from Yayasan Universiti Teknologi PETRONAS fund under project number: 0153AA-E92.

References

1. Wachel JC, Tison JD (1994)Vibrations in reciprocating machinery and piping systems. Proceedings of the 23rd turbomachinery symposium (Texas: Texas A&M University Turbomachinery Laboratories) pp 243–272
2. Sekacheva AA, Pastukhova LG, Alekhin VN, Noskov AS (2019) IOP conference series: materials science and engineering, vol 481 (IOP Publishing Ltd) p 012019
3. Jiang Y, Zhu L (2018) Vibroengineering PROCEDIA vol 21 (JVE International) pp 42–47
4. Liu Z, Wang J, Gho WM, Liu X, Yu X (2017) Shock and vibration
5. Kudus SA, Suzuki Y, Matsumura M, Sugiura K (2018) Jurnal Teknologi 80:37
6. Sutar S, Madabhushi R, Babu P (2016) Int J Mech Eng Autom 3:96–106
7. Sharma J (2019) Engineering vibration. Commun Inf Process 478:177–186
8. Ashrafizadeh H, Karimi M, Ashrafizadeh F (2013) Eng Fail Anal 32:188–201
9. Hussain M, Naeem MN (2018) Advanced engineering testing, p 47
10. Ashri M, Karuppanan S, Patil S, Ibrahim I (2014) MATEC web of conferences, vol 13 (EDP Sciences) p 04030
11. Wachel JC, Morton SJ, Atkins KE (1990) Proceedings of the 19th turbomachinery symposium Texas A&M University. Turbomachinery Laboratories
12. Sunil K, Raghunandana K (2014) Int J Emerg Technol Adv Eng 4(9):1–6
13. Liu J, He X, Liu Q, Naibin J, Chen H (2014) Comput Model New Tech 18(5):14–18
14. Al-Hashimy Z (2009) Comp Model New Tech 18(5):14–18

Effect of Hole Shape and Location on the Natural Frequency and Deformation of Composite Beams

Amar Salim Hamid and Marwah Sabah Fakhri

Abstract Using the composite material is due to its excellent properties and weight reduction. It is used in many applications; one of such applications is the prosthetic foot in which the weight reduction is very important. In this work, glass fiber with epoxy resin was used to manufacture the composite beams. The effect of hole shapes (square, or circular)along with hole location on the natural frequency and deformation of composite cantilever beams were studied. The hole locations were (0.25L, 0.5L, 0.75L) from the fixed end, and it seems that the effect on the first natural frequency is not more than 7% theoretically and 9% experimentally, were the second mode is not more than 4% theoretically and about 12%experimentally, while there is a slight effect in the third mode theoretically and experimentally. The beams natural frequency was theoretically computed using ANSYS program and the conclusion compared with the experimental work, in addition to the maximum deformation was studied theoretically by ANSYS program.

1 Introduction

The phenomenon of the hole shape and location has change into one of the important subject in engineering applications, especially in the artificial limbs, which can make openings in order to reduce weight and other uses, for example, building structure. In this work the effect of holes with different shapes and location on natural frequency for composite beam was studied theoretically and experimentally in addition to the maximum deformation was calculated theoretically by ANSYS software. The composite materials are known by their good combination of low weight with high structural stiffness. The isotropic materials have important advantages, among these is their ability to be used for specific applications by perfect design parameters,

A. Salim Hamid
Department of Mechanical Engineering, University of Technology, Baghdad, Iraq

M. Sabah Fakhri (✉)
Mechanical Engineering, Re-Construction and Projects Directorate, Ministry of Higher Education & Scientific Research, Baghdad, Iraq
e-mail: 20086@uotechnology.edu.iq

F. Ahmad et al. (eds.), *ICPER 2020*, Lecture Notes in Mechanical Engineering,
https://doi.org/10.1007/978-981-19-1939-8_44

such as mass sequence, ply orientation and performance base [1]. Different of web openings shapes namely (rectangular, hexagonal and elliptical) were considered in partially and fully encased with concrete studies [2]. The obtained results showed that the elliptical web opening is efficient for the better behaviour. The shape of the web opening is decided according to the structural planner and the purpose of the opening. Moreover, it is simple to select the shape of openings from uniform shapes whether it is rectangular or circular [3]. Many parameters affected on the natural frequency such as the cut-out orientation, fiber orientation angle, and cut-out shape, in this study [4] it was concluded that the natural frequency of composite plate is affected by the cut-out location and fiber orientation. But it is not influenced by cut-out shape.

Vibration test is very important from this point of view. This test explains the dynamic behaviour of the elements in any structure under sever working. The data collected from the vibration test assist the designer to redesign the structure to avoid the failures by vibration in resonance condition which leads to a maximum amplitude [5]. The openings in the web will be built depending upon the choice of the designer and the purpose of it. There are no rules or difficulties for choosing the opening shapes. Circular or rectangular opening are almost selected depending upon the designer acceptation. Web openings cause reduction in beam stiffness and the deflection will be larger than the no web openings beam. The plastic deformations of the beams with openings caused by moment and shear at the openings, gives an indication to the strength of the beam [6]. The openings in beams are most possible to be circular or rectangular, and may be as single openings or as a serial group along the beam [8].

Natural frequency is the one of the most important parameters that we can get from vibration, which is classified as free or forced, damped or undammed and linear or nonlinear vibration [9]. In structures, reducing the vibration is one of the important requirements of engineers. To reduce the vibration, there are different methods to move the natural frequency far away from the frequency of the excited force. The amplitude of a system that vibrates with a frequency to natural frequency grows rapidly as very low input energy to the system [10]. When a defect or crack occurs in a shaft or beam, then the stiffness is reduced. Measuring these differences can help to characterize a crack that holes may be affected the dynamics characteristics of the beam depend on their location and sizes [11]. The circular holes' effects on the beam natural frequency was studied, and shows the natural frequency value for the composite material with circular shape is more than natural frequency value of composite material with other cut of shape [12].

2 Numerical Simulation of Composite Beams

The ANSYS FEM software was used to simulate the composite beam in this work as shown in Fig. 1.

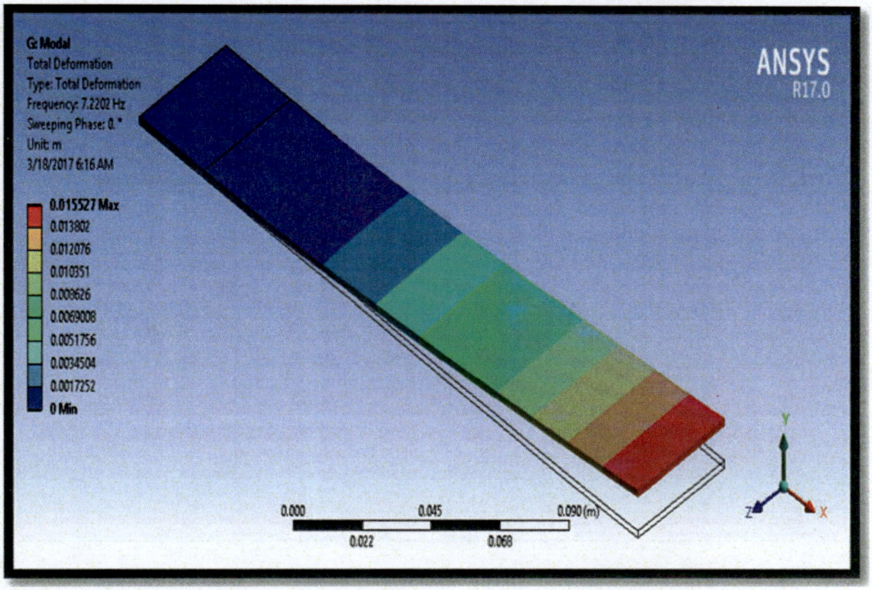

Fig. 1 Composite beams with ANSYS software

Three cases were studied for composite beams; one of them is without holes and the other beams with circular and square holes in a different position. In all cases the natural frequency was theoretically measured with three modes. The boundary condition was applied as a cantilever beams. One side was fixed on a shaker and the other side was free as shown in the Table 1. Maximum deformation was also measured in all composite beams cases by ANSYS software.

3 Experimental Work

Main steps for this work include manufacturing the composite material and carrying out laboratory tests to determine the mechanical properties for composite material, along with preparing the composite beams by making the circular and square holes to calculate the natural frequency and maximum deformation.

3.1 Composite Material Preparing

Composite material can be defined as a combination of two or more materials that resulting better properties than those of the component used alone. In this work, the

Table 1 Three modes of the natural frequency in ANSYS

1st mode	2nd mode	3rd mode
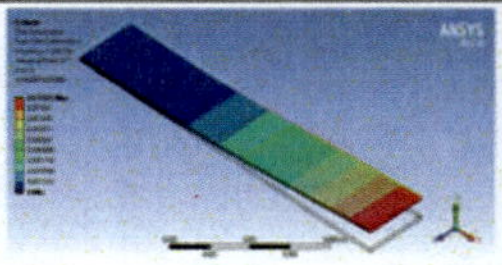	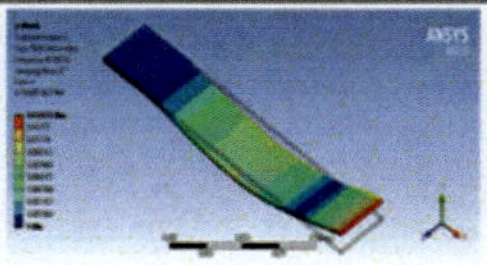	
	Beam; B	
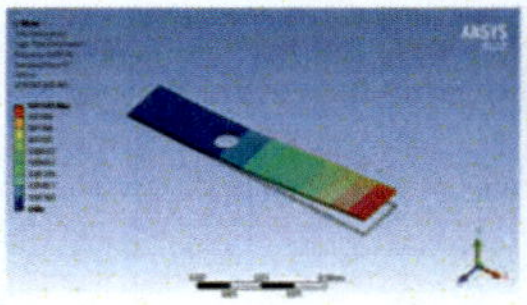		
	Beam; 0.25 Lc	
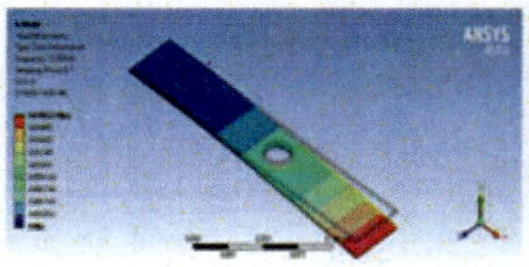		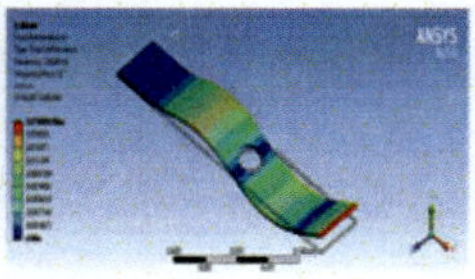
	Beam; 0.5 Lc	
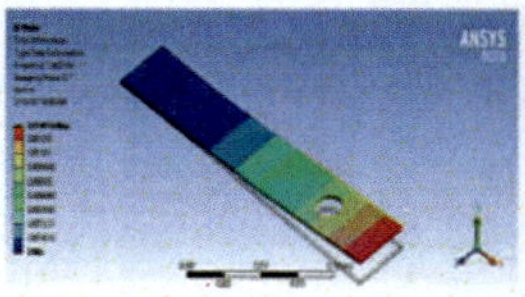	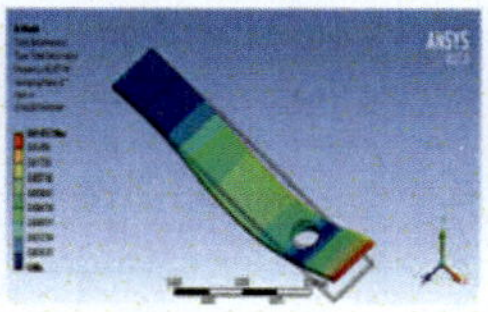	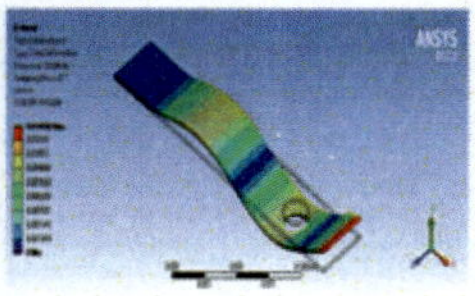
	Beam; 0.75 Lc	
	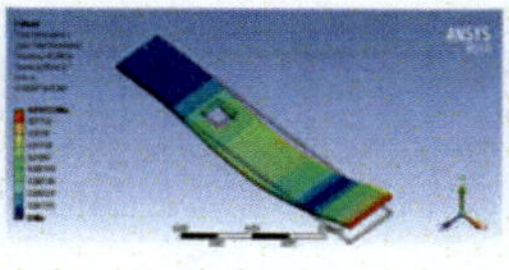	
	Beam; 0.25 Ls	
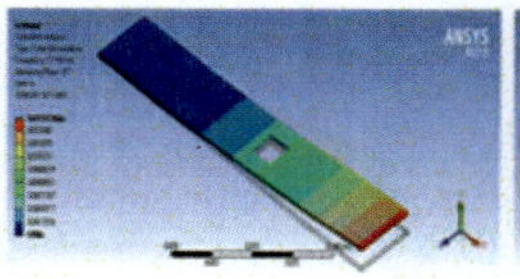	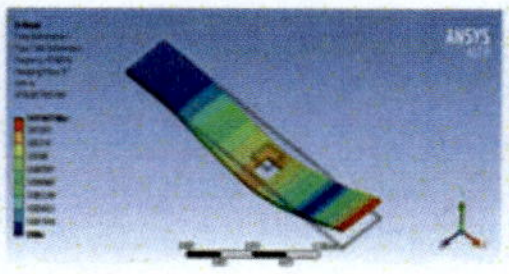	
	Beam; 0.5 Ls	
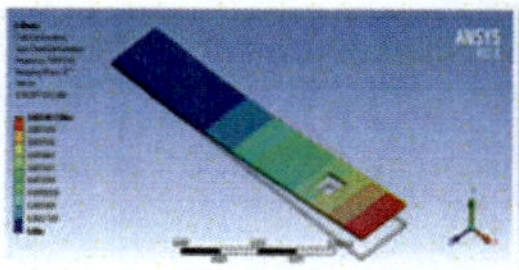	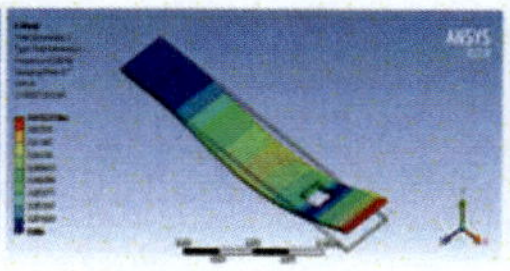	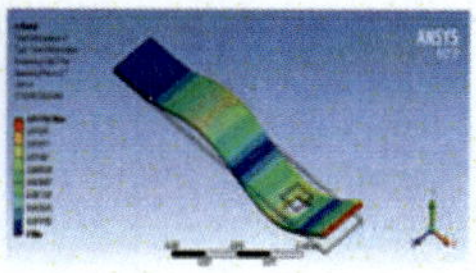
	Beam; 0.75 Ls	

Fig. 2 **a** Fibber glass, **b** Epoxy

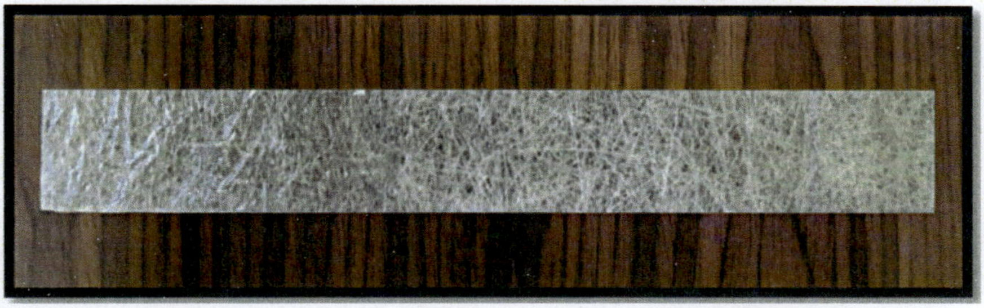

Fig. 3 Composite material in this work

materials (fibber glass layer with epoxy) were used to prepare the composite material as shown in Fig. 2.

The composite material was constructed using a layer of fibber glass, after that, an amount of epoxy was poured over it in a mold, and then it was necessary to wait for a period of time for the mixture to become firm or solid, as shown in Fig. 3.

3.2 Tensile Test

To determining the mechanical properties of materials such as strength, toughness, ductility, strain hardening, and elastic modulus, the tensile test is the most common mechanical test. The sample was tested for tensile test, then it was prepared by CNC machine as shown in Fig. 3, to obtain the exact profile with accurate required dimensions for ASTM D3039 [13] (Fig. 4).

The profile for that standard is rectangular cross section with dimensions 25 mm wide and 250 mm long as shown in the Fig. 5.

The tests were done in Universal Testing Machine (UTM) with strain speed of 10 mm/min, see Fig. 6. Stress—strain diagram was created automatically from the

Fig. 4 Test specimen cutting by CNC machine

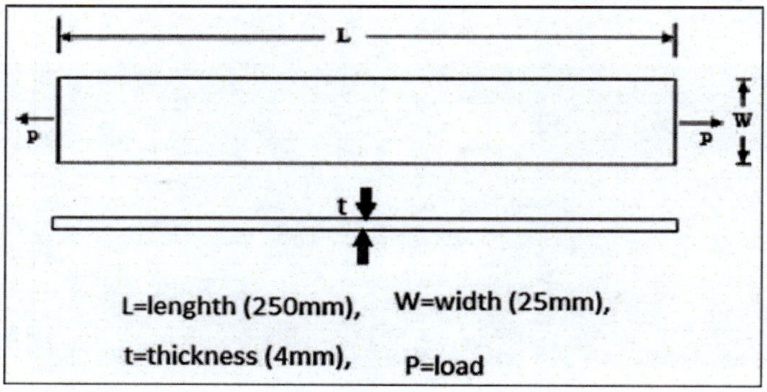

Fig. 5 Tensile test specimen dimensions for ASTM D3039

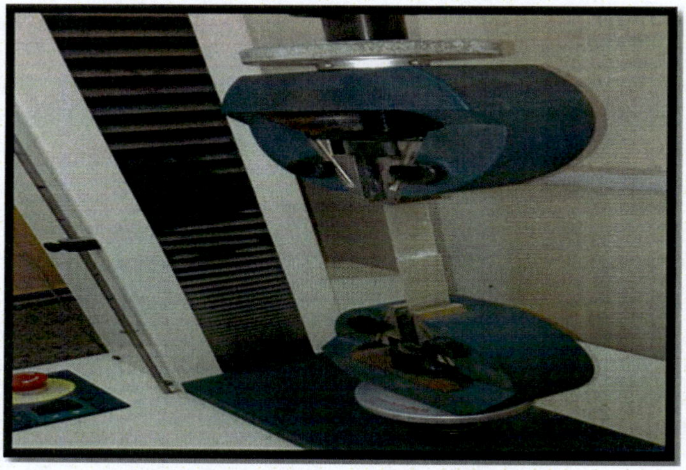

Fig. 6 Tensile test device (UTM)

Table 2 Properties composite material tensile test

Ultimate stress (MPa)	Ultimate force (N)	σy (MPa)	Modulus (GPa)
49.24	3940	35	20

UTM instrument and the mechanical behaviour of the composites was state approve to the diagram.

From this testing the properties for the sample tested results are shown in Table 2.

3.3 Holes Shape and Location in Composite Beam

Two holes (circular and square) with three locations (0.25L, 0.5L,0.75L) prepared by CNC machine to get the accurate profile with exact required dimensions were studied in this work. Figures 7, 8 and 9 show the cantilever beam with holes' shape and locations.

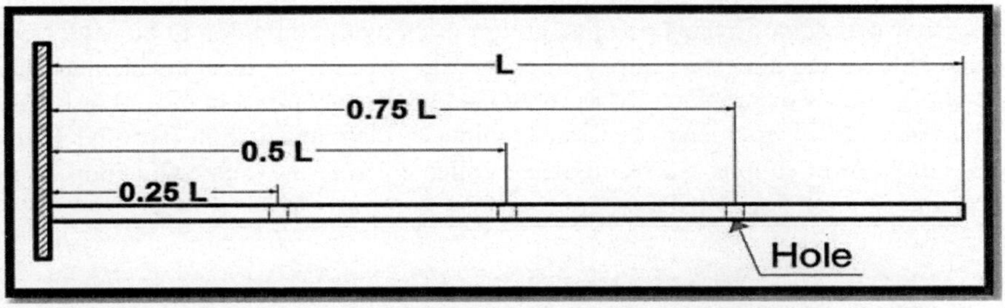

Fig. 7 Cantilever beam with the square holes' and this location

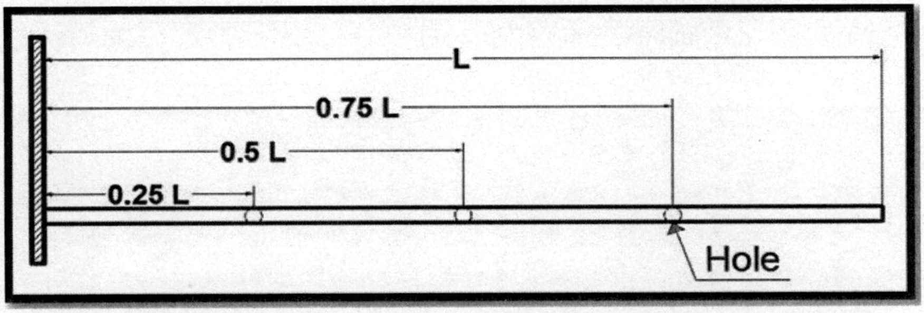

Fig. 8 Cantilever beam with the circler holes' and this locations

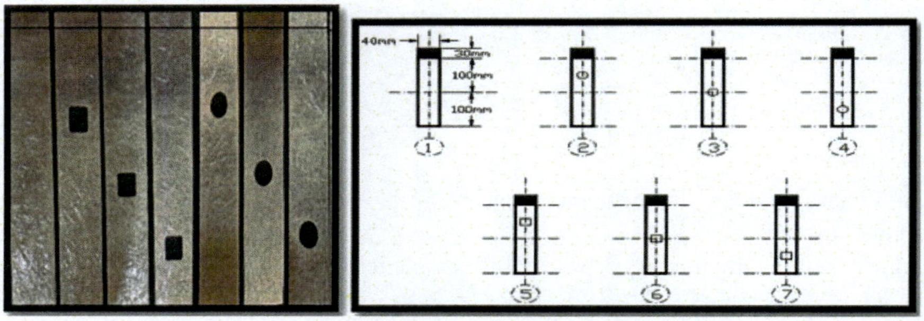

Fig. 9 Holes' shapes and locations in cantilever beam

3.4 Natural Frequency

In this work, seven cases were studied to calculate the natural frequency of the beams with dimensions (200 mm active long, and 40 mm width) with 30 mm extra-long for fixing. CNC machine made all holes in composite beams. The specimen studies as beams comprise a subset of shells and plates whose widths are small compared with lengths. This permits motion in the width direction to be neglected which reduces the dimensionality of models. The system was used to calculate the natural frequency experimentally, as shown in Fig. 10, it consists mainly of a shaker (type B&K 4808), frequency generator, amplifier, data acquisition (type NI USB 4431) and power supply. All results were collected from the data acquisition by a computer through LABVIEW program.

4 Theoretical Analysis

In this work, the properties of the used material are shown in the Table 2, Rule of mixture was used to find the composite modules of elasticity by Eq. 1 below [14]:

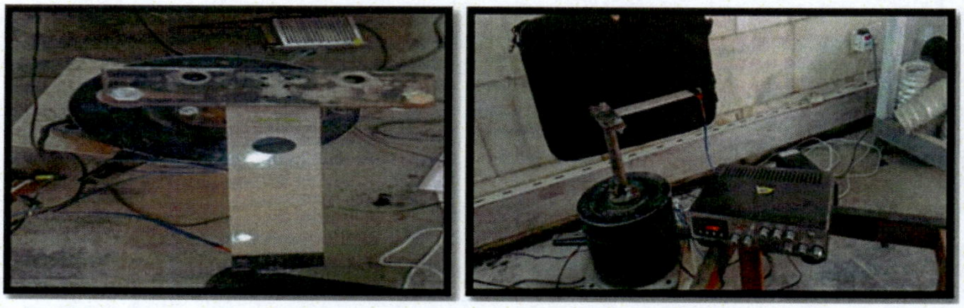

Fig. 10 Vibration device

Table 3 Natural frequency for all cases for beams

	1st mode Hz		2nd mode Hz		3rd mode Hz	
Holes location	ANSYS	experimental	ANSYS	Holes location	ANSYS	experimental
With out	7.22	8.13	45.195	With out	7.22	8.13
0.25 Lc	6.69	7.42	45.596	0.25 Lc	6.69	7.42
0.5 Lc	7.123	7.95	43.58	0.5 Lc	7.123	7.95
0.75 Lc	7.463	8.01	43.857	0.75 Lc	7.463	8.01
0.25 Ls	6.693	7.44	45.506	0.25 Ls	6.693	7.44
0.5 Ls	7.1185	7.88	43.429	0.5 Ls	7.1185	7.88
0.75 Ls	7.441	7.98	43.835	0.75 Ls	7.441	7.98

$$E_c = E_e V_e + E_f V_f \tag{1}$$

where: $V_f = 22.5\%$, $V_e = 77.5\%$, and $E_f = 74$ Gpa, $E_e = 4$ Gpa; $E_c = 16.65 + 3.1 = 19.75$ Gpa.

The composite module of elasticity by Eq. 1 was very near with the magnitude in the Table 3 from tensile test result (experimentally). In this work, for the same area of circle and square holes the composite beams were put along with assuming the radius of the circle to be equal 10 mm, the length of square was found by equations below:

$$A_c = \pi \cdot r^2 \tag{2}$$

$$A_s = L^2 = A_c \tag{3}$$

$$\pi \cdot r^2 = L^2 = 3.14 \times (10)^2 \tag{4}$$

$$L = 17.72\text{mm}$$

5 Results and Discussion

The results cover of the first three natural frequencies for each specimen, were calculated experimentally and theoretically. In addition, the maximum deformation of each specimen was calculated theoretically.

5.1 Natural Frequency

The experimental and theoretical results for circular and square holes in different locations with three modes of natural frequency were calculated as shown in Table 3.

As the hole is far from the support, the first natural frequency will be increased for the circular and square hole, which means that the stiffness of the specimen is also affected. Also it is noticed that the first natural frequencies were not affected by the same area hole shape which was removed. Experimentally for the second natural frequency, the results showed that the frequency increased by about (8–10) % for both shape holes at (0.25L) than the specimen without hole because of the shape that the specimen has taken, and then the frequency will be decreased as the hole would be far from the fixed line.

Theoretically, the frequency increased not more than (2%) for the hole at the (0.25L) which it means that the effect is only from the mass difference, but when the hole was at (0.5 L or o.75L) the frequency decreased by about 5% due to change in mass and stiffness. In the third mode, the effect was about 5% on third natural frequency appeared where the two shapes hole in the 0.75L, theoretically and experimentally. Figures 11 and 12 below show the (1st, 2nd, and 3rd) modes of natural frequency for all cases of beams by diagram.

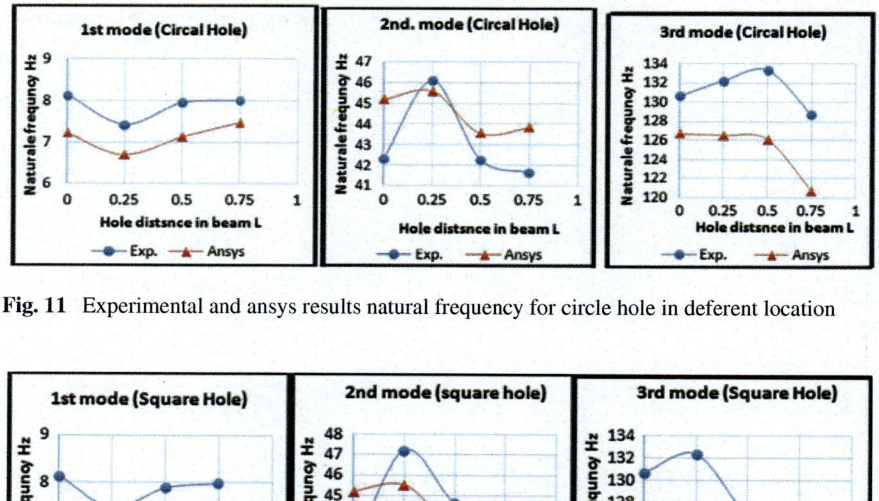

Fig. 11 Experimental and ansys results natural frequency for circle hole in deferent location

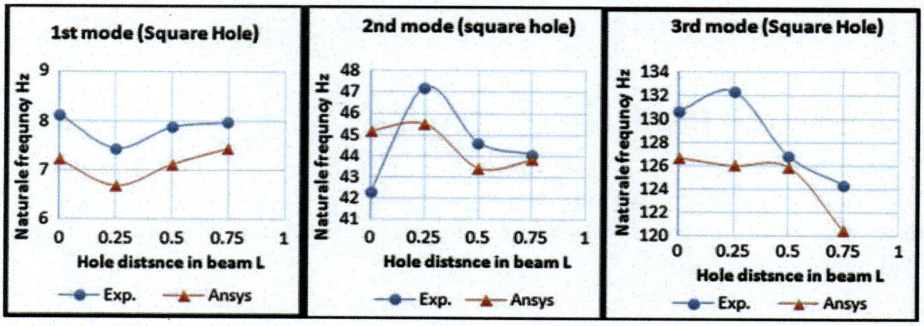

Fig. 12 Experimental and ansys results natural frequency for square hole in deferent location

Table 4 Theoretically result for maximum deformation

Holes location	Max. def. 1st mode (mm)	Max. def. 2nd mode (mm)	Max. def. 3rd mode (mm)
With out	15.52	14.37	20.4
0.25 LC	15.3	16.3	21.86
0.5 LC	18.5	18.67	16.8
0.75LC	14.95	14.57	14.1
0.25 LS	12.92	19.7	21.5
0.5 LS	15.53	15.67	17.4
0.75LS	13.758	15.2	17.19

From the above results, the shape has an effect mainly on the second mode due to the profile of the specimen in that mode.

5.2 *Maximum Deformation*

The maximum deformation for all specimens was theoretically calculated by using (ANSYS) program.

As seen in the Table 4 for first mode, the deflection of beams has the same behaviour for the two shapes of hole, but the deflection for the circular holes was more than at square holes, in spite of that the same mass removed which means that the stiffness between the hole edge and the beam edge at that line has a large effect. For the second mode, the maximum deflection seems to be between (0.25L–0.5L) due to the shape of the hole. For the third mode, the effect for the maximum deflection seems also when the hole was at (0.5L- 0.75L) far from the fixed end.

Figure 13 below shows the maximum deformation (1st, 2nd, and 3rd) modes for all cases of beams by diagrams.

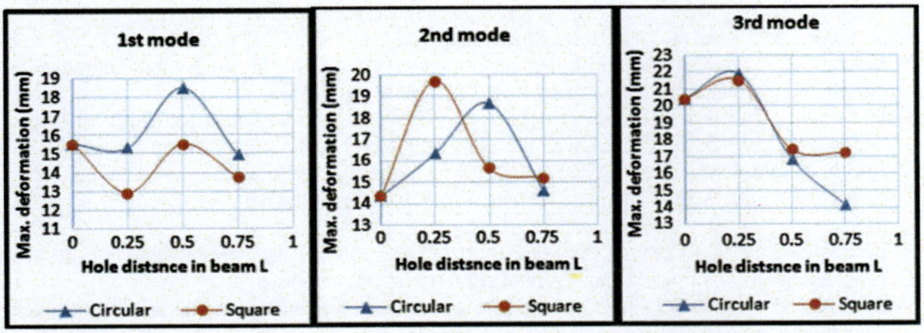

Fig. 13 Maximum deformation for all cases of beams

6 Conclusions

From the results above we can conclude that:

- The fundamental natural frequency decreases as the location distance from the fixed end increases.
- The effect of circular hole on the deflection of composite beam is more than the effect of square holes for 1st mode, and the 2nd 3rd deflection value with square holes more than circular hole.
- As application use, we can choose the best shape and location holes as designer decided.
- The effect of the hole shape and position on the first natural frequency related to the mass removed. While, for the third mode frequency it seems when the hole at third quarter of the beams.
- The effect of the hole shape and position on the second mode frequency seems to be reduced when the holes were between (0.25L and 0.5L) so we can remove the material where the beams have low stress intensity to reduce the weight which is important as in prosthetic limbs.
- The maximum deformation due to vibration seems to be at the hole of 0.5L in the first mode, while at third mode be slightly reduced between 0.5 and 0.75L.
- At the second mode, the maximum deformation is affected by the hole shape as seen in Fig. 9 it was seen that it is maximum for square hole at (0.25L) while it is maximum for circular hole at (0.5L), that because of profile of the beam in the mode and hole shape which affected on the stiffness of the beam.

Acknowledgements The author is obliged to the University of Technology, Baghdad in Iraq, for providing the Department of Mechanical Engineering.

References

1. Priyadarsi MP, Amrit D (2019) Inter J Res Appl Sci Eng Tech 6:2321–2325
2. Serene KT, Aswathy P (2019) Finite element analysis of composite beams and columns with castellated members. Int J Sci & Eng Res 10(5). ISSN 2229-5518
3. El-Dehemy H (2017) Static and dynamic analysis web opening of steel beams. World J Eng Technol 5:275–285. ISSN online: 2331-4249, ISSN print: 2331-4222
4. Erkliğ A, Bulut M, Yeter E (2013) Effects of cutouts on natural frequency of laminated composite plates. SciEng Compos Mater 20(2):179–185. https://doi.org/10.1515/secm-2012-0120
5. Dive V, Bhosale M, Chavan V, Durugkar N (2017) Analysis of natural frequencies of cantilever beam using ansys. Int Res J Eng Technol (IRJET) 04(05)
6. Callioglu H, Atlihan G (2011) Vabration analysis of delaminated composite beam using analytical and FEM models. Indian J Eng Mater Sci 18:7–14
7. Thakare S, Damale A (2013) Free vibration analysis of circular plates with holes and cutouts. IOSR J Mech Civ Eng (IOSR-JMCE) 8(2)
8. Steel Beams with Web Openings (2015) Chapter 28. Version II

9. Cameira Neto TM (2014) Analysis and design of composite beams with web openings. T. M. CameiraNeto/Instituto Superior Técnico
10. Gawali AL, Kumawat S (2011) Vibration analysis of beam. World Res J Civ Eng 1(1):15–29. ISSN: 2277-5986 & E-ISSN: 2277-5994
11. Abdilrazzaq T (2013) The effect of circular holes on the natural frequencies of a beam using transverse vibration measurements method. Al-Rafidain Eng 21(2)
12. Abdullah Al-Shammari M (2015) Effect of hole shape on free vibration characteristic of unidirectional composite plates. Iraqi J Mech Mater Eng 15(1)
13. ASTM D3039 standard test method for tensile properties of polymer matrix composite materials
14. Pavan S, Sudhabindu B, Ajay S, Yaswanth Reddy G, Singh K (2019) A Basic Approach to Find Elastic Constants by Rule of Mixtures on Jute/E-Glass Composite Sheet. Int J Recent Technol Eng (IJRTE) 8(1). ISSN: 2277-3878

Analysis of Stress Distribution of Custom Fitted Femoral Component Knee Implant for Asian Anatomy

Rosdayanti Fua-Nizan, Ahmad Majdi Abdul-Rani, Mohamad Yazid Din, and Suresh Chopra

Abstract Total Knee Replacement is a corrective medical procedure which is performed to remove the damaged bones and cartilage of knee with osteoarthritis, and replaced with implant components. The disparity in anatomy between western and Asian on another hand has created some issues regarding the fit of the knee implant and affected the living quality of the patients during post Total Knee Replacement stage. In conjunction with the development of patient specified prostheses for several medical applications, a custom fitted femoral component was developed by utilizing Digital Imaging and Communications in Medicine images of MRI scanned knee. Stress distribution analysis at the maximum flexion angle was done on the femoral component to analyse the stress value and distribution in comparison with the existing femoral component used by a local hospital. The results showed that the Von Misses stress value recorded for the existing femoral component is 81.83% lower than custom fitted femoral component, due the design features of the existing femoral component. The result recorded that successfully highlighted the critical areas of the custom fitted femoral component and it was concluded that the femoral component requires design enhancement to be deemed as feasible for Total Knee Replacement.

R. Fua-Nizan (✉) · A. Majdi Abdul-Rani
Mechanical Engineering Department, Universiti Teknologi Petronas, 32610 Bandar Seri Iskandar, Perak, Malaysia
e-mail: rosdayanti_g03569@utp.edu.my

M. Yazid Din
Orthopaedic Department, Hospital Tuanku Fauziah, Pusat Bandar Kangar, 01000 Kangar, Perlis, Malaysia

S. Chopra
Orthopaedic Department, Hospital Sultanah Bahiyah, Alor Setar, Perlis, Malaysia

F. Ahmad et al. (eds.), *ICPER 2020*, Lecture Notes in Mechanical Engineering,
https://doi.org/10.1007/978-981-19-1939-8_45

1 Introduction

Human joints, specifically knees, are prone to various illness such as rheumatoid arthritis, traumatic arthritis and osteoarthritis (OA) cause by variety of factors. Amongst these arthritis, osteoarthritis is the most common especially among elderly with obesity problem. The illness causes the joint to degenerate through the thinning of the menisci cartilages, which function to protect the bones between the knee joint from friction. Upon the final stage of osteoarthritis, the menisci cartilages would already disappear and the femur and tibia bones would be deformed and completely damaged. The illness would cause excruciating pain for the patient and hinder their ability to stand and walk normally.

Commonly, this illness was treated by a series of medications including injections to subdue the pain. However, Total Knee Replacement (TKR) procedure would be the best option when the illness has reached the final stage. Damaged cartilage and bone surfaces are resected and replaced with a set of knee implants, commonly made from CoCr alloy or Ti6Al4V4 alloy. The outside surface of the metal components shall be coated with biocompatible coating to suit the surrounding environment while the inside surface shall possess the ability of osteointegration to accelerate the healing process [1]. Particularly for femoral component, the outside surface shall possess the functionality in terms of shape and surface properties [2] in order to facilitate the flexion and extension of the knee joint. The procedure shall eliminate the pain, restore legs alignment, the patellar position, and the quadriceps angle (q-angle) as well as balancing the soft tissues. The knee implant for TKR is consist of three main components, the femoral component, the tibial insert and tibial component. The knee implants components are commonly designed and manufactured by the Western manufacturer which focuses on Caucasian as target market. Although TKR has become common worldwide, due to the difference in the anatomy of Caucasian and Asian [3, 4], the sizing of the knee implant to Asian patient has become an issue which has reduced the life-cycle of the implants and subsequently reduce the living quality of the patients [5, 6]. Patient specified prosthesis have been utilized via reversed engineering and additive manufacturing for several application such as cranial [7] and dental [8]. Hence, this research was initiated in order to analyse the feasibility of custom fitted femoral component knee implant through finite element analysis at maximum flexion angle, in comparison with the existing femoral component used in the market.

2 Methodology

Magnetic Resonance Imaging (MRI) scanned images of a local patient with osteoarthritic knee condition were acquired from a local hospital. The MRI consist of a series of 2D images in Digital Imaging and Communications in Medicine (DICOM) format which require a specified software which shall enable the user to read and

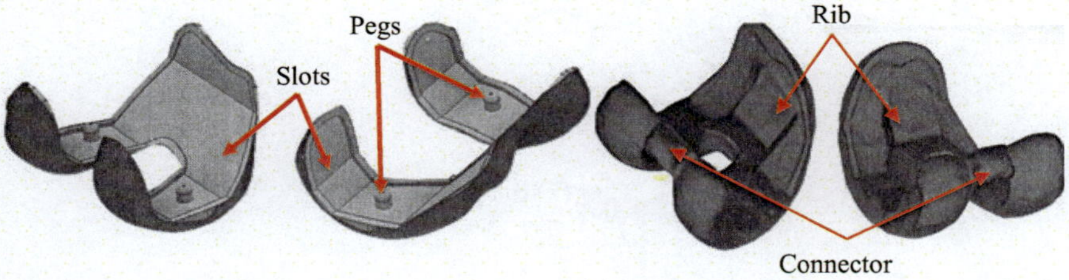

Fig. 1 Custom fitted (left) and existing femoral component (right) [10]

manipulate the image series. ITK –Snap is a medical image processing system which allows the reconstruction of the medical images into 3D [9]. The system was utilized to enhanced the DICOM images quality and redefined the information contained inside the imaged. Via edge attraction method [10], the bones and the soft tissues were differentiated for the east of reconstructing the bones into 3D image. Considering the scope of the study, femur was extracted from the reconstructed 3D image and exported for design modification. Computer Aided Design (CAD) system of Autodesk Inventor was utilized for the modification of the femur bone [10] into custom fitted femoral component knee implant based on standard TKR method [11] for ideal resection angle and segment thickness. The existing femoral component knee implant, taken out from a patient during TKR revision surgery was acquired from the same local hospital and was scanned in order to obtain the 3D image, for the purpose of analysis and comparative study. The custom fitted and existing femoral components are illustrated in Fig. 1.

Both femoral components were simulated at 150° flexion angle, which is the maximum angle required by the Asian Muslim [12] to perform daily activities via Finite Element Analysis (FEA), ANSYS.

The boundary condition of the simulation consist of the tibiofemoral forces in vertical (FTFV) and horizontal direction (FTFH), patellofemoral force (FPF) which is normal to the contact area between the patellar and femur and moment about the knee joint (M) [13, 14], as illustrated in Fig. 2.

Previous researchers have successfully conducted studies to determine the force components acting on the knee during various flexion angle. However, those research were conducted separately for the forces acting on the tibiofemoral (TF) joint and patellofemoral (PF) joint [14–17]. There was a research which had studied the force components for both TF and PF joint [13]. However, the reliability of the technology used for the research is questionable, taking into account the year of the research was conducted. Hence, the value of the forces and moment applied for the simulation were tabulated via meta-analysis method [10] based on more recent research [14–17], and tabulated with respect to body weight (BW) and height (H), as listed in the following Table 1.

However, the personal information of the patient, specifically the body weight and height were not provided along with the DICOM images. Hence, the information were acquired from a study of weight and height for Malaysian [18].

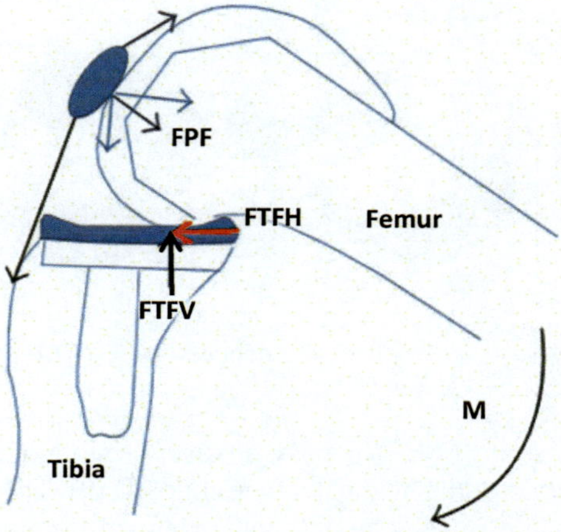

Fig. 2 Forces component applied for simulation during knee flexion [14]

Table 1 Value of forces and moment about the knee joint at 150° flexion [10]

Properties	Value
FTFV, N	7.10BW
FTFH, N	4.39BW
FPF, N	1.93BW
M, N.m	11.30BWxH

Table 2 The mechanical properties of CoCr F75 [19]

Properties	Value
Tensile strength, MPa	979
Yield strength, MPa	752
Density, g/cm3	8.6
Young's modulus, MPa	7000

On the other hand, the material used for the simulation is CoCr-F75 metal, a material commonly used for metal prosthesis such as knee implant components. The mechanical properties of the material are listed in the following Table 2.

3 Result

Based on the simulated custom-fitted femoral component knee implant at 150° flexion, it showed that the stress was not evenly distributed and was more concentrated around the edge [10] of the anterior flange and the femoral condyle. The maximum

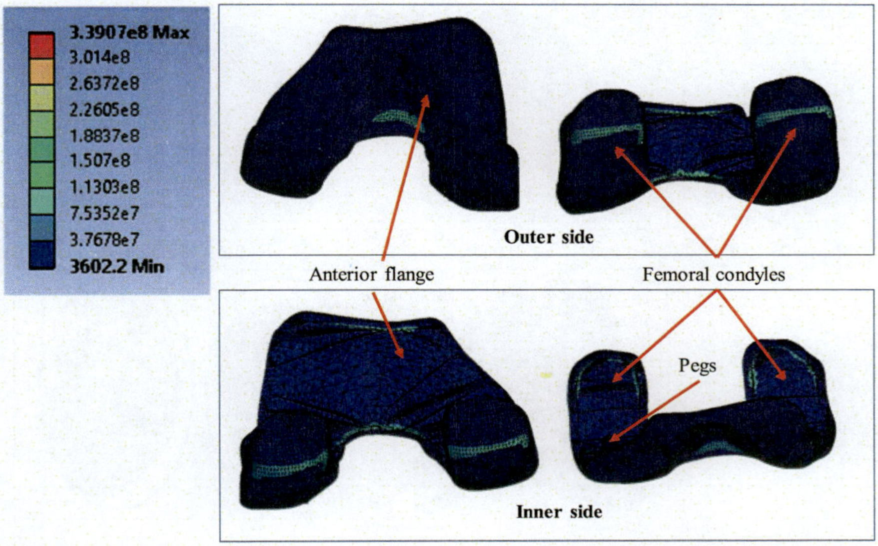

Fig. 3 Stress distribution of the custom-fitted femoral component knee implant at 150° flexion [10]

Von Misses stress value recorded for the femoral component was 339.1 MPa [10], as illustrated in Fig. 3.

On the other hand, the maximum Von Misses stress value recorded for the existing femoral component is 61.6 MPa [10].

Based on the stress distribution, the stress was concentrated at the inner side of the anterior flange, specifically around the rib area. The stress at the femoral condyle was observed to be uniformly distributed around the edges of both condyle as illustrated in Fig. 4.

4 Discussion

Following previous study conducted to analyse the force components on knee joint initiated by Dahlkvist, there have been researches conducted to analyse the stress distribution of knee implant during high flexion. A research conducted by Zelle (2009) to analyse the stress distribution on high flexion type knee implant at 120° flexion of released posterior cruciate ligament (PCL) condition and recorded approximately between 10 to 15 Mpa stress [20] and Anuar (2014) which studied the sheer stress of mobile bearing type knee implant also recorded similar result [21]. Although these research have provided comprehensive methodology and outcomes for the study of stress distribution on tibial insert component, there have been much lesser study conducted for femoral component knee implant. Considering this limitation,

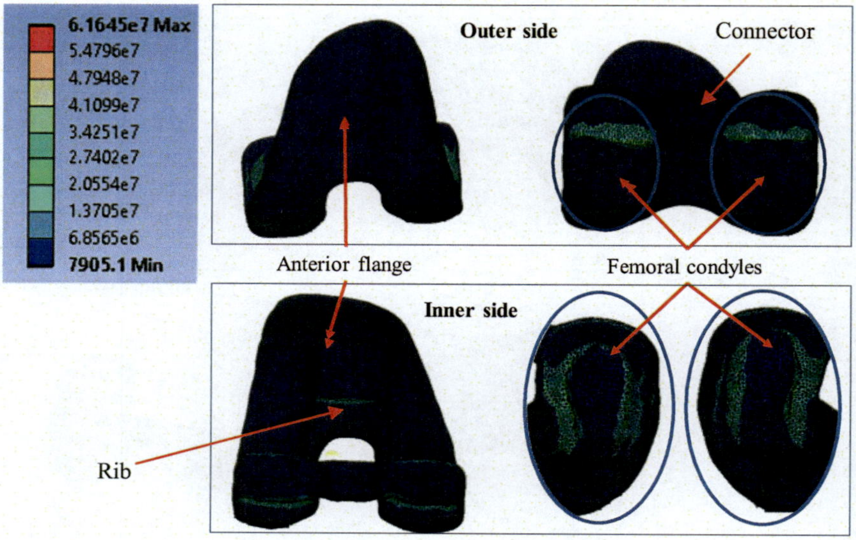

Fig. 4 Stress distribution of the existing femoral component at 150° flexion [10]

this research was initiated with purpose to analyse the stress distribution of femoral component knee implant, specifically during high flexion.

The maximum Von Misses stress values which were documented for both femoral components has showed that the existing femoral component recorded significantly lower stress value, by approximately 81.83% compared with the custom fitted femoral component. The stress distribution of the existing femoral component showed that the rib and connector had played a crucial role in reducing the stress value of the existing femoral component. The connector had provided support and had evenly distributed the stress at both femoral condyle. The rib on the other hand, which is located in the inside of the anterior flange had provided additional thickness [10] which had supported the area in between the implant with the bone by shifting the stress concentrated at the rib, instead of the anterior flange. Based on the stress distribution pattern, unlike the existing femoral component knee, the custom fitted femoral component was lacking such additional design features to act as support, specifically at high flexion with higher force exerted on the implant.

Similar with previous researches conducted for tibial insert knee implant, the simulation conducted for this research has successfully highlighted the critical areas of the custom fitted femoral component during flexion and indicated the limitation of the component which is at the inside edges of the anterior flange, as well as the femoral condyle.

5 Conclusion

The research was initiated to analyse the stress distribution of the custom fitted femoral component knee implant, in comparison with the existing femoral component used in the market. Based on the stress value and stress distribution, the existing femoral component possess the design features which had provided support for the femoral component and had subsequently reduce the stress value significantly, in comparison with the custom fitted femoral component. The high stress value of the custom fitted femoral component would affect the performance of the femoral component in the long run, especially for high flexion needs and consequently increase the risk of revision surgery within less than 10 years after TKR [22]. It was concluded that in order for the custom fitted femoral component to be deemed feasible for TKR, it would require several design enhancements to ensure better performance as well as successful TKR. The methodology to design and develop novel custom-fitted knee implant could be further developed as well and implemented for the design of tibial component and tibial insert as well.

Acknowledgements The authors would like to acknowledge the financial support provided by Ministry of Education Malaysia through grant No. FRGS/1/2014/TK01/UTP/02/08.

References

1. Campanelli SL, Contuzzi N, Ludovico AD, Caiazzo F, Cardaropoli F, Sergi V (2014) Manufacturing and characterization of Ti_6Al_4V lattice components manufactured by selective laser melting. Mater (Basel) 7(6):4803–4822
2. Yamaguchi H, Graziano AA (2014) Surface finishing of cobalt chromium alloy femoral knee components. CIRP Ann—Manuf Technol 63(1):309–312
3. Yue B, Vadarajan KM, Ai S, Tang T, Rubash HE, Li G (2011) Differences of knee anthropometry between chinese and white men and women. J Arthroplasty 26(1):124–130
4. Chaichankul C, Tanavalee A, Itiravivong P (2011) Anthropometric measurements of knee joints in Thai population: correlation to the sizing of current knee prostheses. Knee 18(1):5–10
5. Ko Y et al (2011) Health-related quality of life after total knee replacement or unicompartmental knee arthroplasty in an urban asian population. Value Heal 14(2):322–328
6. Xie F et al (2010) Evaluation of health outcomes in osteoarthritis patients after total knee replacement: a two-year follow-up. Heal Qual Life Outcomes 8:87
7. Jardini AL et al (2014) Cranial reconstruction: 3D biomodel and custom-built implant created using additive manufacturing. J Cranio-Maxillofac Surg 42(8):1877–1884
8. Ojeda CJ, Castro MB, Céspedes JA, Chumán VE, Arriola EA, Díaz-Lantada A (2017) Design and biomechanical simulation of custom dental implants. Biomed Eng (NY) 147–153
9. Yushkevich PA et al (2006) User-guided 3D active contour segmentation of anatomical structures: significantly improved efficiency and reliability. Neuroimage 31(3):1116–1128
10. Fua-Nizan R (2020) The study of custom-fitted femoral component knee implant for high flexion. University Technology Petronas
11. Vail TP, Lang JE, Van Sikes C, Chapter 105 surgical techniques and instrumentation in total knee arthroplasty. Musculoskelet. Key

12. Acker SM, Cockburn RA, Krevolin J, Li RM, Tarabichi S, Wyss UP (2011) Knee kinematics of high-flexion activities of daily living performed by male muslims in the middle east. J Arthroplasty 26(2):319–327
13. Dahlkvist NJ, Mayo P, Seedhom BB (1982) Forces during squatting and rising from a deep squat. Eng Med 11(012632852):69–76
14. Thambyah A, Fernandez J (2014) Squatting-related tibiofemoral shear reaction forces and a biomechanical rationale for femoral component loosening. Sci World J 2014
15. Thambyah A (2008) How critical are the tibiofemoral joint reaction forces during frequent squatting in Asian populations? Knee 15(4):286–294
16. Freedman BR, Sheehan FT, Lerner AL (2015) MRI-based analysis of patellofemoral cartilage contact, thickness, and alignment in extension, and during moderate and deep flexion. Knee 22(5):405–410
17. Sharma A, Leszko F, Komistek RD, Scuderi GR, Cates HE, Liu F (2008) In vivo patellofemoral forces in high flexion total knee arthroplasty. J Biomech 41(3):642–648
18. Azmi MY et al (2009) Body mass index (BMI) of adults: findings of the Malaysian adult nutrition survey (MANS). Malays J Nutr 15(2):97–119
19. Lu Y et al (2015) Investigation on the microstructure, mechanical property and corrosion behavior of the selective laser melted CoCrW alloy for dental application. Mater Sci Eng C 49:517–525
20. Zelle J, Heesterbeek PJC, De Waal Malefijt M, Verdonschot N (2010) Numerical analysis of variations in posterior cruciate ligament properties and balancing techniques on total knee arthroplasty loading. Med Eng Phys 32(7):700–707
21. Afzan M, Anuar M, Todo M, Nagamine R, Hirokawa S (2014) Dynamic finite element analysis of mobile bearing type knee prosthesis under deep flexional motion. Sci World J 2014:6
22. Dalury DF, Pomeroy DL, Gorab RS, Adams MJ (2013) Why are total knee arthroplasties being revised? J Arthroplasty 28(8 SUPPL):120–121

Evaluation of NARX Network Performance on the Maintenance Application of Rotating Machines

N. A. Samsuri, S. A. Raman, and T. M. Y. S. Tuan Ya

Abstract Rotating machines are widely used in many industries even though the maintenance of this equipment is high. Most of the industries employ preventive maintenance to avoid expensive failures and shutdown. From the view of maintenance application, a predictive approach utilizing real time data provide a better understanding of the system degradation. This paper presents the evaluation of ANN technique for a data driven model which is nonlinear autoregressive neural network design with exogenous input (NARX) for a maintenance application for rotating machines. The data used for the research are obtained from NASA Data Depository. Analysis was done to understand the data better. Before model training, a feature selection for the inputs have been made allowing dimensionality reduction narrowing the parameters consisting features that could make an impact on detecting failure. NARX neural network model from the neural net time series is chosen as it can adequately handle a time series data in a forecasting model network. NARX network is first trained in open-loop mode before being trained in closed-loop mode. Results show that the network model is suitable for time series data prediction.

Keywords Artificial neural network · Conditioned-based maintenance

Abbreviations

ANN	Artificial Neural Network
C-MAPSS	Commercial Modular Aero Propulsion System Simulation
MLP	Multi-Layer Perceptron
NARX	Nonlinear Autoregressive Neural Network with eXogenus inputs

N. A. Samsuri (✉) · S. A. Raman · T. M. Y. S. Tuan Ya (✉)
Department of Mechanical Engineering, Universiti Teknologi PETRONAS, 32610 Seri Iskandar, Perak, Malaysia
e-mail: nurain_20000265@utp.edu.my

T. M. Y. S. Tuan Ya
e-mail: tyusoff.ty@utp.edu.my

F. Ahmad et al. (eds.), *ICPER 2020*, Lecture Notes in Mechanical Engineering,
https://doi.org/10.1007/978-981-19-1939-8_46

RMS	Root Mean Square
TDL	Time Delay Line

1 Introduction

A lot of industries nowadays try to optimize the cost of operation and production and this can be help by doing predictive maintenance. The standard maintenance which is corrective, and preventive is not as effective as predictive maintenance as these two types of maintenance method cannot predict an upcoming failure thus resulting in loss of both productions and reputations [1]. A catastrophic failure of machines may cause an escalating machine downtime and overhauling expenses that can cause an economic shortfall of the industries. This has leads to the advancement of the industry maintenance strategies in which better methods are instigated necessarily ensuring a smooth performance of the rotating machineries. In recent years, the usage of condition-based maintenance has been widely applied to the industry as it has evolved into a more capable maintenance technique [2].

Condition-based maintenance allows maintenance plans to be arranged before any machinery failure could occur that could simultaneously lessen the risks of major breakdowns thus the unnecessary maintenance expenditures could be avoided [3]. The said maintenance technique works with a decision-making approach by monitoring the historical conditions statistics that allows predictions of the systems future health [4]. The capability to predict makes condition-based maintenance as an attractive choice for systems maintenance, reliability and in-service expenses [5]. The mechanism of an effective condition-based maintenance can be layered into three essential steps: data acquisition, data processing and decision making [5] that can be further explained in Fig. 1.

The predictions of the machine health conditions are extremely beneficial for maintenance scheduling and planning especially for demanding applications, in which the maintenance shall be done carefully safe and economic wise for the whole lifetime [4]. Diagnostic can be inferred as the evaluation of the available (and past) system conditions based on examined symptoms. It covers the task of discovering

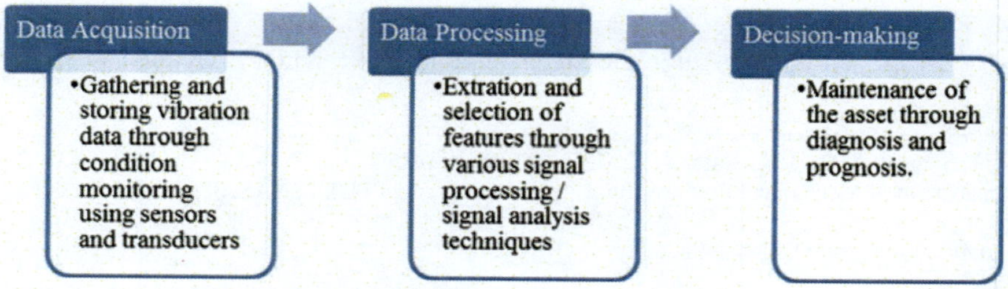

Fig. 1 Key steps of condition-based maintenance for rotating machines

errors and faults, seclusions, and identification once fault ensues [2]. On the other hands, prognostics involved with the assessment of the machine's future conditions and remaining time before any breakdown events provided the existing machine age and historical working applications are known [2].

Diagnosis of the approaching failures and asset health status prognostic are ensured by acquiring periodic data from statistics of the systems [6]. Prognostics on data collection basis are far more effective methods in rotating machine health prediction applications due to the ease in finding related data and consistency in intricate processes [7]. Data-driven prognostics are also significant as it has the capability to assimilate innovative and conventional methods by producing an inclusive predictions approach over a broad dataset [4]. Despite the facts that the information and literature of rotating machines maintenance are heavily available, the study that revolves around prognostic is arising starting to become a trend in the coming years [2].

In their studies, Bednarz et al. [8] outlines the diagnostics of rotating machines utilizing neural network approach that are considered as more accurate methods. Bektas and Jones [4] presents a new method of failure predictions for rotating machinery, also with neural network approach. Another study was made where the rotating machinery health conditions used in industries are monitored with the prospects of neural network [9]. Prognostic models are designed by the basis of Nonlinear Autoregressive Neural Network with eXogenus inputs (NARX) to support the research [2, 4, 8–10]. There are also few researchers provides with some noteworthy evaluations of rotating malfunctions [11, 12].

Neural network can be considered as an impactful approach to imitate engineering tasks with extensive features of nonlinear dynamic systems, data trimming models, nonlinear regression and conditional models [13]. The monitorization of the systems that are illustrated from complex engineering applications may be provided with lack of precise data, at the same time, the input values may not be directly linked to the preferred results. For such issues, neural network is useful to handle with the missing and imprecise data as it can build a model without knowing the exact relationship between input and output data [14].

Artificial Neural Networks (ANN) is one of the most common method used for a data driven prognostic [15]. Among all the available prognostic methods, usually a multi-step long term prediction with dynamic modelling are chosen to solve the complex system of prognostic algorithm as they are easier to calculate and much faster [4]. ANN can be quite challenging for multi-step predictions as little knowledge about time series is available thus not knowing about the failure point when it comes in longer term. Therefore, Nonlinear Autoregressive Neural Network with eXogenus inputs (NARX) is used for this paper as it is suitable to predict a time series data using operational past values [4]. In a time-series modelling, a NARX model is a recurrent dynamic network that uses past values of time series to do prediction [4]. The model deduced predictions of a time series data made up from some parameters that are basing of their performance in the system. The goal of this paper is to evaluate the performance of NARX network on the application of rotating machines for maintenance purpose.

2 Methodology

2.1 Problem Description

This project uses data from National Aeronautics and Space Administration (NASA) Prognostics Center of Excellence Data Repository (http://ti.arc.nasa.gov/tech/dash/pcoe/prognostic-data-repository), which consist of multiple multivariate time series data [16].

This dataset is damage modelling propagation that was done by using Commercial Modular Aero Propulsion System Simulation (C-MAPSS). The simulation data accommodates prognostic diagnostics of the engines. The C-MAPSS simulation possess the ability that allow input variations of health-related parameters and recording of the ensued output sensor measurements [16].

There are 4 sets of data with different type of operational settings and failure mode [16]. Each dataset contains 100 machine id with multivariate time steps. The data gained from the simulation contains 21 sensors measurement with each have different impact on the gas turbine engine performance. Table 1 shows the operational settings of the engine simulations when simulating atmospheric model, while Table 2 shows the sensor parameters of the dataset used to measure the system response [16].

Table 1 Operational parameters of the engine simulation [16]

S. no	Description	Unit
1	Altitude	ft
2	Mach number	M
3	Sea-level temperature	°F

Table 2 Sensor parameters used throughout the simulation. The output data are collected will be used as variables in measuring the system response [16]

S. no	Descriptions	Units	S. no	Descriptions	Units
1	Total temperature at fan inlet	°R	12	Ratio of fuel ow to Ps30	pps/psi
2	Total temperature at LPC outlet	°R	13	Corrected fan speed	rpm
3	Total temperature at HPC outlet	°R	14	Corrected core speed	rpm
4	Total temperature at LPT outlet	°R	15	Bypass Ratio	–
5	Pressure at fan inlet	psia	16	Burner fuel–air ratio	–
6	Total pressure in bypass-duct	psia	17	Bleed Enthalpy	–
7	Total pressure at HPC outlet	psia	18	Demanded fan speed	rpm
8	Physical fan speed	rpm	19	Demanded corrected fan speed	rpm
9	Physical core speed	rpm	20	HPT coolant bleed	lbm/s
10	Engine pressure ratio	–	21	LPT coolant bleed	lbm/s
11	Static pressure at HPC outlet	psia			

Table 3 List of dataset from C-MAPSS

Dataset	Training	Testing	Regimes
FD0001	100	100	1
FD0002	260	259	6
FD0003	100	100	1
FD0004	248	249	1

The C-MAPSS dataset is shown in Table 3. Dataset of FD0001 was chosen as it was the simplest data to work with. At the beginning of operation, the engine was operating normally. At some point of the time series, the engine starts to degrade. This is actually known as the exponential degradation of the gas turbine engine operation. The data then will be divided into two subsets which is training data and testing data [16].

2.2 *Data Pre-processing*

The data is contaminated with noise and therefore a data cleansing must be done in order to produce an accurate result. Not all the features contributed to the research variables as some are constant throughout the time series [17, 18]. This step is important as it minimizes the dimensions of the unrefined data acquired. The data pre-processing process normally considers relationship of the data with the analysis, time propagation and linearity of the data.

The data pre-processing done was by doing a feature selection that was selected based on numerous research paper. Thus, only 9 features are selected as input altogether with 3 operational settings and unit number. The selected features are assigned based on a prognostic parameter-choosing measures in which considering the "Monotonicity", "Prognosability" and "Trendability" of the options [17]. The sum of the measures can determine the fitness of a candidate prognostic parameter.

$$fitness = monotonicity + prognosability + trendabilityy$$

The result of the measures eliminated signals that does not have any large impact on the degradation pattern. Various analysis from numbers of study on the dataset analysis has supported the findings where the most common selected sensors are as Table 4. Meanwhile, the output of the network is determined by median of input variables at each of the time step which is according to the order statistic [4];

$$x_{f1} = \min_j \tilde{y}_{tj}, x_{f2}, , x_{fN-1}, x_{fN} = \max_j \tilde{y}_{tj} \tag{1}$$

The formula for median input variables at time step j is as follows;

Table 4 Features selected as the input for the network training. The features are basing on the overall sensor measurements of the engine simulation

S. no	Description no
2	Total temperature at LPC outlet (°R)
3	Total temperature at HPC outlet (°R)
4	Total temperature at LPT outlet (°R)
7	Total pressure at HPC outlet (psia)
11	Static pressure at HPC outlet (psia)
12	Ratio of fuel flow to Ps30 (pps/psi)
15	Bypass Ratio
20	HPT coolant bleed (lbm/s)
21	LPT coolant bleed (lbm/s)

$$y_{tj} = \begin{cases} x_{f(N+\frac{1}{2})} & \text{if N is odd} \\ \frac{1}{2}\left(x_{fN/2} + x_{f1+N/2}\right) & \text{if N is even} \end{cases} \tag{2}$$

2.3 Feature Extraction

The diagnosis efficiency can be increased by having suitable elements extracted and selected from the data. For the data of vibrations, there are three main aspects of elements that can be extracted from the data: frequency domain, time domain and time–frequency domain. Time domain features can be portrayed as the statistical analysis of the data that consist of mean, Root Mean Square (RMS), value of maximum and minimum, standard deviation, skewness, and crest factor [19].

Standard deviation can be described as the scattered data from its respective mean. Greater value of the standard deviation in every dataset will illustrate a more scattered data. The mean and standard deviation are listed in Eqs. (3) and (4) respectively. From the equations, K represents the number of data points while x(k) is the series of signals [20].

$$xm = \frac{1}{K}\sum_{K=1}^{K} x(k) \tag{3}$$

$$xstd = \sqrt{\frac{\sum_{K=1}^{K}[x(k) - xm]^2}{K-1}} \tag{4}$$

Besides, another crucial parameter for features extraction is the RMS. This is due to its function to assess the signal strength. RMS can also be identified as the quadratic mean, also representing as square root of the mean square. Nevertheless,

RMS is not an important aspect in fault detection. Listed below is Eq. (5) where formulated equation for RMS is shown.

$$xrms = \sqrt{\frac{\sum_{K=1}^{K}[x(k)]^2}{K}} \tag{5}$$

Crest factor can be deduced by maximum and minimum value from the time domain features. The value of crest factor can be used to determine the change of the impulse vibration signals in the failures of machine. It can be determined from the ratio of the data maximum value to RMS value, as shown in Eq. (6).

$$CF = \frac{x\max}{xrms} \tag{6}$$

The time domain features normally identify the state of the machine during short period of time. However, it can hardly be notified especially on the initial stage of breakdowns. This shows that during diagnosis stage, frequency domain features are required as well. Lei et al. [21] claimed that frequency domain features can be used to identify the signs of machine breakdowns.

Discreet Fourier Transformation (DFT) are used to convert time domain features to frequency domain features. The DFT signals are then determined by utilizing the Fast Fourier Transforms (FFT). Irrelevant frequencies are eliminated using High Frequency Resonance Techniques (HFRT) [19].

2.4 Artificial Neural Network and NARX

Artificial Neural Network (ANN) are one of the various methods of machine learning that can be used in a broad field of cases such as clustering, identification, optimization, pattern categorization, function approximation and prediction [22–24]. ANN is a machine learning algorithm that imitates the neural network of human brain. It has the impactful functionalities that are able to mimics the human brain mechanism: to learn, store and recall information [25].

The system structures are designed based on the biological neural structure in which every unit or network node is a neuron that receives input [26]. Each of the data processing neurons connects with the numerous units by computing the output value using input data. A neural input–output network model is recommended for the applications where the priori knowledge of the structure and nonlinearities is unknown [8, 25]. In prognostic, the key factor of a system to be impactful is the prediction tools especially when the data variable revolves around time series. Many researchers [2, 4, 8, 25] have claimed that NARX neural network have a good reputation in time series predictions, thus it has been selected for this study.

The model deduced predictions of a time series data made up from some parameters that are basing of their performance in the system. It consists sets of variables

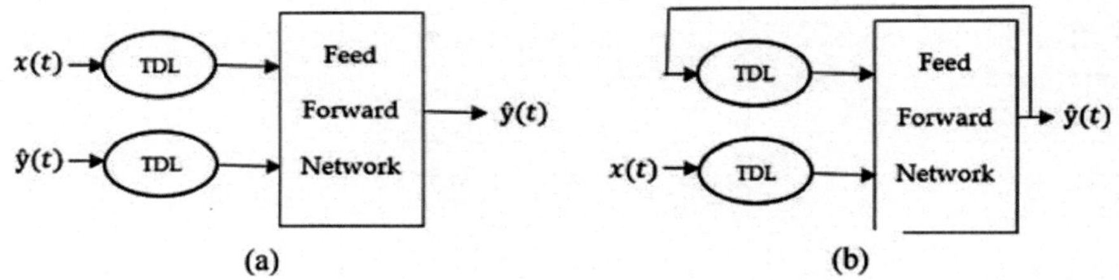

Fig. 2 The structural design of NARX network where **a** is the series–parallel structure and **b** shows the parallel structure

known as weights and biases [8]. Korbicz et al. [27] has presented various architectures of neural networks in their study. NARX network is designed from a Multilayer Perception (MLP) where input state variables are used as a window of past input and the current output are figured out by the output values [4]. The MLP is one of the popular structures constructed in a neural network application [8]. Below is the equation of operation for NARX model:

$$y(t+1) = f\left(y_{t-1}, y_{t-2}, \ldots y_{t-n_y}, u_{t-1}, u_{t-2}, \ldots, u_{t-n_y}\right) \tag{7}$$

In this equation, un and yn is the externally defined training input and target variables that are to be expressed and predicted, while t is the discrete time step [10]. The equation follows the direction of feedforward neural network. The functions are approximated when the previous values of both exogeneous input and output signal regressed together attaining the following y(t) values [4].

NARX neural network model are established with two different architectures: series–parallel architecture and parallel architecture. The network architecture of the network is depicted by Fig. 2 where (a) illustrates the design structure for a series–parallel structure while the parallel structure is shown by (b). This study applies the series–parallel concept structure during training phase as the true past values of the time series are readily accessible. The constructed network that are totally feedforward would allow the application of Multi-Layer Perceptron (MLP) to the model [25].

Equations (8) and (9) portrayed the series–parallel network architecture and parallel network architecture respectively.

$$\hat{y}(t+1) = F\left(\begin{array}{c} y(t), y(t-1), \ldots, y\left(t-n_y\right), x(t+1), \\ x(t), x(t-1), \ldots, x(t-n_x) \end{array}\right) \tag{8}$$

$$\hat{y}(t+1) = F\left(\begin{array}{c} \hat{y}(t), \hat{y}(t-1), \ldots, \hat{y}\left(t-n_y\right), x(t+1), \\ x(t), x(t-1), \ldots, x(t-n_x) \end{array}\right) \tag{9}$$

From the equation, $F(\cdot)$ signifies the mapping functions of the network, $\hat{y}(t-1)$ is NARX output value at time t for the time $t + 1$ at the same time signifies the expected

y values for the time $t + 1$. Apart from that, $\hat{y}(t), \hat{y}(t-1), \ldots, \hat{y}(t-n_y)$ are the previous outputs of the network. For the true previous outputs of the NARX network, it was represented by $y(t), y(t-1), \ldots, y(t-n_y)$ where it can also be identified as the desired output values. The input values of the time series are presented by $x(t), x(t-1), \ldots, x(t-n_y)$ where the n_x is used for input delays and n_y for output delays [25].

2.4.1 Series–Parallel Network Mode

Similarly known as Open Loop mode, where it can only be operated with the actual values of the targeted series to develop the target series regressor. Usually, the network training process between for target elements and main components (input series) uses this mode [4]. The equation for series–parallel network is can be found at Eq. (8).

2.4.2 Parallel Network Mode

Can also be identified as Close Loop mode. This mode uses response from the time series regressor to predict future values. The training of both input and target series must be done first in the series–parallel network mode prior to the task of multi-step time series forecasting that later can be proceeded in parallel mode [4]. The equation for parallel network is can be found at Eq. (9).

2.4.3 Multi-layers Perception

Multi-Layers Perceptron (MLP) is one of the core items that allows approximations of an unknown functions for the duration of training process of forecasting outputs [25]. A basic schematic of MLP are structured with three layers which is the input, hidden and output layers, illustrated by Fig. 3. In a NARX model that are constructed with MLP, the input value elements are used as a window of previous input and output values and calculating the current output. The fundamentals built of an MLP structure are the neurons where it assesses and evaluates all the input state variables [4]. Apart from that, MLP also consists of activation functions and weights [25]. Figure 4 further elaborates the detail structure of a neuron.

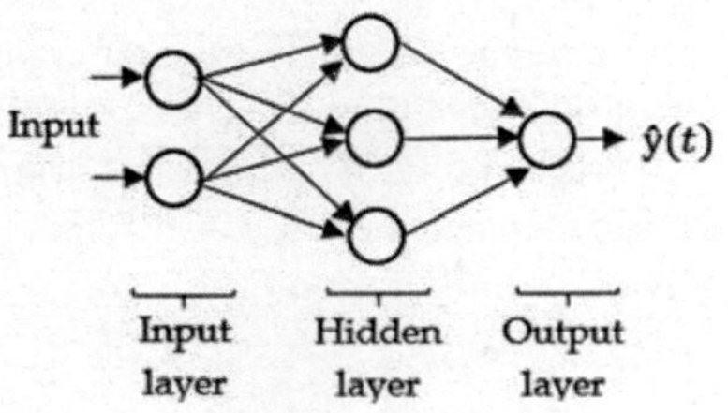

Fig. 3 The architecture of a Multi-Layered Perceptron network where the structure formed with an input, hidden and output layer [27]

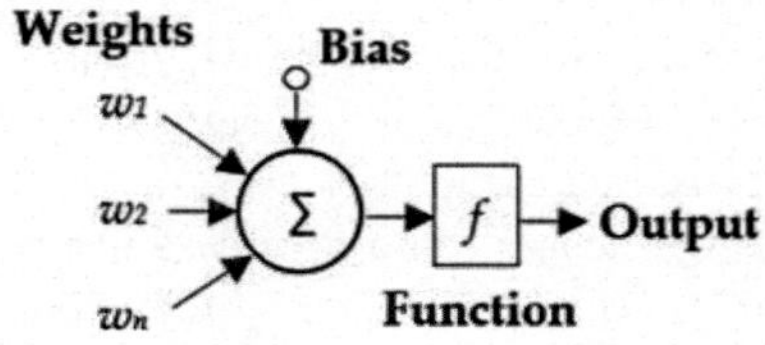

Fig. 4 A detailed structure of a neuron where it consists of weights, bias, a function that correlates with each other to create an output

Table 5 Some of the available examples of activation functions

Function	Definition
Linear	x
Sigmoid (Logsig)	$\frac{1}{1+e^{-1}}$
Hyperbolic tangent (Tansig)	$\frac{e^x-e^{-x}}{e^x+e^{-1}}$

Nodes in hidden layer operates with the function of:

$$y = \sigma\left(\sum_{i=1}^{n} w_i a_i + b\right) \quad (10)$$

The neuron's activations (y) are determined by Eq. (11) where $w_i$is the predetermined real-valued weights, a_i stands for state variables and b is the bias. σ signifies the nonlinear activation functions of a neuron [28, 29]. Table 5 further exhibits the various type of the available activation functions. For the application to this research, y can be identified as the assumed values of the machine performance, while u corresponds to the input dataset that contains several multivariate time series originated from a specified engine type. Apart from that, x can be denoted as the state variables of input values at the nodes processing [4]. However, errors presented in data training and predictions will cause the predicted current value of the output variables to be less accurate using historical data [4].

2.5 *Network Configuration and Training*

The network was trained by using neural net time series with Levenberg Marquadt (LM). MATLAB (Version: 8.0.0.783 (R2012b)) neural network toolbox is used for the training of ANN model. The configuration of ANN model uses tansig (Hyperbolic tangent sigmoid) transfer function in its hidden and output layer. Bayesian regularisation helps to reduce combination of squared errors and weights during network training. This network training style is highly beneficial for the generalization steps especially for difficult and noisy datasets. The table below shows the network configuration (Table 6).

Table 6 Network configuration of the NARX network

Architecture	Details
Training function	Bayesian regularization backpropagation
Input and feedback delays	1:4
Size of hidden layers	10
Data division	Random division of data
Performance function	Mean Square Error (MSE)

The network architecture allows us to decide the number of hidden neurons and number of delays. All the dataset should be segregated into 3 parts which is training, testing and validation sets prior the configuration. With the intention to minimize error numbers concerning the trained outputs and targets, numbers of open-loop networks are constructed in a double loop structure over the increasing number of hidden nodes of an outer loop and an arbitrary weight initialization of an inner loop [30].

Figures 5 and 6 shows the schematic of the open and closed loop mode of the NARX network architecture. Variables for such, the input series x and target values y are introduced to the network as the first stage of training. Both input and target

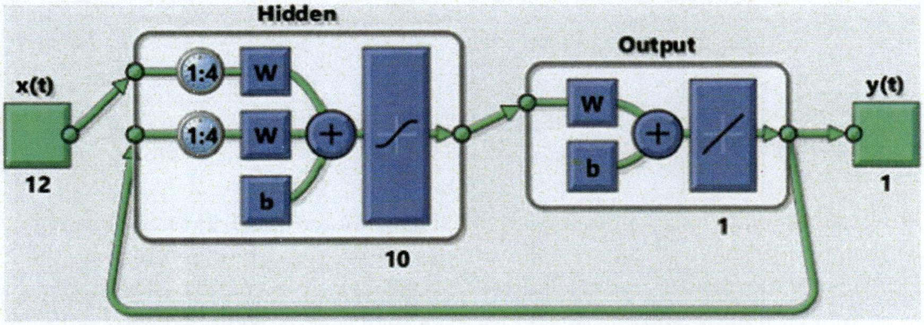

Fig. 5 Design of closed loop NARX for data training

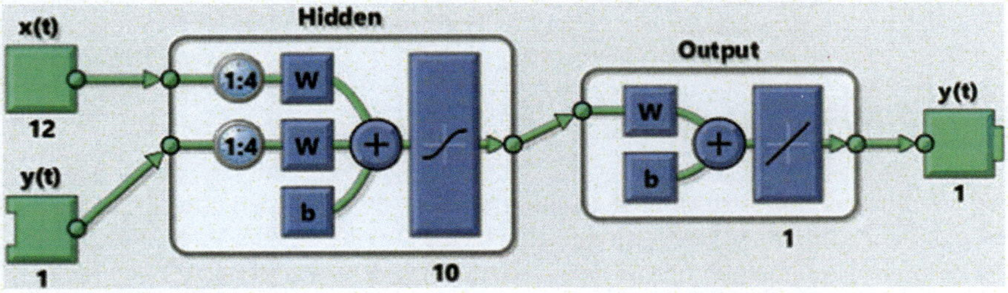

Fig. 6 Design of open loop NARX for data training

start the training with a pre-set value of four delays (1:4) of the trapped delay lines. Due to the difference in initial weight (w) and bias (b) values, variation of solutions is produced in every neural network training [4].

2.6 Mean Square Error (MSE) Validation

MSE is the Mean Square Error and its function is to calculate the performance of the network. The loops will stop after the minimum preferred error has occurred. It is the average squared difference between achieved output values from open loop mode and initial target values.

Stated below is the equation of MSE, where $\left(\breve{y}_{r(i)} - y_{r(i)}\right)$ is called residual, which is the difference in true and estimated output and the goal of MSE loss function is to minimize the residual sum of squares.

$$MSE = \frac{1}{t}\sum_{i=1}^{t}\left(\breve{y}_{r(i)} - y_{r(i)}\right)^2 \quad (11)$$

For validation, a prearranged MSE target is set prior selecting the best open loop design mode [4].

3 Result and Discussion

Various tests were done to choose the database structure and the neural network configuration. Prior to the start of the simulation, datasets are classified into three sections: training, testing and validation sets. During training period, the training set are used in computing the weights and bias of the network and proceeded with the test set, in which used to test the arrangement. After completing with both phases, the validation set are applied to model the network at the same time, evaluating its performance.

3.1 MSE Performance

Table 7 illustrate the performance of the trained NARX network. The MSE for test set is 0.0165 while the MSE for train set is 0.0095.

During learning period, one of the issues that might occurred for ANN network is overfitting where ANN are not able to provide an accurate data prediction with unseen data as compared to the seen data [31]. It can be detected when the expected prediction values deviate drastically when a new unseen data is used. From the result, it shows

Table 7 Score performance of the network training

Item	Performance score (MSE)
Overall (open-loop/one-step ahead)	0.0105
Train set	0.0095
Test set	0.0165
Closed-loop	0.0105
Multistep	0.0084

that the MSE for test set is not far off from the MSE value of the trained set. This condition signifies that there is no overfitting problem occurred. MSE assessment results are shown in Fig. 8. The best performance for MSE is 0.0094376.

3.2 Model Evaluation

The performance of the network is plotted before acquiring an MSE plot of the train and test, presented in Fig. 7. The result performance then undergoes a regression analysis. The equation of the analysis are as follows:

$$\hat{Y} = a + b_1X_1 + b_2X_2 + \dots + b_iX_i \tag{12}$$

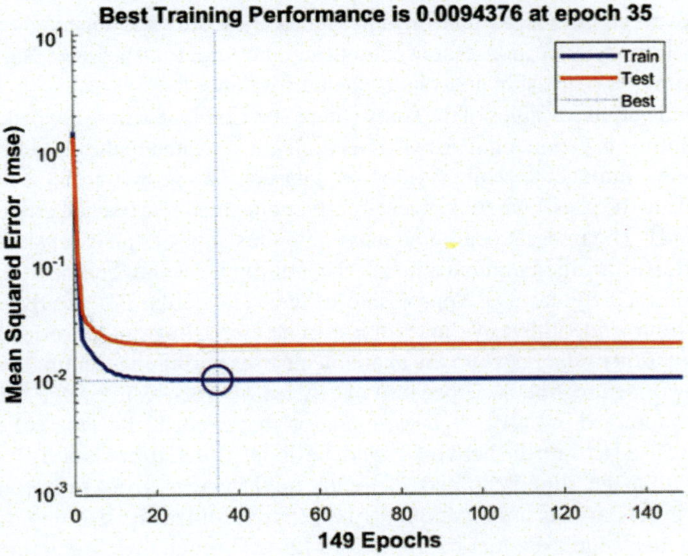

Fig. 7 MSE performance of the trained network

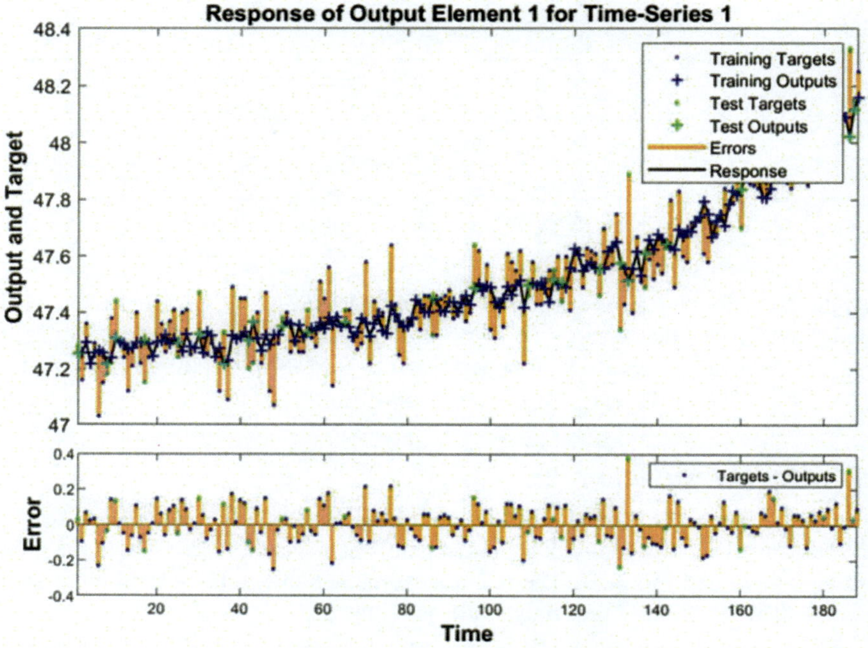

Fig. 8 Time series response of the network performance

where $\widehat{Y}$ denotes the studied score on the dependent variables, a portray the intercept, b_i indicates slope of independent variables and X_i is score on independent variables [32]. The time series response between output and target is shown in Fig. 8 while Fig. 9 exhibit the regression analysis of the performance.

Input and output variables of the study can be seen in Fig. 8 where the pattern of the graph is hinting an exponential growth. Generally, it is related to the conditions of the main engine components of the simulation datasets. The growth trend is somewhat comparable to the characteristics of an aircraft engine modules like the efficiency and flow level [4]. The interval beginning and end points signifies an extensive working period within a specified point at a justified condition level and the final limit, which can be identified by the user appropriate to their standards. For the research, the threshold limit is identified by the resulting value of training output series [4].

Extrapolation of data beyond the available dataset during the training period of a neural network cannot be done because of few factors such as the value for weights that are randomized initially. Any extrapolation that exceeds the range of collected data that are used to train the network cannot be done. [26]. Extrapolation of a datasets that are within the slightly imperfect trends might cause errors to be magnified. However, if the datasets are incomplete, the network utilizes its learning trends and patterns to make and excellent interpolations [26]. Through this, predictions can be made from series of available data by learning how variables affects the machines conditions.

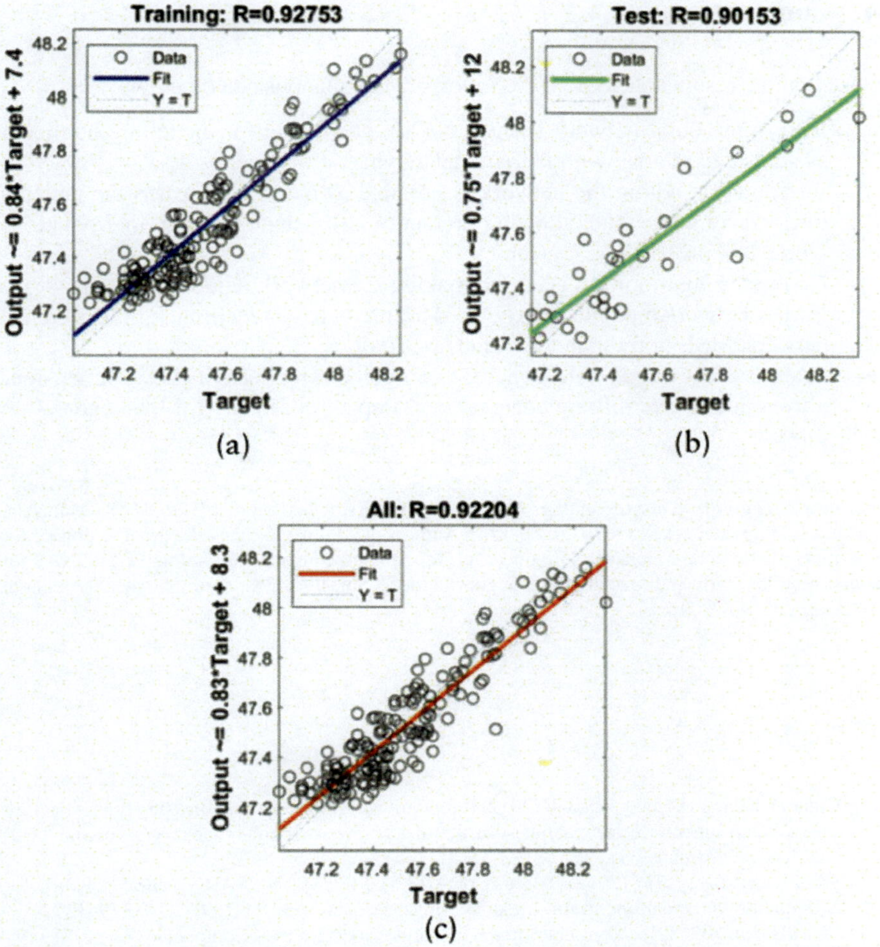

Fig. 9 Regression analysis of the network performance, where **a** is the training set, **b** is the test set and the overall set is shown by (**c**)

Apart from that, overtraining may be another issue faced by neural network. This happens due to the network's capacity that are overly enhanced as either it is too large, or it has undergone a large number of training cycles. The network adapts by allowing outliers to be accommodated causing the simple pattern of understanding prognostic mechanism to be lost [26].

The data scatter of the network regression can be seen in Fig. 9. The graph with blue-coloured fit line (graph (a)) is the training regression data, the green-coloured fit (graph (b)) data represents the test regression data and red-coloured fit line (graph (c)) is the overall regression data analysis.

4 Conclusion

Based on the results discussion, the following conclusions can be drawn:

a. Recurrent neural network such as NARX is suitable for modelling nonlinear systems and relatively good to model a dynamic time series data.
b. Interpolation within the network is possible as the system learns the patterns of the incomplete data, allowing prediction to be made through interpolations with a low probability of error.
c. No overfitting problems occurred making the network reliable.
d. NARX network modelled can be used for predictive modelling as the difference between MSE of train and test data are small.
e. Feature selection has many benefits to a predictive model as it allows dimensionality reduction permitting adequate and impactful sensors for the degradation model.

Acknowledgements This research is funded by Universiti Teknologi PETRONAS, Malaysia, under YUTP grant, cost centre (015LC0-124). Authors would like to thank the grant's sponsor for providing the funding, Mechanical Engineering Department, Universiti Teknologi PETRONAS for supporting this study and special gratitude goes to the technical staff for their guidance throughout the research.

References

1. Galar D, Kumar U, Lee J, Zhao W (2012) Remaining useful life estimation using time trajectory tracking and support vector machines. In: Journal of physics: conference series, vol 364, no 1, p 012063. IOP Publishing
2. Rai A, Upadhyay SH (2017) The use of MD-CUMSUM and NARX neural network for anticipating the remaining useful life of bearings. Measurement 111(397–410):119–127
3. Shin JH, Jun HB (2015) On condition based maintenance policy. J Comput Des Eng 2(2):119–127
4. Bektas O, Jones JA (2016) NARX time series model for remaining useful life estimation of gas turbine engines. In: Proceedings of third european conference of the prognostics and health management society
5. Jardine AK, Lin D, Banjevic D (2006) A review on machinery diagnostics and prognostics implementing condition-based maintenance. Mech Syst Signal Process 20(7):1483–1510
6. Peng Y, Dong M, Zuo MJ (2010) Current status of machine prognostics in condition-based maintenance: a review. Int J Adv Manuf Technol 50(1–4):297–313
7. Heng A, Zhang S, Tan AC, Mathew J (2009) Rotating machinery prognostics: State of the art, challenges and opportunities. Mech Syst Signal Process 23(3):724–739
8. Bednarz J, Barszcz T, Uhl T (2011) NARX model in rotating machinery diagnostics. Mech Control 30:41–52
9. Santoso GB, Prahasto T, Widodo A (2013) Prognosis of bearing damage performance to industrial system using nonlinear autoregressive with exogenous (NARX). In: 2013 International conference on QiR, pp 60–63
10. Siegelmann HT, Horne BG, Giles CL (1997) Computational capabilities of recurrent NARX neural networks. IEEE Trans Syst, Man, Cybern, Part B (Cybern) 27(2):208–215

11. Einsenmann RC (1997) Machinery Malfunction Diagnosis and Correlation. Hewlett Packard Professional Books
12. Kicinski J (2005) Rotordynamics, Wydawnictwo IMP PAN, Gdansk
13. Sarle WS (1994) Neural networks and statistical models
14. Murata N, Yoshizawa S, Amari SI (1994) Network information criterion-determining the number of hidden units for an artificial neural network model. IEEE Trans Neural Netw 5(6):865–872
15. Schwabacher M, Goebel K (2007) A survey of artificial intelligence for prognostics. In: AAAI fall symposium, pp 107–114
16. Saxena A, Goebel K, Simon D, Eklund, N (2008) Damage propagation modelling for aircraft engine run-to-failure simulation. In: 2008 International conference on prognostics and health management, pp 1–9. IEEE
17. Coble J, Hines JW (2009) Identifying optimal prognostic parameters from data: a genetic algorithms approach. In: Annual conference of the prognostics and health management society. san diego, ca, september
18. Guyon I, Elisseeff A (2003) An introduction to variable and feature selection. J Mach Learn Res 3(Mar):1157–1182
19. Mahamad AK, Hiyama T (2008) Development of artificial neural network based fault diagnosis of induction motor dearing. In: Paper presented at the 2008 IEEE 2nd international power and energy conference
20. Mohamad AKB (2010) Diagnosis of rotary machinery using AI
21. Lei Y, He Z, Zi Y (2008) A new approach to intelligent fault diagnosis of rotating machinery. Expert Syst Appl 35(4):1593–1600
22. Araújo RA, Oliveira ALI, Meira S (2017) A morphological neural network for binary classification problems. Eng Appl Artif Intell 65:12–28
23. Gong T, Fan T, Guo J, Cai Z (2017) GPU-based parallel optimization of immune convolutional neural network and embedded system. Eng Appl Artif Intell 62:384–395
24. Sánchez D, Melin P, Castillo O (2017) Optimization of modular granular neural networks using a firefly algorithm for human recognition. Eng Appl Artif Intell 64:172–186
25. Boussaada Z, Curea O, Remaci A, Camblong H, Mrabet Bellaaj N (2018) A nonlinear autoregressive exogenous (NARX) neural network model for the prediction of the daily direct solar radiation. Energies 11(3):620
26. Pleune TT, Chopra OK (2000) Using artificial neural networks to predict the fatigue life of carbon and low-alloy steels. Nucl Eng Des 197(1–2):1–12
27. Korbicz J, Obuchowicz A, Uciński D (1994) Artificial neural network. Akademicka Oficyna Wydawnicza PLJ
28. Barad SG, Ramaiah P, Giridhar R, Krishnaiah G (2012) Neural network approach for a combined performance and mechanical health monitoring of a gas turbine engine. Mech Syst Signal Process 27:729–742
29. Krenker A, Kos A, Bêster J (2011) Introduction to the artificial neural networks. INTECH Open Access Publisher
30. Heath G (2015) Narnet tutorial on multistep ahead predictions. https://uk.mathworks.com/matlabcentral/newsreader/viewthread/338508. (Online forum comment)
31. Durodola JF, Ramachandra S, Gerguri S, Fellows NA (2018) Artificial neural network for random fatigue loading analysis including the effect of mean stress. Int J Fatigue 111:321–332
32. Multiple regression, vol 2019. NCSS Statistical Software. https://ncss-wpengine.netdna-ssl.com/wp-content/themes/ncss/pdf/Procedures/NCSS/Multiple_Regression.pdf

A Case Study to Predict Structural Health of a Gasoline Pipeline Using ANN and GPR Approaches

Nagoor Basha Shaik, Srinivasa Rao Pedapati, Abdul Rahim Othman, and Faizul Azly B. A. Dzubir

Abstract Almost all fluid transportation in oil and gas facilities from one point to another are mainly using simple and complex pipelines system. These pipelines system is vital as it generates significant cash revenue to the facilities. However, the structural health of pipeline deteriorates over time due to various damage mechanism. Any Loss of Primary Containment (LOPC) to the pipelines can lead to heavy business losses. Facility engineers will normally apply the most common approach to mitigate the LOPC by providing schedule maintenance, perform repair activities as well as system replacement. These activities are mostly controlled manually by the engineers. This paper will focus on the prediction of structural health monitoring of gasoline pipeline located in one of the facilities in Malaysia, using historical inspection reports by means Artificial Neural Networks (ANN) and Gaussian Process Regression (GPR) regression techniques. The collected data was analyzed and applied on these two methods to determine the best fitness and performance. It is noticed that both Bayesian-regularization ANN and exponential GPR model demonstrate better performance compared to the others when assessed based on the highest R^2 value. The results were deemed satisfactory as the R^2 value for both methods were close to 1.0. The results also showed that both ANN and GPR models were almost equal in predicting the structural health of a pipeline with an accuracy of 99% and 97.3%, respectively. This work may help in controlling/monitoring the inspection cost and to preplan the maintenance scheduling for gasoline pipeline network in oil and gas industry.

N. B. Shaik (✉) · S. R. Pedapati · A. R. Othman
Mechanical Engineering Department, Universiti Teknologi PETRONAS, 32610 Bandar Seri Iskandar, Perak, Malaysia
e-mail: nagoor_16000473@utp.edu.my

F. A. B. A. Dzubir
Mechanical Department, Group Technical Solutions, Project Delivery and Technology Division, Petroliam Nasional Berhad, 50050 Kuala Lampur, Malaysia

N. B. Shaik
Artificial Intelligence and Machine Learning Research Unit, Faculty of Engineering, Chulalongkorn University, Bangkok 10330, Thailand

F. Ahmad et al. (eds.), *ICPER 2020*, Lecture Notes in Mechanical Engineering,
https://doi.org/10.1007/978-981-19-1939-8_47

Keywords Artificial neural networks · Regression · Prediction · Pipeline · Oil and gas

1 Introduction

Pipelines are the safest mode of transportation for all chemical fluids such as petrol, gasoline and crude oil in oil and gas or petrochemical facilities. Most of the chemical or oil-based products are highly toxic and flammable. These pipelines are susceptible to Loss Of Primary Containment (LOPC) due to various damage mechanism such as corrosion, natural hazards, environmental conditions and biological terms [1]. Hence, majority of oil operators employed sophisticated safety guidelines in handling these types of fluids. Over the years, oil operators will implement various techniques to avoid LOPC such as scheduled inspection, periodic maintenance etc. Oil operators also uses various detection signals using sensors to provide early detection of the pipeline failures. These expensive methods of analysis are currently reliable to be applied to avoid any pipeline leakages that leads to LOPC. Decades ago, Ahamed initiated a deterministic model that can predict the remaining strength of corroded pipeline [2]. After few years, life prediction estimation of a pipeline subjected to corrosion was done using reliability analysis by Castaneda et al. [3]. At a later stage, this type of studies become popular with some researchers have conducted several studies to predict the structural health of the oil pipelines in corrosive environments [4–6]. In 2008, Rabih et al. developed a finite element (FE) model to assess the fatigue life of a submarine gas pipeline [7]. Whilst Velazquez et al. proposed a predictive model for pitting corrosion in oil and gas pipelines [8]. This was followed by Shuxin et al. whereby they developed a mechanical probabilistic model to predict the remaining useful life (RUL) of a pipeline [9]. Cheng et al. developed new method of the residual life prediction based on the corrosion rate of a buried oil pipeline [10]. Subsequently, Senouci et al. developed models to predict the failure of pipelines using artificial neural networks and regression analysis subjected to various conditions such as physical, mechanical and operational factors [11]. Again, the RUL of a pipeline was predicted using ANN by Zahra et al. [12]. Zhou et al. predicted RUL of buried gas pipelines using an iterative method under alternating loads and corrosion with field data [13]. In recent years, few researchers have used ANN to predict the failure and RUL of a pipeline for different conditions such as failure pressure prediction, apparent viscosity prediction, wax deposition prediction and failure sources prediction [14–17]. Nagoor et al. recently introduced ANN approaches for predicting the life condition of oil and gas pipelines [18], including condition evaluation scales [19], and developed an intelligent model that could predict both the life condition and metal loss categorization of a pipeline in one study [20].

However, there is lack of technical paper available which reported the application of ANN or a combination of ANN and GPR methodology for prediction of a specific gasoline pipeline. The main objective of this paper is to predict the structural health of a gasoline pipeline which is located in one of the PETRONAS facility in Malaysia using Artificial Neural Networks (ANN) and Gaussian Process Regression (GPR) techniques. For this purpose, the criterion is given based on E.R.F. value subjected to factors such as length (l), width (w), depth (d), wall thickness (wt) and pressure (p). The representation of a gasoline pipeline with the selected factors for the present work are shown in Fig. 1. The contribution of this research work would be helpful to PETRONAS or to other oil operators worldwide in making prior actions or decisions to reduce the unnecessary periodical inspection and maintenance cost and also indirectly reducing failure rate frequencies for their gasoline pipelines.

In this study, the gasoline pipeline is located at one of the PETRONAS onshore facilities. The inspection reports have been provided for years 2001, 2004, 2010, 2013 and 2015. The historical data and pipeline characteristics of the gas pipeline are specified in Table 1.

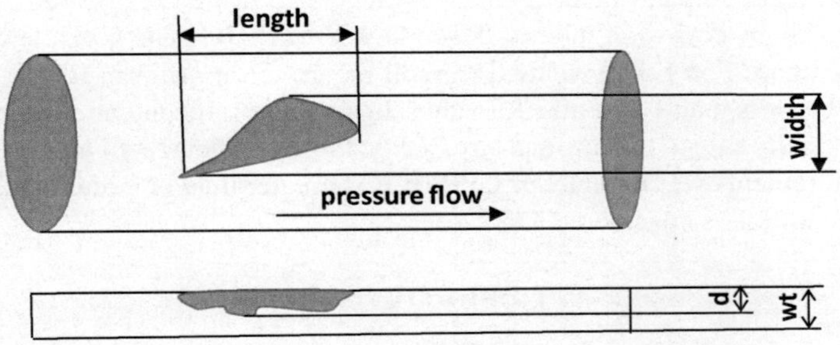

Fig. 1 Pipeline representation with defects

Table 1 Pipeline characteristics

Pipeline length	15 km
Material grade	API 5L X65
MAOP	15.3 Mpa
Product type	Gas
Year of construction	1998
Years of inspection	2001, 2004, 2010, 2013, 2015
Nominal diameter	10 in.
Pipe type	Seamless
Inspection location	Onshore
SMYS (specified minimum yield strength)	448 Mpa
Nominal wall thickness	8.7 mm

2 Methodology

A complete historical inspection data record was obtained from one of the PETRONAS facilities. Each data was carefully selected and multiplied with specified to reflect the present study. Due to limitation of available data, only five factors (length, width, depth, wall thickness and pressure) were considered in this study. All the selected data were then analyzed and normalized in a scale of 0–1.0 and divided into training, testing and validation data sets in ANN model development; whereas the analyzed data is divided into training and validation datasets in GPR model development. The inputs included for ANN and GPR were basically the selected factors, whereas the pipeline condition (which is based on E.R.F) is considered as an output for ANN and responses for GPR. The detailed framework of this methodology is shown in Fig. 2.

The Estimated Repair Factor (ERF) was calculated using Eq. 1 for all metal loss defects caused by corrosion in accordance to ASME B31G. The pipeline condition is evaluated based on E.R.F. value. From all the criterion given in E.R.F, the ERF value shall be less than 1. For all ERF values below than 1, the intended pipeline will be automatically safe to operate and any ERF value more than 1 will not be complied to the requirements sets in ANN or GPR. If it is greater than or equal to 1, then the pipeline condition is considered a risk zone.

$$\mathrm{ERF} = \mathrm{P_m}/\mathrm{P_s} \tag{1}$$

where P_m is maximum operating pressure and Ps is safe operating pressure.

Based on the input data provided, the ANN and GPR models are developed as per Fig. 2. These models represent the condition of a pipeline which is estimated from a scale 0–1. This scale indicates the present structural health of the pipeline. Upon obtaining the results for both models, the results will be compared with the actual pipeline conditions based on PETRONAS Risk Based Inspection (PRBI) data.

2.1 ANN Model Development

Under the ANN model development, there are five parameters have been selected as an inputs for this study i.e. length (mm), width(mm), depth (%), wall thickness (mm) and pressure (bar). From these inputs, a series of hidden and output determinations of algorithm will be applied to generate E.R.F. (pipeline condition) which is the output of ANN model as shown in Fig. 3.

In this study, a built in algorithm package in MATLAB software has been used. Three infamous algorithms have been selected during the training process of ANN model development in order to obtain the best coefficient of determination (R^2) value and mean square error (MSE) as per Eqs. 2 and 3. The three algorithms are Levenberg–Marquardt (LM), Baysian Regularization (BR) and Sacled conjugate

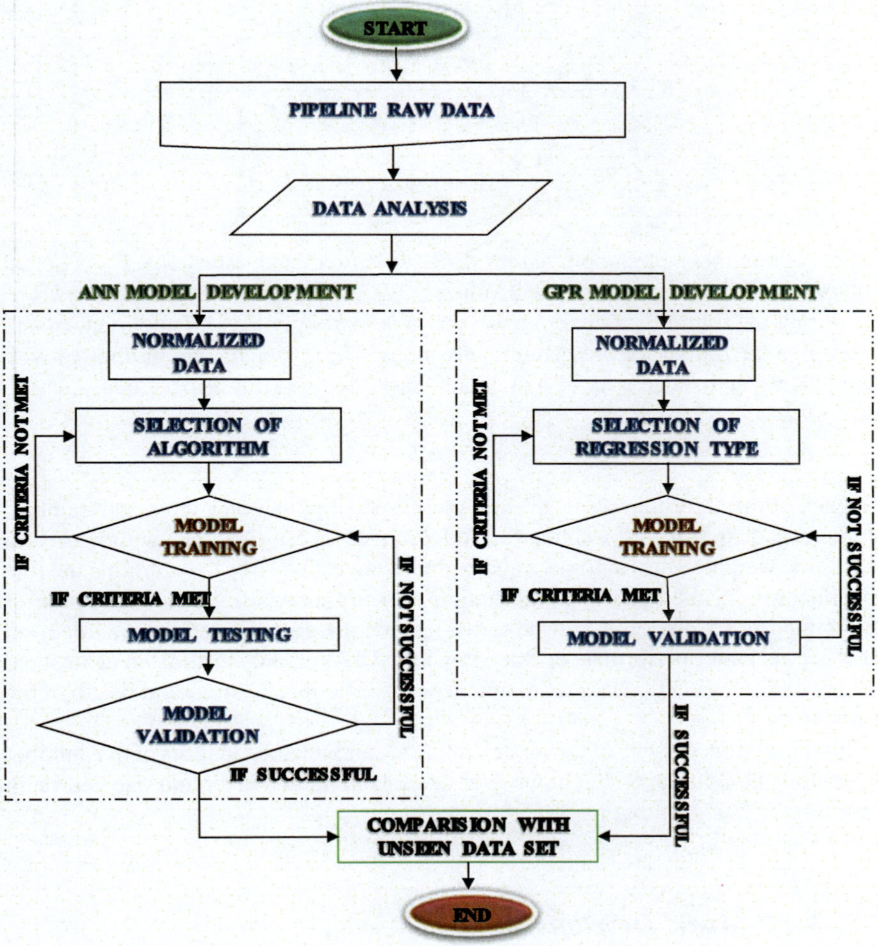

Fig. 2 Framework workflow

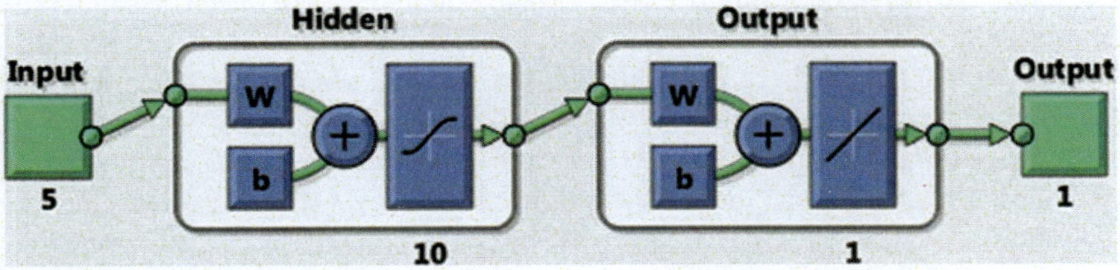

Fig. 3 Artificial neural network architecture

(SC).

$$R^2 = 1 - \sum_{i=1}^{n}\left[\frac{\left(E.R.F._p^i - E.R.F._a^i\right)^2}{\left(E.R.F._{avg,a} - E.R.F._a^i\right)^2}\right] \tag{2}$$

$$MSE = \frac{1}{n}\sum\left(E.R.F._p^i - E.R.F._a^i\right)^2 \tag{3}$$

where n is number of sample values, $E.R.F._p$ is predicted value, $E.R.F._a$ is actual value, $E.R.F._{avg}$ is average of actual values.

During the training process, the weight values strongly play an important role to determine the output for the given input values. These weights are multiplied with input values and added with bias to obtain least possible error as per Eq. 4.

$$x_{new} = wx + b \tag{4}$$

Each of the algorithm package comes with specific function address as 'trainlm', 'trainbr' and 'trainsc' for training the ANN model. These functions mainly used to adjust the weight values automatically and minimizes the error to determine the true combination for the better generaliazation of neural network. Upon completion of the trained ANN model, it will be tested against the given divided testing data set to check its better performance. Once it is tested, the model is validated against the given divided validation data set to obtain new predictions of structural health of the pipeline.

For the complete cycle of the model, the ANN architecture consists of five number of neurons in input layer, ten number of neurons in hidden layer and one neuron in output layer as shown in Fig. 3.

2.2 GPR Model Development

GPR model uses five input parameters similarly like ANN to generate E.R.F. (pipeline condition). However, GPR produces Root Mean Square Error (RMSE), Mean Absolute Error (MAE), Mean Square Error (MSE) and R^2 values instead of only R^2 value compared to ANN model. These additional RMSE, MAE, MSE values are obtained from Eqs. 3, 5 and 6.

Using similar package in MATLAB software, it has four built in different statistical techniques where each technique will provide RMSE, MAE, MSE and R^2 values accordingly. The four built in techniques are Exponential GPR, Squared Exponential GPR, Matern GPR and Rational Quadratic GPR. Each of the technique has been trained using divided training dataset. From the results obtained, the finest technique is selected in accordance with the largest R^2 value.

$$RMSE = \sqrt{\sum_{i=1}^{n} \frac{\left(E.R.F._p^i - E.R.F._a^i\right)^2}{n}} \tag{5}$$

$$MAE = \frac{1}{n}\sum_{i=1}^{n}\left|\left(E.R.F._p^i - E.R.F._a^i\right)\right| \tag{6}$$

where n is number of sample values, $E.R.F._p$ is predicted value, $E.R.F._a$ is actual value.

Each of the GPR technique also has its own built in training process that is similar to ANN but using kernel functions instead of function (trainbr, trainlm, trainsc) that was used in ANN. This additional 'kernel' functions in all GPR techniques are used to minimise the error to define the covariance of a prior distribution over the output functions. The MSE, RMSE, R^2 and MAE values helps to recognize the best fit GPR model. Upon completion of the selection of trained GPR model, it will be validated against the given divided validation data set for further predictions.

3 Results and Discussion

3.1 ANN Model Training

From the three algorithms in MATLAB software ANN tool package i.e. 'trainlm', 'trainbr' and 'trainsc', the results of R^2 values obtained from Eq. 2 are illustrated in Table 1. From the results, it shows that R^2 value from 'trainbr' in BR approach provides the most accurate results close to 1.0 with the value of 0.99596 compared to the other two approaches i.e. LM and SC with values of 0.98386 and 0.93959 respectively. All these R^2 value results in training stages are graphically represented in Fig. 4a, d, and g.

Based on the R^2 values in the training stages for all algorithms, the tested results show that BR approach provides the most accurate results with R^2 value of 0.99018 while the other two approaches i.e. LM and SC with values of 0.98213 and 0.96795 respectively. All these R^2 value results in testing stages are graphically represented in Fig. 4b, e, and h.

From both training and testing R^2 results, the overall R^2 value for BR approach is found to be the most value that is close to 1.0 with the value of 0.99442; while the other two LM and SC approaches provide overall R^2 values of 0.98214 and 0.94601 respectively. All these overall R^2 value results are graphically represented in Fig. 4c, f, and i.

The variations in MSE and R^2 values in training and testing stages are observed in all the algorithms. It is noted that BR based training process is found to be more better as the R^2 value is recorded highest and MSE value recorded lowest among the

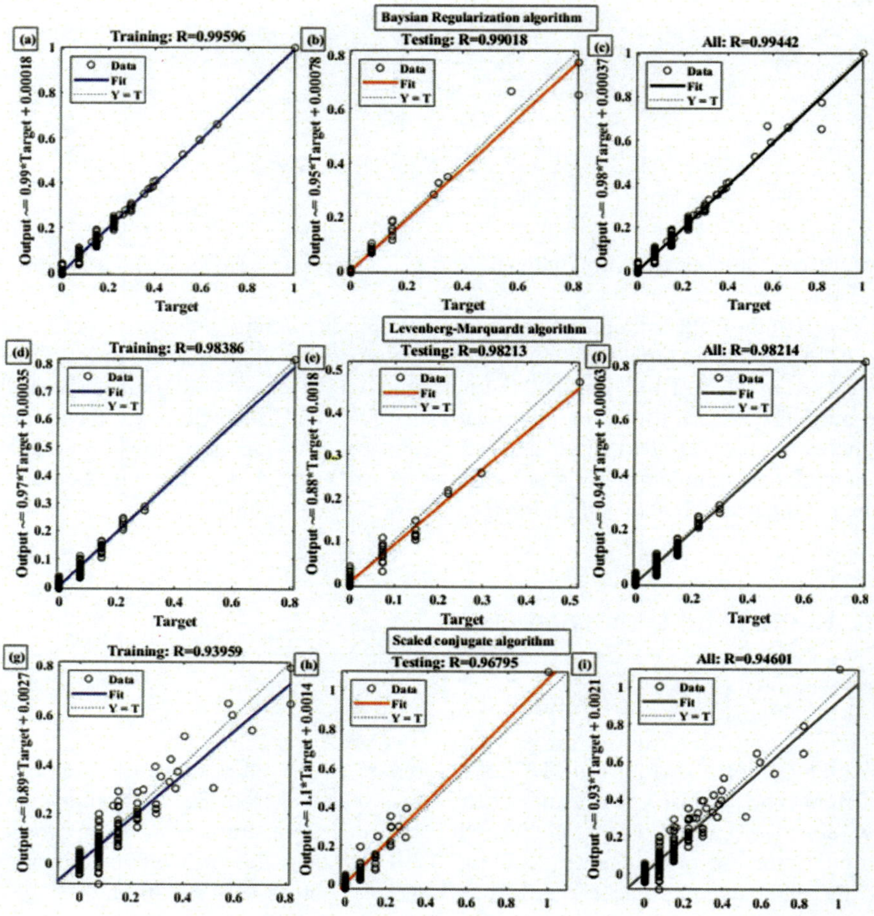

Fig. 4 R^2 values for three algorithms

other LM and SC algorithm based training processes. The summary of all R^2 and MSE values are illustrated in Table 1.

The plot between MSE and number of epochs to represent the best training performance is obtained at epoch 702 reaching its minimum MSE value of $4.699e^{-5}$ as shown in Fig. 5.

3.1.1 ANN Model Validation

Upon the completion of model training and testing, the model is then validated using validation dataset for new predictions. During the ANN validation, the model is given only input parameter values and allowed to predict new E.R.F. values. The results

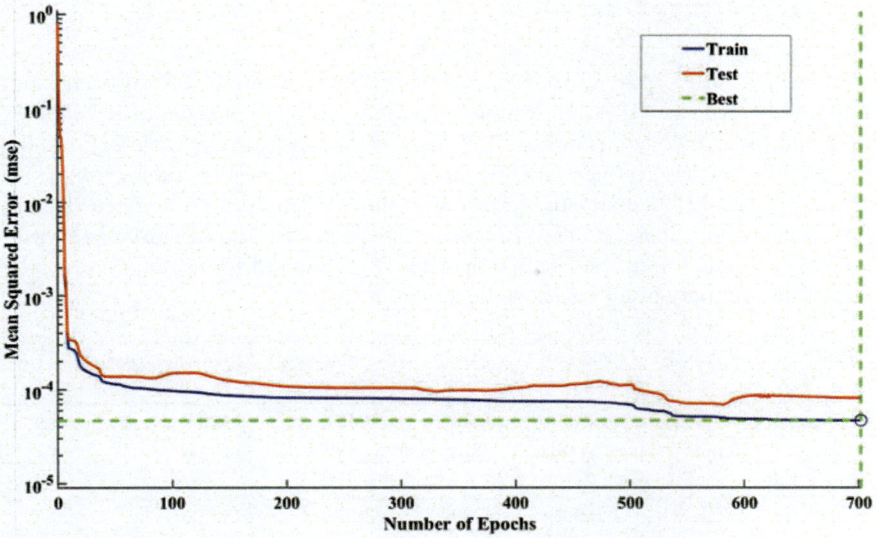

Fig. 5 Training performance plot

found that the BR based ANN model has been predicted accurately with a prediction accuracy of 98.6%. The plot of actual vs predicted values against validation data set is shown in Fig. 6.

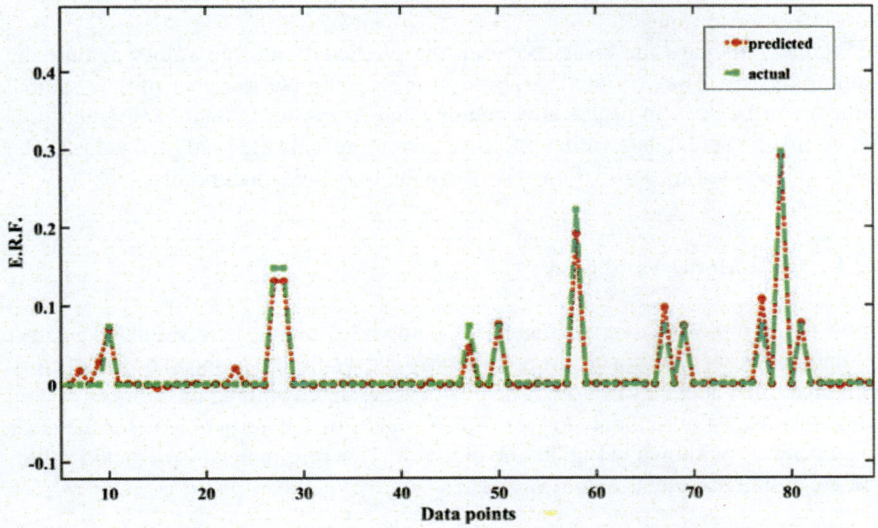

Fig. 6 ANN predicted versus actual values

3.2 GPR Model Training

From the four GPR types in MATLAB software i.e. exponential GPR, squared exponential GPR, Matern GPR and Rational quadratic GPR; the results of R^2, RMSE, MAE and MSE obtained from Eqs. 2, 3, 5 and 6 are illustrated in Table 2. From the results, it shows that exponential GPR model provides highest R^2 value close to 1.0 with the value of 0.97 compared to the other three GPR types i.e. squared, matern and rational with values of 0.58, 0.72 and 0.93 respectively. The variations in RMSE, R^2, MAE, training time and prediction speed are observed for all the GPR model types. The summary of all values is illustrated in Table 2.

GPR type	RMSE	MSE	MAE	Training time (sec)	Prediction speed (obs./sec)	R^2
Exponential	**0.00893**	**7.9901e^{-5}**	**0.0025**	**63.11**	**19,000**	**0.97**
Squared exponential	0.03506	0.03506	0.0083	84.17	18,000	0.58
Matern	0.0284	0.0284	0.0049	62.64	15,000	0.72
Rational quadratic	0.0138	0.0138	0.0029	131.1	9700	0.93

The performance of exponential GPR can be checked using actual vs predicted response plot. Figure 7 shows the scattered plot of predicted response against true response generated by exponential GPR model. The scattered data points that are near to the diagonal line reflects that the model is good for further predictions. It can be seen from the Fig. 7 that the proposed model is able to make predictions well for various response values.

Figure 8 displays the errors between the predicted and true values against the training dataset. The predicted response vs record number (number of data sample values) can be seen in a plot after training the regression model. The green dots represents the actual data values and blue dots represents the predicted values. The errors are denoted by vertical lines between the true and actual points.

3.2.1 GPR Model Validation

Upon the completion of exponential GPR model training, it is validated against validation dataset to perform new predictions. The results are found to be satisfactory with prediction accuracy of 96%. The plot is generated by taking the number of data sample values on horizontal axis and E.R.F. values on the vertical axis for the clear understanding as shown in Fig. 9. The blue color line represents the predicted values and green line represents actual values.

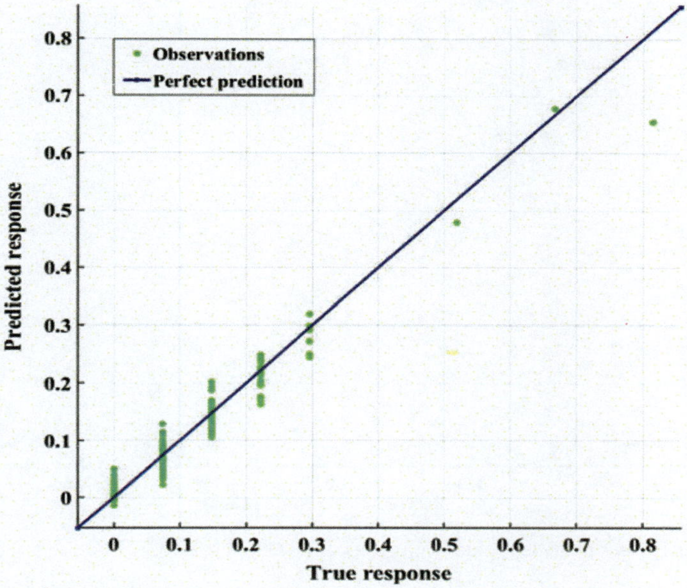

Fig. 7 Exponential GPR predicted against actual response plot

Table 2 Variations of MSE and R^2 values in training and testing stages

ANN algorithm	Stage	MSE	R^2
Levenberg–marquardt	Training	$1.23058e^{-4}$	$9.8386e^{-1}$
	Testing	$3.83056e^{-4}$	$9.8213e^{-1}$
Baysian Regularization	**Training**	$\mathbf{4.96136e^{-5}}$	$\mathbf{9.9596e^{-1}}$
	Testing	$\mathbf{1.163282e^{-4}}$	$\mathbf{9.9018e^{-1}}$
Scaled conjucate	Training	$2.71367e^{-4}$	$9.3959e^{-1}$
	Testing	$1.00258e^{-3}$	$9.6795e^{-1}$

3.3 Comparison of ANN with Regression

It is important to compare the both approaches to consider the best prediction model for unknown datasets in future. Hence, the BR based ANN and exponential GPR models are given some random set of values for the new predictions of pipeline condition. It is found that BR based ANN model almost predicted the pipeline condition very accurately with 99 percent accuracy and exponential GPR model predicted with a prediction accuracy of 96.7%. It is observed that BR based ANN model performed slightly better than exponential GPR model. The comparison plot of ANN vs Regression vs Actual data is generated graphically for the better understanding as shown in Fig. 10. The green colour scattered dotted line with square shaped markers represents

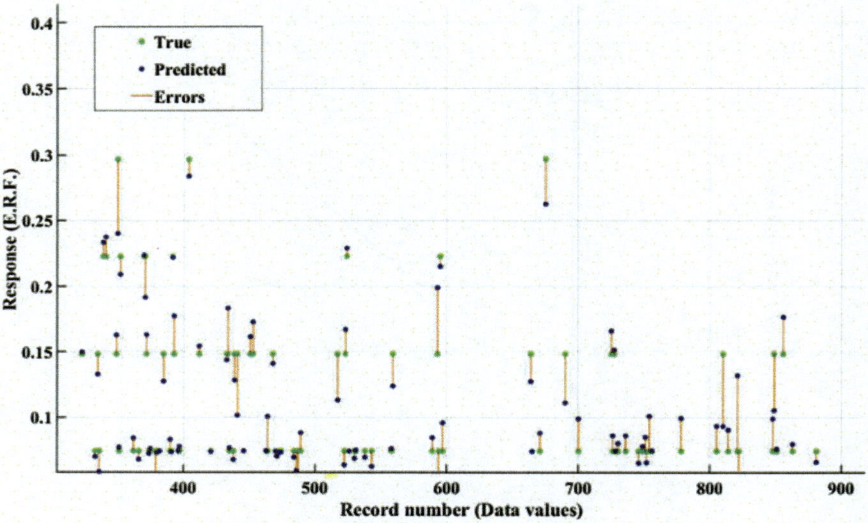

Fig. 8 Actual versus predicted values with errors using GPR

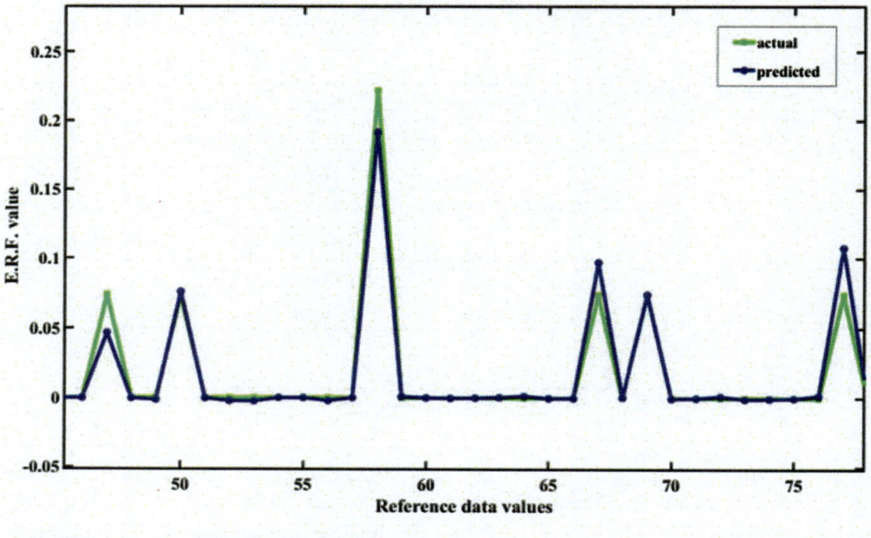

Fig. 9 Regression predicted values versus actual values

the ANN predicted line; The blue colour scattered dotted line with diamond markers represents the GPR predicted line; whereas the red colour scattered dotted line with circle markers represents the actual data line as represented in Fig. 10.

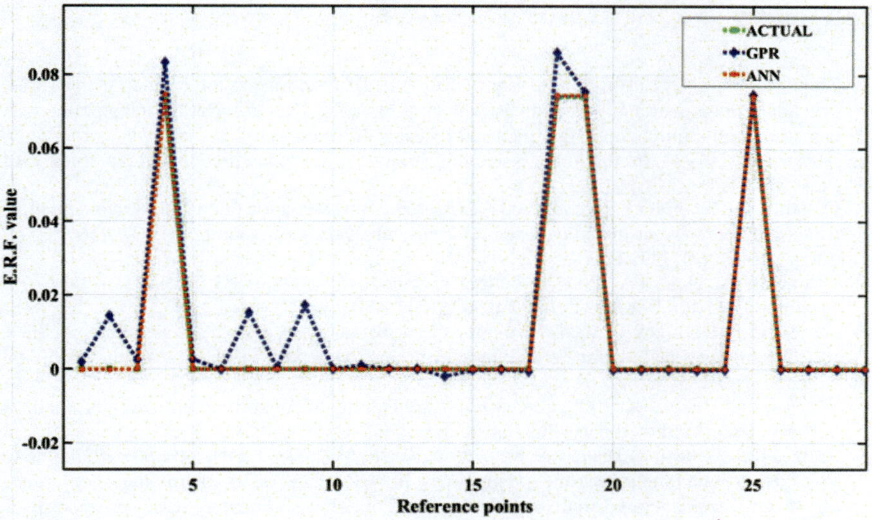

Fig. 10 ANN versus GPR versus actual values comparison plot

4 Conclusion

This study presents the development of ANN and GPR models to predict the structural health condition of a gasoline pipeline subjected to the factors such as length, width, depth, wall thickness and pressure. The ANN and GPR models are developed by considering the selected factors as inputs and E.R.F (pipe condition) as output based on historical inspection reports. The data is divided subsequently to train and validate the models. Each of the technique has been trained in both ANN and GPR approaches using divided training dataset; and the finest technique is selected in accordance with the largest R^2 value. The Bayesian regularization ANN approach and exponential GPR approach are found to be more suitable in terms of structural health prediction of a gasoline pipeline with an accuracy of 99% and 96% respectively. This study helps in estimating the life of a gasoline pipeline and to helps in controlling the inspection budget and to plan the maintenance schedule accordingly.

Acknowledgements The authors are grateful to Universiti Teknologi PETRONAS, Malaysia for the facilities provided to conduct this research.

Conflicts of Interest The authors declare no conflict of interest.

References

1. Zhou M, Zhang Q, Liu Y, Sun X, Cai Y, Pan H (2019) An integration method using kernel principal component analysis and cascade support vector data description for pipeline leak detection with multiple operating modes. Processes 7(10):648
2. Ahammed M (1997) Prediction of remaining strength of corroded pressurised pipelines. Int J Press Vessels Pip 71(3):213–217
3. Castaneda H, Alamilla J, Perez R (2004) Life prediction estimation of an underground pipeline using alternate current impedance and reliability analysis. Corrosion 60(5):429–436. https://doi.org/10.5006/1.3299238
4. Priambudi A, Nafis PA (2004) Lifetime prediction of pressurized pipelines in corrosive environments Alfin Priambudi. In: CORROSION 2004: NACE international
5. Fenyvesi L, Lu H, Jack T (2004) Prediction of corrosion defect growth on operating pipelines. In: 2004 International pipeline conference: American society of mechanical engineers digital collection, pp 225–230
6. Wu M-l, Hao D, Liu J-K (2008) Forecast of the remaining life of corrosive oil and gas pipelines based on Grey-Markov theory. Pipeline Tech Equip 5
7. Kansao R, Casanova E, Blanco A, Kenyery F, Rivero M (2008) Fatigue life prediction due to slug flow in extra long submarine gas pipelines. In: ASME 2008 27th international conference on offshore mechanics and arctic engineering, vol 3: pipeline and riser technology; ocean space utilization, pp 685–692. https://doi.org/10.1115/OMAE2008-58005
8. Velázquez JC, Caleyo F, Valor A, Hallen JM (2009) Predictive model for pitting corrosion in buried oil and gas pipelines. Corrosion 65(5):332–342. https://doi.org/10.5006/1.3319138
9. Li S-X, Yu S-R, Zeng H-L, Li J-H, Liang R (2009) Predicting corrosion remaining life of underground pipelines with a mechanically-based probabilistic model. J Petrol Sci Eng 65(3–4):162–166
10. Cheng Y-P, Bai Y, Li X-Y, Fan H (2009) Residual life prediction method of corroded buried oil pipeline. Press Vessel Technol 2
11. Senouci A, Elabbasy M, Elwakil E, Abdrabou B, Zayed T (2014) A model for predicting failure of oil pipelines. Struct Infrastruct Eng 10(3):375–387
12. Zangenehmadar Z, Moselhi O (2016) Assessment of remaining useful life of pipelines using different artificial neural networks models. J Perform Constr Facil 30(5):04016032
13. Weiguo Z, Dongjing L, Hai W, Xinxin P (2015) Remaining-life prediction and reliability assessment of buried gas pipelines under corrosion and alternating loads. J Pipeline Syst Eng Pract 6(1):05014002
14. Liu X, Xia M, Bolati D, Liu J, Zheng Q, Zhang H (2020) An ANN-based failure pressure prediction method for buried high-strength pipes with stray current corrosion defect. Energy Sci & Eng 8(1):248–259
15. Zhang F, Mukhtar YMF, Liu B, Li J (2019) Application of ANN to predict the apparent viscosity of waxy crude oil. Fuel 254:115669
16. Alnaimat F, Ziauddin M (2019) Wax deposition and prediction in petroleum pipelines. J Pet Sci Eng 106385
17. Zakikhani K, Zayed T, Abdrabou B, Senouci A (2020) Modeling failure of oil pipelines. J Perform Constr Facil 34(1):04019088
18. Shaik NB, Pedapati SR, Taqvi SAA, Othman AR, Dzubir FAA (2020) A feed-forward back propagation neural network approach to predict the life condition of crude oil pipeline. Processes 8(6):661
19. Shaik NB, Pedapati SR, Dzubir FAB (2021) Remaining useful life prediction of a piping system using artificial neural networks: a case study. Ain Shams Eng J
20. Shaik NB, Pedapati SR, Othman AR, Bingi K, Dzubir FAA (2021) An intelligent model to predict the life condition of crude oil pipelines using artificial neural networks. Neural Comput Appl 1–22

Characteristics of Dynamics Electromagnetic Force Acting on a Moving Metal Rod at Very Close Distance

S. A. A. Ghani, M. B. Baharom, and M. H. H. Hassan

Abstract At very close distance, most of the theoretical formula used to compute electromagnetic forces are not valid because the results approach infinity. Past researches have shown that under static condition, measured and computed electromagnetic forces versus position at a very close distance is almost linear. Here, we studied the characteristics of dynamic electromagnetic forces at very close distances of less 0.6 mm. Experiments were conducted to measure accelerations of a metal rod under attraction forces by an electromagnet, and the forces were computed from the accelerations. It was found that the applied dynamics forces were constant for each experiment, indicating that the findings differed from the results of the static condition.

1 Introduction

In recent years, the demand for electronic products has increased worldwide. This demand requires the accuracy in measuring properties of electromagnets to improve the design and function of the devices. One of the important measures that can be improved is to precisely predict how electromagnetic forces behave at close distances.

The experiments of attraction forces between permanent magnets of various shapes in close distance under static conditions have been conducted by Vokoun [1] and [2], and the authors showed that the attraction force could be calculated analytically. Beleggia [3] presented an analytical analysis between the permanent magnet and soft magnetic plate. Research from previous studies showed that there was a drastic increment of magnetic forces of a permanent magnet at a very close distance [1]. Previous studies by Beleggia [4] also presented a new method to compute forces of magneto-statics pairs of uniformly magnetized shapes using theoretical and numerical methods. Based on the literature in [1], it can be observed that the measured static electromagnetic force versus distance at less than 2.0 mm is almost linear and the behaviour was verified based on the experimental results.

S. A. A. Ghani (✉) · M. B. Baharom · M. H. H. Hassan
Mechanical Department, Universiti Teknologi PETRONAS, 32610 Seri Iskandar, Perak, Malaysia
e-mail: syifatul_17006033@utp.edu.my

F. Ahmad et al. (eds.), *ICPER 2020*, Lecture Notes in Mechanical Engineering,
https://doi.org/10.1007/978-981-19-1939-8_48

Based on past literatures, it can be concluded that most of the work done on the prediction of electromagnetic forces at close distance have been conducted under static conditions only. The behaviours of electromagnetic forces when acting on a moving object having magnetic materials are unknown and need to be investigated for verification. Hence, it is questionable whether the behaviours of electromagnetic forces under static and dynamics forces are the same.

The purpose of this study is to predict the characteristics of electromagnet forces acting on a moving metal rod under dynamic force at very close distances. We wanted to find out whether the dynamic electromagnetic force was linear as shown in the static force.

Experiments have been conducted using electromagnets and a ferrous metal rod that have the same diameter and thickness as the electromagnets. The distance between electromagnets and steel rods are varied between 0.1 to 0.6 mm. The drag force acting on the surface of the ferrous metal and friction are assumed to be negligible. The force characteristics found in this research is required for the modelling of our new invented motor called 'Pulsating Electromagnetic Motor' which works based on the principle of face-to-face contacts between electromagnets and steel rods.

2 Verification of Results from Past Researches

This section discusses on an example of a common model called 'Gilbert Model' used to compute electromagnetic forces, and verification of a past research on the attraction forces between permanent magnets at close distances.

2.1 Gilbert Model

The force between well-defined poles of electromagnets can be found using 'Gilbert Model' represented in following equation:

$$F = \frac{\mu q_A q_B}{4\pi r^2} \tag{1}$$

The Gilbert formula for electromagnetic forces were found to be invalid for close-distance cases because the force approaches infinity as the distance approaches zero [5]. It is very common for most equations used to compute electromagnetic forces to include the term $\frac{1}{r^2}$. Hence, when r is less than 1, the term approaches to infinity as r approches to zero or at a very close distance. However, this is not the case for real life application because the maximum attraction forces between electromagnets and magnetic materials are finite and can be experimentally measured. Vokoun [1] conducted experiments at static conditions to compute the attraction forces between

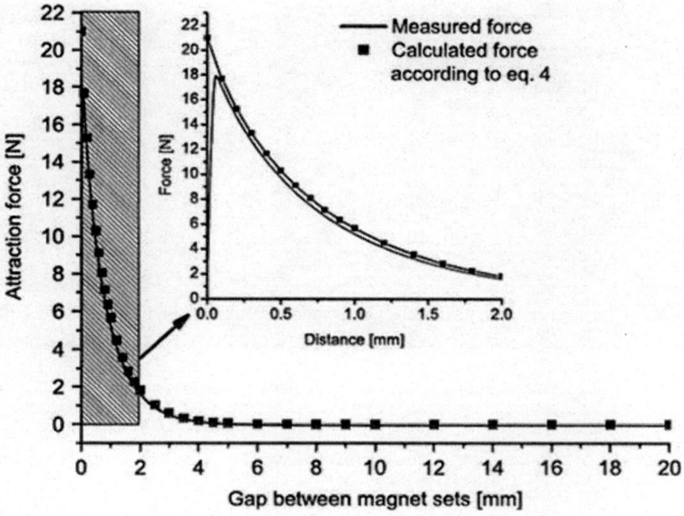

Fig. 1 Experimental and calculation data for attraction force [1]

two permanent magnets of different shapes and proposed a computational formula to calculate the forces at close distances. The computational and experimental results are presented in Fig. 1. Based on their results, it can be observed that the magnetic force at a distance from 0 to 0.5 mm is almost linear, depending on expected accuracy.

In this research, our main intention is to determine the characteristics of electromagnetic forces when acting on a moving magnetic material at very close distance. However, prior to performing the dynamics experiments, we wanted to clarify the reported previous results because there were not many literatures available in this field. A simple experiment has been conducted to determine the actual value of electromagnetic force between two identical electromagnets at very close distance. The experimental apparatus used in this experiment is shown in Fig. 2.

The right electromagnet and the load cell are both fixed to the base. The left electromagnet is moveable and it is connected to a slider. As the right electromagnet attracts the left electromagnet, the slider attached to the left electromagnet also moves to the right and it presses against the load cell. The magnitude of the electromagnetic forces was measured using the calibrated load cell, where the force acting on the load cell connected to the left electromagnet is proportional to the voltage reading.

It can be observed from Fig. 2 that the electromagnets are aligned together and in face to face contact. One of the electromagnets is adjusted at a certain distance. In this experiment, the distance is varied from 1.0 mm to the direct contact position. The attraction forces cause the electromagnets to be close to each other. Both electromagnets have a diameter of 20.5 mm and length of 25.0 mm, and a measured magnetic field at the center of $B = 31.1$ mT. The experimental results for the experiment under static force condition is shown in Fig. 3.

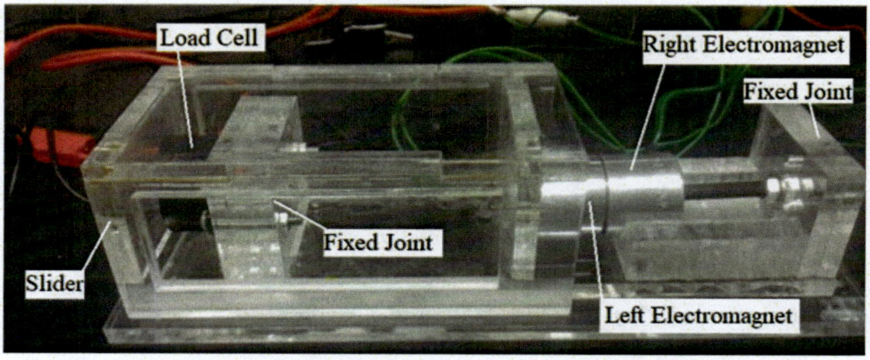

Fig. 2 Experiment of electromagnetic force between electromagnets

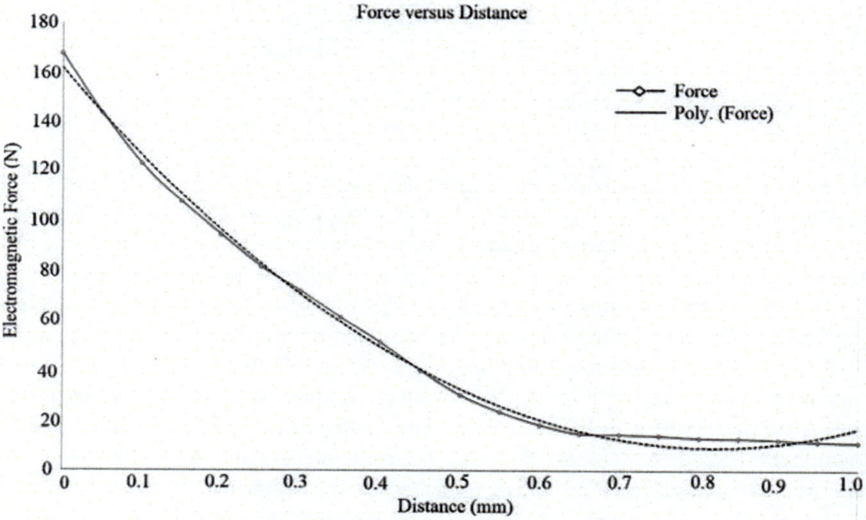

Fig. 3 Attraction force versus position of electromagnet

It can be observed from Fig. 3 that the characteristics of electromagnetic force versus position at a close distance is similar to the previous results presented in past literature [2]. The force versus position is almost linear when the distance is less than 0.5 mm. The maximum recorded force is 167.87 N at face to face contact, while based on manufacturer's data, the maximum force is around 140 N. The results can be considered acceptable owing to the errors in measurements and sensitivity of measuring apparatus. The force is computed using the following equation:

$$F = ma$$

where the acceleration of a constant mass is directly proportional to the force.

3 Methodology

This sections illustrates the experimental setups and methodology for measuring the dynamics force characteristics.

3.1 Experimental Works

The experimental devices used for the experiments is shown in Fig. 4. Two electromagnets having a magnetic field of B = 31.1 mT, similar specifications as in statics experiment, are installed on the left and right side of the device. In between the two magnets is a cylindrical and moveable ferrous metal having magnetic properties of $200N \cdot A^{-2}$. The applied current for each electromagnet is 0.3 A while the voltage is 12 V. Each electromagnet takes turn to be energized at a switching frequency of 2 Hz, controlled by a MOSFET driver. The mass of the ferrous metal rod measured is 0.47 g. The close-up photo of the electromagnets and the ferrous metal is shown in Fig. 5. Figure 6 shows the experimental setup that consists of a power supply, frequency generator and high-speed camera.

The alternate attraction force from one of the electromagnets tends to move the ferrous metal rod back-and-forth within the electromagnets. The images of the moving metal rod were captured using a high-speed camera (Phantom Miro M310) with the setting of 30,000 frames per second. The images were then processed to

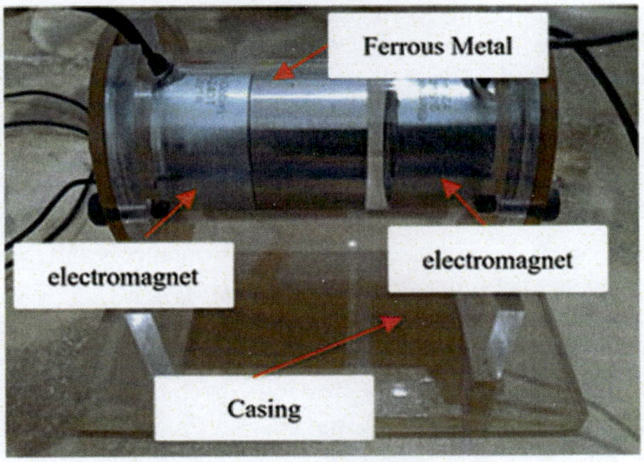

Fig. 4 Experimental device

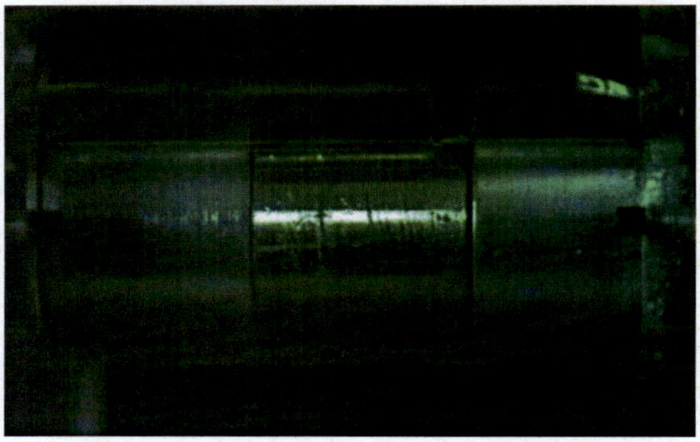

Fig. 5 Close-up image

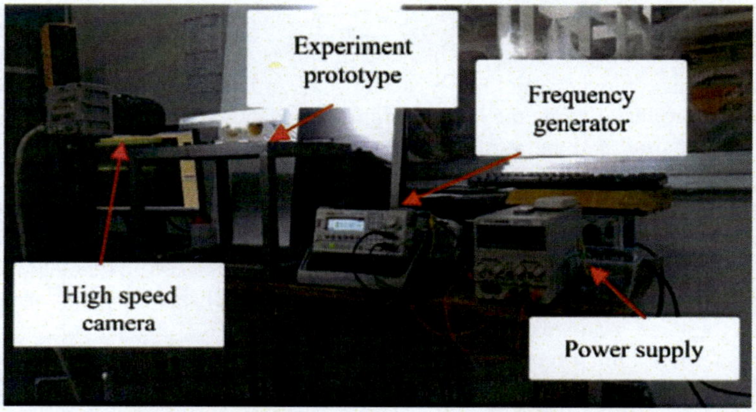

Fig. 6 Experimental setup

obtain the positions of the metal rod versus time. The results were only analyzed when the ferrous metal rod travelled from right to left. The purpose of the right electromagnet is to relocate the metal into its original position.

The maximum gap designed between electromagnet and iron metal is 5 mm but for this experiment, we only conducted experiments at very close distances between 0.1 to 0.6 mm. Six samples of repetitive experimental testing with the incremental value of 0.1 mm for each experiment have been performed. For analysis purposes, the images of the distance between electromagnets and ferrous metal have been rescaled to the actual dimension as in the experiment.

The data for the plots of positions versus time have been obtained from the experiments. We tested a few curved fitting functions to the data and determined the best function based on the regression correlation square values. We found that polynomials of degree two have the best fits compared to all other functions. Based on the selected curve fitting function, the velocities and accelerations of the moving metal rod are computed for further analysis.

4 Results and Discussions

The plots for the positions verses time for cases of 0.1–0.6 mm have been obtained and shown in Fig. 7.

It can be observed from Fig. 7 that polynomials of degree 2 fit the data very well with the squares of regression coefficients, r^2 having the values close to 1. This implies that the function is able to predict the behaviour of the data very well.

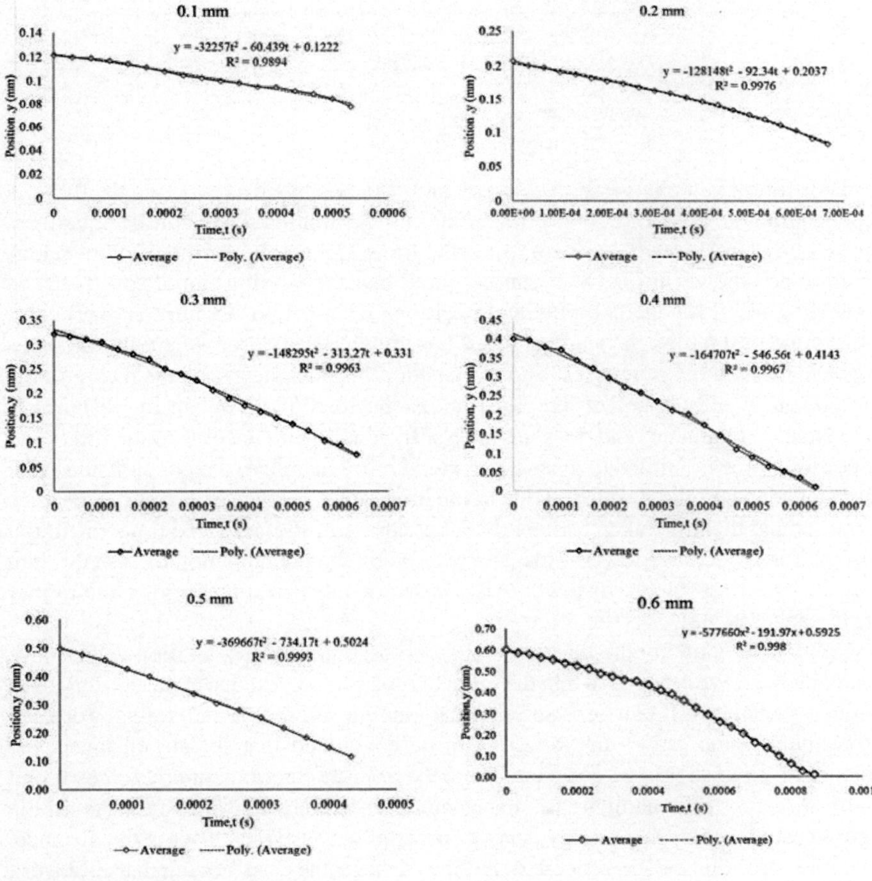

Fig. 7 Graph of distance versus time

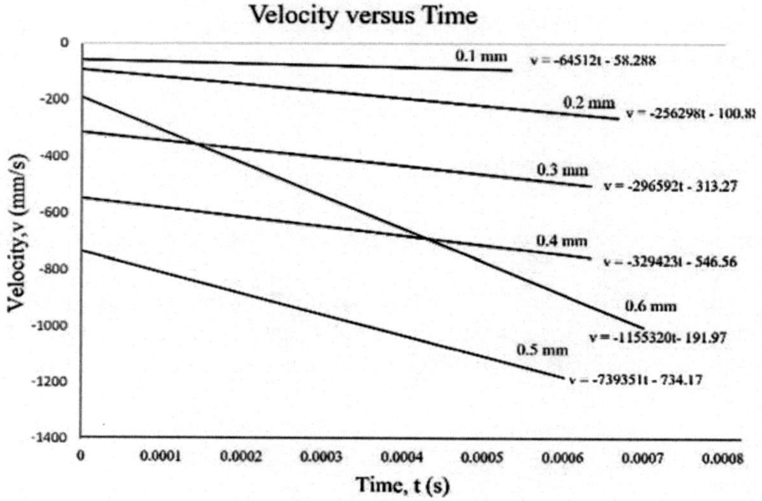

Fig. 8 Graph of velocity versus time

The plots of velocity verses time were then obtained by differentiating the formula representing the positions versus time. After differentiating the quadratic equations, the velocity curves are linear with respect to time. The results for the velocity versus position are shown in Fig. 8. It can be noted that the starting initial speeds are in increasing trend for the experimental positions from 0.1 to 0.5 mm. However, for the sample of 0.6 mm, the starting speed has tremendously drop where its value lies between 0.1 and 0.3 mm cases. The explanation for this phenomena is owing to the time taken to fully energize the electromagnetic force/field. When the distance is very small, a magnetic field does not have sufficient time to develop to the full force, hence less force is applied to the moving metal rod. The longer the position, the more force that can be fully developed due to the longer time. However, when the distance is too far and the time taken is more than the required fully-activated time, the initial applied force becomes lower. This is explained by the phenomenon of case 0.6 mm where its starting velocity is lower. Also, based on the statics conditions, the farther the distance, the lower is the force (Fig. 9).

Upon differentiating the linear velocity curves, constant acceleration values have been obtained. We concluded that the trend of results is correct due to the conservation of energy analysis. It can be noted that the amount of current and voltage for each experiment is the same, therefore, it can be concluded that for any distance, the amount of force should be constant. The behaviour of electromagnetic forces found in this research for dynamics case makes more sense than the statics case. As for the statics case, it seems that energy is not conserved because the closer is the distance, the more force that can be generated. Although this is the case, not much useful work can be cultivated from a very close distance, and the phenomenon might be due to the

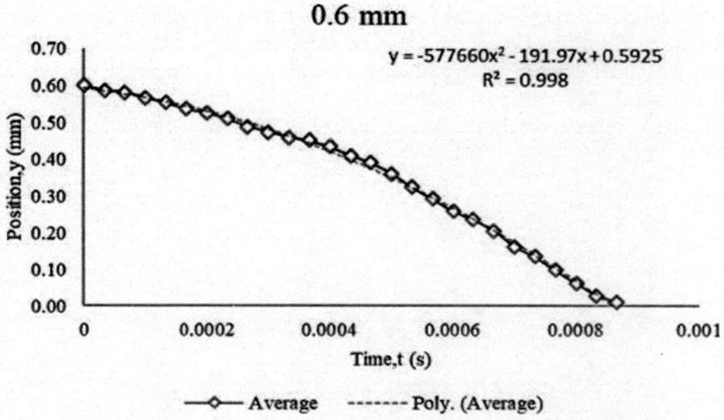

Fig. 9 Graph of acceleration versus position

nature of magnetic fields at close distance which knowledge has not been discovered yet by researchers.

5 Conclusion

Based on experimental results, we concluded that at close distance, when a moving ferrous steel rod is subjected to dynamics loading of electromagnetic forces, the position varies with polynomial of order 2 with time, the velocity varies linearly with time, while the acceleration is constant. The behaviour of electromagnetic forces under dynamics loading differs from the statics case. In brief, for close distance, the behaviour for statics case is almost liner while for the dynamics case is constant. The experimental results found in this research are very sound due to the explanations of conservation of energy.

The closer is the position, the less is the force being generated due to the shorter time for the magnetic force to fully activate. When a distance yields more time than the required time to fully active, the initial starting force is lower, but a moving metal rod starts to gain momentum and accelerates faster than the ones at closer distances.

References

1. Vokoun D, Tomassetti G, Beleggia M, Stachiv I (2011) Magnetic forces between arrays of cylindrical permanent magnets. J Magn Magn Mater 323(1):55–60. https://doi.org/10.1016/j.jmmm.2010.08.029
2. Vokoun D, Beleggia M, Heller L, Šittner P (2009) Magnetostatic interactions and forces between cylindrical permanent magnets. J Magn Magn Mater 321(22):3758–3763. https://doi.org/10.1016/j.jmmm.2009.07.030

3. Beleggia M, Vokoun D, De Graef M (2012) Forces between a permanent magnet and a soft magnetic plate. IEEE Magn Lett 3:2–5. https://doi.org/10.1109/LMAG.2012.2214027
4. Beleggia M, De Graef M (2005) General magnetostatic shape—shape interactions, vol 285, pp 1–10. https://doi.org/10.1016/j.jmmm.2004.09.004
5. Mushtaq S (2016) Experimental validation of dual output stator-less motor. Universiti Teknologi PETRONAS

Strategic Planning Tools to Sustaining a Synergistic Industrial Advisory Panel and Academic Collaboration

Zahraa Alzubaidi and Thar Mohammed Badri Albarody

Abstract Recruitment delay or shortage happen nowadays is due to the gap between academic qualification and industrial expectation which ultimately influence graduates' skill and experience. This paper proposes strategic planning tools as a starting point for systematizing industrial advisory panel and academic collaboration in a strategic manner. Industrial advisors point of view shared communication, assist academic to evaluate recruitment environment, expect future trends in higher education, find solutions to solve recruitment problems in diverse forms which helps to elevate awareness, decrease the risks affects students intake and finally deliver a framework for assessing the relative prominence of various academic program portfolios. Thus, this paper introduces several strategic planning tools to sustain a synergistic industrial advisory panel and academic collaboration, to divulge the spirit of strategic planning and its tools, to single out the successful collaboration mechanism.

1 Introduction

One of the reasons in recruitment delay or shortage is the gap between academic qualification and industrial field which ultimately influence graduates' skill and experience. This involve of several faculties such as engineering, science and pharmacy, supply the industrial sectors. Although graduates' academic qualification has an important role in acquiring the job, however personal leadership, subject knowledge, and business sharpness are important to apply for it. As skills mismatching between fresh graduates and employers' demands consider one of the unemployment main reason. Despite of that Soft skills have an enormous influence of entry-level graduates, but most universities show deficiency in those areas. In addition, some companies and educational institutes do not always match graduating students with their

Z. Alzubaidi
Faculty of Pharmacy, Universiti Kebangsaan Malaysia, 50300 Bangi, Malaysia

T. M. B. Albarody (✉)
Department of Mechanical Engineering, Universiti Teknologi PETRONAS, 32610 Seri Iskandar, Malaysia
e-mail: dher.albarody@utp.edu.my

F. Ahmad et al. (eds.), *ICPER 2020*, Lecture Notes in Mechanical Engineering,
https://doi.org/10.1007/978-981-19-1939-8_49

entry level prospects at the initial stages of recruitment. Thus, career service centres are underutilised and mostly do not identify companies' requirements. Furthermore, the Talent Corp-World Bank assessment stated that only half of the companies that responded to the form bargains structured internship programmes, could aid expose university students to a selection of career paths and assistance them develop essential soft skills for any position [1]. Moreover, any collaboration between public education institutions in companies or non-governmental organizations with the absence of industrial training will eventually produce low competitive graduates as compare to graduates who get industrial training from their institutions [2]. Furthermore, the presence of student guide or role models is very vital in helping students to improve their career focus, this is because of immense career opportunities range in which the students are unaware of [3, 4]. As a result, the World Bank recommended that universities need to improve the youth skills employability via detect skill development mismatch and create a performance feedback tool between the educational institutions and the industry [5]. Lastly, In order to construct an integrated human workforce, three primary strategies need to conduct; educational system restriction in order to raise the students' achievements, advancing the graduates' skills in order to raise the graduates marketability and reorganisation the labour market system in order to elevate the income [6]. Thus, it's important to synergistic a collaboration between academic and industry by set a regular meeting between academic staff and professional industrial representors; so called Industrial Advisory Panel (IAP) to help produce quality talent that will bring the industry forward.

2 Academic Challenges from Industrial Perspective

There are several concerns surrounding graduate employability. Chang [7] defined that the graduates' wrong degree is the main reason for unemployment. As some graduate qualification degree are already plentiful in the market, whereas other degree like engineering and other science graduates are still in great demand. Another reason is that graduates with a degree no more consider as qualify for their job. Instead, graduates who own the greatest information and skills in their field of study are the first to be employed [7]. Depend on a graduate tracer study in 2006, 30.7% of graduates stayed jobless for six months after their convocation [8]. Hillage and Pollard indicated that the graduates' knowledge, skills and aptitudes is the key factor for the graduate employability [9]. Currently employers prefer a graduate who is set with interactive, personal [10] and generic skills [11]. These graduates' skills are more significant in the staffing process than the graduates' academic concerts [10]. This outcome has also been reinforced by Purcell et al. who have shown that for some employers, a degree may now not symbolise anything more than a minimum requirement. The lack of graduate skills and personal qualities required by the employers despite of their high academic qualifications, lead to fail in securing the job which eventually raise unemployed graduates alarming numbers [12]. Hence, many higher education institutions (HEIs) have tried to embed skills into the curriculum [13].

Despite these findings relate to global graduate employability, business schools need to take a step from this shocking statistic and yield proactive actions to address this concern. Thus to increase attractiveness between graduates and employers, it is key for the graduates to proof of having the ability to deal with uncertainty, the capability to work under pressure, exhibit action-planning skills, communication skills, information technology skills, team work, an eagerness to explore and create opportunities, self-confidence, self-management skills and a willingness to learn [14]. Moreover, managers are preferring to hire individuals who can add to team success than others. Many of the communication and inter-personal skills are hunted by the employers, such as; English fluency and the capability to present ideas, to describe issues and problems, to express in a constructive manner, to resolve complications, to understand company's problems and to come up with practical solutions [7]. The harmonisation of education structures and qualification systems could rise international student movement and create a true labour market [15]. Furthermore, several variation have found when classified success attributes in the retailing field between industry members and educators, as the outcome revealed that affective skills like "leadership" and "decision making" are more contribute to retail success than cognitive skills [16]. To find a solution for graduate unemployment, internship programs or industrial training have been established which include hands-on or practical experiences for undergraduates. Hence, better job application will be accepted through feedback corrective measures from host organisation towards their interns or trainees. Finally, a collaboration between HEIs and industry will maximize graduate performance yielding in increase the employability [17].

3 Industry Advisory Panel (IAP)

An industry advisory panel is a group of engineering/pharmacist/scientist professionals and key opinion leaders (KOL) from the industry who has wide experience and expertise in his/her field, headed by a chair person who directs the meeting to discuss the available literature according to a predefined agenda and to come up with a consensus or guidelines and/or an agenda for the next meeting. The meetings are normally conducted in an informal and flexible manner and issues that were not included in the initial agenda may also be discussed.

3.1 Purpose of IAP Meetings

AIP design and review to guarantee the continuous development of the academic programme quality. AIP review and remark of both program objectives and curriculum fulfilled depending to current and future requirements of the industries. To assess and comment on the teaching and learning standards approaches. This include the delivery implemented methods, the content and tools used for the delivery of the lectures, the industrial implementation training programme and the laboratory facilities applied for the practical courses. AIP also deliver an industry perspective on the graduates quality to ensure that they are: of impact to the profession; steady with industry & management objectives; in accordance with industry needs; up-to-date with technology changes; encourage research and progress in areas that are relate to the profession, the university development and the nation follow.

3.2 Common IAP Shortcomings

Several drawbacks may happen during IAP meeting such as: Deviating from the IAP mission whenever the mission is not the focus point of the IAP meeting, then meeting will not gain the profitable purpose. Complacency, the active participation of IAP members is vital, otherwise any action of miss meetings, not performing the assigned tasks or even not join in the discussion is consider complacency. Multiple Voices, despite of IAP member's different opinions, however, their final decision must be of a total agreement and should not disclosed to media. Limitless terms, IAP should be open to new ideas or perspectives, as term limit aid an IAP avoid unproductivity. Lack of self-assessment, periodic self-assessment is important if followed by self-improvement, i.e., corrective actions [18].

3.3 Ways to Improve IAP

IAP need a strategic guidance to deliver best outcome, several practices such as: Be Objective: Setting define goals and objectives is vital for a well-run IAP. It is equally beneficial for both the participants and the organization, as it aids organizations choose the right participants as well as to fulfils participants' requests. Moreover, a defined objective serves as a check to guide them if IAP lose sight of goals. Aim for Right Participants: to collect a broad spectrum IAP opinions and perspectives, it's recommended to combine both national and regional members. Meetings Frequency: IAP meeting need to set twice a year and IAP officers must participate with the academic administrator to organise it. Keep up the Communication Flow: one of the best practices to achieve well-run advisory panel is to communicate regularly. Assign a great meeting moderator (in healthcare sector called medical writer): his responsibility is to steer energetic of discussions which eventually rise the meeting successful. Clarify mutual expectations: IAP members need to collaborate with the moderator to come up with agenda that reflect both side expectations about end-result. Convert meeting results into action steps: it's important for a member to follow up the advisors feedback to deliver updates and future steps [18–20].

3.4 IAP Work of Plan

The IAP work of plan includes the IAP Outcomes determination. To ensure best IAP work of plan, it need to know several things, such as academic program, IAP resources, the accreditation standards, the time required to achieve each IAP Outcome and set IAP priority. IAP best practice is summarize in four categories:

3.4.1 IAP Operating Procedures

The IAP Bylaws should reviewed and update periodically need a strategic plan, concomitant action plans, and work of plan. The high-impact IAP meetings should established, supply regularly scheduled and recorded and staffed by active IAP members.

3.4.2 IAP Curricula Involvement and Program

AIP must evaluate assessment program goals and objectives and compare it with the program accomplishments to suggest employability skills to be incorporated in the curriculum. Curriculum should undergo review committees to compare actual occupation duties and responsibilities with the program content "evaluating course using its syllabus". AIP must meet seniors graduates to talk over the academic program as well as to need to assess graduate's quantity and quality of graduates and to talk over job placements. AIP can serve as classroom visitor lecturers or adjunct faculty to serve as skilled technicians to balance instructor's experience by delivering instructional material offer "real-world" projects. AIP could provide whatever need for course instructors to elucidate industrial methods concerns and to encourage to arrange for "site visits" and "field trips" to facilitate the knowledge delivery of specific course lessons.

3.4.3 IAP Industrial Awareness and Recognition

AIP can assist in engaging possible industry with academic by offering student internships opportunities or make sure faculty internships available (at worksites). AIP can contribute to lead improvement workshops for both academics and students to develop a plan for the academic program marketing or a plan for participating in occupation and carriers' fairs. AIP can conduct employer/student presentations or seminars to help in developing student interview skills as well as to inform instructors of job vacancy for students or indirectly allow the AIP to select and hire graduates of the academic program. Utilizing AIP in developing competitive aptitudes events or even judge competitive aptitudes events to support student association activities.

It is recommended to arrange AIP meet with academic administrators regularly to connect with legislators about academic program needs, endorse legislator's tours and endorse and encourage legislation for academic education and occupation development. On the other hand, academic administrators' advice to present awards for exceptional AIP members. Academic administrators advised to enforce producing an IAP newsletter regularly.

3.4.4 IAP Funding Tools and Deliverables

Faculties must have an IAP fees structure, if possible or at least establish interior fund raising, to support student/department events, more importantly AIP meeting. Investment in said event can support student scholarships or lead and support determinations to start endowments (for faculty positions and scholarships). AIP can offer support the faculty research or postgraduate program by contributing to fund several academic program activities in return it help industrial sectors to claim research joint venture with universities.

In general, whenever the work of plan has been demonstrated and the best practices have been recognised, the responsible parties should consign to each best practice. For each best practice, a clear information must include; like the tasks required, the anticipated results, the deadlines, the evaluation plan and corrective actions timeframe [18].

3.5 IAP Quality Systems

IAP Quality Systems comprise of three measures:

3.5.1 IAP Self-assessing and Evaluation Performance

The importance of this assessment is to ensure the effective and dynamic IAP meetings. Table 1 shows the IAP possible self-assessment that could be filled individually or cooperatively at an IAP meeting. However, it is suggested that the IAP self-assessment be filled individually and anonymously. This is because the cooperatively filled might affected by group pressure or the impression of not being member of the group.

3.5.2 Quality of IAP Outcomes (Best Practices)

The IAP Outcomes are recognised and assigned to selected parties; for example, academic and faculty can help in collecting and evaluate data due to their expertise. The data collected must be relevant and reliable. Then it must go through evaluation process by compare it to the standard of the desired performance. There is two

Table1 IAP self-assessment

IAP self-assessment		
Assessment item	Response	
1. The IAP held at least two meetings during the calendar or academic year	Yes	No
2. IAP members are notified of meetings in a timely manner	Yes	No
3. The IAP agendas are prepared and distributed in advance of each meeting	Yes	No
4. The IAP delivered high impact meetings	Yes	No
5. Percentage of IAP members who attended all meetings (from the IAP secretary)	Percent	
6. The IAP meeting are minutes to maintained as a permanent record and are distributed to the IAP members and appropriate administrators	Yes	No
7. The IAP elects' officers on a regular systematic basic (yearly)	Yes	No
8. The IAP reports formally to the academic administrator	Yes	No
9. The IAP recruit "active" members	Yes	No
10. The IAP membership is divers in gender and ethnicity, and is composed of individuals who represent a broad cross-section of the construction industry	Yes	No
11. IAP members are public recognized for their service	Yes	No
12. IAP members are aware of IAP mission	Yes	No
13. The IAP utilizes up to date Bylaws to govern its operations	Yes	No
14. The IAP has an up to date Strategic of Plan	Yes	No
15. The IAP develops and delivers an annual Plan of Work	Yes	No
16. The IAP uses quality system to assess and evaluate their services and programs	Yes	No
17. The IAP takes corrective action to address problems and deficiencies	Yes	No
18. IAP members are knowledgeable about accreditation standards	Yes	No
19. IAP members are familiar with the curriculum of the academic program	Yes	No
20. The IAP has reviewed the current academic program and curriculum to ensure the employment needs of business and industry are endorsed	Yes	No
21. The IAP reviews outcome data from the program, including: student competences, placement rates, etc	Yes	No
22. The IAP promotes and advocates for the academic program	Yes	No

evaluation process of data source; which is direct and indirect. Direct data sources involve actual observations and written performance reviews, while indirect data sources comprise surveys and opinions view. On the others hand, AIP may prefer another approach like meeting with the graduating seniors, go to their capstone demonstrations, observe internships performance and evaluation the performance assessments of the internship supervisors.

3.5.3 IAP Benchmarking

Once the quality of IAP outcomes accomplished, IAP benchmarking can proceed. The benchmarking numerical ranking is as shown in Fig. 1.

4 Strategic Planning Tools

Several researchers had mentioned and defined strategic planning tools, as Ansoff was the first to put forward in his book "From Strategic Planning to Strategic Management". After that he describes strategic management as an action related with determining goals and objectives of the organization, keep of relationships between the organization and its adjacent that benefit it to reach its goals, link to its internal abilities and let it remain liable to external demands [21]. Hence, strategic management can be defined as the process of running an organization depend on its potential, that guides its activities to consumer needs, offers stretchy management and timely fluctuations in the organization, run into the challenges of the surrounds and allows to generate and maintain competitive benefits, which make the organization able to survive and accomplish its objectives in the long time period. Even though the academics show a great interest in strategic management, the strategic management tools have not yet define. Knott explained the term "strategy tool" as embrace the full range of, ideas, concepts, methods and approaches that assembly or influence strategic thinking, and strategy implementation action. Strategy tools display a starting point for organising strategic management activity [22]. On the other hand, Stenfors and Tanner offer another definition: "strategic management tools and techniques—heterogeneous group of products designed to support organizations in dealing with the complex demands of competitive markets and the quest to create and maintain strategic advantages" [23].

The prominence impact of strategic planning tools is due to various point of view offers, help individual and shared communication, assist executives to make decisions, evaluate environment, know relations, expect future, find solutions to solve problems, organise information and display it in diverse forms which helps to elevate

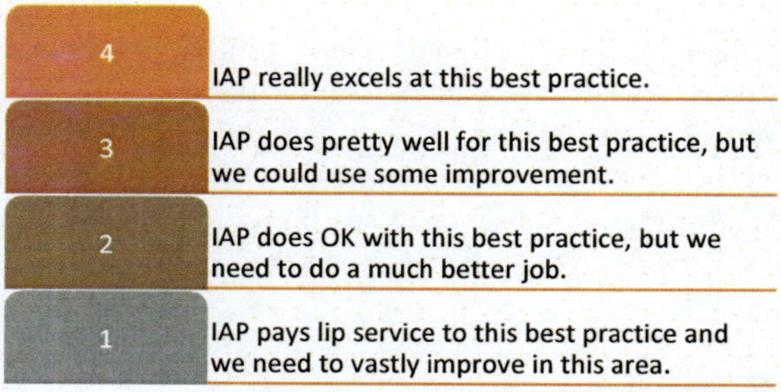

Fig. 1 IAP Benchmarking from 4 to 1

awareness, decrease the risk relate in decision making and finally deliver a framework for assessing the relative prominence of various business portfolios.

4.1 SWOT Analysis Leading to a TOWS Analysis

SWOT is one of the old of strategic analysis tool, as it has shown in the marketing textbook since 1972. It promoters say that it can be used to measure the degree of "fit" between the academy's educational strategies and its environment, i.e., Engineering program outcomes and technical skills demand of nowadays industry see Table 2. It is made to suggest ways in which the academic organization can benefit from strengths and opportunities and identify PEOs weaknesses and threats.

4.1.1 Strengths

Determine an Program's strong points. This could be from faculty members and IAP Prospective. A strength is an advantageous source relative to the university competitors from public or private universities and the needs of the industry that are served by programs offered, or those that it expects to offer in the university. It is a distinct competence when it offers the programs given by the university a comparative advantage in the industry place.

4.1.2 Weaknesses

Determine an organisation's weaknesses, not only from faculty members point of view, but also more importantly, from IAP. Although it may be difficult for an organisation to acknowledge its weaknesses it is best to handle the bitter reality without procrastination. A weakness is a "limitation or deficiency in one or more resources or competencies relative to competitors that impedes a university's effective performance".

Table 2 TOWS matrix

External factor	Internal factor	
	Strength(S)	Weakness(W)
Opportunities(O)	SO strategies	WO strategies
Threats(T)	ST strategies	WT strategies

4.1.3 Opportunities

Another major factor is to determine how engineering educational organisations can continue to grow within the industryplace. After all, opportunities are everywhere, such as the changes in technology, government policy, social patterns, and so on.

4.1.4 Threats

No one likes to think about threats, but we still have to face them, despite the fact that they are external factors that are out of our control, for example, the recent economic slump in Asia. It is vital to be prepared and face threats even during turbulent times. A threat is a main unfavourable condition in an engineering educational's environment. Threats are basic impairments to the university's current or desired situation. The entry of new competitors, low student intake, technological alterations, and new or revised rules could characterize threats to a university's success.

4.1.5 Maxi-Maxi (S/O)

This grouping displays the institution's strengths and opportunities. An institution should attempt to maximise its strengths to achieve new opportunities.

4.1.6 Maxi-Mini (S/T)

This grouping shows the institution's strengths in respect of threats, e.g., from competitors. an institution should attempt to use its strengths to minimise threats.

4.1.7 Mini-Maxi (W/O)

This grouping express the institution's weaknesses in tandem with opportunities. It is an effort to overcome the institution's weaknesses by making of any new opportunities.

4.1.8 Mini-Mini (W/T)

This grouping shows the institution's weaknesses by comparison with the recent external threats. This consider the most defensive scheme, to minimise an institution's internal weaknesses and evade external threats [24].

4.2 PEST Analysis

Is a type of strategic tool that explore/expect external stimuli of organisation or department from numerous perspectives. It may be used to review a strategy, the organisation direction, a marketing scheme, or product development and future business initiatives. To run a PEST with best result, a lot of managers are need to enrol and the following points need to do:

- Collecting data about political, economic, social and technological changes.
- Detecting which of the PEST features characterise as opportunities or threats.

For example: discussion of weak job market in the meeting consider one of the economic factor components due to challenging in finding a well-paid job in the industrial field after completing their degree [25, 26].

4.3 Balanced Scorecard (BSC)

Is another type of strategic planning tools used to: Clarify and bring up to date vision and strategic path; link strategic objectives and methods throughout the organization; Align goals of both department and individual with the institution's vision and strategy; Connect strategic objectives to extended time goals and annual budgets; Recognise and arrange strategic initiatives; set intermittent performance reviews to study about and advance strategy; and Get feedback to know about and develop strategy [27]. A scorecard building logic for an academic institution is depending upon customers (student) requirements to market opportunities. The BSC is a performance measurement structure and it is a multiple time activity. It consist of two phases which is the scorecard building and its implementation as shown in Table 3. It is a continuous process that starts with the first phase of planning the scorecard by taking into account the short and long-term strategies of the institution followed by determining what business procedures are critical to accomplishing these objectives in the implementation phase [28].

The BSC has four measurement perspectives assist to collect as much data as possible for assessment and discussion between AIP and academic; these are financial perspective, customer perspective, internal process perspective and learning and growth perspective. In the below table is the scorecard development structure model for institutions of higher education [29].

Table 3 Balanced Scorecard (BSC)

Perspective	Objectives	Measures
Customer perspective	• Highly valued program • Quality academic advising • Flexible course scheduling • Quality instruction Effective student placemen	Alumni evaluation, recruiter evaluation, average starting salaries, No. of companies recruiting on campus
	• Alumni Knowledge updating with	Alumni feedback
	• passage of time Knowledge reinforcement	Alumni satisfaction survey
Internal business perspective	• Resources to improve new product and service development	Number of new products and services introduced i.e. new courses, syllabi, programs and curriculum changes
	• Internship program	Number of internships available, number of companies available, student evaluation
	• Unique or specialized curriculum	Number of faculty in specialized area, number of schools offering the same program
Innovation and learning perspective	• Incorporating technology into teaching	Number of courses incorporating new technology
	• Curriculum innovation	Number of curriculum revisions in last five years; number of new courses offered in last five years
	• Partnering with corporations for campus recruitment	Number of firms involved; number of joint activities
	• Resource management	Number of campus partnerships; entrepreneurial initiatives; trends in energy use
Financial perspective	• Maximize asset utilization	More efficient and effective use of facilities, space, services, systems and resources as measured by various usage studies and statistics

4.4 Academic Program Portfolio Model (APPM)

Is a type of strategic analysis tool adapted for universities. It is modified from the General Electric McKinsey Product Portfolio Model which have been used widely in business [30].

Table 4 Summary of APPM criteria

Program and institution capabilities	Program marketplace attractiveness
Market share	Student demand for degree
Share growth	Growth rate of student demand for degree
Quality of degree program	Employer demand for graduates
Reputation of degree program	Growth in employer demand for graduates
Market access to degree program	Number of Universities offering degree and extent of competition
Promotional effectiveness	Technological requirements necessary to offer degree
Graduation rate and time	Sensitivity of demand to economic conditions
Per student costs	Demand for intellectual capital of the program
Access to tangible resources	Social trends influencing market and employer demand
Research of faculty	Legal issues
Quality of faculty	Political issues

The APPM's has two dimensions which are the attractiveness marketplace of the academic and capabilities of the program and institution, review external conditions outside the academic program control and internal factors to the program which directly impact its success in long-term, respectively. The APPM's systematically analyse the academic administrators guides depending on strategic track, resource allocation, and performance anticipations for each academic program [31]. At the beginning, APPM need criteria identification to aid the graduate and institution capabilities, then it need to identify criteria to evaluate the program's marketplace attractiveness as shown in Table 4. Moreover, it requires the identification of the criteria necessary to evaluate the program's marketplace attractiveness of each program's marketplace. In addition to the identification of the criteria, judgment is also used to decide the importance of each criterion to the total capabilities assessment (relative weights totalling 1.00) and to the total attractiveness assessment of marketplace (relative weights totalling 1.00). The criterion rank of the academic program is determined by scale (1 to 5 scales, 1 meaning weak and 5 meaning strong). After that, the resultant value multiplied by the relative weight importance of each criterion to get the weighted score as shown in Table 5. The total of these weighted scores deter-

Table 5 The academic program portfolio model

	Program and institution capabilities			
		Weak	Moderate	Strong
Attractiveness of program market	• Highly attractive	• Withdraw from academic area • Invest to address program issues	• Build selectively on program strengths • Invest to build on strengths	• Invest to grow academic program • Maintain program strengths
	• Moderately attractive	• Control risk of offering program • Reduce academic program	• Protect program • Invest in select programs and select market segments	• Invest in select market segments • Invest in academic programs in niche areas
	• Unattractive	• Eliminate program • Cut costs and investments	• Concentrate on few market segments • Minimize investments • Reduce weak program area	• Concentrate on a few viable market segments • Focus on redirecting strengths

mine a composite amount of capabilities and attractiveness, and are the projection that detect the graduate program in the APPM matrix. At the end of this calculation, these scores composite should fall between 1 and 5. Several factors may influence rating score, such as former resource allocations, the faculty research activity, trends in regional industry, and the competition strength effect the capabilities and graduate program marketplace in the sciences differentially. The ranking of each program on each criterion is an image of these differences. As a final point, A summary of the composite marks for each of program is compare with percentage of the overall graduate students number to conclude which field appeal more student against market attractiveness [31].

4.5 Importance-Performance Analysis (IPA)

The importance-performance concept is based on multi-attribute models. This technique identifies the performance of an attribute that can be changed without affecting its importance [32]. According to Nale et al. a particular application of the technique starts with the identification of the attributes that are relevant to the situation chosen to be investigated [33]. This approach, also known as quadrant analysis, was introduced by Martilla and James [34]. In their study of customer satisfaction they focused on pinpointing those quality and service elements that:

- Are very crucial to customers and/or are probably make the big contribution to total customer satisfaction and loyalty.
- Are in need of enhancement because customers' assessments of the company's performance on these features are unfavourable.

The study was carried out in two phases. Phase 1 involved a focus group session of ten managers who, through a brainstorming session, provided information regarding employer expectations. This session focused on the growing concern among employers about the relevance of the HEI's management curriculum in the face of developments in the real world today. The researchers, being the moderators of the focus group, sought unprompted discussion of the issues contributing to graduate unemployment in Malaysia. Using the KSAP dimensions as a basis for the study, attributes were derived from the focus group interview and specific issues put forward by the group. In Phase 2, the dominant theme(s) emerging from the focus group study were used to establish a questionnaire pertaining to the KSAP dimensions and were pre tested through a pilot study. The questionnaire was then distributed personally to the managers of all the companies in industries that employed graduates from this HEI either during the AIP meeting or before it then discuss the result during the meeting. This exercise enabled the examination of the gap between the perceptions of managers towards the important characteristics of these graduates and their actual performance in terms of their knowledge, skills, abilities and personality (KSAP) [34]. Several factors can investigated during questionnaires for example explicit and tacit knowledge, hard and soft skills, intellectual and physical abilities, conscientiousness, agreeableness, emotional stability, openness, extroversion and others. The number attributes used to represent each factor were either four or five as shown in Fig. 2. This is in keeping with the work of Hair et al. (2006), who indicated that a minimum of four attributes is recommended to represent a factor; for example attributes of soft skills factor include (1) able to solve situational problems; (2) present strong leadership skills; (3) able to establish good relationship with co-workers; (4) able to establish good relationships with top managers; and (5) able to establish good relationships with senior staff members. The company's managers of different discipline (Manufacturing, Construction, Service, Information technology, Heavy industry and Others) need to judge the important characteristics of graduates, where 1 represented "extremely unimportant" and 7 represented "extremely important", followed by a second evaluation of actual performance of these graduates, where 1 represented "extremely bad" and 7 "extremely good" [35]. According to important

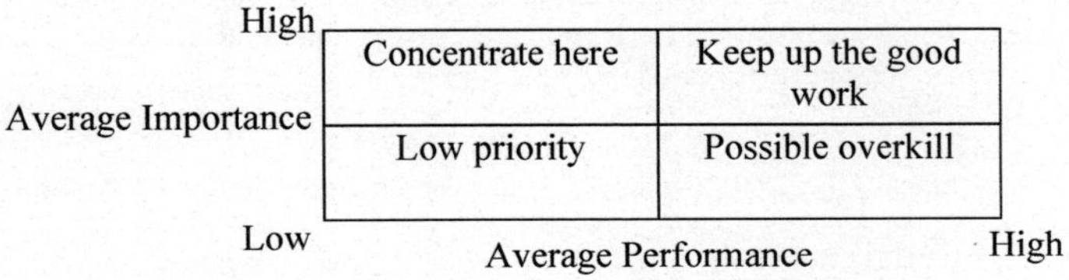

Fig. 2 Importance-performance analysis

and performance scale subtract from company's managers, an IPA matrix can draw resembling the figure below [36]. If the factor or its attributes relay on the Keep up the good work (high importance, high performance), this mean these elements or attributes are assumed to be key drivers of respondent satisfaction/preference, and the management's job is to ensure that the organization continues to deliver/perform well in these areas. However if factor or its attributes relay Concentrate here (high importance, low performance). These elements or attributes are also assumed to be key drivers of respondent satisfaction/preference and should be viewed as critical performance shortfalls. It is management's responsibility to ensure that adequate resources are allocated to improve performance in these areas.

These areas are priorities for improvement. On the other hand, the elements or attributes relay on Low priority (low importance, low performance) assumed to be relatively unimportant, and as such they should not be given a great deal of priority or attention by management. Lastly, the elements or attributes relay on Possible overkill (low importance, high performance) assumed to be relatively unimportant, should be viewed as an area of performance "overkill" and management may want to redirect resources from these elements to high-priority areas in need of improved performance [37].

5 Conclusions

Programs such as engineering scientist and pharmacy are running under turbulent environment due to instability in financial markets effecting federal and state funding fluctuating student demand. Synergistic industrial advisory panel and academic collaboration may help to allocate and align of limited resources so that the university serves its mission and meets its objectives as well as help student to get recruited. Therefore, a strategic tool propose herein to facilitate synergistic engagement with industrial advisors, establishing STEM educations are able to sustain a clear and meaningful differential advantage relative to competition and to admit program objectives consistent with institution's mission.

References

1. Hossain MI, Yagamaran KSA, Afrin T, Limon N, Nasiruzzaman M, Karim AM (2018) Factors influencing unemployment among fresh graduates: a case study in Klang Valley, Malaysia. Int J Acad Res Bus Soc Sci 8(9):1494–1507
2. Omar CMZC, Rajoo S (2016) Unemployment among graduates in Malaysia. Int J Econ Commerce Manag 4(8):367–374
3. Baxter A, Waldock J (2012) Developing employer engagement in STEM through career mentoring
4. Begum H, Ye J (2013) Nurturing talent–mentoring that makes a difference
5. Aina N (2017) More than 1.7 million foreign workers in Malaysia; majority from Indonesia. The New Straits Time

6. Leong PC (2011) Technical education and vocational training (TEVT) in Malaysia: revisiting the TEVT framework. Malaysia Prince Hotel, Kuala Lumpur
7. Chang M (2004) Why some graduates are more marketable than others: Employers' perspective. In: 2004, paper presented to the workshop on enhancing graduate employability in a globalised economy. Economic Planning Unit: Kuala Lumpur
8. Khaled N (2009) Issues, concerns and the way forward, in speech by the minister of higher education Malaysia to the seminar on enhancing graduate employability
9. Hillage J, Pollard E (1998) Employability: developing a framework for policy analysis. Labour Market Trends 107:83–84
10. Harvey L (2000) New realities: the relationship between higher education and employment. Tert Educ Manag 6(1):3–17
11. Hager P, Holland S, Beckett D (2002) Enhancing the learning and employability of graduates: the role of generic skills. Business/Higher Education Round Table, Melbourne, p 16
12. Purcell K, Morley M, Rowley G (2002) Employers in the new graduate labour market: recruiting from a wider spectrum of graduates. Council for Industry & Higher, London
13. Atlay M, Harris R (2000) An institutional approach to developing students' "transferable" skills. Innov Educ Train Int 37(1):76–84
14. Raybould J, Sheedy V (2005) Are graduates equipped with the right skills in the employability stakes? Ind Commer Train 37(5):259–263
15. Jurse M, Tominc P (2008) Professional competences of graduates as a labour market mechanism for aligning business school curriculum reform with the Bologna declaration principles. Management 13:17–36
16. Nicholson A, Cushman L (2000) Developing successful employees: perceptions of industry leaders and academicians. Educ Train 42(6):366–371
17. Morley L (2001) Producing new workers: quality, equality and employability in higher education. Qual High Educ 7:131–138
18. McIntyre C, Fox P (2014) Developing a "High-Impact" industry advisory board
19. Singh N, Sharma R (2016) Covering a medical advisory board meeting and creating the report or publication: the role of the professional medical writer. Med Writing 25:37–44
20. Pugh S, Grove M (2014) Establishing industrial advisory boards using a practice transfer model. New Direct Teach Phys Sci 10(1)
21. Strategicheskoye upravleniye IA (1989)
22. Knott P (2006) A typology of strategy tool applications. Manag Decis 44:1090
23. Stenfors S, Tanner L, Haapalinna I (2004) Executive use of strategy tools: building shared understanding through boundary objects. Front E-Business Res
24. Yigit A, Al-Ansary M, Al-Najem N (2003) SWOT analysis and strategic planning as an effective tool for improving engineering education. Kuwait University
25. Thompson J, Martin F (2010) Strategic management: awareness & change, 6th edn. Cengage Learning EMEA, England
26. David FR (2012) Strategic management: concepts and cases. McGraw-Hill Educat, London
27. Kaplan RS, Norton DP (1996) Using the balanced scorecard as a strategic management system. Harv Bus Rev 74:75–85
28. Patro A (2016) Using Balance scorecard in educational institutions. Int J Bus Manag Invent 5(11):70–77
29. Kaplan RS, Norton DP (2001) Strategy focused organization :how balanced scorecard companies thrive in the new business environment. In: Schmalenbach business review. Harvard Business School Press, Boston
30. YIP GS (1984) Market selection and direction: role of product portfolio planning. Harvard Business School Publishing, Boston
31. Wells R, Wells C (2011) Academic program portfolio model for universities: guiding strategic decisions and resource allocations. Res High Educ J 11:46–54
32. Kitcharoen K (2004) The importance-performance analysis of service quality in administrative departments of private universities in Thailand. ABAC J 24(3):20–46

33. Nale R, Rauch D, Wathen S, Barr P (2000) An exploratory look at the use of importance performance analysis as a curricular assessment tool in a school of business. J Workplace Learn Employee Counsel Today 12(4):139–145
34. Martilla JA, James JC (1977) Importance-performance analysis. J Market
35. Hair JF, Black WC, Babin BJ, Anderson RE, Tatham RL (2006) Multivariate data analysis. Upper Saddle River Pearson Prentice Hall
36. Eskildsen J, Kristensen K (2006) Enhancing importance-performance analysis. Int J Product Perform Manag 55(1):40–60
37. O'Neill Martin A (2004) Importance-performance analysis: a useful tool for directing continuous quality improvement in higher education. Qual Assur Educ 12(1):39

The Design and Simulation of an Exhaust Manifold for Free-Piston Linear Generator Engine

M. Muzani Masri, A. R. A. Aziz, Z. Ezrann, A. Zainal, and Salah E. Mohammed

Abstract Free-piston linear generator is single axis mechanism operating engine which much simple compare to conventional. This paper focus more on free-piston linear generator gas exchange simulation. A computational study has been conducted under the design of exhaust manifold and simulation of the effect of exhaust manifold on gas exchange process. The engine already had intake port and exhaust port design, but the design of exhaust port was multi hole exhaust port which give a backlash of difficulty of collecting the gas exhaust residual and reduce the combustion pressure performance. Based on study, smaller and compact manifold volume will increase the back pressure but the velocity of air exchange is increase. The back pressure also should less than atmospheric pressure so that the residual still can exit from exhaust port.

Keywords Free-piston linear generator · Exhaust manifold · Gas exchange · Scavenging

1 Introduction

Vehicle transportation and industrial consumption of fossil fuel causing greenhouse emission and other pollutants being produced. Due to higher demand for fossil fuel, the emission has increased exponentially [1]. An alternative fuel cell has been investigated and has been implementation in vehicle technology such as hybrid car, electric car and Natural Gas Vehicle (NGV) [2, 3]. One of the alternated is free-piston

M. M. Masri · A. R. A. Aziz (✉) · Z. Ezrann · A. Zainal · S. E. Mohammed
Centre for Automotive Research and Electric Mobility (CAREM), Research and Innovation, Universiti Teknologi PETRONAS, 32610 Seri Iskandar, Perak, Malaysia
e-mail: rashid@utp.edu.my

M. M. Masri
e-mail: muzani.masri@utp.edu.my

A. R. A. Aziz · Z. Ezrann · A. Zainal · S. E. Mohammed
Department of Mechanical Engineering, Faculty of Engineering, Universiti Teknologi PETRONAS, 32610 Seri Iskandar, Perak, Malaysia

F. Ahmad et al. (eds.), *ICPER 2020*, Lecture Notes in Mechanical Engineering,
https://doi.org/10.1007/978-981-19-1939-8_50

linear generator (FPLG). FPLG was a simple and compact compare to conventional engine, where the movement of translator was single axis which can produce a faster expansion stroke, reduce friction, variable compression ratio [4, 5].

FPLG has several improvement over conventional engines as an electrical power generator due to its simple design with great controllability, low maintenance costs, also capable to obtain homogeneous charge compression ignition (HCCI) [4, 6]. Linear alternator or linear generator has several type of design which one of the design placing the magnet in the middle of translator and the coil will surround the middle of the engine, in a different design, the magnet was around the engine body meanwhile the coil was in between the translator. Thus design objective is to produce electricity by using concept of motoring alternator where the coil will cut the magnetic field then the electric will generate [7, 8]. The rate of compression ratio was flexible depend on energy requisite in FPLG design, so when generator required higher energy demand, the compression ratio can be change because it depend on the stroke length so the piston can travel further and the translator velocity can travel much quicker due to higher compression ratio can obtain higher pressure [9]. The exhaust port and manifold can influence on engine performance under scavenging process, but the compression ratio of variable stroke also may have effects on scavenging performance. Scavenging process depend on intake air and exhaust gas residual exchange in-cylinder [8, 10].

2 Design and Meshing of Exhaust Manifold

The design of FPLG engine consist of variable of parameter such as intake port, exhaust port, injector, spark plug and sensor. For two stroke engine, an air intake and exhaust gas outlet process simultaneously in one stroke process before going for combustion stroke process. Based on Chenheng Yuan study, the different between conventional engine with FPLG was the gas exchange in FPLG are shorter gas exchange duration and more residual burned products [11]. The design in UTP FPLG engine consists of one side of combustion chamber with another side of bounce chamber, with 56 mm bore, 1.5 and 2.0 aspect ratio of stroke length. UTP FPLG engine using force scavenging process design in order to improve the gas exchange process where the duration was short due to the design. UTP FPLG design the multiple hole exhaust port at bottom dead center cylinder linear. The design of multi hole was to assist the flushing process during gas exchange. Exhaust manifold was to concentrate and collect all release gas into one pipeline.

The design for exhaust manifold will affect the combustion performance and gas exchange. The structure of UTP FPLG consist of multi hole exhaust port to assist the gas exchange but it reduces the combustion pressure performance and back pressure. Furthermore, the structure also gives a difficult on collecting the exhaust residual releasing from the port and measure it. So the design of exhaust manifold should cover all the exhaust port hole and have one an exit of exhaust pipeline. In order to study the gas exchange process in a FPLG, a multidimensional simulation was set up using ANSYS Fluent R19.1 under UTP Academic Licenses. The simulation required an internal volume of FPLG in 3-dimensional geometry, to conduct the gas exchange process. Figure 1 shows the meshing for first design of exhaust manifold internally, the design of exhaust manifold was quite big in order to avoid high back pressure in-cylinder. Meanwhile, Fig. 2 shows the meshing for second design of exhaust manifold from side, the design of exhaust manifold was quite smaller and compact compare to first design.

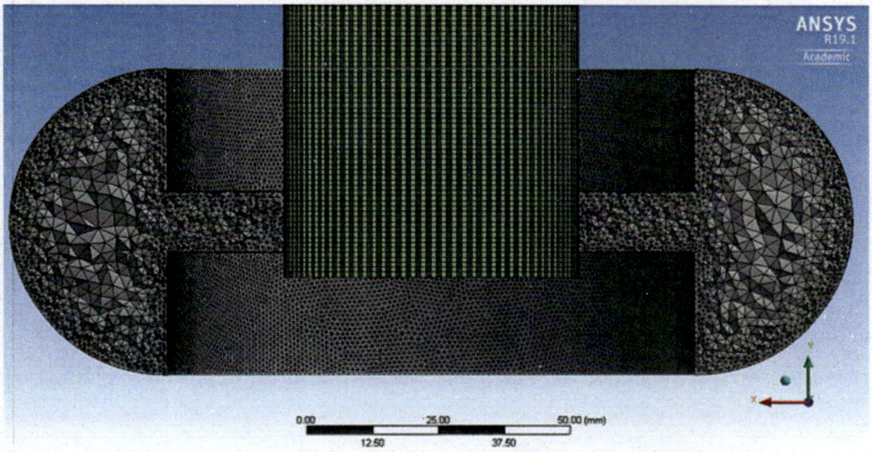

Fig. 1 First design exhaust manifold meshing cross-section

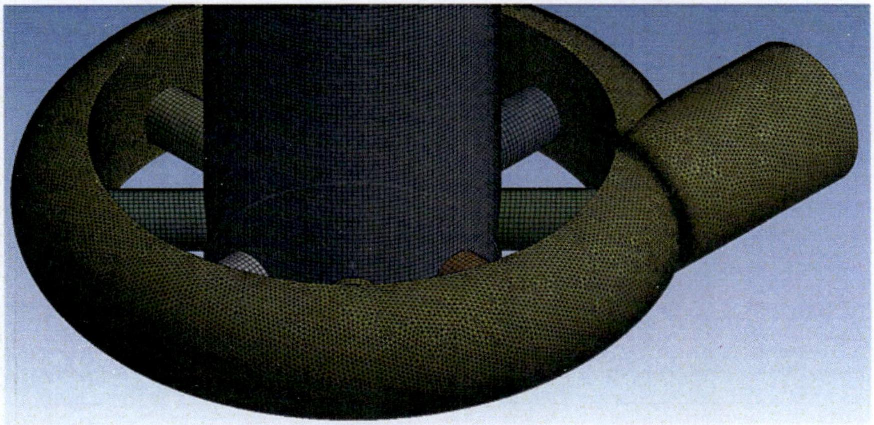

Fig. 2 Meshing for second design

3 Simulation Set up

Turbulence was modelled with a swirl dominated flow RNG $k - \epsilon$ and scalable wall function. The boundary condition for mass flow inlet is 0.0204 kg/s, meanwhile, for pressure outlet boundary condition being set to atmospheric pressure condition. The condition where the end of exhaust manifold still not yet connected with exhaust pipeline. The exhaust pipeline will further change the result of gas exchange and combustion parameter which is not ideal for the purpose to study the effect of exhaust manifold. The case setup uses a coupled scheme to solve a pressure–velocity coupling case. The case also using a standard initialize for pre-calculation before starting to solve the calculation. The case study using transient with 10 ms final time output which indicated the air intake time for flushing and replacing a fresh air only 10 ms.

4 Result and Discussion

Based on the result obtain above, we can observer that the first design can easily release from exhaust port while the second design has a restriction to release form exhaust port, both result can be observer in Fig. 3 and Fig. 6. Due to flow restriction

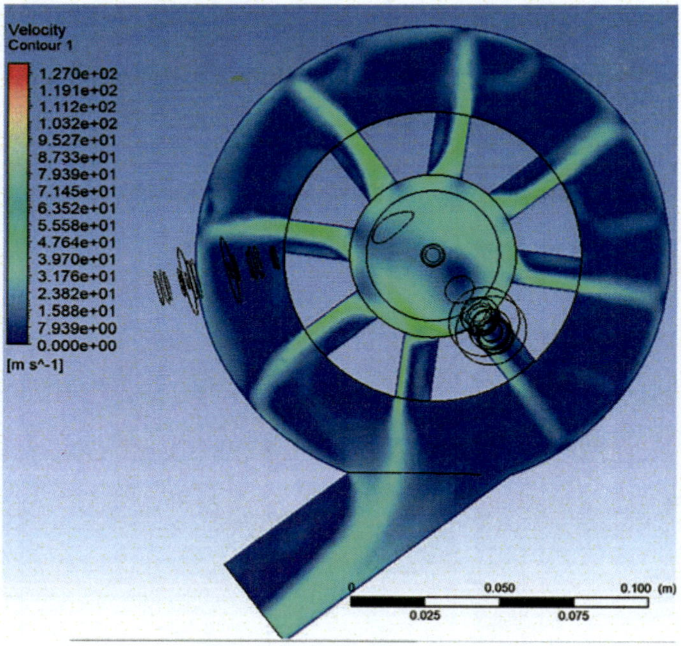

Fig. 3 Velocity contour for first design of exhaust manifold

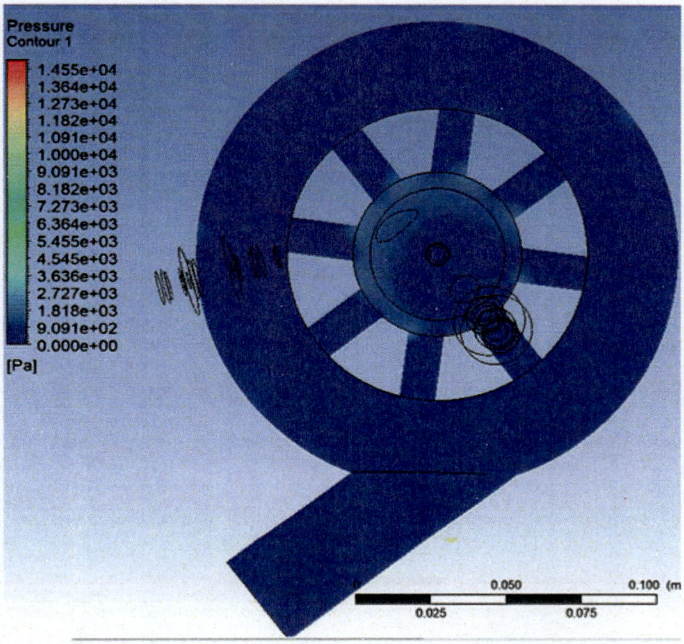

Fig. 4 Pressure contour for first design of exhaust manifold

in second design, the pressure is increase as shown in Fig. 6 compare to Fig. 4, where the pressure inside the chamber and manifold is low. Even though the second design restriction on air flow, the velocity out from the exhaust manifold is higher than first design. Thus we can conclude the second design should have better flushing design but have higher pressure increase inside the chamber. The back pressure inside the cylinder can help increase the compression pressure and improve the combustion performance Figs. 3, 4, 5.

5 Conclusion

The size of exhaust manifold effect the gas exchange flow and pressure inside the cylinder linear. The back pressure should help increase the compression pressure and improve the combustion performance but to high back pressure may cause gas exchange not complete and trapping the residual exhaust gas inside the combustion chamber and exhaust manifold. The trapping residual gas will reduce the possible for combustion to occur inside the combustion chamber. The trapping residual inside the exhaust manifold also have high possible to block the new residual gas to exit the exhaust port. For future study, simulation for exhaust pipeline study together with

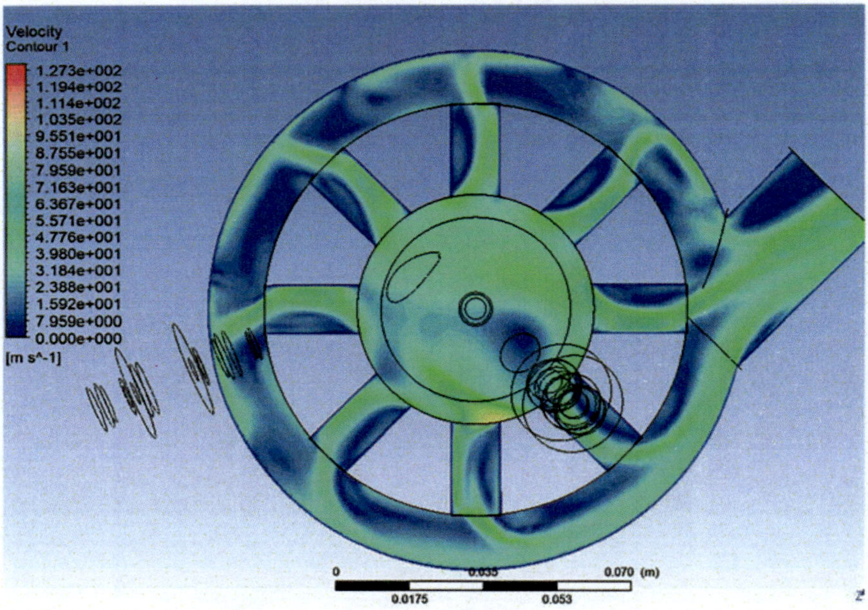

Fig. 5 Velocity Contour for second design of exhaust manifold

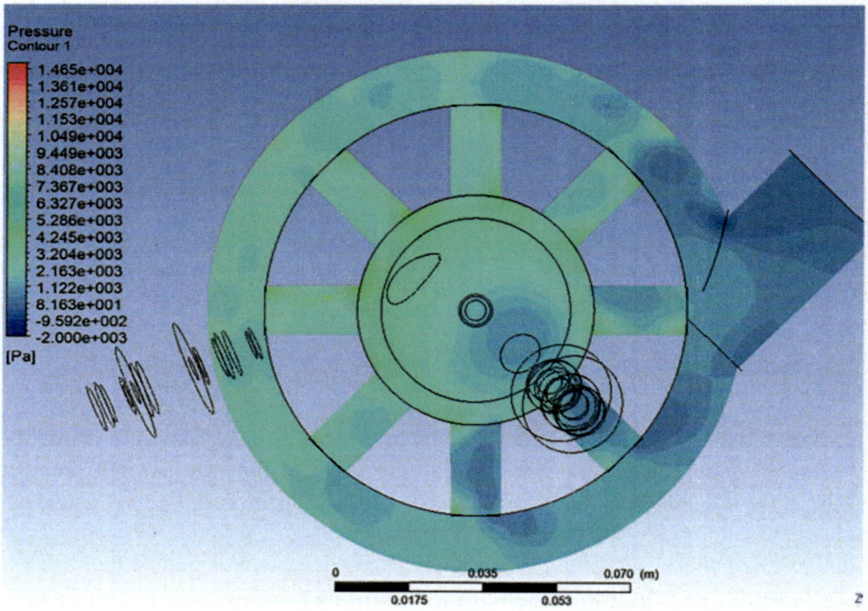

Fig. 6 Pressure Contour for second design of exhaust manifold

exhaust manifold can be run to get a much better combustion performance and CO_2 release for pipeline.

Acknowledgements The authors would like to thank Universiti Teknologi Petronas and Petroleum Research Fund (PRF) for their financial support under grant PRF 0153AB-A34 and facilities for the research.

References

1. Ezrann Z et al (2015) Effect of motoring voltage on compression ratio of a free-piston linear generator engine. J Mech Eng Sci 8:1393–1400
2. Flynn PC (2002) Commercializing an alternate vehicle fuel: lessons learned from natural gas for vehicles. Energy Policy 30(7):613–619
3. Kenneth Holmberg AE (2019) The impact of tribology on energy use and CO_2 emission globally an in combustion engine and electric cars. Tribol Int 135:389–396
4. Hanipah MR, Mikalsen R, Roskilly AP (2015) Recent commercial free-piston engine developments for automotive applications. Appl Therm Eng 75:493–503
5. Mikalsen R, Roskilly AP (2007) A review of free-piston engine history and applications. Appl Therm Eng 27(14):2339–2352
6. Mikalsen R, Roskilly AP (2008) Performance simulation of a spark ignited free-piston engine generator. Appl Therm Eng 28(14):1726–1733
7. Mao J et al (2011) Multi-dimensional scavenging analysis of a free-piston linear alternator based on numerical simulation. Appl Energy 88(4):1140–1152
8. Mikalsen R, Roskilly AP (2008) The design and simulation of a two-stroke free-piston compresion ignition engine for electrical power generation. Appl Therm Eng 28(5–6):589–600
9. Jia B, Smallbone A, Zuo Z, Feng H, Roskilly AP (2016) Design and simulation of a two- or four-stroke free-piston engine generator for range extender application. Energy Convers Manage 111:289–298
10. Jia B, Wang Y, Smallbone A, Roskilly AP (2018) Analysis of the scavenging process of a two-stroke free-piston engine based on the selection of scavenging ports or valves. Energies 11(324):1–14
11. Yuan C, Ren H, Xu J (2017) Comparison of the gas exchange of a loop scavenged free-piston engine alternator and the conventional engine. Appl Therm Eng 127:638–649

Probabilistic Hydrocarbon Release Leading to Fire and Explosion Incorporating System and Human Reliability Analysis: A Comparison of Onshore and Offshore Facilities

Hamizah Rozuhan, Masdi Muhammad, Usama Muhammad Niazi, and Ainul Akmar Mokhtar

Abstract Human support and the inevitable interaction necessarily play a major part in daily production and maintenance activities at onshore and offshore facilities. The impact of human reliability on performance of workers at offshore facility is higher compared to onshore facility, owing to the harsh offshore working environment. This study presents an integrated probability model that combines human and system reliabilities and utilises a Bayesian network (BN). A comparison was made between onshore and offshore facilities in terms of human reliability and probability of hydrocarbon release leading to fire and explosion in order to investigate the influence of human performance shaping factors (PSFs) of human error at each working environment. Particularly, this research uses probability values obtained from expert opinions in fire and explosion scenarios caused by hydrocarbon release in onshore and offshore facilities. The proposed integrated approach is applied to a case study on hydrocarbon storage tank at onshore and offshore facilities. Accordingly, three possible consequences were evaluated (i.e., probability of safe condition, hydrocarbon release and high pressure) that could lead to fire and explosion. The probabilities of consequences are compared between onshore and offshore working conditions. It has been observed that the probability of hydrocarbon release of at onshore is slightly higher compared to offshore working environment with a percentage difference of 0.032% and this can be neglected due to the small difference. Meanwhile, the influence of PSFs of human error are higher at offshore working condition. Therefore, it can be concluded that there is a higher influence of human PSF of stress, task complexity, training, experience and atmospheric factor of human error at offshore compared to onshore due to the harsh working environment.

Keywords Hydrocarbon release · Fire and explosion · Human reliability analysis · Onshore facility · Offshore facility and performance shaping factors

---

H. Rozuhan · M. Muhammad (✉) · U. Muhammad Niazi · A. Akmar Mokhtar
Mechanical Engineering Department, Universiti Teknologi Petronas, 32610 Bandar Seri Iskandar, Perak Darul Ridzuan, Malaysia
e-mail: masdimuhammad@utp.edu.my

F. Ahmad et al. (eds.), *ICPER 2020*, Lecture Notes in Mechanical Engineering,
https://doi.org/10.1007/978-981-19-1939-8_52

1 Introduction

In oil and gas field, the major cause of accidents is human error, contributing to over 70% of total number of accidents. At onshore and offshore facilities, hydrocarbon is stored and processed. A small mishap may lead to the release of hydrocarbon fuels, which can immediately escalate to major accidents, such as fire and explosion [1–5]. The Health and Safety Executive (HSE) indicted that over 70% of accidents on offshore facilities are caused by hydrocarbon releases [4]. Unsurprisingly, 75% of fire and explosion accidents are caused by human error [6–10].

Workers at offshore facility are exposed to tough environment and challenging conditions, physically and psychosocially. This working environment consists of long working hours with environmental and physical stressors at specific working conditions [11]. As an example, workers at offshore facilities are possibly exposed to confined working area, bad quality of air, hazardous chemicals, unpredictable vessel movement and vibration, erratic weather conditions, noise, and heavy workload [12–14]. These stressors have a significant influence in human reliability of workers to perform their job and to manage emergencies [15, 16]. Additionally, these stressors are likely to interrupt activities during relaxation time, reduce sleep quality and disrupt fatigue recovery [14, 17–19].

Apart from that, offshore workers are exposed to psychosocial stressors including inadequate activities during leisure time, remote living environment apart from family and friends, and limited privacy due to the need of sharing accommodation [11]. The quality and type of the activities done during relaxation time are important in resting and recovering period from working [20], and inadequate leisure activities after an extended working time might cause the accumulation of fatigue. Furthermore, the work-related stressors can possibly increase the impact of extended working time in a harsh working environment such as offshore facility. Owing to the harsh working environment at offshore facility, the probability of performing human error is higher compared to onshore facility due to physical and psychosocial human PSFs.

Several fire and explosion accidents at offshore work environment have occurred in the past, which resulted in numerous fatalities and caused significant pollution and damage to the environment [21–24]. The Piper Alpha disaster that occurred in 1988 caused 167 deaths owing to hydrocarbon release that led to fire and explosion [21]. In 2001, a series of explosion accidents occurred on the Petrobas P-36 facility, which resulted in several deaths [22]. In 2010, a fire and explosion accident occurred on the BP Deepwater Horizon semi-submersible mobile offshore drilling unit (MODU), which was caused by a blowout that resulted in 11 deaths and also caused over $61 billion in damage [23]. Amongst these offshore accidents, the Piper Alpha and Deepwater Horizon accidents are known as the most disastrous explosion accidents.

Sklet et al. [25] used the released offshore facility statistics as bases to classify the factors contributing to hydrocarbon releases that lead to fire and explosion accidents. The aforementioned study identified five main causes of hydrocarbon release: technical failures, design failures, external events, process upsets and human errors.

Hydrocarbon releases are generally not triggered by a single failure but by the confluence of mistakes. On offshore platforms, one of the primary reasons for hydrocarbon releases is the combination of process upsets and human errors [6].

Several surveys have been conducted and concluded that 75% of hydrocarbon releases that lead to fire and explosion accidents are due to human errors [6–10]. This percentage is attributed to the unpredictability of human behaviour because of the involvement of high uncertainties in human performance. Human errors play an important role when changes occur in working environments (e.g., harsh weather conditions) on offshore facilities. Accordingly, human reliability analysis (HRA) should be performed to minimise human errors and obviate system degradation. HRA focuses on the interaction between human and machine and integrates human factors into system safety analysis [26]. The tasks are related to how system performance is influenced by human performance, such as in operating any system or equipment, maintenance and repair of equipment, tasks related to safety procedures and analysis.

Some HRA methods that have been previously proposed and the majority of the methods are based on expert judgement techniques. The first-generation methods include the technique for human error rate prediction (THERP) by Swain and Guttmann [26] and success likelihood index methodology (SLIM) by Kirwan [27]. The second-generation of human reliability analysis methods include a technique for human error analysis (ATHEANA) by Cooper et al. [28] and cognitive reliability and error analysis method (CREAM) by Hollnagel [29]. THERP is used to estimate human reliability in quantitative and qualitative studies research [26] and SLIM is used in managing human error data deficiency [27]. Furthermore, ATHEANA and CREAM were developed by considering human–machine systems extended to human factors (i.e., psychologically, emotionally and behaviourally) in the human reliability research [28, 29]. The third-generation methods include the latest human reliability analysis approach, which is BN developed by Pearl [30]. In the 1980s, the BN approach was developed to ease the process of prediction in artificial intelligence analysis [30]. Pearl [31] and Nielsen and Jensen [32] presented a thorough conceptual background of BN.

Significant research has been conducted to quantify the probability of human error by employing the first-, second- and third-generation methods and applied in different areas, such as crude oil tanker, liquified petroleum gas (LPG) cargo, maintenance and shutdown inspection and offshore conditions. Akyuz et al. [33] presented a risk-based approach to quantify human error probability during the gas inerting operations of crude oil tankers. Akyuz et al. [34] predicted human error potential using CREAM for designated tasks during loading and discharge operations of LPG cargo that presents a significant hazard. Hameed et al. [35] proposed a novel risk-based methodology by introducing human error modeling during maintenance and shutdown inspection whilst modeling system failure.

Several recent studies have used third-generation methods (e.g., BN and dynamic Bayesian network (DBN)) to estimate the probability of accidents that integrate human and system reliabilities [36–43]. This causal network-based technique is commonly agreed to be a powerful tool to estimate the probability of an accident given the evidence of another accident in the same network [43]. Musharraf et al. [36]

proposed a quantitative approach to HRA for offshore emergency conditions using the BN approach, in which it represents dependency between human factors and related actions during muster action and the results were compared with SLIM. Cai et al. [37] proposed a quantitative risk assessment (QRA) of human errors owing to offshore blowouts by applying pseudo-fault tree analysis and DBN. Arzaghi et al. [39] developed a dynamic risk-based approach methodology for maintenance planning by monitoring the growth of fatigue cracks of subsea pipelines using an extension of BN to an influence diagram. BahooToroody et al. [40] developed a framework of an assessment of asset integrity, which incorporates a risk-based optimisation approach for maintenance scheduling as regards process variables (PVs) using an integration of DBN, Damage Modelling and sensitivity analysis. Abaei et al. [41] presented a safety framework using hierarchical Bayesian analysis to evaluate the likelihood of a vessel touching the seabed in shallow waters by considering the obtained results of dynamic under keel clearance. Abaei et al. [43] proposed a framework to predict human fatigue over time in offshore platforms by using DBN.

Significant research has been conducted to quantify the probability of human error in different areas by using BN. However, hydrocarbon releases leading to fire and explosion could be caused by the combination of process system reliability and human reliability. Past studies have addressed these issues separately. Therefore, these two factors should be integrated to establish a realistic probabilistic estimation of fire and explosion scenarios on onshore and offshore facilities. Hence, the current study proposes an integrated probabilistic framework by using BN for two major factors that lead to hydrocarbon release and subsequent fire and explosion scenario considering onshore and offshore facilities. A comparison is conducted between onshore and offshore facilities in order to investigate the influence of human PSFs of human error at each working environment.

2 Methodology

This study proposed a methodology to estimate process system reliability and integrate it with human reliability in the hydrocarbon release scenario leading to fire and explosion at onshore and offshore facilities. This proposed framework utilises BN to model the integration of process system and human reliabilities, including its PSF in an event of a hydrocarbon release.

2.1 Process System Reliability

Several safety systems are attached in hydrocarbon storage tank to prevent the release of hydrocarbon fuel. In case of high flow of hydrocarbon entering the unit, failure of safety systems could result in several potential accident scenarios, such as hydrocarbon release and high pressure in the system, which can lead to fire and explosion. BN is used to relate an initial abnormal event, such as high flow entering the unit to its different consequences (safe condition, hydrocarbon release and high pressure).

2.2 *Human Reliability*

In the first stage of human reliability analysis, the task that involves human intervention is identified. For example, if hydrocarbon overflows in a hydrocarbon storage tank, then the alarm goes off to notify the operator in charge to close the valve manually. In this case, human intervention happens when the operator operates the valve manually and human error is a possibility. The current study considers the maloperation of the valve during manual operations is adapted from Sklet [25]. For this task, a set of PSFs that influence human performance is identified. The determination of five PSFs is adapted from DiMattia [44]. PSFs determined in this study include stress (stress level whilst operating MV), task complexity (complexity or difficulty of operating MV), training (training level of the operator in charge), experience (experience level of the operator in operating MV), time available (the available time for the operator to complete MV operation) and atmospheric factor (the condition of working environment including weather and noise).

After the related PSFs are identified, two questionnaires were developed to collect data on the influence of each PSF and conditional probabilities of PSFs in a hydrocarbon release scenario owing to the maloperation of valves during manual operation. A total of 18 participants were asked to rate the influence (in terms of probability) of each PSF and conditional probabilities of PSFs in a hydrocarbon release scenario owing to the maloperation of valves during manual operation. The rating used a Likert scale that ranges from 1 to 10, where 1 represents the minimum influence and 10 represents the maximum influence. Islam et al. used the same technique to collect conditional probabilities by using expert opinion [45]. The questionnaire data were used as bases to calculate the average weight using Eq. (1). Given that the Likert scale data are from 1 to 10, the average weight was transformed into probabilities by changing it into a range of 0 to 1 by using Eq. (2). Weightage was given to the data based on years of working experience.

$$\text{Average weight} = \frac{\sum Likert\,scale\,data}{Number\ of\ participants} \tag{1}$$

$$\text{Probability} = \frac{Average\,weight}{10} \tag{2}$$

2.3 Integrated Approach of Process System Reliability and Human Reliability

Given that human intervention is involved in a process system, an integrated approach is implemented by combining human reliability with process system reliability using BN to compute overall reliability of the unit.

3 Results and Discussion

This section applies the proposed integrated approach of process system reliability and human reliability to a case study that considers a hydrocarbon storage tank with the related safety systems.

3.1 Application of the Proposed Methodology: Case Study

The current study adopts a case study of a storage tank of hazardous chemicals that was developed by Kalantarnia et al. [46] and utilised as a hydrocarbon storage tank at onshore and offshore facilities. The safety system of the tank comprises safety systems, such as basic process control (BPC), bypass line, high-level alarm (HLA), pressure safety valve (PSV) and manual valve (MV). In the process, the major risk is likely to be the excess flow of hydrocarbon fuel through input pipes. Figure 1 shows the illustration of the storage tank with a safety system.

In this process, BPC will detect the overflow of hydrocarbon coming into the system and it will open the bypass valve. Thus, the overflow will be transferred to the bypass line to exit the system. In case BPC fails to activate, the system triggers HLA to indicate that the operator in charge should close MV. Accordingly, closing MV will ensure that the excess flow of hydrocarbon is stopped from entering the storage tank. If the operator fails to close MV, PSV will be activated to release the

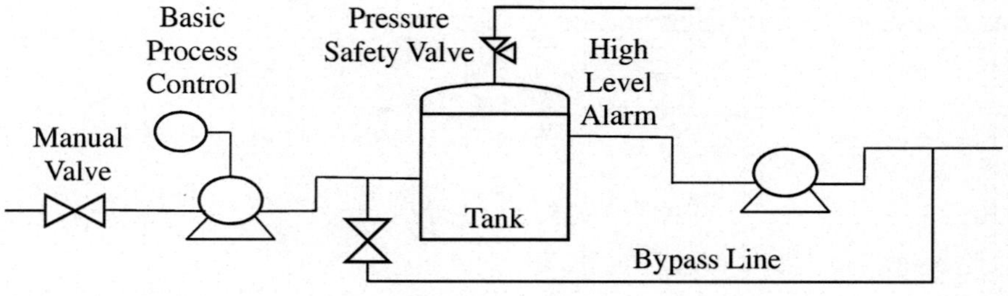

Fig. 1 Safety system in a hydrocarbon storage tank [46]

Table 1 Probability of consequence

Consequence	Probability of consequence
Safe condition	0.98732
Hydrocarbon release	0.01211
High pressure	0.00057

excess pressure. However, if PSV fails to activate, then high pressure will build up in the unit, thereby possibly resulting in the rupture of storage tanks and leading to fire and explosion.

The high flow rate of hydrocarbon entering the storage tank is set as the initial abnormal event. Three types of consequences of this event, namely, safe condition, hydrocarbon release and high pressure in the hydrocarbon storage tank, are plausible if the safety system fails to operate. Amongst these consequences, the release of hydrocarbon and high pressure are acknowledged as accident scenarios that could lead to fire and explosion.

3.1.1 Process Safety System Reliability of Storage Tank

Table 1 presents the calculated probabilities of consequence using BN. It shows that the probability of the system to be in a safe condition is 0.98732. Meanwhile, the probability of hydrocarbon release in the system is 0.01211 and the probability of high pressure in the tank is 0.00057. The results indicate that a high probability of a safe condition is obtained whenever all safety systems are activated. Thus, the probability of accidents, such as hydrocarbon release and high pressure, is low.

3.1.2 Human Reliability

The probabilities of internal PSFs (i.e., stress, experience and training) and external PSFs (i.e., time available, task complexity and atmospheric factors) that can influence the success or failure of human reliability in MV operation, as well as conditional probabilities are obtained using expert opinions by the means of questionnaires and are tabulated in Table 2, 3, 4, 5 and 6.

Based on Table 2, in terms of internal PSF, insufficient experience has the highest influence in terms of probability of failure of human reliability in operating MV with a probability of 0.582 (onshore) and 0.599 (offshore). Meanwhile, internal PSF high

Table 2 Probability of failure of the internal PSF of human reliability in operating manual valve

PSF	Stress	Experience	Training
Status	High	Insufficient	Insufficient
Probability (Onshore)	0.441	0.582	0.491
Probability (Offshore)	0.582	0.599	0.575

Table 3 Probability of failure of the external PSF of human reliability in operating manual valve

PSF	Time available	Task complexity	Atmospheric factor
Status	Low	High	Bad
Probability (Onshore)	0.643	0.341	0.262
Probability (Offshore)	0.604	0.489	0.438

stress has the lowest influence at onshore with a probability of 0.441 and internal PSF insufficient training has the lowest influence at offshore working condition with a probability of 0.575.

Based on Table 3, in terms of external PSF, low time available has the highest influence in terms of probability of failure of human reliability in operating MV with a probability of 0.643 (onshore) and 0.604 (offshore). Meanwhile, PSF bad atmospheric factor has the lowest influence at both onshore and offshore working environment with a probability of 0.262 and 0.438 respectively.

Based on Tables 2 and 3, it can be observed that the influence of PSFs such as high stress, insufficient experience, insufficient training, high task complexity and bad atmospheric factor are much higher at offshore compared to onshore working environment.

In using BN, CPT should be defined for the particular task of MV operation by consulting experts. CPT values are obtained from a questionnaire participated in by experts with working experiences of 10 years and above. Three CPTs are used, namely, CPT of internal PSF, CPT of external PSF and CPT of MV operation. Tables 4, 5 and 6 show CPTs obtained. Table 4 shows the CPT in terms of dependency of overall internal PSF on each internal PSF, whereas Table 5 shows the CPT in terms of dependency of overall external PSF on each external PSF. Table 6 shows the CPT in terms of dependency of human reliability on internal and external PSFs.

Table 4 shows that a maximum conditional probability value of 0.962 of poor internal PSF is obtained when training is insufficient, experience is insufficient, and stress is high. Table 5 presents a maximum conditional probability value of 0.912 for poor external PSF when atmospheric factor is bad, task complexity is high and time available is low.

Table 4 Conditional probability table of the internal PSF of human reliability in operating manual valve

Training	Insufficient				Sufficient			
Experience	Insufficient		Sufficient		Insufficient		Sufficient	
Stress	Low	High	Low	Low	Low	Low	Low	Low
Internal factors (poor)	0.787	0.962	0.625	0.737	0.662	0.812	0.287	0.562
Internal factors (good)	0.212	0.037	0.375	0.262	0.337	0.187	0.712	0.437

Table 5 Conditional probability table of the external PSF of human reliability in operating manual valve

Atmospheric factors	Bad				Good			
Task complexity	Low		High		Low		High	
Time available	Low	High	Low	High	Low	High	Low	High
External factors (poor)	0.712	0.637	0.912	0.77	0.562	0.387	0.762	0.6
External factors (good)	0.287	0.362	0.087	0.22	0.437	0.612	0.237	0.4

Table 6 CPT of human reliability in operating manual valve

Internal factors	Bad		Good	
External factors	Bad	Good	Bad	Good
Human reliability (failure)	0.900	0.800	0.650	0.362
Human reliability (success)	0.100	0.200	0.350	0.638

Table 6 shows that a maximum value of human reliability failure probability of 0.900 is obtained when the internal and external PSF are bad. Meanwhile, a human reliability success probability of 0.638 is obtained when internal and external PSF are good.

3.1.3 Integrated Bayesian Network of Process System Reliability and Human Reliability

Given that human intervention directly affects the operation of MV, it is incorporated in the process system using BN and probabilities of consequences are updated. Figures 2 and 3 present the integrated BN at onshore and offshore respectively.

Figures 2 and 3 show that the nodes of the internal and external PSFs are linked to the human reliability node. In terms of probability of failure and success, human reliability can be calculated. To integrate human reliability into safety system reliability, the human reliability node is linked to the MV node. By linking these nodes together, the probability of the MV node is updated and a new probability of consequences can be estimated.

Based on Figs. 2 and 3, it can be observed that there is a higher probability of hydrocarbon release of 0.03146 at onshore compared to 0.03114 at offshore working environment with a percentage difference of 0.032%. Table 7 shows a comparison between the probability of consequence (safe condition, hydrocarbon release and high pressure) in process system reliability and the probability of consequence of onshore and offshore integrated approach reliabilities.

Table 7 presents the comparison of the results of safety system reliability without human reliability from the past study by Kalantarnia [46] with onshore and offshore integrated approach reliabilities. Based on onshore integrated reliability, the probability of having a safe condition has decreased to 0.96706 (2.024% decrement) after

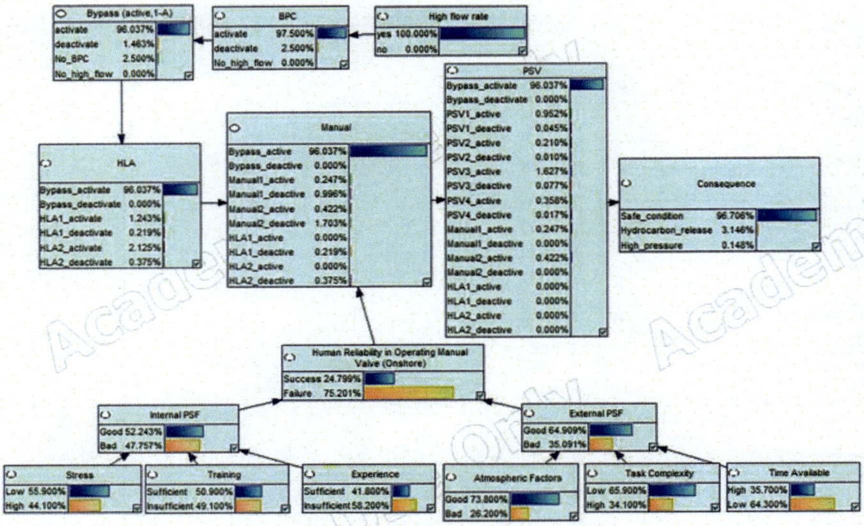

Fig. 2 BN of the integrated approach (Onshore)

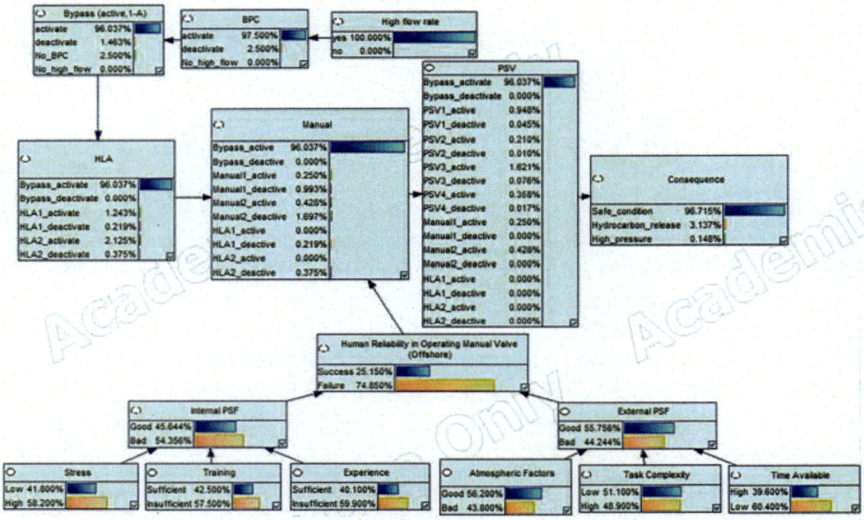

Fig. 3 BN of the integrated approach (Offshore)

Table 7 Comparison of the probability of consequence of process system reliability with onshore and offshore integrated approach reliabilities

Consequence	Probability of consequence (process system reliability without human reliability)	Probability of consequence (onshore integrated approach reliability)	Probability of consequence (offshore integrated approach reliability)
Safe condition	0.9873	0.96706	0.96715
Hydrocarbon release	0.0121	0.03146	0.03114
High pressure	0.0006	0.00148	0.00148

human reliability is considered in the MV operation. Meanwhile, the probability of having hydrocarbon release and high pressure has increased to 0.03146 (1.936% increment) and 0.00148 (0.088% increment), respectively.

Based on offshore integrated reliability, the probability of having a safe condition has decreased to 0.96715 (2.015% decrement) after human reliability is considered in the MV operation. Meanwhile, the probability of having hydrocarbon release and high pressure has increased to 0.03114 (1.904% increment) and 0.00148 (0.088% increment), respectively. It can be observed that there is only a small difference of 0.032% in the probability of hydrocarbon release of onshore and offshore integrated approach reliabilities, and thus, it can be neglected.

Integrating human reliability into system reliability provides a significant effect by contributing 2% of decrement of safe condition probability and increasing the probability of hydrocarbon release scenario by 1.9%. Thus, human reliability with its related PSFs in system reliability analysis should be considered to obtain a better reliability estimation of hydrocarbon release scenario that can result in an event of fire and explosion.

4 Conclusion

This study presents an integrated approach that relates process system reliability with human reliability for fire and explosion scenario at onshore and offshore facility. The probability of process upsets is high and safety system failure can contribute to hydrocarbon release in an oil and gas facility. Human reliability also contributes to hydrocarbon release.

The application of the proposed approach is demonstrated using a case study of a hydrocarbon storage tank at onshore and offshore facility and considering the high flow of hydrocarbon in the tank. Three consequences (i.e., safe condition, hydrocarbon release and high pressure) are estimated using BN. A comparison of the proposed approach of integrating human and safety system reliability with process system reliability shows a significant reduction of 2% of probability of a safe condition, an increase of 1.9% in hydrocarbon release and an increase of 0.088% in high pressure after human reliability is introduced into the process system reliability analysis. However, there is only a small difference of 0.032% in the probability of hydrocarbon release of onshore and offshore work environment, and thus, it can be neglected.

An integrated BN approach that combines process system reliability and human reliability ensures a better estimation of fire and explosion accidents that could help in minimising the occurrence of such accidents, thereby preventing fire and explosion due to hydrocarbon release at onshore and offshore facilities. It can be observed that the influence of human PSFs such as high stress, insufficient experience, insufficient training, high task complexity and bad atmospheric factor are much higher at offshore compared to onshore working environment. Therefore, it can be concluded that human PSFs has a significant influence on the human reliability especially at a harsh working environment such as an offshore facility.

Acknowledgements The authors would like to acknowledge the financial support provided by the Universiti Teknologi PETRONAS.

Funding This study was supported by Universiti Teknologi PETRONAS through its YUTP Grant Scheme (Grant Number: 0153AA-E53).

References

1. Das B, Weinberg M (2012) Improving flammable mass estimation for vapor cloud explosion modeling in an offshore QRA. Saf Sci 50:1218–1227
2. Kang K-Y, Choi K-H, Choi J, Ryu Y, Lee J-M (2016) Dynamic response of structural models according to characteristics of gas explosion on topside platform. Ocean Eng 113:174–190
3. Paik JK, Czujko J, Kim BJ, Seo JK, Ryu HS, Ha YC et al (2011) Quantitative assessment of hydrocarbon explosion and fire risks in offshore installations. Mar Struct 24:73–96
4. HSE (2005) Accident Statistics for Floating Offshore Units on the UK Continental Shelf (1980–2003)
5. Marsh GC (2018) The 100 largest losses 1978–2017–large property damage losses in the hydrocarbon industry. London
6. Ren J, Jenkinson I, Wang J, Xu D, Yang J (2008) A methodology to model causal relationships on offshore safety assessment focusing on human and organizational factors. J Safety Res 39:87–100
7. Bryant D (1991) The human element in shipping casualties. Report prepared for the Dept. of Transport, Marine Directorate, United Kingdom
8. Rothblum AM (2000) Human error and marine safety. National Safety Council Congress and Expo, Orlando, FL
9. Islam R, Khan F, Abbassi R, Garaniya V (2018) Human error assessment during maintenance operations of marine systems–what are the effective environmental factors? Saf Sci 107:85–98
10. Cormier PJ (1994) Towing vessel safety: analysis of congressional and coast guard investigative response to operator involvement in casualties where a presumption of negligence exist
11. Parkes K Human factors, health and safety in the offshore oil/gas industry
12. Haward BM, Lewis CH, Griffin MJ (2009) Motions and crew responses on an offshore oil production and storage vessel. Appl Ergon 40:904–914
13. Gardner R (2003) Overview and characteristics of some occupational exposures and health risks on offshore oil and gas installations. Ann Occup Hyg 47:201–210

14. Oldenburg M, Hogan B, Jensen H-J (2013) Systematic review of maritime field studies about stress and strain in seafaring. Int Arch Occup Environ Health 86:1–15
15. Stevens SC, Parsons MG (2002) Effects of motion at sea on crew performance: a survey. Mar Technol 39:29–47
16. Pisula P, Lewis C, Bridger R (2012) Vessel motion thresholds for maintaining physical and cognitive performance: a study of naval personnel at sea. Ergonomics 55:636–649
17. Hansen JH, Holmen IM (2011) Sleep disturbances among offshore fleet workers. a questionnaire-based survey. Int Marit Health 62:123–130
18. Lauridsen Ø, Tronsmoen S, Berland J, Gitlesen J, Ringstad A, Pedersen T, et al (1991) Shift-work and health. Phillips Petroleum Company/Rogaland Research, Stavanger, Norway
19. Oldenburg M, Baur X, Schlaich C (2010) Occupational risks and challenges of seafaring. J Occup Health 1007160136–1007160136
20. Sonnentag S, Zijlstra FR (2006) Job characteristics and off-job activities as predictors of need for recovery, well-being, and fatigue. J Appl Psychol 91:330
21. Paté-Cornell ME (1993) Learning from the piper alpha accident: a postmortem analysis of technical and organizational factors. Risk Anal 13:215–232
22. Atherton J, Gil F (2008) Incidents that define process safety. Wiley Online Library
23. Dadashzadeh M, Abbassi R, Khan F, Hawboldt K (2013) Explosion modeling and analysis of BP deepwater horizon accident. Saf Sci 57:150–160
24. PGJ (2015) FPSO explosion off brazil claims 5 lives. Pipeline Gas J 14–14
25. Sklet S (2006) Hydrocarbon releases on oil and gas production platforms: release scenarios and safety barriers. J Loss Prev Process Ind 19:481–493
26. Swain AD, Guttmann HE (1983) Handbook of human-reliability analysis with emphasis on nuclear power plant applications. Final report. Sandia National Labs
27. Kirwan B (1994) A guide to practical human reliability assessment. CRC Press
28. Cooper S, Ramey-Smith A, Wreathall J, Parry G (1996) A technique for human error analysis (ATHEANA). Nucl Regulat Commission
29. Hollnagel E (1998) Cognitive reliability and error analysis method (CREAM). Elsevier
30. Pearl J (1998) Bayesian networks. In: The handbook of brain theory and neural networks. pp 149–153
31. Pearl J (1988) Probabilistic reasoning in intelligent systems: networks of plausible inference. Morgan Kaufmann
32. Nielsen T, Jensen F (2009) Bayesian networks and decision graphs. Springer
33. Akyuz E (2015) Quantification of human error probability towards the gas inerting process on-board crude oil tankers. Saf Sci 80:77–86
34. Akyuz E, Celik M (2015) Application of CREAM human reliability model to cargo loading process of LPG tankers. J Loss Prev Process Ind 34:39–48
35. Hameed A, Khan F, Ahmed S (2016) A risk-based shutdown inspection and maintenance interval estimation considering human error. Process Saf Environ Prot 100:9–21
36. Musharraf M, Hassan J, Khan F, Veitch B, MacKinnon S, Imtiaz S (2013) Human reliability assessment during offshore emergency conditions. Saf Sci 59:19–27
37. Cai B, Liu Y, Zhang Y, Fan Q, Liu Z, Tian X (2013) A dynamic Bayesian networks modeling of human factors on offshore blowouts. J Loss Prev Process Ind 26:639–649
38. Bhandari J, Abbassi R, Garaniya V, Khan F (2015) Risk analysis of deepwater drilling operations using Bayesian network. J Loss Prev Process Ind 38:11–23
39. Arzaghi E, Abaei MM, Abbassi R, Garaniya V, Chin C, Khan F (2017) Risk-based maintenance planning of subsea pipelines through fatigue crack growth monitoring. Eng Fail Anal 79:928–939
40. BahooToroody A, Abaei MM, Arzaghi E, BahooToroody F, De Carlo F, Abbassi R (2019) Multi-level optimization of maintenance plan for natural gas system exposed to deterioration process. J Hazard Mater 362:412–423
41. Abaei MM, Arzaghi E, Abbassi R, Garaniya V, Javanmardi M, Chai S (2018) Dynamic reliability assessment of ship grounding using Bayesian Inference. Ocean Eng 159:47–55

42. Leoni L, BahooToroody A, De Carlo F, Paltrinieri N (2019) Developing a risk-based maintenance model for a natural gas regulating and metering station using Bayesian network. J Loss Prev Process Ind 57:17–24
43. Abaei MM, Abbassi R, Garaniya V, Arzaghi E, Toroody AB (2019) A dynamic human reliability model for marine and offshore operations in harsh environments. Ocean Eng 173:90–97
44. DiMattia D (2004) Human error probability index for offshore platform musters. PhD Thesis, Dalhousie University, Halifax, Nova Scotia. See electronic files PRAC# 88_AppendixA1. doc and PRAC# 88_AppendixA2. doc," Development of a human error probability index for offshore platform evacuations
45. Islam R, Khan F, Abbassi R, Garaniya V (2018) Human error probability assessment during maintenance activities of marine systems. Saf Health Work 9:42–52
46. Kalantarnia M, Khan F, Hawboldt K (2009) Dynamic risk assessment using failure assessment and Bayesian theory. J Loss Prev Process Ind 22:600–606

Comparative Analysis of Lubrication Oil Age Prediction Model

Najat Mohammad Nazari and Masdi Muhammad

Abstract Quality of lubrication oil will impact the performance of equipment. Lubrication oil properties are being monitored periodically to ensure the quality of the oil is always good. Currently, oil change activity is conducted in time-based manner based on engine manufacturer recommendation. Therefore, lubrication oil will be discarded even though it is still useful. The idea of this paper is to consider multiple variables to assess the quality of lubrication oil as a higher number of variables are expected to give a more accurate prediction. In this study, multiple regression and artificial neural network (ANN) model were compared by assessing the R squared value and prediction error when predicting lubrication oil age. Spearman's correlation was applied to the lubrication oil analysis data to assess the relationship between lubrication oil age with oil analysis parameters and identify the parameters that are highly correlated with oil age. Total base number (TBN), oxidation, iron (Fe), lead (Pb) and zinc (Zn) were identified as parameters that were strongly correlated with oil age. Multiple regression and ANN were applied to predict the oil age using these parameters as the predictor variables. Both models were compared based on its R squared value and prediction error namely mean square error (MSE) and mean absolute deviation (MAD). Multiple regression presented a better prediction accuracy with higher R squared value of 0.9249 and lower prediction error. However, the P value for the model were more than 0.05 which may be due to the multicollinearity that exist between the independent variables. The R squared value for ANN is considerably high with value of 0.758, which proved its ability to predict the desired oil age.

N. Mohammad Nazari (✉) · M. Muhammad
Mechanical Engineering Department, Universiti Teknologi PETRONAS, 32610 Seri Iskandar, Perak, Malaysia
e-mail: najat_19001016@utp.edu.my

F. Ahmad et al. (eds.), *ICPER 2020*, Lecture Notes in Mechanical Engineering,
https://doi.org/10.1007/978-981-19-1939-8_53

1 Introduction

Lubricant condition monitoring (LCM) is one of the important condition monitoring techniques applied to a scheduled or unscheduled maintenance activity. It reveals the particles in the fluid representing the mechanical wear of the machine. When implemented correctly, it can provide great information about the machine in operation. Lubricant reduces or eliminate wear between rotating members and aid the reduction of heat and friction which is essential to ensure proper operability of equipment. LCM plays a vital role in maintaining the operating conditions of machinery and plants. Maintenance personnel will have access to gather information on equipment and lubricant's condition and state through this monitoring activity [1]. The importance of LCM in maintenance decision support validate the importance for an accurate and reliable remaining useful life (RUL) framework to be established [2].

A substantial amount of oil is wasted annually because of premature and unnecessary oil changes conducted. Oil condition should always be known for optimization and extension of oil drain interval without sacrificing the life of the equipment. Users or maintenance personnel are usually interested in two output from predictive maintenance. They are the condition of the oil and the time that oil requires changes. Current practice is lack of practicality and usefulness when implemented in the industry [3]. The Engine manufacturers (OEM) pre-determined oil change intervals were designed to deliver maximum engine protection when operating under a wide variety of conditions. Despite the majority of equipment owners that comply with these guidelines, there is a rise in trend to extend the oil change intervals beyond OEM recommendation [4]. Oil change based on a standard interval is a conservative approach and results in inefficient usage of oil [5]. Oil condition measured by mileage and time intervals does not reflect the existing condition of oil or accurate level of decay [6]. The assessment of remaining useful life (RUL) on previous researches were based on a few or a combination of limited lubricant characteristics [5].

Correlation analysis is a technique used extensively in identifying relationships in data [7]. Regression correlation assessment among variables and several traditional clustering methods were implemented in the selection of Fe and Pb applicability as material wear indicators to estimate soft failure time point and optimize coefficient for hard time preventive maintenance [8]. In [9], the authors stated that statistical method specifically correlation-regression analysis to be a relevant approach to study the process taking place in lubricating oil due to the complexity of physiochemical processes of oil and their interactions. Statistical analysis was applied for different purposes including development of predictive maintenance algorithm to classify analysis, multiple regression analysis and factor analysis [10]. Comparison between the ability of multiple regression and ANN was conducted in [11] to predict the spreadable quality of Gouda cheese where ANN models were proven to be more accurate. Linear regression (LR) and ANN were applied in [12] to predict engine performance. The authors proved that LR modelling approach was able to predict the desired parameters, however ANN gave a more accurate result.

This study aims to assess multiple regression and artificial neural network (ANN) models by comparing the prediction error of predicted oil age from both models. The data used in this research were from the available oil analysis report from the routine lubrication oil analysis activities of rotating equipment installed at offshore. The machine learning application for predictive model namely the statistical approach and ANN were applied. Correlation analysis was conducted to identify the relationship between oil age with the oil analysis test parameters and to identify the parameters that have high correlation with oil age. Multiple regression and ANN were applied for prediction of the lubrication oil age and the results were compared. The models were evaluated based on its coefficient of determination (R squared) and prediction error expressed in terms of mean square error (MSE) and mean absolute deviation (MAD).

2 Research Methodology

2.1 Data Collection

In this study, the prediction of the lubrication oil age for gas engine that is driving the pump was conducted. Data was taken from the available lubrication oil analysis reports where the analysis were conducted every 2 months. Among the information that were extracted from the oil analysis report are the kinematic viscosity (KV) @ 100 °C, water content (WC), total base number (TBN), soot, oxidation (Oxi), glycol (Gly), contaminants and elemental analysis which include wear metal elements and additive metals. Figure 1 shows the sample of oil analysis report.

2.2 Correlation Analysis

A Spearman's correlation analysis was conducted to evaluate the relationship between oil age and lubrication oil analysis test parameters [13]. The type of correlation can be categorized into three types whose values range from −1 to 1. A positive correlation indicates the variables are changing in the same direction, negative correlation indicates the variables are changing in the opposite direction and no correlation indicates there is no linear relationship between the variables [7, 14]. A correlation coefficient of zero indicates that there is no relationship between variables, values between 0 and 0.3 implies a weak relationship, values of 0.3–0.7 suggests moderate relationship and correlation coefficient of 0.7–1 signifies strong relationship [15].

DIAGNOSTIC & RECOMMENDATIONS | **EQUIPMENT PRIORITY LEVEL**

No abnormality. Oil is fit for further use until next sampling (Aug '17).

SAMPLE DETAILS						
Sample Rating		1	2	3	4	5
Sampling Date		16/12/16	25/2/17	23/3/17	22/4/17	26/6/17
Testing Date		27/12/16	3/3/2017	6/4/2017	28/4/17	7/5/2017
Date of Last Oil Change			6/2/2017			3/24/2017

TEST RESULTS							
Criteria	Units	1	2	3	4	5	Fresh Oil
Appearance	-	Dark	Dark	Dark	Dark	Dark	Clear
Kinematic viscosity @100°C	cSt	13.7	14.0	14.3	14.3	14.2	14.5
Water Content	ppm	647.0	485.0	881.0	532.0	694.0	366.0
Total Base Number (TBN)	mgKOH/g	12.5	12.4	11.8	12.4	10.6	13.5
Insolube Test (Soot)	%	0.0	0.0	0.0	0.0	0.0	
Oxidation	abs/0.1 m	8.8	9.1	10.8	9.1	14.0	6.5
Glycol		0.8	0.8	0.7	0.5	0.4	3.9

CONTAMINANTS						
Test	Elements	1	2	3	4	5
Contaminant (ppm)	Sodium	0	0	0	0	1
	Potassium	0	0	0	0	0
	Silicon	1	0	1	0	4
	Boron	91	81	92	86	63
	Vanadium	0	0	0	0	0

WEAR METALS ELEMENTS						
Test	Elements	1	2	3	4	5
Elemental Analysis (ppm)	Iron	1	2	2	2	31
	Copper	2	1	1	1	43
	Aluminum	1	1	1	0	4
	Tin	0	0	0	0	0
	Lead	0	1	1	0	5
	Antimony	3	11	3	6	5
	Chromium	0	0	0	0	1
	Nickel	0	0	0	0	0
	Moly	0	2	1	3	0

ADDITIVES METALS						
Elemental Analysis (ppm)	Zinc	240	237	252	248	282
	Magnesium	2	2	2	2	2
	Calcium	313	318	376	307	377
	Barium	0	0	0	0	0
	Phosporus	206	238	213	257	215

Fig. 1 Oil analysis report

2.3 Regression Analysis

Regression analysis enables us to find the line and corresponding equation that offers the best fit to our dataset. Oil age was set as the dependent variable ($\hat{Y}$) and the lubrication oil analysis test parameters that are strongly correlated with oil age were assigned as the independent variables. The regression equation is given by Eq. (1):

$$\hat{Y} = a + b_1 X_1 + b_2 X_2 + \cdots + b_i X_i \tag{1}$$

In Eq. (1),

$\hat{Y}$ Observed score on the dependent variable

a Intercept

b_i Slope of independent variables

X_i Score on independent variables.

The values that were observed were the R squared value and P value. The goodness of fit of the regression model was measured by the results of R squared which range from 0 to 1. It is the proportion of variation of $\hat{Y}$ variable that is explained by the predictors [15, 16]. Predictor with low P value shows that predictor selected is significant towards the model. Changes in predictor's value are related to changes in the response variable.

2.4 *Artificial Neural Network*

ANN comprised of a few simple and highly interconnected processing elements just like neurons in the brain. It can be interpreted as a black box that consist series of equations to calculate the output with the given inputs [12]. ANN is a useful tool for function approximation, classification and prediction [17]. A regular neural network comprised of n neurons in the input layer, m neurons in the hidden layer and i neurons in output layer [18].

Bayesian Regularization backpropagation algorithm was used in this analysis. Backpropagation neural network is a supervised learning technique and has been widely used. It is the most popular, effective and easy to learn model [19]. The residual sum of squares and sum of squared weights of Bayesian Regularization (BR) minimized estimation errors to reach a good generalized model. The model also obtains a lower MSE compared to other algorithms [20]. Figure 2 shows sample of ANN model which comprised of input layer, hidden layer and output layer.

Fig. 2 ANN model

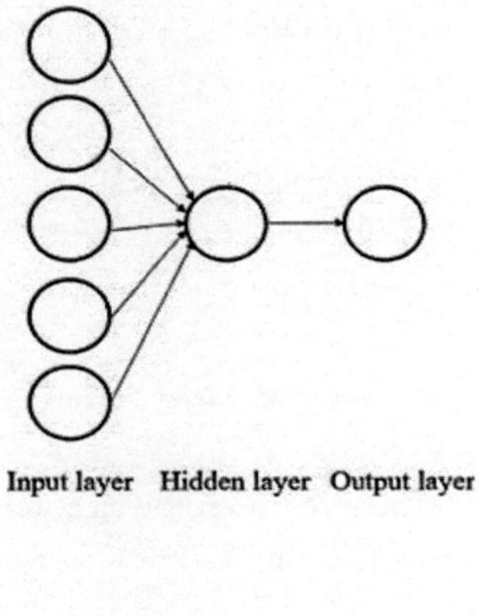

2.5 Model Evaluation

The goodness of fit for the multiple regression and ANN model were evaluated using performance indicators. Performance indicators used to compare both models were the coefficient of determination (R squared), mean square error (MSE) and mean absolute deviation (MAD) [12, 17, 21]. The equation for MSE and MAD are as in Eqs. (2) and (3) respectively.

$$\mathrm{MSE} = \frac{\sum_{i=1}^{n}(\mathrm{Y}_{\mathrm{actual}} - \mathrm{Y}_{\mathrm{predicted}})^2}{\mathrm{n}} \quad (2)$$

$$\mathrm{MAD} = \frac{\sum_{i=1}^{n}|\mathrm{Y}_{\mathrm{actual}} - Y_{\mathrm{predicted}}|}{\mathrm{n}} \quad (3)$$

In Eqs. (2) and (3),
Y_{actual} Actual oil age
$\mathrm{Y}_{\mathrm{predicted}}$ Predicted oil age
n Number of samples.

3 Results and Discussion

3.1 Correlation Coefficient

The longest duration of oil used for this equipment was 244 Days. Correlation between oil age with parameters from oil analysis was conducted to observe the relationship between them and identify lubrication oil parameters that have high correlation with oil age. The results of the correlation between the parameters is shown in Table 1.

Based on the guidelines and referring to the results presented in Table 1, high correlation between oil age with TBN, oxidation, iron (Fe), lead (Pb) and zinc (Zn) can be can be observed with correlation coefficient of −0.944, 0.922, 0.758, 0.705 and 0.751 respectively. These parameters will be included in the models of subsequent analysis as predictors for the oil age prediction. The results also presented that high correlation exist between the predictors variable selected.

3.2 Regression Analysis

Multiple regression was conducted between the oil age with TBN, oxidation, Fe, Pb and Zn. Table 2 shows the results of multiple regression between these parameters.

Table 1 Correlation between oil age oil analysis parameters

Variable	Oil age	KV	WC	TBN	Oxi	Gly	AOD	Na	K	Si	B	V	Fe	Cu
Oil age	1	0.505	0.341	−0.94	0.922	−0.47	0.617	0.419	0.266	0.631	−0.5	0.044	0.758	0.611
KV	0.505	1	0.386	−0.5	0.419	−0.04	0.334	0.186	−0.12	0.359	−0.32	0.127	0.327	0.106
WC	0.341	0.386	1	−0.2	0.217	0.139	0.045	−0.31	−0.64	0.672	−0.15	−0.38	0.279	−0.01
TBN	−0.944	−0.5	−0.2	1	−0.98	0.523	−0.67	−0.51	−0.41	−0.61	0.602	−0.07	−0.81	−0.72
Oxi	0.922	0.419	0.217	−0.98	1	−0.48	0.705	0.505	0.408	0.592	−0.71	0.07	0.823	0.792
Gly	−0.471	−0.04	0.139	0.523	−0.48	1	−0.43	−0.75	−0.55	−0.16	0.187	0.381	−0.27	−0.55
AOD	0.617	0.334	0.045	−0.67	0.705	−0.43	1	0.636	0.615	0.072	−0.4	0.134	0.32	0.578
Na	0.419	0.186	−0.31	−0.51	0.505	−0.75	0.636	1	0.739	0.019	−0.39	0.036	0.316	0.67
K	0.266	−0.12	−0.64	−0.41	0.408	−0.55	0.615	0.739	1	−0.29	−0.21	0.334	0.116	0.442
Si	0.631	0.359	0.672	−0.61	0.592	−0.16	0.072	0.019	−0.29	1	−0.39	−0.21	0.768	0.435
B	−0.5	−0.32	−0.15	0.602	−0.71	0.187	−0.4	−0.39	−0.21	−0.39	1	−0.2	−0.62	−0.74
V	0.044	0.127	−0.38	−0.07	0.07	0.381	0.134	0.036	0.334	−0.21	−0.2	1	−0.11	−0.07
Fe	0.758	0.327	0.279	−0.81	0.823	−0.27	0.32	0.316	0.116	0.768	−0.62	−0.11	1	0.728
Cu	0.611	0.106	−0.01	−0.72	0.792	−0.55	0.578	0.67	0.442	0.435	−0.74	−0.07	0.728	1
Al	0.359	0.037	−0.05	−0.51	0.596	−0.21	0.503	0.547	0.461	0.351	−0.62	0.059	0.667	0.815
Pb	0.705	0.256	−0.11	−0.8	0.817	−0.69	0.784	0.885	0.678	0.324	−0.56	0.044	0.633	0.863
Sb	−0.192	−0.29	−0.72	0.184	−0.15	0.183	0.053	0.26	0.435	−0.53	−0.01	0.628	−0.2	0.105
Cr	−0.262	−0.4	−0.14	0.175	−0.11	0.381	−0.52	−0.31	−0.15	0.18	−0.26	0.319	0.106	0.053

(continued)

Table 1 (continued)

Variable	Oil age	KV	WC	TBN	Oxi	Gly	AOD	Na	K	Si	B	V	Fe	Cu
Ni	−0.446	0.127	−0.02	0.359	−0.36	0.536	−0.34	−0.31	−0.15	−0.21	−0.02	0.319	−0.27	−0.54
Mo	−0.066	−0.28	−0.27	0.197	−0.19	0.029	−0.08	−0.31	0.108	−0.54	0.259	0.23	−0.51	−0.39
Zn	0.751	0.297	0.085	−0.76	0.713	−0.56	0.412	0.533	0.436	0.551	−0.36	−0.04	0.728	0.644
Mg	−0.374	−0.17	−0.18	0.134	−0.1	0.309	−0.02	−0.21	0.076	−0.13	−0.03	−0.03	0.028	−0.11
Ca	0.099	−0.13	0.234	−0.09	0.047	0.301	−0.36	−0.44	−0.26	0.468	0.182	−0.16	0.459	−0.15
P	0.126	−0.35	−0.08	−0.06	0.066	0.254	−0.32	−0.52	−0.12	0.041	0.093	0.026	0.223	−0.15

Variable	Al	Pb	Sb	Cr	Ni	Mo	Zn	Mg	Ca	P
Oil age	0.359	0.705	−0.19	−0.26	−0.45	−0.07	0.751	−0.37	0.099	0.126
KV	0.037	0.256	−0.29	−0.4	0.127	−0.28	0.297	−0.17	−0.13	−0.35
WC	−0.05	−0.11	−0.72	−0.14	−0.02	−0.27	0.085	−0.18	0.234	−0.08
TBN	−0.51	−0.8	0.184	0.175	0.359	0.197	−0.76	0.134	−0.09	−0.06
Oxi	0.596	0.817	−0.15	−0.11	−0.36	−0.19	0.713	−0.1	0.047	0.066
Gly	−0.21	−0.69	0.183	0.381	0.536	0.029	−0.56	0.309	0.301	0.254
AOD	0.503	0.784	0.053	−0.52	−0.34	−0.08	0.412	−0.02	−0.36	−0.32
Na	0.547	0.885	0.26	−0.31	−0.31	−0.31	0.533	−0.21	−0.44	−0.52
K	0.461	0.678	0.435	−0.15	−0.15	0.108	0.436	0.076	−0.26	−0.12
Si	0.351	0.324	−0.53	0.18	−0.21	−0.54	0.551	−0.13	0.468	0.041
B	−0.62	−0.56	−0.01	−0.26	−0.02	0.259	−0.36	−0.03	0.182	0.093
V	0.059	0.044	0.628	0.319	0.319	0.23	−0.04	−0.03	−0.16	0.026
Fe	0.667	0.633	−0.2	0.106	−0.27	−0.51	0.728	0.028	0.459	0.223

(continued)

Table 1 (continued)

Variable	Al	Pb	Sb	Cr	Ni	Mo	Zn	Mg	Ca	P
Cu	0.815	0.863	0.105	0.053	−0.54	−0.39	0.644	−0.11	−0.15	−0.15
Al	1	0.711	0.2	0.217	−0.16	−0.61	0.577	0.214	0.121	−0.16
Pb	0.711	1	0.138	−0.23	−0.45	−0.36	0.677	−0.13	−0.22	−0.3
Sb	0.2	0.138	1	0.287	−0.05	0.216	−0.1	−0.15	−0.31	0.053
Cr	0.217	−0.23	0.287	1	0.319	−0.11	−0.13	0.192	0.297	0.332
Ni	−0.16	−0.45	−0.05	0.319	1	−0.11	−0.44	0.536	0.175	−0.04
Mo	−0.61	−0.36	0.216	−0.11	−0.11	1	−0.26	−0.31	−0.25	0.473
Zn	0.577	0.677	−0.1	−0.13	−0.44	−0.26	1	−0.34	0.263	0.044
Mg	0.214	−0.13	−0.15	0.192	0.536	−0.31	−0.34	1	0.331	0.09
Ca	0.121	−0.22	−0.31	0.297	0.175	−0.25	0.263	0.331	1	0.627
P	−0.16	−0.3	0.053	0.332	−0.04	0.473	0.044	0.09	0.627	1

Table 2 Multiple regression analysis between oil age with TBN, oxidation, Fe, Pb and Zn

	Estimate	SE	tStat	pValue
(Intercept)	1388.2	681.98	2.0355	0.081249
TBN	−104.77	37.718	−2.7777	0.027387
Oxidation	−22.059	13.529	−1.6306	0.14701
Fe	−0.66959	1.1597	−0.5774	0.58176
Pb	2.8756	3.2815	0.87629	0.40993
Zn	0.47975	0.65044	0.73758	0.48476

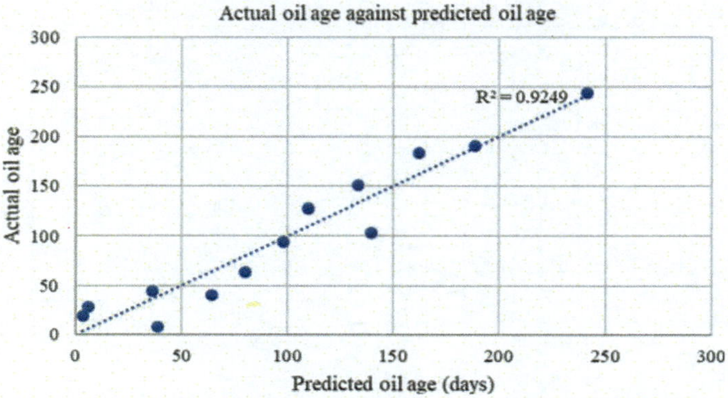

Fig. 3 Graph of actual oil age against predicted oil age from multiple regression analysis

The model has an R squared of 0.925, adjusted R squared of 0.871 and P value of more than 0.05. Equation (4) shows the regression equation for this multiple regression analysis constructed using the coefficients from Table 2. Figure 3 shows the graph of the actual oil age against predicted oil age based on the multiple regression analysis.

$$\hat{Y} = 1388.2 - 104.77X_1 - 22.0591X_2 - 0.66959X_3 + 2.8756X_4 + 0.47975X_5 \quad (4)$$

In Eq. (4),
$\hat{Y}$ Oil age
X_1 Score on TBN
X_2 Score on oxidation
X_3 Score on Fe
X_4 Score on Pb
X_5 Score on Zn.

3.3 Artificial Neural Network Model

The neural network comprised of five neurons (TBN, oxidation, Fe, Pb, Zn) in the input layer and one neuron (oil age) in the output layer. The number of neurons in the hidden layer was selected using try and error method with a forward approach. The training and testing of the neural network were conducted starting with a small number of neurons in the hidden layer. The procedure was repeated by increasing the number of neurons until training and testing is improved [22, 23]. Figure 4 shows the plot of R value against the number of neurons. Based on the graph, it can be observed that the R value is maximum with two neurons in the hidden layer. Therefore, the optimum number of neurons in the hidden layer for this model were two with an R value of 0.87061. Figures 5 and 6 shows the final ANN model and the graph of the actual oil age against the predicted oil age with an R squared value of 0.758 respectively.

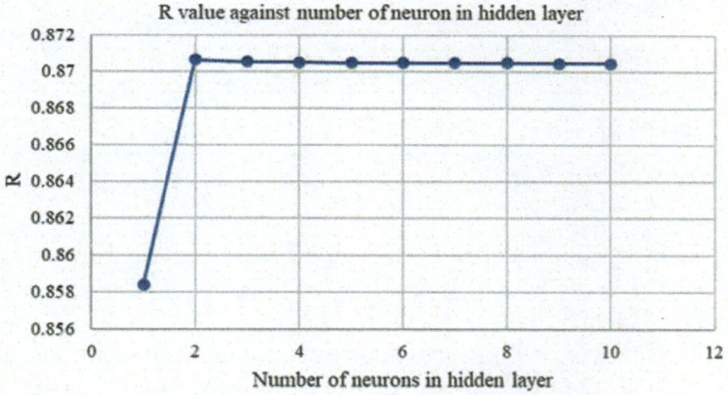

Fig. 4 Graph of R value against number of neurons in hidden layer

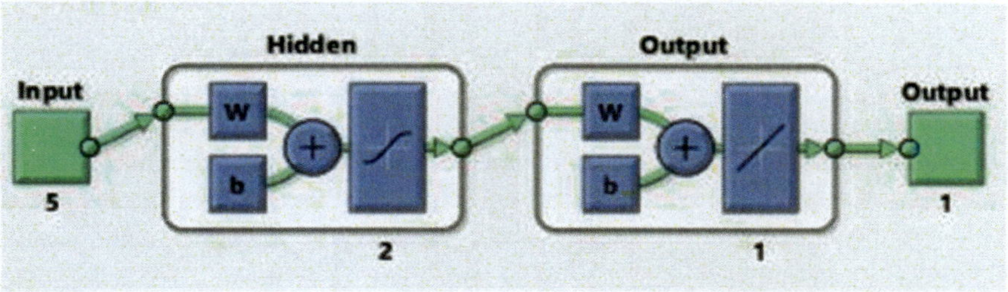

Fig. 5 Final ANN model

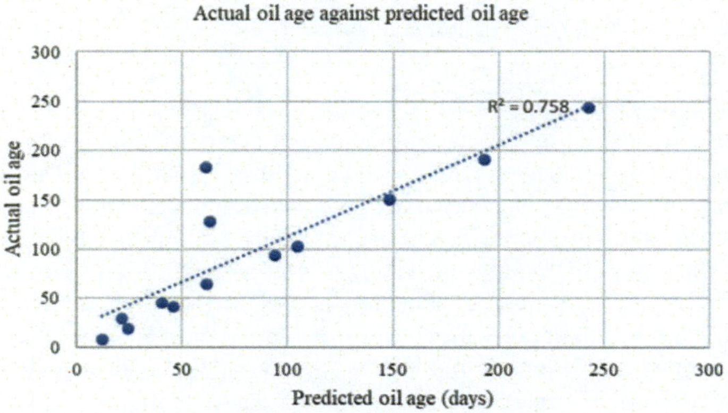

Fig. 6 Graph of actual oil age against predicted oil age from ANN model

Table 3 Model evaluation

Analysis	R squared	MSE	MAD
Multiple regression	0.9249	388.17	16.91
ANN	0.758	1481.52	17.15

3.4 Model Evaluation

The multiple regression model and ANN model were evaluated by comparing its R squared value and prediction error expressed in terms of MSE and MAD using Eqs. (2) and (3). Table 3 summarizes the results.

The results show that multiple regression has a higher R squared value of 0.9249, lower MSE and MAD of 388.17 and 16.91 respectively. The R squared value for ANN is considerably high with value of more than 0.7. Therefore, ANN has proven to be able to predict the desired lubrication oil age.

4 Conclusion

The primary aim of this paper is to compare the prediction between multiple regression and ANN when predicting the oil age. Spearman's correlation analysis was applied to the data to evaluate the relationship between the parameters and identify lubrication oil parameters that have high correlation with oil age. Five parameters of lubrication oil namely TBN, oxidation, Fe, Pb and Zn were identified to have a high correlation with oil age. Multiple regression and ANN analysis were applied to predict the oil age using these parameters and the models were compared based on its R squared, MSE and MAD. Multiple regression presented a better prediction accuracy with higher R squared value and lower prediction error. However, the R squared value for ANN is considerably high which proved that it is able to predict the desired oil age.

Moving forward, the analysis on prediction of RUL for lubrication oil will be conducted. Preliminary results of this study suggest further research is required in this domain for the development of RUL prediction model.

Acknowledgements The authors would like to acknowledge the financial support provided by the Universiti Teknologi PETRONAS.

References

1. Wakiru J, Pintelon L, Chemweno P, Muchiri P (2017) Analysis of lubrication oil contamination by fuel dilution with application of cluster analysis. XVII international science conference industrial system (Serbia), pp 252–257
2. Tanwar M, Raghavan N (2020) Lubrication oil remaining useful life prediction using multi-output Gaussian process regression IEEE. Access 8:128897–128907
3. Tič V, Tašner T, Lovrec D (2014) Enhanced lubricant management to reduce costs and minimise environmental impact. Energy 77:108–116
4. Extended Oil Drain Intervals In: Filter manufacturers council 2010
5. Sharma BC, Gandhi OP (2006) RUL assessment of lube oil using AHP and vector projection approach industrial lubrication and tribology 58(4):187–194
6. Hashim H, Haron MA, Osman FN, Al Junid SAM, Md Idros MF, Wan Nawang WN (2010) Artificial neural networks based algorithm for identifying engine oil parameters. 2nd international conference on computational intelligence, communication systems and Networks (Liverpool), pp 244–249
7. Kumar S, Chong I (2018) Correlation analysis to identify the effective data in machine learning: prediction of depressive disorder and emotion states. Int J Environ Res Public Health 15(12):2907
8. Vališ D, Žák L, Pokora O (2015) System condition estimation based on selected Tribodiagnostic data qual. Reliab. Engng Int 32:635–645
9. Sharma GK, Chawla OP (1988) Modelling of lubricant oil alkalinity in diesel engines Tribology Int 21(5):269–274
10. Idros MF, Abdul Razak AH, Al Junid SAM, Suliman AI, Halim AK (2017) FPGA implementation of predictive degradation model for engine oil lifetime IOP conference series: material science and engineering, vol 341, p 012024
11. Stangierski J, Weiss D, Kaczmarek A (2019) Multiple regression models and artificial neural network (ANN) as prediction tools of changes in overall quality during the storage of spreadable processed gouda cheese European food research and technology, vol 245, pp 2439–47
12. Tosun E, Aydin K, Bilgili M (2016) Comparison of linear regression and artificial neural network model of a diesel engine fuelled with biodiesel-alcohol mixtures. Alex Eng J 55:3081–3089
13. Wakiru J, Pintelon L, Muchiti P, Chemweno P (2020) A Data mining approach for lubricant-based fault diagnosis. J Qual Maint Eng
14. Gogtay NJ, Thatte UM (2017) Principles of correlation analysis. J Assoc Physicians India 65:78–81
15. Ratner B (2009) The correlation coefficient: its values range between +1/−1, or Do they? J Target Meas Anal Mark 17:139–142

16. Multiple regression NCSS Statistical Software. https://ncss-wpengine.netdna-ssl.com/wp-content/themes/ncss/pdf/Procedures/NCSS/Multiple_Regression.pdf
17. Ul-Sufie AZ, Yahya AS, Ramli NA, Hamid HA (2011) Comparison between multiple linear regression and feedforward back propagation neural network models for predicting PM10 concentration level based on gaseous and meteorological parameters Int. J Appl Sci Technol 1(4):42–49
18. Souza AVD, Neto AB, Piazentin JC, Dainese Junior BJ, Gomes EP, Bonini CDSB, Putti FF (2019) Artificial neural network modelling in the prediction of Bananas' harvest. Sci Hortic 257:108724
19. Sapna S, Tamilasari A, Kumar MP (2012) Backpropagation learning algorithm based on Levenberg Marquardt algorithm. The 4th international conference workshop on computer networks & communications, pp 393–98
20. Kayri M (2016) Predictive abilities of Bayesian regularization and Levenberg-Marquardt algorithms in artificial neural networks: a comparative empirical study on social data. Math Comput Appl 21(20):1–11
21. Sheela KG, Deepa SN (2013) Review on methods to fix number of hidden neurons in neural networks mathematical problems in engineering
22. Panchal FS, Panchal M (2014) Review on methods of selecting number of hidden nodes in artificial neural network Int. J Comput Sci Mobile Comput 3(11):455–464
23. Abd Majid MA, Soomro Afzal A, Akmar A (2019) Artificial neural network modelling approach for assessment of stratified thermal energy storage tank ARPN. J Eng Appl Sci 14(19):3371–3378

Growth of NGVs and Comparative Study of Cylinder Material for CNG Storage

Mohammad Azeem, Hamdan Haji Ya, Mohammad Azad Alam, Md Rehan Sadique, Mazli B Mustapha, Ainul Akmar Bin Mokhtar, Tauseef Ahmed, Mohamed Thariq Hameed Sultan, and Rehan Khan

Abstract In the present paper, the growth of Natural Gas Vehicles (NGVs) and the various pressure vessels used for storing compressed natural gas (CNG) is discussed. Also, the comparison of Von-mises stress in pressure vessels (PVs) due to internal pressure with different materials has been investigated numerically. Substantial use of natural gas is recorded as an alternative fuel, and the high-pressure storage of natural gas has resulted in using various types of cylinders in automobiles. Metal cylinders are extensively used in NGVs and are fabricated with different materials to satisfy services, safety and related factors. Metals and their alloys are often employed to make type I cylinders. It is important to calculate the failure pressure limit for a particular geometry of the tank against the yield strength of the material in case of metallic cylinders. An axisymmetric finite element model was made using the ABAQUS finite element platform to mimic the stress levels for two materials, viz steel alloy, and an aluminium alloy. The internal test pressure and geometric details of the cylinder were kept constant. The yield stress of material and numerical results have been compared. It was found that the steel alloy has outperformed the aluminium alloy for the given thickness of the cylinder. In contrast, the aluminium alloy failed to sustain the test pressure of 25 MPa.

M. Azeem (✉) · H. Haji Ya · M. Azad Alam · M. B. Mustapha · A. Akmar Bin Mokhtar
Mechanical Engineering Department, Universiti Teknologi PETRONAS, Seri Iskandar, Malaysia
e-mail: mazeem.me@zhcet.ac.in

M. R. Sadique
Department of Civil Engineering, Aligarh Muslim University, Aligarh, India

M. T. H. Sultan
Aerospace Manufacturing Research Centre (AMRC), Faculty of Engineering, University Putra Malaysia, Selangor, Malaysia

T. Ahmed
Özyeğin University, Mechanical Engineering Department, Orman Street, Turkey

R. Khan
Department of Mechanical Engineering, College of Electrical and Mechanical Engineering, National University of Sciences and Technology, Islamabad, Pakistan

F. Ahmad et al. (eds.), *ICPER 2020*, Lecture Notes in Mechanical Engineering,
https://doi.org/10.1007/978-981-19-1939-8_54

1 Introduction

High-pressure gas storage was always a significant concern with increasing industrialisation and technical advancements. Metallic pressure vessels have been increasingly used for storing and transporting gases at high pressure. Metal cylinders are extensively used in fuel storage in automobiles like NGVs and Liquid Petroleum Gas (LPG) storage. A rapid increment of using cylinders in Compressed Natural Gas (CNG) powered vehicles for storing gas is reported. CNG has presently qualified as a suitable replacement for gasoline and Diesel fuel in terms of availability, cost, greenhouse gas emissions, and pollution. The pricing of CNG is about 50% less than petrol/gasoline and Diesel, and is nontoxic and lighter than air. It is around 20–29% less CO_2 emissions in CNG-powered vehicles than petrol or Diesel combustion engines [1–3]. Type I CNG storage tanks are the most common in CNG-powered automobiles in emerging economies. However, given the total growth rate for future NGV sales, the expected growth rate of 7% for vessels of type II, III, and IV indicates a rising market for composite, manufacturing, and process technologies. The composite pressure vessels eliminate gasoline fuel systems and lighten the vehicle. For example, type III and IV containers might lower the gas containment system's weight for standard transit buses or commercial vehicles by over 1000 lb (454 kg). This weight decrease would enhance fuel-efficiency, load transfer capability, and operational benefits. For type III and type IV, the weight reduction reaches 0.45 kg/Liter. However, type III and type IV are nearly twice as expensive as type II and 3.5 times as expensive as type I [4].

1.1 Background

Several methods and policies need to be implemented to achieve the objective of reducing pollutants and greenhouse gas emissions. The introduction of low CO_2 emitting fuels such as hydrogen and biofuels, as well as the introduction of electric automobiles in an environment where power plants create electricity from renewable sources such as solar and wind generators, will play a critical role. However, due to several technological and economic restrictions, solutions will only help to reduce pollution in the medium to long term, whereas it is critical to establish efforts that can deliver immediate impacts. As a result, researchers believe that CNG is one of the finest short-term alternatives [5]. NGVs have proven to be a clean and cheaper alternative to petrol and Diesel powered vehicles, with 15–20% reduced CO_2 [6]. In addition, the NGVs reported minor wear in the engine than the gasoline-driven engines. Therefore, various nations and companies are promoting the use of NGVs. High strength materials like steel alloys, aluminium alloys, and filament wound composite cylinders are commonly used for making CNG storage tanks. Steel cylinders have been used mainly in NGVs since the 1940s as type I cylinders. Steel alloy tanks are cost-effective and have required strength but are heavier than their other material counterparts. Steel has a good record in bonfire testing and melting. Tanks made of aluminium alloys deliver lightweight and easy to manufacture with required mechanical properties [7]. The metallic type I tanks are, however, prone to corrosion and fatigue.

1.2 Distribution

Many firms are now manufacturing high-pressure gas storage tanks and distribution systems. The CNG distribution to the consumers is done via various modules like underground pipe networks, tube trailers, steel-based gas containment systems, metallic and FRP vessels. Interstate or Intrastate pipe network is one of the efficient methods of distributing fuel gas. Trailers equipped with multiple large tanks, also known as virtual pipeline trailers, can be used during interruptions and maintenance of underground pipelines. In addition, plenty of onboard CNG fuel systems are available that are equipped with lightweight CNG tanks, well suited for cars, trucks and buses etc.

1.3 Types of CNG Tanks

We now have several types of CNG storage tanks starting from the earlier metallic tanks to composite overwrapped pressure vessels depending upon the material and construction configuration. CNG cylinders come in various sizes and weights to suit a variety of applications. The worldwide CNG tank market may be divided into type I, type II, type III, and type IV tank variants. Metals, mainly steel and aluminium, are used to construct type I tanks. The type II tank has a thinner steel liner that is overwrapped by fibre in the hoop direction. Only the middle of the tank's metal liner is wound by composite materials like glass or carbon fibre with a matrix, in type II. In type III tanks, a metal liner is fully wrapped/encased by the fibre resin system in the hoop and helical winding configurations. In the structural architecture of type IV tanks, a plastic liner is used fully overwrapped by fibre-resin layers, which reduces its weight to a greater extent. A new variant type V CNG tank would be released soon for CNG vehicles. Generally, cylinder prices rise as cylinder weight lowers. Table 1 shows the classification of different types of CNG tanks commercially available in the market. CNG has gained acceptance as a automotive fuel alternatives due to its availability, cost features and various advantages, Fig. 1. The all metal tank, type I is shown in Fig. 2 and the installation of CNG tanks on the vehicle chassis is shown in Fig. 3.

Table 1 Types of CNG cylinders

Type	Material/configuration
Type I	Full metal construction made of Steel/Aluminium/alloys
Type II	Metal liner hoop wrapped with Carbon/Glass fibre etc.
Type III	Metal liner fully wrapped with Carbon/Glass fibre etc.
Type IV	Plastic liner fully wrapped with Carbon/Glass fibre etc.

Fig. 1 Features of CNG as a fuel [2]

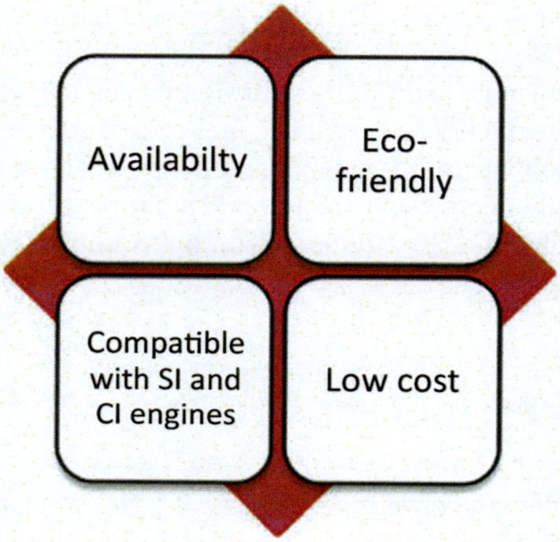

Fig. 2 All metal tanks [8]

Fig. 3 Tank installation
Source Galus Australis

2 CNG Statistics

Several countries are adopting CNG as an alternative to conventional fuels. As a result, an increasing number of vehicles are now manufactured are CNG driven. All medium and heavy-duty trucks, buses, and transport trailers now have CNG kit installation in various capacities viz 30–450 Liters for type I to IV. Type I usually are of low capacity, while as we go towards type IV, the volume and operating pressures increases. Generally, the rated pressure for CNG tanks is around 3600 psi or 248 bar. The following data refers to the NGV statistics. Tables 2 and 3 show the global and regional growth of NGVs.

There is an increasing demand for CNG-driven vehicles due to their availability, cost, eco-friendliness etc. We now have more than 28 million NGVs worldwide

Table 2 Worldwide NGV statistics[a] [9]

NGVs/fueling stations	Number
Natural Gas Vehicles	28,540,819
Natural Gas Fueling Stations	33,383

[a]*Data updated 31 December 2019*

Table 3 Regional growth of NGVs and refuelling stations[a] [9]

Region	NGVs	Gas Fueling Stations
Asia–Pacific	20,473,673	20,275
Europe	2,062,621	5,194
North America	224,500	1,856
Latin America	5,484,676	5,848
Africa	295,349	210

[a]*Data updated 31 December 2019*

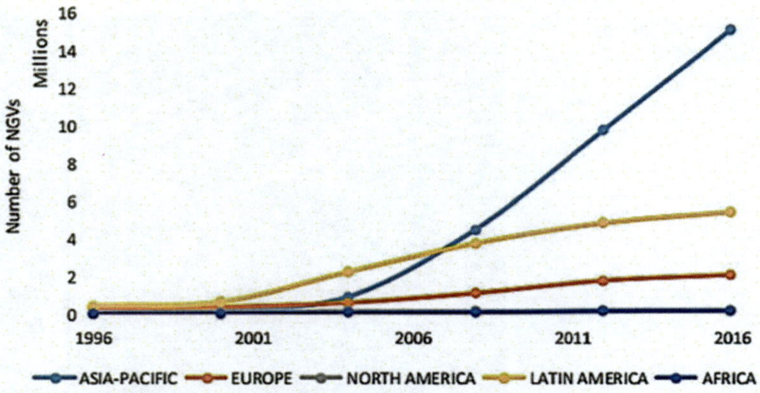

Fig. 4 Regional NGVs development [13]

with more than 33 thousand gas refuelling stations. Alternative energy buses are being extensively promoted worldwide, with several advantageous policies being issued. For example, the European Union has granted several favorable tax policies to promote alternative fuel buses, resulting in NG buses replacing a huge number of regular diesel buses [10]. Furthermore, buses that run on natural gas have been aggressively marketed in China, and they already make up a significant fraction of the fleet in several cities [11]. Beginning in the late 1990s, several Asian nations, including India, China, and Pakistan, marked considerable NGV growth [12]. Asia-Pacific alone comprises more than 20 million NGVs recorded until 2019. Figure 4 shows the regional NGV growth globally (until 2016). The Asia–Pacific region shows an enormous surge in NGV rise since 2006 while Africa and North America each added more than 0.2 million vehicles. Latin America accounts for more than 5 million NGVs, while Europe has revealed more than 2 million NGVs. European statistics are shown in Fig. 5. Italy is a major NGV handler than the other five countries shown in the graph.

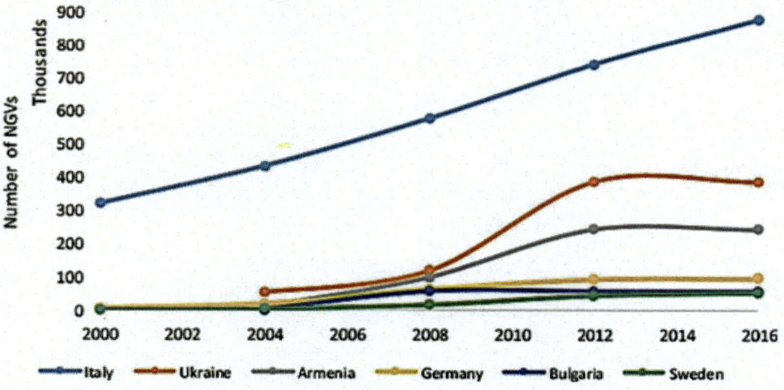

Fig. 5 NGVs rise in six European countries [13]

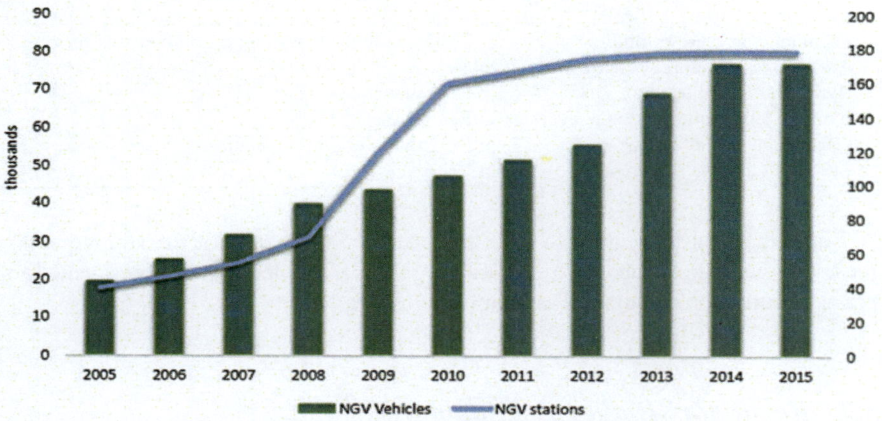

Fig. 6 Malaysian Statistics for NGVs and refueling stations [14]

2.1 *CNG Malaysia*

In Malaysia, CNG was introduced in the late 1990s for taxicabs and airport limousines, and now the road users are showing interest in switching to CNG power. Malaysia has revealed a steady increment in accepting NGVs, as in Fig. 6. Until 2017 Malaysia constitutes around 77,000 NGVs on-road, majorly taxis, with 178 stations for refueling. The public bus transport in Putrajaya is CNG-powered [14]. The national brand Proton in Malaysia provides CNG fittings in some of its models. Hyundai distributor in Malaysia is also offering CNG-powered vehicles. Other worldwide manufacturers like Ford Motor, Volkswagen, General Motors, Honda, and Nissan are the top five CNG-powered vehicle market vendors. The government of Malaysia is encouraging NGV program by providing incentives, reducing road tax, and relaxation in conversion kits [1].

3 Tests and Standards

3.1 *Tests for CNG Tanks*

Various tests are formulated for pressure vessel testing to ensure its durability in different loadings and environments. In addition, some review tests on pressure tanks are also performed after a few years, which are likely to be used multiple times. Following is the list of the tests done on CNG storage vessels [15].

· Pneumatic pressure cycling · Ambient temperature pressure cycling · Hydrostatic burst testing · Extreme temperature pressure cycling · Permeation or leak testing · Accelerated stress rupture	· Damage tolerance testing: drop/impact damage and flaw · Exposure to chemicals · Fire testing · Penetration or gunfire testing · UV exposure

Some material tests are also done to determine the health of the storage tanks; these tests include Sulphide Stress Cracking (SSC), characteristics of permeation of plastics, corrosion calculation of coatings etc. [8].

3.2 Regulatory Bodies and Standards

Various governments have devised regulations for using and transporting high-pressure cylinders. Multiple tests are recommended to assess the reliability of the pressure vessel. The names of the regulating agencies are listed below. Apart from national rules, independent testing agencies such as Underwriters Laboratories (UL), a global safety certification company in Illinois, also provide approved safety testing services. Table 4 enlists various standards and approval agencies

ISO states that the standards define only the minimum requirements for CNG tanks for on-board storage as a fuel. The unforeseen external loadings such as impact and damage due to vehicle crash are not included [16].

4 Finite Element Analysis

The numerical modelling and analysis are presented in this section, and the results are discussed. Finite element software ABAQUS/Standard (SE) is employed to find

Table 4 The common standards followed in major countries for CNG tanks [8]

Country	Standards	Approval authority/agency
Australia	AS 2030.1, NZS 5454	Work Cover Authority
Canada	CSA B51 P.2	Technical Standards & Safety Authority
European Union Countries	ECE/ONU Reg.110	Competent Authorities
India	ISO 11439	P.E.S.O. of India (Formerly Dept. of Explosives)
USA	FMVSS, ANSI NGV 2–2007	Independent Inspection Authorities

out the stress variation in the cylinder wall of the internally pressurised pressure vessel. Two different materials are considered for comparison.

4.1 Geometry

The geometry of the model in the analysis corresponds to the real CNG storage tank. A 60 Liters metallic CNG storage cylinder is considered, as shown in Fig. 7.

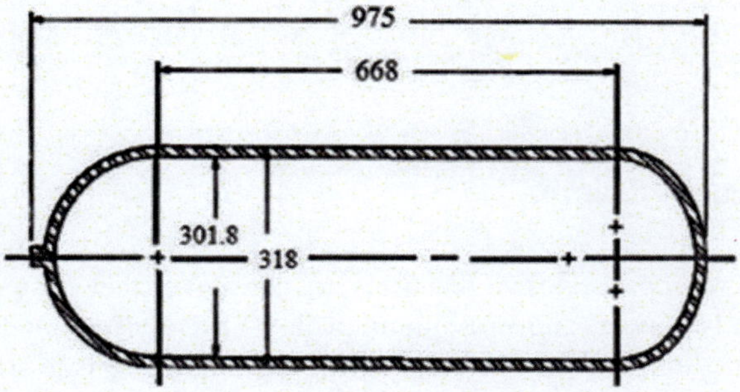

Fig. 7 Faber™ all metal cylinder [17]

Italian faber™ firm manufactures such cylinders. CNG storage vessels typically have a cylindrical shape with semi-spherical domes covering both ends. This geometry ensures the equal distribution of stresses safeguarding maximum possible safety. One-fourth part is made in the analysis due to symmetry (Fig. 11).

Equation (1) shows the circumferential stress or hoop stress σ_θ, and the longitudinal stress σ_z. The radial stress being nullified according to thin vessel theory [18], where p is the internal pressure, D is the internal diameter and t is the wall thickness of the tank. Figure 8 shows the elemental representation and the effective stresses on it.

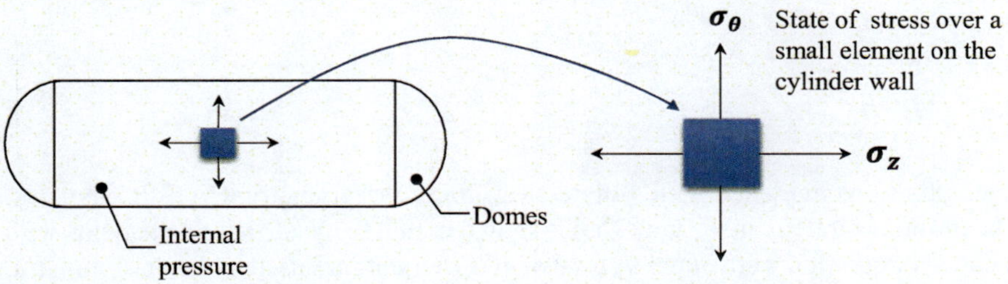

Fig. 8 Biaxial state of stress on an element in an internally pressurized thin cylinder

Table 5 Elastic properties of the materials under analysis

S.no	Material	Youngs modulus E (GPa)	Poissons ratio υ	Yield stress σ_y (MPa)
1	4340 Alloy Steel [19]	200	0.29	710
2	Al206.0-T7 Aluminium alloy [20]	70	0.33	350

$$\begin{aligned}\sigma_\theta &= \sigma_1 = pD/2t \\ \sigma_z &= \sigma_2 = pD/4t \\ \sigma_r &= \sigma_3 \sim 0\end{aligned} \tag{1}$$

4.2 Material Definition

For stress levels comparison, two materials are chosen: a steel alloy generally used for making type I cylinders and an aluminium alloy. Steel 4340 is a nickel-chromium-molybdenum medium-carbon steel alloy with outstanding tensile strength, fatigue resistance, and toughness. It has a range of applications in the military, including ground vehicles and military aircraft, commercial aircraft landing gear, automotive applications, rocket motor casings, pressure vessels etc [21]. Table 5 shows the elastic properties of the materials.

4.3 Pressure Load

Generally, the pressure rating of CNG tanks is around 25 MPa, hence a uniform 25 MPa (250 bar) is applied as a test pressure to the inner surface of the cylinder in the LOAD module, as in Fig. 9. The pressure is assumed to be uniform. The non-linearity is not considered in this analysis.

4.4 Mesh

The cylinder is modelled with 4-node biquadratic axisymmetric CAX4R elements. The meshing strategy in the tank wall is shown in Fig. 9. Quadrilateral elements with reduced integration were opted. A total of 400 elements are used in the axisymmetric 1/4 part.

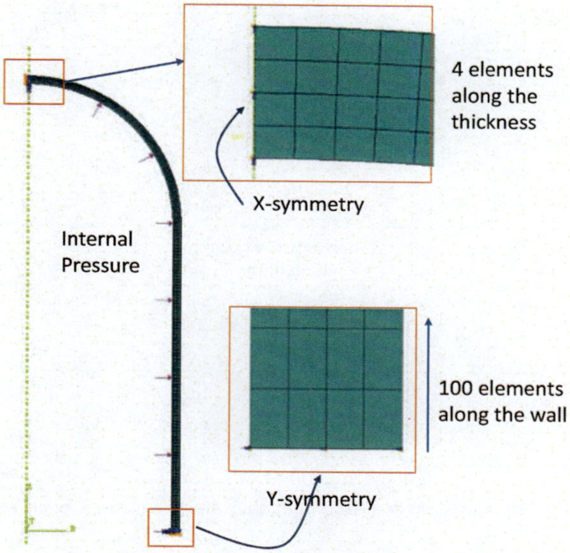

Fig. 9 Axisymmetric ¼ part model of CNG tank in FE solver, internal pressure on the inner surface, and meshing strategy

5 Results and Discussion

This section compares the FE results and theoretical stress levels reached in the cylinder wall.

5.1 *Hoop and longitudinal stress*

The theoretical values of the hoop and longitudinal stresses can be computed from equation (1) as $\sigma_1 = pD/2t = 465.7$Mpa and $\sigma_2 = pD/4t = 232.8$Mpa. Correspondingly, the probe values on an element on the surface of PV for the two stresses are picked, as shown in Fig. 10. The numerical results value is in good correlation with the theoretical calculations.

5.2 *Comparison on yield criterion*

Von-mises stress distribution was considered for the comparison and failure criteria for the materials under analysis. The yielding happens when the Von-mises stress, σ_v equals the yielding stress of the material, according to the Von-mises theory of failure. The theoretical Von-mises stress is shown by Eq. (2) and (3) defined for triaxial and biaxial stress loading. In the case of the internally pressurised thin cylinder, plane stress conditions exist, making negligible stress in the third direction i.e. across the

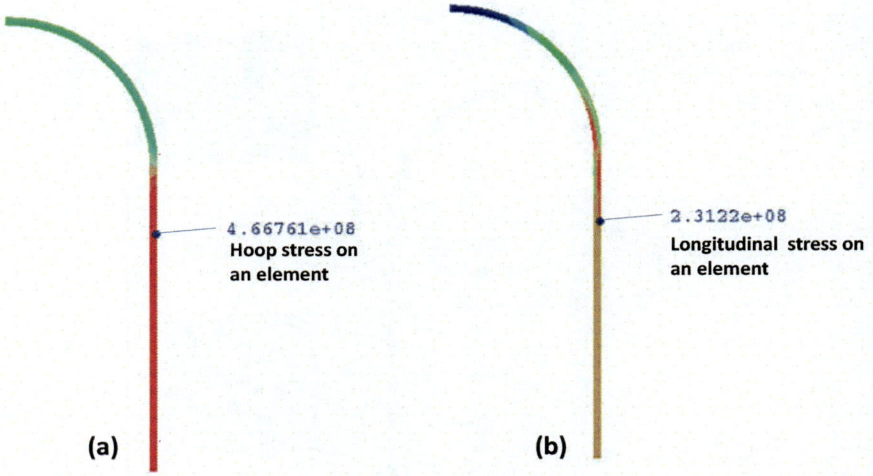

Fig. 10 (a) Hoop and (b) longitudinal stresses on an element on the surface of the pressure vessel

wall thickness of the cylinder the stress is insignificant ($\sigma_3 \sim 0$). The material's yield stress values are given in Table 6. The Von-mises failure region is shown in Fig. 11.

for the triaxial state of stress,

$$\sigma_v = \sqrt{\frac{1}{2}\left[(\sigma_1 - \sigma_2)^2 + (\sigma_2 - \sigma_3)^2 + (\sigma_3 - \sigma_1)^2\right]} \tag{2}$$

for the biaxial state of stress,

$$\sigma_v = \sqrt{{\sigma_1}^2 - \sigma_2\ \sigma_2 + {\sigma_2}^2} = 403.3\text{Mpa} \tag{3}$$

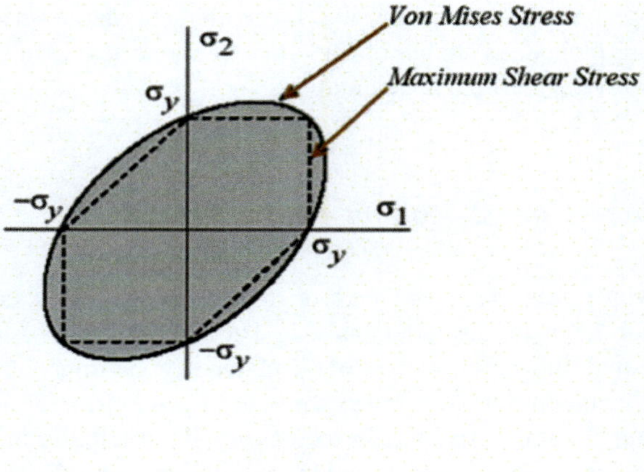

Fig. 11 Von mises failure region

Table 6 Stress comparison of materials under applied pressure of 25 MPa

S. no	Material	Yield stress σ_y (MPa)	Numerical Von mises stress σ_s' (MPa)
1	4340 Alloy Steel [19]	710	403.3
2	Al206.0-T7 Aluminium alloy [20]	350	403.3

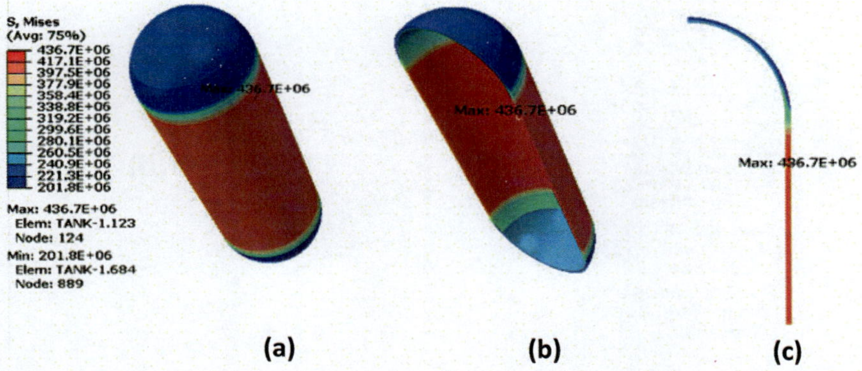

Fig. 12 Von-mises stress distribution in the cylinder wall

Figure 12 (a), (b), (c) shows the Von-mises stress distribution for the given geometry and pressure load with various viewing schemes. The Von-mises stress's numerical or the theoretical value from Eq. (2) and (3) is independent of material. Fig. 12(a) is the stress contour for the 360o surface of the pressure vessel. The highest stress is seen in the cylindrical section of the pressure vessel, as in circumferential or hoop direction, the stress is two times higher than in the longitudinal direction. The stress value in the dome sections is nearly half of the stress in the cylindrical section. Fig. 12(b) shows the internal surface. There is a slight difference in stress from the internal to external surface of pressure vessel; however, thin vessel theory assumes it as nearly constant, which is an appropriate assumption for thin shell designs.

In Table 6 the yield stress in 4340 steel is 710 MPa, and the Von mises stress value from the numerical result shows 436.7 MPa (Fig. 12), which is well below the yield value of the material. While for Al206.0-T7 Aluminium alloy the Von-mises stress reached exceeds the yield value of 350 MPa, leading to pressure vessel failure.

5.3 *Stress transition*

The change in the stress levels in a uniform thickness pressure vessel is not abrupt. The stress contours show that the transition occurs in a distributed manner from high to lower stress levels along and across the wall. In thin shells, this change is not considerable across the thickness. Figure 13 (a) shows the stress distribution in the

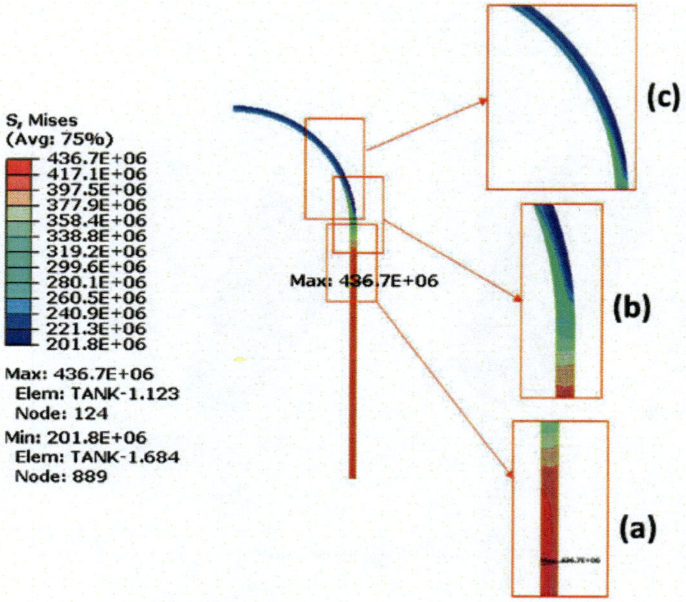

Fig. 13 Stress transitions in the cylinder wall

cylindrical section, (b) shows the start of the dome section, and finally, in (c) the mid dome section is shown. Along the wall, the Von-mises stress contour from the cylindrical section to the dome section shows a higher stress shift where the dome section starts.

6 Conclusion

This paper presented the growth and statistics of NGVs as an alternative to conventional gasoline/Diesel powered vehicles. Also, a comparative numerical study of the stress distribution in the wall of the metallic cylinders with different materials typically used for making type I tanks is done. Two materials (steel and aluminum alloys) are considered for comparative analysis.

The conclusions of the study are stated as follows:

1. There is an increasing demand for NGVs due to the availability, cost, eco-friendliness etc of compressed natural gas. More than 28 million NGVs we now have worldwide with more than 33 thousand gas refuelling stations. Asia alone

comprises more than 20 million NGVs recorded until 2019. The Asia–Pacific region shows a huge increase in NGV rise since 2006 while Africa and North America each contribute to more than 0.2 million CNG powered vehicles. Latin America records more than 5 million NGVs while Europe has shown more than 2 million NGVs. Within Europe, Italy is the biggest NGV user, after Italy, Ukraine, Armenia, Germany, Bulgaria, and Sweden come in a sequence of having CNG-powered vehicles.

2. Malaysia has shown a regular increment in adopting NGVs since the 1990s. More than 77 thousand NGVs are running in Malaysia and still rising. The government encourages consumers to have a CNG-powered vehicle via relaxation in taxes and related aspects. Considerable number of road vehicles are now CNG-powered, and its easy to convert the gasoline/Diesel driven engines to CNG power by installing the CNG kits, which are available with various capacities of tanks for storing gas at high pressures.
3. Several tests have been devised to assess the performance of CNG tanks against various loading and environmental factors. Regulatory bodies and agencies define the standards for testing and certification of pressure vessels.
4. The on-board vehicle storage is an important issue, as the gas needs to be stored at high pressures. Steel cylinders have been utilised due to their excellent pressure storage service, but these contribute to the weight of the vehicle. Aluminium cylinders are lightweight than steel but cannot be used at larger pressures as steel can withstand. Filament wound composite cylinders are a good replacement for storing higher pressure gas. Various types of cylinders are now available named as type I to type IV. Many companies came into existence that fabricates high-pressure storage cylinders for various applications.
5. The test pressure of 25 MPa is chosen for two materials in comparison, viz. 4340 Alloy steel and Al206.0-T7 Aluminium alloy. The thickness and other geometry details were kept same. Von- mises stress distribution was considered for the comparison and failure criteria for the materials under analysis. Steel alloy shows the 439 MPa Von-mises stress in the cylinder wall, which is below it's yield value of 710 MPa. So, 4340 alloys would perform without failure at this test pressure without failure, for the given geometric details.
6. The second alloy in question i.e., Al206.0-T7 has exceeded theoretical and numerical Von-mises value and this value is more than the yield strength of the material (350 MPa). Hence the material would fail for the given geometry and pressure load.

Acknowledgements This work is sponsored by FRGS, Fundamental Research Grant Scheme, FRGS/1/2019/TK08/UTP/02/1. We extend our gratefulness to the Ministry of Higher Education (MOHE), Malaysia, and Universiti Teknologi PETRONAS for the support. We are thankful to ICPER (2020) team & editors for the vital execution of the event (ESTCON) and Springer team for facilitating the publication process.

References

1. G. S. LLC, Gas Souths CNG solution (2020). https://www.gas-south.com/business/compressed-natural-gas.aspx
2. Khan MI, Yasmin T, Shakoor A (2015) Technical overview of compressed natural gas (CNG) as a transportation fuel. Renew Sustain Energy Rev 51:785–797. https://doi.org/10.1016/j.rser.2015.06.053
3. Idris S, Abu Bakar R (2009) An overview of compressed natural gas as an alternative fuel and Malaysian scenario. Eur J Sci Res 34(1):6–15
4. Azeem M et al (2022) Application of filament winding technology in composite pressure vessels and challenges: a review. J Energy Storage 49:103468. https://doi.org/10.1016/j.est.2021.103468
5. Stocchetti A, Volpato G (2010) In quest for a sustainable motorisation: the CNG opportunity. Int J Automot Technol Manag 10(1):13. https://doi.org/10.1504/IJATM.2010.031454
6. Frick M, Axhausen KW, Carle G, Wokaun A (2007) Optimization of the distribution of compressed natural gas (CNG) refueling stations: Swiss case studies. Transp Res Part D Transp Environ 12(1):10–22. https://doi.org/10.1016/j.trd.2006.10.002
7. Maida J (2020) Business Wire Inc. A Berkshire Hathaway Company. https://www.businesswire.com/news/home/20170530005719/en/Top-5-Vendors-Global-Compressed-Natural-Gas
8. F. I. SPA, Faber Cylinders. http://www.faber-italy.com/eng-quality-approvals.html
9. NGV Global, (2019). http://www.iangv.org/current-ngv-stats/
10. Rosero F, Fonseca N, López J-M, Casanova J, (2021) Effects of passenger load, road grade, and congestion level on real-world fuel consumption and emissions from compressed natural gas and diesel urban buses. Appl Energy 282:116195. https://doi.org/10.1016/j.apenergy.2020.116195
11. Yue T et al (2016) Gaseous emissions from compressed natural gas buses in urban road and highway tests in China. J Environ Sci 48:193–199. https://doi.org/10.1016/j.jes.2016.01.028
12. Yeh S (2007) An empirical analysis on the adoption of alternative fuel vehicles: the case of natural gas vehicles. Energy Policy 35(11):5865–5875. https://doi.org/10.1016/j.enpol.2007.06.012
13. Hagos DA, Ahlgren E (2017) A state-of-the art review on the development of CNG/LNG infrastructure and natural gas vehicles (NGVs). Chalmers Univ. Technol
14. Natural Gas—the fuel of choice towards clean mobility working committee 5 . 3 on Natural Gas for Transport Contents
15. P. L. I. Canada, "CNG Tank Testing," Powertech. https://www.powertechlabs.com/services-all/cng-tank-testing
16. The International Organization for Standardization, "ISO 11439:2013(en) Gas cylinders." https://www.iso.org/obp/ui/#iso:std:iso:11439:ed-2:v1:en
17. Ribarits SG (1992) Assessment of inspection criteria and techniques for recertification of natural gas vehicle (NGV). The University of British Columbia
18. Choi JC, Kim C, Jung SY (2004) Development of an automated design system of a CNG composite vessel using a steel liner manufactured using the DDI process. Int J Adv Manuf Technol 24(11–12):781–788. https://doi.org/10.1007/s00170-003-1798-4
19. Ariatapeh MY, Mashayekhi M, Ziaei-Rad S (2014) Prediction of all-steel CNG cylinder fracture under impact using a damage mechanics approach. Sci Iran 21(3):609–619
20. Manjunath DS (2017) Frontal crash analysis of a conformable CNG tank using finite element analysis
21. "Steel 4340 | Tech Steel & Materials." https://www.techsteel.net/alloy/steel/4340 (accessed May 19, 2022).

Residual Strength Prediction of a Pipeline with Interacting Corrosion Defects Subjected to Combined Loading Using Artificial Neural Networks

Prasshanth Perumal, Saravanan Karuppanan, and Mark Ovinis

Abstract Predicting the failure pressure of corroded pipelines has long been a topic of great interest for researchers all around the world. There are several methods and guidelines available to estimate the failure pressures of pipelines with the most commonly used being DNV-RP-F101. However, despite being the most comprehensive method, neither DNV-RP-F101 nor any other widely used corrosion assessment methods consider interacting defects subjected to both internal pressure and longitudinal compressive stress, despite this scenario being extremely common in the real world. In this work, the relationship between interacting corrosion defects and applied loadings (internal pressure and longitudinal compressive stress) with the failure pressure of pipeline is investigated by predicting the failure pressure using an artificial neural network (ANN). Data regarding the failure pressure of pipelines with interacting defects and combined loading is collected from FEA and full-scale burst tests. These data are then fed to an artificial neural network, allowing it to provide appropriate results for new cases, i.e., for other defects, corrosion defect parameters and loadings. This research will provide a gateway to help predict the failure pressure of pipes with interacting corrosion defects subjected to both internal pressure and longitudinal compressive stress using ANN. The greatest advantage of utilizing an artificial neural network is its ability and ease to improve failure prediction accuracy. Over time, more failure data can be fed into the artificial neural network, allowing it to "learn", thus being able to provide better failure pressure predictions.

Keywords Failure pressure · Artificial neural network · Pipeline · Interacting corrosion defects

---

P. Perumal (✉) · S. Karuppanan · M. Ovinis
Mechanical Engineering Department, Universiti Teknologi PETRONAS, 32610 Seri Iskandar, Perak, Malaysia
e-mail: prasshanth_18001237@utp.edu.my

F. Ahmad et al. (eds.), *ICPER 2020*, Lecture Notes in Mechanical Engineering,
https://doi.org/10.1007/978-981-19-1939-8_55

1 Introduction

One of the major concerns in industries that heavily utilise pipelines such as the oil and gas industry is the failure of pipelines due to corrosion. In pipeline-dependent industries, the highest risk factor of pipeline failure includes internal corrosion, internal erosion, external impacts, external corrosion, free span and on-bottom stability. At present, a vast majority of the pipelines in Malaysia have been in operation for two or three decades and have gone into the accident-prone period. Once the accident happens, it will be a threat to personal safety and damage the ecological environment.

When analysing corrosion in pipelines, the most common international guideline referred to is the DNV-RP-F101 guideline. According to DNV documents, external longitudinal compressive stress has been proven to be instrumental in influencing the failure behaviour of corroded pipes. Although the DNV-RP-F101 provides a comprehensive guideline for various scenarios related to pipeline corrosion, the analysis of interacting defects subjected to both internal pressure and longitudinal compressive stress is not covered by the DNV code even though this scenario is very common in real-world application. Therefore, an accurate assessment of corroded pipelines with interacting defects and combined loadingis required.

This study is focused on the failure pressure estimation of corroded pipelines with interacting defects subjected to internal pressure and longitudinal compressive stress using an artificial neural network. An improved failure pressure prediction method will be formulated and this would help the engineers to estimate the failure pressure of corroded pipelines with interacting defects subjected to combined loadings accurately and efficiently.

Benjamin et al. [1] has proven that interacting defects cause a greater reduction in failure pressure in comparison to single defects. In addition, compressive stresses caused by the Poisson's effect, variation in transport fluid temperature, and ground movement are also detrimental to pipeline integrity [2]. Some of the detrimental effects of longitudinal compressive stress on pipeline include buckling and wrinkling [3]. Amongst the most widely used corrosion assessment methods, the DNV-RP-F101 can be considered the most comprehensive, as it provides assessments for single defects subjected to internal pressure, single defects subjected to internal pressure and longitudinal compressive stresses and also interacting defects subjected to internal pressure only. Despite its comprehensiveness, neither the DNV-RP-F101 nor any other widely used corrosion assessment methods take into account interacting defects subjected to both internal pressure and longitudinal compressive stress which is a more common real world scenario [4].

In this paper, a new failure pressure prediction method to determine the failure pressure of pipes with interacting defects subjected to both internal pressure and longitudinal compressive stress using an artificial neural network is described. In addition, the effects of defect geometries on the failure pressure of pipes with interacting defects subjected to both internal pressure and longitudinal compressive stress using artificial neural networks is discussed.

2 Literature Review

2.1 Interacting Defects

Defects in pipes come in various shapes and sizes. According to Teran et al. [5], "there are several studies of corrosion defects with different geometries, such as oval-shaped defects, elliptical-shaped defects, ellipsoid-shaped defects, holes-shaped defects, groove-shaped defects, long blunt grooves, rectangular-shaped defects, semi-spherical shaped defects, and pitting". When it comes to assessment of failure behaviours, interacting defects have a higher degree of complexity due to the interaction among adjacent defects [6].

Li et al. [7] found that adjacent defects introduce disturbances in the stress and strain fields of the pipeline that spread further out than the original border of the defects. The regions where these disturbances are prevalent are called areas of influence of the defect. In cases where the areas of influence of adjacent defects overlap, the failure pressure of the multiple defects have a high probability of being lower in comparison to failure pressure of single or isolated defects [1]. As the distance between the interacting defects increases, the size of the overlapping area of influence decreases until eventually the interaction vanishes and the corrosion defect is then treated as single and isolated defect with the failure pressure being governed by the most severe corrosion defect [8] However, if the depth of the defects is less than 20% of the thickness of the pipe walls, the distance between the defects and its interactions have a negligible effect on the failure pressure of the pipe [9].

Analyses from both the FEA and the full-scale burst test proved that the failure pressure of corrosion with multiple defects were significantly lower than single defects. This fact was highlighted by Lee et al. [10] who found that for defects in a fixed area, when the number of defects, in this case pits, increases, the burst pressure of the pipe decreases. Most of the failure behaviours of multiple interacting corrosion defects included the region that contained the deepest sections of the defects, and this has been observed by other researchers such as Silva et al. [8]. This behaviour was expected and subsequently proven by FEA conducted by Chen et al. [11]. Mondal and Dhar [12] further verified that pipes with corrosion defects that are close enough to interact with each other significantly reduce the burst pressure of pipes, as compared to independent or standalone defects.

In-depth analyses proved that the corrosion depth increases the stress field concentrations and thus caused a decrease in failure pressure of interacting defects. Another FEA research, conducted by Chauhan et al. [2] mentioned that a reduction of 26% of the failure pressure was obtained by decreasing the spacing between the interacting defects from 6 to $1t$ where t represents the thickness of the pipe. However, when it comes to predicting the failure pressure of these pipes with corrosion defects, there are some commercially available codes and guides. One such example is the DNV-RP-F101. However, Belachew et al. [13] found that available codes for pipeline failure pressure estimation are generally conservative. That is to say, failure pressure estimates obtained from these codes are usually lower than the actual failure pressure of the pipe in question. Lee et al. [14] agreed that among all the codes, the DNV-RP-F101 is the most accurate in predicting the failure pressure of pipes with corrosion defects, although it still has a certain level of conservatism to its results.

2.2 Combined Loadings

Pipelines are often subjected to a combination of different loads. These loads include internal pressure, longitudinal compressive and tensile stresses, as well as axial and bending loads [15]. Extensive research has been carried out to investigate the behaviour of pipelines under various combinations of stresses and loads, to help provide a better understanding of its mechanics and thus help avoid untoward incidents in the future. One such research was conducted by Smith and Waldhart [16], which investigated the parameters that would lead a corroded pipe to rupture, or buckle without rupturing while being subjected to internal pressure, bending and axial thermal stress. The research concluded that "if rupture and wrinkling of the corroded pipe are to be predicted accurately, sufficient information describing the in-service pipe's operating loads and load history, defect dimensions, and well-defined material properties must be obtained".

Bjornoy [4] found that external longitudinal loads cause a higher reduction to the failure pressure of pipe in comparison to pipes subjected to internal pressure only. This is in line with real life corrosion scenarios where pipelines are subjected to both interacting corrosion defects as well as external longitudinal compressive stresses. For instance, internal pressure due to fluid transport results in a circumferential expansion of the pipeline which causes a contraction in the longitudinal direction due to the Poisson's effect [17]. Besides that, increase in transport fluid temperature as well as ground movement due to mining subsidence, landslides and earthquakes have been shown to have the aforementioned effect to pipelines [11].

Although a limited amount of research has been conducted to specifically study the effects of pure longitudinal compressive stress on corroded pipes with internal pressure, it was found that longitudinal compression stress in general results in adverse effects on the structural integrity of a pipe. FEA study conducted by Chauhan et al. [2] revealed that application of axial compressive stress and internal pressure caused a buckling behaviour which is similar to results obtained by Li et al. [7]. A study conducted by Peng et al. (2011) revealed that under compressional loads, circumferentially oriented defects showed stronger interacting behaviour in comparison to longitudinally oriented defects. Barbosa et al. [18] stated that the presence of internal pressure in a pipe has a detrimental effect on the bending moment capacity of a pipe and that this effect is more significant as the level of applied initial pressure increases.

Xu and Cheng et al. [3] carried out a study to determine the effects of tensile and compressive strain on corrosion-affected pipelines. In general, results showed that reduction in failure pressure occurred regardless of the presence of tensile or compressive strain as supported by other researches such as the one by Cosham and Hopkins (2002). In addition, failure pressure of pipeline further decreases as the intensity of the tensile or compressive force increases [19]. Full scale burst test conducted on corroded pipe with single defect subjected to both internal pressure and longitudinal compressive forces revealed that under compressive stress, axially oriented corrosion defects have a lower failure pressure in comparison to pipes with circumferentially oriented corrosion defects, despite the circumferential corrosion encompassing the entire diameter of the pipe [4].

2.3 *Artificial Neural Networks and Failure Pressure of Pipes*

Some of the more common machine learning models include Decision Trees, Bayesian Networks, Artificial Neural Networks and Genetic Algorithms. To effectively select the most appropriate model, it is first important to determine the type of problem that is to be solved.

There are several types of problems that can be solved using machine learning applications. They include classification problems, clustering, regression, feature learning and reinforcement learning problems, to name a few [20]. Depending on the problem being tackled each has its own appropriate process of solving. For example, a classification problem is a problem where a new observation needs to be identified and categorized, such as gender, colour or other pattern recognition [21]. A regression problem is when it is necessary to approximate the relationship between multiple different variables. It is commonly used for predicting and forecasting a dependent variable given the independent variables. A reinforcement learning problem on the other hand, is when the software needs to evaluate the environment and adjust its behaviour to obtain the desired output reward. Thus, this type of software utilises dynamic programming techniques.

It is evident that the task of predicting the failure pressure of a pipe given parameters such as defect depth, defect length, spacing between defects and longitudinal compressive stress represents a regression problem. The main task is prediction or estimation of a dependent variable (failure pressure) given the independent variables. Also, in this case, the dependent variable is continuous [22]. As this is a regression problem, the process of supervised learning will be utilised.

One of the most common machine learning methods is known as Supervised Learning. Supervised learning algorithms are "algorithms that reason from externally supplied instances to produce a general hypothesis, which then make predictions about future instances" [23]. According to LeCun et al. [24], supervised learning makes it easier and more efficient for the neural network to modify its internal adjustable parameters to reduce its output error. Since this application involves known inputs and corresponding correct outputs, a supervised learning algorithm should be used [25].

Having identified the most appropriate machine learning process, it is noted that there are several different machine learning algorithms that fall under the supervised learning method, such as Naïve Bayes, Artificial Neural Networks, Support Vector Machines, Decision Trees and Ensembles such as Boosting and Random Forests.

Naïve Bayes models are known for their simplicity and thus computational efficiency. However, this simplicity comes with its drawbacks. Its simplicity is caused by the high degree of independence assumption that it takes on and is thus not suitable when using multiple inputs together [26]. It is possible to enhance the system's accuracy by enhancing the Naïve Bayes' rule or even using it in tandem with other known techniques, however this causes the model to be overly complicated compared to simply using a different model altogether [27].

Decision Trees, on the other hand, are more robust and is thus able to handle combinations of terms to produce acceptable results. However, the process of training a Decision Tree is somewhat complex as the number of nodes required increases exponentially as the number of parameters increase [28]. This makes using Decision Tree models not very practical as interpreting the model with the required number of parameters will be tedious. Random Forests are an extension of the Decision Tree algorithm. It is made up of multiple Decision Trees. Random Forests are able to form a strong classifier from a large number of weak or weakly-correlated classifiers [29]. However, as they are fundamentally made up of Decision Trees, the limiting criteria in terms complexity associated with increasing of number of parameters remains relevant.

A Support Vector Machine (SVM) is a type of supervised learning algorithm that is suitable to be used to solve classification and regression problems. According to Parikh and Shah [30], SVM is a kernel based supervised learning algorithm used predominantly for binary classification problems. Although they can be applied to regression problems, its learning ability is better suited for classification problems, being able to classify highly non-linear data using robust mathematical theory and kernel function.

Another example of a machine learning algorithm that utilizes a supervised learning algorithm is known as a Neural Network. Neural Networks are extremely "powerful techniques for representing complex relationships between inputs and outputs" [26]. Although Neural Networks also have a high degree of complexity, it has an added advantage over other machine learning models in that it is possible to convert the Neural Network into a set or multiple sets of rules for the purpose of determining what exactly the Neural Network has learnt [31]. Besides that, Neural Networks are extremely efficient in terms of utilising computing resources which makes it very attractive and suitable for large datasets and seemingly uncorrelated pieces of data [32]. From this, it is evident that for the task of predicting the failure pressure of pipes based on the available data, an Artificial Neural Network would be the most apt choice of a supervised learning algorithm.

Artificial Neural Networks typically have a structure which begins with an input layer for the parameters that are to be fed to it. This data then goes through one or more hidden layers, depending on the requirements of the specific task the Neural Network is designed to perform. Following that, the relevant output from the Neural Network is produced at the output layer [33]. The idea of using artificial neural networks (ANNs) for forecasting and prediction is not new. The first such application dates back to the mid-1960s. Artificial neural networks are made up of a number of simple interconnected processing elements called nodes or neurons. Each node receives an input signal from either other nodes or external stimuli, processes it locally through an activation or transfer function and produces a transformed output signal to other nodes or external outputs [34]. Although each individual neuron implements its function rather slowly and imperfectly, collectively a network can perform a surprising number of tasks quite efficiently [35].

One way an ANN differs from a conventional computer is although computers have the ability to solve complex mathematical problems involving operations such as subtraction, multiplication and fundamental logic elements, ANNs have the ability to also solve complex problems, non-linear or stochastic problems and mathematically ill-defined problems [36]. This is an extremely important feature of ANNs, as it learns from the data that is provided to it, forming patterns and identifying relationships between the individual pieces of data and establishing connections that help it come to a conclusion and produce an accurate output when presented with similar data in the future.

For the past few years, the utilization of artificial neural networks (ANN) has shown great promise in predicting the failure pressure of corroded pipes. The main advantage of an artificial neural network model is its ability to improve its output accuracy over time, as it is fed with additional data, thus becoming better at failure pressure prediction. It has an outstanding ability to learn information and patterns from samples. An example is research that was carried out by Zangenehmadar and Moselhi [37] to determine the residual life of pipes using an ANN. Results obtained from the ANN model showed a coefficient of determination of greater than 98%. Xu et al. [38] observed that "an insignificant error existed between experimental values and results obtained from trained ANN, whereby the results obtained by the ANN model showed better correlation when compared to the burst pressures predicted by ASME B31G and DNV-RP-F101 codes". Therefore, it is clear that an ANN is an extremely powerful computational tool that can prove to be a viable option to develop an accurate failure pressure prediction model of pipes with interacting defects while being subjected to both internal pressure and longitudinal compressive stress.

3 Methodology

As discussed in the previous section, there are many different machine learning methods and models. For the problem of predicting the failure pressure of a pipeline, the best approach would be to employ a Supervised Learning approach, using an Artificial Neural Network.

Firstly, an ANN model was developed using Python programming language. The performance criterion of the ANN is the accuracy of the prediction, rather than speed or other performance criteria.

The next step would be to source data that is to be used to train and validate the ANN. Training samples for the "learning" phases were obtained from previous extensive parametric studies that have been conducted by past researchers at Universiti Teknologi PETRONAS. This was followed by data pre-processing which involves data cleaning and formatting to make it as efficient as possible to be interpreted by the ANN.

Once the architecture is determined, the stopping criteria and optimization method has to be identified. This step involves evaluating the training error and ensuring that the convergence criterion is met. For this application, two model evaluation methods were used. Firstly this was carried out by evaluating the root-mean-squared error (RMSE) of the model. The second method was by evaluating the coefficient of determination (r^2) of the model.

The failure pressure results obtained from the developed ANN model were then compared with those obtained using DNV-RP-F101 and full-scale burst tests. A good correlation amongst these values suggests that the ANN model is able to accurately estimate the failure pressure. After that, the following parameters were varied (with random parameters) to determine the failure pressure of pipes with interacting defects:

a. Pipe defect parameters (i.e. defect depth, length, width, longitudinal and circumferential spacing).
b. Longitudinal compressive stress.

From the parametric study, parameters that have negligible effects on the failure pressure were eliminated. The remaining parameters were analysed to develop an improved failure pressure prediction method, thus allowing for an accurate estimation of the failure pressure caused by interacting defects subject to internal pressure and longitudinal compressive stress. Results obtained from the new failure pressure prediction method were compared against training samples, DNV-RP-F101 predictions, and actual burst test results.

An Artificial Neural Network (ANN) was developed, with its learning data sourced from results of parametric studies that were available. Data that was fed to the ANN involves the failure pressure of pipes that were subjected to a range of compressive stresses, with varying defect depths and lengths, as well as varied spacing between the defects, with approximately 1440 data points in total for API 5L X52 steel pipes. 80% of the failure pressure data available was fed to the ANN to train it while the remaining 20% of the available data that was not fed to the ANN will be used to check the accuracy of the failure pressure prediction output from the ANN.

The output of the ANN was validated via several methods. It is important to verify if the relationship between the variables that was obtained is an acceptable description of the data. The first method of validating the output would be to analyse the root-mean-square error (RMSE) of the results obtained. The RMSE measures the deviation or difference between the failure pressure values predicted by the model and the available data. The closer the RMSE is to "0" the more accurate the predicted value is deemed to be.

Another method of validating the output of the ANN is to analyse the coefficient of determination, or R-squared. This measure of goodness-of-fit measures how close the predicted values are to the fitted regression line. The R-squared value is obtained by dividing the total variation with the explained variation. The closer the R-squared value is to 100%, the more accurate the ANN is deemed to be.

Once this is achieved, the failure pressure estimations produced by the ANN are deemed acceptable. The results obtained from the ANN were compared with results from estimates using equations for single corrosion defects as found in the DNV-RP-F101 code, as well as results from full scale experimental burst test that have been conducted by fellow researchers. This comparison was done to establish a correlation between the results of the ANN and the other two sources of results. A strong correlation between these results would suggest that the results produced by the ANN is indeed accurate and can be used as a reliable source for the estimation of the failure pressure of pipes with interacting defects subject to internal pressure and longitudinal compressive stress.

Further studies using the ANN was then conducted by varying the conditions and parameters applied to the pipe such as defect depth, length as well as spacing between the interacting defects. This will provide a greater understanding of how the pipe would behave under these various circumstances. It would also be a good way to verify that the ANN is accurate and behaving as it should be when tasked to predict the failure pressure of a pipe with these varying parameters.

With this, from the data gathered from the ANN, a new failure pressure prediction method to help predict the failure pressure of pipes with interacting defects subject to internal pressure and longitudinal compressive stress can be developed. This new assessment method can then be verified by comparing its results with results obtained from past researches. The accuracy and reliability of the new assessment method can then be determined and improved if necessary.

4 Results and Discussion

An ANN was successfully developed using Python programming language. Figure 1 shows the flowchart of the ANN that was developed, while the pseudocode is shown in Fig. 2. The model was trained using existing failure pressure data. 80% of the data was used to train the ANN while the remaining 20% was used to validate the results produced by the ANN. This allocation is in line with generally accepted machine learning and training practices.

The ANN model was successfully trained and the accuracy of its output was validated by assessing the Root Mean Square Error (RMSE) of the outputs and also by assessing the R-squared value of the ANN model. The RMSE of the ANN is <0.05. This value is very close to "0" and shows that the model is producing accurate results. The R-squared of the ANN model is >95%. This value is very close to 100% and again reflects the accuracy and validity of the results produced.

The results produced by the ANN were compared with predictions obtained using DNV under various interacting corrosion defect length subjected to internal pressure and axial compressive stress. This was done in four separate cases, each by varying the defect length, defect depth, defect spacing and magnitude of compressive stress. Table 1 summarizes the default values used for each of these parameters when performing this comparison.

Fig. 1 ANN model flowchart

```
import packages pandas, numpy, matplotlib.pyplot

read file analysis_data

declare variables defect_type, compressive_stress, defect_depth, defect_length, spacing, pressure


train and test :

 import package math
  import sklearn.impute, sklearn.preprocessing, sklearn.pipeline, sklearn.model_selection, sklearn.neural_network, sklearn

 imputer, simpleimputer, strategy=mean
 scaler, standardscaler
 model, MLPRegressor

 calculate rmse
 print (rmse)
 calculate R-squared
 print (R-squared)


user interface for prediction :

 input data compressive_stress, defect_depth, defect_length, spacing, defect_type

 output data pressure
 print ("The failure pressure of the pipe is", pressure)
```

Fig. 2 ANN model pseudocode

Table 1 Geometric parameters for parametric study

Input parameters	Default values
Normalized defect depth, d/t	0.5
Normalized defect length, l/D	1.0
Normalized defect spacing, $s_l/\sqrt{Dt}$	1.0
Normalized axial compressive stress, $\sigma c/\sigma y$	0.5

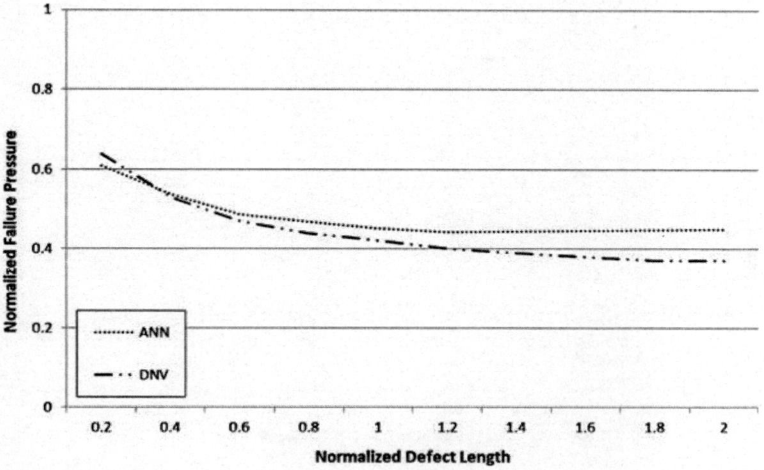

Fig. 3 Normalized failure pressure predictions of ANN under various interacting corrosion defect length subjected to internal pressure and axial compressive stress

Figure 3 shows the results obtained for interacting longitudinal defects with varying defect lengths subjected to internal pressure and axial compressive stress. The normalized defect length is varied from 0.2 to 2 while the normalized defect depth, normalized defect spacing and normalized axial compressive stress are fixed to the default values shown in Table 1. From the curves, it is seen that the predictions from the DNV produces more conservative results as compared to the failure pressures predicted by the ANN.

Figure 4 illustrates results obtained for interacting defects with varying defect depths subjected to internal pressure and axial compressive stress. The normalized defect depth is varied from 0.2 to 0.8 while the normalized defect length, normalized defect spacing and normalized axial compressive stress are fixed to the default values shown in Table 1.

Again, it is observed that both failure pressure predictions have similar trend lines, and that the predictions obtained using DNV is more conservative.

Figure 5 details the results obtained for interacting defects with varying defect spacing subjected to internal pressure and axial compressive stress. The normalized defect spacing is varied from 0 to 4 while the normalized defect depth, normalized defect length and normalized axial compressive stress are fixed to the default values shown in Table 1. With increasing distance, the size of the overlapping area of influence decreases until eventually the interaction vanishes, and the corrosion defect is then treated as single and isolated defect with the failure pressure being governed by the most severe corrosion defect [8].

Figure 6 shows failure pressures obtained for interacting defects with varying axial compressive stress subjected to internal pressure. The normalized axial compressive

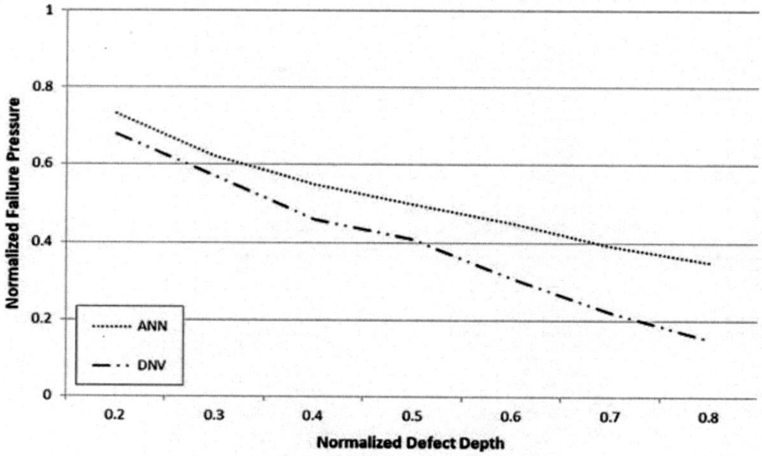

Fig. 4 Normalized failure pressure predictions of ANN under various interacting corrosion defect depth subjected to internal pressure and axial compressive stress

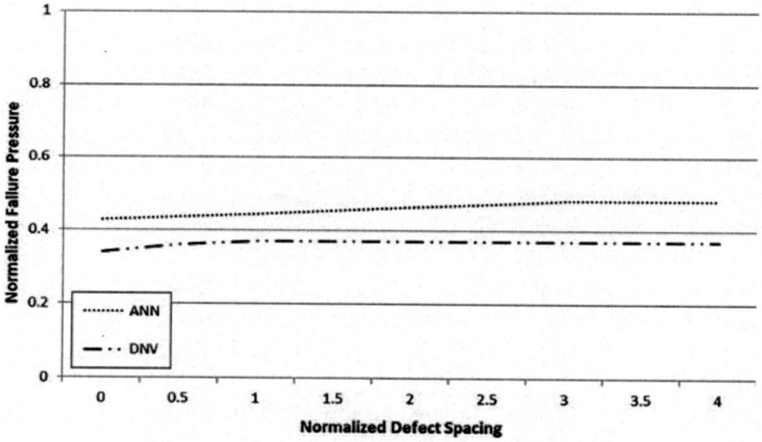

Fig. 5 Normalized failure pressure predictions of ANN under various interacting corrosion defect spacing subjected to internal pressure and axial compressive stress

stress is varied from 0 to 0.8 while the normalized defect depth, normalized defect length and normalized defect spacing are fixed to the default values shown in Table 1.

It is observed that the two trendlines start apart, again with the DNV results being the more conservative of the two, but the trendlines converge as the normalized axial compressive stress increases.

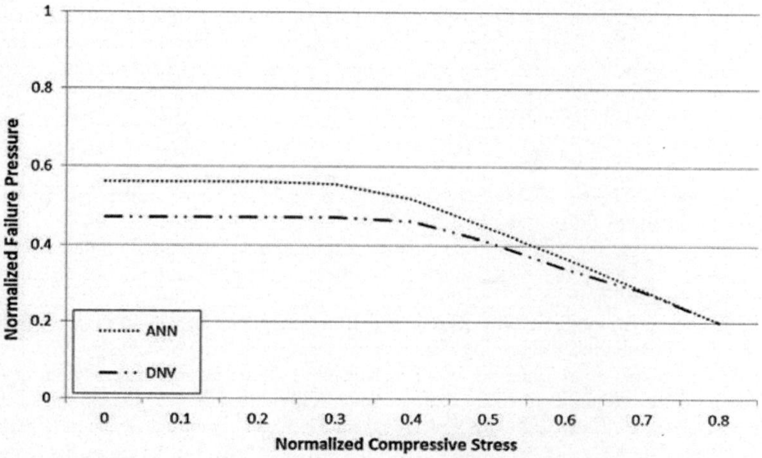

Fig. 6 Normalized failure pressure predictions of ANN under various normalized axial compressive stress subjected to internal pressure and with interacting corrosion defects

5 Conclusion and Recommendation

A new method to predict the failure pressure of pipes that have interacting corrosion defects and are subjected to combined loading was developed using an Artificial Neural Network. Corrosion assessment methods currently available do not cater to scenarios where pipes have interacting defects and are subjected to internal pressure and axial compressive stress, although this is the most common scenario in real world application. This is a major hurdle when trying to estimate the failure pressure of a pipe subjected to the aforementioned conditions. The development a novel failure pressure prediction method using an Artificial Neural Network helps address this major shortcoming in current pipeline corrosion assessment methods.

Acknowledgements This work was supported by Yayasan Universiti Teknologi PETRONAS, Malaysia (Grant number 015LC0-110).

References

1. Benjamin A, Freire J, Vieira R (2007) Part 6: analysis of pipeline containing interacting corrosion defects. Exp Tech 31(3):74–82
2. Chauhan V, Swankie TD, Espiner R, Wood I (2009) Developments in methods for assessing the remaining strength of corroded pipelines. NACE international
3. Xu L, Cheng Y (2012) Reliability and failure pressure prediction of various grades of pipeline steel in the presence of corrosion defects and pre-strain. Int J Press Vessels Pip 89:75–84
4. Bjørnøy OH, Sigurdsson G, Marley MJ (2004) Updated DNV-RP-F101 for corroded pipelines, 2004. 2004 international pipeline conference, vol 1, 2 and 3

5. Terán G, Capula-Colindres S, Velázquez J, Fernández-Cueto M, Angeles-Herrera D, Herrera-Hernández H (2017) Failure pressure estimations for pipes with combined corrosion defects on the external surface: a comparative study. Int J Electroch Sci, 10152–10176
6. Andrade EQ, Benjamin AC, Machado PR, Pereira LC, Jacob BP, Carneiro EG, ... Noronha DB (2006) Finite element modeling of the failure behavior of pipelines containing interacting corrosion defects. Vol 4: Terry Jones Pipeline Technology; Ocean Space Utilization; CFD and VIV Symposium
7. Li X, Chen Y, Zhou J (2010) Plastic interaction relations for corroded steel pipes under combined loadings. Earth and Space 2010
8. Silva R, Guerreiro J, Loula A (2007) A study of pipe interacting corrosion defects using the FEM and neural networks. Proceedings of the fourth international conference on engineering computational technology
9. Karuppanan S, Aminudin A, Wahab A (2012) Burst pressure estimation of corroded pipeline with interacting defects using finite element analysis. J Appl Sci 12(24):2626–2630
10. Lee Y, Kim YP, Moon M, Bang WH, Oh KH, Kim WS (2005) The prediction of failure pressure of gas pipeline with multi corroded region. Mater Sci Forum 475–479:3323–3326
11. Chen Y, Zhang H, Zhang J, Zheng W, Liu X, Liang L (2015) Failure analysis of X90 pipeline with circumferentially aligned interacting corrosion defects. Adv Energy Equip Sci Eng, 2141–2145
12. Mondal B, Dhar A (2017) Interaction of multiple corrosion defects on burst pressure of pipelines. Can J Civ Eng 44(8):589–597
13. Belachew CT, Ismail MC, Karuppanan S (2011) Burst strength analysis of corroded pipelines by finite element method. J Appl Sci 11(10):1845–1850
14. Lee GH, Pouraria H, Seo JK, Paik JK (2015) Burst strength behaviour of an aging subsea gas pipeline elbow in different external and internal corrosion-damaged positions. Int J Naval Archit Ocean Eng 7(3):435–451
15. Zhou T, Tian Y, Cassidy MJ (2018) Effect of tension on the combined loading failure envelope of a pipeline on soft clay seabed. Int J Geomech 18(10)
16. Smith MQ, Waldhart CJ (2000) Combined loading tests of large diameter corroded pipelines. vol 2: Integrity and Corrosion; Offshore Issues; Pipeline Automation and Measurement; Rotating Equipment
17. Ahammed M, Melchers R (1997) Probabilistic analysis of underground pipelines subject to combined stresses and corrosion. Eng Struct 19(12):988–994
18. Barbosa A, Teixeira A, Soares C (2017) Strength analysis of corroded pipelines subjected to internal pressure and bending moment. Progress in the analysis and design of marine structures
19. Selig E, DiFrancesco L, McGrath T (1994) Laboratory test of buried pipe in hoop compression. Buried plastic pipe technology: 2nd vol
20. Schank R, Kass A (1990) Explanations, machine learning, and creativity. Mach Learn, 31–48
21. Alpaydin E (2014) Introduction to machine learning. Mit Press, Cambridge, MA
22. Sen A (2012) Regression analysis: theory, methods, and applications. Springer
23. Osisanwo F, Akinsola J, Awodele O, Hinmikaiye JO, Olakanmi O, Akinjobi J (2017) Supervised machine learning algorithms: classification and comparison. Int J Comput Trends Technol 48(3):128–138
24. LeCun Y, Bengio Y, Hinton G (2015) Deep learning. Nature 521(7553):436–444
25. Kotsiantis S (2007) Supervised machine learning: a review of classification techniques. Informatica 31
26. Xhemali D, Hinde C, Stone R (2009) Naive Bayes vs. decision trees vs. Neural networks in the classification of training web pages. International J Comput Sci Issues 4
27. Addin O, Sapuan S, Mahdi E, Othman M (2007) A Naïve-Bayes classifier for damage detection in engineering materials. Mater Des 28(8):2379–2386
28. Russell S, Norvig P (2003) Artificial intelligence: a modern approach. Prentice Hall, London
29. Mao W, Wang F-Y (2012) Cultural modeling for behavior analysis and prediction. Adv Intell Secur Inform, 91–102

30. Parikh KS, Shah TP (2016) Support vector machine—a large margin classifier to diagnose skin illnesses. Procedia Technol 23:369–375
31. Towell G, Shavlik J (1993) Extracting refined rules from knowledge-based neural networks. Mach Learn 13(1):71–101
32. Liu Z, Zhang Y (2001) A competitive neural network approach to web-page categorization. Int J Uncertainty Fuzziness Knowl Syst 9:731–741
33. Tu JV (1996) Advantages and disadvantages of using artificial neural networks versus logistic regression for predicting medical outcomes. J Clin Epidemiol 50(11):1309–1310
34. Zhang G, Eddy Patuwo B, Hu MY (1998) Forecasting with artificial neural networks: the State of the Art. Int J Forecast 14(1):35–62
35. Reilly DL, Cooper LN (1990) An overview of neural networks: early models to real world systems. In: Zornetzer SF, Davis JL, Lau C (eds) An Introduction to Neural and Electronic Networks. Academic Press, New York, pp 227–248
36. Graupe D (2013) Principles of artificial neural networks. World Scientific, Singapore
37. Zangenehmadar Z, Moselhi O (2016) Assessment of remaining useful life of pipelines using different artificial neural networks models. J Perform Constr Facil 30(5):04016032
38. Xu W, Li CB, Choung J, Lee J (2017) Corroded pipeline failure analysis using artificial neural network scheme. Adv Eng Softw 112:255–266

Manufacturing of a Low Cost Flexible Lightweight Walker for Fractured Foot

Ali S. Khazaal, Sadiq J. Azez, and Ahmed A. Shandookh

Abstract In this paper, plastic materials (polypropylene) were used to manufacture ankle foot orthosis (AFO). They were fabricated using vacuum moulding technique and added steel joint to make it adjustable. The data collecting of gait cycle Ground Reaction Force (GRF) and centre of pressure (COP) by using Dynamic gait analysis on the treadmill and pressure distribution by using (F-socket) were collected via the young patient, and his age was about (26 years) with height (168 cm) and weight (69 kg). The patient was wearing the model orthosis type AFO with left lower suffering from an ankle joint injury. While for a healthy young man the same steps of collect data was applied, the patient age was about (20–30 years) with height (160–170 cm) and weight (65–75 kg). The results of the gait cycle (GRF, Pressure distribution, centre of Pressure (COP), Candace, Stride length and foot print analysis) will show the major difference between the pathological case and healthy case, who wearing ankle foot orthosis (AFO) compare with a healthy case. Measuring the interference (pressure and force) between the legs with AFO area of contact for all cases then used this data for analysis the mathematical model by used (ANSYS V.15) software. In addition, the analysis of the AFO model will be illustrated here to know the results of the safety factor of fatigue and the analysis of the (Von-Mises) stress. Where it shows that the highest stress concentration in the ankle joint region, where the stress reached to (418.9 MPa), safety factor (2.4) for adjustable hinge AFO. Also, the analysis of the kinematics data will indicate the differences of the behaviour for the gait cycle between the healthy and pathological case by used (tracker) software.

1 Introduction

The word of "Orthosis" is referring to a device that used externally to limit or facilities movement via supporting, aligning and protecting body parts. Orthotics can improve the body regions duty that moves to prevent the deformity or make the deformation

A. S. Khazaal (✉) · S. J. Azez · A. A. Shandookh
Mechanical Engineering Department, University of Technology, Baghdad, Iraq
e-mail: 20182@uotechnology.edu.iq

F. Ahmad et al. (eds.), *ICPER 2020*, Lecture Notes in Mechanical Engineering,
https://doi.org/10.1007/978-981-19-1939-8_56

straightening. The orthosis is a device usually used for medical applications, this device applied externally in order to enhance or repair the functional and/or structural features of the Musculo-skeletal and/or nervous systems. Oftentimes it is designed to support, increase, and/or supplement the function of an extremity's limbs and/or body part, usually it is applied and/or fixed to the parts of lower limbs to improve its functions by implementing support through controlling motion, correcting flexible deformities and stabilizing gait in order to reducing pain through devolve load to another area, and preventing proceeding of fixed deformities [1].

Broken ankles are usually meaning "ankle fractures", which refers to that one or more of the bones constituents the ankle joint is broken. Fractured ankles range from simple breaks of a single broken bone, which might not prevent from being able to walk, to multiple fractures, which can pull the ankle out of place and might cause to prevent placing weight on the ankle for a few months. The more broken bones, lead to more unstable ankle will become. Ligaments could be damaged as well. The function of ligaments is to hold ankle and the joints with bones in a suitable correct position. People of all ages can be influenced by the broken ankles. During the past three or four decades, doctors have found an increasing number of broken ankles, especially when it comes to active, older people [2]. The motion of the foot is governed by two joints, the ankle joint and the subtalar joint. The first one is the joint lies between the talus and the tibia, and the last one is lies between the talus and the calcareous [3].

2 Experimental Implementation

2.1 Case Study

The data of kinetics and kinematics were collected by a young person, who was about (26 years) with (168 cm) height and (69 kg) weight. He was wearing the AFO with left lower part. While, the data collected from the same product were done on the healthy young man, who was about (24 years) with (168 cm) height and (62 kg) weight. Figure 1 shows three different bones that form the ankle joint, the shin bone (tibia), the small bone nestled between the heel bone and the fibula and the tibia (talus) and the smaller bone within the lower leg (fibula). Doctors are often assorting the fracture based upon the area where occurred the break. when a fracture occurs two joints are often involved the ankle joint, which is where the talus, the tibia and fibula meet, and the syndesmosis joint, which is lies between the tibia and the fibula and held together with ligaments. to ensure the ankle joint is stable takes multiple ligaments [2].

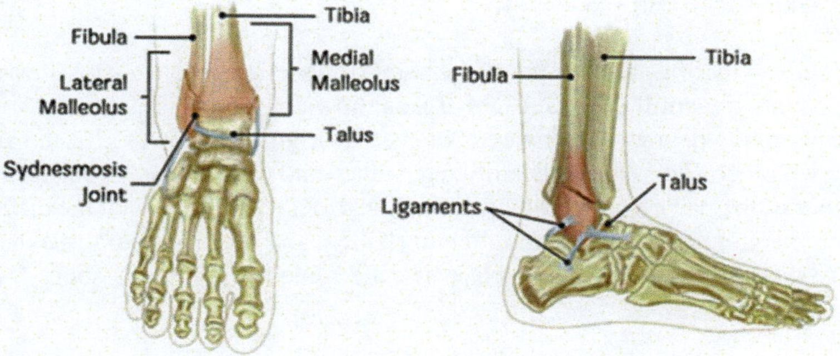

Fig. 1 Case study the Anatomic image [2]

2.2 Materials

The AFO materials needed for this study are as following:

- Jepson's Materials.
- Polypropylene.
- Perlon stockinet white (Ottobock health care 623T3). Which using for covering the Jepson.
- steel hinge.

2.3 Manufacturing Procedure

Measurement, manufacturing, and alignment for the partial foot orthosis for the amputated patient's foot were done according to ICRC's physical rehabilitation centers [4] and the Fig. 2 shows the steps of manufacturing procedure.

2.4 Testing the Orthosis

This section includes two tests: gait cycle test with using Dynamic gait analysis on the treadmill and the second test used F-socket system to measure the interference pressure. The tests are classified into two cases (healthy and pathological).

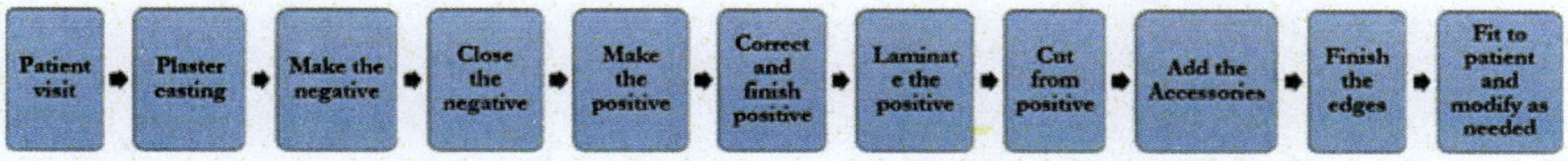

Fig. 2 Flow chart showing Manufacturing steps of AFO model

2.4.1 Gait Cycle and GRF Test

At the University of Babylon/Sport Education College, gait cycle analysis was done on force plate treadmill (Zebris, FDM–T), as shown in the Fig. 3.

The obtaining data were compared between the right, which was affected and the left leg which considered as a healthy leg, stance and swing phase were giving the behaviour of the gait cycle, as shown in the Fig. 4. The GRF exhibited the difference between the gait pattern for both right and left leg, and this added to these data, the pressure distribution under the insole when the subject wearing his shoes for both feet.

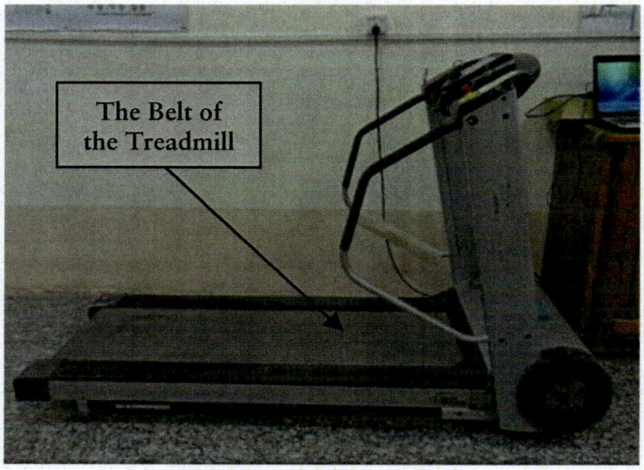

Fig. 3 The force plate treadmill

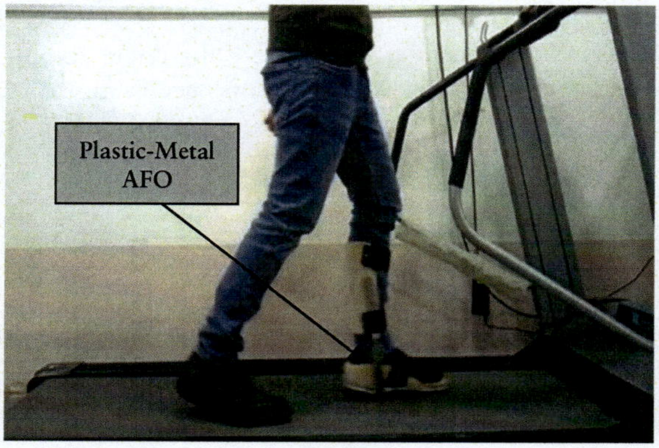

Fig. 4 Patient is walking on a force plate

2.4.2 Interface Pressure Test

The alternating load between the calf and patient's leg, for the man who was wearing the AFO model, was measured as force and pressure. The sensor type of (Mat-scan) is more agreeable for this type of dynamic load. The interface pressure was gained by recording the sensor's output signal through an instrument of multi-meter which is an interface with the computer and recording the data with the time illustrate in Fig. 5.

2.4.3 Kinematic Analysis

The kinematic analysis was done in University of Babylon/Sport Education College by using the tracker software (version 5.1.3) to detect the angle between the thigh and shank with the horizontal lines, as presented in the Fig. 6 via recording a digital video for a young man wearing different model of ankle foot orthosis.

He was walking in his normal speed, after putting markers on his leg to be as a reference point for drawing the main lines. The analysis was done according to the full gait cycle staring from heel strike, mid stance, toe off and the swing phase until repeating to heel strike for the second step. The obtaining data (the angles) were collected according to the time of the gait cycle.

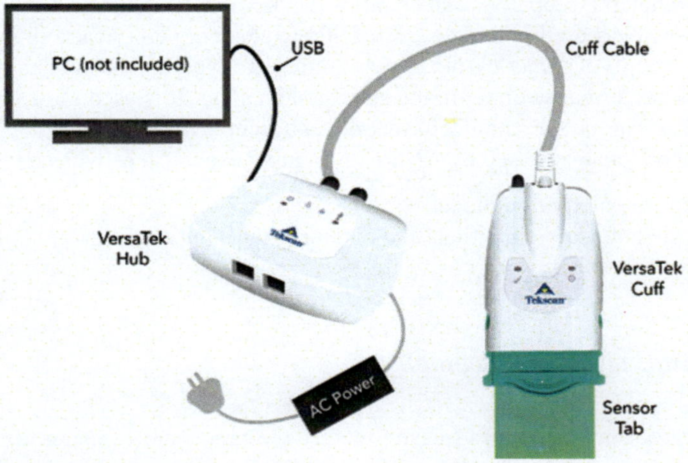

Fig. 5 The F-Socket pressure system

Fig. 6 The angles between the thigh and shank with the horizontal lines

3 Computational Modeling

The modern application of finite element method (FEM) nowadays, is extremely implemented for different fields in engineering and/or other science application. Taking into consideration the privilege of rapid development in the filed of digital computers with large capability, and huge computation. This method is known as one of the most important numerical methods because of its capabilities which include non-linear material properties and complex geometrical boundaries. The FEM software (Ansys) was used in this research as finite element numerical tool in order to predict the results of numerical models to simulate the experimental models to explain the fatigue performance effect in the structural element. It is used to determine the behaviour of fatigue life, total deformation, and safety factor [5].

The general analysis used by ANSYS has three distinct steps which are:

- Build the geometry as a model.
- Apply the boundary conditions and obtaining the solution.
- Review the results.

3.1 Building up the Geometry

The ankle foot orthosis types (plastic metal), the procedure of using the ANSYS Workbench program, for the plastic—metal model, will be used in this chapter to simplify the demonstrating of the main ANSYS process that is to say, modelling, meshing method, applying loads, etc.

This model was drawn by using CAD system (Solid-work version 2016) is illustrated in a Fig. 7, which fabricated according to an original prototype in three dimensions. Most of the small details in this model were taken into account in drawing.

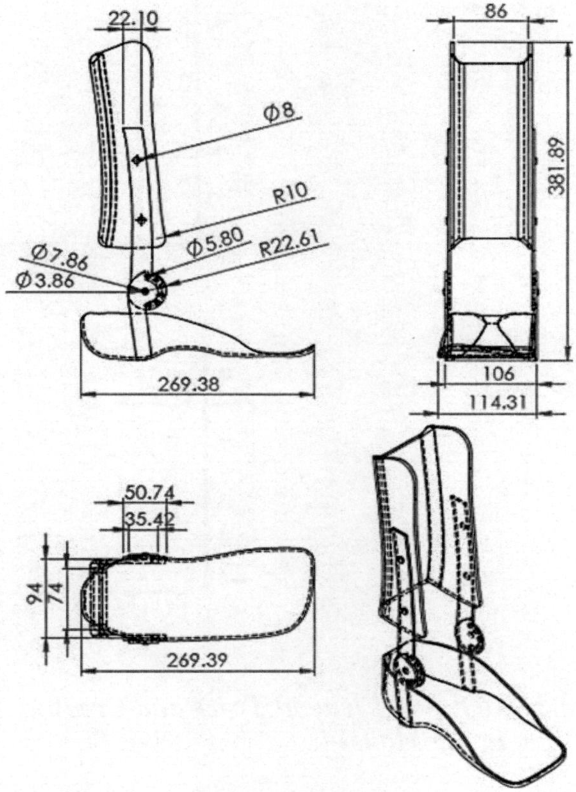

Fig. 7 Solid-work model of AFO

3.2 Determination of the Geometry

In this paper, ankle foot orthosis (AFO) model (are used as it will be explained in Table 1 and Fig. 8. The model of ankle foot orthosis that used in ANSYS Workbench V.15 is illustrated in the Fig. 9.

This model was drawn by using Solid-work program which fabricated according to an original prototype in three dimensions. Every detail was taken into consideration when drawing the model.

Table 1 The mechanical properties of selected materials used in numerical models [11]

Material	σ_y (MPa)	σ_{ult} (MPa)	E (GPa)
Polypropylene	27.20	37.30	1.24
Stainless Steel	792	820	192

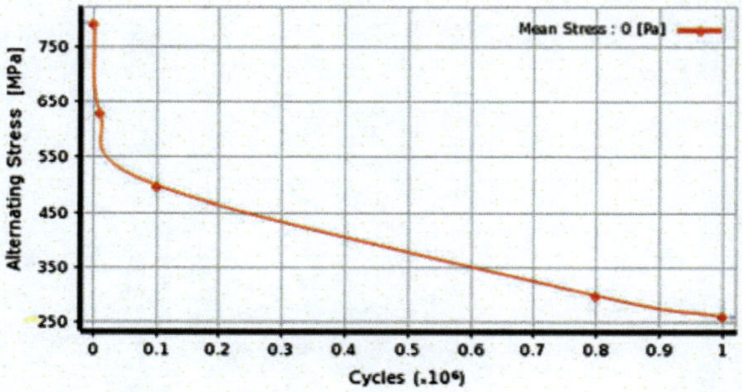

Fig. 8 S–N curve for stainless steel [7]

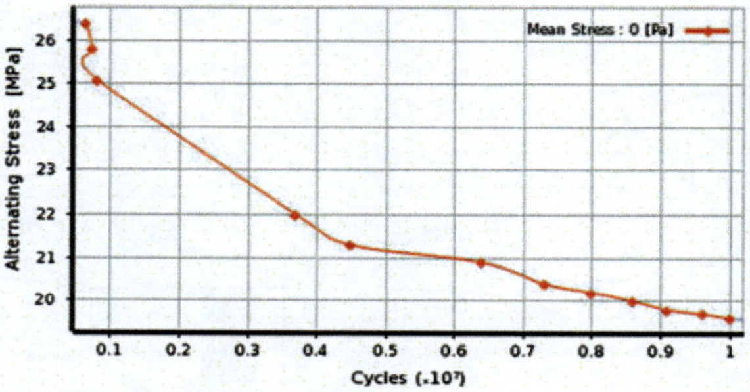

Fig. 9 S–N curve for polypropylene [8]

3.3 Defining Element Types and Creating the Mesh in the Model

For the FEM numerical models, the selected element of (Solid, Brick 8 node 45) was implemented. Usually, Solid 45 is applied for most of the 3D modeling of solid structures. The element is known by eight nodes having three degrees of freedom at each one translation in the nodal x, y, z directions [6]. For the meshing steps, a free mesh has been applied by choosing the volume, then the shape of elements were selected as tetrahedron (Automatic meshing), as shown in Figs. 10 and 11.

3.4 Defining the Analysis Type and Applying Load

The expression load" includes the boundary conditions (supports, constraints, and boundary field specification), as well as other internally and externally applied loads. For the applied loads used in the FEM program (Ansys) software will be (Fixed Support) at the side of the PFO heel segment [9]. The interface pressure was distributed at particular positions in a value taken from the interface pressure test for adjustable hinged AFO started from (zero to 277 kPa)) as shown in Fig. 12.

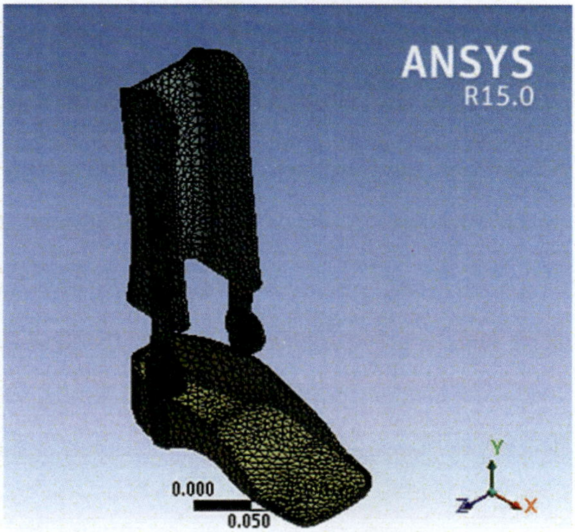

Fig. 10 The Mesh of the FEM Model

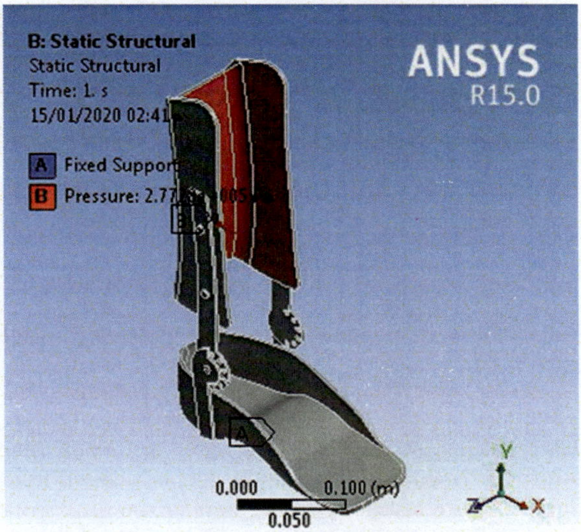

Fig. 11 Load Applied for the FEM Model

For the fatigue condition, fatigue tool was implemented to FEM model to find the life at particular loads, equivalent stress, and safety factor. The other solutions include deformation, stress, strain, etc.

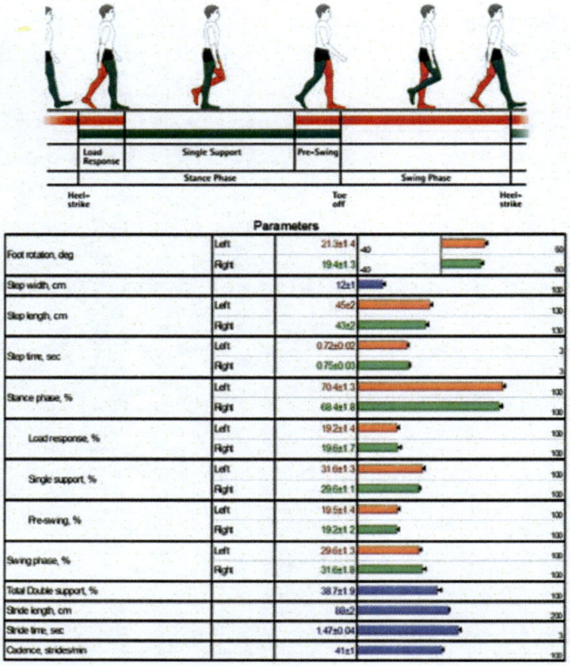

Fig. 12 The parameters of the gait cycle for the normal subject

4 Results and Discussion

Orthosis testing, after testing the patient in the Dynamic gait analysis on the treadmill and F-Socket, walking software will analyse the results and turn them into tables and curves.

4.1 Gait Cycle and GRF Test

The obtained data from the gait cycle test were compared between the two cases, normal and pathological, to recognize the major differences for the parameters of the left and the right leg. From the comparison results obtained for the normal subject, as shown in the Fig. 12 and the pathological case study shown in Fig. 13 and Table 2 shown the difference between the left and right feet for the two cases. The difference can be detected as in the stance phase percent for the left leg equal to (68.4%) and the right (72.0%) for the adjustable hinged AFO, while the normal case was (70.4%) and (68.4%), respectively. From Table 2 clarified a Large difference between both cases, the difference between the left and right foot of change stance phase, change load response, change pre-swing, change swing phase, total double support, stride length, stride time and cadence was (2, 0.4, 2, 0.3, 2 and 38.7) % respectively in the

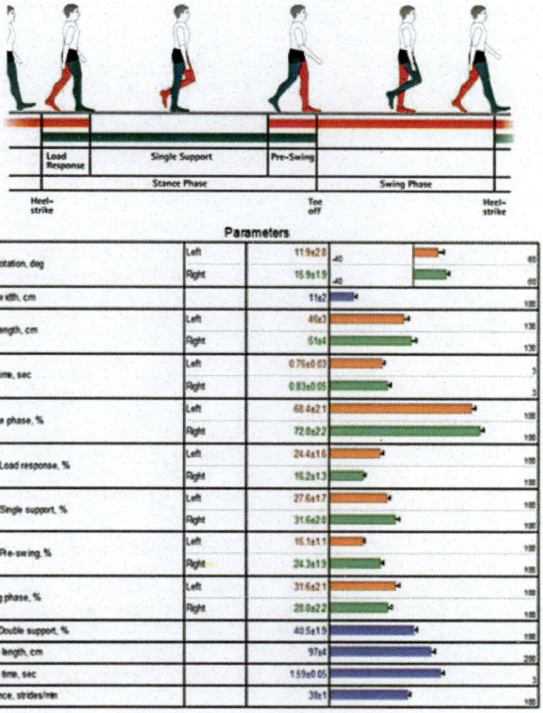

Fig. 13 The parameters of the gait cycle for the pathological subject wearing adjustable hinged AFO on his left leg

normal subject, but in the pathological subject of wearing adjustable hinged AFO, the results were (3.6, 8.2, 4, 8.3, 3.6 and 40.5) %.

The ground reaction force (GRF) for the normal and pathological subject for the left leg, also we see the effect of wearing the AFO on the right foot, illustrated in the Figs. 14, 15 and 16, respectively. Regardless of the magnitude of the GRF, which it is depending on the subject's weight, in general, the discussion will be supported by the behaviour of the GRF curve. To understand this curve, it should be divided into three parts (segments). The first segment is from the starting until the first peak which represents the heel strike, the second segment of the curve is the stage of mid stance and distributes the load of the body for the whole area of the foot, and the third segment is considered from the initial of the second peak which is equivalent to the toe off a stage. All these segments (GRF) occurred through the stance phase.

The pressure distribution for normal and pathological case for the left and right lower limb is represented in the Figs. 17, 18, and 19, 20 and 21 respectively. There is a sufficient difference at the initial of the curve as a sudden valley in the normal subject who is due to the distribution of the load under the subject's heel of his shoe, while it's not clear at the pathological subject and that backs to the using of the patient's heel in a gradual manner. The path of the centre of pressure (COP) for the normal subject is presented in the Figs. 22 and 23, for right, left foot and the butterfly form. This data is showing the behaviour of the COP under the foot which

Table 2 The difference result between left and right foot on gait analysis

	Normal subject	Adjustable hinged AFO
Change foot rotation (°)	1.9	4
Step width (cm)	12	11
Change step length (cm)	2	5
Change step time (s)	0.03	0.07
Change stance phase (%)	2	3.6
Change load response (%)	0.4	8.2
Change single support (%)	2	4
Change pre-swing (%)	0.3	8.3
Change swing phase (%)	2	3.6
Total double support (%)	38.7	40.5
Stride length (cm)	88	97
Stride time (s)	1.47	1.59
Cadence (strides/min)	41	38.4

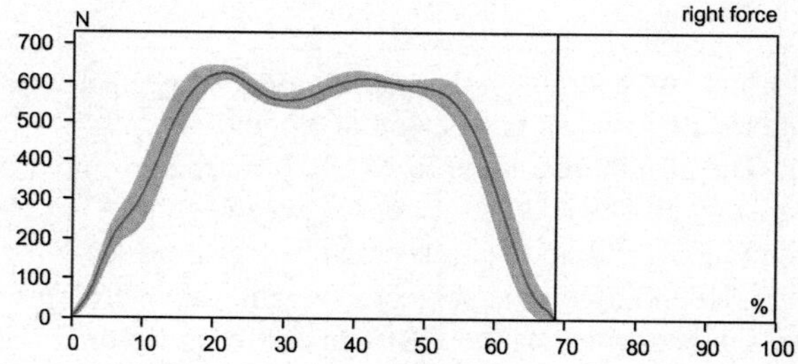

Fig. 14 The GRF (right) for the normal subject

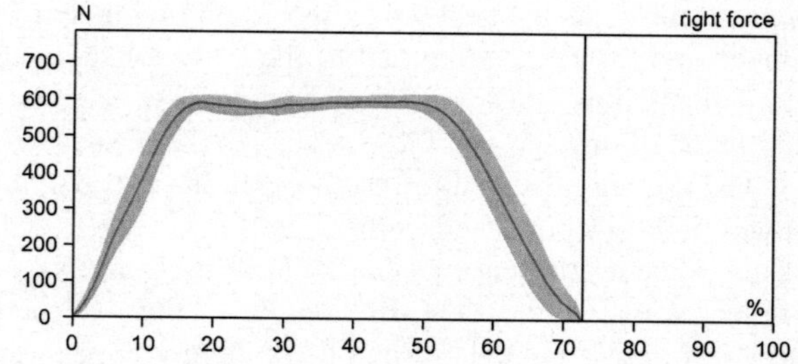

Fig. 15 The GRF (right) for pathological subject wearing adjustable hinged AFO

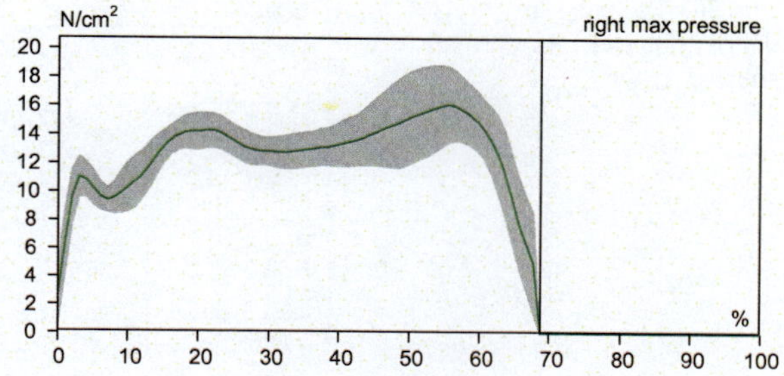

Fig. 16 The pressure distribution (right) for normal subject

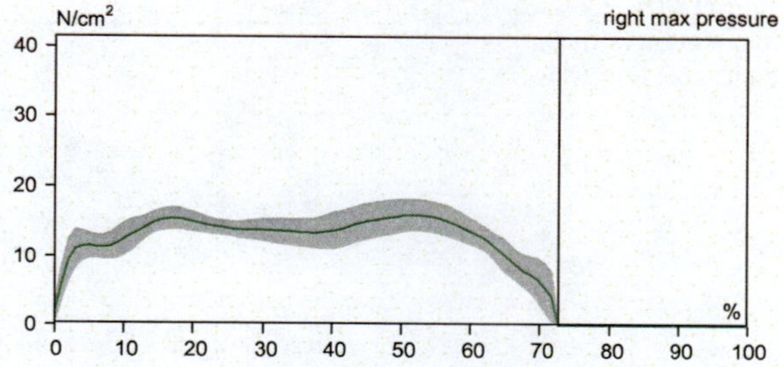

Fig. 17 The pressure distribution (right) for adjustable hinged AFO

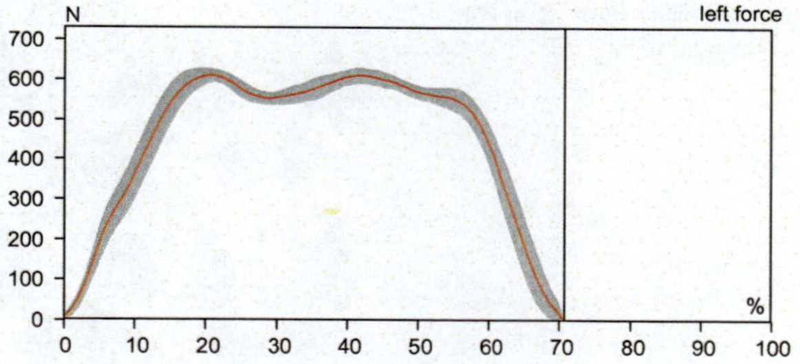

Fig. 18 The GRF (left) for normal subject

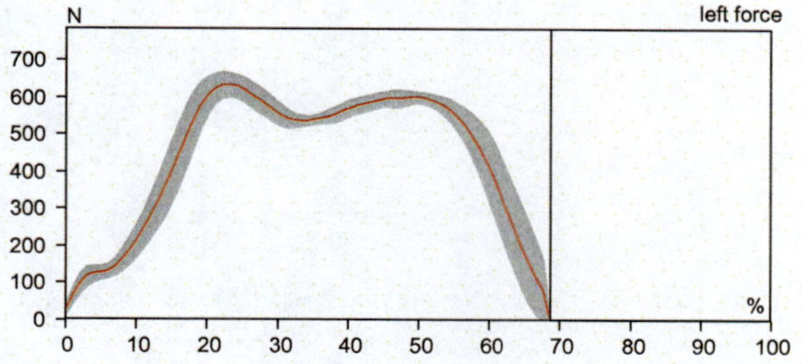

Fig. 19 The GRF (left) for pathological subject wearing adjustable hinged AFO

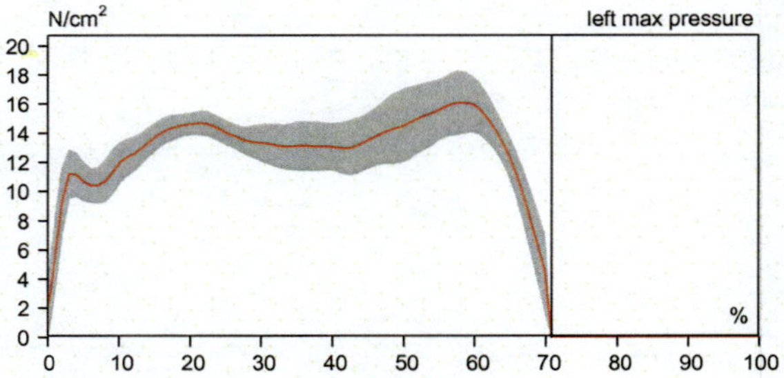

Fig. 20 The pressure distribution (left) for normal subject

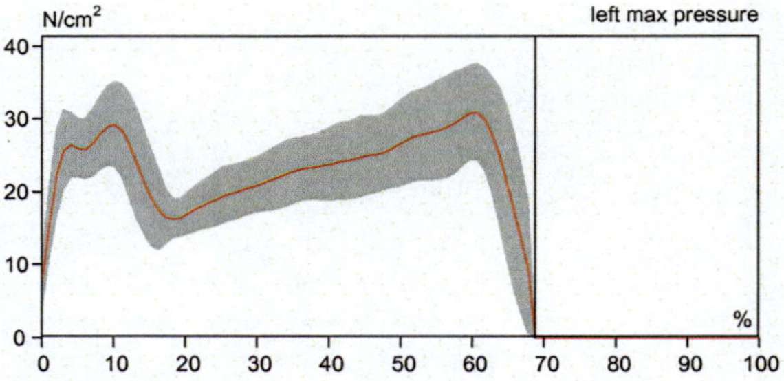

Fig. 21 The pressure distribution (left) for adjustable hinged AFO

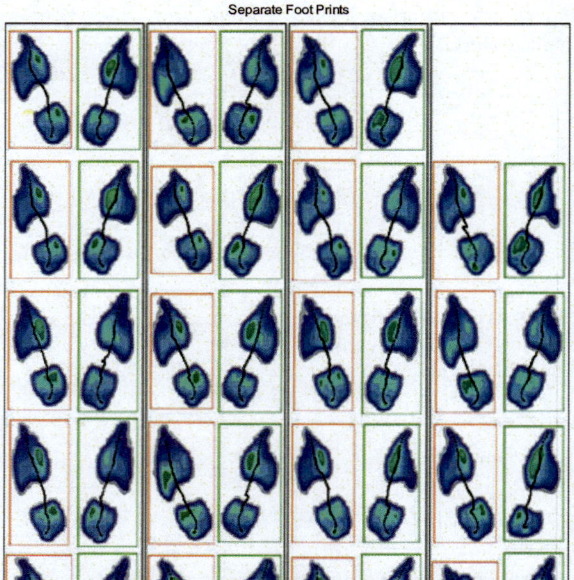

Fig. 22 The Foot Print for normal subject

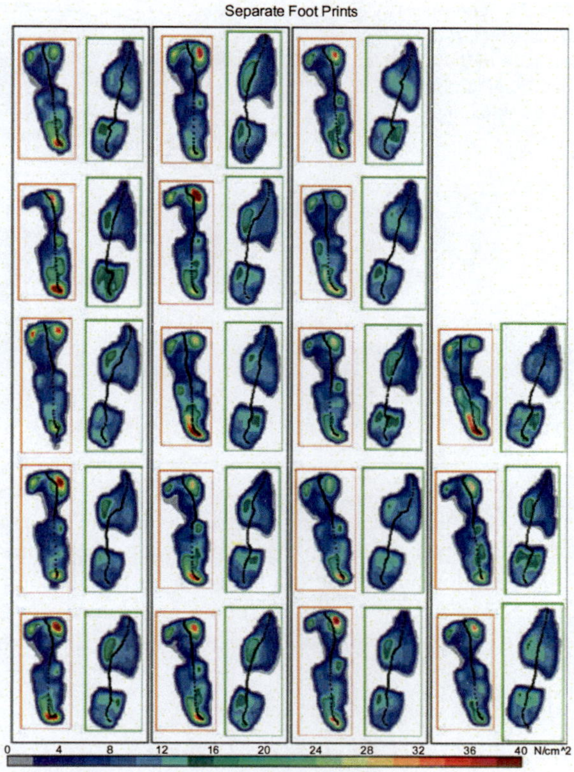

Fig. 23 The Foot Print for pathological subject adjustable hinged AFO

Butterfly Parameters

	15-02-2009 Gait Analysis	
	Left	Right
Gait line length, mm	256±13	268±10
Single support line, mm	93±16	94±17
Ant/post position, mm	135	
Ant/post variability, mm	5	
Lateral symmetry, mm	-2	
Lateral variability, mm	10	

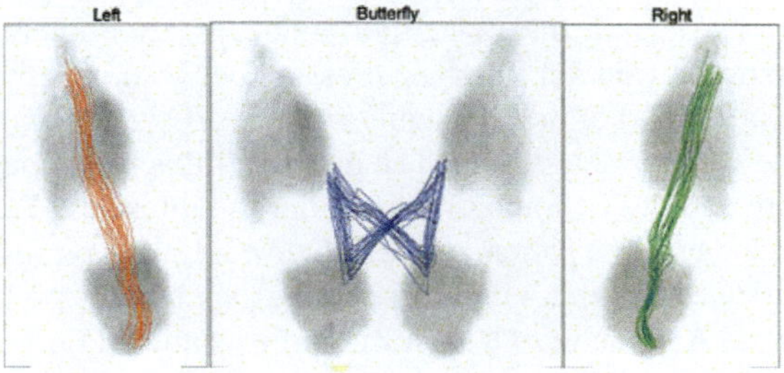

Fig. 24 The Butterfly parameters of the gait cycle for path of the centre of pressure for normal subject

Fig. 25 The Butterfly parameters of the gait cycle for path of the COP for pathological Subject adjustable hinge

Butterfly Parameters

	15-02-2009 Gait Analysis 2	
	Left	Right
Gait line length, mm	266±13	289±12
Single support line, mm	148±19	153±22
Ant/post position, mm	152	
Ant/post variability, mm	19	
Lateral symmetry, mm	14	
Lateral variability, mm	18	

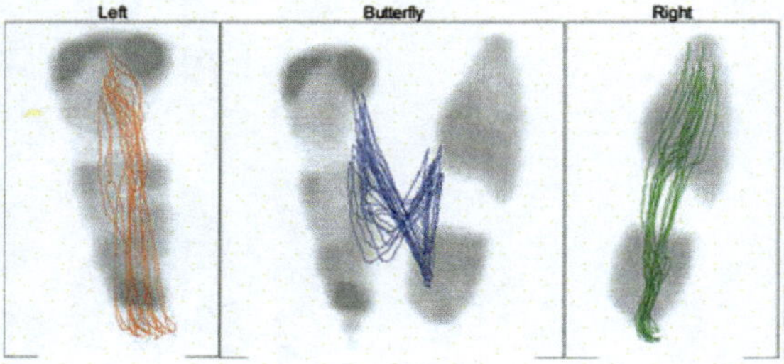

started from the heel strike, mid stance and toe off a stage. Therefore, through each of these stages, there is a maximum value of pressure. The butterfly form refers to the abduction and adduction between the feet through the ambulation, as explained in the Figs. 24 and 25, which represent the contour of pressure distribution in three dimensions for the normal and pathological subject, respectively.

4.2 F-Socket Results and Discussion:

The pressures were only represented over the gait cycle by the use of contact method between the AFO and the patients at the calf reign. The data are normalized to 100 percent of the gait cycle [9]. The pressure for subjects is different at weight acceptance from one patient to another.

The interference force and pressure between calf and the leg inside orthosis in this test were calculated, for this case study, the experimental part with the fracture damage of ankle due to the sport injure, the experimental results arrived state that the maximum value of interface pressure between patients' leg and AFO is recorded at the upper part of the adjustable hinged AFO (277 kPa) and decreased in the direction of lower part of the AFO. The reason for this behaviour is that the calf muscles have more activity as comparing with the foot muscles. The pressure distribution with a percentage for a patient of the right lower limb is represented in Fig. 26. Also, Fig. 27 illustrated the maximum force peak is applied on the leg is (111.2 N) for adjustable hinged AFO and (79.8 N) for modified static AFO.

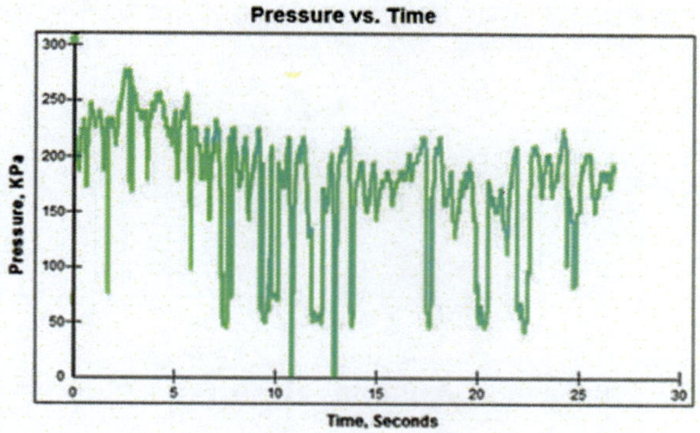

Fig. 26 The Interference Force Between the Left Leg and AFO Model

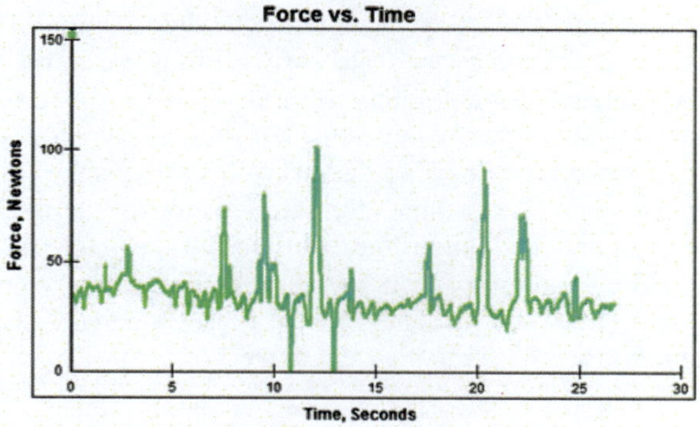

Fig. 27 The Interference Force Between the Right Leg and AFO Model

4.3 Kinematic Analysis Result

For this result we took five readings and then averaged this result to obtain an appropriate behaviour.

the normal condition shown in a Fig. 28 in black line the maximum angle is (102.12). in a Fig. 29. The ankle angle for adjustable hinged AFO is (91.79).

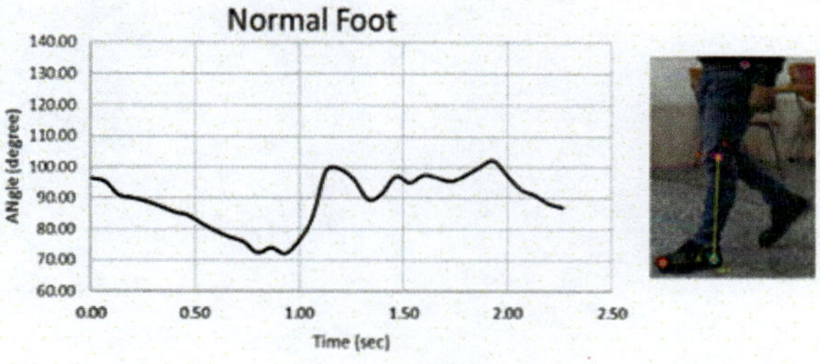

Fig. 28 Ankle angle analysis for normal foot

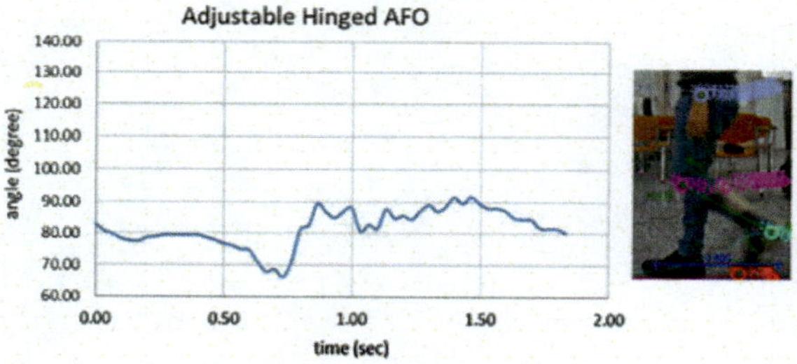

Fig. 29 Ankle angle analysis for adjustable hinged AFO

5 Numerical Analysis Results

The AFO analysis for the model was settled by the FEM software (Ansys) in order to compute the safety factor and total deformation of fatigue. AFO model analysis was represented by the FEM software (Ansys) in order to compute the (Von-Mises) equivalent stress and safety factor of the fatigue condition. According to the principle of Von-Mises theory that considers the yield stress as criteria; ($\sigma_e < \sigma_y$, safe), ($\sigma_e = \sigma_y$, critical) and ($\sigma_e > \sigma_y$, failed). Where (σ_e) is the equivalent stress, and (σ_y) is the yield stress. That's mean the safety factor for the fatigue condition will be safe for design only and only if the safety factor is equal or more than (1.25) [10].

The contours of Von-Mises stress and safety factor for (Polypropylene, steel) AFO model are presented in the Figs. 30 and 31 respectively, the maximum von mises stress 418.9 in the steel hinge.

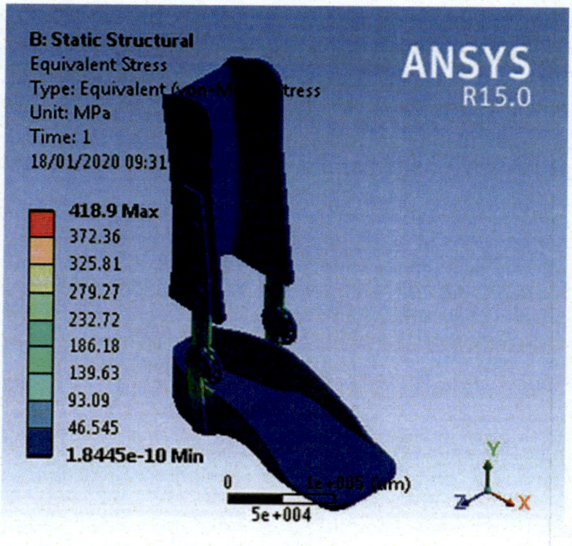

Fig. 30 Equivalent Stress (Von-Mises) (adjustable hinged AFO)

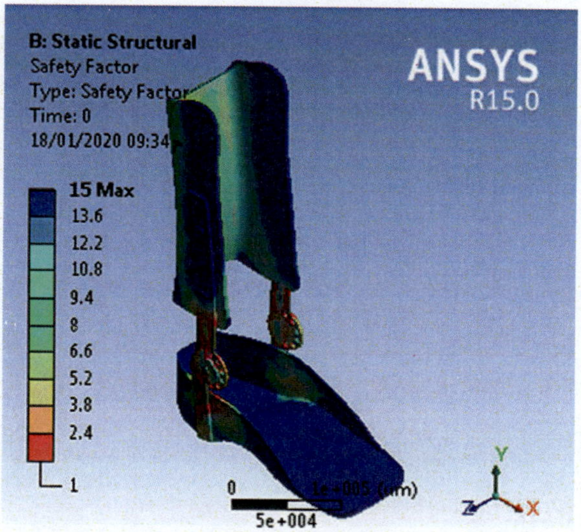

Fig. 31 The Safety Factor for Fatigue (adjustable hinged AFO)

5.1 Adjustable Hinged AFO Model

The Von-Mises stress & safety factor for adjustable hinged AFO model is depicted in Figs. 30 and 31 respectively.

6 Conclusions

The following conclusions may be drawn from the results obtained in this work:

1. The comparison process between the pathological subject and the normal subject in the gait cycle test showed major differences between the two subjects and that back to the circumduction.
2. For the results of gait cycle test, it shows a large difference between both models. As example, the difference between the right and left foot of the change stance phase, changing load response, changing swing phase, changing pre-swing, stride length, total double support, stride time and cadence was (2, 0.4, 2, 0.3, 2 and 38.7)% respectively in the normal subject, but in the pathological subject of wearing adjustable hinged AFO, the results were (3.6, 8.2, 4, 8.3, 3.6 and 40.5)%.
3. The Mat-Scan sensor, which is used to measure the interface pressure, was suitable for the alternating load between the calf and the patient's leg.
4. The max interface pressure result is (277 kPa) for adjustable hinged AFO.
5. The modified static AFO type polypropylene exhibited failure in fatigue life design when the value of the safety factor was about (2.4).

6. The adjustable hinged AFO gave good results in equivalent Von-Mises stress and the safety factor for fatigue, and this led to the longer life design. This type of the AFO is commended to a limited kind of the patients, depending on the weight, because of the concentration of the forces between the regions of the stainless steel hinge and the patient's foot, and that causes uncomforting for the patient.
7. Through a(tracker) software using kinematic analysis, it was found that the pathological condition is closer to the normal state when using this type of AFO.

Acknowledgements This work was carried out under the supervision of Asst. Prof. Dr. Sadiq J. Aziz and Asst. Prof. Dr. Ahmed A. Shandookh to whom the author wishes to express his sincere appreciation for his help, discussion and criticism during the execution of the work described in this article. I would like to express my thanks to Dr. Ali Jawad and MT. Masoud Ali Akbar for giving me a helping hand.

References

1. Mohammed YA (2008) The role of new technology in designing and fabrication of orthosis and prosthesis. Thesis, Zagazig University, M.Sc
2. https://www.rehabmypatient.com/ankle/ankle-fracture
3. O&P Chen J, Siegler S, Schneck CD (1988) The three dimensional kinematics and flexibility characteristics of the human ankle and subtalar joint part II: flexibility characteristics. J Biomech Eng 110:374–85
4. International Committee of the Red Cross (2006) Manufacturing guideless. http://www.icrc.org/en/doc/assets/files/other/eng-afo.pdf
5. Resan KK, Ali H, Hilli A, Muhsin Ali M (2011) Design and analysis of a new prosthetic foot for people of special needs. Iraqi J Mech Mater Eng 11(2)
6. ANSYS Program Help (Version 15)
7. Lampman S (1996) Fatigue and fracture, properties of stainless steels. ASM Int Handbook 19:1811
8. Maier C, Calafut T (1998) Polypropylene—the definitive user's guide and databook. In: Woishnis W (eds) Plastics design library a division of William Andrew Inc. New York
9. Takhakh AM, Salih Hussain H (2017) Manufacturing, testing, and numerical modeling a lower limb orthosis for a patient that has a partial foot amputation. Int J Energy Environ 8(4):347–356
10. Hassan SS, Resan KK, Mahdi AZ (2013) Design and analysis of knee ankle foot orthosis (KAFO) for paraplegia person. Eng Tech J 31, Part (A), No. 8
11. Shandookh AA (2010) The effect of heat sink fins, length and material on its performance. Diyala J Eng Sci 3:1–11

Leakage Quantification of Bolted Flange Joint Subjected to Different Bolt Sizes and External Load

Ir. Alzakri Ekhwan, A. R. Othman, Andre Franzen, M. F. Othman, and Azliza Ishak

Abstract The reliability of bolted connections is largely controlled by the level of initial clamping force and by the stability of the clamping force over time. For many types of structural systems, a failure of a bolted connection is often due to an improper installation, which may lead to reduction in joint tightness and subsequently lead to flange leaks. Further load loss is progressively prolonged due to non-linear behavior during service, reaching below the minimum required clamping load. The study is therefore aimed to analyze the effect of bolt size and external load due to channel head weight on the gasket contact pressure for the quantification of the flange leakage. The analysis is conducted on the tube sheet flange of feed exchanger at a petrochemical plant. The flange was subjected to re-boring prior to commissioning of the tube sheet flange bolt holes. Leak has manifested immediately after startup at 12 o'clock position due to load loss caused by channel head weight. Finite element analysis has been developed to quantify the sensitivity of channel head weight and bolt sizes on the flange integrity. The model comprises of the channel head flange, shell flange, tube sheet, bolt, but and gasket. Series of bolt sizes, i.e. M24, M30 and M36 were introduced for sensitivity study. The external loads were varied, ranging between 380 and 2000 N. Bolt loads were calculated using standard WRC 538 with 50 and 60% bolt yields. The bolt root area and tightening stress were obtained by referring to standard ASME PCC-1 Appendix H. The analysis has been validated with literature and site findings. The stresses at bolt and flange hub are quantified, and the correlation with contact pressure is subsequently assessed. Early analysis has shown that the leak occurred due to loss in bolt stresses subjected to channel head weight. Higher longitudinal/axial stress concentration is observed at lower section

Ir. Alzakri Ekhwan (✉) · M. F. Othman · A. Ishak
Engineering Department, Group Technical Solutions Project Delivery & Technology (PD&T), Petroliam Nasional Berhad (Petronas), 50088 Kuala Lumpur, Malaysia
e-mail: alzakri@petronas.com

A. R. Othman
Department of Mechanical Engineering, Universiti Teknologi PETRONAS, 32610 Seri Iskandar Perak, Malaysia

A. Franzen
PETRONAS Research Sdn Bhd, Group Research & Technology (GR&T), Jalan Ayer Itam, Kawasan Institusi Bangi, 43000 Bandar Baru Bangi, Selangor, Malaysia

F. Ahmad et al. (eds.), *ICPER 2020*, Lecture Notes in Mechanical Engineering,
https://doi.org/10.1007/978-981-19-1939-8_57

of the bolt due to hub rotation. In addition, higher stress values are observed at flange hub due to a concentrated stress region. Contact pressure was found insufficient for the minimum clamping force as a result of loss of bolt tension (small bolt size), to sustain the max stress at bolt.

Keywords Bolted flange joint · Flange leakage · Finite element analysis

1 Introduction

The reliability of bolted connections is largely controlled by the level of initial clamping force and by the stability of the clamping force over time. For many structural systems, a failure of a bolted connection is often due to an improper installation, which may lead to reduction in joint tightness and subsequently lead to flange leaks. Further load loss is progressively increased due to non-linear behavior during service, thus getting below the minimum required clamping load.

While failure of a single non-critical fastener may not be significant to overall structural stability and serviceability, an improper level of clamping load of bolts in a connection may result in excessive vibration or insufficient stiffness of a member. In addition, even the bolt flanges are properly installed with the initial bolt load (i.e. within the maximum allowable clamping load), there will be load loss during flange assembly process following the plant start up [1]. The load loss is progressively extended due to non-linear service behaviour, reaching below the minimum required clamping load, as seen in Fig. 1. The condition subsequently, results in the bolt removal, reducing the joint tightness and lead to flange leaks.

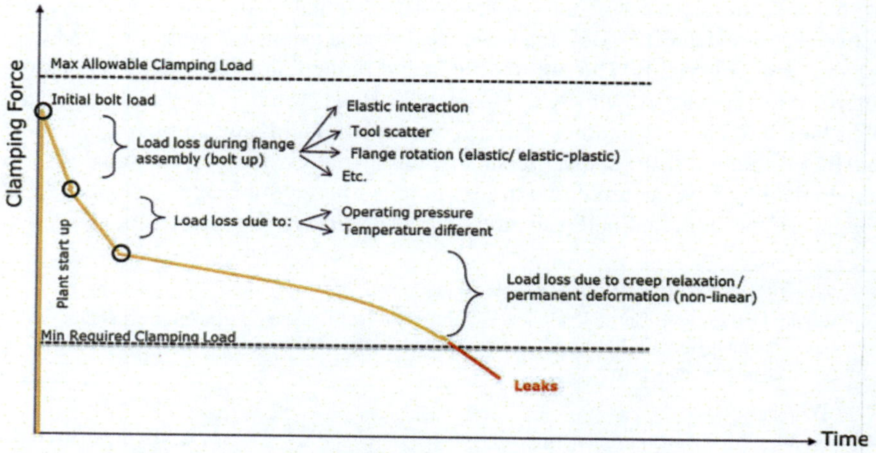

Fig. 1 Summary of non-linear in-service behaviour that may cause load loss and reduce the joint tightness

The sealing performance of the pipe flange connection has been evaluated through series of experimental and numerical analyses [2]. It has been suggested that the contact gasket stress, which is a function of the bolt preload, determines the leakage. The changes in contact gasket stress at tension and compression sides were not symmetrical when the external bending moment was introduced. The reduction in the contact gasket stress of tension side was larger than that of compression side due to the non-linear stress–strain behaviour of the gasket. Similarly, Sawa et al. [3] has conducted an analysis using elastoplastic (EP) finite element method to evaluate the contact gasket stress distribution in the connection with a spiral wound gasket subjected to internal pressure and bending moments. The effects of initial clamping bolt force (bolt preload), the nominal pressure and the nominal diameter of pipe flange and the equivalent pressure when the leakage occurs were characterized. By using the results of the leakage tests and the calculated contact gasket stress distributions, it was found that the sealing performance of the connection can be estimated, that is, when the contact gasket stress becomes zero, the leakage occurs.

In a more recent study, a detailed three-dimensional nonlinear finite element analysis of bolted gasketed flange joint was performed, to study its performance under combined internal (pressure) and external (axial and bending) load [4]. It has been concluded that the internal pressure plus axial loading has significant effect on the bolt behaviour. It was observed in combined loading that axial loading has a more prominent impact on the axial displacement and also on the sealing capacity, providing chances of leakage. The addition of bending load slightly increased the bolt stresses, but has very insignificant effect on the joint strength.

An established standard of WRC Bulletin 538 has been prepared in ASME PCC-1-2010 Appendix O [5], to assist in the understanding of the joint integrity and to improve the practicality and reliability of the joint. It determines the flange limits to prevent any permanent damage to the joint during assembly or operation. However, the reference focussed only on the procedure of correct bolt load design for acceptable bolt stresses that can be applied to flanged joints without damaging them.

In literature, limited data are available on the methods to statistically quantify the possible root cause of leaks due operational loadings. Currently, there is no means to develop an explicit analysis to compute the load losses and hence joint tightening healthiness. The load loss currently embedded in the current work processes and tools are estimated figures. The study is therefore aimed to quantify the effect of bolt size and external load due to channel head weight on the gasket contact pressure for the quantification of the flange leakage.

2 Methodology

2.1 Finite Element Model

The pressurized vessel of a stripper reboiler consists of tube bundles and shells that are joined using bolted flanges, known as FJ-101. Generally, there are three flanges that are tied together, namely channel head flange, tube sheet flange and shell flange. Detail specifications are given in Table 1. The flange was subjected to re-boring prior to commissioning of the tube sheet flange bolt holes. Leakage was manifested on the shell side of the flange at the 12 o'clock position upon the start-up and even after retightening of the bolts.

A finite element (FE) model has been developed to quantify the sensitivity of channel head weights and bolt sizes on the flange integrity. The model comprised of the channel head flange, shell flange, tube sheet, bolt, and gasket. All the geometrical dimensions were acquired as per equipment drawing from the plant. Some simplifications have been made to the model where only flanges were included, disregarding

Table 1 Specification of the bolted flange FJ-101 at the pressurized vessel

Specification			
1	Code and standard applied		(i) WRC 538 (ii) ASME PCC-1
2	Joints location		Channel head-tubesheet-shell
3	Flange size/rating		Non-standard
4	Flange material		A105
5	No. of bolt		68
6	Bolt material		A194-2H
7	Bolt size		M24
8	Type gasket		Metal grooved gasket
9	Operating medium	Shell	C4
10		Tube	H20
11	Design pressure (kPa)	Shell	3309
		Tube	2209
12	Operating pressure (kPa)	Shell	3010
		Tube	380
13	Design temperature (°C)	Shell	150
		Tube	150
14	Operating temperature (°C)	Shell	110
		Tube	142
15	Nominal thickness (mm)	Shell	17
		Tube	2.11

the shell column and channel head geometries. Half bolt model assembly with gasket thickness was modelled based on symmetry. Figure 2 shows the mesh model of flange, gasket, and bolt of flange FJ-101.

The FE analysis was performed using commercial software called ABAQUS. Quadratic 3D elements with reduced integration (C3D20R) were used in the current study for computing stresses at flange, bolts, and gasket. Contact was defined as surface-to-surface between the parts' surface. In this model, a simplification on the material was considered to include elastic-perfectly plastic model (EPP). Table 2 tabulates the material properties of flange, tube sheet, bolt, nut, and gasket used in the current study, taken from ASME Section II Part D [6].

A series of bolt sizes, i.e. M24, M30 and M36 were introduced for a sensitivity study. Bolt loads were calculated at a bolt yield of 50% based on WRC Bulletin 538 [5], as shown in Table 3. ASME PCC-1 Appendix H [6] was used as a reference to

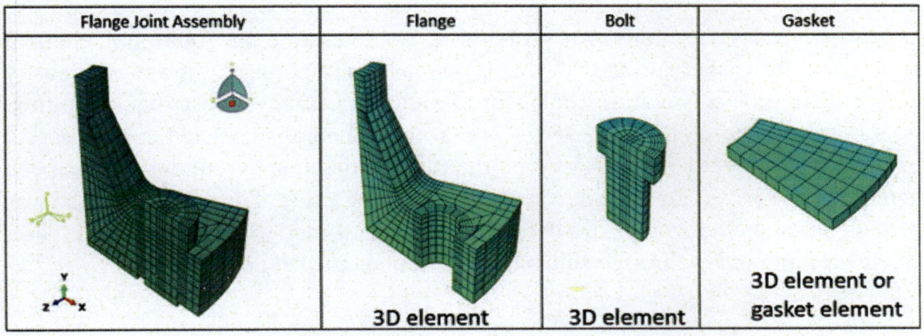

Fig. 2 Mesh model of the flange joint assembly

Table 2 Material properties of the flange components

Component	Material	Young modulus (MPa)	Poisson ratio
Channel head flange	A105	190,985	0.3
Shell flange	A105	190,985	0.3
Tube sheet	A266	186,158	0.3
Bolt	A193	206,843	0.3
Nut	A194	206,843	0.3
Gasket	NA filler	19,510	0.2

Table 3 Bolt load calculation

Bolt size	FJ-101 (50% bolt yield) (N)
M24	57,510
M30	96,968
M36	146,685

obtain the bolt root region and the bolt pressure for tightening. The extruded side of the model was constrained to allow movement only in axial and radial direction. The pin support was introduced at the end of the shell flange. The bolts were constrained in the radial direction, thus giving a better response axially along the bolt. In order to achieve the required bolt-preload, a displacement constraint was applied in the y-direction at the bottom of the bolt shank. After application of bolt pre-load, the external load representing the channel head weight was applied at the end of the shell flange.

3 Model Validation

3.1 Element Mesh Sensitivity

A mesh convergence study was conducted to determine the optimum element numbers at the gasket surface. This analysis involved increasing the mesh density until it achieved satisfactorily balances in output accuracy and computing time. Figure 3 show the plot of *Von Mises stress* against number of elements at gasket for the flange model FJ-101. Based on the plot, as the element numbers have been increased, *Von Mises stress* tends to converge at some value. The optimum mesh was selected based on the critical size before stress value started to converge. In this case, 5 element numbers in through-thickness direction at gasket are selected.

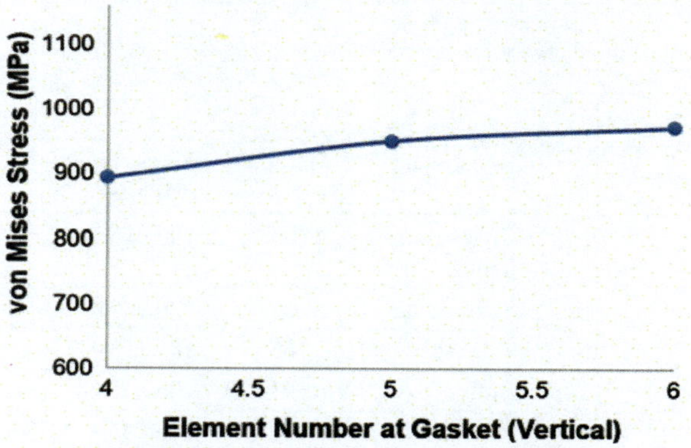

Fig. 3 *Von Mises stress* versus element numbers for FJ-101 analysis

Table 4 Results from analytical calculation

Calculation	Value
Bolt stiffness constant, K_B	155.11 kN/mm
Joint stiffness constant, K_J	112.69 kN/mm
Elongation of bolt, ΔL	1280 mm
Change in bolt load, ΔF_B	741 N

3.2 FE Validation with Bickford [8]

The half bolt model was validated with the theoretical calculation based on the Ref. [8]. The governing equations to calculate the stiffness and spring constants are as follow:

$$\text{For bolt load,}\ \ K_B = F_P/\Delta L; \qquad \text{For joint,}\ d,\ K_J = F_P/\Delta T$$
$$\text{By trigonometry,}\ \ \Delta F_B = \left[\frac{K_B}{K_B+K_J}\right] L_x$$

The joint will experience separation when the external load reaches the critical value defined by $L_{x,crit} = F_P\left(1 + \frac{K_B}{K_J}\right)$, in which the load factor, $\varnothing = \frac{\Delta F_B}{L_x} = \left(\frac{K_B}{K_B+K_J}\right)$. Therefore, the change in bolt load is calculated as $\Delta F_B = \varnothing_K L_x$. The results from the analytical equations are tabulated in Table 4. A comparison with the FE results shows a very close agreement at 0.36%.

Result from FE analysis indicated the maximum reaction forces R_F after and before the external load are 1746 N and 2485 N, respectively. Thus, the change in bolt load $\Delta F_B = 739$ N. The percentage difference between analytical equation and FE results:

$$\%diff = \frac{741 - 739}{741} \times 100\% = 0.36\%.$$

4 Results and Discussion

Figure 4 presents the overall deformation of FJ-101 joint at 12 o'clock position when subjected to loading from the channel head weight. The analysis indicates excessive deformation at the flange hub of the column side due to the external weight. This causes loss in bolt stresses, and as a result, contact pressure is insufficient for the minimum clamping force at the gasket surface. The FE results agreed with the site finding as shown in Fig. 5. The leak consistently occurred at the column side, at the 12 o'clock position when the channel head weight was loaded. It is anticipated that this was due to insufficient bolt size, after re-boring during the repair work.

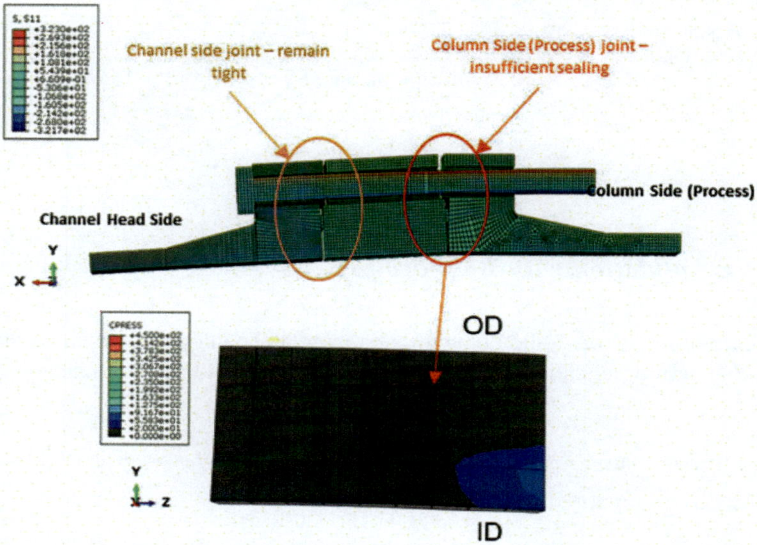

Fig. 4 Deformation of FJ-101 joint under channel head weight loading

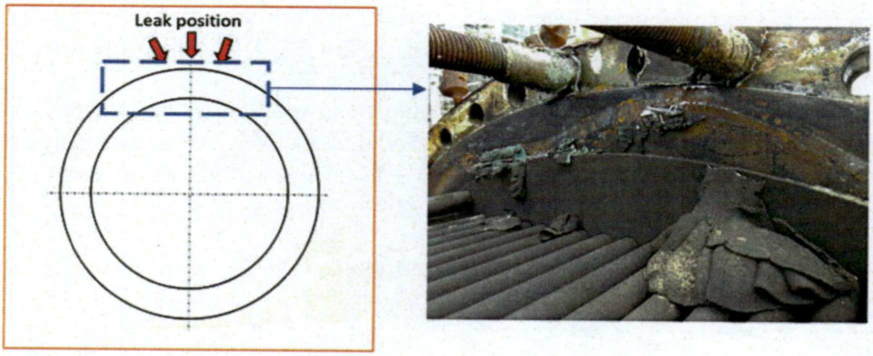

Fig. 5 Site observation and findings

In addition, the FE analysis has shown that the bolt thread experiences plastic deformation immediately after the start-up when subjected to channel head weight. Figure 6 indicates that the stress concentration on the tapped hole causes plastic deformation on the bolt thread, and these lead to maintenance problems (i.e. difficult to remove and insert any bolt). Further analysis has shown that an increase of the bolt load may sufficiently address channel head weight issue.

Figure 7 provides the contour of the contact pressure at the gasket for three different cases. At the bolt load condition prior to the applying the external weight (a), the gasket experience full and sufficient sealing for M24 bolt size. However,

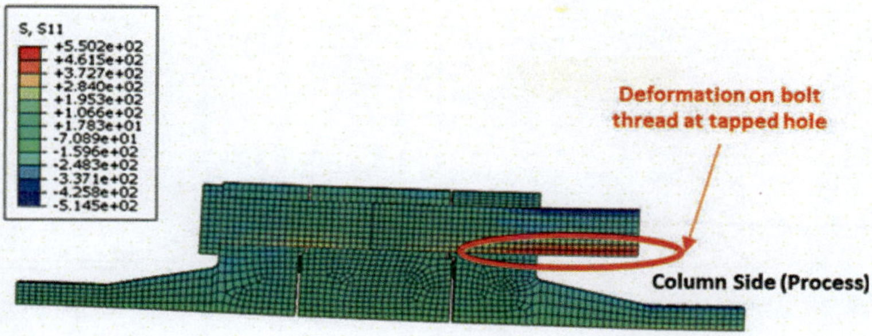

Fig. 6 Deformation of bolt thread subjected to channel head weight loading

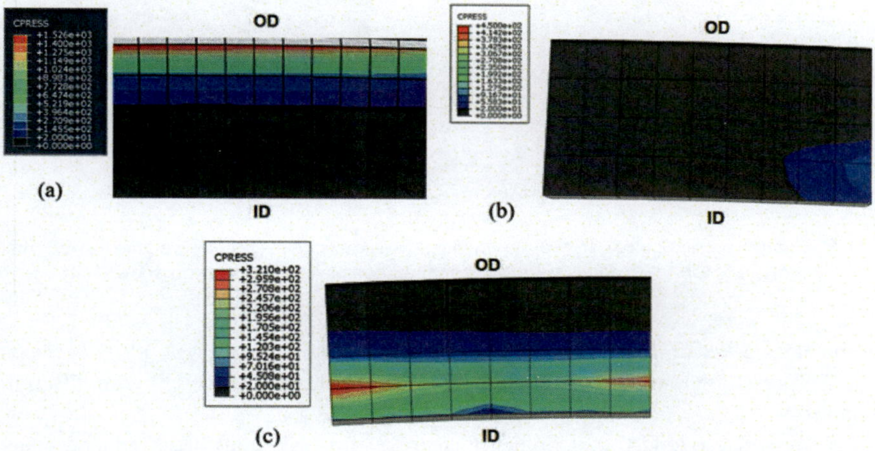

Fig. 7 Contour of the gasket contact pressure: **a** using M24 bolt before and **b** after applying CH weight **c** with M36 bolt

after the external weight was applied (b), the sealing capacity has significantly been reduced. Zero contact pressure of FJ-101 at the gasket surface of 12 o'clock position was apparent, causing the leak at the flange. It is anticipated that the additional channel head weight can be addressed by increasing bolt load, hence bigger bolt size required. An analysis based on M36 bolt size has shown that the use of a bigger bolt can sufficiently address channel head weight. The gasket contact pressures at the 12 o'clock position under the channel head weight loading in (c) indicates sufficient sealing was achieved.

Additional loading from the channel head weight certainly effects the gasket contact pressure. The weight creates a moment force and pulls the channel head downwards. The unsupported weight will cause the insufficient sealing at the gasket area, hence causing the flange to leak. A sensitivity study has been conducted further to look at the effect of the weight with different bolt sizes on the contact pressure

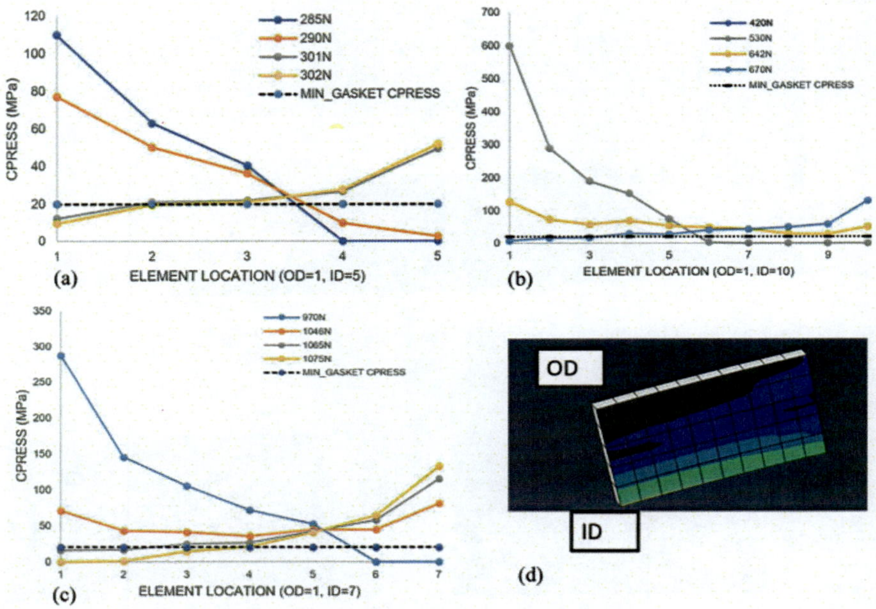

Fig. 8 Contact pressure along the gasket elements at various channel head weights for different bolt sizes of **a** M24, **b** M30, and **c** M36. The configuration of gasket element is shown in **d**

along the gasket area of FJ-101 flange model. Figure 8a–c shows plots between contact pressure along the gasket after the addition of channel weight with different bolt sizes.

For bolt size of M24, the contact pressure at the gasket has been reduced when external loading of 301 N was applied. At this point, the gasket starts to pry due to the counterweight from the channel head. As a result, the bolt experiences relaxation, and thus the bolt starts to lose its tightening force. The loss of the preload in the pretensioned bolt has been discussed in [8]. This condition may cause leaks at the gasket. When larger bolts were used, the threshold limit to sustain the external load without causing the loss in gasket sealing was increased to 670 and 1065 N for M30 and M36, respectively. This is apparent from the plots in Fig. 8b and c as the contact pressure is reduced below the minimum gasket limit.

Basically, the larger bolt was tightened with a larger bolt load, and thus a strong compressive force will prevent the gasket from leaking. Other than that, a larger bolt requires a bigger flange due to the bolt spacing. The relationship of bolt size and channel weight is expected to be linear, as depicted in Fig. 9. It is shown that in order to overcome the channel weight, a bigger bolt is needed to achieve the required sealing. From this analysis, it has been shown that larger bolt sizes will provide a good sealing at gasket and perhaps can withstand larger channel weight.

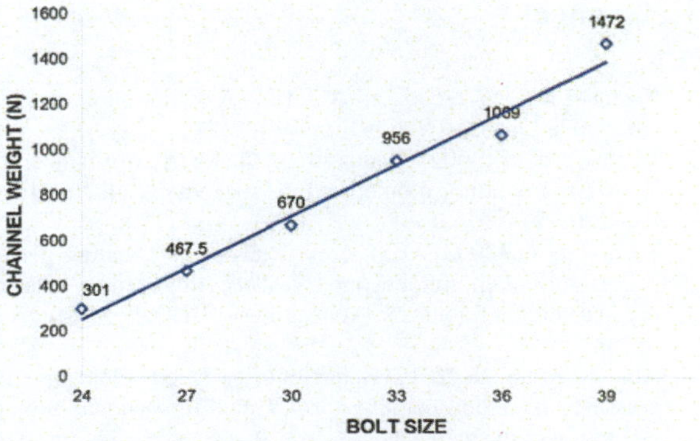

Fig. 9 Correlation of channel head weight versus bolt sizes for flange joint FJ-101

5 Conclusion

Finite element analysis has been developed to quantify the possible root cause of flange leaks experienced in operational loadings. The analysis included the variation of channel head weight and also the variation of bolt sizes for computing the load losses and gasket contact pressure, hence the joint integrity. The following are the main findings from the study:

- The analysis has confirmed the site findings, the leak consistently occurred at the Column side, at 12 o'clock position when the channel head weight is loaded. This results into zero contact pressure of the FJ-101 at the gasket surface of 12 o'clock position, causing the leak at the flange. It is anticipated that this was due to an insufficient bolt size, after the re-boring work.
- Early assessments have indicated that the leak occurred due to loss in bolt stresses caused by channel head weight. Higher longitudinal/axial stress concentration is observed at the lower section of the bolt, causing plastic deformation due to hub rotation.
- Insufficient sealing capacity has been found due to low contact pressure at the gasket surface when subjected to external channel head weight.
- The use of a larger bolt (i.e. M36) increases the threshold limit to sustain the external load without causing any loss in the gasket sealing. The additional channel head weight can be addressed by increasing the bolt load/size. Linear correlation analysis between channel head weight and bolt sizes has provided indicative of bolt size required to withstand the specific channel head weight to ensure a sufficient sealing capacity at the gasket surface.

Acknowledgements The authors would like to thank PETRONAS GR&T and Universiti Teknologi PETRONAS for the financial support through the Master Research Agreement (MRA) Funding of 015MD0-023 for this research work.

References

1. Nikravesha SMY, Goudarzib M (2017) A review paper on looseness detection methods in bolted structures. Lat Am J Solids Struct 14(12)
2. Takagi Y et al (2009) Evaluation of sealing performance of pipe flange connection subjected to external bending moments. In: Proceedings of ASME pressure vessels and piping 2009, PVP2009-77494
3. Sawa T et al (2008) FEM stress analysis and sealing performance evaluation in pipe flange connections with spiral wound gaskets subjected to external bending moments: case where internal fluid is liquid. In: Proceedings of ASME pressure vessels and piping 2008, PVP2008-61168
4. Abid M, Khan NB (2017) Behavior of gasketed bolted pipe flange joint under combined internal pressure, axial, and bending load: Three-dimensional numerical study. In: Proceedings of the institution of mechanical engineers, Part E: Journal of process mechanical engineering, vol 232, no 3, pp 314–322
5. Brown W (2014) PVRC-WRC bulletin 538: determination of pressure boundary joint assembly bolt loads
6. The American Society of Mechanical Engineers (2017) ASME boiler and pressure vessel code, Section II, Part D Properties
7. PCC-1: Guidelines for Pressure Boundary Bolted Flange Joint Assembly (2019) The American society of mechanical engineers. ISBN 9780791872635
8. Bickford JH (2018) An introduction to the design and behavior of bolted joints, 3rd edn. Revised and Expanded. CRC Press, Routledge

Decision Support Tools: Machine Learning Application in Smart Planner

Muhammad Azri Ahmad Baharom, Mohd Shahrizan Abd Rahman, Ahmad Rashidi Sabudin, and Mohd Faizairi Mohd Nor

Abstract Immaculate Project Planning and Execution (PPE) is capital to edge over competitors, decrease costs and honour delivery dates. Project Management Information System (PMIS) is necessary towards an improved and efficient quality of any project. Machine Learning (ML) Algorithms enabled learned the date of experience to develop insights into various associations between data and outcomes. A defined set of rules prescribed by the analysts makes the probability of the fault possible. In this paper, Regression Model compute across all viable sectors expending the tool for Downstream Business and other Facilities Upstream, including Resource Estimation Schedule Generation. Extending structured information into a reliable database allows super users to define the data structures and completely configurable the setting's dynamics. The model used to decrease the approximation error and measure the closest possible outcome. This subset of artificial intelligence has tremendous potential in improving schedule generation configuration to develop Project Planning timely and financially smartly. This paper aims to share standard protocols and methods applied in ML-aided as a tool in PPE decision making. Additionally, the abundant used data resources devoted to implementing ML are outlined. Finally, ML success as a Decision Support tool in project management by having a Smart Planner in supporting project recommendation accelerates the decision process, increases stakeholder confidence, and minimizes uncertainty; results are reviewed and analyzed where gaps and potential improvement for future projects are being noted and highlighted.

M. A. A. Baharom (✉) · M. S. A. Rahman · A. R. Sabudin
PETRONAS, Level 19, Tower 3, KLCC, 50088 Kuala Lumpur, Malaysia
e-mail: muhammadazri@petronas.com.my

M. S. A. Rahman
e-mail: msharizan.ar@petronas.com.my

A. R. Sabudin
e-mail: ahmadrashidi.sabud@petronas.com.my

M. F. M. Nor
University Teknologi PETRONAS, Seri Iskandar, 32610 Tronoh Perak, Malaysia
e-mail: mfaizairi_mnor@utp.edu.my

F. Ahmad et al. (eds.), *ICPER 2020*, Lecture Notes in Mechanical Engineering,
https://doi.org/10.1007/978-981-19-1939-8_58

Keywords Machine learning · Smart planner · Regression model · Project control · Project Management Information System (PMIS) · Agile · Benchmarking

1 Introductıon

Since the inception of PETRONAS back in 1974, the national oil company has embarked on thousands of projects from one spectrum of complexity to another. Many valuable data were captured from business idea to feasibility studies and further to execution and later decommissioning and abandonment. Some are used now as we continue to develop the oil fields. Some data, however, were resided in some subject matter expert's machine that sometimes were not cascaded over the new generation. However, most of these cost and schedule data are being tracked in Microsoft Office applications. There is an initiative to collect, process, and automate using quite an advanced visual basic back in 2017. A standalone application was developed to collect raw data, process it, and later become benchmarking values for new projects.

A Project Management Information System (PMIS) inevitably is a choice to provide manageable project information systematically [1], which could provide according to Gartner, the probability of a successful project increased to 75% [2]. With PMIS, systemizing the two works on operational and project management as mentioned by Project Management Institute (PMI), systematic processes PMIS improved project performance effectiveness. With the application of data science and Artificial Intelligence (AI) powered solutions, PETRONAS can predict future outcomes, make actionable decisions, minimize risks, determine the best possible managing strategies, and control the work of numerous project.

The investment grows more complex and extensive through the years, and more data becomes available and traditional tools no longer bring sufficient insights. Hence, more than a dozen Machine Learning (ML) algorithms were explored, and the paradox is that they don't ease the choice. Understanding the implementation and mathematical intuition behind several supervised and unsupervised ML Algorithms, PETRONAS come across the Regression model as a suitable choice. Phase 1 of the Smart Planner was developed in 2019 and was tied to a static table structure for data input in which any customization will rely solely on the developer. This restricted users from expanding the tool for Downstream Business and other Facilities Upstream. Phase 2, having a more dynamic table structure, people can do the future expansion in-house. This enhancement allows super users to define the data structures and completely configurable the settings dynamically. PMI on Principles of Project Management [3] concerning PMBOK Guide provides the categories and function of PMIS on Scheduling software, Cost Control Software, Resource Management Software and Information (document) Management and all four can be integrated via a simple visual display in Phase 2 of this ML decision tool. Retnowardhani and Suroso [4] wrote a review on PMIS effectiveness and produced a comparison table [4] on the use of PMIS on a number of a project across board.

a. Case 1: Iran—Decision by project managers on graphical form eases informed decision makings.
b. Case 2: Canada—Assisting project managers on planning, scheduling and project control.
c. Case 3: Kenya—Utilization of information in decision making.
d. Case 4: Latvia—Use a communication module more frequently to improve project performance.
e. Case 5: Slovenia—Culture adoption to the system upon implementation

The overall review indicates aspects of user-friendliness and overburden on project managers where information are limited to small but essential decision-makers and shareholders. Limitation to current PMIS is its display, limited content of information integration from clients and owners, multiple references and knowledge contents are scattered.

In 2020, despite the challenges of working with the pandemic environment, Phase 2 of the Smart Planner was developed using the Agile approach. Having two more modules: Resource Estimation and Schedule Generation onboard, gave more advantage in adapting high accuracy producing manhours and initial cost and standard project planning based on the scope of work. The applied could compare the norms in productivity and build a notional schedule in Excel or Primavera format. The core objective of schedule generation is to develop Initial Project Planning without waiting for the users' essential requirement to be signed off. Working in parallel has been evidenced in UAT completion within a planned sprint of 2–3 months. This paper will describe ML success as a Decision Support tool in project management by having a Smart Planner in supporting project recommendation accelerate the decision process, increase stakeholder confidence and minimize uncertainty.

2 Methodology

The development of the Smart Planner combined data and information of various projects integrating project interviews from suppliers and contractors, business partners including internal stakeholders. Abundant information individuals and interactions over processes and tools, software and documentation, benchmarking reports, contract negotiations, communications, procurements, risk management as on all responded change over following a planned action. In other words, the Smart Planner is agile. The use of .NET as an open-source developer platform whereas Microsoft SQL Server (MSSQL) as a data platform. Essential features capable of being extracted for Smart Planner, however, maintain the following details as PMIS tools (Fig. 1).

- Scheduling and planning, i.e. the critical path
- Resource Management, i.e. allocation
- Budget and cost, i.e. associate cost
- Control and Performance, i.e. plan against actual inclusive changes

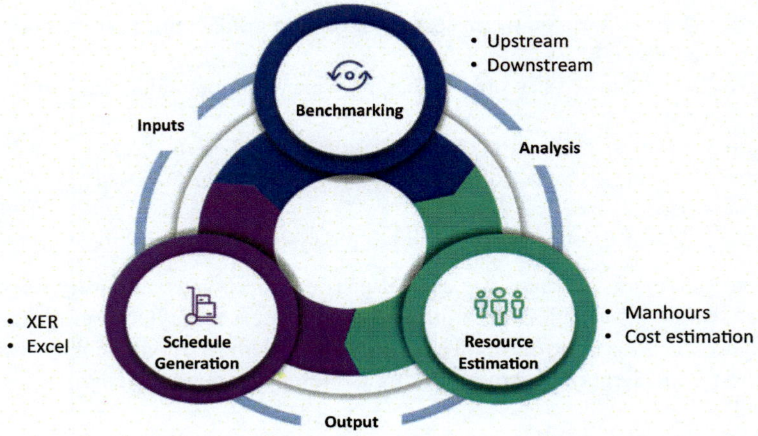

Fig. 1 Smart planner features

- Reporting, i.e. graphs and charts
- Integration, i.e. inventory

A standard methodology for data mining, analytics, and data science projects, CRoss Industry Standards Process for Data Mining (CRISP-DM) Life Cycle framework, is applied in the process. CRISP-DM phases help the team in planning, organizing, and implementing the machine learning progress [5]. Closed-loop deployment enhances the system continuous improvement from Phase 2 onwards. The developer team have a few guardrails in creating the platform for Smart Planner, namely:

- Improve schedule predictability/Schedule forecasting
- Enhance productivity
- Improve schedule control fraternity capabilities to have a better prediction of schedule
- Improve productivity by enhancing visibility, planning and controls of projects, program and portfolio.

The next step after the first-order concept in producing a Smart Planner is on the Exploratory Data Analysis (EDA). This critical stage performs initial investigations on data, identifying patterns and abnormalities. Much of the statistical calculations filters the information and data according to the required inputs benchmarked during evaluation and review processes. Dissecting data and the way the team look at the data, the way the team look at the data and how data is interpreted put EDA as a more appreciated graphical data than statistical graphics.

The statistical concept behind Smart Planner is a result of Regression Analysis on the relationship of any variables chosen by the interface user expressed by a logarithmic equation. The least Square Method provides the best-fit curve line depending on the different relationship displayed as selected dependent variables. At this stage, a compilation of data, knowledge and statistical probability offers an overview of a 'what-if' scenario for stakeholders to review and negotiate upon concerning risk assessment.

3 Result and Discussion

Digitization, automation and information technology is the niche in Industry 4.0. Big data and new digital technologies are essential in projects. With simultaneous hundreds of projects running in a year and with a handful of Project Managers (PM) available, a digital dashboard could help make decisions. Rapid advances hold a promise to assist PM's multiple, complex projects. Smart Planner derived from comprehensive PMIS eases the manual review of project reports, lack of quality live data, and cross-linked information, which is time-consuming. Smart Planner provides a unified conceptual model that can resolve the matter of fragmented data sources in projects. Linking Project scheduling, resource management, cost control, risk analysis, drawing interface, document management and many other future data, Smart Planner is the platform which is worth to be used as forward looking usage and decision making.

The following is a caption on a result on Topsides projects dependent variable Duration against its independent variable of Dry weight. Plotted data of various past projects enable the data analysis to provide scatter plots that have a trend. Limits of upper and lower data dictated the outliers while those closer to Average are numerous reference. The best fit graph is visually easy to interpret. Smart Planner provided users with the flexibility to choose either R squared or RMSE for generating the best-fit graphs among the 4 types of regression models (linear, power, exponential and logarithmic) depending on the data pattern (Fig. 2).

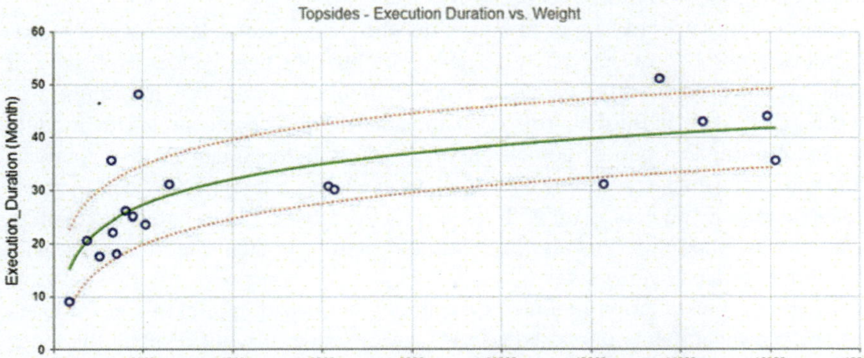

Fig. 2 Exploratory data analysis—topsides

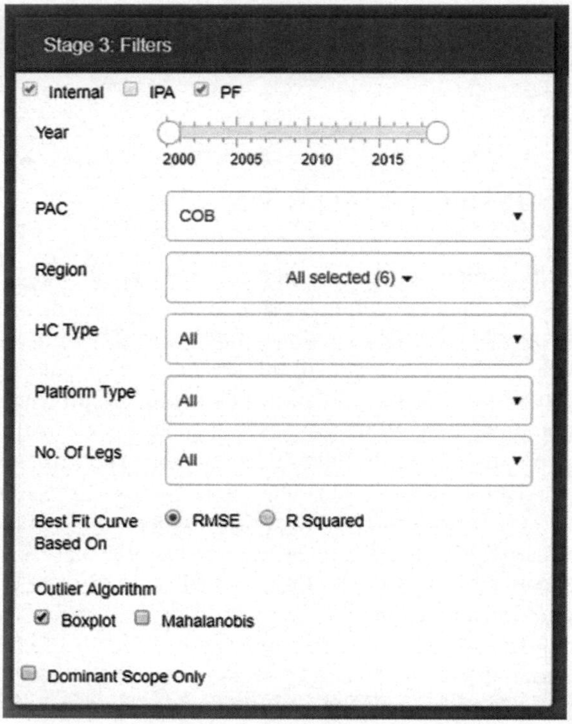

Fig. 3 Sample of stage 3 filter

The user interface provides flexibility in choosing information to display and categorize data for review and analysis. The digital Filters consist of 3 stage selection (at the moment), and it is a platform where all information compilation is being stored. User-defined service is applied on a see-fit basis still applies to gathered knowledge and understanding of each project (Fig. 3).

Smart Planner uses two types of outlier detection algorithms, Boxplot and Mahalanobis distance [6], to detect the outliers. The boxplot is a standardized way of displaying the distribution of data based on five numbers summary—minimum, first quartile (Q1), median, third quartile (Q3) and maximum. The benchmarking module can also plotting 3rd parties best-fit curves, such as Independent Project Analysis (IPA) and Performance Forum (PF), which we can harvest insight provided by these analysts company and compare with the internal database. It can boost stakeholder confidence in making a business decision and getting trust in the analysis model algorithm deployed [7] (Fig. 4).

The resource estimation module concentrates on providing a platform for users to develop project manhours based on the project scope of work. Having the latest and accurate activity norms, the user can have total project manhours quickly, and Smart Planner can generate the final report to Microsoft Excel. It can guarantee less rework, error and having an accurate result. The total manhours can be translated

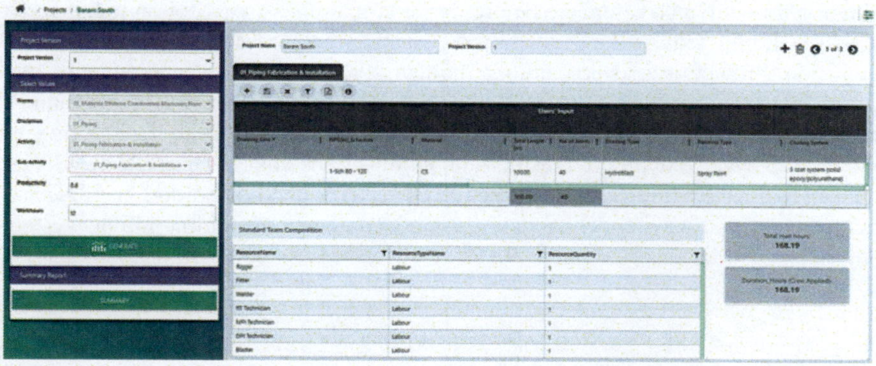

Fig. 4 Resource estimation tools User Interface (UI)

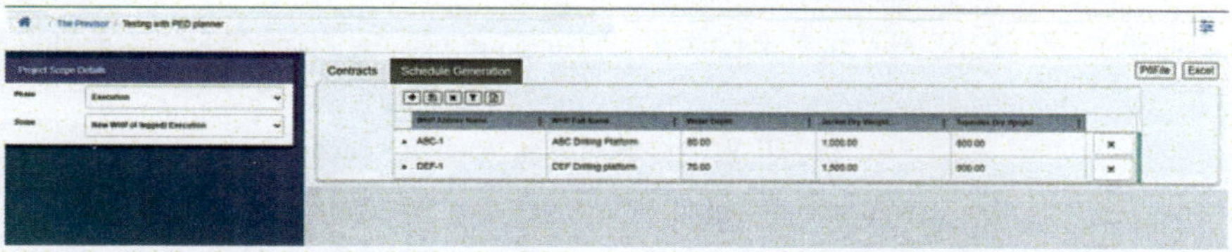

Fig. 5 Schedule generation User Interface (UI)

into the total project cost. Standardization on formulizing any calculation made is essential for stakeholder to make a proper decision (Fig. 5).

Meanwhile, the Schedule Generation function focuses on furnishing user a standard Project Planning in Microsoft Excel file format (XLSX) and Primavera P6 file format (XER). Each activity inserted will be calculated based on norms, 3 points estimation based on the benchmarking data, and some study data from IPA or PF to generate the activity duration. The project team can have initial insight based on the work scope; either it can meet the stakeholders' objectives. Having said intuition and prediction project team can refine the planning to ensure they can deliver the project as per stakeholder requirements.

4 Conclusions

In conclusion, this paper's primary motivation comes from the need to have a new data analysis and representative developed under PETRONAS own umbrella. Problems identified are common in many sectors apart from oil and gas, the lack of data centre, knowledge bridging and process data generation to make decision-making faster and more accurate. This paper presents a finding on detecting deviations in project execution, optimizing data-driven decision-making and project execution patterns for study and reference before decision-making. An exception to the result is that it should address the needs of those outside the numerical reference. Human inclusion, protection and precautions, human behaviour, ethical assessments, and a few humanistic factors of project management are only a few of the humanistic aspects to consider. The system still requires a person or a team to heavily review, interact, analyze, and interpret results and findings. This Smart Planner, evolved with understanding, genre, and interest, needs a corrective so-called borderline guide and guided users and AI is understandably potential advantage addition into the system currently established.

References

1. Jaafari A, Manivong K (1998) Towards a smart project management information system. Int J Project Manag 16(4):249–265
2. Light M, Rosser B, Hayward S (2005) Realizing the benefits of projects and portfolio management
3. Newton P. Principles of project management—project skills. ISBN 978-1-62620-958-9
4. Retnowardhani A, Suroso JS (2019) Project management information systems for project management effectiveness: comparison of case studies. In: Proceedings of ICOMTEE, pp 160–164, October 2019
5. Rivo et al (2012) Cross-industry standard process for data mining is applicable to the lung cancer surgery domain, improving decision making as well as knowledge and quality management
6. Maesschalck et al (2000) The Mahalanobis distance
7. Yang M et al (2020) Deep learning algorithms and multicriteria decision-making used in big data: a systematic literature review

Numerical Investigation of PDC Cutter—Rock Interaction in Drilling Operation Using SPH

Nanthar Kugarn N. Paramananthan, Tamiru Alemu Lemma, Mokhtar Awang, and Mas Irfan Purbawanto Hidayat

Abstract Polycrystalline diamond compact (PDC) bit is one of the most widely used drill bits in drilling operation. The interaction model between the PDC cutters with the rock is essential in optimizing the rate of penetration (ROP), and tool life. There are several factors which affect the performance and structural integrity of the drill bit. Some of these factors include rock lithology, cutter material, number of cutters, cutter profile, cutter orientation, weight on bit (WOB), depth of cut, rotational speed, drilling fluid property (e.g. viscosity and specific heat), working condition (pressure and temperature). The interactions can be studied by using a numerical model. This research is aimed at developing a new numerical model by using smooth particle hydrodynamics (SPH). From the simulation results, empirical correlation will be developed using response surface analysis. Furthermore, a numerical simulation model will be developed and validated using the experimental data as a benchmark case study. A series of simulation studies will be performed based on design of experiment. It is hoped that the result from the study will be used to determine a set of drilling parameters to produce the optimum ROP and attain higher efficiency in drilling operation.

1 Introduction

PDC bit is fixed cutter type, where there is no moving part, but drilling occurs due to shearing, scraping or abrasion of the rock. The interaction model between the PDC cutters with the rock is crucial in predicting the ROP, increase tool life, and to optimize the drilling operation [1–3].

N. K. N. Paramananthan · T. A. Lemma (✉) · M. Awang
Department of Mechanical Engineering, Universiti Teknologi PETRONAS, Seri Iskandar, Perak, Malaysia
e-mail: tamiru.lemma@utp.edu.my

M. I. P. Hidayat
Departemen Teknik Material dan Metalurgi-FTIRS, Institut Teknologi Sepuluh Nopember (ITS), Kampus ITS, Surabaya, Keputih Sukolilo 60111, Indonesia

F. Ahmad et al. (eds.), *ICPER 2020*, Lecture Notes in Mechanical Engineering,
https://doi.org/10.1007/978-981-19-1939-8_59

Rock cutter interactions can be studied either by experimental investigation or numerical modelling [4]. Experimental investigation of rock cutter interaction can help understand the effects of different parameters. However, the high cost, the difficulty to exactly represent the down hole conditions, time to conduct many experiments, difficulty to directly observe the rock fragmentation, and removal process leads to using numerical simulation as a means to study the process at early stage. Since simulation is cheaper and can be done in shorter time, it is far more ideal compared to experimental investigation. On the other hand, numerical simulations require verification and validation with experimental results. Hence, it is always advisable to conduct both numerical and experimental analysis.

There are several factors which affect the performance of drill bit. Some of these parameters include WOB, depth of cut, cutting velocity, rock hardness and bit design features such as cutter material, number of cutters, cutter profile shape, and cutter characteristics (shape, type, and orientation). It is essential to determine a set of drilling parameters to produce the optimum ROP, extend useful life, and attain higher efficiency [5] during drilling operation. In addition, determining the optimum design features from such analysis helps drill bit designers to incorporate these features in their future design. Hence, cutter rock interaction model became a critical feature in the design process as well [2].

A number of simulation studies have been conducted with respect to numerical analysis of rock cutter interactions. Che et al. [1] used finite element modelling of the rock cutting process to investigate the chip formation and force response in 2D orthogonal cutting of rock using explicit finite element code in LS-DYNA. They also pointed out the challenges of FEA modelling resulting from complex material properties of rock, rock fracture and chip formation phenomena and large force oscillations during the dominant brittle cutting mode. In addition, various numerical models such as finite element method (FEM) by Liu et al. [6], discrete element method (DEM) by Rojek et al. [7], and boundary element method (BEM) by Chen et al. [8] have been used in the past for modelling of rock–cutter interaction problems. None of these models have included the WOB in their model which is one of the main factors in drilling operation. The numerical model to be developed in this research can be used to optimize the cutting efficiency considering the WOB and other parameters such as drill bit rotational speed, rock lithology and back-rake angle. The findings from this study will be useful in designing drill bits and determining the optimum drilling parameters. Furthermore, the result can be used as input to further study on drill string vibration and bit wear. There is also a huge gap in the fluid-structure interaction where drilling fluids are introduced into the model. For simulation studies, the effect of drilling fluid can be assimilated as hydrostatic pressure and lower frictional coefficient while for experimental study, actual drilling fluid may be injected as what was done by Rafatian et al. [9]. Most of the researchers have used different rock lithologies to study the difference in effect on the cutter-rock interaction model. Not many studies have been conducted to optimize the drilling operation, in terms of MSE or the force produced. Many researchers focus on the cutting forces produced and not on the MSE which adds significance to this study. The cutting force may represent the trend of the MSE but the correlation exists only when the cutting area is constant.

2 Smooth Particle Hydrodynamic (SPH)

The SPH model is a Lagrangian meshless method first introduced by Qianhong Yang et al. [10]. The SPH model was initially used in the astrophysics field, before it gained popularity and after that it was widely used in multiphase flow, aerodynamics, explosion and fluid-structure interaction. Andrea Amicarellith et al. [11] states that although the SPH model is a meshless Computational Fluid Dynamics (CFD) model used for free surface and multi-phase flow modelling, the SPH model can also be used to deal with multiple body dynamics such as in astrophysics and solid mechanics.

The SPH model is basically a particle method, which means each particle is considered as a computational node and each particle has its own mass, velocity, momentum, temperature, concentration, and other hydrodynamic properties. Liu et al. [12] have stated in their research in 2003 that the combination of a meshless and Lagrangian model has given the SPH model some advantages over the Eulerian models such as Volume of Fluid model which are that the SPH model can handle large deformations, large inhomogeneities and tracing free surface even in the extreme transient explosion process. He also claims that the SPH model has been extensively studied in the previous decade and developed to be able to handle dynamic response with material strength, like in high velocity impact cases. The SPH model also has conceptual advantages compared to the other grid based Eulerian formulation [13]. In the SPH model also, material interfaces are shown self adaptively without having complex interface capturing or front tracking algorithms.

According to Ma et al. [14], the SPH model is an interpolation theory in essence. A kernel function is introduced as the interpolating function which gives the kernel estimate of the field variables involved at a point. Along this process, the properties of each particle are calculated by either the integrals or sums over each value of its neighboring particles. This can be shown in Fig. 1 where Ma et al. [14] considers a domain which is discretized by particles. From Fig. 1, W is the kernel function of which it is assumed to have a compact supporting domain and a radius of kh. Using these assumptions, the approximation of the functions $f(x)$ and its differential form $\langle \nabla f(x) \rangle$ at particle i is shown in Eqs. (1) and (2) where the summation is divided by the total number of particles inside the domain including the particle i.

$$f(x)_{x=x_i} = \sum_{j=1}^{N} \frac{m_i}{p_j} f(x_j) W_{x_i}(x_i - x_j, h) \tag{1}$$

$$\langle \nabla f(x) \rangle_{x=x_i} = -\sum_{j}^{N} \frac{m_i}{p_j} (f(x_i - x_j)) \nabla [W_{x_i}(x_i - x_j, h)] \tag{2}$$

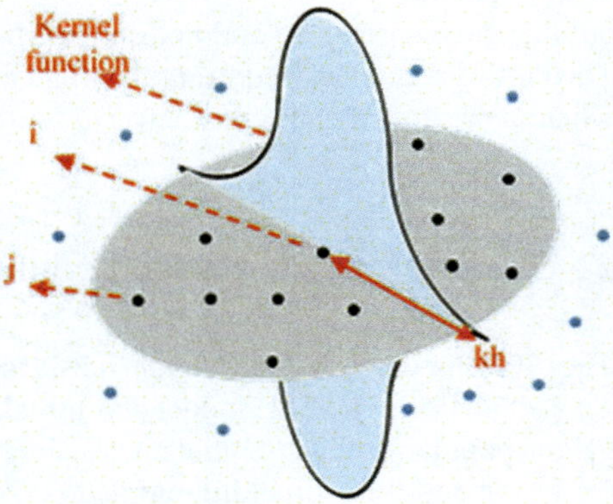

Fig. 1 The kernel function as described by Ma et al. [14]

The j term is the other neighboring particles to particle i which are also influenced by particle i and the h term refers to the smoothing length which defines the particles' supporting domain and lastly $W_{x_i}(x_i - x_j, h)$ is the smoothing kernel function.

3 Continuous Surface Cap Model Failure Criterion

The Continuous Surface Cap Model is a relatively new model developed by Schwer and Murray [15]. This model has its own set of conditions it must meet before the rock undergoes failure and is unique in the sense that it combines the shear failure and compaction (cap) surfaces in a continuous surface, hence the name. This can be seen in Fig. 2.

From Fig. 2, this model is made up of the two surfaces which are the shear failure surface and cap surface. The shear failure surface equation is as shown in Eq. (3). The parameters J_1 and $J_2^{'}$ are the first and second invariants of the deviatoric stress

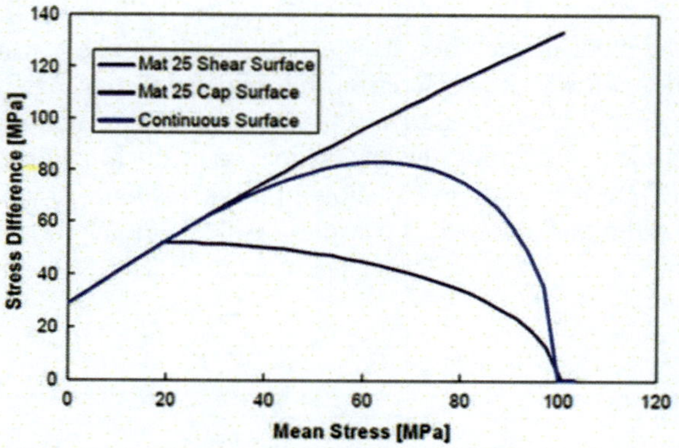

Fig. 2 The combination of the CSCM model to form a continuous surface from the shear surface and compaction surface as shown by Schwer and Murray [15]

tensor, respectively, while the other symbol in Eq. (3) are the material parameters.

$$\sqrt{J_2'} = F_e(J_1) = \alpha - \gamma.exp(-\beta J_1) + \theta J_1 \tag{3}$$

The compaction surface on the other hand is shown in Eq. (4) which is a modification made to the cap surface developed by Sandler and Rubin

$$\sqrt{J_2'} = F_c(J_1, \mathfrak{K}) = 1 - \frac{[J_1 - L(\mathfrak{K})][|J_1 - L(\mathfrak{K})| + J_1 - L(\mathfrak{K})]}{2[X(\mathfrak{K}) - L(\mathfrak{K})]^2} \tag{4}$$

From Eq. (4), $\mathfrak{K}$ is a hardening parameter which causes the expansion and contraction of the cap surface, $X(\mathfrak{K})$ is where the outer edge of the cap surface intersects the J_1 axis and $L(\mathfrak{K})$ is the position in which the cap surface intersects the shear failure surface on the J_1 axis.

A continuous shear and cap surface model can be achieved by multiplying together the functional forms of the shear failure surface and cap surface, as shown in Eq. (5).

$$\sqrt{J_2'} = F_f(J_1, \mathfrak{K}) = F_e\sqrt{F_f} \tag{5}$$

Equations (3)–(5) can be seen in Fig. 2 where F_e is the shear failure surface, F_C is the cap surface and the continuous blending of those two surfaces gives birth to the F_f which is the continuous surface.

The three invariant formulations being implemented in the CSCM model is a general and easy to use model as it can be calibrated to a higher range of experimental data compared to the simple linear fit given by the Mohr-Coulomb model. In the case of insufficient experimental data, Schwer and Murray [15] recommends using the Mohr-Coulomb form of the Rubin's three invariant formulation.

4 Mechanical Specific Energy (MSE)

The MSE is a common parameter used in rock cutting studies to measure the efficiency of the rock cutting process. According to Teale [16], the MSE is a useful parameter which is the work done per unit volume of rock excavated. It is the minimum energy required to remove a unit volume of rock during the cutting process. Apart from the force response considered in this study, the MSE is also considered since one of the objective of this study is to study the interaction between the rock and cutter. Rafatian et al. [9] used the MSE parameter to experimentally find the interaction between PDC cutter and rock and found that as the confining pressure increases, the MSE increases as well, which means more energy is required to drill at higher confining pressure. For single cutter application, Rafatian et al. [9] states the MSE can be simplified into work done in cutting action, divided by the volume of the rock extracted as shown in Eq. (8). The work done, W in cutting action is shown in Eq. (6) where the work done, or the energy needed to cut the rock can be obtained by integrating the cutting force with respect to the displacement of the cutter. The W can also be expressed as the area under the curve for the force vs displacement graph. For the volume of the rock extracted, V_R, instead of calculating the volume of cut, the mass removed can be obtained from the software, which is then used to find V_R using Eq. (7).

$$W = \int (F_x)dx \tag{6}$$

$$V_R = \frac{Mass\ of\ rock\ removed}{Density\ of\ rock} \tag{7}$$

$$MSE = \frac{W}{Volume\ of\ rock\ extracted} \tag{8}$$

As stated by Rafatian et al. [9] and as shown in Eqs. (6)–(8), the mechanical work done by the cutter can be obtained by finding the area under the force vs displacement curve

5 Model Set-Up

This study will commence starting from developing a 3-dimensional, rock cutting, numerical simulation model in LS-Dyna. The model is then validated by comparing it to experimental results from previous work results. The validated model is then used to run a set of experiments with different varying parameters. Several experimental runs will be conducted using the DOE developed to investigate the effects of WOB, back rake angles, drill bit rotational velocity, depth of cut, confining pressure, and rock lithology on the cutting efficiency by observing the responses involved.

Data acquisition system will be used to record the real-time bit weight, torque and displacement data which will be resulted in the analysis of the responses related to the drilling operation. The database generated will be developed into empirical correlation using response surface analysis (Fig. 3).

The numerical study consists of two parts. The first part consists of model with only varying back rake angle and cutting depth. Cutting velocity is kept constant while WOB and confining pressure is not accounted in the first part. Most of the studies on single PDC cutter is done in this way. Therefore, the second part of the numerical consists of five varying parameters where WOB is applied to the single PDC cutter along with confining pressure and varying cutting velocity.

Fig. 3 Flow diagram for numerical simulation study

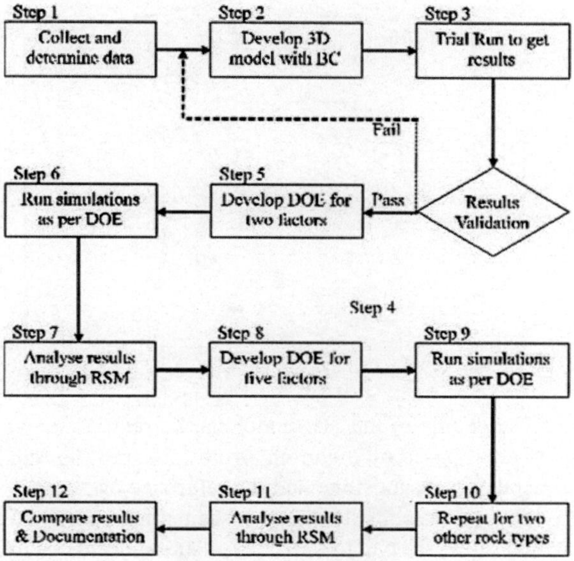

5.1 Model Set-Up

5.1.1 First Part

The isometric view of the 3-Dimensional model is shown in Fig. 4 where the cutter moves at a constant velocity along the negative X-axis and the WOB is applied as a uniform force acting on the negative Z-axis. From Fig. 5, d shows the depth of cut while α is the back-rake angle.

The rock is a SPH model generated into a box shape. It has dimensions of 12 mm width, 10 mm length and 10 mm height.

Fig. 4 Isometric view of 3-D model for first part in LS-DYNA

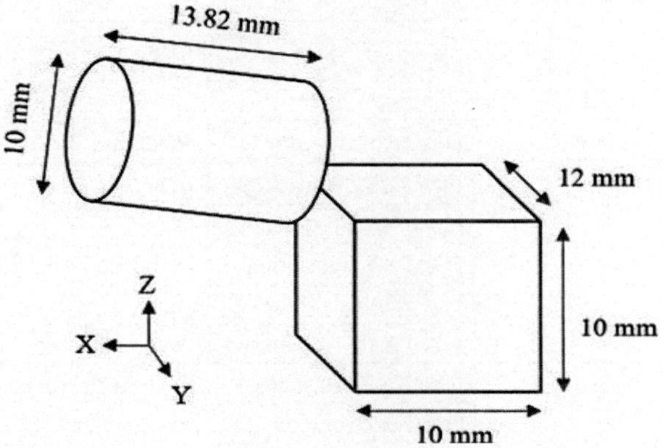

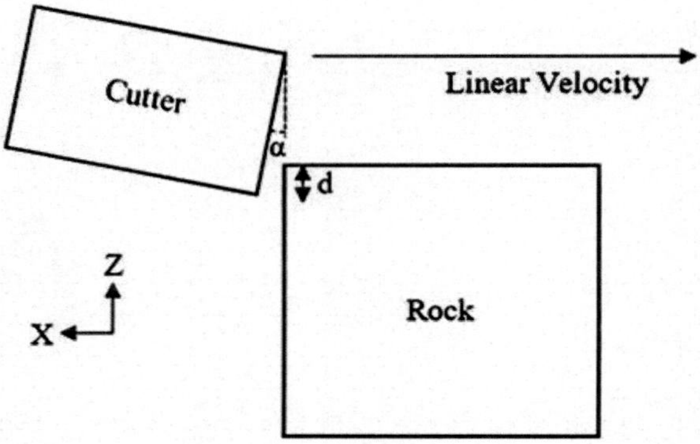

Fig. 5 Side view of the model for first part in LS-DYNA

The details of the 3D model can be seen as shown in Fig. 4. The SPH model is generated based on the grain size for each of the rock. The Table 1 shows the grain size taken for sandstone and the reference used.

The generated SPH model is then constrained in a few parts as shown in Table 2 which shows the part of the rock constrained and in which manner they are constrained as well as the purpose of the constraints. Fig. 6 shows the constrained parts of the rock.

The rock material model used in the LS-Dyna is the Continuous Surface Cap Model (CSCM) which is chosen based on our literature review where the CSCM model was proven to show good results although it was initially designed for concrete models.

The modelling of the single cutter is done by creating a cylindrical solid in LS-DYNA. The dimensions for the single cutter and for the rock model is as shown in Fig. 4. The PDC single cutter is modelled as a RIGID model which is constrained

Table 1 Grain size for each rock type

Rock references used		
Rock type	Grain size (mm)	Reference
Vosges sandstone	0.15	Jaime [17]

Table 2 Rock constraints

Section of rock	Degree of freedom	Function
Bottom	Fixed XYZ translation	To hold the rock in place during the cutting process
	Fixed XYZ rotation	
Sides	Fixed XY translation	To induce plane strain conditions
	Free Z translation	
	Fixed XYZ rotation	
Top	Free XYZ translation	To allow the rock nodes to be cut
	Free XYZ rotation	

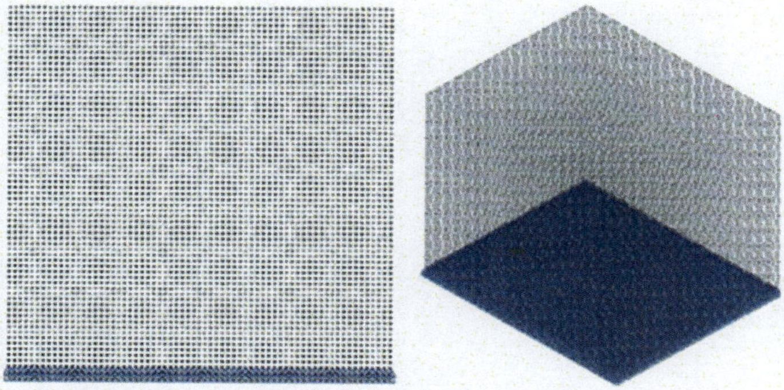

Fig. 6 Right and Isometric view of the constrained nodes of the rock model

in all degree of freedom except for the translational X direction in which the cutting velocity is applied.

5.1.2 Second Part

The second part consists of WOB acting upon the single PDC cutter which will cause the cutter to move downward in the Z-direction. Therefore, the dimension of the rock model must be modified to create a higher rock model. The WOB range determined to be used in this study was taken from an actual case study where the WOB was taken from an actual drilling report. Since the maximum WOB used is too high, the rock model is forced to be made with higher height and the termination time is also limited to make the simulation time acceptable. The dimensions of rock used to study the second part is shown in Fig. 7.

Figure 7 shows isometric view of the full rock cutting model. Just like in the first part, the rock is constrained as shown in Table 2.

The cutter model is similar to the one used in the first part except for the degree of freedom. In the second part, the cutter body is free to move translationally in the X and Z direction. X direction motion is for the cutter velocity while Z direction motion is an effect of the WOB on the cutter. Apart from this small modification, everything else about the cutter is identical to the first part where the cutter is modelled as a rigid material.

5.2 Boundary Conditions

The first part and second part have different terms of the boundary condition as the first part only includes rake angle and cutting depth, while the second part includes

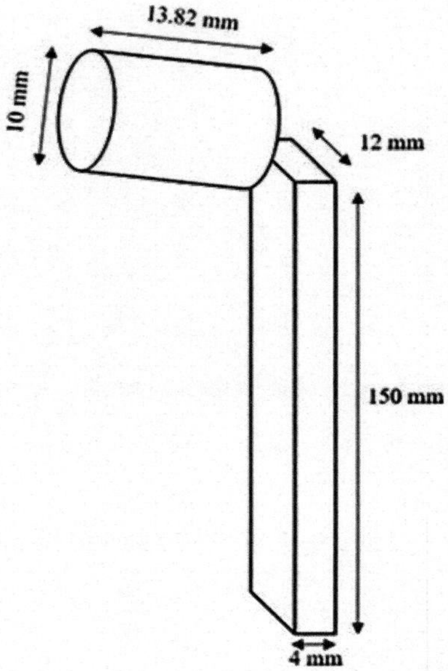

Fig. 7 Isometric view of the second part

those two parameters as well as confining pressure due to the high depth of drilling, WOB and cutting velocity and all five of these parameters were studied.

5.2.1 First Part

The two factors which are varied in this phase of the study are the rake angle and the depth of cut while the linear velocity is kept constant. The range of values for some of these parameters are taken from actual drilling report and are shown in Table 3.

The angular velocity of the PDC drill bit is used to calculate the linear velocity of the single cutter using Eq. (9).

$$Linear\ Velocity = RPM.2\pi.R_{bit}.\frac{1\ \text{min}}{60\ \text{s}} \tag{9}$$

Table 3 Rnage of tool parameters

Parameters	Range
Cutting depth	1–5 mm
Back rake angle	5°–25°
Cutting velocity[a]	60–120 RPM
	The maximum value is taken to be constant throughout all nine runs

[a]Value taken from actual drilling report

Table 4 Assimilation of boundary conditions for first part

Parameters	Values	Methods of assimilation in model	Keyword used In LS-Dyna
Cutting velocity	Constant linear velocity of 1.915 mm/ms throughout all nine runs	To set the cutter to move along a constant velocity along the X-axis	BOUNDARY_PRESCRIBED_MOTION_RIGID & DEFINE_CURVE
		The bottom nodes of the rock are constrained to be rigid to prevent the rock from shifting	
Rake angle	As per the DOE developed	By rotating the cutter along the Y-axis	Element Tools_Transform function
Depth of cut	As per the DOE developed	By translating the cutter along the Z-axis	Element Tools_Transform function

From Eq. (9), R_{bit} is the radius of the bit used which is obtained from the drilling report to be 6 in. or 152.4 mm. For 120 RPM and 6 in. bit radius, we get linear velocity of 1.915113 mm/ms which is used as a constant throughout all nine runs.

These parameters are then assimilated into the 3D numerical model as shown in Table 4. From the many ways of incorporating these parameters into the model, the ones shown in Table 4 are chosen for the simplicity in incorporating them as well as these methods have also been used by many other researchers and have proven to give good results.

5.2.2 Second Part

The second part consists of five varying parameters across three types of rock. The range of these five parameters are taken from actual drilling report from a real case study. The parameters which are taken from actual drilling report are shown in Table 5.

Table 5 Range determined for two factor study

Parameters	Range
Weight on Bit[a]	44.48–155.69 kN
Rotational speed[a]	60–120 r/min
Back rake angle	5–25°
Depth of cut	1–5 mm
Confining pressure[a]	0–50 MPa

The best way of assimilation of these parameters into the models were looked upon in the literature review part. The method of assimilation of these boundary conditions into the numerical simulation model in LS-Dyna are as shown in Table 6. Just as in the first part, careful analysis was done to choose the best way to assimilate these parameters into the model. Based on careful considerations, these methods in Table 6 have been chosen as they have been proven to give good results by researchers who have previously done study on PDC bit drilling operation using the LS-Dyna software.

5.3 *Design of Experiment*

The DOE was developed taking into consideration the simulation time and the total time to complete all the runs to ensure this study can be completed within the given period of time, but also for it to be as accurate and useful as possible. The type of DOE chose are different for both parts.

5.3.1 First Part

For the first part, a three-level factorial design has been developed. The two varying factors are the rake angle and the depth of cut. The first part contains a total of nine runs based on the DOE generated.

5.3.2 Second Part

A Central Composite Design (CCD) was used to develop the Design of Experiment for the five factors across the three rock types. The design has an α value of 1 and consists of 52 runs with 10 centre points, 10 axial points, and 32 cube points.

5.4 *Validation of Results*

The validity of our model developed is first tested to check its validity in giving accurate results. The model has been perfected until it gives values close to those values from previous experimental works done. Once the model is proven to give results close to the experimental results, only then we change the boundary conditions as per the DOE and run the simulation. Two types of validation method were used, which are using mathematical models and previous experimental results. The experimental work used to validate our model is from Glowka [6]. His study involves conducting experiments on rock cutting in high temperatures and hard-rock drilling environment which makes it suitable to be used as a source of validation in our study. Figure 8

Table 6 Assimilation of boundary conditions for second part

Parameters	Values	Methods of assimilation in model	Keyword used in LS-Dyna
Cutting velocity	As per the DOE developed	To set the cutter to move along a constant velocity along the X-axis	BOUNDARY_PRESCRIBED_MOTION_RIGID & DEFINE_CURVE
		The bottom nodes of the rock are constrained to be rigid to prevent the rock from shifting	
Rake angle	As per the DOE developed	By rotating the cutter along the Y-axis	Element Tools_Transform function
Depth of cut	As per the DOE developed	By translating the cutter along the Z-axis	Element Tools_Transform function
Weight on Bit (WOB)	As per the DOE developed	By applying load on the single cutter model. The WOB was evenly distributed into 50, assuming a PDC bit consists of 50 evenly distributed single cutter	LOAD_RIGID_BODY & DEFINE_CURVE
Confining pressure	As per the DOE developed	By using the density versus depth function in LS-Dyna which initiates hydrostatic pressure on the model as an effect of gravity acting on an object	LOAD_DENSITY_DEPTH & DEFINE_CURVE

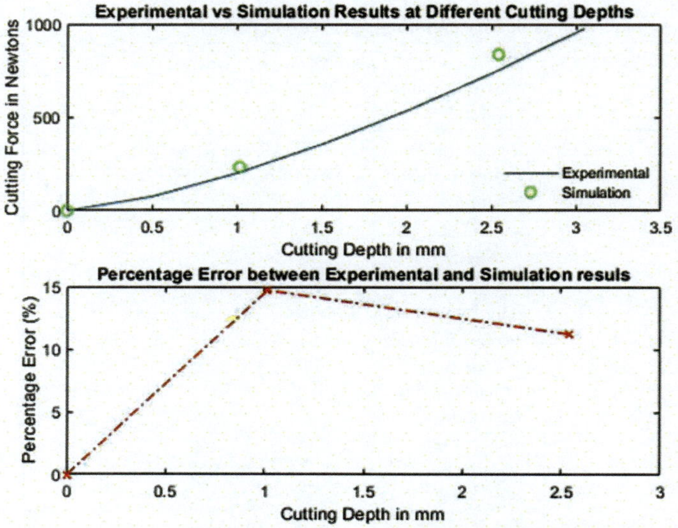

Fig. 8 Simulation results compared to experimental results by Glowka [18] at different depths and the percentage error compared between them

shows the simulation results compared to the experimental results as stated above. It can be seen that the simulation results are close to the experimental work with a maximum error margin of 14.71%. Therefore, the model is validated and can be used to run simulations.

Four types of mathematical models were used to validate the model, out of which the Nishimatsu and Modified Nishimatsu have given results close to the results in the first part as shown in Fig. 9. It can be seen that for lower depth, the results are closer to the Nishimatsu model while at higher depths, the results are closer to the modified Nishimatsu model.

6 Results

The results were collected as four types of responses which are the average cutting force, average Z-force, average X-force and the MSE for the rock cutting simulation. In the paper, we focus only on the cutting force and the MSE (Fig. 10).

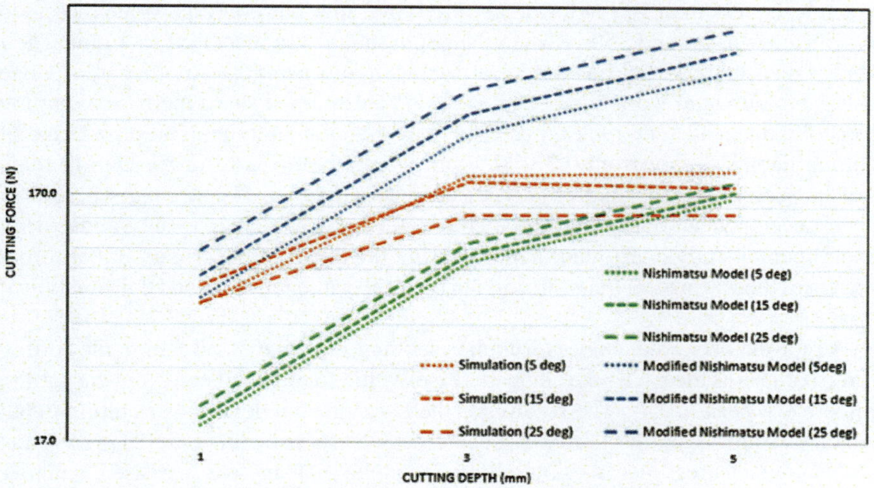

Fig. 9 Simulation results compared to the Nishimatsu and modified Nishimatsu model

Fig. 10 Perturbation plot for the cutting force at different deviations from reference point

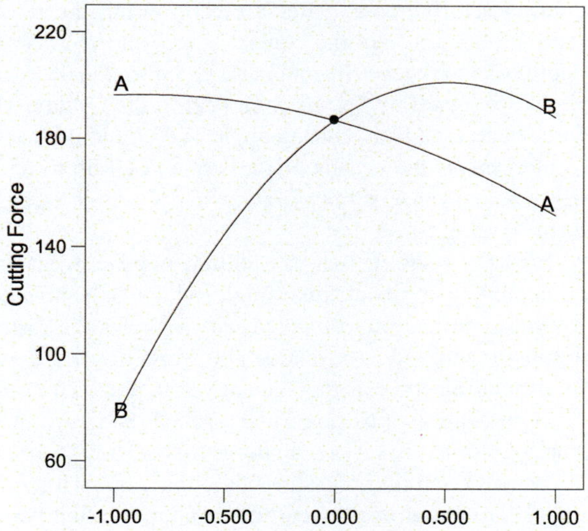

6.1 *First Part*

6.1.1 Cutting Force

The cutting force increases significantly with a peak trend and then reduces slightly when the cutting depth is increased from 1 mm to 5 mm. Menezes et al. [4] obesrved somewhat similar results where for velocity value 50 mm/s, the cutting force was noted to be less than 100 N for depth of cut of 1 mm but the force value increased significantly to around 350 N when the depth of cut was increased to 4 mm. Their study consists of a 2D FEM model of rock cutting operation. Another study done using experimental work by Richard et al. [19] also noted sharp increase in cutting force for the sandstone rock type. Their experimental study is done on a range of cutting depth varying from 0.2 to 0.9 mm where the increase in the cutting force can be seen clearly. Cheng et al. [20] has also gotten similar finding where they have developed an analytical model for PDC cutter which is to be validated by experimental study and found that increasing the depth of cut from 0.3 to 1.5 mm increases the cutting force significantly in both experimental work and their analytical model.

As for the back rake angle, for higher depth of cuts like 3 and 5 mm, the cutting force reduces as the back rake angle increases while for depth of cut 1 mm the cutting force has a peak at 15°. The results for the back rake angle at lower cutting depth which is 1 mm is similar to that of observed by Rostamsowlat et al. [21] where the scaled contact stress increases initially as the inclination angle is increased, but after a certain inclination angle the scaled contact stress begins to fall, giving a 'peak' trend. It is shown in their study that the stress increases as the rock experiences an elastic regime and slowly moves toward elastoplastic regime where the stress continues to increase. After a certain point, the rock begins experiencing plastic regime, this is where the contact stress starts to decrease. For the results at higher depth of cut however like at 3 and at 5 mm, the decreasing trend observed is similar to that observed by Samui et al. [22] in their study. They observed that the cutting force decreases from cutting angle of 5° up to 45°, and then the cutting force starts to increase as the cutting angle is increased up to 85°. This trend is similar to our result as our range is only from 5° to 25° which would explain the decreasing trend of the cutting force.

Figure 11 shows that the cutting depth affects the cutting force by 86.858% compared to the back rake angle which only affects 13.412%. Figure 12 is the response surface plot obtained which shows the distribution of forces at different cutting depths and back rake angles while Fig. 13 shows the trend of each parameter on the cutting force as they deviate from the centre point.

Optimization of cutting force for the first part would result in as shown in Fig. 12. The conditions are set to be in range for back rake angle, cutting depth, linear velocity, WOB and confining pressure while the cutting force is set to be maximized. From Fig. 12, to maximize the cutting force in the first part, the back rake angle is to be 7.78° and the cutting depth is 4.29 mm.

6.1.2 MSE

The MSE results from the first part of the study shows a decreasing trend when both depth of cut and the back rake angle is increased. The result for the depth of cut is in line with that of Balci and Tumac [23] where they have observed that increasing the depth of cut from 3 to 7 mm for Arcose rock specimen using V-type disc cutter

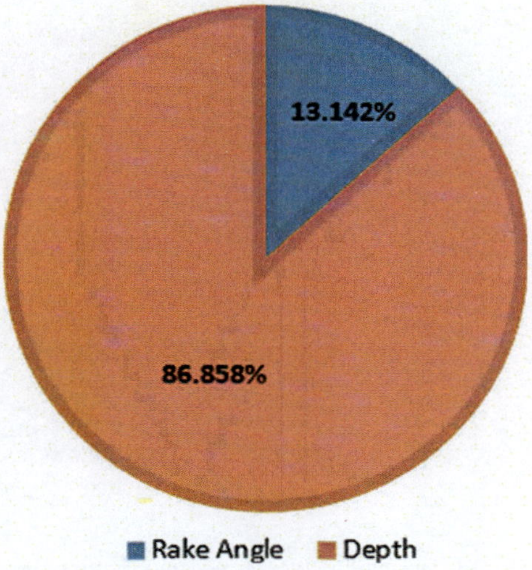

Fig. 11 Combinatorial effect on the two parameters on the cutting force

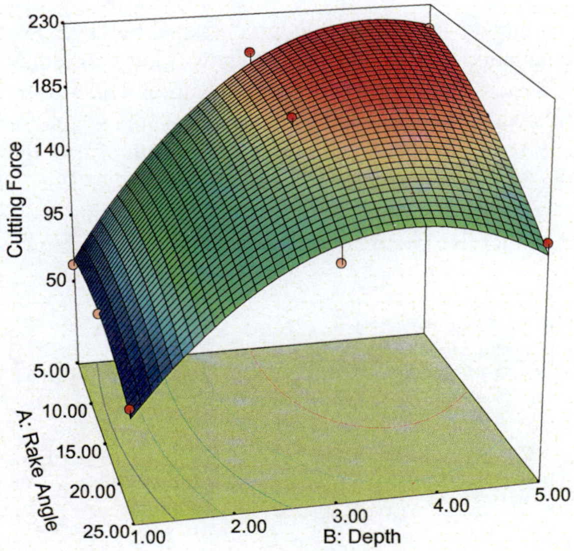

Fig. 12 Response surface plot for cutting depth versus back rake angle on the cutting force

significantly reduces the specific energy in terms of kWh/m^3. The lower MSE is preferred as it would mean lower energy is required to cut a unit volume of rock. As for the back rake angle, no previous research was found relating the MSE to the back rake angle but it can concluded that the higher cross sectional area of cut resulting from the higher back rake angle helps reduce the energy required to cut a unit volume of rock [24]. This is because higher sectional area of cut results in higher friction force generated as stated by Doshvarpassand et al. [25], where they

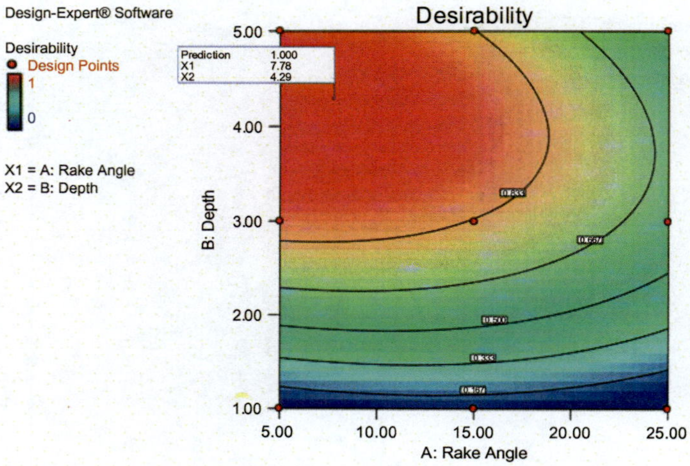

Fig. 13 Desirability plot for optimum cutting force (first part)

assume the cutting force is proportional to the sectional area of cut. Therefore, it is concluded that increasing the depth of cut will reduce the MSE which would mean lower cost incurred for drilling operation while for the back rake angle, higher back rake angle would give higher sectional area which in return results in lower MSE.

Figure 14 shows that the cutting depth affects the MSE by 84.9% compared to the back rake angle which only affects 15.1% (Figs. 15 and 16).

Optimisation for the MSE is done by setting the MSE to be minimum while all the other five parameters are set in range. The optimized desirability plot is as shown

Fig. 14 Combinatorial effect on the two parameters on the MSE

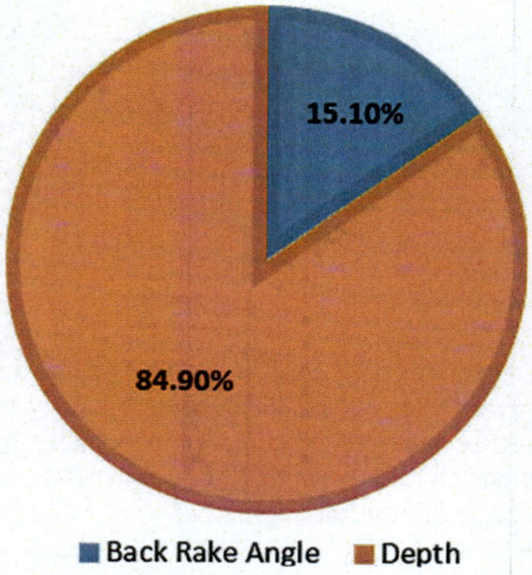

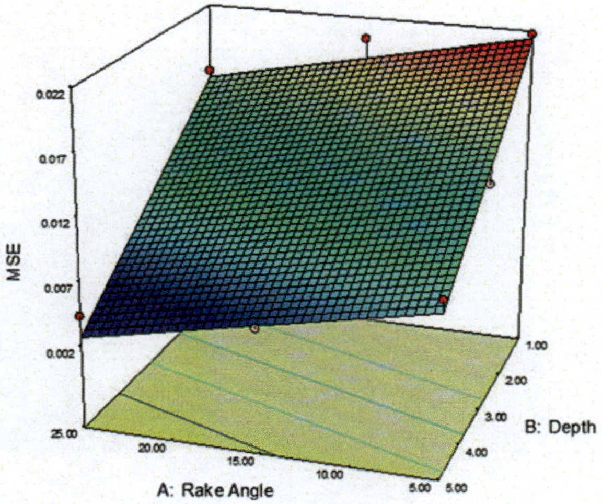

Fig. 15 Response surface plot for cutting depth versus back rake angle on the MSE

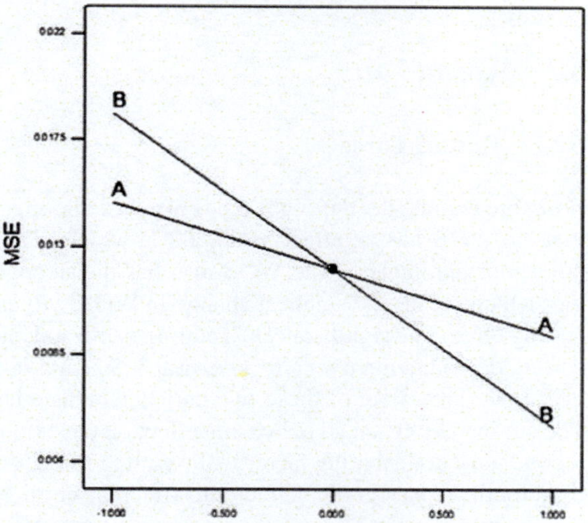

Fig. 16 Perturbation plot for the MSE at different deviations from reference point

in Fig. 17 where the back rake angle is set to be 22.78° and cutting depth at 4.97 mm. These values will give the minimum MSE which means maximum drilling efficiency can be reached as less energy is consumed to remove per unit volume of rock.

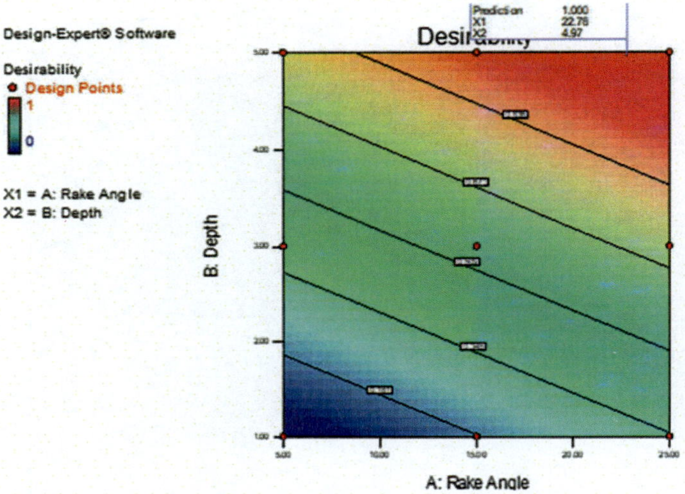

Fig. 17 Desirability plot for optimum MSE (first part)

6.2 *Second Part*

6.2.1 Cutting Force

From the results, the cutting force is affected significantly by the WOB. It is shown that the WOB has an effect of 80.202%, which is a huge portion compared to the other four parameters. The WOB also has the steepest gradient in the perturbation plot which means the smallest change in WOB will cause the biggest change in the cutting force. The significances of cutting depth and rake angle are close enough with cutting depth having a significance of 9.379% and rake angle having a significance of 8.35%. The effect of these two parameters have been discussed in the first part. The cutting depth had an increasing effect on the cutting force for both parts. It is the second most contributing factor to the cutting force. The back rake angle comes in at third for the most significant factor to affect the cutting force. Higher back rake angle results in lower cutting force as shown in the perturbation plot. The linear velocity which represents the rotational velocity can be seen from the perturbation plot to reduce the cutting force as it increases. Although its effect is not as significant as the WOB, nevertheless, it is shown to be significant from the ANOVA study carried out. This result for linear velocity is contradicting that of Menezes [3] where it was found that higher velocity produces higher forces. However, the range of linear velocity used in their study is from 0.1 to 100 mm/s. Since the range of linear velocity used in this experiment is very small, it may seem like the velocity has a lower significance.

From Fig. 18, it can be seen that the WOB has 80.202% influence toward the cutting force, while the confining pressure has influence close to zero. The cutting depth and back rake angle are in the top three highest contributors to the cutting force

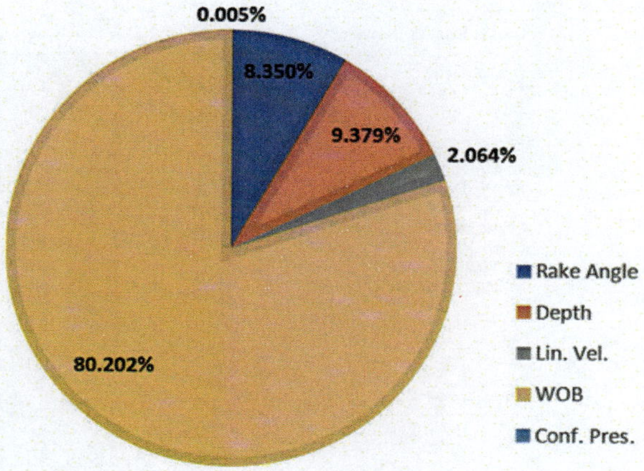

Fig. 18 Combinatorial effect of all five parameters on the cutting force for sandstone rock type

with an influence of 9.379% and 8.350%, respectively. The linear velocity only has an effect of 2.064% on the cutting force. As for Fig. 19, the trend of the cutting force as each parameter is changed can be observed and it is shown that the WOB has the highest slope which points out that the cutting force will have the biggest change for the same amount as deviation compared to other parameters.

The desirability plot is as shown in Fig. 20 where it can be seen that at back rake angle is 5°, cutting depth is 5 mm, linear velocity and WOB at their maximum values and confining pressure at 35.64 MPa.

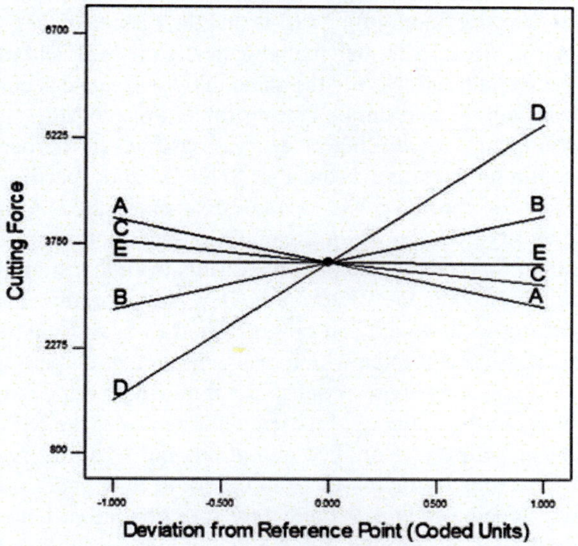

Fig. 19 Perturbation plot of all five parameters for the cutting force for sandstone rock type

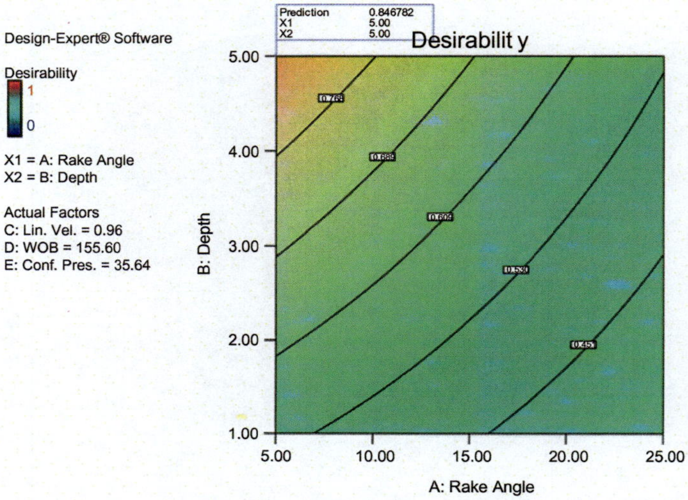

Fig. 20 Desirability plot for optimum cutting force (second part)

6.2.2 MSE

It is observed that the cutting depth and WOB have an upward trend, meaning the MSE increases as the values of these two parameters increase while the back rake angle and the linear velocity have a downward trend. Generally, a lower MSE would be preferred since it means lower energy is required to remove a unit mass of rock.

The key finding for MSE is that the results are very different from the results for cutting force. Although mathematically the MSE and the cutting force are correlated, the result trend was not the same in both first and second part. This shows that MSE does not depend on the cutting force. For the MSE, almost all the five parameters show either increasing or decreasing trend in the perturbation plot except for the confining pressure. From the ANOVA study performed, the WOB has the most effect on the MSE with a 48.63 % as measured in this study. WOB was also the majority contributor for the cutting force. Not many studies have been done on the effect of WOB to compare the results from this study. Zhang et al. [26] have found that increasing the WOB on the PDC cutter tends to increase the temperature of the cutting teeth which can be correlated to having higher force as this will generate more frictional force which results in higher temperature.

The second biggest contributor is the linear velocity with a contribution percentage of 34.85 %. Chen et al. [24] have found that the MSE will not be affected by the linear velocity at drained conditions but will undergo significant reduction when linear velocity increases from 0.2 to 40 in./s under transition condition. They also developed a method to predict the rock conditions between undrained, transition and drained. It was found that for undrained and undrained rock conditions, the MSE will not be affected by the linear velocity. Only in the transition rock condition will the results be same as the results from this study. The back rake angle and cutting depth both have a significance of 8.895 and 7.606% and their effect on the MSE has been discussed in first part. As for the confining pressure, no significant affect was observed on the MSE. Even the ANOVA study showed that the confining pressure is not significant. More research on the effect of confining pressure is suggested in the future.

It can be seen from Fig. 21 that the WOB is the highest contributing factor with an angle of 48.630% followed by the linear velocity which has an effect of 34.85%. The back rake angle and cutting depth have an effect of 8.895% and 7.606%, respectively. The confining pressure has very little effect of 0.019%. Figure 22 shows the trend of

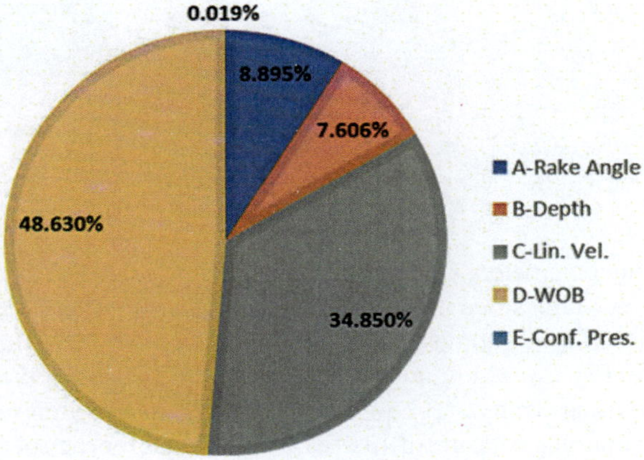

Fig. 21 Combinatorial effect of all five parameters on the MSE for sandstone rock type

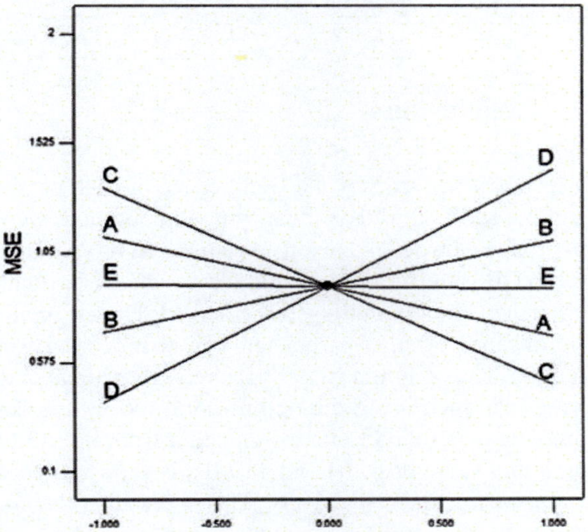

Fig. 22 Perturbation plot of all five parameters for the MSE for sandstone rock type

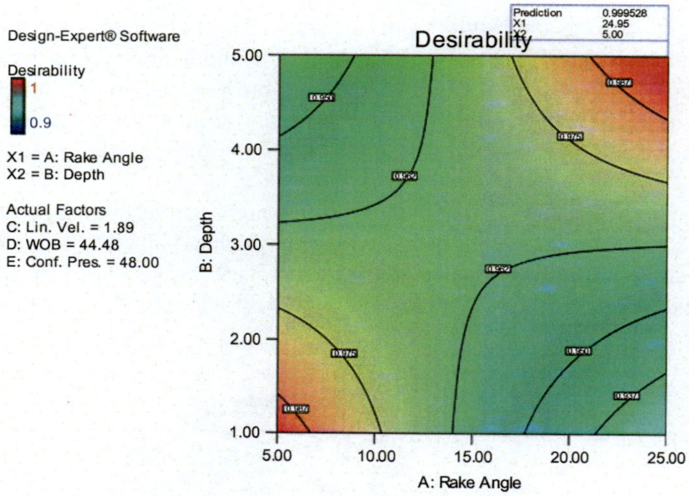

Fig. 23 Desirability plot for optimum MSE (second part)

the parameters as it deviates from low to high. Linear velocity and WOB both have very high slope in the graph which means it affects the MSE the most.

The optimum values to get minimum MSE was calculated statistically and shown in Fig. 23. Minimum MSE value is achieved when the back rake angle is at 24.95°, cutting depth at 5 mm, linear velocity at 1.89 mm/ms, WOB of 44.48 kN and at confining pressure of 48 MPa. These values for each of the parameter are the optimum values as these values gives the smallest MSE, or also the highest efficiency as smaller MSE value means lower energy is used to remove the saame unit volume of rock.

7 Conclusions

This study investigates the significance of each parameter toward the cutting force and the MSE as well as identifying the optimum values for each of the parameters to produce the desired results. From the ANOVA study carried out, it was found that the WOB is the most influential factor which is critical to achieve maximum drilling efficiency as both cutting force and MSE was greatly affected by the WOB. The significance of other parameters can also be seen from the pie chart in the results section. The optimum values to achieve maximum cutting force and minimum MSE were identified using the optimization tool in the design software. The confining pressure was found to be insignificant from the ANOVA study carried out, but an optimum value of 35.64 and 48 MPa is shown to produce the highest cutting force and lowest MSE respectively. This shows that the confining pressure does help in increasing the drilling efficiency but only up to a certain extent. Depending on the goal, whether it is to produce higher forces or to minimize energy during drilling, this research has highlighted the optimum value for both conditions. It is hoped that this finding will help make a difference in the drilling industry where efficiency plays a huge role.

It is also highlighted that the SPH model developed is acceptable based on the experimental and numerical model validation carried out. More modifications is hoped to be done in this model to further improve the accuracy and for it to be used in future works.

Acknowledgments The research is supported by YUTP grants with cost centers: 0153AA-H17 and 015ME0-148. The authors are also thankful to the support provided by Universiti Teknologi PETRONAS.

References

1. Che D, Han P, Peng B, Ehmann K (2014) Finite element study on chip formation and force response in two-dimensional orthogonal cutting of rock, V002T02A038
2. Gerbaud L, Menand S, Sellami H (2006) PDC bits: all comes from the cutter/rock interaction. Presented at the IADC/SPE drilling conference, Miami, Florida, USA. https://doi.org/10.2118/98988-MS. Accessed 1 Jan 2006
3. Menezes PL, Lovell MR, Avdeev IV, Higgs CF (2014) Studies on the formation of discontinuous rock fragments during cutting operation. Int J Rock Mech Min Sci 71:131–142
4. Menezes PL, Lovell MR, Avdeev IV, Lin J-S, Higgs CF (2013) Studies on the formation of discontinuous chips during rock cutting using an explicit finite element model. Int J Adv Manuf Technol 70(1–4):635–648
5. Munoz H, Taheri A, Chanda E (2016) Rock cutting characteristics on soft-to-hard rocks under different cutter inclinations. Int J Rock Mech Min Sci 87:85–89
6. Liu H, Kou S, Lindqvist P (2008) Numerical studies on bit-rock fragmentation mechanisms, Int J Geomech 8, 1 January 2008
7. Rojek J, Oñate E, Labra C, Kargl H (2011) Discrete element modelling of rock cutting. In: Particle-based methods (Computational methods in applied sciences), pp 247–267
8. Chen C-S, Pan E, Amadei B (1998) Fracture mechanics analysis of cracked discs of anisotropic rock using the boundary element method. Int J Rock Mech Min Sci 35(2):195–218, 3 December 1998
9. Rafatian N, Miska SZ, Ledgerwood LW, Yu M, Ahmed R, Takach NE (2010) Experimental study of MSE of a single PDC cutter interacting with rock under simulated pressurized conditions. In: SPE drilling & completion, vol 25, no 1, pp 10–18, 1 March 2010
10. Qianhong Yang JY, Huang Z, Asif M (2019) A comprehensive SPH model for three-dimensional multiphase interface simulation. Comput Fluids 187. Elsevier Ltd.
11. Andrea Amicarellith RA, Mirauda D, Agate G, Sole A, Guandalini R (2015) A smoothed particle hydrodynamics model for 3D solid body transport in free surface flows. Comput Fluids 116. Elsevier Ltd.
12. Liu GRLMB, Zong Z, Lam KY (2003) Computer simulation of high explosive explosion using smoothed particle hydrodynamics methodology. Comput Fluids 32. Elsevier Science Ltd.
13. Adami XYHNAAS (2010) A new surface-tension formulation for multi-phase SPH using a reproducing divergence approximation. J Comput Phys 229. Elsevier Inc.
14. Ma GW, Wang XJ, Ren F (2011) Numerical simulation of compressive failure of heterogeneous rock-like materials using SPH method. Int J Rock Mech Min Sci 48(3):353–363

15. Schwer L, Murray Y (2002) Continuous surface cap model for geomaterial modeling: a new LS-DYNA material type. In: Proceedings, 7th international LS-DYNA users conference, vol 16, 1 January 2002
16. Teale R (1965) The concept of specific energy in rock drilling. Int J Rock Mech Min Sci Geomech Abst 2(1):57–73, 1 March 1965
17. Jaime MC (2011) Numerical modeling of rock cutting and its associated fragmentation process using the finite element method. Ph.D., Graduate Faculty of the Swanson School of Engineering, University of Pittsburgh
18. Glowka DA (1989) Use of single-cutter data in the analysis of PDC bit designs: Part 1—Development of a PDC cutting force model. J Petrol Technol 41(8):797–849, 1 August 1989
19. Richard T, Dagrain F, Poyol E, Detournay E (2012) Rock strength determination from scratch tests. Eng Geol 147–148:91–100
20. Cheng Z, Li G, Huang Z, Sheng M, Wu X, Yang J (2019) Analytical modelling of rock cutting force and failure surface in linear cutting test by single PDC cutter. J Petrol Sci Eng 177:306–316
21. Rostamsowlat I, Akbari B, Evans B (2018) Analysis of rock cutting process with a blunt PDC cutter under different wear flat inclination angles. J Petrol Sci Eng 171:771–783
22. Samui P, Kumar R, Kurup P (2016) Determination of optimum tool for efficient rock cutting. Geotech Geol Eng 34(4):1257–1265
23. Balci C, Tumaç D (2012) Investigation into the effects of different rocks on rock cuttability by a V-type disc cutter. Tunnelling and underground space technology, 30:183–193
24. Chen P et al (2019) Modeling of PDC single cutter—Poroelastic effects in rock cutting process. J Petrol Sci Eng 183
25. Doshvarpassand S, Richard T, Mostofi M (2017) Effect of groove geometry and cutting edge in rock cutting. J Petrol Sci Eng, 151:1–12
26. Zhang Z, Zhao D, Zhao Y, Zhou Y, Tang Q, Han J (2020) Simulation and experimental study on temperature and stress field of full-sized PDC bits in rock breaking process. J Petrol Sci Eng 186

Materials

Improving Knowledge of Ductile Fracture Control in CO_2 Transportation Pipelines

A. Mohammed Nor, M. T. Sallehud-Din, M. F. Suhor, A. Z. Abas, E. Bertelli, M. Di Biagio, and E. Mecozzi

Abstract A specific activity was carried out to overcome knowledge gaps concerning fracture control on pipelines for CO_2 transportation. Special focus was given to the Battelle Two-Curve Method (BTCM) for the prediction of the minimum toughness requirements for fracture arrest and gas decompression phenomenon. Mid-scale testing on a pipeline section pressurized with CO_2 mixture was carried out in onshore and offshore conditions. A correction function that significantly reduces the mis-prediction of BTCM in terms of arrest and propagation has been developed, and the knowledge about backfill effects in fracture propagation has been improved, specifically on the effect of external overpressure in offshore conditions.

Keywords Fracture propagation · Pipeline · CO_2 · BTCM · Backfill · Correction function · Offshore

1 Introduction

South East Asia has about 182 Tcf of undeveloped hydrocarbon gas reserves that reside in high CO_2 fields [1]. The CO_2 content can be as high as 81 mol% as in the case of Kuala Langsa field in Aceh [2]. High CO_2 content renders the monetization of such hydrocarbon gas fields technically challenging; this is because proven CO_2 management technology that will allow for the economic development may not be readily available yet. On the other hand, the use of conventional technology may not be applicable or may lead to an uneconomical development cost. The technology should encompass capture, transportation, utilization, and storage. One of the strategies for evacuating hydrocarbon gas in high CO_2 fields is to transport full well stream onshore for CO_2 separation and then transport the dense-phase CO_2 to sequestration sites. In terms of transportation, the challenges arise in the forms of

A. Mohammed Nor (✉) · M. T. Sallehud-Din · M. F. Suhor · A. Z. Abas
PETRONAS Research Sdn. Bhd, Bandar Baru Bangi, Malaysia
e-mail: azminr@petronas.com.my

E. Bertelli · M. Di Biagio · E. Mecozzi
RINA Consulting—CSM S.p.A, Rome, Italy

F. Ahmad et al. (eds.), *ICPER 2020*, Lecture Notes in Mechanical Engineering,
https://doi.org/10.1007/978-981-19-1939-8_60

potentially high corrosivity of full-well stream fluids [3] and running ductile fracture associated with a high internal pressure required to transport the dense-phase CO_2 to sequestration sites [4]. Like high fluid corrosivity that requires a more accurate CO_2 prediction model, an effective corrosion control, and a suitable corrosion monitoring, running ductile fracture also calls for a better prediction model to predict the minimum fracture toughness required and clamps to arrest the crack propagation.

While PETRONAS has successfully developed its corrosion management technology to address transportation of dense-phase CO_2, it still needs to address the problem of running ductile fracture (RDF). To prevent Running Ductile Fracture, CO_2 pipeline standards such as BS ISO 27913 and DNVGL-RP-F104 stipulate that dense-phase CO_2 pipelines must be designed with an adequate fracture toughness. For lean natural gas pipelines, Battelle Two Curve Model (BTCM) has been widely employed to determine the minimum fracture toughness required. However, the model cannot be extended to dense-phase CO_2 unless a suitable decompression model that covers for dense-phase CO_2 with some impurities is established. The model must also be enhanced to account for high strength steel as the original model was developed for a low strength steel. Furthermore, as the dense-phase CO_2 pipelines will traverse subsea, the enhanced model must also consider the effects of seawater as the backfill.

2 Methodology

The approach used to improve the accuracy of the prediction of arrest/ propagation behavior for dense-phase CO_2 pipelines is schematically reported in Fig. 1:

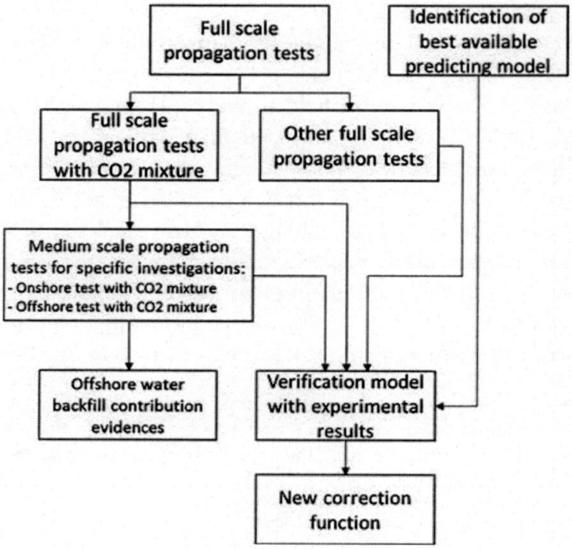

Fig. 1 The flow chart for the enhancement of BTCM

The first step was the identification of the best prediction model for the description of the arrest/propagation condition in a pressurized pipeline. In this regard, the much-gained experience in the natural gas transportation system suggests the use of the Battelle Two Curve model developed by [5]. This model is actually used in updated Standards for the definition of minimum toughness requirement for both natural gas and CO_2 mixture transportation pipelines. The model is based on an uncoupled approach that allows an independent modeling of a complex decompression behavior of rich natural gas and CO2 mixtures during pipeline decompressions.

The second step is the collection of the full-scale propagation test data carried out in the last 40 years, using natural gas and CO_2 mixtures as the pressurizing medium. The number of CO_2 mixture full-scale propagation tests is limited. For this reason, the full-scale database was extended to all the available tests.

With the aim of specifically investigating the decompression behavior of the CO_2 mixtures pipeline, two medium-scale tests were performed: one in onshore conditions while the other was in open seawater conditions. For the latter, the offshore test was carried out at 20 m water depth. The selected depth value was necessary to permit for a realistic development of overpressure around the pipe and its evolution up to the free sea surface.

To conclude the activity, a new Battelle Two Curve correction methodology was developed using a statistical approach focused to identify a correction function based on geometrical and mechanical properties of the pipe.

2.1 *Full Scale Test Data*

Full scale propagation test results were collected, including data from CSM proprietary database that comprised the results of around 150 full-scale burst tests, corresponding to more than 600 pipes tested in the last 40 years. Table 1 presents the information about the data collected in CSM database, including 25 full scale tests performed with CO_2 mixtures.

Table 1 CSM database information

Number of tests (pipes)	>150 (>600)
Number of published test 1 (pipes)	>100 (>350)
Range of diameters	16–56 in.
Range of wall thicknesses	5–30, 5 mm
Range of grades	X52–X120
Range of test pressures	45–255 bar
Type of backfill	Air, soil and seawater backfill
Type of pressurizing mixture	Air, nitrogen, natural gas, rich natural gas and CO_2 mixture
Period of tests	1972–2018

352 data were considered in the statistical analysis aimed at developing the correction function for the BTCM.

2.2 *Statistical Analysis*

A sensitivity analysis was carried out to identify the most significant variables involved in the process. The following parameters, related to pipe geometry and steel characteristics, were selected as the ones playing a crucial role in the BTCM prediction:

- Charpy V-Notch Energy (CV).
- Backfill effect.
- Yield stress of the material.
- Ultimate tensile stress.
- DWTT total energy value.
- Pipe diameter.
- Pipe thickness.

A multiple regression procedure was applied in which the accuracy of prediction was determined as a balance between the following factors:

- For all the points where the corrected BTC failed in its prediction, the sum of residuals between the CVN (actual BTC) with the new expression (in Joules) was calculated and different weights were applied in case of false arrests and false propagations (4:1 ratio).
- Residual = [weight * abs(CVN actual – CVN new)]$^{0.5}$.
- The average of all residuals was minimized in order to find the coefficients of the correction factor.

Moreover, during the optimization procedure, different mathematical formulations of the correction factor were tested in which the most promising one was optimized.

Figure 2 shows the workflow to achieve the reliable correction function:

Two different types of error were considered:

- False propagation: the model predicts a propagation of the fracture while the full-scale test exhibited an arrest.
- False arrest: the model predicts an arrest of the fracture while the full-scale test exhibited propagation of the fracture.

The developed equation, an outcome of the analysis, is an equation with a sum of different contributions; each of them consists of a power of non-dimensional ratio between similar variables related to three different properties:

- $\frac{Cv}{DWTT}$ related to toughness properties of the material.

Fig. 2 Workflow to identify the correction formula

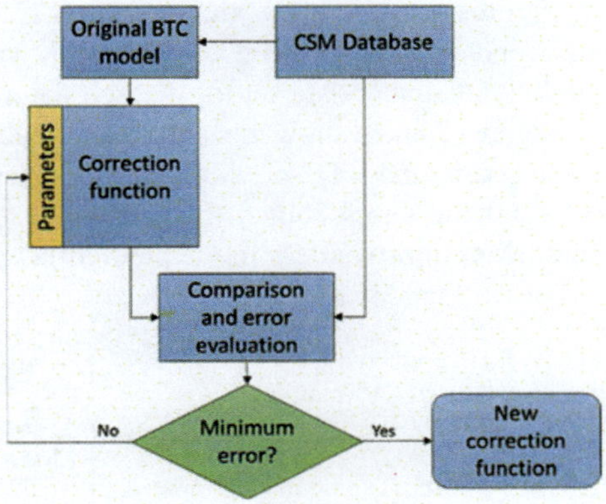

- $\frac{Y_S}{R_m}$ related to tensile characteristics of the material.
- $\frac{D}{t}$ related to geometric properties of the pipe.

$$BTC_{Corr} = BTC\left(A + B\left(\frac{Cv}{DWTT}\right)^{a} + C\left(\frac{Y_S}{R_m}\right)^{\beta} + D\left(\frac{D}{t}\right)^{\gamma}\right) \quad (1)$$

where:

BTC_{corr} = Corrected BTC.
BTC = Battelle Two-Curve (J).
Cv = Charpy V-notch energy (J/cm^2).
$DWTT$ = Drop Weight Tear Test (J/cm^2).
Y_s = yield stress (MPa).
R_m = ultimate tensile strength (MPa).
D = pipe diameter (mm).
T = pipe thickness (mm).

BTC, corrected through the application of the correction function significantly reduced the number of errors in fracture arrest prediction (from 40.1 to 19.3% on a set of 352 pipes). In particular, a significant reduction in the prediction of the most critical and unsafe type of error has been achieved; that is, the fracture propagation predicted as an arrest while the full-scale test exhibited propagation of the fracture (126 vs. 12). On the other hand, there is an increment of the fracture arrest predicted as propagation (15 vs. 56).

As for the fracture propagation tests performed with CO_2 mixtures (25 pipes), the BTC, corrected through the correction function, also allowed a significant error reduction (60 vs. 16% in general), and from 15 to 4% for fracture propagation predicted as an arrest.

Figures 3 and 4 show the measured CV values vs original BTC-predicted CV and versus corrected BTC-predicted CV values, respectively considering only the publicly available data set for the rich natural gas and CO_2 mixtures.

Figures 5 and 6 show measured CV values versus original BTC-predicted CV and versus corrected BTC-predicted CV, respectively for CO_2 mixtures only.

In all the cases above, the use of the developed correction function allows a significant improvement in the prediction capability and accuracy.

Fig. 3 BTC-predicted CV versus measured CV (public data)

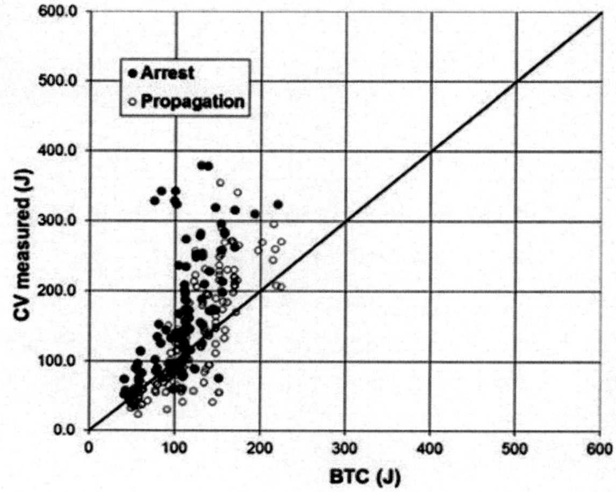

Fig. 4 Corrected BTC-predicted CV versus measured CV (public data)

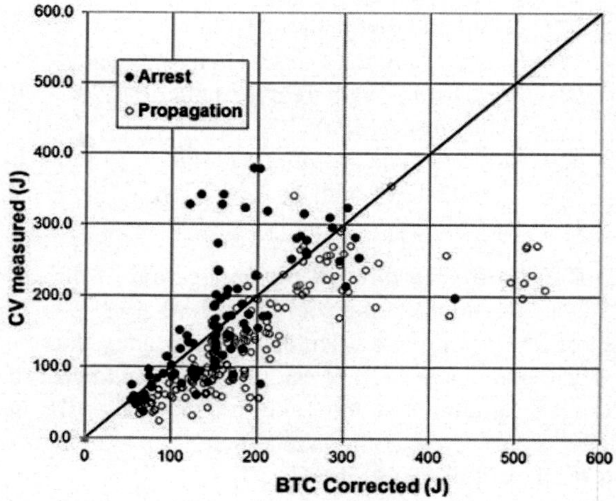

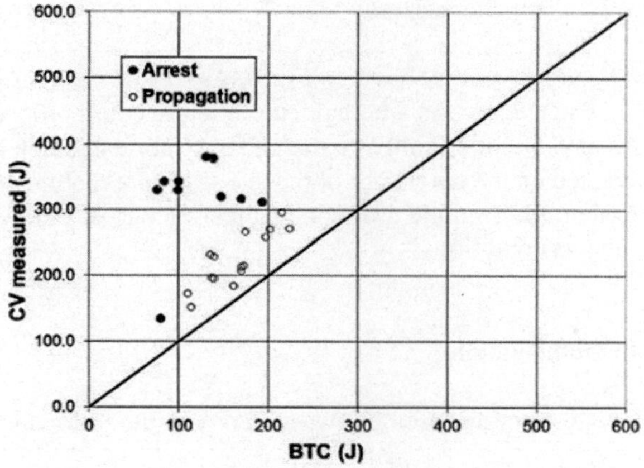

Fig. 5 BTC-predicted CV versus measured CV values for CO_2 mixtures

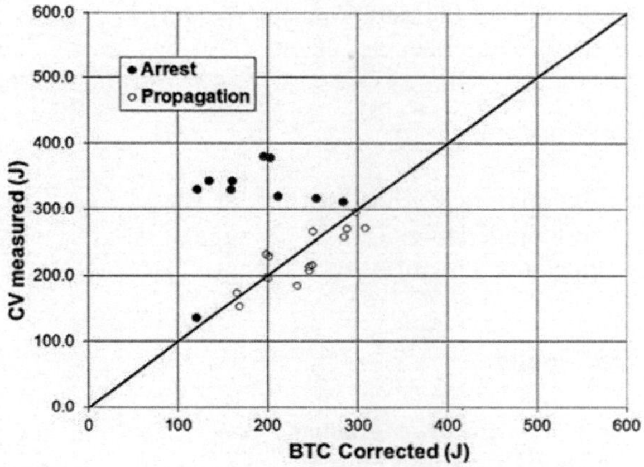

Fig. 6 Corrected BTC-predicted CV versus measured CV values for CO_2 mixtures

2.3 *Field Testing*

Two medium-scale field tests were carried out, one in onshore and the other in offshore conditions, using CO_2 mixture as the pressurization fluid in dense/supercritical phase.

The aim of the tests was to collect valuable information and data to investigate the effect of the external hydrostatic pressure on the gas decompression evolution and to investigate the backfill coefficient to be used in the BTC model.

Tests were carried out using X42 nominal steel grade pipes with geometry 141.3 mm outer diameter (OD), 4.78 mm wall thickness (WT).

The onshore test was carried out in CSM full-scale laboratory in Castel Romano (Roma) whereas the offshore propagation test was carried out in the offshore testing laboratory in Reggio Calabria (South of Italy).

2.3.1 Onshore Tests

The test layout is shown in Fig. 7.

The test section where ductile fracture could propagate was at the central section located about 3.9 m from the ends. A fabricated defect (crack initiation point) was created on the outer wall of the test section. Fracture was initiated a by a mechanical system to avoid the use of an explosive charge. Below are the instruments used and the test conditions:

Instrumentation

The test piping assembly was fitted with the following sensors:

- No. 44 Timing Wires (TW).
- No. 6 Pressure Transducers (PT).
- No. 6 Blast Pressure Transducers (BTP).
- No. 6 Thermocouples (TC).

Test Conditions

- Gas mixture: Carbon dioxide: 97.7%, Nitrogen: 2.3%.
- Test temperature: 47.3 °C (average of all TCs).
- Pipe burst pressure: 165.5 bar (average of all PTs).

Test Results

The test section was gradually pressurized until the pipe burst, starting from the **artificially machined notch** when reaching 165 bars. The fracture propagated in a fully ductile manner along the entire path and the rupture surface presented the typical 45° inclination of ductile propagation.

Figure 8 presents the sequence of pictures taken during the test while Fig. 9 shows schematically the fracture path. Figure 10 shows further the details of the arrest on both test lines.

On side A, the fracture ran at a roughly constant speed in the range between 140 and 120 m/s with a slight decrement. The arrest occurred in correspondence of the

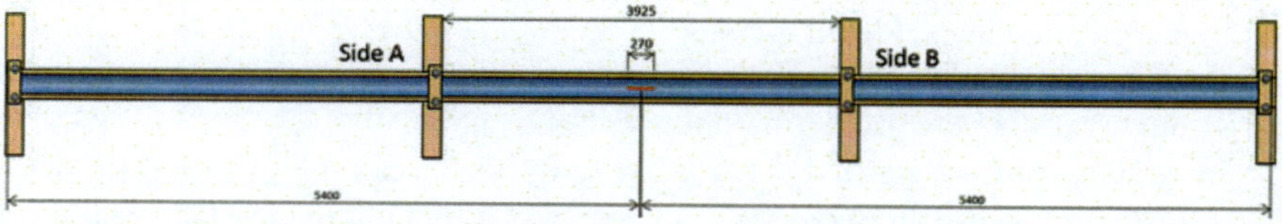

Fig. 7 The test layout showing a fabricated defect in the middle

Fig. 8 A sequence of pictures during the test

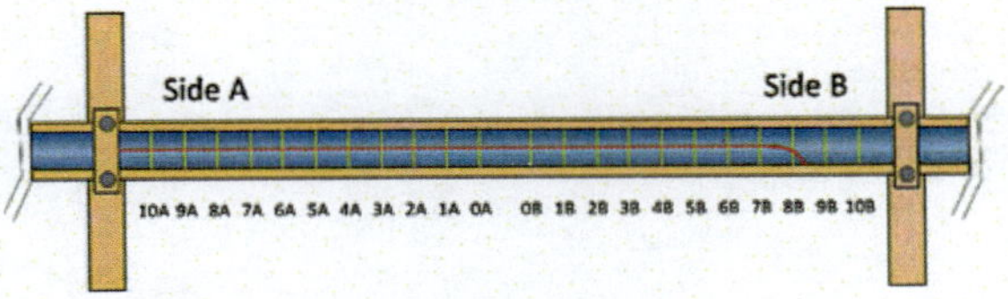

Fig. 9 Fracture path

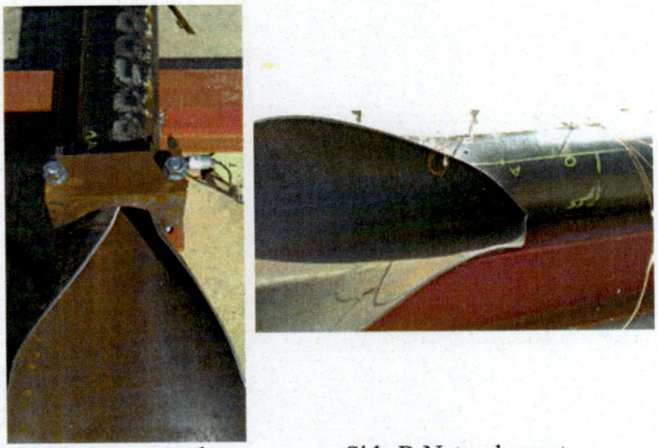

Side A-Arrest on clamp

Side B-Natural arrest

Fig. 10 Details of the arrest

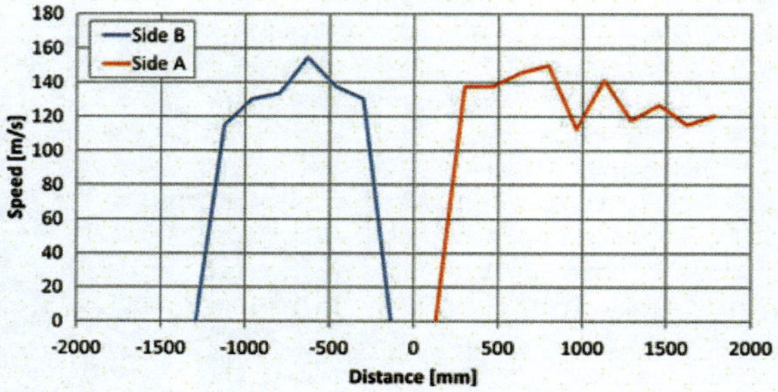

Fig. 11 Crack speed diagram

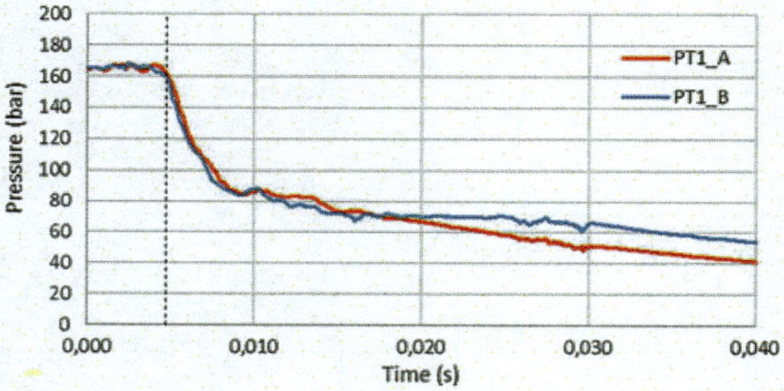

Fig. 12 Decompression data acquired from PT1_A and PT1_B

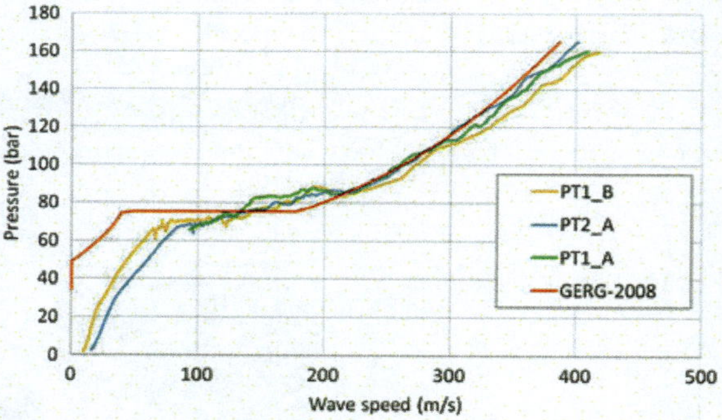

Fig. 13 Comparison between experimental decompression curve and GERG2008 equation

clamp, with fracture entering at 120 m/s. On side B, the fracture initiated at the same speed (around 140 m/s), but after one meter, the velocity started decreasing where the fracture was rapidly arrested after a few centimeters; the arrest was due to the material toughness.

From the data acquired by the TWs, the crack speed diagram as displayed in Fig. 11 was plotted.

The pressure profile versus time was successfully recorded via the pressure transducers installed on the test pipe. An example is shown in Fig. 12.

The experimental depressurization curves after bursting were compared with the curves modelled using the GERG 2008 equation as shown in Fig. 13.

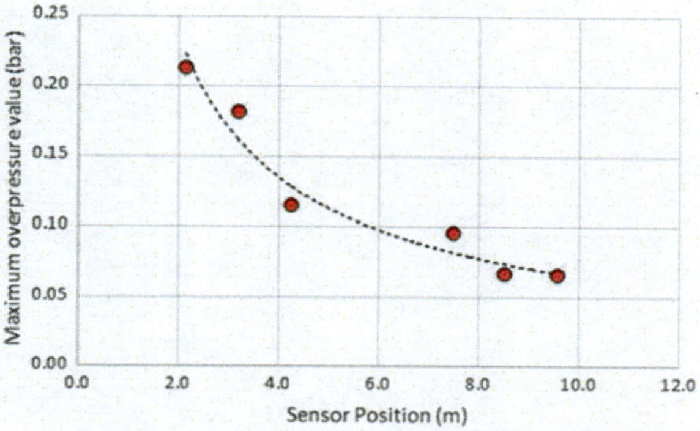

Fig. 14 A decrease in the maximum blast pressure with sensor distance

The experimental decompression results in dense phase exhibited good agreement with GERG 2008 equation prediction. The shape of the plateau is not flat as expected. However, its average value is quite in line with the model prediction. Additionally, all the Blast Pressure Sensors installed outside and around the pipe successfully acquired the data as shown in Fig. 14.

2.3.2 Offshore Test

The offshore test location was 50 m away from shoreline with 20 m water depth (see Fig. 15). The test layout, the instrumentation installed, and the initiation device were the same as the ones used for the onshore test.

Fig. 15 Laying phase

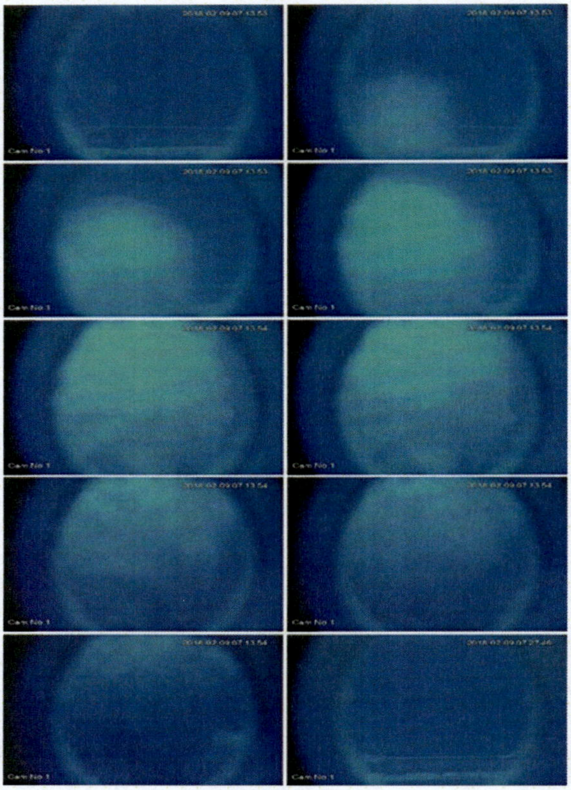

Fig. 16 Sequence of pictures during the test

Test Conditions

- Gas mixture: Carbon dioxide: 97.5%, Nitrogen: 2.5%.
- Test temperature: 16.5 °C (average of all TCs).
- Pipe burst pressure: 137.8 bar (average of all PTs).

Test Results

The test section was gradually pressurized until the pipe burst at 137.8 bar, starting from the artificially machined notch. Even if the burst pressure was lower than the target (150 bar), the CO_2 mixture condition was already in dense phase. Figure 16 presents the sequence of pictures taken during the test where it was evident the growth of spherical bubble of gas around the pipe while Fig. 17 shows the fracture path.

Fracture started regularly from the notch tips and was arrested on both sides (Side A and Side B) for a very short length after a typical spiral path (see Figs. 18 and 19).

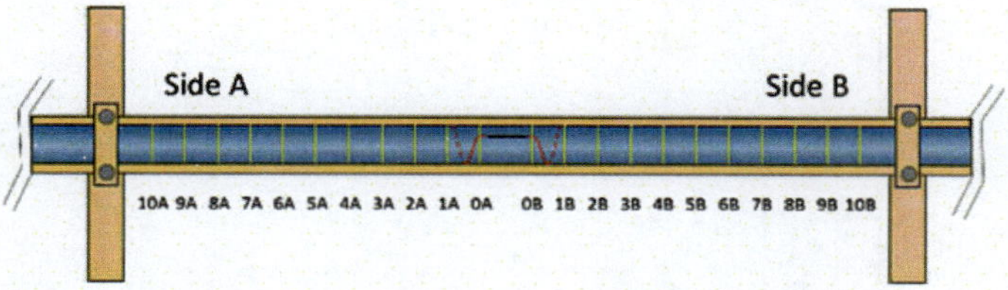

Fig. 17 Fracture path

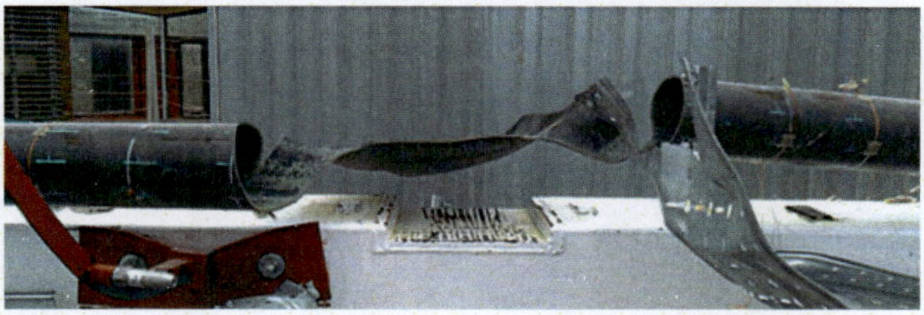

Fig. 18 Details of the test section

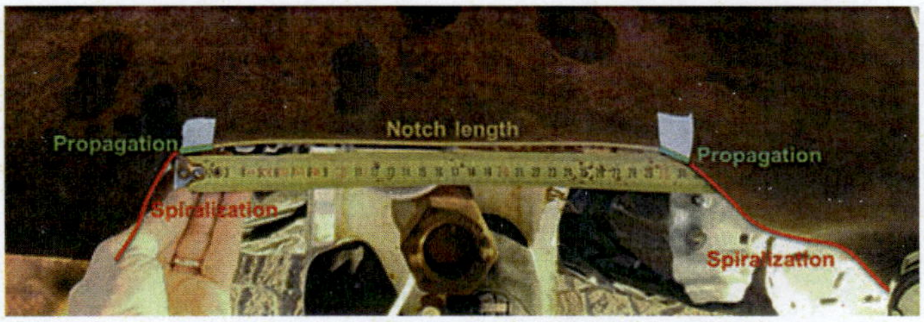

Fig. 19 Details of onset, propagation and arrest path

Although the signals from the TWs were captured, it was not possible to identify a fracture speed diagram, as the fracture propagation length was shorter than the distance between the first TWs.

The pressure profile versus time was successfully recorded on one side of the test pipe while the transducers installed on the opposite side failed during the test. Nevertheless, the data acquired were still enough to verify the decompression behavior of the pipe during the test as shown in Fig. 20.

The Blast Pressure Sensors correctly acquired the data. Considering the sound speed in offshore environment (about 1500 m/s) there is a very good correlation between the first wave arrival time with the expected one (see Fig. 21). Figure 22 shows the data acquired by the oxygen cell sensors as an indicator of CO_2 concentration installed 1 m over the sea level.

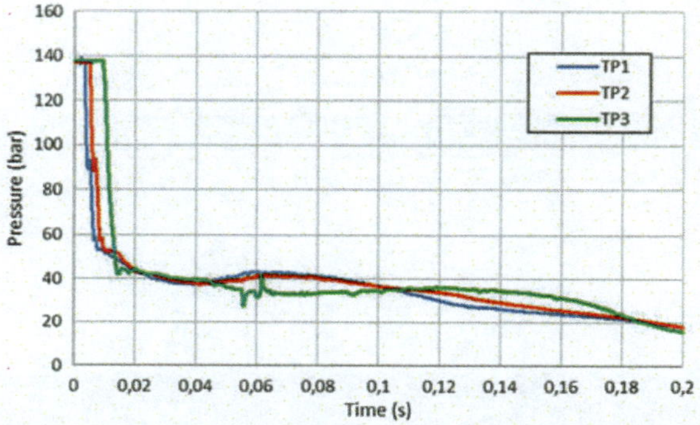

Fig. 20 Decompression curves acquired

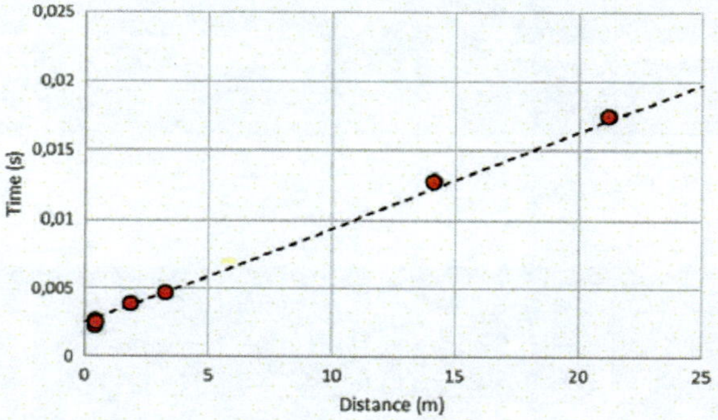

Fig. 21 Analysis of first wave arrival time at sensor position

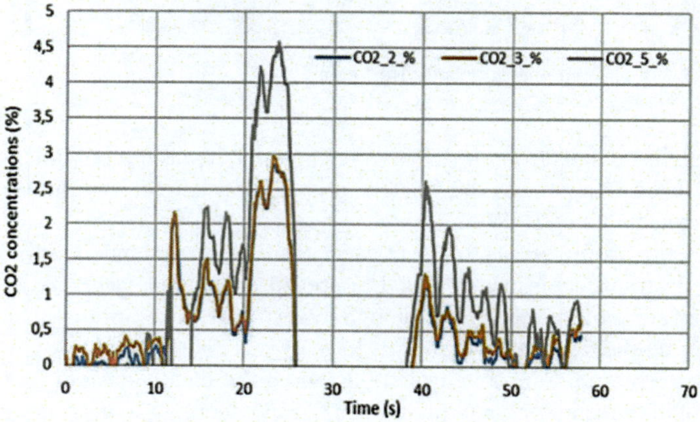

Fig. 22 Oxygen cells—CO_2 concentration 1 m over the sea level

Fig. 23 Schematic description of bubble formation around the pipe

Offshore Water Backfill

Generally, pipelines can be either buried (soil backfill), unburied (air backfill, meaning that the pipeline is not covered by soil), or laid on seabed in offshore conditions (water backfill).

Air backfill is limited to the inertia of the pipe steel to the rapid opening. The reaction of air is very limited with respect to the internal pressure of the line. In this case, the contribution of air backfill to the resistance forces is quite limited.

Soil backfill results in a higher contribution to the resistance force due to the sum of the steel inertia effect with the soil inertia. That is, there are different actions inside the trench during the pipe-opening event. In particular, the soil that is located on the pipe upper is immediately ejected from the trench, due to the pressure of gas inside the line. On the contrary, part of the soil in contact with the flaps is ejected by the flap opening generating a strong inertia, and part is compacted inside the trench.

A specific explanation is required for the water backfill behavior. In fact, in this case, there is the inertia effect of pipe wall and of water around flaps, but the stronger resistance effect is generated by the external overpressure that is generated around the pipe during fracture.

Furthermore, due to the presence of water around the pipe (i.e. a fluid with very low compressibility), the pipe opening produces a "bubble" of overpressure centered in the pipe rupture point and with a radius that increases with time. The overpressure in this area can be added to the external hydrostatic pressure and produces a force on external pipe surface, countering the internal pressure effect.

Figure 23 displays the schematic description of the bubble formation around the pipe, as explained above.

In the '80s the first studies on sea water conditions were conducted by Maxey [6, 7]. Such studies mainly consisted of lab-scale tests designed to ascertain whether aqueous backfill had any influence on propagation and whether it constrained the discharge of gas from the pressure line.

Afterwards, a set of five full scale propagation tests have been carried out in offshore conditions, i.e. open sea water in a range between 20 and 30 m depth [7, 8]. The two tests at higher water depth were carried out by CSM. The results of all the

Table 2 Full scale offshore tests

O.D (in.)	Thickness (mm)	Length (m)	Depth (m)	Pressure (bar)	Medium
22	6.5	45	12	100	Nitrogen
42	15	53	12	116	Nitrogen
30	9.5	57	12	63	Nitrogen
20	20	100	30	255	Air
36	20.8	108	30	157	Air

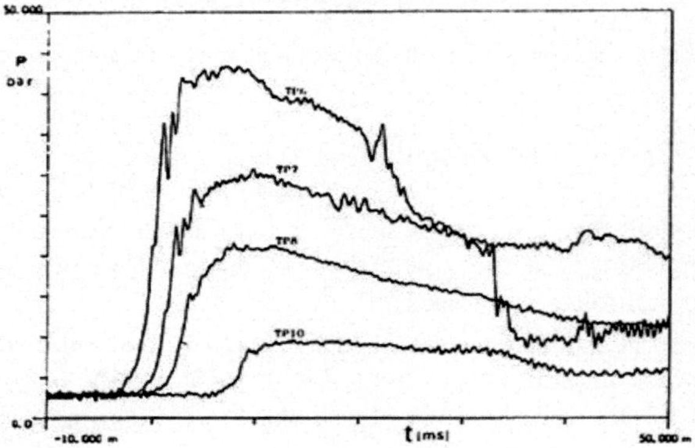

Fig. 24 External underwater overpressure during the propagation test [8]

tests allowed to identify a backfill coefficient value to be applied to the BTC model in case of offshore conditions.

Table 2 lists the main parameters of the offshore tests carried out.

The main scope of these tests was to confirm the hypotheses that are the basis of the explanation of water backfill action on fracture propagation event, particularly:

- The higher resistance effect of water with respect to the soil.
- The formation of the overpressure bubble.
- The monitoring of the cycle of bubble from the formation to the ascent up to the sea level.

These tests allowed to verify the formation of the overpressure bubble around the pipe. In addition, it is clear from Fig. 24 that the external overpressure level is strongly higher than hydrostatic pressure confirming the relevant effect of bubble overpressure on pipe flaps.

The studies about the effect of marine backfill containment were based on purely empirical-statistical processing of experimental data, and on the schematization of the fluid dynamics phenomenon associated with flap opening and evolution.

In practice, the study was based on the scheme developed for onshore tests (see Fig. 25) that consisted of the empirical analysis of experimental results with the aim to determine the link among fracture velocity, steel toughness and nominal crack tip pressure. These values reported in the chart above show a correlation function

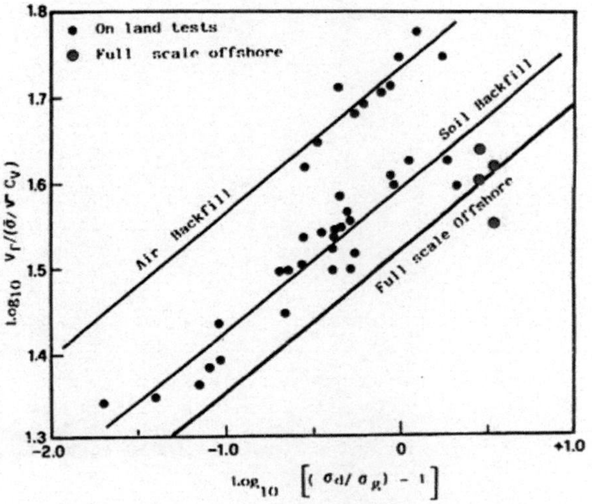

Fig. 25 Ductile fracture velocity correlations [8]

of the backfill condition. As seen in the BTC model description, this correlation is presented with the following equation.

$$V_f = A \times \frac{\tilde{\sigma}}{\sqrt{R}} \times \left(\frac{\sigma_d}{\sigma_a} - 1\right)^{\frac{1}{6}}$$

where:

V_f: fracture velocity
$\tilde{\sigma}$: flow stress of the steel
σ_d: nominal hoop stress of the pipe
σ_a: level of hoop stress at fracture arrest
R: fracture strength of the steel
A: constant

The constant A, a function of the backfill, is determined as indicated in Fig. 25 scheme, which also includes data on the test performed without backfill (only inertia of flaps is involved) and with earth backfill (buried pipe) for the sake of comparison.

Starting from these considerations, aqueous backfill exerts a real constraining effect on propagation; this is greater than that exerted by earth backfill. The main mechanism involved deals with the overpressure field created outside the pipe, which reduces the overall driving force.

To confirm the experimental evidences a dedicate CFD model was developed for the evaluation of external pressure during the fracture propagation event. In particular, the model aimed at identifying the profile and duration of external pressure and the dense CO_2 mixture depressurization behavior. A scheme of the model that represents the pipe laid in seabed without soil covering is shown in Fig. 26.

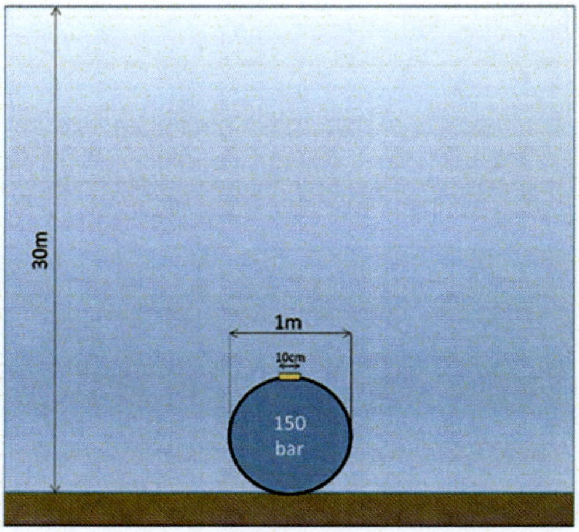

Fig. 26 CFD model scheme

Scope of the hereby presented activity was to evaluate the external overpressure in different scenarios during the pipe decompression. For this purpose, a set of cases were performed based on the following scenarios:

- Case #1: N_2 versus air ambient.
- Case #2: N_2 versus 30 m depth water ambient.
- Case #3: N_2 versus 500 m depth water ambient.
- Case #4: CO_2 versus 30 m depth water ambient.

The comparison between case #1 and case #2 was performed to verify if and how the water backfill affects the generation of the overpressure bubble around the pipe.

The comparison between case #2 and case #3 was performed to verify the effect of sea water depth.

The comparison between case #2 and case #4 was performed to verify the effect of the decompression of dense CO_2 mixture on the overpressure bubble formation.

The results obtained from case #1 (see Fig. 27) show that when onshore unburied pipelines are considered (i.e. no soil material as backfill), the external overpressure around the pipe is almost negligible.

On the contrary, the simulation of case #2 performed in the same operating conditions as that of case #1 (i.e. same geometry, pressure and temperature), but in offshore scenario, shows that sea water backfill produces, after the rupture, a strong overpressure around the pipe. Figure 28 shows that, in this case, the pressure around the pipe results in a range 30–50 bar, that is much higher than in case #1. This external pressure cannot be considered a negligible contribution to the resistance forces, as it acts contrasting the internal pressure action in flaps opening. Moreover, it must be noted that the formation of bubble is very rapid and is maintained constant for several milliseconds.

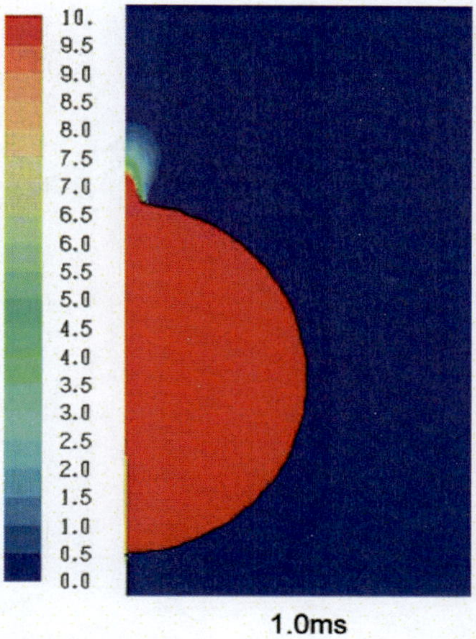

Fig. 27 Pressure contour of case #1

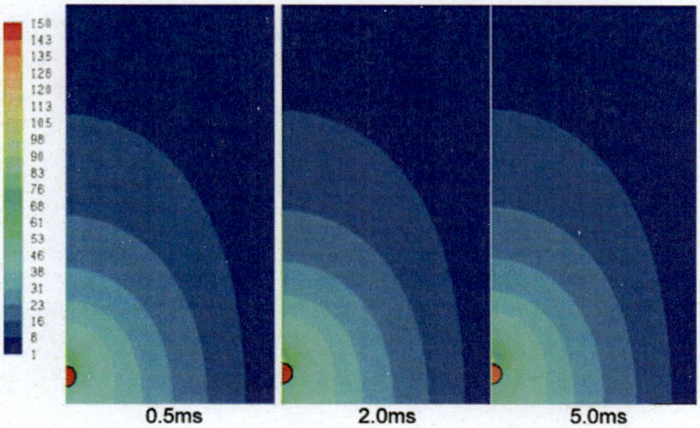

Fig. 28 Pressure contour of case #2

To validate the developed CFD model in offshore conditions, its results were compared to the offshore full-scale test carried out by CSM in '80s [8]. The full-scale test was carried at 30 m water depth using a pipeline with outside diameter equal to 36in, pressurized with air at 157 bar.

The test was instrumented with external pressure transducers positioned around the pipe.

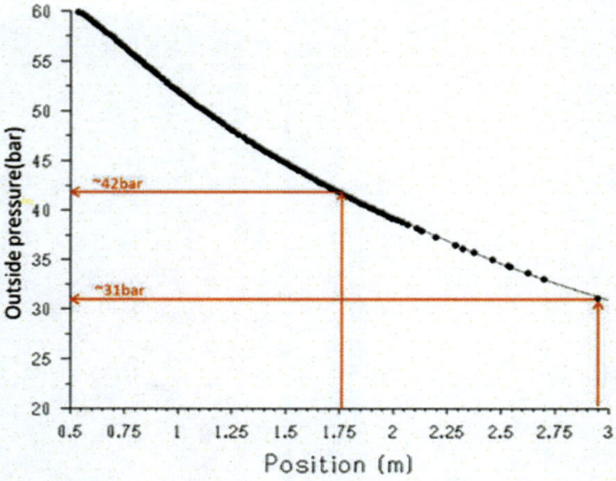

Fig. 29 External overpressure (CFD simulation)

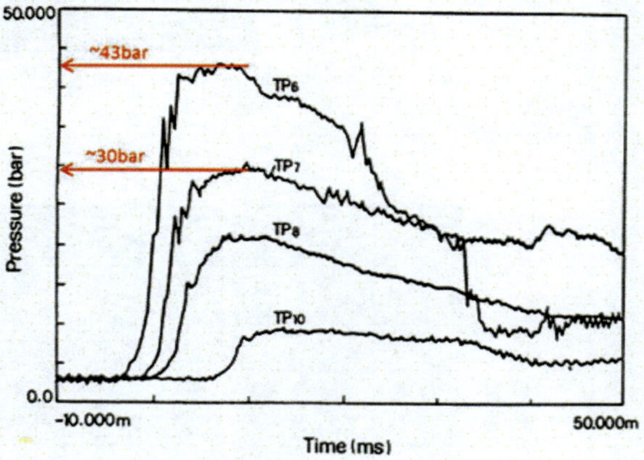

Fig. 30 External overpressure (experimental verification)

The comparison between the CFD prediction of overpressure values for two pressure transducers (see Fig. 29), positioned at different distance from pipe, with those monitored in the experimental test (see Fig. 30), shows a very good agreement confirming the reliability of the CFD model.

All these evidences confirm the formation of the overpressure bubble around the pipe in offshore conditions. The overpressure results were lower than the internal pressure, but of the same order of magnitude. This means that this external pressure gives an important contribution in resisting the fracture driving force (internal pressure on flaps).

Figure 31 shows clearly that the internal pressure starts to decrease from the initial value (i.e. 150 bar) near the area of pipe rupture while the external pressure rapidly increases in a large external volume around the pipe. It is therefore evident that as

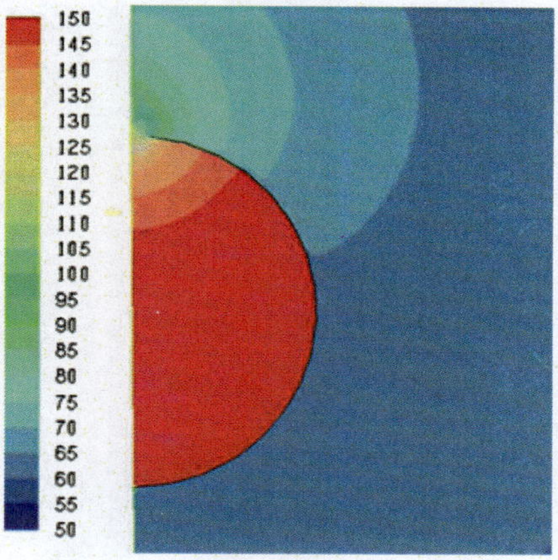

Fig. 31 Details of decompression around the pipe (CFD simulation)

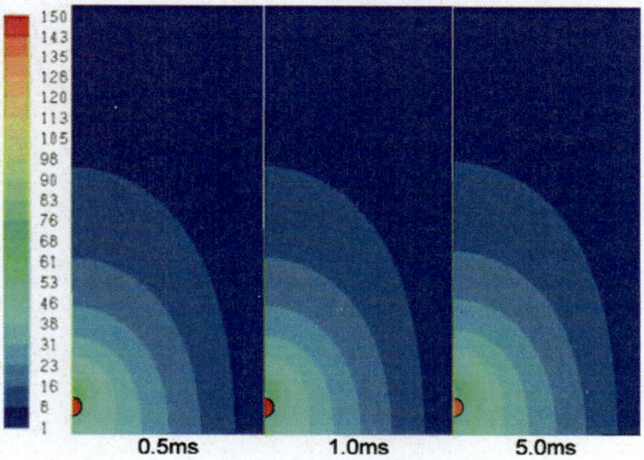

Fig. 32 Pressure contour of case #2

the pipe flaps are subjected internally to the internal fluid pressure, it externally is subjected to a significantly considerable pressure as well.

To evaluate the influence of sea water depth on the phenomenon under investigation, a further simulation was performed with a higher water depth, i.e. 500 m (case #3). Comparing Case #2 (shallow water) with Case #3 (deep water) reveals that despite the different hydrostatic pressures, the shape and the pressure magnitude of the overpressure bubble are very similar for the two cases. In other words, the pressure increment over the hydrostatic pressure in two cases is very similar. Figures 32 and 33 show the same shape of the pressure contour around the pipe taking into account that the scale starts from the hydrostatic pressure. This means that the overpressure bubble is formed as an additional pressure over the hydrostatic values.

CFD simulation of the scenario identified as case #4 has been carried out using pure CO_2 in dense phase as pressurizing medium. The scope was to verify the effect

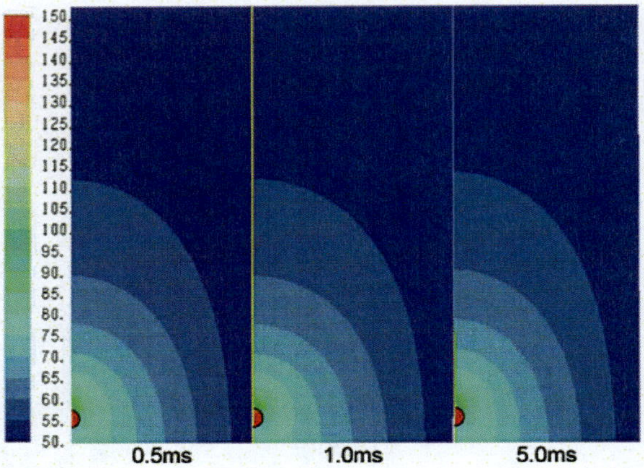

Fig. 33 Pressure contour of case #3

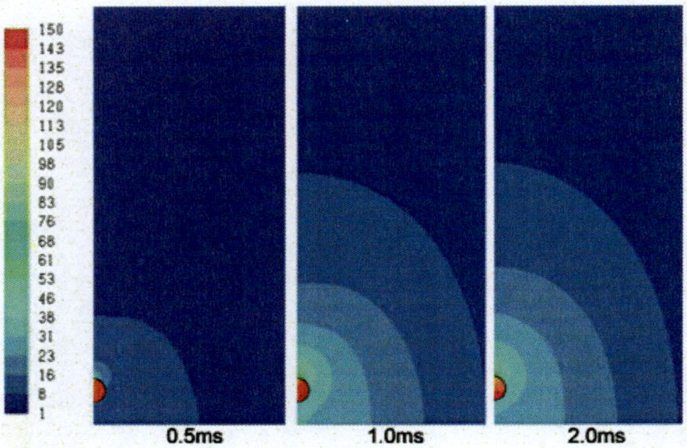

Fig. 34 Pressure contour of case #4

of a different internal medium on the formation of the external overpressure as in the operating conditions object of this project. Case #4 presents the same operating conditions (i.e. pressure, temperature), same geometrical properties and same water depth as those of case #2. The difference is in the internal medium in which for the case #4, the medium is dense-phase pure CO_2.

Figure 34 shows similar shape of overpressure bubble around the pipe as seen in case #2. The differences between nitrogen and CO_2 mixture decompression in offshore condition is mainly due to the following aspects:

- At the beginning of pipe decompression (case #4) the magnitude of overpressure is a bit lower than that of case #2. This effect is probably due to the dense state of CO_2 mixture in the first stage of decompression.
- During the whole decompression, the overpressure bubble is slightly lower than that of case #2.

This means that the decompression of CO_2 mixture produces an external overpressure around the pipe. Due to the nature of dense fluid, the effect of decompression

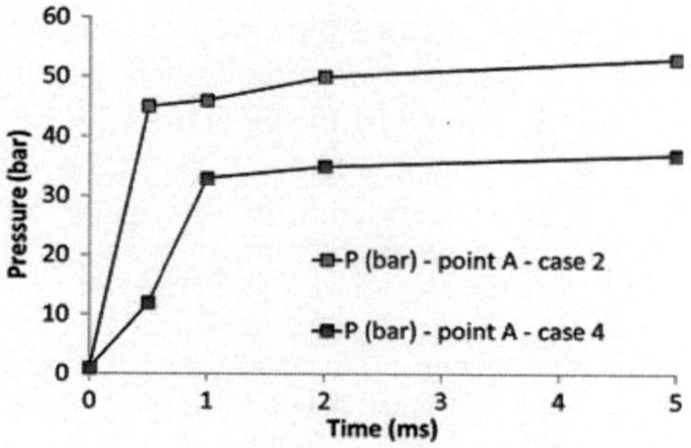

Fig. 35 External overpressure comparison between case #2 and case #4

is strong, but lower than that in the case of nitrogen decompression (see Fig. 35). On the contrary, once the bubble is formed, the dimension and trend are very similar.

3 Conclusions

The following are the conclusions from this study:

(a) In terms of the enhancement of existing BTCM, the comparison between the BTC in the original formulation, and BTC corrected through the correction function shows, for all of the analyzed cases, a significant capacity in reducing the total number of errors in prediction (from 40.1 to 19.3%) and, in particular, the number of propagation predicted as arrest (i.e. the most critical case) from 126 to 12.

(b) With regard to the backfill effect on fracture propagation phenomenon in terms of backfill coefficient reliability in offshore conditions, the study shows that the model is able to correctly predict the significant overpressure bubble forming around the pipe at the point of the rupture in offshore conditions. This phenomenon is not present in onshore conditions for an unburied pipe (i.e. air backfill) where the external overpressure is low and can be considered almost negligible.

(c) The medium-scale propagation tests carried out onshore (aboveground installation that is air backfill) validated the air backfill coefficient currently used for the BTC model even when CO_2 mixtures are used as internal pressurizing medium.

(d) The second test performed in offshore condition highlights that:

- The backfill effect in sea water was confirmed to be stronger than that in aboveground installation. In fact, the fracture in the offshore test arrested very quickly while in the onshore test (aboveground installation) a long propagation was observed.

- The external pipe pressure monitoring allowed to confirm the bubble overpressure formation around the pipe with a magnitude similar to what was predicted by the CFD model.
- CFD model showed that the overpressure bubble forms regardless of the sea water depth. Bubble shapes as well as pressure values are very similar (in terms of increment of the hydrostatic pressure) at various water depths.

Acknowledgements The authors would like to acknowledge PETRONAS for funding the research project and the Reggio Calabria University and AMBITECH s.r.l. and in particular to the Professor Pasquale Filianoti for the technical support and the availability of offshore site.

References

1. Veron J (2007) An overview of the non-developed gas reserves in Southeast Asia. PetroMin 3:22–28
2. Sitawati S, Nugroho SM (2013) An example of factors affecting monetization efforts for high CO_2 gas field development. In: Indonesian petroleum association thirty-seventh annual convention & exhibition, May 2013
3. Suhor MF, Mohamed MF, Mohammed Nor A, Singer M, Nesic S (2011) CorrosionTests of mild steel in high pressure CO_2 environments, Corrosion/2011
4. Cosham A, Eiber RJ (2008) Fracture propagation of CO_2 pipelines. J Pipeline Eng 281–292
5. Maxey WA (1974) Fracture initiation, propagation and arrest. In: 5th symposium on line pipe research, paper J, November 1974, Houston, TX
6. Maxey WA, Wilkowski GM, Kiefner JF (1978) Fracture initiation and propagation in underwater pipeline. In: Tenth annual offshore technology conference, Houston, 8 May 1978
7. Maxey WA (1984) Fracture propagation in underwater gas pipelines. In: ASME third annual pipeline engineering symposium, New Orleans, Louisiana, 12–16 February 1984. American Gas Association NG-18 paper no. 83-1
8. Demofonti G, Maresca A, Buzzichelli G (1988) Ductile fracture propagation and arrest in offshore pipelines. ASME Appl Mech Rev 41(2):85–95. https://doi.org/10.1115/1.3151883

A Comparative Review of Surface Modification Technique Between Physical Vapor Deposition (PVD) and Plasma Electrolytic Oxidation (PEO) for Biomedical Applications

I. Ahmed, A. Ahmad, A. M. Abdul-Rani, N. Tasnim, and M. Al-Amin

Abstract Whether the product is a surgical instrument, an orthopaedic implant or a heart valve, it is crucial to meet the criteria of hygiene, durability and protection that are extremely high. Nevertheless, it is quite unlikely for a metal to fully demonstrate all the desired properties. Recently, researchers have focused on coating bio-friendly materials over these tools in order to achieve desired durability and further ensure their safety and reliability. Among the numerous prevailing coating techniques, Physical Vapor Deposition (PVD) and Plasma Electrolytic Oxidation (PEO) coating techniques attain most of the requirements which are desirable for bioimplants and medical instruments. The aim of this review is to show the constructive comparisons between PVD and PEO coatings based on their tribomechanical properties, corrosion resistance and antibacterial characteristics leading to compatibility in biomedical applications. These coatings have positive effects on the biocompatibility and tribological properties of implant surfaces. These coatings have been characterized by scanning electron microscope (SEM), wear track morphology, potentiodynamic polarization test and antimicrobial tests.

1 Introduction

Ti6Al4V is the most widely used titanium alloy for orthopaedic applications, bone screws and plates for knee and hip prosthesis. This alloy is commonly used as hard tissue substitutes in artificial bones and joints due to its outstanding characteristics such as high strength, low density, resistance to corrosion, total stiffness of the body system, improved mobility, relatively low Young's Modulus and high capacity to join with bone or other tissues [1]. However, the most critical concern of this and other titanium alloys is their poor tribological character, defined by a high

I. Ahmed (✉) · A. Ahmad · A. M. Abdul-Rani · M. Al-Amin
Mechanical Engineering Department, Universiti Teknologi PETRONAS (UTP), 32610 Seri Iskandar, Perak, Malaysia
e-mail: ahmed_19001670@utp.edu.my

N. Tasnim
Pharmacy Department, Varendra University, 529/1, Kazla, Motihar, 6204 Rajshahi, Bangladesh

F. Ahmad et al. (eds.), *ICPER 2020*, Lecture Notes in Mechanical Engineering,
https://doi.org/10.1007/978-981-19-1939-8_61

coefficient of friction and low wear resistance [2]. Titanium medical applications research and development focus on new alloys, processing processes and surface treatments that improve biocompatibility and avoid fretting fatigue. Presently, the relatively weak tribological properties and possible corrosion problems of titanium alloy orthopaedic devices have led to the production of adequate surface treatments to effectively improve near-surface power, enhance durability and abrasive wear resistance, thereby reducing the coefficient of friction and thus improving the biocide ability and osseointegration of implants. Mechanical, electrochemical, and physical processes can be some of the surface treatment methods used to achieve these objectives. Anodic oxidation, cathodic deposition, and plasma electrolytic oxidation (PEO) provide electrochemical methods to improve corrosion resistance, biocompatibility, and bioactivity. Physical techniques include thermal spray, physical vapor deposition (PVD), plasma treatments for ion implantation and glow discharge, primarily to improve wear and corrosion resistance.

1.1 Physical Vapor Deposition (PVD) Technologies

The highest biocompatibility, corrosion resistance and basic strength (ratio of tensile strength to density) between different biomaterials are exhibited by titanium and its alloys [1]. They exhibit low wear resistance and high coefficients of friction although. In the region of the implants, poor abrasive wear resistance results in wear debris forming, causing metal ion release, inflammation and pain. Moreover, human life expectancy, which involves the creation of a new generation of implants of greater longevity, is increasingly growing. Applying protective hard coatings with PVD techniques characterized by high strength and durability, low elasticity modulus, low friction coefficient and improved wear and corrosion resistance in body fluids, along with better biocompatibility, may achieve substantial improvements in this field. In terms of coating material selection, PVD techniques are very versatile, as any type of inorganic material can be deposited. Physical vapor deposition (PVD) technology refers to a range of techniques for deposition of thin film where in a high vacuum setting a solid material (target) is vaporized and then transported to the substrate where it condenses to form a film [2, 3].

Physical vapor deposition (PVD) technologies are widely used today in the biomedical sector for the growth of protective hard coatings on various titanium components. Among them are the most common techniques for cathodic arc evaporation (CAE), cathodic cage plasma nitriding (CCPN) and magnetron sputtering (MS). Due to their excellent adhesion, high strength, low friction coefficient and increased wear and corrosion resistance in body fluids, such techniques have proven to be excellent candidates for improving the efficiency and durability of titanium and titanium alloys in biomedical applications. Furthermore, to manufacture biomedical protective coatings with personalized properties, the parameters of the PVD process can be modified. In addition, in comparison to chemical and galvanic surface treatment methods, PVD is an eco-friendly technology. It is clean and dry and does not produce toxic waste or water contamination, with no hazardous materials involved. A broad range of coating materials can be stored in PVD technologies, including alloys, metals, ceramics, and semiconductors. Therefore, PVD coatings are an attractive substitute to improve titanium and its alloys' surface characteristics while increasing its longevity. PVD techniques can deposit protective biocompatible coatings with high hardness, smooth surface, high corrosion and wear resistance and low friction coefficients [4, 5]. This can also act as effective barriers to ion release minimization due to tribocorrosion. In addition, solving the problem of colonizing bacteria that exists on implant products, antibacterial agents such as silver or copper can be introduced into the coating matrix.

1.2 Plasma Electrolytic Oxidation (PEO)

Plasma electrolytic oxidation (PEO) is an electrochemical oxidation process which produces a ceramic oxide coating on the surface of light alloys such as titanium, aluminum, or magnesium, also known as micro arc oxidation (MAO). The oxide layers produced by this method are characterized by high hardness, adhesion and improved wear and corrosion resistance [3–7]. The samples are anodically polarized to high voltages that surpass the dielectric breakdown voltage in this process, creating discrete short-lived plasma micro-discharges on the surface [3–6]. PEO is a high temperature process that results in the formation of a crystalline structured oxide consisting of rutile or anatase [8]. The improved wear resistance of the PEO layers created is due to the influence of high-temperature oxides and complex compounds, in particular rutile, formed during rapid melting and cooling by micro-discharges in interaction with the refrigerated electrolyte [2]. These oxide layers have excellent adhesive bonding to the base and elevated hardness. The layers are developed in the PEO method by polarizing the titanium-based portion to the dielectric breakdown voltage in a sufficient electrolyte. The PEO layer formed by titanium alloys typically consists of two crystallographic phases i.e., anatase and rutile. Although anatase can pose poor resistance to corrosion against certain acids and metal oxide solutions, rutile generally shows better protective properties. The use of rutile in orthopedic implants is required for all these purposes to increase the tribological and corrosion properties of titanium alloys. The PEO coating microstructure relies on the substrate material, the composition and temperature of the electrolyte, the dielectric breakdown, and the parameters of the process [6–8]. A porous microstructure resulting from the gas trapped in the material and the discharge channels formed during the process are the properties of these coatings. In the anodic oxide layer, the higher discharge energy produced by greater spark voltages results in larger pores. The oxide layer created by the PEO technique typically consists of two or three different porosity layers i.e. an external porous layer, an intermediate layer, and an inner layer that is denser.

The use of alkaline electrolytes makes the PEO process sustainable and environmentally friendly. In addition, to verify the structure of the resulting oxide layer, the chemical composition of the electrolyte can be precisely regulated and formulated. For example, some previous studies on the development of PEO coatings on titanium alloys for biomedical implants have incorporated into their composition bioactive elements such as phosphorus and calcium to enhance their osseointegration and promote cell growth [9, 10]. Finally, the simple and quick operation and its ecological friendliness make the PEO process a promising method for industrial applications related to surface modification of titanium alloy [11].

1.3 Titanium Based Biomedical Implants Using PEO Technique

Several investigations have already corroborated the improvement of the tribological properties of titanium alloys, as well as the osseointegration and bactericidal properties obtained through the PEO technique. The results of some studies are summarized and described briefly in this section.

The tribological output in dry abrasion conditions of a PEO-treated Ti6Al4V alloy was investigated by Ceschini et al. [12]. They discovered that the treatment improved the behavior of the alloy, reducing both the degree of wear and the coefficient of friction, even under high loads of up to 35 N. This enhancement was due to the high hardness and thickness of the coating, which as long as the oxide layer was present, could sufficiently support the load applied to protect the substrate. The biological response in vivo of PEO treated titanium alloys was confirmed in a study conducted by Ravanetti et al. [13]. The anodic oxide film was observed to promote in vivo early osteoblast adhesion and osseointegrative properties. To produce anodic oxide films for dental implants, Alves et al. [10] used the PEO technique. The effect of process parameters on the properties of the coating was assessed, i.e. morphology, thickness, roughness, elemental composition, etc. Artificial saliva (AS) for corrosion and tribocorrosion tests showed an improvement in the efficiency of Titanium after PEO treatment in terms of improved corrosion and resistance to wear. In order to have the best results, with improved mechanical resistance and corrosion, the AS corrosion and tribocorrosion tests revealed the sample with a higher rutile and anatase ratio. Finally, tests of cell-material interaction with human osteosarcoma cells showed an increase in the cell proliferation of PEO coating relative to the substrate. Saenz de Viteri et al. [14] developed TiO_2 coatings on the biomedical Ti6Al4V alloy used by the PEO technique in hip joint implants. In addition to Ca and P, they also incorporated iodine into the composition of the oxide layer to obtain antibacterial properties to produce sufficient surface characteristics for cell adhesion and boost osseointegration. Iodine has previously been reported to be strongly antibacterial, biocompatible and without cytotoxicity [15]. The Scanning Electron Microscopy (SEM) image presents a typical morphology of the topmost layer of the TiO_2 layer, showing a high porosity sufficient for cell proliferation.

2 Analysis of Results

2.1 Surface Morphology

The surface morphology of the treated sample is depicted in Fig. 1a, b and the magnified surface images are shown in the inset of Fig. 1a, b. The PVD-TiN coated sample surface in (Fig. 1a) indicate the existence of voids, pores, and pinholes that are caused by cooling strains and oxygen absorbed through the pores and grain structure [16]. The cathodic cage plasma nitriding (CCPN) processed sample shown in (Fig. 1b) confirms the existence of uniformly distributed particles apparently sputtered from the cathodic cage [17]. This sample of PVD-CCPN (Fig. 1b) shows no voids and pinholes on the surface, while under similar conditions, ceramic TiN was deposited by PVD. Although the coating is not strong, the diffusion from the base also strengthens the adhesion. The CCPN-treated PVD-coated sample indicates that the particles sputtered from CC (Cathodic Cage) complete the process [18]. It is clear that hardness is determined by the oxygen content. The oxygen absorbed decreases the amount of nitrogen absorption sites that are present on the surface. The residual gas is simply oxygen; it serves as a barrier to the diffusion or absorption of nitrogen. The higher the residual oxygen gas, the greater the absorption of oxygen, which limits the absorption of nitrogen and thus impacts the hardness.

Normally, the PEO coating shows a significantly higher thickness ($18 \pm 4\ \mu m$) than PVD coatings and a different morphology, which is seen more clearly in the SEM image in (Fig. 2). The outer part of the coating is porous, from the point of view of hardness, this attribute is unacceptable but can be useful in lubricated contacts for oil retention. By shifting towards the interface with the substrate, the coating becomes increasingly compact. This form of morphology results in a relatively high surface roughness; average R_a values of $1.3\ \mu m$ [12] are given. Figure 2 demonstrates the surface morphologies of the coatings developed in the concentrated aluminate electrolyte for 4 min. The coatings display irregularly formed nodules and micro

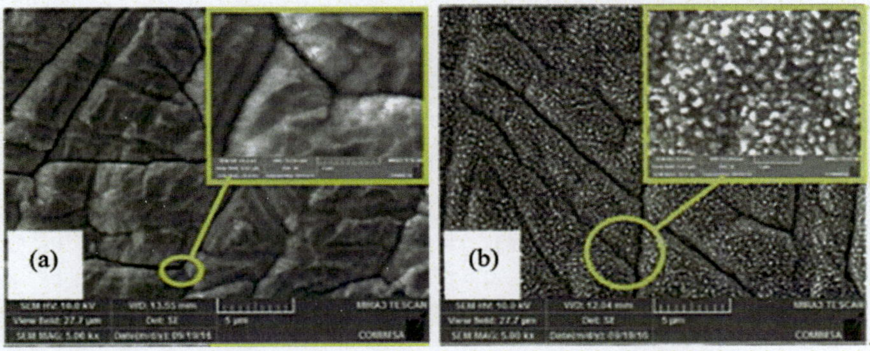

Fig. 1 SEM surface images of **a** PVD-TiN coated **b** PVD-CCPN treated sample [18]

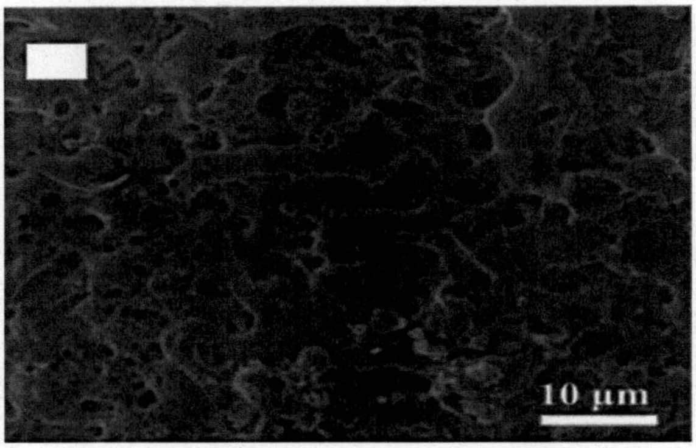

Fig. 2 SEM image of the PEO coatings formed in 32 g/L $NaAlO_2$ electrolyte [19]

pores, but the absence of pancakes is a significant notification, indicating that the penetrating discharges have been largely removed during the process of coating formation. The aluminum electrolyte coatings show an increased growth rate and enhanced uniformity. The aluminate electrolyte can be observed in different porous and "fused and re-solidified" regions (Fig. 2).

2.2 *Wear Track Morphology*

Using the ball-on-disc tribometer, the sliding wear activity of untreated and handled samples is carried out. Extreme adhesive wear, which typically occurs in soft substrates, is shown by substantial variations in friction plots for untreated samples and CCPN treated aluminum samples. First of all, there are negligible variations in the friction curve for the TiN coated sample. Variations gradually increase at a greater distance and later stop because of the sudden rise in the coefficient of friction. For PVD-TiN coated surface, high-resolution SEM photograph reveals that this occurs due to an increase in ball contact area and higher wearing of the coating. There are negligible variations on the surface of PVD-CCPN treated samples and they display an almost smooth curve. Although the CCPN coating has a slightly greater coefficient of friction, the coating is hard and has a remarkably low wear rate. The PVD-CCPN SEM image (Fig. 3a) also displays minimal wear track. This also indicates that the CCPN treatment produces cleaner, tougher, smoother, and adhesive coatings. The smallest friction coefficient and smoother friction curve are present in the PVD-CCPN treated sample. It has low wear rates as well. This is due to the diffusion of aluminum from the substrate to the PVD-CCPN sheet, which makes the layer more adhesive. Wear rates clearly indicate a substantial decrease in wear rates for treated samples, with the smallest value observed for the CPPN sample of 4.691 $\times 10^{-7}$ mm^3/Nm [18]. The contrast of worn surfaces indicates that CCPN processing significantly reduces the wear rate.

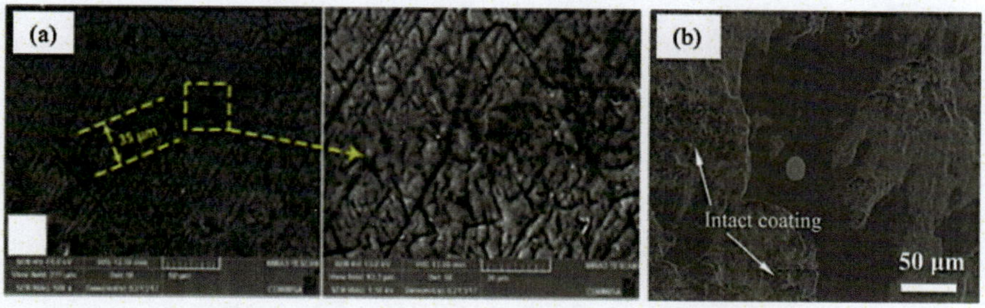

Fig. 3 SEM-images of wear tracks **a** normal and high-resolution images of selected regions of PVD-CCPN [18]. **b** PEO coated substrate in the aluminate electrolyte [19]

Figure 3b demonstrates the wear track on the coating shaped for 4 min in the aluminate electrolyte after the dry sliding time of 1800s. The wear track on the silicate electrolyte-formed coating is narrower than that. It is clear that the wear track includes tiny patches of intact coating. There are also large patches of dark materials present. The result reveals that the so-called "transfer layer" [20] or "tribolayer" [21] are the dark patches. The transfer layer can be created by the steel counterface tribo-oxidation, under the combined action of mechanical stress and frictional heating. Asperities on the surface of the PEO coating are assumed to provide mechanical interlocking for the adherence of the transfer layer. It is evident that the coating continues to maintain its original thickness, protected by a thin layer of light material that is the transfer layer referred to above. The previous view that the wear track profile only depicts the undulation of the transfer layer is confirmed by the SEM cross section [19]. The appearance of plate-shaped wear debris (Fig. 3a) indicates that extreme wear occurred by delamination. The thin PVD coatings, therefore do not provide good sliding wear resistance under these conditions, despite their high hardness, because the substrate does not provide sufficient mechanical support for the thin film [12]. On the contrary, in the case of PEO-treated samples, even under the more extreme sliding conditions, there was still a moderate wear regime. During tests, the worn surfaces appeared smooth, without signs of grooving and fine wear debris (Fig. 3b).

2.3 *Tribomechanical Properties*

In comparison with that of the substrate content, Vickers microhardness values of PVD and PEO coated samples are shown in (Fig. 4). All coatings improve the Ti alloy's surface hardness, most notably PVD TiN coatings. The penetration depth of the diamond indenter does not meet the requirement to be many times (typically 10) smaller than the coating thickness, despite the low indentation load (25 gm). The calculated microhardness values are therefore representative of the

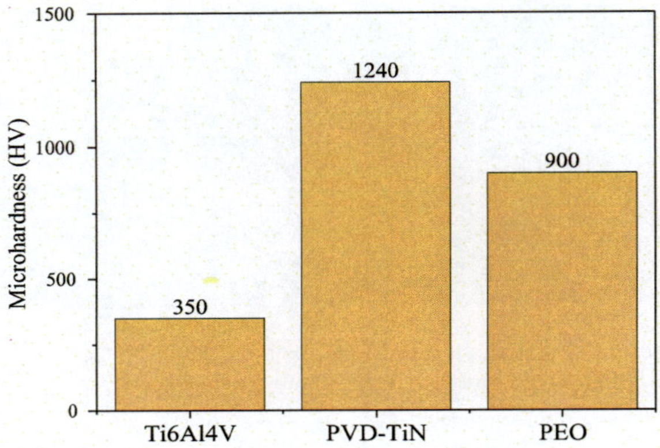

Fig. 4 Vickers microhardness values of surface modified samples

composite coating/substrate microhardness, because they are affected by the deformation of the substrate below the coating [22]. This effect especially stigmatizes the thinnest coating, i.e. the superlattice, which, thanks to coherence stresses between the nanolayers, can achieve intrinsic hardness values of at least 3000 HV. In order to be able to reliably calculate the dimensions of the indentation on the rough surface, the hardness of the PEO coating is measured by a higher applied load (100 gm) [12]. This treatment also gives a significant improvement in the hardness of the Ti alloy. In this case, the coating thickness is high enough to allow the essential hardness to be determined.

The PEO treated samples show the best tribological behavior, both in terms of average friction coefficient and high wear resistance (Fig. 5a and b). In this case, the coefficient of friction remains low (approximately 0.2) and the linear wear is negligible for the duration of the test, even under the maximum applied load of 15 N, suggesting that this coating is capable of shielding the titanium alloy from extreme

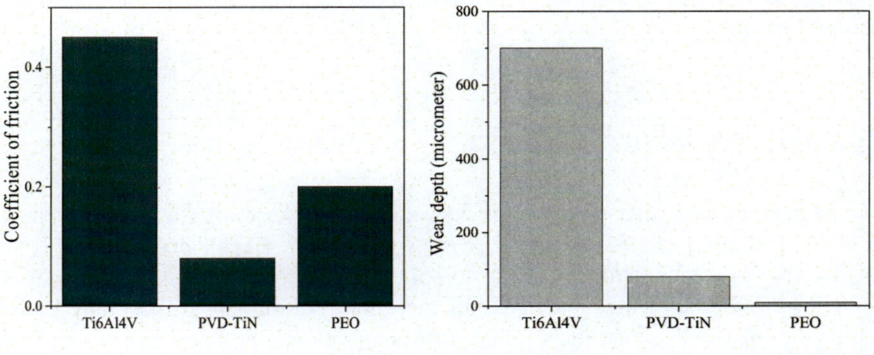

Fig. 5 By applying load 15 N **a** Average values of the coefficient of friction for different samples. **b** Wear depth of different surfaces

wear, thereby also reducing the coefficient of friction significantly. The average friction coefficient values determined under the 15 N loads applied are summarized in (Fig. 5a) [23]. For PVD-TiN coatings, it should be noted that the lowest values of the friction coefficient are shown for up to 15 N TiN coatings. At 15 N applied loads, the friction coefficient of the PVD-TiN coated alloy is 0.1. The wear damage can be measured on both the fixed sliders and the rotating ceramic counter material by calculating the maximum wear scar depth. The better wear resistance of the PVD-TiN coating is shown between the coatings, possibly due to its higher composite microhardness. Owing to the poor load carrying capacity of these thin hard coatings, deposited on a much softer substrate, with consequent coating fracture and delamination, all other PVD hard coatings are worn through and stripped from the substrate when the usual load rises. Indeed, it is well known that the layer under the thin hard coating must be as hard as possible in order to: (i) prevent deformation from fracturing the brittle coating; (ii) increase the load support; and (iii) minimize the actual contact area [24–26]. As the coating is weakened and micro-fractured, the "third body" [12], which includes base material chips and hard wear debris from the coating, forms a coefficient of friction and wear damage during sliding with a consequent rise.

The PEO-treated samples show the highest wear resistance, with marginal wear loss even under the highest load applied of 15 N. This good tribological behavior should be related to the higher thickness of this coating, which can better support the load applied. It is important to note that negligible wear damage to the counter material is also created when the coating is not affected. On the contrary, due to the development of an abrasive "third-body", extreme wear damage to the coatings often contributes to high wear of the counter content. It is also noteworthy that, even with its lower hardness, the Ti6Al4V alloy caused substantial wear damage to the ceramic counter material. Dong and Bell [27] also found a similar activity and explained it based on tribochemical reactions between the face of the alumina counter and Ti and Al transferred from the Ti6Al4V slider.

2.4 Corrosion Resistance

From corrosion current density (i_{corr}) and current potential (E_{corr}) value of potentiodynamic polarization curve, it can be easily inferred the corrosion resistivity of any object. If the i_{corr} parameter shows less value, then the corrosion resistance will be higher and vice-versa. Over time, the uncoated Ti-6Al-4 V alloy potential becomes more optimistic. This behavior indicates that a naturally formed passivation film based on TiO_2 is thickened, rendering titanium alloys spontaneously passive in NaCl media. The responses to this corrosion process are defined as follows [28, 29]:

$$O_2 + 2H_2O + 4e^- = 4OH^-(\text{Cathodic reaction}) \quad (1)$$

Table 1 Polarization and open circuit potential electrochemical parameters

Sample	E_{corr}	i_{corr} ($\mu A/cm^2$)	References
Ti6Al4V	$-0.10\ V_{SCE}$	0.37	[32]
PVD-TiN	$-0.09\ V_{SCE}$	0.63	[32]
PEO	46.30 mV	0.01	[42]

$$Ti + 4OH^- - 4e^- = TiO_2 + 2H_2O \text{ (Anodic reaction)} \quad (2)$$

The cathodic reaction involves oxygen reduction, whereas anodic reactions form a TiO_2-based oxide on the surface that slows down the corrosive method. Although the variance in open circuit potential (OCP) for coated alloys does not follow the same pattern, their final OCPs (called E_{corr}) are similar to the value calculated for uncoated alloys after 3 h exposure [30]. This is indicative that the key compound on the three materials produced in the passive film is almost the same. In the TiO_2 stability area, all measured E_{corr} are recorded in (Table 1). Pohrelyuk et al. [31] stated that the passive behavior of such chloride solution coatings is consistent with TiN oxidation, leading to the formation of TiN_xO_y oxynitride and the active dissolution of this oxynitride to Ti^{3+} followed by Ti^{3+} oxidation to TiO_x oxide [32]. The passive conduct of the TiN coated alloy in the NaCl solution is presumed to control these reactions. It is a difficult task to compare the E_{corr} values with published data, as E_{corr} is highly dependent on the microstructure of the coatings, which is related to the process and conditions of deposition. The corrosion activity of plasma altered Ti-6Al-4 V alloy in 3.5 wt.% NaCl solution at room temperature has been studied by Amaya-Vazquez et al. [33] and Savonov et al. [34]. Atapour et al. [35] measured a corrosion potential of 0.380 V_{SCE} for Ti-6Al-4 V alloy in 0.9% NaCl solution at 37 °C. The key electrochemical parameters acquired from the potentiodynamic polarization curves are presented in Table 1.

The corrosion current density (i_{corr}) values for the uncoated and coated alloy are in the same magnitude value order and are very low, indicating strong corrosion resistance [36, 37]. For such coatings in similar solutions, several authors [38–40] recorded corrosion current density (i_{corr}) values between 10^{-8} and 10^{-6} A/cm^2. The corrosion current density values of approximately 10^{-7} A/cm^2 recorded in this current study have the same magnitude of the i_{corr} values obtained by Li et al. [36], Cunha et al. [38], and Chung [40]. The current density of uncoated Ti-6Al-4 V alloy in 3.5 wt.% NaCl is 0.37 $\mu A/cm^2$ [32]. Martin and colleagues [41] who researched the impact of microstructure and texture on the corrosion behavior of Ti-6Al-4 V alloy in 1% NaCl found values of the same order, between 4.51 and 5.90 $\mu A/cm^2$ depending on the various microstructures and preferential crystallographic plane alignment. It can be concluded that at all temperatures, the TiN coated Ti-6Al-4 V alloy demonstrated superior corrosion resistance in the NaCl solution. The coated alloy based on PVD-TiN demonstrates the best corrosion behavior, bearing i_{corr} values at 0.63 $\mu A/cm^2$ and maintaining its structural integrity on the surface at 25, 60 and 80 °C during polarization.

Potentiodynamic polarization curves of the PEO coatings are generally obtained over a potential range of ±600 mV in 3.5% NaCl solution. It is found that the corrosion current density of the coating is reduced as the NaF concentration increased from 0.15 to 0.25 mol/L and it shows the value of i_{corr} is 0.01 μA/cm^2 (Table 1). Moreover, the coating formed in 0.25 mol/L NaF electrolyte has the best corrosion resistance, which can be attributed to the thicker coating. To further confirm the influence of F ions on the corrosion resistance of the PEO coating, the AC impedances of Ti alloy and PEO-treated samples are normally measured. Studies have shown that that the reactance arc in the high frequency region reveals the properties of the solution/film interface, and the reactance arc in the low frequency region indicates the properties of the oxide film itself [42, 43]. ZHU et al. [42] found that there is a direct relationship between arc radius and corrosion rates. Compared with the Ti alloy substrate, the corrosion resistance of the PEO-treated sample is significantly improved. As the NaF concentration increased from 0.15 to 0.25 mol/L, the radius of the capacitance loop and |Z| (Z is the impedance of coated samples) decreased, which indicates an increase in the corrosion resistance of the oxide film. Therefore, the incorporation of F ions (using NaF electrolyte with a concentration of ≤0.25 mol/L) can promote the growth of the PEO coating and improve the corrosion resistance of the sample. However, as the NaF concentration increased to 0.50 mol/L [42], the discharge became increasingly violent, and coating ablation occurred, leading to inferior corrosion resistance of the PEO coating.

3 Biocompatibility

Mainly, antibacterial activity depends on the hydrophilicity and contact angle measurement. Another factor that can influence this antibacterial property is surface roughness. Higher surface roughness and lower contact angle leads to higher antibacterial activity. Surface energy plays an important role in the concept of bioactivity. In comparison to hydrophobic surfaces, hydrophilic surfaces allow better cell attachment and proliferation [44–46]. A well-established reality is the impact of conformational changes in adhesive proteins on altered biological properties [47, 48]. As blood coagulation is directly associated with fibrinogen adsorption [49, 50] or surgical instruments [51], bio inert materials are favored for blood interaction. Consequently, for bone implants, certain materials are suboptimal. However, an association with integrin cell surface receptors is observed in addition to protein interaction with surfaces through conformational modifications, again with hydrophobic surfaces showing a detrimental effect on bioactivity [52, 53]. Oses et al. [54] showed that in PVD coating techniques possess higher surface roughness value of sample exhibits lower contact angle. The wettability of the implanted coatings with respect to the non-implanted ones was observed about the contact angle measurements, leading to an improvement in the contact angle. For the non-implanted samples, the measured angle increased from 79° to 95° after the implantation procedure [54]. The results obtained in the test for antimicrobial activity and efficacy showed that the ion implantation treatments had a proper antibacterial impact. Oses et al. [54] also found that the PVD coating has a higher rate of *Escherichia Coli* removal than the bacteria *Staphylococcus Aureus*.

In terms of porosity, thickness, and composition, the morphology of the intermediate and outer layers can be adjusted to monitor the process parameters and the composition of the electrolyte. In the outer porous layer, the pore diameter varies from 3 to 8 μm in diameter, resulting in relatively high surface roughness [56]. The rough and porous TiO_2 surface formed by the PEO process makes these coatings suitable for cell adhesion [9]. By comparing the different PEO treated surfaces containing silver particles with two other surfaces, antibacterial properties can be assessed: one untreated and one PEO treated without silver particles. In order to ensure good reproducibility, the experiments are normally conducted in triplicate. A gram-positive (*Staphylococcus Aureus*) and a gram-negative (*Escherichia Coli*) are used to compare two different types of bacteria with which silver demonstrates its action [56, 57]. The effects of counting the slides of the *Escherichia Coli* and *Staphylococcus Aureus* colonies after 3 h of exposure can be observed in (Fig. 6). It can be observed from the results that contact with the surfaces of the untreated and conventionally anodized samples does not produce any significant effect on the decrease in bacterial colony density, in reality only a decrease comparable to the experimental error is reported [58]. Action with the silver-free PEO coating potentially results in a decrease of about one order of magnitude in the density of bacterial colonies due to the alkaline atmosphere produced when the bacteria's broth meets the treated surface. All the samples treated in the electrolyte with silver particles demonstrate a very powerful antimicrobial activity. This is in line with the high antimicrobial effect of silver on this bacterium [57], which however, appeared to be substantially higher in these experiments than in the literature [59], due to the high amount of silver present in the coating. However, owing to the better corrosion resistance previously

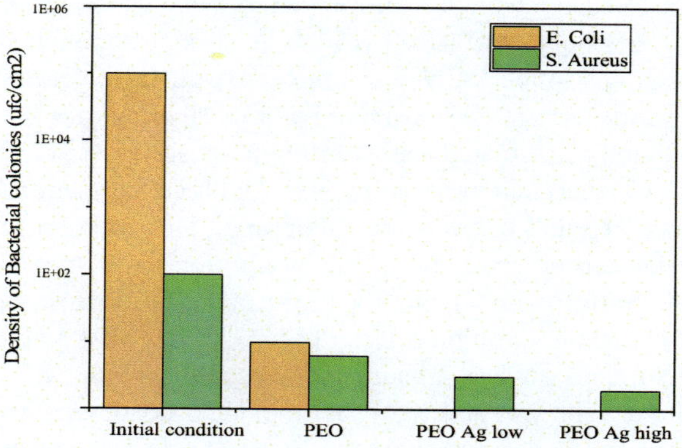

Fig. 6 Counting of *E. coli* and *S. Aureus* bacteria colonies after 3 h of exposition to the different surfaces

checked, sealed samples should be preferred. More recently, Chen et al. [60], in a KOH solution, formed PEO coatings on a Ti-39Nb-6Zr alloy. The corrosion and wear efficiency of the coated alloy in PBS was evaluated and the biocompatibility of osteoblast cultured cells was investigated. After the treatment, the corrosion and wear performance of the alloy was considerably improved, showing higher corrosion resistance and lower loss of material. The oxide layer morphology was found to be beneficial for protein adsorption, and the porous and rough surface also enhanced cell adhesion and osteoblast cell proliferation.

The effect of the PEO coating on the activity of *S. Aureus* was evaluated in the same way as for *E. Coli* and the results of the counting and the slides with the bacteria after 3 h of exposition can be observed in (Fig. 6). The results substantially confirmed the ones obtained for *E. Coli.* In fact, contact with the untreated sample (initial condition) does not produce an antimicrobial effect, whereas contact with the surface treated with PEO without silver particles results in a reduction of the density of the bacterial colonies by about one order of magnitude. For *S. Aureus,* the presence of silver produced a major increase in the antimicrobial effect generated by the surfaces. In addition, certain variations can be found between the various samples containing silver. It can be demonstrated that in case of *S. Aureus* bacteria, density of bacterial colonies get lower with the increment of silver(Ag) contents [61, 62] and exhibit higher antibacterial activity.

4 Conclusions

Additionally, PVD technologies are highly versatile and are distinguished for the production of well-adherent, rough, compact and highly durable films. Under low temperature deposition processes a wide range of coating materials can be produced by PVD, made of different compounds or in multilayer structures. It must be highlighted:

- PVD techniques deposit very low friction coefficient and wear resistance carbon-based films to shield Titanium from mechanical degradation. High corrosion and wear resistance transition metal nitride films are grown using PVD techniques that significantly improve Titanium's tribocorrosion efficiency.
- Without affecting their biocompatibility, the introduction of silver into either of these coating matrixes will increase the antibacterial property of the films.

PEO techniques are among the most promising among the different electrochemical surface modification technologies available today to improve the characteristics of titanium-based materials for biomedical applications, for the following reasons:

- To boost the different properties, the composition of the electrolytes can be tailored to successfully integrate many elements into the oxide film composition. By

improving osseointegration and avoiding infection at the same time, the enrichment of the oxide film structure with bioactive and antimicrobial agents can induce new bio-selective functional surfaces for osseointegrated titanium implants.
- The growth of an oxide film on Titanium surfaces in simulated body fluids, such as artificial saliva, will significantly improve their corrosion and tribocorrosion properties. In the case of PEO, which typically causes the growth of an oxide film with a crystalline structure, the improved performance of TiO_2 films in corrosion and tribocorrosion has been reported to occur for greater rutile and anatase ratios.

In conclusion, this study focused on promising electrochemical and physical developments to alter the surface characteristics of titanium-based materials in order to improve their functionality for applications in biomedical implants. Therefore, the most adequate technique between PVD and PEO for surface treatment should be selected considering the required surface features of the implant, namely surface morphology, wear track morphology, microhardness, as well as tribocorrosion processes. Based on this approach, implants may be constructed with single or multiple functionalities, including osseointegrable implants with antimicrobial properties, showing enhanced corrosion and tribomechanical properties at the same time.

References

1. Niinomi M, Nakai M, Hieda J (2012) Development of new metallic alloys for biomedical applications. Acta Biomater 8(11):3888–3903
2. Mattox DM (2010) Handbook of physical vapor deposition (PVD) processing. William Andrew
3. Martin PM (2009) Handbook of deposition technologies for films and coatings: science, applications and technology. William Andrew
4. Antunes RA, De Oliveira MCL (2009) Corrosion processes of physical vapor deposition-coated metallic implants. Crit Rev Biomed Eng 37(6)
5. Ching HA, Choudhury D, Nine MJ, Osman NAA (2014) Effects of surface coating on reducing friction and wear of orthopaedic implants. Sci Technol Adv Mater 15(1):014402
6. Rapheal G, Kumar S, Scharnagl N, Blawert C (2016) Effect of current density on the microstructure and corrosion properties of plasma electrolytic oxidation (PEO) coatings on AM50 Mg alloy produced in an electrolyte containing clay additives. Surf Coat Technol 289:150–164
7. Hussein R, Nie X, Northwood D (2016) Production of high quality coatings on light alloys using plasma electrolytic oxidation (PEO). WIT Trans Built Environ 166:439–454
8. Dehnavi V, Shoesmith DW, Luan BL, Yari M, Liu XY, Rohani S (2015) Corrosion properties of plasma electrolytic oxidation coatings on an aluminium alloy–The effect of the PEO process stage. Mater Chem Phys 161:49–58
9. Alves S, Bayón R, Igartua A, Saénz de Viteri V, Rocha L (2014) Tribocorrosion behaviour of anodic titanium oxide films produced by plasma electrolytic oxidation for dental implants. Lubr Sci 26(7–8):500–513
10. Alves S et al (2015) Tribocorrosion behavior of calcium-and phosphorous-enriched titanium oxide films and study of osteoblast interactions for dental implants. J Bio Tribo-Corros 1(3):23
11. Yetim A (2010) Investigation of wear behavior of titanium oxide films, produced by anodic oxidation, on commercially pure titanium in vacuum conditions. Surf Coat Technol 205(6):1757–1763

12. Ceschini L, Lanzoni E, Martini C, Prandstraller D, Sambogna G (2008) Comparison of dry sliding friction and wear of Ti6Al4V alloy treated by plasma electrolytic oxidation and PVD coating. Wear 264(1–2):86–95
13. Ravanetti F et al (2010) In vitro cellular response and in vivo primary osteointegration of electrochemically modified titanium. Acta Biomater 6(3):1014–1024
14. Kuromoto NK, Simão RA, Soares GA (2007) Titanium oxide films produced on commercially pure titanium by anodic oxidation with different voltages. Mater Charact 58(2):114–121
15. Shirai T, Shimizu T, Ohtani K, Zen Y, Takaya M, Tsuchiya H (2011) Antibacterial iodine-supported titanium implants. Acta Biomater 7(4):1928–1933
16. McIntyre D, Greene J, Håkansson G, Sundgren J, Münz W (1990) Oxidation of metastable single-phase polycrystalline Ti0. 5Al0. 5N films: kinetics and mechanisms. J Appl Phys 67(3):1542–1553
17. Gallo SC, Dong H (2009) On the fundamental mechanisms of active screen plasma nitriding. Vacuum 84(2):321–325
18. Bashir M et al (2017) Enhanced surface properties of aluminum by PVD-TiN coating combined with cathodic cage plasma nitriding. Surf Coat Technol 327:59–65
19. Kang S, Tu W, Han J, Li Z, Cheng Y (2019) A significant improvement of the wear resistance of Ti6Al4V alloy by a combined method of magnetron sputtering and plasma electrolytic oxidation (PEO). Surf Coat Technol 358:879–890
20. Martini C, Ceschini L, Tarterini F, Paillard J, Curran J (2010) PEO layers obtained from mixed aluminate–phosphate baths on Ti–6Al–4V: dry sliding behaviour and influence of a PTFE topcoat. Wear 269(11–12):747–756
21. Arrabal R, Mohedano M, Matykina E, Pardo A, Mingo B, Merino M (2015) Characterization and wear behaviour of PEO coatings on 6082–T6 aluminium alloy with incorporated α-Al_2O_3 particles. Surf Coat Technol 269:64–73
22. Bull SJ, Rickerby DS (2001) Characterization of hard coatings. Handb Hard Coat Park Ridge Noyes Publ, 181–228
23. Hutchings I, Tribology EA (1992) A division of hodder headline PLC
24. Holmberg K, Ronkainen H, Matthews A (2000) Tribology of thin coatings. Ceram Int 26(7):787–795
25. Hogmark S, Jacobson S, Larsson M (2000) Design and evaluation of tribological coatings. Wear 246(1–2):20–33
26. Podgornik B et al (2001) Tribological properties of plasma nitrided and hard coated AISI 4140 steel. Wear 249(3–4):254–259
27. Dong H, Bell T (1999) Tribological behaviour of alumina sliding against Ti6Al4V in unlubricated contact. Wear 225:874–884
28. Pjescic J, Mentus S, Blagojevic N (2002) Investigation of titanium corrosion in concentrated NaOH solutions. Mater Corros-Werkst Korros 53(1):44–50
29. McCafferty E (2005) Validation of corrosion rates measured by the Tafel extrapolation method. Corros Sci 47(12):3202–3215
30. Pourbaix M (1974) Atlas of electrochemical equilibria in aqueous solution. NACE 307
31. Pohrelyuk I, Fedirko V, Tkachuk O, Proskurnyak R (2013) Corrosion resistance of Ti–6Al–4V alloy with nitride coatings in Ringer's solution. Corros Sci 66:392–398
32. Oliveira V, Vazquez A, Aguiar C, Robin A, Barboza M (2015) Protective effect of plasma-assisted PVD deposited coatings on Ti–6Al–4V alloy in NaCl solutions. Mater Des 88:1334–1341
33. Amaya-Vazquez M, Sánchez-Amaya J, Boukha Z, Botana F (2012) Microstructure, microhardness and corrosion resistance of remelted TiG2 and Ti6Al4V by a high power diode laser. Corros Sci 56:36–48
34. Savonov G, Ueda M, Oliveira R, Otani C (2011) Electrochemical behavior of the Ti6Al4V alloy implanted by nitrogen PIII. Surf Coat Technol 206(7):2017–2020
35. Atapour M, Pilchak A, Shamanian M, Fathi M (2011) Corrosion behavior of Ti–8Al–1Mo–1V alloy compared to Ti–6Al–4V. Mater Des 32(3):1692–1696
36. Li J, Li S, Hao Y, Yang R (2014) Electrochemical characterization of nanostructured Ti–24Nb–4Zr–8Sn alloy in 3.5% NaCl solution. Int J Hydrog Energy 39(30):17452–17459

37. Sivakumar B, Singh R, Pathak LC (2015) Corrosion behavior of titanium boride composite coating fabricated on commercially pure titanium in Ringer's solution for bioimplant applications. Mater Sci Eng C 48:243–255
38. Cunha L, Andritschky M, Rebouta L, Silva R (1998) Corrosion of TiN,(TiAl) N and CrN hard coatings produced by magnetron sputtering. Thin Solid Films 317(1–2):351–355
39. Ahn S, Hong J, Kim J, Han J (2007) Effect of microstructure on corrosion behavior of TiN hard coatings produced by a modified two-grid attachment magnetron sputtering process. Thin Solid Films 515(17):6878–6883
40. Chung K, Liu G, Duh J, Wang J (2004) Biocompatibility of a titanium–aluminum nitride film coating on a dental alloy. Surf Coat Technol 188:745–749
41. Martin E, Azzi M, Salishchev G, Szpunar J (2010) Influence of microstructure and texture on the corrosion and tribocorrosion behavior of Ti–6Al–4V. Tribol Int 43(5–6):918–924
42. Zhai D, Feng K (2019) Preparation of micro/nano-structured ceramic coatings on Ti6Al4V alloy by plasma electrolytic oxidation process. Trans Nonferrous Met Soc China 29(12):2546–2555
43. Yerokhin A, Nie X, Leyland A, Matthews A, Dowey S (1999) Plasma electrolysis for surface engineering. Surf Coat Technol 122(2–3):73–93
44. Van Wachem P, Beugeling T, Feijen J, Bantjes A, Detmers J, Van Aken W (1985) Interaction of cultured human endothelial cells with polymeric surfaces of different wettabilities. Biomaterials 6(6):403–408
45. Groth T, Zlatanov I, Altankov G (1995) Adhesion of human peripheral lymphocytes on biomaterials preadsorbed with fibronectin and vitronectin. J Biomater Sci Polym Ed 6(8):729–739
46. Lim JY, Liu X, Vogler EA, Donahue HJ (2004) Systematic variation in osteoblast adhesion and phenotype with substratum surface characteristics. J Biomed Mater Res Part Off J Soc Biomater Jpn Soc Biomater Aust Soc Biomater Korean Soc Biomater68(3):504–512
47. Grinnell F (1987) Fibronectin adsorption on material surfaces a. Ann N Y Acad Sci 516(1):280–290
48. Iuliano DJ, Saavedra SS, Truskey GA (1993) Effect of the conformation and orientation of adsorbed fibronectin on endothelial cell spreading and the strength of adhesion. J Biomed Mater Res 27(8):1103–1113
49. Sawyer PN, Pate JW (1953) Bio-electric phenomena as etiological agents in intravascular thrombosis. Surgery 34(3):491–500
50. Baurschmidt P, Schaldach M (1977) Electrochemical aspects of thrombogenicity of a material. J Bioeng 1(4):261–278
51. Cloud A, Kumar S, Kavdia M, Abu-Safe H, Gordon M (2008) Protein adsorption on low temperature alpha alumina films for surgical instruments. Surf Coat Technol 203(5–7):913–917
52. Groth T, Altankov G, Kostadinova A, Krasteva N, Albrecht W, Paul D (1999) Altered vitronectin receptor (αv integrin) function in fibroblasts adhering on hydrophobic glass. J Biomed Mater Res Off J Soc Biomater Jpn Soc Biomater Aust Soc Biomater 44(3):341–351
53. PeŠŠková V, Kubies D, Hulejova H, Himmlova L (2007) The influence of implant surface properties on cell adhesion and proliferation. J Mater Sci Mater Med 18(3):465–473
54. Osés J, Palacio J, Kulkarni S, Medrano A, García J, Rodríguez R (2014) Antibacterial PVD coatings doped with silver by ion implantation. Appl Surf Sci 310:56–61
55. de Viteri VS, Fuentes E (2013) Titanium and titanium alloys as biomaterials. Tribol Adv InTech, 155–181
56. Panáček A et al (2006) Silver colloid nanoparticles: synthesis, characterization, and their antibacterial activity. J Phys Chem B 110(33):16248–16253
57. Ruparelia JP, Chatterjee AK, Duttagupta SP, Mukherji S (2008) Strain specificity in antimicrobial activity of silver and copper nanoparticles. Acta Biomater 4(3):707–716
58. Cerchier P et al (2017) Antibacterial effect of PEO coating with silver on AA7075. Mater Sci Eng C 75:554–564
59. Sondi I, Salopek-Sondi B (2004) Silver nanoparticles as antimicrobial agent: a case study on E. Coli as a model for Gram-negative bacteria. J Colloid Interface Sci 275(1):177–182

60. Chen L et al (2018) Characterization of plasma electrolytic oxidation film on biomedical high niobium-containing β-titanium alloy. Surf Coat Technol 352:295–301
61. Jung WK, Koo HC, Kim KW, Shin S, Kim SH, Park YH (2008) Antibacterial activity and mechanism of action of the silver ion in Staphylococcus aureus and Escherichia coli. Appl Env Microbiol 74(7):2171–2178
62. Chowdhury NR, MacGregor-Ramiasa M, Zilm P, Majewski P, Vasilev K (2016) 'Chocolate'silver nanoparticles: synthesis, antibacterial activity and cytotoxicity. J Colloid Interface Sci 482:151–158

Stress Corrosion Cracking and Oxide Layer Growth of P91 Steel in High Temperature Steam Environment

Nizam Yusoff and Mokhtar Che Ismail

Abstract The cracking of ferritic alloy steel SA213 P91 materials has been analysed. Environmental factors such as chloride, pH, temperature and dissolved oxygen content in steam may contribute to the propagation of corrosion and SCC cracks on ferritic alloy steel SA213 P91 materials. The SCC crack propagation on P91 alloy will increase with an increase of high chloride content, low pH value, high steam temperature and low oxygen concentration. The remaining steel life also was increase with an increase of oxygen level in test environment.

Keywords Environmental factors · Steam corrosion · SCC cracks

1 Introduction

Stress corrosion cracking in ferritic alloy steel SA213 P91 is a significant problem pertaining to the boiler tubes in the power generation industry. In accordance with an article from [1] the most common causes of SCC in steam environment are chloride contents, temperature, pH and dissolved oxygen gases level. The topic for discussion in this paper is the environmental factors that contribute specifically to the ferritic alloy steel failure of SA213 P91 (UNS K91560) or 9Cr 1Mo materials in steam systems; these include chloride, pH, temperature and the oxygen content present in steam.

N. Yusoff (✉) · M. C. Ismail (✉)
Universiti Teknologi PETRONAS, Seri Iskandar, Malaysia
e-mail: mohd_g03294@utp.edu.my

M. C. Ismail
e-mail: mokhtis@petronas.com.my

F. Ahmad et al. (eds.), *ICPER 2020*, Lecture Notes in Mechanical Engineering,
https://doi.org/10.1007/978-981-19-1939-8_62

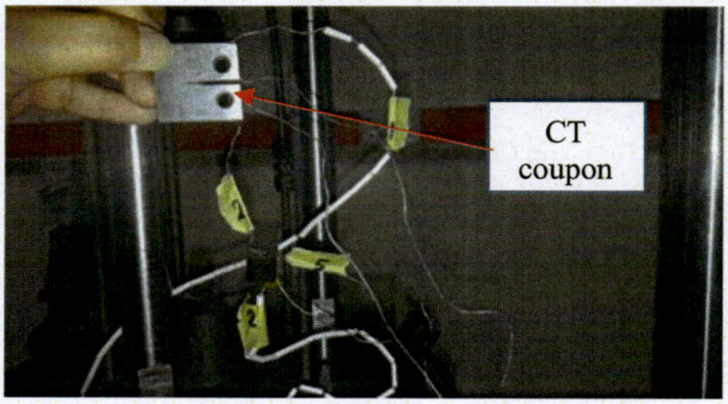

Fig. 1 Stress corrosion cracking (SCC) CT coupons prior test

Table 1 Steam environment for stress corrosion cracking test

Steam parameters	74 h	1344 h
Tensile stress (KN)	11	6.53
Oxygen (ppb)	7.0	5500
pH	9.29	9.50
Temperature (°C)	612	568
Chloride (ppm)	Non detected	Non detected

2 Method

Stress corrosion cracking test was conducted using compact tension (CT) types coupon and exposed to simulated steam environment in SCC test rig. The coupons was manufactured by international overseas coupons company located in USA and in according with ASTM G4-01 [2] standard. All the coupons was mounted in such way so that it securely held and was electrically isolated from contact with all other metals. Figure 1 shows SCC compact tension coupons prior to high temperature steam test. The coupons was exposed to simulated high temperature steam environment for a period of 74 h and 1344 h which has been shown in the Table 1 and Fig. 2.

3 Test Result

Figures 3, 4, 5 and 6 show the physical condition of ferritic alloy steel of SA213 P91 (UNS K91560) or 9Cr 1Mo after 74 h high temperature steam exposure. It was found that corrosion product were observed on the ferritic alloy steel SA213 P91 (UNS K91560) coupons which indicate metal loss had occurred to the steel. Oxide layer (Fe_3O_4) thickness on the ferritic alloy steel SA213 P91 (UNS K91560) or 9Cr 1Mo coupons after 74 h exposure were recorded as 8–9 μ. Meanwhile, Figs. 7, 8, 9 and 10 show the physical condition of ferritic alloy steel of SA213 P91 (UNS

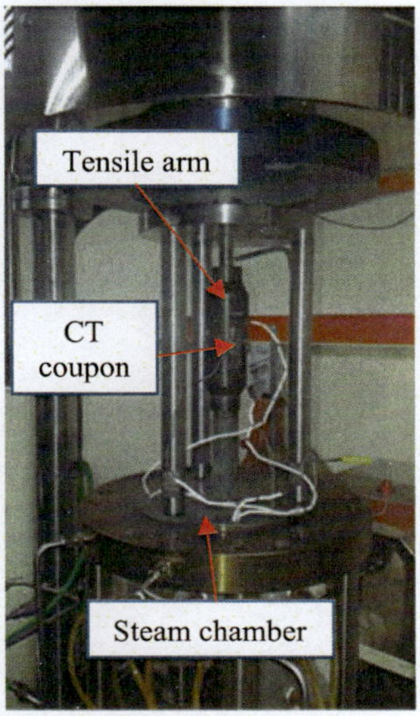

Fig. 2 Stress corrosion cracking (SCC) test in steam environment

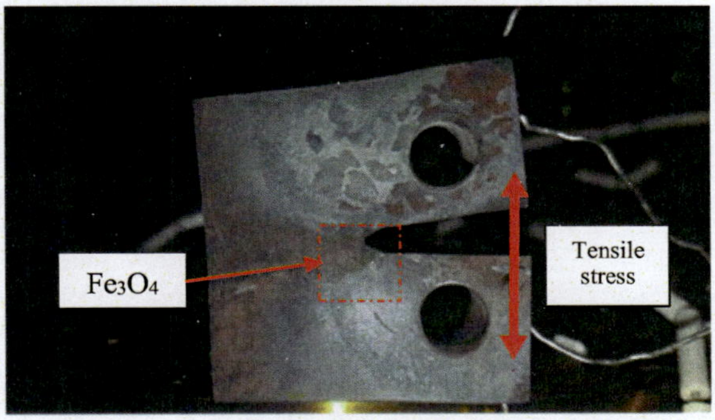

Fig. 3 CT sample after 74 hours of SCC test

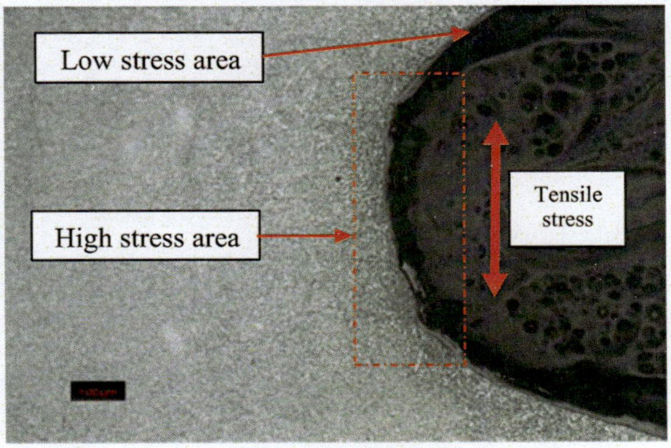

Fig. 4 Close view of oxide layer at notch area of CT coupon

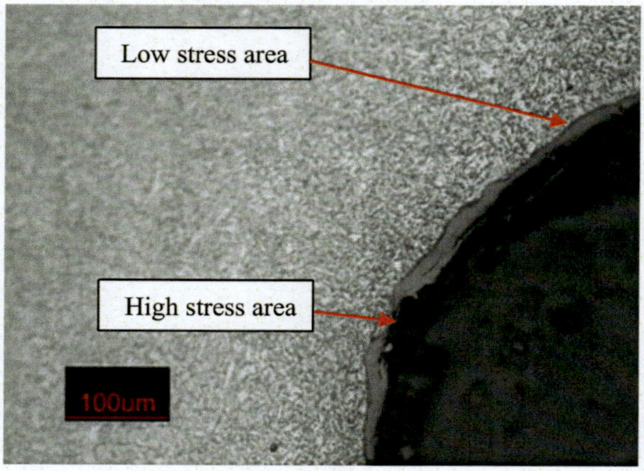

Fig. 5 Close view of oxide layer at high & low stress area on coupon

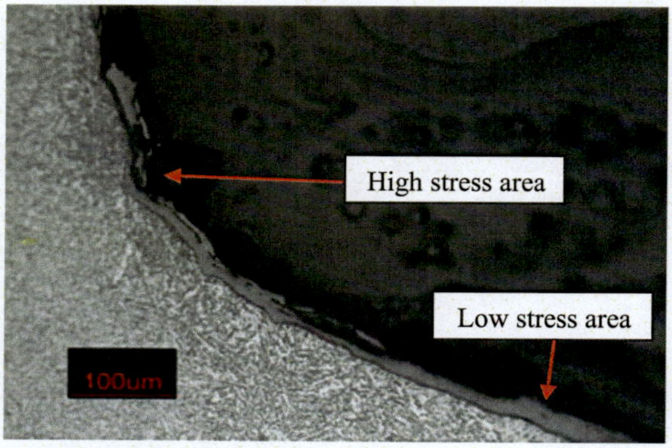

Fig. 6 Close view of high and low stress area on coupon

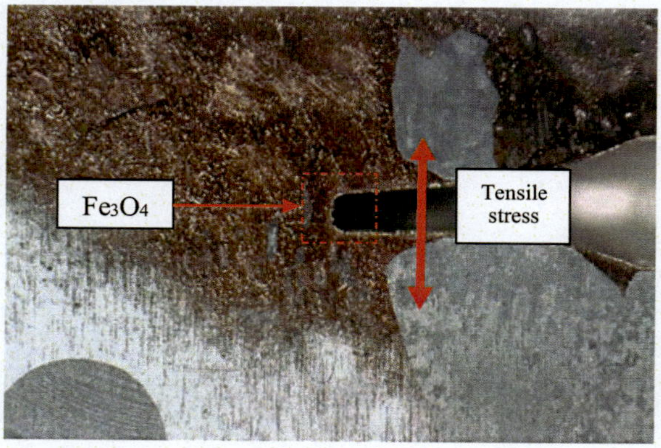

Fig. 7 Stress corrosion cracking (SCC) pre crack CT coupon after 1344 h test

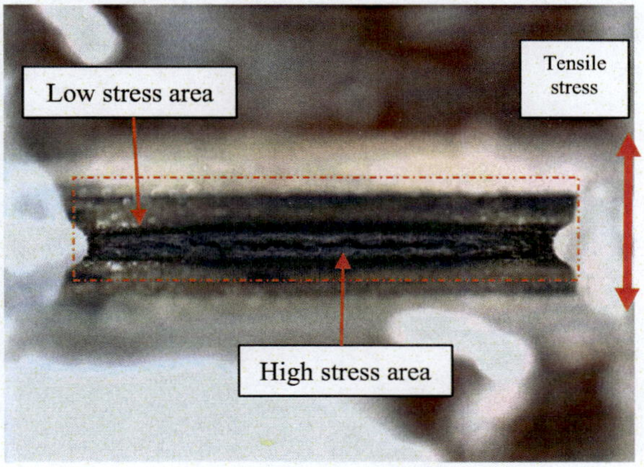

Fig. 8 Close view of crack on coupons after 1344 hr test

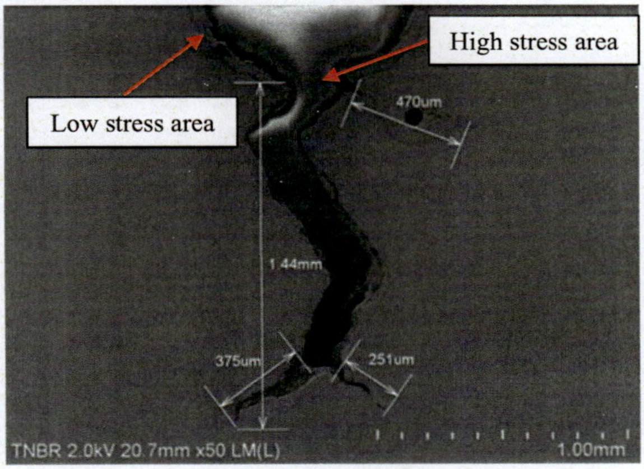

Fig. 9 Close view of oxide layer at high & low stress area of CT coupon

K91560) or 9Cr 1Mo after 1344 h high temperature steam exposure. It was found that corrosion product were observed on the ferritic alloy steel SA213 P91 (UNS K91560) coupons which indicate metal loss had occurred to the steel.

The thickness of oxide layer (Fe_3O_4) recorded as 730 μ.

The microstructure of SA213 P91 (UNS K91560) after 74 h exposure was revealed using high magnification microscope. However, no cracks were detected on metal surface. The surface of SA213 P91 (UNS K91560) materials were protected with black corrosion product known as magnetite (Fe_3O_4). Cracks with length of 1.44 mm were detected on metal surface after 1344 h exposure. The surface of SA213 P91 (UNS K91560) materials were also protected with black corrosion product known as magnetite (Fe_3O_4).

Fig. 10 Oxide layer composition on CT coupons

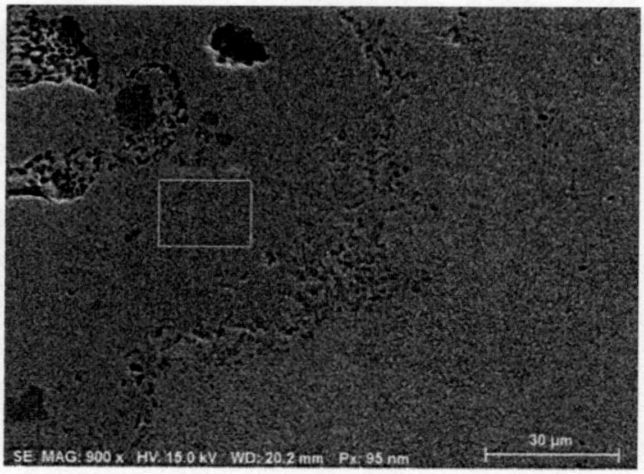

El	AN	Series	unn. C [wt.%]	norm. C [wt.%]
Fe	26	K-series	35.85	54.17
O	8	K-series	19.85	29.99
Cr	24	K-series	8.16	12.33
C	6	K-series	2.32	3.51
		Total:	66.19	100.00

4 Discussion

Steam parameters such as chloride contents, temperature, oxygen concentration and pH level play critical roles in the development of general corrosion and stress corrosion cracking of SA213 P91 (UNS K91560) sample. Our previous investigation on austenitic stainless steel 304 (UNS S30400) materials has revealed, SCC was occurred after 2 month of seawater test, with chloride recorded as 17,480 ppm and temperature of 28.5 °C. However, during a period of 74 h testing and with no chloride was recorded in steam solution, the cracks due to SCC could not be revealed on SA213 P91 (UNS K91560) type samples. Oxide layer on metal surface had tendency to slow down the cracks initiation process on CT samples.

Meanwhile, in another investigation on P91 materials, SCC was occurred after 59 h of high temperature steam test, with oxygen level recorded as 7 ppb and temperature of 568 °C. For this test, tensile stress was recorded as 6.53KN and pH in steam solution is 9.5. The cracks due to SCC have been revealed on SA213 P91 (UNS K91560) after 59 h of testing period. However, under similar setup condition and with high oxygen level concentration of 5500 ppb, the failure time has been increased to 1344 h due to an effect of oxide layer.

According to Logan [3], the oxide layer or protective film likely to repair itself at constant stress. The process called as film-rupture mechanism, is continuously repeated which result the steel surface been protected from environment attacked, and stress corrosion cracking may ceased. However, when stress is readjusted or high enough to cause plastic deformation, the formed protective film will be broken and the exposed metal will be electronegative with respect to filmed materials. The stress corrosion will proceed discontinuously to failure. Figures 4, 5 and 6 show close view of broken oxide layer under high stress condition.

The rates of SCC initiation and propagation is faster with increasing temperature, stress and chloride concentration. As stated in the literature [4], a concentration of approximately 400 ppm chloride is required for SCC initiation at 20 °C and 100 ppm at 100 °C. In addition, the role of oxygen appears to be important for SCC initiation and propagation. Without oxygen, SCC will not occurred even at high chloride concentrations [4].

Meanwhile, pH in steam solution can contribute to the development of SCC initiation and propagation on SA213 P91 (UNS K91560) materials. Various researchers in their experiments have demonstrated at low pH and high chloride solutions, cracks can occur at temperature below 500 °C [5]. However, at pH more than 7 can significantly prevent SCC [6]. Our investigation result showed no evidence of SCC cracks on SA213 P91 (UNS K91560) materials at steam solution of pH 9.29 during 74 h test.

The result however showed an evidence of SCC cracks on SA213 P91 (UNS K91560) materials at steam solution of pH 9.5 for longer test exposure 1344 h. As previously discussed, high chloride level and oxygen content in steam solution were a main contributor to the SCC cracks initiation for SA213 P91 (UNS K91560) steel materials. Since no chloride detected in steam system, the oxygen concentration in steam for this case had play a critical roles for retarding and accelerating crack initiation process on SA213 P91 (UNS K91560) steel materials.

5 Conclusion

The factors influencing the stress corrosion cracking of SA213 P91 (UNS K9156 0) or 9Cr 1Mo materials have been extensively studied by previous researchers. The results of this investigation indicate that steam solution pH, chloride concentration, oxygen, temperature, stress and corrosion products play critical roles for crack initiation and propagation on SA213 P91 (UNS K91560) steel materials.

References

1. Parrott R, Pitts H (2011) Chloride stress corrosion cracking in austenitic stainless steel. Health and Safety Executive Books
2. ASTM G4–01 Conducting corrosion coupon tests in plant equipment. American Society for Testing and Materials, Philadelphia, PA
3. Logan HL (1952) Film-rupture mechanism of stress corrosion. J Res National Bureau of Standard
4. McIntyre DR, Dillon CP (1985) MTI Publication No. 15, Guidelines for preventing stress corrosion cracking in the chemical process industries
5. Fielder JW, Hobson BA, Pickett ML (1998) The stress corrosion cracking behaviour of stainless steel at temperatures below 50°. European Commission, Technical Steel Research, EUR 18001
6. Platts N, Adams R Review of SCC in low chloride environments relevant to HAST operation. Serco Assurance SA/SIS/14319/R001, Commercial in Confidence

Corrosion Resistance Enhancement of Stainless Steel Powder Metallurgy in Artificial Body Fluids by Reinforcement of Boron and Niobium

Sri Hastuty, Fandika Reza Ardi Saputra, Byan Wahyu Ryandwita, Muhammad Awwaluddin, Maman Kartaman, Yanlinastuti, Ika Dyah Widharyanti, Ayu Dahliyanti, Haryo Satriya Oktaviano, Sadaqat Ali, Abdul' Azeez Abdu Aliyu, Ahmad Majdi Abdul-Rani, and Amar Prasad Yadav

Abstract The usage of powder metallurgy technology has been extended until nowadays due its flexibility and practical use in mixing with other alloying elements. This research investigate the analysis of the corrosion behavior of powder metallurgy alloy AISI 316L stainless steel with variations in the elemental content of Niobium in human body simulation fluid (SBF) at 25 °C. The various elements of niobium are

---

S. Hastuty (✉) · F. R. A. Saputra · B. W. Ryandwita
Department of Mechanical Engineering, Universitas Pertamina, Jakarta 12220, Indonesia
e-mail: sri.hastuty@gmail.com; sri.hastuty@universitaspertamina.ac.id

M. Awwaluddin
Faculty of Engineering, Department of Mechanical Engineering, Universitas Indonesia, Depok 1642, Indonesia

M. Kartaman · Yanlinastuti
Center for Nuclear Fuel Technology, National Nuclear Energy Agency of Indonesia, Jakarta 12710, Indonesia

I. D. Widharyanti · A. Dahliyanti
Department of Chemical Engineering, Universitas Pertamina, Jakarta 12220, Indonesia

H. S. Oktaviano
Research and Technology Innovation, PT Pertamina (Persero), Mega Kuningan Barat III, Sopo Del Building, 51st Fl. Jl., Jakarta Pusat, Indonesia

S. Ali
School of Mechanical and Manufacturing Engineering (SMME), National University of Sciences and Technology (NUST), H12, Islamabad, Pakistan

A. A. A. Aliyu · A. M. Abdul-Rani
Department of Mechanical Engineering, Bayero University Kano, Gwarzo Road, Kano, Nigeria

A. A. A. Aliyu
Department of Mechanical Engineering, Universiti Teknologi PETRONAS, 32610 Bandar Seri Iskandar, Perak, Malaysia

A. P. Yadav
Central Department of Chemistry, Tribhuvan University, Kathmandu, Nepal

F. Ahmad et al. (eds.), *ICPER 2020*, Lecture Notes in Mechanical Engineering,
https://doi.org/10.1007/978-981-19-1939-8_63

0.5, 1.0, 1.5, 2.0 wt%. After mixing and compaction, the green compacted samples were sintered in nitrogen atmosphere at 1150 °C. The corrosion testing used in this research were electrochemical methods consisting of open circuit potential and anodic polarization. In addition, characterization testing was also carried out in this study using an optical microscope and an scanning-electron microscope (SEM–EDS). The results obtained from this test are based on the corrosion rate, the general corrosion resistance of the alloy increases with the variation of niobium from 0.5, 1.0 and 1.5 wt Nb. However, it decreased in the variation of 2.0%wt Nb. Each sample are still showed susceptibility to pitting corrosion. Based on its repassivation capacities, the increase of Nb content tend to cause detrimental effect on alloys. SEM–EDS observations predict a decrease in corrosion resistance in AISI 316L alloy with the addition of 2.0%wt Nb due to the concentration of the Nb element at a certain point so that there is a failure in preventing the depletion of chromium elements at the grain boundaries resulting in chromium carbide at grain boundaries.

Keywords Powder metallurgy · 316L · Niobium · Corrosion rate · Pitting corrosion · SBF

1 Introduction

Metal is a material whose scope of use is quite extensive. Almost in all fields such as engineering, metal is very popular or mostly used for research, construction and so on. No exception, metal is also used in the medical field. The use of metal material in the medical field can vary in form, one of them is used for orthopedic applications as a bioimplant. The purpose of using implants is to interact with biological systems in the human body with the aim of evaluating, treating, augment or replace tissue, organs or body functions that are damaged or malfunctioned. The use of implants in the human body began in 1660, and the use of implants in the modern medical field began in 1960 [1]. The level of popularity of metal as an implant material is increasing every time. Bioimplants are taking increasingly more importance to improve the lives of human beings. Noted just in the United States the use of implants, especially in the total hip ararthroplasty (THA) procedure, increased in the decade 2002–2012 by 2.1% [2].

The metals use for the construction of bioimplant must have several requirements; among of them is biocompatible for human body, sufficient mechanical strength and sufficient corrosion resistant ability. Biocompatible for human body means that implanted material should not cause adverse effect such as allergy, inflammation, toxicity and other bad problem that can interfere metabolic mechanism. Material should posses sufficient mechanical strength to withstand the forces to which they are subjected without deform. Corrosion resistant ability means that the material should not be degraded in the human body because it can adversely effect both biocompatibility and mechanical strength [3].

There are several conventional material that can be used for bioimplant application such as titanium and its alloy, cobalt and its alloy, and stainless steel [4]. That material are essential for orthopaedic implants, bone fixator, artificial joints, etc., since it can subtitute for function of hard tissues in orthopaedic. Among these all, titanium and its alloy are most corrosion resistant. However, the drawbacks are their high cost, low formability and high susceptibility to friction and wear compare to Co-based and stainless steel [3]. Compare to Ti-based and stainless steel, Co-based possesses greater wear resistance but as same as Ti based Co-based is still high cost and low formability. As an alternative the other metals that suitable for implant application is stainless steels. Compare to Ti-based and Co-based, large amounts of stainless steels are often used for implant device because of easy to form and less expensive [5–7]. Among stainless steel types, austenitic 316L SS is most suitable and is often used as an implant in human bones.

Bulk austenitic 316L SS has much higher elastic modules than human bones. The ultimate high tensile strength and fracture toughness guarantee satisfactory mechanical performance, because the material can bear significant loads and experience sufficient plastic deformation before damage [8]. That reason is also a consideration why stainless steel is suitable for use as an bone implant.

Altough the stainless steels are good material as bone implant in terms of popularity, formability, mechanical properties and relatively low cost material, but there are inferior in terms of corrosion resistance. Not only that, but also there is an obstacle due to the mechanical mismatch between bone elastic modules and stainless steel elastic modules. This will lead to condition that called "*stress shielding effect*". This effect implies that bone insufficiently loaded which eventually leads to bone resorption [9]. One way to alleviate these problem is to reduce elastic modulus of stainless steel to be the same as possible with human bone by introducing pores [10]. One method of making porous material is powder metallurgy (PM). In the medical field purposes, especially orthopaedic applications, the materials that manufactured by powder metallurgy are very promising for the following reason [11]: (1) they involve nearly waste-free net-shape forming; (2) they offer precise choice of chemical composition through the use of high-melting or hard-to machine alloys to improve implant biocompatibility; (3) they offer porosity of appropriate pore geometry. This improves bone tissue in-growth, thus enhancing endoprosthesis stability.

However, there is also limitation with the PM technique. Not only reduce the ductility and strenght of th stainless steel, this PM technique stainless steel is also has corrosion resistance problem. The PM technique stainless steel is less corrosion resistance compared to the other conventional one such as cast and wrought [11, 12]. This low corrosion resistance is due to porosity. Porosity can make stainless steel susceptible to pitting corrosion [12]. It was reported that 90% failure of 316L stainless steel that implanted in human body were due to corrosion attacks due to pitting [13]. This is very hazardous for a long-term application because that corrosion would not only degrade the implant mechanical function but also be likely to release ions such as iron, chromium and nickel to human body. Ni ion is typical high-risk element from perspective of metal toxicity problems [14]. For those reasons, therefore development in 316L SS is needed in order to overcome this problem.

Several investigation have been carried out to improve corrosion properties of 316L SS. One of them is alloying the stainless steel with rare earth material, Nb. Huang et al. observed that addition of Nb to 22Cr25NiWCoCu promote Cr_2O_3 formation and increasing the niobium content (0–0.86 wt%) will increase the oxide layer thickness after 100 h oxidation with the form of Fe_2O_3 as outer layer and (Cr, Fe, $Mn)_3O_4$ as inner layer [15]. Wu et al. investigate that increasing Nb content (0, 0.25, 0.75, 1.25 wt%) to as–built novel stainless steel can decrease the current density and increase the charge transfer resistance in 3.5 wt% NaCl environment [16]. Prifiharni et al. show that increasing level of Nb in the matrix of Fe25Ni15Cr stainless steel can increase the pitting corrosion resistance on 3.5 wt% chloride containing solution. Elemental niobium most likely formed niobium oxides, which protect the surface. Protection by the oxide layer formed by Nb is stronger than the protection of chromium oxides [17].

There is also a particular study conducted by Ali et al. [13]. They invesigate that addition Nb in PM 316L SS matrix can increase the corrosion resistance. They use nitrogen atmosphere in sintering process to maintain austenitic structure of 316L SS so that can minimize the Ni ion leaching. Therefore by combining the process between the addition of niobium and the use of powder metallurgy as the manufacturing process seems to be an allternative choice to make a good bioimplants.

This present work was aimed to investigate the effect of increasing Nb content on pitting corrosion behaviour of PM 316L stainless steel in simulated body fluid condition (Ringer's Lactate).

2 Experiment

2.1 *Sample Preparation and Electrolyte*

Five AISI 316L stainless steel with different composition were prepared. The samples produced by powder metallurgy that provided by Universiti Teknologi Petronas (UTP), Malaysia. The composition of each samples as follows (Table 1).

The addition of 0.25%wt boron was used for maximum densification for powder metallurgy proccess [13]. All the samples were disc shape with ±4 mm in thickness and 30 mm in diameter. Before experiment, each sample were grinded and polished

Table 1 Amount of B and Nb Modified 316L Stainless Steel (wt%)

Sample no	B	Nb
1	0.25	0.5
2	0.25	1.0
3	0.25	1.5
4	0.25	2.0

with SiC paper (120, 240, 400, 600, 800, 1000 grit), degreased with alcohol, washed with distilated water and dried with tissue paper and hair dryer.

The simulated body fluid used in this study was Ringer's Lactate (RL). The volume of solution is about 100 mL and maintained at ambient temperature open to air.

2.2 Electrochemical Equipments and Experiments

MetroOhm Autolab PGSTAT302N was used as potentiostat and a three-electrode system with a silver/silver chloride (Ag/AgCl) as a reference electrode and platinum wire as counter electrode was used to conduct the electrochemical test. The surface area of samples was 132.73 mm^2. Before test, all samples were exposed to solution for about 10 min to make its open-circuit potential (OCP). The potentiodynamic polarization were performed with scan rate of 10 mV/s from the potential of 1000 mV below the OCP to 1000 mV above the OCP. The value of corrosion potential (E_{corr}) and corrosion current density (i_{corr}) were automatically known from the polarization curve by ECLab software through Tafel slope extrapolation. The pitting potential E_{pit} was determined from anodic polarization curves at the point where a stable increase in current density occurred. The protect potential E_{prot} was determined from the reverse polarization at the intersection point with the forward polarization curve.

Backscattered scanning electron microscopy was used for assessment of the surface morphology and Energy-Dispersive X-ray Spectroscopy (EDS) was used for chemical characterization of surface layer of the sample after corrosion studies.

3 Result and Discussion

3.1 Density and Porosity

The density of each samples are shown in Table 2. The green density of sample 1 is 6.370 g/cm^3 which is the highest value. Sample 1 is also achieved maximum sintered density of 7.411 g/cm^3. The green density of sample 2, 3, and 4 is 6.340 g/cm^3,

Table 2 Densities of all samples [13]

Sample no	Green density (g/cm^3)	Standard deviation	Sintered density (g/cm^3)	Standard deviation	Relative density (%)
1	6.370	0.0432	7.411	0.023	93.81
2	6.340	0.057	7.0.367	0.051	93.25
3	6.160	0.035	7.285	0.092	92.21
4	6.080	0.029	7.190	0.028	91.01

Table 3 Porosity estimation of each sample

Sample no	1	2	3	4
Porosity	5.68	6.20	7.36	8.60

6.160 g/cm^3, and 6.080 g/cm^3, respectively. The sintered density of sample 2, 3, and 4 is 7.367 g/cm^3, 7.285 g/cm^3 and 7.190 g/cm^3, respectively. There is a decrease in the density value, not only the green density but also the sintered density, as the niobium content increased. This indicates that boron and niobium particles tried to induce themselves in the SS matrix [13].

Porosity is related to density of alloys. The lower density of alloy causes the porosity of alloy much greater. Porosity can be estimated with the following equation [13]:

$$Porosity = 1 - \frac{\rho_a}{\rho_{th}} \quad (1)$$

with ρ_a is sintered denity and ρ_{th} is theoretical density. The estimation of the pores present in the sintered samples is presented in Table 3. It can be shown that the increase of Nb content leads to increasethe porosity of alloys. Sample 4 (SS316L + 2.0%wt Nb) has the greatest porosity with the value of 8.60.

3.2 Potentiodynamic Polarization Curves

Tafel extrapolation was undertaken to investigate the influence of niobium addition on 316L stainless steel on corrosion resistance behaviour. As shown on Fig. 1, the behaviour of each sample can be seen clearly. All samples behave a similar manner which it didn't show active–passive transition. This phenomenon reveal that passive film on all samples form spontaneosly. This phenomenon indicate that there are dissolved oxygen in environment, especially Ringer's Lactate that was used in this research [3]. All samples also show relatively the same passive region, ranging from −450 to 0 mV.

The corrosion potential value (E_{cor}) of sample 1 (SS316L + 0.5%wt Nb), sample 2 (SS316L + 1.0%wt Nb), sample 3 (SS316L + 1.5%wt Nb) and sample 4 (SS316L + 2.0%wt Nb) is −614.72 mV, −600.83 mV, −568.27 mV, and −601.67 mV, respectively. This indicates that the increase of Nb content tends to decrease the corrosion potential value of the alloys. Based on positive logarithmic of current density value, sample 4 (316L + 2.0 5%wt Nb) has the highest value of 19.37, followed by sample 1 (SS316L + 0.5%wt Nb) with the value of 18.21, then sample 2 (SS316L + 1.0%wt Nb) with the value of 17.49 and the lowest value is belongs to sample 3 (SS316L + 1.5%wt Nb) with the value of 10.52. The higher current density indicates corrosion rate occurs at a higher rate. This result indicates that sample 4 (SS316L + 2.0%wt

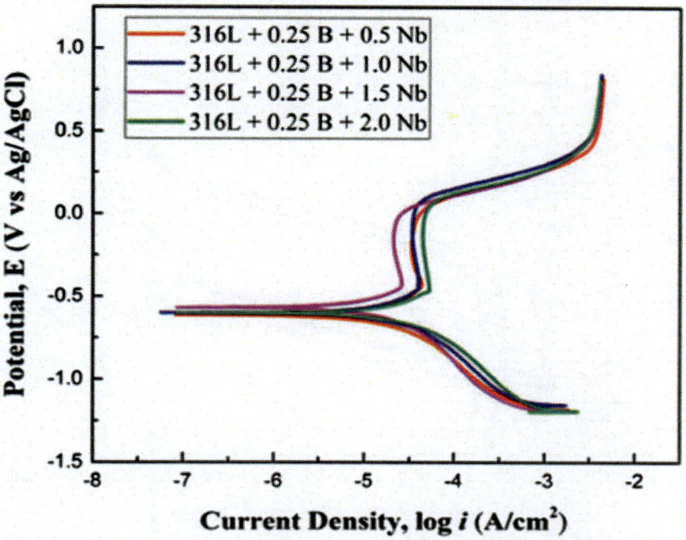

Fig. 1 Potentiodynamic Polarization Curves of Each Samples in Ringer's Lactate Solution

Table 4 Electrochemical parameter result of each samples in Ringer's Lactate Solution

Sample no	E_{cor} (mV)	E_{pit} (mV)	E_{rep} (mV)	i_{cor} (log Ii/mAI)	$E_{pit} - E_{rep}$ (mV)	CR (mpy)
1	−614.72	54.626	−217.315	18,21	271.942	6.08
2	−600.83	90.209	−214.904	17,49	305.113	5.88
3	−568.27	57.861	−261.077	10,52	318.939	3.50
4	−601.67	81.634	−246.917	19,37	328.551	6.67

Nb) has the highest corrosion rate value while sample 3 (SS316L + 1.5%wt Nb) has the lowest corrosion rate value. Based on increment of Nb content from 0.5%wt to 1.5%wt, the increase of Nb content tends to increase the general corrosion resistance of alloys. This phenomenon is due to role of Nb that stabilize the alloys so there are chromium-rich zone that can lead to formation of strong and stable chromium oxide passive film on its surface [15]. The result of polarization studies are collected in Table 4.

3.3 Cyclic Polarization Curves

From Fig. 2, the pitting potential (E_{pit}) value of sample 1 (SS316L + 0.5%wt Nb), sample 2 (SS316L + 1.0%wt Nb), sample 3 (SS316L + 1.5%wt Nb) and sample

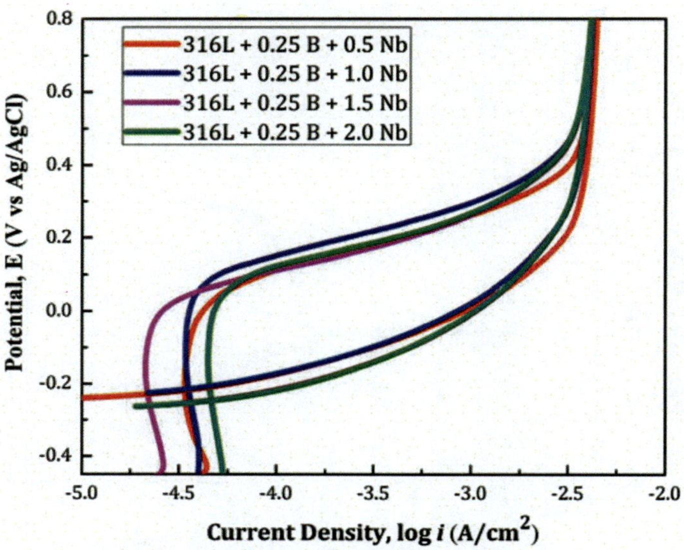

Fig. 2 Cyclic polarization curve of each samples in Ringer's Lactate Solution

4 (SS316L + 2.0%wt Nb) is 54.626 mV, 90.209 mV, 57.861 mV and 81.634 mV, respectively. The pitting corrosion behaviour of all samples didn't show clear effect with respect to increasing Nb content. From a sample with 0.5%wt to a sample with 1.0%wt Nb showed an increase in E_{pit} but an increase in the Nb content from 1.0%wt to 1.5%wt makes E_{pit} decrease and increase again on the sample with 2.0%wt Nb. It can be seen that sample with the addition of 1.0%wt Nb has highest pitting potential with the value of 90.209 mV and sample with the addition of 0.5%wt Nb has lowest pitting potential with the value of 54.626 mV (Fig. 3).

During reverse scan, there is a hysteresis loop occur in all samples. This indicates pitting corrosion that developed in the sample during the test. The repasivation capacities is indicated from $E_{pit} - E_{rep}$ value. The low $E_{pit} - E_{rep}$ differences indicate high repassivation capacities in all cases. The increase in Nb content makes the difference in $E_{pit} - E_{rep}$ value even greater. From Table 3, the highest $E_{pit} - E_{rep}$ value belongs to sample with addition of 2.0%wt Nb with the value of 328.551 mV and the lowest $E_{pit} - E_{rep}$ value belongs to sample with addition of 0.5%wt Nb with the value of 271.942 mV.

3.4 *Effect of Nb on Pitting Corrosion Resistance*

In this research, the influence of Nb content as alloying element in powder metallurgy 316L stainless steel on pitting corrosion resistance is not well defined yet. Based on potentiodynamic polarization test, all samples are still susceptible to pitting corrosion

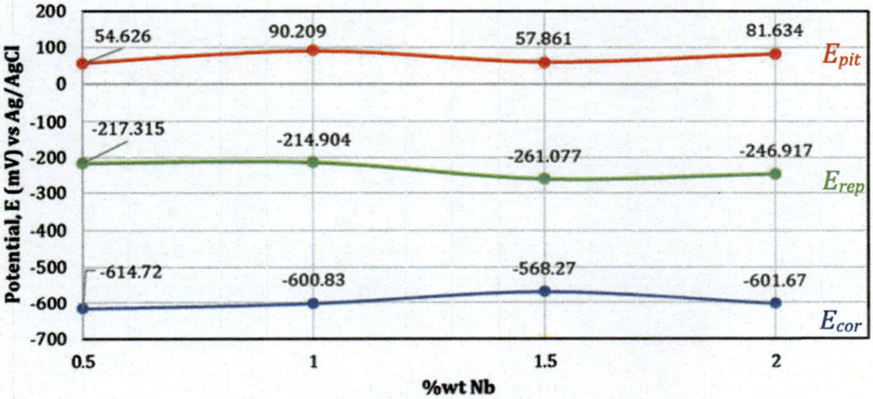

Fig. 3 Comparison of E_{pit}, E_{rep}, and E_{cor} of all samples

which is indicated by presence of hysteresis loop [18]. Repasivation capacities of each sample are relatively low. Even, based on $E_{pit} - E_{rep}$ value, the increase of Nb content has detrimental effect on pitting corrosion resistance in powder metallurgy 316L stainless steel.

As shown on Figs. 1, 2 and Table 4, the sample with the greatest niobium addition with the value of 2.0%wt actually has the worst effect in terms of corrosion resistance both in general and pitting corrosion resistance. This can be validated by looking at Figs. 4 and 5. As shown on the pictures, corrosion tends to start at the porosity of the sample. Sample with the addition of 2%wt Nb has relatively more porosity than sample with the addition of 1.5%wt Nb (best sample in term of general corrosion resistance). It looks that corrosion that occurs in sample with the addition of 2%wt Nb is more severe than in sample with the addition of 1.5%wt Nb. From this examination it can be concluded that porosity in powder metallurgical material is a weak point in terms of corrosion resistance.

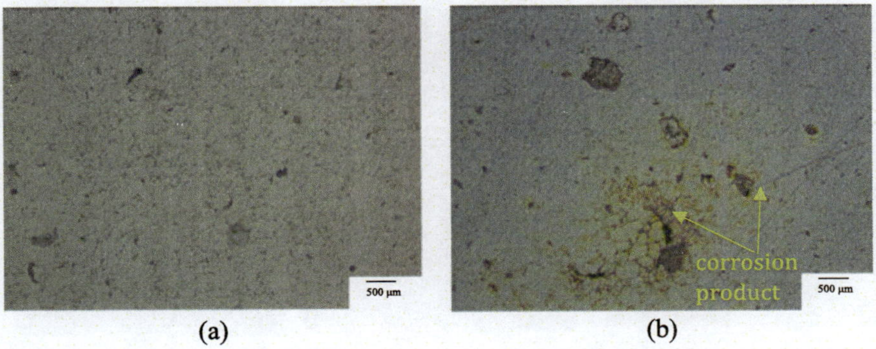

Fig. 4 Sample with 1.5%wt Nb **a** before corrosion **b** after corrosion

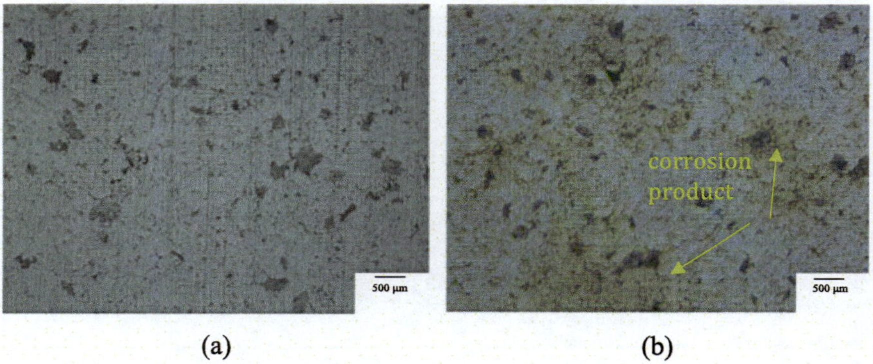

Fig. 5 Sample with 1.5%wt Nb **a** before corrosion **b** after corrosion

The corrosion resistance of a passive metal is highly dependent on its chemical composition. In stainless steel the content of chromium plays a very important role in corrosion resistance by forming a very protective layer. From SEM–EDS studies on Fig. 6, there is constituent that contain a lot of molybdenum and niobium (red

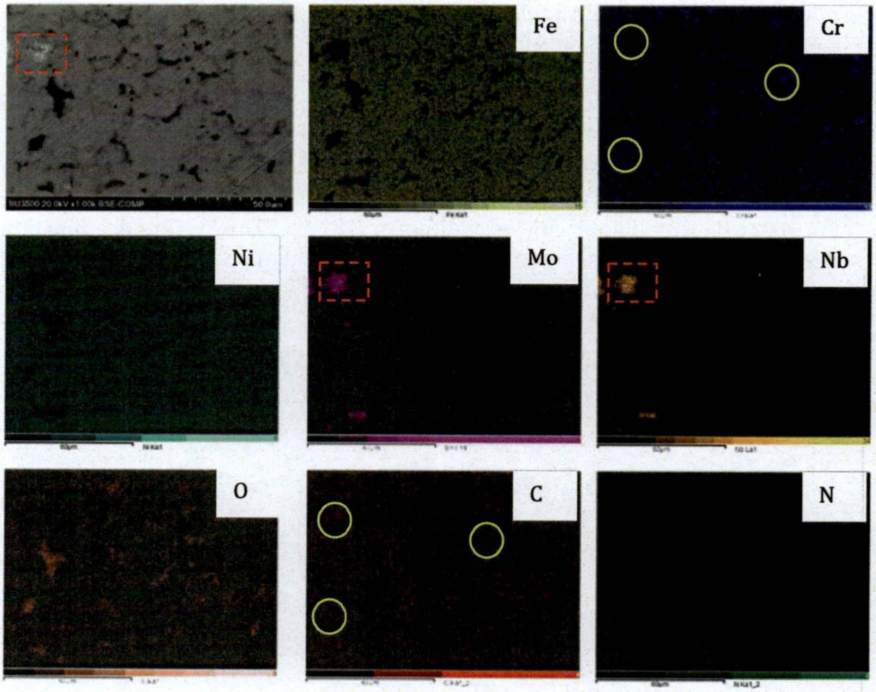

Fig. 6 SEM–EDS Analysis of Sample with The Addition of 2.0%wt Nb in Ringer Solution

dashed-square marks). Niobium which acts as an alloy stabilizer is concentrated at a particular point. This causes the element chromium to be depleted towards the grain boundaries. This fact is shown by the increase in the color contrast of the chromium element shown by the yellow circle marks in Fig. 6. This increase in color contrast is also accompanied by an increase in the color contrast of the carbon element at the same point. This indicates that the chromium carbide bonds are formed at the grain boundaries [19]. It is known that the formation of chromium carbide compounds can cause a decrease in the quality of the passive layer formed from chromium elements, such as chromium oxide, because the chromium content to form the passive layer is reduced. So it might be the reason why sample with the highest niobium content has lower corrosion resistance than the others.

4 Conclusions

Research on the influence of niobium content variation on corrosion resistance in 316L stainless steel has been conducted. Th results showed that an increase in the amount of niobium content ranging from 0.5%wt, 1.0%wt and 1.5%wt led to an increase in corrosion resistance in general. This is characterized by an increase in the corrosion potential value and a decrease in the value of the corrosion rate. However, there was a decrease in corrosion resistance in samples with a content of 2.0%wt Nb, this was indicated by the potential value of corrosion decreased and the value of the corrosion rate increased relative to other samples.

In addition, based on observations of polarization diagrams, all samples still showed susceptibility to pitting corrosion. This is characterized by the formation of hysteresis loops on the polarization diagrams of each sample. Based on the $E_{pit} - E_{rep}$, the samples showed low repasivation capacities and the increase on niobium content causing detrimental effects. In this case the sample with the addition of 2%wt Nb had the worst effect indicated by it has biggest $E_{pit} - E_{rep}$ value difference.

Detrimental effect on the sample with addition of 2.0%wt Nb observed by microscopic observations. Based on microscopic examination, porosity in the sample with addition of 2%wt Nb is relatively high. This porosity becomes a weak point and becomes where the first corrosion process taking place. In addition, SEM–EDS observations indicate that there is an indication of a bond that occurs between molybdenum and niobium concentrated at particular point. This resulted in niobium not being spread evenly in the sample so that the stability of the alloy became disturbed. Disturbed stability of the alloy is characterized by the depletion of chromium elements at the boundaries of the grain and tends to bond with carbon. This condition predicts the formation of a passive layer of chromium oxide is relatively lower than other samples. It might be the reason that can answer why the sample with the addition of 2.0%wt Nb has a low general corrosion resistance and low pitting corrosion resistance.

References

1. Naci K, Yavuz S, Bunyamin C, Hayrettin A (2012) Production of 316L stainless steel implant material by powder metallurgy and investigation of their wear properties. Chinese Sci Bullet 1873–1878
2. Bozic ML, Kevin J (2014) Trends in total hip arthroplasty implant utilization in the United States. J Arthroplasty 29(10):1915–1918
3. Eliaz N (2019) Corrosion of metallic biomaterials: a review. Materials 12(3)
4. Prasad K, Bazaka O, Chua M, Rochford M, Fedrick L, Spoor J (2017) Metallic biomaterial: current challenges and opportunities. Materials 10(8)
5. Sivakumar M, Mudali UK, Rajeswari S (1994) In vitro electrochemical investigations of advanced stainless steels for applications as Orthopaedic implants. J Mater Eng Perform 3(6):744–753
6. Talha M, Behera CK, Sinha OP (2012) Potentiodynamic polarization study of Type 316L and 316LVM stainless steels for surgical implants in simulated body fluids. J Chem Pharm Res 4(1):203–208
7. Brooks EK, Brooks RP, Ehrensberger MT (2017) Effect of simulated inflammation on the corrosion of 316L stainless steel. Mater Sci Eng C 71:200–205
8. Fang C-WK, Zhou F (2018) State of the art of bioimplants manufacturing: part I. Adv Manuf 6:20–40
9. Mariotto SdFF, Guido V, Cho LY (2011) Porous stainless steel for biomedical applications. Mater Res 14(2):146–154
10. Dewidar MM, Khalil KA, Lim JK (2007) Processing and mechanical properties of porous 316L stainless steel for biomedical application. Trans Nonferrous Metals Soc China 17:468–473
11. Dewidar MM, Yoon H-C, Lim JK (2006) Mechanical properties of metals for biomedical application using powder metallurgy process: a review. Met Mater Int 12(3):193–206
12. Garcia C, Martin F, Tiedra Pd, Cambronero LG (2007) Pitting corrosion behaviour of PM austenitic stainless steels sintered in Nitrogen–Hydrogen atmosphere. Corros Sci 49(4):1718–1736
13. Ali S, Rani AMA, Mufti RA, Azzam FI, Hastuty S, Baig Z, Hussain M, Shehzad N (2019) The influence of nitrogen absorption on microstructure, properties and cytotoxicity assessment of 316L stainless steel alloy reinforced with boron and niobium. MDPI
14. Lane J, Mait J, Unnanuntana A, Hirsch B, Shaffer A, Shonuga O (2011) Materials in fracture fixation. Comprehens Biomater 6:219–235
15. Huang Y-S, Tsai W-T, Pan Y-T, Kuo J-C, Chen H-W, Lin D-Y (2019) Effect of niobium addition on the high-temperature oxidation behaviour of 22Cr25NiWCoCu stainless steel in air. MDPI 9(9):975
16. Wu SQ, Zhang CH, Zhang S, Wang Q, Liu Y, Abdullah A (2018) Effect of Nb addition on microstructure and corrosion resistance of novel stainless steels fabricated by direct laser metal deposition. Mater Res Express 5(3)
17. Prifiharni S, Anwar MS, Nikitasari A, Mabruri E (2018) The hardness, microstructure, and pitting resistance of austenitic stainless steel Fe25Ni15Cr with the addition of tungsten, niobium, and vanadium. in AIP conference proceedings, Banten
18. Gurappa (2002) Characterization of different materials for corrosion resistance under simulated body fluid conditions. Mater Characteriza 49:73–79
19. Mishra S, Datta R (2001) Embrittlement of steel. in Encyclopedia of materials: science and technology (Second Edition). Oxford, Elsevier, pp 2761–2768

Mechanical Properties and Pitting Corrosion Behavior of Zr-Mo-Al-Ti-Y in Simulated Human Body Fluids

Muhammad Awwaluddin, Djoko Hadi Prajitno, Sri Hastuty, and Tresna P. Soemardi

Abstract The mechanical properties and pitting corrosion resistance of Zr-Mo-Al-Ti-(x)Y in simulated human body fluids was investigated. The amount of Yttrium added into Zr alloys was 2, 3, and 4 wt%. From the Vickers hardness test, the hardness value was increased by the increasing of Yttrium content. In other hand, the pitting corrosion behaviour of Zr alloys was carried out by cyclic polarization in Ringer's Lactate Solution at room temperature. The study found out that the pitting resistance reached its optimum value by the addition of 3 wt% of Y.

Keywords Hardness · Pitting corrosion · Zr Alloy · Simulated human body fluids · Zr-Mo-Al-Ti-Y

1 Introduction

Zr alloys has withdrawn a great attention for application as structural biomaterial because it has high toughness to fracture, non-toxic, and good biocompatibility [1, 2]. Despite of all its advantages, this type of material has several back draws such as inadequate corrosion resistance and low ultimate tensile strength, as the compensation of having high toughness. The required modulus of elasticity for natural human bone is ranging 15–30 GPa, while in case of Zr alloys only up to 33.16 GPa [3]. Bio implants are prone to be corroded as they are exposed to human body fluid which contained chloride and other minerals. Corrosion could take place in various forms, one of them is pitting corrosion, which is considered have a detrimental effect on

M. Awwaluddin (✉) · T. P. Soemardi
Faculty of Engineering, Department of Mechanical Engineering, Universitas Indonesia, Depok 16424, Indonesia
e-mail: muhammad.awwaluddin91@ui.ac.id

D. Hadi Prajitno
Center for Applied Nuclear Science and Technology, National Nuclear Energy Agency of Indonesia, South Jakarta, Indonesia

S. Hastuty
Department of Mechanical Engineering, Universitas Pertamina, Jakarta 12220, Indonesia

F. Ahmad et al. (eds.), *ICPER 2020*, Lecture Notes in Mechanical Engineering,
https://doi.org/10.1007/978-981-19-1939-8_64

structural integrity of bio implants. In addition, Zr metal is very susceptible for pitting to occur among metals for biomaterial; for example, titanium, niobium and nickel [4]. To overcome this problem, researchers investigated bulk metallic glass of Zr alloys [5–10]. However, its manufacturing process is complicated and then produce defects, segregation of chemical composition, and short range ordered structure.

Among methods to improve mechanical hardness and strength and also corrosion resistance of Zr alloys significantly, alloying is still the best option [11–13]. From other point of view, such as manufacturing process, casting is an established process, yet a low cost one. Yttrium is chosen as an alloying element of Zr alloys to enhance these properties [14]. Therefore, this research was focused on the effect of Ytrrium on the hardness and pitting corrosion behaviour of Zr-6Mo-6Al–Ti alloys. The advantage of this research alloy is that it has good corrosion resistance in simulated Human body fluid Linger‘s Lactate, but the hardness obtained is still high so that further research is needed, for example annealing treatment.

2 Experimental Methods

The alloy ingots with nominal compositions of Zr-6Mo-6Al–Ti-xY, with x (2–4) were prepared by arc melting furnace of the mixture of pure Zr (zirconium sponge with (99%wt)), Mo (molybdenum rod with (99.8%wt) purity), Al (aluminium with (99.9%wt) purity), Ti (titanium) and Y (yttrium) with (99.9%wt) purity. The alloys were remelted four times in order to obtain a homogeneous composition. Alloy with diameter of 20 mm and width 6 mm was fabricated by copper mold casting method under high-purity argon atmosphere.

A standard procedure for hardness tests i.e., grinding until grade 1000 were prepared. The hardness test of the Zr-6Mo-6Al–Ti-xY samples were examined by SHIMADZU HSV-310. In terms of corrosion testing, ASTM G59 was referred as the standard whereas the specimens were cathodically and anodically polarized in ringer's lactate solution at room temperature (25 °C). The testing employed three electrodes configuration consists of working, counter and reference electrodes. Working electrode was the specimens, counter electrode was platinum wire of 0.5 mm in diameter, and reference electrode was silver—silver chloride (SSE) or Ag–AgCl. Corrosion resistance measurements were performed with an Autolab Potentiostat PGSTAT 302 N. The potentiodynamic polarization curves were recorded between start potential -1 V_{OCP} and stop potential 1 V_{OCP} with potential sweep scan rate 10 mV/s and step 1 mV in anodic direction after monitoring the open circuit potential (OCP) for 10 min.

Table 1 shows the chemical composition of Zr based alloys that used in this study. The number before the element indicates the amount of that element, %wt as the unit. The Zr is the balance element. In this experiment, the amount of Yttrium acts as the variable factor.

Table 1 The chemical composition of Zr Based Alloy (In %Wt)

Composition alloy (As cast)	Zr		Mo		Al		Ti		Y	
	%	g	%	g	%	g	%	g	%	g
Zr-6Mo-6Al-1Ti-2Y	85	12.76	6	0.9	6	0.9	1	0.15	2	0.3
Zr-6Mo-6Al-1Ti-3Y	84	12.58	6	0.9	6	0.9	1	0.15	3	0.45
Zr-6Mo-6Al-1Ti-4Y	83	12.45	6	0.9	6	0.9	1	0.15	4	0.6

Table 2 Chemical composition of Ringer's Lactate medium

Reagent	Composition
Calcium Chloride $2H_2O$	0.10 g
Potassium Chloride	0.15 g
Sodium Chloride	3.00 g
Sodium	130 mEq/L
Potassium	4 mEq/L
Calcium	2.7 mEq/L
Sodium Lactate	28 mEq/L
Water for Injection	1.55 g
Chloride	ad 500 mL
Lactate (HCO_3^-)	108.7 mEq/L
Ca^{++}	109.4 mEq/L
Cl^-	2.7 mEq/L
Osmolarity	273 mOsm/L
Na^+	130.3 mEq/L
K^+	4 mEq/L

The chemical composition of Ringer's Lactate medium was displayed in Table 2. It presents as simulated human body fluid. Sodium chloride was the main component. Because of the chloride ions, most of the metals including biomaterials occurre of corrosion.

3 Results and Discussion

Figure 1 was the result of Vickers hardness test of Zr based alloys. It is clearly showed that the increasing of Yttrium content in the alloys will increase the hardness. Alloy with 2 wt% Yttrium had a hardness value of 467 HV, while 3 and 4 wt% Yttrium had 501 and 505 HV, respectively. There was increment of 6.7% from alloy of 2 wt% Y to alloy of 3 wt% Y and. In case of the change of hardness from alloy with 2 to

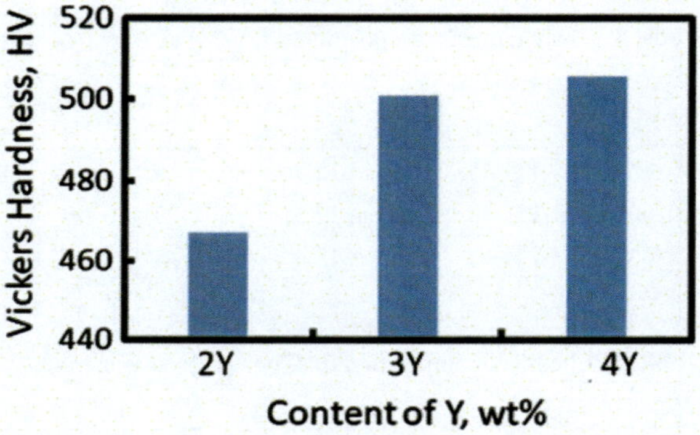

Fig. 1 The hardness of Zr based alloys

4 wt%, the increment reached 7.5%. This change is quite significant by addition of Yttrium content up to 2 wt%.

Figure 2 demonstrates the change of potential during immersion test of Zr based alloys in Ringer's Lactate Solution for 10 min. In the test, all three alloys show the potential stability, which indicate the passivation of Zr based alloys in the chloride ions containing-solution. The difference between them is alloy with 3%wt Y had a higher corrosion potential compare to other alloys. It is thought that the addition of 3%wt Y has reached its optimum value in order to enhance its corrosion resistance.

In Fig. 3, the pitting performance of Zr based alloy in Ringer's Lactate solution was measured by cyclic polarization. Polarization started from corrosion potential (E_{corr}) towards anodic region followed by the increasing of current density. When the potential reached pitting potential (E_{pit}), the current density decreased gradually, and made intersection with the forward scan. This point is called pitting protection potential (E_{prot}). The result was Zr based alloy with 3%wt of Yttrium showed the pitting potential in the forward scan began at higher potential and lower current density, which means lower corrosion rate compare to other alloys. However, E_{prot}

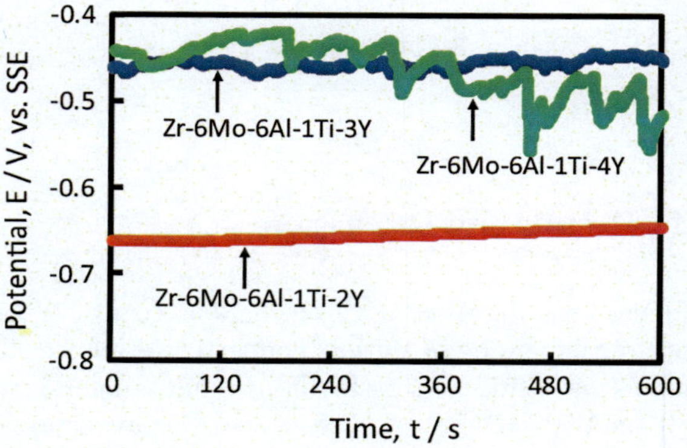

Fig. 2 Measurement of open circuit potential of Zr based alloys in Ringer's Lactate Solution

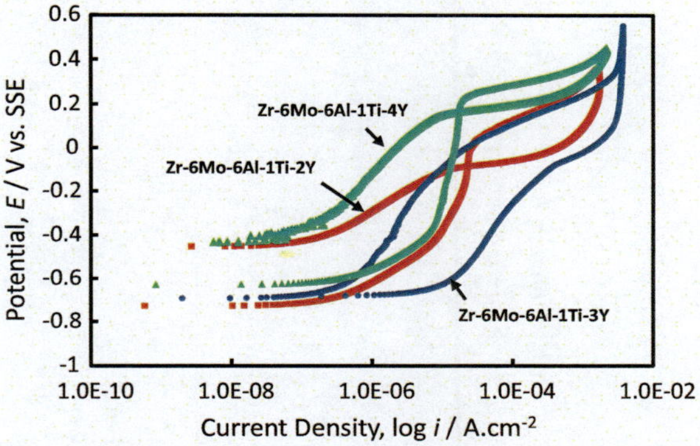

Fig. 3 Cyclic polarization curve of Zr based alloys in Ringer's Lactate Solution

occurred at lower potential compare to the original OCP, indicated that the repassivation process did not take place easily. It is thought that the increasing of Yttrium into Zr alloys affected the ability of this alloy to resist the pitting occurrence and delayed the pitting repassivation.

The data of cyclic polarization of Zr based alloys in Ringer's Lactate medium i.e., Epit, Ecorr, was displayed in Table 3.

In Fig. 4 plot of corrosion, pitting, and protection potential of Zr based alloys added with Yttrium in Ringer Lactate's solution at 25 °C is shown. Ecorr of Zr alloys after 10 min immersion is plotted in x–y coordinates. It revealed there was a fluctuates in Ecorr measurement. However, a trend can be still withdrawn from the graph. It showed an increasing trend of Ecorr by increasing the addition of Yttrium content. A similar trend was also applied for Epit and Eprot. The pitting corrosion resistance of Zr alloys was increased by the addition of Y from 2 to 4%wt. It is probably due to the enrichment of Y oxide in the whole oxide covered the surface. In case of Eprot, their value is also increased by increasing the amount of Y weight percentage. It means the ability of alloys to repassivate was significantly improved.

Figure 5 displayed the microstructure of Zr-6Mo-6Al-1Ti-xY (x = 2, 3, and 4) alloys after corrosion testing in ringer's lactate solution. They showed that all composition of Zr based alloys had suffered from pitting corrosion. The alloy with 2%wt of Y (Fig. 5a) had suffered the most severe pitting corrosion compare to other two alloys. It is shown by the larger and round shape of pits which has diameter about 300 μm at the largest pit. These pits are considered as defects in growth stage of pitting corrosion. As for Fig. 5b and c, the alloys with addition of higher content

Table 3 The data of Cyclic polarization of Zr based alloys in Ringer's Lactate Solution

Alloy	E_{pit} (V)	E_{corr} (V)	E_{prot} (V)
2Y	0.3548	0.0016	−0.1
3Y	0.5499	0.0038	−0.7
4Y	0.4517	0.0022	0.2

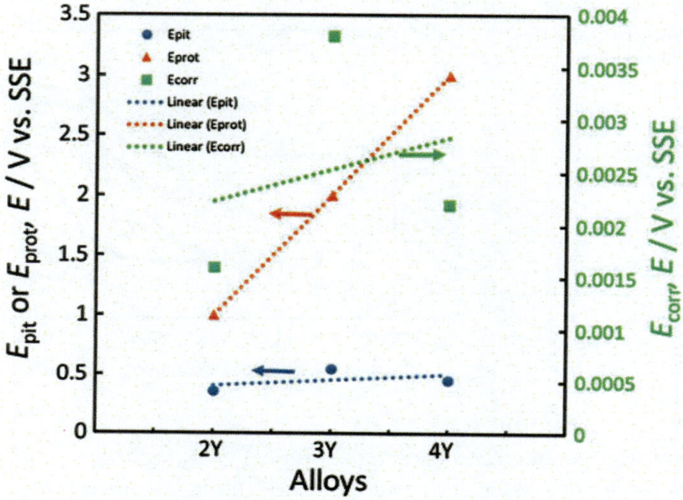

Fig. 4 Plot of Corrosion, Pitting, and Protection Potential of Zr based alloys added with Yttrium in Ringer Lactate's solution at 25 °C

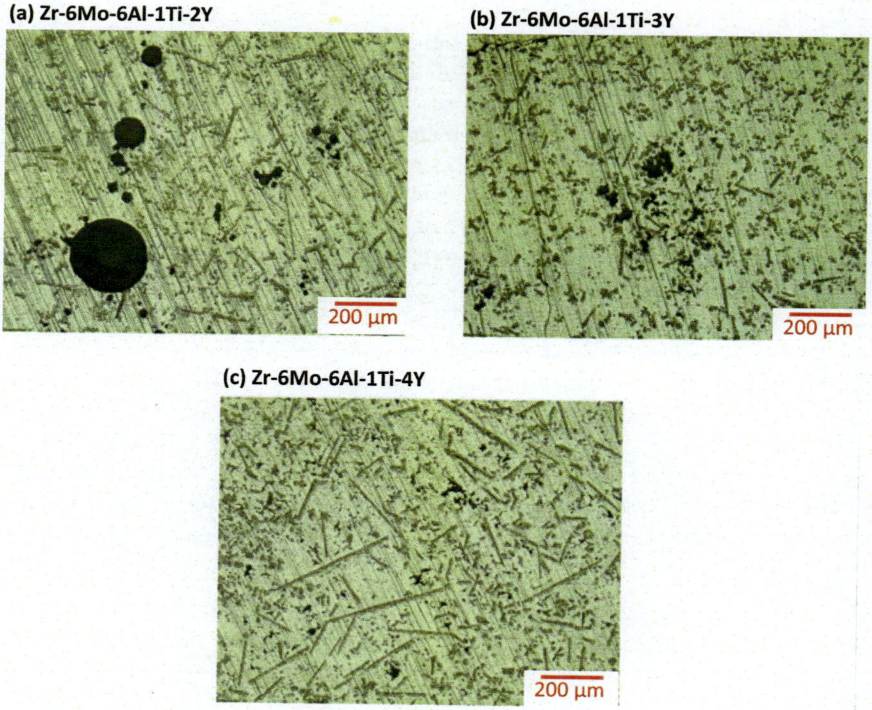

Fig. 5 Microstructure of Corroded Samples Zr-6Mo-6Al-1Ti-xY (x = 2, 3, and 4) after Cylic Polarization in Ringer's Lactate Solution

of Y showed fewer and smaller pits. In alloy with 3%wt of Y, a cluster of pits were formed in range of 10 to 100 μm, which can be categorized as micro pits. In case of alloy with 4%wt (Fig. 5c), it has much smaller pits and the maximum size is around 20 μm and scattered. The shape of pits for these two alloys are irregular, different with alloys with 2%wt of Y. The irregular shape is suggested included in the initiation stage of pitting corrosion. After they occurred, growth stages followed and formed a larger and round shape of pits. From this evidence, increasing Yttrium content into Zr-6Mo-6Al-1Ti alloys have a beneficial effect on pitting corrosion resistance [15].

According to Fig. 6, the resistance of the material to pitting corrosion is linear with the addition of Y element in the Zr-6Mo-6Al-Ti material. The large pitting corrosion occurred on the surface of the Zr-6Mo-6Al-Ti material with the addition of Y element

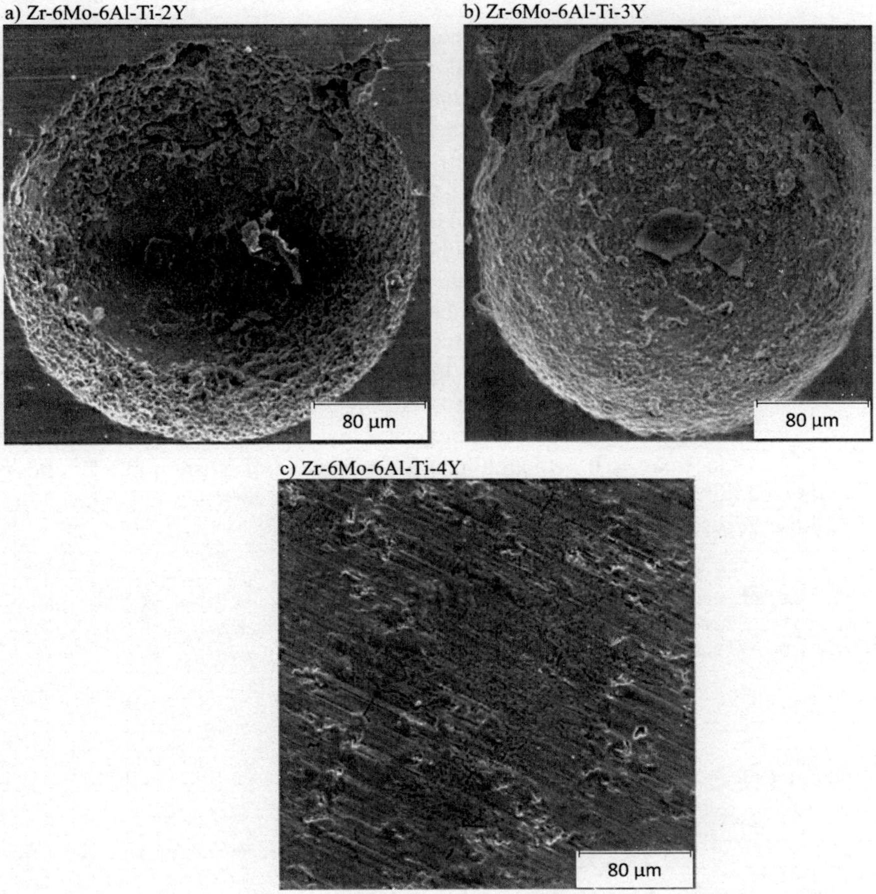

Fig. 6 Typical secondary electron micrographs (SEM) of pitting corrosion of the samples Zr-6Mo-6Al-1Ti-xY (x = 2, 3, and 4) after Cylic Polarization in Ringer's Lactate Solution

2%wt. this can be seen in Fig. 6a. When viewed from the surface roughness of the pitting corrosion results, it looks rougher than the other two materials. Zr-6Mo-6Al-Ti material with the addition of Y element 3%wt has a smoother pitting corrosion surface than 3%wt addition of Y (Fig. 6b). At the addition of 4%wt of element Y shows the best results with a smooth surface and has a very small pitting corrosion (Fig. 6c). From this evidence, increasing Yttrium content into Zr-6Mo-6Al-1Ti alloys have a beneficial effect on pitting corrosion resistance.

4 Conclusions

Based on the experimental results, these conclusions were withdrawn as shown below.

1. The Vickers hardness test showed that the increasing of Yttrium content in the alloys will increase the hardness. Alloy with 2%wt Yttrium had a hardness value of 467 HV, while 3 and 4%wt Yttrium had 501 and 505 HV, respectively. The percentage of increment could reached up to 7.5 from alloy of 2%wt Y to alloy of 4%wt Y.
2. In the immersion test, all three alloys show the potential stability, which indicate the passivation of Zr based alloys in the chloride ions containing-solution. The difference between them is alloy with 3%wt Y had a higher corrosion potential compare to other alloys. It is thought that the addition of 3%wt Y has reached its optimum value in order to enhance its corrosion resistance.
3. Zr based alloy with 3%wt of Yttrium showed the pitting potential in the forward scan began at higher potential and lower current density, which means lower corrosion rate compare to other alloys. However, E_{prot} occurred at lower potential compare to the original OCP, indicated that the repassivation process did not take place easily. It is thought that the increasing of Yttrium into Zr alloys affected the ability of this alloy to resist the pitting occurrence and delayed the pitting repassivation.

Acknowledgements This research was financially supported by Saintek Scholarship the Ministry of Research, Technology and Higher Education of Indonesia or Saintek Scholarship the Ministry of Research, Technology and BRIN through Doctoral Scholarship program.

References

1. Bell B et al (2016) The influence of alloying elements on the corrosion of Zr alloys. Corros Sci 105:36–43
2. Kim HG, Jeong YH, Kim TH (2004) Effect of isothermal annealing on the corrosion behavior of Zr–xNb alloys. J Nucl Mater 326(2–3):125–131
3. Chui P (2017) Near β-type Zr-Nb-Ti biomedical alloys with high strength and low modulus. Vacuum 143:54–58

4. Revie RW (2011) Uhlig's corrosion handbook. John Wiley & Sons
5. Mudali UK, Baunack S, Eckert J, Schultz L, Gebert A (2004) Pitting corrosion of bulk glass-forming zirconium-based alloys. J Alloy Compd 377(1–2):290–297
6. Wang D, Wang S, Wang J (2012) Relationship between amorphous structure and corrosion behaviour in a Zr–Ni metallic glass. Corros Sci 59:88–95
7. Kawashima A, Ohmura K, Yokoyama Y, Inoue A (2011) The corrosion behaviour of Zr-based bulk metallic glasses in 0.5 M NaCl solution. Corros Sci 53(9):2778–2784
8. Mondal K, Murty B, Chatterjee U (2006) Electrochemical behavior of multicomponent amorphous and nanocrystalline Zr-based alloys in different environments. Corros Sci 48(8):2212–2225
9. Nie X et al (2011) The effect of oxidation on the corrosion resistance and mechanical properties of a Zr-based metallic glass. Corros Sci 53(11):3557–3565
10. Kondo R, Nomura N, Tsutsumi Y, Doi H, Hanawa T (2011) Microstructure and mechanical properties of as-cast Zr–Nb alloys. Acta Biomater 7(12):4278–4284
11. Wang Z, Ma Y, Zhang J, Hou W, Chang X, Wang J (2008) Influence of yttrium as a minority alloying element on the corrosion behavior in Fe-based bulk metallic glasses. Electrochim Acta 54(2):261–269
12. Awwaluddin M, Prajitno DH, Adi WA, Kartaman M, Soemardi TP (2020) Mechanical properties and corrosion behavior of novel β-type biomaterial Zr-6Mo-4Ti-xY alloys in simulated body fluid Ringer's lactate solution for implant applications. AIMS Mater Sci 7(6):887–901
13. Nie L, Zhan Y, Liu H, Tang C (2014) Novel β-type Zr–Mo–Ti alloys for biological hard tissue replacements. Mater Des 53:8–12
14. Huang L et al (2009) Bio-corrosion study on zirconium-based bulk-metallic glasses. Intermetallics 17(4):195–199
15. Hastuty S, Nishikata A, Tsuru T (2010) Pitting corrosion of Type 430 stainless steel under chloride solution droplet. Corros Sci 52(6):2035–2043

The Application of Knowledge Management Practices in Corrosion Management for Oil and Gas Industry: A Review

A. P. Alwi Mohd Yunus, Mohd Ridwan Kamarulzaman, and A. P. Ir Mokhtar Che Ismail

Abstract The oil and gas industry is a largest capital-intensive industry with assets ranging from wells, risers, drilling rigs, and offshore platforms in the upstream segment, to pipelines, liquefied natural gas (LNG) terminals, and refineries in the midstream and downstream segments. The integrity of these assets can be compromised by corrosion and the impact of various corrosion management (CM) is implemented to reduce the cost of corrosion (COC). Therefore, the critical step in reducing COC is not only dependent on the implementation of effective technical corrosion control but also incorporating the knowledge management (KM) domain. KM practices such as knowledge creation, acquisition, sharing, application, knowledge storage and retrieval, and communities of practice on CM programs and supported by organizational culture are expected to contributing in COC reduction. This study examines the recent development of KM practice applications in CM for the oil and gas industry, how the KM element could be embedded in Corrosion Management Program for reducing cost of corrosion and the likely Corrosion Knowledge Management framework suitable for that purpose. The results are expected to contribute to the establishement of corrosion knowledge management with proposed framework for reducing COC.

Keywords Knowledge management · Corrosion management · Corrosion knowledge management · Oil and gas · Industry · Cost of corrosion · KM Practices

A. P. Alwi Mohd Yunus (✉) · M. Ridwan Kamarulzaman
Universiti Teknologi MARA, Shah Alam, Malaysia
e-mail: alwiyunus@uitm.edu.my

M. Ridwan Kamarulzaman
e-mail: mridwan@uitm.edu.my

A. P. Ir Mokhtar Che Ismail
Universiti Teknologi Petronas, Seri Iskandar, Malaysia
e-mail: mokhtis@utp.edu.my

F. Ahmad et al. (eds.), *ICPER 2020*, Lecture Notes in Mechanical Engineering,
https://doi.org/10.1007/978-981-19-1939-8_65

1 Introductıon

Since the 1950s several countries considered the economic consequences of corrosion. Studies conducted during this time indicated that the cost of corrosion (COC) to society was significant. The global COC is estimated to be US$2.5 trillion, which is equivalent to 3.4% of the global GDP (2013). By using available corrosion control practices, it is estimated that savings of between 15 and 35% of the COC could be realized; i.e., between US$375 billion and US$875 billion annually on a global basis. These costs typically do not include individual safety or environmental consequences. Through near misses, incidents, forced shutdowns (outages), accidents, etc., several industries have come to realize that lack of corrosion management (CM) which can be very costly and that, through proper corrosion management, significant cost savings can be achieved over the lifetime of an asset. To relate the COC studies to a global COC, a relationship between economic sectors and corrosion costs is needed. Furthermore, the GDP of the economic sectors by country must be known to permit the use of the "percent cost of corrosion by economic sector" within the extrapolation to global corrosion costs.

Corrosion adversely affects the lives of people and organizations because of the high cost of direct (indirect and indirect) substitution, substitution and environmental damage [1]. Indirect costs are difficult to measure from the category of hardship, loss of business and ultimately loss of life while direct costs are associated with actual costs of replacing steel [2, 3]. Zero-tolerance leaks occur due to the fluid properties of the product, this was stated by Mohammad and Hosseini [4]. Corrosion enhances strategic asset maintenance and downtime and increases the likelihood of process security incidents [5].

Based on the failure, approximately 25% of the associated corrosion has resulted in a loss of 20–30% of production in the oil and gas industry [6, 7], and more than 50% of leaks are due to aging assets [6]. Unfortunately, BP, Shell, ConocoPhillips, Exxon, Car estimated total assets of $100 billion are corrosion and metals [8]. Corrosion costs in the United Kingdom increased to $1 trillion in 2012, accounting for 6.2% of GDP [7]. Shell's corrosion costs alone amounted to $400 million in 1995, while 6% as reported by BP which was equivalent to its net asset value. Corrosion contributes to about 85% of all unforeseen damages [5] in the oil and gas industry.

Corrosion is a major threat affecting the integrity of the industry. The COC has been a standard measure that demonstrates the effectiveness of erosion control and management practices. The issue of COC improvement despite many controls and management of erosion required detailed analysis of the root causes. As CM not only involves technological aspects, but also the human factors, it states that human factors are important for efforts to reduce COC. According to Kpch et al. [9], who said that the difficulty in promoting CM is that cost savings from corrosion control are difficult to measure,that is, (i) a decrease in maintenance costs; (ii) supervision or inspection of cost or inspection costs often increases; (iii) less failure to save time on production and/or loss of product, reduce injury, reduce property damage, reduce environmental liberation, improve public relations; and (iv) the expansion of the asset's assets may go directly downwards and/or withholding capital expenditure. All of this can be included in the business case for CMs to be enhanced.

The discipline of KM has emerged in both academic research and in practice for many years [10, 11]. It addresses knowledge management issues as a source of business. Although there are many different definitions of KM, it can be said to enhance organizational performance by enabling individuals to recognize, take, share and apply their collective knowledge to make the best decisions. Organizations must have the latest technology for global development, especially in today's multinational business environment. However, to achieve bottom line success, attention must be given to other issues as well. Experience in many organizations has shown that no more than one-third of the knowledge management budget should be devoted to KM technology [12]. One of the key factors in KM's success is developing an understanding of existing connections and mechanisms and then developing ways to enable people to improve existing processes [13]. Bixler [14] said that implementing KM framework can be complex and dynamic, no matter how well planned and advanced the organization is. This is because it involves people and factors from other organizations.

The oil and gas industry has taken advantage of the development of KM [15, 16]. To date, as the industry has undergone rapid transformation, with rapid technological advances, advanced offshore drilling, multiple acquisitions, growing reliance on foreign oil resources, and a focus on environmental issues, KM initiatives have played a key role in operations, more efficient and effective [15].

KM is a key aspect of organizational capability in driving the best resources available [17]. It is important to note that every individual, society or organization needs knowledge to make real and continuous progress over time. The KM team provides support through the transfer of technology and knowledge, as well as asset management as oil and gas organizations deal with new technologies, outsourcing, new partnerships and government regulation. When business issues involve capacity management, cost reduction and the environment, KM practices play a role through forecasting/scheduling and process and technical innovation. Recently, a great deal of research emphasis has placed KM elsewhere in the world [18]. Considering the recent existence of KM and organizational theory, there is evidence that several local oil and gas organizations involved in project management have placed an emphasis on KM practices [19, 20]. This study reviews existing strategies in the development of knowledge management, examining future challenges and proposed actions to address these issues to facilitate improvement in local oil and gas industry. It is also coordinated with CM's research to understand the development path and its effectiveness from the management aspect which is more towards reducing the COC in local view.

The research aims to expand the literature on the relationship between KM practices that show the Four Pillars in KM, CM and the role of organizational culture support in managing COCs in the oil and gas industry. The resulting research framework comprises strategies and practices that can be used to address the issue of COC reduction in the local oil and gas industry, as well as promote the effectiveness of managing CMs based on the best practices of KM. The objective of the research is to apply the current practice of KM for CM in the practice of COC reduction in the local oil and gas industry. This research will be based on a systematic approach that investigates current CMs and recommends a framework based on KM practices. The framework is expected to contribute to corrosion knowledge management with the aim of reducing COCs in the local oil and gas industry.

The remaining sections of this article discuss the KM practices, Four Pillars of KM, CM, and organizational culture followed by discussion and implications, and conclusions.

2 Knowledge Management

Human factor domain is not only related to the measurement of required skill sets but also driven on structured framework in relation to technology or knowledge transfer within the organization. Malaysia oil and gas industry is now looking at the current COC and also current CM including human factors involve in the corrosion field. According to Dalkir [21], KM was beneficial for individual employees, communities of practice and the organizations. KM provides individuals to helps people do their jobs, better decision making, problem solving, challenges and opportunities to contribute. In community of practice, KM develops professional skills, promotes peer-to-peer mentoring, facilitate more effective networking and collaboration. Meanwhile, in the organization, KM helps drive strategy, solves problems quickly, diffuses best practices, improves knowledge embedded in products and services, builds organizational/corporate memory and enables the organization to stay ahead of the competition better. KM plays a role in forecasting/scheduling and processes and also the techniques in innovation [16]. KM is crucial in reducing costs so that properly implemented KM can lead to cost reduction [22].

KM practitioners have been practicing "KM on the fly". There have been many empirical studies, lesson learnt and best practices gained through the experience of individuals and groups of people across various organizations. A key role of the KM model is to ensure a certain level of in-depth equipment in the effective implementation of KM and to provide long-term benefits to individuals and organizations. The coherent model found in KM on knowledge-driven processes is critical to KM's ability to overcome strategic successes even if it only partially. This is not to say that KM is a silver bullet or that it will solve all organizational problems, but that it has the potential to be a platform for improvement in performance [21].

The Four Pillars of KM model involves four elements that influence the environment, namely: Political, Economic, Government and Social (Fig. 1). The four pillars of the framework are: Leadership, Organization, Technology and Learning. These four pillars are domains that demonstrate the potential to encompass all aspects of effective knowledge sharing and collaborative culture [23]. Leadership promotes decision making and strategic alignment for KM initiatives with business objectives. Organizational pillars emphasize the design and strategic coordination of operational processes and procedures to ensure the success of KM initiatives within the

Fig. 1 Four pillars of knowledge management [23]

organization. The technology pillars define the importance of a technology infrastructure that enables the implementation of KM in an organization and without the use of KM applications in any organization, it is impossible. Finally, The Learning in this context, is defined as the acquisition of knowledge or skills through study, experience or teaching and emphasizes the fact that organizations must address approaches to facilitate KM such as enhancing internal communication, encouraging cross-functional teams and creating learning communities. This Four Pillars of KM model may be manipulated by organizational culture to reduce COC in CM activities/program through KM practices.

Choo [24] described a model of knowledge management that emphasizes direction (mainly based on [25]), knowledge creation (based on [26]) and decision-making (based on, among other concepts [27]) as shown in Fig. 2. This KM Choo's model focuses on how information elements are selected and how they are incorporated into the actions of an organization. Organizational actions result from focusing and absorbing information from the external environment into each successive cycle, as shown in Fig. 2. Each phase, sense making, knowledge creation, and decision making, has an external stimulus or trigger. Choo [24] applied the core to the model of Nonaka's model [26] for the foundations of KM model. The strength of the KM Choo's model is a holistic treatment of the KM cycle process that includes organizational decision-making, where there is a lack of other KM theories. This makes the Choo model one of the more "realistic" or KM-worthy models, as it represents the actions of "high fidelity" organizations. The KM Choo model is ideal for simulation and hypothesis or scenario testing applications [21]. The relationship between the

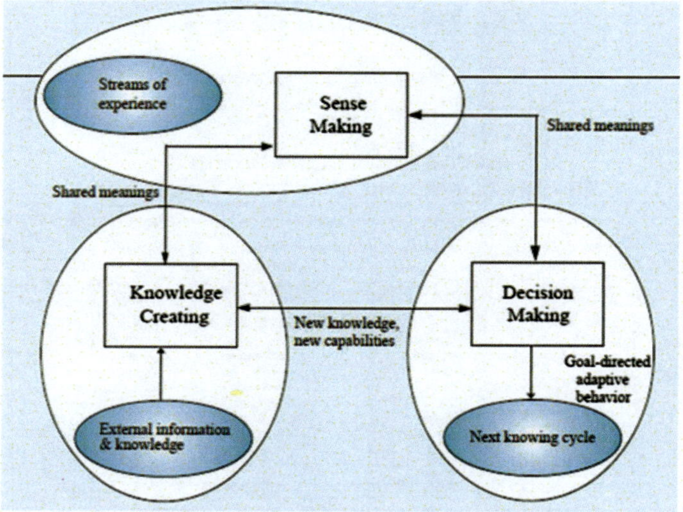

Fig. 2 Overview of Choo's (1998) KM model [24]

elements of sense making, knowledge creating and decision making is well-applied in the CM activities for oil and gas industry. The KM Choo model complies with the findings of the integrated KM framework whereby organizations are strongly influenced by organizational culture support in COC reduction.

2.1 *Knowledge Creation*

Knowledge creation is the result of the interaction between knowledge and the act of knowing, which is done through practices, actions, and ways of interacting with others [28]. It is important for an organization to allocate sufficient resources to create new knowledge as it will accelerate the development of innovation and increase the production of new technologies [29], which in turn will enable the organization to achieve sustainability as a whole [30]. Dynamic organizations facilitate the knowledge creation environment by encouraging employees to share their knowledge [31], providing systems, such as infrastructure and information, enabling them to practice new knowledge creation, and offering rewards to employees who introduce new ideas or solutions [32]. Knowledge-based organizations aim to promote efficient use of resources and tend to be environmentally friendly [33]. Such organizations not only encourage and facilitate the process of green product development but also constantly consider the impact of their operations on the environment [34].

2.2 Knowledge Acquisition

Knowledge acquisition refers to the activities of organizations acquiring, extracting and organizing knowledge from different sources [35]. The purpose of acquiring knowledge is to understand customer needs and their experience through organizational products and services. Qasrawi et al. [36] state that the majority of workers acquire knowledge from internal sources either from team members or other partners. This shows that the more employees know each other, the greater the likelihood that their productivity will improve. Acquisition of knowledge from external sources refers to knowledge gained from customers, competitors, suppliers, partners, and external experts' expertise [37]. In doing so, organizations make relevant changes to achieve customer satisfaction, leading to enhanced economic sustainability [38]. Shahzad et al. [39] explain that an organization's ability to acquire and absorb knowledge positively affects the financial performance of the organization.

2.3 Knowledge Sharing

Knowledge sharing is the process by which knowledge is communicated explicit or tacit between individuals or groups of people [31]. This is one of the social interaction methods in the organization, which enables employees to solve problems in a more creative way [35] and provides excellent support for strategic planning, decision making and the learning environment [40]. Knowledge sharing significantly improves explicit and tacit knowledge among workers, resulting in reduced errors and errors, and increased operational and economic sustainability [28]. Dynamic organizations consider knowledge sharing as a social responsibility and regularly participate in social awareness programs to reduce knowledge sharing gaps [41].

2.4 Knowledge Application

The use or integration of knowledge gained in designing or delivering organizational products and services is referred to as knowledge application [37]. It is also regarded as a timely response to changes in operations through technology and strategies as well as the ability to benefit them to produce and design new products and services [42]. Knowledge applications are also a key resource for creating new core competencies for organizations and improving their economic performance [43]. By using knowledge applications, firms can uncover new processes that can significantly improve their performance. Stakeholders are aware that dynamic organizations follow environmental practices and integrate existing and new knowledge into research and development activities to introduce new processes and technologies

[33]. This enables the firm to produce high quality products by taking the least resources, which not only benefit the environment, but the organization itself [44].

2.5 Knowledge Storage and Retrieval

Knowledge storage in an organization is an organizational mechanism that captures, stores, and disseminates knowledge gained from previous experience used in organizational decision making [45]. Chirumalla and Parida [46] state that many organizations consider reuse and for effective business performance as key to KM. Watson and Hewett [47] suggest that firms can leverage knowledge by using existing knowledge codified in the knowledge repository to gain competitive advantage. According to Jennex and Croasdell [48], organizational knowledge storage can be defined as the way organizations retain organizational knowledge and apply it to carry out organizational activities. Therefore, companies with strong storage knowledge and capabilities for strong retrieval can enhance their business performance and growth by adopting coding strategies. Literature has shown that knowledge storage and retrieval in organizations is positively associated with KM [48–50]. Each organization must have its own process of storing and retrieving knowledge to ensure that previously acquired knowledge is available for current use or future reference. It is important for organizations to have knowledge storage and acquisition processes, as expert knowledge storage will not be lost and enable the retrieval of relevant and systematic knowledge as needed. This storage process ensures that knowledge is stored properly, not lost or corrupted and does not have unauthorized access to organizational knowledge storage. The process of storing knowledge is only for explicit knowledge where it is documented even when organizational knowledge contains its own and tangible information [51].

2.6 Communities of Practice

Communities consist of people and not technology [52]. Community members may take an active role by contributing to discussions or providing assistance to other members. This is referred to as "participation." Communities of practice (CoP) initiative was implemented with the idea of starting effective KM practices across the entire organization to better facilitate the transfer of information between the various divisions in a swift and effective manner. A leader or a champion is chosen among each CoP team based on their performance criteria. Regular monitoring of the CoP operations and impact of the same are undertaken. These leaders are also encouraged to share success stories on an online CoP newsletter. To encourage enthusiastic participation from the staff, the KM team is also currently promoting a reward system wherein the reports from the various CoP's would be reviewed and the best performing team recognized and awarded [53]. This CoP approach has shown significant positive results so far with effective information exchange between various operating units already visible across the organization. This would also significantly improve the training for younger generation staff as they fast replace the retiring older generation [53].

This study emphasizes the role of KM as the key thinking and management change in CM among the oil and gas industry. In the oil and gas sector, KM serves as a practitioner link to information across various fields of exploration and production. This extension includes the processes and incentives needed for practitioners to not only trust and use the available data but to contribute to the global data repository. Donegan et al. [54] states that intellectual capital is rapidly replacing physical capital as competition. Investment in KM solutions brings the power of intellectual assets to the forefront and translates them into value. In their findings, Donegan et al. [54] suggest that about 80 percent of the world's largest companies have some kind of KM-based, efficient, and highly efficient business. The oil and gas industry has shown significant policy savings by adopting and fostering the collection and sharing of knowledge. Evidence from this shows that since the early 1990s, KM's tendency has been to accelerate the pace of technological and market changes that has led to innovation and learning becoming increasingly important for business success and with rapid advances in information and communications technology (ICT). It offers greater opportunities to exploit existing knowledge in organizations [15]. As stated from the literature studied, the oil and gas sector has been at the forefront of both the development and use of KM techniques due to several factors [15]:

- Rapid advances in information and communication technologies (ICT)
- Technological and market changes in the petroleum sector became increasingly intense during the 1990s and first decade of the twenty-first century.
- Global oil and gas organizations have undergone a major change in their dominant logic.
- Conditions specific to the oil and gas industry further suggest the potential of KM to provide solutions for critical problems.

This study uses the concept of Four Pillar of KM model contain leadership, organization, technology, and learning combined with KM practices which involves knowledge creation, knowledge acquisition, knowledge sharing, knowledge application, knowledge storage and retrieval, and community of practice (CoP).

3 Corrosion Management

Corrosion is a major expense in operating oil and gas facilities. Therefore, most oil and gas companies have some type of corrosion control or level of management program depending on the size, geographical location, and culture within the company. To explain the differences across the oil and gas industry, NACE International Impact 2016 reports; The CM's practices for international, national, intermediate and unconventional oil and gas companies are regularly reviewed. Lack of CM practices in the organization can also lead to inefficient use of resources as corrosion control activities are not prioritized based on ROI (all costs include financial and safety risks and immediate environment). In all cases, a robust CM program is essential as a driver to realize / maximize the ROI of corrosion mitigation activities.

3.1 International Oil Companies (IOC)

The IOCs have adopted corrosion control, different levels of corrosion management depending on the project portfolio, risk tolerance, and management culture. IOC benchmarks suggest significant differences in management approaches, and in-depth examination of selected IOCs to highlight the differences in CM practices. For all companies, the comparison between the two major IOCs and the aggregate oil and gas scores shows significant achievement. This shows different CM approaches to management. According to the NACE International Impact (2016) report, the results of Company A (Middle East) survey suggest that CM are mostly social. For example, in one region where regulations are strong, CM has been developed and implemented, whereas in other regions with weak regulations, CMs are limited to corrosion control tasks, and only to meet local minimum requirements. On the contrary, Company B (Asia) reports that strong centralized CM programs have been implemented equally across all parts of the organization including all regions. Company B considers itself the closest to "best practice". However, with a strong centralized CM program, it is often difficult to launch and implement the program into different parts and territories of the company, due to differences in operating culture and company practices.

3.2 National Oil Companies (NOC)

A number of additional businesses are often associated with NOCs, including electricity generation, chemicals, minerals, and all types of infrastructure. According to a 2016 Nace International Impact report, the NOC in Asia operates a general retail store as these are the channels through which petroleum products are distributed in the country. The results of three NOC surveys in the Middle East, Asia, and South America, along with the oil and gas aggregate results. Companies A and B show a similar trend in aggregate results, where the low scores are for the elements of Performance, Policy, Stakeholder Integration, Organization, and Accountability. Low scores indicate little commitment from management to the CM program (i.e., low baseline scores), and no follow-up practices are shown by low performance scores.

Further examination of the company's CM practices explains some of the findings from the survey. Without clear management commitment to the form of CM, other elements regardless of scoring are not optimized. Even with the full support of management, effective CMs often do not. With this commitment, the company has developed a comprehensive CM approach and implementation plan. However, the implementation of this plan has many obstacles. In this case, the relationship between having a CM and its implementation plan is lost. Therefore, before launching the plan, the company should spend time educating its own staff in the field as well as contractors on the use of related plans and instilling a buying interest. However, significant overall trends were not observed and the implementation of CMs in both oil and gas companies (IOC and NOC) but did not reduce COCs.

CM is comprehensive management and the process of stopping or reducing corrosion rates from occurring [4]. It includes maintenance activities, as part of asset integrity, often associated with business management system operators with higher management domain exclusions [20]. It involves the development, implementation, review, and maintenance of the overall management of corrosion policy [7]. Implications for mismanagement of CM include high probability of failure, increased cost of repair and downtime [55–57]. Identifying and designing corrosion as a threat at the design level increases the asset's reliability at the operational level [57]. Corrosion and control management is achieved through the application of appropriate corrosion management system (CMS) strategies and achieving a balance between the material and the environment [7]. CM strategies are guidance documents that ensure the design, operation and maintenance activities are carried out in such a way that they protect assets against corrosion [1, 57, 58]. CM techniques are categorized into two approaches, the technology approach and the management approach [55]. Technological approaches representing specific erosion include design, material selection, use of protective coating, cathode retention and protection, procedures and other design. However, Hansson [1] benchmarked the management approach into nine domains of management systems: policies, strategies and objectives; stakeholder integration: organization; accountability; resources; communication; integration of corrosion management practices; continuous improvement; and performance measures as shown in Fig. 3.

Management approaches include inspection, modeling and flow [4, 7]. Implementing corrosion measures can be costly and inefficient [59]. Hansson [1] show that by incorporating corrosion control in existing management systems and mitigating the effects, corrosion-related expenses can be optimized using corrosion management through a probabilistic and deterministic approach. This approach outlines each asset's maintenance strategy. Probabilistic approaches provide more risk-oriented maintenance while deterministic approaches suggest that assets are likely to fail based on their service time and thus lead to time-based maintenance [60]. Management approach requires competent staff for effective implementation [61]. Corrosion management emphasizes metrics and the establishment of key performance indices and cost capture that are not normally captured in operations [1].

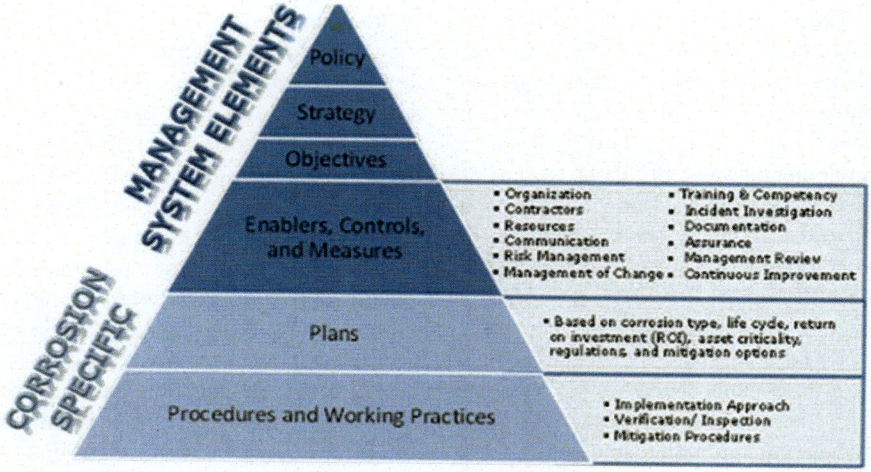

Fig. 3 Hierarchy of General and Corrosion-Specific Management Elements (*Adopted from* NACE International Impact Report, 2016)

4 Organızational Culture

According to Davenport and Prusak [62], organizational culture is the determination of the values and beliefs that a person chooses and absorbs is crucial to the effectiveness and performance of an organization. Wellman [63] basically explains that culture is "the way it is around here". Organizational culture classified into four types which are communal, mercenary, networked and fragmented [21]. A communal culture can give its members an idea, even if it is also guided by their obligations. Leaders of this culture are usually very inspiring and charismatic. The major negative is that they often influence and other members rarely talk. In a networked culture, members are considered friends and family. They are said to have a close relationship with each other and love each other. They are willing to help and share information. The weakness of this culture is that people are so good to each other that they refuse to show and criticize bad performance. A mercenary culture focuses on strict goals. Members are expected to meet their goals and to get the job done quickly. Because everyone is focused on goals and objectives, there is no room for political class. The negative is that people with poor performance can be treated with inhumanity. In a fragmented culture, belonging to and identity with an organization is usually very weak. Individuals are organizations, and their priorities are given to members and work assignments. Weaknesses are especially evident due to lack of cooperation between members and teams.

Organizational culture influences the level of management efficiency, cost savings and restructuring. The various types of organizational culture are explained with the aim of understanding the key dimensions of the microculture evolving within the organization. The enablers and barriers of shared knowledge sharing, along with discussions on how to set up organizational change are required to better and efficiently manage knowledge management. Implementation of knowledge management is closely linked to cultural change—if not complete transformation, at least to the extent that cultural tweaks exist to promote a culture of collaboration and knowledge sharing.

Corporate culture is an important component of ensuring critical knowledge and information flow within the organization. The strength and commitment of corporate culture is more important than communication technology to promote knowledge sharing. The communication system within an organization can be regarded as a cultural diffuser [64]. Sveiby [65] show that the climate of cooperation is one of the major factors affecting the effectiveness of knowledge work. Gruber and Duxbury [66, 66] have conducted in-depth studies on the relationship between the research department and the development of a high-technology company. They examine the relationship between organizational culture and knowledge sharing and use variables of trust, openness, top management support, and organizational reward structure to explain correlations. Gladwell [67] discuss the preparation and features of offices that may influence innovation and knowledge sharing.

5 Application of KM Practices in CM

KM practices have many advantageous properties and can enforce applications in various fields and various other applications. In general, KM practices applications can be classified as in Table 1. There has been a lot of KM-related research and

Table 1 Application of KM practices in CM

Category	Applications	References
KM practices	Knowledge creation Knowledge acquisition Knowledge sharing Knowledge application Knowledge storage and retrieval Communities of practice	Maravilhas [28, 31], Albort et al. [33], Mothe et al. [37], Wijethilake [38], Shahzad [39] Attia [35], Bolisani et al. [40], Albort et al. [33], Mulhim [43] Watson [47], Wang et al. [50], Vij [49], Cook [52], Nickols [68]
Four pillars of KM	Leadership Organization Technology Learning	Skyrme [69], Coleman [70] Argyis [71], Gruber [66], Hofstede [72] Kilmann et al. [73] Argyis et al. [71]
Corrosion management (CM)	CM strategy and skilled technical team CM techniques involved technological and management approach	Njomane [74] Al-Arada et al. [55]

publications that have taken the organizational perspective and given little attention to the effects of KM exploitation on the oil and gas industry. These studies typically focus on how an organization's knowledge and capabilities are managed through the use of KM practices with the overall goal of gaining competitive advantage over potential competitors [69, 75, 76]. In this regard, KM needs to move beyond the organizational level through a more robust approach to developing push-pull strategies. This strategy is for integration of personal, organizational, project and industry interface knowledge to maintain existing knowledge and create new knowledge.

The strengthening of the Four Pillars of KM which has been the catalyst for the formation and implementation of the KM practice, based on a recent study of Njomane and Telukdarie [74] the four pillars are:

Leadership: Develops a strategic, objective, and strategic vision of KM in line with business vision and mission; Encouraging employee advocacy and staff empowerment to reduce reliance on other people or organizations and encourage initiatives by providing development training and ensuring adequate resourcing; and Promote an open, encouraging and supportive culture. *Organizations*: to ensure organizational and technological infrastructure flexibility to meet changing market demands; and Establish formal and informal networks where knowledge can flow smoothly without restrictions.

Technology: Ensures that systems are managed for the authenticity and reliability of information and are up-to-date; and Ensure controlled access to the organization's information system during operation. *Learning*: Encouraging employee training and development; and Promoting Practice Communities and cross teams.

Design practices often fail to consider erosion; therefore, the cost of corrosion can be avoided. There are two aspects to the problem. First, design engineers, generally mechanical, chemical, and electrical engineers, have limited understanding of materials/environmental interactions and various corrosion modes. Second, life cycle costs or total cost of ownership are often not considered in the design phase. CM is able to change the design paradigm based on KM practices is the best approach in overcoming these two aspects. The provision of current corrosion control technology "best practices" should be provided to designers. These include corrosion performance in design criteria, and promoting the life cycle and total cost of ownership analysis. Provides designers with an understanding of corrosion engineers with databases and design tools for optimal corrosion management. Educates design engineers, corrosion engineers, and current maintenance managers on the importance and potential savings from proper corrosion management. This requires knowledge sharing methods. A corrosion engineer needs to do more than assess that "corrosion management pays off." Corrosion engineers must be prepared to demonstrate attractive investment returns to design corrosion management, as well as improve service and reliability. This requires knowledge of engineering economics that enables rust engineers to perform economic calculations (such as life cycle cost analysis and cost/benefit analysis). Corrosion engineers must be able to determine design and maintenance practices that reduce corrosion, as well as evaluate these practices to inform decision makers.

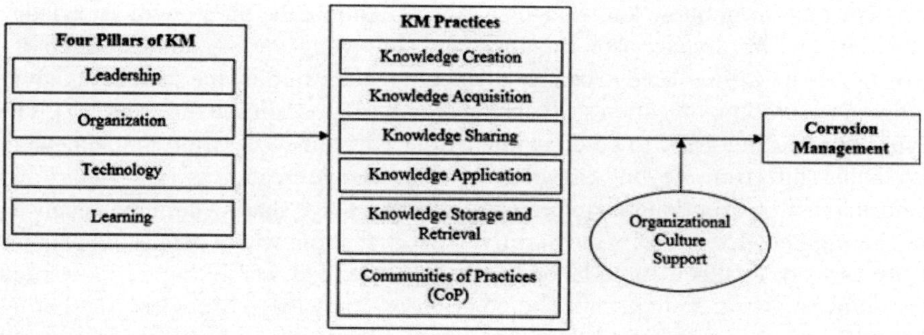

Fig. 4 Conceptual Framework of KM Practices Application in CM

Organizational culture support intervening the Four Pillars of KM contain leadership, organization, technology and learning are related to KM practices which are knowledge creation, knowledge acquisition, knowledge sharing, knowledge application, knowledge storage and retrieval, and communities of practices (CoP). Both independent variables (Four pillars of KM and KM Practices) give impacts on the CM activities/program to reduce the COC in the oil and gas industry.

The following conceptual framework was developed and proposed from the previous discussion. Figure 4 shows the conceptual framework of KM practices application in CM for oil and gas industry.

6 Discussion and Implications

KM practice is important in the collective competition and development of the performance of oil and gas organizations. It launches an effective reuse of (limited) resources such as cost savings, especially in high technology organizations. KM practices help promote collaborative collaboration between teams in various disciplines. This can help reduce millions of dollars in operating and cost control by helping employees avoid unnecessary delays and recurring mistakes. KM practices involving the community of practice show good acceptance among staff to maintain optimal level of management. As more staff are encouraged to participate in KM activities, more KM practices and collaboration platforms can be created. With the practice of KM practices, the knowledge gap can be reduced between the expertise and the new learner. These activities will help in realizing the integration of the organization's overall process and business alignment. Continued support from management is important despite the limited time and resources allocated to improving KM practices to generate large productivity gains over the long term.

Organizational culture support is an important component of ensuring critical knowledge and information flow smoothly. The strength and commitment of corporate culture is almost always more important than the communication technology implemented to promote knowledge sharing. Traditionally, the flow of knowledge was vertical, from supervisor to supervisor (following the organizational chart). Today's organizations need to change their culture to rewarding the stream of knowledge as well. Communication systems can be regarded as cultural diffusers [77]. The types of ideas that need to be disseminated for KM are successfully implemented, including shifts from viewing knowledge and knowledge creation as a proprietary and solo acquisition to collaborative perceptions. Knowledge sharing culture is one way of sharing knowledge that is the norm, not the exception, where people are encouraged to work together, to collaborate and to share, and where they are rewarded for doing so. The paradigm shift should go from "knowledge is power" to "sharing knowledge is more powerful" and this culture will determine what you can and will do with the organization's knowledge assets. Sveiby and Simons [78] suggest that the climate of cooperation is one of the major factors affecting the effectiveness of knowledge work. The extent to which organizational culture can be assessed, and this provides a good indication of the success of KM. Gruber and Duxbury [66, 66] conduct in-depth studies of the research and development department of a high-technology company. They look at the relationship between organizational culture and knowledge sharing and use variables of trust, openness, top management support, and organizational reward structure to try to explain any correlation. Some of the factors that make it difficult to share implicit knowledge include the attitude that knowledge is power, not knowing who the member is, not knowing whether that knowledge exists, and losing knowledge when people leave the company. An ideal knowledge sharing culture is one in which communication and coordination between groups is emphasized, whereby experts will not actively control their knowledge, and where knowledge sharing will be active and clearly encouraged at all levels of the hierarchy through recognition and reward sharing of knowledge.

The current study examines the multidimensional relationship between the Four Pillars of KM, KM practices, CM and influenced by organizational culture support. The importance of developed framework and contributions to theoretical and practical as well as implications are as follows.

6.1 Importance of the Developed KM Practices Framework for CM

There is a tendency for research to become obsolete over the course of the research project for a number of reasons e.g. changes to the research status quo due to new findings, or due to the research problem being solved already elsewhere. This research, however, has been found to be valid and relevant to date as COC remains endemic. Recent industry report [79, 80], have re-iterated the need to reduce the COC across the oil and gas industry. It can therefore not be over-stated that the industry needs improvement in this area in which the research presents a solution and offers real practical application for reducing COC. Continuous updates on literature review also ensure that there is no duplication or work with other ongoing research.

Corrosion knowledge management is considered one of the latest methods of reducing corrosion. Therefore, minimal material management can reduce losses and save costs. Corrosion and non-corrosion experts can apply the principles of corrosion knowledge management, especially to managers with little or no background in rust science to find the best approach in resource management [81]. The impact of an application of KM based Four Pillars of KM in CM activities/program supported by organizational culture become the CKM and it suit for oil and gas industry. The research presents a CKM model for harnessing and integrating knowledge across personal, organizational, project and industry boundaries which adds to the body of knowledge both for academics and industry practitioners. The future CKM application for reducing COC in the oil and gas industry (on-shore pipeline) which has never before been presented in academia and industry (local oil and gas industry scope of study).

6.2 Theoretical and Practical Contributions and Implications

The current study has multiple theoretical implications. Firstly, the research enriches the limited literature on the correlation between KM practices and CM based on KM theory and CM theory and explains how KM practices facilitate CM activities in the oil and gas industry. Secondly, it highlights the important role of organization culture in the relationship between KM practices and CM and has never been studied in Malaysia oil and gas industry. This study also highlights the importance of implementing all dimensions of KM practices to enhance CM capabilities in managing COC.

The output from the research contributes to the body of knowledge in the area of KM practices and CM particularly in understanding the impact of KM practices in reducing COC in the oil and gas industry. Previous studies have not covered this area therefore this study provides a new valuable insight for industry practitioners and academics. The research presents a new definition of 'cost reduction' in the CM context which adds to the body of knowledge in COC management. It also presents a new definition of corrosion knowledge management in the oil and gas industry which contributes to KM body of knowledge. The output from the study contributes to industry knowledge in the area of KM best practices and continuous improvement on CM in leading the COC in Malaysia oil and gas industry.

7 Conclusions

Malaysia oil and gas industry is looking at the current of CM and also COC involving human factor (KM practices) in the corrosion field. Oil and gas organizations had been struggled to combat voluntary and involuntary turnover and the resulting loss of knowledge. The limitation of this study focuses on examining the extent to human factor approaches. It is worth mentioning that the findings of the study provide CM practitioners and local oil and gas organizations with an integrated KM practices approach in leading the COC reduction. Further research needs to focus on CKM processes to ensure the findings also taking into consideration different level of workers' position (top management level and middle management level) in CM activities/program which are vital to KM practices. Research can collect primary data through surveys among technical workers and interviews among managers to increase the reliability and validity of this review study.

References

1. Hansson C (2011) Metallugrical and materials transaction: impact of corrosion on society. Struct Mater Am 42(10):2952–2962
2. National Research Council (2009) Assessment of corrosion education. The National Academic Press, Washington, D.C.
3. The Institute of Corrosion, Corrosion Management (2013) J Inst Corros 112
4. Mohammad S, Hosseini K (2017) Robust prediction of CO_2 corrosion rate in extraction and production hydrocarbon industry. Anti-Corros Meth Mater 64(1):36–42
5. Ritchie D (2011) The role of asset integrity and life extension in major accident prevention, Anerdeen
6. Horrocks P, Mansfield D, Parker K, Atkinson T, Worsley J (2012) Managing ageing plant
7. Ossai C (2012) Advances in asset management techniques: an overview of corrosion mechanisms and mitigation strategies for oil and gas pipelines. ISRN Corrosion
8. Nicholson R, Feblowitz J, Madden C, Bigliani R (2010) The role of predictive analytics in asset optimization for the oil and gas industry-white paper
9. Koch G, Varney J, Thompson N, Moghissi O, Gould M, Payer J (2016) International measures of prevention, application, and economics of corrosion technologies study. NACE International, pp 1–216
10. Lang JC (2001) Managerial concerns in knowledge management. J Knowl Manage 5(1):43–59
11. Leseure MJ, Brookes NJ, N.J. (2004) Knowledge management benchmarks for project management. J Knowl Manage 8(1):103–116
12. Odell C, Hasanali F, Hubert C (2000) Knowledge management: a guide for your journey to best-practice processes. American Productivity and Quality Centre, Houston, Texas, USA
13. Gilbert M, Leibold M, Probost G (2002) Five styles of customer knowledge management and how smart companies use them to create value. Eur Manage J 20(5):459–469
14. Bixler CH (2002) Applying the four pillars of knowledge management. Knowl Manage World 11(1)
15. Grant RM (2013) The development of knowledge management in the oil and gas industry. Universia Bus Rev 92–125
16. Leavitt P (2002) Applying knowledge management to oil and gas industry challenges. Am Product Qual Center 1–6

17. Rabiu M (2009) Knowledge management in oil and gas industry: theoretical frameworks, practical challenges and opportunities. Petrol Technol Dev J 1:1–10
18. Pawlowski J, Bick M (2012) The global knowledge management framework: towards a theory for knowledge management in globally distributed settings. Electron J Knowl Manage 10(1):92–108
19. Ekemena EH (2011) Knowledge management in the Nigeria public sector. Unpublished PhD Thesis, Aberystwth University
20. Nwafor M, Salau S (2009) Knowledge management and development targets in Nigeria. Report no. NSSP 002, International Food Policy Research Institute
21. Dalkir K (2013) Knowledge management in theory and practice. Knowledge management in theory and practice
22. Hooshyar B (2010) Strategic knowledge management and its effects on cost management in projects. Advanced manufacturing engineering, quality and production systems, pp 177–182
23. Stankosky MF, Calabrese F, Baldanza C (2003) A systems approach to engineering a knowledge management system. Management Concepts Press, Washington, DC
24. Choo C (1998) The knowing organization. Oxford University Press, New York
25. Weick K (2001) Making sense of the organization. Basil Blackwell, Malden, MA
26. Nonaka I, Takeuchi H (1995) The knowledge-creating company: how Japanese companies create the dynamics of innovation. Oxford University Press, New York
27. Simon H (1957) Models of man: social and rational. John Wiley & Sons, New York
28. Maravilhas S, Martins J (2019) Strategic knowledge management a digital environment: tacit and explicit knowledge in Fab Labs. J Bus Res
29. Habib M, Abbas J, Noman R (2019) Are human capital, intellectual property rights, and research and development expenditures really important for total factor productivity? An empirical analysis. Int J Soc Econ
30. UNGC (2018) United nations global compact. United Nations, New York
31. Jarrahi MH (2018) Social media, social capital, and knowledge sharing in enterprise, vol 20. IT Professional, pp 37–45
32. Chatzoudes D, Chatzoglou P, Vraimaki E (2015) The central role of knowledge management in business operations: developing a new conceptual framework. Bus Process Manage J 21:1117–1139
33. Albort-Morant G, Leal-Millan A, Cepeda-Carrion G (2016) The antecedents of green innovation performance: a model of learning and capabilities. J Bus Res 69:4912–4917
34. Tseng SM (2014) The impact of knowledge management capabilities and supplier relationship management on corporate performance. Int J Prod Econ 154:39–47
35. Attia A, Salama I (2018) Knowledge management capability and supply chain management practices in the Saudi food industry. Bus Process Manage J 24:459–477
36. Qasrawi BT, Almahamid SM, Qasrawi ST (2017) The impact of TQM practices and KM processes on organizational performance: an empirical investigation. Int J Qual Reliab Manage 34:1034–1055
37. Mothe C, Nguyen-Thi UT, Triguero A (2017) Innovative products and services with environmental benefits: design of search strategies for external knowledge and absorptive capacity. J Environ Plan Manage 1–21:0568
38. Wijethilake C (2017) Proactive sustainability strategy and corporate sustainability performance: the mediating effect of sustainability control systems. J Environ Manag 196:569–582
39. Shahzad M, Ying Q, Ur Rehman S, Zafar A, Ding X, Abbas J (2019) Impact of knowledge absorptive capacity on corporate sustainability with mediating role of CSR: analysis from the Asian context. J Environ Plan Manage
40. Bolisani E, Bratianu C (2018) The elusive definition of knowledge. In: Emergent knowledge strategies: strategic thinking in knowledge management. Springer International Publishing, Cham, pp 1e22
41. Khodadadi M, Feizi M (2015) The impact of knowledge management on social responsibility at social security organization in East Azarbaijan province. Kuwait Chap Arab J Bus Manage Rev 4:165–169

42. Bar~ao A, de Vasconcelos JB, Rocha A, Pereira R (2017) A knowledge management approach to capture organizational learning networks. Int J Inf Manage 37:735–740
43. Mulhim AF (2017) The effects of knowledge creation process on organizational performance: evidence from Saudi banking sector. Int J Manage Sci Bus Res 6:1–12
44. Mardani A, Nikoosokhan S, Moradi M, Doustar M (2018) The relationship between knowledge management and innovation performance. J High Technol Manage Res 29:12–26
45. Wang CL, Ahmed PK, Rafiq M (2008) Knowledge management orientation: construct development and empirical validation. Eur J Inf Syst 17(3):219–235
46. Chirumalla K, Parida A (2016) Understanding knowledge reuse process: a case study in a production maintenance organization. IUP J Comput Sci 10(1):25–42
47. Watson S, Hewett K (2006) A multi-theoretical model of knowledge transfer in organizations: determinants of knowledge contribution and knowledge reuse. J Manage Stud 43(2):141–173
48. Jennex M, Croasdell DT (2004) Knowledge management, organizational memory and organizational learning cluster
49. Vij S, Farooq R (2016) Moderating effect of firm size on the relationship between IT orientation and business performance. IUP J Knowl Manage 14(4):1–19
50. Wang LC, Hult MTGJ, Ketchen JD, Ahmed KP (2009) Knowledge management orientation, market orientation, and firm performance: an integration and empirical examination. J Strateg Mark 17(2):99–122
51. Mostert JC, Snyman MMM (2007) Knowledge management framework for the development of an effective knowledge management strategy. South African J Inf Manage 9(2)
52. Cook P (1999) I heard it through the grapevine: making knowledge management work by learning to share knowledge, skills and experience. Ind Commer Train 31(3):101–105
53. Shariff M (2008) PETRONAS: engaging knowledge worker communities to stimulate innovation and build corporate capability, Viewed Jan 23rd 2012
54. Donegan J, Burgoyne M, Day P (2001) Oilfield review spring 2001—Building a knowledge-sharing culture. Oilfield Review
55. Al-Arada M, Al-Refai F, Joshi S, Patil S (2015) Effective corrosion management for enhancing mechanical integrity. NACE international
56. Debruyn H, Al-Ghamdi S (2016) Corrosion management implementation in large oil and gas projects. NAÇE International
57. Morshed A (2007) Offshore assets: from corrosion engineering to corrosion management. Chem Treatment 46:33–48
58. Tino A, Wiryolukito I, Abduh M (2008) Corrosion: technical and economic driver for Indonesia oil and gas industry, Petroenergy Magazine Edition
59. Nyborg R (2010) CO_2 corrosion models for oil and gas production systems, Houston
60. Lawson K (2005) Pipeline corrosion risk analysis—an assessment of deterministic and probabilistic methods. Anti-Corros Meth Mater 52(1):3–10
61. Su X, Yin Z, Cheng Y (2012) Corrosion of 16Mn line pipe steel in a simulated soil solution and the implication on its long-term corrosion behavior. J Maer Eng Perform 22:498–504
62. Davenport TH, Prusak L (2000) Working knowledge: how organizations manage what they know. Harvard Business School Press, Boston, MA
63. Wellman JL (2009) Organizational learning: how companies and institutions manage and apply knowledge. Palgrave Macmillian
64. Bloom B, Mesia B, Krathwohl R (1964) Taxonomy of educational objectives. Volumes 1 and 2. David McKay, New York
65. Sveiby K (2001) Methods for measuring intangible assets
66. Gruber H, Duxbury L (2001) Does organizational culture affect the sharing of knowledge? Presentation Health Canada
67. Gladwell M (2000) Designs for working. The New Yorker
68. Nickols F (2000) Community of practice start up kit. The distance consulting company
69. Skyrme D (2011) A knowledge-enriching culture
70. Coleman J (1994) Social capital in the creation of human capital. Am J Sociol ('94 Supplement):97

71. Argyris C, Schon D (1978) Organizational learning: a theory of action perspective. Addison-Wesley, Reading, MA
72. Hofstede G, Neuijen B, Daval Ohayv D, Sanders G (1990) Measuring organization cultures: a qualitative and quantitative study across twenty cases. Adm Sci Q 35:286–316
73. Kilmann RH, Saxton MJ, Serpa R (1986) Issues in understanding and changing culture. Calif Manage Rev 28:87–94
74. Njomane L, Telukdarie A (2018) Corrosion management: A case study on South African oil and gas company. In: Proceedings of the international conference on industrial engineering and operations management, pp 581–595
75. Garstenauer A, Blackburn T, Olson B (2014) A knowledge management based approach to quality management for large manufacturing organizations. Eng Manage J 26(4):47–58
76. Newell S, Robertson M, Scarbrough H et al (2009) Managing knowledge work and innovation. Palgrave MacMillan, Basingstoke
77. Bloom H (2000) The global brain. John Wiley and Sons, New York
78. Sveiby K, Simons R (2002) Collaborative climate and effectiveness of knowledge work—an empirical study. J Knowl Manage 6(5):420–433
79. BIS (Department for Business, Innovation and Skills) (2013) UK construction: an economic analysis of the sector. London UK
80. HM Government (2013) Construction 2025. Industrial Strategy—government and industry in partnership. HM Government, London
81. Javaherdashti R (2006) Corrosion knowledge management: an example of an interface for more efficient managerial communication. Mater Corros 57(12):945–950
82. Tompang N, Yunus AM (2017) The effectiveness of knowledge sharing practices in government agency. Int J Acad Res Bus Soc Sci 7(12). https://doi.org/10.6007/IJARBSS/v7-i12/3749
83. Rambeli SA, Yunus AM (2018) The knowledge management practices at PTAR in promoting creativity and innovation among staff. Int J Acad Res Bus Soc Sci 8(9). https://doi.org/10.6007/IJARBSS/v8-i9/4598
84. Malic NHA, Malik A, Yunus AM, Nik NA (2018) Organizational development of knowledge management: an analysis on the consultancy company. Int J Acad Res Bus Soc Sci 8(9). https://doi.org/10.6007/IJARBSS/v8-i9/4651

Rapid CO_2 Gas Decompression Performance of Epoxy Coatings Containing Graphene Oxide-Zinc Oxide Hybrids

Othman Nurul Husna, Che Ismail Mokhtar, and Mustapha Mazli

Abstract Carbon steel pipelines are widely used in oil and gas industry. To ensure the reliability and long-term performance of the pipelines, coatings tailored for corrosion protection of metallic substrates are of the utmost relevance. Although there is a strong emphasis on the development of advanced functional and smart coatings, the need for a more advanced coating for conventional applications has yet to be fulfilled. Recently, graphene attracted significant attention due to its robust and inert properties. In this study, we report on the development of graphene oxide-zinc oxide hybrids (GO-ZnO) applied as fillers to reduce the permeability of epoxy coatings applied in high pressure carbon dioxide gas conditions. The structural stability of GO-ZnO nanosheets were compared with GO sheets using Raman Spectroscopy. The identification of mechanical strength of nanocomposites after subjected to rapid carbon dioxide gas decompression conditions at pressures 1, 25 and 65 bars were also obtained and discussed. Finally, the mechanism of protection against CO_2 gas rapid decompression conditions was established.

1 Introduction

Crude oil and natural gas are important ingredients to produce various petroleum products such as petrol, diesel, liquefied petroleum gas (LPG) and several other hydrocarbons [1]. Obtaining these poses a big challenge as oil exploration requires transporting the ingredients several miles away from the sea to production refineries through underground pipelines. Taking into account their inflammable properties, any leakage could cause severe accidents that could lead to massive explosion. Therefore, the dependency on pipelines for the past 40–50 years demands for innovative and robust techniques to improve its corrosion inhibition efficiency. Up until now,

O. Nurul Husna (✉) · C. Ismail Mokhtar · M. Mazli
Centre for Corrosion Research, Universiti Teknologi PETRONAS, 32150 Tronoh, Perak Darul Ridzuan, Malaysia
e-mail: mokhtis@utp.edu.my

Mechanical Engineering Department, Universiti Teknologi PETRONAS, 32150 Tronoh, Perak Darul Ridzuan, Malaysia

F. Ahmad et al. (eds.), *ICPER 2020*, Lecture Notes in Mechanical Engineering,
https://doi.org/10.1007/978-981-19-1939-8_66

different coating materials such as nanocomposite coating, organic-inorganic hybrid and hydrophobic coatings have been proven to increase the life cycle of carbon steel pipelines [2–4]. Still, problems continue in the long-term protection of metal against aggressive environments [5].

Graphene has attracted global attention due to its unique properties such as chemically and thermally inert, high strength and ductility and impermeable to discrete molecules. Previous works [6, 7] have demonstrated the ability of graphene to provide short term corrosion and barrier protection for the substrate underneath. It is very challenging to transfer CVD grown graphene as the structure defects of graphene are formed during the transfer. To harness graphene's unique properties, chemically synthesized graphene oxide (GO) based nanosheets can be utilized with organic coatings such as polymer and epoxy to provide impermeability to corrosive ions. The very high surface area and nanometric thickness of graphene oxide sheets provides great potential to be served as additives to enhance the barrier properties of organic coatings [8].

However, GO sheets are prone to form severe aggregations within the polymer matrix, arising from their intrinsic van der waals' interaction and high surface area. This can cause poor distribution of GO within the polymeric matrix and limit further application. However, GO contains abundant oxygen functional groups that can be easily functionalize with organic components. Li et. al. functionalized graphene oxide with tetraethoxysilane to form organofunctional silane composite coatings. The composite coatings exhibited significant improvement in protective properties compared to neat silane coatings [9]. In a study done by Sheng et al., GO was modified with p-phenylenediamine to improve the dispersion stability in acidic conditions and compatibility with polyaniline. The results shown that the GO is highly exfoliated and intercalated among the polymeric matrix and the corrosion rate obtained by the nanocomposite was lower compared to individual polyaniline coatings [10].

Recently, studies reported the decoration of GO with nanoparticles (NP) such as titanium dioxide [11], silica [12], alumina [13] and iron oxide [14]. This approach has attracted considerable attention as the advantages of both GO and NPs can be utilized. Zinc oxide is one of the nanoparticles that is commonly used in coatings [15]. ZnO can be called a multifunctional material thanks to its unique physical and chemical properties. Besides used as UV blocking agents, ZnO is relatively easy to be dispersed compared to other nanoparticles such as titanium oxide and alumina. This is due to the stable surface of ZnO that is able to maintain its lattice parameter at any condition, allowing room for crystallization and nucleation of polymer. The anti-bacterial and anti-fungal properties of ZnO make it useful for surface coating, textiles and plastics. Studies have shown that the heat, scratch, abrasion resistance and thermal stability properties improved with the incorporation of ZnO nanoparticles in polymer matrix.

To fully utilize the impermeability, strength and stability of Graphene oxide sheets, Graphene Oxide-Zinc Oxide (GO-ZnO) hybrids were developed to function as barriers in against corrosive ions in an epoxy coating system and provide the best mechanical support as internal pipeline liners. The performance of the coating under rapid decompression in CO_2 gas was evaluated. In the offshore petroleum industry, polymer liners are often used in certain areas of pipeline. Within such a framework, the liners are required to resist high pressures (up to 100 MPa) in aggressive chemical fluids such as hydrogen sulphide and carbon dioxide. Some liners will exhibit blistering due to the high decompression rate in the pipeline [16]. In addition, CO_2 gas tend to saturate in the polymer sheath of the pipe. Within a sudden operating interruption, the gas present in the polymer sheath tend to diffuse outside the material and leads to the damage of polymer liners [17]. The presence of homogeneous GO-ZnO fillers in epoxy is believed to prevent the mechanical damage of coating, as CO_2 is unable to enter the polymer sheath due to the impermeability of graphene. This speculation was evaluated in this paper scope. Hence, the stability of GO-ZnO in high pressure CO_2 is evaluated to ensure that the conditions are unable to damage the impermeability structure of hybrids.

2 Materials and Methods

2.1 *Materials*

Graphene Oxide powder was obtained from Graphenea. 50 nm sized Zinc Oxide nanoparticles was used to be decorated onto GO sheets with the aid of 3-(aminopropyl)triethoxysilane (APTES). Epoxy resin of commercial-grade, diglycidyl ether of bisphenol A (DGEBA, EPON 828) was used with a modified cycloaliohatic amine type of hardener known as EPIKURE curing agent F205. The mixing ratio was 1:0.58 of epoxy to hardener. All other reagents and solvents were purchased from commercial suppliers and used as received.

2.2 *Preparation of Graphene Oxide-Zinc Oxide (GO-ZnO) Hybrids*

ZnO dispersion was prepared by mixing 10 g ZnO powder and 50 g toluene in a round bottomed flask by stirring at room temperature using a magnetic stirrer for 5 min. The suspension was then treated in an ultrasonic bath for 10 min. 1 g (0.0055 mol) of APTES was added into the suspension and it was stirred at room temperature. After a yellow-transparent dispersion was obtained, the reaction mixtures were then refluxed for a further 24 h. The solvents in the dispersion were evaporated to dryness by using a rotation evaporator [18]. After that, APTES modified ZnO powder was washed to remove unreacted APTES molecules with ethanol for three times. The powder was dried at 50 °C in the vacuum oven for 2 h. Solvent exchange process was carried out to transfer Graphene Oxide sheets into DMF from water suspension. To carry out this method, 30 ml of Graphene Oxide (GO) in water dispersion (0.12 g GO) was mixed with 100 ml DMF and ultrasonicated in a water bath (75 °C) for 1 h. After that, water was removed from the solution by heating the mixture up to 90 °C for 3 h. To produce GO-ZnO with a mixing ratio of GO:ZnO OF 1:5, 0.05 g of functionalized-ZnO was added into the suspension and ultrasonication was continued for another 30 min. The reaction process was started at 105 ºC for 5 h. The final suspension was subjected to vacuum filtration recover GO-ZnO hybrids. The recovered product was washed with ethanol thrice. Finally, the procedure was repeated to produce other mixing ratios of GO:ZnO OF 1:3, 1:4 and 1:6.

2.3 *Preparation of Nanocomposite Samples*

The composite samples were prepared by directly mixing the vacuum filtered GO-ZnO hybrids in hardener and stirred for 1 h at 1100 rpm. The mixture was then ultrasonicated for 10 min. The mixture of hybrids and hardener were added with epoxy according to ratio 1:0.58. The resulting mixture were poured into a mould and was left in the vacuum oven for 1 h to remove air bubbles. The mixture was then post cured for 24 h at room temperature and 1 h at 100 ºC. Various weight percentage of GO-ZnO and GO (0.1 wt%, 0.2 wt%, 0.3 wt%, and 0.4 wt%) was applied in epoxy to proceed with rapid decompression tests. The best mixture based on the rapid decompression results are selected for the high-pressure permeability tests. Examples of the prepared dog-bone samples are shown in Fig. 1.

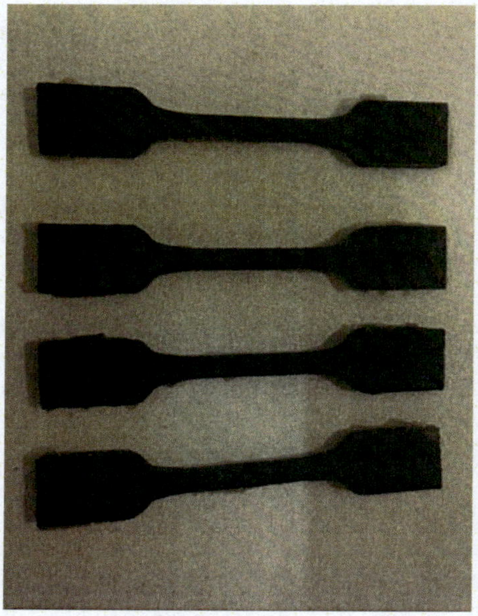

Fig. 1 Dog-bone samples for rapid decompression tests

2.4 *Rapid Decompression Tests*

The tests were performed according to TM0192. Tensile dog bone sample were exposed to 1, 25 and 65 bar of CO_2 gas (99.99 purity) for 72 h at room temperature in a high-pressure autoclave. The pressure released from the test vessel was controlled within ≤1 min. The dog bone samples were then immediately subjected to tensile tests to identify the effect of rapid decompression on the properties of nanocomposite. At the same time, powder samples of GO and GO-ZnO of ratios 1:3,1:4, 1:5 and 1:6 were exposed to similar conditions and was further analysed using Raman Spectroscopy.

2.5 *Characterization*

The tensile properties were studied using Universal Testing Machine equipped with a 10 kN load cell at a displacement rate of 5 mm/min at room temperature according to ASTM D638. The cross section of composite coatings was observed with SEM. Raman spectroscopy was also used to analyse the stability of GO and GO-ZnO sheets under high pressure CO_2 gas. TEM images were captured to confirm the mechanism of barrier protection of nanocomposites under similar condition.

3 Results and Discussions

3.1 *Stability of Graphene Oxide (GO) and GO-ZnO Under High-Pressure Carbon Dioxide Gas Conditions*

GO sheets and different ratios of GO:ZnO were subjected to different CO_2 gas pressure. The structural stability of the sheets is studied using Raman Spectroscopy. Raman spectroscopy is known to be very informative technique to characterize different carbon materials [19, 20]. The most intensive bands in Raman spectra of sp^2 carbon networks are so-called G band (doubly degenerate phonon mode at the Brillouin zone center), D band (disorder-induced phonon mode corresponding to scattering on the K-point phonon), 2D band (overtone of the D band) and S3 band attributed to the combination of D and G peaks (i.e. G+D band) [6, 19, 21]. The G band is considerably broadened in graphene oxide and there is also a strong D band that is not present in exfoliated graphene. The appearance of the D band is a clear indication of the presence of sp^3 bonding in the graphene oxide although other features of the spectrum indicate defects in the structure [22].

Typical Raman spectra of the nanocomposite coatings before and after exposure to high pressure carbon dioxide gas are presented in Fig. 3. In all spectra, two prominent Raman features at ~1349 and ~1591 cm^{-1} can be seen, corresponding to the D and G bands. In addition to the G and D bands, two weak peaks at ~2930 (S3-band) and

~2690 cm^{-1} (2D band) were also observed for GO and GO-ZnO Raman spectra. As the pressure of carbon dioxide increases, I band at ~1180 became more prominent in Fig. 2b and c suggesting that the structure of C–C, C = C or sp^2-sp^3 carbon bonds has become heavily disordered [23]. After multiple-peak fittings of the broadened D, G and high order bands, the Raman spectra for GO and all ratios of GO:ZnO showed some evidence of disorder with increasing carbon dioxide pressure conditions. For graphene oxide sheets, the gas pressure increments to 65 bars resulted an apparent broadening of the full widths at half maximum (FWHM) of D band by 27% with respect to the FWHM of D at the control pressure, as depicted in Fig. 3. This indicates that there is an increase of defects or disorder due to the expansion of the GO sheets with sudden pressure rise [24]. In contrast, the attachment of ZnO onto GO sheets managed to minimize the change of FWHM of D band which suggests that GO:ZnO is more stable with pressure increment as compared to GO sheets [25].

The intensity ratio of the D and G bands (ID/IG) can serve as a standard to measure the defects in GO and GO-ZnO with increasing carbon dioxide pressure environment [26, 27]. In general, it can be seen in Fig. 4 that the ID/IG increases with increasing CO_2 gas pressure. This shows that the stability GO and GO-ZnO is

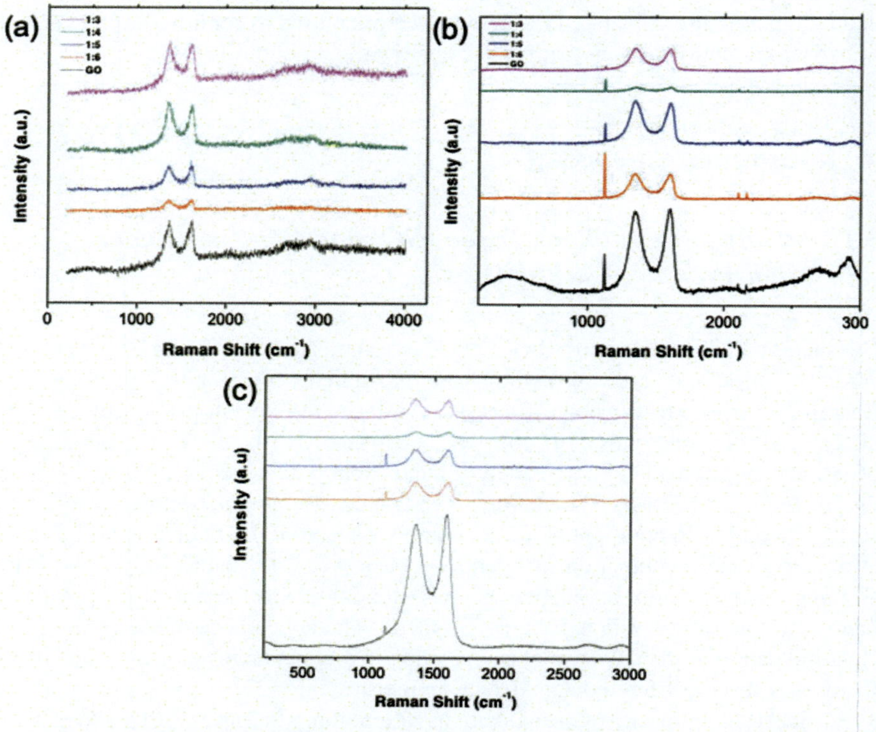

Fig. 2 Raman spectra of GO and GO-ZnO of different ratios (1:3, 1:4, 1:5 and 1:6) at **a** 1 bar, **b** 25 bar, **c** 65 bar carbon dioxide gas condition

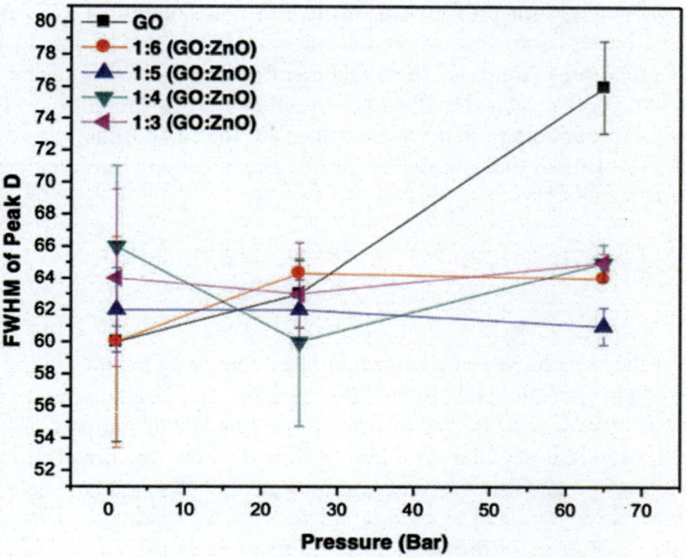

Fig. 3 Full width height maximum (FWHM) of GO and GO-ZnO powder of different ratios with respect to carbon dioxide pressure rise

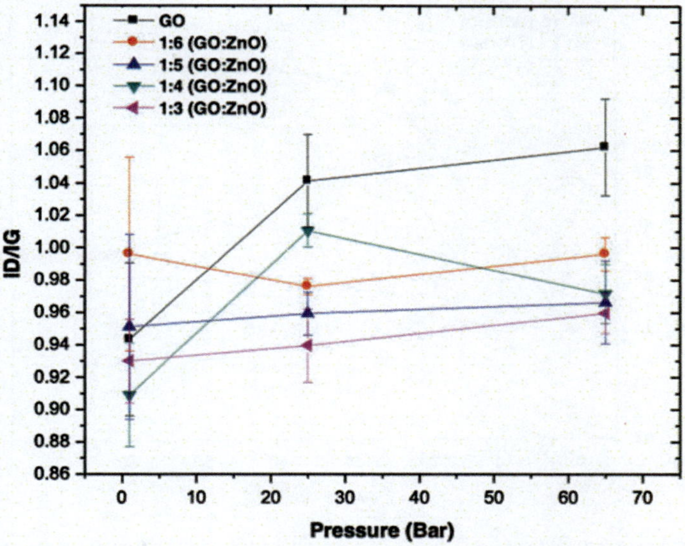

Fig. 4 Change of ID/IG of GO and different ratios of GO-ZnO with carbon dioxide gas pressure increment

disturbed with the presence of high-pressure CO_2 gas conditions. GO showed the highest defectiveness at 65 bars where the value of ID/IG increased up to 12% from the control pressure. In contrast, the attachment of ZnO onto GO sheets is proven to minimize the defect caused by the high-pressure CO_2 gas conditions. The defect ratio variation throughout pressure were below 7% for all GO:ZnO ratios.

The defect density of the nanosheets can be calculated with the equation obtained from Sharma et al. [28]:

$$n_D = \frac{2.4 \times 10^{22}}{\lambda_L^4} \times \frac{I_D}{I_G} \tag{1}$$

where λ_L is the wavelength of the Raman laser source and ID/IG is the intensity ratio of the D and G bands. The calculated values are plotted against the gas pressure and is shown in Fig. 5. It can be seen from the figure that ratios of GO to ZnO, ratio 1:5 showed the best stability in high pressure among the other ratios studied. Unlike the other GO:ZnO ratios, there was only a small change in the defect density over pressure increment. Hence, ratio of 1:5 for GO:ZnO ratio was selected for the preparations of samples for the rapid decompression tests.

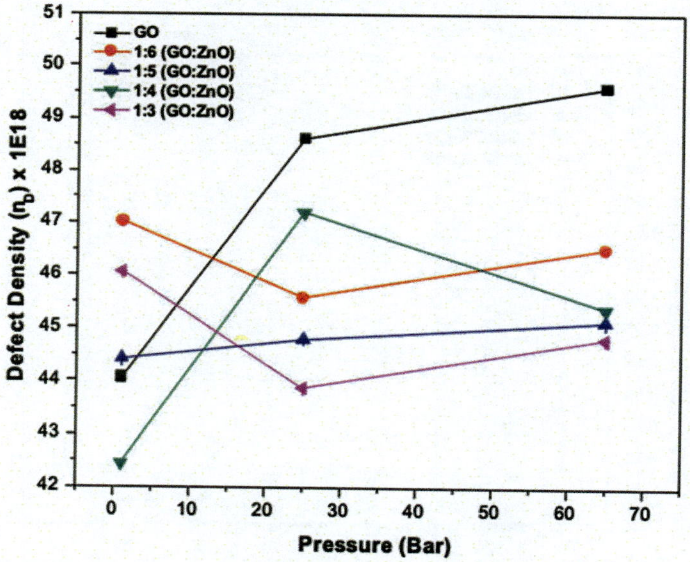

Fig. 5 Defect ratio of GO and different ratios of GO-ZnO with carbon dioxide gas pressure increment

3.2 Mechanical Strength of Bare Epoxy, GO/Epoxy, GO-ZnO/Epoxy Under Rapid Decompression of Carbon Dioxide Gas Conditions

Rapid decompression measurements were undergone at pressure 1, 25 and 65 bar of CO_2 gas. The results are displayed in Fig. 6.

Based on Fig. 6a and b, it can be seen that the sudden pressure drop affected the mechanical properties of GO-ZnO/Epoxy and GO/Epoxy except for concentration of 0.1 wt% of nanofillers. The long period of CO_2 gas exposure would result in the sorption of gas molecules within the polymer chains due to the high solubility of CO_2 gas. As the pressure was rapidly reduced, the trapped gas molecules were forced to exit the nanocomposite material, resulting in the decrease in the tensile strength of the material. Although the rapid pressure drop resulted in a more brittle nanocomposite due to the icing effect of CO_2, there was only a slight decrease in tensile strength after 65 bar CO_2 gas pressure drop for GO-ZnO/Epoxy and GO/Epoxy containing 0.1 wt% of nanofillers (~6% and ~8% respectively). However, GO-ZnO/Epoxy still overpowered GO/Epoxy by having a larger tensile strength after rapid decompression conditions, as shown in Fig. 7c. The robustness of GO-ZnO/EP is undeniably excellent as the tensile strength of the coating only reduced by 24% (Bare Epoxy: 50%; GO/EP: 55%) under 65 bars of CO_2 gas of depressurization conditions. This was due to the excellent epoxy chain anchor on the ZnO NPs surfaces. In the case of GO/EP, the very small CO_2 molecules are forced to enter the coating when the system pressure is high, causing the air-filled gaps between the poor GO/EP interface being filled with CO_2 gas. This is opposite for GO-ZnO/EP, since GO-ZnO/EP have very strong matrix/filler adhesion due to the presence of ZnO NPs in the GO sheets. The ZnO NPs can readily form a dense layer of tightly bound polymer chains on the ZnO NPs owing to its stable surface [163, 164]. On the other hand, bare epoxy shows the worse resistance against rapid decompression conditions with 49% drop of tensile strength at 65 bar CO_2 depressurization. This is due to the increase in brittleness especially at pressure conditions above 25 bar.

In terms of filler concentration, the tensile strength for both GO/EP and GO-ZnO/EP having a filler content of 0.1 wt%. portrayed the best value. After this concentration level, the addition of more GO and GO-ZnO sheets resulted in a decrease in the tensile strength. The addition of 0.1 wt% of fillers improved the tensile strength of bare epoxy by 52% and 30% for GO-ZnO/EP and GO/EP respectively. A further content increase (0.2 wt% to 0.4 wt%) resulted in poor tensile strength of the composite coating. This behaviour has been observed for other composites in other studies [138, 249] and is associated with the poor dispersion of the fillers. The relationship between tensile strength and concentration of filler is not a linear function [250]. After a certain critical concentration, a further rise in filler content would result in a decrease in tensile strength. Zhao and coworkers considered this as a 'mechanical percolation', where graphene can be homogeneously exfoliated in the polymer matrix and exhibits effective reinforcement when the lower than this

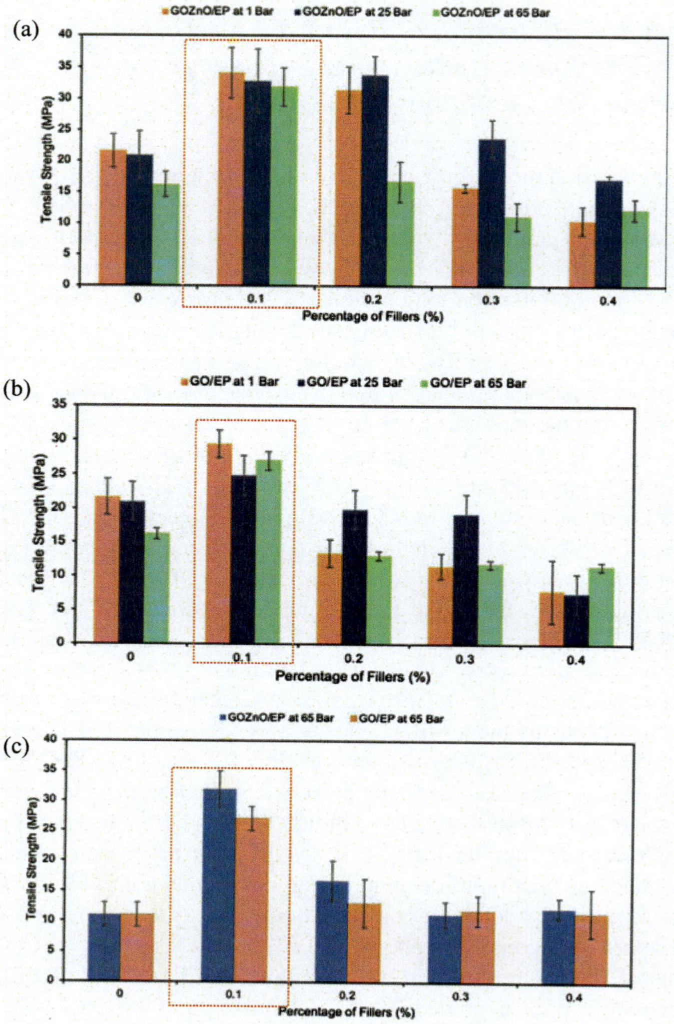

Fig. 6 **a**, **b** are the compilation of tensile strengths after exposed to rapid decompression pressures of 1, 25 and 65 bars of carbon dioxide gas for GO-ZnO/EP and GO/EP respectively. **c** Compares the tensile strength between GO/EP and GO-ZnO/EP after being exposed to rapid gas decompression at 65 bars of CO_2 gas. The best nanocomposite filler percentage that conveys the highest tensile strength is marked by the red-lined box

content [251]. Further loading would result in the stacking of graphene sheets due to the strong van der Waals forces between them.

Mechanical percolation can be explained by the matrix/filler theory [252]. When a relatively small volume of fillers is present, the fillers would influence the surrounding matrix by acting as reinforcements. Nevertheless, there is enough matrix to support

Fig. 7 FESEM images of the fracture surface **a** 0.4 wt% and **b** 0.1 wt% loading content of GO-ZnO/EP composite

the composite as they are not subjected to interactions with the filler surface. This makes the tensile strength to be close to bare epoxy. As the concentration of fillers become close to the mechanical percolation, the mobility of the epoxy chains in the composite will be affected. A fracture would lead to various crack pinning and crack deflection from the embedding of fillers that restrict the movement of cracks. As more fillers being incorporated, the entire matrix would become influenced by the filler, limiting the reinforcement form of epoxy chains. More epoxy chains would participate in bonding with the fillers, leaving little space for mechanical reinforcements from the epoxy chains. Although the presence of the filler limits to the matrix plastic flow during the material fracture process, the high number of fillers could also contribute to internal defects that could initiate fracture [250]. This would lead to the reduction in tensile strength of the composite coating. This theory is further justified by the FE-SEM images depicted in Fig. 8. It can be observed in Fig. 7a that at high filler loading (0.4 wt%), the GO-ZnO hybrids occupied most of the parts in the epoxy matrix due to agglomeration. Therefore, the fracture surface displayed a brittle morphology which is associated with the low fracture resistance. There was no indication of sheet debonding as the fillers themselves fractured along with

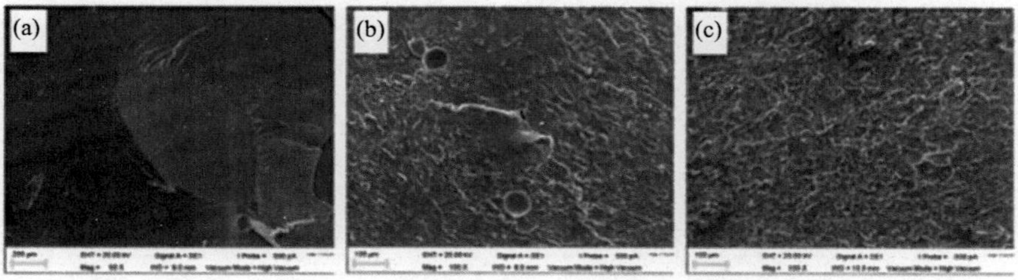

Fig. 8 SEM images of cross section of **a** bare epoxy, **b** GO/epoxy (0.1 wt%) and **c** GO-ZnO/epoxy (0.1 wt%) after being subjected to rapid decompression of 65 bar of carbon dioxide gas

the undeformed epoxy matrix. In contrast, dimples were present in 0.1 wt% GO-ZnO/EP (Fig. 7b) which shows that the fracture path deviated due to the presence of reinforcements from the GO-ZnO hybrids [250].

The results from the rapid gas decompression tests are further emphasised in the SEM images in Fig. 8. Figure 8a depicts the presence of brittle fracture at the cross section of Bare Epoxy after the test. Brittle fracture is conveyed by the smooth surface of fracture with feather markings on the studied SEM image. Meanwhile, most of the cross-section area in Fig. 8b and c consist of ductile fracture, consisting of a relatively rougher fracture surface having multiplane features. These observations are consistent with the obtained tensile strength of the nanocomposite, where the fillers act as reinforcing points that divert the cracks when fracture is occurring.

Figure 9 depicts the TEM micrographs of epoxy matrix containing 0.1 wt% of GO-ZnO and GO respectively. As shown in Fig. 9c, the pristine GO sheets cluster with each other and some of them debond from the epoxy matrix indicated by the red arrows. When the cluster is magnified, as shown in Fig. 9d, voids can be clearly seen as a result of the high stress during the preparation of TEM specimen [29]. The aggregates of GO sheets delaminate or slip by each other due to weak sheet/sheet and sheet/matrix adhesion. In the case of GO-ZnO/Epoxy, GO-ZnO seems to disperse and exfoliate homogeneously in epoxy matrix, although some small-scale aggregation still exists (Fig. 9a). Higher magnification in Fig. 9b shows that GO-ZnO and epoxy matrix have good interfacial quality. Although the sheets are wrinkled in

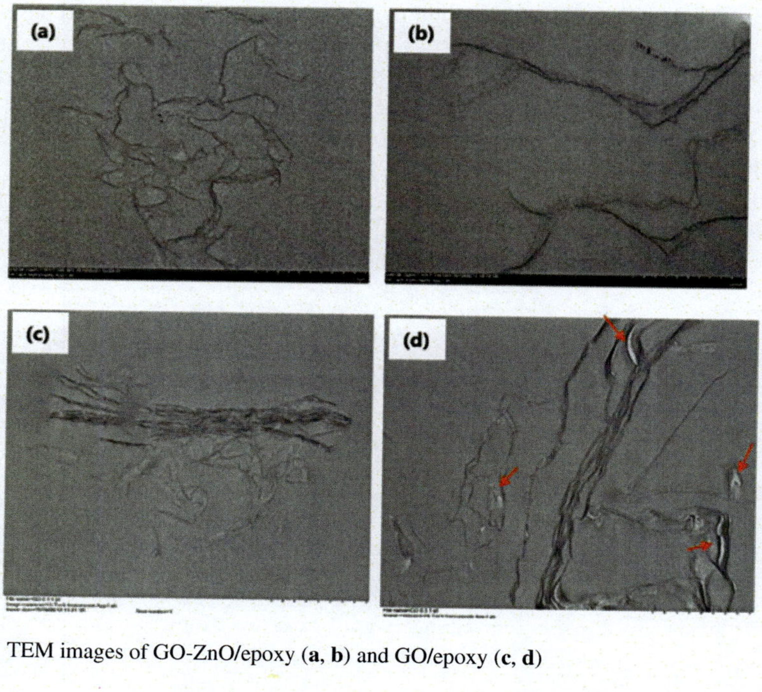

Fig. 9 TEM images of GO-ZnO/epoxy (**a**, **b**) and GO/epoxy (**c**, **d**)

both systems, the improvement of barrier protection against CO_2 permeation were contributed by the tortuous pathway that they create. GO-ZnO/Epoxy coating have small scale of aggregation, which further creates a convoluted system for gas to enter.

4 Conclusions

Incorporating GO-ZnO in epoxy has proven to effectively prevent the damage caused by high pressure carbon dioxide gas. The structural stability GO-ZnO is better as the defectiveness shown by ID/IG is less than 7% compared to 12% for GO sheets with respect to the ID/IG ratio at the control pressure. Based on the defective ratios, it can be concluded that GO:ZnO ratio of 1:5 showed the lowest change over pressure increment among the other ratios tested. Under 65 bar of CO_2 gas rapid decompression conditions, the tensile strength drop from the control pressure (1 bar) for Bare Epoxy, 0.1 wt% GO-ZnO/Epoxy and 0.1 wt% GO/Epoxy are 50%, 24%, 55% respectively. These results indicate that GO-ZnO/Epoxy at with 0.1 wt% loading can sustain in a high-pressure CO_2 condition.

Acknowledgements The authors would like to express profound gratitude to Yayasan Universiti Teknologi PETRONAS (YUTP-FRG) cost center: 0153AA-E67 and Centre for Corrosion Research, Universiti Teknologi PETRONAS, Malaysia for providing financial support and services throughout this research.

References

1. Khanna AS (2016) Corrosion protection of underground gas/crude pipelines. In: Khanna AS (ed) Corrosion and materials in the oil and gas industries, pp 185–196
2. ur Rahman O, Kashif M, Ahmad S (2015) Nanoferrite dispersed waterborne epoxy-acrylate: anticorrosive nanocomposite coatings. Prog Org Coatings 80:77–86. https://doi.org/10.1016/j.porgcoat.2014.11.023
3. Presuel-Moreno F, Jakab MA, Tailleart N, Goldman M, Scully JR (2008) Corrosion-resistant metallic coatings. Mater Today 11(10):14–23. https://doi.org/10.1016/S1369-7021(08)70203-7
4. Ding S et al (2017) Fabrication of self-cleaning super-hydrophobic nickel/graphene hybrid film with improved corrosion resistance on mild steel. Mater Des 117:280–288. https://doi.org/10.1016/j.matdes.2016.12.084
5. Sørensen PA, Kiil S (2009) Anticorrosive coatings: a review. J Coatings Technol Res 6(2):135–176. https://doi.org/10.1007/s11998-008-9144-2
6. Yamada T, Ishihara M, Hasegawa M (2013) Large area coating of graphene at low temperature using a roll-to-roll microwave plasma chemical vapor deposition. Thin Solid Films 532:89–93. https://doi.org/10.1016/j.tsf.2012.12.102
7. Batzill M (2012) The surface science of graphene: metal interfaces, CVD synthesis, nanoribbons, chemical modifications, and defects. Surf Sci Rep 67(3–4):83–115. https://doi.org/10.1016/j.surfrep.2011.12.001

8. Ramezanzadeh B, Ghasemi E, Mahdavian M, Changizi E, Mohamadzadeh Moghadam MH (2015) Covalently-grafted graphene oxide nanosheets to improve barrier and corrosion protection properties of polyurethane coatings. Carbon NY 93:555–573. https://doi.org/10.1016/j.carbon.2015.05.094
9. Li J, Cui J, Yang J, Li Y, Qiu H, Yang J (2016) Reinforcement of graphene and its derivatives on the anticorrosive properties of waterborne polyurethane coatings. Compos Sci Technol 129:30–37. https://doi.org/10.1016/j.compscitech.2016.04.017
10. Sheng X, Cai W, Zhong L, Xie D, Zhang X (2016) Synthesis of functionalized graphene/polyaniline nanocomposites with effective synergistic reinforcement on anticorrosion. Ind Eng Chem Res 55(31):8576–8585. https://doi.org/10.1021/acs.iecr.6b01975
11. Liu J et al (2018) Silane modification of titanium dioxide-decorated graphene oxide nanocomposite for enhancing anticorrosion performance of epoxy coatings on AA-2024. J Alloys Compd 744:728–739. https://doi.org/10.1016/j.jallcom.2018.01.267
12. Chen L et al (2012) Enhanced epoxy/silica composites mechanical properties by introducing graphene oxide to the interface. ACS Appl Mater Interf 4(8):4398–4404. https://doi.org/10.1021/am3010576
13. Yu Z et al (2015) Fabrication of graphene oxide-alumina hybrids to reinforce the anti-corrosion performance of composite epoxy coatings. Appl Surf Sci 351:986–996. https://doi.org/10.1016/j.apsusc.2015.06.026
14. Vermisoglou EC et al (2014) Reduced graphene oxide/iron carbide nanocomposites for magnetic and supercapacitor applications. J Alloys Compd 590:102–109. https://doi.org/10.1016/j.jallcom.2013.11.087
15. Dhoke SK, Rajgopalan N, Khanna AS, Used Atlas AM, Hyv M (2013) Effect of nano-zinc oxide particles on the performance behavior of waterborne polyurethane composite coatings. Int J Mater Sci 53:47–55. https://doi.org/10.1017/CBO9781107415324.004
16. Mason JF (2000) Effect of rapid decompression conditions on liner materials. Corros 2000, no 00782
17. Grandidier J, Cange L (2011) A damage model for the blistering of polyvinylidene fluoride subjected to carbon dioxide decompression. J Mech Phys Solids 59:1909–1926. https://doi.org/10.1016/j.jmps.2011.04.010
18. Othman NH, Zaireen W, Yahya N, Ismail MC, Koi ZK (2019) Highly dispersed graphene oxide–zinc oxide nanohybrids in epoxy coating with improved water barrier properties and corrosion resistance. J Coatings Technol Res 10(4). https://doi.org/10.1007/s11998-019-00245-y
19. Asghar A, Naderi R, Ramezanzadeh B (2019) Epoxy-polyamide nanocomposite coating with graphene oxide as cerium nanocontainer generating effective dual active/barrier corrosion protection. Compos Part B 172(January):363–375. https://doi.org/10.1016/j.compositesb.2019.05.055
20. Bondarenko EA, Mazanik AV, Streltsov EA, Kulak AI, Korolik OV (2015) SnO_2/reduced graphene oxide composite films for electrochemical applications. Mater Sci Eng B 202:61–67. https://doi.org/10.1016/j.mseb.2015.10.002
21. Li Z et al (2013) Control of the functionality of graphene oxide for its application in epoxy nanocomposites. Polym (United Kingdom) 54(23):6437–6446. https://doi.org/10.1016/j.polymer.2013.09.054
22. Young RJ, Kinloch IA, Gong L, Novoselov KS (2012) The mechanics of graphene nanocomposites: a review. Compos Sci Technol 72(12):1459–1476. https://doi.org/10.1016/j.compscitech.2012.05.005
23. Sadezky A, Muckenhuber H, Grothe H, Niessner R (2005) Raman microspectroscopy of soot and related carbonaceous materials: spectral analysis and structural information. Carbon NY 43:1731–1742. https://doi.org/10.1016/j.carbon.2005.02.018
24. Huda N, Bakar A, Ali GAM, Ismail J, Algarni H, Chong KF (2019) Size-dependent corrosion behavior of graphene oxide coating. Prog Org Coatings 134(February):272–280. https://doi.org/10.1016/j.porgcoat.2019.05.011

25. Yoon SM et al (2012) Synthesis of multilayer graphene balls by carbon segregation from nickel nanoparticles. ACS Nano 6(8):6803–6811. https://doi.org/10.1021/nn301546z
26. Liu D, Zhao W, Liu S, Cen Q, Xue Q (2016) Comparative tribological and corrosion resistance properties of epoxy composite coatings reinforced with functionalized fullerene C60 and graphene. Surf Coatings Technol 286:354–364. https://doi.org/10.1016/j.surfcoat.2015.12.056
27. Lin X et al (2012) Fabrication of highly-aligned, conductive, and strong graphene papers using ultralarge graphene oxide sheets. ACS Nano 6(12):10708–10719. https://doi.org/10.1021/nn303904z
28. Sharma R, Chadha N, Saini P (2018) Determination of defect density , crystallite size and number of graphene layers in graphene analogues using X-ray diffraction and Raman spectroscopy. Indian J Pure Appl Phys 55:625–29
29. Guan L et al (2014) Toward effective and tunable interphases in graphene oxide/epoxy composites by grafting different chain lengths of polyetheramine onto graphene oxide. J Mater Chem A Mater Energy Sustain 2(36):15058–15069. https://doi.org/10.1039/C4TA02429J

Numerical Failure Analysis of Corroded Pipeline Elbow with Longitudinally Aligned Interacting Corrosion Defects

Thibankumar Arumugam, Saravanan Karuppanan, and Mark Ovinis

Abstract This paper describes the application of finite element method (FEM) to predict the failure pressure of pipeline elbow with longitudinally aligned interacting corrosion defects subjected to internal pressure. The FEM incorporated material nonlinearity and utilized the true UTS to determine the point of failure. Finite element analysis (FEA) results were used to determine the failure pressure as well as the presence of interaction. FEA revealed that interaction occurs and is most significant when medium depth corrosion defect is at Location 1 (i.e. 45° degree at the pipe bend). DNV-RP-F101's guideline for interaction limit was not applicable for medium depth corrosion defect at Location 1, while is applicable for deep corrosion defect at Location 1.

1 Introduction

1.1 Overview of Corrosion

Pipeline is the primary means of transporting fluid especially in the oil and gas industry. Pipelines made of carbon steel are commonly used due to their excellent material properties and relatively low cost [1, 2]. However, one of the most common issue pertaining steel pipelines is corrosion. Corrosion is defined as the degradation of a material when it reacts with the environment through the means of electrochemical reactions. One of the most common classification of corrosion is uniform general corrosion. Uniform general corrosion occurs due to dissolution of metallic component into its respective metallic ions or due to chemical attacks [3]. For pipeline with bends, one of the most common corrosion is erosion corrosion. Erosion corrosion occurs when mechanical erosion force impinging the surface accelerates the corrosion rate. Several studies have revealed that pipe elbows are critical pipe components

T. Arumugam (✉) · S. Karuppanan · M. Ovinis
Mechanical Engineering Department, Universiti Teknologi PETRONAS, 32610 Bandar Seri Iskandar, Perak Darul Ridzuan, Malaysia
e-mail: thibankumar_g03585@utp.edu.my

F. Ahmad et al. (eds.), *ICPER 2020*, Lecture Notes in Mechanical Engineering,
https://doi.org/10.1007/978-981-19-1939-8_67

that often result in high reduction in structural integrity especially when subjected to corrosion [4–6].

1.2 Parameters of Corrosion

The severity of a corrosion defect is often determined based on the parameters of the defect. For single corrosion defect, the major parameters that are often considered for failure pressure predictions are depth, d, length, l and width, w of defect. A burst test study conducted by Netto et al. [7] proved that the most detrimental parameter is the defect depth. The burst test revealed a 18% failure pressure drop when the corrosion defect depth was increased by 0.20 t (pipe wall thickness, t). In comparison, only a 14% failure drop was observed when the defect length was doubled from 0.5 D to D (nominal pipe outer diameter, D). This result was also supported by Ahammed and Melchers [8] who stated that the increase in corrosion depth decreases the reliability of the pipe due to reduced wall thickness to resist the hoop stress development. In cases involving interacting corrosion defects, corrosion defect spacing must be taken into consideration. This is because the stress-strain disturbances of adjacent interacting corrosion defects coalesce resulting in high probability of lower failure pressure in comparison to single and isolated corrosion defects [2, 9–12]. With increasing corrosion defect spacing, stress-strain coalescence decreases until eventually the interaction vanished and the corrosion defect is then treated as single and isolated corrosion defect [9].

1.3 General Overview of Corrosion Assessment Method

Among the various corrosion assessment methods, DNV-RP-F101 [13] (DNV) is one of the most comprehensive and accurate corrosion assessment method. DNV was developed by considering results of multiple full-scale burst test results and finite element analysis (FEA). DNV has been proven to be the least conservative and gives the closest failure pressure prediction in predicting the failure pressure for single corrosion defect [11, 14, 15] and interacting corrosion defects [15]. DNV also provides interaction limit guideline for interacting corrosion defects in longitudinal, $(s_l)_{lim}$ and circumferential, $(s_c)_{lim}$ direction, as presented by Eqs. (1) and (2), respectively. However, DNV was developed for straight pipes and not for pipe bends. Hence, the application of DNV on pipe bends would result in inaccurate failure pressure predictions as pipe bends generally undergo higher stress due to the Bourdon effect [16]. The Bourdon effect occurs as the applied internal pressure tends to straighten the pipe due to the difference between the intrados and extrados surfaces of the pipe bend.

$$(s_l)_{lim} = 2.0\sqrt{Dt} \tag{1}$$

$$(s_c)_{lim} = \pi\sqrt{Dt} \tag{2}$$

1.4 Application of Finite Element Methods for Pipeline Corrosion Assessment

Finite element method (FEM) has become a popular method in determining the failure behaviour as well as predicting the failure pressure of pipelines with corrosion defects. Study conducted by Lee et al. [4] successfully utilized both fluid and structural FEM to determine the failure behaviour of pipe elbow with corrosion defect. Hence, in this study, a detailed structural FEM to simulate and predict the failure pressure of pipe elbow with interacting corrosion defects subjected to internal pressure is presented. The developed FEM was verified against full-scale burst test results obtained from Benjamin et al. [17]. Based on the developed FEM, parametric study was conducted to determine the failure behaviour of pipe elbows with longitudinally aligned interacting corrosion defects subjected to internal pressure.

2 Methodology

2.1 Overview of Corrosion Geometric Values

Parametric study was conducted based on the geometric parameters as presented in Table 1. The design specifications of the pipe elbow were based on ASME/ANSI B16.9 [18]. Note that the corrosion defect parameters varied in this study were the depth, d and longitudinal spacing, s_l, and they are presented as normalized values. The normalized longitudinal spacings, $s_l/\sqrt{Dt}$ were varied till a value of 2 as per DNV's guideline for longitudinal interaction limit.

Table 1 Geometric parameters for FEA parametric study

Input parameters	Values
Outer diameter of pipe, D (mm)	100
Wall thickness, t (mm)	5
Pipe bend radius, R (mm)	130
Straight pipe length, L (mm)	300
Normalized defect length, l/D	0.5
Normalized defect width, w/D	0.5
Normalized defect depth, d/t	0.1, 0.2, 0.4, 0.6
Normalized longitudinal spacing, $s_l/\sqrt{Dt}$	0.5, 1.0, 2.0

2.2 Development of Finite Element Method

The interacting corrosion defects were located at two locations namely Location 1 and Location 2 as illustrated in Fig. 1. The corrosion at Location 1 was modelled at 45° of the pipe elbows as Lee et al. [4] determined that angles ranging from 40° to 50° had the highest amount of erosion. Location 2 was modelled based on the longitudinal spacing from the corrosion defect at Location 1. The corrosion defects were modelled at the exterior instead of interior of the pipe after preliminary analyses revealed no difference in the failure pressure prediction via finite element analysis (FEA).

The corrosion defects were assumed to be rectangular as this idealization has been successfully utilized by various researchers as it gives a reasonably accurate failure pressure prediction [7, 13]. Next, a smaller model was utilized in this study as it allows for reduced computational resources without compromising the results' accuracy. A symmetrical half pipe elbow as illustrated in Fig. 2 was designed to represent the actual pipe elbow geometry. In addition, the straight pipe lengths at both ends of the pipe elbow were designed with a length of 300 mm to prevent end points from influencing the corrosion defect regions during FEA.

ANSYS 16.1 Structural Product of Mechanical ANSYS Parametric Design Language (ANSYS) contains a large range of elements to be used for meshing corroded pipes. Due to the thickness of pressurised pipes, solid elements allow for more accurate element meshing in comparison to shell elements [19]. Hence, as illustrated in Fig. 3, tetrahedral SOLID186 elements were used to mesh the corrosion defects (region bounded in yellow) as well as the longitudinal spacing between the defects (region bounded in red). Meanwhile, hexahedral SOLID185 elements were used to mesh the rest of the pipe elbow (region bounded in green). Both SOLID185 and SOLID186 have the ability to accommodate high stress stiffening, plasticity, creep, welling, large deflection and large strain [20]. The difference between the elements is SOLID186 has 20 nodes that can tolerate irregular shapes such as the corrosion defect region.

It is recommended by the British Standard Institute [21] that at least 3 to 4 layers of elements to be used at the corrosion defect region to ensure accurate prediction

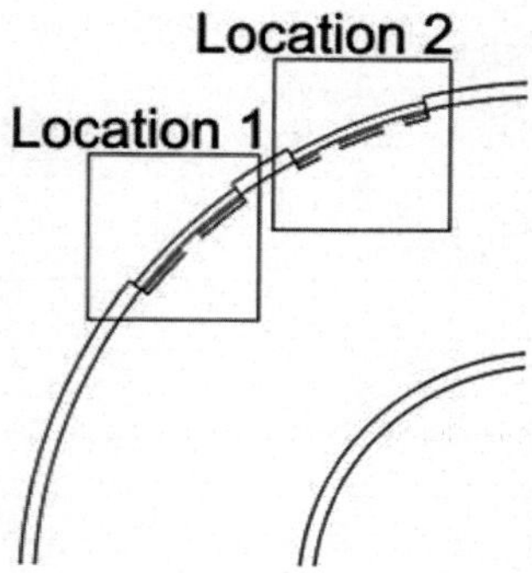

Fig. 1 Region bounded in squares highlight corrosion defects in Location 1 and Location 2

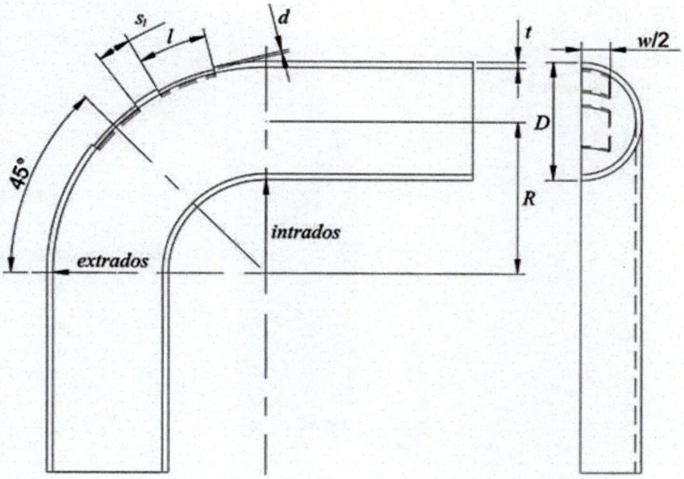

Fig. 2 Annotated half model of pipe elbow with longitudinally aligned interacting corrosion defects

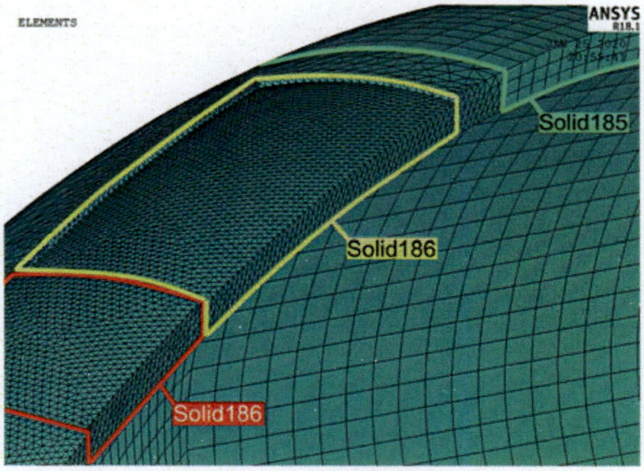

Fig. 3 Application of SOLID185 and SOLID186 in meshing the pipe elbow with interacting corrosion defect

using FEM. Thus, as seen in Fig. 3, a minimum of 3 layers of element were used at the corrosion defect region (region bounded in yellow) as well as longitudinal spacing between the defects (region bounded in red). The element lengths were set to be 1 mm for both regions.

With the usage of half pipe elbow models, appropriate boundary conditions were applied at the necessary areas of the model so that the half pipe elbow model would behave as a whole pipe elbow model during the FEA. Figure 4 illustrates the region in red where symmetrical boundary conditions were applied. The degrees of freedom (DOF) of the pipe were constrained to prevent unwanted rigid body motion. Referring to the blue markings in Fig. 4, the half model was constrained in the x, y and z

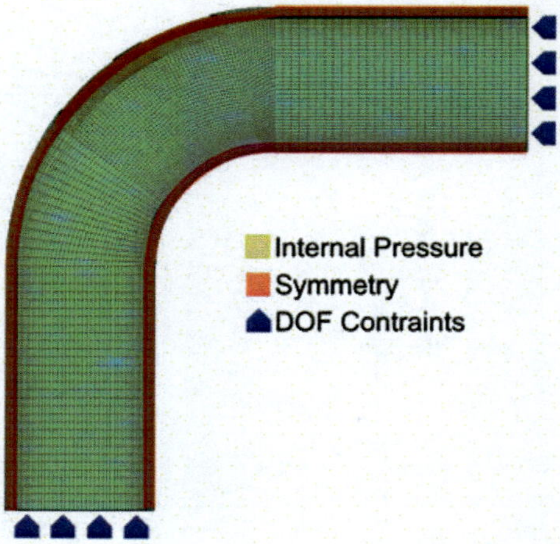

Fig. 4 Application of internal pressure, symmetrical boundary conditions, internal pressure, and DOF constraints for half pipe elbow model

directions. Internal pressure was applied on the inner surface of the half pipe model as depicted by the region in yellow.

2.3 Material Properties

A non-linear structural analysis was used for this FEM due to the non-linear stress–strain relationship for the material. The material properties of the pipe modelled in the FEM were based on the non-linear true stress-strain curve of API 5L X52 steel, as per Fig. 5. Note that the true UTS, σ_{UTS}^{*} is 612 MPa at a strain value of 0.1745. To consider high non-linearity during the FEA, large strains and displacements as

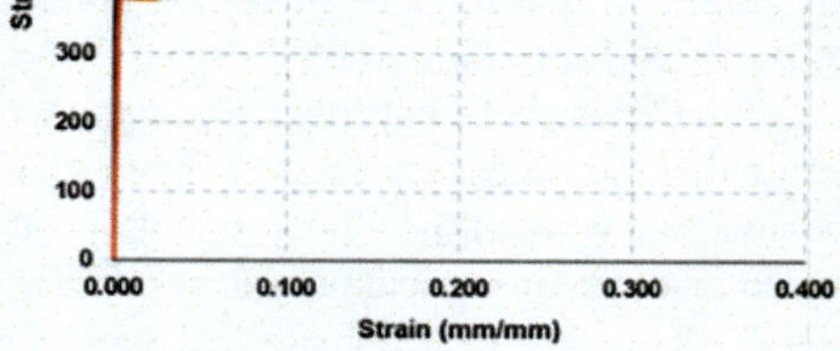

Fig. 5 True stress-strain curve for API 5L X52 steel [22]

well as material stress stiffening were considered. The FEM utilizes the full Newton-Raphson method, which involve iterative linear steps taken to evaluate the out-of-balance load vector to obtain a closer approximation to achieve exact solution for convergence. In addition, rate-independent plasticity model that uses von Mises yield criterion and isotropic hardening rule were employed. The modulus of elasticity and Poisson's ratio used were 200 GPa and 0.3, respectively.

2.4 Failure Criterion

In this study, the failure criterion for the corroded pipe elbow takes into consideration of the local stress at the defect area. Failure is determined when the local stress at the defect area exceeds the specified stress, as per DNV guidelines [13]. Based on this information, von Mises theory was applied in ANSYS in order to analyse the stress distribution of the corroded pipe. When the von Mises stress reaches the σ_{UTS}^{*} throughout the entire wall thickness, the corresponding applied internal pressure can be defined as the failure pressure. The prediction of failure pressure based on σ_{UTS}^{*} provides more accurate results and reasonable sensitivity on defect geometries compared to yield stress and flow strength because the failure mechanism is controlled by plastic collapse [23].

3 Methodology Verification

Simulation validation is important to ensure that the developed FEM produces accurate results. In general, FEM should provide good results (percentage error within 10%) when compared to actual real-life results. Experimental study conducted by Duan and Shen [24] utilized pipe elbow fabricated from 20 g steel material. Table 2 lists the dimensions of the handcraft polished corrosion defect at Location 1 (as per experiment). The failure criterion utilized by the experiment was twice-elastic-slope which was subsequently applied in this validation. The pipe was quarter modelled (due to simpler pipe geometry), and the loads as well as necessary boundary conditions were applied as per developed FEM. Table 2 also lists the failure pressure from the experiment, the FEA failure pressure predictions, and the absolute percentage

Table 2 Tabulation of corrosion defect failures obtained using developed FEM and full-scale burst test conducted by Duan et al. [24]

Specimen	l (mm)	d (mm)	w (mm)	Experiment failure pressure (MPa)	FEA failure pressure (MPa)	Absolute percentage difference (%)
1	40.06	1.25	42.41	28.58	31.50	10.22
2	80.11	6.00	84.82	17.67	18.50	4.70

difference between both failure pressure values. For Specimen 1, although the absolute percentage difference is relatively high at 10.22%, the FEA managed to accurately determine the location of failure which was at the intrados. It is also possible that the relatively high percentage difference is due to inaccuracies in the experimental study. For Specimen 2, the absolute percentage difference is relatively small at 4.70%.

4 Results and Discussion

4.1 Preliminary Analyses

Before analysing the failure behaviour trends of pipe elbows with interacting corrosion defects, it is important to understand the adverse effects of corrosion defects on pipes. In general, when corrosion is present, it means that there is certain degree of thinning of the pipe wall thickness. In other words, as the corrosion defect depth increases, the thickness of pipe wall decreases. The hoop stress distribution is adversely affected due to the decrease in pipe wall thickness. Figures 6 and 7 illustrate the changes in the hoop stress distribution due to the introduction of corrosion defects. The corrosion defects prevent the hoop stress from being evenly distributed throughout the pipe and causing them to be concentrated at the corrosion defect region. Therefore, due to the continuous increases of hoop stress, the pipe eventually fails at the corrosion defect region [8, 25, 26]. In Fig. 7, it is noticeable that the failure

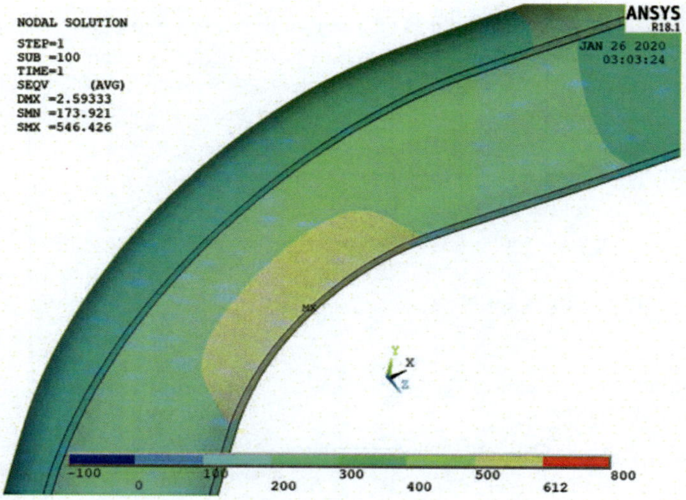

Fig. 6 Stress contour plot of an intact pipe elbow when subject to an internal pressure of 47.5 MPa

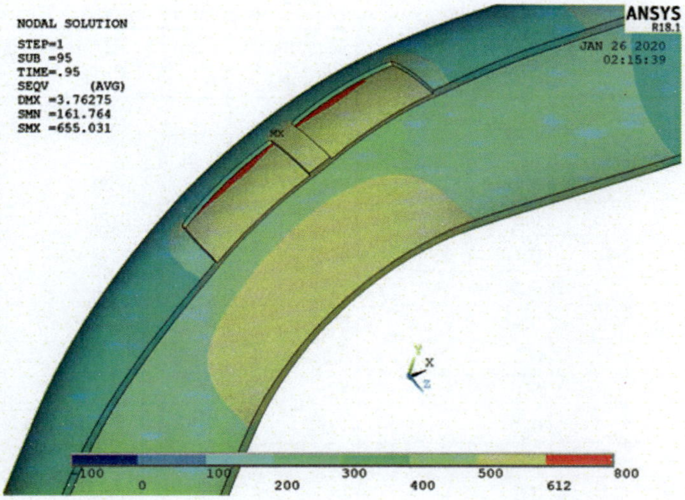

Fig. 7 Stress contour plot of pipe elbow with longitudinal aligned corrosion defects subjected to internal pressure of 47.5 MPa

of pipe elbow with interacting corrosion defect occurs at the defect region (red stress contour) when the σ_{UTS}^{*} has been exceeded. The corrosion defect region also undergoes bulging, a phenomenon that has been previously observed by past researchers investigating pipe with single corrosion defect subjected to internal pressure [25, 27].

As a generally note, shallow corrosion defect is defined as corrosion defect with $d/t \leq 0.2$, while medium depth corrosion defect is defined as corrosion defect with $0.2 < d/t < 0.6$, and deep corrosion defect is defined as corrosion defect with $0.6 \leq d/t \leq 0.8$. It is necessary to determine the failure pressure, P_f of single corrosion defect so that interaction can be determined when it occurs. As mentioned in the Introduction (Sect. 1.2), interaction results in a lower failure pressure compared to that of a single isolated corrosion. Table 3 lists the dimension of the single corrosion defects as well as the respective failure pressure in their normalized form. The failure pressures were normalized by the failure pressure of intact pipe elbow. Intact pipe elbow generally fails at intrados as it is the region that undergoes the highest amount of deformation and hoops stress when subjected to internal pressure [16]. For this pipe specimen, the failure pressure of intact pipe elbow, P_i is 50.05 MPa. From Table 3, for this study, it is clear that the failure pressure for single corrosion defect is the same regardless of the location at the pipe elbow with the exception for deep corrosion defect (d/t =0.6) at Location 1 of the pipe elbow.

Table 4 details the results of the parametric study of pipe elbows with longitudinal interacting corrosion defects subjected to internal pressure. These results are explained in graph forms in subsequent sections.

Table 3 Failure pressure of pipe elbow with single corrosion defect subjected to internal pressure

Location of corrosion defect	Corrosion defect depth, d/t	Longitudinal defect spacing, $s_l/\sqrt{Dt}$	FEA failure pressure, P_f/P_i
1	0.1	–	1
1	0.2	–	1
1	0.4	–	0.99
1	0.6	–	0.79
2	0.1	0.5	1
2	0.2	0.5	1
2	0.4	0.5	0.99
2	0.6	0.5	0.80
2	0.1	1.0	1
2	0.2	1.0	1
2	0.4	1.0	0.99
2	0.6	1.0	0.80
2	0.1	2.0	1
2	0.2	2.0	1
2	0.4	2.0	0.99
2	0.6	2.0	0.80

4.2 *Shallow Corrosion Defect at Location 1*

When a shallow corrosion defect at Location 1 is adjacent to a shallow corrosion defect at Location 2, corrosion interaction does not occur regardless of the longitudinal spacing. This is illustrated by the graph in Fig. 8. However, by referring to Fig. 9, when the shallow corrosion defect at Location 1 is adjacent to medium depth corrosion defect at Location 2, interaction occurs even up to $s_l/\sqrt{Dt} = 2$, which is not in line with DNV's guideline for interaction limit. This is expected since DNV's interaction limit guideline was recommended for straight pipe instead of pipe with bends. Note that if interaction is not present, the normalized failure pressure would be as indicated by the horizontal line across the graph. Next, when the shallow corrosion at Location 1 is adjacent to a deep corrosion defect at Location 2, interaction occurs but only when $s_l/\sqrt{Dt} < 1$. This interaction behaviour is illustrated in Fig. 10. Notice that interaction does not occur for shallow ($d/t = 0.1$) corrosion at Location 1 adjacent to deep corrosion defect at Location 2. This can be attributed to the weak stress-strain disturbance by corrosion defect with $d/t = 0.1$, as such not being much of an influence on the neighbouring corrosion defect.

Table 4 Failure pressure of pipe elbows with longitudinal aligned interacting corrosion defects subjected to internal pressure

Corrosion defect depth, d/t (Location 1)	Corrosion defect depth, d/t (Location 2)	Longitudinal spacing,$s_l/\sqrt{Dt}$	FEA failure pressure prediction, P_f/P_i
0.1	0.1	0.5	1.00
0.1	0.2	0.5	1.00
0.1	0.4	0.5	0.98
0.1	0.6	0.5	0.80
0.2	0.1	0.5	1.00
0.2	0.2	0.5	1.00
0.2	0.4	0.5	0.98
0.2	0.6	0.5	0.79
0.4	0.1	0.5	0.98
0.4	0.2	0.5	0.98
0.4	0.4	0.5	0.95
0.4	0.6	0.5	0.78
0.6	0.1	0.5	0.79
0.6	0.2	0.5	0.79
0.6	0.4	0.5	0.78
0.6	0.6	0.5	0.75
0.1	0.1	1.0	1.00
0.1	0.2	1.0	1.00
0.1	0.4	1.0	0.98
0.1	0.6	1.0	0.80
0.2	0.1	1.0	1.00
0.2	0.2	1.0	1.00
0.2	0.4	1.0	0.98
0.2	0.6	1.0	0.80
0.4	0.1	1.0	0.98
0.4	0.2	1.0	0.98
0.4	0.4	1.0	0.97
0.4	0.6	1.0	0.79
0.6	0.1	1.0	0.79
0.6	0.2	1.0	0.79
0.6	0.4	1.0	0.79
0.6	0.6	1.0	0.78
0.1	0.1	2.0	1.00
0.1	0.2	2.0	1.00

(continued)

Table 4 (continued)

Corrosion defect depth, d/t (Location 1)	Corrosion defect depth, d/t (Location 2)	Longitudinal spacing, $s_l/\sqrt{Dt}$	FEA failure pressure prediction, P_f/P_i
0.1	0.4	2.0	0.98
0.1	0.6	2.0	0.80
0.2	0.1	2.0	1.00
0.2	0.2	2.0	1.00
0.2	0.4	2.0	0.98
0.2	0.6	2.0	0.80
0.4	0.1	2.0	0.98
0.4	0.2	2.0	0.98
0.4	0.4	2.0	0.98
0.4	0.6	2.0	0.79
0.6	0.1	2.0	0.79
0.6	0.2	2.0	0.79
0.6	0.4	2.0	0.79
0.6	0.6	2.0	0.79

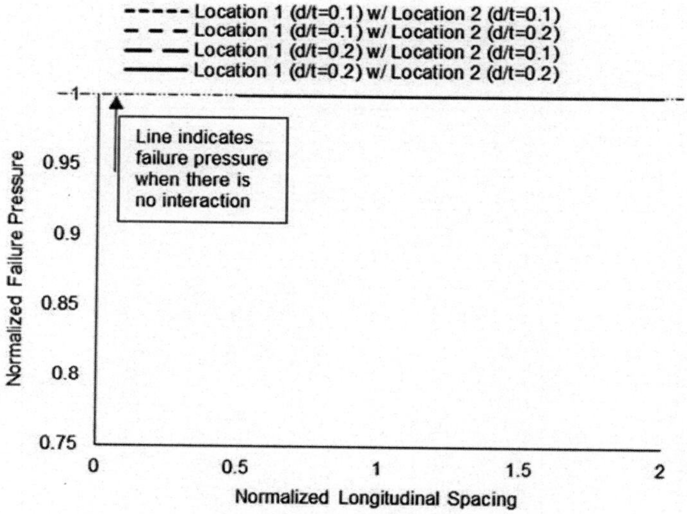

Fig. 8 Failure trend line of shallow corrosion defect at Location 1 adjacent to shallow corrosion defect at Location 2

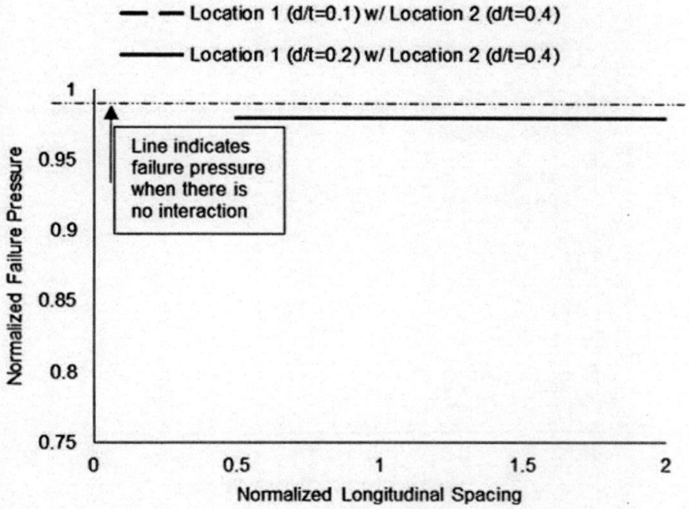

Fig. 9 Failure trend line of shallow corrosion defect at Location 1 adjacent to medium depth corrosion defect at Location 2

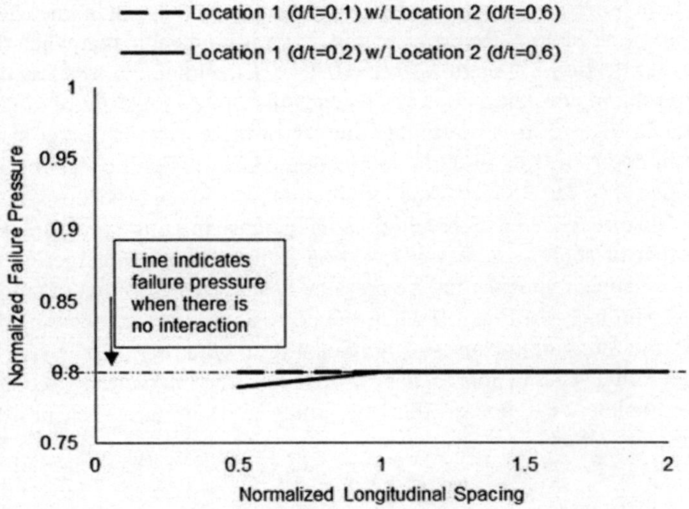

Fig. 10 Failure trend line of shallow corrosion defect at Location 1 adjacent to deep corrosion defect at Location 2

4.3 Medium Corrosion Defect at Location 1

When a medium depth corrosion defect at Location 1 is adjacent to corrosion defect of any depth at Location 2 (i.e. regardless if it is shallow, medium, deep corrosion

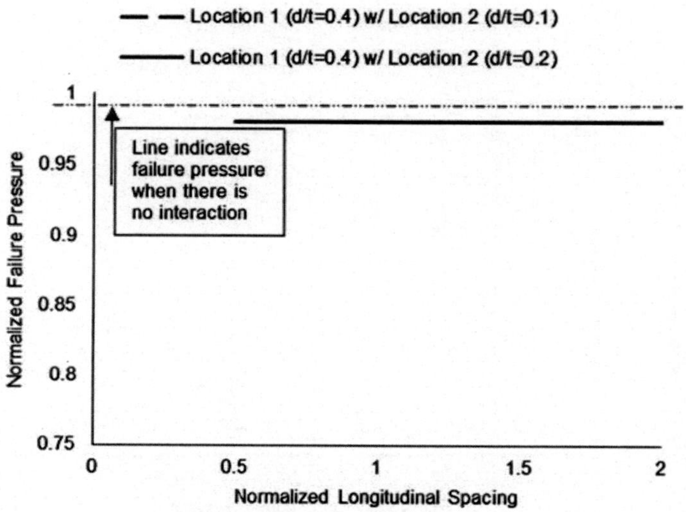

Fig. 11 Failure trend line of medium depth corrosion defect at Location 1 adjacent to shallow corrosion defect at Location 2

defect), interaction occurs even beyond $s_l/\sqrt{Dt} = 2$, which is not in line with DNV's interaction limit guideline. Referring to Fig. 11, it is noticeable that when the corrosion defect at Location 2 is shallow ($d/t = 0.1, 0.2$), the interaction causes the failure pressure to remain constant and never reaches the failure pressure of single corrosion defect. In Fig. 12, it is noticeable that when the corrosion defect at Location 2 is medium depth ($d/t = 0.4$), the interaction causes a significant drop in failure pressure when $s_l/\sqrt{Dt} = 0.5$. As the longitudinal spacing approaches $s_l/\sqrt{Dt} = 2$, the interaction effect decreases resulting in an increase in failure pressure. However, similar to that of shallow defect at Location 2, the effect of interaction does not completely disappear, thus the failure pressure never reaches to that of single corrosion defect. Similarly, in Fig. 13, when the corrosion defect at Location 1 is deep (d/t = 0.6), the interaction causes a significant drop when $s_l/\sqrt{Dt} = 0.5$, however as the longitudinal spacing approaches $s_l/\sqrt{Dt} = 2$, the interaction effect decreases but never completely disappears, thus the failure pressure never reaches to that of single corrosion defect.

4.4 Deep Corrosion Defect at Location 1

When a deep corrosion defect at Location 1 is adjacent to a shallow corrosion defect at Location 2, corrosion defect interaction does not occur regardless of the longitudinal spacing, as illustrated by Fig. 14. Note that in this observation, interaction behaviour is different from shallow corrosion defect at Location 1 adjacent to deep

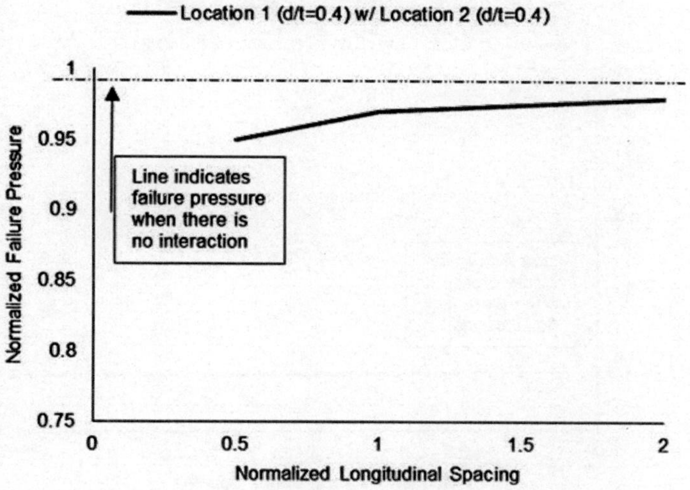

Fig. 12 Failure trend line of medium depth corrosion defect at Location 1 adjacent to medium depth corrosion defect at Location 2

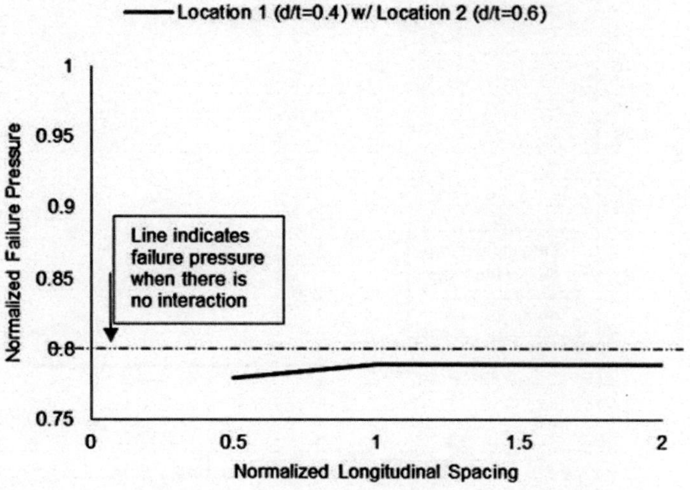

Fig. 13 Failure trend line of medium depth corrosion defect at Location 1 adjacent to deep corrosion defect at Location 2

corrosion defect at Location 2 as there is an instance when this pair of corrosion undergoes interaction (i.e. shallow corrosion defect at Location 1 ($d/t = 0.2$) adjacent to deep corrosion ($d/t = 0.6$) with $s_l/\sqrt{Dt} = 0.5$). This proves that the location of the corrosion in the longitudinal axis plays an important role in interaction. Next, referring to Fig. 15, when a deep corrosion defect at Location 2 is adjacent to a

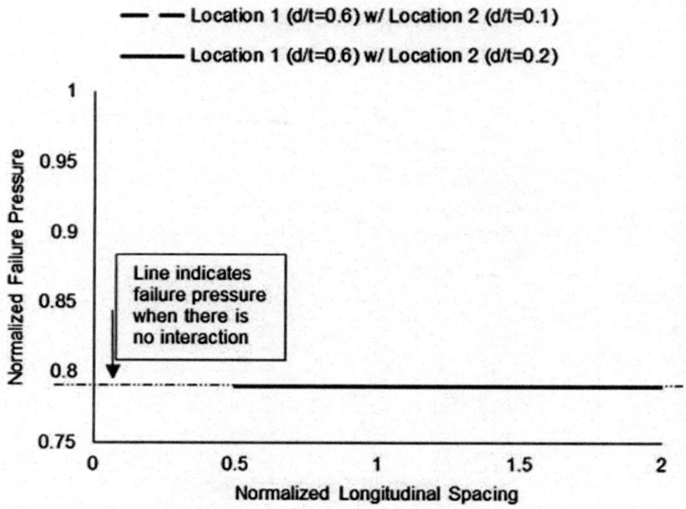

Fig. 14 Failure trend line of deep corrosion defect at Location 1 adjacent to shallow corrosion defect at Location 2

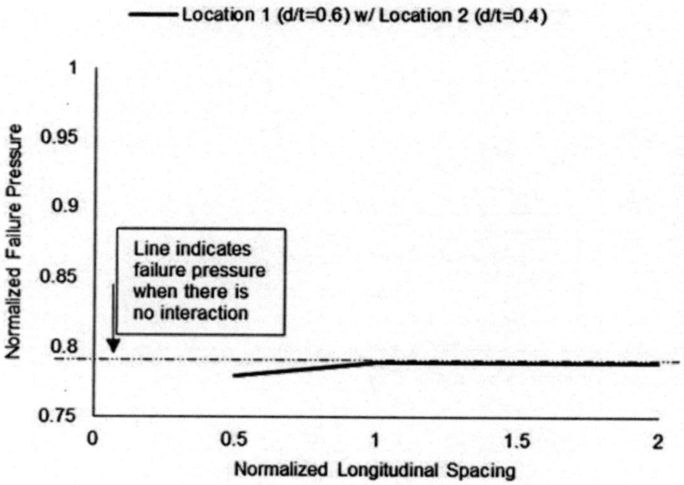

Fig. 15 Failure trend line of deep corrosion defect at Location 1 adjacent to medium depth corrosion defect at Location 2

medium depth corrosion defect at Location 2, corrosion interaction does occur, but only when $s_l/\sqrt{Dt} < 1$. Finally, referring to Fig. 16, when a deep corrosion defect at Location 1 is adjacent to deep corrosion defect at Location 2, corrosion interaction does occur, but when $s_l/\sqrt{Dt} < 2$. Based on these results, DNV's guideline may be appliable when deep corrosion defect is at Location 1.

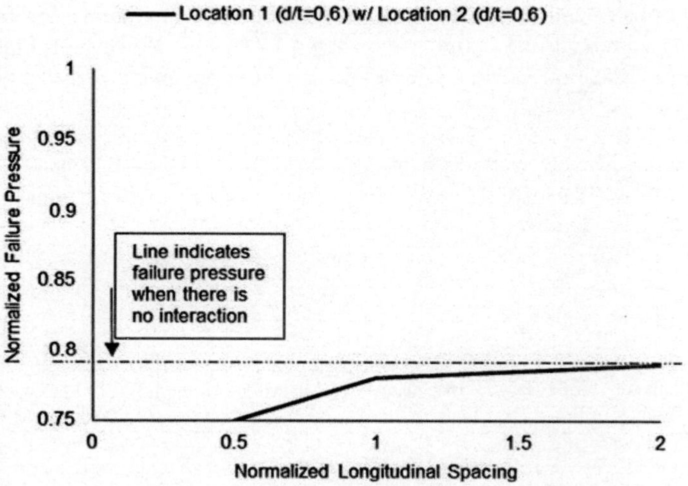

Fig. 16 Failure trend line of deep corrosion defect at Location 1 adjacent to deep corrosion defect at Location 2

4.5 Recommendations for Future Studies

Future studies involving interacting corrosion defects at pipe elbows should vary several parameters as to determine their exact influence on the failure pressure. For instance, the variation for longitudinal spacing should be increased beyond $s_l/\sqrt{Dt} = 2$ because in this study, it was observed that interaction does occur up till $s_l/\sqrt{Dt} = 2$ for certain instances. Furthermore, even more corrosion defect depth (i.e. including d/t = 0.3, 0.5, 0.7, 0.8) should be analysed so that accurate and comprehensive failure behaviour could be determined. In addition, corrosion defect length should be varied as well as it has been proven in past researches to play a significant role in failure pressure prediction [7, 14, 28]. Next, the circumferential spacing and positions should be studied as well as Khalajestani et al. [6] determined that circumferential position of corrosion defect on pipe bends is important and adversely affects the failure pressure the closer it is to the intrados. In terms of improving the FEM, the corrosion defect region could be meshed in hexahedral (brick) mesh as it would allow quicker computation and lower computing resource utilization.

5 Conclusion

In this study, von Mises stress failure criterion was used to predict the failure pressure of API 5L X52 pipe elbow with interacting corrosion defects subjected to internal pressure. The FEA revealed that interaction does occur and is most significant when medium depth corrosion defect is at Location 1. It also revealed that for medium corrosion defect depth at Location 1, DNV's guideline for interaction limit is not applicable as interaction occurred even when $s_l/\sqrt{Dt} = 2$. However, for deep corrosion defect at Location 1, DNV's guideline may be applicable as interaction was not found when $s_l/\sqrt{Dt} = 2$.

Acknowledgements This work was supported by Yayasan Universiti Teknologi PETRONAS, Malaysia [015LC0-304 and 015LC0-110].

References

1. Al-Sabagh A, Abd-El-Bary H, El-Ghazawy R, Mishrif M, Hussein B (2011) Corrosion inhibition efficiency of linear alkyl benzene derivatives for carbon steel pipelines in 1M HCl. Egypt J Pet 20(2):33–45
2. Benjamin AC, Freire JLF, Vieira RD, Cunha DJ (2016) Interaction of corrosion defects in pipelines—Part 1: fundamentals. Int J Pres Ves Pip 144:56–62
3. Schweitzer PA (2009) Fundamentals of corrosion: mechanisms, causes, and preventative methods. CRC Press
4. Lee GH, Pouraria H, Seo JK, Paik JK (2015) Burst strength behaviour of an aging subsea gas pipeline elbow in different external and internal corrosion-damaged positions. Int J Nav Arch Ocean 7(3):435–451
5. Khalajestani MK, Bahaari MR (2014) Investigation of pressurized elbows containing interacting corrosion defects. Int J Pres Ves Pip 123:77–85
6. Khalajestani MK, Bahaari MR, Salehi A, Shahbazi S (2015) Predicting the limit pressure capacity of pipe elbows containing single defects. Appl Ocean Res 53:15–22
7. Netto T, Ferraz U, Estefen S (2005) The effect of corrosion defects on the burst pressure of pipelines. J Constr Steel Res 61(8):1185–204
8. Ahammed M, Melchers RE (1996) 1996 Reliability estimation of pressurised pipelines subject to localised corrosion defects. Int J Pres Ves Pip 69(3):267–272
9. Silva R, Guerreiro J, Loula A (2007) A study of pipe interacting corrosion defects using the FEM and neural networks. Adv Eng Softw 38(11):868–875
10. Benjamin A, Freire J, Vieira R (2007) Part 6: analysis of pipeline containing interacting corrosion defects. Exp Tech 31(3):74–82
11. de Andrade EQ, Benjamin AC, Machado PR, Pereira LC, Jacob BP, Carneiro EG, Guerreiro JNC, Silva RCC, Noronha DB (2006) Finite element modeling of the failure behavior of pipelines containing interacting corrosion defects. In: Proceedings of 25th international conference on offshore mechanics and arctic engineering, American Society of Mechanical Engineers
12. Benjamin AC, de Andrade EQ, Jacob BP, Pereira LC, Machado PR (2006) Failure behavior of colonies of corrosion defects composed of symmetrically arranged defects. In: Proceedings of international pipeline conference, American Society of Mechanical Engineers
13. Det Norske Veritas (2015) Recommended practice DNV-RP-F101 corroded pipelines, DNV GL AS
14. Alang N, Razak N (2013) Finite element analysis on burst pressure of steel pipes with corrosion defects. In: Proceedings of the 13th international conference on fracture
15. Benjamin AC, Freire JLF, Vieira RD, Cunha DJ (2016) Interaction of corrosion defects in pipelines–Part 2: MTI JIP database of corroded pipe tests. Int J Pres Ves Pip 145:41–59
16. Abdulhameed D, Adeeb S, Cheng R, Martens M (2016) The influence of the bourdon effect on pipe elbow. In: Proceedings of 11th international pipeline conference, American Society of Mechanical Engineers Digital Collection

17. Benjamin AC, Freire JLF, Vieira RD, Diniz JL, de Andrade EQ (2005) Burst tests on pipeline containing interacting corrosion defects. In: Proceedings of 24th international conference on offshore mechanics and arctic engineering, American Society of Mechanical Engineers
18. American Society of Mechanical Engineers (2018) ASME B16.9: Factory Made Wrought Steel Buttwelding Fittings, American Society of Mechanical Engineers
19. Wang Y-L, Li C-M, Chang R-R, Huang H-R (2016) State evaluation of a corroded pipeline. J Mar Eng Technol 15(2):88–96
20. ANSYS (2014) ANSYS release 15.0 documentation. SAS IP, Inc., Canonsburg, PA
21. British Standard Institute (1999) Guide on methods for assessing the acceptability of flaws in metallic structures, British Standard Institution
22. Belachew CT, Mokhtar CI, Karuppanan S (2016) Strength assessment of a corroded pipeline through the burst test: case study. J Pip Syst Eng 2016:04016003
23. Choi J, Goo B, Kim J, Kim Y, Kim W (2003) Development of limit load solutions for corroded gas pipelines. Int J Pres Ves Pip 80(2):121–128
24. Duan Z-X, Shen S-M (2006) Analysis and experiments on the plastic limit pressure of elbows. Int J Pres Ves Pip 83(10):707–713
25. Han C, Zhang H, Zhang J (2016) Failure pressure analysis of the pipe with inner corrosion defects by FEM. Int J Electrochem 11:5046–5062
26. Oh C-K, Kim Y-J, Park C-Y (2009) 2009 Effects of local wall thinning on net-section limit loads for pipes under combined pressure and bending Nucl. Eng Des 239(2):261–273
27. Cosham A, Hopkins P (2003) The assessment of corrosion in pipelines—guidance in the pipeline defect assessment manual (PDAM). In: Proceedings of international colloquium reliability of high pressure steel pipelines
28. Suzuki T, Meshii T (2008) Failure pressure of straight pipe with wall-thinning under internal pressure. Int J Pres Ves Pip 85(9):628–634

Assessment of Internal Tubular Coating System for High CO_2 Application

Ahmad Shalabi Md Sauri and Muzdalifah Zakaria

Abstract Assessment of three identified internal tubular coating candidates, namely Phenolic Epoxy Coating (PE), Fusion Bonded Epoxy coating (FBE) and Fluoropolymer coating (FP) had been conducted in simulated static operational condition of both CO_2 Producer (5,000 psi, 135 °C, 60–80% CO_2, 200–400 ppm H_2S, synthetic brine) and CO_2 Injector (5,000 psi, 160 °C, 100% CO_2, 50–100 ppm H_2S) wells in accordance to NACE TM0185 (N. TM0185, "Standard Test Method Evaluation of Internal Plastic Coatings for Corrosion Control of Tubular Goods by Autoclave Testing," 2006.). All the coating candidates underwent the autoclave test with the exposure period of 30 days. Upon the completion, visual inspection, adhesion, hardness, microscopic analysis, EIS, impact resistance and linear abrasion test were done on the test section to further evaluate the coating performance post exposure. PE demonstrated a tremendous performance at both CO_2 Producer and Injector conditions with the adhesion strength between 14 and 16 MPa as compared to the other coating candidates. Impedance value of 10^{10} Ω/cm^2 is higher than acceptance limit of 10^6 Ω/cm^2. All the obtained results indicated that PEC can be an alternative solution to the Corrosion Resistant Alloys (CRA) for high CO_2 applications.

1 Introduction

Development of high CO_2 field poses significant challenges compared to conventional natural gas field due to high CO_2 content, higher production and injection rate. One of the main challenges need to be considered is corrosion issues on the carbon steel (CS) alloy commonly used as downhole tubing materials. In the presence of free water, CO_2 will partially dissolve and form carbonic acid that can accelerate the degradation rate on the CS material. Application of CRA is the current practice to

A. S. Md Sauri (✉) · M. Zakaria
Gas Sustainable Technology (GaSTech), Group Research & Technology (GR&T), Project Delivery & Technology (PD&T), PETRONAS Research Sdn. Bhd, Kuala Lumpur, Malaysia
e-mail: shalabi.sauri@petronas.com.my

M. Zakaria
e-mail: muzdalifah@petronas.com.my

F. Ahmad et al. (eds.), *ICPER 2020*, Lecture Notes in Mechanical Engineering,
https://doi.org/10.1007/978-981-19-1939-8_68

minimize the corrosion impacts toward the well completion assembly in high CO_2 service, but the development cost (CAPEX) becomes less economical. An alternative solution to the CRA material is the internal tubular coating which will still allow the use of CS tubing hence enable CAPEX cost to be kept relatively low. The assessment of physical and chemical properties of the identified internal tubular coating candidates at the aggressive environment of replicated high CO_2 field, including the mechanical testing (abrasion and impact) were furthered discussed in this paper.

2 Methodology

API 5CT L80 carbon steel test specimens with dimension of 250mm (L) × 70mm (W) × 5mm (T) were initially abrasive blast cleaned to Sa 2 ½ as per ISO 8501-1:2007 and were coated as per manufacturer recommendation. All the test specimens were holiday tested in accordance to NACE TM0384:2016 [1] using Elcometer 270 Pinhole tester to ensure defect-free specimens. Dry film thickness of the coated specimens were measured at 8 location of each specimen side in accordance to SSPC PA 2 [2] and the average thickness for PE, FBE and FP are 400, 350 and 60 microns, respectively.

Autoclave tests were conducted in simulated static operation conditions as followed:

CO_2 Producer

- 5,000 psi, 135 °C, 60–80% CO_2, 200–400 ppm H_2S, synthetic brine (wet)

CO_2 Injector

- 5,000 psi, 160 °C, 100% CO_2, 50–100 ppm H_2S (dry).

The tests were initiated by inserting all the specimen in the autoclave, followed by overnight pressure tested using oxygen free N_2 to ensure leakage free test setup. The autoclave was heated to the required test temperature and pressurized to the required test pressure using the gas mixture. The gas mixture used, is based on the abovementioned fluid composition. After 720 h, the heater was switched off and the autoclave was allowed to cool to 93 °C before depressurizing in accordance to NACE TM0185 [3] at approximately 10 bar/min. The exposed specimens were then evaluated in accordance to below post analysis test methods and acceptance criteria, as highlighted in Table 1.

Table 1 Acceptance criteria for coating post analysis

No.	Test	Standard	Acceptance criteria
1	Visual (blistering)	ASTM D714	• No blisters formation observed after the autoclave test
2	Pencil hardness	ASTM D3363	• Hardness grade after exposure must not be less than one grade from the baseline sample hardness grade
3	Knife adhesion	ASTM D6677	• Rating A (coating is extremely difficult to remove)
4	Pull-off strength adhesion	ISO 4624	• Pull-off strength $\geq$12 MPa
5	Electrochemical impedance spectroscopy	ISO 16773	• $Z_{0.1\ Hz} > 10^8 \Omega cm^2$
6	Abrasion resistance	ASTM D4060 (modified)	• Mass loss <100 mg at 1,000 cycle using CS-17 abradant
7	Impact resistance	ISO 6272 (modified)	• Impact resistance >5 J

3 Results

The static condition is chosen to evaluate the resistivity of coating candidates towards the solvent effect of supercritical CO_2 and the acidic condition of carbonic acid due to higher contact time. In addition, supercritical CO_2 is potentially became main threat to polymeric coating as it capable to dissolve and decompose organic compounds by removing low molecular weight components from their molecules structure [4]. At CO_2 Producer condition, half of the specimens' length were immersed in the bulk water (saturated with CO_2) and the rest of the length were exposed to CO_2 gaseous. From the visual inspection, visible blistering was only observed on FBE and FPC coated specimens at both liquid and gas phase. Figure 1 showed comparison of the coated specimens before and after exposure for all three coating candidates. The formation of blisters were also observed on FBE coated specimen after 30 days exposure at CO_2 Injector condition and no significant physical and chemical degradation. The unaffected exposed coated specimens underwent adhesion and hardness tests to gauge their performance after the exposure tests. The results of both adhesion and hardness tests are summarized in Table 2.

PE showed no significant change after exposure on the pull-off adhesion tests for CO_2 Producer condition. However, a significant change in pull-off adhesion strength at CO_2 Injector condition was observed. It was due to formation additional crosslinking at the active site of novolac backbone after the post heat curing at 160 °C, hence, improving the mechanical and chemical resistance of the coating [5]. Pull-off adhesion test is not applicable for FPC coated specimen due to its non-sticky properties. Therefore, the performance of FP at CO_2 Injector condition were only evaluated via V-cut adhesion test and pencil hardness test. Based on the observation, no significant change after exposure on the V-cut adhesion test and pencil hardness test for FP coated specimen as compared to the control specimen. In addition, similar

Fig. 1 Comparison of the coated specimen before and after exposure

Table 2 Summary of the adhesion and hardness tests of the unaffected coated specimens at both CO_2 Producer and CO_2 Injector conditions

Coating	Specimen	Results		
		V-cut adhesion	Pencil hardness	Pull-off adhesion (MPa)
FP	Control	A	2B	N/A
	Exposed (CO_2 Injector)	A	2B	N/A
PE	Control	A	2H	14.60
	Exposed (CO_2 Producer)	A	2H	14.01
	Exposed (CO_2 Injector)	A	2H	16.80

observation were noted on PE coated specimen at both CO_2 Producer and CO_2 Injector conditions.

EIS were conducted on both control and exposed coated specimens in accordance to ISO 16773 [6]. A two electrode system (working and counter electrodes) was used and the measurement were conducted at a frequency range of 100 kHz to 10 mHz with sinusoidal voltage perturbation signal of 200 mV in 3% NaCl electrolyte. The specimens were initially immersed in 50 mm depth of 3% NaCl electrolyte and left to saturate for 24 h prior to the EIS measurement. The resulted impedance spectra was assessed for quantifying the barrier properties of the coating candidates. The total coating impedance (Ω cm^2) at an applied signal frequency of 0.1 Hz ($Z_{0.1Hz}$) is widely used as the indicator for determining barrier properties of coating. Coating with an impedance $Z_{0.1Hz} > 10^6$ Ω cm^2 provides corrosion protection, while the excellent protective barrier is achieved at $Z_{0.1Hz} > 10^8$ Ω cm^2 [7, 8].

Impedance results for PE coated specimens at CO_2 Producer condition indicates some reduction in barrier properties after the exposure but still within the excellent protective barrier range. The rest of the impedance results do not show any reduction. FP coated specimen at CO_2 Injector also shows similar pattern of impedance result as PE at CO_2 Producer condition. EIS results indicated no significant chemical degradation on the abovementioned coating candidates after the exposure at the respective test conditions. Due to proprietary issue, both Nyquist and Tafel plots are not allowed to be published. Only nett charge transfer resistance (R_{ct}) are provided as the performance indicator for the investigated coating candidates (Table 3).

Abrasion and impact resistance test are widely used to measure the resistivity level of coating towards the mechanical damage during wireline intervention. The abrasion tests were conducted in accordance to ASTM D4060 using a Taber Linear Abraser fitted with CS-17 Wearasers under 1 kg load. The stroke length was set to 101.6 mm (4 inches) and the speed was 60 cycles/min. According to PETRONAS Technical Standard [9], the weight loss of the coated specimens must less than 150 mg after 1,000 cycles under 1 kg load. Based on the below results (Fig. 2), all the coating candidates has passed the criteria with the weight loss less than 60 mg after 5,000 cycle. Among them, PE demonstrated a better abrasion resistance at both test conditions. On the other side, the impact resistance tests were designed to evaluate the

Table 3 Summary of EIS tests at both CO_2 Producer and CO_2 injector conditions

Coating	Impedance (Ω cm^2) at $Z_{0.1 Hz}$		
	Control	CO_2 Producer	CO_2 Injector
FP	1.69×10^{10}	N/A	8.68×10^9
PE	5.67×10^{10}	2.54×10^9	6.30×10^{10}

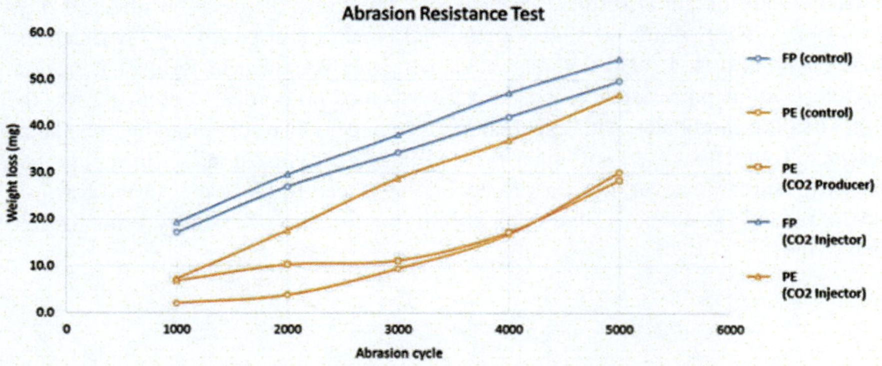

Fig. 2 Summary of abrasion tests for all coating candidates

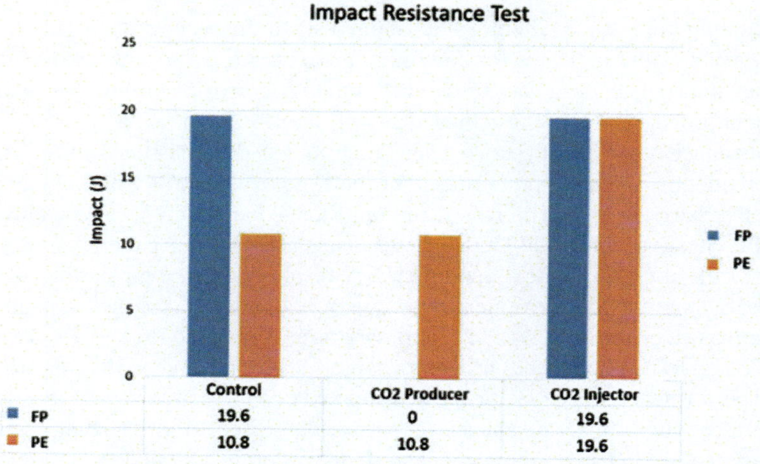

Fig. 3 Summary of impact resistance tests for all coating candidates

resistance of the coating to cracking or peeling from the substrate when it is subjected to a deformation caused by a falling weight under standard condition. In this study, a 20 mm diameter spherical indenter was used to assess the performance of the coating candidate in accordance to ISO 6272. From the obtained results as shown in Fig. 3, no change of impact resistance values observed for the exposed coated specimens as compared to the control specimens, except for PE coated specimen at CO_2 Injector condition. Similar to pull-adhesion strength, the increase of impact resistance value is due to the formation of additional crosslinking at the novolac backbone.

4 Conclusion

Based on the obtained results, PE does not show any degradation after 30 days exposure at both simulated static operational condition at both CO_2 Producer and Injector condition. It also demonstrated better performance in term of adhesion, hardness, abrasion resistance and impact resistance properties as compared to the other coating candidates. This excellent performance is also supported by impedance test result which showed sufficient protection against corrosion is still maintained post exposure. However, this coating technology need to be further tested for fluid erosion, sand erosion, wireline compatibility and coating flexibility prior to field deployment.

References

1. N T 0384 (2016) Detection of internal tubular coatings of less than 250 μm (10 mils) dry-film thickness
2. S P 2 (2015) Procedure for determining conformance to dry coating thickness requirements
3. N TM0185 (2006) Standard Test method evaluation of internal plastic coatings for corrosion control of tubular goods by autoclave testing
4. Fukushima Y (1999) Applications of supercritical fluids. R&D Review of Toyota CRDL
5. Munger CG (1999) Corrosion prevention by protective coatings, NACE
6. I 16773 (2016) Electrochemical impedance spectroscopy (EIS) on coated and uncoated metallic specimens. ISO
7. Bacon RC, Smith JJ, Rugg FM (1948) Electrolytic resistance in evaluating protective merit of coatings on metals. ACS
8. Niglia J, Hoijman N, Marino M (2011) Design of an internal onshore system qualification program for tubular goods. NACE
9. P 30.48.00.31 (2012) Protective coatings and linings

Investigating the Mechanical Performance of Intumescent Coating Enhanced with Magnesium Oxide (MgO) for Structural Steel Application

Y. Azmi, F. Ahmad, S. Kabir, Y. X. Lee, A. Zulfiqar, G. H. Yeoh, A. Qaiser, and Patrick J. Masset

Abstract Intumescent fire-retardant coating is a passive type of protection against fire, design to reduce the heat transfer from the source of fire to the steel substrate in a duration of time. This research investigates the effect of Magnesium Oxide as a filler in expandable graphite-based intumescent fire-retardant coatings. The coating was developed by varying weight percentage (0.5–2.5wt%) of magnesium oxide towards binder in the coating in addition to ammonium polyphosphate, melamine, boric acid, zinc borate, and expandable graphite. The samples were subjected to fire tests to investigate their performance. The heat shielding fire test shows a decrease of back substrate temperature when increasing the amount of magnesium oxide in the formulation up to 242 °C for 1 h. The samples were subjected to a fire control test to investigate the intumescent factor (IF) at 600 °C for 1 h in a carbolite furnace. MG1.5 shown the highest with the expansion of 8.33 Intumescent factor. Thermal gravimetric analysis, water immersion, and adhesion tests were also performed to study the sample's thermal stability and adhesion strength towards the substrate. The result showed that adding magnesium oxide in the formulation of intumescent coating improved the coating's thermal performance, thus increasing the protection time of steel substrate.

Y. Azmi · F. Ahmad (✉) · S. Kabir · Y. X. Lee
Department of Mechanical, Universiti Teknologi PETRONAS, Bandar Seri Iskandar, Perak, Malaysia
e-mail: faizahmad@utp.edu.my

A. Zulfiqar
Institute of Chemical Engineering and Technology, University of Punjab, Lahore, Pakistan

G. H. Yeoh
School of Mechanical and Manufacturing Engineering, University of New South Wales, Kensington, NSW, Australia

A. Qaiser
Department of Polymer and Process Engineering, University of Engineering and Technology Lahore, Lahore, Pakistan

P. J. Masset
Koszalin University of Technology, ul.Sniadeckich 2, 75-453 Koszalin, Poland

F. Ahmad et al. (eds.), *ICPER 2020*, Lecture Notes in Mechanical Engineering,
https://doi.org/10.1007/978-981-19-1939-8_69

Keywords Magnesium oxide · Intumescent coating · Fire retardant · Structural steel

Abbreviations

IC	Intumescent Coating
IF	Intumescent factor
MgO	Magnesium Oxide
ZB	Zinc Borate
BA	Boric Acid
APP	Ammonium polyphosphate
EG	Expandable graphite
Mel	Melamine
DGEBA	Bisphenol A diglycidyl ether
EEW	Equivalent Weight

1 Introduction

When exposed to fire, structure steel significantly weaken, with unprotected members fail after only 5–10 min of direct exposure to an intense fire [1]. According to the American Iron and Steel Institute (AISI), in general, steel strength remains unchanged until 600°F (315 °C). It retains about 50% of its strength both in tension and compression at 1100°F (593 °C), where an unprotected steel start showing a visible sign of deform, buckle and twist before loses all its strength at 2700 °F (1482 °C).

Intumescent coating (IC) is a type of passive fire protection (PFP) to protect the structural steel against fire. Under the action of heat, this coating will expand and form a thick porous charred layer to insulate the substance. The substance is protected against the high-temperature rise and exposure to oxygen within a period [2, 3].

Vandersall, in 1971 published in his compressive review paper, divided intumescent coating into three critical main components and are widely agreed by the researcher [3–6]. The classical three components involved acid source, carbon source, and blowing agent. The addition of binder, addictive, and filler is introduced later to improve the existing fire-retardant properties of intumescent coating [7]. Alongi et al. (2014) mention in their journal, an additional filler may help to modify the chemical (reactivity of the filler versus the ingredient of the intumescent system) and physical (expansion, char strength, and thermophysical properties) behaviours of the intumescent coating when exposed to heat [8]. Metal elements as a filler in the intumescent coating reported enhanced carbonization, the thermal stability of char, and promoting dehydrogenation of the polymer, thus providing effective insulation [9–12]. Duquesne (2013) found, adding metal element material such as titanium dioxide will actively react with ammonium polyphosphate (blowing agent) at high temperatures leading to the development of thermally stable phosphate species (TiP_2O_7), consequently prolong the protection time for the substrate [13].

Magnesium oxide is a stable compound in the form of white hygroscopic. It has high melting and boiling point and is well-known flame-retardant material. Most scholars reported magnesium oxide as a high potential candidate to improve the performance of flame-retardant composite material such as polycarbonate but barely was reported to be used as filler in coating application [12, 14, 15]. In addition, a various scholar has reported a unique trait of magnesium oxide having a high adhesion strength when combining with other composite formulation.

This study investigates the influence of magnesium oxide as a flame retardant additive on the mechanical performance of intumescent fire-retardant coatings.

2 Experimental

In this section, the material and procedure to develop the intumescent coating in the study is explained.

2.1 *Materials*

Bisphenol A diglycidyl ether (DGEBA) with Equivalent Weight (EEW) 188, Expendable Graphite (EG), and Zinc Borate (ZB) purchased from Mc-Growth Chemical Sdn. Bhd. Boric Acid (BA), melamine (MEL) obtained from Sigma-Aldrich (M) Sdn. Bhd. Malaysia. Ammonium polyphosphate (APP) supplied by Clariant (Malaysia) Sdn. Bhd and magnesium oxide obtained from Guangzhou Jiechunag Trading Co. Ltd. Polyamide amine brought from Mc-Growth Chemical Sdn Bhd. Malaysia.

2.2 *Sample Preparation for Intumescent Coating*

The experiment started by weighing all materials according to their respective weight percentage as per Table 1. Zinc borate, melamine, boric acid, and ammonium polyphosphate are ground for 1 min before mixed into a beaker together with bisphenol A diglycidyl ether at 30 rpm for 15 min using a shear mixer. A hardener (polyamide amine) was added to cure the mixture and mixed for another 10 min. The coating was applied to the ASTM A36 steel substrate with a dimension of 100 mm × 100 mm × 1.5 mm and 50 mm × 50 mm × 1.5 mm before leaving to cure at room temperature for 24 h. The thickness of the coating is measured to calculate the intumescent factor (IF). The cured samples were also subjected to characterization and testing as per Sect. 3.

Table 1 Formulation for Intumescent coating enhanced with MgO

Samples	APP wt%	MEL wt%	ZB wt%	BA wt%	EG wt%	MGO wt%	Binder and curing agent (wt%)
MG0.0	11.23	5.5	5	6.11	5.5	0.0	66.66
MG0.5	11.23	5.5	5	6.11	5.5	0.5	66.16
MG1.0	11.23	5.5	5	6.11	5.5	1.0	65.66
MG1.5	11.23	5.5	5	6.11	5.5	1.5	65.16
MG2.0	11.23	5.5	5	6.11	5.5	2.0	64.66
MG2.5	11.23	5.5	5	6.11	5.5	2.5	64.16

3 Characterization and Testing

The characterization and testing performed on the cured intumescent coating describe as per the following subsection. The 10 × 10 × 1.5 cm cured samples were subjected to a heat-shielding test while 5 × 5 × 1.5 samples were exposed to the control fire test, immersion test, and adhesion test.

3.1 Heat Shielding Test

An imitation of a real fire test for a lab scales sample was conducted to gauge the back temperature of the steel structure for one hour.

A butane cartridge equips with a torch was set up in a rectangular heat shielding box. The intumescent samples were placed 7 cm from the tip of the torch to the coating sample. Three thermocouples were attached at the back of the substrate, and temperature is recorded using Anritsu Compact data logger. The schematic diagram for the equipment setup is shown in Fig. 1.

3.2 Control Fire Test

The test was performed to evaluate the char properties of the coating. The cured samples were burned to 600°C with a dwelling time of 1 h to ensure complete burning in *Carbolite Gero RWF 1100*. The samples were then left cooled for a day before measuring to avoid thermal shock to the sample. The height of expanded char produced was recorded for intumescent factor calculation as per Eq. (1) below and sent for characterization.

$$IF = D_2 - D_0 / D_1 - D_0 \quad (1)$$

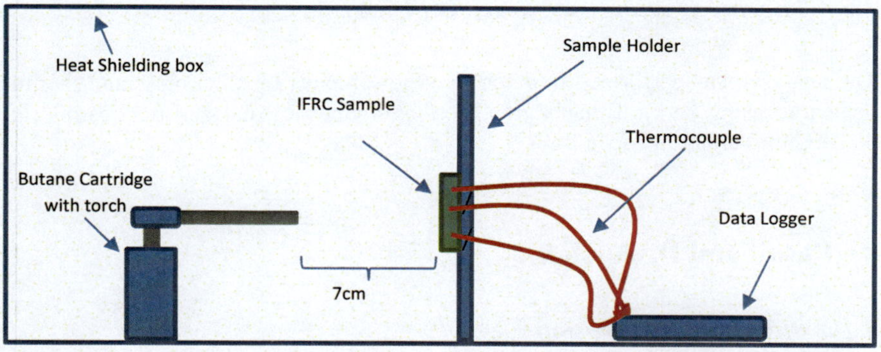

Fig. 1 Heat shielding equipment setup

D_2	*the thickness of the char with the substrate,*
D_1	*the thickness of coating with a substrate,*
D_0	*the thickness of the steel substrate.*

3.3 Water Immersion Test (ASTM D 870)

The cured coating is weighted before immersed in seawater acquired from Teluk Batik, Perak, at room temperature. After 168 h, the samples are dry for 5 min before weighting the sample [16]. The rate of weight change, R, is then calculated as per Eq. (2), with m_1 represent coating weight before immersing and m_2 signify the weight of coating after immersion.

$$R(\%) = \left|\left({m_1 - m_2}/{m_1}\right) * 100\right| \quad (2)$$

3.4 Adhesion Test (ASTM D 4541)

The test was performed according to the Test Method for Pull-Off Strength of Coatings Using Portable Adhesion Testers, ASTM 4541 [17]. A Dolly was fixed into each coating surface sample using metal-type glue and cured at room temperature for at least 24 h before being pulled using Elcometer 108.

3.5 *Thermogravimetric Analysis (TGA)*

The analysis was performed to determine the sample's residual mass and thermal behaviour using Perkin Elmer Pyris Thermogravimetric Analyser at a heating rate of 10°C/min under nitrogen flow.

4 Result and Discussion

4.1 *Physical Observation*

All the cured sample was exposed to control fire environment in the Carbolite Furnace to investigate the sample's intumescent factor (IF) and char properties. The IC samples enhanced with magnesium oxide shown in Fig. 2, together with a cross-section of the char after burned. The char expanded nonhomogeneous with plenty of worms that look like structures protruding from the surface. This structure may be due to agglomeration from the uneven distribution of expandable graphite flakes, leading to the uneven distribution of cell size in the char layer [7]. Plenty of residues was observed at the bottom of substrate steel after a part of the char was cut. This trait might be an indication of suitable adhesion properties that come from magnesium oxide. This finding is supported by the evidence of high adhesion strength attained from the adhesion test discussed in Sect. 4.4.

4.2 *Char Expansion Analysis*

The Intumescent Factor (IF) value calculated from the char expansion is deemed one of the key parameters to evaluate the performance of intumescent coating but does not necessarily lead to the most satisfactory performance. The intumescent formulation aims to gain a multicellular char layer to reduce the heat transfer steel plate [18].

Figure 3 shown the IF value of 6 samples, including control formulation. The IF values of char expansion as follow: MG0.0 (5.52) > MG0.5 (6.2) > MG1.0 (6.59) > MG1.5 (8.33) > MG 2.0 (7.93) > MG 2.5 (6.59). As observed, the trend shows the char expansion increases with increasing the amount of magnesium oxide up to 1.5%wt in the formulation before it decreases. Sample MG1.5 achieved the highest char expansion with 8.33 times. The decrease of expansion when more than 1.5%wt MgO added may vary due to excessive loading of MgO into the formulation.

Fig. 2 The intumescent coating enhances with MgO and their char after burned

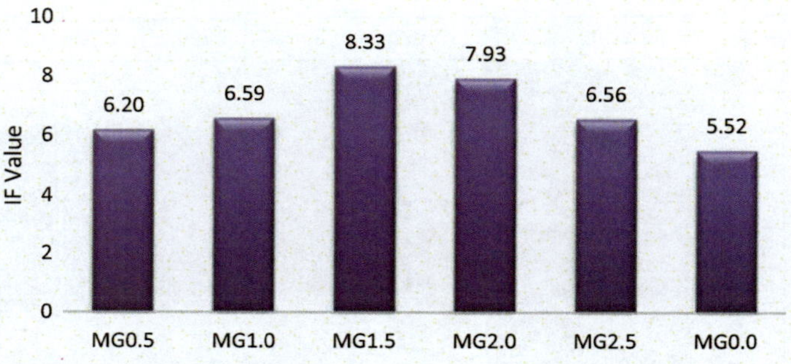

Fig. 3 Intumescent Factor (IF) of intumescent coating enhance with MgO

4.3 Fire Test Analysis

An imitation of the actual fire test in the lab-scale range was performed to evaluate the heat penetration of fire to the steel substrate for each intumescent formulation. When exposed to hydrocarbon-type fire, the MG0.0 formulation (without magnesium oxide) increases slowly compared to the other five formulations shown in Fig. 4 but achieved the highest back steel temperature at 367°C when approaching 40 min and retaining the temperature until the end of the test. In contrast, when Magnesium oxide was added to the formulation, the temperature rises rapidly in the first 5 min before reaching a constantly stable temperature up to 1 h with Mg2.5 obtain the lowest back steel temperature of 242°C. This suggested MgO has a positive effect on flame retardant performance, thus improving the time for the protection to the steel substrate.

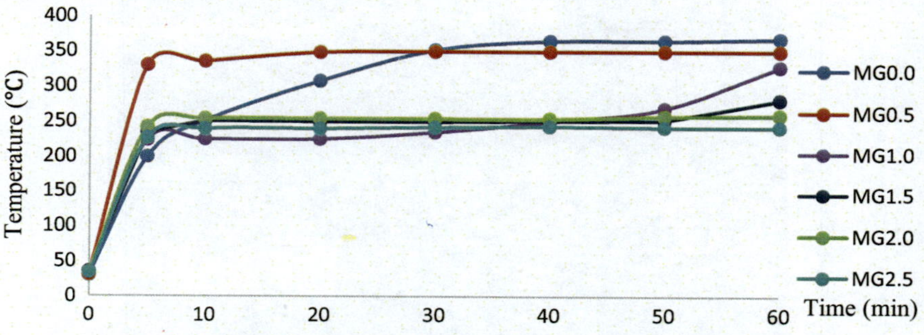

Fig. 4 Temperature versus Time, Fire Test graph

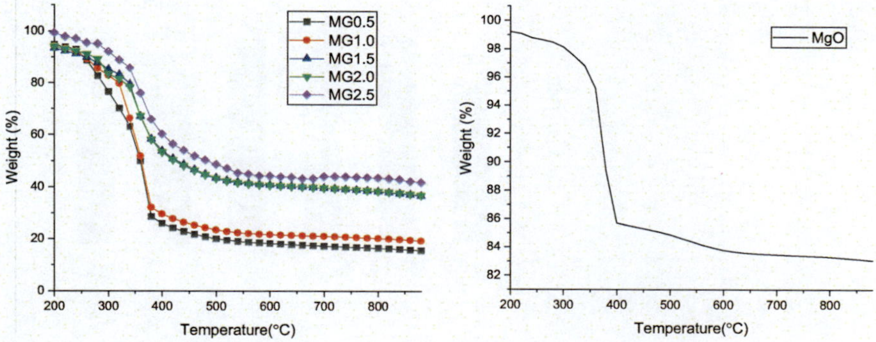

Fig. 5 TGA analysis for IC samples (left) and raw MgO (right)

4.4 Thermal Stability Analysis

The residual mass trends for the influence of magnesium oxide on IC by varying weight percentages are shown in the thermograph of Fig. 5. The graph observed that increasing magnesium oxide loading in the coating formulation increasing the onset temperature and the residual weight of the samples. A similar result also achieved in the experiment performed by K.W. Soo et al. (2016) using magnesium oxide to increase the adhesion properties of natural rubber [19]. Sample MG0.5 and MG1.0 were displaying higher weight loss with 15.52% and 19.28%, respectively.

The coating thermal stability surges up when MgO loading of 1.5%wt, 2.0% wt, and 2.5wt % of magnesium oxide used in the formulation with 36.09%, 37.01549%, and 41.9019% residual weight left respectively at the end of the analysis. This comparable result was also achieved in the fire test discussed earlier. The fascinating phenomenon may result from magnesium oxide's excellent thermal properties itself, shown in the graph below. The MgO shows a drop of only 18% total weight loss at a temperature of 880°C, thus explaining the excellent thermal performance of the coating.

4.5 Adhesion Test

The adhesion test was conducted following the international standard ASTM 4541. The recorded strength for each sample is presented in Fig. 6. MG1.5 exhibits the highest peel strength with 5.4 MPa, followed by Mg2.0, MG2.5, MG1.0, and MG0.5 with 5.18Mpa, 4.9Mpa, 4.59 Mpa, and 4.14 Mpa, respectively. The coating adhesive strength rises when 0.5% to 1.5% of MgO added to the formulation but started to lose strength after 2.0%wt. This same arrangement was also observed by Poh et al. [20, 21] on the effect of magnesium oxide on adhesion properties of epoxidized natural rubber. In addition, the high adhesion strength achieved for all the samples with

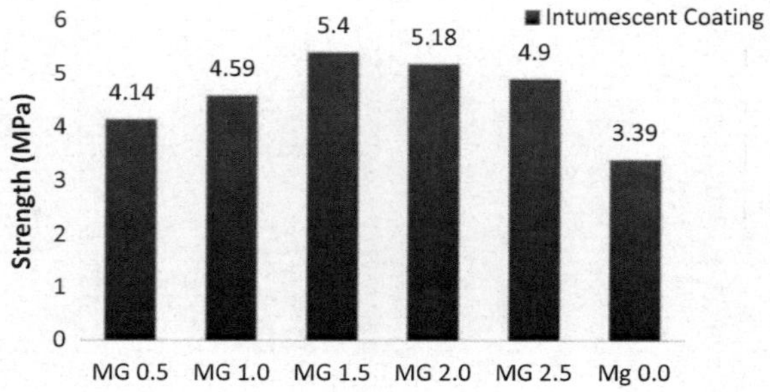

Fig. 6 Intumescent coating's adhesion test result

MgO compare without it (3.18Mpa) exceeded the minimum requirement set by the industrial standard of 2.07 MPa for ambient and 3.45 MPa for splash zone [22]. This excellent adhesive strength of IFRC with the metal surface was due to magnesium oxide characteristics having an adequate interfacial bonding [7].

4.6 Water Resistance of Coatings

All the sample was immersed in seawater with pH level of 8.10 at room temperature for seven days and their weight before and after being recorded. The rate of weight change, R for the coating are calculated and arranged as Fig. 7. The chart showed an increase of weight trend after immersion, with MG0.5 is the lowest and MG2.5 is the highest with 0.2% and 1.62% weight gain indicating a permeation process occurred. In the permeation process, water and corrosive ions could infiltrate into the pore structure of the coating, leading to an increase in the coating weight [23]. The tendency of the coating sample to gain weight when more MgO is introduced was due to its hygroscopic nature.

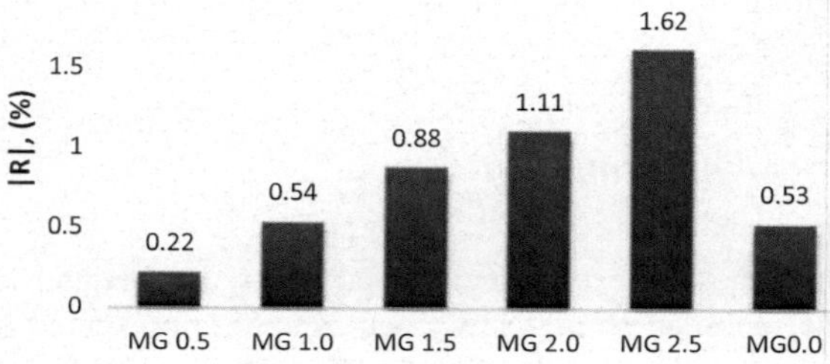

Fig. 7 Intumescent coating's Water Resistance Test result

5 Conclusion

Intumescent coating enhanced with Magnesium Oxide, suggesting a positive effect for improving the performance of flame-retardant coating for application on structural steel. MG1.5 and MG2.0, with the addition of 1.5 and 2%wt MgO, respectively, show excellent adhesion to the substrate, thermal stability, and lower heat transfer, thus prolonging the substrate's protection time. Adding more than 2.0%wt MgO will result in excessive loading, thus decreasing the performance of the coating.

Acknowledgements The authors would like to express their gratitude to the Department of Mechanical Engineering, Universiti Teknologi PETRONAS (UTP), and Advanced Functional Materials (AFM), providing all facilities required for experimentation and testing. The authors also want to appreciate the financial support given through the FRGS-015MA0-081 research grant by MOHE, Malaysia.

References

1. Fitzgeral R (1986) Fire protection handbook. In: Structural integrity during fire, national fire protection association, Section 7
2. Jimenez M, Duquesne S, Bourbigot S (2006) Characterization of the performance of an intumescent fire protective coating. Surf Coatings Technol 201:979–987
3. Yasir M, Ahmad F, Yusoff P S M M, Ullah S, Jimenez M (2019) Latest trends for structural steel protection by using intumescent fire protective coatings: a review. Surf Eng 0:1–30
4. Papaspyrides CD, Kiliaris P (2014) Polymer green flame retardants, 1st edn. Elsevier
5. Wang JQ, Chow WK (2005) A brief review on fire retardants for polymeric foams. J Appl Polym Sci 97:366–376
6. Weil ED, Levchik S (2004) A review of current flame retardant systems for epoxy resins. J Fire Sci 22:25–40
7. Mariappan T (2016) Recent developments of intumescent fire protection coatings for structural steel: a review. J Fire Sci 34:120–163
8. Alongi J, Han Z, Bourbigot S (2015) Intumescence: TRADITION versus novelty. A comprehensive review. Prog Polym Sci 51:28–73
9. Ahmad F, Ullah S, Mohamad WF (2014) Non-toxic mineral based intumescent fire retardant coating for structural applications
10. Aziz H, Ahmad F (2016) Effects from nano-titanium oxide on the thermal resistance of an intumescent fire retardant coating for structural applications. Prog Org Coatings 101:431–439
11. Amir N et al (2017) Synergistic effects of titanium dioxide and zinc borate on thermal degradation and water resistance of epoxy based intumescent fire retardant coatings, vol 740, pp 41–47
12. Zhang Y, Peng R qun, Zhou G da, Fang Z ping, Li X nan (2015) Flammability characterization and effects of magnesium oxide in halogen-free flame-retardant EVA blends. Chin J Polym Sci , English edn, vol 33, pp 1683–1690
13. Duquesne S, Bachelet P, Bellayer S, Bourbigot S, Mertens W (2013) Influence of inorganic fillers on the fire protection of intumescent coatings. J Fire Sci 31:258–275
14. Dong Q et al (2011) A polycarbonate/magnesium oxide nanocomposite with high flame retardancy, pp 2–10
15. Wang C, Wang Y, Han Z (2018) Enhanced flame retardancy of polyethylene/magnesium hydroxide with polycarbosilane. Sci Rep 8:1–10

16. ASTM International "ASTM D870–09, Standard Practice for Testing Water Resistance of Coatings Using Water Immersion." West Conshohocken, PA, 2009.
17. ASTM International (2017) ASTM D4541–17, Standard test method for pull-off strength of coatings using portable adhesion testers. West Conshohocken, PA
18. Jimenez M, Duquesne S, Bourbigot S (2006) Multiscale experimental approach for developing high-performance intumescent coatings. Ind Eng Chem Res 45:4500–4508
19. Soo KW, Azahari B, Poh BT (2016) Effect of magnesium oxide loading on adhesion properties of ENR 25/NBR blend adhesives in the presence of petro resin and gum rosin tackifiers. J Polym Environ 24:334–342
20. Poh BT, Suid NH (2016) Effect of magnesium oxide and the testing rate on the viscosity and adhesion properties of epoxidized natural rubber/acrylonitrile-butadiene rubber blend based adhesives. J Adhes 92:282–294
21. Poh BT, Gan CF (2010) Viscosity and peel strength of magnesium oxide-filled adhesive prepared from epoxidized natural rubber (ENR 25). Polym Plast Technol Eng 49(2):191–194
22. PETRONAS (2019) PETRONAS technical standards: protective coatings and lining
23. Wang G, Yang J (2010) Influences of binder on fire protection and anticorrosion properties of intumescent fire resistive coating for steel structure. Surf Coatings Technol 204:1186–1192

Application of Artificial Neural Network for Failure Pressure Prediction of Pipeline with Circumferential Groove Corrosion Defect

Suria Devi Vijaya Kumar, Saravanan Karuppanan, and Mark Ovinis

Abstract This paper describes the application of artificial neural network (ANN) to develop a corrosion assessment equation for predicting the failure pressure of pipeline with circumferential groove corrosion defect. Finite element analysis (FEA) was utilised to obtain the failure pressure of pipeline for various defect depths and defect length. The FEA results were used to train the ANN model that consisted of three inputs that are true ultimate tensile strength, normalised defect depth, and normalised defect length while the output of the model was the normalised failure pressure of the pipeline. The weights and biases of the ANN model was used to develop a new equation to predict the failure pressure of a pipe with circumferential groove corrosion defect subjected to internal pressure only.

1 Introduction

1.1 Overview of Corrosion

Pipelines being the largest form of transport for oil and gas are prone to corrosion due to their exposure to various corrosive environments and substances [1]. Corrosion is the result of pipe degradation when it undergoes electrochemical reactions that cause the thinning of the pipe wall [2, 3]. Corrosion has always been a severe problem in the industry and recent reports state that the global cost of corrosion has been evaluated to be around USD 2.5 trillion [4]. During operations, pipelines are subjected to internal pressure due to the fluid flow through the pipeline. The internal pressure contributes to the circumferential expansion of the pipe. In the presence of a corrosion defect, the operating pressure of the pipeline is reduced. The hoop stress distribution across the pipe results in strain buildup at the region of defect where the deepest point of the corrosion defect experiences the largest stress [5]. When this occurs, the pipeline

S. Devi Vijaya Kumar · S. Karuppanan (✉) · M. Ovinis
Mechanical Engineering Department, Universiti Teknologi PETRONAS, 32610 Bandar Seri Iskandar, Perak Darul Ridzuan, Malaysia
e-mail: saravanan_karuppanan@utp.edu.my

F. Ahmad et al. (eds.), *ICPER 2020*, Lecture Notes in Mechanical Engineering,
https://doi.org/10.1007/978-981-19-1939-8_70

will fail at a pressure lower than the operating pressure. Among the common types of corrosion that occurs in steel pipes is circumferential groove corrosion. This type of corrosion is described as the thinning of a pipe wall along the circumference of the pipe. Groove corrosion usually occurs at seam welded areas, collars and brackets.

1.2 General Overview of Pipeline Corrosion Assessment Method

Over the years, a variety of methods have been developed to assess the residual strength of corrosion affected pipelines. This effort was driven by the need for an accurate residual strength prediction method in the industry. Fitness-for-service analysis of pipelines used in the oil and gas industry requires a detailed technical assessment of a defect to ensure that the structure can serve its purpose as long as the failure conditions are not reached [6]. By relating the defect geometry, pipe properties, and stresses acting on the structure, the residual strength of a pipeline can be predicted. ASME B31G, DNV-RP-F101 (DNV), and RSTRENG are some of the industrial guidelines or codes that were developed to assess metal loss due to corrosion [7–9].

However, in reality, the conventional corrosion defect assessment procedures result in predictions that are conservative [10]. Although the conservativeness of the predictions represents a certain amount of intrinsic safety factor, unnecessary maintenance of pipelines may arise at a large cost for the business operators. Besides, the predictions from the existing methods differ from one another due to the assumptions made during its development. In comparison to other corrosion assessment methods, the most comprehensive code is noted to be DNV [11].

Despite its comprehensiveness, the DNV code does not provide a thorough assessment method or equation for circumferential groove corrosion defect. The existing equation for circumferential corrosion defect subjected to internal pressure as presented in Eq. (1) is only applicable for circumferential groove corrosion defect that has a length of less than 1.5t.

$$P_{f,grov} = \gamma_{mc} \frac{2t\sigma_{UTS}}{D - t} \tag{1}$$

1.3 Application of Artificial Neural Network (ANN) for Pipeline Corrosion Assessment

Since the development of artificial intelligence, machine learning has proven to be useful in various fields of study. Its ability to recognise and infer from patterns without needing explicit instructions has made it suitable to be applied as a prediction tool. In recent researches, ANN is increasingly being utilised to predict the failure pressure of corroded pipelines.

ANN has the ability to learn from complex non-linear data and produce accurate outcomes enabling researchers to utilise ANN to predict various occurrences in their specific fields provided that the ANN has been trained sufficiently [12]. To achieve this, an ANN mimics the synapses in a human brain by employing weights and biases to the neurons in the hidden layers of the neural network [13]. The input and output data are analysed by neural network and appropriate weights and biases are allocated to each neuron until the neural network produces accurate results. When this is achieved, the neural network is said to be trained. Hence, it is crucial that the training dataset consists of sufficient number of inputs and corresponding outputs to minimize percentage error in the output results of the neural network.

Zangenehmadar and Moselhi's research in 2016 on utilising the back-propagation method to train their ANN model not only proved that the accuracy of ANN models can be applied in predicting the remaining life of pipelines but also the failure pressure of corroded pipelines subjected to various conditions [14].

The application of ANN in determining the failure pressure of corroded pipelines is not limited to pipelines with single corrosion defects only but also interacting defects. Xu et al. modelled an ANN that incorporated the longitudinal and circumferential spacing of corrosion defects in addition to corrosion defect length and depth. The failure pressure predictions using ANN when compared with the experimental burst pressure results proved to have an insignificant error between them. Besides, the failure pressure prediction using ANN had a strong correlation with the failure pressures obtained from the ASME-B31G and DNV's corrosion assessment codes [15]. Upon training and verifying the ANN model, the weights and biases from the model can also be used to form an empirical equation for the prediction of failure pressure in addition to the existing equations.

2 Methodology

2.1 Overview of Corrosion Geometric Values

To obtain training data for the ANN, finite element analysis (FEA) was conducted based on the parameters shown in Table 1. In this study, the manipulated parameters are the length of defect and the depth of defect. These parameters are presented as normalised values.

Table 1 Geometric parameters for FEA parametric study

Input parameters	Values
Outer diameter of pipe, D (mm)	300
Wall thickness, t (mm)	10
Straight pipe length, L (mm)	2000
Normalized defect length, l/D	0.05, 0.1, 0.2, 0.4, 0.6, 0.8
Normalized defect depth, d/t	0.2, 0.4, 0.6, 0.8

2.2 *Development of Finite Element Method*

Prior to conducting the FEA, several assumptions and simplifications were made to the model. The assumptions and simplifications are as follows:

- Rectangular-shaped corrosion defect.
- A quarter model was used to reduce computational time without compromising the accuracy of the results.
- The pipe length was designed to be 2000 mm to prevent the influence of end cap conditions on the area of corrosion defect.

ANSYS 16.1 Structural Product of Mechanical ANSYS Parametric Design Language (ANSYS) offers a wide range of element types for meshing. In this study, hexahedral SOLID185 element type was used as it accommodates for high stress stiffening, creep, large deflection, plasticity, swelling and large strain [16].

To ensure that the quarter pipe model was treated as a whole model, boundary conditions were applied to the model during the FEA using ANSYS. to prevent undesired rigid body movements, degree of freedom (DOF) constraints in the x, y, and z directions were applied at the end of the model. As this study only takes in to account internal pressure acting on the pipe, the internal pressure was applied on the inner surface of the pipe incrementally known as ramped loading.

2.3 *Development of ANN Model*

A feed forward neural network using back-propagation algorithm was developed using MathWorks MATLAB R2019b. Levenberg-Marquardt training function was used to optimise the weight and biases of the ANN model as it is recommended for supervised learning algorithms and requires less computation time. When applying this function, performance is assumed as the mean or sum of squared errors. Hence, mean square error (MSE) performance function was used in this study. A sigmoidal function as presented in Eq. (2) was used as the activation function for the nodes at the hidden layers as the output values are all in the range of positive numbers and a linear function as presented in Eq. (3) was used as the activation function at the output node.

$$a(x) = \frac{1}{1 + e^{-x}} \tag{2}$$

$$f(x) = x \tag{3}$$

2.4 Material Properties

In this study, API 5L X52, API 5L X65 and API 5L X80 was used. The non-linear true stress-strain curves for each of the material were used to determine the material properties of the pipes modelled as in Fig. 1. The true ultimate tensile strength values used for API 5L X52, API 5L X65 and API 5L X80 were 455 MPa, 530 MPa, and

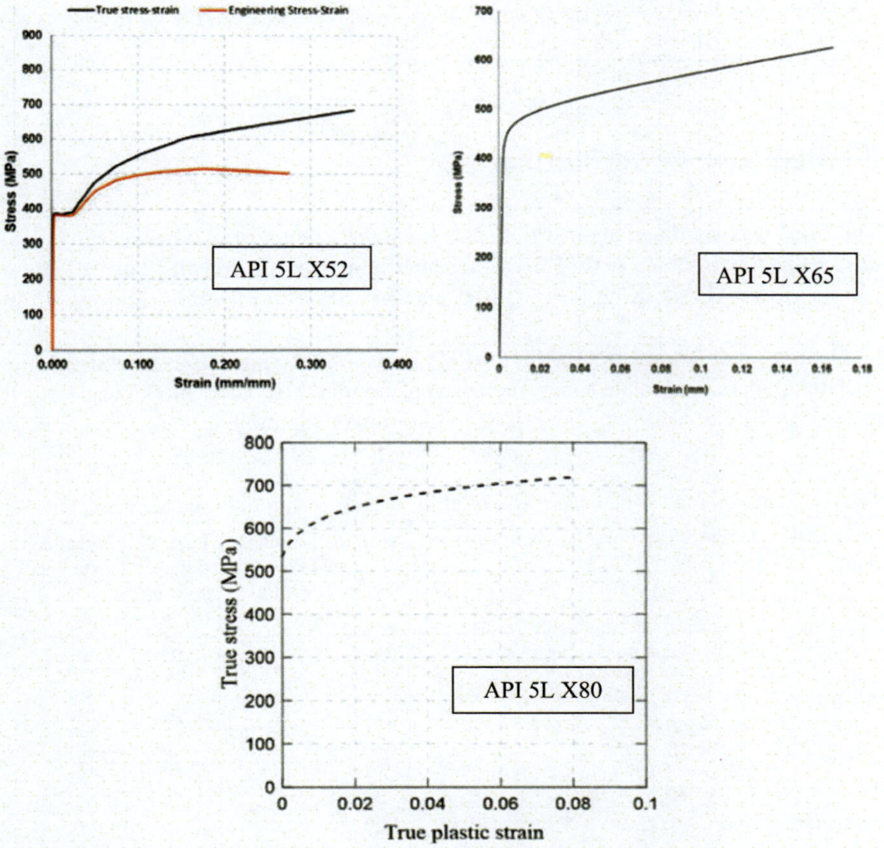

Fig. 1 True stress-strain curve for API 5L X52, API 5L X65 and API 5L X80 [18]

718.2 MPa respectively. The Poisson's ratio and modulus of elasticity were set at a value of 0.3 and 200 GPa respectively. Material stress stiffening and large strain as well as displacements were considered during the FEA to account for high material nonlinearity.

2.5 *Failure Criterion*

In this study, stress-based criterion was utilized to determine the point of failure. The pipe is said to have failed when the localized stress at the region of defect exceeds the true ultimate tensile strength of the pipe. In ANSYS, the stress distribution along the corroded pipe is analyzed using von Mises theory where the effective stress is calculated as a function of hoop, radial and axial stress as shown in Eq. (4) [17].

$$\sigma_e = \sqrt{\frac{1}{2}[(\sigma_h - \sigma_r)^2 + (\sigma_h - \sigma_l)^2 + (\sigma_r - \sigma_l)^2} \tag{4}$$

3 Methodology Verification

The FEM was validated against full scale burst test results from other researches to ensure that the results from the FEA represented actual real-life results. The validation results of the API 5L X52, API 5L X65 and API 5L X80 FEM are summarised in Table 2.

Quarter models were created for the FEA based on the dimensions provided in the research papers for each test specimen. To ensure that the quarter model was treated as a whole model necessary boundary condition were applied.

Table 2 Summary of FEA failure pressure results validation against full scale burst tests [19–21]

Author(s), year	Grade	Specimen	*d* (mm)	*l* (mm)	*w* (mm)	Burst Pressure (MPa)	FEA failure pressure (MPa)	Percentage difference
Bjornoy and Sigurdsson (2000)	X52	Test no. 1	5.15	243	154.5	23.20	22.95	1.08
Ma et al. (2013)	X65	Test no. 61	4.4	200	50	24.11	23.25	3.57
Benjamin et al. (2005)	X80	IDTS 2	5.39	39.6	31.9	22.68	22.40	1.23

Table 3 Categories of corrosion defect based on depth

Category	Range
Shallow corrosion defect	$d/t \leq 0.2$
Medium depth corrosion defect	$0.2 < d/t < 0.6$
Deep corrosion defect	$0.6 \leq d/t \leq 0.8$

4 Results and Discussion

4.1 Preliminary Analyses

In the presence of a circumferential groove corrosion defect, the thickness of the pipe is reduced in the circumferential direction. Generally, corrosion defects can be categorised based on their depths as shown in Table 3.

The thinning of the pipe wall causes a disturbance in the hoop stress distribution of the pipe. In a defect free pipe, the hoop stress is distributed evenly along the pipe. When a corrosion defect is present, there is a variation in the hoop stress distribution and it causes stress build up in specific regions. The stress concentration continues to increase in that region and eventually the pipe fails at the region of corrosion defect [22, 23]. This occurs when the magnitude of stress in the region of defect exceeds the σ^*_{UTS} of the pipe.

4.2 Generation of ANN Training Data Using FEA

In this study, FEA was used to predict the failure pressure of three different pipe grades that are API 5L X52 (low toughness), API 5L X65 (medium toughness) and API 5L X80 (high toughness). Table 4 lists the dimensions of the circumferential groove corrosion defect and the respective normalised failure pressures for each grade when subjected to internal pressure only. The failure pressure data for API 5L X65 was obtained from previous studies conducted by Arumugam et. al. [17]. The failure pressures obtained using FEA were normalised by the intact pressures of the pipes. The failure pressure of the intact pipe was calculated using Eq. (5).

$$P_i = \frac{\sigma_{UTS} t}{r_i} \tag{5}$$

The failure pressure of the intact pipe for API 5L X52, API 5L X65, and API 5L X80 are 32.50 MPa, 37.92 MPa, and 51.30 MPa respectively. This data was used as the training data for the ANN model to predict the failure pressure of a pipe with circumferential groove defect subjected to internal pressure only.

Upon comparing the results obtained from the FEA, the normalised failure pressures of the pipe was plot for the grades as shown in Figs. 2, 3, 4 and 5. The graphs

Table 4 Failure pressure of circumferential corrosion defect subjected to internal pressure

Corrosion defect depth, d/t	Corrosion defect length, l/D	Normalised FEA failure pressure, P_f/P_i		
		X52	X65	X80
0.2	0.05	0.98	1.00	0.98
0.2	0.10	0.95	0.99	0.96
0.2	0.20	0.92	0.97	0.94
0.2	0.40	0.88	0.92	0.90
0.2	0.60	0.85	0.89	0.86
0.2	0.80	0.84	0.86	0.84
0.2	1.00	0.83	0.85	0.82
0.2	1.20	0.83	0.85	0.81
0.4	0.05	0.92	0.98	0.96
0.4	0.10	0.89	0.94	0.93
0.4	0.20	0.81	0.85	0.85
0.4	0.40	0.71	0.74	0.75
0.4	0.60	0.66	0.69	0.69
0.4	0.80	0.64	0.66	0.65
0.4	1.00	0.64	0.65	0.63
0.4	1.20	0.63	0.64	0.62
0.6	0.05	0.87	0.92	0.88
0.6	0.10	0.80	0.84	0.83
0.6	0.20	0.67	0.70	0.71
0.6	0.40	0.52	0.54	0.55
0.6	0.60	0.46	0.47	0.48
0.6	0.80	0.44	0.45	0.44
0.6	1.00	0.43	0.44	0.43
0.6	1.20	0.43	0.43	0.42
0.8	0.05	0.76	0.82	0.78
0.8	0.10	0.64	0.70	0.66
0.8	0.20	0.45	0.47	0.46
0.8	0.40	0.27	0.28	0.29
0.8	0.60	0.24	0.24	0.24
0.8	0.80	0.22	0.23	0.23
0.8	1.00	0.22	0.22	0.22
0.8	1.20	0.21	0.22	0.21

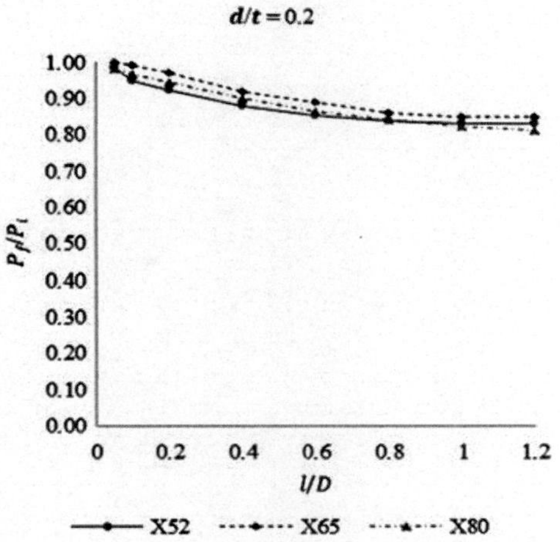

Fig. 2 Normalised FEA failure pressure against various normalised defect length at constant normalised corrosion defect depth of 0.2

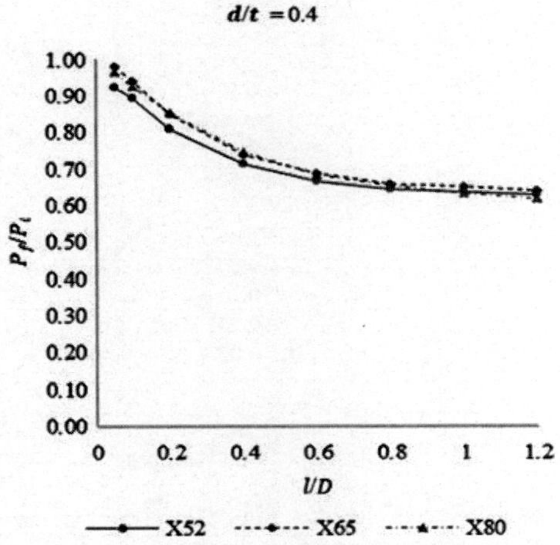

Fig. 3 Normalised FEA failure pressure against various normalised defect length at constant normalised corrosion defect depth of 0.4

of the normalised failure pressure for the three pipe grades closely matched with a highest percentage difference of 7.41%.

4.3 Development of the ANN Model

The ANN model was developed using MathWorks MATLAB R2019b. A feed forward neural network using back-propagation algorithm with two hidden layers and two nodes in each hidden layer was used as illustrated in Fig. 6.

The ANN was developed to produce one output based on three input variables. The input variables are the ultimate tensile strength of the pipe, normalised defect

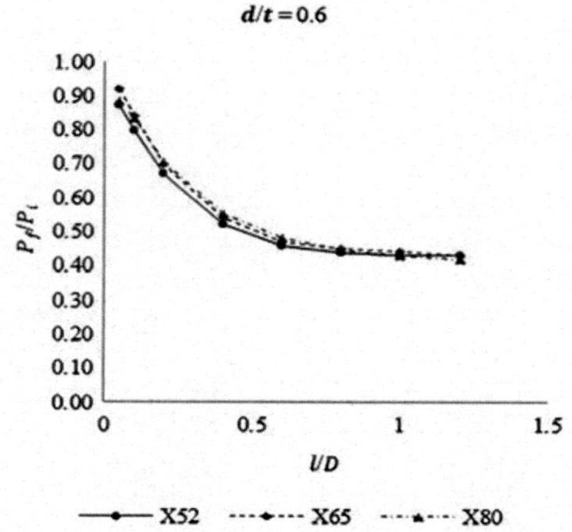

Fig. 4 Normalised FEA failure pressure against various normalised defect length at constant normalised corrosion defect depth of 0.6

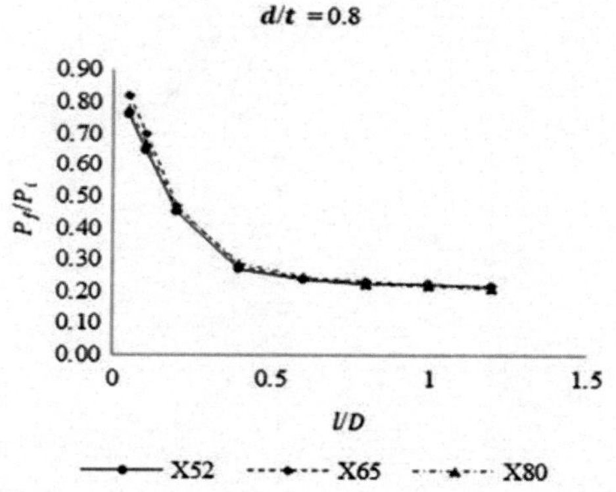

Fig. 5 Normalised FEA failure pressure against various normalised defect length at constant normalised corrosion defect depth of 0.8

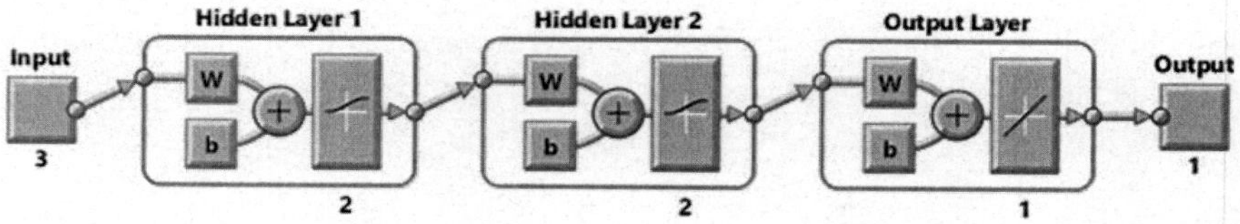

Fig. 6 Architecture of the ANN model extracted from MathWorks MATLAB R2019b

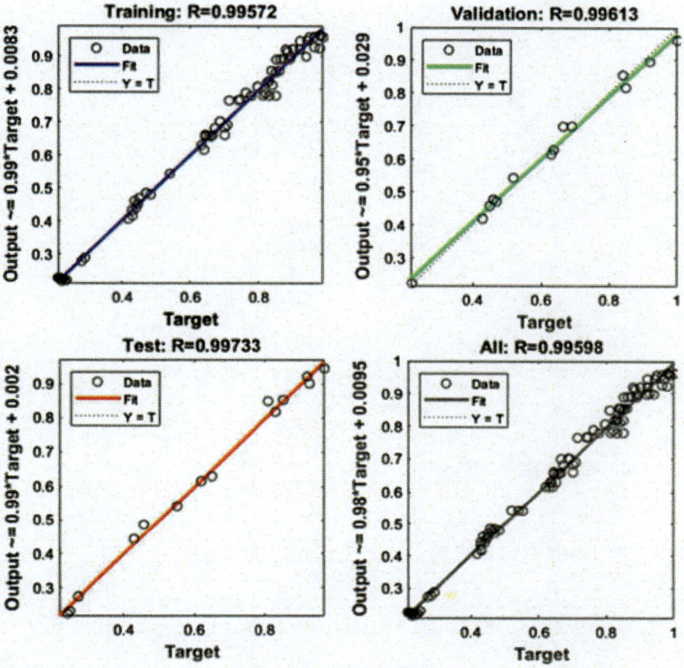

Fig. 7 Regression plot extracted from MathWorks MATLAB R2019b

depth, and normalised defect length. The output of the ANN model is the normalised failure pressure of the pipe. A total number of 96 datasets generated by FEA was used to train the ANN model. The performance of the ANN model was evaluated using a regression plot and a mean squared error performance function as shown in Fig. 7.

4.4 Development of a New Pipeline Failure Pressure Prediction Method

The input values of the ANN that are the ultimate tensile strength of the pipe, normalised defect depth, and normalised defect length are normalised by the ANN model so that the normalised input value, i_n, falls within the range of -1 to 1 before the values are transferred to the first hidden layer. Hence, the inputs need to be normalised using Eq. (6) before being used in the equation. Likewise, the output values that were used to train the ANN model were normalised to range between the values -1 and 1. Hence, the final output, o, has to be denormalized using Eq. (7) to produce the failure pressure of the corroded pipe.

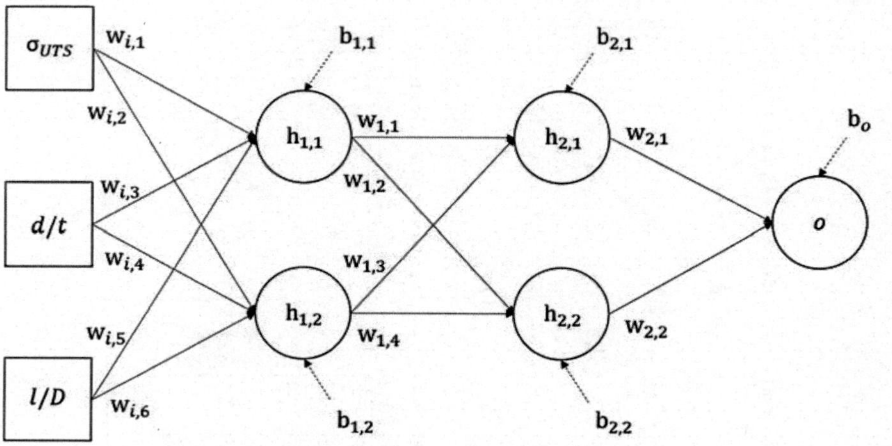

Fig. 8 Architecture of the ANN model

$$i_n = \frac{(i_{n,max} - i_{n,min})(i - i_{min})}{(i_{max} - i_{min})} + i_{n,min} \tag{6}$$

$$o = \frac{(i_n - i_{n,min})(i_{max} - i_{min})}{(i_{n,max} - i_{n,min})} + i_{min} \tag{7}$$

In this study, the weights and biases of the ANN model was used to develop an empirical equation to predict the failure pressure of a pipe with circumferential groove corrosion. The weights are biases of the trained ANN model can be represented in matrix form as presented in Eqs. (8)–(13) based on the illustration of the ANN model in Fig. 8.

$$\begin{bmatrix} \mathrm{h}_{1,1} \\ \mathrm{h}_{1,2} \end{bmatrix} = \begin{bmatrix} \mathrm{w}_{i,1} & \mathrm{w}_{i,3} & \mathrm{w}_{i,5} \\ \mathrm{w}_{i,2} & \mathrm{w}_{i,4} & \mathrm{w}_{i,6} \end{bmatrix} \begin{bmatrix} \sigma_{UTS_n} \\ (d/t)_n \\ (l/D)_n \end{bmatrix} + \begin{bmatrix} \mathrm{b}_{1,1} \\ \mathrm{b}_{1,2} \end{bmatrix} \tag{8}$$

$$\begin{bmatrix} \mathrm{h}_{2,1} \\ \mathrm{h}_{2,2} \end{bmatrix} = \begin{bmatrix} \mathrm{w}_{1,1} & \mathrm{w}_{1,3} \\ \mathrm{w}_{1,2} & \mathrm{w}_{1,4} \end{bmatrix} \begin{bmatrix} a(\mathrm{h}_{1,1}) \\ a(\mathrm{h}_{1,2}) \end{bmatrix} + \begin{bmatrix} \mathrm{b}_{2,1} \\ \mathrm{b}_{2,2} \end{bmatrix} \tag{9}$$

$$o_n = f\left(\begin{bmatrix} a(\mathrm{h}_{2,1}) \\ a(\mathrm{h}_{2,2}) \end{bmatrix} \begin{bmatrix} \mathrm{w}_{2,1} & \mathrm{w}_{2,2} \end{bmatrix} + [b_o] \right) \tag{10}$$

Upon training the ANN model, specific constant values were obtained for the weights and biases as presented below.

$$\begin{bmatrix} h_{1,1} \\ h_{1,2} \end{bmatrix} = \begin{bmatrix} -0.00738 & 6.3214 & 0.88069 \\ 0.01927 & -3.209 & -1.4889 \end{bmatrix} \begin{bmatrix} \sigma_{UTS_n} \\ (d/t)_n \\ (l/D)_n \end{bmatrix} + \begin{bmatrix} -4.4684 \\ 2.629 \end{bmatrix} \tag{11}$$

$$\begin{bmatrix} h_{2,1} \\ h_{2,2} \end{bmatrix} = \begin{bmatrix} -64.4473 & -55.0117 \\ -6.6915 & -1.9911 \end{bmatrix} \begin{bmatrix} a(h_{1,1}) \\ a(h_{1,2}) \end{bmatrix} + \begin{bmatrix} 58.0681 \\ 5.327 \end{bmatrix} \tag{12}$$

$$o_n = f\left(\begin{bmatrix} a(h_{2,1}) \\ a(h_{2,2}) \end{bmatrix} [-23.4996 \; 33.3951] + [-8.8604]\right) \tag{13}$$

The equations used to predict the failure pressure of corroded pipeline with circumferentially grooved corrosion defect in this study is summarised in the following steps:

Step 1: Normalisation of input parameters

$$\sigma_{UTSn} = \frac{2\sigma_{UTS} - 910}{263.2} - 1 \tag{14}$$

$$(d/t)_n = \frac{2(d/t)_i - 0.4}{0.6} - 1 \tag{15}$$

$$(l/D)_n = \frac{2(l/D)_i - 0.1}{1.15} - 1 \tag{16}$$

Step 2: Calculation of the normalised output value

$$a(h_{1,1}) = \frac{1}{1 + e^{0.00738\sigma_{UTSn} - 6.3214(d/t)_n - 0.88069(l/D)_n + 4.4684}} \tag{17}$$

$$a(h_{1,2}) = \frac{1}{1 + e^{-0.01927\sigma_{UTSn} + 3.209(d/t)_n + 1.4889(l/D)_n - 2.629}} \tag{18}$$

$$a(h_{2,1}) = \frac{1}{1 + e^{64.4473 * a(h_{1,1}) + 55.0117 * a(h_{1,2}) - 58.0681}} \tag{19}$$

$$a(h_{2,2}) = \frac{1}{1 + e^{6.6915 * a(h_{1,1}) + 1.9911 * a(h_{1,2}) - 5.327}} \tag{20}$$

$$f(o_n) = 23.4996 * a(h_{2,1}) + 33.3951 * a(h_{2,2}) - 8.8604 \tag{21}$$

$$o_n = f(o_n) \tag{22}$$

Step 3: Denormalization of the output value

$$P_f/P_i = \frac{0.789474o_n + 0.789474}{2} + 0.210526 \tag{23}$$

Step 4: Calculation of the failure pressure

$$P_i = \frac{\sigma_{UTS} t}{r_i} \tag{24}$$

$$P_f = P_f/P_i * P_i \tag{25}$$

4.5 Performance Evaluation of the New Corrosion Assessment Method

Upon comparing the failure pressure prediction results obtained using the newly developed equations with the results obtained from the burst test, it was observed that results obtained from both the methods show close agreement. The percentage errors range from −9.63 to 9.64% with a standard deviation of 2.5. Figure 9 depicts the probability distribution of the percentage error obtained using the newly developed failure pressure prediction method for circumferential groove corrosion defect subjected to internal pressure. Conservative failure pressure is indicated by the negative values while over-estimation is indicated by the positive values. Since the error

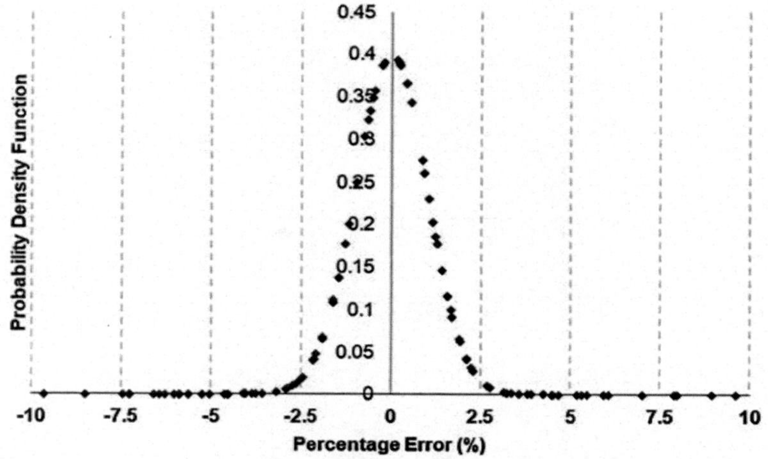

Fig. 9 Probability distribution of the percentage error obtained using the newly developed failure pressure prediction method for circumferential groove corrosion defect subjected to internal pressure

percentage is within the accuracy limit of 10%, these equations can be incorporated with risk-based safety factors to ensure efficient pipeline operations.

4.6 Recommendations for Future Studies

Future studies should consider using a greater number of ANN training dataset that consists of more types of material and varying parameters of corrosion defects to increase the accuracy of the ANN model as well as the equation. Besides, axial compressive stress should be taken into account during the FEA. In reality, the pipelines are subjected to both internal pressure and axial compressive stress due to ground movement and various other occurrences [19]. The presence of axial compressive stress greatly effects the failure pressure of a pipeline. By taking into account the axial compressive stress acting on the pipe, more realistic applications of the failure pressure prediction can be utilised.

5 Conclusion

In this study, FEA was used to generate training data using three materials that are API 5L X52, API 5L X65 and API 5L X80 to develop an ANN model for the prediction of failure pressure of a pipe with circumferentially grooved corrosion defect. The weights and biases of the ANN model was then used to develop an empirical equation for the prediction of failure pressure of a pipe with circumferentially grooved corrosion defect as a function of true ultimate tensile strength, normalised defect depth and normalised defect length.

Acknowledgements This work was supported by Yayasan Universiti Teknologi PETRONAS, Malaysia [015LC0-110].

References

1. Ren L, Jiang T, guang Jia Z, sheng Li D, lin Yuan C, nan Li H (2018) Pipeline corrosion and leakage monitoring based on the distributed optical fiber sensing technology. Meas J Int Meas Confed 122:57–65
2. Cosham A, Integrity P, Palmer A, Hopkins P (2004) The assessment of corrosion in pipelines-guidance in the pipeline defect assessment manual (PDAM)
3. Belachew CT, Ismail MC, Karuppanan S (2011) Burst strength analysis of corroded pipelines by finite element method. J Appl Sci 11(10):1845–1850
4. Koch G, Thompson N, Moghissi O, Gould M, Payer J, Varney J. International measures of prevention, application , and economics of corrosion technologies study
5. Fekete G, Varga L (2012) The effect of the width to length ratios of corrosion defects on the burst pressures of transmission pipelines. Eng Fail Anal 21:21–30

6. Cosham A, Hopkins P, Macdonald A (2007) Best practice for the assessment of defects in pipelines—corrosion, vol 14, pp 1245–1265
7. American Society of Mechanical Engineers (2015) Manual for determining the remaining strength of corroded pipelines: supplement to ASME B31 code for pressure piping
8. DNV (2017) Recommended practice DNV-RP-F101
9. Kiefner JF, Vieth PH (1989) A modified criterion for evaluating the remaining strength of corroded pipe. Pip Res Comm
10. Chiodo MSG, Ruggieri C (2009) Failure assessments of corroded pipelines with axial defects using stress-based criteria: numerical studies and verification analyses, pp 164–176
11. Arumugam T, Karuppanan S, Ovinis M (2019) Finite element analyses of corroded pipeline with single defect subjected to internal pressure and axial compressive stress
12. Song H, Zhang Y, Yin X, Wang W, Xia F (2011) How to improve the prediction accuracy of the ANN model to underground water content effectively. In: 2011 International symposium on water resources environment protection, vol 1, pp 350–353
13. Haque ME, Sudhakar KV (2001) Prediction of corrosion–fatigue behavior of DP steel through artificial neural network. Int J Fatigue 23:1–4
14. Zangenehmadar Z, Moselhi O, Ph D, Eng P (2016) Assessment of remaining useful life of pipelines using different artificial neural networks models, vol 30, no 2007, pp 1–7
15. Xu W-Z, Chun BL, Choung J, Lee J-M (2017) Corroded pipeline failure analysis using artificial neural network scheme. Adv Eng Softw 255–266
16. ANSYS. ANSYS theory reference. ANSYS Inc., Canonsburg
17. Arumugam T, Karuppanan S, Ovinis M (2019) Residual strength analysis of pipeline with circumferential groove corrosion subjected to internal pressure. In: Mater Today Proc 1–19
18. De Andrade EQ et al (2006) Finite element modeling of the failure behavior of pipelines containing interacting corrosion defects. In: Proceedings of the international conference on offshore mechanics and arctic engineering—OMAE, vol 2006
19. Bjørnøy OH, Sigurdsson G, Cramer E (2000) Residual strength of corroded pipelines, DNV test results, vol II
20. Ma B, Shuai J, Liu D, Xu K (2013) Assessment on failure pressure of high strength pipeline with corrosion defects, pp 209–219
21. Benjamin AC, Freire JLF, Vieira RD, Diniz JLC, de Andrade EQ (2005) Burst tests on pipeline containing interacting corrosion defects. In: 24th International conference on offshore mechanics and arctic engineering (OMAE 2005), no Omae 2005, pp 1–15
22. Ahammed M, Melchers RE (1997) Probabilistic analysis of underground pipelines subject to combined stresses and corrosion. Eng Struct 0296(12)
23. Chen Y, Zhang H, Zhang J, Li X, Zhou J (2015) Failure analysis of high strength pipeline with single and multiple corrosions. Mater Des 67:552–557

Optimisation of Fe-Al Diffusion Bonding in Air

Asmawi Ismail, Muhammad Al'Hapis Abdul Razak, Mazli Mustapha, M. Zaki Abdullah, Patthi Hussain, Asmalina Mohamed Saat, Fauzuddin Ayub, and Muhammad Muaz bin Mohd Faisal

Abstract Joining of dissimilar metal is challenging due to formation of oxide layer on the intermetallic compound at the joining area. Aluminium and steel bars with length of 28 mm and 47 mm, respectively, are joined via diffusion bonding in featuring gallium at the joint interface to ensure grain boundary attack especially on the embrittlement of aluminium. Experimentation is designed in accordance to Response Surface Methodology (RSM). The holistic approach of optimization technique is exploited to optimise the diffusion bonding parameters to obtain the maximum bonding strength of the dissimilar Al–Fe joint. Impact mechanical properties were evaluated at constant pressure with various heating temperature, soaking time and surface roughness. A good bonding intensifies at the early heating process of the successful bonding boundary. A small toughness value of Izod test is probably due to an incomplete bonding process, at which, air entrapment is present along the joining interface. After a complete 15 cycles of trialling, the most optimal parameters are approximately, 550 °C bonding temperature, 60 min holding time and P900 grit surface roughness.

1 Introduction

A joining of metals such as stainless steel, aluminium alloy, even titanium still be an interesting problem to solve. Metals of matrix composites and/or super alloy are amongst advanced materials that face a greater challenge to having a good quality

A. Ismail · M. Mustapha (✉) · M. Zaki Abdullah · P. Hussain
Mechanical Engineering Department, Universiti Teknologi PETRONAS, 32610 Seri Iskandar, Perak, Malaysia
e-mail: mazli.mustapha@utp.edu.my

M. Al'Hapis Abdul Razak
Manufacturing Section, Universiti Kuala Lumpur Malaysian Spanish Institute, 09000 Kulim, Kedah, Malaysia

A. Mohamed Saat · F. Ayub · M. Muaz bin Mohd Faisal
Universiti Kuala Lumpur, Malaysian Institute of Marine Engineering Technology, 32200 Lumut, Perak, Malaysia

F. Ahmad et al. (eds.), *ICPER 2020*, Lecture Notes in Mechanical Engineering,
https://doi.org/10.1007/978-981-19-1939-8_71

join. Diffusion bonding is an alternative method to a common fusion-arc welding, especially for joining dissimilar metals [1].

A process of diffusion bonding between materials takes place in a solid-state phase, when the pressure is applied on the faying surfaces. Even at an upraised temperature and with no melting of materials, the macroscopic deformation occurs during heating [2]. Diffusion bonding enables a joining between materials even though they are difficult to join using a conventional fusion welding, hence, this is the main reason for it being a preferred technique to joining complex parts or precision production articles [3, 4].

Joining dissimilar metal such as aluminium alloy to mild steel, titanium to stainless steel, and metal composite to polymer are getting more popular. The main variables in joining metal by diffusion bonding are welding temperature, pressure, holding time and surface roughness [2, 5]. An ideal condition for diffusion bonding is in vacuum environment to ensure protection from contamination or oxidation on the mating surfaces and providing a clean surfaces during dissolution or sublimation of oxides that diffuse into the materials [2].

This study focused on joining dissimilar metal of aluminium (AA 5083) with mild steel (A36) by diffusion bonding. The optimum parameter were determine from impact Izod testing by evaluation of heating temperature, soaking time and surface roughness. Design of experiment were conducted to select experiment setting and analysed the optimize parameter that contribute to highest toughness value.

2 Materials and Methodology

The impact specimens of bars were machined to dimension of 10 mm × 10 mm according to Izod Impact testing standard. The mechanical properties of the bonding on the diffusion bonded samples, for impact test were carried out in accordance to ASTM E23-16b [8]. Faying surfaces were polished using abrasive SiC paper with three different roughness, P400, P800 and P1200 grit. Gallium was used as an interlayer between the mating surfaces. Ground surfaces were rinsed in acetone, dried in hot air, and then, applied a thin layer of gallium before joining. Mixed Boron nitride and acetone was stirred and coated on the part of fixture that was in contact with the specimen, to prevent the specimen from any chemical reaction. Figure 1 represents the schematic diagram of the fixture assembly prior to heating the pre-pressed specimens into the atmospheric heating furnace.

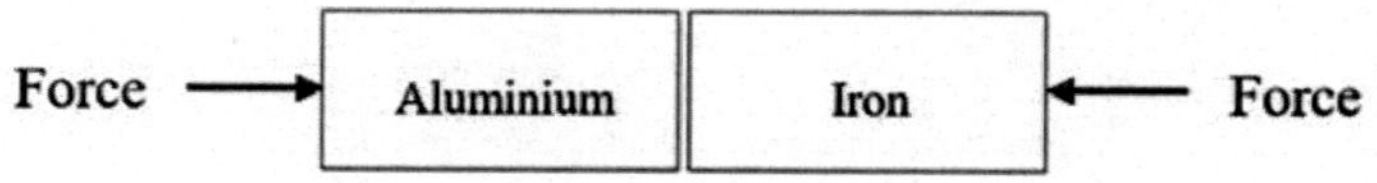

Fig. 1 Diagram of specimens assembly [6]

Table 1 Level of parameters

Factor/Parameter	Unit	Notation	Level		
			−1	0	1
Temperature heating	°C	T	500	525	550
Holding time	min	t	60	120	180
Surface roughness	grit	R	400	800	1200

Table 2 Composition of A36 mild steel and AA5083 aluminium alloy

	Composition specification (wt.%)							
	Al	Mg	Mn	Zn	Cr	Si	Fe	C
AA5083	92	4.9	0.4	0.2	0.2	0.4	–	–
A36	–	–	1.03	–	–	0.28	98	0.29

The specimens from the Al–Fe diffusion bonded joints were tested for toughness using the Ceast Instron impact tester machine. Pressure is an important factor, after temperature, that contributes the most on blocking the oxidation from occurring at the bonding interface [7]. Aluminium and steel were assembled using a G-clamp fixture to ensure the mating surfaces undergo a pre-pressed pressure. The specimens were heated, soaked, and then, cooled in the furnace. The heating rate of the furnace was about 30 °C/min. Furnace cooling was recommended for reducing oxidation from affecting the quality at join.

Bonding parameters were selected based upon the literature and preliminary experiment conducted at an earlier stage. The joining process was conducted in air under atmospheric condition. The level or range of factors is shown in Table 1. Table 2 shows composition of A36 mild steel and AA5083 aluminium alloy used.

3 Results and Discussion

There are two Energy (E) being discussed in this paper, E_1 and E_2. E_1 represents a general Impact Energy produced by the pendulum and E_2 represents the individual energy difference of each 15 runs as to rank them based on their individual bonding strength. The conditions during Izod testing, including the temperature of specimens, Impact Velocity of the pendulum and a general Impact Energy (E_1) of the pendulum are summarised in Table 3. Three factors and three levels are set, and the number of experiment were limited to 15 sets as *designed* by the orthogonal array matrix of *Response Surface Methodology released 12* (RSM R12) software. T represents heating temperature, t represents holding time, R represents surface roughness, and E_2 represents an individual energy for each of 15 runs calculated by the Izod testing machine. Summary of inputs for response 1 (E_2) and the parameters or design matrix for all 15 successful experiments is shown in Table 4. Experimentation and modelling

Table 3 Total energy, Joule (J) from Izod test of A36 mild steel and AA5083 aluminium alloy

Specimen temperature (°C)	Impact velocity (m/s)	Impact energy (E_1), Joule (J)
23.000	3.460	22.000

Table 4 Design matrix and experimental results

		Factor 1	Factor 2	Factor 3	Response 1
Std	Run	A:T heating temperature °C	B:t holding time min	C:R surface roughness grit	E_2 individual energy Joule
10	1	550	120	800	0.312
8	2	550	180	1200	0.035
15	3	525	120	800	0.039
1	4	500	60	400	0.069
5	5	500	60	1200	0.165
4	6	550	180	400	0.068
12	7	525	180	800	0.040
11	8	525	60	800	0.213
13	9	525	120	400	0.008
2	10	550	60	400	0.095
7	11	500	180	1200	0.009
14	12	525	120	1200	0.058
3	13	500	180	400	0.145
6	14	550	60	1200	0.204
9	15	500	120	800	0.148

are run based upon experiment designed by RSM R12. The optimum factor levels for every results and response are established as a result of the interaction and effect plots. Out of 15 experiments that have undergone the Izod test, the maximum toughness (E_2) of 0.312 J was obtained from Run #1, with the conditions of bonding temperature 550 °C, holding time 120 min and surface roughness P800 grit. The value of 0.312 J is used as to initiate a calculation of the optimisation process.

Inputs of all 15 E_2 were decoded for simulating a model. Figure 2 represents a 2D contour model of graphical optimization for E_2 viewing from C:R versus B:t at A:T = 549.985. The overshot 0.325085 peak value of toughness is very high beyond the upper limit of T and the lower limit of t, but at the selective limit of R.

The flowability of metal at a low temperature is irrefutable due to the high yield strength of parent metals, at which, the mating surfaces still at the incomplete coalescence stage [9]. Relatively, the toughness of the joint is high with the increase of temperature, where the atoms from one parent metal diffuse to the other parent metal at the opposite end; it happens more easily with the occurrence of high temperature as to manifesting a good chemical bonding [10]. At an increased temperature mainly

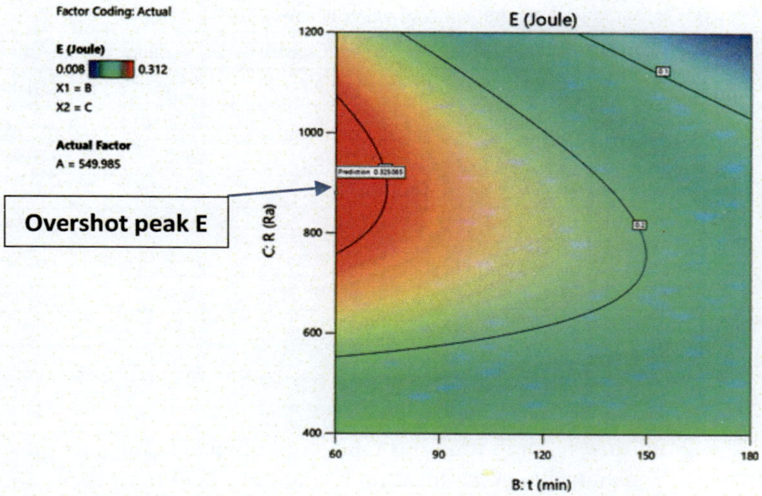

Fig. 2 2D Contour model of graphical optimization for E viewing from C:R versus B:t at A:T = 549.985

about 550 °C, the bonding strength also increase as indicated in Fig. 2, where the peak point of highest energy (red zone, 0.325085) is detected on the contour model due to the movement of elements (mass transfer) across the interface.

Figure 2 shows a red region on the left hand side of the 2D contour model concentrates at the region on X2 = C (C = R, roughness) axis, ranging from 750 to 1080 grit, with the peak value of energy E prediction (0.325085) is detected, exactly at R = 879.422 grit. This peak of the red region occurs on the left hand side of X1 = B (B = t, time) axis. It can be concluded that, the highest toughness value, which is represented by the highest energy, was observed at the very minimum holding time, 60 min. The longer the holding time, the more thickness of interlayer developed at the joining area [6]. A brittle intermetallic compound of the interdiffusion layer does not influence the strength of a diffusion bonded joint because a good bonding inflates at the early heating process of the successful bonding boundary [11].

A study on the effect of surface roughness on the diffusion-bonded join is also discussed. According to Zuruzi, 1999, a rougher surface granted a greater strength than a smoother surface even though there was considered as incomplete joining due to air entrapment along the joining line [11]; moreover, the oxide layer occurs on the rougher surface, hence, performs a metallurgical bonding on the metal–metal contact. Singh and Alpas highlighted that severe wear occurs at a temperature above 150 °C [12], because the rougher surface contributes to a more wear of the oxide layer, which leads to a greater strength. Failures from Izod tests happened at the bond line interface. From the study (Fig. 2), at a combination of high joining temperature

Table 5 ANOVA for reduced quadratic model of surface response

Source	Sum of squares	df	Mean square	F-value	p-value
Model	0.2112	6	0.0352	5.04	0.0199
A-T	0.0056	1	0.0056	0.8049	0.3958
B-t	0.0588	1	0.0588	8.42	0.0198
C-R	0.0006	1	0.0006	0.0888	0.7733
BC	0.0523	1	0.0523	7.48	0.0256
A^2	0.0656	1	0.0656	9.40	0.0155
C^2	0.0659	1	0.0659	9.44	0.0153
Residual	0.0559	8	0.0070		
Cor Total	0.2671	14			

(549.985 °C) and low holding time (60 min), it is observed that a rough surface (P400) does not guarantee a good diffusion bonding ($E_{max} = 0.325085$). The E_{max} observably occurs at the middle segment of the upper and lower roughness limits.

ANOVA for the reduced quadratic model of a surface, with significant p-values, at response E (Energy), transform Square Root, and constant "0" is shown in Table 5. ANOVA is significant when P-value is less than 0.1 and is not significant when P-value is more than 0.1.

Factor coding is *Coded*. Sum of squares is *Type III—Partial.*

The *Model F-value* of 5.04 implies the model is significant. There is only a 1.99% chance that an F-value this large could occur due to noise.

P-values less than 0.0500 indicate model terms are significant. In this case B, BC, A^2, C^2 are significant model terms. Values greater than 0.1000 indicate the model terms are not significant. If there are many insignificant model terms (not counting those required to support hierarchy), model reduction may improve the model. Those insignificant measurements such as AB, AC and B^2 were withdrawn from ANOVA reduced quadratic analysis in obtaining the formula on parameters input to controlling the impact energy or toughness of the joint.

Subsequent to the elaboration of a basic quadratic mathematical model in Eq. 1, the final equation model be summarised as Eqs. 2 (Coded Factors) and 3 (Actual Factors), respectively; where E represents toughness, the total energy absorbed by the joint during Izod test.

$$E = \beta_0 + \beta_A A + \beta_B B + \beta_C C + \beta_{BC}(B)(C) + \beta_{AA} A^2 + \beta_{CC} C^2 \quad (1)$$

$$\begin{aligned} E = {} & 0.299228 + (0.023706 \times A) + (-0.0766776 \times B) + (0.00787499 \times C) \\ & + (-0.0808215 \times BC) + (0.153071 \times A^2) + (-0.15343 \times C^2) \end{aligned} \quad (2)$$

$$E = 66.5065 + (-0.256212 \times T) + (0.00141609 \times t) + (0.00195809 \times R)$$

$$+ (-3.3675610^{-6} \times t \times R) + (0.000244914 \times T^2) + (-9.5893710^{-7} \times R^2) \quad (3)$$

The coded factors equation is used to predict the response for given levels of each factor. The high levels of the factors are coded as + 1 and the low levels are coded as −1. This coded equation is useful for identifying the relative impact of the factors by comparing the factor coefficients. The actual factors equation is used to predict the response for given levels of each factor. The levels are specified in the original units for each factor. This equation is not used to determine the relative impact of each factor because the coefficients are scaled to accommodate the units of each factor and the interception is not at the center of the design space.

The equation indicates the energy absorbed by the joint during Izod test is forecastable due to the changing of diffusion bonding parameters, provided that, it is within the upper and lower level or limit. The most significant interaction between the three parameters, is *BC:tR* (i.e.: $-3.3675610^{-6} \times t \times R$); and the discussion of this paper is solely based upon BC:tR. As seen in Fig. 4a, the base axis of the 3D surface of selected solution are represented by X1 = B and X2 = C, being that B represents t (time) and C represents R (roughness). From Eq. 3, the final equation of actual factors is simulated in the 3D Surface (Fig. 4a) and on graphs of Impact Energy over bonding temperature, heating time, and surface roughness (Fig. 4b, c and d, respectively). The effect of holding time and surface roughness is more significant than bonding temperature. Knowing the diffusion theory, diffusion bonding takes place at between 50 and 75% of the melting temperature of a material, i.e.: between 330 and 495 C for aluminium [2]. The optimisation of the parameters would maximise the bonding quality of the AA5083 and A36 joint by means of analyzing combined desirable criteria. The value of one (1) represents an ideal case in searching for a desirable result. As shown in Fig. 3, of 30 suggested solutions listed in RSM, the recommended values for optimal solution are indicated by red points in each of the parameters range. The earlier input of maximum E_2 value 0.312, is indicated by the blue point. It is postulated that the best optimised factors, with the intention of maximising the energy E_2 (projected to 0.318), as recommended by RSM is summarized in Table 6.

After the analysis, using earlier inputs of Response 1 (Table 4) and referring Fig. 4a, with input design points of E_2 between 0.008 and 0.312, it is predicted that the highest energy, E_{max} could go up to 0.325085. This peak E_{max} only exists at the optimum value of parameters. The optimum parameters for this simulation of diffusion bonding are at a holding time B:t = 60.026, surface roughness C:R = 879.422, and bonding temperature A:T = 549.985 °C. A confirmation test of another Al–Fe diffusion join has to be carried out as to verify the equation and E_{max}, via applying the above mentioned values of B:t = 60.026, C:R = 879.422 and A:T = 549.985. It is considerably significant, should the prediction results from the confirmation test be vastly above 90% similarity to the actual values of impact energy [13].

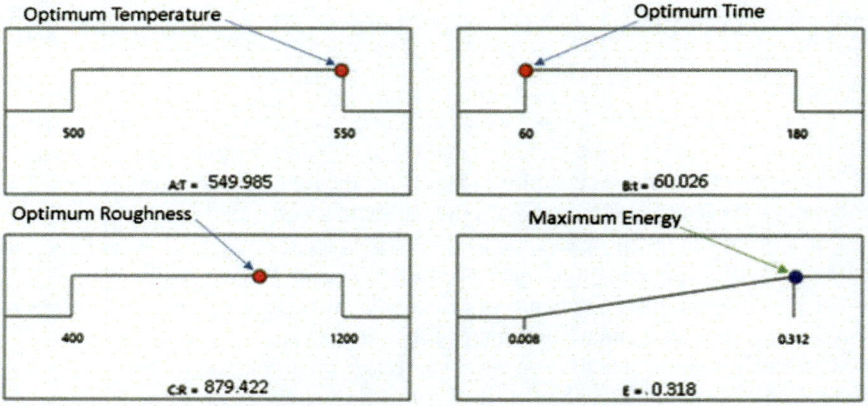

Fig. 3 Optimal factors of optimisational model

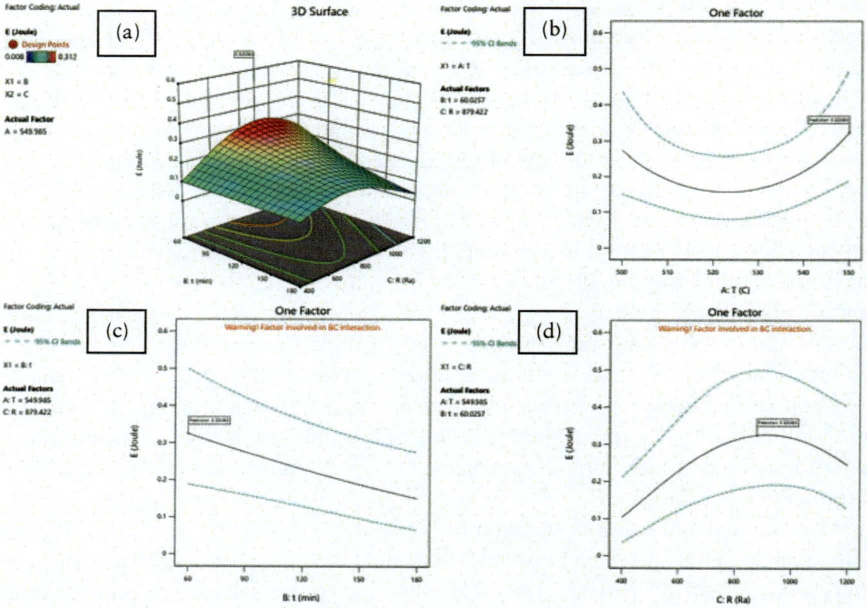

Fig. 4 **a** 3D surface of selected solution; **b** Graph of impact energy over bonding temperature; **c** Graph of impact energy over bonding time; **d** Rate of impact energy over surface roughness

Table 6 Data summary of optimisational model

A:T	B:t	C:R	E	Desirability
549.985	60.026	879.422	0.318	1.000

4 Conclusion

The optimisation of bonding parameters can maximise the bonding quality of the AA5083 and A36 joint by means of analysing combined desirable criteria. It can be concluded that:

1. A small toughness value of Izod test is probably due to an incomplete bonding process, at which, air entrapment is present along the joining interface. It is observed that a rough surface does not guarantee a good diffusion bonding.
2. The energy absorbed by the joint during Izod test is forecastable, at which, a different combination of parameters input is resulting in a different output value. It is reckoned that the most significant interaction is tR, hence, the effect of holding time and surface roughness is more significant than bonding temperature. The interactive parameters of tR, T^2 and R^2 are the significant model terms that influence the toughness.
3. Of all the running cycles conducted recommended by the matrix of RSM, the most optimal parameters are approximately, at the bonding temperature of 550 °C, holding time of 60 min and surface roughness is at slightly below P900 grit.
4. High bonding temperature improves bonding quality, to specifically, altering the toughness or absorbed energy that associates during impact testing. The flowability of a metal does not only facilitate the migration of atoms to diffuse at a distinctive rate, also to manifesting a good chemical bonding due to the movement of elements (mass transfer) across the interface.
5. The formula modelled by RSM is regulated to magnify the quality of bonding via manipulation of temperature, timing and roughness. In the near future, the confirmation test is to be executed as to verify the equation and energy; and it is expected to be above 90% similarity to the actual value.

Acknowledgements The authors are grateful and acknowledge the financial support provided by The Ministry of Higher Education–Malaysia (MOHE) under the Fundamental Research Grant (FRGS - Grant No. FRGS/1/2019/TK03/UTP/02/6). The authors would like to thank Universiti Teknologi PETRONAS for providing the resources to perform this research. The authors also would like to thank Ts. Dr. Al'Hapis for his contribution bestowed following the encouragement and input on the discussion of DOE/RSM in the paper.

References

1. Kundu S, Ghosh M, Laik A, Bhanumurthy K, Kale G, Chatterjee S (2005) Diffusion bonding of commercially pure titanium to 304 stainless steel using copper interlayer. Mater Sci Eng A 407:154–160
2. Kazakov N (1985) An outline of diffusion bonding in vacuum. In: Diffusion bonding of materials. Pergamon Press, pp 10–16

3. Kolařík L, Janovec J, Kolaříková M, Nachtnebl P (2015) Influence of diffusion welding time on homogenous steel joints. Proc Eng 100:1678–1685
4. Wang A, Ohashi O, Ueno K (2006) Effect of surface asperity on diffusion bonding. Mater Trans 47:179–184
5. Tsakiris V, Lucaci M, Leonat L, Alecu G (2007) Materials bonding by diffusion welding technology. In: 2007 1st international conference innovative technologies for joining advanced materials
6. Ismail A, Hussain P, Mustapha M, Nuruddin MF, Saat AM, Abdullah A et al (2016) Fe-Al diffusion bonding: effect of reaction time on the interlayer thickness. J Mech Eng 13:10–20
7. Momono FAT, Katayama HG, Sakura I, Matsuo K (1990) Diffusion bonding of cast iron to steel under atmospheric pressure. Casting Science and Technology, The Japan Foundry men Society
8. A. E23-16b (2018) Standard test methods for notched bar impact testing of metallic materials. An American National Standard, p 26
9. Ureña A, De Salazar JG, Quiñones J (1992) Interface reactions and bonding strength in aluminium alloy (AA-7075)-alumina diffusion bonds. J Mater Sci 27:5291–5296
10. Venugopal S, Mahendran G, Nadu T (2018) Optimization of the diffusion bonding parameters on aa5083 aluminium alloy using Taguchi's technique
11. Zuruzi A, Li H, Dong G (1999) Effects of surface roughness on the diffusion bonding of Al alloy 6061 in air. Mater Sci Eng A 270:244–248
12. Singh J, Alpas A (1996) High-temperature wear and deformation processes in metal matrix composites. Metall Mater Trans A 27:3135–3148
13. Razak M, Rani A, Saad N, Littlefair G, Aliyu A (2018) Controlling corrosion rate of Magnesium alloy using powder mixed electrical discharge machining. In: IOP conference series: materials science and engineering, 2018, p 012010

Review of Current Development of Knee Rehabilitation Device Using Series Elastic Actuator (SEA)

Krishnan Subramanim, Sriraman, Victor Amirtham, A. M. A. Rani, Sadaqat Ali, and Thinagaran Perumal

Abstract The purpose of this study is to review about the Series Elastic Actuator (SEA), which analyses the theoretical and experimental results of impedance control, stiffness, torque control, energy consumes to generate the actuator. Normally the torque control of SEA depends on the spring and the amount of current giving in from the motor to generate the actuator. The stiffness in SEA also depends on the spring where many customize the spring according to the experimental needs. Impedance control is important because of the importance of interactions with the human body. This SEA largely used in robotic rehabilitation field since it has one of the best actuation techniques where the manufacturer can customize them according to their needs. SEAs are easy to manufacture, light, low cost and can be safely used in human operations without any major safety requirements. A new design was introduced to improve the range of motion and to have better torque control.

Keywords Torque control · Stiffness · Energy consumptions

K. Subramanim (✉) · Sriraman
Department of Mechanical Engineering, Manipal International Universiti (MIU), Nilai, Negeri Sembilan, Malaysia
e-mail: Krishnan.subramaniam@miu.edu.my

V. Amirtham
Politeknik Ungku Omar, Ipoh, Perak, Malaysia

A. M. A. Rani
Mechanical Engineering Department, Universiti Teknologi PETRONAS (UTP), 32610 Bandar Seri Iskandar, Perak, Malaysia

S. Ali
National University of Sciences and Technology, Islamabad, Pakistan

T. Perumal
Faculty of Computer Science and Information Technology, Department of Computer Science, Universiti Putra Malaysia (UPM), Seri Kembangan, Malaysia

F. Ahmad et al. (eds.), *ICPER 2020*, Lecture Notes in Mechanical Engineering,
https://doi.org/10.1007/978-981-19-1939-8_72

1 Introduction

The first SEA was created for a robot with the name of Spring Flamingo [1]. It was used on planar bipedal walking robot. This SEA can drive through 6 degrees of freedom mechanism, 3 each on the leg, hip knee and ankle to generate movement like a human. It can walk after programmed at 1.25 m/s on flat surface. It can also walk uphill and downhill surfaces up to 15° in slope. The Spring Flamingo was then decided to be upgraded and its upgraded version called M2 which was a 3-D bipedal walking robot [2]. It has 12 degrees of freedom with six on each leg. M2 has 3 degrees of freedom on the hip, 2 degrees of freedom on the ankle and 1 on the knee. Since to upgrade the Spring Flamingo, M2 became much more complicated for a 3-D movement. Another robot was developed with SEA as for the lower limb which was named a Corndog [3]. This robot is a planar running robot as it only represents half of a dog and it was then employed with electro-mechanical SEA. A humanoid robot was created with applying series elastic technology on upper limb of its robot [4]. It has 6 degrees of freedom just at its arm with SEA operation. This humanoid robot has the ability to do hammering and turning a crank. After wide studies on using SEA in robots, it was then brought into rehabilitation field where the first SEA used in rehabilitation field was compact rotary SEA for knee joint assistive exoskeleton [5]. A worm gear was used to reduce the compactness of the motor but it increases the friction while functioning and makes it more difficult to control. A robust control algorithm was used to control the system. Series elastic actuated joint for robotic gait rehabilitation was designed [6]. This robotic gait was designed to help impaired patients to train their assisted-as- need approach with providing actuated torque to help the patient to complete the gait cycle. SEAs was then brought into prosthesis devices where a prosthetic knee was developed with stiffness adjustments being described to achieve optimum values for energy consumptions and peak power for running and walking [7]. An actuator is a device that moves or controls some mechanism. An actuator turns a control signal into mechanical action such as an electric motor. Actuators may be based on hydraulic, pneumatic, electric, thermal or mechanical means, but are increasingly being driven by software. In this paper three types of actuators are discussed, which are Shape Memory Alloy (SMA) [8–10], Pneumatic Artificial Muscle [11–14] and Silk Pneumatics Artificial Muscle (SPAM) [15].

2 Characteristics of SEAs

SEAs have been successfully applied in many of the applications for almost around 20 years [16]. As much as SEA widely being used and it as well provides many benefits in force control of robotics. SEAs have an elastic element attached to the mechanical output and with this SEA has advantage over many actuators around which mostly are rigid actuators with low impedance, tolerances to impacts, increased peak power output and passive mechanical energy storage. These properties that have been mentioned are the reason for SEAs to be widely in used in legged actuation systems and human orthosis for rehabilitation [17].

3 Applications and Performance Evaluation

There are various kind of knee rehabilitation device which are already being used in the rehabilitation center. Each device will be powered by an actuator and the dimension of the knee rehabilitation device will be dependent on the actuator. A bigger actuator makes the device even bigger. Series Elastic Actuators (SEA) provides torque and impedance control which can be used to adjust the level according to the treatment being received by the patient [18]. For example, the torque and impedance can be controlled using the DC motor where the amount of current being produce changes the output of the torque and impedance. SEA also have high range of motion where it can be customized accordingly to the requirement needed [19].

A rotational SEA for knee restoration have been introduced by Dos Santos. It comprises of Maxon Motor RE 40, graphite brushes, 150 Watt DC engine, worm gear set with decrease proportion of 150:1, modified torsion spring, precise contact orientation, magneto-safe steady encoder, and opto- electronic gradual encoder. The torsion spring was modified and structured through a simulation program on finite element method to satisfy the necessities of the trial as far as acceptable admissible peak load, low stiffness and a compact and lightweight design yet the spring constant decided during the test was lower than which got by finite element analysis and the explanation was on the grounds because the material used and some imperfection in the model made the differences. Figure 1 shows the cross section of rotary SEA [20].

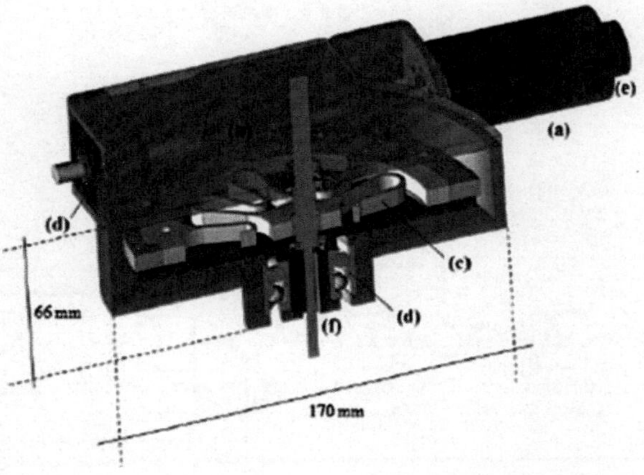

Fig. 1 Cross section of the rotary SEA [20]

When the customized spring get compressed shown in Fig. 2, the stiffness is around 106 Nm/rad and when the torsional spring gets extended, the stiffness is around 103 Nm/rad [20]. To find the desired torque, a block diagram was used in term of impedance control loop showed in Fig. 3. From the block diagram, the desired torque can be expressed by:

$$\tau_{L,d} = k_v(\theta_{L,d} - \theta_L) + b_v(\theta_{L,d} - \theta_L) \tag{1}$$

kv and bv are the virtual stiffness and damping parameters.

The same author came up with an active knee orthosis driven by a rotary SEA to have a better control over the torque using the same design of torsional spring. In this research being proposed that to satisfy a partially supported the knee joint for extension and flexion during physical movement or rehabilitation in individuals with low motor range. Robust torque and impedance controllers are designed to ensure the interaction with the patient is secured and the device has a good system performance. The spring constant achieved was not as planned but this SEA device can be used while walking with the performance controllers [21].

An active knee orthotics was designed, controlled and prototyped by using SEA. In this study, the author compared different type of actuators to see which type of it gives the best performance. The actuators compared are Direct drive, Parallel Elastic

Fig. 2 Customized torsional spring [20]

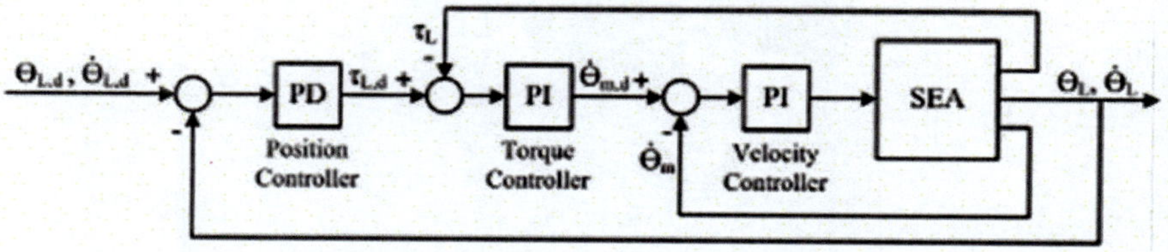

Fig. 3 Block diagram of the impedance control [20]

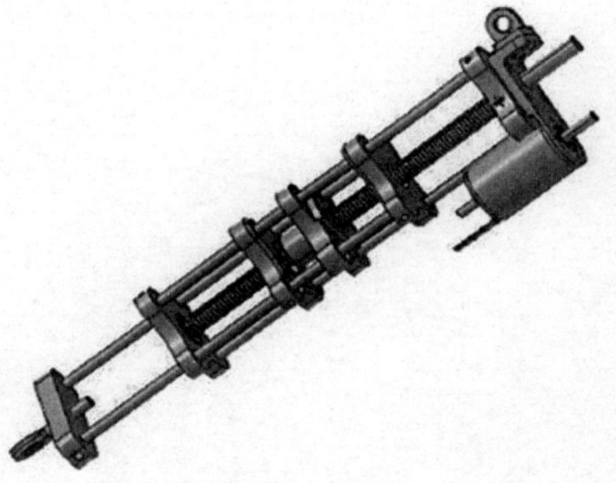

Fig. 4 CAD model of SEA actuator [22]

Actuator and finally SEA. The power required to function the actuator and the energy needed to move the actuator were compared among the 3 actuators [22]. The orthotics energy consumption of each type of actuator goes with the formula:

$$E = \int |P_M| dt \tag{2}$$

It turns out that SEA being the best choice to construct an active knee orthosis for daily usage and eventually an actuator with a mass less than 1kg was developed that provides the ability to produce the required knee joint force and speed. The series elastic actuator (SEA) design was shown in Fig. 4.

A research with the point Design and Characterization of a Compact Rotary SEA for Knee Assistance During over ground walking was done. This paper commonly introduced about offering little help to the knee during walking. The prerequisite was to be wearable and having a decent torque control. The plan incorporates a rotating optical gradual encoder, level DC engine, Harmonic Drive, torsion spring, magnetic incremental encoder and a yield shaft. The torsional spring was modified because there was no torsional spring in the market that could give the necessary result. The manufactured component has lower stiffness than the necessary ones which was presumably because of the real material's elastic modulus being somewhat not quite the same as the one utilized for the finite element simulation and furthermore accuracy in fabricating it. The customized SEA weighs around 1.8 kg. in this way it can permit maker to straightforwardly introduce the actuator on the edge of a knee orthosis which is wearable [18]. Figure 5 shows the cross section of the rotary SEA that was used in this experiment.

In this version of SEA, the torsional spring was customized to fulfil the stiffness of 10 Nm and torque of 150 Nm/rad as shown in Fig. 6. A test bed was used to hold the components which are the torque sensor, the flexible coupling and the SEA. Figure 7 shows the test bed [18].

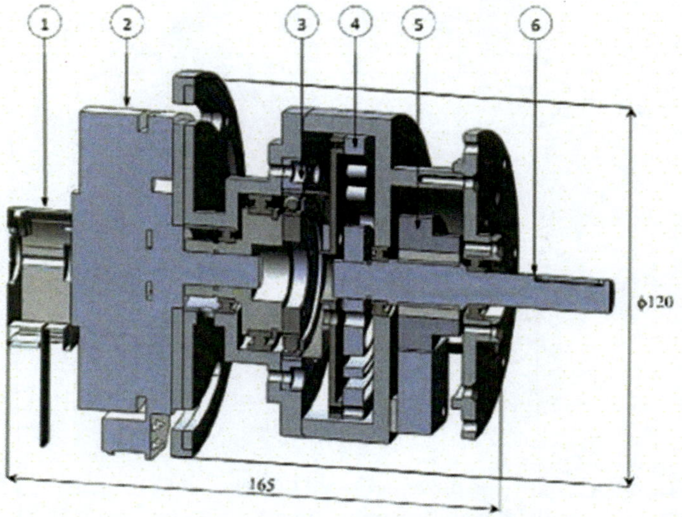

Fig. 5 Cross section of rotary SEA [18]

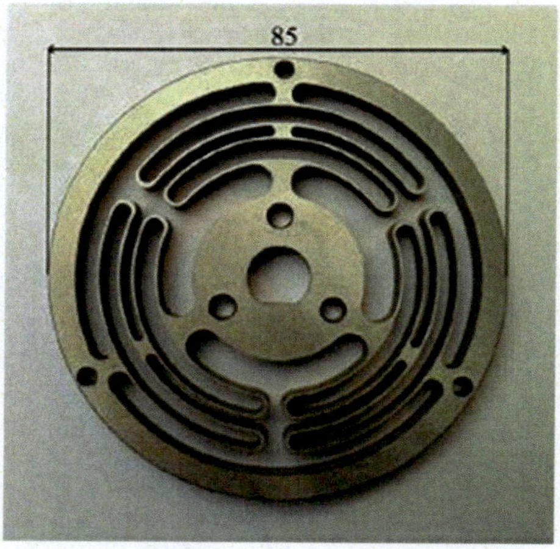

Fig. 6 Customized torsional spring [18]

A research on design and performance evaluation of Rotary SEA was made. The spring deflection is the main factor which influences the torque of the SEA. Without following the original spring deflection, the author embedded a load cell to assess whether the torque prediction from spring deflection is valid and analyze system performance. The linearity of the torsion spring was also examined through some experiments and comparing it with desired torque data [23]. Figure 8 shows the schematic representation of rotary SEA.

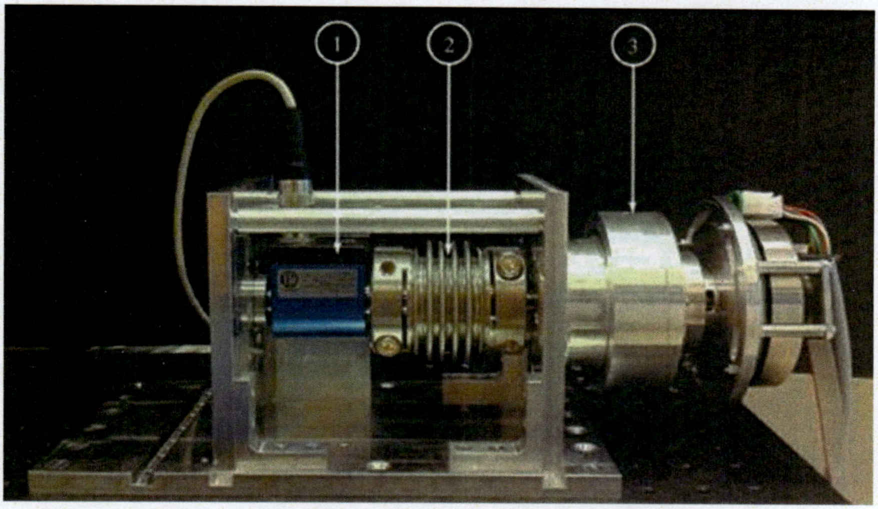

Fig. 7 Test bed for spring characterization [18]

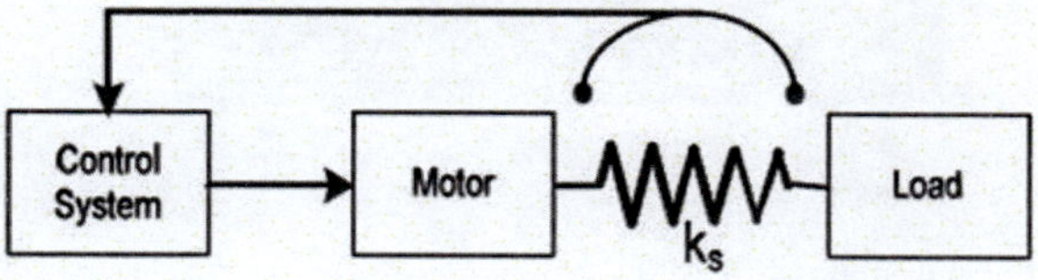

Fig. 8 Schematic of rotary SEA [23]

This research was done to improve the reliability of SEA by not having a significant size. It was tested on with a motor and human to check the actual torque being given by the SEA based on the components that set up to design this SEA. Figure 9 shows the experimental set up of SEA to evaluate the performance [17]. The actual torque of the SEA was calculated with the following formula after testing:

$$\tau = k_{spring} * (\theta_{motor} - \theta_{human}) \tag{3}$$

The main reason for the load cell being included was to measure actual time output torque of the Rotary SEA. It was the discovered that the Rotary SEA can precisely follow required point contrasts with least measure of mistake and less vibration. The spring assumes a major job in the Rotary SEA. A few issues, for example, the weaknesses of the torsion spring dependability, where the spring rate is not generally straight, the obscure properties of the energy included for the deformation of the spring makes a greater hole among actual and needed torque were faced to get the precise outcomes in this test [23].

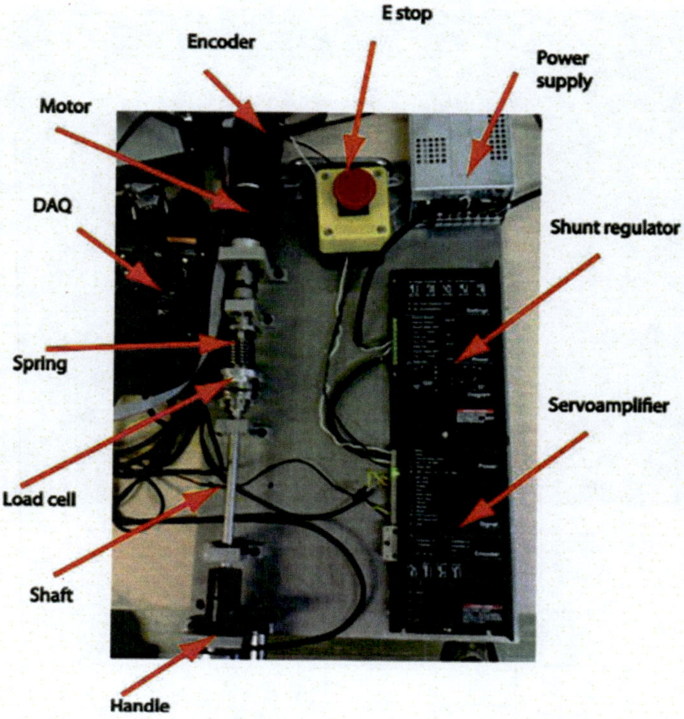

Fig. 9 The experimental set up of SEA [23]

A Gait Phase-Based Control for a Rotary Series Elastic Actuator Assisting the Knee Joint was designed for rehabilitation and assistive system purpose to obtain the desired torque accurately with including the non-linear resistive factors. It was said that to design a non-linear control algorithms are much easier and straightforward than the linear robust controllers. A sliding mode control was used to get better robustness compared to the feedback control. A boundary layer was applied around the sliding surface to decrease the chattering effect and it helps in getting an effective tracking performance. The research was based on human movement by installing the rotary SEA at the knee joint and observe the 2 main gait phases which are swing and stance. The performance of the rotary SEA was verified with human walking motions. Figure 10 shows the rotary SEA was used in this experiment [24].

Normally the torque will be generated by the geared motor but since this is a non-linear resistive torque, the effort from human is needed to overcome the resistive force. Therefore, a normal dc motor was used for this actuator by balancing it. Figure 11 shows the schematic plot of human joint and rotary SEA [24].

The sliding factor mainly depends on the spring where there is sliding mode control to deal with this problem. The spring deflection was calculated with the error

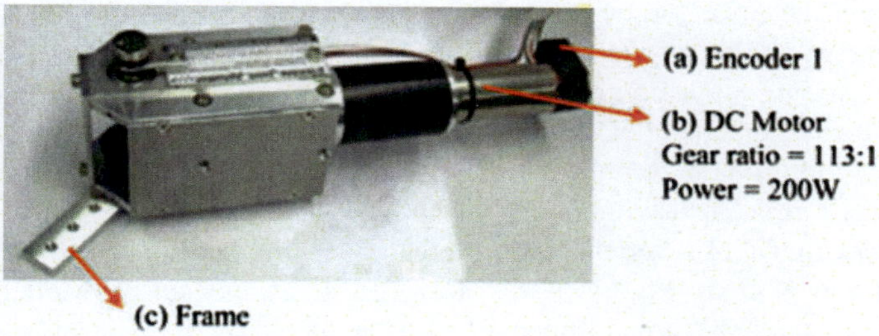

Fig. 10 Model of rotary SEA [24]

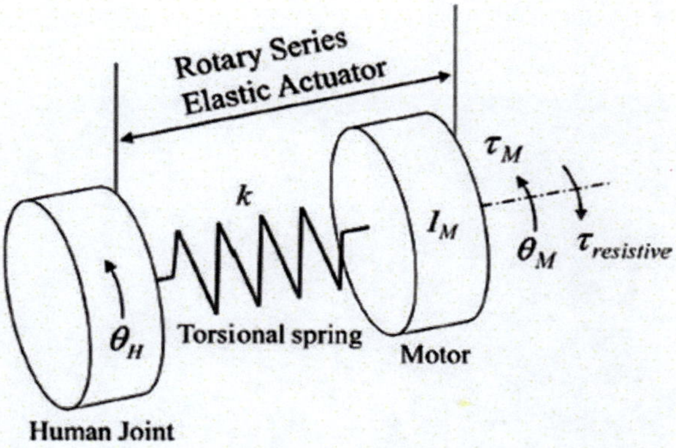

Fig. 11 Schematic plot of human joint and rotary SEA [24]

and actual desired deflection and finally to get the sliding surface value. Figure 12 shows the smart shoes installed in orthosis with rotary SEA.

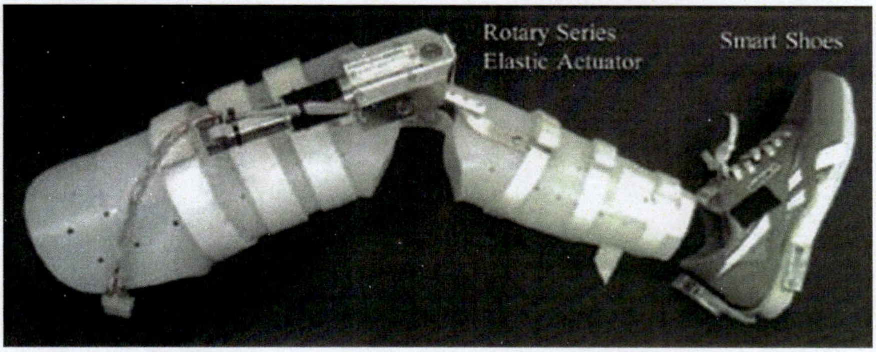

Fig. 12 Rotary SEA with smart shoes installed in orthosis [25]

With this the knee torque was obtained with torque sensor by varying to the human motion angle and the angle produced by the motor. For the robustness for the model, sliding mode control was introduced but it produces chattering which is not desirable to human-robot interactions. The boundary layer was installed to have a smooth sliding mode control and it is varied for swing and stance of human to decrease the torque error without chattering [25].

A Force Control of a Series Elastic Actuator for Impedance Control of an Ankle Rehabilitation Robotic Platform was designed and analyzed. The effects of the dynamic variations of human impedance was analyzed on the control performance during the rehabilitation process. An optimal H1- based force controller implemented in this impedance control strategy for this ankle rehabilitation platform [25]. Figure 13 shows the actual platform for ankle rehabilitation with SEA based and Fig. 14 show the model of SEA based ankle rehabilitation platform.

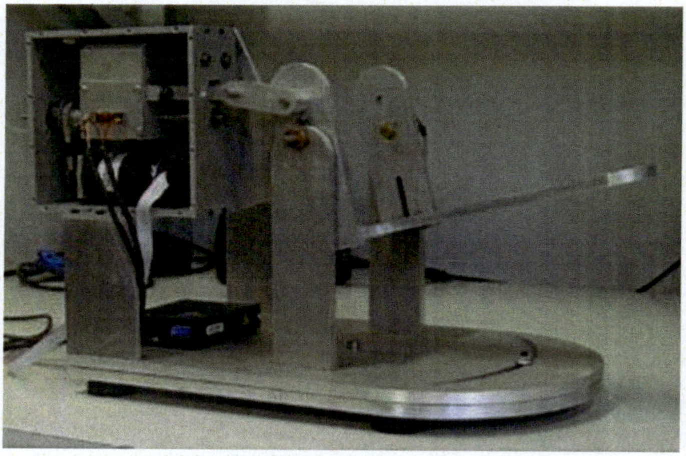

Fig. 13 SEA-based platform for ankle rehabilitation [25]

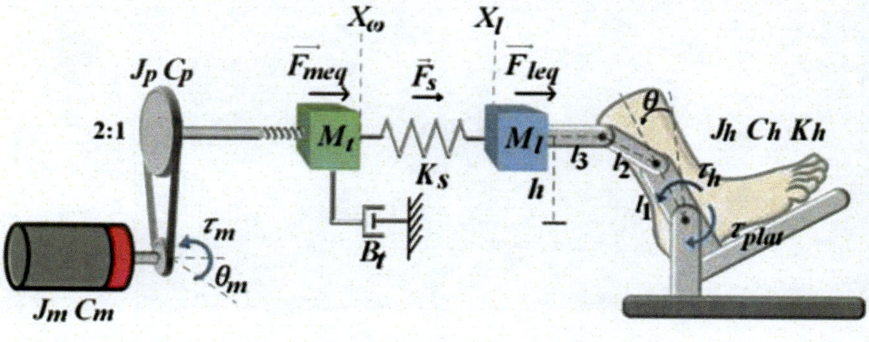

Fig. 14 Model of SEA based ankle rehabilitation platform [25]

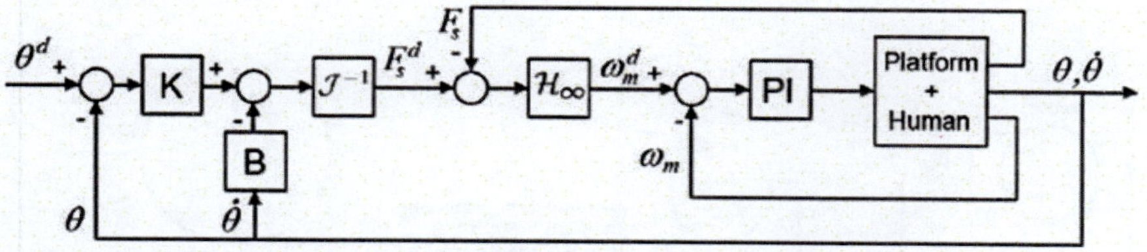

Fig. 15 Architecture of the controller [25]

The active and passive movement of human was experimented in this ankle rehabilitation. The human active motion is when the actuator is attached to the human leg while the person is giving out high stiffness about the neutral position. Human passive motion is that the actuator attached to the human leg and the person does not apply any stiffness as the person goes with the flow of the movement. The force control design shown in Fig. 15 by having the human and robot interaction as disturbance. A mixed sensitivity approach was done to shape the sensitivity, S and control the sensitivity, KcS. To ensure the robust tracking performance, to reject the disturbance and to limit the control signals, the S should be shaped accordingly [25].

A Clutchable series-elastic actuator (CSEA) was implicated to the prosthetic knee design to biologically imitate accurate human knee movement. This is a new mechanism being proposed. Firstly, this device does its work like the standard SEA where it has force control and impedance control for shock impact. Then, the CSEA comes up to store the elastic energy in the customized spring at low electrical power consumption. A small and low mass clutch placed on the motor mainly to reduce the electrical power consumption as the spring being reacted elastically to produce the torque. The main motive of this research is to mechanical design a knee prosthesis which provides stances and flexion of the knee by using less amount of electrical energy [26]. The Fig. 16 shows the model of CSEA.

When the motor runs as the current generates to move the SEA, the clutch remains off. The clutch turns on when the current loop across the clutch and the next round loop runs with the clutch. Clutch plays a vital role on reducing the power usage and consumption [18]. Figure 18 shows the views of the CSEA with translucent protective cosmetic.

By using CSEA, it was able to support a wide range in clinical field for traditional passive prosthetic knees. It seems to be lighter in weight, showed an accurate torque

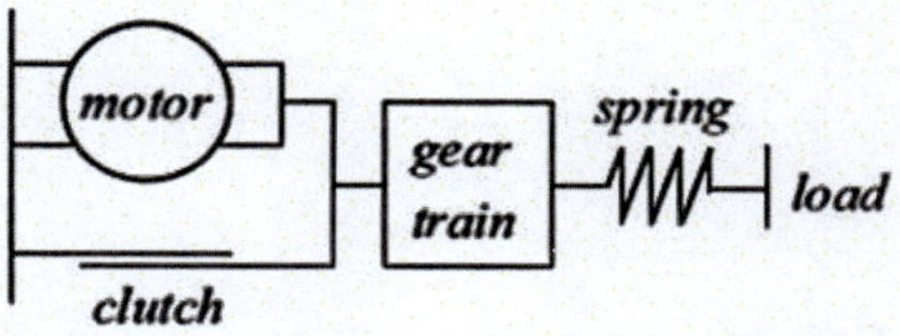

Fig. 16 Model of CSEA [26]

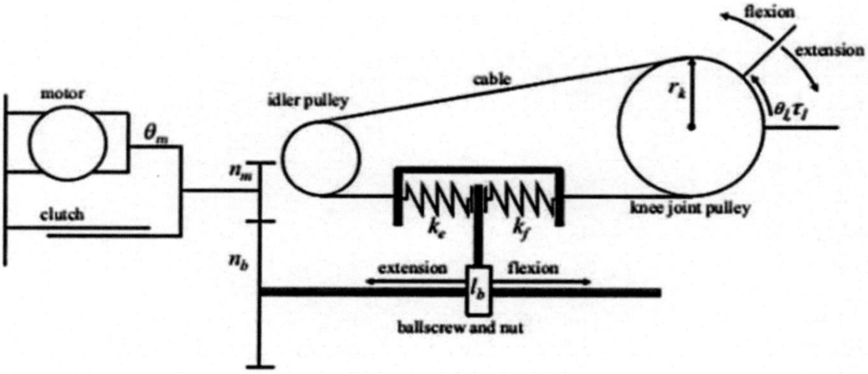

Fig. 17 The geometric configuration of CSEA of knee [25]

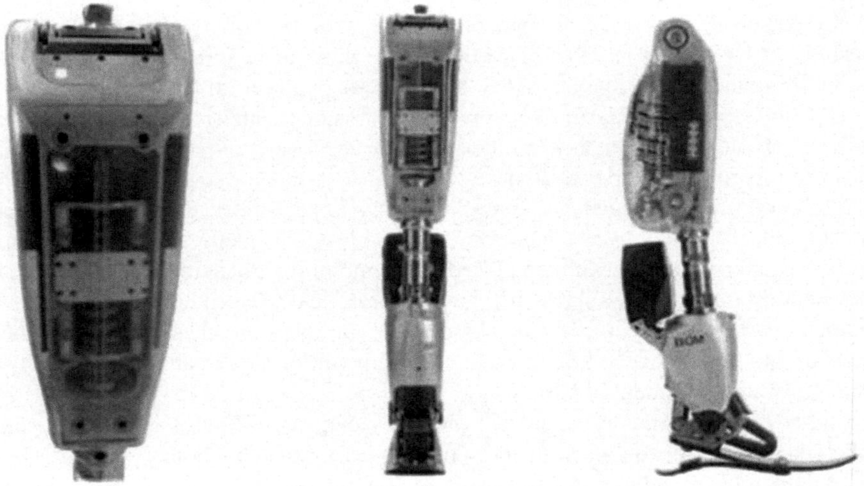

Fig. 18 Anterior view of the clutchable (CSEA) knee shown with translucent protective cosmetic [26]

behaviour and high flexion angle produced. This work can as well include in other biomechanical field which will help to conserve the power consumption [26].

4 Proposed Design

Figure 20 shows the Series Elastic Actuator (SEA) used in knee orthotics rehabilitation device. Figure 19 is the CAD design of the modified version of the knee rehabilitation device using SEA. The modifications will be done on increasing the range of motion as well as a better torque control/resistive force in the device. The existing SEA knee rehabilitation device has less range of motion due to the actuator placed between the thighs and calves. It limits the amount flexion can be done using the existing knee rehabilitation device. In Fig. 19, the actuator will be placed at the side of the thighs and a rotary device with bearing together with the springs, which will be controlled by the pulley generated by the motor. There will not be any hindrance in getting full flexion of the knee, which will allow the user to be more comfortable with the device. The springs play a roll on controlling the torque whereby as we increase the number of springs, it will increase the stiffness of the device. The torque control data for the previous research done with a limit around 0.15Nm. According to the Hooke's Law formula F=kx, k is the spring constant, when the spring constant increases the force increases as well. Since the previous research was done with 4 springs in a place, so there would 4k which will increase the force. Therefore, to reduce the stiffness for a better torque control, the springs are separated 2 at each side different from the existing device where all 4 springs are actuated from the same side (Fig. 20).

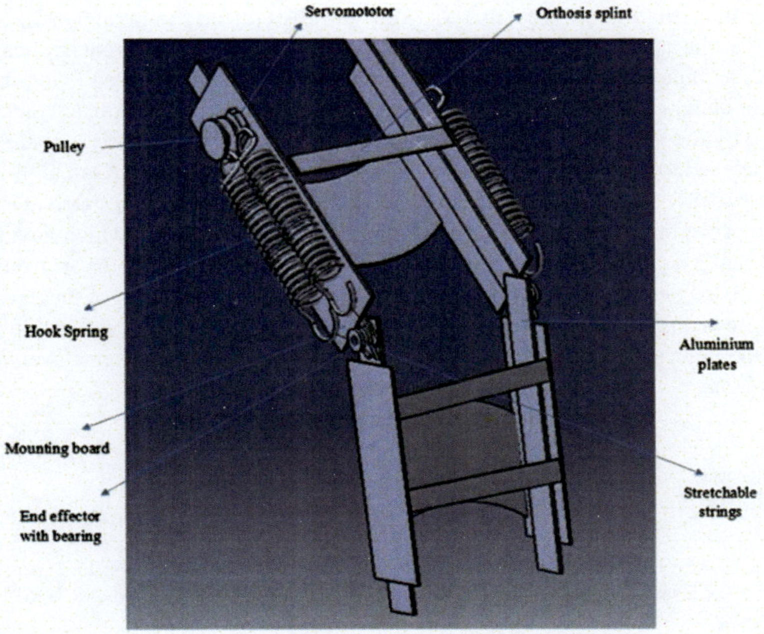

Fig. 19 Side view CATIA design of SEA knee rehabilitation device

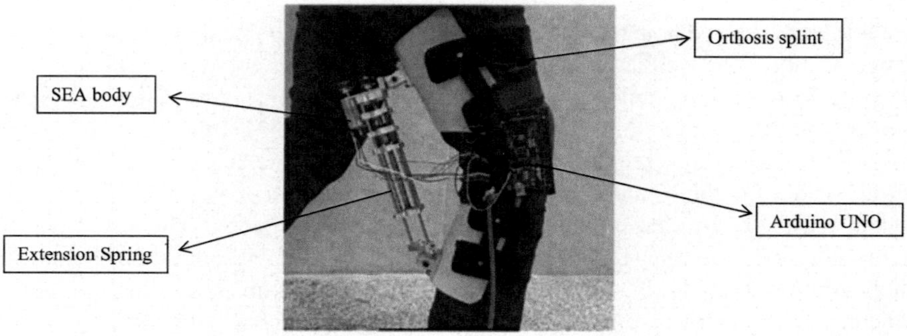

Fig. 20 Active knee orthosis device using SEA

5 Conclusion

The SEAs can be used widely on many fields such as robotics, rehabilitation and also prosthetic. There are various fabrication method and designs for SEAs, which have been stated by the authors of the journals through experiments and calculations. Furthermore, the authors have also come up with customized component in the SEAs, which has the ability to increase the efficiencies by relating it to different applications. Various techniques such as Rotary, Linear and Clutchable can be used to modify the SEA for better-desired results according to the situations being used in and outcomes desired by the researchers. Mainly the torque, impedance and energy consumption was analysed for a further and better control and performance of SEAs in biomedical field especially related to the joints of the body. The purposed design stated in Fig. 19 is to improve the resistive force for the rehabilitation of the knee, which having the torque being controlled from the stiffness in Newtons of the springs. Apart from that, the proposed design have been done to increase the range of motion of the device so that there will be more flexibility so that the user will feel more comfortable when using the device as rehabilitation is a process of recovery for a better movement of body parts.

References

1. Pratt J, Pratt G (1998) Intuitive control of a planar bipedal walking robot. In: Proceedings. 1998 IEEE international conference on robotics and automation (Cat. No. 98CH36146), pp 2014–2021
2. Robinson DW, Pratt JE, Paluska DJ, Pratt GA (1999) Series elastic actuator development for a biomimetic walking robot. In: 1999 IEEE/ASME international conference on advanced intelligent mechatronics (Cat. No. 99TH8399), pp 561–568
3. Krupp BT (2000) Design and control of a planar robot to study quadrupedal locomotion. Massachusetts Institute of Technology

4. Brooks RA, Breazeal C, Marjanović M, Scassellati B, Williamson MM (1998) The Cog project: building a humanoid robot. In: International workshop on computation for metaphors, analogy, and agents, pp 52–87
5. Kong K, Bae J, Tomizuka M (2010) A compact rotary series elastic actuator for knee joint assistive system. In: IEEE international conference on robotics and automation, pp 2940–2945
6. Lagoda C, Schouten AC, Stienen AH, Hekman EE, van der Kooij H (2010) Design of an electric series elastic actuated joint for robotic gait rehabilitation training. In: 2010 3rd IEEE RAS & EMBS international conference on biomedical robotics and biomechatronics, pp 21–26
7. Au SK, Weber J, Herr H (2007) Biomechanical design of a powered ankle-foot prosthesis. In: 2007 IEEE 10th international conference on rehabilitation robotics, 2007, pp 298–303
8. Abdul-Rani AM, Krishnan S, Nagarajan T, Rao T, Ramiah R, Ambaraj W (2017) Early mobilization using heat treated shape memory alloy (SMA) for rehabilitation phases. In: Key engineering materials, pp 112–116
9. Abdul-Rani AM, Krishnan S, Nagarajan T, Rao T (2017) Inverse kinematics-stewart platform actuated by shape memory alloy for immobilization of ankle-foot rehabilitation. In: Key engineering materials, pp 105–111
10. Krishnan S, Nagarajan T, Rani A, Ambaraj W, Ramiah R (2016) Rehabilitation for foot/ankle-continuous passive motion (CPM) using shape memory alloy (sma) actuated Stewart platform. ARPN J Eng Appl Sci 11:22
11. Wise KD, Bhatti PT, Wang J, Friedrich CR (2008) High-density cochlear implants with position sensing and control. Hear Res 242:22–30
12. Amirtham V, Nagarajan T, Krishnan S, Hashim FM (2013) SMA actuation for wrist motion with split-tube flexures. Asian J Sci Res 6:615
13. Krishna S, Nagarajan T, Rani A (2011) Review of current development of pneumatic artificial muscle. J Appl Sci 11:1749–1755
14. Nagarajan T, Krishnan S, Amirtham V, Abdul-Raniand AM, Rao T (2013) Experimental investigation-natural fiber braided sleeve for pneumatic artificial muscles actuation. Asian J Sci Res 6:596
15. Krishnan S, Nagarajan T, Rani AMA, Rao T (2012) Silk pneumatic artificial muscle (SPAM) construction for bio-medical engineering application. In: Business engineering and industrial applications colloquium (BEIAC), 2012 IEEE, pp 302–306
16. Pratt GA, Williamson MM (1995) Series elastic actuators. In: Proceedings 1995 IEEE/RSJ international conference on intelligent robots and systems. human robot interaction and cooperative robots, pp 399–406
17. Pestana J, Bombín R, Arevalo J, Garcia E (2010) Characterization of emerging actuators for empowering legged robots. In: Emerging trends in mobile robotics. World Scientific, pp 997–1004
18. Sergi F, Accoto D, Carpino G, Tagliamonte NL, Guglielmelli E (2012) Design and characterization of a compact rotary series elastic actuator for knee assistance during overground walking. In: 2012 4th IEEE RAS & EMBS international conference on biomedical robotics and biomechatronics (BioRob), pp 1931–1936
19. Junior AL, de Andrade RM, Bento Filho A (2016) Series elastic actuator: design analysis and comparison. In: Recent advance in robotic systems, vol 1, pp 203–234
20. Santos dos WM, Siqueira AA (2014) Impedance control of a rotary series elastic actuator for knee rehabilitation. In: IFAC proceedings volumes, vol 47, pp 4801–4806
21. dos Santos WM, Caurin GA, Siqueira AA (2017) Design and control of an active knee orthosis driven by a rotary series elastic actuator. Control Eng Pract 58:307–318
22. Vadiati A, Bagheri A, Mahjoob M, Sadigh M (2018) Design, control, and prototyping of a series elastic actuator for an active knee orthosis. Eng Solid Mech 6:241–252
23. Donald M,. Li Q (2012) Design and performance evaluation of a rotary series elastic actuator. In: International conference on intelligent robotics and applications, pp 555–564
24. Bae J, Kong K, Tomizuka M (2011) Gait phase-based control for a rotary series elastic actuator assisting the knee joint. J Med Devices 5

25. Pérez-Ibarra JC, Alarcón ALJ, Jaimes JC, Ortega FME, Terra MH, Siqueira AA (2017) Design and analysis of $\mathscr{H}\infty$ force control of a series elastic actuator for impedance control of an ankle rehabilitation robotic platform. In: American control conference (ACC), pp 2423–2428
26. Rouse EJ, Mooney LM, Herr HM (2014) Clutchable series-elastic actuator: implications for prosthetic knee design. Int J Robot Res 33:1611–1625

The Study of Pseudo-ductile Polymer Nano Composite Reinforced with Non-treated Inorganic Nanofillers

Tauseef Ahmed, H. H. Ya, Mohd Amin A. Majid, Mian Imran, Rehan Khan, M. Azeem, M. Azad Alam, H. U. Khalid, and Aamir Mehmood

Abstract The outright understanding of tensile test can furnish details about the pseudo-ductility of High Density polyethylene nano composite. Non-rigid filler particles render a soft composite limiting the applications of a polymer composite while non-treated rigid filler particles compromises the uniform morphology resulting deficient mechanical properties. Therefore, tensile testing of hybrid high density polyethylene was performed using ASTM D638 type I in order to study the pseudo-ductility. In this study, four samples of different weight percentages 0,5,10,30 were tested after fabrication via melt compounding. The phenomenon of stress softening, debonding of filler/matrix were discussed for the hybrid composites after the tensile test. The results obtained indicated that both non-treated fillers can act in synergy, not only to increase the strength but also retain the ductility to a substantial amount in pure polymer. Reinforcement of polymer with the combination of non-treated inorganic fillers of different microstructural lay out proved to be an unprecedented approach for high strength and reasonable ductility. Additionally, morphological analysis indicated that an artful combination of fillers having different microstructural layout can render the function of solid compatiblizers. Moreover, the inhomogeneity can be minimized by application of nano clay.

T. Ahmed (✉) · H. H. Ya · M. A. A. Majid · M. Azeem · M. Azad Alam · H. U. Khalid
Mechanical Engineering Department, Universiti Teknologi PETRONAS, 32610 Seri Iskandar, Malaysia
e-mail: tauseef.17007229@gmail.com

T. Ahmed
Research Center for Nano Materials and Energy Technology, Sunway University, 47500 Selangor, Malaysia

M. Imran
Department of Mechanical Engineering, Institute of Space Technology Islamabad, Islamabad 44000, Pakistan

R. Khan
College of Mechanical Engineering, National University of Sciences and Technology, Islamabad 44000, Pakistan

A. Mehmood
Department of Physics, Air University Islamabad, Islamabad 44000, Pakistan

F. Ahmad et al. (eds.), *ICPER 2020*, Lecture Notes in Mechanical Engineering,
https://doi.org/10.1007/978-981-19-1939-8_73

Keywords Pseudo-ductility · Toughness · Solid compatiblizers · Interfacial adhesion

1 Introduction

High density polyethylene (HDPE) Clay polymer nano composite is fairly a new class of polymer composites having nano sized layered silicate structured nano fillers. It is a class of nano composites that is reinforced by layered structure material called nano clay [1]. Moreover, it is conditional that the dimension of reinforcements/fillers in the clay polymer nano composite to be in nano meter range [2]. That is to say, that the nano metric size of fillers not only increase the properties at a very low fillers loading but also retain the toughness of a polymer nano composite. [3]. The improvement in performance is attributed to the spacious surface area with dimensions in nano meters, for example a fillers bearing dimensions in nano meters possess surface area of as large as large as 15–25 m^2/g [3, 4].

Pseudo-ductility is the mechanical property of a functional material to possess the ability of brittle (high strength) and ductile material (high value of elongation) at the same time [5]. According to the most accepted theory, it is possible for a material to undergo pseudo-ductile behaviour, if the deformation starts before the yield point in a tensile test [6]. Therefore, the outright knowledge of various aspects of tensile testing (a measure of hardness, a measure of onset of viscoelastic deformation, the area under the curve and slope of the stress–strain curve of materials, strength) are of great importance [7].

It is an established fact that polymer such as HDPE overtakes to become purposelessly brittle, when reinforced with rigid filler particles. It is because (a) inorganic nano fillers generate microstructural flaws as a result of inhomogeneity in phase morphology [8], and (b) generation of cavities around rigid filler particle, due to difference of elastic moduli [9]. On the other hand, HDPE acquires ductility (low strength) when reinforced with rubber filler particles. Furthermore, rubber particles cannot withstand the compressive and shear load in wear process. Additionally, the rubber filler particles are more vulnerable to thermal oxidative degradation [10]. The agreement found in the studies is that, if there is an adequate bonding between the polymer-filler interfaces in case of rigid filler, properties (mechanical, morphological) can be improved [3]. Thus, the polymer-filler interfacial interactions are of prime importance. It can be well postulated that a combination of nano fillers can be chosen so that both of the filler possess different microstructural layout such as layered silicates bentonite (nano clay) and spherical Calcium Carbonate particles. By doing so, this combination gives rise a stable morphology and compatibilizing effect of nano clay reduces the inhomogeneity of the fillers resulting the required pseudo-ductility [11, 12].

2 State of the Art

That being the case, in light of literature review we can report that there is no work conducted on the application of solid compatiblizers resulting from two non-treated inorganic nano filler particles. Thus, this work explains the state of the art concept of achieving a pseudo-ductility, uniform morphology, and the resulting mechanical properties with the help of non-treated inorganic nano fillers.

3 Aim of Investigation

The aim of the composite fabrication and testing is mainly to investigate, if a tensile strength can be enhanced with the ductility being retained. The nano composite was first fabricated via melt compounding in the form of solid pellets and then tested for bulk mechanical. The filler loadings were chosen in order to investigate the variation in cases where $CaCO_3$ or bentonite was used. Moreover, it was proposed the $CaCO_3$ being a stiffer material will enhance the strength and the layered bentonite will prompt the ductility by sliding mechanisms of the layers over one another [13]. Consequently, both the fillers will act in synergy to impede the slip of polymer chain over one another and increase the strength.

4 Materials and Experimental Procedures

4.1 Materials

HDPE Polymer was supplied in the form of pellets from titanvene Malaysia, with the trade name HD5218EA. $CaCO_3$ and bentonite nano clay were purchased from Alfa Aesar with the trade names 33,295 Calcium carbonate and A15795 sodium bentonite, respectively.

4.2 Composite and Sample Fabrication

To remove moisture from the $CaCO_3$ and trapped water molecules from bentonite both the nano fillers were dried in a vacuum oven for 48 h. Formulations selected were on the base of weight percent of fillers and were melt compounded with HDPE resin via twin screw extruder with. Temperature for fabrication of melt compounding of composite and fabrication of samples selected was 120 °C. In a sample labelled as BC-x/y, B and C indicates bentonite and $CaCO_3$ while the x and y indicates the

Table 1 Mechanical properties resulting from tensile testing

Sample	Mechanical properties			
	Tensile strength (Mpa)	Elastic limit (Mpa)	Y.M (Mpa)	Elongation @ break (mm)
BC-0/0 (HDPE)	19.06 ± 0.48	10.65 ± 2.31	596 ± 58.43	17.91 ± 0.48
BC-0/5	23.06 ± 2.16	13.24 ± 4.55	788 ± 63.79	4.11 ± 2.43
BC-5/0	20 ± 1.44	12.97 ± 3.11	772 ± 45.57	7.85 ± 1.05
BC-5/5	21.85 ± 1.85	13.81 ± 1.51	448 ± 42.18	10.43 ± 1.31
BC-15/15	19.69 ± 2.36	11.47 ± 2.50	323 ± 68.56	3.99 ± 1.11

quantity of weight % in the corresponding sample as presented in Table 1. Moreover, the rpm for melt compounding chosen was14 such that to have proper mixing and to avoid unwanted degradation of polymer chains as a result of excessive shear.

5 Results

The Tensile test of the composites was investigated by analysis of three different mechanical properties such as tensile strength, Youngs modulus and elongation at break. The intent of fabricating a hybrid composite is to achieve a composite which has a higher strength than the pure polymer and additionally having ductility in agreement with the pure polymer because tensile strength and Youngs modulus indicates the stiffness and strength of the composites while the elongation at break indicates the ductility induced in the composites. Additionally, electron microscopic image analysis adds further to the microstructural phenomenon resulting the pseudo-ductility.

5.1 Tensile Strength

The summary of the tensile testing shows variation of tensile strength with variation of weight percent in the samples (Table 1). The table presents the variation of tensile strength with the weight percent (wt%) of hybrid fillers $CaCO_3$ and bentonite nano clay included in the samples. The sample BC-0/5 exhibited the highest tensile strength of the value 23.06 Mpa. The stiff behaviour of $CaCO_3$ induced brittle nature in pure polymer resin initially having a tensile strength value of 19.06 Mpa as noticed in the sample BC-0/0. This brittle behaviour can be well attributed to the phenomenon of polymer chains entanglement with the particulate $CaCO_3$, impeding the sliding of HDPE chains over one another which increased the strength of BC-0/5 by almost 21% [14]. The sample BC-5/0 displayed tensile strength of 20 Mpa. The inclusion of same amount of bentonite as $CaCO_3$ into pure HDPE hints to the sliding

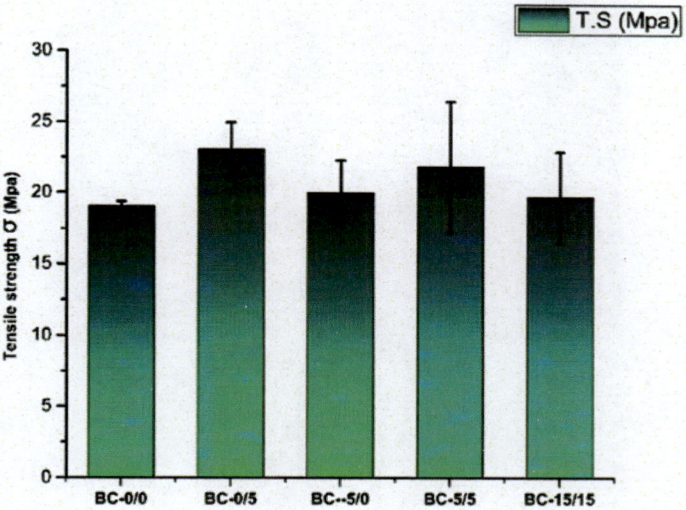

Fig. 1 Variation of tensile strength of the sample with respect to weight percent

phenomenon of bentonite layers over one another instead of absolute stiffness, which possibly induced the element of more ductility as compared to BC-0/5 (Table 1) [13]. Amongst the hybrid samples BC-5/5 and BC-15/15 with same comparative wt% of reinforcements, the sample BC-5/5 showed an extra ordinary behaviour of enhanced tensile strength (Fig. 1).

The fracture of BC-5/5 indicates necking during the final fracture as compared to the fracture sample BC-0/5 which is a typical brittle fracture without necking (Fig. 2a). Worth noting is the uneven edges of fracture in sample BC-15/15 which does not resembles sample BC-5/5 or BC-0/5 (Fig. 2c). The elongation of the sample BC-15/15 is even lesser which gives indication of agglomeration of filler particles due to excessive loading (30 wt %).

5.2 *Youngs Modulus*

The fundamental concept of tensile behaviour of a materials instructs that generally the more the tensile strength, the more would be the value of Youngs modulus. Moreover, Youngs modulus is the measure of resistance to deformation which is caused by bonding strength between atoms in microstructure of the material. The values of Youngs modulus obtained in the tensile test performed, exhibited the results which agreed to the fundamental concept i-e the samples with higher values of tensile strength showed higher values of Youngs modulus such as BC-0/5 which is 788 Mpa. Moreover, lower the values of tensile strength for BC-0/0 lower the value of YM i-e 596 Mpa (Fig. 3). However, cases of observation were noted for the samples BC-

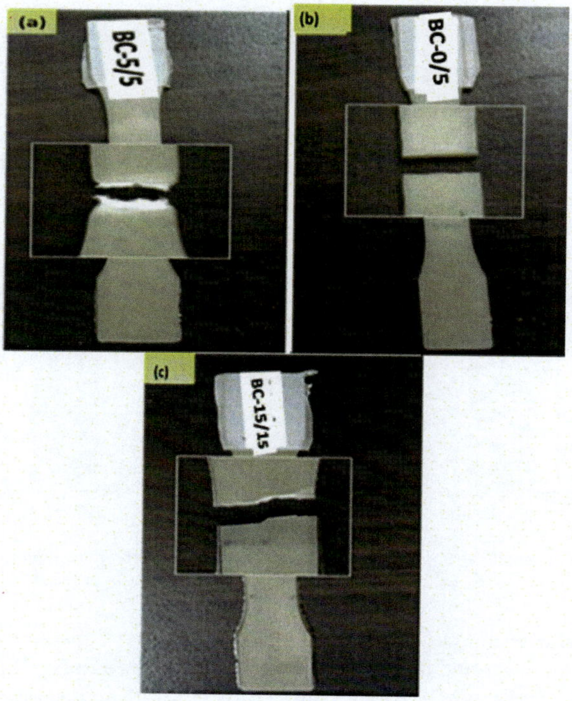

Fig. 2 Fractures in tensile test **a** ductile fracture, **b** brittle fracture, **c** fracture with most deficient behaviour

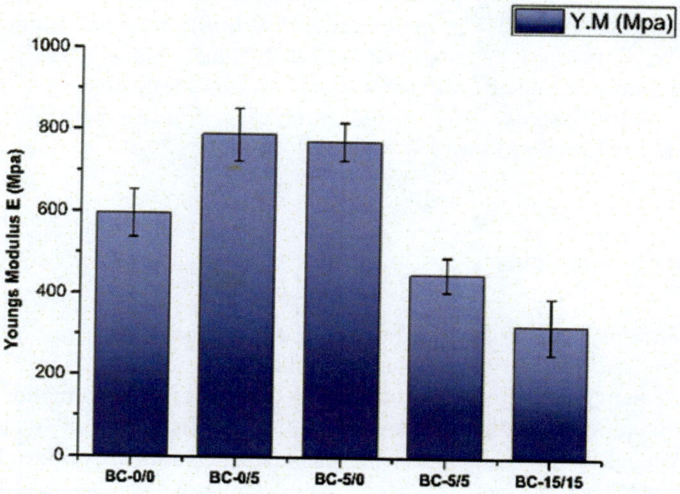

Fig. 3 Comparison of modulus of elasticity of the specimen

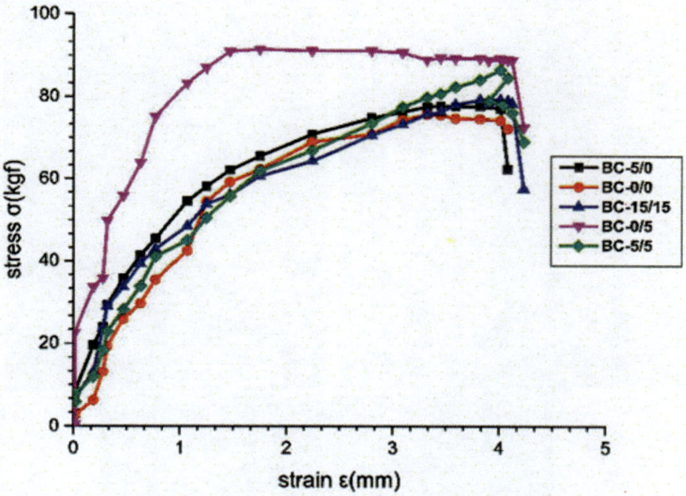

Fig. 4 Stress–Strain curves indicating area under the curves

5/5 and BC-5/0 which indicates the phenomenon of stress softening [3, 5, 15]. In a stress softened sample there is enough atomic bonding to render high values of tensile strength and elongation but low values of Youngs modulus due to toughness jump. The toughness is what responsible for the phenomenon of pseudo-ductility is a composite material. This lowering phenomena of elastic modulus in hybrid is attributed to the toughening mechanism that takes place during the crack propagation [16, 17]. The toughness or the toughness jump of the samples can be calculated by the area under the curves in stress strain curve in Fig. 4.

Furthermore, the premature fracture in BC-5/5 was prevented by the compatibilizing effect of the clay [11, 12] resulting in adequate polymer/filler adherence to one another. The role of a solid compatiblizers of nano clay can be seen by the cavities around the filler particles in SEM images (Fig. 5).

There is negligible cavity around the filler particles noted for the sample BC-5/5 (Fig. 5a) and BC-5/0 as opposed to visible cavities for sample BC-15/15 (Fig. 5c) due to incompatibility and agglomeration. On the other hand, non-treated fillers deboned lately in the process of elongation due to absence of any other external chemical bonding to render toughness or low Youngs Modulus.

5.3 Elongation at Break

The value of elongation at break gives indication of ductility in polymer. As it was the intent of the developing a hybrid composite to achieve a composite with fine tune property of enhanced strength and a retained valued of ductility at the same time. To

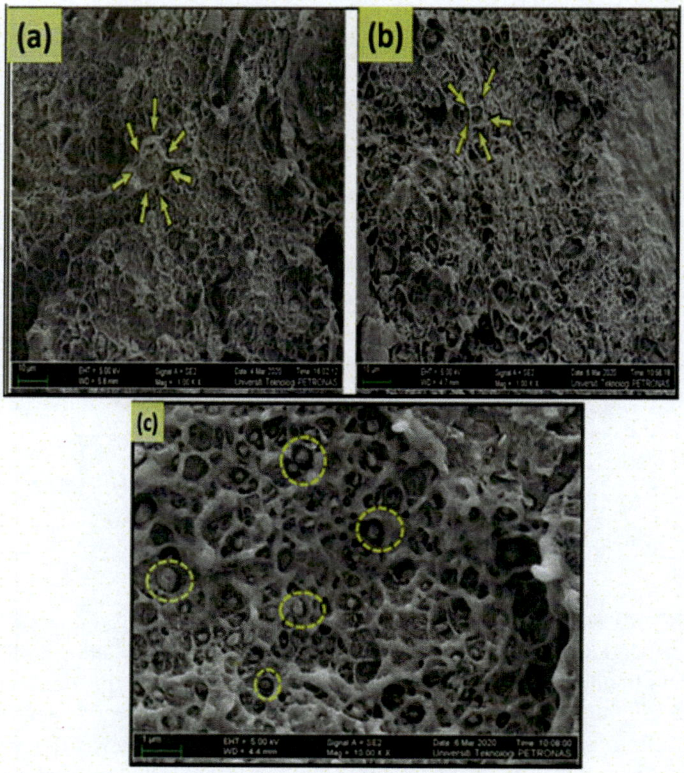

Fig. 5 Morphological analysis, **a** homogenous morphology, **b** negligible cavities, **c** visible cavities around the particles

illustrate this, in Fig. 6 gives elongation of different ranges as discussed. The "elongation at break" of the samples BC-5/0 and BC-15/15 are closely comparable having values of 4.11 and 3.99 mm, respectively. Truly, the phenomenon of agglomeration in both samples could have possibly resulted this deficient behavior. This behaviour can be well attribute to the inadequate breaking of particle clusters by the shearing force during the fabrication of [18, 19] and excessive loading of fillers in BC-15/15 [20].

The higher value of elongation in sample BC-5/5 is due to the fact that the fillers contents were dominated by sliding bentonite layers [13]. Moreover, the final fracture is controlled by the imperfection caused by the inhomogeneity in the morphology and the agglomerated particles.

The agglomeration is evaluative in reduction of elongation at break. The more weight % of fillers the more is chances of agglomeration due to the chance encounter of the propagating neck with such an agglomerate. The debonding of

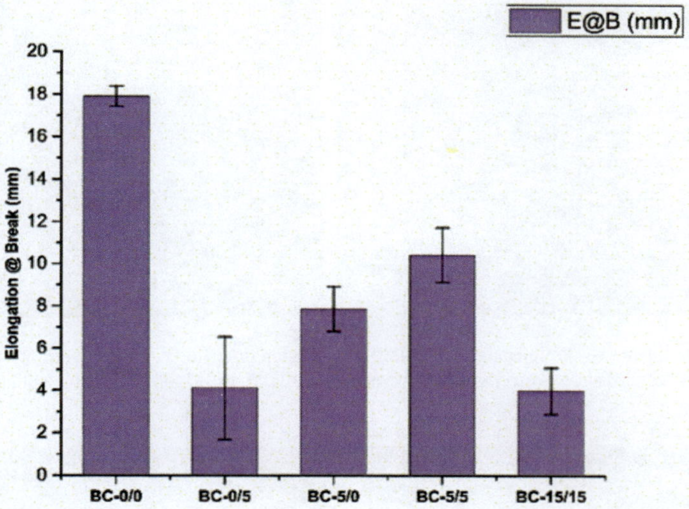

Fig. 6 Elongation of the specimens

polymer/filler occurring during the later period of the test allowed large unhindered plastic extensions for BC-5/5 and BC-5/0 which was the final outcome of the toughness induced [21].

6 Discussion

An additional point for further discussion is dissimilar features of micro cracks resulting from the hybridization of filler particles. The comparative weight percentage of both fillers seed different mode of fracture; therefore, it gave rise to unique features of micro cracks. For illustrations, micro cracks of tensile tested composites are shown in the Fig. 7a and b. Moreover, Fig. 7a appears to possess dissimilar pattern of failure in the inner edges of microcrack than Fig. 7b. For the purpose of demonstration, dotted lines are drawn showing a regular and even edges of micro cracks indicating a brittle mode of failure in the sample BC-0/5 in Fig. 7a. This microcrack possessing glassy fracture can be regarded as to have been resulted from premature polymer/filler debonding (one is somehow regular closed circle and the other is uneven edges at the microcrack) [17, 21]. By this occurrence there is no pinning of the particles during elongation and micro crack propagates easily leaving out no material "pull out" [16].

On the other hand, due to compatibilization effect, the nano clay adhered well to the polymer. The $CaCO_3$ particle could have been possibly penetrated into the clay galleries [17] The contact of $CaCO_3$ particles was reduced to minor, by the morphology resulting from this penetration. Therefore, the material appeared to be "pulled out" during the fracture as shown in Fig. 7b which is a microcrack that

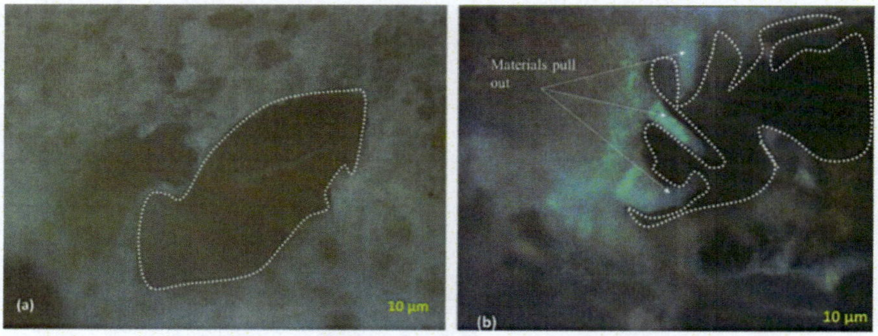

Fig. 7 Optical microscopic images of micro-cracks generated in the cross sectional area of the specimen **a** very fine and even edges of the cracks of BC-0/5, **b** BC-5/5 indicating a ductile fracture

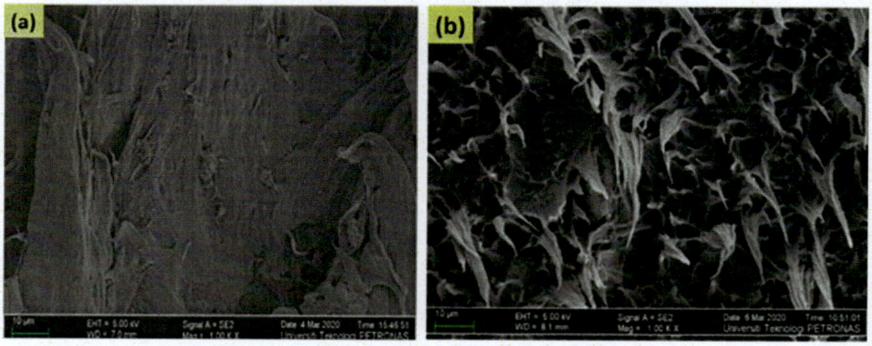

Fig. 8 **a** Smooth patterns indication brittle fracture, **b** ductile fracture with rough patterns

resulted a ductile fracture in BC-5/5. Further investigation of the fractures via fractographic analysis of the fractures in SEM indicates smooth patterns of BC-0/5 as compared rough patterns BC-5/5 (Fig. 8). The roughness explains the crack pinning phenomenon inducing toughness while the smooth patterns explains the easy crack propagation resulting a ductile and brittle fractures simultaneously.

7 Conclusion

The tensile testing performed for the hybrid composite was a profound insight into microstructural characteristics of inorganic nano fillers. Moreover, it was evident from the results that the particulate fillers $CaCO_3$ and the layered bentonite assisted each other to perform well for strength and ductility. It was concluded that surface treatment is not always needed for effective stress transfer between polymer/filler.

Moreover, application of surfactants can hinder the process of polymer/filler debonding which reduces the toughness.

An additional conclusion drawn is application of rubber filler for toughness cannot enhance the young modulus therefore nano clays can render the function of compatibilization and the approach of combining rigid fillers of different microstructural layout lead to rigid and tough composite.

Acknowledgements Authors would like to acknowledge sponsorship of Newton Link Grant No 015ME0-015 for the material acquisition and technical support.

References

1. Anderson J (1982) High density and ultra-high molecular weight polyethenes: their wear properties and bearing applications. Tribol Int 15(1):43–47
2. Osman MA, Mittal V, Morbidelli M, Suter UW (2003) Polyurethane adhesive nanocomposites as gas permeation barrier. Macromolecules 36(26):9851–9858
3. Lazzeri A, Zebarjad SM, Pracella M, Cavalier K, Rosa R (2005) Filler toughening of plastics. Part 1—the effect of surface interactions on physico-mechanical properties and rheological behaviour of ultrafine CaCO3/HDPE nanocomposites. Polymer 46(3):827–844
4. Osman MA, Rupp JE, Suter UW (2005) Tensile properties of polyethylene-layered silicate nanocomposites. Polymer 46(5):1653–1660
5. Renner K, Yang MS, Móczó J, Choi HJ, Pukánszky B (2005) Analysis of the debonding process in polypropylene model composites. Eur Polymer J 41(11):2520–2529
6. Prato A, Longana ML, Hussain A, Wisnom MR (2019) Post-impact behaviour of pseudo-ductile thin-ply angle-ply hybrid composites. Materials 12(4):579
7. Wood W, Kumar S, Zhong WH (2010) Synthesis of organosilane-modified carbon nanofibers and influence of silane coating thickness on the performance of polyethylene nanocomposites. Macromol Mater Eng 295(12):1125–1135
8. Lin Y, Chen H, Chan C-M, Wu J (2010) The toughening mechanism of polypropylene/calcium carbonate nanocomposites. Polymer 51(14):3277–3284
9. Zuiderduin W, Westzaan C, Huetink J, Gaymans R (2003) Toughening of polypropylene with calcium carbonate particles. Polymer 44(1):261–275
10. Sahin M et al (2018) Tailoring the interfaces in glass fiber-reinforced photopolymer composites. Polymer 141:221–231
11. Ruamcharoen J, Ratana T, Ruamcharoen P (2014) Bentonite as a reinforcing and compatibilizing filler for natural rubber and polystyrene blends in latex stage. Polym Eng Sci 54(6):1436–1443
12. Tiwari RR, Paul DR (2011) Effect of organoclay on the morphology, phase stability and mechanical properties of polypropylene/polystyrene blends. Polymer 52(4):1141–1154
13. Liu T, Wang Y, Eyler A, Zhong W-H (2014) Synergistic effects of hybrid graphitic nanofillers on simultaneously enhanced wear and mechanical properties of polymer nanocomposites. Eur Polym J 55:210–221
14. Thio Y, Argon A, Cohen R, Weinberg M (2002) Polymer (Guildf) 43:3661. (https://doi.org/10.1016, S0032–3861 (02), pp. 00193–3)
15. Ahmed T, Ya HH, Khan R, Hidayat Syah Lubis AM, Mahadzir S (2020) Pseudo-ductility, morphology and fractography resulting from the synergistic effect of CaCO3 and bentonite in HDPE polymer nano composite. Materials 13(15):3333
16. Kinloch A, Maxwell D, Young R (1985) The fracture of hybrid-particulate composites. J Mater Sci 20(11):4169–4184

17. Wang K, Chen L, Wu J, Toh ML, He C, Yee AF (2005) Epoxy nanocomposites with highly exfoliated clay: mechanical properties and fracture mechanisms. Macromolecules 38(3):788–800
18. Fawaz J, Mittal V (2015) Synthesis of polymer nanocomposites: review of various techniques. In: Synthesis techniques for polymer Nanocomposites, pp 992–1057
19. Paul DR, Robeson LM (2008) Polymer nanotechnology: nanocomposites. Polymer 49(15):3187–3204
20. Eiras D, Pessan LA (2009) Mechanical properties of polypropylene/calcium carbonate nanocomposites. Mater Res 12(4):517–522
21. Bartczak Z, Argon A, Cohen R, Weinberg M (1999) Toughness mechanism in semi-crystalline polymer blends: II. High-density polyethylene toughened with calcium carbonate filler particles. Polymer 40(9):2347–2365

Manufacturing

A Study on the Effect of Structural Properties and Configuration of Low-Stiffness-Resilient-Shaft (LSRS) on Torsional Behavior

M. B. Baharom, Salah E. Mohammed, Najaf Hussain, A. Rashid A. Aziz, Z. Ezrann A. Zainal, and Mhadi A. Ismael

Abstract The use of the Low-Stiffness-Resilient-Shaft (LSRS) in the Semi-Active-Steering (SAS) system helps remove the drawbacks of the conventional and Steer-By-Wire (SBW) systems. However, the underlying structure of suitable configurations of LSRS which provide equal stiffness in both directions of twists have never been proposed before. This paper dealt with the modelling of LSRS using wire ropes and analysis of its torsional stiffness. Two potential design configurations of LSRS designated as config.1 and config.2 have been proposed. Finite Element Analysis (FEA) was carried out to determine their torsional stiffnesses. An experimental test rig (Torsion test machine) was also designed and fabricated and then it was used for the experimental validation of the simulation models. It was observed that config.2 had equal torsional stiffness in both directions while config.1 resulted in a slight error of 0.9%. In comparing simulation with experimental data, it was found that the maximum percentage error was about 12.41%, indicating the validity of the established model. The model and findings achieved from this research can be used to design an LSRS of required torsional properties which allows the steering to have equal torsional stiffness in both directions.

Keywords Low-Stiffness-Resilient-Shaft · Semi-Active-Steering · Steer-by-Wire · Torsional behavior

1 Introduction

The architectures of steering systems of passenger cars have evolved from the basic manual rack and pinion steering concept to the advanced concept like Steer by Wire

M. B. Baharom (✉) · S. E. Mohammed · N. Hussain · A. R. A. Aziz · Z. E. A. Zainal · M. A. Ismael
Center for Automotive Research and Electric Mobility (CAREM), Universiti Teknologi PETRONAS, 32610 Perak, Seri Iskandar, Malaysia
e-mail: Masrib@utp.edu.my

S. E. Mohammed · M. A. Ismael
College of Engineering and Architecture, Agricultural Engineering Department, University of Bahri, Khartoum 1660, Sudan

F. Ahmad et al. (eds.), *ICPER 2020*, Lecture Notes in Mechanical Engineering,
https://doi.org/10.1007/978-981-19-1939-8_74

(SBW) which eliminates the use of steering shaft [1]. To satisfy the conditions for safety and ride comfort, several researchers came up with the ideas of steering systems which fall in between the previously mentioned concepts [2, 3]. Whatever the case, the basic configuration of the vehicle steering system has not modified much since the first modern automobile was manufactured many years ago. The driver input from the steering wheel is transmitted to universal joints and gearboxes of the front tires through mechanical or non-mechanical connections.

As vehicle velocities increased, more force reactions occurred at road wheels which then caused an increase in steering rack loads. Due to this, a large force is required by drivers to turn the steering wheel [1]. To counter this problem, an inventor named Delphi Saginaw develop a new hydraulic power assisted steering system (HPAS). This system was employed by General Motors in its automobiles in 1953 was the first ever system with hydraulic assistance [1].

The hydraulic steering though very helpful, has some disadvantages as there might be fluid leakage in the hydraulic system due to accidents or corrosion. Furthermore, it puts a load on the engine as it is operated by a hydraulic pump which is powered by the engine and finally it requires frequent maintenance of the power steering oil [2, 3]. Due to these issues, vehicle manufacturers have moved their interest to electric power steering (EPS) systems. In EPS, an electric motor is used rather than a hydraulic pump which means that all the torque from the motor is efficiently transferred to the steering wheel [2, 4–6]. It is also noted that by using an EPS over HPAS, there is a reduced fuel usage of 5% to 15% since the engine load is reduced [7]. The latest and most recent advancement in the steering systems is the invention of the Steer-By-Wire (SBW) system and Semi-Active-Steering (SAS) [8, 9].

The SBW system is like an electric power steering system, with the main difference being the exclusion of the permanent steering shaft connecting the steering wheel and the rack and pinion with actuators [8]. This gives the steering system several advantages over the conventional systems such as active steering technology, safety to driver in an accident, more packaging space, freedom of interior design (as the steering wheel can be placed on both sides), decreasing noise and vibration and increased fuel economy [2, 8, 9]. Although, the removal of the rigid steering shaft gives the SBW its advantages over the conventional systems, it is still not accepted by some automotive industries due to safety reasons. The removal of the steering shaft makes the SBW a completely electronically controlled system with no physical connection between the driver and the road, and in the case of an electronic system failure, the driver will not be able to control the vehicle. This concern led to the conceptualization of the Semi-Active-Steering (SAS) system.

The SAS system bridges the gap of the SBW and the conventional systems by introducing a flexible steering shaft between the steering wheel and the rack and pinion. This flexible shaft is called Low-Stiffness-Resilient-Shaft (LSRS) and it allows the SBW to be operational without interfering with the steering system because it is not torsionally rigid. Furthermore, in the case of an electronic system failure, the LSRS would then take over and the driver can drive the vehicle to safety. Similarly, in the case of any frontal collision, the LSRS buckles because of its axial flexibility, and the forces from the front wheels will not be transferred to the steering wheel.

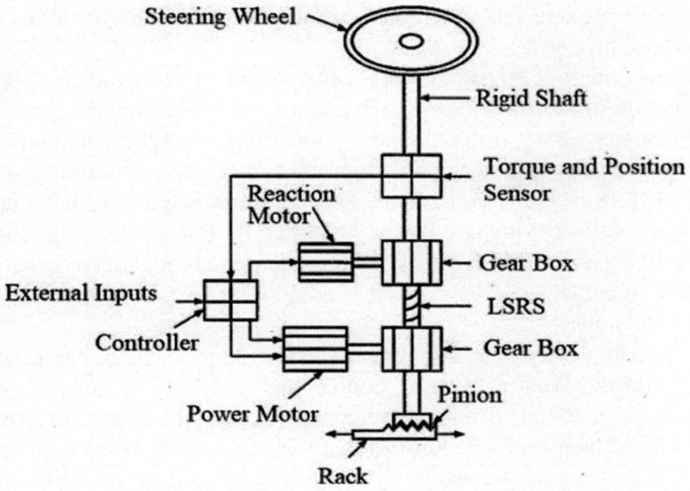

Fig. 1 Basic configuration of the SAS [2]

Hence, it will save the driver from injuries which are usually caused by the steering wheel. Figure 1 shows the basic architecture of the SAS, where the LSRS is located between the two gear boxes attached to the reaction and power motors.

The Low-Stiffness-Resilient-Shaft is the important component of the Semi-Active-Steering system that distinguishes it from other steering systems. In the SBW system with backup [10], the rigid shaft is engaged with the help of a clutch mechanism in the event of failure of the electronic system. But in SAS, the LSRS is a fundamental part and is always coupled to the system permanently. The LSRS needs to have certain properties for the SAS system to function effectively, which are:

- It needs to be soft and supple along the axis so that it buckles in the case of any collision [9, 11, 12].
- It needs to have a compromise of torsional stiffness, so that when the SBW system is intact it does not interfere with the steering capabilities but when the SBW system fails it should have enough torsional stiffness to steer the vehicle to safety [9, 11, 12].

One of the most important criteria for a vehicle system to be successfully commercialized is that it must provide consumer with a safety confidence level as high as possible. Although the fly-by-wire can be commercially accepted by most people, it is very difficult for a SBW system on a car to be commercially accepted by most consumers as they are car drivers themselves which makes it skeptical for them to find out that the vehicle they are driving does not have any mechanical connection between the steering wheel and steered road wheels. Through the presence of the LSRS, the SAS is hoped to have advantages in terms of gaining consumers' confidence [9]. Because the LSRS needs to have certain specific properties, certain materials need to be selected for its construction, so that the materials allow the LSRS to exhibit those properties.

The author Baharom [9] discusses the possible choices of products to be used for the construction of the LSRS. One such example is a flexible drill cable. Flexible shafts are constructed with wire wound tightly around a central wire and each layer of wire increases the diameter of the shaft, also increasing its torque transmission capability [13]. It generally consists of several coils which are wound alternatively in different directions. Although it was suggested by Baharom [9] that the flexible shaft can be used as the LSRS for the SAS system, the literature suggests that even bi-directional flexible shafts have variable torsional stiffness in clockwise and anti-clockwise directions.

Due to the complexity of the structure of flexible shafts and their nature of unequal torsional stiffness, inventor [9] used double springs which have similar torsional stiffness in both directions to carry out the experiments to find out the effectiveness of the LSRS. Although double springs were used to perform the experiments by Baharom [9], they cannot be used for the creation of the LSRS. The problem with the double springs is that they are axially rigid which means that they do not satisfy the requirements for the construction of the LSRS.

A wire rope is made from individual wires twisted into strands. Several strands are then twisted into ropes according to specific lay patterns [13–15]. Due to the high axial strength and stiffness of the wire ropes along with flexibility in bending, they have been employed extensively in many engineering applications [16, 17].

Due to the vast advantages of wire ropes, a lot of research has been done to predict the behavior of them. Most of the analytical studies rely on the solutions of equilibrium equations without frictional effects due to the complexity of the theory [18]. With advances in computer technology and finite element software, it has become easier to predict the behavior of wire ropes as compared to the analytical studies. Similarly, the effect of frictional contacts on the wire ropes can also be considered using finite element analysis. Therefore, researchers have been developing advanced analytical techniques for the design and behavioral analysis of wire strands and wire ropes [19]. In this paper, several finite element computer models have been discussed to better predict the behavior of the wire ropes.

Stanova et al. [20] derived and discussed parametric equations to model multi-layered wire ropes. These parametric equations were then used to create computer models in CATIA V5 and implemented in a finite element software to observe the behavior of the wire rope under tensile loading. They considered a frictional value of 0.2 for the interwire connections and then compared the results obtained to the theoretical values [19]. Similarly, Imrak and Erdonmez [18] developed a new method to model the Independent Wire Rope Core and then analyzed it in an FEA environment with frictional consideration to compare the results with the analytical theory of Costello [21]. Since the theory of Costello does not consider the frictional contacts and it is only valid in the elastic region, the results found were similar in the elastic region but differed in the plastic region due to the excessive stress. The same result was obtained and published by Kastratovic et al. [16].

There is some research done on the torsional behavior of wire ropes, but that too is to observe the torque produced in wire ropes because of the axial loading due to its lay pattern [22–24]. Therefore, experimental testing needed to be done to study the behavior of wire ropes under a purely torsional environment. It should be noted that the stress in wire ropes in previous research was in tension, while in a purely torsional environment, the wire ropes will be in the state of compression.

Aziz Ameen et al. [25] discussed the effects of annealing on carbon steel. Torsion test experiments and finite element analysis results were then compared to see the difference in the results. It should be noted that for the experiment, the specimen was clamped on both ends in the torsion test machine. Similar boundary conditions can be seen in [26], where FEM investigation is done to see the results of heat treatment on alloy steel. One end is completely fixed and on the other end a torque of some magnitude is applied.

The aim of this paper is to present the modelling of LSRS made of wire ropes and to study its torsional stiffness using Finite Element Analysis (FEA).

2 Materials and Methods

In this section, the methodology used to analyze the LSRS model will be introduced. Firstly, the layout pattern of the LSRS will be presented. This will be followed by a demonstration of how the model of the LSRS with different configuration was designed, and how the material was selected. The simulation technique using FEA was used to analyze the behavior of the LSRS models as well as the experimental test procedure will be presented.

2.1 Configuration Conceptualization of LSRS

The LSRS need to have similar torsional stiffness in both CW and ACW directions for the SAS system to function effectively. It was noted that the single strand wire rope due to its unidirectional lay structure would have unequal torsional stiffness in CW and ACW directions. Hence, a certain wire rope configuration needed to be created, so that it could be utilized as the base layout for the LSRS. Two separate configurations namely config.1 and config.2 were chosen, based on the idea that coupling two sets (4 strand wire rope) of single strand wire ropes having the same structural properties but opposite direction. Figure 2 shows the wire rope sets arrangement of LSRS.

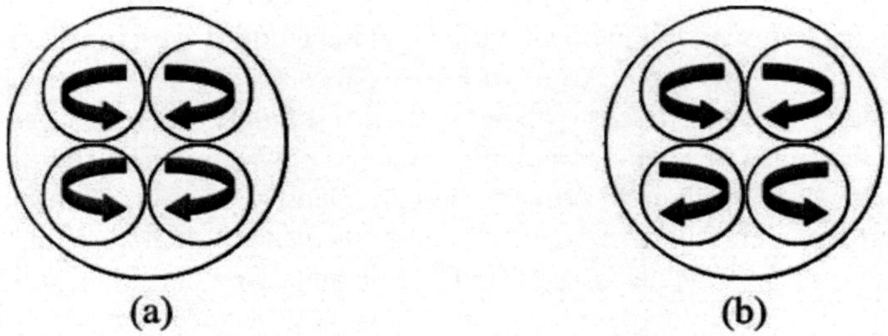

Fig. 2 Wire ropes arrangement for **a** config.1 **b** config.2 [27]

2.2 Modelling of config.1 and config.2

To perform simulations to observe the torsional behavior of the wire rope, 3D solid modeling was done in Solidworks™. The first model created was of a single strand 1 × 7 wire rope and a wire diameter of 1.2 mm was chosen. The main goal was only to observe the angular deflection of the wire rope in the clockwise and anticlockwise direction. The dimensions of the wire rope to be modelled are expressed in Table 1.

To make it easier to apply boundary conditions to the solid model, torque handles ware modeled similar to those for 1 × 7 wire rope strand. Figure 3 shows the solid

Table 1 Parameters of the (1 × 7) strand wire rope

Geometric parameters	Values
Wire diameter	1.2 mm
Strand diameter	3.54 mm
Pitch length	50 mm
Total length	100 m

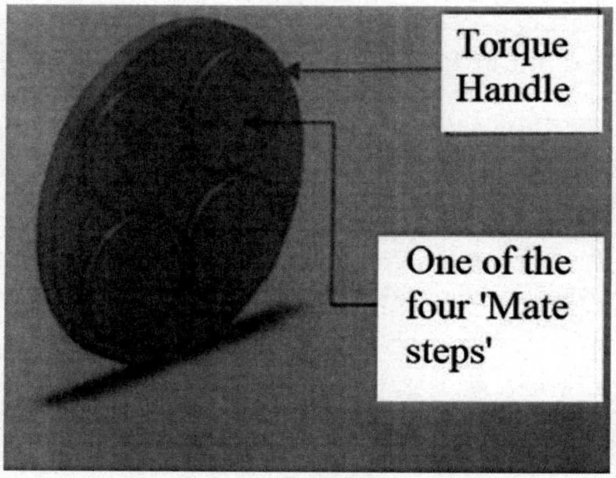

Fig. 3 Solid models of the torque handle for config.1 and config.2

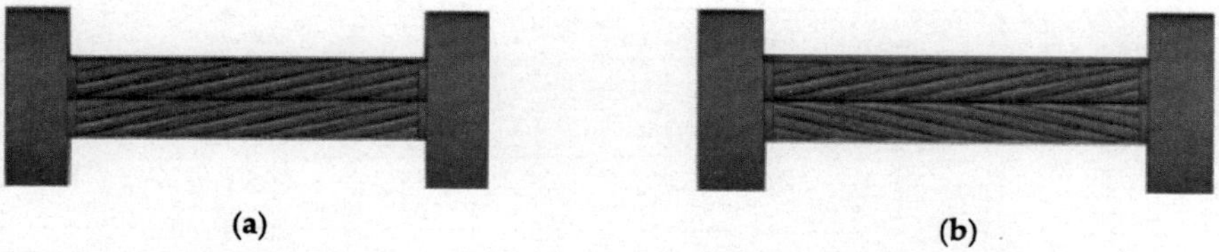

Fig. 4 Finalized solid models of the **a** config.1 **b** config.2

model of the torque handle for config.1 and config.2.

The assembly of solid models of config.1 and 2, along with two instances of the torque handles are shown in Fig. 4. The only difference in config.1 and 2 was the arrangement of the wire ropes as evident from Fig. 4. As discussed earlier, in config.1, the two adjacent wire ropes are in the same direction either right or left lay while in config.2, the two alternative wire ropes are in the same direction.

2.3 *Finite Element Analysis of Variation 1 and Variation 2*

The purpose of doing the analysis was to obtain the angular deflection and to determine the arrangement pattern of wire ropes within the LSRS so that it has similar torsional properties in both CW and ACC directions. The solid modeling of wire rope was done in Solidworks™ and the finite element analysis was done in ANSYS™ Workbench 15.0. To import the created solid model, it was firstly saved in IGES format and then imported into ANSYS™. After importing the model, suitable boundary conditions were applied. One end of the wire rope was fixed while on the other end, a certain magnitude of torque was applied. The angular deflection obtained was then compared to the experimental tests performed.

2.3.1 Material Properties

The selection of material model and the accuracy of the material data have a significant impact on FEA accuracy. Assigning a material to the model is a crucial step to get realistic and appropriate results. The material properties applied to the model were that of ASTM A525 Galvanized Steel which is commercially used are shown in Table 2. The reason for choosing the galvanized steel was because the galvanized steel wires were later used to create wire ropes and then tested experimentally.

2.3.2 Defining Contact Surfaces

Once the model was imported and the material was assigned, the next step was to define the contact regions. This is also a crucial step as it determines the behavior as

Table 2 Assigned material properties for galvanized steel

Specifications	Value
Young's modulus	200 GPa
Poisson's ratio	0.29
Density	7850 kg.m^3
Tensile yield strength	203.9 MPa
Compressive yield strength	250 MPa
Tensile ultimate strength	356.9 MPa

to how different parts of the assembly are interacting with each other. The software automatically detected all the connections between different parts by determining the faces which were interacting with each other. However, the contact conditions still needed to be altered to represent the wire rope in real world application.

For a single wire rope strand, there were a total of 26 connections altogether. Since the torque handles were only created for the applications of boundary conditions to the wire rope, they were bonded with every individual wire. The reason for the torque handles being bonded to the wires was because they transmit the torque directly to the wire rope. There was a total of 56 bonded contacts altogether (single wire rope has 14 bonded contacts). The formulation was set as 'Multi Point Constraint (MPC)' and the detection method was set as 'Nodal-Normal to Target'. There is a reason as to why 'MPC' was preferred over the 'normal formulation' for bonded contacts. The bonded contacts are a linear form of contact-based connection. For a linear contact connection, there must be contact elements on one body and target elements on the other, which lie on the outer surface of each body's solid elements. At the beginning of each load increment for 'normal formulation', the software requests the contact elements to search for the target elements in the allowable range. The contact elements have a stiffness in normal direction called the 'normal contact stiffness', which means that despite the connection being 'bonded', there is some flexibility in the connection between two bodies. On the other hand, for an MPC® connection, the software does not calculate any stiffness for the connection. It uses rigid constraint equations between the solid elements on the contact and target faces for a real bonded connection.

Similarly, the settings were also changed for the frictional contacts to ensure convergence. There was a total of 48 frictional contacts in config.1 and config.2 which were established for inter-wire connections (single wire rope has 12 frictional contacts). Usually, for the construction of wire ropes, the manufacturing companies use a lubricant to reduce the inter-wire friction. However, since no lubricant was used for the fabrication of wire ropes in this research, the coefficient of friction chosen for the inter-wire contact was 0.74 [28]. The formulation method for the frictional contact (which is a nonlinear contact) was changed from 'Programme Detected' to 'Augmented Lagrange'. Augmented Lagrange is a penalty-based contact formulation which is recommended for general frictionless or frictional contact in large deformational problems [29, 30]. It makes sure that the penetration of the contact element within the target element is not too much due to a low contact stiffness. Due to this reason, 'Augmented Lagrange' takes more time for the simulation to converge. The detection method was again chosen to be 'Nodal Normal to Target' with the reason being that the nodal detection could handle contacts at edges better. The normal contact stiffness was changed from 'automatic' to a 'manual' value of 0.1. The normal contact stiffness is the most important parameter affecting both accuracy and convergence behavior. A large value of stiffness gives better accuracy but for non-linear contacts it becomes difficult for the solution to converge. Therefore, the normal contact stiffness is usually kept low for frictionless and frictional contacts (i.e. both are non-linear contacts). For bending dominated problems, a contact stiffness value of either 0.1 or 0.01 is recommended. The interface treatment was changed to 'Adjust to Touch' for the contact. By changing to 'Adjust to Touch' the software assumes that the two faces are in contact with each other at the start of simulation. Due to errors associated with complex CAD models, a few gaps exist between the contact faces. When the 'Adjust to touch' is selected the software would make sure that the faces are in contact so that the simulation does not have any problem to converge.

2.3.3 Defining Remote Point to Measure Angular Deflection

The main objective of performing the FEA using torsion test simulations was to calculate the angular deflection of the wire ropes at the application of torque. Since ANSYS™ Workbench 15.0 does not have an option to calculate angular deflection directly, a 'Remote point®' was used to calculate the rotation of a point on the wire rope (SSWR). The first step in measuring the rotation of a single strand wire rope was to associate a Remote Point® with the geometry of one of the wires in the wire rope. The purpose of associating a Remote Point® with the geometry was that this point would act as a marker and it would be used to trace the movement when the whole body rotates. Since the whole geometry rotates an equal amount upon application of torque, this point was placed on one of the outer wires. The behavior of the associated geometry was set to 'Deformable®', so that the FEA model behavior was not affected by the 'Remote Point®'. Figure 5 indicates the position of the Remote Point® on the SSWR which was used to measure the rotation of the wire rope.

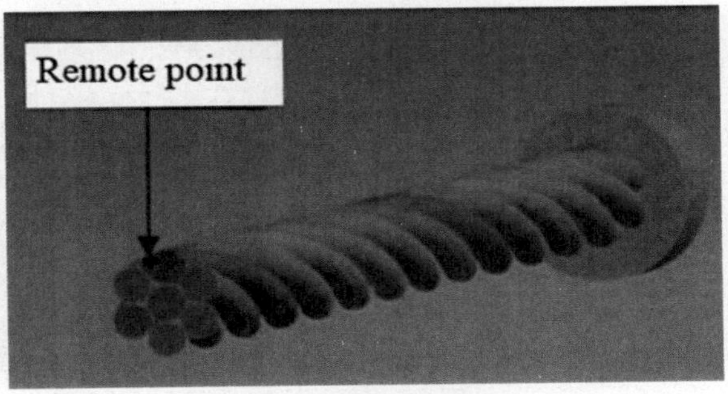

Fig. 5 Indication of the Remote Point on SSWR

In ANSYS™ Workbench, whenever a 'Remote Point®' is placed in a geometry, it is associated with a parameter '_npilot®'. A command called "Measure_pilot = _npilot" was introduced, which tells the software to measure all of the properties associated with the 'Remote Point®'. A second Command Object® was then inserted at the post-processor in the 'Solution®' area. This command then asks the software to calculate the rotation of 'Remote Point®' in all three axes. The APDL® code written in the Remote Point® section helps assigning a name to the remote point and enables the measurement of its rotation.

2.3.4 Boundary Conditions

The boundary conditions applied to the single strand wire rope model are as follows:

- One end of the single strand wire rope called 'End 1' was fixed. The 'Fixed support®' boundary condition was applied to the torque handle on 'End 1'. This makes sure that it will not translate or rotate in any direction.
- On 'End 2' meanwhile, the torque handle was applied with 'Frictionless support®' so that it would be easier for that end to rotate freely along the axis of the wire rope, but not in any other direction.
- On 'End 2' a clockwise and counter-clockwise 'Moment®' of 0.25–1.5 N.m at an interval of 0.25 N.m was applied.
- The angular deflection of the assigned Remote Point® was then observed after the solution had converged successfully to calculate the wire ropes torsional stiffness.

This whole setup of boundary condition simulates a torsion test in which one end of the wire rope is fixed while the other end is allowed to freely rotate under a torque of 0.25–1.5 N.m and is highlighted in Fig. 6.

The boundary conditions chosen for the solid models of config. 1 and config. 2 were similar to the single strand wire rope model. Both models were 'fixed' at one end with the help of the torque handle and on the other end a 'frictionless support' was applied. A 'Moment' value of 1–5 N.m was applied on the same faces as the 'frictionless support' at an interval of 1 N.m in the clockwise and anti-clockwise direction. The angular deflection in both the directions was then obtained using the remote point assigned earlier.

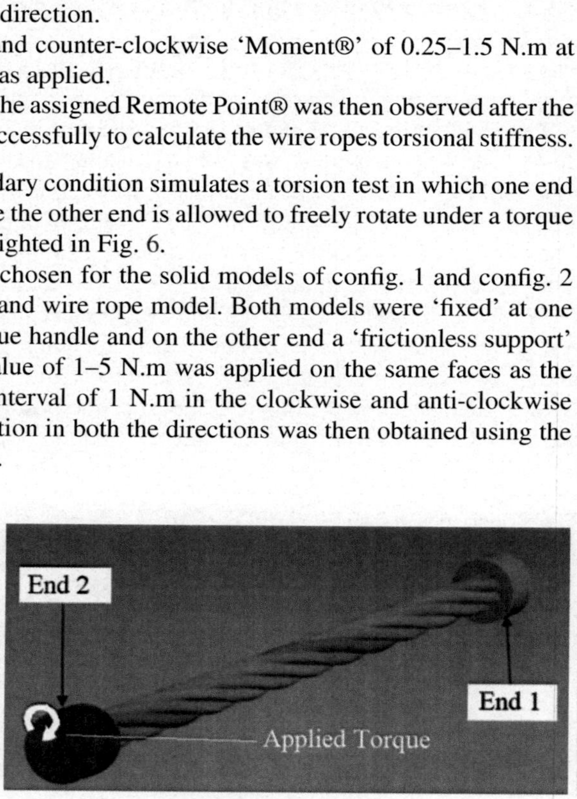

Fig. 6 Clockwise moment is applied at End 2

2.3.5 Experimental Validation of the FEA

To validate the simulation results, a few experiments were carried out on wire ropes under similar boundary conditions. Therefore, a Wire Rope Maker Machine (WRMM) was designed and manufactured to fabricate the wire ropes. Left- and Right-hand lay wire ropes (similar to solid models created for FEA) were fabricated using the WRMM. To conduct torsion testing on the wire ropes, an experimental testing was setup. Hence, a Torsion Testing Machine (TTM) was then designed and manufactured so that the model could be experimentally validated. Torsion testing was then done for the LSRS in CW and ACW directions and the simulated models were validated by comparing the experimental results with the simulation results.

The experimental setup contains 4 major steps described below:

- Design and fabrication of the "Wire Rope Making Machine" (WRMM).
- Manufacturing of wire ropes out of individual wires using the WRMM.
- Design and fabrication of the torsion test machine to calculate the angular deflection.
- Using the torsion test machine to conduct the experiment and verify the FEA.

(a) Fabrication of the Wire Rope Making Machine (WRMM)

The WRMM was fabricated using mild steel. CNC machining was done for most of the parts. It should be noted that the machine created was for the creation of 1 × 7 wire rope strands with the individual wire diameter being 1.2 mm, so that the wire ropes for the validation of simulation could be manufactured.

(b) Fabrication of the Wire Rope

The wire ropes were manufactured using the WRMM. 1.2 mm-galvanized steel wires were used to create the wire ropes. The material was chosen based on its availability as it is readily available. Furthermore, galvanized steel wire ropes are normally used for manufacturing of wire ropes as it protects against rusting and galvanizing and reduces the break strength of the rope by 10 percent [31]. That is why all the Finite Element Analysis was carried out based on the raw material that was going to be used for the experimentation. Right and left-hand lay lx7 strand wire ropes of 50 mm lay length was created using the WRMM.

(c) Fabrication of the torsion test machine

To test the validity of the FEA, experiments were carried out on a test rig called the Torsion Test Machine (TTM) which was designed and fabricated. The TTM was able to hold the wire rope at one end and on the other end a controlled torque could be applied. The machine was able to measure the angular deflection made by the wire rope and it had a sliding mechanism so that different lengths of wire ropes or LSRS could be tested.

(d) Torsion tests of the wire ropes

To validate the FEA results obtained earlier, torsion tests were carried out on the wire rope samples manufactured earlier. The wire ropes were analyzed for

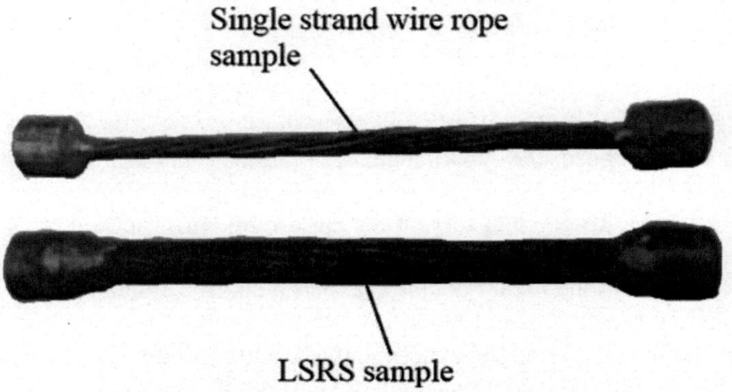

Fig. 7 Manufactured single strand wire rope and the LSRS sample with torque handles

irregularities and covered in heat-shrink sleeves so that the wire ropes did not unwind when they were cut in smaller pieces. A mini metal cutter machine was used to cut a total of 15 wire rope sections having a length of 100 mm each. For the single strand torsion test, 3 wire rope sections in the same direction were prepared. While for the analysis of the LSRS, the rest of the 12 Sects. (4 for each LSRS) were used. Torque handles were machined out of mild steel cylinders and then connected to the wire rope sections using copper brazing to create test samples. Figure 7 shows the samples of the single strand wire rope and the LSRS right next to each other.

The angular deflection was measured using a potentiometer. The gear on the solid shaft of the TTM was meshed with the gear in the potentiometer assembly.

The procedure used to perform the torsion test was as follows.

- The torque handles of the sample were fixed within the wire rope holders on both ends of the torsion test machine.
- To apply torque to the wire rope, the pulley was gradually loaded with 0.1–1 kg of mass. This equated to a torque of 0.098–0.981 N.m for each increment, since the radius of the pulley was 0.1 m.
- After loading the pulley with loads, the wire rope would stop twisting and deforming because of the internal torsional stiffness. The readings of the resistance were recorded from a multimeter, as the pulley was loaded at every interval.
- For all cases, the samples were loaded in both, CW and ACW directions.

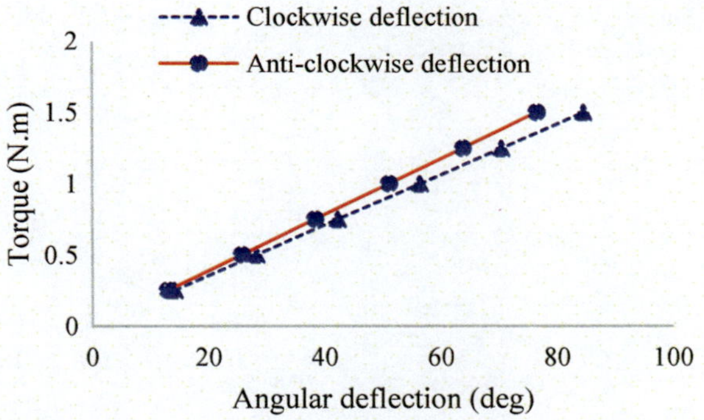

Fig. 8 Torque verses angular deflection curve of a single strand wire rope

3 Results and Discussions

This section presents the results and discussion of the finite element analysis using two configurations of the LSRS. The results from the experiments were discussed and the data was used to validate the simulation model.

3.1 Simulation Results of SSWR

To verify the torsional stiffness of a SSWR, used to fabricate the LSRS, in the CW and ACW directions, a FE analysis was conducted. Figure 8 shows torque verses angular deflection curve of a single strand wire rope in CW and ACW directions. The difference of slopes in the CW and ACW directions can be clearly observed with maximum error of 10.33%, indicating different values of torsional stiffness. The results confirmed the assumptions made in the introduction.

3.2 Simulation Results of config.1

This configuration was modeled as having 2 adjacent wire ropes in left lay direction and the other 2 adjacent in right lay direction as shown in Fig. 4a. A turning moment of 1–5 N.m with an interval of 1 N.m was applied to one end of the configuration, while keeping the other end fixed, in order to observe its torsional behavior in CW and ACW direction. Figure 8 shows the torque verses angular deflection curve of config.1 in CW and ACW directions. As the torque is increased, the angular deflection also increases in a linear relationship. The torsional stiffness values are calculated from the slope of the curves from Fig. 9. From the simulation results, the torsional stiffness value obtained for the CW deflection is 4.40 N.m/rad while for the ACW deflection is 4.44 N.m/rad. The values are almost identical with a percentage difference of only

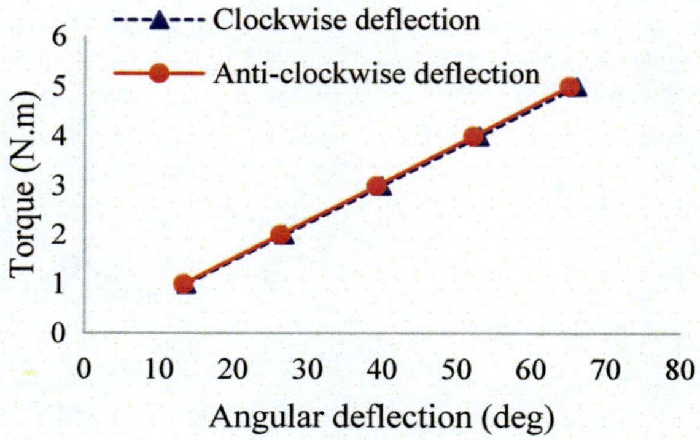

Fig. 9 Torque verses angular deflection curve of config.1

0.90%. It can be clearly observed that the difference between the CW and ACW deflection is negligible and it can be concluded that both directions have yielded almost the same stiffness values.

3.3 Simulation Results of config.2

Configuration 2 is a bit different than configuration 1 because the 2 alternate wire ropes are in right lay direction while the other 2 alternate wire ropes are in the left lay direction which can be seen from the Fig. 4b. The boundary conditions applied were like config.1 with a turning moment of 1 to 5 N.m. The results of torque verses angular deflection curve obtained are shown in Fig. 10. Like config.1, the relationship between the applied torque and an angular deflection is also linear. It can be observed that the CW and ACW deflection curves are overlapping each other and there is no significant difference between the two. The results show that there is almost no difference in the CW and the ACW angular deflection values. Both directional torsion stiffness (obtained from the slope of the curves in Fig. 10) are

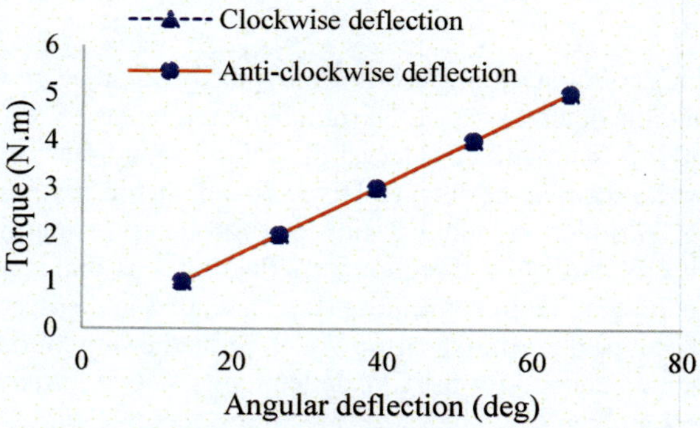

Fig. 10 Angular deflection curve of config.2

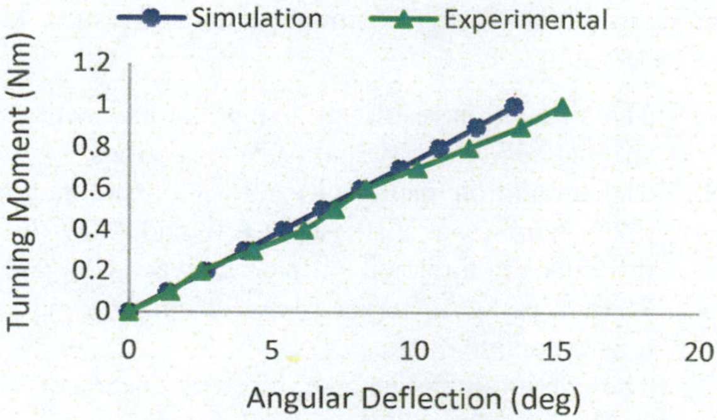

Fig. 11 Anti-clockwise LSRS Angular deflection Curve of simulation and experimental results

found to be 4.44 N.m/rad, indicating that config.2 behaves identically in both CW and ACW directions.

3.4 Experimental Validation of the Wire Rope Configuration (LSRS Layout)

The LSRS sample made for the experimental verification were based on the LSRS model used for simulations. The wire ropes in the LSRS had a lay length of 50 mm, the individual wire diameter was 1.2 mm and the total length was 100 mm. 4 wire ropes were coupled together using config.2 as the base layout like the simulated models and welded to the torque handles using copper blazing.

Figure 11 shows the angular deflection of the LSRS in anti-clockwise direction for the simulation and the experimental results. It can be observed from the figure that there was a good accordance between the simulated and experimental results. It was found that the maximum error was about 12.41%. The presence of error is also attributed to numerous reasons, including the assumptions made in the simulation, and limitations of the experiment due to equipment machining accuracy. In can be concluded that the developed model was valid to be used to design LSRS and analyze its properties.

4 Conclusion

This paper dealt with the modelling and Finite Element (FE) analysis in torsional behavior of two configurations of Low Stiffness Resilience Shafts (LSRS) constructed using combinations of single-strand wire ropes. Two configurations of LSRS have been proposed, i.e. config.1 and config.2 using the combinations of 4 single-strand wire ropes, and a finite element model was established for them. The model was then validated using experimental results. The findings can be summarized as follows:

1. The simulation results of a single-strand wire rope clearly indicated different stiffnesses in the CW and ACW directions.
2. The simulation results have shown that config.1 had a percentage difference of 0.9% in torsional stiffness for CW and ACW directions, while config.2 had no difference in torsional stiffness between CW and ACW directions.
3. The percentage difference between config.1 and config.2 are very low and within the acceptable range.
4. The maximum percentage difference between the simulated and experimental results was about 12.41% at all intervals.

Conclusively, the percentage difference between config.1 and config.2 are very low and within the acceptable range. Either one of these two can be selected as the base layout structure of the LSRS. However, config.2 was chosen as the base layout for all future simulations that were conducted beyond this point due to better confidence level. The model and results from the analysis will be used to design the LSRS for the Semi-Active Steering (SAS) system.

References

1. Holt DJ (2002) Electric steering, a revolution in steering technology: society of automotive engineers
2. Maharun M, Baharom MB, Mohd MS (2015) Ride comfort simulation of a vehicle equipped with semi-active steering system. Int J Automot Mech Eng (IJAME) 11:2495–2503
3. Chang S (2009) Implementation and control logic design of intelligent electric power steering system. World Electric Veh J 3:1–8
4. Markel G (2006) A brief history of the steering wheel & power steering. http://www.brakeandfrontend.com/a-brief-history-of-the-steeringwheel-powersteering
5. Rodriguez-Angeles A, Garcia-Antonio JA (2014) Active disturbance rejection control in steering by wire haptic systems. SA Trans 53:939–946
6. Baharom MB, Hussain K, Day AJ (2013) Design of full electric power steering with enhanced performance over that of hydraulic power-assisted steering. Proc Inst Mech Eng Part D J Autom Eng
7. Haggag S, Alstrom D, Cetinkunt S, Egelja A (2005) Modeling, control, and validation of an electro hydraulic Steer-By-Wire system for articulated vehicle applications. IEEE/ASME Trans Mechatron 10
8. Tavoosi V, Kazemi R, Hosseini SM (2014) Vehicle handling improvement with Steer-By-Wire system using hardware in the loop method. J Appl Res Technol 12:769–781
9. Baharom MB (2008) Design of a semi-active steering system for a passenger car. PhD, School of Engineering, Design and Technology, University of Bradford
10. Husain M, Daugherty B, Oynolan J (2004) Motor vehicle steering system. United States Patent
11. Hussain N, Maharun M, Baharom MB (2015) Analysis to determine the best configuration of wire rope for the development of the Low-Stiffness Resilient-Shaft (LSRS) of the Semi-Active-Steering (SAS) system. In: 3rd international conference on mechanical engineering research, Pahang, Malaysia
12. Hussain N, Maharun M, Baharom MB (2016) The effect of total length of wire ropes on the torsional properties of Low-Stiffness-Resilient-Shaft. J Eng Appl Sci

13. Broder G, Boehm M (2013) Flexible options with flexible shafts. Power Trans Eng 40–42
14. Milburn DA, Rendler NJ (1972) Methods of measuring the mechanical behavior of wire rope. Naval Research Laboratory, Washington, D.C
15. Xu B, Dong P, Zhang J, Zhang S (2014) A three dimensional approach to model steel wire ropes used in high energy absorber apparatus. Int J Solids Struct 2014(51):4280–4293
16. Kastratovic G, Vidanovic N, Bakic V, Rasuo B (2014) On finite element analysis of sling wire rope subjected to axial loading. Ocean Eng 88:480–487
17. Peterka P, Kresak J, Kropuch S, Fedorko G, Molnar V, Vojtko M (2014) Failure analysis of hoisting steel wire rope. Eng Fail Anal 45:96–105
18. Imrak CE, Erdonmez C (2010) On the problem of wire rope model generation with axial loading. Math Comput Appl 15:259–268
19. Stanova E, Fedorko G, Fabian M, Kmet S (2011) Computer modelling of wire strands and ropes part II: finite element-based applications. Adv Eng Softw 42:322–331
20. Stanova E, Fedorko G, Fabian M, Kmet S (2011) Computer modelling of wire strands and ropes part I: theory and computer implementation. Adv Eng Softw 42:305–315
21. Costello GA (1997) Theory of wire rope, vol 2. Springer, New York
22. Wang D, Zhang D, Wang S, Ge S (2013) Finite element analysis of hoisting rope and fretting wear evolution and fatigue life estimation of steel wires. Eng Fail Anal 27:173–193
23. Xiang L, Wang HY, Chen Y, Guan YJ, Wang YL, Dai LH (2015) Modeling of multi-strand wire ropes subjected to axial tension and torsion loads. Int J Solids Struct 58:233–246
24. Argatov I (2011) Response of a wire rope strand to axial and torsional loads: asymptotic modeling of the effect of interwire contact deformations. Int J Solids Struct 48:1413–1423
25. Aziz Ameen H, Salman Hassan K, Mahdi Salah M (2012) Experimental and finite element investigation of annealing on the torsional aspects of carbon steel St35. J Mech Eng Autom 2:135–139
26. Mahasneh MM, Al-Qawabah SMA, Al-Qawabeha UF (2010) Experimental and FEM investigation of heat treatment on the torsional aspects of D2 alloy steel. Res J Appl Sci Eng Technol 2:552–557
27. Hussain N, Baharom MB, Maharun M (2016) Analysis of optimum wire rope configuration for equal unidirectional torsional stiffness for flexible steering shaft. In: MATEC web of conferences, vol 74, pp 00017
28. Carbide Depot (2017) Coefficient for static friction of steel. http://www.carbidedepot.com/formulas-frictioncoefficient.htm. Accessed 17 Apr 2017
29. Imaoka S (2009) Contact analysis tips. In: ANSYS release: 11.0. ANSYS
30. Stefancu A-I, Melenciuc S-C, Budescu M (2011) Penalty based algorithms for frictional contact problems. Gheorghe Asachi Technical University of Iasi, Faculty of Civil Engineering and Building Services
31. Karwa R (2006) A text book of machine design. Laxmi Publications (P) LTD, New Delhi

Effect of Valve Opening Frequency on the Performance of Free Piston Linear Generator

Farith Ikmal Harith, Abdul Rashid Abdul Aziz, Masri Baharom, Ezrann Zharif Zainal Abidin, Salah Eldin Elfakki, Wasiu Babatunde Ayandotun, and Wan Nor Azleen Wan Nadhari

Abstract Combustion is considered as the main method of power generation of the free piston linear generator (FPLG), however other existing method such as expansion of the FPLG using compressed air will also be applicable in generating electricity. In this article, the effect of varying the valve opening frequency onto the performance of the FPLG was experimentally investigated whereby the generator unit of the FPLG was assembled with metal plates, and a constant compressed air pressure of 5 bar was supplied throughout the investigation. From the investigation, by increasing the valve opening frequency, the translator speed increased which led to an overall increase in both, the average power output and the peak power output. But, due to a higher valve opening frequency which reflects on the decrement of injection duration, the stroke of the piston decreased with increasing frequency.

List of Nomenclatures

FPLG	Free Piston Linear Generator
CO	Carbon Monoxide
CO_2	Carbon Dioxide
GDP	Growth Domestic Product
VOF	Valve Opening Frequency
FPE-LG	Free Piston Engine-Linear Generator

F. Ikmal Harith (✉) · A. Rashid Abdul Aziz · M. Baharom · E. Zharif Zainal Abidin · S. Eldin Elfakki · W. Babatunde Ayandotun · W. Nor Azleen Wan Nadhari
Centre for Automotive Research and Electric Mobility, Universiti Teknologi PETRONAS, Persiaran UTP, 32610 Seri Iskandar, Perak, Malaysia
e-mail: farith.harith@utp.edu.my

F. Ahmad et al. (eds.), *ICPER 2020*, Lecture Notes in Mechanical Engineering,
https://doi.org/10.1007/978-981-19-1939-8_75

1 Introduction

The demand of power generation via clean renewable energy is vastly growing amidst the continuous emission of greenhouse gasses due to the burning of fossil fuels to generate electricity. Since 1978, the use of natural gas as one of the many existing fuels to generate electricity has increased by 3083.73% whereby in 2016, 62,723 ktoe of natural gas, was reported used in Malaysia's electricity generation [1]. Therefore, the emission of greenhouse gasses (GHG) such as carbon monoxide (CO), and carbon dioxide (CO_2) would have increased since then whereby a total of 1.5 and 3,558 tons per year per thousand households of CO and CO_2 were believed to be emitted during the process of power generation [2]. In line with the nation's commitment to reduce the emission of GHG while meeting the county's growing power demand in sync with the increment of the Growth Domestic Product (GDP), significant technological advancement will contribute in meeting the required demands [3].

To meet the said demands, various research bodies continue to develop and improve the output of the FPLG by means of simulation, and experimentation. In general, the FPLG may be operated with various types of fuel in order to achieve combustion whereby it leads to the production of electricity. Hydrogen is one of the many fuels in the market, which is used in FPLG, whereby it would technically achieve just about zero percent emission while generating electricity [4].

Not only that, besides operating the FPLG as an engine whereby it combusts, it can also be operated as an expander whereby the movement of the translator can be performed by injecting compressed air or by harvesting waste heat from a different process of power generation [5–7].

The idea of injecting compressed air or harvesting waste heat into the FPLG as a substitute to fuels as a combustible element will ultimately lead towards the absence of GHG emission and this would certainly serve the same purpose in generating clean energy to be supplied to the end users. For instance, in a conventional internal combustion engine, the injection timing can be adjusted according to the crank angle whereby it effects the engine's overall power output at different engine speeds [8]. The aim of this study was to investigate the effect of changing VOF to the output of the FPLG in terms of the piston velocity, stroke length, average power, and peak power output.

2 Experimental Setup

The principle of operation of the FPLG is similar to a free piston expander-linear generator (FPE-LG) whereby it applies the concept of using compressed air as the working fluid to move the translator [9–11]. In this study, the experiment was carried out on the FPLG whereby a constant compressed air pressure of 5 bar was injected via a set of solenoid valves according to the respective VOFs. Throughout the experiment, the FPLG was tested at room temperature, and no cooling nor heating systems were connected to the FPLG, therefore, to ensure that the FPLG's temperature maintained throughout the experiment, a 10 min-gap of cooling down period between each experiment with different VOFs was scheduled. A buffer tank was used in order to support the continuous demand of the FPLG to continue operating on compressed air. The power generated by the FPLG was due to the cutting of the magnets on the translator passing through the stator which supports the copper coils inside of the FPLG's generator unit. The generator unit was connected to a 10Ω-load bank and the current passing through the load bank was recorded with a current clamp in order to calculate the power output of the FPLG. Furthermore, the case holding the copper coils in place were made up of ferromagnetic material which was galvanized steel thus The overall experimental layout can be observed in Fig. 1.

The VOF which represents the system's injection duration for compressed air can be manually adjusted via the control and data acquisition system which is based on National Instruments' hardware and software. The VOF was calculated from the injection duration and tabulated as according Table 1.

The experiment was executed and the registered data that were recorded by our control and data acquisition system which was then processed further in order to extract the required results. The results were than represented in proper graphs and figures for further explanation and clarification.

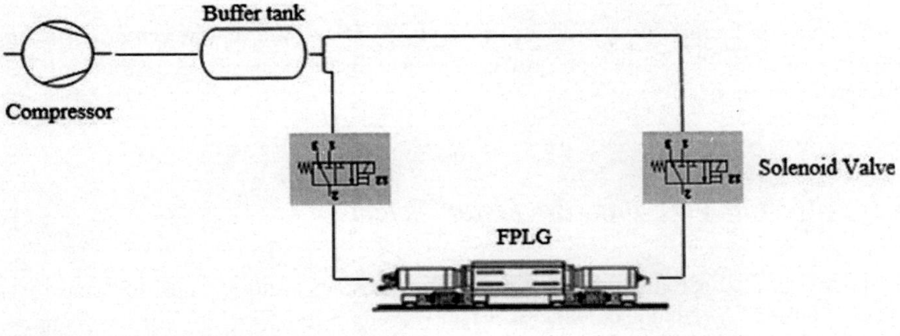

Fig. 1 Overall experimental layout of the FPLG

Table 1 Valve opening frequency for every injection duration of compressed air

Injection duration (ms)	Valve opening frequency (Hz)
200	5
100	10
60	16.67
30	33.3

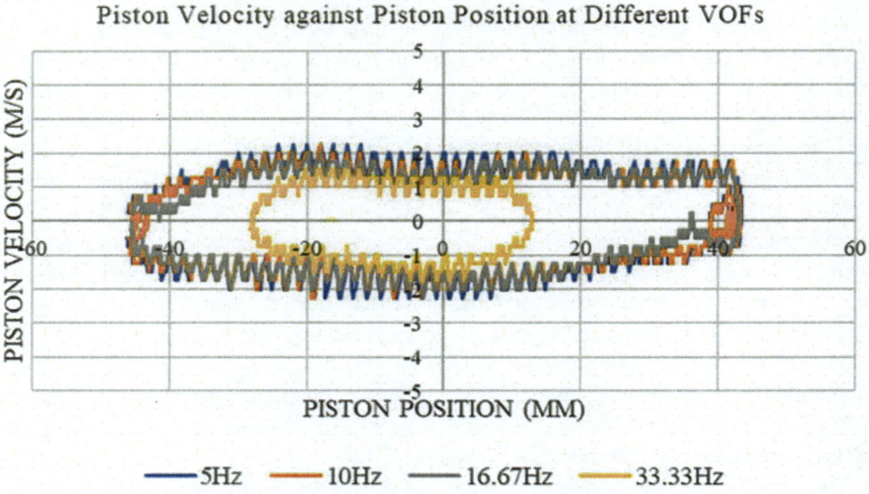

Fig. 2 Plot of piston velocity against piston position at different VOFs

3 Results and Discussion

In this part of the article, the effect of varying VOF to the piston velocity, piston stroke, and the FPLG's average power output and the peak power output will be discussed in detail.

3.1 *Effect of VOF onto the Piston Velocity*

Based on Fig. 2, due to the higher VOF which represents a shorter injection duration, the piston velocity reduced considerably from 2.5 ms^{-1} to 1.5 ms^{-1} while the distance travelled by the piston which also represents the stroke decreases with increasing VOF in which it will be further discussed in Sect. 3.2.

3.2 *Effect of VOF onto the Piston Stroke*

According to Fig. 3, and as highlighted previously in Sect. 3.1, the piston stroke decreased with increasing VOF. According to a past study, the stroke length ultimately decreased with increasing natural frequency of the free piston engine due to the decrement in expansion volume and the membrane stiffness [12]. However, in this case, this was mainly due to the shorter injection duration that allowed a considerably

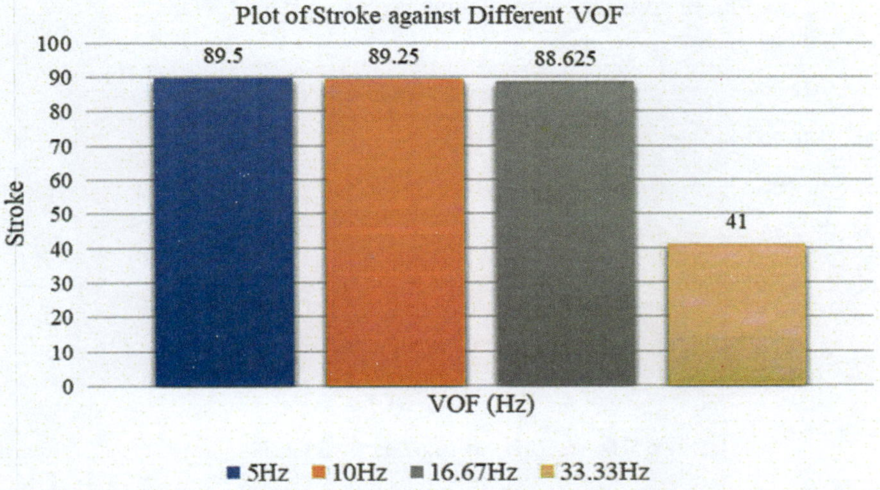

Fig. 3 Plot of stroke length against different VOFs

lower amount of compressed air to be injected into the FPLG thus limiting the travel distance which is also the stroke of the FPLG.

3.3 *Effect of VOF onto the Average Power Output*

In relation with Fig. 4, the average power output was calculated from the voltage recorded via the control and data acquisition system. It can be observed that there is a considerably steady increment in the average power output of the FPLG with increasing VOFs. In other studies, whereby the operating frequency was varied due to the piston mass or motor setting, a similar trend was achieved whereby with increasing operating frequency, the power output would also increase [13, 14]. From this calculated data, the peak power output was also determined using a set of equations and discussed in Sect. 3.4.

3.4 *Effect of VOF onto the Peak Power Output*

Referring to Fig. 5, the peak power output of the FPLG increased steadily with an increment in the VOF, however, the peak power output experienced a steep decrement when the VOF was set to 33.33 Hz. The reduction of the power output was mainly due to the reduction of the stroke length passing through the generator unit [15], whereby in this study, a shorter stroke length indicates that the magnets on the translator did not fully pass through the coils on the stator. Nevertheless, from the experiment

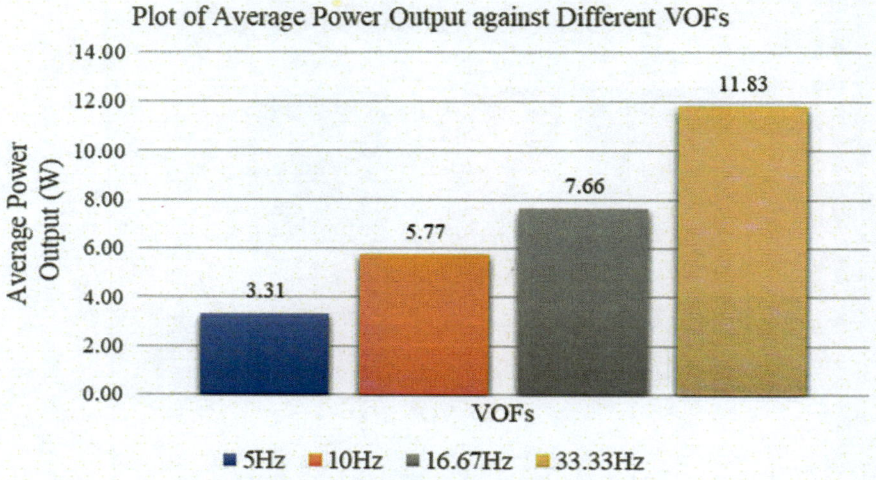

Fig. 4 Plot of average power output at different VOFs

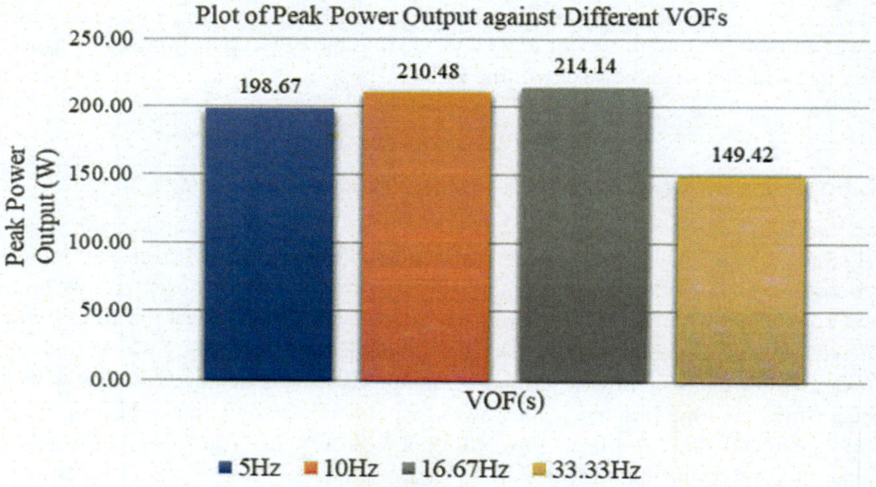

Fig. 5 Plot of the peak power output against different VOFs

carried out on the FPLG, the increased VOFs led to a reduction of piston stroke which contributed in an increase in the power output [16]. Therefore, this provides an indication that the VOF does not possess a linear relationship with the power output since increasing the VOF may reduce the stroke length as stated in Sect. 3.2, which will affect the average power output and the peak power output stated in Sects. 3.3, and 3.4.

4 Conclusion

In conclusion towards the article, it can be said that the method of power generation via expansion mode on the FPLG is promising in delivering a clean renewable energy whereby there are no GHG emissions to the atmosphere, thus meeting with the nation's drive in clean energy production. With the growing demand of energy supply amidst the continuous advancement of technology, the scene of energy production via FPLG will continue to develop whereby numerous improvements can be made in order to further improve the power generation in order to provide a sustainable yet clean renewable energy.

Acknowledgements This study was managed to be completed due to the support of fellow research officers in Centre for Automotive Research and Electric Mobility (CAREM) in Universiti Teknologi PETRONAS, with the financial assistance under the Petroleum Research Fund (PRF) Grant (0153AB-A34) for the project entitled, "Development of a concept-demonstrator free piston linear generator powered by flexi-fuel".

References

1. Malaysia Energy Statistics Handbook (2018) Suruhanjaya Tenaga, Putrajaya, pp 18–19, ISSN No.: 2289-6953
2. Suruhanjaya Tenaga, Energy Malaysia, 2019, pp 1828, ISSN No.: 2289-6953
3. Tan CS, Maragantham K, Leong YP (2013) Electricity energy outlook in Malaysia
4. Woo Y, Lee Y, Lee Y (2009) The performance characteristics of a hydrogen-fuelled free piston internal combustion engine and linear generator system. Int J Low-Carbon Technol 4(1):36–41. https://doi.org/10.1093/ijlct/ctp003. (Oxford University Press (OUP), 1 Mar 2009)
5. Champagne C, Weiss L (2013) Design and optimization of free piston expander for energy harvesting. In: Volume 6A: energy. American Society of Mechanical Engineers. https://doi.org/10.1115/imece2013-64798. (15 Nov 2013)
6. Li G, Zhang H, Yang F, Song S, Chang Y, Yu F, et al (2016) Preliminary development of a free piston expander-linear generator for Small-Scale Organic Rankine Cycle (ORC) waste heat recovery system. In: Energies, vol 9, no 4. MDPI AG, p 300. https://doi.org/10.3390/en9040300. (20 Apr 2016)
7. Hou X, Zhang H (2018) Performance analysis of free piston expander-linear generator for organic rankine cycle waste heat recovery system. Int J Electron Electric Eng 6(2):16–21. https://doi.org/10.18178/ijeee.6.2.16-21. (EJournal Publishing)
8. Rahman MM, Kamil M, Bakar RA (2011) Engine performance and optimum injection timing for 4-cylinder direct injection hydrogen fueled engine. Simul Modell Practice Theory 19(2):734–751. https://doi.org/10.1016/j.simpat.2010.10.006. (Elsevier BV, Feb 2011)
9. Wang Y, Chen L, Jia B, Roskilly AP (2017) Experimental study of the operation characteristics of an air-driven free-piston linear expander. In: Applied energy, vol 195. Elsevier BV, pp. 93–99. https://doi.org/10.1016/j.apenergy.2017.03.032. (Jun 2017)
10. Tenissara N, Thepa S, Monyakul V (2017) Performance study of a small-single piston expander using compressed air as working fluid. In: Energy procedia, vol 138. Elsevier BV, pp 610–615. https://doi.org/10.1016/j.egypro.2017.10.169. (Oct 2017)
11. Zhang H, Yu F, Li G, Hou X, Liu H, Tian Y (2017) Experiment of free-piston expander-linear generator driven by compressed air (Sept 2017)

12. Weiss LW (2010) Study of a MEMS-based free piston expander for energy sustainability. J Mech Design 132(9). https://doi.org/10.1115/1.4002083. (ASME International, 30 Aug 2010)
13. Champagne C, Weiss L (2013) Performance analysis of a miniature free piston expander for waste heat energy harvesting. In: Energy conversion and management, vol 76. Elsevier BV, pp 883–892. https://doi.org/10.1016/j.enconman.2013.08.045. (Dec 2013)
14. Feng H, Song Y, Zuo Z, Shang J, Wang Y, Roskilly A (2015) Stable operation and electricity generating characteristics of a single-cylinder free piston engine linear generator: simulation and experiments. In: Energies, vol 8, no 2. MDPI AG, pp 765–85. https://doi.org/10.3390/en8020765. (23 Jan 2015)
15. Hong Y, Eriksson M, Boström C, Waters R (2016) Impact of generator stroke length on energy production for a direct drive wave energy converter. In: Energies, vol 9, no 9. MDPI AG, p 730. https://doi.org/10.3390/en9090730. (9 Sept 2016)
16. Preetham BS, Weiss L (2016) Investigations of a new free piston expander engine cycle. In: Energy, vol 106. Elsevier BV, pp 535–545. https://doi.org/10.1016/j.energy.2016.03.082. (Jul 2016)